Mechanics of Liquids and Gases

6th edition

L. G. Loitsyanskiy
St. Petersburg State Technical University, Russia

English Edition Editor:
Robert H. Nunn
formerly with the US Naval Postgraduate School Monterey, California

begell house, inc.
New York • Wallingford (UK)

Translators: Greta R. Malyavskaya*
Natalia K. Shveyeva*

Translation Editor: Oleg G. Martynenko*

* A.V. Luikov Heat and Mass Transfer Institute
Minsk, Belarus

Library of Congress Cataloging-in-Publication Data

Loitsyanskiy , L.G. (Lev Gerasimovich)
(Mekhanika zhidkosti I gaza. English)
Mechanics of liquids and gases /L.G. Loitsyanskiy.-- 6th ed
p. cm.
Includes indexes.
ISBN 1-56700-042-8 (hardcover)
1. Hydrodynamics. 2. Aerodynamics. I. Title.
QA911.L613 1995
532'.05--dc20 95-25679
CIP

Contents

7. Modification of the Laminar Boundary Layer Equations to Equations for an Incompressible Fluid 830

Author's Preface

This publication is the translation of the 6th edition (1987, Nauka Press, Moscow) of my monographical textbook "Mechanics of Liquids and Gases" which was extensively revised and augmented as compared with the 2nd edition (1957, Nauka Press, Moscow) translated into English and published by Pergamon Press in 1966. The 6th edition incorporates significant advances towards the solution of the problems of the boundary layer theory, dynamics of viscous fluids and theory of turbulence. Moreover, this English edition is augmented with the 16th Chapter on computational hydrogasdynamics written by my colleagues at the Department of Hydroaerodynamics of the Leningrad Polytechnic Institute: Prof. Yu. P. Golovachev and Ass. Prof. S. B. Koleshko to whom, taking this opportunity, I wish to express my gratitude. I also wish to thank the Joint Soviet-American Enterprize "TEMPO" in the persons of its founders Mr. W. Begell and Corresponding Member of the Byelorussian Academy of Sciences O. G. Martynenko and also staff-members of the USSR Editorial Office of the International Journal of Heat and Mass Transfer G. R. Malyavskaya and N. K. Shveyeva who has taken the trouble to translate my book into English.

The textbook is intended for students, postgraduate students, engineers, and research workers specializing in the field of mechanics.

Author's Biography

Born on 26 December 1900 in St. Petersburg, L. G. Loitsyanskiy graduated in 1921 from the Physico-Mathematical Department of the University with the speciality "pure mathematics". From 1922 to 1934, L. G. Loitsyanskiy is Assistant Lecturer, then Assistant Professor, thereafter Professor at the Department of Theoretical Mechanics of the Petrograd (Leningrad) Polytechnical Institute and thereupon he accepted the chairmanship of the Hydroaerodynamic Department of the same Institute.

L. G. Loitsyanskiy is Doctor of Physical and Mathematical Sciences (1935), State Prize Winner (1946), Merited Worker of Science and Technology (1968), Full Member of the International Astronautical Academy (1970), Foreign Member of the Polish Mathematical Society (1972), Honored Member of the Yugoslavian Society of Mechanics (1982).

L. G. Loitsyanskiy has authored more than 120 scientific works in the fields of applied mathematics, dynamics of viscous fluids, theory of boundary layer and turbulence. Among these works are textbooks, monographs and scientific reviews, *viz.* "Theoretical Mechanics" in 3 volumes (1933) and "A Course in Theoretical Mechanics" in 2 volumes (1934) co-authored with A. I. Luriye (Corresponding Member of the USSR Academy of Sciences); "Fundamentals of the Mechanics of Viscous Fluids" (in two parts, 1933), "Aerodynamics of the Boundary Layer" (1941), "Laminar Boundary Layer" (1962), "Mechanics of Liquids and Gases" (1950), a review "The Development of Boundary Layer Theory in the USSR" (1970, Ann. Review of Fluid Mechanics, Palo Alto, California, USA).

Preface to the 6th Edition

The present edition is the outcome of an extensive revision made, on the one hand, to bring its content closer to that of a textbook and, on the other, to update it by adding some new present-day problems.

Since it was impossible to extend any further so bulky a book as it was, some special sections dealing with the mechanics of liquids and gases but not fulfilling the objectives of the present general consideration had to be sacrificed. Thus, the problem of motion of multiphase media was dropped from the dynamics of inhomogeneous fluids, and only the derivation of the equations for multicomponent gas flows was retained as being closely associated with the description of the phenomena of dissociation and ionization of molecules in supersonic gas flows. Also omitted was the section dealing with the macroscopic approach to calculations of dust-laden gas flows which gave way to the methods of the kinetic theory of molecular motions. This was also the reason for dropping the section which discussed the shock wave from the viewpoint of continuum mechanics as well as some other problems dealt with in the previous edition of the book.

As distinct from previous editions where only vector and tensor calculus formulae were given, the present one provides a short background discussion on this area of mathematics. The section on the elements of similarity theory is supplemented with a presentation of dimensionality theory and with the proof of the main theorem in this theory.

Newly written are three sections dealing with some general methods for numerical integration of differential equations and their application to Navier-Stokes viscous fluid dynamics equations. Extensive revision of the chapter on turbulence culminated in the appearance of a new chapter dealing specially with the techniques for calculating turbulent boundary layers. Methodological blunders noticed in many places throughout the book have been eliminated.

Professor G. Yu. Stepanov, Dr. Sc. (Physics and Mathematics) read the manuscript with an exceptional accuracy and made a great number of scientific and methodological comments important for the quality of the book.

Professor Yu. V. Lapin, Dr. Sc. (Physics and Mathematics) read the manuscript of Chapter XIV and made a number of useful comments. An invaluable help in the work on the manuscript and in proof-reading was lent by Assistant Professor at the Chair of Hydroaerodynamics of Leningrad Polytechnic Institute S. B. Koleshko who not only undertook the scientific editing of the manuscript and made many useful remarks, but also wrote three sections (102–104) concerned with numerical solution of viscous fluid dynamics problems. The author regards it his duty to acknowledge cordially their assistance for this.

Preface to the English Edition

Many of us who study and teach "fluid mechanics" have come to think of it as some sort of counterpart, or alternative, to "solid mechanics." Indeed, western university curricula typically address the two subjects with specialized faculty using texts that do little to reveal their common ground. Fluid mechanics to Loitsyanskiy, as his title proclaims, springs naturally from the vast realm of mechanics itself. Thus his technique is to use, over and over again, the general formulations of mechanics as the foundation of theories describing the behaviors of fluids--viscous and inviscid, liquids and gases.

The result may be somewhat daunting to the pragmatic student of fluid mechanics, for the practical results are often achieved by Loitsyanskiy only after substantial mathematics gyrations of the most general kind. Readers should take heart, however, because if the material seems abstract there are often repeated explanations with very helpful nuances. For the fluid mechanics devotee (especially non-Russian readers, I suspect) Loitsyanskiy offers words that are exquisitely chosen; and many new ways of looking at the wonderful world of fluids are waiting for those who persevere. As one of many examples, I recall how airfoil theory is almost completely expounded by Loitsyanskiy without even defining the famous Joukowsky transformation.

As the editor of the translation, I must compliment the translators whose work makes it clear that they have been alert to the significance of the document and the danger of losing even the smallest part of its "personality" in translation. I, also, have done my best not to interfere. Occasionally, I have even left some "fuzzy" passages in their as translated form for fear of losing the beauty of the narrative for the sake of technical precision. In these instances, readers will have to reach their own conclusions.

To be asked to edit the English translation of *Mechanics of Liquids and Gases* was, to me, like being invited to a private viewing prior to the unveiling of a great masterpiece. *The Mechanics of Liquids and Gases* is huge--in coverage and in quality--and its availability in English is a great new learning opportunity.

Robert H. Nunn, Ph.D.
formerly with the US Naval Postgraduate School
Monterey, California

the mechanics of fluids and ensuing general integrals or theorems. In order application of these equations may be fully definitive, it is necessary that the following additional *phenomenological laws* be taken into account: state equation, Newton's law for viscosity, Fourier's law for heat conduction, Fick's law for diffusion, etc.

2. Some Information on the Molecular Structure of Substance

Depending on the quantitative relationship between the kinetic energy of the motion of molecules and potential energy of intermolecular dynamic interaction, there originate different molecular structures and types of internal motion of molecules.

Essential in *solid bodies* is the molecular energy of the interaction of molecules due to which the latter are arranged in regular crystal lattices with stable equilibrium positions at lattice nodes. Thermal motions in a solid body consist of small vibrations of molecules about lattice nodes with a high frequency (of order 10^{12} Hz) and amplitude proportional to spacings between these nodes. Both "short-range" and "long-range" orders are effected in the molecular structure of a solid body. To solid bodies also belong substances in amorphous state which do not have crystalline structure but which possess "short-range" order closely resembling that in liquids (see below). Amorphous states are not very stable and change easily to crystalline states.

In contrast to a solid body, both "short-range" and "long-range" orders are absent in gases. The molecules of a gas move in a random motion, with their interaction being reduced only to collisions. The interaction of molecules in the intervals between collisions is neglected, and this corresponds to the smallness of the potential energy of the dynamic interaction of molecules as compared with the kinetic energy of their random motion. The mean distance between two consecutive collisions of molecules determines the "free path length". The "free path" velocity of molecules is commensurable with the speed of propagation of small disturbances (speed of sound) in this state of gas.

As to their molecular structure and thermal motion, liquid bodies occupy an intermediate position between solid and gaseous bodies. According to the current views, a certain molecule, acting as a central one, collects around itself a group of neighboring molecules that slightly vibrate with a frequency close to that mentioned earlier for vibrations of solid body molecules in a lattice and with an amplitude of the order of mean distance between molecules. The central molecule either remains stationary (in a liquid at rest), or migrates at a speed coinciding in magnitude and direction with the local mean velocity of the macroscopic motion of liquid. In the molecular structure of a liquid the potential energy of molecular interaction is comparable in order with the kinetic energy of their thermal motion, with the "short-range" order being present and the "long-range" order not. The evidence for the vibrations of molecules in liquids is provided by the well-known "Brownian motion" of tiny solid particles introduced into liquid. Vibrations of these particles can be easily seen under a microscope and may be looked upon as the result of collision of solid particles with the molecules of liquid.

The difference between molecular structures and thermal motions of solid, liquid and gaseous bodies reveals itself clearly in the phenomenon of diffusion consisting in the propagation of one substance (inclusion) into the other (carrier). Diffusion of one gas in another (for example, the propagation of odor in air) due to intensive molecular motion leads to rapid penetration of odor into the farthest corners in a room. Conversely, the diffusion of a liquid in a liquid occurs much more slowly because of weak migration of central molecules with groups of molecules bound around them. The best example is provided by the historical experiment of Reynolds who introduced a thin jet of dye into a (laminar) water flow slowly moving through a cylindrical pipe. The jet remained nearly the same in thickness over almost *the mechanics of fluids and ensuing* general integrals or theorems. In order application of these equations may be fully definitive, it is necessary that the following additional *phenomenological laws* be taken into account: state equation, Newton's law for viscosity, Fourier's law for heat conduction, Fick's law for diffusion, etc.

2. Some Information on the Molecular Structure of Substance

Depending on the quantitative relationship between the kinetic energy of the motion of molecules and potential energy of intermolecular dynamic interaction, there originate different molecular structures and types of internal motion of molecules.

Essential in *solid bodies* is the molecular energy of the interaction of molecules due to which the latter are arranged in regular crystal lattices with stable equilibrium positions at lattice nodes. Thermal motions in a solid body consist of small vibrations of molecules about lattice nodes with a high frequency (of order 10^{12} Hz) and amplitude proportional to spacings between these nodes. Both "short-range" and "long-range" orders are effected in the molecular structure of a solid body. To solid bodies also belong substances in amorphous state which do not have crystalline structure but which possess "short-range" order closely resembling that in liquids (see below). Amorphous states are not very stable and change easily to crystalline states.

In contrast to a solid body, both "short-range" and "long-range" orders are absent in gases. The molecules of a gas move in a random motion, with their interaction being reduced only to collisions. The interaction of molecules in the intervals between collisions is neglected, and this corresponds to the smallness of the potential energy of the dynamic interaction of molecules as compared with the kinetic energy of their random motion. The mean distance between two consecutive collisions of molecules determines the "free path length". The "free path" velocity of molecules is commensurable with the speed of propagation of small disturbances (speed of sound) in this state of gas.

As to their molecular structure and thermal motion, liquid bodies occupy an intermediate position between solid and gaseous bodies. According to the current views, a certain molecule, acting as a central one, collects around itself a group of neighboring molecules that slightly vibrate with a frequency close to that mentioned earlier for vibrations of solid body molecules in a lattice and with an amplitude of the order of mean distance between molecules. The central molecule either remains stationary (in a liquid at rest), or migrates at a speed coinciding in magnitude and direction with the local mean velocity of the macroscopic motion of liquid. In the molecular structure of a liquid the potential energy of molecular interaction is comparable in order with the kinetic energy of their thermal motion, with the "short-range" order being present and the "long-range" order not. The evidence for the vibrations of molecules in liquids is provided by the well-known "Brownian motion" of tiny solid particles introduced into liquid. Vibrations of these particles can be easily seen under a microscope and may be looked upon as the result of collision of solid particles with the molecules of liquid.

The difference between molecular structures and thermal motions of solid, liquid and gaseous bodies reveals itself clearly in the phenomenon of diffusion consisting in the propagation of one substance (inclusion) into the other (carrier). Diffusion of one gas in another (for example, the propagation of odor in air) due to intensive molecular motion leads to rapid penetration of odor into the farthest corners in a room. Conversely, the diffusion of a liquid in a liquid occurs much more slowly because of weak migration of central molecules with groups of molecules bound around them. The best example is provided by the historical experiment of Reynolds who introduced a thin jet of dye into a (laminar) water flow slowly moving through a cylindrical pipe. The jet remained nearly the same in thickness over almost the entire working section of the pipe, thus testifying to a slow molecular diffusion typical of laminar motion of water. On transition to a turbulent mode of flow, when the molecular mechanism of diffusion gives place to turbulent mixing of finite macrovolumes of liquid, there arises an intensive turbulent diffusion and the dye rapidly fills the entire flow.

It should be noted that the phenomenon of diffusion is also observed in solid bodies, although it is much weaker here than in liquids. Mutual penetration of molecules can be seen in two specimens of different metals tightly fitted to each other after the lapse of a long period.

The problems of astrophysics associated with the study of ionosphere, "star clouds" and other astronomical objects and especially various physico-technical problems connected with the design of thermonuclear reactors and magnetohydrodynamic generators for direct conversion of thermal into electrical energy spurred a considerable upsurge of interest in the dynamics of ionized gases (plasmas).

In contrast to ordinary electrically neutral gases in which randomly moving molecules display dynamic interaction only ion only on their mutual collision, much more substantial Coulomb interactions originate in plasma due to a high concentration of charged particles. This imparts specific properties to plasma, as e.g. higher electrical conductivity, showing up most vividly on exposure of plasma flows external electrical and magnetic fields.

The exceptional, in their physical importance and applied possibilities, and at the same time very specific properties of plasmas suggested the idea of the fourth (after solid, liquid and gaseous) aggregated state of matter.[1]

---

[1]A detailed survey of the properties of plasma and phenomena occurring in it can be found in the paper by B. B. Kadomtsev "Plasma" (*Physical Encyclopedic Dictionary*, Vol. 4, Moscow, Sov. Entsiklopediya Press, 1965, pp. 15–24) and also in Sects. 1.3–1.4 of the book by Lukiyanov G. A. *Sverkhzvukovye strui plazmy* (*Supersonic Plasma Jets*). Leningrad, Mashinostroeniye Press, 1985, pp. 13–21 and the list of references in it.

1

The Field of a Physical Quantity and Conditions for the Physical Objectivity of Quantities Specified Analytically: Basic Opcrations of the Field

1. Scalar and Vector Fields: Conditions for the Physical Objectivity of Scalar and Vector Quantities Specified Analytically

The *field* of a physical quantity (velocity, acceleration, pressure, density, temperature, etc.) is the name given to a set of the values of this quantity unambiguously determined at each point in some part of space. There may exist *scalar, vector,* and, as seen later, *tensor fields,* depending on the kind of the quantity, the distribution of which is prescribed by the field.

If scalar, vector, or tensor quantities have the same values at all the points of the field, the field is called *uniform*; otherwise they are *nonuniform*. The fields of physical quantities may show no variation with time, i.e., they may be *stationary,* and may vary with time, i.e., be *nonstationary.*

The field of the scalar quantity φ (e.g., temperature, pressure, density) is specified analytically as a function of the coordinates of field points and time ($\boldsymbol{r}$ is the instantaneous radius-vector of the point M of space)

$$\varphi = \varphi(x, y, z; t) = \varphi(M; t) = \varphi(\boldsymbol{r}; t) \tag{1}$$

Here, time is considered to be a parameter that denotes separate instants. The fields pertain to these when they are nonstationary and can be omitted when the fields are stationary.

The field of the vector quantity $\boldsymbol{a}$ (velocity, acceleration, force, etc.) is prescribed by the vector function of the coordinates of points M and time

$$\boldsymbol{a} = \boldsymbol{a}(x, y, z; t) = \boldsymbol{a}(M; t) = \boldsymbol{a}(\boldsymbol{r}; t) \tag{2}$$

or by three projections of the vector $\boldsymbol{a}$

$$a_x = a_x(x, y, z; t) \qquad a_y = a_y(x, y, z; t) \qquad a_z = a_z(x, y, z; t) \tag{3}$$

To shorten the amount of formulation, letter indexing of coordinates and vector projections is often replaced by *numerical indexing* assuming that

$$x = x_1 \quad y = x_2 \quad z = x_3 \qquad a_x = a_1 \quad a_y = a_2 \quad a_z = a_3$$

and Equation 3 will be replaced by

$$a_i = a_i(x_1, x_2, x_3; t) \qquad (i = 1, 2, 3)$$

Furthermore, proceeding as advised by Einstein, omit the summation sign $\sum$ in front of monomials being summed up if the index, over which the summation is performed, appears twice in the monomial. Thus, for example,

$$c_k = \sum_{l=1}^{3} a_l b_{lk} = a_l b_{lk} \qquad (k = 1, 2, 3)$$

$$f_{pq} = \sum_{l=1}^{3} \sum_{m=1}^{3} a_{pl} b_{qm} c_{lm} = a_{pl} b_{qm} c_{lm} \qquad (p, q = 1, 2, 3)$$

The indices that appear twice and over which summation is carried out are called "dummy indices"; they can be denoted arbitrarily in the process of calculations:

$$c_k = a_l b_{kl} = a_m b_{km} \qquad (k = 1, 2, 3)$$

The rest of the indices (as, for example, k in the previous formula) remain unchanged in each given process of calculation and are called "free indices".

Let us recall the analytical geometry formulas, which will be needed later, having expressed them in the notation just adopted.

Consider two systems of *rectangular* Cartesian coordinates: $0x_1x_2x_3$ and $0x_1'x_2'x_3'$ (Figure 1), which have the common origin at point 0, but which are rotated arbitrarily with respect to one another. We shall denote the *cosines* of the angles between "new" (primed) and "old" axes by α_{kl} according to the matrix

$$\begin{array}{cccc} & x_1 & x_2 & x_3 \\ x_1' & \alpha_{11} & \alpha_{12} & \alpha_{13} \\ x_2' & \alpha_{21} & \alpha_{22} & \alpha_{23} \\ x_3' & \alpha_{31} & \alpha_{32} & \alpha_{33} \end{array} \tag{I}$$

In each coordinate system we refer to its *set of unit vectors directed along the coordinate axes* as the *basis* of the coordinate system. In the present case there are two such bases: $\boldsymbol{i}_1$, $\boldsymbol{i}_2$, $\boldsymbol{i}_3$ and $\boldsymbol{i}_1'$, $\boldsymbol{i}_2'$, $\boldsymbol{i}_3'$. The radius-vector $\boldsymbol{r}$ of the point M can be expressed in terms of these bases as

$$\boldsymbol{r} = x_1\boldsymbol{i}_1 + x_2\boldsymbol{i}_2 + x_3\boldsymbol{i}_3 = x_1'\boldsymbol{i}_1' + x_2'\boldsymbol{i}_2' + x_3'\boldsymbol{i}_3' \tag{4}$$

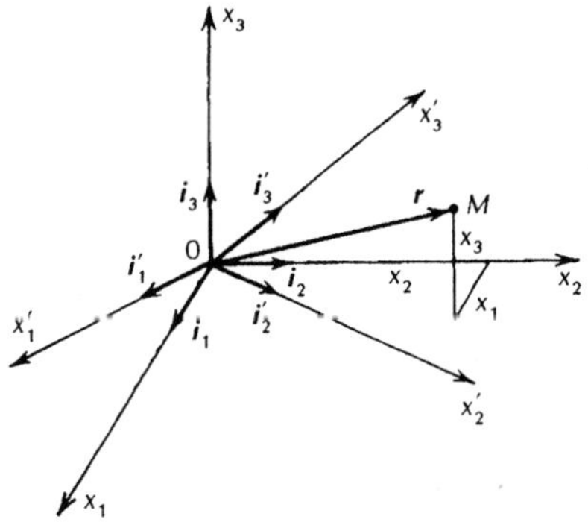

Fig. 1

Form the scalar product of both sides of this equality successively with $\boldsymbol{i}_1$, $\boldsymbol{i}_2$, $\boldsymbol{i}_3$ and $\boldsymbol{i}'_1$, $\boldsymbol{i}'_2$, $\boldsymbol{i}'_3$ by recalling that

$$\boldsymbol{i}_k \cdot \boldsymbol{i}_l = \boldsymbol{i}'_k \cdot \boldsymbol{i}'_l = \delta_{kl} = \begin{cases} 1 & k = l \\ 2 & k \neq l \end{cases} \tag{5}$$

and noting that according to the matrix (I) above:

$$\boldsymbol{i}'_k \cdot \boldsymbol{i}_l = \alpha_{kl} \tag{6}$$

Thus, we shall obtain the following formulas for transforming the "old" coordinates x_1, x_2, x_3 into "new" coordinates x'_1, x'_2, x'_3:

$$x'_k = \alpha_{k1}x_1 + \alpha_{k2}x_2 + \alpha_{k3}x_3 = \alpha_{kl}x_l \quad (k = 1, 2, 3) \tag{7}$$

and vice versa:

$$x_k = \alpha_{1k}x'_1 + \alpha_{2k}x'_2 + \alpha_{3k}x'_3 = \alpha_{lk}x'_l \quad (k = 1, 2, 3) \tag{8}$$

Taking into account that the projections of the vector $\boldsymbol{a}$ on the coordinate axes a_1, a_2, a_3 or a'_1, a'_2, a'_3 are given by the differences between the coordinates that locate the end and beginning of the segment that represents the vector, it can be concluded that the formulas for the vector projections in conversion from one coordinate system to the other will be analogous in form

$$a'_k = \alpha_{kl}a_l \qquad a_k = \alpha_{lk}a'_l \quad (k = 1, 2, 3) \tag{9}$$

Note that Equation 9 is valid only in orthogonal coordinate systems that will be employed below almost without exception.

Let us agree that *physically objective* will be the name given to those quantities which, by virtue of their physical essence, *are independent of the choice of the coordinate system* in which they are represented analytically. Thus, for example, the physical scalar (temperature, pressure, density, etc.) prescribed by its field, i.e., by the function of the coordinates $\varphi(x_1, x_2, x_3)$, should be an *invariant* which satisfies the equality

$$\varphi(x_1, x_2, x_3) = \varphi(x'_1, x'_2, x'_3) = \text{inv}$$

where it is presumed that the sets of "old" and "new" coordinates correspond to the same point of space.

Analogously, a vector prescribed by its three projections $a_k(x_1, x_2, x_3)$ $(k = 1, 2, 3)$, or, in another coordinate system, by the projections $a'_k(x'_1, x'_2, x'_3)$ $(k = 1, 2, 3)$ should be independent of the coordinate axes for it to be capable of representing a physical quantity.

In order to find the conditions for the physical objectivity of the vector function $\boldsymbol{a}$, we shall first prove that the scalar product $\boldsymbol{a} \cdot \boldsymbol{b} = a_k b_k$, where a_k and b_k satisfy the conversion formulas of Equation 9, is independent of the directions chosen for the coordinate axes, in which these multiples were determined (i.e., where it is an *invariant*).

According to Equation 9, we have (k is a dummy index)

$$a_k b_k = \alpha_{lk} a'_l \alpha_{mk} b'_m = \alpha_{lk} \alpha_{mk} a'_l b'_m \tag{10}$$

To conform with the well-known theorems of analytical geometry, the cofactor $\alpha_{lk}\alpha_{mk}$ on the right side of Equation 10 is the expression of the cosine of the angle between the axes $0x'_l$ and $0x'_m$ in terms of the cosines of the angles between these axes and those of the system $0x_1x_2x_3$. In compliance with the orthogonality property of the coordinate axes

$$\cos(\widehat{x'_l, x'_m}) = \alpha_{lk}\alpha_{mk} = \begin{cases} 1 & \text{when} \quad m = l \\ 0 & \text{when} \quad m \neq l \end{cases} \tag{11}$$

Analogously, expressing the cosine of the angle between the axes $0x_l$ and $0x_m$ in terms of the cosines of the angles about the axes of the system $0x'_1x'_2x'_3$, we shall obtain

$$\cos(\widehat{x_l, x_m}) = \alpha_{kl}\alpha_{km} = \begin{cases} 1 & \text{when} \quad m = l \\ 0 & \text{when} \quad m \neq l \end{cases} \tag{12}$$

Taking into account Equation 11, we shall obtain the equality that proves the *invariance of the scalar product* $\boldsymbol{a} \cdot \boldsymbol{b}$

$$a'_k b'_k = a_l b_l = \text{inv} \tag{13}$$

Assuming that in this equality $\boldsymbol{b} = \boldsymbol{a}$, we may note the *invariance of the sum of the squared projections*

$$a'_k a'_k = a_l a_l = \text{inv} \tag{14}$$

i.e., squared length (absolute value) of the vector and, consequently, of the very length (absolute value) of the vector. Bearing in mind that

$$a_k b_k = \boldsymbol{a} \cdot \boldsymbol{b} = ab\cos(\widehat{\boldsymbol{a}, \boldsymbol{b}})$$

where $\boldsymbol{a}$ and $\boldsymbol{b}$ are invariants, and specifying different vectors $\boldsymbol{b}$, we shall also note the invariance of the location of the vector $\boldsymbol{a}$ relative to the points of space.

The role of Equation 9 as conditions for the physical objectivity in the analytical determination of the vector is clear. As regards the formulas of conversion from one coordinate system to the other (Equation 7 and 8), they constitute the specific case of Equation 9 in application to the radius-vector $\boldsymbol{r}$ of point M and, consequently, they establish the physical objectivity, in analytical determination of the radius-vector of the point, with the aid of its coordinates.

Calling attention to the identity of the formulas of conversion from one coordinate system to the other for the Cartesian coordinates themselves (Equations 7 and 8) and for vector projections (Equation 9), we can formulate the conditions for the physical objectivity in the analytical representation of vectors: In converting from one rectangular Cartesian coordinate system to the other, the projections of a physical vector must be transformed in the same way as the coordinates.

2. A Second-Rank Tensor: Conditions for the Physical Objectivity of Its Analytical Specification – A Dyad and a Rotation Tensor

Along with scalar quantities, the field of which is determined by one numerical function, and vector quantities with the field specified by three numerical functions (vector projections onto the coordinate axes), the mechanics of continua also employ more mathematically complex physical quantities, i.e., *second-rank tensors.*[1]

The second-rank tensor can be defined by two independent techniques leading to identical results.

Generalizing the notions of the scalar and vector and the conditions for their physical objectivity, we shall determine the second-rank tensor T at each point of a three-dimensional space as a set of nine quantities T_{pq} $(p, q = 1, 2, 3)$ which are the tensor components specified in a certain rectangular Cartesian coordinate system and which, when converting to another coordinate system, are transformed by the formulas (r and s are dummy indices)

$$T'_{pq} = \alpha_{pr}\alpha_{qs}T_{rs} \qquad T_{pq} = \alpha_{rp}\alpha_{sq}T'_{rs} \qquad (p, q = 1, 2, 3) \tag{15}$$

These equalities can be interpreted as the formulae of transition from one rectangular Cartesian coordinate system to the other from the product of two coordinates. In fact,

$$x'_p x'_q = \alpha_{pr}x_r\alpha_{qs}x_s = \alpha_{pr}\alpha_{qs}x_r x_s$$

$$x_p x_q = \alpha_{rp}x'_r\alpha_{sq}x'_s = \alpha_{rp}\alpha_{sq}x'_r x'_s$$

in agreement with Equations 15.

Thus, according to the first definition, the second-rank tensor represents the set of nine quantities (tensor components) which are transformed as a *product of two coordinates* when converting from one coordinate system to the other. The second definition of the tensor is based on the concepts of *linear vector-function* or *linear transformation of the vector.* Both are expressed analytically as the equalities (l is a dummy index, k is a free index)

$$b_k = T_{kl}a_l \qquad (k = 1, 2, 3) \tag{16}$$

where a_i and b_i are the coordinates of the vectors $\boldsymbol{a}$ and $\boldsymbol{b}$; T_{kl} are the coefficients of the linear vector-function or of linear transformation.

The second definition of the second-rank tensor states: if in the linear transformation (linear vector-function; Equation 16) to the physical vector $\boldsymbol{a}$, the physical vector $\boldsymbol{b}$ also corresponds, then the set of the transformation coefficients T_{kl} $(k, l = 1, 2, 3)$ represents a physically objective quantity, i.e., the second rank tensor.

[1] Sometimes "order" rather than "rank" is used in this Context, i.e., "second-order tensor".

In what follows, the rank of the tensor will not be referred to because as a rule only second-rank tensors will be encountered. The single exception is the Levi-Civitas third-order tensor (Equation 34). The tensors are denoted by capital italic letters (*P*, *Q*, *S*, etc.), and their components are designated by the same letter notation and sometimes also by lower-case symbols with two subscripts. Thus, a tensor that represents a set of linear transformation coefficients (Equation 16) can be denoted with the capital letter T and its components can be designated as T_{kl}.

To the tensor T there corresponds its *matrix* (*table*) in which the first subscript usually denotes the number of the row in the table, and the second subscript denotes the number of the column. A tensor whose matrix is formed by means of *transposing* (i.e., by interchanging subscripts) the matrix of the initial tensor T is called a *conjugated tensor* or the transpose of the tensor T and is designated by T^*. These two matrices are such as below:

$$T\begin{pmatrix} T_{11} & T_{12} & T_{13} \\ T_{21} & T_{22} & T_{23} \\ T_{31} & T_{32} & T_{33} \end{pmatrix} \qquad T^*\begin{pmatrix} T_{11} & T_{21} & T_{31} \\ T_{12} & T_{22} & T_{32} \\ T_{13} & T_{23} & T_{33} \end{pmatrix} \tag{17}$$

The conditions for the physical objectivity in analytical specification of a tensor are incorporated in both of its definitions. According to the first definition, in converting from one coordinate system to another the tensor components are required to be transformed as products of the coordinates. According to the second definition (Equation 16), the tensor components should be the coefficients of linear transformation of one physical vector into another physical vector. We shall demonstrate that these requirements follow from each other.

According to Equations 9 and 16:

$$b_s' = \alpha_{sk}b_k = \alpha_{sk}T_{kl}a_l = \alpha_{sk}T_{kl}\alpha_{pl}a_p' = \alpha_{sk}\alpha_{pl}T_{kl}a_p'$$

but to conform with the same equality (Equation 16) applied to a system with primed quantities, we will have

$$b_s' = T_{sp}'a_p'$$

Comparing the above with the previous expression will give

$$T_{sp}' = \alpha_{sk}\alpha_{pl}T_{kl}$$

i.e., Relation 15. Thus, starting with the second definition of the tensor T (Equation 16) and the conditions for the physical objectivity of vectors $\boldsymbol{a}$ and $\boldsymbol{b}$ (Equation 9), we have arrived at the first definition of the tensor in Equation 15, thereby proving equivalence between the two definitions.

The definition adopted for the tensor is *physical*, as attested to by the conditions of its physical objectivity. There is also the definition of the tensor as a *mathematical object* with prescribed properties relative either to the transformation of coordinates or to the expression of the linear vector-function. In the present book, tensors are considered only in *rectangular* Cartesian or curvilinear coordinate systems, thus making it possible to use only lower indices for denoting tensor components. Let us consider two typical examples for each of the definitions.

Using the projections a_k ($k = 1, 2, 3$) and b_l ($l = 1, 2, 3$) of two physical vectors $\boldsymbol{a}$ and $\boldsymbol{b}$, make up the set of nine quantities $d_{kl} = a_kb_l$. With the aid of Equation 7 one can easily show the physical objectivity of these products of vector projections taken in pairs. In fact, according to these formulas

$$d'_{kl} = a'_k b'_l = \alpha_{km} a_m \alpha_{ln} b_n = \alpha_{km} \alpha_{ln} a_m b_n = \alpha_{km} \alpha_{ln} d_{mn}$$

These equalities coincide with the conditions for the physical objectivity of tensor D prescribed analytically by its components $d_{mn} = a_m b_n$. The tensor D is called a *dyad* or a *multiplicative tensor* which is formed from two physical vectors $\boldsymbol{a}$ and $\boldsymbol{b}$; the formation of the dyad is represented as:

$$D = \boldsymbol{ab} \tag{18}$$

The tables (matrices) of the tensor D and of the conjugate to its tensor D^* are

$$D \begin{pmatrix} a_1 b_1 & a_1 b_2 & a_1 b_3 \\ a_2 b_1 & a_2 b_2 & a_2 b_3 \\ a_3 b_1 & a_3 b_2 & a_3 b_3 \end{pmatrix} \qquad D^* \begin{pmatrix} a_1 b_1 & a_2 b_1 & a_3 b_1 \\ a_1 b_2 & a_2 b_2 & a_3 b_2 \\ a_1 b_3 & a_2 b_3 & a_3 b_3 \end{pmatrix} \tag{19}$$

Sometimes the expression on the right side of Equation 18 is also called the *dyad product* of two vectors; note that the scalar product $\boldsymbol{a} \cdot \boldsymbol{b} = \boldsymbol{b} \cdot \boldsymbol{a}$ is a scalar, the vector product $\boldsymbol{a} \times \boldsymbol{b} = -\boldsymbol{b} \times \boldsymbol{a}$ is a vector, the dyad product $\boldsymbol{ab}$ is a tensor with

$$D^* = (\boldsymbol{ab})^* = \boldsymbol{ba} \tag{20}$$

The simplest examples of a dyad can be provided by *coordinate dyads* which are dyad products of single coordinate vectors, i.e., unit vectors $\boldsymbol{i}_k$, $\boldsymbol{i}_l$. Just as the vector $\boldsymbol{a}$ can be represented by the expansion $\boldsymbol{a} = a_k \boldsymbol{i}_k$, so the tensor T can be expanded into coordinate dyads $T = T_{kl}\, \boldsymbol{i}_k \boldsymbol{i}_l$ which can be easily verified by writing out the components of the both sides of this equality and noting, for example, that

$$\boldsymbol{i}_1 \boldsymbol{i}_2 \begin{pmatrix} 0 & 1 & 0 \\ 0 & 0 & 0 \\ 0 & 0 & 0 \end{pmatrix} \qquad T_{12} = (T_{kl}\, \boldsymbol{i}_k \boldsymbol{i}_l)_{12} = T_{12}$$

and the like. This is an example of the expansion of the dyad of the general form into coordinate dyads.

As the other example illustrates the second tensor definition, let us consider the tensor $\mathfrak{A}$ with a matrix composed of the cosines of angles between two different coordinate systems

$$\mathfrak{A} \begin{pmatrix} \alpha_{11} & \alpha_{12} & \alpha_{13} \\ \alpha_{21} & \alpha_{22} & \alpha_{23} \\ \alpha_{31} & \alpha_{32} & \alpha_{33} \end{pmatrix} \tag{21}$$

There is no need to prove the physical objectivity of the analytical assignment of the tensor $\mathfrak{A}$ using its components α_{kl} because the latter are the coefficients of the transformations (Equations 7 and 8) of the physical vector $\boldsymbol{r}$ into another (also physical) vector $\boldsymbol{r}'$, or of transformations (Equation 9) of any physical vector $\boldsymbol{a}$ into the physical vector $\boldsymbol{a}'$.

The system of coordinates $0x'_1 x'_2 x'_3$ can be considered a result of rotation of the system $0x_1 x_2 x_3$ such that the "old" and "new" axes coincide. In this case, any vector fixed in the system $0x_1x_2x_3$ will rotate with it in space. Let us show that the product of the tensor $\mathfrak{A}$ (Equation 21) by the arbitrary vector $\boldsymbol{a}$, fixed with respect to $0x_1x_2x_3$, expresses the rotation of this vector in space together with the system $0x_1x_2x_3$. According to Equation 21 (the definition of the product of the tensor by the vector is given further in Section 3), we have

$$(\mathfrak{A}\boldsymbol{a})_i = \mathfrak{A}_{ij}a_j = \alpha_{ij}a_j$$

or, according to the first equality of Equation 9,

$$(\mathfrak{A}\boldsymbol{a})_i = a_i'$$

This equality, the left side of which contains the ith projection of the vector $\boldsymbol{a}$ in the system $0x_1x_2x_3$ having been multiplied by $\mathfrak{A}$, and the right side of which involves an analogous projection a_i' of the initial vector $\boldsymbol{a}$ in the coordinate system $0x_1'x_2'x_3'$, indicates that multiplication of the vector $\boldsymbol{a}$ by the tensor $\mathfrak{A}$ yields the same vector $\boldsymbol{a}$, but it has been rotated in space as an entity with the coordinate system $0x_1x_2x_3$. This interpretation of the vector $\mathfrak{A}\boldsymbol{a}$ allows the physical tensor $\mathfrak{A}$ to be called the *rotation tensor.*

In Section 3, different tensors will be encountered, and their definition is connected in the majority of cases with the notion of linear vector-function (or linear transformation) of one physical vector one into the other, i.e., corresponding to the second means of defining the physical tensor.

3. Basic Operations of Tensor Algebra

The simplest operations with tensors are summation (subtraction) scalar multiplication, and vector multiplication. The summation (subtraction) and scalar multiplication are defined by the equalities

$$(P \pm Q \pm R \pm \cdots)_{mn} = P_{mn} \pm Q_{mn} \pm R_{mn} \pm \cdots$$
$$(\lambda P)_{mn} = (P\lambda)_{mn} = \lambda P_{mn} \qquad (22)$$

A somewhat more complex operation is that of multiplication of a tensor by a vector and the somewhat different (in the case of an arbitrary tensor) operation of multiplication of a vector by a tensor. These operations respectively define the vectors $\boldsymbol{Pa}$ and $\boldsymbol{aP}$ with the projections (n is a dummy index)

$$(\boldsymbol{Pa})_m = P_{mn}a_n \qquad (\boldsymbol{aP})_m = a_nP_{nm} \qquad (m = 1, 2, 3) \qquad (23)$$

Recalling the definition given in Section II for the linear vector function (linear transformation) (Equation 16) and comparing the right sides of equality (Equation 16) with the first system of equalities (Equation 23), we shall note the equivalence of the operation of tensor multiplication by the vector, $\boldsymbol{Pa}$, to the linear transformation of the physical vector $\boldsymbol{a}$ into another physical vector. It is evident that multiplication of the vector by the tensor, $\boldsymbol{aP}$, is equivalent to linear transformation of the physical vector $\boldsymbol{a}$ by tensor P^* according to Equation 16.

The operation of tensor by vector multiplication defined by the first equality of Equation 23 is also called the *multiplication of the tensor by the vector on the right,* and that defined by the second equality of Equation 23 is called the *multiplication of the tensor by the vector on the left.*

Note that regardless of the sequence of multiples, the dummy subscripts on the tensor and the vector are always adjacent. It is also informative to note the identities

$$Pa = aP^* \qquad P^*a = aP \tag{24}$$

A tensor whose components are independent of the order of subscripts, so that

$$S_{mn} = S_{nm} \qquad (m, n = 1, 2, 3) \tag{25}$$

is called *symmetrical.* In the matrix of Equation 29, the components, which are mirror images about the main diagonal from the left upper corner to the right lower corner, are identical. The symmetrical tensor has only six components which all differ from one another.

The simplest example of a symmetrical tensor is the so-called unit tensor E, the components of which represent the already familiar (see equalities in Equation 5) Kronecker delta symbols δ_{kl} and the matrix

$$E \begin{pmatrix} 1 & 0 & 0 \\ 0 & 1 & 0 \\ 0 & 0 & 1 \end{pmatrix} \tag{26}$$

in all coordinate systems.

In effect, it follows from the first technique of tensor definition (see Equation 15) that if the matrix represented in Equation 26 is expressed in the old coordinates, then in new coordinates the E components will take on the form

$$\delta'_{kl} = \alpha_{km}\alpha_{ln}\delta_{mn} = \alpha_{km}\alpha_{lm}$$

or, recalling Equation 12 (m is a dummy index),

$$\delta'_{kl} = \begin{cases} 1 & \text{when} \quad l = k \\ 0 & \text{when} \quad l \neq k \end{cases}$$

which coincides with Equation 5 ($\delta'_{kl} = \delta_{kl}$).

The unit tensor owes its name to the obvious relation

$$Ea = a \tag{27}$$

The tensor A, whose components undergo sign reversal on interchanging the subscripts, is called *antisymmetrical.*

We have

$$A^* = -A \qquad A_{mn} = -A_{nm} \qquad A_{mm} = 0 \quad (m, n = 1, 2, 3) \tag{28}$$

The antisymmetric tensor has only three components that differ in absolute value. Here are the tables of the symmetric and antisymmetric tensors:

$$S \begin{pmatrix} S_{11} & S_{12} = S_{21} & S_{13} = S_{31} \\ S_{21} = S_{12} & S_{22} & S_{23} = S_{32} \\ S_{31} = S_{13} & S_{32} = S_{23} & S_{33} \end{pmatrix} \tag{29}$$

$$A\begin{pmatrix} 0 & A_{12}=-A_{21} & A_{13}=-A_{31} \\ A_{21}=-A_{12} & 0 & A_{23}=-A_{32} \\ A_{31}=-A_{13} & A_{32}=-A_{23} & 0 \end{pmatrix} \tag{30}$$

By defining the tensor × vector multiplication operations we find for the symmetric tensor S and antisymmetric tensor A

$$S\boldsymbol{a} = \boldsymbol{a}S = S^{*}\boldsymbol{a} = \boldsymbol{a}S^{*}$$

$$A\boldsymbol{a} = \boldsymbol{a}A^{*} = -\boldsymbol{a}A \tag{31}$$

Let us consider in more detail the multiplication of the antisymmetric tensor A by the vector $\boldsymbol{a}$ on the right. It will be proved that the set of three components of an antisymmetric tensor, which are placed in a certain order, may yield the so-called A-accompanying vector $\boldsymbol{c}$ with the components

$$c_1 = A_{23} \qquad c_2 = A_{31} \qquad c_3 = A_{12} \tag{32}$$

Note that in this definition of the accompanying vector the values of the subscripts on the left and right sides of each of these equalities obey the rule of cyclical permutation ($1 \to 2 \to 3 \to 1 \cdots$).

"Accompanying" underscores that though the components of the vector $\boldsymbol{c}$ coincide with three of the nine components of the tensor A (including the zero values of A_{ii}), the equality of the vector $\boldsymbol{c}$ to the tensor A is, of course, impossible because these have different quantities.

Equalities 32 can be replaced by

$$c_r = \varepsilon_{rst} A_{st} \tag{33}$$

where ($r \to s \to t$, according to the cyclical permutation rule; $r, s, t = 1, 2, 3$; and no summation should be made over the indices s and t). We now introduce into consideration the Levi-Civitas symbols. When the symbols are taken together, they represent the third-order tensor

$$\varepsilon_{rst} = \boldsymbol{i}_r \cdot (\boldsymbol{i}_s \times \boldsymbol{i}_t) \tag{34}$$

where $\boldsymbol{i}_r, \boldsymbol{i}_s, \boldsymbol{i}_t$ are the unit vectors of rectangular coordinate axes. According to Equation 34 ε_{rst} is equal to zero if there are at least two identical indices: +1 if the indices form a cyclical permutation and −1 if cyclical permutation is violated. An example of the application of the Levi-Civitas symbols is provided by the general formula for the components of the vector product of two vectors:

$$(\boldsymbol{a} \times \boldsymbol{b})_r = \varepsilon_{rst} a_s b_t \tag{35}$$

which presupposes the summation over repeated "dummy" indices. We have

$$(\boldsymbol{a} \times \boldsymbol{b})_1 = \varepsilon_{1st} a_s b_t = \varepsilon_{123} a_2 b_3 + \varepsilon_{132} a_3 b_2 = a_2 b_3 - a_3 b_2 \cdots$$

Let us prove that three quantities c_k ($k = 1, 2, 3$), expressed in terms of the components of the physical antisymmetric tensor A by Equations 32 and 33, form the physical vector $\boldsymbol{c}$. In other words, let us prove that if the components A_{ln} ($l, n = 1, 2, 3$) satisfy the conditions of the physical objectivity of a tensor, then the quantities c_k ($k = 1, 2, 3$) satisfy the conditions of the physical objectivity of a vector (Equation 9).

According to Equation 15, the following equality holds in the primed system

$$A'_{23} = c'_1 = \alpha_{2r}\alpha_{3s}A_{rs} \tag{36}$$

or, in expanded form,

$$c'_1 = A'_{23} = (\alpha_{22}\alpha_{33} - \alpha_{23}\alpha_{32})A_{23} + (\alpha_{23}\alpha_{31} - \alpha_{21}\alpha_{33})A_{31} + (\alpha_{21}\alpha_{32} - \alpha_{22}\alpha_{31})A_{12} =$$

$$= (\alpha_{22}\alpha_{33} - \alpha_{23}\alpha_{32})c_1 + (\alpha_{23}\alpha_{31} - \alpha_{21}\alpha_{33})c_2 + (\alpha_{21}\alpha_{32} - \alpha_{22}\alpha_{31})c_3 \tag{37}$$

Squaring the determinant of the table of cosines (I) by the multiplication rule for determinants [2] and making use of Equation 12 we shall see that the squared determinant (det) is equal to unity

$$[\det(\alpha_{kl})]^2 = \det(\delta_{mn}) = \begin{vmatrix} 1 & 0 & 0 \\ 0 & 1 & 0 \\ 0 & 0 & 1 \end{vmatrix} = 1$$

and, consequently, the determinant is equal to

$$\det(\alpha_{kl}) = \pm 1 \tag{38}$$

where the plus corresponds to the case of *codirected* coordinate systems (both systems are right or are left handed), whereas the minus means *differently directed* systems (one is right handed and the other is left handed).

Assuming in Equation 11 that $l = 1, 2, 3$ at $m = 1$, we shall obtain a system of equalities (the conditions of orthogonality of coordinate systems):

$$\begin{aligned} \alpha_{11}\alpha_{11} + \alpha_{12}\alpha_{12} + \alpha_{13}\alpha_{13} &= 1 \\ \alpha_{21}\alpha_{11} + \alpha_{22}\alpha_{12} + \alpha_{23}\alpha_{13} &= 0 \\ \alpha_{31}\alpha_{11} + \alpha_{32}\alpha_{12} + \alpha_{33}\alpha_{13} &= 0 \end{aligned} \tag{39}$$

which can be considered to be a linear nonhomogeneous system of equations in the unknowns α_{11}, α_{12}, α_{13}. Noting that according to Equation 38 the determinant of this system is equal to unity (the minus sign is omitted because only codirected, *viz.*, right handed systems will be used in the followings) we will find its solution:

$$\begin{aligned} \alpha_{11} &= \alpha_{22}\alpha_{33} - \alpha_{23}\alpha_{32} \\ \alpha_{12} &= \alpha_{23}\alpha_{31} - \alpha_{21}\alpha_{33} \\ \alpha_{13} &= \alpha_{21}\alpha_{32} - \alpha_{22}\alpha_{31} \end{aligned} \tag{40}$$

On the right there are *minors* of the elements in the determinant of the table of cosines α_{11}, α_{12}, α_{13}. It can be shown analogously that any of the elements in the determinant of

---

[2] $\det(a_{ij}) \cdot \det(b_{ij}) = \det(c_{mn})$, $c_{mn} = a_{mk}b_{kn}$.

the table of cosines $\det(\alpha_{kl})$ of the angles between two coordinate axes is equal to its minor. Determinants with this property are called reciprocal.

Returning to equality (Equation 137), we shall rewrite it, according to Equation 40, in the form

$$c_1' = \alpha_{11}c_1 + \alpha_{12}c_2 + \alpha_{13}c_3 \tag{41}$$

Analogously, using the proven property of reciprocity for the determinant of the table of cosines, we shall express the rest of the projections c_2' and c_3' as

$$c_2' = \alpha_{21}c_1 + \alpha_{22}c_2 + \alpha_{23}c_3$$

$$c_3' = \alpha_{31}c_1 + \alpha_{32}c_2 + \alpha_{33}c_3 \tag{41a}$$

It is the set of equalities (Equations 41 and 41a) that evidently proves the physical objectivity of the vector $\boldsymbol{c}$ which accompanies the antisymmetric tensor A.

In order to show the main application of the vector $\boldsymbol{c}$ that accompanies the tensor A, let us compose the product of the tensor A by any vector $\boldsymbol{a}$, for example, on the right. According to Equation 32, we find

$$(A\boldsymbol{a})_1 = A_{11}a_1 + A_{12}a_2 + A_{13}a_3 = a_2c_3 - a_3c_2 = (\boldsymbol{a}\times\boldsymbol{c})_1$$

$$(A\boldsymbol{a})_2 = A_{21}a_1 + A_{22}a_2 + A_{23}a_3 = a_3c_1 - a_1c_3 = (\boldsymbol{a}\times\boldsymbol{c})_2$$

$$(A\boldsymbol{a})_3 = A_{31}a_1 + A_{32}a_2 + A_{33}a_3 = a_1c_2 - a_2c_1 = (\boldsymbol{a}\times\boldsymbol{c})_3$$

or using vector representation,

$$A\boldsymbol{a} = \boldsymbol{a}\times\boldsymbol{c} \tag{42}$$

This means that when the antisymmetric tensor A is multiplied by vector $\boldsymbol{a}$ on the right, the set of its components (Equation 32) behaves as the accompanying vector $\boldsymbol{c}$ in the sense that the result is the vector product of the vector $\boldsymbol{a}$ by the accompanying vector $\boldsymbol{c}$. This equivalence between the antisymmetric tensor and the accompanying vector in the operation of multiplication does not, of course, mean that the tensor A is equal to the vector $\boldsymbol{c}$.

Equation 42 can be supplemented with an analogous formula for the product of the antisymmetric tensor by the vector on the left

$$\boldsymbol{a}A = \boldsymbol{c}\times\boldsymbol{a} \tag{42a}$$

By introducing into consideration the conjugate vector $\boldsymbol{c}^* = -\boldsymbol{c}$ which accompanies the conjugate tensor $A^* = -A$, then Equation 32 will go over into the formulas

$$c_1^* = A_{23}^* = -A_{23} \qquad c_2^* = A_{31}^* = -A_{31} \qquad c_3^* = A_{12}^* = -A_{12}$$

with the preservation of the cyclical permutation of indices but reversal of sign in front of the tensor A.

The following formulas are valid

$$A^*\boldsymbol{a} = \boldsymbol{c}\times\boldsymbol{a} \qquad \boldsymbol{a}A^* = \boldsymbol{a}\times\boldsymbol{c} \tag{43}$$

The adopted operation of vector multiplication by tensor, $\boldsymbol{a}T$, or tensor multiplication by vector, $T\boldsymbol{a}$, does not exhaust the variety of possible operations of this kind. There are also the following operations, but in somewhat different notation than is contained in the present text:

1. The external product $\boldsymbol{a}T$ which leads to the third-order tensor represented by the expansion into coordinate "triads" $\boldsymbol{i}_k\,\boldsymbol{i}_l\,\boldsymbol{i}_m$ (further, the rule of summation over the indices from 1 to 3 that occur twice in monomials is retained):

$$\boldsymbol{a}T = \boldsymbol{i}_k \boldsymbol{i}_l \boldsymbol{i}_m a_k T_{lm} \tag{44}$$

2. The internal product $\boldsymbol{a} \cdot T$ which gives a vector whose expansion into single vectors $\boldsymbol{i}_m$ has the form

$$\boldsymbol{a} \cdot T = a_k T_{km} \boldsymbol{i}_m \tag{45}$$

which coincides with $\boldsymbol{a}T$;

3. The vector product $\boldsymbol{a} \times T$ which leads to the second-rank tensor with the following expansion into coordinate dyads:

$$\boldsymbol{a} \times T = (\boldsymbol{i}_k \times \boldsymbol{i}_l)\, \boldsymbol{i}_n a_k T_{lm} = \boldsymbol{i}_n \boldsymbol{i}_m \varepsilon_{kln} a_k T_{lm} \tag{46}$$

4. Expansion of a Tensor into Symmetric and Antisymmetric Parts. Invariants: Expansion of a Tensor into Spherical and Deviatoric Parts

Using the simplest operations with tensors given in the previous section, we shall show that the second-rank tensor P general form can be expanded into symmetric and antisymmetric parts. This follows directly from the identity

$$P \equiv \frac{1}{2}(P + P^*) + \frac{1}{2}(P - P^*) = S + A \tag{47}$$

where

$$S = \frac{1}{2}(P + P^*) = \frac{1}{2}(P^* + P) = S^* \tag{48}$$

represents the symmetric and

$$A = \frac{1}{2}(P - P^*) = -\frac{1}{2}(P^* - P) = -A^* \tag{49}$$

antisymmetric parts of the tensor P.

Let us write the matrices of the symmetric and antisymmetric parts of the tensor P:

$$S\begin{pmatrix} P_{11} & \frac{1}{2}(P_{12} + P_{21}) & \frac{1}{2}(P_{13} + P_{31}) \\ \frac{1}{2}(P_{21} + P_{12}) & P_{22} & \frac{1}{2}(P_{23} + P_{32}) \\ \frac{1}{2}(P_{31} + P_{13}) & \frac{1}{2}(P_{32} + P_{23}) & P_{33} \end{pmatrix}$$

$$A\begin{pmatrix} 0 & \frac{1}{2}(P_{12}-P_{21}) & \frac{1}{2}(P_{13}-P_{31}) \\ \frac{1}{2}(P_{21}-P_{12}) & 0 & \frac{1}{2}(P_{23}-P_{32}) \\ \frac{1}{2}(P_{31}-P_{13}) & \frac{1}{2}(P_{32}-P_{23}) & 0 \end{pmatrix}$$

The vectors $\boldsymbol{c}$ and $\boldsymbol{c}^*$ which accompany the antisymmetric part of the tensor P will have the components

$$c_r = \frac{1}{2}\varepsilon_{rst}(P_{st}-P_{ts}) \qquad c_r^* = \frac{1}{2}\varepsilon_{rst}(P_{ts}-P_{st})$$

or, in expanded form,

$$c_1 = \frac{1}{2}(P_{23}-P_{32}) \qquad c_2 = \frac{1}{2}(P_{31}-P_{13}) \qquad c_3 = \frac{1}{2}(P_{12}-P_{21})$$

$$c_1^* = \frac{1}{2}(P_{32}-P_{23}) \qquad c_2^* = \frac{1}{2}(P_{13}-P_{31}) \qquad c_3^* = \frac{1}{2}(P_{21}-P_{12})$$

If the tensor P represents the dyad $\boldsymbol{ab}$ (this case will be dealt with in the following), the projections of the accompanying vector $\boldsymbol{c}$ for the antisymmetric part in the expansion of the dyad $\boldsymbol{ab}$ will be equal to

$$c_r = \varepsilon_{rst}a_s b_t \quad (r = 1, 2, 3)$$

$$c_1 = \frac{1}{2}(a_2b_3-a_3b_2) \qquad c_2 = \frac{1}{2}(a_3b_1-a_1b_3) \qquad c_3 = \frac{1}{2}(a_1b_2-a_2b_1)$$

Hence, it follows that

$$\boldsymbol{c} = \frac{1}{2}(\boldsymbol{a}\times\boldsymbol{b}) \tag{50}$$

In other words, the accompanying vector of the antisymmetric part of the dyad is equal to half the vector product of the vectors that form the dyad.

Using the components of any tensor P, it is possible to compose three scalar expressions which will be invariant to the change of coordinates. They are called the *invariants of a tensor.*

The first invariant J_1 ("spur") represents the *sum of the diagonal elements of the tensor matrix*

$$J_1 = P_{11}+P_{22}+P_{33} = P_{mm} \tag{51}$$

The proof of the invariance is straightforward. In the new coordinate system we have

$$J_1' = P_{kk}' = \alpha_{kl}\alpha_{km}P_{lm} = \delta_{lm}P_{lm} = P_{ll}$$

Here the foregoing familiar equality is applied

$$\alpha_{kl}\alpha_{km} = \delta_{lm} = \begin{cases} 1 & \text{when} \quad m = l \\ 0 & \text{when} \quad m \neq l \end{cases}$$

The second invariant J_2 of the tensor P is the name given to a *scalar product of the tensor into itself*, conventionally denoted as a squared tensor and equal to the sum of squares of all tensor components

$$J_2 = P_{mn}P_{mn} = P^2 \tag{52}$$

The third invariant J_3 of the tensor is the determinant of its matrix

$$J_3 = \begin{vmatrix} P_{11} & P_{12} & P_{13} \\ P_{21} & P_{22} & P_{23} \\ P_{31} & P_{32} & P_{33} \end{vmatrix} = \det(P_{mn})$$

Using the first invariant, it is possible to compose the identity

$$P \equiv J_1 E + (P - J_1 E) = P^{(s)} + P^{(d)} \tag{53}$$

which is the expansion of the tensor P into the *spherical*, $P^{(s)} = J_1 E$, and *deviatoric*, $P^{(d)} = P - J_1 E$, parts. The matrices of these parts have the form

$$P^{(s)} = \begin{pmatrix} J_1 & 0 & 0 \\ 0 & J_1 & 0 \\ 0 & 0 & J_1 \end{pmatrix} \tag{54}$$

$$P^{(d)} = \begin{pmatrix} P_{11} - J_1 & P_{12} & P_{13} \\ P_{21} & P_{22} - J_1 & P_{23} \\ P_{31} & P_{32} & P_{33} - J_1 \end{pmatrix} \tag{55}$$

The tensor $P^{(s)}$ which possesses spherical symmetry (isotropy) is called a spherical tensor.

Of the two means of tensor expansion into the components given above, of greatest interest for subsequent discussion is the expansion of the tensor into symmetric and antisymmetric parts. The vector product by the antisymmetric part of the tensor on the right or on the left is expressed (recall Equations 42 and 42a) in terms of the vector product of this tensor into the vector $\boldsymbol{c}$ accompanying the antisymmetric tensor so that

$$\begin{aligned} Pa &= Sa + Aa = Sa + a \times c \\ aP &= aS + aA = Sa + c \times a \end{aligned} \tag{56}$$

5. Principal Axes and Principal Values of the Symmetric Tensor

Among all kinds of coordinate systems, there are those called the *principal axes* of the symmetric tensor, in which the vectors located along these axes preserve their directions in space on linear transformation corresponding to this tensor.

Let us denote the unit vector of some principal axis of the given symmetric tensor S by the symbol $\boldsymbol{e}$ and subject it to the condition of direction conservation after the transformation (λ is a scalar, E is a unit tensor). This condition can be written in the form of either of the two equivalent equalities

$$S\boldsymbol{e} = \lambda \boldsymbol{e} \qquad (S - \lambda E)\boldsymbol{e} = 0 \tag{57}$$

which when expanded gives the following system of equations in e_1, e_2, e_3:

$$\begin{aligned}(S_{11} - \lambda)\, e_1 + S_{12}e_2 + S_{13}e_3 &= 0\\ S_{12}e_1 + (S_{22} - \lambda)\, e_2 + S_{23}e_3 &= 0\\ S_{13}e_1 + S_{23}e_2 + (S_{33} - \lambda)\, e_3 &= 0\end{aligned} \tag{58}$$

This system with the unknowns e_1, e_2, e_3 is *homogeneous* and has a *trivial zero* solution which does not agree with the obvious condition

$$e_1^2 + e_2^2 + e_3^2 = 1 \tag{59}$$

For the existence of *nontrivial solutions* of the system of equations (Equation 58), it is known that the condition of the zero determinant of this system must be fulfilled:

$$\begin{vmatrix} S_{11} - \lambda & S_{12} & S_{13} \\ S_{12} & S_{22} - \lambda & S_{23} \\ S_{13} & S_{23} & S_{33} - \lambda \end{vmatrix} = 0 \tag{60}$$

Cubic Equation 60 is called *characteristic* and its roots are called the eigenvalues of the matrix S. Generally, it has three roots $\lambda^{(1)}$, $\lambda^{(2)}$, and $\lambda^{(3)}$ which are not equal. To each of these which there their values of unit vectors: $\boldsymbol{e}^{(1)}$, $\boldsymbol{e}^{(2)}$, $\boldsymbol{e}^{(3)}$ correspond.

Let us first prove that these vectors are mutually *orthogonal*. Resorting to Equation 57, we find the equalities

$$(S - \lambda^{(1)}E)\boldsymbol{e}^{(1)} = 0 \qquad (S - \lambda^{(2)}E)\boldsymbol{e}^{(2)} = 0 \qquad (S - \lambda^{(3)}E)\boldsymbol{e}^{(3)} = 0 \tag{61}$$

With scalar multiplication of both parts of the first two equalities, respectively, by $\boldsymbol{e}^{(2)}$ and $\boldsymbol{e}^{(1)}$, we find

$$\boldsymbol{e}^{(2)} \cdot (S\boldsymbol{e}^{(1)}) = \lambda^{(1)}\boldsymbol{e}^{(1)} \cdot \boldsymbol{e}^{(2)} \qquad \boldsymbol{e}^{(1)} \cdot (S\boldsymbol{e}^{(2)}) = \lambda^{(2)}\boldsymbol{e}^{(1)} \cdot \boldsymbol{e}^{(2)} \tag{62}$$

Noting that by symmetry of the tensor S

$$\boldsymbol{e}^{(2)} \cdot (S\boldsymbol{e}^{(1)}) = e_m^{(2)} S_{mn} e_n^{(1)} = e_m^{(1)} S_{mn} e_n^{(2)} = \boldsymbol{e}^{(1)} \cdot (S\boldsymbol{e}^{(2)})$$

and subtracting term-by-term the left and right sides of equalities in Equation 62, we obtain

$$0 = (\lambda^{(1)} - \lambda^{(2)})\boldsymbol{e}^{(1)} \cdot \boldsymbol{e}^{(2)}$$

From this, by virtue of the adopted condition of nonequal roots of the characteristic equation $\lambda^{(1)}$ and $\lambda^{(2)}$, the orthogonality of the vectors $\boldsymbol{e}^{(1)}$ and $\boldsymbol{e}^{(2)}$ follows. The orthogonality of

$e^{(2)}$ and $e^{(3)}$ is shown analogously, and thus the mutual orthogonality of all three vectors $e^{(1)}, e^{(2)}, e^{(3)}$.

We shall prove now that all the nonequal roots are real. For instance, let $\lambda^{(1)}$ and $\lambda^{(2)}$ be complex and, more to the point, *conjugate numbers*. Then, the projections $e_m^{(1)}$ and $e_m^{(2)}$ corresponding to them, where $m = 1, 2, 3$, will have the form of conjugate complex numbers

$$e_m^{(1)} = \mu_m + \nu_m i \qquad e_m^{(2)} = \mu_m - \nu_m i$$

and the sum of their products (summation over m)

$$e_m^{(1)} e_m^{(2)} = \mu_m \mu_m + \nu_m \nu_m$$

as the sum of squares, will be greater than zero; however, this sum is nothing more than the scalar product $e^{(1)} \cdot e^{(2)}$ which, as has just been proved, is equal to zero. Thus, the reality of all three roots $\lambda^{(1)}, \lambda^{(2)}, \lambda^{(3)}$ is proved.

Finally, we shall show that by adopting, as coordinate axes, the three principal axes corresponding to the unit vectors $e^{(1)}$, $e^{(2)}$, and $e^{(3)}$, the nondiagonal components of the symmetric tensor S will be equal to zero and the diagonal components will be *the roots of the characteristic equation* $\lambda^{(1)}, \lambda^{(2)}, \lambda^{(3)}$. These are the *principal values* of the symmetric tensor S.

It follows from equalities (57) (no summation over m) that

$$Se^{(m)} = \lambda^{(m)} e^{(m)}$$

$$(Se^{(m)}) \cdot e^{(n)} = \lambda^{(m)} e^{(m)} \cdot e^{(n)} = \lambda^{(m)} \delta_{mn}$$

where δ_{mn} are the components of the unit tensor. By priming the components of the symmetric tensor S in principal axes, we obtain, based on the last equality,

$$S'_{mn} = \lambda^{(m)} \delta_{mn} = \begin{cases} \lambda^{(m)} & \text{when} \quad n = m \\ 0 & \text{when} \quad n \neq m \end{cases} \tag{63}$$

which proves the above statement that the nondiagonal terms are equal to zero and diagonal terms are equal to principal values.

In principal axes the matrix of the symmetric tensor has the form

$$S' \begin{pmatrix} \lambda^{(1)} & 0 & 0 \\ 0 & \lambda^{(2)} & 0 \\ 0 & 0 & \lambda^{(2)} \end{pmatrix} \tag{64}$$

If the roots of the characteristic equation are repeated for example, $\lambda^{(1)} = \lambda^{(2)}$, only the third principal axis that corresponds to the unit vector $e^{(3)}$ will have a certain direction in space; the others will lie in the plane perpendicular to the third axis, and their location in space will be determined to within an *arbitrary rotation* around the third axis.

When all three roots are equal ($\lambda^{(1)} = \lambda^{(2)} = \lambda^{(3)} = \lambda$), the symmetric tensor S will have the property of *isotropy* of diagonal components, i.e., they will be independent of the directions of coordinate axes. As already mentioned, such tensors are called spherical. It is defined as the product of λ by the unit tensor, and its matrix has the form

$$\begin{pmatrix} \lambda & 0 & 0 \\ 0 & \lambda & 0 \\ 0 & 0 & \lambda \end{pmatrix} = \lambda \begin{pmatrix} 1 & 0 & 0 \\ 0 & 1 & 0 \\ 0 & 0 & 1 \end{pmatrix} = \lambda E \tag{65}$$

The first invariant of the symmetric tensor S in the principal axes is expressed as

$$J_1 = \lambda^{(1)} + \lambda^{(2)} + \lambda^{(3)} \tag{66}$$

the second as

$$J_2 = (\lambda^{(1)})^2 + (\lambda^{(2)})^2 + (\lambda^{(3)})^2 \tag{67}$$

and the third as

$$J_3 = \lambda^{(1)} \lambda^{(2)} \lambda^{(3)} \tag{68}$$

Along with very popular operations of summation (subtraction) of tensors, their multiplication by a scalar quantity or by a vector on the right or on the left, the *operation of multiplication of two tensors by each other* must be dealt with; though rare, we have separated these operations from the previous and shall give them at the end of this section. According to the kind of result obtained (scalar, vector, or tensor), these operations (and there are three of them) are defined and called:

1. Scalar product (contraction) of two tensors:

$$P \cdot Q = P_{ij} Q_{ij} \tag{69}$$

(i and j are "dummy" indices)

2. Vector product of two tensors:

$$(P \times Q)_i = \varepsilon_{imn} P_{ms} Q_{sn} = P_{js} Q_{sk} - P_{ks} Q_{sj} \tag{70}$$

Here, m, n, s are "dummy" indices; j and k follow i in the order of cyclical permutation $i \to j \to k \to i \to \cdots$

3. Tensor product of two tensors:

$$(PQ)_{ij} = P_{is} Q_{sj} \tag{71}$$

(s is a "dummy" index; i and $j = 1, 2, 3$).

Let us note some specific cases, but first the formula

$$P \cdot E = E \cdot P = P_{ii} = P_{11} + P_{22} + P_{33} \tag{72}$$

The right side is the *first invariant* of tensor P. By taking the scalar product of the tensor P by itself, we obtain

$$P \cdot P = P_{ij} P_{ij} = P^2 = |P|^2 \tag{73}$$

i.e., the squared tensor P. This is the *second invariant* of tensor P.

The second operation ($i \to j \to k \to \cdots$; summation over the subscript s) yields

$$(P \times E)_i = P_{js} E_{sk} - P_{ks} E_{sj} = P_{jk} - P_{kj} \tag{74}$$

We conclude from this equality that the condition for the symmetry of the tensor $P = S$ can be written in the following form, independent of coordinates:

$$P \times E = 0 \tag{75}$$

6. Derivative with Respect to the Prescribed Direction. Space Derivatives in Scalar, Vector, and Tensor Fields

Because a detailed coverage of the foundations of vector analysis is provided by a number of general textbooks, we shall give only the amount of information required. First, we shall define the *derivative with respect to the prescribed direction* and then the *general space derivative*.

Let us assign a certain axis l in space and introduce a rectangular Cartesian coordinate system. Let the orientation of the axis, and simultaneously, the positive direction on it be defined by the unit vector $\boldsymbol{l}(l_1, l_2, l_3)$ where l_1, l_2, l_3 are the direction cosines of the axis in the given coordinate system (Figure 2). Henceforth we shall have to deal often with the concept of the derivative of a scalar, vector, or tensor function. This concept can be put in an easily memorizable form by employing the symbolic vector-operator ∇ ("nabla")

$$\nabla\left(\frac{\partial}{\partial x_1}, \frac{\partial}{\partial x_2}, \frac{\partial}{\partial x_3}\right) = i_k \frac{\partial}{\partial x_k}$$

The derivative operation of the quantity Ψ with the direction along the axis l at the given point M will be determined as an ordinary derivative by the limit

$$\frac{d\Psi}{dl} = \lim_{\overline{MM'} \to 0} \frac{\Psi(M') - \Psi(M)}{\overline{MM'}} = \lim_{\Delta l \to 0} \frac{\Delta\Psi}{\Delta l}$$

where Ψ is any scalar, vector or tensor function for which this limit exists. The derivative $d\Psi/dl$ can be calculated by the formula of the derivative of a complex function:

$$\frac{d\Psi}{dl} = \frac{\partial\Psi}{\partial x_1}\frac{dx_1}{dl} + \frac{\partial\Psi}{\partial x_2}\frac{dx_2}{dl} + \frac{\partial\Psi}{\partial x_3}\frac{dx_3}{dl}$$

or, by taking into account that

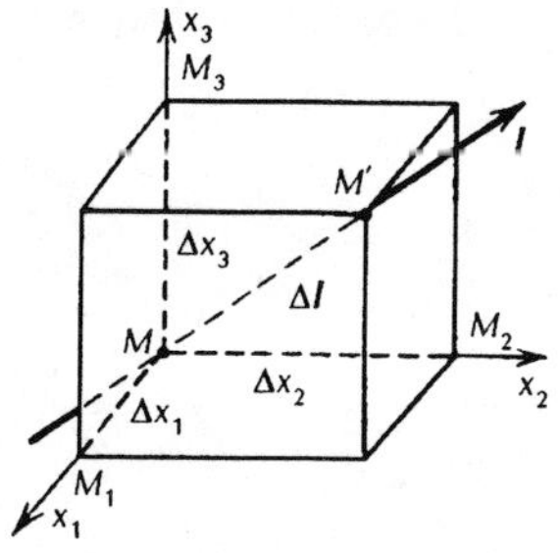

Fig. 2

$$\frac{dx_1}{dl} = \cos(\widehat{l, x_1}) = l_1 \qquad \frac{dx_2}{dl} = \cos(\widehat{l, x_2}) = l_2 \qquad \frac{dx_3}{dl} = \cos(\widehat{l, x_3}) = l_3$$

by the formula

$$\frac{d\Psi}{dl} = \frac{\partial \Psi}{\partial x_1} l_1 + \frac{\partial \Psi}{\partial x_2} l_2 + \frac{\partial \Psi}{\partial x_3} l_3$$

Because the form of the function Ψ is not needed to define derivative operation, we shall obtain, after omitting Ψ, the following symbolic formula

$$\frac{d}{dl} = \boldsymbol{l} \cdot \nabla = l_1 \frac{\partial}{\partial x_1} + l_2 \frac{\partial}{\partial x_2} + l_3 \frac{\partial}{\partial x_3} \tag{76}$$

The use of the derivative operation along the direction to the scalar φ, vector $\boldsymbol{a}$, and tensor T gives the following expressions:

$$\frac{d\varphi}{dl} = (\boldsymbol{l} \cdot \nabla)\,\varphi = \boldsymbol{l} \cdot \operatorname{grad} \varphi = l_i \frac{\partial \varphi}{\partial x_i}$$

$$\frac{d\boldsymbol{a}}{dl} = (\boldsymbol{l} \cdot \nabla)\,\boldsymbol{a} = l_i \frac{\partial \boldsymbol{a}}{\partial x_i} \qquad \frac{dT}{dl} = (\boldsymbol{l} \cdot \nabla)\,T = l_i \frac{\partial T}{\partial x_i} \tag{77}$$

An example can be provided by the derivative of the vector $\boldsymbol{b}$ along the direction of the vector $\boldsymbol{a}$ which is frequently met with

$$(\hat{\boldsymbol{a}} \cdot \nabla)\,\boldsymbol{b} = \frac{1}{a}(\boldsymbol{a} \cdot \nabla)\,\boldsymbol{b} \tag{78}$$

where $\hat{\boldsymbol{a}} = \boldsymbol{a}/a$ is the unit vector directed along the vector $\boldsymbol{a}$.

Among all kinds of directions in the field of the scalar quantity φ we note one characteristic direction – that of the *normal to the level surface* of the function φ (Figure 3)

$$\varphi(x_1, x_2, x_3) = C = \text{const}$$

and prescribe it by the unit vector $\boldsymbol{n}$, having directed it to the side of an increasing φ so that $d\varphi/dn > 0$. As is seen from Figure 3, for the same increment ΔC of the function φ on transition from one level surface ($\varphi = C$) to the other ($\varphi = C + \Delta C$, $\Delta C > 0$), different increments $\Delta l = \Delta n/\cos\alpha$ correspond to different directions l so that $\Delta l \geq \Delta n$ and

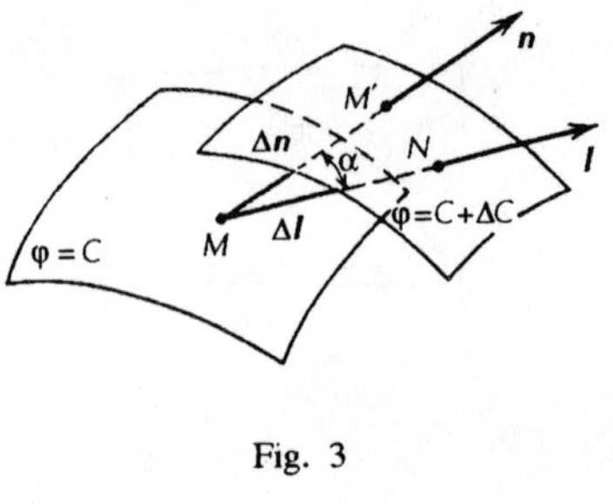

Fig. 3

$$\frac{d\varphi}{dl} = \frac{d\varphi}{dn}\frac{dn}{dl} = \frac{d\varphi}{dn}\cos\alpha \le \frac{d\varphi}{dn}$$

The gradient of the scalar function φ is defined as the vector

$$\text{grad}\,\varphi \equiv \nabla\varphi = \frac{d\varphi}{dn}\boldsymbol{n}$$

equal in magnitude to the derivative of the scalar function φ along the normal to the level surface and directed to the side of the greatest change of φ at the given point.

The general space derivative of scalar, vector, or tensor functions is defined as the limiting integral operation (the existence of the limit is postulated)

$$\lim_{\tau\to 0}\frac{1}{\tau}\int_{\sigma}\boldsymbol{n}\cdots\delta\sigma = \nabla\cdots \tag{79}$$

where τ is the volume enclosing the given point of the field, σ is the surface bounding the volume, $\boldsymbol{n}\,\delta\,\sigma$ is the oriented element of the surface σ ($\boldsymbol{n}$ is the unit vector of the outward normal to the surface σ) and the ellipses in Equation 79 stand for the place of the scalar, vector, or tensor with corresponding symbols to form the products of $\boldsymbol{n}$ being integrated.

These space derivatives are

$$\lim_{\tau\to 0}\frac{1}{\tau}\int_{\sigma}\boldsymbol{n}\varphi\delta\sigma = \text{grad}\,\varphi$$

$$\lim_{\tau\to 0}\frac{1}{\tau}\int_{\sigma}\boldsymbol{n}\cdot\boldsymbol{a}\;\delta\sigma = \text{div}\,\boldsymbol{a}$$

$$\lim_{\tau\to 0}\frac{1}{\tau}\int_{\sigma}\boldsymbol{n}\times\boldsymbol{a}\;\delta\sigma = \text{rot}\,\boldsymbol{a}$$

$$\lim_{\tau\to 0}\frac{1}{\tau}\int_{\sigma}\boldsymbol{n}\boldsymbol{a}\;\delta\sigma = \text{Grad}\,\boldsymbol{a}$$

$$\lim_{\tau\to 0}\frac{1}{\tau}\int_{\sigma}\boldsymbol{n}T\;\delta\sigma = \text{Div}\,T$$

$$\lim_{\tau\to 0}\frac{1}{\tau}\int_{\sigma}\boldsymbol{n}\times T\;\delta\sigma = \text{Rot}\,T \tag{80}$$

Using the surface of the coordinate rectangular parallelepiped as σ and applying the above-mentioned limiting transition, we obtain analytical expressions of the aforementioned space differentiation operations (summation over twice-repeated indices):

$$(\text{grad}\,\varphi)_i = \frac{\partial\varphi}{\partial x_i} \quad (i = 1, 2, 3) \qquad \text{div}\,\boldsymbol{a} = \frac{\partial a_i}{\partial x_i}$$

$$(\text{rot}\,\boldsymbol{a})_p = \varepsilon_{pqr}\frac{\partial a_r}{\partial x_q} = \frac{\partial a_r}{\partial x_q} - \frac{\partial a_q}{\partial x_r} \qquad (p \to q \to r \to p \cdots)$$

$$(\text{Grad}\,\boldsymbol{a})_{ij} = \frac{\partial a_j}{\partial x_i}$$

$$(\text{Div}\,T)_i = \frac{\partial T_{ji}}{\partial x_j}$$

$$(\text{Rot}\,T)_{nm} = \varepsilon_{kln}\frac{\partial T_{lm}}{\partial x_k} \tag{81}$$

(compare with Equation 46).

The same results can be arrived at by using symbolic expressions of operations with the vector ∇:[3]

$$\text{grad}\,\varphi = \nabla\varphi = \frac{\partial \varphi}{\partial x_k}\boldsymbol{i}_k$$

$$\text{div}\,\boldsymbol{a} = \nabla \cdot \boldsymbol{a} = \frac{\partial a_k}{\partial x_k}$$

$$\text{rot}\,\boldsymbol{a} = \nabla \times \boldsymbol{a} = \begin{vmatrix} \boldsymbol{i}_1 & \boldsymbol{i}_2 & \boldsymbol{i}_3 \\ \partial/\partial x_1 & \partial/\partial x_2 & \partial/\partial x_3 \\ a_1 & a_2 & a_3 \end{vmatrix}$$

$$\text{Grad}\,\boldsymbol{a} = \nabla\boldsymbol{a} = \boldsymbol{i}_k\boldsymbol{i}_l\frac{\partial a_l}{\partial x_k}$$

$$\text{Div}\,T = \nabla T = \frac{\partial T_{jk}}{\partial x_j}\boldsymbol{i}_k$$

$$\text{Rot}\,T = \nabla \times T = \boldsymbol{i}_n\boldsymbol{i}_m\varepsilon_{kln}\frac{\partial T_{lm}}{\partial x_k} \tag{82}$$

(compare with Equation 46).

Analogous expressions in the system of *orthogonal curvilinear* coordinates appear to be more complex. Let us introduce the so-called *Lamé coefficients* (q_i are curvilinear coordinates):

$$H_i = \sqrt{\left(\frac{\partial x_1}{\partial q_i}\right)^2 + \left(\frac{\partial x_2}{\partial q_i}\right)^2 + \left(\frac{\partial x_3}{\partial q_i}\right)^2} \tag{83}$$

where the specific form of the system of curvilinear coordinates q_1, q_2, q_3 is given by the equalities

[3]The term "curl ( )" is often used instead of "rot ( )".

$$x_k = x_k(q_1, q_2, q_3) \qquad (k = 1, 2, 3)$$

Note that each of the quantities H_i possesses the meaning of the proportionality factor in the equality, which couples the elementary increment ds_i of the segment length along the ith coordinate line with the increment of the corresponding curvilinear coordinate dq_i:

$$ds_i = H_i dq_i$$

(do not sum over i). Then, the analytical expressions of the space derivatives in curvilinear coordinates will have the following form (do not sum over repeated indices):

$$(\operatorname{grad}\varphi)_{q_i} = \frac{1}{H_i}\frac{\partial\varphi}{\partial q_i} \qquad (i = 1, 2, 3)$$

$$\operatorname{div}\boldsymbol{a} = \frac{1}{H_1H_2H_3}\left[\frac{\partial}{\partial q_1}(a_{q_1}H_2H_3) + \frac{\partial}{\partial q_2}(a_{q_2}H_3H_1) + \frac{\partial}{\partial q_3}(a_{q_3}H_1H_2)\right]$$

$$\operatorname{rot}_{q_1}\boldsymbol{a} = \frac{1}{H_2H_3}\left[\frac{\partial(a_{q_3}H_3)}{\partial q_2} - \frac{\partial(a_{q_2}H_2)}{\partial q_3}\right]$$

$$\operatorname{rot}_{q_2}\boldsymbol{a} = \frac{1}{H_3H_1}\left[\frac{\partial(a_{q_1}H_1)}{\partial q_3} - \frac{\partial(a_{q_3}H_3)}{\partial q_1}\right]$$

$$\operatorname{rot}_{q_3}\boldsymbol{a} = \frac{1}{H_1H_2}\left[\frac{\partial(a_{q_2}H_2)}{\partial q_1} - \frac{\partial(a_{q_1}H_1)}{\partial q_2}\right]$$

$$\nabla^2\varphi = \operatorname{div}\operatorname{grad}\varphi =$$

$$= \frac{1}{H_1H_2H_3}\left[\frac{\partial}{\partial q_1}\left(\frac{H_2H_3}{H_1}\frac{\partial\varphi}{\partial q_1}\right) + \frac{\partial}{\partial q_2}\left(\frac{H_3H_1}{H_2}\frac{\partial\varphi}{\partial q_2}\right) + \frac{\partial}{\partial q_3}\left(\frac{H_1H_2}{H_3}\frac{\partial\varphi}{\partial q_3}\right)\right]$$

$$(\operatorname{Div}T)_{q_1} = \frac{1}{H_1H_2H_3}\left[\frac{\partial}{\partial q_1}(H_2H_3T_{q_1q_1}) + \frac{\partial}{\partial q_2}(H_3H_1T_{q_2q_1}) + \frac{\partial}{\partial q_3}(H_1H_2T_{q_3q_1})\right] +$$

$$+ \frac{T_{q_1q_2}}{H_1H_2}\frac{\partial H_1}{\partial q_2} + \frac{T_{q_1q_3}}{H_1H_3}\frac{\partial H_1}{\partial q_3} - \frac{T_{q_2q_2}}{H_1H_2}\frac{\partial H_2}{\partial q_1} - \frac{T_{q_3q_3}}{H_1H_3}\frac{\partial H_3}{\partial q_1}$$

$$(\operatorname{Div}T)_{q_2} = \frac{1}{H_1H_2H_3}\left[\frac{\partial}{\partial q_1}(H_2H_3T_{q_1q_2}) + \frac{\partial}{\partial q_2}(H_3H_1T_{q_2q_2}) + \frac{\partial}{\partial q_3}(H_1H_2T_{q_3q_2})\right] +$$

$$+ \frac{T_{q_2q_1}}{H_1H_2}\frac{\partial H_2}{\partial q_1} + \frac{T_{q_2q_3}}{H_2H_3}\frac{\partial H_2}{\partial q_3} - \frac{T_{q_1q_1}}{H_1H_2}\frac{\partial H_1}{\partial q_2} - \frac{T_{q_3q_3}}{H_2H_3}\frac{\partial H_3}{\partial q_2}$$

$$(\operatorname{Div}T)_{q_3} = \frac{1}{H_1H_2H_3}\left[\frac{\partial}{\partial q_1}(H_2H_3T_{q_1q_3}) + \frac{\partial}{\partial q_2}(H_3H_1T_{q_2q_3}) + \frac{\partial}{\partial q_3}(H_1H_2T_{q_3q_3})\right] +$$

$$+ \frac{T_{q_3q_1}}{H_1H_3}\frac{\partial H_3}{\partial q_1} + \frac{T_{q_3q_2}}{H_2H_3}\frac{\partial H_3}{\partial q_2} - \frac{T_{q_1q_1}}{H_1H_3}\frac{\partial H_1}{\partial q_3} - \frac{T_{q_2q_2}}{H_2H_3}\frac{\partial H_2}{\partial q_3} \qquad (84)$$

For reference purposes give the formulas in the most popular systems of curvilinear coordinates: cylindrical and spherical. The cylindrical (including the *polar*) system r, ε, z:

$$x = r\cos\varepsilon \quad y = r\sin\varepsilon \quad z = z \quad r = \sqrt{x^2+y^2} \quad \mathrm{tg}\,\varepsilon = y/x$$

$$(0 \le r < \infty \quad 0 \le \varepsilon \le 2\pi \quad -\infty < z < \infty)$$

$$H_r = 1 \quad H_\varepsilon = r \quad H_z = 1$$

$$ds^2 = dr^2 + r^2 d\varepsilon^2 + dz^2$$

$$d\sigma_r = r\,d\varepsilon\,dz \quad d\sigma_\varepsilon = dr\,dz \quad d\sigma_z = r\,dr\,d\varepsilon \quad d\tau = r\,dr\,d\varepsilon\,dz$$

$$\mathrm{grad}_r\varphi = \frac{\partial\varphi}{\partial r} \quad \mathrm{grad}_\varepsilon\varphi = \frac{1}{r}\frac{\partial\varphi}{\partial\varepsilon} \quad \mathrm{grad}_z\varphi = \frac{\partial\varphi}{\partial z}$$

$$\mathrm{div}\,\boldsymbol{a} = \frac{1}{r}\frac{\partial(ra_r)}{\partial r} + \frac{1}{r}\frac{\partial a_\varepsilon}{\partial\varepsilon} + \frac{\partial a_z}{\partial z}$$

$$\mathrm{rot}_r\boldsymbol{a} = \frac{1}{r}\frac{\partial a_z}{\partial\varepsilon} - \frac{a_\varepsilon}{\partial z} \quad \mathrm{rot}_\varepsilon\boldsymbol{a} = \frac{\partial a_r}{\partial z} - \frac{\partial a_z}{\partial z} \quad \mathrm{rot}_z\boldsymbol{a} = \frac{1}{r}\left[\frac{\partial(ra_\varepsilon)}{\partial r} - \frac{\partial a_r}{\partial\varepsilon}\right]$$

$$\nabla^2\varphi = \mathrm{div\,grad}\,\varphi = \frac{1}{r}\frac{\partial\left(r\frac{\partial\varphi}{\partial r}\right)}{\partial r} + \frac{1}{r^2}\frac{\partial^2\varphi}{\partial\varepsilon^2} + \frac{\partial^2\varphi}{\partial z^2}$$

$$(\mathrm{Div}T)_r = \frac{\partial T_{rr}}{\partial r} + \frac{1}{r}\frac{\partial T_{\varepsilon r}}{\partial\varepsilon} + \frac{\partial T_{zr}}{\partial z} + \frac{T_{rr} - T_{\varepsilon\varepsilon}}{r}$$

$$(\mathrm{Div}T)_\varepsilon = \frac{\partial T_{r\varepsilon}}{\partial r} + \frac{1}{r}\frac{\partial T_{\varepsilon\varepsilon}}{\partial\varepsilon} + \frac{\partial T_{z\varepsilon}}{\partial z} + \frac{T_{r\varepsilon} + T_{\varepsilon r}}{r}$$

$$(\mathrm{Div}T)_z = \frac{\partial T_{rz}}{\partial r} + \frac{1}{r}\frac{\partial T_{\varepsilon z}}{\partial\varepsilon} + \frac{\partial T_{zz}}{\partial z} + \frac{\partial T_{rz}}{r} \qquad (85)$$

The spherical system R, θ, ε:

$$x = R\sin\theta\cos\varepsilon \quad y = R\sin\theta\sin\varepsilon \quad z = R\cos\theta$$

$$(0 \le R < \infty \quad 0 \le \theta \le \pi \quad 0 \le \varepsilon \le 2\pi)$$

$$H_R = 1 \quad H_\theta = R \quad H_\varepsilon = R\sin\theta$$

$$ds^2 = dR^2 + R^2 d\theta^2 + R^2\sin^2\theta\,d\varepsilon^2$$

$$d\sigma_R = R^2\sin\theta\,d\theta\,d\varepsilon \quad d\sigma_\theta = R\sin\theta\,dr\,d\varepsilon \quad d\sigma_\varepsilon = R\,dR\,d\theta$$

$$d\tau = R^2\sin\theta\,dR\,d\theta\,d\varepsilon$$

$$\mathrm{grad}_R\varphi = \frac{\partial\varphi}{\partial R} \quad \mathrm{grad}_\theta\varphi = \frac{1}{R}\frac{\partial\varphi}{\partial\theta} \quad \mathrm{grad}_\varepsilon\varphi = \frac{1}{R\sin\theta}\frac{\partial\varphi}{\partial\varepsilon}$$

$$\operatorname{div} \boldsymbol{a} = \frac{1}{R^2}\frac{\partial (R^2 a_R)}{\partial R} + \frac{1}{R\sin\theta}\frac{\partial (a_\theta \sin\theta)}{\partial\theta} + \frac{1}{R\sin\theta}\frac{\partial a_\varepsilon}{\partial\varepsilon}$$

$$\operatorname{rot}_R \boldsymbol{a} = \frac{1}{R\sin\theta}\left[\frac{\partial (a_\varepsilon \sin\theta)}{\partial\theta} - \frac{\partial a_\theta}{\partial\varepsilon}\right]$$

$$\operatorname{rot}_\theta \boldsymbol{a} = \frac{1}{R}\left[\frac{1}{\sin\theta}\frac{\partial a_R}{\partial\varepsilon} - \frac{\partial (R a_\varepsilon)}{\partial R}\right]$$

$$\operatorname{rot}_\varepsilon \boldsymbol{a} = \frac{1}{R}\left[\frac{\partial (R a_\theta)}{\partial R} - \frac{\partial a_R}{\partial\theta}\right]$$

$$\nabla^2\varphi = \frac{1}{R^2}\left[\frac{\partial\left(R^2\dfrac{\partial\varphi}{\partial R}\right)}{\partial R} + \frac{1}{\sin\theta}\frac{\partial\left(\sin\theta\dfrac{\partial\varphi}{\partial\theta}\right)}{\partial\theta} + \frac{1}{\sin^2\theta}\frac{\partial^2\varphi}{\partial\varepsilon^2}\right]$$

$$(\operatorname{Div} T)_R = \frac{\partial T_{RR}}{\partial R} + \frac{1}{R}\frac{\partial T_{\theta R}}{\partial\theta} + \frac{1}{R\sin\theta}\frac{\partial T_{\varepsilon R}}{\partial\varepsilon} + \frac{1}{R}(2T_{RR} + T_{\theta R}\operatorname{ctg}\theta - T_{\theta\theta} - T_{\varepsilon\varepsilon})$$

$$(\operatorname{Div} T)_\theta = \frac{\partial T_{R\theta}}{\partial R} + \frac{1}{R}\frac{\partial T_{\theta\theta}}{\partial\theta} + \frac{1}{R\sin\theta}\frac{\partial T_{\varepsilon\theta}}{\partial\varepsilon} + \frac{1}{R}[2T_{R\theta} + T_{\theta R} + (T_{\theta\theta} - T_{\varepsilon\varepsilon})\operatorname{ctg}\theta]$$

$$(\operatorname{Div} T)_\varepsilon = \frac{\partial T_{R\varepsilon}}{\partial R} + \frac{1}{R}\frac{\partial T_{\theta\varepsilon}}{\partial\theta} + \frac{1}{R\sin\theta}\frac{\partial T_{\varepsilon\varepsilon}}{\partial\varepsilon} + \frac{1}{R}[2T_{R\varepsilon} + T_{\varepsilon R} + (T_{\theta\varepsilon} + T_{\varepsilon\theta})\operatorname{ctg}\theta] \quad (86)$$

The symbolic operator ∇ was prescribed in the *rectilinear orthogonal* coordinate system by its projections $\partial/\partial x_i$ $(i = 1, 2, 3)$. Note that this operator also can be extended to any orthogonal curvilinear system. The general expressions of different space derivatives (Equation 80) of scalar, vector, and tensor functions in terms of the symbolic operator ∇ were given above and filled in for the case of the rectilinear orthogonal coordinate system. The use of the operator ∇ is expedient not only for the easy to remember formulae; it also can be used to derive expressions for the space derivatives of complex scalar, vector, and tensor functions. Recalling the ordinary rule for the differentiation of the product of two functions

$$(\varphi\psi)' = \varphi\psi' + \psi\varphi' \quad (87)$$

we shall extend it to scalar, vector, and tensor operations with the symbolic vector ∇. For this purpose, we shall compose, according to the vector algebra rules, different combinations of ∇ and differentiable quantities including only those combinations that involve the operator ∇ to the left of these quantities or which can be reduced to such a form, and then we will sum up these combinations.

By employing this technique, we shall find

$$\operatorname{grad}(\varphi\psi) = \nabla(\varphi\psi) = \varphi\nabla\psi + \psi\nabla\varphi = \varphi\operatorname{grad}\psi + \psi\operatorname{grad}\varphi$$

$$\operatorname{div}(\varphi\boldsymbol{a}) = \nabla\cdot(\varphi\boldsymbol{a}) = \varphi(\nabla\cdot\boldsymbol{a}) + \boldsymbol{a}\cdot\nabla\varphi = \varphi\operatorname{div}\boldsymbol{a} + \boldsymbol{a}\cdot\operatorname{grad}\varphi$$

$$\operatorname{rot}(\varphi\boldsymbol{a}) = \nabla\times(\varphi\boldsymbol{a}) = \varphi(\nabla\times\boldsymbol{a}) + (\nabla\varphi)\cdot\boldsymbol{a} = \varphi\operatorname{rot}\boldsymbol{a} + \operatorname{grad}\varphi\times\boldsymbol{a}$$

$$\operatorname{div}(\boldsymbol{a}\times\boldsymbol{b}) = \nabla\cdot(\boldsymbol{a}\times\boldsymbol{b}) = \boldsymbol{b}\cdot(\nabla\times\boldsymbol{a}) - \boldsymbol{a}\cdot(\nabla\times\boldsymbol{b}) = \boldsymbol{b}\cdot\operatorname{rot}\boldsymbol{a} - \boldsymbol{a}\cdot\operatorname{rot}\boldsymbol{b}$$

$$\operatorname{rot}(\boldsymbol{a}\times\boldsymbol{b}) = \nabla\times(\boldsymbol{a}\times\boldsymbol{b}) = (\boldsymbol{b}\cdot\nabla)\boldsymbol{a} - (\boldsymbol{a}\cdot\nabla)\boldsymbol{b} + \boldsymbol{a}(\nabla\cdot\boldsymbol{b}) - \boldsymbol{b}(\nabla\cdot\boldsymbol{a}) =$$

$$= (\boldsymbol{b}\cdot\nabla)\boldsymbol{a} - (\boldsymbol{a}\cdot\nabla)\boldsymbol{b} + \boldsymbol{a}\operatorname{div}\boldsymbol{b} - \boldsymbol{b}\operatorname{div}\boldsymbol{a}$$

$$\operatorname{grad}(\boldsymbol{a}\cdot\boldsymbol{b}) = (\boldsymbol{a}\cdot\nabla)\boldsymbol{b} + (\boldsymbol{b}\cdot\nabla)\boldsymbol{a} + \boldsymbol{a}\times(\nabla\times\boldsymbol{b}) + \boldsymbol{b}\times(\nabla\times\boldsymbol{a}) =$$

$$= (\boldsymbol{a}\cdot\nabla)\boldsymbol{b} + (\boldsymbol{b}\cdot\nabla)\boldsymbol{a} + \boldsymbol{a}\times\operatorname{rot}\boldsymbol{b} + \boldsymbol{b}\times\operatorname{rot}\boldsymbol{a}$$

$$\operatorname{grad}(\boldsymbol{a}^2) = 2(\boldsymbol{a}\cdot\nabla)\boldsymbol{a} + 2\boldsymbol{a}\times\operatorname{rot}\boldsymbol{a}$$

$$\operatorname{div}\operatorname{grad}\varphi = \nabla^2\varphi \quad (\text{Laplacian } \varphi)$$

$$\operatorname{rot}\operatorname{grad}\varphi \equiv 0 \qquad \operatorname{div}\operatorname{rot}\boldsymbol{a} \equiv 0$$

$$\operatorname{grad}\operatorname{div}\boldsymbol{a} = \nabla(\nabla\cdot\boldsymbol{a}) = \nabla\times(\nabla\times\boldsymbol{a}) + \nabla^2\boldsymbol{a} = \operatorname{rot}\operatorname{rot}\boldsymbol{a} + \nabla^2\boldsymbol{a} \tag{88}$$

It is not advisable to present analytical coordinate expressions of Equation 88, because the expressions are cumbersome and will not be needed later.

7. A Measure of the Vector Field Nonuniformity – A Differential Dyad and Its Components: Deformation and Rotation of the Vector Field

The differential dyad $D = \nabla\boldsymbol{a}$, which was introduced in Section 6, with the components $D_{ij} = \nabla_i a_j = \partial a_j/\partial x_i$ and the matrix table

$$D\begin{pmatrix} \frac{\partial a_1}{\partial x_1} & \frac{\partial a_2}{\partial x_1} & \frac{\partial a_3}{\partial x_1} \\ \frac{\partial a_1}{\partial x_2} & \frac{\partial a_2}{\partial x_2} & \frac{\partial a_3}{\partial x_2} \\ \frac{\partial a_1}{\partial x_3} & \frac{\partial a_2}{\partial x_3} & \frac{\partial a_3}{\partial x_3} \end{pmatrix} \tag{89}$$

may serve as a *measure of the vector field nonuniformity.* Only its equality to zero at all points indicates the *vector field uniformity.*

The conjugate dyad D^* with the components $D^*_{ij} = D_{ji}$ has the matrix table

$$D^*\begin{pmatrix} \frac{\partial a_1}{\partial x_1} & \frac{\partial a_1}{\partial x_2} & \frac{\partial a_1}{\partial x_3} \\ \frac{\partial a_2}{\partial x_1} & \frac{\partial a_2}{\partial x_2} & \frac{\partial a_2}{\partial x_3} \\ \frac{\partial a_3}{\partial x_1} & \frac{\partial a_3}{\partial x_2} & \frac{\partial a_3}{\partial x_3} \end{pmatrix} \tag{90}$$

A simple, useful example can be provided by a *differential dyad of vector-radii* $\boldsymbol{r}(x_1, x_2, x_3)$ *of the points in the field* $\nabla \boldsymbol{r}$ with the matrix table

$$\begin{pmatrix} \frac{\partial x_1}{\partial x_1} & \frac{\partial x_2}{\partial x_1} & \frac{\partial x_3}{\partial x_1} \\ \frac{\partial x_1}{\partial x_2} & \frac{\partial x_2}{\partial x_2} & \frac{\partial x_3}{\partial x_2} \\ \frac{\partial x_1}{\partial x_3} & \frac{\partial x_2}{\partial x_3} & \frac{\partial x_3}{\partial x_3} \end{pmatrix} = \begin{pmatrix} 1 & 0 & 0 \\ 0 & 1 & 0 \\ 0 & 0 & 1 \end{pmatrix}$$

which expresses the validity of the equality

$$\nabla \boldsymbol{r} = E \tag{91}$$

The condition of symmetry (Equation 75) of the tensor P is now reduced to the form

$$P \times \nabla \boldsymbol{r} = 0 \tag{92}$$

The tensor D, the columns of which contain the elements of the *gradients of projections* of the vector $\boldsymbol{a}$, is designated as

$$D = \nabla \boldsymbol{a} = \text{Grad}\, \boldsymbol{a} \tag{93}$$

(see also Equations 80 and 89), where the "vector gradient" operator, in contrast to the scalar function gradient, is capitalized.

The expression for the *total differential* of the vector $\boldsymbol{a}$ can be represented by the equalities

$$(d\boldsymbol{a})_i = \frac{\partial a_i}{\partial x_j} dx_j = D_{ji} dx_j = D^*_{ij} dx_j = (D^* d\boldsymbol{r})_i$$

or, in the form independent of coordinates, by the equality

$$d\boldsymbol{a} = D^* d\boldsymbol{r} \tag{94}$$

Let us introduce the operation "the derivative of the vector $\boldsymbol{a}$ with respect to the vector $\boldsymbol{b}$" and designate it by the symbol $d\boldsymbol{a}/d\boldsymbol{b}$; then

$$D^* = \frac{d\boldsymbol{a}}{d\boldsymbol{r}} \tag{95}$$

The measure for the nonuniformity of the vector field of general form, tensor D can be split into two components: *deformation* and *rotation* of the field. For this purpose, we shall avail ourselves of the earlier tensor separation into the symmetric and antisymmetric parts:

$$D \equiv \frac{1}{2}(D + D^*) + \frac{1}{2}(D - D^*) = S + A$$

$$D^* \equiv \frac{1}{2}(D^* + D) + \frac{1}{2}(D^* - D) = S + A^* \tag{96}$$

The symmetric part S equal to

$$S = \frac{1}{2}\left(\operatorname{Grad}\boldsymbol{a} + \frac{d\boldsymbol{a}}{d\boldsymbol{r}}\right) = \operatorname{def}\boldsymbol{a} \tag{97}$$

will be called the *measure of deformation*, or more concisely, the *deformation of the vector field*. In a rectilinear orthogonal Cartesian coordinate system its components are

$$(\operatorname{def}\boldsymbol{a})_{ij} = \frac{1}{2}\left(\frac{\partial a_j}{\partial x_i} + \frac{\partial a_i}{\partial x_j}\right) \quad (i, j = 1, 2, 3) \tag{98}$$

The components of the tensor def $\boldsymbol{a}$ in orthogonal curvilinear coordinates are

$$(\operatorname{def}\boldsymbol{a})_{q_r q_s} =$$

$$= \begin{cases} \dfrac{1}{2}\left(\dfrac{1}{H_r}\dfrac{\partial a_{q_s}}{\partial q_r} + \dfrac{1}{H_s}\dfrac{\partial a_{q_r}}{\partial q_s} - \dfrac{a_{q_r}}{H_r H_s}\dfrac{\partial H_r}{\partial q_s} - \dfrac{a_{q_s}}{H_r H_s}\dfrac{\partial H_s}{\partial q_r}\right) & \text{under the condition } r \neq s \\ \dfrac{1}{H_s}\dfrac{\partial a_{q_s}}{\partial q_s} - \dfrac{a_{q_s}}{H_s^2}\dfrac{\partial H_s}{\partial q_s} + \dfrac{a_{q_1}}{H_1 H_s}\dfrac{\partial H_s}{\partial q_1} + \dfrac{a_{q_2}}{H_2 H_s}\dfrac{\partial H_s}{\partial q_2} + \dfrac{a_{q_3}}{H_3 H_s}\dfrac{\partial H_s}{\partial q_3} & \text{under the condition } r = s \end{cases}$$

$$(\operatorname{def}\boldsymbol{a})_{q_1 q_1} = \frac{1}{H_1}\frac{\partial a_{q_1}}{\partial q_1} + \frac{a_{q_2}}{H_2 H_1}\frac{\partial H_1}{\partial q_2} + \frac{a_{q_3}}{H_3 H_1}\frac{\partial H_1}{\partial q_3}$$

$$(\operatorname{def}\boldsymbol{a})_{q_1 q_2} = \frac{1}{2}\left(\frac{1}{H_1}\frac{\partial a_{q_2}}{\partial q_1} + \frac{1}{H_2}\frac{\partial a_{q_1}}{\partial q_2} - \frac{a_{q_1}}{H_1 H_2}\frac{\partial H_1}{\partial q_2} - \frac{a_{q_2}}{H_1 H_2}\frac{\partial H_2}{\partial q_1}\right)$$

$$(\operatorname{def}\boldsymbol{a})_{q_1 q_3} = \frac{1}{2}\left(\frac{1}{H_1}\frac{\partial a_{q_3}}{\partial q_1} + \frac{1}{H_3}\frac{\partial a_{q_1}}{\partial q_3} - \frac{a_{q_1}}{H_1 H_3}\frac{\partial H_1}{\partial q_3} - \frac{a_{q_3}}{H_1 H_3}\frac{\partial H_3}{\partial q_1}\right)$$

$$(\operatorname{def}\boldsymbol{a})_{q_2 q_2} = \frac{1}{H_2}\frac{\partial a_{q_2}}{\partial q_2} + \frac{a_{q_1}}{H_1 H_2}\frac{\partial H_2}{\partial q_1} + \frac{a_{q_3}}{H_2 H_3}\frac{\partial H_2}{\partial q_3}$$

$$(\operatorname{def}\boldsymbol{a})_{q_2 q_3} = \frac{1}{2}\left(\frac{1}{H_2}\frac{\partial a_{q_3}}{\partial q_2} + \frac{1}{H_3}\frac{\partial a_{q_2}}{\partial q_3} - \frac{a_{q_2}}{H_2 H_3}\frac{\partial H_2}{\partial q_3} - \frac{a_{q_3}}{H_2 H_3}\frac{\partial H_3}{\partial q_2}\right)$$

$$(\operatorname{def}\boldsymbol{a})_{q_3 q_3} = \frac{1}{H_3}\frac{\partial a_{q_3}}{\partial q_3} + \frac{a_{q_1}}{H_1 H_3}\frac{\partial H_3}{\partial q_1} + \frac{a_{q_2}}{H_2 H_3}\frac{\partial H_3}{\partial q_2}$$

$$(\operatorname{def}\boldsymbol{a})_{q_2 q_1} = (\operatorname{def}\boldsymbol{a})_{q_1 q_2} \qquad (\operatorname{def}\boldsymbol{a})_{q_3 q_1} = (\operatorname{def}\boldsymbol{a})_{q_1 q_3}$$

$$(\operatorname{def}\boldsymbol{a})_{q_3 q_2} = (\operatorname{def}\boldsymbol{a})_{q_2 q_3} \tag{99}$$

Thus, in the *cylindrical coordinate system* r, ε, z ($H_r = 1$, $H_\varepsilon = r$, $H_z = 1$) we have

$$(\mathrm{def}\,\boldsymbol{a})_{rr} = \frac{\partial a_r}{\partial r} \qquad (\mathrm{def}\,\boldsymbol{a})_{\varepsilon\varepsilon} = \frac{1}{r}\frac{\partial a_\varepsilon}{\partial \varepsilon} + \frac{a_r}{r} \qquad (\mathrm{def}\,\boldsymbol{a})_{zz} = \frac{\partial a_z}{\partial z}$$

$$(\mathrm{def}\,\boldsymbol{a})_{\varepsilon z} = \frac{1}{2}\left(\frac{1}{r}\frac{\partial a_z}{\partial \varepsilon} + \frac{\partial a_\varepsilon}{\partial z}\right) \qquad (\mathrm{def}\,\boldsymbol{a})_{rz} = \frac{1}{2}\left(\frac{\partial a_r}{\partial z} + \frac{\partial a_z}{\partial r}\right)$$

$$(\mathrm{def}\,\boldsymbol{a})_{r\varepsilon} = \frac{1}{2}\left(\frac{\partial a_\varepsilon}{\partial r} - \frac{a_\varepsilon}{r} + \frac{1}{r}\frac{\partial a_r}{\partial \varepsilon}\right) \tag{100}$$

In the spherical system R, θ, ε ($H_R = 1$, $H_\theta = R$, $H_\varepsilon = R\sin\theta$)

$$(\mathrm{def}\,\boldsymbol{a})_{RR} = \frac{\partial a_R}{\partial R} \qquad (\mathrm{def}\,\boldsymbol{a})_{\theta\theta} = \frac{1}{R}\frac{\partial a_\theta}{\partial \theta} + \frac{a_R}{R}$$

$$(\mathrm{def}\,\boldsymbol{a})_{\varepsilon\varepsilon} = \frac{1}{R\sin\theta}\frac{\partial a_\varepsilon}{\partial \varepsilon} + \frac{a_\theta}{R}\,\mathrm{ctg}\,\theta + \frac{a_R}{R}$$

$$(\mathrm{def}\,\boldsymbol{a})_{\theta\varepsilon} = (\mathrm{def}\,\boldsymbol{a})_{\varepsilon\theta} = \frac{1}{2R}\left(\frac{\partial a_\varepsilon}{\partial \theta} - a_3\,\mathrm{ctg}\,\theta\right) + \frac{1}{2R\sin\theta}\frac{\partial a_\theta}{\partial \varepsilon}$$

$$(\mathrm{def}\,\boldsymbol{a})_{R\theta} = (\mathrm{def}\,\boldsymbol{a})_{\theta R} = \frac{1}{2}\left(\frac{\partial a_\theta}{\partial R} - \frac{a_\theta}{R} + \frac{1}{R}\frac{\partial a_R}{\partial \theta}\right)$$

$$(\mathrm{def}\,\boldsymbol{a})_{\varepsilon R} = (\mathrm{def}\,\boldsymbol{a})_{R\varepsilon} = \frac{1}{2}\left(\frac{1}{R\sin\theta}\frac{\partial a_R}{\partial \varepsilon} + \frac{\partial a_\varepsilon}{\partial R} - \frac{a_\varepsilon}{R}\right) \tag{101}$$

The antisymmetrical part

$$A = \frac{1}{2}\left(\mathrm{Grad}\,\boldsymbol{a} - \frac{d\boldsymbol{a}}{d\boldsymbol{r}}\right) \tag{102}$$

with the components in the rectangular Cartesian system

$$A_{ij} = \frac{1}{2}\left(\frac{\partial a_j}{\partial x_i} - \frac{\partial a_i}{\partial x_j}\right) \tag{103}$$

which for the "accompanying" vector has the vector

$$c_k = A_{ij} = \frac{1}{2}\left(\frac{\partial a_j}{\partial x_i} - \frac{\partial a_i}{\partial x_j}\right) = \frac{1}{2}\mathrm{rot}_k\boldsymbol{a} \qquad (k = 1, 2, 3;\; i \to j \to k \to \cdots) \tag{104}$$

i.e., the vector

$$\boldsymbol{c} = \frac{1}{2}\,\mathrm{rot}\,\boldsymbol{a} \tag{105}$$

is called the rotation of the vector field.[4]

---

[4]The origin of the terms *deformation* and *rotation* is associated with their application to the fields of *displacements* or *velocities* of the points of a continuous medium (see Sections 12 and 13 in Chapter 2).

According to Equation 56:

$$\boldsymbol{b}D = D^{*}\boldsymbol{b} = \boldsymbol{b}\,\mathrm{def}\,\boldsymbol{a} + \frac{1}{2}\,\mathrm{rot}\,\boldsymbol{a} \times \boldsymbol{b} \tag{106}$$

8. Basic Integral Formulas of the Field: The Gauss-Ostrogradskiy and Stokes, Theorems

Using the definition (Equation 79) of the space derivative, it is possible to derive integral formulas of the field that are of fundamental importance for fluid mechanics.

Rewrite the limiting definitions of the space derivatives (Equation 80) in an approximate form which is accurate to the small fourth-order quantities with respect to the characteristic linear dimension h of the elementary region. The derivation will then be considered using as an example the first of the formulas (Equation 80). For the small volume $\delta\tau$, introduce

$$\int_{\sigma} \boldsymbol{n}\varphi\,\delta\sigma \approx \mathrm{grad}\,\varphi\,\delta\tau + O(h^4) \tag{107}$$

Sum both sides of this equality over all the elementary volumes $\delta\tau$ that constitute the final volume τ. Note that on the common boundary $\delta\sigma$ between any two adjacent elementary volumes, the unit vectors $\boldsymbol{n}$ will have opposing directions for each of these volumes, and this, with the scalar function φ being continuous, will bring about cancellation of the elements of the surface integral in pairs on the left side of equality (Equation 107). This cancellation will probably affect only the surfaces of volume elements not on the outer boundary of the surface σ which bounds the volume τ.

The sum of such noncompensated elements leads to the integral $\int_{\sigma} \boldsymbol{n}\varphi\,\delta\sigma$ for the surface of the final volume.

Summation of the first terms on the right side yields the integral $\int_{\tau} \mathrm{grad}\,\varphi\,\delta\tau$. Thus, we shall arrive at the exact formula

$$\int_{\sigma} \boldsymbol{n}\varphi\,\delta\sigma = \int_{\tau} \mathrm{grad}\,\varphi\,\delta\tau \tag{108}$$

since the summation over the volume of small fourth-order quantities will accumulate the first-order error, which vanishes in the limit as $h \to 0$.

Analogously, according to Equation 80 we shall obtain a system of integral formulas

$$\int_{\sigma} \boldsymbol{n} \cdot \boldsymbol{a}\,\delta\sigma = \int_{\tau} \mathrm{div}\,\boldsymbol{a}\,\delta\tau$$

$$\int_{\sigma} \boldsymbol{n} \times \boldsymbol{a}\,\delta\sigma = \int_{\tau} \mathrm{rot}\,\boldsymbol{a}\,\delta\tau$$

$$\int_{\sigma} \boldsymbol{n}T\,\delta\sigma = \int_{\tau} \mathrm{Div}\,T\,\delta\tau \tag{109}$$

the set of which can be presented in a symbolic form convenient for memorization:

$$\int_\sigma n \cdots \delta\sigma = \int_\tau \nabla \cdots \delta\tau \tag{110}$$

Equation 110 allows the conversion from surface to volume integral via the replacement of the normal $\boldsymbol{n}$ in the first integral by ∇ in the second. The ellipses on both sides stand for the integrable quantity with the pertinent multiplication sign.

The first formula in Equation 109, which in expanded form (the summation over i) can be also written as

$$\int_\sigma \boldsymbol{n} \cdot \boldsymbol{a}\, \delta\sigma = \int_\sigma a_n \delta\sigma = \int_\sigma (a_i n_i)\, \delta\sigma = \int_\tau \operatorname{div} \boldsymbol{a}\, \delta\tau = \int_\tau \frac{\partial a_i}{\partial x_i} \delta\tau \tag{111}$$

expresses the Gauss-Ostrogradskiy theorem.

Let us introduce the concept of the flux of the vector $\boldsymbol{a}$ through the surface σ:

$$F_\sigma(\boldsymbol{a}) = \int_\sigma a_n \delta\sigma = \int_\sigma \boldsymbol{a} \cdot \boldsymbol{n}\, \delta\sigma \tag{112}$$

where $a_n = \boldsymbol{a} \cdot \boldsymbol{n}$ is the projection of the vector $\boldsymbol{a}$ onto the normal $\boldsymbol{n}$ to the surface σ at the given point on it. The definition (Equation 112) by itself has meaning for both closed and open surfaces, with the sign of $F_\sigma(\boldsymbol{a})$ obviously dependent on the choice of the direction for the normal $\boldsymbol{n}$ to one side or the other. Equation 111 contains the closed surface and the outward normal. Thus, the Gauss-Ostrogradskiy theorem can be formulated as follows: the vector flux through a closed surface which bounds a certain volume is equal to this volume integral of the vector divergence.

To prove another important theorem Stokes theorem, we will need the concept of the circulation of the vector $\boldsymbol{a}$ along the arbitrary contour C (Figure 4). This circulation is defined by the curvilinear integral along the contour C

$$\Gamma_C(\boldsymbol{a}) = \int_{A\,(C)}^{B} \boldsymbol{a} \cdot \delta\boldsymbol{r} = \int_{A\,(C)}^{B} a_i \delta x_i \tag{113}$$

In the case of the closed ($B \to A$) contour C, which does not intersect itself, the circulation is designated by the symbol

$$\Gamma_C(\boldsymbol{a}) = \oint_{(C)} \boldsymbol{a} \cdot \delta\boldsymbol{r} = \oint_{(C)} a_i \delta x_i \tag{114}$$

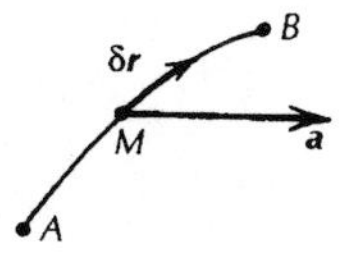

Fig. 4

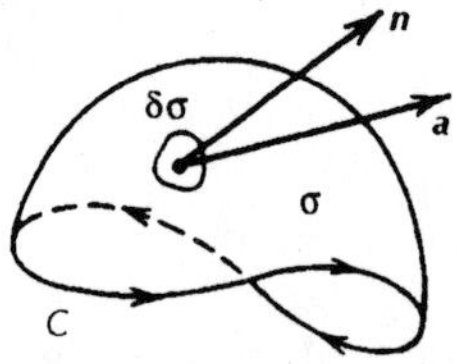

Fig. 5

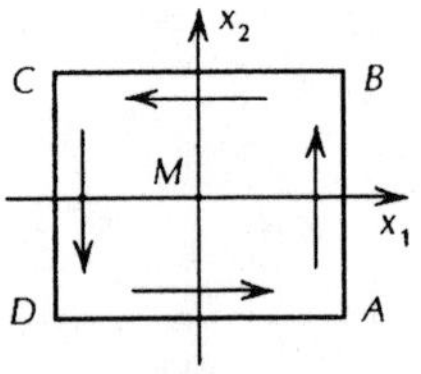

Fig. 6

The direction of positive circulation around the contour C is selected so that the surface σ, bounded by the contour, remains on the left.

The Stokes, theorem states that the *circulation of a vector along the closed contour C on which the open surface σ rests* (Figure 5) *is equal to the flux of the vector* rot $\boldsymbol{a}$ *through the surface σ*:

$$\Gamma_C(\boldsymbol{a}) = F_\sigma(\mathrm{rot}\,\boldsymbol{a}) \tag{115}$$

and the Stokes, formula, corresponding to this theorem, has the form

$$\oint_C \boldsymbol{a}\cdot\delta\boldsymbol{r} = \oint_\sigma \mathrm{rot}_n\boldsymbol{a}\,\delta\sigma \tag{116}$$

The direction of the normal should correspond to the direction of circulation around the contour C; i.e., on the side of the normal the circulation should be represented in the positive (counterclockwise) direction. Let us first prove the Stokes, theorem for an elementary rectangle $ABCD$ (Figure 6), which is located in the plane x_3 = const (the $0x_3$ axis is directed perpendicularly to the plane of the figure toward the reader). This rectangle has sides parallel to the axes $0x_1$ and $0x_2$, and at the center the point M is placed with the coordinates x_1, x_2, x_3 and with the vector $\boldsymbol{a}(x_1, x_2, x_3)$ prescribed at this point. The values of the vector $\boldsymbol{a}$ in the middle of the sides will be expressed as accurate to $O(\delta x^2)$ by the equalities

$$\text{on } \overline{AB}: \boldsymbol{a} + \frac{\partial a}{\partial x_1}\delta x_1$$

$$\text{on } \overline{BC}: \boldsymbol{a} + \frac{\partial a}{\partial x_2}\delta x_2$$

$$\text{on } \overline{CD}: \boldsymbol{a} - \frac{\partial a}{\partial x_1}\delta x_1$$

$$\text{on } \overline{DA}: \boldsymbol{a} - \frac{\partial a}{\partial x_2}\delta x_2$$

The circulation along the contour of the rectangle will be obtained as the sum of circulations around separate segments $\overline{AB}$, $\overline{BC}$, $\overline{CD}$, and $\overline{DA}$ separately equal, with regard for the direction of circulation, to

$$\Gamma_{AB} = \left(a_2 + \frac{\partial a_2}{\partial x_1}\delta x_1\right)|AB| = \left(a_2 + \frac{\partial a_2}{\partial x_1}\delta x_1\right)2\delta x_2$$

$$\Gamma_{BC} = -\left(a_1 + \frac{\partial a_1}{\partial x_2}\delta x_2\right)|BC| = -\left(a_1 + \frac{\partial a_1}{\partial x_2}\delta x_2\right)2\delta x_1$$

$$\Gamma_{CD} = -\left(a_2 - \frac{\partial a_2}{\partial x_1}\delta x_1\right)|CD| = -\left(a_2 - \frac{\partial a_2}{\partial x_1}\delta x_1\right)2\delta x_2$$

$$\Gamma_{DA} = \left(a_1 - \frac{\partial a_1}{\partial x_2}\delta x_2\right)|DA| = \left(a_1 - \frac{\partial a_1}{\partial x_2}\delta x_2\right)2\delta x_1$$

The summation will yield

$$\Gamma_{ABCD} = \left(\frac{\partial a_2}{\partial x_1} - \frac{\partial a_1}{\partial x_2}\right)4\delta x_1\delta x_2 = \text{rot}_3\,\boldsymbol{a}\,\delta\sigma \tag{117}$$

where $\delta\sigma$ denotes the area of the rectangle $ABCD$ $\delta\sigma = 2\delta x_1 \cdot 2\delta x_2 = 4\delta x_1\delta x_2$. Equality (117) shows that the circulation of the vector $\boldsymbol{a}$ along the contour of the rectangle is equal to the flux of the vector rot $\boldsymbol{a}$ through the rectangle area. This formula can be applied to any element $\delta\sigma$ of the surface σ in the form

$$\delta\Gamma = \text{rot}_n\boldsymbol{a}\,\delta\sigma \tag{118}$$

Summing up both sides of Equation 118 over all the elements of the surface σ, note that circulations along the common sides of adjacent cells cancel out and only circulations along those sides of the cells remain that constitute the contour C. We shall obtain in the limit

$$\sum\delta\Gamma = \oint_C \delta\Gamma = \oint_C \boldsymbol{a}\cdot\delta\boldsymbol{r}$$

The sum of the right sides in Equation 118 will be reduced to the *total flux of the vector* rot $\boldsymbol{a}$ *through the surface* σ. This proves the validity of Stokes, theorem.

The basic concepts and formulas of vector and tensor analysis that have been given in this chapter will suffice for understanding standing the subject matter of the book as a whole. For a more extensive account of the foundations of vector and tensor calculus, reference may be made to the fòllowing special textbooks:

1. Kochin, N. Ye., *Vektornoye ischisleniye i nachala tenzornogo ischisleniya (Vector Calculus and the Laws of Tensor Calculus)*, Nauka Press, Moscow, 1965.
2. Sedov, L. I., *Mekhanika sploshnoi sredy (Mechanics of a Continuous Medium)*, Part 1, Nauka Press, Moscow, 1983; *A Course in Continuum Mechanics*, Vol. 1 to 4, Wolters-Nordhoff Publishing, Amsterdam, 1971 to 1972.
3. Luriye, A. I., *Teoriya uprugosti (The Elasticity Theory)*, Nauka Press, Moscow, 1970 (Appendices I, II, III, pp. 799 to 870).
4. Mase, G. E., *Teoriya i zadachi mekhaniki sploshnykh sred (The Theory and Problems of the Mechanics of Continua)*, Mir Press, Moscow, 1974.

The previous (5th) edition (Nauka Press, Moscow, 1978) of the present book contains the summary of the most popular formulas in the vector and tensor calculi.

2

The Kinematics of a Continuous Medium

1. Specification of the Position and Motion of a Continuous Medium. Streamlines and Trajectories. Stream Tubes and Jets

There are two means for assigning the position and motion of a continuous medium, in particular, of a liquid or a gas.

The first, associated with Lagrange, consists of assigning instantaneous values of the coordinates for the medium particles x_1, x_2, x_3 as functions of time; i.e., it coincides with the system of point particles applied in the kinematics. That these particles could be considered individually requires that the values of the coordinates of each of them (a_1, a_2, a_3) be specified additionally at a certain initial time instant $t = t_0$. These coordinates must be not only rectangular Cartesian Coordinates, but also curvilinear. Because an infinite number of particles are in a continuous medium, the quantities a_k themselves form a continuous field. The quantities a_k and t are called the *Lagrange variables*.

The kinematic equations of motion are written in vector or coordinate form:

$$x_1 = x_1(a_1, a_2, a_3; t) \qquad x_2 = x_2(a_1, a_2, a_3; t) \qquad x_3 = x_3(a_1, a_2, a_3; t) \tag{1}$$

or

$$x_i = x_i(a_1, a_2, a_3; t) \quad (i = 1, 2, 3)$$

or

$$\boldsymbol{r} = \boldsymbol{r}(a_1, a_2, a_3; t) = \boldsymbol{r}(\boldsymbol{a}, t)$$

The other method, that of the Euler equation, which is much more popular than the Lagrange technique, consists of analytical or vector specification of the *velocity field* of the velocity vector $\boldsymbol{V}$ with the projections V_1, V_2, V_3

$$V_1 = V_1(x_1, x_2, x_3; t) \qquad V_2 = V_2(x_1, x_2, x_3; t) \qquad V_3 = V_3(x_1, x_2, x_3; t) \qquad (2)$$

or

$$V_i = V_i(x_1, x_2, x_3; t) \qquad (i = 1, 2, 3)$$

or

$$\boldsymbol{V} = \boldsymbol{V}(x_1, x_2, x_3; t) = \boldsymbol{V}(\boldsymbol{r}, t)$$

In this case the variables x_1, x_2, x_3; t are called the *Euler variables.*

It should be emphasized that in the Lagrange technique the coordinates x_1, x_2, x_3 relate to moving particles of the medium, whereas in the Euler equation, the coordinates relate to the points in space with which these particles coincide at the given time instant. This event allows rewriting of the system of equations (2) in the form of one vector equation, or as a system of three ordinary first-order differential equations:

$$V_1 = \frac{dx_1}{dt} = V_1(x_1, x_2, x_3; t) \qquad V_2 = \frac{dx_2}{dt} = V_2(x_1, x_2, x_3; t)$$

$$V_3 = \frac{dx_3}{dt} = V_3(x_1, x_2, x_3; t) \qquad (3)$$

or

$$V_i = dx_i/dt = V_i(x_1, x_2, x_3; t) \qquad (i = 1, 2, 3)$$

or

$$\boldsymbol{V} = d\boldsymbol{r}/dt = \boldsymbol{V}(x_1, x_2, x_3; t)$$

in the unknowns $\boldsymbol{r}(t)$ or $x_1(t)$, $x_2(t)$, $x_3(t)$.

Three integration constants, C_1, C_2, C_3, which appear in this case and which will enter into the integrals of the system in Equation 3

$$x_1 = x_1(C_1, C_2, C_3; t) \qquad x_2 = x_2(C_1, C_2, C_3; t)$$

$$x_3 = x_3(C_1, C_2, C_3; t) \qquad (4)$$

can be eliminated with the aid of initial conditions at $t = t_0$

$$a_1 = x_1(C_1, C_2, C_3; t_0) \qquad a_2 = x_2(C_1, C_2, C_3; t_0)$$

$$a_3 = x_3(C_1, C_2, C_3; t_0) \qquad (5)$$

This will ultimately lead to the Lagrange form of assignment (Equation 1). Later, it will be more convenient to use the Euler variables and to prescribe motion in the form of Equation 2.

The velocity field is *uniform* when *velocity vectors* at all points of the field are identical. The velocity field is *stationary* if it does not vary with time ($\partial \boldsymbol{V}/\partial t = 0$), otherwise, it is

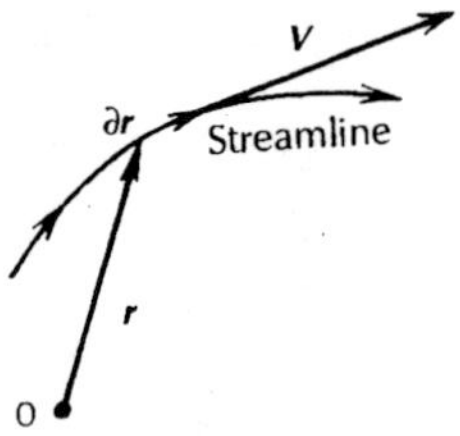

Fig. 7

nonstationary.[1] In the section on the kinematics of a continuous medium the velocity field is regarded as given.

The fields of physical quantities can not be readily surveyed directly. To regulate their representation, special techniques have been developed. The scalar field is split by the *level surfaces* of the scalar function; for example, isotherms, which are the surfaces of identical temperatures and isopotential surfaces in the dynamics on which the values of the force field potential are retained as constant.

In vector fields, vector lines are considered at each point at which the vector is directed along the tangent to this line. In the following, the first to be encountered will be the *streamlines* (Figure 7). The velocity vector of the medium is directed along the tangent to individual points along the streamline. Time is a fixed parameter in streamlines. (Later, the streamlines will be differentiated from other vector lines by arrows showing motion.)

In general, at the given time instant one can draw a streamline through each point of the velocity field, with this streamline being the only one. However, there are also points in the velocity field, *singular points*, through which an infinite number of streamlines may pass, or through which no streamlines may pass. Examples of such singular points are given in Chapter 7.

The flow of a liquid can be visualized, for example, by introducing into it fine solid particles (powders of lycopodium and aluminum or polystyrene spheres) or small gas bubbles (usually hydrogen), and photographing them in reflected light with short exposure times or by making a motion picture. The streamlines pictorially demonstrate the general character of motion. Such patterns of liquid motion are called *flow spectra.* Similarly *smoke spectra* are also used to visualize air flow. Professor van Dyke of Stanford University (Stanford, CA) composed an excellent atlas (as to content and design) of the spectra of water and air flow and also of optical pictures of sub- and supersonic flow. This atlas is a valuable textbook on the mechanics of liquids and gases.[2]

Having written the condition that the velocity vector $\boldsymbol{V}$, which at the given point is directed along the tangent to the streamline, and the differential $\delta\boldsymbol{r}$ (Figure 7) of the radius-vector $\boldsymbol{r}$ of this point both have 2 common direction, we shall obtain the vector equality which expresses the condition of the parallelism of these two vectors

$$\boldsymbol{V} \times \delta\boldsymbol{r} = 0 \tag{6}$$

or in projections

$$\frac{\delta x_1}{V_1(x_1, x_2, x_3; t)} = \frac{\delta x_2}{V_2(x_1, x_2, x_3; t)} = \frac{\delta x_3}{V_3(x_1, x_2, x_3; t)} \tag{7}$$

[1] The terms "steady" and "unsteady" are also frequently used in these contexts.

[2] Van Dyke, M., *An Album of Fluid Motion*, Parabolic Press, Stanford, CA, 1982. The book was translated into Russian and published by Mir Press, Moscow, in 1986.

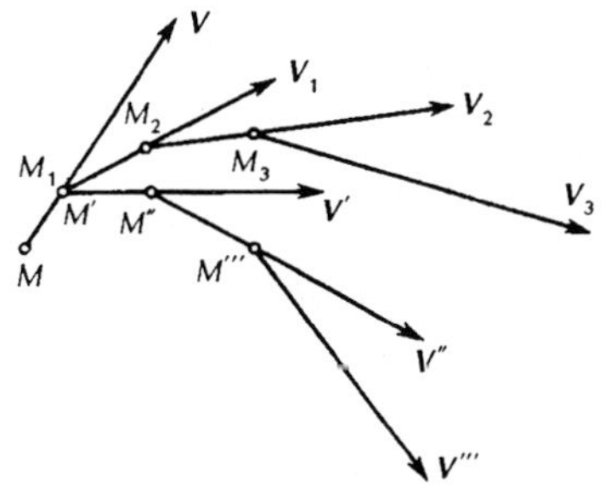

Fig. 8

It should be emphasized that the use of the Greek letter δ to denote differentials indicates an infinitely small increment of the quantity in space when the field of the given quantity is fixed at the given time instant. The italic letter *d* is preserved to designate the differentials of quantities as infinitely small increments in time during medium motion.

In equalities (7) V_1, V_2, V_3 are the given functions of position and time. Regarding t to be a parameter that takes on any fixed values, we may consider the set of equalities (7) as a system of two ordinary differential equations (the third equation results from the set.) The integrals of these equations will represent the final relationships between coordinates which are equations of two surfaces; their intersection will give the sought-after streamline. In what follows, we shall not use such a direct, but a very laborious, method for obtaining streamlines.

The streamlines should be differentiated from the *trajectories* of motion of the continuous medium particles defined as the locus of successive locations of separate particles in space at time instants following one another. At each given time instant t the differential $\boldsymbol{dr}$ of the radius-vector $\boldsymbol{r}$ of a moving particle coincides in direction to the velocity vector so that the differential equations of trajectories will be analogous to equations (Equation 7) of streamlines, but with the essential difference that now the time t will be an independent variable as are x_1, x_2, x_3 and the trajectory equation must be written in the form

$$\frac{dx_1}{V_1(x_1, x_2, x_3; t)} = \frac{dx_2}{V_2(x_1, x_2, x_3; t)} = \frac{dx_3}{V_3(x_1, x_2, x_3; t)} = dt \tag{8}$$

If the velocity field is stationary, the time t does not enter into velocity projections. It also disappears from Equations 8 and will coincide with Equations 7. From this it follows that *in stationary velocity fields the trajectories and streamlines coincide at any time instant.*

The same result can be also derived on the basis of the following simple geometrical arguments. To construct a streamline passing through the point M (Figure 8), we shall lay off an arbitrary small segment $\overline{MM_1}$ along the velocity vector. At the same time instant t (t is a fixed parameter), due to the nonuniformity of the velocity field, the velocity at the point M_1 will be different, say V_1. Laying off the segment M_1M_2 along the direction of V_1, we obtain the next point M_2, and so on. Figure 8 shows the polygon $MM_1M_2M_3 \cdots$. When the segments are infinitely small, this polygon will convert to a continuous curve – a streamline.

The trajectory of the point is plotted in another fashion. After the lapse of a small interval of time Δt the point M will be displaced by $\overline{MM'} = \boldsymbol{V}\Delta t$ and will pass over into the point M' which can be mated with M_1 by the corresponding selection of Δt. From this location, after a new interval of time Δt, the point M' will displace along the velocity vector $\boldsymbol{V}'$, which corresponds to the point M' and to the time instant $t + \Delta t$, which will in turn displace into the point M'' and then to the point M''' along the velocity vector $\boldsymbol{V}''$, etc.

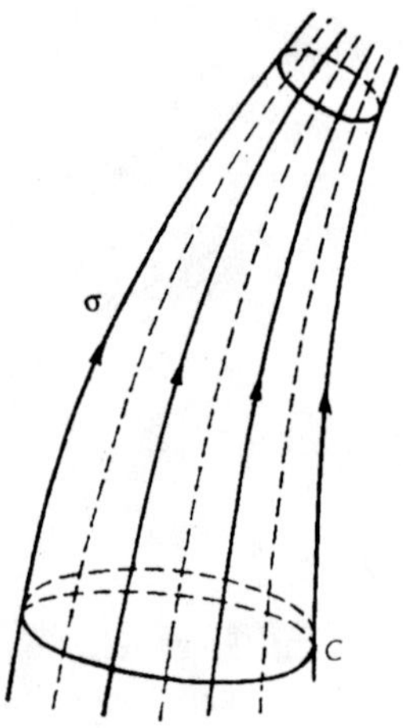

Fig. 9

In the general case of the nonstationary velocity field the polygon $MM'M''M'''\cdots$ will obviously differ from the polygon $MM_1M_2\cdots$. The limiting position of the points of the polygon $MM'M''\cdots$ when $\Delta t \to 0$ will determine the trajectory of the moving point of the medium M which, in the case of the nonstationary velocity field, will not be coincident with the streamline. The streamline and the trajectory, which pass through the given point M, will have at this point the common tangent directed along the velocity vector $\boldsymbol{V}$.

In the case of the stationary flow, by selecting the arbitrary points M_1, $M_2,\cdots$, to be coincident with M', $M'',\cdots$, and noting that according to the stationary condition the velocity vectors $\boldsymbol{V}_1$, $\boldsymbol{V}_2,\cdots$ coincide at any time instant with $\boldsymbol{V}'$, $\boldsymbol{V}'',\cdots$, we ensure that the streamline will also coincide with the trajectory.

By drawing streamlines at the given time instant through the points of the closed contour C (Figure 9), we shall obtain the *stream surface*, which encloses a portion of liquid called the stream tube. By considering further the contour C as a liquid, i.e., *at any time instant consisting of the same particles of liquid*, the set of the trajectories of the points of this contour will limit a jet. In the case of the stationary velocity field the *jet coincides with the stream tube*. At any point of the stream surface the velocity vector $\boldsymbol{V}$ is directed along the tangent to the streamline so that the medium does not penetrate through the stream surface. Therefore, if the medium is incompressible (its density is the same everywhere) and, moreover, there are no sources or sinks inside the stream tube, then the rate of volume discharge of it, Q (i.e., the velocity vector flux through the cross-section of the tube), is the same at the given instant for any cross-section of the tube:

$$Q = \int_\sigma V_n \delta\sigma = \text{const} \tag{9}$$

where $\boldsymbol{n}$ is the normal to the tube cross-section σ at the given point.

Elementary (plotted on an infinitely small contour) *stream tubes*, whatever the velocity field, admit the drawing of cross-sections normal to them, with these sections considered to be plane accurate to higher-order small quantities. It is different with stream tubes of a finite size. For such tubes to have the sections normal to the streamlines inside the tubes and on their surfaces, it is necessary that the *condition for the existence of surfaces normal to the streamlines be fulfilled.*

Consider a set of surfaces $\varphi(x_1, x_2, x_3) = \text{const}$ that intersects the streamlines. The condition for the orthogonality of the streamlines to these surfaces is (λ is a scalar function):

$$\text{grad}\ \varphi = \lambda \boldsymbol{V}$$

Operating with rot on both sides of this vector equality, we obtain zero on the left, whereas on the right, according to the third equality of Equation 88 in Chapter 1, we have the relation

$$\text{rot}\ (\lambda \boldsymbol{V}) = \lambda \text{rot}\ \boldsymbol{V} + \text{grad}\ \lambda \times \boldsymbol{V}$$

Performing scalar multiplication of the both sides of the aforementioned equality

$$0 = \lambda \text{rot}\ \boldsymbol{V} + \text{grad}\ \lambda \times \boldsymbol{V}$$

by $\boldsymbol{V}$ and taking into account that the second term on the right is perpendicular to the vector $\boldsymbol{V}$, we see that the condition for the existence of the surfaces orthogonal to streamlines is

$$\boldsymbol{V} \cdot \text{rot}\ \boldsymbol{V} = 0 \tag{10}$$

Such is the limitation imposed on the velocity field. Without fulfilling it, the existence of the sections of stream tubes normal to the streamlines is impossible.

2. Velocity Distribution in an Elementary Volume of the Medium. Helmholtz's First Theorem

The kinematics of a perfect solid body, which represents the simplest example of a continuous medium, give rise to the Euler theorem of splitting the velocity vector $\boldsymbol{V}$ of any point of an arbitrarily sized body into the component $\boldsymbol{V}_0$ of translational motion, together with about an arbitrarily selected axis 0 the rotational component of angular velocity vector $\boldsymbol{\omega}$. A similar theorem extended to include deformations of the medium and continued, when necessary, to an infinitely small volume, is established for any continuous medium, including a liquid or a gas.

Recall that in the general case of motion of a perfect solid body, the velocity distribution was given by the formula

$$\boldsymbol{V} = \boldsymbol{V}_0 + \boldsymbol{\omega} \times (\boldsymbol{r} - \boldsymbol{r}_0) \tag{11}$$

where $\boldsymbol{V}$ is the velocity vector of any point M of the body; $\boldsymbol{V}_0$ is the velocity of the point M_0 taken to be the axis; $\boldsymbol{\omega}$ is the angular velocity vector of body rotation identical for any point of the body; $\boldsymbol{r}$ and $\boldsymbol{r}_0$ are the vector-radii of points M and M_0. In projections on the coordinate axis equality (11) will take the form

$$V_1 = V_{10} + \omega_2 (x_3 - x_{30}) - \omega_3 (x_2 - x_{20})$$

$$V_2 = V_{20} + \omega_3 (x_1 - x_{10}) - \omega_1 (x_3 - x_{30})$$

$$V_3 = V_{30} + \omega_1 (x_2 - x_{20}) - \omega_2 (x_1 - x_{10}) \tag{12}$$

Bearing in mind, that the quantities $x_{10}, x_{20}, x_{30}; V_{10}, V_{20}, V_{30}; \omega_1, \omega_2, \omega_3$ are identical at the given time instant t for all points of the body, we conclude that equality (11) and its expanded form (Equation 12) represent the linear vector-function of the vector-radius $\boldsymbol{r}$ of the point M or, equivalently, of the coordinates x_1, x_2, and x_3.

In the kinematics of a perfect solid body, Equations 11 or 12 express the velocity vector $\boldsymbol{V}$ of any point of the body as a given function of coordinates. Let the inverse problem be posed, which is usually not treated in Volumes on theoretical mechanics, and which concerns the search for the expression of the angular velocity vector of body rotation in terms of the field of the velocity vector $\boldsymbol{V}$ specified by equalities (11) or (12). For this purpose, using equalities (12), we shall find the following differences of the derivatives of velocity projections V_1, V_2, V_3 with respect to the coordinates x_1, x_2, x_3:

$$\frac{\partial V_3}{\partial x_2} - \frac{\partial V_2}{\partial x_3} = \omega_1 - (-\omega_1) = 2\omega_1$$

$$\frac{\partial V_1}{\partial x_3} - \frac{\partial V_3}{\partial x_1} = \omega_2 - (-\omega_2) = 2\omega_2$$

$$\frac{\partial V_2}{\partial x_1} - \frac{\partial V_1}{\partial x_2} = \omega_3 - (-\omega_3) = 2\omega_3 \qquad (13)$$

The differences on the left sides of these equalities represent the projections of the space derivative of the vector $\boldsymbol{V}$, called the *rotor*, or the *curl*, of the velocity vector and is denoted by the symbol rot $\boldsymbol{V}$ (see Chapter 1 Equations 80 and 81).

Equalities (13) yield

$$\boldsymbol{\omega} = \frac{1}{2}\text{rot}\,\boldsymbol{V} \qquad (14)$$

which is the solution to the posed inverse problem in the kinematics of a perfect solid body.

In general motion of a perfect solid body, Equations 12 represented the components of the velocity vector $\boldsymbol{V}$ in terms of the linear functions of position by virtue of which the magnitude and direction of the velocity curl, rot $\boldsymbol{V}$, and angular velocity vectors, $\boldsymbol{\omega}$, were identical throughout the entire space.

In the case of a *deformable medium*, in particular of a liquid or a gas, the velocity field will not be linear with respect to the position; however, by confining our consideration at the given time instant to an infinitely small volume, the velocity field inside it can be considered linear accurate to infinitely small higher-order quantities, and Equation 13 can be regarded to be valid. Having imagined the volume element of the medium as being instantly solidified, we shall conclude that the angular velocity of its rotation at the given instant will be *equal to half the velocity vector curl* which can be calculated in a small region of this instantly solidified nondeformable liquid.

In the case of a perfect solid body Equation 14 is valid globally (i.e., for the entire body), whereas in the case of a deformable medium it is valid only locally in infinitely close proximity to the given point of the medium.

Later, the angular velocity vector of an elementary liquid volume will be given a more direct interpretation, but will not be based on the imagined solidification of the volume.

Let us consider the motion of an elementary liquid volume having isolated with in it two arbitrary points M and M_1 (Figure 10) with the velocities $\boldsymbol{V}$ and $\boldsymbol{V}^{(1)}$. The relative location of the points M_1 and M is dictated by the vector-radius $\delta\boldsymbol{r} = \boldsymbol{r}_1 - \boldsymbol{r}_0$ where $\boldsymbol{r}_1$ and $\boldsymbol{r}_0$ are the vector-radii of the points M_1 and M, respectively.

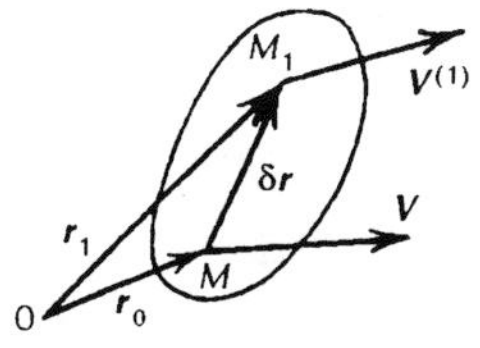

Fig. 10

Availing ourselves of the continuous nature of the velocity field and assuming the existence of the first derivatives of velocity projections $\boldsymbol{V}$ with respect to the coordinates, let us compare the velocities $\boldsymbol{V}^{(1)}$ and $\boldsymbol{V}$. Employing for this purpose the symbolic representations for the total differential of the vector $\boldsymbol{V}$ in space and of the total derivative of the vector $\boldsymbol{V}$ with respect to the vector-radius $\boldsymbol{r}$, we obtain (according to Equations 94 to 96 of Chapter 1):

$$\delta \boldsymbol{V} = \boldsymbol{V}^{(1)} - \boldsymbol{V} = \left(\frac{\delta \boldsymbol{V}}{\delta \boldsymbol{r}} \delta \boldsymbol{r}\right) = \dot{D}^* \delta \boldsymbol{r} = \dot{S} \delta \boldsymbol{r} + \dot{A}^* \delta \boldsymbol{r} \tag{15}$$

where the bracketed expression represents the product of the tensor $\delta \boldsymbol{V}/\delta \boldsymbol{r} = \dot{D}^*$ by the vector $\delta \boldsymbol{r}$ on the "right".[3] Now, using equality (106) of Chapter 1 with $\boldsymbol{a} = \boldsymbol{V}$, $\boldsymbol{b} = \delta \boldsymbol{r}$, we shall obtain

$$\boldsymbol{V}^{(1)} = \boldsymbol{V} + \frac{1}{2} \operatorname{rot} \boldsymbol{V} \times \delta \boldsymbol{r} + (\operatorname{def} \boldsymbol{V})\, \delta \boldsymbol{r} \tag{16}$$

Having reduced this expression, according to Equation 14, to the form

$$\boldsymbol{V}^{(1)} = \boldsymbol{V} + \boldsymbol{\omega} \times \delta \boldsymbol{r} + (\operatorname{def} \boldsymbol{V})\, \delta \boldsymbol{r} \tag{17}$$

we shall compare it with equality (11) at $\boldsymbol{r} - \boldsymbol{r}_0 = \delta \boldsymbol{r}$. The first term on the right side can be considered the velocity of the pole M; the second can be considered the rotational velocity of the points of the volume element with the relative vector-radii $\delta \boldsymbol{r}$ and angular velocity $\boldsymbol{\omega}$, and the sum of these two terms as the velocity $\boldsymbol{V}_{\text{qs}}$ in quasisolid motion of the volume element. The last term will be called the *rate of strain* and will be designated by $\boldsymbol{V}_{\text{str}}$. Then, equality (17), rewritten as

$$\boldsymbol{V}^{(1)} = \boldsymbol{V}_{\text{qs}} + \boldsymbol{V}_{\text{str}} \tag{18}$$

where

$$\boldsymbol{V}_{\text{qs}} = \boldsymbol{V} + \boldsymbol{\omega} \times \delta \boldsymbol{r} = \boldsymbol{V} + \frac{1}{2} \operatorname{rot} \boldsymbol{V} \times \delta \boldsymbol{r}$$

$$\boldsymbol{V}_{\text{str}} = (\operatorname{def} \boldsymbol{V})\, \delta \boldsymbol{r} = \dot{S} \delta \boldsymbol{r} \tag{19}$$

will represent Helmholtz's first theorem, which is a direct extension of the Euler theorem about velocity distribution in motion of a perfect solid body. Helmholtz's first theorem states: *the velocities of the volume element points of a continuous medium are composed of the velocities of quasisolid and deformational motions of it.*

---

[3]The dots in the notation $\dot{D}$, $\dot{S}$, and $\dot{A}^*$, which are usually used to designate time derivatives, are used here to distinguish these velocity field tensors from the tensors D, S, A^* for the field of displacements.

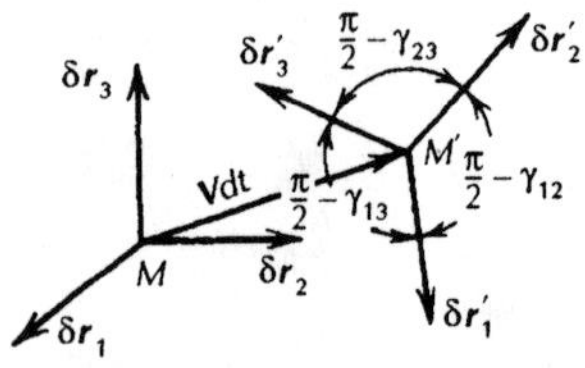

Fig. 11

The quasisolid motion, more precisely the general case of motion of a perfect solid body, is considered in detail in the author's book on theoretical mechanics.[4] Let us consider next the description of the deformational velocity component.

3. Distorting Motion of a Liquid, Rate of Strain Tensor and the Kinematic Sense of Its Components. The Principal Axes of the Rate of Strain Tensor

Figure 11 is an arbitrarily oriented system of rectangular coordinates $Mx_1x_2x_3$ at a certain time instant at the point selected. The set of three elementary liquid segments $\delta\boldsymbol{r}_1$, $\delta\boldsymbol{r}_2$, $\delta\boldsymbol{r}_3$ is now considered with the projections onto these axes:

$$\delta\boldsymbol{r}_1\,(\delta x_1, 0, 0) \qquad \delta\boldsymbol{r}_2\,(0, \delta x_2, 0) \qquad \delta\boldsymbol{r}_3\,(0, 0, \delta x_3) \tag{20}$$

After the lapse of an infinitely small period of time *dt*, the liquid coordinate segments (Equation 20) will displace into the adjacent location $M'x_1'x_2'x_3'$, having performed quasi-solid and deformational displacement (the second equality of system 19). The new values the coordinate segments (20) and of their projections will be equal (we shall preserve the brief notation $\dot{S}$ for def $\boldsymbol{V}$)

$$\delta\boldsymbol{r}_i' = \delta\boldsymbol{r}_i + \dot{S}\delta\boldsymbol{r}_i dt$$

$$\delta x_i' = \delta x_i + (\dot{S}\delta\boldsymbol{r}_i)_i dt = \delta x_i + \dot{S}_{ij}(\delta\boldsymbol{r}_i)_j dt = \delta x_i + \dot{S}_{ii}\delta x_i dt \tag{21}$$

because, according to Equation 20, $(\delta\boldsymbol{r}_i)_j = 0$ if $j \neq i$ ($i = 1, 2, 3$; in Equations 21 to 23, do not sum over *i*).

According to Equation 21, the relative elongations of the magnitudes of these segments will be

$$\varepsilon_i = \frac{\delta x_i' - \delta x_i}{\delta x_i} = \dot{S}_{ii}\,dt \tag{22}$$

and, consequently, *the diagonal components of the rate of strain tensor* $\dot{S}_{ii}$ *are equal to the rates of relative elongations* ε_i *of the coordinate segments* $\delta\boldsymbol{r}_i$:

$$\dot{S}_{ii} = \frac{\varepsilon_i}{dt} = \dot{\varepsilon}_i \tag{23}$$

---

[4]Loitsyanskiy, L. G. and Luriye, A. I., *Kurs teoreticheskoi mekhaniki* (*The Course of Theoretical Mechanics*), Vol. 1, Nauka Press, Moscow, 1982, pp. 283 to 285.

For the time dt the coordinate vectors $\delta \boldsymbol{r}_1$, $\delta \boldsymbol{r}_2$, $\delta \boldsymbol{r}_3$ will displace into the positions $\delta \boldsymbol{r}_1'$, $\delta \boldsymbol{r}_2'$, $\delta \boldsymbol{r}_3'$, and will be deformed (elongated or shortened), whereas the (initially) right angles between them will cease to be (they will become "skewed").

The new values of the angles between $\delta \boldsymbol{r}_1'$, $\delta \boldsymbol{r}_2'$, and $\delta \boldsymbol{r}_3'$ will be designated as

$$\widehat{\delta \boldsymbol{r}_1, \delta \boldsymbol{r}_2} = \frac{\pi}{2} - \gamma_{12} \qquad \widehat{\delta \boldsymbol{r}_2, \delta \boldsymbol{r}_3} = \frac{\pi}{2} - \gamma_{23} \qquad \widehat{\delta \boldsymbol{r}_1, \delta \boldsymbol{r}_3} = \frac{\pi}{2} - \gamma_{13}$$

We shall agree to regard the "skewness" of the right angles γ_{12}, γ_{23}, γ_{13} as positive when the angles become smaller.

To determine the skewed angles, rewrite the equalities ($\approx$ is the sign of approximate equality)

$$\cos(\widehat{\delta \boldsymbol{r}_1', \delta \boldsymbol{r}_2'}) = \frac{\delta \boldsymbol{r}_1'}{\delta x_1'} \cdot \frac{\delta \boldsymbol{r}_2'}{\delta x_2'} = \cos\left(\frac{\pi}{2} - \gamma_{12}\right) = \sin \gamma_{12} \approx \gamma_{12}$$

$$\cos(\widehat{\delta \boldsymbol{r}_2', \delta \boldsymbol{r}_3'}) = \frac{\delta \boldsymbol{r}_2'}{\delta x_2'} \cdot \frac{\delta \boldsymbol{r}_3'}{\delta x_3'} = \cos\left(\frac{\pi}{2} - \gamma_{23}\right) = \sin \gamma_{23} \approx \gamma_{23}$$

$$\cos(\widehat{\delta \boldsymbol{r}_1', \delta \boldsymbol{r}_3'}) = \frac{\delta \boldsymbol{r}_1'}{\delta x_1'} \cdot \frac{\delta \boldsymbol{r}_3'}{\delta x_3'} = \cos\left(\frac{\pi}{2} - \gamma_{13}\right) = \sin \gamma_{13} \approx \gamma_{13} \tag{24}$$

Using Equation 21 and recalling that in the initial position $\delta \boldsymbol{r}_1 \cdot \delta \boldsymbol{r}_2 = 0$, obtain an expression for the skewness of the coordinate angle $\widehat{\delta \boldsymbol{r}_1, \delta \boldsymbol{r}_2}$

$$\gamma_{12} = \frac{\delta \boldsymbol{r}_1'}{\delta x_1'} \cdot \frac{\delta \boldsymbol{r}_2'}{\delta x_2'} = \frac{\delta \boldsymbol{r}_1 + (\dot{S}\delta \boldsymbol{r}_1)dt}{\delta x_1 + \dot{S}_{11}\delta x_1 dt} \cdot \frac{\delta \boldsymbol{r}_2 + (\dot{S}\delta \boldsymbol{r}_2)dt}{\delta x_2 + \dot{S}_{22}\delta x_2 dt} =$$

$$= \frac{(\dot{S}\delta \boldsymbol{r}_1) \cdot \delta \boldsymbol{r}_2 dt + (\dot{S}\delta \boldsymbol{r}_2) \cdot \delta \boldsymbol{r}_1 dt + \cdots}{\delta x_1 \delta x_2 + \cdots} \approx 2\dot{S}_{12} \frac{\delta x_1 \delta x_2 dt}{\delta x_1 \delta x_2} = 2\dot{S}_{12} dt$$

The skewness of the other coordinate angles will be found analogously:

$$\gamma_{23} = 2\dot{S}_{23} dt \qquad \gamma_{13} = 2\dot{S}_{13} dt$$

whereas the skewness rates[5] of the coordinate angles will be equal to

$$\dot{\gamma}_{12} = 2\dot{S}_{12} \qquad \dot{\gamma}_{23} = 2\dot{S}_{23} \qquad \dot{\gamma}_{13} = 2\dot{S}_{13} \tag{25}$$

It follows that the nondiagonal rate of strain tensor components are halves of the rates of skewness of the coordinate angles. One other characteristic of deformational motion is important – the rate of relative volumetric expansion of the medium at the given point $\dot{\theta}$ defined by the equality

[5] "Angular strain" is often used instead of "skewness". Thus, $\dot{\gamma}_{ij}$ are angular strain rates.

$$\dot{\theta} = \frac{\delta\tau' - \delta\tau}{\delta\tau dt} = \frac{1}{\delta\tau}\frac{d}{dt}(\delta\tau) \tag{26}$$

as the ratio of the increment in the elementary liquid volume $\delta\tau$ to the initial value of this volume and to the deformation time dt. Assuming that

$$\delta\tau = \delta x_1 \delta x_2 \delta x_3, \quad \delta\tau' = \delta x_1' \delta x_2' \delta x_3' = (1 + \dot{\varepsilon}_1 dt)\delta x_1 \cdot (1 + \dot{\varepsilon}_2 dt)\delta x_2 \cdot (1 + \dot{\varepsilon}_3 dt)\delta x_3$$

and omitting the infinitely small quantities in the final version, the following relation will be obtained, which corresponds to the limiting transition:

$$\dot{\theta} = \dot{\varepsilon}_1 + \dot{\varepsilon}_2 + \dot{\varepsilon}_3 = \frac{\partial V_1}{\partial x_1} + \frac{\partial V_2}{\partial x_2} + \frac{\partial V_3}{\partial x_3} = \operatorname{div} V \tag{27}$$

Thus, it follows from the definition of $\dot{\theta}$ (Equation 26) and from equality (27) that *the divergence of the velocity vector* **V** *is the rate of relative change in the elementary liquid volume at the given point of the velocity field.*

In the model of an incompressible liquid, the rate of relative volumetric expansion in the absence of distributed sources is equal to zero by definition. Irrespective of all the other properties of the liquid (density, viscosity, etc.), the projections of its velocity turn out to be connected by a purely kinematic equality, the incompressibility equation, which is also called less definitely the *continuity equation*

$$\operatorname{div} V = \frac{\partial V_1}{\partial x_1} + \frac{\partial V_2}{\partial x_2} + \frac{\partial V_3}{\partial x_3} = 0 \tag{28}$$

Recall (see Chapter 1) that for any symmetric tensor it is possible to indicate such a rectangular coordinate system (its axes are called principal axes) in which all the nondiagonal components are equal to zero and the tensor table is reduced to the principal values of the tensor which are located on the main diagonal. As applied to the rate of strain tensor $\dot{S}$ this means that in its principal axes the rates of skewness of coordinate angles are equal to zero, whereas the rates of the relative elongation of infinitely small coordinate segments are equal to the main values $\dot{\lambda}^{(1)}$, $\dot{\lambda}^{(2)}$, $\dot{\lambda}^{(3)}$ of the rate of the strain tensor. Using primes on the principal coordinates and tensor components in the principal axes gives

$$\begin{aligned} \dot{S}'_{11} &= \dot{\lambda}^{(1)} & \dot{S}'_{12} &= 0 & \dot{S}'_{13} &= 0 \\ \dot{S}'_{21} &= 0 & \dot{S}'_{22} &= \dot{\lambda}^{(2)} & \dot{S}'_{23} &= 0 \\ \dot{S}'_{31} &= \dot{\lambda}^{(1)} & \dot{S}'_{32} &= 0 & \dot{S}'_{33} &= \dot{\lambda}^{(3)} \end{aligned} \tag{29}$$

The fact that the principal coordinate axes do not skew at each given instant at the given point of the medium allows the vector $\boldsymbol{\omega}$ to be defined as *the vector of instantaneous angular velocity of the trihedron of the principal axes*, which is an instantaneously rigid (in the sense given by the direction of the axes) "skeleton" of the liquid volume element. It must not be forgotten that at the next time instant the trihedron of the principal axes will already pass through other particles of the medium.

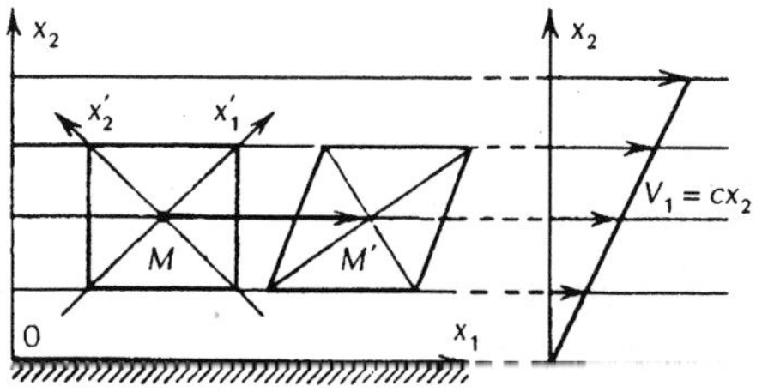

a) b)

Fig. 12

As the simplest example, consider the motion of *pure shear* in which the distribution of velocities has the form (Figure 12b)

$$V_1 = cx_2 \qquad V_2 = 0 \qquad V_3 = 0 \tag{30}$$

Among the shear rate tensor components in the coordinates x_1, x_2 only $\dot{S}_{12} = \dot{S}_{21} = \frac{1}{2}\frac{\partial V_1}{\partial x_2} = \frac{c}{2}$ differ from zero. The characteristic Equation 60 of Chapter 1 will take the form

$$\begin{vmatrix} -\dot{\lambda} & \frac{c}{2} & 0 \\ \frac{c}{2} & -\dot{\lambda} & 0 \\ 0 & 0 & -\dot{\lambda} \end{vmatrix} = -\dot{\lambda}\left(\dot{\lambda}^2 - \frac{c^2}{4}\right) = 0 \tag{31}$$

which will have the roots, the main values of the shear rate tensor,

$$\dot{\lambda}^{(1)} = \frac{c}{2} \qquad \dot{\lambda}^{(2)} = -\frac{c}{2} \qquad \dot{\lambda}^{(3)} = 0 \tag{32}$$

By isolating at the initial instant a liquid contour in the form of an infinitely small square with the center M, it will pass into another position after time dt and will turn into a rhomb (accurate to small second-order quantities). It is clear from physical considerations that the rate of relative elongation of the axis Mx_1' (Figure 12a) is positive, and the rate of the axis Mx_2' is negative; i.e., as time progresses the segments of the axis Mx_1' become longer and the segments of the axis Mx_2' shorter. The right angle between the diagonals of the square Mx_1' and Mx_2' is preserved after small deformation, having become an angle between the diagonals of the rhomb. This indicates that the axes Mx_1' and Mx_2' are principal axes. The same conclusion can be arrived at analytically.

Equations 58 and 59 of Chapter 1 are reduced to two independent equations in the projections $e_1^{(1)}$, $e_2^{(1)}$ of the unit vector $e^{(1)}$ of one of the principal axes:

$$-\dot{\lambda}^{(1)} e_1^{(1)} + \dot{\lambda}^{(1)} e_2^{(1)} = 0 \qquad e_1^{(1)2} + e_2^{(1)2} = 1$$

when, according to Equation 32,

$$e_1^{(1)} = e_2^{(1)} = \frac{\sqrt{2}}{2} \tag{33}$$

Analogously, for $e_1^{(2)}$ and $e_2^{(2)}$:

$$e_1^{(2)} = -\frac{\sqrt{2}}{2} \qquad e_2^{(2)} = \frac{\sqrt{2}}{2} \tag{34}$$

and for $e_1^{(3)}$ and $e_2^{(3)}$, the zero values in conformity with the two-dimensional character of flow.

At the same time the projections of the unit vectors of the principal axes $\boldsymbol{e}^{(1)}$, $\boldsymbol{e}^{(2)}$, $\boldsymbol{e}^{(3)}$ on the axis of the coordinate system $0x_1x_2x_3$ are equal to the corresponding cosines of the angles between the principal and ordinary axes. It follows from Equations 33 and 34 that the angles between the first principal axis and the axes $0x_1$, $0x_2$ are equal to 45°, and between the second principal axis and the same axes to 135° and 45°, respectively. From this, the first two principal axes of the rate of strain tensor at point M are located along the diagonals of the square (Figure 12); i.e., they are just the axes Mx_1' and Mx_2'. The third principal axis forms an angle of 90° with the plane of the figure, and this was previously evident the motion considered is parallel to the plane $0x_1x_2$.

The displacement $\overline{MM'}$ (Figure 12) characterizes the translatory displacement $\boldsymbol{V}dt$ of the volume element for the time interval dt. It can be noted that for this very time interval the system of the principal axes, which (as above) is the "skeleton" of the liquid volume, rotated clockwise through the angle $\omega_3 dt = \frac{1}{2}\left(\frac{\partial V_2}{\partial x_1} - \frac{\partial V_1}{\partial x_2}\right)dt = -\frac{c}{2}dt$. The minus sign arises because the axis $0x_3$ is directed perpendicularly to the plane of the figure so that from the side of this axis the rotation is seen to be clockwise from the axis $0x_2'$ to the axis $0x_1'$, i.e., in the negative direction.

In liquid and gas mechanics, the deformation component of motion is not as important as in the theory of a solid body (elastic, plastic, etc.). The strain rate tensor components will occur only in the dynamics of viscous liquids and gases where they will be connected with the stress tensor components (the generalized Newton law).

In subsequent chapters, the quasisolid component of motion will be of prime importance, but in this case one should not forget the presence of the strain component.

4. A Vortex, a Vortex Line, and a Vortex Tube. Helmholtz's Second Theorem

It follows from the previous section that the quasisolid motion of a volume element is composed of the translatory motion with the velocity V, determined by any point of this element and taken to be the pole, and of the rotational motion around an instantaneous axis with the angular velocity vector $\boldsymbol{\omega} = \frac{1}{2}\text{rot } V$. The liquid motion accompanied by the rotation of its separate volume elements is called rotational and the rotating element itself (sometimes its angular velocity) is called a vortex. When the motion of the volume element of the medium is reduced to only translational and deformational motion, it is called irrotational.

Generally, to each point M of the velocity field of a moving medium there two vectors are assigned: the velocity $\boldsymbol{V}$ and the angular velocity $\boldsymbol{\omega}$ (Figure 13). To each of these two vectors correspond their own vector lines: streamlines to the vector $\boldsymbol{V}$ and the vortex lines to the vectors $\boldsymbol{\omega} = \frac{1}{2}\text{rot } \boldsymbol{V}$ or $\boldsymbol{\Omega} = \text{rot } V$, i.e., the lines at each point of which the tangent coincides in

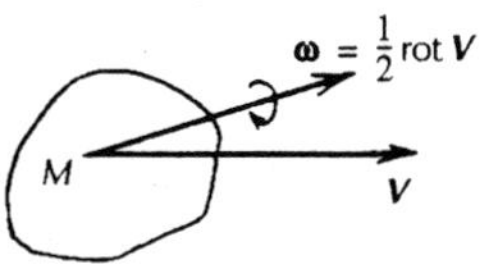

Fig. 13

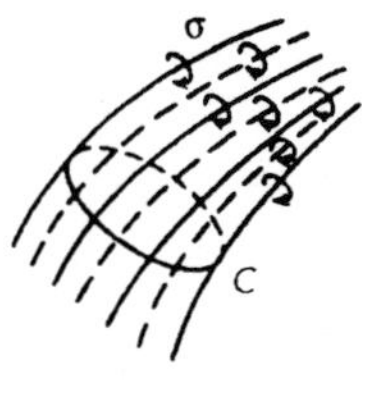

Fig. 14

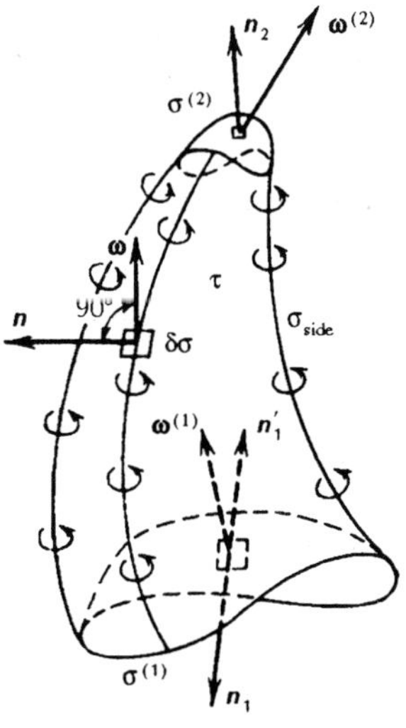

Fig. 15

direction with the vector $\boldsymbol{\Omega}$ or $\boldsymbol{\omega}$. As outlined below the vortex lines often are denoted by rounded arrows representing the rotation of volume elements around tangents to vortex lines.

The stream surfaces were constructed (see earlier in the Chapter), therefore, the vortex surface σ is obtained by drawing vortex lines through the points of a certain closed contour C (Figure 14), whereas the portion of liquid bounded by it and participating in rotation is called the vortex tube.

Analogously to the rate of volume flow, which is equal to the velocity vector flux through any cross-section of the stream tube, and is the same for an incompressible liquid along the entire stream tube, the flux of the vector $\boldsymbol{\omega}$ or of the vortex $\boldsymbol{\Omega} = \text{rot}\, \boldsymbol{V}$ will also be the same at the given time instant along the vortex tube.

Let us prove the following second theorem of Helmholtz: the flux of the velocity curl $F_\sigma(\boldsymbol{\Omega})$ (just as the flux of the angular velocity vector $F_\sigma(\boldsymbol{\omega})$) through any cross-section of the vortex tube is the same at the given time instant along the entire tube.

Let us isolate in the vortex tube the volume τ enclosed (Figure 15) between the side surface σ_{side} of the tube and its two arbitrary cross-sections $\sigma^{(1)}$ and $\sigma^{(2)}$. Let us apply to this volume the Gauss-Ostrogradskiy formula in the form of Equation 111 of Chapter 1 at $\boldsymbol{a} = \boldsymbol{\omega}$:

$$\int_\sigma \omega_n \,\delta\sigma = \int_\tau \text{div}\, \boldsymbol{\omega}\, \delta\tau \tag{35}$$

where the surface σ bounding the volume of the vortex tube τ consists of three surfaces: $\sigma^{(1)}$, $\sigma^{(2)}$, and σ_{side}. According to the vortex line definition, $\omega_n = 0$ on the side surface, and the integrand expression on the right side of Equation 35 is identically equal to zero ($\text{div}\,\text{rot}\, \boldsymbol{V} \equiv 0$). Equality (35) will be reduced to the form $\int_{\sigma^{(1)}} \omega_n \delta\sigma + \int_{\sigma^{(2)}} \omega_n \delta\sigma = 0$. By reversing the direction of the unit vector $\boldsymbol{n}_1$ to the opposite $\boldsymbol{n}_1'$ so that the cosine of the angle between $\boldsymbol{\omega}^{(1)}$ and $\boldsymbol{n}_1'$ can be positive, we shall obtain

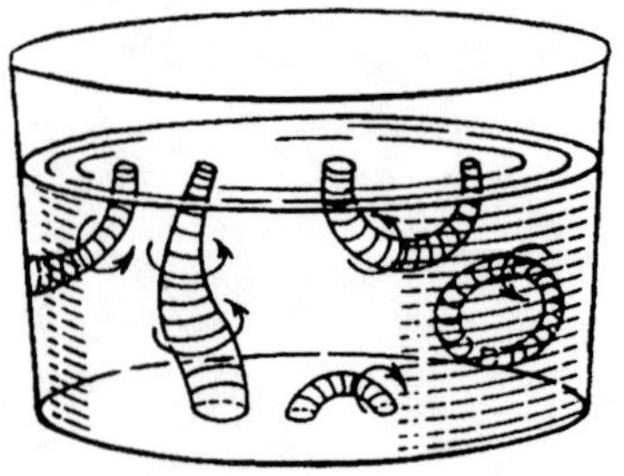

Fig. 16

$$\int_{\sigma^{(1)}} \omega_{n'} \, \delta\sigma = \int_{\sigma^{(2)}} \omega_n \, \delta\sigma. \quad (36)$$

Because the cross-sections $\sigma^{(1)}$ and $\sigma^{(2)}$ were selected arbitrarily, it is equality (36) that proves the second theorem of Helmholtz.

It follows from the second theorem of Helmholtz that the flux of the vector $\mathbf{\Omega}$ (or $\boldsymbol{\omega}$), identical at the given time instant for any cross-sections of the vortex tube, can be taken as a measure of the strength of the vortex tube *i* as a whole. It can also be concluded from the second theorem that vortex tubes cannot terminate inside the medium because then ($\sigma^{(1,2)} = 0$) the quantity ω or Ω should have become infinite. The experiment confirms this fact. In nature, either closed vortex rings are observed (Figure 16) (for example, smoke rising from a chimney), or vortex tubes starting and terminating at the interfaces between a solid surface and a liquid or a liquid and a gaseous media. These are whirlpools which are frequently observed in rivers and which are vortex tubes on one side resting on the river bed, and on the other, on the river-free surface that separates water from the atmosphere, or only on the bed or only on the free surface. Vortex tubes can be observed by the naked eye in a glass of milk or coffee when a spoon is taken out of it quickly. Gigantic whirlpools are observed in oceans behind islands (e.g., Jamaica) which are surrounded by rapid streams. The most formidable vortex tubes seem to be waterspouts. Their lower end rests on the ocean surface and their upper end spirals off into the layer of overhanging clouds. The reason for their origin are not well established, but seem to have a bearing on thermal phenomena at the interface between the ocean and atmosphere. Cyclones and anticyclones in the atmosphere do not behave not as vigorously.

Special attention is given to a vortex tube formed when water is discharged from a reservoir with a small hole at the bottom. Rotating together with the Earth, the water-filled reservoir has a small angular velocity which increases sharply when water passes through a narrow hole. Due to a sharply increased vortex (see Chapter 4), the liquid quickly rotates and forms a rarefaction region into which the air breaks out and creates an air cavity (the "hollow vortex"). The same phenomenon also occurs in waterspouts, at the bottom where water is collected from the body of water, and in the upper portion, where steam is collected from the clouds. Tornadoes originating in deserts entrain sand. For their description these phenomena require the use of the dynamics of a liquid or of a gas. The results of present-day investigations into different vortical flows in nature and engineering equipment can be found in a book by Albring.[6]

[6]Albring, W., *Elementarvorgänge Fluider Wirbelbewegungen*, Academie Verlag, Berlin, 1981, Section 305.

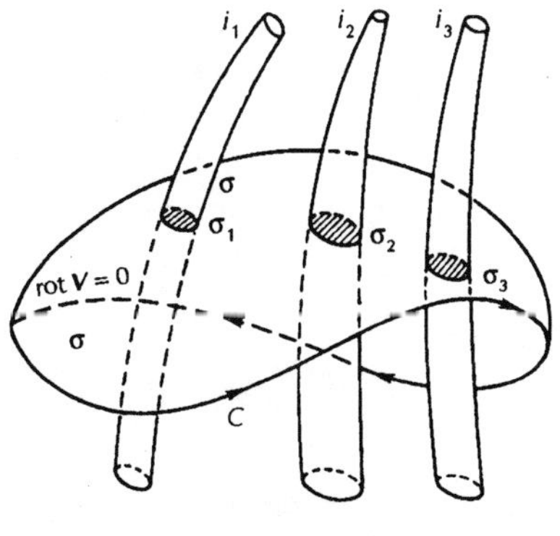

Fig. 17

5. The Stokes, Theorem on the Relationship between the Strength of the Vortex Tube and Velocity Circulation Around the Encircling Contour

Direct observation of a vortex tube in a transparent liquid that has no admixture or that is not specially heated for flow visualization is practically impossible. There are also special devices for measuring the strength of vortex tubes; however, this strength can be calculated by finding the velocity distribution on a corresponding contour with the aid of an appropriate instrument: revolving-vane, hot-wire, and laser anemometers and pressure tubes. For this calculation to be made, the use of the following Stokes, theorem is sufficient: *the strength of a vortex tube is equal to velocity circulation along the contour which is located on the surface of the vortex tube and which encircles it once.*

To prove the Stokes, theorem, use will be made of the Stokes, formula (Equation 116 Chapter 1), assuming that $\boldsymbol{a} = \boldsymbol{V}$ and having determined the strength of the vortex tube i, for example, as the flux of the velocity vortex vector $\boldsymbol{\Omega} = \text{rot } \boldsymbol{V}$.

This will give (σ is an arbitrary tube section resting on the contour C)

$$i = \int_{\sigma} \Omega_n \delta\sigma = \oint_{C} \boldsymbol{V} \cdot \delta \boldsymbol{r} \tag{37}$$

According to this theorem, the determination of the vortex tube strength is reduced to the integration of the elements of circulation $\boldsymbol{V} \cdot \delta\boldsymbol{r}$ along the contour which encircles the tube.[7]

It is common practice use the terms "strength" and "circulation" interchangeably when referring to vortices.

The vortex tube, which encases the agitated portions of liquid, may be surrounded by liquid whose particles do not rotate about their axes, but move translationally along closed (also round) trajectories. This irrotational "envelope" is caused (induced) by a vortex tube. In order to determine the strength of the vortex tube, there is no need in this case to calculate the circulation along the contour that encircles the tube once, but it is sufficient to calculate the velocity circulation around any contour encircling the tube once. In the case of several vortex tubes separated by irrotational flows of strength $i_1, i_2, \cdots$ (Figure 17), their total strength equal to the algebraic sum of the strengths of separate tubes will be determined by velocity circulation along the contour C which encircles the entire set of the vortex tubes.

[7] It is common practice use the terms "strength" and "circulation" interchangeably when referring to vortices.

The measured circulation along the closed contour in the velocity field makes it possible to not judge the presence or absence of vortex tubes inside it, but only their overall strength. The zero velocity circulation along the closed contour does not yet allow a conclusion about the absence of vortex tubes because the contour may contain vortex tubes with different directions of rotation to which the strengths that differ in sign correspond.

6. Acceleration of the Medium Particle. Local and Convective Components of Acceleration. Total Acceleration

The acceleration of a liquid particle, as with the acceleration of a point particle in general, is defined by *the time derivative of the velocity vector of this particle*

$$\dot{\mathbf{V}} = \frac{d\mathbf{V}}{dt} \tag{38}$$

In the Lagrange representation of motion (Equation 1) the velocity vector $\mathbf{V}$ is defined as

$$\mathbf{V} = \frac{d}{dt}\mathbf{r}(a, b, c; t) \tag{39}$$

and the acceleration vector $\dot{\mathbf{V}}$ as

$$\dot{\mathbf{V}} = \frac{d^2}{dt^2}\mathbf{r}(a, b, c; t) \tag{40}$$

(a, b, c are the parameters that separate the given particle from the multitude of other particles – they are the coordinates of the particle at the initial instant).

In the Euler representation, the velocity vector is prescribed by equalities (2), whereas for calculating the acceleration, i.e., the vector derivative of $\mathbf{V}$ with respect to time, it is necessary to take into consideration the following two circumstances that are for the variation in time of the velocity vector $\mathbf{V}$:

1. The local variation of velocity in time because of unsteadiness in the field of flow
2. Convective variation of velocity due to the *displacement of the liquid particle in a non-uniform velocity field*

The first, local acceleration, is defined as

$$\dot{\mathbf{V}}_{\text{loc}} = \frac{\partial \mathbf{V}}{\partial t} \tag{41}$$

To calculate the second, convective acceleration, it is necessary to find the increment of the velocity vector of the liquid particle due only to its displacement (convection) into the adjacent point of space in a "frozen" velocity field and relate it to the corresponding time interval. With the use of the Euler method, this requires differentiation of $\mathbf{V}(x_1, x_2, x_3; t)$ with respect to time as a multivariable function when the fourth argument t is fixed and $x_1(t)$, $x_2(t)$, $x_3(t)$ are variable coordinates of the liquid particle. As a result, the derivative of $\mathbf{V}$ in the direction of the trajectory is multiplied by the derivative of the path s with respect to time; i.e., by the velocity magnitude. According to Equation 78 of Chapter 1, we shall have ($\hat{\mathbf{V}} = \mathbf{V}/V$)

$$\dot{V}_{\text{conv}} = \frac{\partial V}{\partial s}\frac{ds}{dt} = (\hat{V}\cdot\nabla)VV = \left(\frac{V}{V}\cdot\nabla\right)VV = (V\cdot\nabla)V \tag{42}$$

Summing up the local and convective components of acceleration, obtain the expression for the total acceleration:

$$\dot{V} = \frac{\partial V}{\partial t} + (V\cdot\nabla)V \tag{43}$$

or in projections on the coordinate axes

$$\dot{V}_1 = \frac{\partial V_1}{\partial t} + V_1\frac{\partial V_1}{\partial x_1} + V_2\frac{\partial V_1}{\partial x_2} + V_3\frac{\partial V_1}{\partial x_3}$$

$$\dot{V}_2 = \frac{\partial V_2}{\partial t} + V_1\frac{\partial V_2}{\partial x_1} + V_2\frac{\partial V_2}{\partial x_2} + V_3\frac{\partial V_2}{\partial x_3}$$

$$\dot{V}_3 = \frac{\partial V_3}{\partial t} + V_1\frac{\partial V_3}{\partial x_1} + V_2\frac{\partial V_3}{\partial x_2} + V_3\frac{\partial V_3}{\partial x_3} \tag{44}$$

or

$$\dot{V}_i = \frac{\partial V_i}{\partial t} + V_k\frac{\partial V_i}{\partial x_k} \qquad (i = 1, 2, 3)$$

The local acceleration is equal to zero if the velocity field is stationary. The convective acceleration is absent when the velocity field is uniform. For example, in the case of body impact against a water surface at rest, first the local acceleration of liquid originates and the convective acceleration can be regarded as being equal to zero at the initial time instant; the velocity field nonuniformity and convective acceleration appear later. An analogous situation also takes place at the start of the motion of a body immersed in a motionless liquid.

The operator $V\cdot\nabla$ is called the *convective derivative*, the derivative $\partial/\partial t$ is called the *local derivative*, and the total derivative is written as d/dt, which is equal to[8]

$$\frac{d}{dt} = \frac{\partial}{\partial t} + (V\cdot\nabla) \tag{45}$$

which takes into account the total change in some physical quantity associated with the motion of an individual particle of the medium in the field of this quantity. In a more general case, it is associated in the field of any substance. This is referred to as *the ordinary (Lagrangian) or substantial derivative.*

The total acceleration of the particle in the Euler representation of its motion, which is equal to the sum of the local and conresponsible vective accelerations, is defined as an *ordinary (substantial) derivative of the velocity vector with respect to time*. For this derivative the notation d/dt will be preserved.

The symbolic formula (Equation 42) of convective acceleration can be replaced by another formula that does not contain the symbolic operator ∇. We shall first give the deriva-

[8] d/dt is often used for the total derivative ("derivative moving with the flow").

tion of one important formula of vector analysis which expresses convective acceleration in terms of the spatial derivatives grad and rot.

Recalling the formula of the double vector product

$$\boldsymbol{a} \times (\boldsymbol{b} \times \boldsymbol{c}) = \boldsymbol{b}(\boldsymbol{a} \cdot \boldsymbol{c}) - \boldsymbol{c}(\boldsymbol{a} \cdot \boldsymbol{b}) = \boldsymbol{b}(\boldsymbol{a} \cdot \boldsymbol{c}) - (\boldsymbol{a} \cdot \boldsymbol{b})\boldsymbol{c} \qquad (46)$$

and assume in it $\boldsymbol{b} = \nabla$. Thus,

$$\nabla(\boldsymbol{a} \cdot \boldsymbol{c}) = (\boldsymbol{a} \cdot \nabla)\boldsymbol{c} + \boldsymbol{a} \times (\nabla \times \boldsymbol{c})$$

Having exchanged $\boldsymbol{a}$ and $\boldsymbol{c}$,

$$\nabla(\boldsymbol{c} \cdot \boldsymbol{a}) = (\boldsymbol{c} \cdot \nabla)\boldsymbol{a} + \boldsymbol{c} \times (\nabla \times \boldsymbol{a})$$

In the first of these equalities the operation of differentiation of ∇ relates only to $\boldsymbol{c}$ with constant $\boldsymbol{a}$. In the second case, conversely, ∇ relates to $\boldsymbol{a}$ with constant $\boldsymbol{c}$. By the rule of symbol-forming space derivatives of complex vector expressions (see Chapter 1), combine the two expressions for the same quantity $\nabla(\boldsymbol{a} \cdot \boldsymbol{c})$ and obtain the general result

$$\nabla(\boldsymbol{a} \cdot \boldsymbol{c}) = (\boldsymbol{a} \cdot \nabla)\boldsymbol{c} + (\boldsymbol{c} \cdot \nabla)\boldsymbol{a} + \boldsymbol{a} \times (\nabla \times \boldsymbol{c}) + \boldsymbol{c} \times (\nabla \times \boldsymbol{a})$$

Converting partially from the symbolic notation to the operations grad and rot, we shall obtain the formula of vector analysis

$$\operatorname{grad}(\boldsymbol{a} \cdot \boldsymbol{c}) = (\boldsymbol{a} \cdot \nabla)\boldsymbol{c} + (\boldsymbol{c} \cdot \nabla)\boldsymbol{a} + \boldsymbol{a} \times \operatorname{rot} \boldsymbol{c} + \boldsymbol{c} \times \operatorname{rot} \boldsymbol{a} \qquad (47)$$

given earlier (Equation 88, Chapter 1) without derivation.

Let us assume in this equality that $\boldsymbol{a} = \boldsymbol{c}$, $\boldsymbol{a} \cdot \boldsymbol{c} = a^2$. Then, it will directly follow from vector relation (47) that

$$\operatorname{grad}(a^2) = 2(\boldsymbol{a} \cdot \nabla)\boldsymbol{a} + 2\boldsymbol{a} \times \operatorname{rot} \boldsymbol{a} \qquad (48)$$

This will yield the equality

$$(\boldsymbol{a} \cdot \nabla)\boldsymbol{a} = \operatorname{grad} \frac{a^2}{2} + \operatorname{rot} \boldsymbol{a} \times \boldsymbol{a} \qquad (49)$$

with which $\boldsymbol{a} = \boldsymbol{V}$ will lead to the expression not involving symbolic operators. Representing the convective component of acceleration in terms of the operations grad and rot,

$$\dot{\boldsymbol{V}}_{\text{conv}} = (\boldsymbol{V} \cdot \nabla)\boldsymbol{V} = \operatorname{grad} \frac{V^2}{2} + \operatorname{rot} \boldsymbol{V} \times \boldsymbol{V} \qquad (50)$$

The last expression for $\dot{\boldsymbol{V}}_{\text{conv}}$ will be useful in the dynamics for deriving the Gromeka-Lamb equation. This equality can also be of great help in deriving acceleration projections in the orthogonal curvilinear coordinate system. A direct approach to this derivation would consist of the use of velocity vector expansion into unit vectors $\boldsymbol{i}_r$ of coordinate axes in the curvilinear system

$$\boldsymbol{V} = V_{q_1}\boldsymbol{i}_1 + V_{q_2}\boldsymbol{i}_2 + V_{q_3}\boldsymbol{i}_3 = V_{q_r}\boldsymbol{i}_r \qquad (51)$$

Its subsequent differentiation with respect to time would require calculation of the ordinary derivatives dV_{qr}/dt and $d\boldsymbol{i}_r/dt$, and the process itself is rather laborious. It will suffice to

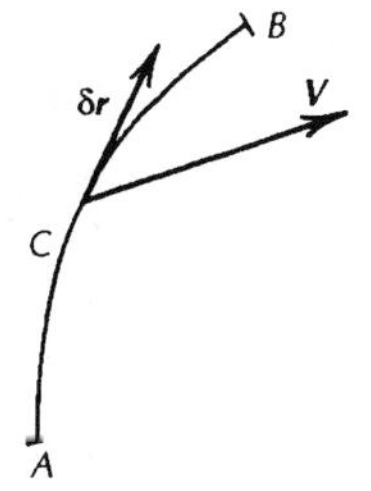

Fig. 18

present the expressions for acceleration projections on the axis of the most popular systems of curvilinear coordinates:

1. Cylindrical system (r, ε, z) (including polar):

$$\dot{V}_r = \frac{\partial V_r}{\partial t} + V_r \frac{\partial V_r}{\partial r} + \frac{V_\varepsilon}{r}\frac{\partial V_r}{\partial \varepsilon} + V_z \frac{\partial V_r}{\partial z} - \frac{V_\varepsilon^2}{r}$$

$$\dot{V}_\varepsilon = \frac{\partial V_\varepsilon}{\partial t} + V_r \frac{\partial V_\varepsilon}{\partial r} + \frac{V_\varepsilon}{r}\frac{\partial V_\varepsilon}{\partial \varepsilon} + V_z \frac{\partial V_\varepsilon}{\partial z} + \frac{V_r V_\varepsilon}{r}$$

$$\dot{V}_z = \frac{\partial V_z}{\partial t} + V_r \frac{\partial V_z}{\partial r} + \frac{V_\varepsilon}{r}\frac{\partial V_z}{\partial \varepsilon} + V_z \frac{\partial V_z}{\partial z} \quad (52)$$

2. Spherical system (R, θ, ε):

$$\dot{V}_R = \frac{\partial V_R}{\partial t} + V_R \frac{\partial V_R}{\partial R} + \frac{V_\theta}{R}\frac{\partial V_R}{\partial \theta} + \frac{V_\varepsilon}{R \sin\theta}\frac{\partial V_R}{\partial \varepsilon} - \frac{V_\theta^2 + V_\varepsilon^2}{R}$$

$$\dot{V}_\theta = \frac{\partial V_\theta}{\partial t} + V_R \frac{\partial V_\theta}{\partial R} + \frac{V_\theta}{R}\frac{\partial V_\theta}{\partial \theta} + \frac{V_\varepsilon}{R \sin\theta}\frac{\partial V_\theta}{\partial \varepsilon} + \frac{V_R V_\theta}{R} - \frac{V_\varepsilon^2}{R} \operatorname{ctg}\theta$$

$$\dot{V}_\varepsilon = \frac{\partial V_\varepsilon}{\partial t} + V_R \frac{\partial V_\varepsilon}{\partial R} + \frac{V_\theta}{R}\frac{\partial V_\varepsilon}{\partial \theta} + \frac{V_\varepsilon}{R \sin\theta}\frac{\partial V_\varepsilon}{\partial \varepsilon} + \frac{V_R V_\varepsilon}{R} + \frac{V_\theta V_\varepsilon}{R} \operatorname{ctg}\theta \quad (53)$$

7. Kelvin Kinematic Theorem

Thus far we have studied the properties of vortex tubes or velocity circulations, equivalent to them by the Stokes' theorem, along the contours encircling these tubes at the given time instant. The forthcoming Kelvin theorem makes it possible, purely kinematically, to predict the behavior of a vortex tube in time.[9] The theorem is formulated as follows: *an ordinary derivative with respect to time of velocity circulation along the closed liquid* (i.e., consisting of the same particles of liquid during the entire time of motion) *contour is equal to the accel-*

[9] Due to a purely kinematic aspect of this theorem not connected with the physical properties of the liquid or any exposure to the action of forces, this theorem is called the kinematic theorem of Kelvin, in contrast to his dynamic theorem, which will be considered later in the dynamics of a fluid.

eration circulation along the same contour. Let us first consider a somewhat more general case: variation in time of velocity circulation along the open liquid contour C (Figure 18).

Let us calculate the ordinary derivative with respect to time of the velocity vector (V) circulation along the open contour C between points A and B. According to the definition of the integral as the limit of the sum, we have

$$\frac{d}{dt}\int_{\substack{A\\(C)}}^{B} V\cdot\delta r = \int_{\substack{A\\(C)}}^{B} \frac{d}{dt}(V\cdot\delta r) = \int_{\substack{A\\(C)}}^{B} \frac{dV}{dt}\cdot\delta r + \int_{\substack{A\\(C)}}^{B} V\cdot\frac{d}{dt}(\delta r)$$

Because the operations d/dt and δ are independent, they can be interchanged and the last integral in the previous equality can be given the form

$$\int_{\substack{A\\(C)}}^{B} V\cdot\frac{d}{dt}\delta r = \int_{\substack{A\\(C)}}^{B} V\cdot\delta\left(\frac{dr}{dt}\right) = \int_{\substack{A\\(C)}}^{B} V\cdot\delta V = \int_{\substack{A\\(C)}}^{B} \delta\left(\frac{V^2}{2}\right)$$

The integral of the differential is equal to the difference between the values of the antiderivative function at the upper and lower limits. Thus,

$$\frac{d}{dt}\int_{\substack{A\\(C)}}^{B} V\cdot\delta r = \int_{\substack{A\\(C)}}^{B} \dot{V}\cdot\delta r + \frac{1}{2}(V_B^2 - V_A^2) \tag{54}$$

Having connected points B and A, we close the curve C and obtain

$$\frac{d}{dt}\oint_{(C)} V\cdot\delta r = \oint_{(C)} \dot{V}\cdot\delta r \tag{55}$$

or, in the earlier notation for the circulation,

$$\frac{d}{dt}\Gamma_C(V) = \Gamma_C(\dot{V}) \tag{56}$$

The Kelvin formula (Equations 55 or 56) expresses the general law of change in time of velocity circulation along the closed liquid contour and thus, according to the Stokes, theorem (Section 5 of this chapter), of the vortex tube strength, with this change being attributed to the circulation of acceleration, i.e., to the volumetric force circulation, which is probably equal to the work of the force along the closed contour.[10] In this resides the specific importance of the Kelvin theorem for the dynamics of a liquid and a gas, as shown in Chapter 7.

---

[10] The authors concept of "volumetric force" is developed further in the following chapter. Among other things, it is actually a force per unit mass and acceleration is an adequate term for present purposes.

3

Distribution of Mass and Force in a Continuous Medium. Cauchy Equalities. Theorem on the Reciprocity of Tangential Stresses

1. The Density of Mass Distribution in a Continuous Medium. Mass Conservation Law. Continuity Equation

In the dynamics of the discrete system of point particles, the dynamic characteristics, including inertia, were ascribed to separate points of the system (masses of the point particles of the system, forces applied to them). Considerations in the case of a perfect solid body were either the overall (but relating to sharply defined regions of a solid body) characteristics (moments of inertia depending on the dimensions, external shape of the body and distribution of masses in it, and vectors of momentum and of moment of momentum) or forces (loads) and the moments of forces concentrated at specific points of the solid body. In the dynamics of continuous media (liquids, gases, elastic, and other "solid" deformable bodies), such an approach is substituted by the assignment of continuous distributions of dynamic and, generally, physical quantities throughout a continuous medium. These distributions are characterized by the density of distribution of these quantities.

The first example of such an assignment is the density of mass distribution as the limit of the ratio between the mass Δm of the small volume $\Delta\tau$, which encloses the given point M, and the volume $\Delta\tau$ when the latter tends to zero as it is shrunk to point M. The ratio $\Delta m/\Delta\tau$ is called the mean density in the volume $\Delta\tau$, whereas the limit of this ratio for $\Delta\tau \to 0$ is called the density of the medium at the given point M and is designated by the letter ρ so that

$$\rho = \lim_{\Delta\rho \to 0} \frac{\Delta m}{\Delta\tau} = \frac{\delta m}{\delta\tau} \tag{1}$$

This equation directly yields expressions for the mass of a volume element δm in terms of the density ρ

$$\delta m = \rho \delta\tau \tag{2}$$

and for the mass m of the designated volume τ isolated in the medium

$$m = \int_\tau \rho \delta\tau \tag{3}$$

The density characterizes the inertial properties of the medium and forms the scalar field $\rho = \rho(x_1, x_2, x_3; t)$, which can be either stationary or nonstationary. The level surfaces in the density field are called *isopycnals*.

The density is expressed in the International System of Units (SI) in kilograms per cubic meter (kg/m^3). Numerical values of the densities of liquids and gases are given in the tables of physical quantities. As before, we shall continue to use the symbol δ for an infinitely small increment in space, and the symbol d will be retained for the increment in time.

In Newtonian mechanics the mass of a liquid volume, i.e., that consisting of the same particles during the duration of motion, retains a constant value, so that according to Equation 2 (d/dt is the symbol for an ordinary derivative with respect to time):

$$\frac{d}{dt}(\delta m) = \frac{d}{dt}(\rho \delta\tau) = \frac{d\rho}{dt}\delta\tau + \rho\frac{d}{dt}(\delta\tau) = 0 \tag{4}$$

Recalling the definition given at the end of Chapter 2, Section 3 for the volumetric expansion rate $\dot{\theta}$, we shall obtain from Equations 26 and 27 of Chapter 2 that

$$\frac{d}{dt}\delta\tau = \operatorname{div} \boldsymbol{V} \delta\tau$$

Consequently, equality (4) will be reduced to

$$\left(\frac{d\rho}{dt} + \rho \operatorname{div} \boldsymbol{V}\right)\delta\tau = 0$$

where, due to the arbitrariness of the quantity $\delta\tau$, it follows that

$$\frac{d\rho}{dt} + \rho \operatorname{div} \boldsymbol{V} = 0 \tag{5}$$

Thus, the mass conservation law in Newtonian mechanics has led to the differential equation that connects density and velocity (Equation 5). It should be emphasized that this connection exists without regard to the forces applied to the medium. Equation 5 is usually called the *continuity equation,* although *mass conservation equation* would have been a better monicker.

Expanding (according to Section 6 of Chapter 2) the ordinary derivative from Equation 5 into the local and convective components:

$$\frac{\partial\rho}{\partial t} + (\boldsymbol{V}\cdot\nabla)\rho + \rho \operatorname{div} \boldsymbol{V} = 0 \tag{6}$$

or, having introduced the scalar quantity ρ within the brackets,

$$\frac{\partial\rho}{\partial t} + \boldsymbol{V}\cdot \operatorname{grad} \rho + \rho \operatorname{div} \boldsymbol{V} = 0$$

Recalling the corresponding formula (88) of Chapter 1 ($\varphi = \rho$, $\boldsymbol{a} = \boldsymbol{V}$):

$$\operatorname{div}(\rho \boldsymbol{V}) = \boldsymbol{V} \cdot \operatorname{grad} \rho + \rho \operatorname{div} \boldsymbol{V}$$

and rewriting equality (6) in the form

$$\frac{\partial \rho}{\partial t} + \operatorname{div}(\rho \boldsymbol{V}) = 0 \tag{7}$$

This is another form of the same continuity equation (5) which has no ordinary derivative of density with respect to time.

When the density field is stationary ($\partial\rho/\partial t = 0$), the continuity equation will take the form

$$\operatorname{div}(\rho \boldsymbol{V}) = 0 \tag{8}$$

or in the velocity projections V_k ($k = 1, 2, 3$) (summation over k)

$$\frac{\partial(\rho V_k)}{\partial x_k} = \frac{\partial(\rho V_1)}{\partial x_1} + \frac{\partial(\rho V_2)}{\partial x_2} + \frac{\partial(\rho V_3)}{\partial x_3} = 0 \tag{9}$$

Finally, in the case of constant density (incompressible homogeneous medium), we obtain the *incompressibility equation* of a fluid

$$\operatorname{div} \boldsymbol{V} = 0 \tag{10}$$

or

$$\frac{\partial V_k}{\partial x_k} = \frac{\partial V_1}{\partial x_1} + \frac{\partial V_2}{\partial x_2} + \frac{\partial V_3}{\partial x_3} = 0$$

Such cases of motion of a continuous medium are discussed later. When along the direction of movement there is the continuous appearance or disappearance of a certain substance due (for example) to the conversion of one of its species into another via chemical reaction or to a change in the phase state of the substance (evaporation of a moving liquid with origination of vapor bubbles in it or, conversely vapor condensation with the appearance of liquid droplets in it, setting of liquid metal, melting of icicles in a water stream, etc.). Here, it is reasonable to refer to the methods of the mechanics of variable mass.[1]

A theoretical model for this kind of phenomena can be a continuous distribution of the sources of mass inflow or sink, which is prescribed *a priori*, determined by the chemical or physical kinetics of the processes occurring in a moving medium, and proceeds with intensity characterized by the mass increment per second and per unit volume at the given point in the flow. We could have easily designated this quantity, which has the dimensions $[M/(L^3T)]$ = density/time, by the symbol $\dot{\rho}$, but to avoid confusion with the ordinary derivative with respect to time $d\rho/dt$, we shall use the symbol J. The connection between the quantities $d\rho/dt$ and J is evident from the relation

[1]Loitsyanskiy, L. G. and Luriye, A. I. *Kurs teoreticheskoi mekhaniki* (*The Course of Theoretical Mechanics*), Vol. 2, Nauka Press, Moscow, 1983, pp. 110 to 114.

$$\frac{d}{dt}(\delta m) = \frac{d}{dt}(\rho\delta\tau) = J\delta\tau \tag{11}$$

which replaces Equation 4, and from the following continuity equations, being derived from Equation 11, but extended to the case of the dynamics of variable mass:

$$\frac{d\rho}{dt} + \rho \operatorname{div} \boldsymbol{V} = J \tag{12}$$

$$\frac{\partial\rho}{\partial t} + \operatorname{div}(\rho\boldsymbol{V}) = J \tag{13}$$

2. Distribution of Forces in a Continuous Medium. Volumetric and Surface Forces. Cauchy Equalities. Stress Tensor

In the dynamics of continuous media there are two classes of forces acting on particles of the media: volumetric (also called body forces) and surface forces. The volumetric forces are meant to be forces that act on the volume elements, for example, the forces of weight, gravity, inertia, electrostatic attraction or repulsion, and the forces of action of a magnetic or electric field on the medium particles. The term "surface" refers to forces which, based on the approach adopted in the mechanics of continuous media, act on the elements of a surface which bounds the volume. These forces include pressure, or, more generally, the forces of flow acting on the surface of the body immersed in it or the response of a body to the flow, and the forces of internal friction (viscosity) in the medium.

It should be noted that this classification of forces is arbitrary because Newton mechanics recognizes only the forces applied to masses, i.e., only volumetric forces. In cases in which the particles onto which the action of forces is concentrated are located in such a thin layer that it can be reduced without great error to a certain "material surface", it is assumed that the forces act on the elements of this surface.

In contrast to the dynamics of a system of discrete points, the dynamics of continuous media do not deal with the forces themselves, but with the densities of their distribution in space. Thus, the density of distribution of volumetric forces (or briefly, the volumetric force) $\boldsymbol{F}$ at the given point M of the medium is meant to be the limit of the ratio between the resultant vector $\Delta\boldsymbol{R}$ of the forces applied to the points of the small volume $\Delta\tau$, which encases the point M and the mass $\Delta m = \rho\Delta\tau$ where ρ is a certain mean value of the density in the volume $\Delta\tau$, when the volume $\Delta\tau$ tends to zero, preserving the point M inside, i.e.,

$$\boldsymbol{F} = \lim_{\Delta m \to 0} \frac{\Delta\boldsymbol{R}}{\Delta m} = \frac{1}{\rho} \lim_{\Delta\tau \to 0} \frac{\Delta\boldsymbol{R}}{\Delta\tau} = \frac{1}{\rho}\frac{\delta\boldsymbol{R}}{\delta\tau} \tag{14}$$

The quantity $\boldsymbol{F}$[2] is measured in the International System of Units (SI) in $\mathrm{N/kg}$ or $\mathrm{m/s^2}$. Thus, it follows that the ordinary Newtonian force $\delta\boldsymbol{R}$ applied to the volume element $\delta\tau$ at point M is defined in terms of the volumetric force $\boldsymbol{F}$ as

$$\delta\boldsymbol{R} = \rho\boldsymbol{F}\delta\tau$$

[2]The symbol $\boldsymbol{F}$ and the term "force" are unfortunate choices for a quantity with the units of acceleration. The reader will need to adopt a "per unit mass" point of view on such occasions.

As an example, it is possible to say that in the case of gravity, $\boldsymbol{F} = \boldsymbol{g}$, where $\boldsymbol{g}$ is the gravity acceleration vector; in the case of the centrifugal force of inertia in a system rotating with the angular velocity $\boldsymbol{\omega}$ the volumetric force $\boldsymbol{F} = \omega^2\boldsymbol{r}$ corresponds, where $\boldsymbol{r}$ is the vector that is equal to the shortest distance between the force point and the axis of rotation and which is directed outward from the axis.

Analogously, the surface forces are prescribed by the stress vector

$$\boldsymbol{p} = \lim_{\Delta\sigma \to 0} \frac{\Delta \boldsymbol{p}'}{\Delta\sigma} = \frac{\delta \boldsymbol{p}'}{\delta\sigma} \tag{15}$$

where $\Delta\boldsymbol{p}'$ is the resultant vector of forces applied from the side of the medium to a small area $\Delta\sigma$ isolated in this medium, and $\delta\boldsymbol{p}'$ and $\delta\sigma$ are their limiting values. The quantity $\boldsymbol{p}$ is measured in N/m^2.

The vector $\delta\boldsymbol{p}$ of the surface force applied to the area $\delta\sigma$ at the given point of space is equal, according to Equation 15, to

$$\delta\boldsymbol{p}' = \boldsymbol{p}\delta\sigma$$

i.e., to the product of the stress vector $\boldsymbol{p}$ by the area element $\delta\sigma$.

Note the basic difference (other than the noncoincidence of dimensions) between the vectors $\boldsymbol{F}$ and $\boldsymbol{p}$: while the vector $\boldsymbol{F}$ is a single-valued vector function of the points of space and time (i.e., forms a vector field), the vector $\boldsymbol{p}$ acquires at each point of space an infinite number of values depending on the orientation of the area to which the stress is applied. Thus, it does not form the vector space.

At the point M of the continuous medium we shall take the area $\delta\sigma$, the orientation of which in space will be determined by the unit vector $\boldsymbol{n}$ of the normal to the area (Figure 19). Let us call one side of the area $\delta\sigma$ the front side, and the other the back side. Draw the unit vector of the normal $\boldsymbol{n}$ on the front side. Then, on the front side of the area imagine the removal of a portion of the medium and replace its action on the area by the force $\boldsymbol{p}_n\delta\sigma$. The subscript n at the stress vector $\boldsymbol{p}_n$ indicates that the force is applied to the front side of the area with the unit vector of the normal $\boldsymbol{n}$. If, conversely, the portion of the medium had been removed from the back of the area, then the force equivalent to the action of the removed medium and applied to the back of the area would have been equal to $-\boldsymbol{p}_n\delta\sigma$, according to the law of action and counteraction. The choice of one of the sides of the area as the front and the other as the back is, of course, quite arbitrary, but it should be mentioned beforehand and fixed in the analysis process.

We shall prove that the stress vector $\boldsymbol{p}_n$ can be represented as the product of the unit vector $\boldsymbol{n}$, which characterizes the orientation of the area $\delta\sigma$ in space by the second-rank tensor P, which is a function of the vector-radius $\boldsymbol{r}$ of the point and time (i.e. forming the tensor field).

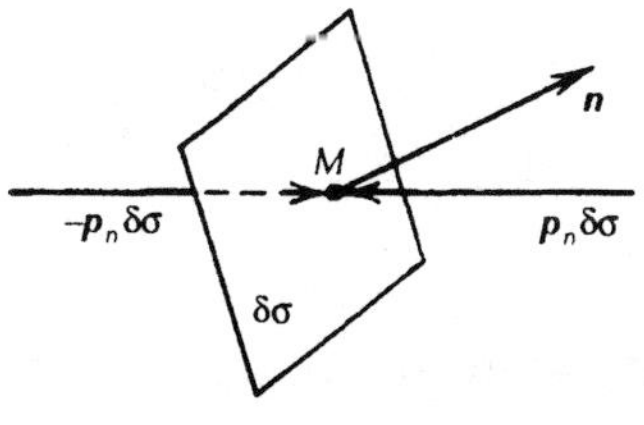

Fig. 19

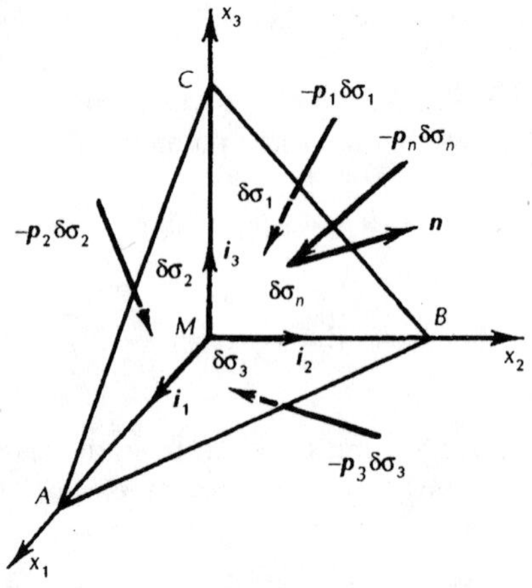

Fig. 20

For this purpose, we shall consider the elementary tetrahedron *MABC* cut out in the medium (Figure 20) and having its vertex at the given point *M*, the bottom (in the form of the triangle *ABC*) formed by the inclined plane intersecting three coordinate planes, and the side faces located in the coordinate planes. The area of the triangle *ABC* will be designated by $\delta\sigma_n$ and the areas of the triangles *BMC*, *AMC*, and *AMB*, which represent the projections of the triangle *ABC* onto the coordinate planes, will be denoted (respectively) by $\delta\sigma_1$, $\delta\sigma_2$, and $\delta\sigma_3$, with the subscripts 1, 2, and 3 at these areas, just as at the stresses $\boldsymbol{p}_1, \boldsymbol{p}_2, \boldsymbol{p}_3$ applied to these areas, indicating the axes which are perpendicular to these areas.

Treating the above, infinitely small tetrahedron as a "liquid" tetrahedron, i.e., consisting of the particles of a moving medium, we write the motion equation for the inertia center of this system of particles, the total mass of which is equal to δm:

$$\dot{\boldsymbol{V}}_C \delta m = \boldsymbol{F}\delta m + \boldsymbol{p}_n\delta\sigma_n - \boldsymbol{p}_1\delta\sigma_1 - \boldsymbol{p}_2\delta\sigma_2 - \boldsymbol{p}_3\delta\sigma_3$$

where $\dot{\boldsymbol{V}}_C$ is the acceleration vector of the inertia center of the tetrahedron; $\boldsymbol{F}$ is the density of distribution of volumetric forces, and $\boldsymbol{p}_n, \boldsymbol{p}_1, \boldsymbol{p}_2, \boldsymbol{p}_3$ are the stress vectors located on the sides of the front faces of the tetrahedron. The front faces are obviously those on which the unit vector $\boldsymbol{n}$ is located on the side, as well as the unit vectors $\boldsymbol{i}_1, \boldsymbol{i}_2, \boldsymbol{i}_3$ (Figure 20). The last three terms on the right side of the above equation contain the minus signs because the stresses $-\boldsymbol{p}_1, -\boldsymbol{p}_2, -\boldsymbol{p}_3$ are applied to the outer, back sides of the lateral faces.

In the equation considered the term on the left side and the first term on the right side can be omitted as the quantities of the third order of smallness (they involve the elements of mass proportional to the volume $\delta\tau$). These are compared to the rest of the terms that contain the elements of the surfaces $\delta\sigma_n$, $\delta\sigma_1$, $\delta\sigma_2$, and $\delta\sigma_3$, which are the quantities of the second order of smallness. Then, retaining only the quantities of the second order of smallness:

$$\boldsymbol{p}_n\delta\sigma_n = \boldsymbol{p}_1\delta\sigma_1 + \boldsymbol{p}_2\delta\sigma_2 + \boldsymbol{p}_3\delta\sigma_3 \qquad (16)$$

Noting that

$$\delta\sigma_1 = \delta\sigma_n \cos(\widehat{\boldsymbol{n}, x_1}) = n_1\delta\sigma_n$$

$$\delta\sigma_2 = \delta\sigma_n \cos(\widehat{\boldsymbol{n}, x_2}) = n_2\delta\sigma_n$$

$$\delta\sigma_3 = \delta\sigma_n \cos(\widehat{n, x_3}) = n_3 \delta\sigma_n$$

where n_1, n_2, n_3 are the cosines of the angles between the unit vector $\boldsymbol{n}$ and the coordinate axes, we shall obtain, after dividing both parts of Equation 16 by $\delta\sigma_n$ and contracting the tetrahedron to its vortex at point M, the exact expression

$$\boldsymbol{p}_n = n_1\boldsymbol{p}_1 + n_2\boldsymbol{p}_2 + n_3\boldsymbol{p}_3 = n_k\boldsymbol{p}_k \tag{17}$$

or, in projections on the axis of Cartesian coordinates,

$$p_{n1} = n_1p_{11} + n_2p_{21} + n_3p_{31} \qquad p_{n2} = n_1p_{12} + n_2p_{22} + n_3p_{32}$$

$$p_{n3} = n_1p_{13} + n_2p_{23} + n_3p_{33} \tag{18}$$

Recalling the definition of the stresses $\boldsymbol{p}_1$, $\boldsymbol{p}_2$, $\boldsymbol{p}_3$, note that in the notation adopted the first subscript at the stresses p_{11}, $p_{12}, \cdots$ denotes the axis perpendicularly to which the area $\delta\sigma$ is oriented, the second subscript denotes the axis onto which this stress is projected; thus, for example, p_{13} denotes the projection onto the axis x_3 of the stress applied to the area perpendicular to the axis x_1. It is evident that $\boldsymbol{p}_r = p_{rs}\boldsymbol{i}_s$ ($r = 1, 2, 3$; summation over s from 1 to 3).

The quantities with identical subscripts p_{11}, p_{22}, p_{33}, which represent the projections of the stress vectors $\boldsymbol{p}_1$, $\boldsymbol{p}_2$, $\boldsymbol{p}_3$ onto the normals to corresponding areas, are called normal stresses, whereas the projections p_{12}, p_{21}, $p_{31}, \cdots$ onto the axes lying in the plane of the areas are called tangential stresses.

The system of equalities (17 or 18) shows that the projections of the physically objective vector $\boldsymbol{p}_n$ onto the coordinate axes are the linear functions of projections onto the same axes of the physically objective vector $\boldsymbol{n}$. The coefficients in this linear connection represent the set of nine quantities

$$\begin{pmatrix} p_{11} & p_{21} & p_{31} \\ p_{12} & p_{22} & p_{32} \\ p_{13} & p_{23} & p_{33} \end{pmatrix} \tag{19}$$

As noted in Section 2 of Chapter 1, the existence of 2 linear connection between the projections of two physical vectors $\boldsymbol{p}_n$ and $\boldsymbol{n}$ indicates the physical objectiveness of the set of quantities (Equation 19) p_{kl} ($k, l = 1, 2, 3$), and thereby the physical objectiveness of the second-rank tensor of Equation 19 with these components.

Equation 19 does not satisfy the adopted order of subscripts: the first subscript should denote the number of the row, the second, of the column. Therefore, the stress tensor is taken to be the tensor P with the table

$$P\begin{pmatrix} p_{11} & p_{12} & p_{13} \\ p_{21} & p_{22} & p_{23} \\ p_{31} & p_{32} & p_{33} \end{pmatrix} \tag{20}$$

whereas corresponding to the tensor with Equation 19 is the tensor P^* conjugated with the stress tensor P. No difference exists between the tensors P and P^* (see below).

Recalling the definition for multiplication of a vector by a tensor on the right (or of a tensor by a vector on the left), we can rewrite Equations 18 in tensor form as

$$p_n = nP = P^* n \tag{21}$$

Equation 17 and equivalent equations 18, which establish a linear dependence between the projections of the stress applied to any area at the given point and the cosines of the angles between the normal to this area and the coordinate axes, are basic to the mechanics of continuous media. They are called the Cauchy equalities, which were initially in 1827. Equation 21 is equivalent to analytical formulas (17) or (18) but has the advantage of being independent of the choice of that or another Cartesian coordinate system, thus emphasizing its physical objectiveness.

3. Theorem on the Reciprocity of Tangential Stresses

According to the theorem of statics, the conditions for the equilibrium state of the tetrahedron *ABCM* should contain, in addition to the requirement of the zero-resultant vector (Equation 16), the requirement of zero-resultant moment of the forces applied.

Therefore, it is necessary to note that the definition of volumetric and surface forces adopted in Section 2 calls for additional definitions of volumetric and surface couples of forces which, for example, can form in ferromagnetic fluids. Being the quantities of the third order of smallness, the former will not enter into the equations of moments, whereas the latter must be taken into account.

Let us move on from this special case and assume that there are no surface pairs of forces in a moving medium.

In order to construct the equilibrium condition of Equation 16, the location of the surface force points is of no importance. The determination of the moments of surface forces requires knowledge of the coordinates of the force points. Because the tetrahedron faces are infinitely small, it is possible to assume that the surface forces are distributed uniformly over them and, consequently, the points of application of the resultants are located at the intersection of the medians in the triangles *ABC*, *AMB*, *BMC*, and *AMC* at the points G, G_1, G_2, G_3 (Figure 21). For the sake of clarity the figure contains only the position of the points G

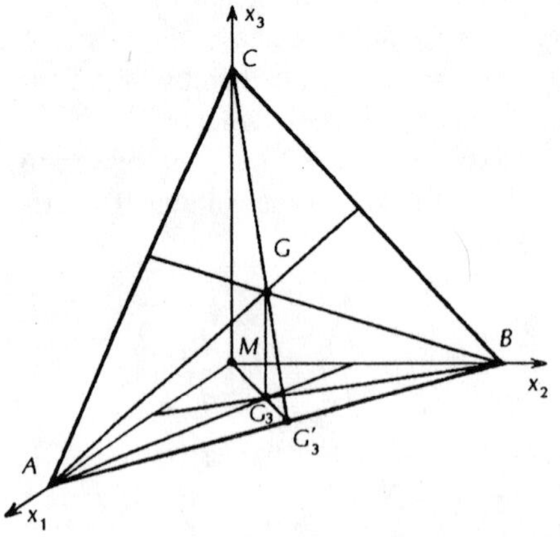

Fig. 21

and G_3 on the face ABC and on its projection AMB. Let us designate the vector-radii of the points of intersection of the medians G, G_1, G_2, G_3 by $\boldsymbol{r}$, $\boldsymbol{r}^{(1)}$, $\boldsymbol{r}^{(2)}$, $\boldsymbol{r}^{(3)}$. Then, the necessary second condition for the equilibrium of the tetrahedron will be the equation of moments

$$\boldsymbol{r} \times \boldsymbol{p}_n \delta\sigma_n = \boldsymbol{r}^{(1)} \times \boldsymbol{p}_1 \delta\sigma_1 + \boldsymbol{r}^{(2)} \times \boldsymbol{p}_2 \delta\sigma_2 + \boldsymbol{r}^{(3)} \times \boldsymbol{p}_3 \delta\sigma_3$$

Replace $\boldsymbol{p}_n$ in this equality by its expansion (Equation 17) and divide by $\delta\sigma_n$:

$$(\boldsymbol{r} - \boldsymbol{r}^{(1)}) \times \boldsymbol{p}_1 n_1 + (\boldsymbol{r} - \boldsymbol{r}^{(2)}) \times \boldsymbol{p}_2 n_2 + (\boldsymbol{r} - \boldsymbol{r}^{(3)}) \times \boldsymbol{p}_3 n_3 = 0 \tag{22}$$

The directed segment $\overrightarrow{G_3G}$ is equal to the difference between the vector-radii $\boldsymbol{r}$ and $\boldsymbol{r}^{(3)}$

$$\overrightarrow{G_3G} = \overrightarrow{MG} - \overrightarrow{MG_3} = \boldsymbol{r} - \boldsymbol{r}^{(3)}$$

and, because the point G_3 is the projection of the point G onto the plane Mx_1x_2, the directed segment $\overrightarrow{G_3G}$ is parallel to the vector $\boldsymbol{i}_3$.

Analogously, the vectors $\boldsymbol{r} - \boldsymbol{r}^{(1)}$ and $\boldsymbol{r} - \boldsymbol{r}^{(2)}$ are parallel to the vectors $\boldsymbol{i}_1$ and $\boldsymbol{i}_2$, so that it is possible to assume ($\lambda^{(k)}$ are scalars) that

$$\boldsymbol{r} - \boldsymbol{r}^{(1)} = \lambda^{(1)} \boldsymbol{i}_1 \qquad \boldsymbol{r} - \boldsymbol{r}^{(2)} = \lambda^{(2)} \boldsymbol{i}_2 \qquad \boldsymbol{r} - \boldsymbol{r}^{(3)} = \lambda^{(3)} \boldsymbol{i}_3 \tag{23}$$

Substituting the values of these differences into Equation 22, we shall form the relation

$$\lambda^{(1)} n_1 (\boldsymbol{i}_1 \times \boldsymbol{p}_1) + \lambda^{(2)} n_2 (\boldsymbol{i}_2 \times \boldsymbol{p}_2) + \lambda^{(3)} n_3 (\boldsymbol{i}_3 \times \boldsymbol{p}_3) = 0 \tag{24}$$

The scalars $\lambda^{(1)}$, $\lambda^{(2)}$, $\lambda^{(3)}$ remain to be determined. For this, we extend the segments $\overrightarrow{MG_1}$, $\overrightarrow{MG_2}$, and $\overrightarrow{MG_3}$ until their intersection with the plane ABC at the points G_1', G_2', G_3' (in the figure this construction is shown only for G_3'). According to the well-known theorem on the intersection of medians in a triangle, we have

$$\overrightarrow{MG_1'} = \frac{3}{2}\overrightarrow{MG_1} = \frac{3}{2}\boldsymbol{r}^{(1)} \qquad \overrightarrow{MG_2'} = \frac{3}{2}\boldsymbol{r}^{(2)} \qquad \overrightarrow{MG_3'} = \frac{3}{2}\boldsymbol{r}^{(3)}$$

Take into account that the ends of the vectors $\overrightarrow{MG_1'}$, $\overrightarrow{MG_2'}$, $\overrightarrow{MG_3'}$, just as of $\overrightarrow{MG}$, lie in the inclined plane ABC. Consequently, designating by h the length of the perpendicular from point M onto the plane ABC, we shall find ($\boldsymbol{n}$ is the unit vector of the normal to ABC)

$$\frac{3}{2}[\boldsymbol{r}^{(1)} \cdot \boldsymbol{n}] = \frac{3}{2}[\boldsymbol{r}^{(2)} \cdot \boldsymbol{n}] = \frac{3}{2}[\boldsymbol{r}^{(3)} \cdot \boldsymbol{n}] = h \tag{25}$$

The scalar multiplication of the both sides of each of the equalities (23) by $\boldsymbol{n}$ will yield, according to Equation 25, that

$$[(\boldsymbol{r} - \boldsymbol{r}^{(k)}) \cdot \boldsymbol{n}] = \boldsymbol{r} \cdot \boldsymbol{n} - \boldsymbol{r}^{(k)} \cdot \boldsymbol{n} = h - \frac{2}{3}h = \frac{1}{3}h = \lambda^{(k)} n_k$$

($k = 1, 2, 3$; no summation over k!).

From this it follows that $\lambda^{(k)} n_k = h/3$. Substituting the resulting values of $\lambda^{(k)}$ into equality (24) and canceling both sides by h, we arrive at the following vector relation

$$\boldsymbol{i}_1 \times \boldsymbol{p}_1 + \boldsymbol{i}_2 \times \boldsymbol{p}_2 + \boldsymbol{i}_3 \times \boldsymbol{p}_3 = 0 \tag{26}$$

Taking into account the equality $\boldsymbol{p}_r = p_{rs}\boldsymbol{i}_s$ (summation over s), we have

$$\boldsymbol{i}_1 \times \boldsymbol{p}_1 = \boldsymbol{i}_1 \times p_{1s}\boldsymbol{i}_s = p_{1s}(\boldsymbol{i}_1 \times \boldsymbol{i}_s)$$

where

$$\boldsymbol{i}_1 \times \boldsymbol{i}_s = \begin{cases} 0 & s = 1 \\ \boldsymbol{i}_3 & s = 2 \\ -\boldsymbol{i}_2 & s = 3 \end{cases}$$

so that

$$\boldsymbol{i}_1 \times \boldsymbol{p}_1 = p_{12}\boldsymbol{i}_3 - p_{13}\boldsymbol{i}_2$$

Analogously,

$$\boldsymbol{i}_2 \times \boldsymbol{p}_2 = -p_{21}\boldsymbol{i}_3 + p_{23}\boldsymbol{i}_1 \qquad \boldsymbol{i}_3 \times \boldsymbol{p}_3 = p_{31}\boldsymbol{i}_2 + p_{32}\boldsymbol{i}_1$$

Substituting these results into equality (26), we obtain

$$\boldsymbol{i}_1 \times \boldsymbol{p}_1 + \boldsymbol{i}_2 \times \boldsymbol{p}_2 + \boldsymbol{i}_3 \times \boldsymbol{p}_3 = (p_{23} - p_{32})\boldsymbol{i}_1 + (p_{31} - p_{13})\boldsymbol{i}_2 + (p_{12} - p_{21})\boldsymbol{i}_3 = 0$$

from which it follows directly that

$$p_{23} = p_{32} \qquad p_{31} = p_{13} \qquad p_{12} = p_{21} \qquad P = P^* \tag{27}$$

These equalities express the theorem on the reciprocity of tangential stresses: if three mutually perpendicular, infinitely small areas are drawn through any point of the medium, the projections of the vector of stress applied to one of the areas onto the normal to the other will be equal for each pair of these areas. Equalities (27) establish that under the adopted condition of the absence of continuous distribution of the surface couples of forces in the medium, the stress tensor components are independent of the order of subscripts, or in other words, the tensor of stresses is symmetric.

The mechanics of continuous media (liquids, gases, elastic bodies) for which the theorem on the reciprocity of tangential stresses is valid (and consequently, the stress tensor is symmetric) is called "symmetric", and otherwise "nonsymmetric". The latter has only recently begun development, has become the subject of a small number of studies, and thus is not discussed in this volume. In the treatment that follows only the symmetric mechanics of liquids and gases is considered.

These essential conclusions from the theorem of the reciprocity of shear stresses should be noted:

1. Of nine unknown components of the stress tensor, only six are different

2. The order of subscripts at the stress tensor components in the Cauchy equalities (18) is interchangeable
3. Equality (21) acquires the form

$$p_n = nP = Pn \tag{28}$$

It should be emphasized that the presence of volumetric or surface couples of forces does not restrict the validity of the Cauchy equalities because the couples do not contribute to forming the resultant vector of forces.

4

General Theorems of the Dynamics of a Continuous Medium

1. The Theorem of Momentum. Equation of Dynamics in Terms of "Stresses"

The general theorem of the change in the resultant vector of the momentum of a system is *the time derivative of the resultant vector of the momentum of a system is equal to the resultant vector of external forces applied to the system.* As applied to the mechanics of continuous flowing media, this formulation undergoes some changes and becomes: *an ordinary derivative of the resultant vector of the momentum of a "liquid" volume is equal to the resultant vector of external volumetric and surface forces applied to particles located in the volume and on the bounding surface, respectively.*

Isolate in the volume an arbitrary finite liquid volume τ (Figure 22). If $\delta\tau$ is the volume element containing the point M with the velocity V, then its momentum vector δK is equal to

$$\delta K = \rho V \delta\tau$$

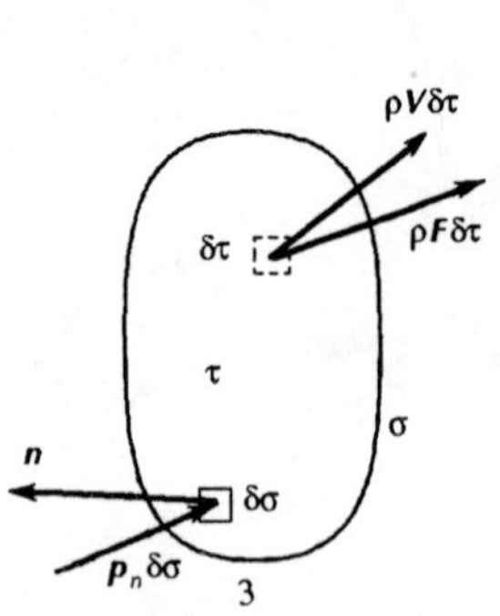

Fig. 22

and the resultant vector of momentum throughout the entire volume τ is equal to the sum or the integral of elementary momenta

$$\boldsymbol{K} = \int_\tau \rho \boldsymbol{V} \delta\tau \tag{1}$$

For the same reasons, the resultant vectors of the external volumetric and surface forces will be equal, respectively, to

$$\int_\tau \rho \boldsymbol{F} \delta\tau \quad \text{and} \quad \int_\sigma \boldsymbol{p}_n \delta\sigma \tag{2}$$

Thus, the theorem dealing with the change in momentum and just formulated in terms of the dynamics of continuous media leads to the equality

$$\frac{d}{dt}\int_\tau \rho \boldsymbol{V} \delta\tau = \int_\tau \rho \boldsymbol{F} \delta\tau + \int_\sigma \boldsymbol{p}_n \delta\sigma \tag{3}$$

The left side of Equation 3 is transformed into

$$\frac{d}{dt}\int_\tau \rho \boldsymbol{V} \delta\tau = \int_\tau \frac{d}{dt}(\rho \boldsymbol{V} \delta\tau) = \int_\tau \rho \frac{d\boldsymbol{V}}{dt} \delta\tau + \int_\tau \boldsymbol{V} \frac{d}{dt}(\rho\delta\tau)$$

According to the mass conservation condition Equation 4 of Chapter 3, the last integral is equal to zero; consequently,

$$\frac{d}{dt}\int_\tau \rho \boldsymbol{V} \delta\tau = \int_\tau \rho \frac{d\boldsymbol{V}}{dt} \delta\tau \tag{4}$$

In the last term on the right side of Equation 3 the quantity $\boldsymbol{p}_n$ will be replaced by expressions (17) or (28) from Chapter 3. According to the Gauss-Ostrogradskiy theorem, we obtain the equivalent results

$$\int_\sigma \boldsymbol{p}_n \delta\sigma = \int_\sigma (n_1 \boldsymbol{p}_1 + n_2 \boldsymbol{p}_2 + n_3 \boldsymbol{p}_3)\, \delta\sigma = \int_\tau \left(\frac{\partial \boldsymbol{p}_1}{\partial x_1} + \frac{\partial \boldsymbol{p}_2}{\partial x_2} + \frac{\partial \boldsymbol{p}_3}{\partial x_3}\right) \delta\tau$$

$$\int_\sigma \boldsymbol{p}_n \delta\sigma = \int_\sigma \boldsymbol{n} P \delta\sigma = \int_\tau \nabla P \delta\tau = \int_\tau \text{Div } P \delta\tau \tag{5}$$

Returning to equality (3), substitute into it the terms that are reduced to volumetric integrals and combine them by transposing to the left side of the equation:

$$\int_\tau \left(\rho \frac{d\boldsymbol{V}}{dt} - \rho \boldsymbol{F} - \frac{\partial \boldsymbol{p}_1}{\partial x_1} - \frac{\partial \boldsymbol{p}_2}{\partial x_2} - \frac{\partial \boldsymbol{p}_3}{\partial x_3}\right) \delta\tau = 0$$

$$\int_{\tau}\left(\rho\frac{d\boldsymbol{V}}{dt}-\rho\boldsymbol{F}-\operatorname{Div}P\right)\delta\tau = 0$$

where, due to the arbitrariness of the volume of integration τ, we derive two possible expressions for the equation of the dynamics of a continuous medium in terms of stresses:

$$\rho\frac{d\boldsymbol{V}}{dt} = \rho\boldsymbol{F}+\frac{\partial \boldsymbol{p}_1}{\partial x_1}+\frac{\partial \boldsymbol{p}_2}{\partial x_2}+\frac{\partial \boldsymbol{p}_3}{\partial x_3} = \rho\boldsymbol{F}+\frac{\partial \boldsymbol{p}_i}{\partial x_i}$$

$$\rho\frac{d\boldsymbol{V}}{dt} = \rho\boldsymbol{F}+\operatorname{Div}P \tag{6}$$

The vectors

$$\frac{\partial \boldsymbol{p}_i}{\partial x_i} = \frac{\partial \boldsymbol{p}_1}{\partial x_1}+\frac{\partial \boldsymbol{p}_2}{\partial x_2}+\frac{\partial \boldsymbol{p}_3}{\partial x_3} = \operatorname{Div}P$$

on the right sides of Equations 6 can be considered, according to equalities (5), as the limits

$$\lim_{\Delta\tau\to 0}\frac{1}{\Delta\tau}\int_{\Delta\sigma}\boldsymbol{p}_n\delta\sigma = \lim_{\Delta\tau\to 0}\frac{1}{\Delta\tau}\int_{\Delta\sigma}\boldsymbol{n}P\delta\sigma \tag{7}$$

of the ratio between the resultant vector of surface forces applied to the surface $\Delta\sigma$, which bounds the volume $\Delta\tau$ containing point M, and this volume, i.e., as the volumetric density of the distribution of the resultant vector of surface forces. This density is called the volumetric action of surface forces.[1]

In the Cartesian system of rectangular coordinates, Equations 6 acquire the form

$$\rho\left(\frac{\partial V_i}{\partial t}+V_k\frac{\partial V_i}{\partial x_k}\right) = \rho F_i+\frac{\partial p_{ki}}{\partial x_k}$$

($i = 1, 2, 3$; summation over k), or, in expanded form,

$$\rho\left(\frac{\partial V_1}{\partial t}+V_1\frac{\partial V_1}{\partial x_1}+V_2\frac{\partial V_1}{\partial x_2}+V_3\frac{\partial V_1}{\partial x_3}\right) = \rho F_1+\frac{\partial p_{11}}{\partial x_1}+\frac{\partial p_{21}}{\partial x_2}+\frac{\partial p_{31}}{\partial x_3}$$

$$\rho\left(\frac{\partial V_2}{\partial t}+V_1\frac{\partial V_2}{\partial x_1}+V_2\frac{\partial V_2}{\partial x_2}+V_3\frac{\partial V_2}{\partial x_3}\right) = \rho F_2+\frac{\partial p_{12}}{\partial x_1}+\frac{\partial p_{22}}{\partial x_2}+\frac{\partial p_{32}}{\partial x_3}$$

$$\rho\left(\frac{\partial V_3}{\partial t}+V_1\frac{\partial V_3}{\partial x_1}+V_2\frac{\partial V_3}{\partial x_2}+V_3\frac{\partial V_3}{\partial x_3}\right) = \rho F_3+\frac{\partial p_{13}}{\partial x_1}+\frac{\partial p_{23}}{\partial x_2}+\frac{\partial p_{33}}{\partial x_3} \tag{8}$$

This system of equations is augmented by the continuity equation (Equation 7) of Chapter 3, which in the above-mentioned coordinates has the form

---

[1] It may be useful to note the units of Equation 6, $[F/L^3]$, and hence the consistency of the "volumetric" terminology in the present context.

$$\frac{\partial \rho}{\partial t} + \frac{\partial}{\partial x_1}(\rho V_1) + \frac{\partial}{\partial x_2}(\rho V_2) + \frac{\partial}{\partial x_3}(\rho V_3) = 0 \qquad (9)$$

Equations 8, together with Equation 9, provide the basis for the dynamics of continuous media. They contain, as specific cases, different forms of equations corresponding to these or additional assumptions concerning the physical properties of certain models of continuous media (ideal, viscous liquid, etc.).

The previous derivation of the equation in terms of stresses related to the case of the absence of internal sources of mass inflow; i.e., according to the end of Section 1 of Chapter 3, it corresponded to the condition $J = 0$. In the presence of sources of this kind ($J \neq 0$) and under motionless conditions, the additional term JV appears on the left side of Equations 6 to represent a change in time of the momentum due to the presence of these sources; thus, instead of Equations 6:

$$\rho \frac{dV}{dt} + JV = \rho F + \text{Div } P$$

or, by transposing the term JV to the right side,

$$\rho \frac{dV}{dt} = \rho F + \text{Div } P - JV \qquad (10)$$

This equation can be considered to be the equation of the dynamics of a continuous medium of variable mass,[2] and the last term on the right may be interpreted as a reactive force based on the volume unit or as the density of distribution of reactive forces.

2. Theorem of Moments and Resulting Theorems on the Reciprocity of Tangential Stresses

The law of change in the resultant moment of momentum of a system of point particles is: *the time derivative of the resultant moment of momentum of a system of point particles is equal to the resultant moment of external forces applied to the system.* In the dynamics of a continuous medium this law is formulated as follows: *an ordinary derivative of the resultant moment of momentum of the finite liquid volume is equal to the resultant moment of external volumetric and surface forces applied to this volume and its bounding surface.*

Using Figure 22, find the moments of the vectors indicated with respect to the arbitrary center 0 (not shown). Designating by r the instantaneous vector-radius of the points in the volume τ and on its bounding surface σ, express the just-formulated theorem by the following equality

$$\frac{d}{dt}\int_\tau r \times \rho V \delta\tau = \int_\tau r \times \rho F \delta\tau + \int_\sigma r \times p_n \delta\sigma \qquad (11)$$

The integral on the left side is transformed as follows:

[2]Loitsyanskiy, L. G. and Luriye, A. I., *Kurs teoreticheskoi mekhaniki (The Course of Theoretical Mechanics)*, Vol. 2, Nauka Press, Moscow, 1983, pp. 110 to 114.

$$\frac{d}{dt}\int_\tau \boldsymbol{r}\times\rho\boldsymbol{V}\delta\tau = \int_\tau \frac{d\boldsymbol{r}}{dt}\times\rho\boldsymbol{V}\delta\tau + \int_\tau \boldsymbol{r}\times\rho\frac{d\boldsymbol{V}}{dt}\delta\tau + \int_\tau \boldsymbol{r}\times\boldsymbol{V}\frac{d}{dt}(\rho\delta\tau)$$

The first of the integrals on the right side is equal to zero because

$$\frac{d\boldsymbol{r}}{dt}\times\rho\boldsymbol{V} = \boldsymbol{V}\times\rho\boldsymbol{V}\equiv 0$$

the third integral is also equal to zero because, according to the mass conservation law, $\frac{d}{dt}(\rho\delta\tau) = 0$, and we arrive at the equality

$$\frac{d}{dt}\int_\tau \boldsymbol{r}\times\rho\boldsymbol{V}\delta\tau = \int_\tau \boldsymbol{r}\times\rho\frac{d\boldsymbol{V}}{dt}\delta\tau$$

Consider the transformation of the surface integral of the right side of Equation 11 into the volumetric one. According to the Cauchy formula (17) of Chapter 3, it can be rewritten in the form

$$\int_\sigma \boldsymbol{r}\times\boldsymbol{p}_n\delta\sigma = \int_\sigma \boldsymbol{r}\times(n_1\boldsymbol{p}_1 + n_2\boldsymbol{p}_2 + n_3\boldsymbol{p}_3)\,\delta\sigma =$$

$$= \int_\sigma [n_1(\boldsymbol{r}\times\boldsymbol{p}_1) + n_2(\boldsymbol{r}\times\boldsymbol{p}_2) + n_3(\boldsymbol{r}\times\boldsymbol{p}_3)]\,\delta\sigma$$

or, using the Gauss-Ostrogradskiy integral relation, as

$$\int_\sigma \boldsymbol{r}\times\boldsymbol{p}_n\delta\sigma = \int_\tau\left[\frac{\partial}{\partial x_1}(\boldsymbol{r}\times\boldsymbol{p}_1) + \frac{\partial}{\partial x_2}(\boldsymbol{r}\times\boldsymbol{p}_2) + \frac{\partial}{\partial x_3}(\boldsymbol{r}\times\boldsymbol{p}_3)\right]\delta\tau =$$

$$= \int_\tau\left[\boldsymbol{r}\times\left(\frac{\partial\boldsymbol{p}_1}{\partial x_1} + \frac{\partial\boldsymbol{p}_2}{\partial x_2} + \frac{\partial\boldsymbol{p}_3}{\partial x_3}\right) + \left(\frac{\partial\boldsymbol{r}}{\partial x_1}\times\boldsymbol{p}_1\right) + \left(\frac{\partial\boldsymbol{r}}{\partial x_2}\times\boldsymbol{p}_2\right) + \left(\frac{\partial\boldsymbol{r}}{\partial x_3}\times\boldsymbol{p}_3\right)\right]\delta\tau \qquad (12)$$

Substituting the resulting expressions into Equation 11 and grouping the terms, we shall arrive at the expression

$$\int_\tau\left[\boldsymbol{r}\times\left(\rho\frac{d\boldsymbol{V}}{dt} - \rho\boldsymbol{F} - \frac{\partial\boldsymbol{p}_1}{\partial x_1} - \frac{\partial\boldsymbol{p}_2}{\partial x_2} - \frac{\partial\boldsymbol{p}_3}{\partial x_3}\right) - \right.$$

$$\left. - \frac{\partial\boldsymbol{r}}{\partial x_1}\times\boldsymbol{p}_1 - \frac{\partial\boldsymbol{r}}{\partial x_2}\times\boldsymbol{p}_2 - \frac{\partial\boldsymbol{r}}{\partial x_3}\times\boldsymbol{p}_3\right]\delta\tau = 0 \qquad (13)$$

Thus, bearing in mind the first equality of Equations 6 and the arbitrariness of the integration volume τ, we obtain the relationship between the stress vectors $\boldsymbol{p}_1, \boldsymbol{p}_2, \boldsymbol{p}_3$:

$$\frac{\partial r}{\partial x_1} \times p_1 + \frac{\partial r}{\partial x_2} \times p_2 + \frac{\partial r}{\partial x_3} \times p_3 = 0$$

which, noting that

$$\frac{\partial r}{\partial x_1} = \frac{\partial}{\partial x_1}(x_1 i_1 + x_2 i_2 + x_3 i_3) = i_1 \qquad \frac{\partial r}{\partial x_2} = i_2 \qquad \frac{\partial r}{\partial x_3} = i_3$$

can be written in the form

$$i_1 \times p_1 + i_2 \times p_2 + i_3 \times p_3 = 0 \tag{14}$$

The last equality coincides with equality (26) of Chapter 3 which, as was shown, leads to the theorem on the reciprocity of tangential stresses and to the symmetry of the stress tensor (Equation 27, Chapter 3).

It should be noted that relation (14) is valid for any continuous medium regardless of the character of forces applied to it and of the presence or absence of the mass inflow from outside, provided that no distributed couples of forces occur in this medium. In other words, the existence of distributed sources (sinks) of mass would not have violated the symmetry of the stress tensor.

It is obvious that if the property of the reciprocity of tangential stresses is considered to be known on the basis of the geometric conclusion of Chapter 3, the corresponding formula (27, Chapter 3) is substituted into equality (11), and the momentum equation (6) is taken into account, then the equation of moments (11) will simply convert into an identity and it will not give any new information. However, if the geometric derivation of the reciprocity theorem for tangential stresses is ignored, and the transformations that lead the equation of the balance of moments to the form of equation (14) are made independently, this will be the second proof of the same theorem. Finally, based on the definition for the operation of vector multiplication of two tensors (Equation 70, Chapter 1), the third (tensor) proof may have been given for the symmetry of the stress tensor in the form of equality (75) of Chapter 1: $P \times E = 0$, which is completely independent of the choice of a coordinate system.

However, because such tensor operations are not needed in the following Section, this version of the proof is omitted.

3. Theorem of the Change in Kinetic Energy and General Law of Energy Conservation

The general formulation of the law of change in the kinetic energy of a system of point particles is as follows: *the time derivative of the kinetic energy of the system of point particles is equal to the sum of powers of both the external forces applied to the system and the forces of internal interactions between the points of the system*. The author emphasizes the difference in the formulations of the theorem of the change in kinetic energy and the theorems of momenta and moments of momenta: in the above formulation the internal forces are taken into account together with the external forces.

The integral expression of the theorem on the change in the kinetic energy of a moving individual volume of a continuous medium is constructed using the formulation for the dynamics of material systems

$$\frac{d}{dt}\int_\tau \frac{\rho V^2}{2}\delta\tau = \int_\tau \rho \boldsymbol{F}\cdot \boldsymbol{V}\delta\tau + \int_\sigma \boldsymbol{p}_n \cdot \boldsymbol{V}\delta\sigma + \int_\tau N_{\text{in}}\delta\tau \qquad (15)$$

where N_{in} is meant to be the power of internal forces related to the unit volume and also called the density of distribution of the power of internal forces.

Going over from the surface to the volumetric integral with the aid of the transformation

$$\int_\sigma \boldsymbol{p}_n \cdot \boldsymbol{V}\delta\sigma = \int_\sigma (\boldsymbol{n}P)\cdot \boldsymbol{V}\delta\sigma = \int_\sigma \boldsymbol{n}\cdot (P\boldsymbol{V})\,\delta\sigma = \int_\tau \text{div}\,(P\boldsymbol{V})\,\delta\tau$$

which uses the easily verifiable identity

$$(\boldsymbol{n}P)\cdot \boldsymbol{V} \equiv \boldsymbol{n}\cdot (P\boldsymbol{V})$$

as well as the Gauss-Ostrogradskiy formula (Equation 111, Chapter 1), and discarding, in the same way as in the previous sections, the integrals, we shall obtain the expression in differential form for the theorem on the change of the kinetic energy:

$$\rho\frac{d}{dt}\left(\frac{V^2}{2}\right) = \rho \boldsymbol{F}\cdot \boldsymbol{V} + \text{div}\,(P\boldsymbol{V}) + N_{\text{in}} \qquad (16)$$

Let us compare this equation with the equality

$$\rho\frac{d\boldsymbol{V}}{dt}\cdot \boldsymbol{V} \equiv \rho\frac{d}{dt}\left(\frac{V^2}{2}\right) = \rho \boldsymbol{F}\cdot \boldsymbol{V} + \boldsymbol{V}\cdot \text{Div}\,P$$

composed of the second equation (Equation 6) by means of scalar multiplication of both its parts by the velocity vector $\boldsymbol{V}$. The term-by-term subtraction of the last two equations yields the expression for the density of distribution of the power of internal forces

$$N_{\text{in}} = \boldsymbol{V}\cdot \text{Div}\,P - \text{div}\,(P\boldsymbol{V}) \qquad (17)$$

This expression can be simplified further by using the following series of differential transformations with the operator ∇:

$$\text{div}\,(P\boldsymbol{V}) = \nabla\cdot (P\boldsymbol{V}) = (\nabla P)\cdot \boldsymbol{V} + P\cdot (\nabla\boldsymbol{V}) = \boldsymbol{V}\cdot \text{Div}\,P + P\cdot (\nabla\boldsymbol{V}) \qquad (18)$$

Here, the last expression on the right side is the scalar product of the stress tensor by the dyad of the velocity field (see Equations 18, 19, and 69, Chapter 1). Using the previous equality, we shall find, according to Equation 17, that

$$N_{\text{in}} = -P\cdot (\nabla\boldsymbol{V})$$

Alternatively, by expanding the dyad of the velocity field $\nabla\boldsymbol{V}$ into the symmetric $\dot{S}$ and antisymmetric $\dot{A}$ parts and noting that $P\cdot \dot{A} \equiv 0$, due to the symmetry of the tensor P, we find the following expression of the power of internal forces as the scalar product (taken with the opposite sign) of the stress tensor by the tensor of the rate of strain:

$$N_{\text{in}} = -P\cdot \dot{S} \qquad (19)$$

The right side of Equation 19, which expresses the density of the distribution of the power of internal forces, resembles in its structure the formula of the force power. However, the difference consists in the fact that in the case of a discrete force the power is defined as the scalar product of the force vectors by the velocity, whereas in a continuous medium the density of the distribution of the power of internal forces is equal to the scalar product of stress tensors by the rates of strain.

In moving continuous media conversions occur of one form of energy into another and, primarily, of mechanical energy into thermal energy. These conversions are calculated from the energy balance equation derived from the general thermodynamic energy conservation law (the first law of thermodynamics), which for the given individual volume of the moving medium is formulated thus: *the arbitrary time derivative of the total energy of the given moving volume of the medium is equal to the sum of the powers of the external body and surface forces applied to the isolated volume and to its surface. These are related to the unit time of the quantity of energy supplied from outside to the volume.* This law is expressed by the integral equality

$$\frac{d}{dt}\int_\tau \rho\left(U+\frac{V^2}{2}\right)\delta\tau = \int_\tau \rho \boldsymbol{F}\cdot \boldsymbol{V}\delta\tau + \int_\sigma \boldsymbol{p}_n \cdot \boldsymbol{V}\delta\sigma + \int_\tau \rho q\,\delta\tau \tag{20}$$

where U is the specific (related to the unit mass) internal energy of the medium involving all possible energies of internal motions of matter (in the problems in liquid and gas mechanics this in first of all the thermal energy), and q is the specific quantity of energy supplied from the outside per unit time to the given particle of the medium. This involves thermal and non-thermal types of energy that differ from the work of macroscopic mechanical forces. The sum of the internal and external (kinetic) energies is called the total energy.

Using the mass conservation law (Equation 4, Chapter 3) the left side of Equation 20 can be transformed:

$$\frac{d}{dt}\int_\tau \rho\left(U+\frac{V^2}{2}\right)\delta\tau = \int_\tau \rho\frac{d}{dt}\left(U+\frac{V^2}{2}\right)\delta\tau +$$

$$+\int_\tau\left(U+\frac{V^2}{2}\right)\frac{d}{dt}(\rho\delta\tau) = \int_\tau \rho\frac{d}{dt}\left(U+\frac{V^2}{2}\right)\delta\tau$$

and the surface integral on the right side of equality (20) can be transformed, as with the derivation of Equation 16, into the volumetric integral

$$\int_\sigma \boldsymbol{p}_n\cdot \boldsymbol{V}\delta\sigma = \int_\tau \operatorname{div}(\boldsymbol{P}\boldsymbol{V})\,\delta\tau$$

Substituting the above expressions into Equation 20 and taking into account (as above) the arbitrariness of the volume τ, we obtain the total energy balance equation in differential form:

$$\rho\frac{d}{dt}\left(U+\frac{V^2}{2}\right) = \rho\boldsymbol{F}\cdot\boldsymbol{V} + \operatorname{div}(\boldsymbol{P}\boldsymbol{V}) + \rho q \tag{21}$$

Comparing the total energy balance equation (Equation 21) with the earlier-derived equation for the change in the kinetic energy (16), we may obtain, by subtracting term-by-term these equations from each other, the following internal energy balance equation:

$$\rho \frac{dU}{dt} = \rho q - N_{\text{in}} = \rho q + P \cdot \dot{S} \tag{22}$$

This equation does not involve either external volumetric forces or velocities in explicit form and represents the relationship between the individual change in time of the internal energy of medium per unit mass, the inflow of internal energy from the outside, and the power of internal forces.

Using the expression obtained for the density of distribution of the power of internal forces N_{in} (Equation 19), we may reduce Equation 16 to one of the two equivalent forms:

$$\rho \frac{d}{dt}\left(\frac{V^2}{2}\right) = \rho \boldsymbol{F} \cdot \boldsymbol{V} + \boldsymbol{V} \cdot \operatorname{Div} P$$

$$\rho \frac{d}{dt}\left(\frac{V^2}{2}\right) = \rho \boldsymbol{F} \cdot \boldsymbol{V} + \operatorname{div}(P\boldsymbol{V}) - P \cdot \dot{S} \tag{23}$$

One of the balance equations of energy (total or internal) in this or another form can be attached to the general system of the equations for the fluid dynamics, which will nevertheless remain indeterminate. In order to make the system determinate, it is necessary either to introduce additional connections between the components of the stress tensor and of the tensor of strain rates (for example, Newton's law of viscous stresses), or to impose simplifying assumptions on the stress tensor components (a model of ideal liquid, i.e., one devoid of internal friction).

In the presence of the mass inflow from continuously distributed sources, additional terms will appear in the previous formulas. Thus, instead of Equation 16, the differential form of the theorem on the change in the kinetic energy will be represented by the equation

$$\rho \frac{d}{dt}\left(\frac{V^2}{2}\right) = \rho \boldsymbol{F} \cdot \boldsymbol{V} + \operatorname{div}(P\boldsymbol{V}) + N_{\text{in}} - \frac{JV^2}{2} \tag{24}$$

The energy balance equations in differential form (21 and 22) will go to the following:

$$\rho \frac{d}{dt}\left(U + \frac{V^2}{2}\right) = \rho \boldsymbol{F} \cdot \boldsymbol{V} + \operatorname{div}(P\boldsymbol{V}) + \rho q - J\left(U + \frac{V^2}{2}\right) \tag{25}$$

$$\rho \frac{dU}{dt} = pq - N_{\text{in}} - JU \tag{26}$$

Finally, the expression for the power of internal forces will acquire the form

$$N_{\text{in}} = -P \cdot \dot{S} - \frac{JV^2}{2} \tag{27}$$

Equations 24 to 27 have been rather easily derived with the use of equality (11) of Chapter 3 instead of equality (4) of that chapter. When deriving Equations 10 and 24 to 27, it was assumed that continuously distributed mass sources were stationary. In the case of sources moving at velocities $\bar{\boldsymbol{V}}$, the change in the momentum density due to the mass inflow from these moving sources is determined by the difference $J(\boldsymbol{V} - \bar{\boldsymbol{V}})$ and the corresponding reactive force will be determined by the difference $J(\bar{\boldsymbol{V}} - \boldsymbol{V})$. The change in kinetic energy is equal to $J(V^2/2-\bar{V}^2/2)$ and the change in the total energy, $E = U + \frac{1}{2}V^2$, is correspondingly equal to $J(E-\bar{E})$, where $\bar{E} = \bar{U}+\frac{1}{2}\bar{V}^2$. This kind of expression is used in a later Section for deriving equations of motion of inhomogeneous multicomponent gas mixtures in which the masses of some mixture components originate and disappear due to physicochemical conversions.

In practice, direct application of the general laws of the dynamics of liquids and gases is sometimes difficult. In view of this, the preference is given, as in hydraulics, to the use of their integral expressions. This is often accompanied by explicit or implicit averaging of velocity fields or other physical quantities. This problem is the subject matter of the next section.

4. Transport of a Physical Quantity by the Medium Flow through a Surface – The Euler Theorem

The integral forms of the laws of the dynamics and energy balance of continuous media (momenta and moments of momenta, Equations 3 and 11, balance of kinetic and total energy, Equations 15 and 20) contain on the left sides ordinary time derivatives of the expressions of the type $\int_\tau \Phi\,\delta\tau$ called the values of the physical quantity Φ in the volume τ. Depending on the form of the equation, the quantity Φ can be a scalar, vector, or tensor quantity. In the cases enumerated it is correspondingly equal to $\rho\boldsymbol{V}$, $\boldsymbol{r}\times\rho\boldsymbol{V}$, $\rho V^2/2$, and $\rho(U+V^2/2)$.

Let us split the ordinary derivative d/dt of $\int_\tau \Phi\,\delta\tau$ into the local $\partial/\partial t$ and convective $(d/dt)_{\text{conv}}$ parts:

$$\frac{d}{dt}\int_\tau \Phi\,\delta\tau = \frac{\partial}{\partial t}\int_\tau \Phi\,\delta\tau + \left(\frac{d}{dt}\right)_{\text{conv}}\int_\tau \Phi\,\delta\tau$$

To clarify the convective part of the derivative, and also to simplify manipulations, we introduce the concept of transfer of the physical quantity by the medium flow through the so-called control surface. Let us agree that the control surface of a moving liquid volume of the medium will be the surface immovable in space which at the given instant bounds the moving volume being considered. Moving in space, the liquid volume flows through the control surface.

Let us take in the space, filled with a moving medium, an elementary area $\delta\sigma$ with the unit vector of the normal $\boldsymbol{n}$ which determines the front of the area. The product $\Phi V_n\delta\sigma$ of the physical quantity Φ by the rate of the medium volume flow $V_n\delta\sigma$ through the area $\delta\sigma$ is the transfer of the quantity Φ through this area and the integral $\int_\sigma \Phi V_n\,\delta\sigma$ represents the transfer of the same quantity through the surface σ of finite extension.

Assuming, for example, that Φ is equal to the momentum vector $\rho\boldsymbol{V}$ related to the unit volume, we obtain the vector of momentum transfer through the surface σ equal to the inte-

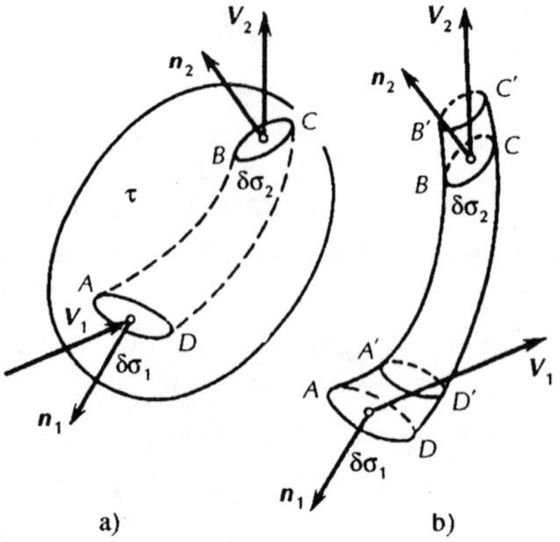

Fig. 23

gral $\int_\sigma \rho V V_n \delta\sigma$. The vector $\int_\sigma \rho r V V_n \delta\sigma$ represents the transfer of the moment of momentum.

The medium mass $\int_\sigma \rho V_n \delta\sigma$ flowing through the surface σ per second can be considered to be the transfer of the density ρ through the surface σ; the scalar quantity $\int_\sigma \rho \frac{V^2}{2} V_n \delta\sigma$ can be considered to be the transfer of the kinetic energy, etc.

Let us prove that the convective time derivative of the integral of a certain quantity, taken over the moving volume, is equal to the transfer of the same quantity through the control surface of the volume.

In order to do this, divide the volume τ (Figure 23a) into elementary stream tubes *ABCD*. For each of these tubes form an expression for an infinitely small convective change in the integral of the quantity Φ over the volume of the elementary tube which occurred for the time *dt* because of the displacement of this volume from the position *ABCD* to the position *A'B'C'D'* (Figure 23b). The convective increment of the quantity $\int_\tau \Phi \delta\tau$ for time *dt* is caused, by definition, only by the change in the position of the integrated volume in the non-uniform field $\Phi(x_1, x_2, x_3; t)$ "frozen" at the initial instant *t*. Consequently, the integral over the common portion of the *A'BCD'* of the new and old volumes will not effect this increment and can be discarded. Then

$$d_{\text{conv}} \int_\tau \Phi \delta\tau = \int_{A'B'C'D'} \Phi(x_1, x_2, x_3; t)\, \delta\tau - \int_{ABCD} \Phi(x_1, x_2, x_3; t)\, \delta\tau =$$

$$= \int_{BB'C'C} \Phi(x_1, x_2, x_3; t)\, \delta\tau - \int_{AA'D'D} \Phi(x_1, x_2, x_3; t)\, \delta\tau$$

Because the volumes are elementary, the following equalities are valid

$$\text{volume} \quad AA'D'D = -V_{1n}\delta\sigma_1 dt$$

$$\text{volume} \quad BB'C'C = V_{2n}\delta\sigma_2 dt$$

Here the subscript n means that the projections are taken in the direction of the external normal $\boldsymbol{n}$ to the surface bounding the volume $ABCD$, which is confined between sections $\delta\sigma_1$ and $\delta\sigma_2$. The sought-after convective change in the integral of Φ in the elementary tube volume is equal to

$$\Phi_1 V_{1n}\delta\sigma_1 dt + \Phi_2 V_{2n}\delta\sigma_2 dt$$

Having made the summation of these quantities over all the elementary tubes into which the volume τ is divided, we obtain an expression for an infinitely small convective change in the considered integral for time dt:

$$d_{\text{conv}} \int_\tau \Phi\delta\tau = \int_\sigma \Phi V_n \delta\sigma dt$$

Dividing by dt, we move to the convective derivative

$$\left(\frac{d}{dt}\right)_{\text{conv}} \int_\tau \Phi\delta\tau = \int_\sigma \Phi V_n \delta\sigma \tag{28}$$

thus proving the earlier statement.

The left side of Equation 28 contains the derivative, whose calculation requires the observation of the volume motion. This is the convective derivative in the Lagrangian representation of the motion of medium (Equation 1, Chapter 2).

The same convective derivative on the right side of Equation 28 requires for its calculation the knowledge of the velocity field, i.e., of the basic element in the Euler variables (Chapter 2, Equation 2). It is possible that the expression on the right side of Equation 28 is the convective derivative in Euler's notation.

The ordinary time derivative of the integral over the moving liquid volume τ of the physical quantity Φ also includes the local part, so that the expanded expression for the ordinary derivative has the form

$$\frac{d}{dt}\int_\tau \Phi\delta\tau = \frac{\partial}{\partial t}\int_\tau \Phi\delta\tau + \int_\sigma \Phi V_n \delta\sigma \tag{29}$$

If the field of the quantity Φ is stationary, the local part can be discarded to give

$$\frac{d}{dt}\int_\tau \Phi\delta\tau = \int_\sigma \Phi V_n \delta\sigma \tag{30}$$

Using Equation 29 and assuming that $\Phi = \rho$, we obtain the mass conservation equation in integral form:

$$\frac{\partial}{\partial t}\int_\tau \rho\delta\tau + \int_\sigma \rho V_n \delta\sigma = 0 \tag{31}$$

This equation is especially simple in cases of a steady flow when the first term on the left is equal to zero. In this case the equation

$$\int_{\sigma} \rho V_n \delta\sigma = 0 \tag{32}$$

states that the mass flow rate of the medium across a stationary closed surface with no mass sources inside is equal to zero.

By applying this equation to the volume of the finite stream tube confined between the sections σ_1 and σ_2, where the velocities and densities are designated by V_1, V_2, ρ_1, ρ_2, respectively, we obtain

$$\int_{\sigma_1} \rho V_{1n} \delta\sigma = \int_{\sigma_2} \rho V_{2n} \delta\sigma \tag{33}$$

This equation expresses the conservation law for the mass flow rate along the stream tube, or, using the terminology of Chapter 1, Section 8, the conservation law for the flux of the momentum vector $\rho \boldsymbol{V}$ along the stream tube.

According to the foregoing, the integral form of the theorem on the change of the medium momentum in the volume τ bounded by the closed surface σ is

$$\int_{\tau} \rho \boldsymbol{F} \delta\tau + \int_{\sigma} \boldsymbol{p}_n \delta\sigma - \frac{\partial}{\partial t} \int_{\tau} \rho \boldsymbol{V} \delta\tau - \int_{\sigma} \rho \boldsymbol{V} V_n \delta\sigma = 0 \tag{34}$$

The last term, including the minus sign, can be interpreted as the momentum transfer across the control surface toward the inside of the volume τ. In fact, the unit vector of the external normal is directed to the outside of the volume, and therefore, if at a certain point of the surface the velocity vector $\boldsymbol{V}$ is also directed to the outside of the volume ($V_n > 0$), the element of the integral $-\rho V_n \boldsymbol{V} \delta\sigma$ is directed into the volume; on the other hand, if the vector $\boldsymbol{V}$ is directed into the volume, then $V_n < 0$ and the element of the integral is directed in the same direction as is the vector $\boldsymbol{V}$, i.e., again toward the inside of the volume.

It is useful to note that the integral forms of the expressions for the mass (31) and momentum (34) conservation laws can easily give, with the use of the Gauss-Ostrogradskiy formula, corresponding differential forms.

In the case of a steady flow, Equation 34 can be interpreted in the following way: the resultant vectors of the external volumetric and surface forces, applied to an isolated liquid volume, together with the vector (taken with the opposite sign) of momentum transfer across the control surface, which corresponds to this liquid volume, form a closed triangle, i.e., the sum of these three vectors is equal to zero. This is Euler's theorem concerning momenta in a continuous medium. This theorem can be applied to analogous formulations for the moment of momenta and mechanical energy which are widely used in the course of hydraulics and applied hydrodynamics. Among the applications, note should be made of the calculations for jet pressure on a plane or curvilinear wall, the determination of the moment of momenta of a liquid moving through a turbine rotor, and the application of the Bernoulli and of the Borda theorems to calculate a spontaneously expanding flow, etc.[3]

---

[3] In addition to the textbooks on hydraulics, see also Loitsyanskiy, L. G. and Luriye, A. I., *Kurs teoreticheskoy mekhaniki* (*A Course of Theoretical Mechanics*), Vol. 2, 6th rev. suppl., Nauka Press, Moscow, 1983, pp. 143 to 147, 191, 192, 245 to 251.

5. Statics of a Flowing Medium. Euler Equations for the Medium Equilibrium

The equations of continuity, momentum, reciprocity of tangential stresses, and the balance of kinetic and total energies constitute the basic system of equations of the mechanics of continuous media. This system is not closed because the number of unknown quantities in it considerably exceeds the number of equations. As a rule, it is impossible to do without additional assumptions concerning the physical properties of the medium. The foundations of these assumptions will be considered in what follows.

At the same time, there is a typical and practically important class of problems when the basic system of equations of the continuous medium can be closed without resorting to new physical concepts. This is statics, i.e., the relative rest and the equilibrium of the medium.

As already noted in the introductory portion of the Chapter, the mechanics of continuous media is based on two assumptions: continuity, i.e., continuous distribution of all the properties in space, and fluidity or ease of mobility. To construct all the sections of kinematics, the first assumption is sufficient, whereas for the dynamics the second assumption is also of great importance.

Recall that except for some special fluids, flowing continuous media, in the absence of nondiagonal components of the tensor of strain rates, i.e., shear rates, also have no tangential components of the stress tensor. Likewise, in a quiescent ($\boldsymbol{V} \equiv 0$) fluid medium the tangential stresses are absent

$$p_{12} = p_{21} = p_{23} = p_{32} = p_{13} = p_{31} = 0 \tag{35}$$

and the expressions of normal stresses have the form

$$\boldsymbol{p}_n = p_n \boldsymbol{n} \qquad \boldsymbol{p}_1 = p_{11}\boldsymbol{i}_1 \qquad \boldsymbol{p}_2 = p_{22}\boldsymbol{i}_2 \qquad \boldsymbol{p}_3 = p_{33}\boldsymbol{i}_3 \tag{36}$$

Using Equations 35, 36, and the Cauchy equations (18, Chapter 3), we find

$$p_{11} = p_{22} = p_{33} = \boldsymbol{p}_n \tag{37}$$

i.e., three normal stresses applied to three mutually perpendicular areas, arbitrarily oriented in space, are all equal. This law of the isotropy of normal stresses at the points of a continuous medium being in equilibrium was discovered in the mid-17th century by Pascal.

The overall value of normal stresses at the given point of the medium, taken with the minus sign, is called pressure, and is designated by the letter p so that

$$p_{11} = p_{22} = p_{33} = -p \tag{38}$$

In order to emphasize that the adopted definition corresponds only to the case of medium in equilibrium, this pressure is called the hydrostatic pressure.

The minus sign in the pressure definition indicates that the normal stress

$$\boldsymbol{p}_n = -p\boldsymbol{n} \tag{39}$$

applied at the points on the surface of a mentally isolated volume is directed to the side opposite to the unit vector of the external normal to this surface, i.e., toward the inside of the volume.

The pressure p represents a physical scalar as do the density, temperature, concentration, and other scalar characteristics. It is measured in Newtons per square meter (Pa), kilograms per square meter (bar).[4]

The pressure in a gas in which surface tensions are impossible is always positive; it vanishes only under vacuum conditions. In a liquid the existence of surface tensions is possible.

In the case of the medium in equilibrium state, the stress tensor in it has, according to Equation 36, the table

$$\begin{pmatrix} -p & 0 & 0 \\ 0 & -p & 0 \\ 0 & 0 & -p \end{pmatrix} = -p\begin{pmatrix} 1 & 0 & 0 \\ 0 & 1 & 0 \\ 0 & 0 & 1 \end{pmatrix}$$

In other words, it possesses the spherical symmetry in conformity with the property of isotropy of normal stresses in a quiescent medium.

It is possible to assume (E is the unit tensor) that

$$\boldsymbol{P} = -p\boldsymbol{E} \tag{40}$$

The equation for the medium equilibrium can be easily obtained by assuming in the momentum equations (8) that $V_1 = V_2 = V_3 = 0$ and, moreover, by taking into account Equations 35 and 38.

Let us find the following Euler equations for the statics of the medium:

$$\rho F_1 = \frac{\partial p}{\partial x_1} \qquad \rho F_2 = \frac{\partial p}{\partial x_2} \qquad \rho F_3 = \frac{\partial p}{\partial x_3} \tag{41}$$

or an equivalent vector equation

$$\rho \boldsymbol{F} = \operatorname{grad} p \tag{42}$$

The same result could be directly obtained from the second equation (6) when noting that in the case considered

$$\operatorname{Div} P = -\operatorname{grad} p \tag{43}$$

The density and pressure in the equilibrium condition (42) can be eliminated. For this purpose, first operate with curl (rot) on the both sides of this equation; then p is eliminated because rot grad $p \equiv 0$:

$$\operatorname{rot}(\rho \boldsymbol{F}) = 0$$

where, expanding according to the familiar rule (the third formula in Equation 88, Chapter 1):

$$\rho \operatorname{rot} \boldsymbol{F} + \operatorname{grad} \rho \times \boldsymbol{F} = 0 \tag{44}$$

Conduct the scalar multiplication of the both parts of this equation by $\boldsymbol{F}$; then, noting that the second term, as a vector product, is perpendicular to its cofactor $\boldsymbol{F}$, find the following general limitation imposed on the class of forces under the action of which the equilibrium state of a liquid or a gas is possible

[4] Autor implies here the use of the kilogram as a unit of force.

$$\boldsymbol{F} \cdot \operatorname{rot} \boldsymbol{F} = 0 \tag{45}$$

In form this coincides with Equation 10, Chapter 2, and expresses the existence of surfaces normal to the force lines of the field.

The volumetric forces that satisfy Equation 45 include, the forces that have the potential Π because for them

$$\boldsymbol{F} = -\operatorname{grad} \Pi \qquad \operatorname{rot} \boldsymbol{F} = 0$$

In this case

$$\operatorname{grad} \rho \times \boldsymbol{F} = -\operatorname{grad} \rho \times \operatorname{grad} \Pi = 0 \tag{46}$$

from which it follows that in the case of medium equilibrium the force lines of the field of potential volumetric forces are orthogonal to isopycnals (the surfaces of identical densities) and that the isopycnals coincide with the isopotential surfaces of the force field.

It follows from Equation 42 that in the case of medium equilibrium the force lines are orthogonal to isobars (the surfaces of identical pressures). Thus, generally, when a liquid or a gas is in equilibrium, the isopotential surfaces of the field coincide with the isobars and isopycnals under the action of the potential field of volumetric forces.

It is also possible to prove the reverse: if the isobars coincide with the isopycnals, the equilibrium state of a liquid or a gas is possible only under the action of the potential field of volumetric forces. In fact, from the condition

$$\operatorname{grad} p \times \operatorname{grad} \rho = 0$$

or, according to Equation 42,

$$\boldsymbol{F} \times \operatorname{grad} \rho = 0$$

On the basis of Equation 44, it follows that

$$\operatorname{rot} \boldsymbol{F} = 0 \qquad \boldsymbol{F} = -\operatorname{grad} \Pi$$

If in a moving or quiescent gas the density is the function of pressure, then the motion or equilibrium is called barotropic. From the above we can see that the barotropic equilibrium state of a gas is possible only in the presence of potential forces, because under the condition $\rho = \rho(p)$ the isobars and isopycnals will obviously coincide; consequently, as has just been shown, the force field should be potential.

Euler equations (41) represent the so-called system in total differentials and have the general solution of the form

$$\int_{p_0}^{p} \frac{dp}{\rho(p)} - \int_{M_0}^{M} \boldsymbol{F} \cdot d\boldsymbol{r} \tag{47}$$

or, in the presence of the potential $\Pi(M)$,

$$\int_{p_0}^{p} \frac{dp}{\rho(p)} = \Pi(M_0) - \Pi(M) \tag{48}$$

where $p_0 = p(M_0)$. In order to make the problem determinate, it is necessary to employ additional assumptions about the form of the dependence of density on pressure.

6. Equilibrium State of an Incompressible Fluid. Archimedes' Law

The equation of equilibrium of an incompressible fluid in the potential field of volumetric forces is

$$-\rho \operatorname{grad} \Pi = \operatorname{grad} p \tag{49}$$

where at $\rho = \text{const}$ it directly follows that

$$p + \rho\Pi = \text{const} \tag{50}$$

Suppose two immiscible fluids of different densities ρ_1 and ρ_2 are in mutual equilibrium. At the interface between these fluids the pressure p and the potential are continuous. The derivative of the left side of Equation 50 along any direction s in the tangent plane to the interface should simultaneously satisfy the following two equalities:

$$\frac{dp}{ds} + \rho_1 \frac{d\Pi}{ds} = 0 \qquad \frac{dp}{ds} + \rho_2 \frac{d\Pi}{ds} = 0$$

from which the subtraction gives

$$(\rho_1 - \rho_2)\frac{d\Pi}{ds} = 0$$

With the adopted condition that $\rho_1 \neq \rho_2$, the latter equation leads to the constancy of the potential of volumetric forces Π at the interface. According to Equation 50, the pressure p in this case will retain its constant value along the interface surface, which is called the free surface. Thus, in the case of equilibrium state of two immiscible incompressible fluids of different densities in the potential field of volumetric forces, the interface between the fluids will simultaneously represent the isopotential surface and the isobar.

When an incompressible fluid is in equilibrium in the gravity field (the z axis is directed vertically downward), Equation 50 gives $p - \rho g z = \text{const}$, or $p - \gamma z = \text{const}$, when the product ρg is replaced by the specific weight γ.

Let us designate the pressure above the free surface of the fluid (usually atmospheric) by p_a; then, placing the coordinate origin at the point on the horizontal free surface, we find $p = p_a + \rho g z = p_a + \gamma z$.

The pressure at the given point at depth z, minus the additional pressure of an air column onto the free surface, i.e., the pressure

$$p' = \gamma z \tag{51}$$

is called the pressure of the fluid, and p' is the excess of pressure in the fluid over the atmospheric pressure onto the free surface.

The interface – the free surface of the fluid – is the horizontal plane $z = 0$; over the entire plane $p' = 0$.

The resultant vector and the resultant moment of the forces of fluid pressure onto a certain surface σ are determined by the integrals ($\boldsymbol{n}$ is the unit vector of the external normal to the surface σ directed toward the inside of the fluid)

$$\boldsymbol{R} = -\int_{\sigma} \boldsymbol{n}p'd\sigma \qquad \boldsymbol{L} = -\int_{\sigma} \boldsymbol{r}\times\boldsymbol{n}p'd\sigma \tag{52}$$

with the surface σ, generally speaking, being nonclosed. In the specific case of fluid weight, with $p' = \gamma z$:

$$\boldsymbol{R} = -\gamma\int_{\sigma} \boldsymbol{n}zd\sigma \qquad \boldsymbol{L} = -\gamma\int_{\sigma} \boldsymbol{r}\times\boldsymbol{n}zd\sigma \tag{53}$$

When the surface σ (Figure 24) represents an arbitrarily inclined plane area, then $\boldsymbol{n}$ = const and the first formula of Equation 53 yields

$$\boldsymbol{R} = -\gamma\boldsymbol{n}z_C\sigma \qquad R = \gamma z_C\sigma \tag{54}$$

where z_C (Figure 24) denotes the vertical coordinate of the inertia center C of the area σ. Equations 54 show that the resultant vector of the forces of fluid pressure on any plane area, arbitrarily inclined to the horizon, is equal in magnitude to the weight of the cylindrical column of the fluid. The plane area is the bottom and the depth of the gravity center of the area under the fluid-free surface is the height of this column.

This law, applied to a vessel filled with liquid, leads to the conclusion that the liquid pressure on the vessel walls is independent of the shape of the vessel. This conclusion was reached by Pascal and called the hydrostatic paradox.

If the surface σ is closed and bounds a certain finite volume τ, then, according to Equation 52, noting that grad p' = grad p:

$$\boldsymbol{R} = -\int_{\sigma} \boldsymbol{n}p'd\sigma = -\int_{\tau} \text{grad}\, p\, d\tau \tag{55}$$

In the case of a heavy liquid, according to the Euler equation:

$$\text{grad}\, p = \rho\boldsymbol{g} \tag{56}$$

where $\boldsymbol{g}$ is the gravity acceleration vector.

Substituting this value of the gradient into Equation 55:

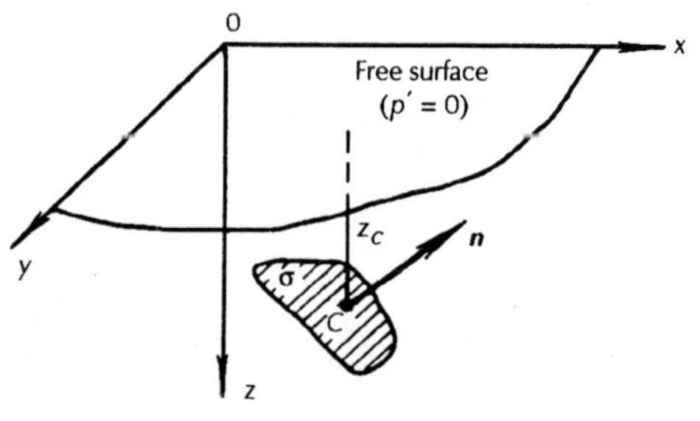

Fig. 24

$$R = -\int_{\tau} \rho g \, d\tau = -G \tag{57}$$

where G is the liquid gravity force vector in the volume of the body immersed into it.

Equation 57 shows that the resultant vector of the forces of the heavy liquid pressure on the surface of the body immersed into it is equal in magnitude to the weight of the liquid in the volume of the body and is directed to the side opposite of the weight force. This is the Archimedes law. The vector R is called the Archimedes force, or the hydrostatic lifting force, signifying that this force tends to push the body out of the liquid, causing it to float to the surface.[5] Any body immersed in fluid loses weight equal to the weight of the fluid displaced by the body.

The Archimedes force can be observed not only in a heavy liquid, i.e., in the liquid under the action of the gravitational forces of the liquid, but in any case when the body is situated in the variable field of pressures. Such a case is the *induction* of the closed working section of a wind tunnel with solid walls after the insertion of an elongated body due to a downstream pressure drop (drop occurred because of the resistance of the working section itself).

The resulting moment of the forces of liquid pressure on the body immersed in it is equal to

$$L = -\int_{\sigma} r \times n p \, d\sigma = \int_{\sigma} n \times p r \, d\sigma = \int_{\tau} \operatorname{rot}(p r) \, d\tau$$

Expanding the expression rot (pr) according to the third formula of Chapter 1, Equation 88 and taking into account that rot $r \equiv 0$:

$$L = -\int_{\tau} r \times \operatorname{grad} p \, d\tau$$

or, according to Equation 56,

$$L = -\int_{\tau} r \times \rho g \, d\tau \tag{58}$$

Moreover, noting that the radius-vector r' of the displaced volume inertia center is equal to

$$r' = \frac{1}{G} \int_{\tau} r \rho g \, d\tau$$

and that evidently

$$g = \frac{G}{G} g$$

we find from Equation 58 that

$$L = -\frac{1}{G} \int_{\tau} r \rho g \, d\tau \times G = -r' \times G = r' \times R \tag{59}$$

[5] Readers will also recognize this as the "buoyancy" force.

This formula shows that the line of action of the resultant vector of liquid pressure on the body immersed in it passes through the gravity center of the liquid volume displaced by the body.

7. Equilibrium State of a Uniformly Rotating Incompressible Fluid. Centrifuging Solid Particles

If an incompressible fluid is relatively at rest with respect to a certain coordinate system uniformly rotating with the angular velocity ω, then, in order to write the condition for the relative equilibrium state of the rotating fluid, to the directly applied forces with the potential Π must be also added the inertial centrifugal force $\boldsymbol{F}^{(c)}$ based on the unit mass and equal to

$$\boldsymbol{F}^{(c)} = \omega^2 \boldsymbol{r} \tag{60}$$

and having the potential

$$\Pi^{(c)} = -\frac{1}{2}\omega^2 r^2 \tag{61}$$

where $\boldsymbol{r}$ is the vector directed along the shortest distance from the axis of rotation to the considered point in the liquid and equal in magnitude to this distance. The equation of relative equilibrium of the rotating liquid is

$$p + \rho\Pi - \frac{1}{2}\rho\omega^2 r^2 = \text{const} \tag{62}$$

and the equation of the free surface of the liquid ($p = \text{const}$) is

$$\Pi - \frac{1}{2}\omega^2 r^2 = \text{const} \tag{63}$$

Let us find the equilibrium configuration for the rotating volume of homogeneous liquid attracted to a motionless center by the force inversely proportional to the distance to the center.

Let the axis z (Figure 25) be the axis of rotation and the coordinate originate origin 0 be the center of gravity. The potential of the gravitational forces based on the liquid unit mass

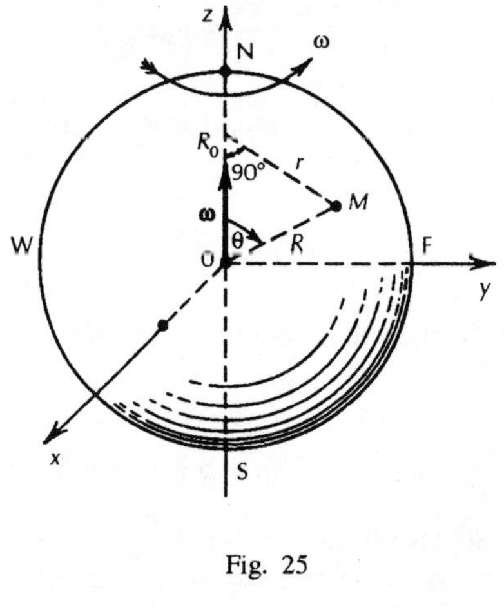

Fig. 25

will be equal to $-C/R$, where C is a certain constant, $R = \sqrt{x^2 + y^2 + z^2}$ is the distance between the liquid particle M and the center of gravity (the coordinate origin 0). The potential of the centrifugal forces based on the liquid unit mass unit (Equation 61) is equal to $\left(-\frac{1}{2}\omega^2 r^2\right)$, where ω is the angular speed of liquid volume rotation, $r = \sqrt{x^2 + y^2}$ is the distance of the liquid particle from the rotation axis $0z$. According to Equation 62, the condition of equilibrium for the rotating liquid with the forces of mutual attraction between the particles being neglected is

$$p - \rho\frac{C}{R} - \frac{1}{2}\rho\omega^2 r^2 = \text{const} \tag{64}$$

and the equation of the free surface which bounds the rotating volume of liquid is

$$\frac{C}{R} + \frac{\omega^2 r^2}{2} = \text{const} \tag{65}$$

This equation describes the sought-after shape of the surface for the equilibrium configuration of the rotating mass of liquid. In order to give a certain, if qualitative, idea about the application of the above formula to the problem of the shape of the Earth, roughly representing a rotating homogeneous liquid tending to the center, we shall prescribe the acceleration g_0 of the mass gravitation at the pole located at the distance R_0 from the Earth's center. Then,

$$\frac{C}{R_0^2} = g_0 \qquad C = g_0 R_0^2$$

and the equation of the equilibrium configuration surface is

$$\frac{g_0 R_0^2}{R} + \frac{\omega^2 r^2}{2} = \text{const}$$

where the constant is determined from the requirement that at the pole $R = R_0$, $r = 0$. It follows that $g_0 R_0 = \text{const}$. The final form of the equation for the free surface is

$$\frac{g_0 R_0^2}{R} + \frac{\omega^2 r^2}{2} = g_0 R_0$$

or, by introducing the polar angle θ and assuming that $r = R \sin\theta$, we shall obtain

$$\frac{g_0 R_0^2}{R} + \frac{\omega^2 R^2 \sin^2\theta}{2} = g_0 R_0 \tag{66}$$

If the Earth did not rotate ($\omega = 0$), the free surface equation would reduce to the equality $R = R_0$, and the equilibrium configuration would be a sphere. Due to a very slow rotation made by the Earth ($\omega \approx \frac{1}{13,700}\ \text{s}^{-1}$), the equilibrium configuration is represented by the body of revolution, which is a sphere slightly flattened at the poles, i.e., a spheroid whose surface equation, due to the smallness of the dimensionless quantity,

$$\frac{\omega^2 R_0}{g_0} = \left(\frac{2\pi}{24 \cdot 60 \cdot 60}\right)^2 \cdot 6.37 \cdot 10^6 \cdot \frac{1}{9.83} \approx 0.0034 \approx \frac{1}{300}$$

can be approximately presented as

$$R = R_0\left(1 + \frac{1}{2}\frac{\omega^2 R_0}{g_0}\sin^2\theta\right)$$

Thus, it is easy to find the relative degree of flattening of the Earth:

$$\varepsilon = \frac{R_{max} - R_{min}}{R_{min}} = \frac{1}{2}\frac{\omega^2 R_0}{g_0} \approx \frac{1}{600}$$

Geodesic measurements give twice as large a value. This discrepancy between theory and experiment is attributable to the roughness of the adopted assumption about the homogeneity of the Earth and to the disregard of the mutual attraction of particles which alters the law of attraction to the center. In fact, the law of attraction depends on the shape of the liquid body, the equilibrium of which is considered, and this makes the rigorous solution of the above problem very difficult.

8. Barotropic Equilibrium of a Gas

The equilibrium of a gas is called barotropic if the gas density can be considered to be the function of pressure alone ($\rho = \rho(p)$). Otherwise, ($\rho = \rho(p, T)$) the equilibrium is called baroclinic.

The main specific features of the barotropic equilibrium are

1. The gas is incompressible, i.e., it has identical density everywhere ($\rho = \text{const} = \rho_0$); this case has already been considered
2. The process is isothermal; the temperature is identical everywhere ($T = \text{const} = T_0$) and it follows from the Clapeyron state equation that

$$\rho = \frac{p}{RT_0} = \text{const} \cdot p = \frac{\rho_0}{p_0} p \qquad (67)$$

3. The process is adiabatic (there is no heat supply from the outside); then there is the adiabatic

$$p = \text{const} \cdot \rho^k = p_0\left(\frac{\rho}{\rho_0}\right)^k \qquad (68)$$

where k is the ratio of specific heats of the gas at a constant pressure c_p and constant volume c_v; for air $k \approx 1.4$. The values p_0, ρ_0, T_0 are related to some reference point in a quiescent gas.

In the case of gas equilibrium in the potential force field the equilibrium equation takes the form of Equation 49.

Regardless of the question of whether there is the equilibrium state or motion, if only the process is barotropic, the pressure function is introduced:

$$\mathcal{P}(p) = \int_{p_0}^{p} \frac{dp}{\rho(p)} \tag{69}$$

the gradient of which is equal to

$$\operatorname{grad} \mathcal{P} = \frac{d\mathcal{P}}{dp} \operatorname{grad} p = \frac{1}{\rho} \operatorname{grad} p \tag{70}$$

According to the considerations presented in Sections 1 and 5, the quantity $(-\frac{1}{\rho}\ \operatorname{grad} p)$ can be interpreted as the resultant vector of pressure forces per unit mass at the given point, or the vector of volumetric action of these forces. Thus, the pressure function $\mathcal{P}$, which satisfies Equation 70, represents the potential of the volumetric action of pressure forces.

The equilibrium equation (49) can now be rewritten in the form $\operatorname{grad}(\Pi + \mathcal{P}) = 0$. Thus, in the case of the barotropic equilibrium of the medium the following equality holds at all of its points

$$\Pi + \mathcal{P} = \text{const} \tag{71}$$

Simple calculations of the integrals in Equation 69 lead to the following values of the pressure function in the above-mentioned specific cases of barotropic processes:

1. Incompressible fluid ($\rho = \text{const}$)

$$\mathcal{P}(p) = \frac{p - p_0}{\rho} \tag{72}$$

2. Isothermal process $\left(\rho = \frac{\rho_0}{p_0} p\right)$

$$\mathcal{P}(p) = \frac{p_0}{\rho_0} \ln \frac{p}{p_0} \tag{73}$$

3. Adiabatic process $\left[p = p_0 = \left(\frac{\rho}{\rho_0}\right)^k\right]$

$$\mathcal{P}(p) = -\frac{k}{k-1} \frac{p_0}{\rho_0} \left[1 - \left(\frac{p}{p_0}\right)^{\frac{k-1}{k}}\right] \tag{74}$$

Let us be satisfied with these brief concepts from the field of the statics of liquids and gases. Those interested in further information are referred to textbooks on hydraulics mentioned in the footnotes, of which a considerable portion concerns the sections on hydrostatics and aerostatics.

5

The Dynamics of an Ideal Medium: Basic Equations and Theorems

1. The Equations of Euler, Gromeka-Lamb, and Helmholtz-Fridman. Theorem of Helmholtz

The simplest model of a continuous flowing medium is an ideal medium which, when moving, possesses the property of ideal fluidity, which is typical, as elucidated in Chapter 4, Section 5, of the majority of continuous media when at rest. This property manifests itself in the absence of tangential stress tensor components due to null tangential components – shear rates – of the rate of strain tensor. For an ideal medium, the absence of tangential stresses is postulated regardless of whether the medium is at rest or in motion. This assumption is equivalent to the requirement of the absence of internal friction in an ideal medium.

In a moving ideal medium the Pascal law (Equations 35 to 40, Chapter 4) is fulfilled

$$p_{12} = p_{21} = p_{23} = p_{32} = p_{31} = p_{13} = 0$$

$$p_{11} = p_{22} = p_{33} = -p \qquad \boldsymbol{p}_n = -p\boldsymbol{n}$$

$$P = -pE \tag{1}$$

which is derived in the same way as shown in Chapter 4, Section 5.

In contrast to the hydrostatic pressure introduced in the previous chapter, the scalar quantity p will be called the hydrodynamic pressure or simply the pressure at the given point of the flow; as in the case of equilibrium, the minus sign emphasizes that the direction of the normal stress vector $\boldsymbol{p}_n$ is opposite of the direction of the unit vector of the normal $\boldsymbol{n}$ to the front side of the area.

Ignoring in the model of an ideal medium the quantitative effect of the internal molecular exchange showing up in real fluids as friction and heat conduction, we shall preserve the qualitative consequence of this exchange, i.e., continuity in the distribution of velocities and other physical quantities. This accounts for the applicability of the theory of an ideal fluid in revealing, for example, the general picture of smooth flow past bodies, pressure distribution over the surfaces of such bodies, formation of the lifting force, and many other practical, important problems.

The continuity principle must be given up only in some cases: at the interface between two ideal media, on the solid boundary of a body in a flow, and also on certain special surfaces of discontinuity. In the first two cases free relative fluid sliding is admitted as well as the fluid sliding along the solid surface, with the only requirement that there be no mutual penetration of fluids or fluid flow through the solid surface (the impermeability condition).

The real medium does not admit discontinuities either inside a moving flow, or on its boundaries with a solid body. In fact, a liquid or a gas cannot slide along a solid surface; the velocities of those particles that contact the solid wall equal zero. (It is as though the medium adheres to the surface of the body.) However, as a rule, this velocity increases steeply at a distance from the surface and at the outer edge of the boundary layer, which is very thin as compared to the body dimensions. The layer attains the values corresponding to the scheme of free sliding of an ideal medium. This is the other reason explaining the possibility for using the ideal medium scheme in calculating the flow past bodies of smooth prolate shape (a wing, fuselage, a blade of a turbomachine, etc.). In the case of a bluff body, the boundary layer separates from the body surface and considerably distorts the pattern of continuous flow of the ideal medium around the body.

The basic differential equations of the dynamics of an ideal medium follow directly from the general equations of motion derived in Chapter 4, either due to the simplification of the latter equations on the basis of Equation 1 or because of the evident equality

$$\operatorname{Div} P = -\operatorname{grad} p \tag{2}$$

The continuity equation (Equations 5 and 7, Chapter 3), as the one not containing stresses, will preserve its general form.

The momentum equation (Equations 6 and 8, Chapter 4) will be considerably simplified and will acquire the Eulerian form

$$\frac{d\boldsymbol{V}}{dt} \equiv \frac{\partial \boldsymbol{V}}{\partial t} + (\boldsymbol{V} \cdot \nabla)\,\boldsymbol{V} = \boldsymbol{F} - \frac{1}{\rho}\operatorname{grad} p \tag{3}$$

or, in projections on the axes of Cartesian rectangular coordinates,

$$\frac{\partial V_1}{\partial t} + V_1\frac{\partial V_1}{\partial x_1} + V_2\frac{\partial V_1}{\partial x_2} + V_3\frac{\partial V_1}{\partial x_3} = F_1 - \frac{1}{\rho}\frac{\partial p}{\partial x_1}$$

$$\frac{\partial V_2}{\partial t} + V_1\frac{\partial V_2}{\partial x_1} + V_2\frac{\partial V_2}{\partial x_2} + V_3\frac{\partial V_2}{\partial x_3} = F_2 - \frac{1}{\rho}\frac{\partial p}{\partial x_2}$$

$$\frac{\partial V_3}{\partial t} + V_1\frac{\partial V_3}{\partial x_1} + V_2\frac{\partial V_3}{\partial x_2} + V_3\frac{\partial V_3}{\partial x_3} = F_3 - \frac{1}{\rho}\frac{\partial p}{\partial x_3} \tag{4}$$

or

$$\frac{\partial V_i}{\partial t} + V_k\frac{\partial V_i}{\partial x_k} = F_i - \frac{1}{\rho}\frac{\partial p}{\partial x_i} \qquad (i = 1, 2, 3)$$

Using the expressions for the acceleration projections on the axis of cylindrical and spherical coordinates (Equations 52 and 53, Chapter 2), write the Euler equations in these two systems:

Cylindrical coordinates:

$$\frac{\partial V_r}{\partial t} + V_r \frac{\partial V_r}{\partial r} + \frac{V_\varepsilon}{r}\frac{\partial V_r}{\partial \varepsilon} + V_z \frac{\partial V_r}{\partial z} - \frac{V_\varepsilon^2}{r} = F_r - \frac{1}{\rho}\frac{\partial p}{\partial r}$$

$$\frac{\partial V_\varepsilon}{\partial t} + V_r \frac{\partial V_\varepsilon}{\partial r} + \frac{V_\varepsilon}{r}\frac{\partial V_\varepsilon}{\partial \varepsilon} + V_z \frac{\partial V_\varepsilon}{\partial z} + \frac{V_r V_\varepsilon}{r} = F_\varepsilon - \frac{1}{\rho r}\frac{\partial p}{\partial \varepsilon}$$

$$\frac{\partial V_z}{\partial t} + V_r \frac{\partial V_z}{\partial r} + \frac{V_\varepsilon}{r}\frac{\partial V_z}{\partial \varepsilon} + V_z \frac{\partial V_z}{\partial z} = F_z - \frac{1}{\rho}\frac{\partial p}{\partial z} \tag{4a}$$

Spherical coordinates

$$\frac{\partial V_R}{\partial t} + V_R \frac{\partial V_R}{\partial R} + \frac{V_\theta}{R}\frac{\partial V_R}{\partial \theta} + \frac{V_\varepsilon}{R\sin\theta}\frac{\partial V_R}{\partial \varepsilon} - \frac{V_\theta^2 + V_\varepsilon^2}{R} = F_R - \frac{1}{\rho}\frac{\partial p}{\partial R}$$

$$\frac{\partial V_\theta}{\partial t} + V_R \frac{\partial V_\theta}{\partial R} + \frac{V_\theta}{R}\frac{\partial V_\theta}{\partial \theta} + \frac{V_\varepsilon}{R\sin\theta}\frac{\partial V_\theta}{\partial \varepsilon} + \frac{V_R V_\theta}{R} - \frac{V_\varepsilon^2}{R}\operatorname{ctg}\theta = F_\theta - \frac{1}{\rho R}\frac{\partial p}{\partial \theta}$$

$$\frac{\partial V_\varepsilon}{\partial t} + V_R \frac{\partial V_\varepsilon}{\partial R} + \frac{V_\theta}{R}\frac{\partial V_\varepsilon}{\partial \theta} + \frac{V_\varepsilon}{R\sin\theta}\frac{\partial V_\varepsilon}{\partial \varepsilon} + \frac{V_R V_\varepsilon}{R} + \frac{V_\theta V_\varepsilon}{R}\operatorname{ctg}\theta = F_\varepsilon - \frac{1}{\rho R\sin\theta}\frac{\partial p}{\partial \varepsilon} \tag{4b}$$

The Euler equation (3) can be rid of the symbolic combination $(\boldsymbol{V}\cdot\nabla)\boldsymbol{V}$, and at the same time the gradient term can be isolated on the left side of it.

For this, recall the earlier established expression for the convective component of acceleration (50) in Chapter 2. Substituting it into the left side of Equation 3, obtain the following modification of the Euler equation as suggested by Gromeka and Lamb

$$\frac{\partial \boldsymbol{V}}{\partial t} + \operatorname{grad}\left(\frac{V^2}{2}\right) + \operatorname{rot}\boldsymbol{V}\times\boldsymbol{V} = \boldsymbol{F} - \frac{1}{\rho}\operatorname{grad} p \tag{5}$$

Of greatest interest is the case in which the volumetric forces have the potential Π, i.e.,

$$\boldsymbol{F} = -\operatorname{grad}\Pi \tag{6}$$

and there exists the pressure function (the motion is barotropic)

$$\mathcal{P}(p) = \int_{p_0}^{p}\frac{dp}{\rho(p)} \tag{7}$$

introduced earlier in Section 8 of Chapter 4. In this case, the Gromeka-Lamb equation will go to

$$\frac{\partial \boldsymbol{V}}{\partial t} + \operatorname{grad}\left(\frac{V^2}{2} + \mathcal{P} + \Pi\right) + \operatorname{rot}\boldsymbol{V}\times\boldsymbol{V} = 0 \tag{8}$$

Let us introduce the notation

$$\frac{V^2}{2} + \mathcal{P} + \Pi = B \tag{9}$$

$$\text{rot}\, V = \mathbf{\Omega} \tag{10}$$

Then, Equation 8 can be represented in the form

$$\frac{\partial V}{\partial t} + \text{grad}\, B + \mathbf{\Omega} \times V = 0 \tag{11}$$

or, in projections onto the Cartesian axes,

$$\frac{\partial V_1}{\partial t} + \frac{\partial B}{\partial x_1} + \Omega_2 V_3 - \Omega_3 V_2 = 0$$

$$\frac{\partial V_2}{\partial t} + \frac{\partial B}{\partial x_2} + \Omega_3 V_1 - \Omega_1 V_3 = 0$$

$$\frac{\partial V_3}{\partial t} + \frac{\partial B}{\partial x_3} + \Omega_1 V_2 - \Omega_2 V_1 = 0 \tag{12}$$

Bernoulli's trinomial (Equation 9) can be interpreted as the total mechanical energy at the given point per unit mass, i.e., according to the conventional terminology, as the total specific mechanical energy. In fact, the first term on the left side represents the specific kinetic energy, and the third term is the specific potential energy of volumetric forces.

As explained in Chapter 4, the pressure function $\mathcal{P}$ has the meaning of the potential (potential energy) of the volumetric action of pressure forces (more precisely, of the mass density of the action of these forces). One should avoid the occasional definition for the pressure function $\mathcal{P}$ (when ρ = const and equal to p/ρ) as the potential of pressures. Pressures form a scalar field for which the concept of the potential or of the potential energy is nonexistent.

Equations 11 or 12 connect purely kinematic quantities V and $\mathbf{\Omega} = \text{rot}\, V$ with the dynamic characteristics Π and $\mathcal{P}$.

Rewriting Equation 11 in the form

$$\frac{\partial V}{\partial t} + \mathbf{\Omega} \times V = -\text{grad}\, B \tag{13}$$

we see that in the case of barotropic motion of an ideal continuous medium under the action of the potential field of volumetric forces, the left, kinematic side of this equality represents the potential vector. Consequently, not every velocity field can be produced in an ideal fluid moving barotropically under the action of the potential field of volumetric forces, but only one that satisfies the equality

$$\text{rot}\left(\frac{\partial V}{\partial t} + \mathbf{\Omega} \times V\right) = 0$$

or

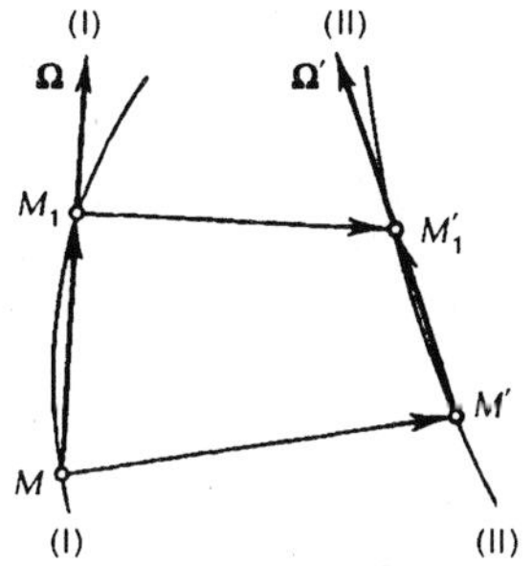

Fig. 26

$$\frac{\partial \mathbf{\Omega}}{\partial t} + \operatorname{rot}(\mathbf{\Omega} \times \boldsymbol{V}) = 0 \tag{14}$$

Using the vector identity (see the fifth equality of Equation 88, Chapter 1):

$$\operatorname{rot}(\mathbf{\Omega} \times \boldsymbol{V}) = (\boldsymbol{V} \cdot \nabla)\mathbf{\Omega} - (\mathbf{\Omega} \cdot \nabla)\boldsymbol{V} + \mathbf{\Omega}\operatorname{div}\boldsymbol{V} - \boldsymbol{V}\operatorname{div}\mathbf{\Omega}$$

and discarding the last term on the right side of this equality as identically equal to zero, rewrite the previous equality as

$$\frac{\partial \mathbf{\Omega}}{\partial t} + (\boldsymbol{V} \cdot \nabla)\mathbf{\Omega} = (\mathbf{\Omega} \cdot \nabla)\boldsymbol{V} - \mathbf{\Omega}\operatorname{div}\boldsymbol{V}$$

or recalling the definition of the ordinary derivative as

$$\frac{d\mathbf{\Omega}}{dt} = (\mathbf{\Omega} \cdot \nabla)\boldsymbol{V} - \mathbf{\Omega}\operatorname{div}\boldsymbol{V} \tag{15}$$

The first term on the right side equal to, according to Equation 106, Chapter 1,

$$(\mathbf{\Omega} \cdot \nabla)\boldsymbol{V} = \mathbf{\Omega}(\nabla \boldsymbol{V}) = \mathbf{\Omega}\dot{S} - \mathbf{\Omega} \times \frac{1}{2}\mathbf{\Omega} = \mathbf{\Omega}\dot{S}$$

expresses the effect of the deformation of the vector $\mathbf{\Omega}$ by the velocity field, the second term ($-\mathbf{\Omega}$ div $\boldsymbol{V}$) determines the influence of compressibility.

Equation 15 is called the equation of the dynamic possibility for the motion of an ideal compressible medium, or the Helmholtz-Fridman equation. It was derived by A. A. Fridman[1] by the extending the Helmholtz equation

$$\frac{d\mathbf{\Omega}}{dt} = (\mathbf{\Omega} \cdot \nabla)\boldsymbol{V} \tag{16}$$

which relates to the case of an incompressible fluid (div $\boldsymbol{V} = 0$).

Let us introduce the concept of the invariance of the vortex lines. Suppose that at a certain time instant there exists in the fluid the vortex line (I, I) (Figure 26), which is the vector

[1]Fridman, A. A., *Opyt gidromelhaniki szhimayemoi zhidkosti* (*An Experiment on the Hydromechanics of a Compressible Fluid*), ONTI Press, Moscow, 1931 (Chapter 1, specifically p. 34); Kochin, N. Ye., Kibel, I. A., and Roze, N. V., *Teoreticheskaya gidromekhanika* (*Theoretical Hydromechanics*), Vol. 1, Gostekhizdat, Moscow, 1948, pp. 150 to 160; Kochin, N. E., Kibel, J. A., and Rose, N. V., *Theoretische Hydromechanik*. Vol. 1 and 2, Berlin, 1954.

line of the vector $\mathbf{\Omega} = \text{rot } V$; let us consider the fluid line (II, II) formed at the instant $t + dt$ by the same fluid particles as is the line (I, I) at the instant t. If the fluid line (II, II), which represents the new position of the line (I, I) by the time instant $t + dt$, is also rotational, we say that the vortex line (I, I) is preserved during medium motion, otherwise it disintegrates.

Let us prove the following theorem of Helmholtz: *in an ideal incompressible fluid moving under the action of the conservative field of volumetric forces the vortex lines are preserved.*

Compare the infinitely small fluid vector $\overline{MM_1}$ and its adjacent position $\overline{M'M_1'}$ (in the case of infinitely small displacements of a fluid, its straight lines remain straight accurate to the higher-order small quantities). From the vector polygon $MM_1M_1'M'$ we have

$$\overline{M'M_1'} = \overline{MM_1} + \overline{M_1M_1'} - \overline{MM'}$$

and, noting that (λ is an arbitrary infinitely small scalar)

$$\overline{MM_1} = \lambda\mathbf{\Omega} \qquad \overline{MM'} = V dt \qquad \overline{M_1M_1'} = [V + (\lambda\mathbf{\Omega} \cdot \nabla) V]\, dt$$

we obtain

$$\overline{M'M_1'} = \lambda\mathbf{\Omega} + V dt + \lambda(\mathbf{\Omega} \cdot \nabla) V dt - V dt = \lambda[\mathbf{\Omega} + (\mathbf{\Omega} \cdot \nabla) V dt] \tag{17}$$

With Equation 16 taken into account, the latter equation leads to the relation

$$\overline{M'M_1'} = \lambda\left(\mathbf{\Omega} + \frac{d\mathbf{\Omega}}{dt}\, dt\right) = \lambda\mathbf{\Omega}' \tag{18}$$

which proves the Helmholtz theorem because the element of the fluid line (II, II) is directed along the vector $\mathbf{\Omega}'$ into which the vector $\mathbf{\Omega}$ will convert during the time interval dt. The above proof is valid only for an ideal incompressible fluid.

As was shown by Fridman (see footnote 1), the theorem also holds in the case of any barotropic motion of an ideal gas.

2. Bernoulli Theorem

Suppose that under the action of the potential field of volumetric forces having the potential Π an ideal fluid undergoes a steady barotropic motion with the pressure function $\mathcal{P}$. The first term in Equation 11 is thus equal to zero, and, performing the scalar multiplication of both parts of Equation 11 by the velocity vector V, we obtain

$$V \cdot \text{grad } B \equiv V\left(\frac{V}{V} \cdot \text{grad } B\right) = 0$$

due to the fact that the last term is perpendicular to the vector V, or, recalling the definition of the derivative with respect to direction,

$$\frac{V}{V} \cdot \text{grad } B = \frac{dB}{ds}$$

where we conclude that

$$\frac{dB}{ds} = 0 \tag{19}$$

where the symbol d/ds denotes the derivative taken along the trajectory or the streamline, which is the same in steady motion. From Equation 19 it follows directly that along the trajectory or the streamline the Bernoulli trinomial B preserves the same value

$$B = \frac{V^2}{2} + \mathcal{G} + \Pi = \text{const} \quad \text{(along the streamline)} \tag{20}$$

This equality can be considered to be the first integral of the Euler equation, being valid in the case of steady motion in the presence of the pressure function (which represents the potential of the volumetric action of surfaces forces) and as the integral of the potential of volumetric forces. This integral, derived by means of scalar multiplication of both parts of Equation 11 by the velocity vector $\boldsymbol{V}$, can be interpreted as the integral of the mechanical energy in the equation of motion for the inertia center of the fluid volume element (the Bernoulli integral).

Equation 20 expresses the following theorem of Bernoulli: *in the case of a steady barotropic motion of an ideal fluid under the action of potential volumetric forces, the sum of the kinetic energy of the unit mass, pressure function, and of the potential of volumetric forces (reduced to the unit mass), preserves its constant value along the streamline (trajectory).*

The Bernoulli integral could be also derived directly from the Euler equation (3) without having been transformed to the Gromeka-Lamb form (Equation 8). In fact, rewriting Equation 3 under the conditions of the theorem as

$$\frac{d\boldsymbol{V}}{dt} = -\text{grad}\ (\Pi + \mathcal{G})$$

and performing scalar multiplication by the vector $\boldsymbol{V}$, we obtain

$$\frac{d}{dt}\left(\frac{V^2}{2}\right) = -\boldsymbol{V} \cdot \text{grad}\ (\Pi + \mathcal{G})$$

but, by the definition of the ordinary time derivative of the scalar function, in the case of steady motion, we have

$$\frac{d}{dt}\left(\frac{V^2}{2}\right) = \boldsymbol{V} \cdot \text{grad}\left(\frac{V^2}{2}\right)$$

so that

$$\boldsymbol{V} \cdot \text{grad}\left(\frac{V^2}{2} + \mathcal{G} + \Pi\right) = 0$$

which, according to the text above leads to Equation 20.

In the case of steady motion, Equation 11 also directly yields the constancy of the value B along any vortex line. Actually, discarding the first term in the case of steady motion, performing scalar multiplication of both parts of Equation 11 by $\boldsymbol{\Omega}$, and using the same sort of reasoning as with Equation 19, we obtain

$$\boldsymbol{\Omega} \cdot \text{grad}\ B \equiv \Omega\left(\frac{\boldsymbol{\Omega}}{\Omega} \cdot \text{grad}\ B\right) = \Omega\frac{dB}{dl} = 0$$

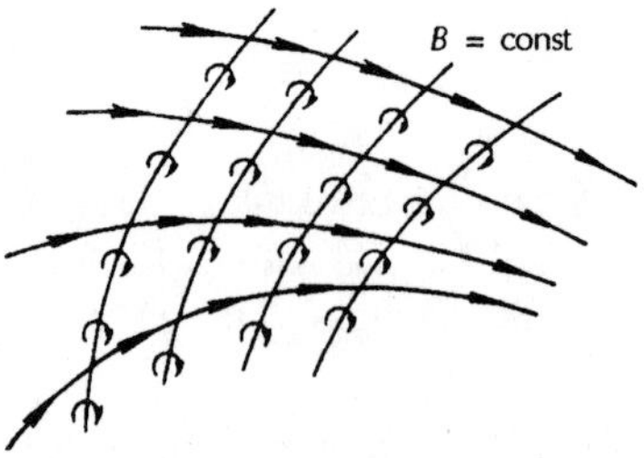

Fig. 27

where d/dl determines differentiation along the arc of the vortex line. From this it follows directly that along the vortex line quantity B has the same value:

$$B = \frac{V^2}{2} + \mathcal{P} + \Pi = \text{const} \quad \text{(along the vortex line)} \tag{21}$$

In the case of steady motion the vector $\boldsymbol{\Omega} \times \boldsymbol{V}$ forms (see Equation 11) the potential vector field with the potential B. In this case, through each point of space one may draw the surface which will be orthogonal to the vector line belonging to the vector field $\boldsymbol{\Omega} \times \boldsymbol{V}$ and passing through this point. These orthogonal surfaces will be the level surfaces of the Bernoulli trinomial, and at the same time of the total specific mechanical energy. The planes tangential to these surfaces contain the vectors $\boldsymbol{\Omega}$ and $\boldsymbol{V}$. The level surfaces can be obtained by taking (Figure 27) any streamline and passing vortex lines through all of its points; these vortex lines form a vortex surface – the level surface which passes through the given streamline. We may also proceed differently: we may take some vortex line and pass streamlines through all of its points. These streamlines will then form the stream surface passed through the given vortex line.

The value of the constant in Equations 20 and 21 is determined by the value of the Bernoulli trinomial at any one point[2] of a streamline or vortex line. Generally, these constants are different for the streamlines or vortex lines that do not lie on the same level surface of the total mechanical energy.

If at all the points of space the following equality is fulfilled

$$\boldsymbol{\Omega} \times \boldsymbol{V} = 0 \tag{22}$$

there are no level surfaces. However, according to Equation 11, in a steady flow

$$\text{grad}\, B = 0 \tag{23}$$

i.e., the Bernoulli trinomial preserves the same value throughout the entire space occupied by the liquid or gas flow.

Equality (22) is fulfilled in the following two cases:

1. $\boldsymbol{\Omega} = 0$: The motion is irrotational; special chapters will be devoted to the detailed analysis of this most important case.
2. $\boldsymbol{\Omega} \parallel \boldsymbol{V}$: The vortex lines coincide with the streamlines; in such a motion the instantly rotating particles revolve about the tangents to the streamlines. Such motion is called helical. The helical motion is dealt with, for example, when considering the socalled free vortices shed from the surface of a wing of finite span.

---

[2]For some reason, this point is a characteristic point, or the value may be given *a priori*.

Let us consider specific cases of the Bernoulli theorem that relate to separate simplest barotropic processes. In the case of a motion of an incompressible fluid (ρ = const) we have

$$\mathcal{P} = \frac{p - p_0}{\rho} = \frac{p}{\rho} + \text{const}$$

Restricting ourselves, among the volumetric forces, only to the gravity forces and directing the vertical axis z upward, we obtain

$$\Pi = gz + \text{const}$$

Then, Equations 20 and 21 will take the form (the symbol const means the preservation of the value of B along both the streamline and the vortex line)

$$B = \frac{V^2}{2} + \frac{p}{\rho} + gz = \text{const} \tag{24}$$

or, on going over from the density ρ to the specific weight $\gamma = \rho g$,

$$\frac{B}{g} = H = \frac{V^2}{2g} + \frac{p}{\gamma} + z = \text{const} \tag{25}$$

The separate terms of equality (25) have the dimensions of length and are respectively called $V^2/(2g)$, the velocity; p/γ, the pressure; and z, the gravity heads. The sum of these heads is called the hydraulic head.

Formula (25) leads to the classical formulation of the Bernoulli theorem: *in steady motion of an ideal incompressible fluid the hydraulic head, equal to the sum of the velocity, pressure, and gravity heads, preserves a constant value along streamlines (trajectory) and vortex lines*. This form of the Bernoulli theorem has fundamental importance in hydraulics and is called the Bernoulli equation.

If we further assume that the volumetric forces[3] can be neglected in comparison with the surface forces (pressure); then the Bernoulli equation (24) will acquire a simpler form

$$p + \frac{\rho V^2}{2} = \text{const} = p_0 \tag{26}$$

The first term on the left side of this equation is called the static pressure, the second is called the velocity pressure or dynamic pressure. Their sum is called the total pressure p_0. The Bernoulli theorem is formulated as follows: in steady motion of an ideal incompressible fluid in the absence of volumetric forces the total pressure, equal to the sum of the static and velocity pressures, preserves its value along streamlines (trajectories) and vortex lines.

Recalling the expressions of the pressure functions $\mathcal{P}$ (Equations 73 and 74, Chapter 4), we obtain the Bernoulli theorem in the following forms:

1. For an isothermal motion (T = const, $p/\rho = p_0/\rho_0$, $p/p_0 = \rho/\rho_0$)

---

[3]The reader is reminded of the term "body-forces" which is more common in English usage than "volumetric forces".

$$\frac{V^2}{2} + \frac{p_0}{\rho_0} \ln \frac{p}{p_0} = \frac{V^2}{2} + \frac{p_0}{\rho_0} \ln \frac{\rho}{\rho_0} = \text{const} = \frac{V_0^2}{2} \tag{27}$$

2. For an adiabatic motion $[p/p_0 = (\rho/\rho_0)^k,\ \rho/\rho_0 = (p/p_0)^{1/k}]$

$$\frac{V^2}{2} - \frac{k}{k-1} \frac{p_0}{\rho_0} \left[1 - \left(\frac{p}{p_0} \right)^{\frac{k-1}{k}} \right] = \text{const} = \frac{V_0^2}{2} \tag{28}$$

$$\frac{V^2}{2} - \frac{k}{k-1} \frac{p_0}{\rho_0} \left[1 - \left(\frac{\rho}{\rho_0} \right)^{k-1} \right] = \text{const} = \frac{V_0^2}{2} \tag{29}$$

Here, the subscript zero relates to the point arbitrarily selected on the streamline (trajectory) or on the vortex line. It also will be applied for the parameters of a gas at rest. If there is no such a point on the given streamline (trajectory) or the vortex line where $V = 0$, it is always possible to imagine a certain continuous adiabatic motion of an ideal gas (it will be shown later that this motion is also isentropic), which takes it from the given position to the "tank" (receiver) of an infinitely large volume in which the gas becomes motionless or, as conventionally said, is in an adiabatic and isentropic stagnation state. In this state the gas parameters are called the parameters of the adiabatic and isentropic stagnation state or the stagnation parameters, and are, respectively, designated by p_0, ρ_0, and T_0. The Bernoulli equations (28 and 29) will acquire in this case one of the following forms (the first equation bears the names of Saint-Venant and Wantzel):

$$V^2 = \frac{2k}{k-1} \frac{p_0}{\rho_0} \left[1 - \left(\frac{p}{p_0} \right)^{\frac{k-1}{k}} \right] \qquad V^2 = \frac{2k}{k-1} \frac{p_0}{\rho_0} \left[1 - \left(\frac{\rho}{\rho_0} \right)^{k-1} \right] \tag{30}$$

In applications, specifically in the analysis of turbomachines, one has to deal with a relative motion of fluid in a certain coordinate system uniformly rotating about a fixed axis. Thus, the absolute flow past the turbine rotor rotating with a constant angular velocity, i.e., the flow relative to the turbine foundation, will probably be unsteady and the Bernoulli theorem cannot be applied to it. In a relative coordinate system moving with the rotating rotor the flow is steady and the Bernoulli theorem is valid for the relative motion.

Denoting by $\boldsymbol{V}_r$ the relative velocity and adding, to the applied volumetric forces, the centrifugal force of inertia with the distribution density $\boldsymbol{F}^{(c)} = \omega^2 \boldsymbol{r}$ and potential $\Pi^{(c)} = -1/2\, r^2 \omega^2$, where r is the distance from the rotation axis, and also the Coriolis force with the distribution density $\boldsymbol{F}^{(cor)} = -2\boldsymbol{\omega} \times \boldsymbol{V}_r$ (which vanishes after the scalar multiplication by $\boldsymbol{V}_r$), we obtain

$$\frac{V_r^2}{2} + \mathcal{P} + \Pi - \frac{r^2 \omega^2}{2} = \text{const} \tag{31}$$

(const only along the relative streamlines or trajectories).

Noting that $r\boldsymbol{\omega} = V_e$ represents the approach (peripheral) velocity, we have the Bernoulli theorem in the final form for a relative motion

$$\frac{V_r^2 - V_e^2}{2} + \mathcal{P} + \Pi = \text{const} \tag{32}$$

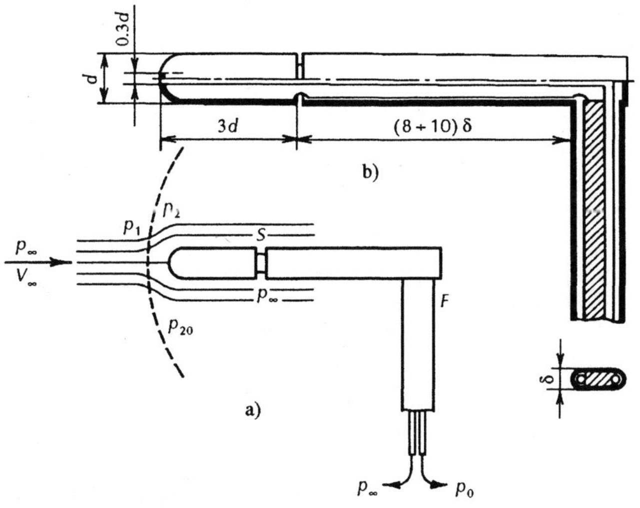

Fig. 28

Usually, in hydraulics and applied hydromechanics courses the Bernoulli theorem is illustrated by reference to numerous examples. To get acquainted with the simultaneous application of the momentum and kinetic energy theorems in the form of the Bernoulli theorem, consider the derivation of Borda's theorem about the shock on sudden flow expansion.[4] The Bernoulli theorem is often referred to in subsequent chapters of this book. An interesting illustration of its application to the motion of an ideal incompressible fluid can be provided by the method of velocity measurements with the aid of the so-called pressure tube, or Prandtl tube, which is very popular in aerodynamic laboratories.[5]

Consider the schematic diagram of the Prandtl tube (Figure 28) used for measuring gas (air) velocities under the conditions when the compressibility effect can be ignored; tubes of this kind are also used for measuring velocities in liquid flows. Along the axis of the tube the flow approaches its nose which has the so-called stagnation hole D (Figure 28a) and passes along the side surface of the tube on which there is a static hole (slit) S. At a sufficient distance of the stem F and nose D of the tube from the static hole (the standard dimensions are shown in Figure 28b), it is possible to assume that the pressure near the hole D is equal (Figure 28a) to the stagnation gas pressure p_0, and near the static slit to the pressure p of the gas passing along the tube. Probably for this reason the pressure p is unjustifiably called static, though this term does not exist in hydrodynamics.

When presenting the theory of viscous fluid motion in a boundary layer on the surface of a body immersed in a flow (Chapter 12), it will be shown that the pressure at any point of the boundary layer section, including those on the surface, coincides with the pressure of the medium at the outer edge of the boundary layer, i.e., in the flow rushing past the static hole of the tube. Thus, if the tube near the slit S is cylindrical and the slit itself is cut flush with the tube wall, so that the fluid passes the slit without being disturbed by the projecting sides of this slit, the pressure in the slit will be equal to the pressure in the fluid not perturbed by the tube and far from it.

---

[4] See the earlier cited Course of Theoretical Mechanics, Vol. 2, pp. 249 to 251, where some applications of the Bernoulli theorem are also given.

[5] "Pitot tube" or "Pitot-static tube" are also used to describe such a device.

Denote the pressure and velocity (p_∞ and V_∞) of the incident uniform flow and bear in mind that the disturbing effect of the nose on the field of the ideal medium flow parameters ceases before the static hole. Thus, V_∞ is the value measured by the tube.

According to the Bernoulli theorem, which for an incompressible fluid is expressed by Equation 26, we have

$$p_\infty + \frac{1}{2}\rho V_\infty^2 = \text{const} = p_0$$

where p_0 is the total pressure measured by the hole D of the tube, and p_∞ is the pressure in the hole S. From this, determine the unknown velocity via

$$V_\infty = \sqrt{\frac{2(p_0 - p_\infty)}{\rho}}$$

If now the Prandtl tube is considered in a nonuniform flow, but it is assumed that the size of the nose is much smaller than the characteristic dimension of the nonuniformity, it is possible to repeat the previous reasoning and obtain an analogous formula:

$$V = \sqrt{\frac{2(p_0 - p)}{\rho}} \tag{33}$$

where p and V indicate the local parameters of the flow not disturbed by the tube.

Measuring the difference of pressures $p_0 - p$ with the aid of a differential manometer and knowing the density of the moving medium, we can find its velocity at the given place in the flow.

In fact, the measured pressure differences can differ somewhat from the actual ones due to the inaccuracies in the fabrication of the tube; to take into account the necessary corrections, a coefficient is introduced into Equation 33 which is close to unity and which is determined by means of tube calibration and its comparison in a flow with a certain standard tube.

The theorems of Helmholtz and Bernoulli are two essential results obtained from the Euler equation, i.e., from the momentum law for an ideal continuous medium. The law of the moments of momenta in the absence of tangential stresses in an ideal medium is fulfilled automatically by virtue of the condition that $P = -pE$. Next in our sequence of presentation, we consider the further laws of mechanics and thermodynamics of continuous media.

3. The Power of Internal Forces. The Equation of Energy Balance

The equation of the kinetic energy balance (Equation 16, Chapter 4) will somewhat simplify for an ideal medium, because the stresses will be reduced to pressure. However, in this form the equation is not of importance by itself, because it can be successfully substituted by the Bernoulli integral, which expresses the law of conservation of the total mechanical energy.

Let us consider the expression for the power of internal forces in a moving ideal medium. In the case of an ideal gas this power corresponds to the work being done by pressure forces to compress the gas. Noting that in the case considered (summation over repeated indices)

$$P = -pE; \quad P \cdot \dot{S} = -pE_{ij}\dot{S}_{ij} = -p\dot{S}_{ii} = -p \operatorname{div} \boldsymbol{V}$$

we obtain, according to Equation 19, Chapter 4, the following expression for the sought-after power

$$N_{\text{in}} = -P \cdot \dot{S} = p \operatorname{div} V \tag{34}$$

Note that according to the continuity equation (5) from Chapter 3

$$\operatorname{div} V = -\frac{1}{\rho}\frac{d\rho}{dt} = \rho\frac{d}{dt}\left(\frac{1}{\rho}\right) = \rho\frac{dv}{dt}$$

where v is the specific volume equal to $1/\rho$ so that Equation 34 is equivalent to

$$N_{\text{in}} = \rho p \frac{dv}{dt} \tag{35}$$

Dividing both parts of this equation by ρ and multiplying by dt yields the expression pdv, which is well known in thermodynamics as the expression of specific elementary work done by the internal pressure forces in an ideal gas (also known as "flow work").

The internal friction (viscosity) in a gas and thermal conductivity represent two aspects of the same process of molecular transfer.

The friction is due to momentum transfer, while the thermal conductivity relates to the transfer of the kinetic energy of molecules. Having assumed in this chapter the scheme of an ideal gas, i.e., the gas devoid of the internal friction, it is only natural to also ignore the thermal conductivity. Neglecting radiation emission, we shall assume that the moving gas is isolated from the inflow of heat from the outside. Such a motion is called adiabatic.[6] Moreover, note that the specific internal energy of a thermodynamically perfect gas is proportional to its absolute temperature and is equal to $U = c_v T$ where c_v is the heat capacity coefficient of the gas at constant volume.

Now consider the equation of the total energy balance (Equation 21, Chapter 4) under the assumptions: the motion is adiabatic ($q = 0$), the gas is ideal ($P = -pE$), the gas is perfect ($p/\rho = RT$, $U = c_v T$):

$$\rho\frac{d}{dt}\left(U + \frac{V^2}{2}\right) = \rho\frac{d}{dt}\left(c_v T + \frac{V^2}{2}\right) = \rho F \cdot V - \operatorname{div}(pV) \tag{36}$$

Along with the internal energy, let us introduce another thermal function, specific enthalpy

$$h = c_p T \tag{37}$$

Recall the derivation of the basic thermodynamic formula

$$c_p - c_v = R \tag{38}$$

which connects the coefficients of the heat capacity of the gas at constant pressure c_p and at constant volume c_v with the gas constant R. Determining the heat capacity coefficients as the quantities that characterize the rate of change in the amount of heat with a temperature rise under the condition of constant pressure or volume

[6]The concept of an adiabatic state was proposed in Chapter 4 to denote a state of the gas for which p/ρ^k = const. As shown later, the two definitions are equivalent for the considered case of an ideal perfect gas.

$$c_p = \left(\frac{\partial Q}{\partial T}\right)_p \qquad c_v = \left(\frac{\partial Q}{\partial T}\right)_v$$

the first law of thermodynamics written in the form ($v = 1/\rho$)

$$dQ = c_v dT + p dv$$

will yield

$$\left(\frac{\partial Q}{\partial T}\right)_p = c_v + p\left(\frac{\partial v}{\partial T}\right)_p \tag{39}$$

and the Clapeyron formula will give

$$\left(\frac{\partial v}{\partial T}\right)_p = \left[\frac{\partial}{\partial T}\left(\frac{RT}{p}\right)\right]_p = \frac{R}{p}$$

In this case Equation 39 is reduced to Equation 38.

Equation 38 will give the relationship between the internal energy and the enthalpy

$$U = c_v T = c_p T - RT = h - \frac{p}{\rho} \tag{40}$$

Substituting this expression of the internal energy into the balance equation (36), we obtain

$$\rho\frac{d}{dt}\left(h + \frac{V^2}{2}\right) = \rho \boldsymbol{F} \cdot \boldsymbol{V} - \operatorname{div}(p\boldsymbol{V}) + \rho\frac{d}{dt}\left(\frac{p}{\rho}\right) \tag{41}$$

By calculating the ordinary derivative and employing the continuity equation in the form of Equation 5, Chapter 3, we can transform the last term on the right side of Equation 41 to

$$\rho\frac{d}{dt}\left(\frac{p}{\rho}\right) = \frac{dp}{dt} - \frac{p}{\rho}\frac{d\rho}{dt} = \frac{dp}{dt} + \frac{p}{\rho}\rho \operatorname{div} \boldsymbol{V} = \frac{dp}{dt} + p \operatorname{div} \boldsymbol{V}$$

Substituting this expression into Equation 41 and making simple transformations, we obtain the energy balance equation in one of the following forms:

$$\left(\frac{dp}{dt} = \frac{\partial p}{\partial t} + \boldsymbol{V} \cdot \operatorname{grad} p\right):$$

$$\rho\frac{d}{dt}\left(h + \frac{V^2}{2}\right) = \rho \boldsymbol{F} \cdot \boldsymbol{V} - \boldsymbol{V} \cdot \operatorname{grad} p + \frac{dp}{dt}$$

$$\rho\frac{d}{dt}\left(h + \frac{V^2}{2}\right) = \rho \boldsymbol{F} \cdot \boldsymbol{V} + \frac{\partial p}{\partial t} \tag{42}$$

Eliminate the volumetric forces and the velocity vector $\boldsymbol{V}$ from the energy balance equation. For this, perform scalar multiplication of both parts of the Euler equation (3) by $\boldsymbol{V}$ and the term-by-term substraction of the resulting equation

$$\rho \boldsymbol{V} \cdot \frac{d\boldsymbol{V}}{dt} \equiv \rho\frac{d}{dt}\left(\frac{V^2}{2}\right) = \rho \boldsymbol{F} \cdot \boldsymbol{V} - \boldsymbol{V} \cdot \operatorname{grad} p$$

from both parts of the first of Equations 42. This will yield the following differential connection between the density, pressure, and enthalpy (i.e., temperature):

$$\rho \frac{dh}{dt} = \frac{dp}{dt} \tag{43}$$

This relation gives the barotropic character of the adiabatic motion of an ideal perfect gas and the corresponding connection between the density and pressure. It follows from Equation 43 and from Clapeyron's law that in the case of differentiation along the trajectory,

$$\frac{dp}{\rho} = dh = c_p dT = \frac{c_p}{R} d(RT) = \frac{c_p}{R} d\left(\frac{p}{\rho}\right) \tag{44}$$

However, according to Equation 38,

$$\frac{c_p}{R}\left(1 - \frac{c_v}{c_p}\right) = 1$$

or, after the introduction of the ratio $c_p / c_v \equiv k$,

$$\frac{c_p}{R} = \frac{k}{k-1} \tag{45}$$

Now Equation 44 can be rewritten in the form

$$\frac{dp}{\rho} = \frac{k}{k-1} d\left(\frac{p}{\rho}\right) \equiv \frac{k}{k-1}\frac{dp}{\rho} - \frac{k}{k-1}\frac{dp}{\rho^2} d\rho$$

or

$$\frac{dp}{p} = k\frac{d\rho}{\rho}$$

Integration gives Poisson's adiabatic

$$\frac{p}{p_0} = \left(\frac{\rho}{\rho_0}\right)^k \qquad \frac{\rho}{\rho_0} = \left(\frac{p}{p_0}\right)^{1/k} \tag{46}$$

where the subscript zero corresponds to the state of the gas at any fixed point on the particle trajectory. Thus, the motion is actually barotropic; the relationship between the density and pressure is determined by the adiabatic (Equation 46).

By virtue of the barotropic character of motion (see above), it follows from Equation 43 that

$$\frac{dh}{dt} = \frac{1}{\rho}\frac{dp}{dt} = \frac{d\mathcal{P}}{dt}$$

i.e., along the trajectory of motion the values of the enthalpy and pressure function differ only by the constant

$$h = \mathcal{P} + \text{const} \tag{47}$$

The expression of the pressure function $\mathcal{P}(p)$ for an adiabatic motion was derived in Chapter 4. Substituting the value of $\mathcal{P}$ from Chapter 3, Equation 74, Equation 47, we shall obtain the adiabatic relationship between the absolute temperature T and pressure p

$$c_p T = \frac{k}{k-1}\frac{p_0}{\rho_0}\left[\left(\frac{p}{p_0}\right)^{\frac{k-1}{k}} - 1\right] + \text{const} = \frac{k}{k-1}RT_0\left[\left(\frac{p}{p_0}\right)^{\frac{k-1}{k}} - 1\right] + c_p T_0$$

or, according to Equation 46,

$$\frac{T}{T_0} = \left(\frac{p}{p_0}\right)^{\frac{k-1}{k}} \qquad \frac{p}{p_0} = \left(\frac{T}{T_0}\right)^{\frac{k}{k-1}} \tag{48}$$

Substituting p by ρ, according to Equation 46, we obtain the adiabatic relationships between the temperature and density

$$\frac{T}{T_0} = \left(\frac{\rho}{\rho_0}\right)^{k-1} \qquad \frac{\rho}{\rho_0} = \left(\frac{T}{T_0}\right)^{\frac{1}{k-1}} \tag{49}$$

A simple connection between the thermodynamic elements of the gas p, ρ, T and the velocity of its motion V can be derived from the energy balance equation on the condition that the motion is stationary and the volumetric forces are conservative, Then we have

$$\frac{d}{dt}\left(h + \frac{V^2}{2}\right) = \boldsymbol{V}\cdot\text{grad}\left(h + \frac{V^2}{2}\right) \qquad \frac{\partial p}{\partial t} = 0 \qquad \boldsymbol{F}\cdot\boldsymbol{V} = -\boldsymbol{V}\cdot\text{grad}\,\Pi$$

and the second of Equations 42 can be reduced to the form

$$\boldsymbol{V}\cdot\text{grad}\left(h + \frac{V^2}{2} + \Pi\right) = 0$$

Noting that the scalar product of the velocity gradient by the scalar function gradient is proportional to the derivative of this function along the trajectory or streamline, we obtain the equation

$$\frac{V^2}{2} + h + \Pi = \text{const} \quad \text{(along the trajectory or streamline)} \tag{50}$$

Combining Equations 50 and 47, we again obtain the Bernoulli equation (20) which was derived from the assumption concerning the barotropic character of motion. The above derivation differs by the fact that the barotropic character of the process was not assumed *a priori*, but resulted from the condition of the adiabatic character of gas motion.

Later, when considering the motions of gas, we shall neglect the effect of volumetric forces (in particular, weight). This effect is substantial only in gas motions within great differences of altitudes above the Earth's surface, for example, in dynamic meteorology. In the absence of volumetric forces ($\Pi = 0$), Equation 50 is reduced to a more simple equation,

$$h + \frac{V^2}{2} \equiv c_p T + \frac{V^2}{2} = \text{const} \tag{51}$$

which states that in the case of an adiabatic motion of an ideal gas there is a certain relationship between its absolute temperature and velocity along the trajectory: with an increase in the velocity, the gas is cooled down, and conversely, with a decrease in the velocity, it becomes hotter.

Let us determine the constant in Equation 51 in terms of the parameters of the gas in the adiabatic stagnation state ($V = 0$, $h = h_0$, $T = T_0$); this will give the following formula which is fundamental to what follows

$$h + \frac{V^2}{2} = h_0 \tag{52}$$

The sum of the specific enthalpy and specific kinetic energy that occurs on the left side of this equation and which is preserved during the adiabatic motion of gas particles along their trajectories (streamlines), is called the total enthalpy or the stagnation enthalpy.

We shall not write out the relationships between the velocity and pressure or between the velocity and density, because Section 5 gives these very formulas in a more symmetric and easy to memorize form.

Along with the state functions, enthalpy h and pressure function $\mathcal{P}$, another state function is introduced, *viz.*, the specific entropy s defined by the differential relation

$$ds = dQ/T \tag{53}$$

where dQ is an elementary inflow of heat per unit mass of gas which, according to the first law of thermodynamics, satisfies the relationship

$$dQ = c_v dT + p dv = c_v dT + \frac{p}{\rho^2} d\rho$$

Substituting this expression into Equation 53:

$$ds = c_v \frac{dT}{T} - \frac{p}{\rho T} \frac{d\rho}{\rho} = c_v \frac{d(p/\rho)}{p/\rho} - R \frac{d\rho}{\rho} = R\left[\frac{c_v}{R} d \ln\left(\frac{p}{\rho}\right) - d \ln \rho\right]$$

or, noting besides that according to Equation 45,

$$\frac{c_v}{R} = \frac{1}{k-1}$$

we find

$$ds = \frac{R}{k-1} d \ln\left(\frac{p}{\rho^k}\right)$$

The integration of this relation will yield the following expression for the specific entropy in the final form:

$$s = \frac{R}{k-1} \ln\left(\frac{p}{\rho^k}\right) + \text{const} \tag{54}$$

The value of the constant is of no consequence because one has to deal only with increments of the entropy rather than with its absolute values.

If the equality $ds = 0$ is fulfilled along the trajectory of motion, i.e., the entropy preserves its value, then such a motion is called isentropic.

It follows from Equation 54 that the adiabatic motion of an ideal gas, which obeys Equation 46, is isentropic. Equation 46 could be called isentropic adiabatic, or more briefly, isentropic. Equations 51 and 52 are also called isentropic.

According to the second law of thermodynamics, the entropy in a closed (adiabatic) material system is a nondecreasing function of time. The growth of entropy in an adiabatic system shows that there are irreversible processes of the conversion of mechanical energy into heat that occur inside the system, which are accompanied by the losses of mechanical energy. An example of the formation of such mechanical losses can be provided by the losses due to internal friction in nonideal liquids and gases. In the next chapter we shall meet with the phenomenon of the loss of mechanical energy by a gas when passing through a shock – the discontinuity surface of kinematic and thermodynamic quantities. In this case, the motion, being adiabatic, will become nonisentropic.

4. Velocity of Propagation of Small Disturbances in an Ideal Gas: Speed of Sound

In order to elucidate the specific features of gas motion, it is very important to compare its velocity with another quantity that is characteristic for the given gas and depends on its thermodynamic state, namely, the velocity of propagation of small disturbances (for example, small compressions) through the gas, or equivalently, the velocity of sound propagation.

For this purpose, consider for simplicity the barotropic flow of an ideal perfect gas all of the streamlines of which are parallel to the x-axis. However, the velocity component u, pressure p, density ρ, and temperature T, are functions only of x and t; the action of volumetric forces will be neglected.

The Euler equation (4) and continuity equation (7) of Chapter 3 are reduced in this case to a nonlinear system of first-order partial differential equations (the letter notation of velocity projection is used, $u = V_1$),

$$\frac{\partial u}{\partial t} + u\frac{\partial u}{\partial x} = -\frac{1}{\rho}\frac{\partial p}{\partial x}$$

$$\frac{\partial \rho}{\partial t} + \frac{\partial}{\partial x}(\rho u) = 0 \qquad (55)$$

with three unknown functions u, p, ρ. In order to make the system determinate, it is necessary in the case of a barotropic motion to add one more equation which will connect ρ and p or, in a more general case, the Clapeyron and energy balance equations. The integrals of the system of equations thus formed must, of course, also satisfy the given initial and boundary conditions.

Consider the solution to the simplest problem of the propagation of small disturbances in a gas, which can be formulated as follows: in a quiescent ideal and perfect gas very small disturbances of velocities, pressures, or density are produced. The resulting motion is one-dimensional, parallel to the x-axis, and barotropic motion depends only on the coordinate x and the time t, it is a requirement to find the elements of the disturbed motion. Let us denote by u, p, and ρ the velocity, pressure, and density of the disturbed motion, and by p_0 and ρ_0, the pressure and density of the gas at rest; the action of volumetric forces is neglected. Then, introducing the notation u', p', ρ' for small disturbances of velocity, pressure, and density:

$$u = u' \qquad p = p_0 + p' \qquad \rho = \rho_0 + \rho'$$

Let us substitute these values of the disturbed elements into the system of equations (55) and discard in them the products of small quantities and their space derivatives as higher-order small quantities. Then, noting that for barotropic motion the following is valid accurate to the first-order small quantities

$$\frac{\partial p}{\partial x} = \frac{dp}{d\rho}\frac{\partial \rho}{\partial x} = \left(\frac{dp}{d\rho}\right)_0 \frac{\partial \rho'}{\partial x}$$

we shall obtain, instead of the nonlinear system (55), the following linear system of two equations with two unknown quantities u' and ρ':

$$\frac{\partial u'}{\partial t} + \frac{1}{\rho_0}\left(\frac{dp}{d\rho}\right)_0 \frac{\partial \rho'}{\partial x} = 0$$

$$\frac{\partial \rho'}{\partial t} + \rho_0 \frac{\partial u'}{\partial x} = 0 \qquad (56)$$

The system of equations (56) can be called linearized as compared to the nonlinear system (55), because it was derived from the latter by means of linearization, which consists of discarding the small quantities of the second and higher orders.

Noting that the quantity $dp/d\rho$ is essentially positive, because the density of a perfect gas grows with pressure, we introduce the notation

$$\left(\frac{dp}{d\rho}\right)_0 = a_0^2 \qquad (57)$$

and rewrite Equations 56 as

$$\rho_0 \frac{\partial u'}{\partial t} = -a_0^2 \frac{\partial \rho'}{\partial x} \qquad \rho_0 \frac{\partial u'}{\partial x} = -\frac{\partial \rho'}{\partial t} \qquad (58)$$

The system of equations (58) can be reduced to one equation. Differentiating both sides of the first equation from system (58) with respect to time t and of the second with respect to x, multiplying both sides of the second equation by a_0^2 and subtracting it term-by-term from the first equation, we obtain the linear, hyperbolic type equation:

$$\frac{\partial^2 u'}{\partial t^2} - a_0^2 \frac{\partial^2 u'}{\partial x^2} = 0 \qquad (59)$$

An analogous equation is found to determine ρ':

$$\frac{\partial^2 \rho'}{\partial t^2} - a_0^2 \frac{\partial^2 \rho'}{\partial x^2} = 0$$

and noting that

$$p' = p - p_0 = \left(\frac{dp}{d\rho}\right)_0 (\rho - \rho_0) = a_0^2 \rho'$$

also to determine p':

$$\frac{\partial^2 p'}{\partial t^2} - a_0^2 \frac{\partial^2 p'}{\partial x^2} = 0$$

The general solution of any of these equations can be presented in the form of the sum, specifically,

$$u' = f_1(x + a_0 t) + f_2(x - a_0 t) \tag{60}$$

the form of the functions f_1 and f_2 depends on the initial conditions of the problem.

Let us introduce the new coordinates ξ_1 and ξ_2 connected to the former coordinates by means of equalities

$$\xi_1 = x + a_0 t \qquad \xi_2 = x - a_0 t \tag{61}$$

This transformation of coordinates has a simple kinematic meaning. The coordinate axis $0_1\xi_1$ is situated along the axis $0x$ and moves translationally to the side of the negative direction of the axis $0x$ with the velocity a_0; in the same way the axis $0_2\xi_2$ moves translationally to the side of the positive direction of the $0x$ axis with the same velocity a_0. In this case, Equation 60 acquires the form

$$u' = f_1(\xi_1) + f_2(\xi_2) \tag{62}$$

Consider each term separately, i.e., the following two specific solutions of Equation 59:

$$u' = f_1(\xi_1) = f_1(x + a_0 t) \qquad u' = f_2(\xi_2) = f_2(x - a_0 t) \tag{63}$$

The function $f_1(\xi_1)$ in the moving system $0_1\xi_1$ represents the time-independent distribution of velocity disturbances. This fixed form of one-dimensional disturbance, prescribed by its initial distribution, moves, according to the first of Equations 63, as an entity along the negative direction of the immovable axis $0x$ with the velocity a_0. Analogously, the function $f_2(\xi_2)$, which characterizes the distribution of disturbance in the moving system $0_2\xi_2$, represents the second fixed form of the disturbance, generally differing from the first and also propagating as an entity in the positive direction of the immovable axis $0x$ with the same velocity a_0.

Assuming in these solutions that ξ_1 = const or ξ_2 = const, we obtain two systems of plane waves:

$$x + a_0 t = \text{const} \qquad x - a_0 t = \text{const} \tag{64}$$

representing two planes that move in opposite directions with velocity a_0 perpendicularly to the $0x$ axis, and each of which carries constant (prescribed by the initial conditions) values of the disturbances of velocity, pressure, density, or temperature; such waves are called simple.

The general solution of Equation 59 and, consequently, of analogous equations for the disturbances of pressure and density, is thus composed of the solutions corresponding to two oppositely propagating simple waves; Equation 59 itself, is a one-dimensional wave equation, as are the analogous equations for density and pressure. Geometrically, the solution obtained can be interpreted as the presence of two families of straight lines (64) in the plane (x, t) which have the angular coefficients $\pm a_0$ and which possess the property that along each of them there persist the constant values of the disturbances prescribed for velocity or other gas parameters by the initial conditions. In the case considered, these two families of straight lines represent two families of the characteristics of wave equation 59. Physically, two such simple waves can be produced by moving a piston with a small amplitude about a certain initial position.

The velocity a_0, common to both waves, is called the velocity of propagation of small disturbances in a gas, and, according to Equation 57, it is defined as

$$a = \sqrt{\frac{dp}{d\rho}} \tag{65}$$

In this formula the subscript zero, which characterizes the undisturbed state of the gas, is discarded, because Equation 65 is also valid for an arbitrarily moving gas, if only the quantity a is understood to be the local velocity of propagation of small disturbances relative to the moving gas at the given point in the flow. Among the most common phenomena of the propagation of small disturbances in liquids and gases is the propagation of sound, consisting, as is known, of the propagation of the waves of small compression and expansion. In view of this, the quantity a is called the speed of sound.

According to Equation 65, the speed of sound depends on the barotropic character of the process of propagation of small disturbances.

If it is assumed that the fluid is incompressible (ρ = const), then, according to Equation 65, $a = \infty$. This means that in the model of an incompressible fluid the disturbances propagate with an infinite velocity; i.e., any change in pressure at the given location in the flow should instantly manifest itself at any other location. In a number of cases this assumption can be employed in calculations with a practically sufficient approximation; in other cases (see below) it must be rejected and the scheme of a compressible medium should be used which has the finite velocity of sound propagation.

Assuming the process of sound propagation to be isothermal and recalling that in an isothermal process (C is a constant)

$$p = C\rho \qquad \frac{dp}{d\rho} = C = \frac{p}{\rho}$$

we obtain the isothermal velocity of sound

$$a_{is} = \sqrt{\frac{p}{\rho}}$$

If it is assumed that the process of sound propagation is so rapid that it is possible to neglect the effect of a comparatively slow process of heat removal and if we consider the process of sound propagation to be adiabatic:

$$p = C\rho^k \qquad \frac{dp}{d\rho} = kC\rho^{k-1} = k\frac{p}{\rho}$$

the adiabatic velocity of sound will be equal to

$$a = \sqrt{k\frac{p}{\rho}} \tag{66}$$

The formula of the isothermal propagation of sound was suggested by Newton. Equation 66 was suggested by Laplace; experiments confirmed the validity of Laplace's formula (Equation 66). In what follows, the speed of sound always means the adiabatic speed of sound, as given by Equation 66.

Table 1 Standard Speeds of Sound at Heights Above Sea Level

H (km)	T (K)	a (m/s)
−1.0	294.5	345
0.0	288.0	341
1.0	281.5	337
2.0	275.0	333
3.0	268.5	329
4.0	262.0	326
5.0	255.5	322
6.0	249.0	317
7.0	242.5	313
8.0	236.0	309
9.0	229.5	306
10.0	223.0	300

Applying the Clapeyron formula, rewrite Equation 66 in the form

$$a = \sqrt{kRT} \tag{67}$$

From this it follows that the velocity of sound propagation in a perfect gas solely depends on the absolute temperature and the physical properties of the gas. Noting that the gas constant R can be expressed in terms of the molecular weight of the gas μ and gravity acceleration g by the formula

$$R = \frac{848\,g}{\mu}\ \frac{\text{m}^2}{\text{s}^2 \cdot \text{K}}$$

we obtain

$$a = \sqrt{\frac{848\,kg}{\mu}}\sqrt{T}\ \text{m/s}$$

For air $k = 1.4$, $\mu = 29$; assuming $g = 9.81\ \text{m/s}^2$, we find that the velocity of sound propagation in air is

$$a = 20.1\sqrt{T}\ \text{m/s}$$

In particular, at $T = 273$ K (0°C) the speed of sound attains the value 332 m/s. The speed of sound in the air atmosphere varies with the height above sea level. Applying the standard atmosphere, we obtain Table 1.

The kinetic theory of gases shows that the speed of sound has the same order as the root mean square (RMS) free path velocity of gas molecules $v_s = \sqrt{\overline{v^2}}$, which is also proportional to the square root of the absolute temperature and is defined by the formula

$$v_s = \sqrt{3RT}$$

Comparing this with Equation 67:

$$a = \sqrt{\frac{k}{3}}\,v_s$$

where $k = c_p/c_v$.

For air ($k = 1.4$), the speed of sound constitutes about 70% of the RMS free path velocity of molecules.

The speed of sound depends substantially on the molecular weight; thus, the speed of sound in argon under normal conditions is smaller than in air, 308 m/s. This speed is still smaller in carbon dioxide, 258 m/s; in gaseous freon-12 the speed of sound at 15°C is 120 m/s.

The description of the phenomenon of propagation of small disturbances in nonuniform media, such as gas-liquid mixtures[7] is still more complex. Let us describe in more detail the propagation of sound waves in water saturated with air bubbles. The process of such mixture formation is called bubbling. The notation α and $(1-\alpha)$, respectively, is introduced for the volumetric concentrations of gas and liquid; the densities of the gas-liquid mixture and gas, and liquid separately are designated by ρ, ρ_g, and ρ_l and the speeds of sound corresponding to them, by a, a_g, and a_l. Then, for the density of the mixture:

$$\rho = \alpha\rho_g + (1-\alpha)\rho_l \tag{68}$$

Assuming that in the volume element the relationship between the gas and liquid masses is preserved:

$$\rho_g = \text{const} \cdot \frac{1-\alpha}{\alpha}\rho_l \tag{69}$$

Let us assume for simplicity that the gas bubbles are completely entrained by the liquid and that the pressure in a bubble p_g coincides with the liquid pressure and, consequently, with the pressure of the mixture at the corresponding point. Moreover, assume that the temperature in a gas bubble is constant and, consequently, the pressure p is proportional to the density ρ_g. Then, according to Equation 69

$$p = \text{const} \cdot \rho_g = \text{const} \cdot \frac{1-\alpha}{\alpha}\rho_l \tag{70}$$

Assuming that the propagation of small disturbances in a gas-liquid mixture is barotropic, take the derivative p from both parts of Equation 68, and in the result obtained,

$$\frac{d\rho}{dp} = \alpha\frac{d\rho_g}{dp} + (1-\alpha)\frac{d\rho_l}{dp} + (\rho_g - \rho_l)\frac{d\alpha}{dp}$$

make the replacement

$$\frac{d\rho}{dp} = \frac{1}{a^2} \qquad \frac{d\rho_g}{dp} = \frac{1}{a_g^2} \qquad \frac{d\rho_l}{dp} = \frac{1}{a_l^2}$$

At the same time note that Equation 70 yields

$$\frac{d\alpha}{dp} = -\frac{\alpha(1-\alpha)}{p} + \frac{\alpha(1-\alpha)}{\rho_l a_l^2}$$

Then,

[7] Van Wijngaarden, L., One-dimensional flow of liquid containing small gas bubbles, *Annu. Rev. Fluid Mech.*, 4, 370, 1972.

$$\frac{1}{a^2} = \frac{\alpha}{a_g^2} + \frac{1-\alpha}{a_l^2} - \frac{\alpha(1-\alpha)\rho_g}{p} + \frac{\alpha(1-\alpha)\rho_g}{\rho_l a_l^2} + \frac{\alpha(1-\alpha)\rho_l}{p} - \frac{\alpha(1-\alpha)}{a_l^2}$$

In the third term on the right side, use should be mad of the isothermicity of the gas bubble compression, and p should be substituted $\rho_g a_g^2$. Then, the previous equation can be rewritten in the form

$$\frac{1}{a^2} = \frac{\alpha}{a_g^2} - \frac{\alpha(1-\alpha)}{a_g^2}\left(1 - \frac{\rho_g a_g^2}{\rho_l a_l^2}\right) + \frac{1-\alpha}{a_l^2} - \frac{\alpha(1-\alpha)}{a_l^2} + \frac{\alpha(1-\alpha)\rho_l}{p}$$

or, noting that $\rho_g \ll \rho_l$, $a_g \ll a_l$, and $\rho_g a_g^2 \ll \rho_l a_l^2$, we finally obtain Wood's formula[8]

$$\frac{1}{a^2} = \frac{\alpha}{a_g^2} + \frac{(1-\alpha)^2}{a_l^2} + \frac{\alpha(1-\alpha)\rho_l}{p}$$

The first two terms on the right are relatively small and can be omitted; in practice it is possible to use the following approximate formula for the velocity of propagation of small disturbances (speed of sound) in a gas-liquid mixture:

$$a^2 = \frac{p}{\alpha(1-\alpha)\rho_l}$$

The minimum of the speed of sound corresponds to the volumetric concentration of gas $\alpha = 1/2$. For water with gas bubbles under ordinary pressure conditions (p = 10 kPa) this minimum is equal to 20 m/s; i.e., it is 17 times smaller than the speed of sound in air (340 m/s) and 75 times smaller than the speed of sound in water (1500 m/s). A substantial difference (a = 50 m/s) is also preserved at a 4% volumetric concentration of air. The cited review by Wijngaarden contains the generalizations of the above formulas of the speed of sound for gas-liquid media which take into account the difference between the velocities of the liquid and the gas bubbles, the effect of the nonisothermicity of the process of bubble compression, the presence of liquid viscosity, the frequency of sonic vibrations, and of other physical details of the process.[9]

5. The Numbers M and λ: Isentropic Formulas

The velocity of propagation of small disturbances or the speed of sound is an important characteristic of the flow of a compressible medium. Depending on whether the velocities of particles are smaller or greater than the speed of sound, the phenomena occurring in the medium will be fundamentally different. This can be demonstrated in the following simple example. Assume that a gas escapes through a converging pipe from a large tank into a certain chamber. The difference in pressures between the tank and the chamber is not great, at the beginning and the velocity of discharge through the pipe is not higher than the speed of sound. Slowly decrease the pressure in the chamber; the discharge velocity will then begin to rise. The disturbances (reductions of pressure) produced in the chamber will propagate upstream,

[8] Wood, A. B., *A Textbook of Sound*, Bell & Sons, Ltd., London, 1941.

[9] A detailed model of these processes can be found in Kogarko, B. S., Motion of the mixture of a liquid with gas bubbles, *Proc. Int. Symp. Nondeveloped Water Flows with High Velocities*, Nauka Press, Moscow, 1973, pp. 243 to 246, and in other papers by the same author.

from the chamber, through the pipe, into the tank, until the velocity at the pipe exit reaches the speed of sound. After this, the pressure disturbances will not be able to penetrate the tank because they will be swept by the flow, which has the same velocity as that of the propagation of disturbances in the gas. A further reduction of pressure in the chamber will not influence the discharge, the velocity of which will remain constant and equal to the speed of sound at the pipe exit. This phenomenon is called flow choking. Later, we shall also encounter other, very peculiar, phenomena in the flows of a compressible medium, i.e., gas.

If somewhere in the gas flow the velocity V becomes equal to the local speed of sound a, then this gas velocity $V = a^*$ is called critical; the corresponding values p^*, ρ^*, and T^* (pressure, density, and temperature) are also critical.

If the discharge velocity in this case attains its critical value at the pipe exit, then in this section (also called critical), the pressure, density, and temperature of the gas will acquire corresponding critical values. In adiabatic motion of gas the critical values of state parameters are identical for all gas particles, and they depend only on its total enthalpy (Section 3) and can be determined, for example, from the "stagnation" values of the parameters in the tank where the gas is regarded to be motionless. The presence of critical phenomena is a characteristic feature of gas flows.

The foregoing example shows that the character of the phenomena developing in the flow has a bearing on the magnitude of the ratio between the velocity at the given point in the flow and the speed of sound or the critical flow velocity.

The ratio between the gas velocity V at the given point in the flow and the local speed of sound corresponding to this point,

$$\mathbf{M} = \frac{V}{a}$$

which shows whether the flow at this point is subsonic ($\mathbf{M} < 1$), sonic ($\mathbf{M} = 1$), or supersonic ($\mathbf{M} > 1$), represents the basic parameter of gas motion, and is called the **M** number (Mach number). The ratio of the flow velocity at the given point to the critical velocity $V/a^* = \lambda$, which is the same for the entire flow, will be called the velocity coefficient. In foreign literature it is denoted by the symbol $\mathbf{M}^*$.

It was indicated at the end of Section 4 that the speed of sound is proportional to the RMS free path velocity of molecules. This allows the squared number **M** to be considered the value that characterizes the ratio of the kinetic energies of the directed (external, studied in the mechanics of liquids and gases) and random, molecular (internal) motions of a gas. In fact, this ratio of energies is equal to

$$\frac{V^2}{v_s^2} = \frac{V^2}{(3/k)\,a^2} = \frac{k}{3}\mathbf{M}^2$$

In subsonic motion ($\mathbf{M} < 1$) the kinetic energy of the directed motion is smaller than the kinetic energy of random motion. This is also preserved in supersonic motion up to $\mathbf{M} = \sqrt{3/k}$ (for air, 1.46): at greater values of the number **M** the kinetic energy of the directed motion exceeds in magnitude the kinetic energy of molecular motion. From the viewpoint of the dynamics of a gas, considered a continuous medium, the characteristic **M** number, which would be critical for the processes of gas motion, is not the value at which the energies of the directed and random motions equilibrate, but rather the value $\mathbf{M} = 1$, which corresponds to the velocity of gas particles being equal to the velocity of propagation of small disturbances at the same point in the gas. It will become clear from what follows that the equations of gas motion themselves differ fundamentally in character: elliptical in a subsonic ($\mathbf{M} < 1$) and hyperbolic in a supersonic ($\mathbf{M} > 1$) flow. This mathematical difference reflects the physical aspects of two basic regimes of gas flow.

Using the **M** and λ numbers, it is possible to form simple, memorizable formulas of connection between the velocity, pressure, density, and temperature in isentropic adiabatic motion.

Having replaced the enthalpy h by its expression in terms of the temperature in Equation 51, we have

$$\frac{V^2}{2} + c_p T = c_p T_0$$

where, as always, the subscript zero indicates that the value is taken at $V = 0$, i.e., in a gas that is in an adiabatic and isentropical stagnation state. Dividing the both parts of the latter equation by $c_p T$, we obtain

$$\frac{T_0}{T} = 1 + \frac{V^2}{2c_p T}$$

Noting that

$$\frac{V^2}{2c_p T} = \frac{V^2}{2\frac{c_p}{kR} \cdot kRT} = \frac{k-1}{2}\frac{V^2}{a^2} = \frac{k-1}{2}\mathbf{M}^2$$

we rewrite the previous formula as

$$\frac{T_0}{T} = 1 + \frac{k-1}{2}\mathbf{M}^2 \qquad \frac{T}{T_0} = \left(1 + \frac{k-1}{2}\mathbf{M}^2\right)^{-1} \tag{71}$$

Recalling that the speed of sound is proportional to the square root of the absolute temperature, we obtain

$$\frac{a_0}{a} = \left(1 + \frac{k-1}{2}\mathbf{M}^2\right)^{1/2} \qquad \frac{a}{a_0} = \left(1 + \frac{k-1}{2}\mathbf{M}^2\right)^{-1/2} \tag{72}$$

With the use of Equations 48 and 49, it is easy to obtain from Equation 71 the isentropic formulas:

For the ratio of densities

$$\frac{\rho_0}{\rho} = \left(1 + \frac{k-1}{2}\mathbf{M}^2\right)^{\frac{1}{k-1}} \qquad \frac{\rho}{\rho_0} = \left(1 + \frac{k-1}{2}\mathbf{M}^2\right)^{-\frac{1}{k-1}} \tag{73}$$

For the ratio of pressures

$$\frac{p_0}{p} = \left(1 + \frac{k-1}{2}\mathbf{M}^2\right)^{\frac{k}{k-1}} \qquad \frac{p}{p_0} = \left(1 + \frac{k-1}{2}\mathbf{M}^2\right)^{-\frac{k}{k-1}} \tag{74}$$

and for the ratio of the flow velocity to the speed of sound in a gas at rest

$$\frac{V}{a_0} = \mathbf{M}\left(1 + \frac{k-1}{2}\mathbf{M}^2\right)^{-1/2} \tag{75}$$

Isentropic equations 71 to 75 establish the parametric connection between temperature, density, pressure, and velocity with the aid of the aid of the parameter **M**. Analogous parametric formulas can be derived that have the velocity coefficient λ as a parameter. For this purpose, note that the enthalpy is related to the speed of sound by the relation

$$h = c_p T = \frac{c_p kRT}{kR} = \frac{a^2}{k-1} \tag{76}$$

Rewrite Equation 51 in the form

$$\frac{V^2}{2} + \frac{a^2}{k-1} = \text{const}$$

where the constant can be determined from both the condition $a = a_0$ at $V = 0$ and the condition $a = a^*$ at $V = a^*$. We obtain one of the following two equalities

$$\frac{V^2}{2} + \frac{a^2}{k-1} = \frac{a_0^2}{k-1} \qquad \frac{V^2}{2} + \frac{a^2}{k-1} = \frac{k+1}{2(k-1)} a^{*2} \tag{77}$$

Dividing both parts of the latter equation by V^2, we obtain the connection between the Mach number **M** and the velocity coefficient λ

$$\frac{1}{\lambda^2} = \frac{k-1}{k+1} + \frac{2}{k+1} \frac{1}{\mathbf{M}^2} \tag{78}$$

which can be easily resolved with respect to λ or **M**. Thus, we obtain

$$\lambda = \sqrt{\frac{k+1}{2}} \frac{\mathbf{M}}{\sqrt{1 + \frac{k-1}{2}\mathbf{M}^2}} \tag{79}$$

and the reverse relation

$$\mathbf{M} = \sqrt{\frac{2}{k+1}} \frac{\lambda}{\sqrt{1 - \frac{k-1}{k+1}\lambda^2}} \tag{80}$$

If $\mathbf{M} = 0$, then $\lambda = 0$; on the other hand, if $\mathbf{M} \to \infty$, then $\lambda \to \lambda_{max} = \sqrt{\frac{k+1}{k-1}}$ (λ_{max} = 2.449 for air at $k = 1.4$).

Note that the binominal $1 + \frac{k-1}{2}\mathbf{M}^2$ entering the isentropic formulas can be expressed in terms of λ by the formula

$$1 + \frac{k-1}{2}\mathbf{M}^2 = \frac{1}{1 - \frac{k-1}{k+1}\lambda^2} \tag{81}$$

so that Equations 71 to 75 can be transformed to give

$$\frac{T}{T_0} = 1 - \frac{k-1}{k+1}\lambda^2 \qquad \frac{a}{a_0} = \left(1 - \frac{k-1}{k+1}\lambda^2\right)^{1/2}$$

$$\frac{\rho}{\rho_0} = \left(1 - \frac{k-1}{k+1}\lambda^2\right)^{\frac{1}{k-1}}$$

$$\frac{p}{p_0} = \left(1 - \frac{k-1}{k+1}\lambda^2\right)^{\frac{k}{k-1}} \qquad \frac{V}{a_0} = \sqrt{\frac{2}{k+1}}\lambda \tag{82}$$

The above formulas, are useful to remember, as are Equations 71 to 75, because they are constantly used in calculations of gas flows.

Let us show that isentropic formulas (73) and (74) contain, as a specific case at $\mathbf{M} = 0$, the formulas of the incompressible fluid

$$\rho = \rho_0 \qquad p + \frac{\rho V^2}{2} = p_0 \tag{83}$$

It is emphasized that in this case the condition $\mathbf{M} = 0$ should be understood as the existence of the infinite speed of sound a, as given by the formula $a = \sqrt{dp/d\rho}$ at $\rho = \text{const}$, i.e., the result of the denominator a in the expression of $\mathbf{M}$ as infinite rather than of the numerator V as vanishing, because the latter would have indicated the absence of flow. Let us expand the right sides of Equations 73 and 74 into power series at small values of $\mathbf{M}$:

$$\frac{p_0 - p}{\frac{1}{2}\rho V^2} = \frac{p}{\frac{1}{2}\rho V^2}\left[\left(1 + \frac{k-1}{2}\mathbf{M}^2\right)^{\frac{k}{k-1}} - 1\right] =$$

$$= \frac{2}{k}\frac{kp/\rho}{V^2}\left\{\left[1 + \frac{k}{2}\mathbf{M}^2 + \frac{\frac{k}{k-1}\left(\frac{k}{k-1} - 1\right)}{2}\frac{(k-1)^2}{4}\mathbf{M}^4 + \cdots\right] - 1\right\} =$$

$$= \frac{2}{k\mathbf{M}^2}\left(\frac{k}{2}\mathbf{M}^2 + \frac{k}{8}\mathbf{M}^4 + \cdots\right) = 1 + \frac{1}{4}\mathbf{M}^2 + \cdots \tag{84}$$

$$\frac{\rho}{\rho_0} = 1 - \frac{1}{2}\mathbf{M}^2 + \cdots \tag{85}$$

Assuming that $\mathbf{M} = 0$, find Equation 83 for an incompressible fluid. Moreover, taking into account the second terms in the resulting expansions, determine the order of the error that appears when the gas moving at small $\mathbf{M}$ is considered as an incompressible fluid. Regarding $\rho = \text{const} = \rho_0$, the terms (the higher of which is equal to $1/2\mathbf{M}^2$) are neglected as compared to unity. If, for example, the 1% relative error appears because of the disregard for gas compressibility, this is equivalent to the condition

$$1/2\mathbf{M}^2 \leq 0.01 \qquad \mathbf{M} \leq 0.14$$

which, for air under normal conditions [$T = (273 + 15)$ K, $a = 340$ m/s] leads to the limitation of the velocity $V < 50$ m/s. At the velocity close to 100 m/s the relative error amounts to 4%. As is seen from Equation 84, the relative error for pressures is twice as small as for densities.

Using the isentropic formulas, find the expression of the critical parameters of the gas T^*, a^*, p^*, ρ^* in terms of the parameters T_0, a_0, p_0, and ρ_0 of the stagnation gas. It suffices to recall that in the critical flow the velocity V is equal to the local speed of sound a, i.e., in this case $\mathbf{M} = 1$. Then, according to equations 71 to 75:

$$\frac{T^*}{T_0} = \frac{2}{k+1} \qquad \frac{a^*}{a_0} = \sqrt{\frac{2}{k+1}} \qquad \frac{p^*}{p_0} = \left(\frac{2}{k+1}\right)^{\frac{k}{k-1}} \qquad \frac{\rho^*}{\rho_0} = \left(\frac{2}{k+1}\right)^{\frac{1}{k-1}} \tag{86}$$

Having formed the isentropic relations for any two points with the numbers $\mathbf{M}_1$ and $\mathbf{M}_2$ or λ_1 and λ_2 of the same flow or of two flows, but with the same parameters of the stagnation flow, and having divided the corresponding relation one by the other:

$$\frac{T_1}{T_2} = \frac{1 + \dfrac{k-1}{2}\mathbf{M}_2^2}{1 + \dfrac{k-1}{2}\mathbf{M}_1^2} = \frac{1 - \dfrac{k-1}{k+1}\lambda_1^2}{1 - \dfrac{k-1}{k+1}\lambda_2^2}$$

$$\frac{a_1}{a_2} = \left(\frac{1 + \dfrac{k-1}{2}\mathbf{M}_2^2}{1 + \dfrac{k-1}{2}\mathbf{M}_1^2}\right)^{1/2} = \left(\frac{1 - \dfrac{k-1}{k+1}\lambda_1^2}{1 - \dfrac{k-1}{k+1}\lambda_2^2}\right)^{1/2}$$

$$\frac{\rho_1}{\rho_2} = \left(\frac{1 + \dfrac{k-1}{2}\mathbf{M}_2^2}{1 + \dfrac{k-1}{2}\mathbf{M}_1^2}\right)^{\frac{1}{k-1}} = \left(\frac{1 - \dfrac{k-1}{k+1}\lambda_1^2}{1 - \dfrac{k-1}{k+1}\lambda_2^2}\right)^{\frac{1}{k-1}}$$

$$\frac{p_1}{p_2} = \left(\frac{1 + \dfrac{k-1}{2}\mathbf{M}_2^2}{1 + \dfrac{k-1}{2}\mathbf{M}_1^2}\right)^{\frac{k}{k-1}} = \left(\frac{1 - \dfrac{k-1}{k+1}\lambda_1^2}{1 - \dfrac{k-1}{k+1}\lambda_2^2}\right)^{\frac{k}{k-1}}$$

$$\frac{V_1}{V_2} = \frac{\mathbf{M}_1}{\mathbf{M}_2}\left(\frac{1 + \dfrac{k-1}{2}\mathbf{M}_2^2}{1 + \dfrac{k-1}{2}\mathbf{M}_1^2}\right)^{1/2} = \frac{\lambda_1}{\lambda_2} \tag{87}$$

6

One-Dimensional Flow of an Ideal Gas

1. One-Dimensional Steady Motion of a Gas through a Tube of Variable Cross-Section

If all the dynamic and thermodynamic parameters of the gas flow are functions of only one coordinate, generally, curvilinear and time, then such a flow is called one-dimensional.

For approximate calculations of gas flows in tubes it is often possible to satisfy ourselves with the following simplified one-dimensional stationary scheme. Assuming the velocity vector at the given section of the tube or a channel to be directed along the axis and the values of the velocity V, pressure p, density ρ, and temperature T to be constant over the section, we consider them to be the values that vary from section to section of the channel, with the law of variation in the cross-sectional area A along the axis considered as given.

Neglecting the effect of the channel axis curvature, we adopt as the independent variable the rectilinear Cartesian coordinate x, reckoned along the channel axis downstream from some initial section. The flow is considered adiabatic and the gas perfect and ideal. Under these conditions, as was already been proven in Chapter 5, the gas motion can be regarded as isentropic.

Using the Euler equation

$$u\frac{du}{dx} = -\frac{1}{\rho}\frac{dp}{dx} \tag{1}$$

and the continuity equation

$$\rho u A = \text{const} \tag{2}$$

it is easy to establish the differential relationship between the variations in the velocity and the cross-sectional area of the tube. For this purpose, transform Equation 1 into

$$u du = -\frac{dp}{\rho} = \frac{dp}{d\rho}\frac{d\rho}{\rho} = -a^2\frac{d\rho}{\rho}$$

and take from both parts of Equation 2 the logarithmic differential. Then,

$$\frac{d\rho}{\rho} = -\frac{du}{u} - \frac{dA}{A}$$

Using this equation to eliminate the density from the previous equation:

$$(u^2 - a^2)\frac{du}{u} = a^2\frac{dA}{A}$$

or, dividing both parts by a^2,

$$(\mathbf{M}^2 - 1)\frac{du}{u} = \frac{dA}{A} \tag{3}$$

From this equation, named for Hugoniot,[1] the corollaries follow:

1. If $\mathbf{M} < 1$, the sign du is opposite to the sign dA, i.e., in subsonic motion of a gas, just as in the case of an incompressible fluid, with an increase of the cross-sectional area of the tube, the velocity of motion decreases and, conversely, with a decrease of the area, the velocity increases.
2. If $\mathbf{M} > 1$, the sign of du is identical to the sign of dA, i.e., in supersonic motion of a gas through a converging tube the motion slows down; in a diverging tube it accelerates. This result, though paradoxical at first sight, is explained by the fact that on gas expansion, its density decreases to such an extent that the product ρA in Equation 2 decreases despite the increase in the area A, and this leads to an increase in the velocity u.
3. If $\mathbf{M} = 1$, then $dA = 0$; the corresponding section of the tube will be critical. The condition $dA = 0$ coincides with the prerequisite condition of the extremum of the cross-sectional area. It can be easily grasped that the critical cross-section will be minimal, because with the approach to the maximum cross-section the subsonic flow slows down, whereas the supersonic flow accelerates. This by no means can lead to the flow with the speed of sound in the critical section.
4. If $dA = 0$ and the section is extremal (maximal or minimal), the either $\mathbf{M} = 1$ (consequently, this section is critical) or $\mathbf{M} \neq 1$ and $du = 0$. In the latter case, whatever the flows – subsonic or supersonic – the velocity in the extremal section also acquires the extremal value: in a subsonic flow of gas it is minimal in the maximal section and maximal in the minimal section; in the maximal section of a supersonic flow, the velocity is maximal, and in the minimal section it is minimal.

Using Equation 2 and the isentropic formulas derived at the end of Chapter 5, we find the relationship between the parameters of a one-dimensional flow and the cross-sectional area prescribed as the function of the coordinate x. In fact, according to Equation 2 of this chapter and Equation 87 of Chapter 5, we have (subscript 1 denotes any fixed tube section)

[1] Hugoniot, P. H., *C. R. Acad. Sci.*, 103, 1178, 1880.

$$\frac{A}{A_1} = \frac{\rho_1 u_1}{\rho u} =$$

$$= \left(\frac{1 + \frac{k-1}{2} \mathbf{M}^2}{1 + \frac{k-1}{2} \mathbf{M}_1^2} \right)^{\frac{1}{k-1}} \frac{\mathbf{M}_1}{\mathbf{M}} \frac{a_1}{a} = \left(\frac{1 + \frac{k-1}{2} \mathbf{M}^2}{1 + \frac{k-1}{2} \mathbf{M}_1^2} \right)^{\frac{1}{k-1}} \frac{\mathbf{M}_1}{\mathbf{M}} \left(\frac{1 + \frac{k-1}{2} \mathbf{M}^2}{1 + \frac{k-1}{2} \mathbf{M}_1^2} \right)^{\frac{1}{2}}$$

or finally

$$\frac{A}{A_1} = \frac{\mathbf{M}_1}{\mathbf{M}} \left(\frac{1 + \frac{k-1}{2} \mathbf{M}^2}{1 + \frac{k-1}{2} \mathbf{M}_1^2} \right)^{\frac{k+1}{2(k-1)}} \tag{4}$$

This relation, together with the isentropic formulas

$$\frac{p}{p_1} = \left(\frac{1 + \frac{k-1}{2} \mathbf{M}_1^2}{1 + \frac{k-1}{2} \mathbf{M}^2} \right)^{\frac{k}{k-1}} \qquad \frac{\rho}{\rho_1} = \left(\frac{1 + \frac{k-1}{2} \mathbf{M}_1^2}{1 + \frac{k-1}{2} \mathbf{M}^2} \right)^{\frac{1}{k-1}}$$

$$\frac{T}{T_1} = \frac{1 + \frac{k-1}{2} \mathbf{M}_1^2}{1 + \frac{k-1}{2} \mathbf{M}^2} \qquad \frac{u}{u_1} = \frac{\mathbf{M}}{\mathbf{M}_1} = \left(\frac{1 + \frac{k-1}{2} \mathbf{M}_1^2}{1 + \frac{k-1}{2} \mathbf{M}^2} \right)^{\frac{1}{2}} \tag{5}$$

gives the parametric solution of the problem of a one-dimensional gas flow in a tube of variable cross-section, with $\mathbf{M}$ playing the role of the parameter. Specifying the function $A(x)$, we shall determine $\mathbf{M}(x)$ from Equation 4 and then the unknown quantities $p(x)$, $\rho(x)$, $T(x)$, and $u(x)$ from Equation 5.

Equation 4 simplifies if it is assume that $\mathbf{M}_1 = 1$; then the section with the area A_1 will be critical ($A_1 = A^*$) and Equation 4 will acquire the form

$$\frac{A}{A^*} = \left(\frac{2}{k+1} \right)^{\frac{k+1}{2(k-1)}} \frac{1}{\mathbf{M}} \left(1 + \frac{k-1}{2} \mathbf{M}^2 \right)^{\frac{k+1}{2(k-1)}} = \Theta^{-1} \tag{6}$$

The plot of this relation for air ($k = 1.4$) is given in Figure 29. The plot confirms the fact revealed earlier: in a subsonic flow ($\mathbf{M} < 1$), the $\mathbf{M}$ number can be increased, thus, it is necessary to decrease A. In a supersonic flow ($\mathbf{M} > 1$), conversely, A should be increased. Thus, for example, it follows from Figure 29 that for $\mathbf{M}$ to be increased from 0.2 to 0.8, the gas must pass through a converging section of the tube whose cross-section was decreased approximately threefold; in order to increase the $\mathbf{M}$ number from 1 in the critical section to 3.2, it is necessary to construct a diverging tube (a supersonic diverging duct), having an the exit area about five times that of the critical section.

Supplementing Equation 6 with the obvious group of formulas

$$\frac{p}{p^*} = \left[\frac{2}{k+1} \left(1 + \frac{k-1}{2} \mathbf{M}^2 \right) \right]^{-\frac{k}{k-1}} \qquad \frac{\rho}{\rho^*} = \left[\frac{2}{k+1} \left(1 + \frac{k-1}{2} \mathbf{M}^2 \right) \right]^{-\frac{1}{k-1}}$$

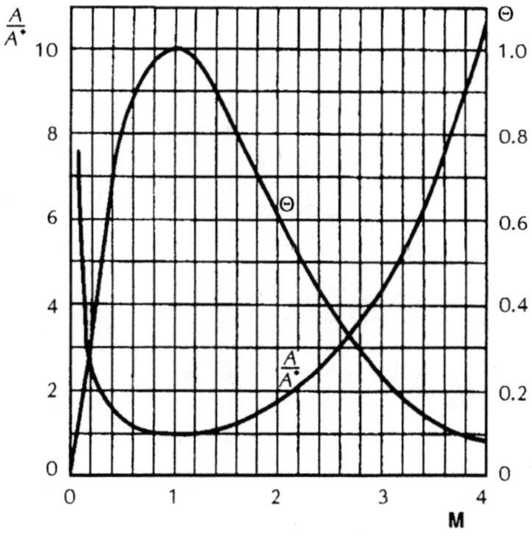

Fig. 29

$$\frac{T}{T^*} = \left[\frac{2}{k+1}\left(1 + \frac{k-1}{2}\mathbf{M}^2\right)\right]^{-1} \qquad \frac{u}{a^*} = \mathbf{M}\left[\frac{2}{k+1}\left(1 + \frac{k-1}{2}\mathbf{M}^2\right)\right]^{-\frac{1}{2}} \tag{7}$$

find the parametric solution involving the critical values A^*, p^*, ρ^*, T^* rather than A_1, p_1, ρ_1, T_1, which refer to an arbitrary section.

Using Equations 80 and 81 of Chapter 5, it is not difficult to write the corresponding formulas expressed in terms of the parameter λ

$$\frac{A}{A^*} = \left(\frac{2}{k+1}\right)^{\frac{1}{k-1}}\left[\lambda\left(1 - \frac{k-1}{k+1}\lambda^2\right)^{\frac{1}{k-1}}\right]^{-1}$$

$$\frac{p}{p^*} = \left(\frac{k+1}{2}\right)^{\frac{k}{k-1}}\left(1 - \frac{k-1}{k+1}\lambda^2\right)^{\frac{k}{k-1}}$$

$$\frac{\rho}{\rho^*} = \left(\frac{k+1}{2}\right)^{\frac{1}{k-1}}\left(1 - \frac{k-1}{k+1}\lambda^2\right)^{\frac{1}{k-1}}$$

$$\frac{T}{T^*} = \frac{k+1}{2}\left(1 - \frac{k-1}{k+1}\lambda^2\right) \qquad \frac{u}{a^*} = \lambda \tag{8}$$

The continuity equation (2) rewritten as

$$\rho u A = \rho^* u^* A^* = \rho^* a^* A^* \tag{9}$$

gives the relation

$$\frac{A^*}{A} = \frac{\rho u}{\rho^* a^*} = \Theta \tag{10}$$

Table 2

M	λ	$\sqrt{\tau}$	p/p^*	p/p_0	ρ/ρ_0	T/T_0	Θ
0	0	0	1.894	1.0000	1.0000	1.0000	0.0000
0.2	0.218	0.089	1.842	0.9725	0.9803	0.9921	0.3374
0.4	0.431	0.176	1.696	0.8956	0.9243	0.9690	0.6288
0.6	0.635	0.259	1.485	0.7840	0.8405	0.9328	0.8416
0.8	0.825	0.337	1.345	0.6560	0.7400	0.8865	0.9632
1.0	1.000	0.408	1.000	0.5283	0.6339	0.8333	1.0000
1.2	1.159	0.474	0.781	0.4124	0.5311	0.7764	0.9705
1.4	1.300	0.531	0.595	0.3142	0.4374	0.7184	0.8969
1.6	1.426	0.582	0.446	0.2353	0.3557	0.6614	0.7999
1.8	1.536	0.627	0.329	0.1740	0.2868	0.6068	0.6949
2.0	1.633	0.667	0.242	0.1278	0.2301	0.5556	0.5926
2.2	1.718	0.701	0.177	0.0935	0.1841	0.5081	0.4988
2.4	1.793	0.732	0.129	0.0684	0.1472	0.4647	0.4161
2.6	1.858	0.758	0.095	0.0501	0.1179	0.4252	0.3453
2.8	1.914	0.781	0.070	0.0368	0.0946	0.3894	0.2857
3.0	1.964	0.802	0.0515	0.0272	0.0762	0.3571	0.2362
3.2	2.008	0.820	0.0383	0.0202	0.0616	0.3281	0.1953
3.4	2.047	0.835	0.0296	0.0151	0.0500	0.3019	0.1617
3.6	2.081	0.849	0.0216	0.0114	0.0409	0.2784	0.1342
3.8	2.112	0.862	0.0163	0.0086	0.0335	0.2572	0.1113
4.0	2.138	0.873	0.0125	0.0066	0.0276	0.2381	0.0933
∞	2.437	1.000	0	0	0	0	0

where the quantity Θ, according to Equation 6 and to the first of Equations 8, at the given k is only the function of **M** or λ

$$\Theta = \left(\frac{2}{k+1}\right)^{-\frac{k+1}{2(k-1)}} \mathbf{M}\left(1+\frac{k-1}{2}\mathbf{M}^2\right)^{-\frac{k+1}{2(k-1)}} =$$

$$= \left(\frac{k+1}{2}\right)^{\frac{1}{k-1}} \lambda\left(1-\frac{k-1}{k+1}\lambda^2\right)^{\frac{1}{k-1}} \tag{11}$$

Figure 29 presents the plot of $\Theta(\mathbf{M})$ for air ($k = 1.4$). For any section A of the tube with the prescribed critical (minimal) section A^*, the value of Θ can be found from Equation 10, **M** or λ from Equation 11, and p, ρ, T, and (directly) velocity u from Equations 7 or 8. In carrying out calculations, it is convenient to use the available tables for air (Table 2).[2] The meaning of the quantity $\sqrt{\tau}$ in the third column of Table 2 is explained in the following section.

2. The Discharge of a Gas through a Nozzle

As the first example showing the application of the above formulas, consider the problem of isentropic gas escape from a reservoir (tank) of very large capacity. First, assume that the nozzle through which the flow escapes has the shape of a convergent duct, i.e., of a duct whose cross-sectional area decreases downstream. Denote by p_0, ρ_0, and T_0 the thermody-

[2]Detailed tables can be found in the books Ferri, A., *Aerodinamika sverkhzvukovykh techeniy* (*Aerodynamics of Supersonic Flows*) (translated from English), Gostekhizdat Press, Moscow, 1953; Abramovich, G. N., *Prikladnaya gazovaya dinamika* (*Applied Gas Dynamics*), Nauka Press, Moscow, 1976.

namic parameters of the gas in the tank, where the gas can be considered quiescent because of the great capacity of the tank ($u = 0$, $\mathbf{M} = 0$); the symbols $\bar{p}$, $\bar{\rho}$, $\bar{T}$, and $\bar{\mathbf{M}}$ will denote the corresponding parameters at the exit with the area $\bar{A}$, and p' will stand for the pressure in the chamber into which the gas escapes; in the discharge theory this pressure p' is called back pressure.

Denote by m the mass of gas flowing per second through any section of the nozzle;

$$m = \rho u A = \bar{\rho}\bar{u}\bar{A}$$

and by m^*, the critical value of this mass corresponding to the $\bar{\mathbf{M}}$ number at the exit equal to unity; we have $m^* = \rho^* a^* \bar{A}$ and, consequently, according to Equation 10, the dimensionless mass flow rate will be equal to

$$\frac{m}{m^*} = \frac{\bar{\rho}\bar{u}}{\rho^* a^*} = \Theta(\bar{\mathbf{M}}) \tag{12}$$

Eliminating $\bar{\mathbf{M}}$ from Equation 12 with the aid of the isentropic relation

$$\frac{\bar{p}}{p_0} = \left(1 + \frac{k-1}{2}\bar{\mathbf{M}}^2\right)^{-\frac{k}{k-1}} \tag{13}$$

derives the expression for the dimensionless flow rate in terms of the ratio of the exit pressure to the pressure in the tank

$$\frac{m}{m^*} = \sqrt{\frac{2}{k-1}\left(\frac{k+1}{2}\right)^{\frac{k+1}{k-1}}\left(\frac{\bar{p}}{p_0}\right)^{\frac{2}{k}}\left[1 - \left(\frac{\bar{p}}{p_0}\right)^{\frac{k-1}{k}}\right]} \tag{14}$$

Figure 30 presents the plot of the dimensionless mass flow rate m/m^* vs. The ratio of the back pressure p' in the chamber to the pressure p_0 in the tank. Until the pressure at the exit from the nozzle $\bar{p}$ becomes equal to the critical pressure p^*, the back pressure p' will coincide with $\bar{p}$, and the curve is plotted from Equation 14 (the right solid branch of the curve). With a further decrease in the back pressure, i.e., when $p' < p^*$, the choking of the exit section will occur, which was described at the beginning of Section 5, Chapter 5. The pres-

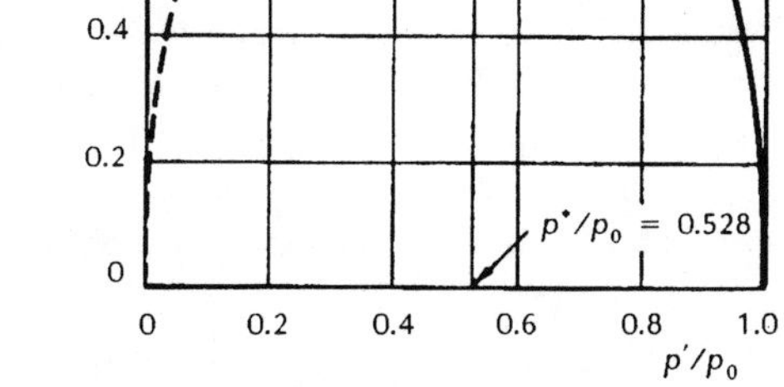

Fig. 30

sure disturbances will not be able to penetrate through the exit section and the discharge will occur with a constant critical velocity in spite of the fact that the back pressure continues to decrease. A portion of the plot in Figure 30 corresponding to the interval $p'/p_0 < p^*/p_0$ (for air $p^*/p_0 = 0.528$) will be represented by a solid horizontal straight line and not by a descending dashed curve, as would follow from Equation 14.

The greatest possible (at the given parameters in the tank) mass flow rate of gas m_{max} through the exit section of the convergent nozzle is equal to $m_{max} = m^* = \rho^* a^* \bar{A}$ and, according to Equation 86 of Chapter 5, can be presented as

$$m_{max} = \left(\frac{2}{k+1}\right)^{\frac{k+1}{2(k-1)}} \sqrt{k p_0 \rho_0} A \tag{15}$$

As was pointed out by Prandtl,[3] the following simple approximate formulas can be used for air discharge

$$G = 0.76\bar{A}\sqrt{\frac{(p_0 - \bar{p})\bar{p}}{T_0}} \quad \text{when} \quad \bar{p} > \frac{1}{2}p_0 \tag{16}$$

$$G_{max} = 0.38\bar{A}\frac{p_0}{\sqrt{T_0}} \quad \text{when} \quad \bar{p} > \frac{1}{2}p_0 \tag{17}$$

where G and G_{max} represent the weight flow rates expressed in Newtons per second in (N/s); the pressures p_0 and $\bar{p}$ should be expressed, respectively, in Pascals times Newtons per square meter (Pa (N/m^2)) or 10 kPa (N/cm^2), depending on whether the discharge area $\bar{A}$ is given square or square centimeters; T_0 is the absolute temperature in the tank.

Equation 16 results in approximate replacement of the right branch of the curve in Figure 30 by the quadrant of an ellipse and of the number 0.528, which represents the ratio of the critical pressure to the pressure in the tank, by the number 0.5. The term-by-term division of Equation 16 by Equation 17 gives the easily memorizable formula

$$\frac{G}{G_{max}} = 2\sqrt{\left(1 - \frac{\bar{p}}{p_0}\right)\frac{\bar{p}}{p_0}} \tag{18}$$

The phenomenon of gas discharge into a chamber with the given back pressure takes place differently if the initial part of the nozzle is convergent and the exit part is divergent. In this case, the gas velocity, having attained its critical value in the section that separates the convergent and divergent parts, can become supersonic in the course of further expansion of the gas in the diffuser part of the nozzle. Such nozzles are called the Laval nozzles, after the Swedish engineer, Laval, who was the first to use them as nozzles for steam turbines.

Let us consider the one-dimensional adiabatic and isentropic gas flow in a Laval nozzle. The course of the change in the area A along the nozzle axis is assigned by the curve in Figure 31a: a corresponding variation of the **M** number is given in the curves of Figure 31b, and the curves of the pressure related to its critical value are presented in Figure 31c.

The **M** and p/p^* curves were plotted from the earlier obtained formulas for an isentropic flow. The form of the curves suggest a conclusion about the phenomena occurring in the Laval nozzle. If in the most narrow cross-section of the nozzle $A = A^*$ the **M** number reached the crit-

[3]Prandtl, L., *Gidroaeromekhanika* (*Hydroaeromechanics*) (translated from German), IL Press, Moscow, 1949, pp. 327, 328.

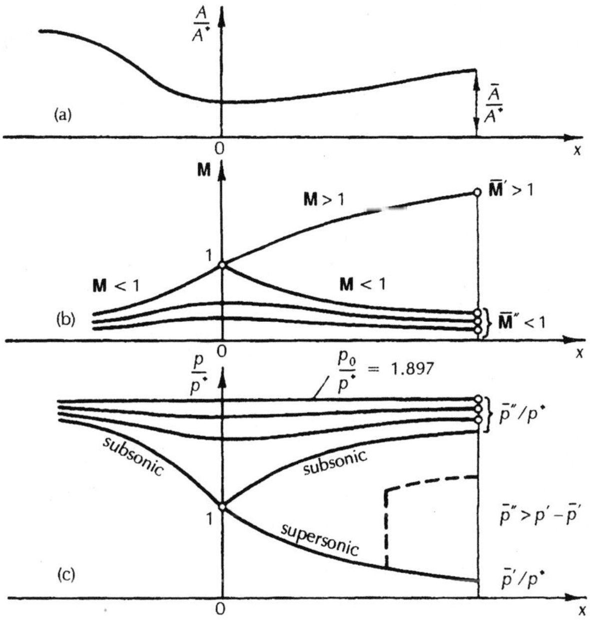

Fig. 31

ical value $\mathbf{M} = 1$, then the further development of the flow may follow the curves that correspond to both $\mathbf{M} > 1$ and $\mathbf{M} < 1$; i.e., the flow may become supersonic or remain subsonic. This alternative is resolved by specifying the back pressure at the nozzle exit.

Using the given ratio $\bar{A}/A^*$ at the Laval nozzle exit in conjunction with Equation 6, or the right ascending branch of the curve in Figure 31, find the exit value $\bar{M}' > 1$. Having substituted it into the right side of the first formula of Equation 7, determine the design ratio of the exit pressure $\bar{p}'$ to the critical pressure p^*. By selecting the back pressure in the chamber equal to this design pressure $\bar{p}'$, the Laval nozzle will operate in the supersonic design regime, the exit velocity will exceed the speed of sound, and will be equal to $\bar{u} = \bar{\mathbf{M}}' \bar{a}$.

At the same value of $\bar{A}/A^*$, but when the right descending branch of the curve in Figure 31 is used, the value $\bar{\mathbf{M}}'' < 1$ can be determined at the nozzle exit and the ratio $\bar{p}''/p^*$ corresponding to it according to Equation 7 can be determined. Selecting the back pressure to be higher than or equal to $\bar{p}''$, obtain different subsonic regimes of discharge from the nozzle exit section.

It should be emphasized that there is an infinite number of subsonic regimes of discharge from the Laval nozzle of given shape, whereas the supersonic regime is unique and can be realized only at the sole value of back pressure equal to $\bar{p}'$.

If the back pressure lies between the design values $\bar{p}'$ and $\bar{p}''$, complex phenomena will originate in or outside the nozzle. In such presence the motion of gas in the nozzle will not be continuously one-dimensional and isentropic everywhere. If the back pressure in the chamber turns out to be smaller than $\bar{p}'$, the gas, on having escaped from the nozzle, will keep expanding continuously and isentropically until it reaches the pressure in the chamber, but its motion outside the nozzle can no longer be considered as being one-dimensional.

The mass flow rate through the Laval nozzle, just as in the case of a purely convergent nozzle, cannot exceed its maximum value equal to that flow rate which would be observed in the nozzle had the local speed of sound be attained in its most narrow section. Unlike a convergent nozzle, the Laval nozzle exit velocity in the supersonic regime exceeds the speed of sound and, by properly selecting the shape and the length of the nozzle, the exit velocity can be made higher with smaller back pressure. It is possible to imagine such an ideal Laval nozzle which will operate in the design regime $\bar{p}' = 0$. This means that there will be an absolute vacuum in the chamber, where along with $\bar{p}'$ the values of ρ' and $\bar{T}'$ will also become zero.

As can be easily concluded from Saint-Venant and Wantzel's formula (30) of Chapter 5, the velocity of such a discharge is maximal at the given gas parameters in the tank and equal to

$$u_{\max} = \sqrt{\frac{2}{k-1}}\sqrt{k\frac{p_0}{\rho_0}} = \sqrt{\frac{2}{k-1}}a_0 = \sqrt{\frac{k+1}{k-1}}a^* \tag{19}$$

Introducing the enthalpy of the gas in the adiabatic stagnation state (total enthalpy) $h_0 = c_pT_0$, it is possible to write

$$u_{\max} = \sqrt{2h_0} \tag{20}$$

which directly follows from Equation 52 of Chapter 5, where $h = 0$.

Note that for air at the absolute temperature $T_0 = 273 + 15 = 288$ K, the maximum velocity will be equal to 757 m/s. Now, also calculate the numbers **M** and λ at the maximum velocity of discharge. Noting that the speed of sound in vacuum is equal to zero and that the maximum discharge velocity is finite, we rind $\mathbf{M}_{\max} = \infty$. The velocity coefficient λ attains in this case its maximum value

$$\lambda_{\max} = \sqrt{\frac{k+1}{k-1}} \tag{21}$$

Along with the numbers **M** and λ, the gas dynamic investigations also employ the Chaplygin parameter τ equal to the squared ratio of the flow velocity to the maximum velocity and expressed, according to Equation 19, in terms of λ as

$$\tau = \left(\frac{u}{u_{\max}}\right)^2 = \frac{k-1}{k+1}\lambda^2 \tag{22}$$

It follows from the very definition of τ that $\tau_{\max} = 1$. The comparison between the values of **M**, λ, and $\sqrt{\tau}$ for air ($k = 1.4$) is given in Table 2.

The above information concerning the motion of gas in nozzles is valid for an ideal gas without internal friction and if the process is fully adiabatic, i.e., in the absence of inflow or outflow of heat in a nozzle. Actually, the motion of gas in a nozzle is much more complex.

The nozzle calculated by the approximate theory may not give the desired increase in the exit **M**; number, and it is only by experimental checks that practically useful results can be attained; more accurate calculations are also employed.

It is often useful for calculations of one-dimensional gas flows to compare the thermodynamic "instantaneous" gas characteristics at any point in the flow with some standard characteristics. In the majority of cases two such states are selected: the gas at rest and the gas in its critical state (the motion with the local speed of sound). These states can be visualized as realized with the aid of one-dimensional adiabatic and isentropic motion of gas through a certain imaginary duct.

If the motion is subsonic, the gas is brought to rest with the aid of a divergent duct – the subsonic diffuser used to convert the kinetic energy of flow into pressure. This process is called the pressure recovery. The higher the degree of pressure recovery, the higher the efficiency of the diffuser. In an ideal adiabatic motion, complete recovery of pressure may occur up to the value of p_0, defined by the already-familiar adiabatic and isentropic formula.

A supersonic flow can also be retarded with the aid of a special supersonic diffuser which is analogous in appearance to the Laval nozzle. In contrast, the flow in the convergent part of the nozzle will be decelerated to a certain velocity, which is smaller than in the inlet section but still supersonic. Later, having passed jumpwise into a subsonic flow, the flow will continue to slow down in the divergent section of the nozzle. Elementary calculations of such a supersonic diffuser will be given in Section 8.

3. An Example of the Nonadiabatic Motion of Gas

Suppose that the adiabatic character of the one-dimensional steady flow of an ideal gas is violated by the supply of heat to the gas from the outside over a certain, very short, section. This causes a change in the gas temperature T_1 by the value $\Delta T = T_2 - T_1$ and in the temperature of the gas in isentropic stagnation state T_{10} by the value $\Delta T_0 = T_{20} - T_{10}$ up to the heated section, downstream of which the adiabatic flow again develops with the temperatures T_2 and T_{20}.

Neglecting the effect of the variable cross-section of the tube over the short heated portion, determine the variation of the Mach number $\mathbf{M}$ over this portion, after which it will not be difficult to also find the variation of the remaining quantities. The governing equations of the problem-set can be easily obtained by taking into account the fact that the inflow of heat does not violate the mass and momentum balance; i.e., when the gas passes the heated section the following two equations remain valid

$$\rho u = \text{const} \qquad p + \rho u^2 = \text{const} \tag{23}$$

Under the adopted conditions of a steady one-dimensional flow through a cylindrical tube ($V_n = V = u$), the first of the above equations follows directly from Equation 33 of Chapter 4. The second equation follows from Equation 34 of the same chapter, provided that both parts of it are projected onto the tube axis; the action of volumetric forces is neglected, the steady character of the flow is taken into account, and it is noted that in the case of the ideal medium $p_{nn} = p_{xx} = -p$ (i.e., provided that the thus-simplified Equation 34 is applied to the volume of the tube confined between two arbitrary plane sections).

Recalling the already-familiar formulas that connect the speed of sound with temperature, pressure, and density of the gas and the definition of the number $\mathbf{M}$:

$$\rho u = k\frac{\rho u}{kp/\rho} = k\frac{\rho u}{a^2} = kp\mathbf{M}\frac{1}{\sqrt{kRT}} = p\mathbf{M}\sqrt{\frac{k}{RT}} = \text{const}$$

$$p + \rho u^2 = p\left(1 + \frac{\rho u^2}{p}\right) = p\left(1 + k\frac{u^2}{a^2}\right) = p\,(1 + k\mathbf{M}^2) = \text{const} \tag{24}$$

By dividing one equation by the other, we obtain the sought-after connection between the Mach number $\mathbf{M}$ and the absolute temperature T or the temperature of the gas T_0 in the adiabatic isentropic stagnation state

$$\frac{1 + k\mathbf{M}^2}{\mathbf{M}}\sqrt{T} = \text{const} \qquad \frac{1 + k\mathbf{M}^2}{\mathbf{M}\sqrt{1 + \dfrac{k-1}{2}\mathbf{M}^2}}\sqrt{T_0} = \text{const} \tag{25}$$

Apply these equations to two flow sections that confine the heated section; this yields:

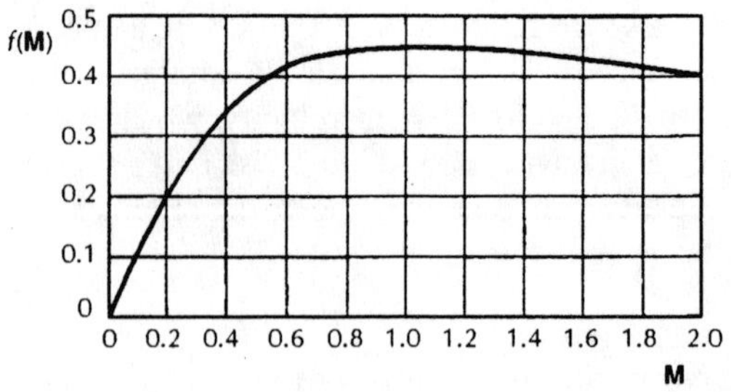

Fig. 32

$$\frac{\mathbf{M}_2}{1+k\mathbf{M}_2^2} = \frac{\mathbf{M}_1}{1+k\mathbf{M}_1^2}\sqrt{\frac{T_2}{T_1}}$$

$$\frac{\mathbf{M}_2\sqrt{1+\frac{k-1}{2}\mathbf{M}_2^2}}{1+k\mathbf{M}_2^2} = \frac{\mathbf{M}_1\sqrt{1+\frac{k-1}{2}\mathbf{M}_1^2}}{1+k\mathbf{M}_1^2}\sqrt{\frac{T_{20}}{T_{10}}} \tag{26}$$

Prescribing the ratio

$$\frac{T_2}{T_1} = 1+\frac{\Delta T}{T_1}$$

and the number $\mathbf{M}_1$ upstream of the heated section, we can find $\mathbf{M}_2$ from Equation 26. Then, from the second of Equations 26, the pressure ratio

$$\frac{p_2}{p_1} = \frac{1+k\mathbf{M}_1^2}{1+k\mathbf{M}_2^2} \tag{27}$$

and all the remaining thermodynamic parameters. Finally, knowing the number $\mathbf{M}_2$ and the temperature T_2, the gas velocity downstream of the heated section can be easily found.

Let us introduce into consideration the function

$$f(\mathbf{M}) = \frac{\mathbf{M}\sqrt{1+\frac{k-1}{2}\mathbf{M}^2}}{1+k\mathbf{M}^2} \tag{28}$$

involved in the second of Equation 26. Having calculated the derivative

$$f'(\mathbf{M}) = \frac{1-\mathbf{M}^2}{(1+k\mathbf{M}^2)^2\left(1+\frac{k-1}{2}\mathbf{M}^2\right)^{1/2}}$$

we shall see that the function $f(\mathbf{M})$ has the maximum at $\mathbf{M} = 1$ and that this maximum is equal to

$$f(1) = \frac{1}{\sqrt{2(k+1)}}$$

In Figure 32 the plot of the function $f(\mathbf{M})$ is given for air ($k = 1.4$). As is seen from the plot and from the second of Equations 26, the heating of gas causes an increase of the function $f(\mathbf{M}_2)$ and also of the number $\mathbf{M}_2$ when $\mathbf{M}_1 < 1$, while in the case of $\mathbf{M}_1 > 1$, conversely, the $\mathbf{M}_2$ number decreases. Consequently, the inflow of heat to a subsonic flow accelerates it, whereas the outflow of heat retards the flow. In the case of a supersonic flow, the inflow of heat retards the flow, whereas the outflow of heat accelerates it. Thus, for example, at $T_{10} = 540$ K and $\mathbf{M}_1 = 0.5$, a 20% increase in temperature leads to the rise of the number $\mathbf{M}$ up to $\mathbf{M}_2 = 0.6$. At the same initial temperature and at $\mathbf{M}_1 = 1.4$, the heating by 7% leads to a decrease of $\mathbf{M}$ to $\mathbf{M}_2 = 1$; the pressure increases in this case by > 50%.

4. The Nonisentropic Motion of a Gas through a Tube in the Presence of Resistance

Let us consider an adiabatic, but not isentropic, motion of a gas through a tube in the presence of frictional resistance, restricting the discussion for simplicity to the tube of constant cross-section.

To sustain a uniform motion of a real fluid in a tube of constant cross-section, it is necessary that the moving pressure gradient Δp be applied to the sections of the tube confining a certain length l. This could balance out the frictional resistance which impedes the fluid motion in the tube. This pressure gradient is called the resistance of the segment of the tube and is represented by the formula

$$\Delta p = \lambda \frac{l}{D} \frac{\rho u_{av}^2}{2} \tag{29}$$

where D is the tube diameter, ρ is the fluid density assumed to be constant, u_{av} is the cross-section-averaged fluid velocity defined by the ratio of the volumetric fluid flow rate through the section and normal to the sectional area. The resistance coefficient λ is a dimensionless number that depends on the physical properties of the medium, density ρ and viscosity μ, and also on the average velocity u_{av} and the tube diameter D. More precisely, λ is a function of a certain complex of these quanties, i.e., of the Reynolds number $\mathbf{Re} = \rho u_{av} D/\mu$.[4]

For the approximate one-dimensional representation of the motion adopted in this section, consider that the average velocity u_{av} coincides with the velocity u of a one-dimensional flow and that the resistance coefficient λ is a constant value. The latter assumption can be justified by noting that λ depends only slightly on **Re**, whereas the **Re** number itself over this segment of the tube usually varies relatively little and can be replaced by its averaged value. Applying the resistance formula (29) to a compressible gas over the segment of length dx:

$$dp = \lambda \frac{dx}{D} \frac{\rho u^2}{2} \tag{30}$$

Form the Euler equation (1) for a one-dimensional steady motion of an ideal gas, taking into account the effect of friction produced by the addition pressure difference (Equation 30); then we have the equation of motion

[4]A more detailed description is given in Chapters 10 and 13.

$$u\frac{du}{dx} = -\frac{1}{\rho}\frac{dp}{dx} - \lambda\frac{u^2}{2D} \tag{31}$$

from which we find

$$\frac{dp}{p} = -\lambda\frac{\rho}{p}\frac{u^2}{2}\frac{dx}{D} - \frac{\rho}{p}udu$$

Making use of Clapeyron's formula and the definition of the speed of sound $a = \sqrt{kp/\rho}$, and having introduced the Mach number $\mathbf{M}$, we obtain

$$\frac{dp}{p} = -\frac{1}{2}\lambda k\mathbf{M}^2\frac{dx}{D} - k\mathbf{M}^2\frac{du}{u} \tag{32}$$

The continuity equation (2) rewritten (A = const) with the aid of Clapeyron's formula as

$$\frac{pu}{T} = \text{const}$$

gives, after logarithmic differentiation,

$$\frac{dp}{p} = \frac{dT}{T} - \frac{du}{u} \tag{33}$$

Comparison of Equations 32 and 33 leads to the relation

$$-\frac{\lambda}{2}k\mathbf{M}^2\frac{dx}{D} + (1 - k\mathbf{M}^2)\frac{du}{u} = \frac{dT}{T} \tag{34}$$

However, by the definition of the Mach number

$$u = \mathbf{M}a = \text{const}\cdot\mathbf{M}\sqrt{T}$$

so that after the logarithmic differentiation

$$\frac{du}{u} = \frac{d\mathbf{M}}{\mathbf{M}} + \frac{1}{2}\frac{dT}{T} \tag{35}$$

Because the motion is adiabatic, i.e., the tube is thermally insulated, write the conservation condition for the total enthalpy h_0, or, equivalently, the gas stagnation temperature

$$T_0 = T\left(1 + \frac{k-1}{2}\mathbf{M}^2\right) = \text{const}$$

then, having again made the logarithmic differentiation, we have

$$\frac{dT}{T} = -(k-1)\frac{\mathbf{M}d\mathbf{M}}{1 + \frac{k-1}{M}\mathbf{M}^2} \tag{36}$$

Expressing du/u and dT/T with the aid of Equations 35 and 36 in terms of the Mach number and its differential $d\mathbf{M}$ and substituting these values into Equation 34, we obtain, after simple transformations, the following basic relation

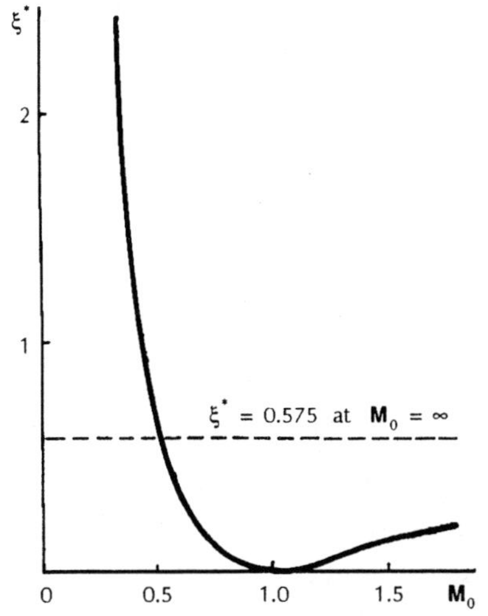

Fig. 33

$$\frac{k\lambda}{2D}dx = d\xi = \frac{1-\mathbf{M}^2}{\mathbf{M}^3\left(1+\frac{k-1}{2}\mathbf{M}^2\right)}d\mathbf{M} \tag{37}$$

Let us assume that $d\xi > 0$; we shall consider the development of motion downstream. It is immediately evident that $d\mathbf{M} > 0$ if $\mathbf{M} < 1$ and that $d\mathbf{M} < 0$ if $\mathbf{M} > 1$. This leads to the following conclusion: in adiabatic motion of a gas through a tube of constant cross-section the presence of friction causes acceleration of a subsonic flow and deceleration of a supersonic flow.

Making use of Equation 37, it is possible to determine the length l^* of the tube segment over which the subsonic flow, having the prescribed value $\mathbf{M}_0 < 1$ at the start, will reach the value $\mathbf{M}_1 = 1$ due to the presence of acceleration. Analogously, it is possible to find the length of the segment over which the supersonic flow will decelerate from the given value $\mathbf{M}_0 > 1$ to $\mathbf{M}_1 = 1$.

Integrating Equation 37 and postulating that $\xi = 0$ in the section with $\mathbf{M} = \mathbf{M}_0$, we have

$$\xi^* = \frac{k\lambda l^*}{2D} = \frac{1}{2\mathbf{M}_0^2} - \frac{1}{2} - \frac{k+1}{4}\ln\frac{1+\frac{k-1}{2}\mathbf{M}_0^2}{\frac{k+1}{2}\mathbf{M}_0^2} \tag{38}$$

Thus, for example, assuming $\lambda = 0.01$ (which corresponds to $\mathbf{Re} = 2.5 \cdot 10^6$), we shall find for air ($k = 1.4$) at $\mathbf{M}_0 = 0.4$ that

$$\frac{l^*}{D} = \frac{2}{1.4 \cdot 0.01} \cdot 1.62 \approx 230$$

and under the same conditions, but at $\mathbf{M}_0 = 1.4$, we shall obtain

$$\frac{l^*}{D} = \frac{2}{1.4 \cdot 0.01} \cdot 0.07 \approx 10$$

The general behavior of the dimensionless parameter ξ^* vs. $\mathbf{M}_0$ is presented in Figure 33. It is possible to note that with the growth of the number $\mathbf{M}_0$ to infinity, the quantity ξ^* tends to the finite value

$$\xi^* \to \frac{k+1}{4} \ln \frac{k+1}{k-1} - \frac{1}{2}$$

In particular, for air ($k = 1.4$) the curve in Figure 33 has a horizontal asymptotic $\xi^* = 0.5751$.

5. Planar Disturbance Waves and a Shock Waves

In contrast to the problem of propagation of small disturbances, the study of the phenomenon of propagation of finite-intensity disturbances presents mathematical difficulties, as it requires integration of nonlinear Equations 55 of Chapter 5. This case is dealt with in Section 11; also considered is Riemann's rigorous explanation for the phenomena of the appearance in an ideal gas of shock waves that represent the surfaces of discontinuities in the parameters of the gas state and its velocity. Let us first consider the elementary theory of shock waves and satisfy ourselves with a simple qualitative explanation of the reason for their origination.

Imagine (Figure 34) a thermally insulated (from the surroundings), infinite cylindrical tube along which a piston moves. The piston and the gas are at first motionless. The piston then instantaneously acquires a certain velocity and moves with this velocity to the left, compressing the gas ahead of it. The disturbance (gas compression) originating in this case will propagate along the tube.

Mentally break down the region of the disturbed gas into a large number of volumes by closely spaced, perpendicular to the tube axis, plane sections. Each has its own corresponding values of the disturbed gas parameters and velocities of propagation with respect to the gas. It can be assumed that the distribution of disturbances along the axis is continuous at each time instant; i.e., in two sufficiently close sections the gas parameters differ little from each other. Then, assuming the gas motion in the given section is relative in the coordinate system, moving translationally and uniformly with the gas velocity in the adjacent section, we can apply to such a Galilean system the theory of propagation of small disturbances. This will allow the statement that the velocity of propagation of disturbances in each section is equal to the local speed of sound.

Thus, the propagation of disturbances produced by the piston can be treated as a sequence of continuous sound waves, with each subsequent wave propagating in the gas disturbed by the previous waves. However, in the adiabatic and isentropic motion considered the compression of the gas is accompanied by its heating, and the propagation velocity of disturbances increases with temperature. From this we conclude that each subsequent wave will move more quickly with respect to the nondisturbed gas than the previous wave. The waves

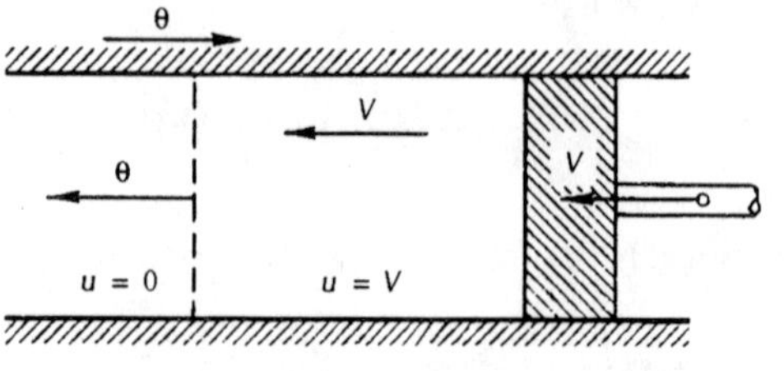

Fig. 34

will drive one another, pile up, and form one compression wave which has the finite intensity and which is called a shock wave.

Note that with the piston motion to the left, rarefaction is formed behind it which will propagate to the right from the piston also in wave fashion. In this case, however, the waves will not drive one another because the subsequent wave will move in the gas cooled by the previous wave, so that the velocity of propagation of the subsequent wave will be smaller than that of the previous one. From this simple reasoning it follows that the rarefaction waves cannot form shock waves. The next section gives an alternative thermodynamic proof of the impossibility of such formations.

It follows from the foregoing description of the development of the compression shock have that as soon as the shock wave has formed (it will be proven in what follows that this occurs in a finite time interval), the gas state parameters and its velocity (absolute or with respect to the moving front) on the both sides of the wave front will have values differing by finite amounts. The shock wave front represents a surface (in the case considered, it is a plane) of the discontinuity in the gas state parameters which moves through the gas and which causes a jumpwise change of these parameters. In the nondisturbed gas upstream the shock wave front has smaller pressure, density, and temperature than that downstream.

The presence of such a jumpwise change in the gas parameters – actually a very sharp change over the segment of length equal by the order of magnitude to the free path length of molecules – shows that here the internal molecular process takes place which is associated with the conversion of the kinetic energy of regular flow of gas into the kinetic energy of irregular thermal motion of molecules. This explains the heating of the gas when it flows from the nondisturbed region ahead of the shock wave front into the region of disturbed motion behind the shock wave front. The growth of the root mean square (RMS) velocity of molecules also causes an increase of pressure and density of the gas when it moves across the shock wave front.

The phenomenon of a one-dimensional propagation of a plane shock wave admits an elementary quantitative calculation. Let V denote the constant velocity of the piston (Figure 34) and let θ denote the velocity of propagation of a relatively nondisturbed gas of the shock wave shown by dashes in Figure 34. Assuming the process of the shock wave origination to be completed, we shall take the gas velocity throughout the entire region between the shock wave and the piston to be the same and equal to the constant velocity of the piston V; in exactly the same way we shall regard the gas parameters also to be constant in this region. Thus, the gas motion and state parameters both on the left from the shock wave (in the region of nondisturbed gas) and on the right remain invariant for all locations of the shock wave. From this it follows that the velocity of shock wave propagation θ is also constant; it is clear from the above reasoning that the shock wave outruns the gas set in motion by the piston, i.e., always $\theta > V$.

The motion of the gas in the coordinate system fixed with respect to the tube will be unsteady because the shock wave, moving along the tube, alters the velocity field in time. Let us reverse the motion by mentally imparting the translational motion to the right with the velocity θ to the entire tube together with the moving gas. In other words, we shall consider the phenomenon occurring in the tube from the viewpoint of the Galilean coordinate system, which moves translationally along the tube axis together with the shock wave front. Then the shock wave will appear as if brought to a stop and as if the gas motion is steady.

The motionless shock wave (the plane is perpendicular to the flow direction) will be called a normal shock wave. Now, the nondisturbed gas is already in motion but approaches the shock wave from the left to right (Figure 35), with the velocity $V_1 = \theta$ and behind the shock it moves with the velocity $V_2 = \theta - V$. It is evident in this case that $V_1 > V_2$; the pressure, density, and temperature in this Galilean system preserve their previous values. We

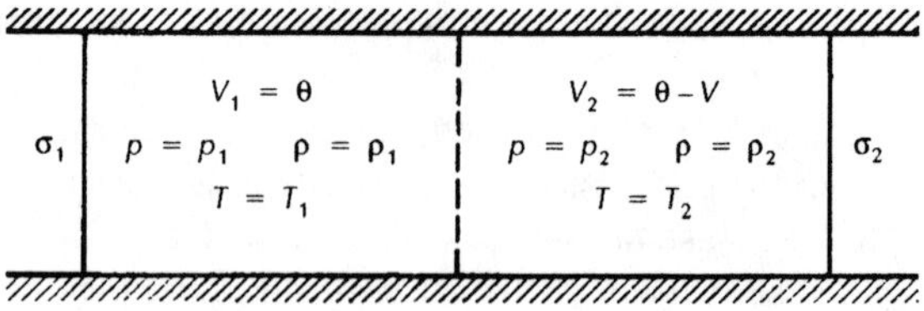

Fig. 35

agree that from this point to use the subscript 1 for the values ahead of the shock and the subscript 2 for the values behind it.

In order to find the connection between V_1, p_1, ρ_1, T_1, and V_2, p_2, ρ_2, T_2, we shall avail ourselves of the steady character of the flow and apply to it the mass, momentum and energy conservation theorem in the Euler form. As follows from the derivations themselves, given in Chapter 4, the Eulerian forms of these theorems can also be applied in the presence of the surfaces of discontinuity in the flow (e.g., of a shock wave). It is only necessary to select the control surface in such a way that those parts of it on which the normal velocity component differs from zero would not coincide or intersect with the surface of discontinuity.

At the control surface select the set of the surface of the cylindrical tube and of two normal sections with identical areas σ_1 and σ_2 (Figure 35). The surface of discontinuity intersects only that part of the control surface where $V_n = 0$. Because the motion has been adapted to be one-dimensional, we regard that in the sections σ_1 and σ_2 the fields of the velocity and of other quantities are uniform. Under these conditions the mass conservation law at $\sigma_1 = \sigma_2$ and $V_n = V_1$ or V_2 will yield directly

$$\rho_1 V_1 = \rho_2 V_2 \tag{39}$$

Assume that the flow is steady and that the volumetric forces are absent ($\boldsymbol{F} = 0$) and the gas is ideal ($\boldsymbol{p}_n = -\boldsymbol{n}p$). Under the conditions of the velocity and pressure fields uniformities in the sections σ_1 and σ_2, the theorem of momentum variation (Equation 34, Chapter 4), after projecting both parts of the equation (34) on the direction of the tube axis, will give the second sought-after equation – the conservation of the total momentum $p + \rho V^2$

$$p_1 + \rho_1 V_1^2 = p_2 + \rho_2 V_2^2 \tag{40}$$

Analogously, the energy balance equation (20) of Chapter 4 under the conditions of an ideal and perfect gas $\boldsymbol{p}_n = -\boldsymbol{n}p$, $U = c_v T$, neglecting the volumetric forces and using the Eulerian form, will be reduced to

$$\int_\sigma \rho\left(c_v T + \frac{V^2}{2}\right) V_n d\sigma = -\int_\sigma p V_n d\sigma$$

Having performed the replacement

$$U = c_v T = h - \frac{p}{\rho}$$

we find

$$\int_\sigma \rho\left(h + \frac{V^2}{2}\right) V_n d\sigma = 0$$

Applying this formula to the selected control surface, we obtain

$$\rho_1 V_1 \sigma_1\left(h_1 + \frac{V_1^2}{2}\right) = \rho_2 V_2 \sigma_2\left(h_2 + \frac{V_2^2}{2}\right)$$

or, by virtue of Equation 39 and equality of the sections σ_1 and σ_2,

$$h_1 + \frac{V_1^2}{2} = h_2 + \frac{V_2^2}{2} \tag{41}$$

This equation represents the conservation law of the total enthalpy $h_0 = h + V_2/2$ of the gas when it moves across the shock wave.

Equations 39, 40, and 41 is a system that can be augmented with Clapeyron's equation and from which we shall have

$$h_1 = c_p T_1 = \frac{c_p}{R}\frac{p_1}{\rho_1} = \frac{k}{k-1}\frac{p_1}{\rho_1}$$

and analogously

$$h_2 = \frac{k}{k-1}\frac{p_2}{\rho_2}$$

after which Equation 41 can be replaced by:

$$\frac{k}{k-1}\frac{p_1}{\rho_1} + \frac{V_1^2}{2} = \frac{k}{k-1}\frac{p_2}{\rho_2} + \frac{V_2^2}{2} \tag{42}$$

Thus, the system of three equations (39, 40, and 42) with three unknown quantities, V_2, p_2, ρ_2 has been formed.

Let us first find the connection between the pressures and densities ahead of and behind the shock wave, having eliminated from consideration the velocities V_1 and V_2. For this, according to Equation 39, we rewrite the equation of momentum variation (40) as

$$p_1 - p_2 = \rho_2 V_2^2 - \rho_1 V_1^2 = \rho_1 V_1 (V_2 - V_1)$$

and multiply the both parts of this equation: on the right by the relation

$$\frac{V_2 + V_1}{\rho_1 V_1}$$

and on the left by the quantity equal to it:

$$\frac{V_2}{\rho_1 V_1} + \frac{1}{\rho_1} = \frac{V_2}{\rho_2 V_2} + \frac{1}{\rho_1} = \frac{1}{\rho_1} + \frac{1}{\rho_2}$$

This will give

$$(p_1 - p_2)\left(\frac{1}{\rho_1} + \frac{1}{\rho_2}\right) = V_2^2 - V_1^2$$

On the other hand, it follows from the energy balance equation (42) that

$$\frac{2k}{k-1}\left(\frac{p_1}{\rho_1}-\frac{p_2}{\rho_2}\right) = V_2^2 - V_1^2$$

so that having set the left sides of the latter two equations equally, we shall find

$$\frac{1}{2}(p_1 - p_2)\left(\frac{1}{\rho_1}+\frac{1}{\rho_2}\right) = \frac{k}{k-1}\left(\frac{p_1}{\rho_1}-\frac{p_2}{\rho_2}\right)$$

Grouping in this equation the terms with p_1 and p_2, we obtain the Hugoniot equation

$$\frac{p_2}{p_1} = \frac{(k+1)\rho_2 - (k-1)\rho_1}{(k+1)\rho_1 - (k-1)\rho_2} = \frac{(k+1)\rho_2/\rho_1 - (k-1)}{k+1-(k-1)\rho_2/\rho_1} \tag{43}$$

Recalling the connection between the pressure and density in adiabatic motion of an ideal gas, defined by the isentropic adiabatic

$$\frac{p_2}{p_1} = \left(\frac{\rho_2}{\rho_1}\right)^k \tag{44}$$

we see that the Hugoniot equation (43) represents an adiabatic which differs from the isentropic; this adiabatic is called the shock adiabatic, or the Hugoniot adiabatic, in contrast to the Poisson isentropic adiabatic (44).

At first sight, the result obtained is at variance with the statement made in Chapter 5 about the isentropic character of the adiabatic motion of an ideal gas. One should not forget, however, that in contrast to the earlier considered continuous (along the tube) stream of motion, the present section considers a discontinuous motion with the finite shock of the gas parameters in a certain section of the tube. From this one has only to draw the conclusion that the passage of an ideal gas across a shock wave is not an isentropic process, but is accompanied by the irreversible conversion of mechanical energy into thermal energy.

In the presence of irreversible processes of mechanical energy conversion into thermal energy, in a closed (adiabatic) system, the entropy of this system must rise. It can be easily verified that in the case considered here of the passage of a gas across the shock wave, the entropy of the gas per unit mass will increase. For this purpose, form the expression for the entropy from Equation 54 of Chapter 5:

$$s_2 - s_1 = \frac{R}{k-1}\left[\ln\left(\frac{p_2}{\rho_2^k}\right) - \ln\left(\frac{p_1}{\rho_1^k}\right)\right] = \frac{R}{k-1}\ln\left[\frac{p_2}{p_1}\left(\frac{\rho_1}{\rho_2}\right)\right] \tag{45}$$

Figure 36 shows the plots of two adiabatics for comparison of isentropic and nonisentropic shock adiabatics. As is seen from this plot, when $\rho_2/\rho_1 > 1$ the shock adiabatic takes a higher location than the isentropic. It follows then that the expression within the squared brackets under the logarithm sign in Equation 45 is greater than unity; the logarithm is positive so that in fact, $s_2 > s_1$.

It follows directly from Equation 45 that the expansion shock cannot exist. Actually, repeating formally all previous arguments regarding the imaginary expansion shock, it could be possible to obtain the same formulas for $p_1 > p_2$ and $\rho_1 > \rho_2$. However, when

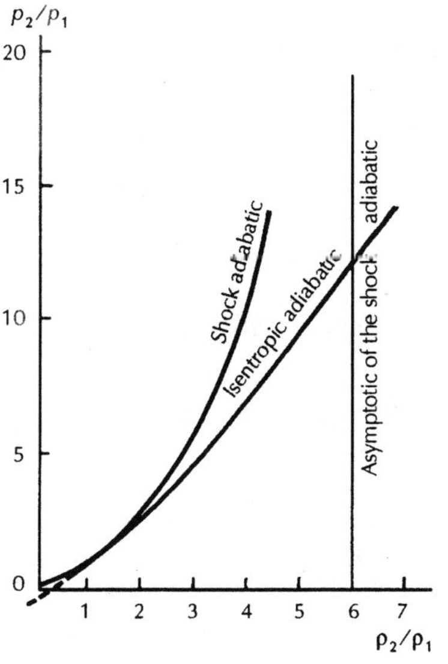

Fig. 36

$\rho_2/\rho_1 < 1$ the curve corresponding to the shock adiabatic lies below the isentropic adiabatic [in Figure 36 this branch is shown by dashes; it intersects the ordinate axis at the point $\left(0, -\frac{k-1}{k+1}\right)$], so that in this case $s_2 < s_1$. This means that when the gas passes across the imaginary expansion shock, the gas entropy per unit mass should have decreased, a contradiction of the second law of thermodynamics. Thus, from general thermodynamic considerations it also follows that the expansion shock is impossible.

Note that the shock adiabatic has the asymptotic

$$\frac{\rho_2}{\rho_1} = \frac{k+1}{k-1}$$

because in this case the ratio of pressures, according to Equation 43, turns out to be infinite. From this it follows that in contrast to the ordinary adiabatic and isentropic compression of the gas (for example, in a thermally insulated cylinder with a piston), no matter how great the compression of the gas p_2/p_1 may be in a shock wave, the compaction of the gas ρ_2/ρ_1 produced by this wave cannot exceed the value $(k+1)/(k-1)$. Thus, for example, on having passed across a shock wave, air cannot increase its density more than six times.

6. Variation of the Velocity and Thermodynamic Parameters of the Gas in Passing across a Normal Shock Wave

According to the Bernoulli theorem, which is valid separately in the regions ahead of and behind the shock wave, we have

$$h_1 + \frac{V_1^2}{2} = h_{10} = c_p T_{10} \qquad h_2 + \frac{V_2^2}{2} = h_{20} = c_p T_{20}$$

Comparing these relations to Equation 41, we conclude that when the gas passes across the shock wave, the total enthalpy h_0 and the adiabatic stagnation temperature of the gas T_0 are preserved; consequently, the quantities a_0, a^*, and T^* are also preserved. Thus, we have

$$h_{10} = h_{20} = h_0 \qquad T_{10} = T_{20} = T_0 \qquad T_1^* = T_2^* = T^*$$

$$a_{10} = a_{20} = a_0 \qquad a_1^* = a_2^* = a^* \tag{46}$$

According to Clapeyron's formula, it is also possible to write

$$\frac{p_{10}}{\rho_{10}} = \frac{p_{20}}{\rho_{20}} \qquad \frac{p_1^*}{\rho_1^*} = \frac{p_2^*}{\rho_2^*} \tag{47}$$

Let us find the connection between the velocities V_1 and V_2 before and after the shock. With this purpose in mind, rewrite the momentum Equation 40 with Equation 39 taken into account. This will give

$$V_1 - V_2 = \frac{p_2}{\rho_2 V_2} - \frac{p_1}{\rho_1 V_1} \tag{48}$$

The Bernoulli equation in the form of the second of Equations 77, Chapter 5, applied before and after the shock, gives

$$\frac{a_1^2}{k-1} \equiv \frac{k}{k-1}\frac{p_1}{\rho_1} = \frac{k+1}{2(k-1)}a^{*2} - \frac{V_1^2}{2}$$

$$\frac{a_1^2}{k-1} \equiv \frac{k}{k-1}\frac{p_2}{\rho_2} = \frac{k+1}{2(k-1)}a^{*2} - \frac{V_2^2}{2}$$

Determining the ratios p_1/ρ_1, p_2/ρ_2 from the above relations, substituting their values into Equation 48, and making simple transformations, we obtain

$$\frac{k+1}{2k}(V_1 - V_2)\left(1 - \frac{a^{*2}}{V_1 V_2}\right) = 0$$

However, according to what was proven above, in the shock wave $\rho_2 > \rho_1$, and, consequently, and to Equation 39, $V_1 > V_2$; therefore the previous equation can be reduced to the relation (suggested by Prandtl):

$$V_1 V_2 = a^{*2} \tag{49}$$

which, on transition to the velocity coefficients λ_1, λ_2, gives

$$\lambda_1 \lambda_2 = 1 \tag{50}$$

When $V_1 > V_2$, the Prandtl formula (49) gives

$$V_1 > a^* > V_2 \qquad \lambda_1 > 1 > \lambda_2$$

i.e., before the shock the motion of the gas is supercritical, and after the shock, it is subcritical.

Table 3

M_1	M_2	p_2/p_1
1.0	1.000	1.000
1.1	0.912	1.246
1.2	0.842	1.520
1.3	0.779	1.824
1.4	0.739	2.120
1.5	0.701	2.455
1.6	0.668	2.820
1.7	0.640	3.200
1.8	0.616	3.604
1.9	0.595	4.043
2.0	0.577	4.500
2.1	0.561	4.970
2.2	0.546	5.480
2.3	0.534	6.000
2.4	0.522	6.550
2.5	0.512	7.400

Making use of Equation 80, Chapter 5, we see that $\mathbf{M}_1 > 1$ when $\lambda_1 > 1$ and $\mathbf{M}_2 < 1$, when $\lambda_2 < 1$. Thus, it is possible to conclude that the motion of a gas before the shock is supersonic and after the shock it is subsonic.

Let us transform in Prandtl's equation (50) from the velocity coefficients to Mach numbers. By relation (79) of Chapter 5 we have

$$\frac{k+1}{2}\frac{\mathbf{M}_1}{\sqrt{1+\frac{k-1}{2}\mathbf{M}_1^2}}\frac{\mathbf{M}_2}{\sqrt{1+\frac{k-1}{2}\mathbf{M}_2^2}} = 1$$

Having solved this equation in $\mathbf{M}_2$, we obtain the connection between $\mathbf{M}_1$ and $\mathbf{M}_2$ before and after the shock:

$$\mathbf{M}_2 = \sqrt{\frac{1+\frac{k-1}{2}\mathbf{M}_1^2}{k\mathbf{M}_1^2 - \frac{k-1}{2}}} \tag{51}$$

To simplify the calculations, set up the derivative $d(\mathbf{M}_2^2)/d(\mathbf{M}_1^2)$ instead of the identical in sign derivative $d\mathbf{M}_2/d\mathbf{M}_1$ and:

$$\frac{d(\mathbf{M}_2^2)}{d(\mathbf{M}_1^2)} = -\frac{(k+1)^2}{4}\frac{1}{\left(k\mathbf{M}_1^2 - \frac{k-1}{2}\right)^2} < 0$$

from which it follows that with an increase of Mach number $\mathbf{M}_1$ of the flow before the shock the Mach number $\mathbf{M}_2$ after the shock decreases monotonically. Table 3 gives the connection between the numbers $\mathbf{M}_1$ and $\mathbf{M}_2$ within the range $1 \le \mathbf{M}_1 \le 2.5$ for air ($k = 1.4$); this table also contains the numerical values of the ratio p_2/p_1 which characterizes the compression of the gas in passage across the shock wave. The formula connecting p_2/p_1 and $\mathbf{M}_1$ is derived later.

As shown by Equation 51, when $\mathbf{M}_1$ grows infinitely, the quantity $\mathbf{M}_2$ tends to its limiting value

$$(\mathbf{M}_2)_{\mathbf{M}_1=\infty} = \sqrt{\frac{k-1}{2k}}$$

equal for air ($k = 1.4$) to 0.378.

Let us determine the relative changes in pressure

$$\frac{\Delta p}{p_1} = \frac{p_2 - p_1}{p_1}$$

density

$$\frac{\Delta\rho}{\rho_1} = \frac{\rho_2 - \rho_1}{\rho_1}$$

and temperature

$$\frac{\Delta T}{T_1} = \frac{T_2 - T_1}{T_1}$$

for the gas passing across the shock wave. For this, we use basic relations (39), (40) and Prandtl's formula (49), having written

$$\frac{p_2 - p_1}{p_1} = \frac{\rho_1 V_1^2 - \rho_2 V_2^2}{p_1} = \frac{\rho_1 V_1^2}{p_1}\left(1 - \frac{V_2}{V_1}\right) = k\frac{V_1^2}{kp_1/\rho_1}\left(1 - \frac{V_1 V_2}{V_1^2}\right) =$$

$$= k\mathbf{M}_1^2\left(1 - \frac{1}{\lambda_1^2}\right)$$

Now use Equations 79 and 80, Chapter 5, for transition from λ_1 to $\mathbf{M}_1$ or from $\mathbf{M}_1$ to λ_1 to obtain the sought-after relations

$$\frac{\Delta p}{p_1} = \frac{2k}{k+1}\frac{\lambda_1^2 - 1}{1 - \frac{k-1}{k+1}\lambda_1^2} \qquad \frac{p_2}{p_1} = 1 + \frac{2k}{k+1}\frac{\lambda_1^2 - 1}{1 - \frac{k-1}{k+1}\lambda_1^2} \tag{52}$$

$$\frac{\Delta p}{p_1} = \frac{2k}{k+1}(\mathbf{M}_1^2 - 1) \qquad \frac{p_2}{p_1} = 1 + \frac{2k}{k+1}(\mathbf{M}_1^2 - 1) = \frac{2k}{k+1}\mathbf{M}_1^2 - \frac{k-1}{k+1} \tag{53}$$

Analogous transformations give

$$\frac{\rho_2 - \rho_1}{\rho_1} = \frac{V_1}{V_2} - 1 = \frac{V_1^2}{a^{*2}} - 1$$

so that

$$\frac{\Delta\rho}{\rho_1} = \lambda_1^2 - 1 \tag{54}$$

$$\frac{\Delta\rho}{\rho_1} = \frac{\mathbf{M}_1^2 - 1}{1 + \frac{k-1}{2}\mathbf{M}_1^2} \qquad \frac{\rho_2}{\rho_1} = \frac{\frac{k+1}{2}\mathbf{M}_1^2}{1 + \frac{k-1}{2}\mathbf{M}_1^2} \tag{55}$$

In order to determine the relative change in the temperature, we use Equation 41, from which it follows that

$$\frac{\Delta T}{T_1} = \frac{h_2 - h_1}{h_1} = \frac{V_1^2 - V_2^2}{2h_1} = \frac{V_1^2}{2a_1^2/(k-1)}\left(\frac{1 - V_2^2}{V_1^2}\right) =$$

$$= \frac{k-1}{2}\mathbf{M}_1^2\left(1 - \frac{a^{*4}}{V_1^4}\right) = \frac{k-1}{2}\mathbf{M}_1^2\left(1 - \frac{1}{\lambda_1^4}\right)$$

Replacing here (as before) $\mathbf{M}_1$ by λ_1 or λ_1 by $\mathbf{M}_1$ we obtain

$$\frac{\Delta T}{T_1} = \frac{k-1}{k+1}\frac{\lambda_1^4 - 1}{\lambda_1^2\left(1 - \frac{k-1}{k+1}\lambda_1^2\right)} \tag{56}$$

$$\frac{\Delta T}{T_1} = \frac{2(k-1)}{(k+1)^2\mathbf{M}_1^2}(\mathbf{M}_1^2 - 1)(1 + k\mathbf{M}_1^2) \tag{57}$$

To estimate the mechanical energy loss by the moving gas that passes across a normal shock wave, we characterize the mechanical energy by the total pressure p_0, i.e., by the adiabatic and isentropic stagnation pressure of the gas. In this case, as the quantitative characteristic of the irreveribility of the process of gas passage across the normal shock wave, we take the value æ of the ratio between the total pressure p_{20} behind the shock and p_{10} before the shock

$$æ = \frac{p_{20}}{p_{10}} \tag{58}$$

Expressing p_{20} and p_{10} by the isentropic formulas in terms of the pressure p_2 and p_1 of the gas after and before the shock, we obtain

$$æ = \frac{p_{20}}{p_{10}} = \frac{p_2\left(1 + \frac{k-1}{2}\mathbf{M}_2^2\right)^{\frac{k}{k-1}}}{p_1\left(1 + \frac{k-1}{2}\mathbf{M}_1^2\right)^{\frac{k}{k-1}}} = \frac{p_2\left(1 - \frac{k-1}{k+1}\lambda_1^2\right)^{\frac{k}{k-1}}}{p_1\left(1 - \frac{k-1}{k+1}\lambda_2^2\right)^{\frac{k}{k-1}}}$$

Replacing here the ratio p_2/p_1, according to Equation 52, by one of the following relations

$$\frac{p_2}{p_1} = \begin{cases} 1 + \dfrac{2k}{k+1}(\mathbf{M}_1^2 - 1) \\[2ex] 1 + \dfrac{2k}{k+1}\dfrac{\lambda_1^2 - 1}{1 - \frac{k-1}{k+1}\lambda_1^2} \end{cases}$$

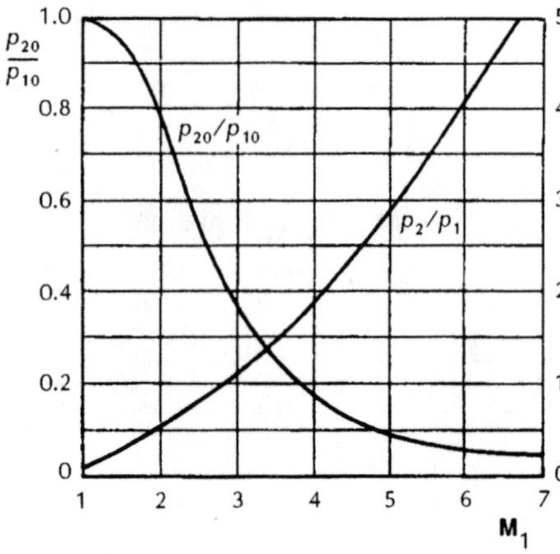

Fig. 37

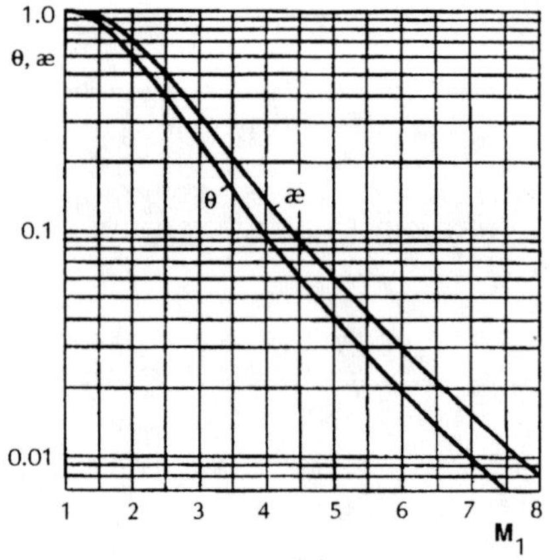

Fig. 38

and using the representation of $\mathbf{M}_2$ and λ_2 in terms of $\mathbf{M}_1$ and λ_1 (Equations 51 and 50, respectively), we obtain the sought-after relation in one of the following forms:

$$
æ = \frac{p_{20}}{p_{10}} = \begin{cases} \lambda_1^{\frac{2k}{k-1}} \left(\dfrac{1 - \dfrac{k-1}{k+1}\lambda_1^2}{\lambda_1^2 - \dfrac{k-1}{k+1}} \right)^{\frac{1}{k-1}} = \lambda_1^2 \left(\dfrac{1 - \dfrac{k-1}{k+1}\lambda_1^2}{1 - \dfrac{k-1}{k+1}\dfrac{1}{\lambda_1^2}} \right)^{\frac{1}{k-1}} \\[2ex] \left(\dfrac{k+1}{2}\right)^{\frac{k+1}{k-1}} \dfrac{\mathbf{M}_1^{\frac{2k}{k-1}}}{\left(1 + \dfrac{k-1}{2}\mathbf{M}_1^2\right)^{\frac{k}{k-1}} \left(k\mathbf{M}_1^2 - \dfrac{k-1}{2}\right)^{\frac{1}{k-1}}} \end{cases} \tag{59}
$$

Recall that, according to Equation 47, these very relations also determine the ratio of the gas stagnation densities before and after the shock.

In Figure 37 the plot of the second relation (59) is given for air ($k = 1.4$); also given is the plot of the compression p_2/p_1 of air in the shock. The plot that expresses the first relation (59) is of the same character, but p_{20}/p_{10} does not vanish when $\mathbf{M}_1 \to \infty$, but at the finite value of the velocity coefficient $\lambda_1 = \sqrt{\dfrac{k+1}{k-1}}$. As is seen from the plot, the higher the values of $\mathbf{M}_1$ or p_2/p_1 that characterize the intensity of the shock, the relatively smaller pressure p_{20} can be obtained at the expense of the energy available from a subsequent adiabatic and isentropic stagnation of the gas that has passed across the shock wave. The reason for this phenomenon was elucidated previously: there is an irreversible conversion of mechanical into thermal energy in the shock wave. The consequence is that the mechanical energy becomes smaller.

The values of the ratio p_{20}/p_{10} at great values of $\mathbf{M}_1$ are more accurately represented by the plot of Figure 38 than Figure 37 because it is constructed there in the logarithmic scale; also plotted is the function $\Theta(\mathbf{M}_1)$ (Equations 10 and 11), which will be needed in Section 8.

At great values of $\mathbf{M}_1$, the second of Equations 59 can be replaced by the approximate asymptotic formula

$$æ = \frac{p_{20}}{p_{10}} \sim \left(\frac{k+1}{2}\right)^{\frac{k+1}{k-1}} \frac{\mathbf{M}_1^{\frac{2k}{k-1}}}{\left(\frac{k-1}{2}\right)^{\frac{k}{k-1}} \mathbf{M}_1^{\frac{2k}{k-1}} k^{\frac{1}{k-1}} \mathbf{M}_1^{\frac{2}{k-1}}} =$$

$$= \left[\frac{(k+1)^{k+1}}{2k(k-1)^k}\right]^{\frac{1}{k-1}} \mathbf{M}_1^{-\frac{2}{k-1}} \tag{60}$$

which gives a simple quantitative estimate for the behavior of the ratio p_{20}/p_{10} at great $\mathbf{M}_1$ numbers. For example, for air (k = 1.4) at great values of $\mathbf{M}_1$ the ratio under consideration is inversely proportional to the number $\mathbf{M}_1$ to the fifth power, thus indicating very appreciable mechanical energy losses by the gas in passing across shocks of high strength. Note that for æ a simple asymptotic formula, which correspond, to this case and which is valid only at great values of $\mathbf{M}_1$, will be $æ \sim 360\ \mathbf{M}_1^{-5}$.

Let us investigate the behavior of the curve in Figure 37 at small values of the difference $\mathbf{M}_1^2 - 1$. Transform the second of Equations 59 as

$$æ = \left[\frac{2k}{k+1} - \frac{k-1}{k+1} + \frac{2k}{k+1}(\mathbf{M}_1^2 - 1)\right]^{-\frac{1}{k-1}} \left[\frac{\frac{k+1}{2} + \frac{k+1}{2}(\mathbf{M}_1^2 - 1)}{1 + \frac{k-1}{2} + \frac{k-1}{2}(\mathbf{M}_1^2 - 1)}\right]^{\frac{k}{k-1}} =$$

$$= \left[1 + \frac{2k}{k+1}(\mathbf{M}_1^2 - 1)\right]^{-\frac{1}{k-1}} \left[\frac{1 + (\mathbf{M}_1^2 - 1)}{1 + \frac{k-1}{k+1}(\mathbf{M}_1^2 - 1)}\right]^{\frac{k}{k-1}}$$

Expanding the powers of the small quantity $\mathbf{M}_1^2 - 1$, we see that the coefficients at $\mathbf{M}_1^2 - 1$ and $(\mathbf{M}_1^2 - 1)^2$ vanish and the expansion of the quantity $æ = p_{20}/p_{10}$ will have the form

$$æ = 1 - \frac{2k}{(k+1)^2} \frac{(\mathbf{M}_1^2 - 1)^3}{3} + \cdots \tag{61}$$

It is seen from the latter expansion that small-intensity shocks do not lead to a marked loss of the total pressure because at $\mathbf{M}_1$ close to unity, p_{20} coincides with p_{10} accurate to a small value:

$$\frac{2k}{(k+1)^2} \frac{(\mathbf{M}_1^2 - 1)^3}{3}$$

Thus, for air (k = 1.4) this value is on the order of $0.16\,(\mathbf{M}_1^2 - 1)^3$ and, for example, will be equal to 0.0015 when the speed of sound is exceeded by 10% ($\mathbf{M}_1$ = 1.1). In the presence of irreversible losses in the adiabatic system its entropy increases. To determine this increase, apply Equation 45 to the parameters of the gas in the adiabatic and isentropic stagnation states. This is admissible because the isentropic stagnation cannot influence the entropy increment. We obtain

$$\frac{s_2 - s_1}{R} = \frac{1}{k-1} \ln\left[\frac{p_{20}}{p_{10}}\left(\frac{\rho_{10}}{\rho_{20}}\right)^k\right]$$

However, as before, $\frac{\rho_{10}}{\rho_{20}} = \frac{p_{10}}{p_{20}}$ and consequently,

$$\frac{s_2 - s_1}{R} = \frac{1}{k-1}\ln\left[\left(\frac{p_{20}}{p_{10}}\right)^{1-k}\right] = -\ln\left(\frac{p_{20}}{p_{10}}\right) = -\ln æ \tag{62}$$

Substituting the ratio p_{20}/p_{10} into this equation by the approximate formula (60), we find the relation for high values of $\mathbf{M}_1$:

$$\frac{s_2 - s_1}{R} \sim \frac{2}{k-1}\ln \mathbf{M}_1 \tag{63}$$

which expresses the asymptotic law of entropy growth during the passage of the gas across shocks of high intensity.

At comparatively small intensity, i.e., when the difference ($\mathbf{M}_1^2 - 1$) is close to zero, we have, according to Equation 61,

$$\frac{s_2 - s_1}{R} \approx \frac{2k}{(k+1)^2}\frac{(\mathbf{M}_1^2 - 1)^3}{3} \tag{64}$$

From this it follows that the small intensity shocks lead to small changes in entropy, i.e., the transonic phenomena can be considered to a sufficient approximation to be isentropic.

7. Velocities of Propagation of a Shock Wave and the Accompanying Flow behind It

Having set up the basic relations for the shock wave, let us return to the analysis or the phenomenon of shock wave propagation in space. Determine the velocity θ of the shock wave propagation with respect to the nondisturbed (quiescent) gas and the velocity V of the disturbed gas motion behind the shock wave; the latter motion could be called the accompanying flow of the gas behind the shock wave. According to Section 38, these velocities are connected with the velocities V_1 and V_2 with respect to the shock wave by the equation

$$\theta = V_1 \qquad V = \theta - V_2 = V_1 - V_2 \tag{65}$$

As the measure of the shock intensity, and consequently, of the shock wave, it is possible to take the number $\mathbf{M}_1$ equal, as before, to

$$\mathbf{M}_1 = \frac{V_1}{a_1} = \frac{\theta}{a_1} \tag{66}$$

or the compression of the gas in the shock p_2/p_1 equal via Equation 53 to

$$\frac{p_2}{p_1} = \frac{2k}{k+1}\mathbf{M}_1^2 - \frac{k-1}{k+1} \tag{67}$$

Then, the velocity θ of the shock wave propagation with respect to the non-disturbed gas will be

$$\theta = \mathbf{M}_1 a_1 = \sqrt{\frac{k-1}{2k} + \frac{k+1}{2k}\frac{p_2}{p_2}} \cdot a_1 \tag{68}$$

Two important corollaries follow from this formula:

1. The velocity of the shock wave propagation with respect to a nondisturbed gas is always higher than the speed of sound in a nondisturbed gas.
2. With a decrease in the shock wave intensity, the velocity of its propagation tends to the speed of sound in a nondisturbed gas

$$\theta = a_1 \quad \text{at} \quad \mathbf{M}_1 = 1 \quad \text{or} \quad p_2 = p_1$$

Thus, the sound wave can be considered as shock wave of very small intensity.

To determine the velocity V of the accompanying flow behind the wave, use is made of the second of Equations 65 and Prandtl's formula (49); this will give

$$V = V_1 - \frac{a^{*2}}{V_1}$$

Having made the replacement

$$a^{*2} = \frac{2}{k+1}\left(1 + \frac{k-1}{2}\mathbf{M}_1^2\right)a_1^2$$

and, simple transformations, we have

$$V = \frac{2}{k+1}\left(\mathbf{M}_1 - \frac{1}{\mathbf{M}_1}\right)a_1 \tag{69}$$

At great values of $\mathbf{M}_1$ it is possible to give an approximate asymptotic formula

$$V \sim \frac{2}{k+1}\mathbf{M}_1 a_1 = \frac{2}{k+1}\theta$$

from which it is seen that the accompanying flow behind the shock wave of very high intensity has a smaller velocity than the velocity of the propagation of the shock wave itself, but rather close to it.

Equation 69 can be easily transformed by Equation 53, to the form

$$V = \sqrt{\frac{2}{k}} \frac{\frac{p_2}{p_1} - 1}{\sqrt{k - 1 + (k+1)\frac{p_2}{p_1}}} a_1 \tag{70}$$

or at very great ratios p_2/p_1

$$V \sim \sqrt{\frac{2}{k(k+1)}} \sqrt{\frac{p_2}{p_1}} a_1 \tag{71}$$

Table 4

M_1	$\Delta p/p_1$	$\Delta\rho/\rho_1$	ΔT (°C)	θ (m/s)	V (m/s)
1	0	0	0	340	0
1.18	0.47	0.30	33	400	93
1.47	1.39	0.81	87	500	224
2.94	9.20	2.77	465	1000	734
4.40	22.20	3.74	1075	1500	1181
5.90	40.3	4.20	1925	2000	1611
8.80	92.3	4.58	5940	3000	2880
11.80	165	4.72	7750	4000	3300
14.70	258	4.78	12100	5000	4135

As can be easily concluded from Equation 70, the velocity of the accompanying flow in the sound wave ($p_2/p_1 \approx 1$) is close to zero. With the growth of the shock wave intensity, the velocity of the accompanying flow increases; at very great intensities this velocity is proportional to the square root of the compression p_2/p_1.

Table 4 displays numerical values of the relative compressions and compactions of the gas by a shock wave propagating in quiescent air ($k = 1.4$) at 15°C ($T = 288$ K) and at the standard atmospheric pressure; the table also contains the values of M_1, θ, V, and the temperature difference ΔT, which correspond to these compressions.

Noted that even comparatively small compressions of air by a shock wave produce a strong accompanying flow. Thus, for example, it can be easily calculated by the previous formulas that the shock wave propagating with the velocity 370 m/s and carrying the relative compression of air $\Delta p/p_1 = 0.22$ could induce an accompanying flow with a velocity of 50 m/s. From this it is evident what negligible compressions of air are produced by ordinary sound waves which hardly displace the air particles.

We give the following data:[5] or a sound which is hundreds or thousands of times more intense than the loudest note play by any orchestra, the amplitude of the change in the air density in the sound wave amounts to only 0.4% of the normal air density; the amplitude of the change in pressure is equal to 0.56% of the atmospheric pressure; the amplitude of the air velocity does not exceed 0.4% of the speed of sound, i.e., is on the order of 1.3 m/s. The amplitude of the displacement of air particles at a frequency of 500 Hz reaches 0.036 cm. The sound produced by a strong siren may cause an accompanying flow – "sound wind" – which is sufficient to put out a candle.

The numerical data presented in Table 4 refer to an ideal gas. The data were calculated under the assumption of the absence of viscosity and thermal conductivity. As the gas is adiabatic, it is also assumed that there is no radiant heat removal, which is appreciable at high temperatures. In fact, the shock wave energy is partially absorbed by the real gas and the wave intensity becomes smaller; at the same time the real processes of the propagation of shock waves are not adiabatic, i.e., their energy is removed to the surroundings.

Note also that Table 4 relates to the propagation of a plane shock wave for which all characteristic quantities are preserved as constant regardless of the distance from the source of disturbance.

In practice one has to deal with spherical shock waves, the process of propagation of which is essentially unsteady, and even in the simplest cases it requires a complex mathematical apparatus for its study.

[5] Zeldovich, Ya. B. and Raizer, Yu. P., *Fizika udarnykh voln i vysokotemperaturnykh gidrodinamicheskikh yavleniy* (*The Physics of Shock Waves and of High-temperature Hydrodynamic Phenomena*), Fizmatgiz Press, Moscow, 1963, pp. 18 and 19.

The difference between the spherical propagation of and plane one propagation can be simply demonstrated using as an example the problem of the propagation of a spherical sound wave. Let us set up the equations of disturbed motion in spherical coordinates with the origin placed into the center of disturbances (a point source of sound). Exact equations will be composed of the motion equation, coinciding with the corresponding equation in the plane case (the first formula of Equations 55 Chapter 5), if x is replaced by the distance r of the point with respect to the source of disturbances and if u is understood to be the radial gas velocity

$$\rho\left(\frac{\partial u}{\partial t} + u\frac{\partial u}{\partial r}\right) = -\frac{\partial p}{\partial r}$$

and of the continuity equation, which in the spherical coordinates can be reduced to the form

$$\frac{\partial \rho}{\partial t} + \frac{1}{r^2}\frac{\partial(r^2 \rho u)}{\partial r} = 0$$

In complete agreement with Section 4 of Chapter 5, using the smallness of the disturbances, we linearize these equations and, satisfying ourselves with the assumption about the adiabatic nature of motion, we come to the following system of equations for small disturbances:

$$\frac{\partial u'}{\partial t} + \frac{a_0^2}{\rho_0}\frac{\partial \rho'}{\partial r} = 0$$

$$\frac{\partial \rho'}{\partial t} + \frac{\rho_0}{r^2}\frac{\partial(r^2 u')}{\partial r} = 0 \tag{72}$$

where, just as in Section 4 of Chapter 5, $a_0^2 = (dp/d\rho)_0$.

The solution of this system in the form of the wave moving out from the center is

$$\rho' = \frac{f(r - a_0 t)}{r} \tag{73}$$

which corresponds to a decrease in the intensity of disturbance by the law $1/r$. As compared to plane waves, the spherical sound waves also have other specific features which will not be considered here.[6]

8. Elementary Theory of a Supersonic Diffuser

A simple example of a one-dimensional flow with a normal shock wave is provided by the flow section of a ramjet engine, the simplified scheme of which is given in Figure 39. The arrangement can be divided into the following parts: the inlet I to III, a supersonic diffuser; the middle part III to IV, a combustion chamber; and the exit IV to V, a Laval nozzle.

The function of the supersonic diffuser is to convert the kinetic energy of the supersonic flow entering the engine into the energy of compression required to increase the combustion intensity of the fuel supplied to the combustion chamber through the injectors. The combus-

[6]See footnote 5 (pages 21 and 22 of the book).

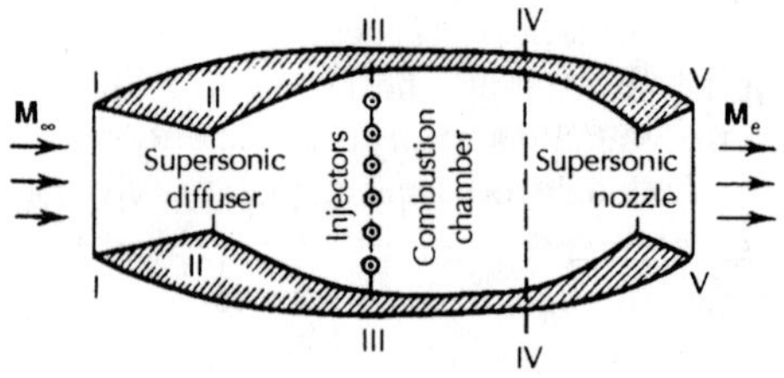

Fig. 39

tion chamber serves to impart to the flow the thermal energy, which is the basic source of gas expansion, and which in the flow-accelerating Laval nozzle IV to V converts into kinetic energy of the jet when exiting nozzle V. The momentum of this jet serves as the source of the reactive force of the engine defined as the product of the gas mass flow rate through the engine exit by the relative rate of the engine exhaust. The simplest calculation of the engine flow section by the one-dimensional theory is elementary and consists in the use, of the isentropic formulas and, of the basic formulas from the normal shock theory. The inflow of heat can be taken into account in this case approximately by the theory analogous to that presented in Section 2.

The calculation is based on the selection of the mode for engine operation. The choice of the regime influences the location, in or beyond the engine flow section, of the normal shock wave unavoidable in supersonic flight. The optimum location is in throat II of the supersonic diffuser or close behind it. In fact, in this case the supersonic flow with $\mathbf{M}_1 > 1$ incident on the engine will begin [recall the corollaries from the Hugoniot equation (Equation 3)] to retard in the convergent duct over sections I to II down to a certain $\mathbf{M}_2 > 1$, although it is smaller than $\mathbf{M}_1$. Then, by means of a comparatively small-intensity shock, it will go over into a subsonic flow and, on having entered into divergent duct II to III, will continue to slow down and recover the pressure. In this case, the operation of entire duct I to III promotes the recovering of the pressure ahead of the combustion chamber beneficially for the engine.

In some regimes of engine operation the shock wave may originate ahead of the engine inlet in the form of the so-called detached shock wave. The presence of such a shock wave at the inlet would have greatly decreased the engine efficiency, as can be judged from the following assessment. Let $V_1 > a_1$ (a_1 is the speed of sound at the given flight altitude H) denote the velocity of the flying vehicle. The pressure in the incident flow is p_1, the pressure in the combustion chamber is p_2'. Assuming first that the bow wave is absent and that the process of air entry into the combustion chamber is isentropic, we shall neglect the air velocity in the chamber which is small compared to the incident flow velocity. Then, we obtain

$$p_2' = p_{20} = p_{10} = p_1\left(1 + \frac{k-1}{2}\mathbf{M}_1^2\right)^{\frac{k}{k-1}}$$

in particular, for air ($k = 1.4$)

$$p_2' = p_1(1 + 0.2\mathbf{M}_1^2)^{3.5}$$

Thus, for example, at $\mathbf{M}_1 = 2$, the compression of air in the chamber is

$$p_2'/p_1 = 1.8^{3.5} \approx 8$$

At a height of $H = 10$ km the pressure p_1 (according to the International Standard Atmosphere) is approximately equal to 26 kPa; in the combustion chamber at $\mathbf{M}_1 = 2$ (at the height

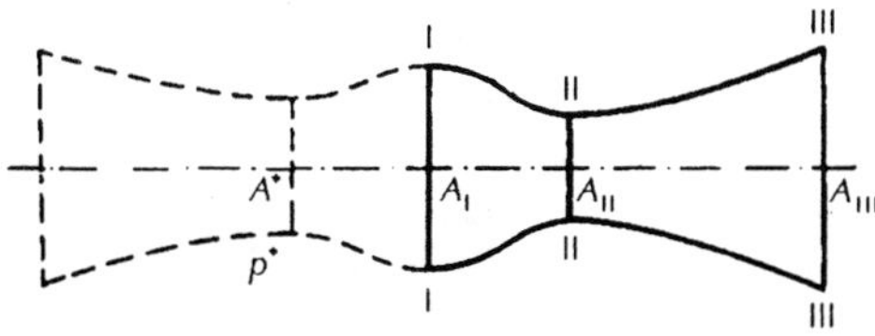

Fig. 40

of 10 km this corresponds to the airplane speed of 600 m/s or 2160 km/h) with isentropic stagnation pressure $p_2' = 200$ kPa. Such an increase in the pressure would have favorably influenced the engine operation. However, because of the presence of the bow shock wave (near the engine inlet the portion of the shock wave can be considered a normal shock), such an isentropic motion is not realized. According to Figure 38, due to the losses of mechanical energy the pressure in the chamber will be equal to

$$p_2' = p_{20} = 0.75\, p_{10} = 0.75 \cdot 200 = 150 \text{ kPa},$$

i.e., 75% of isentropic stagnation pressure. This difference would have been much more significant at great values of $\mathbf{M}_1$. Thus, at $\mathbf{M}_1 = 3$ the pressure in the chamber would have amounted to only 35% of the isentropic stagnation pressure of air, at $\mathbf{M}_1 = 4$, only 15%, at $\mathbf{M}_1 = 5$, only 5%. To preserve the effect of pressure increase in the engine combustion chamber, it is necessary to bring the process of pressure recovery closer to the isentropic one, i.e., to combat the bow wave formed in front of the engine inlet.

One of the means of accomplishing is the use of a supersonic diffuser, because of which the shock, during engine start-up, first "gets on" the entry section and then shifts into the interior of the entry nozzle of the engine, losing its intensity. Another way of accomplishing the bow shock wave consists in its cancellation with the aid of a needle pushed opposite to the supersonic flow at the inlet to the engine. The explanation of the needle effect does not enter into the scope of the one-dimensional flow theory; the corresponding two-dimensional theory is presented later in the volume.

To quantitatively estimate the possibility for pressure recovery in a flow with the aid of a supersonic diffuser, the following, almost practical, assumptions are presented: (1) the flow in the region of the diffuser is adiabatic, (2) the diffuser walls are impenetrable. These two basic assumptions reduce to the requirement of the absence of heat and mass transfer into the diffuser from the outside. Let us use the conservation conditions for the flow mass and total enthalpy or stagnation temperature along the flow, and consequently, of the critical temperature. As indices denote the relation of the quantities to that or other sections of the diffuser, use will be made again of Roman figures[7] (Figure 40) and of conventional designation by an asterisk and null for critical and "stagnation" values of the quantities characteristic for the flow.

Avoiding discussion of particulars of the conditions of external flow entering a supersonic diffuser (secondary for the one-dimensional approach), assume that the flow at the inlet to the diffuser is formed by a certain imaginary ideal Laval nozzle (dashed lines in Figure 40) with the area of the narrow section (throat) A^* and critical characteristics ρ^*, p^*, a^*. The outlet section of this nozzle will be regarded as coinciding with the inlet section of the diffuser.

If the flow in the supersonic diffuser could have been considered as isentropic, neglecting the presence of irreversible processes in it, such as shock waves, friction etc., this diffuser

[7] The arabic numerals 1 and 2 have been used to denote the quantities before and after the shock.

would have represented in turn an ideal Laval nozzle, but only with oppositely directed flow. Such a nozzle is often called the second Laval nozzle.

Sometimes this name is preserved also for a real supersonic diffuser, in which the recovery of pressure is always connected with irreversible nonisentropic processes and especially with the presence of shock waves unavoidable in transition of supersonic into subsonic flows.

Isolate two basic limiting cases among all possible flows in a supersonic diffuser. In the first case the oncoming supersonic flow goes over into the subsonic prior to the entrance into the diffuser after having passed over the detached normal shock wave or through a shock wave in the entry section of the diffuser. Because the flow behind a normal shock is always subsonic, the diffuser operates in this case as a subsonic diffuser. The position of the shock is not stable with respect to small disturbances of the flow and is considered only as a convenient representation for comparison with the second case (optimal for solving the problem of pressure recovery) when the shock wave, on having passed through converging section I to II, will occupy the position in section II or in close proximity behind this section.

To obtain an approximate, comparatively rough, estimate of the gain in the degree of pressure recovery performed by a supersonic diffuser, let us concentrate our attention only on the changes in the total pressure during flow passage across shock waves regarding the flow before and after the shock to be separately isentropic. The quantitative estimate of this recovery will be represented by the coefficient æ, which was been introduced by Equation 58 and which depends on the **M** number in the flow before the shock according to Equations 59, 60, or 61.

Let the ratio of the quantities $æ(\mathbf{M})$ for the second (the shock in the diffuser throat) and first (the shock in the inlet section of the diffuser) cases be called the coefficient of total pressure recovery in a supersonic diffuser *h*. Assume that

$$\eta = \frac{æ(\mathbf{M}_{\mathrm{II}}^{(2)})}{æ(\mathbf{M}_{\mathrm{I}})} \tag{74}$$

where $\mathbf{M}_{\mathrm{I}}$ is the same in both cases compared and $\mathbf{M}_{\mathrm{II}}^{(2)}$ denotes the Mach number in section (II) in the second case. In this case in converging section I to II we have $\mathbf{M}_{\mathrm{I}} > \mathbf{M}_{\mathrm{II}}^{(2)} > 1$ due to the retardation of the supersonic flow.

In the first case (the shock at the entrance) we consider, not losing the generality at all, that the diffuser throat is critical ($A_{\mathrm{II}} = A_{\mathrm{II}}^{*}$) and $\mathbf{M}_{\mathrm{II}}^{(1)} = 1$. To determine η, it remains only to find $\mathbf{M}_{\mathrm{II}}^{(2)}$ at the given $\mathbf{M}_{\mathrm{I}}$ because the function $æ(\mathbf{M})$ itself is already known and, for example, for air it is tabulated.

For this purpose, let us set up the ratios of the critical section areas in the imaginary Laval nozzle and in the diffuser throat A^{*}/A_{II} for the two cases of motion considered and equate them because in both cases the geometric dimensions of the flow region are the same. In the first case we have (the mass conservation law; section II is critical)

$$\rho^{*}a^{*}A^{*} = \rho_{\mathrm{II}}^{*}a_{\mathrm{II}}^{*}A_{\mathrm{II}} \tag{75}$$

or, according to the law of total enthalpy conservation (and consequently of the critical temperature and critical gas velocity),

$$\frac{A^{*}}{A_{\mathrm{II}}} = \frac{\rho_{\mathrm{II}}^{*}}{\rho^{*}}$$

Moreover, let us also use Clapeyron's formula (the gas is perfect)

$$\frac{p^{*}}{\rho^{*}} = RT^{*} = RT_{\mathrm{II}}^{*} = \frac{p_{\mathrm{II}}^{*}}{\rho_{\mathrm{II}}^{*}} \tag{76}$$

Table 5

$\mathbf{M}_I$	1	2	3	4	5	7	10	∞
$\mathbf{M}_{II}^{(2)}/\mathbf{M}_I$	1.000	0.876	0.884	0.891	0.895	0.898	0.901	0.903

Table 6

$\mathbf{M}_I$	1	1.5	2	3	4	6	8	10	∞
η	1.000	1.049	1.157	1.341	1.452	1.558	1.602	1.624	1.666

and, passing from the critical to total pressures, we have for the first case of supersonic diffuser operation

$$\frac{A^*}{A_{II}} = \frac{p_{II_0}}{p_0} = æ(\mathbf{M}_I) \tag{77}$$

Turning our attention to the second case, when the shock is located in throat II of the supersonic diffuser, we recall that in this regime of diffuser operation section II is no longer critical. The $\mathbf{M}_{II}^{(2)}$ number, having become smaller than $\mathbf{M}_I$, will preserve the value $\mathbf{M}_{II}^{(2)} > 1$ corresponding to the supersonic flow in this section.

Let us introduce via Equation 10 the function $\Theta(\mathbf{M})$ which, by virtue of the isentropicity of motion over the section from the throat of the imaginary Laval nozzle to section II, will be determined by Equations 11. According to the mass conservation law, we obtain

$$\frac{\rho_{II} u_{II}}{\rho_* a_*} = \frac{A^*}{A_{II}} = \Theta(\mathbf{M}_{II}^{(2)}) \tag{78}$$

Comparing Equations 77 and 78, we find the sought-after dependence of $\mathbf{M}_{II}^{(2)}$ on $\mathbf{M}_I$

$$\Theta(\mathbf{M}_{II}^{(2)}) = æ(\mathbf{M}_I) \tag{79}$$

This equation admits a simple graphic solution. Using the plots of the functions $\Theta(\mathbf{M})$ and $æ(\mathbf{M})$ in Figure 38 we can, by assigning $\mathbf{M}_I$, the Mach number at the inlet, find $æ(\mathbf{M}_I)$ and, having placed this ordinate onto the curve $\Theta(\mathbf{M})$, determine the corresponding abscissa $\mathbf{M}_{II}^{(2)}$. We can conclude from the shape of the curves in Figure 38 that the ratio $\mathbf{M}_{II}^{(2)}/\mathbf{M}_I$ is close to unity. Table 5 for air ($k = 1.4$) confirms the foregoing in the entire range of $\mathbf{M}_I$ values.

We may assume approximately that

$$\mathbf{M}_{II}^{(2)} = 0.89\,\mathbf{M}_I \quad \text{for} \quad 1.2 \le \mathbf{M}_I \le \infty$$

When $\mathbf{M}_I > 2.5$, the error will not exceed 1%.

The coefficient of total pressure recovery in a supersonic diffuser η will be determined as the function of the Mach number $\mathbf{M}_I$ at the inlet by eliminating $\mathbf{M}_{II}^{(2)}$ from Equations 74 and 79.

Table 6 presents the dependence of η on $\mathbf{M}_I$ for air.

When analyzing Tables 5 and 6, it is possible to note that despite a small difference between the numbers $\mathbf{M}$ at the inlet and in the throat of the supersonic diffuser, the gain in the total pressure recovery is rather appreciable and increases with the number $\mathbf{M}_I$ in the incident flow. This fact finds explanation in the earlier revealed high degree of dependence (the inverse fifth-degree proportionality for air) of the recovery coefficient æ on the Mach number.

The description of various cases of operation of supersonic diffusers and approximate treatment of them from the viewpoint of both one- and two-dimensional theories can be found in special textbooks on the subject.[8]

The elements of the one-dimensional theory of a supersonic diffuser, which are given in this section, very approximately reflect the essence of the real phenomena occurring in it. First, it should be noted that along with normal shock waves, in the flow section of the diffuser and at its inlet there form systems of oblique shocks inclined to the diffuser axis at various angles. These shocks disturb the uniformity of the flow and make it two-dimensional (plane or axisymmetric).

Moreover, the assumption about the ideal character of the gas, i.e., about the absence in it of internal friction, also excessively simplifies the real flow in a diffuser. In reality, the flow deceleration by friction forces produces the nonuniformity of the velocity profile which is not taken into account by the one-dimensional theory. A substantial role in the operation of a supersonic diffuser can be played by the phenomenon of flow (boundary layer) separation from the diffuser walls. This is closely connected with the presence of oblique shocks in the diffuser which fall on and reflect from the diffuser walls. The final chapters of the book are a discussion of this subject.

9. Measurement of Sub- and Supersonic Velocities by the Pneumatic Methods

In contrast to the Prandtl tubes (Figure 28 of Chapter 5) used for small-velocity flows ($\mathbf{M} \ll 1$), the tubes for measuring great supersonic velocities are thin probes with a conical frontal part and static holes for measuring the pressure p_1 in the gas-free stream of density ρ_1. The total pressure p_{10} is measured by a separate tube attached to the device.

Let us assume that the gas moves with great but subsonic velocities ($\mathbf{M}_1 < 1$). In this case it is possible to apply the formula of isentropic motion

$$p_{10} = p_1\left(1 + \frac{k-1}{2}\mathbf{M}_1^2\right)^{\frac{k}{k-1}} \tag{80}$$

Using a micromanometer to record separately the pressure p_{10} in the stagnation and the pressure p_1 in the static holes of a conventional Prandtl tube, we determine the Mach number $\mathbf{M}_1$ of the moving gas and, knowing the gas temperature, we find the speed of sound a_1 in the moving gas and, consequently, the velocity V_1. The temperature can be measured, for example, by a thermocouple or by another thermometer placed at such a point in the pressure tube where the velocity is equal to zero and one can be sure that the isentropic stagnation temperature T_0 of the gas is measured. This place is provided by the nose of the pressure tube where the flow is branched. With T_0 measured directly, we find T_1 from the equation

$$T_0 = T_1\left(1 + \frac{k-1}{2}\mathbf{M}_1^2\right)$$

then the speed of sound

$$a_1 = \sqrt{kRT_1}$$

and the sought-after velocity

[8] See Hermann, R., *Supersonic Inlet Diffusers* (translated from English), Fizmatgiz Press, Moscow, 1960.

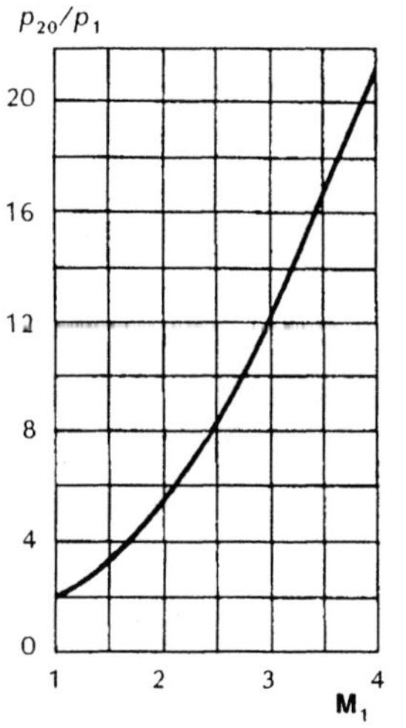

Fig. 41

$$V_1 = \mathbf{M}_1 a_1$$

Reading the pressure p_{10} in the pressure hole D can be considered reliable, but as regards the operation of the static hole, it requires the following remark. At sufficiently great, though smaller than unity, values of $\mathbf{M}_1$ on the spherical surface of the nose the zones of local supersonic velocities may originate. The subsequent transition from the supersonic to subsonic velocities will cause the appearance of shock waves and local pressure distortions on the tube surface in front of the static hole. The smallest value of the number $\mathbf{M}_1 < 1$ of the incident flow at which supersonic zones appear on the surface of the body immersed in the flow (in the present case the measuring tube) is called the critical $\mathbf{M}$ number and it is designated as $\mathbf{M}_{cr}$.[9] If the Mach number $\mathbf{M}_1$ of the incident flow exceeds the critical value $\mathbf{M}_{cr}$, the static hole becomes unreliable and it is necessary to measure the pressure p_1 in the moving gas by some other independent technique, for example, when the gas moves through a cylindrical tube it is possible to measure the pressure on the tube wall in the section close to the pressure tube nose.

The application of the static hole S for measuring velocities in a supersonic flow may also turn out to be unsatisfactory because in this case the bow wave originates in front of the pressure hole and the pressure p_1 will not coincide with the readings of the micromanometer connected with the static hole. Shock waves getting on the portion of the tube surface DS further distort the pressure field in the gas.

Using the readings of p_{20} in the pressure hole behind the bow wave and measuring p_1 by some other technique, we find their ratio p_{20}/p_1. This ratio is connected with the sought-after number $\mathbf{M}_1$ of the incident flow by the Rayleigh equation

$$\frac{p_{20}}{p_1} = \frac{p_{20}}{p_{10}}\frac{p_{10}}{p_1} = \left(\frac{k+1}{2}\right)^{\frac{k+1}{k-1}} \frac{\mathbf{M}_1^{\frac{2k}{k-1}}}{\left(k\mathbf{M}_1^2 - \frac{k-1}{2}\right)^{\frac{1}{k-1}}} \tag{81}$$

which follows from Equation 59.

Figure 41 gives the relationship between p_{20}/p_1 and $\mathbf{M}_1$ for air ($k = 1.4$).

---

[9] This is treated in detail in Chapter 8.

At supercritical subsonic and supersonic velocities the nose of the tube is shaped like a sharp needle (a cone with a small vertex angle) which splits the bow shock wave. This makes it possible to improve the operation of the static hole. Parallel to this needle and at a certain distance from it the total pressure intake is placed. The Mach number is determined by placing the resistance thermometer into the flow.[10]

10. Unsteady One-Dimensional Ideal Gas Flow: Propagation of Finite-Intensity Disturbances

Let us return to the nonlinear system of differential equations of one-dimensional unsteady motion, Equation 55, Chapter 5. In the case of adiabatic motion, by means of the substitution ($\mathcal{P}$ is the pressure function)

$$\frac{dp}{d\rho} = a^2 \qquad \frac{1}{\rho}\frac{\partial p}{\partial x} = \frac{\partial \mathcal{P}}{\partial x} \qquad \frac{\partial \rho}{\partial t} = \frac{d\rho}{dp}\frac{\partial p}{\partial t} = \frac{\rho}{a^2}\frac{\partial \mathcal{P}}{\partial t}$$

$$\frac{\partial \rho}{\partial x} = \frac{d\rho}{dp}\frac{\partial p}{\partial x} = \frac{\rho}{a^2}\frac{\partial \mathcal{P}}{\partial x}$$

can be rewritten in the form

$$\frac{\partial u}{\partial t} + u\frac{\partial u}{\partial x} + \frac{\partial \mathcal{P}}{\partial x} = 0$$

$$\frac{1}{a}\frac{\partial \mathcal{P}}{\partial t} + a\frac{\partial u}{\partial x} + \frac{u}{a}\frac{\partial \mathcal{P}}{\partial x} = 0 \qquad (82)$$

Let us replace the pressure function $\mathcal{P}$ by a new function $\tilde{\mathcal{P}}$ having assumed that

$$d\tilde{\mathcal{P}} = \frac{1}{a}d\mathcal{P} = \frac{1}{\rho a}dp \qquad \tilde{\mathcal{P}} = \int_{p_0}^{p} \frac{dp}{\rho(p)\,a(p)} \qquad (83)$$

Then,

$$\frac{\partial \tilde{\mathcal{P}}}{\partial t} = \frac{1}{a}\frac{\partial \mathcal{P}}{\partial t} \qquad \frac{\partial \tilde{\mathcal{P}}}{\partial x} = \frac{1}{a}\frac{\partial \mathcal{P}}{\partial x}$$

and system (82) will go over into the following

$$\frac{\partial u}{\partial t} + u\frac{\partial u}{\partial x} + a\frac{\partial \tilde{\mathcal{P}}}{\partial x} = 0$$

$$\frac{\partial \tilde{\mathcal{P}}}{\partial t} + a\frac{\partial u}{\partial x} + u\frac{\partial \tilde{\mathcal{P}}}{\partial x} = 0$$

[10] See, for example, Gorlin, S. M., *Eksperimentalnaya aeromekhanika* (*Experimental Aeromechanics*), Vysshaya Shkola Press, Moscow, 1970, pp. 159 to 182.

Term-by-term summation and subtraction will transform this system of equations as

$$\frac{\partial}{\partial t}(\tilde{\mathcal{P}} + u) + (u + a)\frac{\partial}{\partial x}(\tilde{\mathcal{P}} + u) = 0$$

$$\frac{\partial}{\partial t}(\tilde{\mathcal{P}} - u) + (u - a)\frac{\partial}{\partial x}(\tilde{\mathcal{P}} - u) = 0 \tag{84}$$

The left sides of these equations involve one-dimensional substantive derivatives: in the first equation the derivative of the quantity $\tilde{\mathcal{P}} + u$, which is connected to the point moving along the axis $0x$ with the velocity $u + a$, and in the second equation, the derivative of the quantity $\tilde{\mathcal{P}} - u$, connected to the point that moves along the axis $0x$ with the velocity $u - a$. That these substantive derivatives are equal to zero testifies to the conservation of the quantity $\tilde{\mathcal{P}} + u$ at the point moving with the velocity $u + a$ and of the quantity $\tilde{\mathcal{P}} - u$ at the point moving with the velocity $u - a$. The quantities

$$r = \tilde{\mathcal{P}} + u \qquad s = \tilde{\mathcal{P}} - u \tag{85}$$

are called the Riemann invariants.[11]

Noting that in one-dimensional motion along the axis $0x$ to the point with the abscissa x there actually corresponds the plane that is perpendicular to the axis $0x$ and which passes through this point (for example, the plane section of the tube), we may interpret the result obtained as the existence of two families of plane waves propagating along the axis $0x$.

The absolute velocity of the propagation of the waves belonging the first family (C_1) is equal to

$$\frac{dx}{dt} = u + a \tag{86}$$

and the waves of the second family (C_2)

$$\frac{dx}{dt} = u - a \tag{87}$$

Directing the axis $0x$ downstream, i.e., regarding $u > 0$, we may say that the waves of the first family in their relative motion through the gas propagate in the same direction as the gas, whereas the waves of the second family propagate in the opposite direction. The relative velocity of the propagation of waves through the gas is equal to $\pm a$, where a is the local speed of sound; the upper sign relates to the first family, the lower, to the second. The waves of the first family carry the constant value of the invariant r, the waves of the second family of the invariant s.

In the case of a linearized motion of the gas (Section 4 of Chapter 5), an analogous picture of the propagation of disturbances was observed with the only difference that in this case the waves propagated with a constant velocity equal to the speed of sound in a nondisturbed gas and carried constant values of the gas parameters.

Before going over to a more detailed analysis of the problem concerning the propagation of finite disturbances, note that the wave character of equations (84) allows one to propose a

[11] Riemann, B., Über die Fortpflanzung ebener Luftwelleu von endlicher Schwingungsweite, *Abh. Ges. Wiss. Goettingen*, 1860.

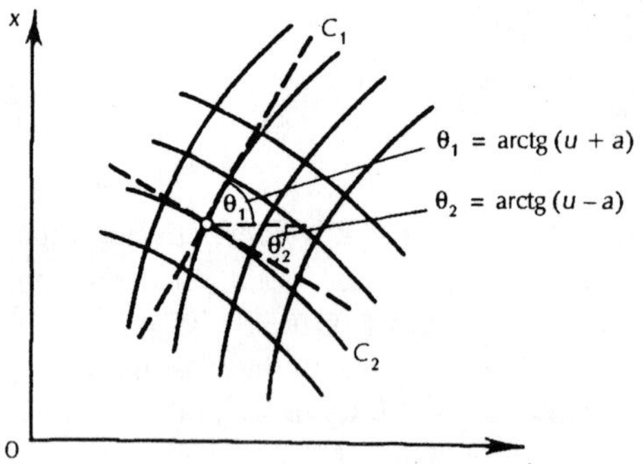

Fig. 42

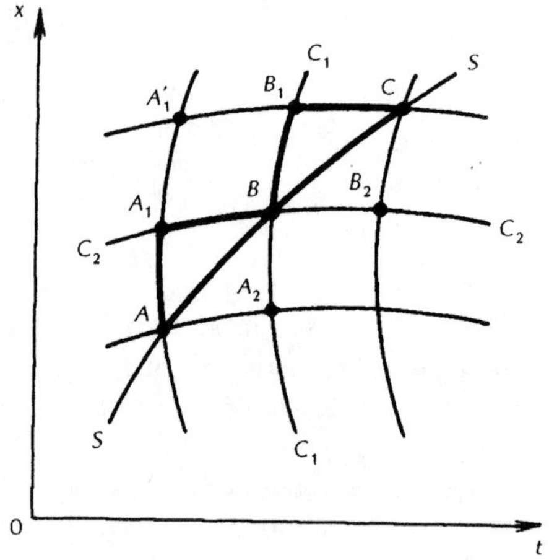

Fig. 43

grapho-analytical method for their integration. Below, the basic idea behind the method is described.

From the geometric point of view, Equations 84 can be interpreted as follows. At each point of the plane (t, x) there are two directions (in Figure 42 they are indicated by dashed lines) with the angular coefficients

$$\text{tg}\,\theta_1 = \left(\frac{dx}{dt}\right)_1 = u + a \qquad \text{tg}\,\theta_2 = \left(\frac{dx}{dt}\right)_2 = u - a \tag{88}$$

which have the property that in the case of an infinitely small displacement along these directions, the corresponding values of the Riemann invariants are retained.

Equations 86 and 87 can be considered to be differential equations for the families of the curves C_1 and C_2. The angular coefficients of the tangents are defined by equalities (88). The equations for these families of the curves cannot be found in closed form prior to the integration of Equations 84 and finding $u(x, t)$ and $a(x, t)$; however, as we shall soon see, the existence of Equations 88 is essential for representing the general character of the processes described by Equations 84, and for integrating these equations.

The curves C_1 and C_2 are the so-called characteristics of system (84) and the directions of the tangents to them (88), the characteristic directions. Recall that in the case of the linearized problem (Section 4 of Chapter 5) the equations of characteristics were known in closed form; these were the families of straight lines

$$x + at = \text{const} \qquad x - at = \text{const}$$

To the mathematical concept of characteristics in the plane (x, t) corresponds the concept of waves as planes moving with the velocities $u + a$ and $u - a$ and carrying the values of the Riemann invariants r and s.

Let us suppose that a certain initial distribution of the velocity $u(s)$ and of the gas parameters $p(s)$, $\rho(s)$, $a(s)$ are given and, consequently, $\tilde{\mathcal{P}}(s)$ is given along a certain curve S (Figure 43), the direction of the tangents does not coincide with the characteristic directions. In particular, this can be the assignment of the initial conditions at $t = 0$, i.e., on the axis $0x$, or of the boundary conditions on the straight lines parallel to the axis $0t$.

Having determined the angular coefficients of the curve C_1 at the point A from Equation 88 and of the curve C_2 at the adjacent point B from the equations

$$\operatorname{tg}\theta_{1A} = \left(\frac{dx}{dt}\right)_{1A} = u_A + a_A \qquad \operatorname{tg}\theta_{2B} = \left(\frac{dx}{dt}\right)_{2B} = u_B - a_B$$

we shall pass the corresponding characteristic directions through these points and construct a small triangle AA_1B.

Over the segment AA_1 of the characteristic C_1 the first invariant r is retained, i.e.,

$$\tilde{\mathcal{P}}_{A_1} + u_{A_1} = \tilde{\mathcal{P}}_A + u_A$$

and over the segment A_1B of the characteristic C_2, the second invariant s, so that

$$\tilde{\mathcal{P}}_{A_1} - u_{A_1} = \tilde{\mathcal{P}}_B - u_B$$

From the resulting system of equations we find

$$\tilde{\mathcal{P}}_{A_1} = \frac{1}{2}(\tilde{\mathcal{P}}_A + u_A + \tilde{\mathcal{P}}_B - u_B)$$

$$u_{A_1} = \frac{1}{2}(\tilde{\mathcal{P}}_A + u_A - \tilde{\mathcal{P}}_B + u_B)$$

and, consequently, we also find P_{A_1}, ρ_{A_1}, and a_{A_1}.

Repeating the same arguments regarding the triangle BB_1C constructed from the angular coefficients of the characteristic C_1 at the point B and of the characteristic C_2 at the point C, we find the values of u_{B_1} and $\tilde{\mathcal{P}}_{B_1}$ at the point B_1 and, consequently, of p_{B_1}, ρ_{B_1}, and a_{B_1}. This will make it possible to envisage the directions of characteristics at the points A_1 and B_1, to construct the triangle $A_1A_1'B_1$ and, as before, to determine the values of u and $\tilde{\mathcal{P}}$ at the point A_1', and then at the points A_2, B_2, and so on. Prescribing a rather thick division of the curve S at the points A, B, C, etc., we shall find the values of the unknown functions u, p, ρ, and a at the points of the plane (x, t) which are spaced as closely as possible.

Let us now consider in more detail the problem of one-dimensional propagation of disturbances of finite intensity in an ideal gas. We shall show that as in the case of small disturbances, the propagation of the disturbances of finite magnitude may also proceed with the aid of simple waves (Section 4, Chapter 5), i.e., the waves traveling with a constant velocity and carrying constant values of the gas parameters. Such propagation of finite-intensity dusturbances takes place if one of the Riemann invariants is constant throughout the entire flow region. For this purpose it seems sufficient that this invariant be constant at the initial time instant (at $t = 0$, along the $0x$ axis). The possibility for this kind of assumption will soon be explained and illustrated in an example.

Let, for the sake of definiteness, the second invariant $s = \tilde{\mathcal{P}} - u$ be constant. Assuming that $p = p_0$ at $u = 0$ and recalling the definition of $\tilde{\mathcal{P}}$ given by the last of Equations 83, we have throughout the entire flow region and at any time instant that

$$\tilde{\mathcal{P}} = u \tag{89}$$

In this case, the second equation of system (84) will be fulfilled identically, whereas the first equation rewritten as

$$\frac{\partial u}{\partial t} + (u + a)\frac{\partial u}{\partial x} = 0 \tag{90}$$

will show that the waves of the family C_1 carry constant values of the velocity u. According to Equation 89, the values of the function $\tilde{\mathcal{P}}$ and, consequently, of the gas parameters p, ρ, and a also will be constant. From this it follows that the absolute velocity of wave propagation $u + a$ and its velocity with respect to the gas "a" preserve constant values in the given wave, i.e., the wave is simple.

We conclude from the adopted condition $u > 0$ and from Equation 89 that $\tilde{\mathcal{P}} > 0$, whereas according to Equation 83 $p > p_0$, i.e., the disturbances carried by the waves of the first family correspond to the compression of the gas and the waves themselves can be called the compression waves.

Similarly, having assumed that the invariant $r = \tilde{\mathcal{P}} + u$ is preserved throughout the entire flow region and in time, i.e., having assumed that

$$\tilde{\mathcal{P}} + u = \text{const} = 0 \tag{91}$$

where the constant is selected to be equal to zero on the same grounds as in the previous case, we see that now the second equation of system (84) will go over into the following

$$\frac{\partial u}{\partial t} + (u-a)\frac{\partial u}{\partial x} = 0 \tag{92}$$

In other words, in the waves of the second family C_2, just as in the waves of the first family, the velocity u preserves its magnitude, and, according to Equation 91, also preserved are $\tilde{\mathcal{P}}$, p, ρ, and a. Thus, then condition (91) is fulfilled, the second family C_2 also represents the set of simple waves that have the absolute velocity of propagation $u - a$ and the velocity a which with respect to the gas is directed opposite to the motion of the gas. According to Equation 91 and to the condition $u > 0$, we conclude that the waves of the second family represent the waves of rarefaction.

In contrast to the case of small disturbances considered in Chapter 5, Section 4, where all the simple waves had the velocities of propagation identical in magnitude and equal to the speed of sound in a nondisturbed gas, in the present case of simple waves carrying the finite-intensity disturbances have the velocities of propagation with respect to the gas, the absolute magnitudes of which are equal to the local speed of sound, but which are not the same for different waves of the given family.

It can be easily verified that in the case of finite-intensity disturbances the absolute and relative velocities of propagation of simple waves of the first family C_1 is the higher, the more intense are the disturbances which are carried by them. For this purpose let us express $\tilde{\mathcal{P}}$ in terms of a, assuming for the sake of definiteness that the motion is isentropic. In this case we have

$$d\tilde{\mathcal{P}} = \frac{1}{a}d\mathcal{P} = \frac{1}{\rho a}dp = \frac{1}{a}\frac{dp}{d\rho}\frac{d\rho}{\rho} = a\frac{d\rho}{\rho} \tag{93}$$

on the other hand,

$$a^2 = k\frac{p}{\rho} = \text{const} \cdot \rho^{k-1}$$

and, calculating the logarithmic differential, we find

$$\frac{d\rho}{\rho} = \frac{2}{k-1}\frac{da}{a}$$

so that

$$d\tilde{\mathcal{P}} = \frac{2}{k-1}da \qquad \tilde{\mathcal{P}} = \frac{2}{k-1}(a-a_0) \tag{94}$$

where the subscript zero, as before, relates to the parameters of the quiescent gas ($u = 0$) and where the constant of integration has been selected in conformity with the definition of $\tilde{\mathcal{P}}$ by Equation 83. Substituting the obtained value of $\tilde{\mathcal{P}}$ into Equation 89, we find

$$a = a_0 + \frac{k-1}{2}u \qquad u + a = a_0 + \frac{k+1}{2}u \tag{95}$$

If follows directly from these equations that for the waves of the first family C_1, the higher the velocity u of the disturbed gas motion, the higher the velocities $u + a$ of the absolute and "a" of the relative propagation of the wave. According to Equation 89, however, to the greater velocity of motion there correspond greater values of $\tilde{\mathcal{P}}$, and along with them, of the pressure p and density ρ of the disturbed gas, because by the definition of the function $\tilde{\mathcal{P}}$

$$\frac{dp}{d\tilde{\mathcal{P}}} - \rho a > 0 \qquad \frac{d\rho}{d\tilde{\mathcal{P}}} = \frac{\rho}{a} > 0$$

Thus, the higher the absolute and relative velocities of propagation of the first family waves (compression waves), the greater the intensities of disturbances carried by them.

For the waves of the second family C_2 (rarefaction waves) we have, according to Equation 91 and 94

$$a = a_0 - \frac{k-1}{2}u \qquad u - a = -a_0 + \frac{k+1}{2}u \tag{96}$$

As the rarefaction increases, i.e., increases the absolute value of the difference $p - p_0$, then, according to Equation 83, the absolute value of the function $\tilde{\mathcal{P}}$ will also increase. Thus, following Equation 91, the gas velocity u will also rise. We conclude from the first equation of system (96) that in this case the velocity a of the propagation of rarefaction waves with respect to the gas will become smaller, whereas the absolute velocity will increase.

From this follows the basic difference between the phenomenon of the propagation of finite disturbances and that of the propagation of small disturbances: the initial form of the distribution of disturbances is not preserved.

In order to get a general idea about the character of deformation of the curve representing the distribution of initial disturbances, we again consider the motion of a piston in a tube (Figure 44). Let the piston velocity u_0 be directed from left to right along the positive direction of the axis $0x$; the gas is compressed to the right from the piston. At the time $t = t_1$ in front of the piston the perturbation of velocities is formed which can be represented in the form of a certain continuous descending curve $u = u_1(x)$. At the next time instant $t = t_2 > t_1$ the velocity distribution curve $u = u_2(x)$ will fall more steeply, because for the time period $t_2 - t_1$ the points of the curve $u_1(x)$ will displace in the horizontal direction by the segments which, according to the theory of propagation of the first family waves (compression waves), are equal to

$$(u_1 + a)(t_2 - t_1) = \left(a_0 + \frac{k+1}{2}u_1\right)(t_2 - t_1) \tag{97}$$

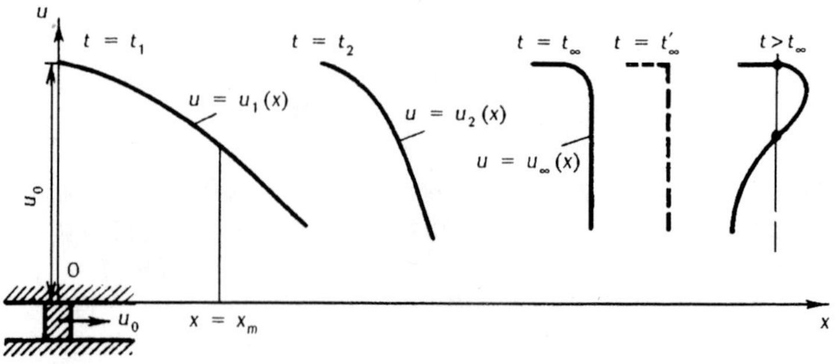

Fig. 44

and, consequently, are the greater the larger are the ordinates u_1 of the curve $u_1(x)$. The slope of the curve will become increasingly steeper with time until at a certain instant $t = t_\infty$ the tangent at one of the points on the curve will become perpendicular to the axis $0x$ and the derivative $u'_\infty(x)$ will become infinite. Henceforth, when $t > t_\infty$, the upper points of the velocity distribution curve should have outrun the lower, and the curve should have acquired the shape of the last curve, as shown at right in Figure 44. Such a shape is not feasible physically, because otherwise to the same tube section at the given instant there would have corresponded several different velocities. Actually, the shape continues to vary only until the tangents become perpendicular to the axis $0x$. The limiting form of the distribution of disturbances is represented by a jumpwise change in the velocity which is shown in Figure 44 by dashes and which probably corresponds to a shock wave.

Thus, from the theory of propagation of the finite-intensity compression waves follows the inevitability of the origination of shock waves in a tube.

Let us estimate the time interval $t_\infty - t_1$ needed for the distribution $u = u_\infty(x)$ to be formed from the initial distribution of velocities with an infinitely great slope at any one point. This time interval can be taken as the measure of time after the lapse of which the distribution, characteristic for the shock wave, forms from the given initial distribution of propagation.

Let $u = u(x)$ be a certain intermediate velocity distribution occurring at time t. The curve $u(x)$ results from the curve $u_1(x)$ by shifting the ordinates of the curve $u_1(x)$ parallel to the axis $0x$ by the distance (97), so that it is possible to write

$$u(\xi) = u_1(x) \tag{98}$$

where it is assumed that

$$\xi = x + \left[a_0 + \frac{k+1}{2}u_1(x)\right](t - t_1) \tag{99}$$

It should be explained that such a displacement of the ordinates of the curve $u_1(x)$ corresponds to the displacement of the first family with constant velocities on the characteristic $u_1(x)$ along rectilinear characteristics in the plane (x, t).

To calculate the derivative at the point x, take the derivative with respect to x of both parts of Equation 98; using Equation 99, we have

$$u'(\xi)\frac{d\xi}{dx} = u'(\xi)\left[\left(1 + \frac{k+1}{2}u'_1(x)(t - t_1)\right)\right] = u'_1(x)$$

where

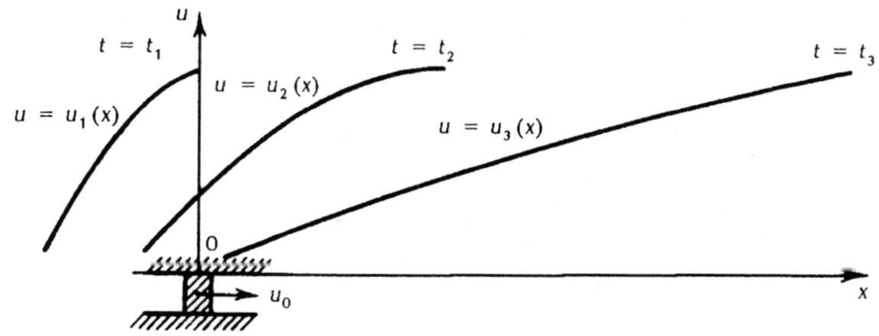

Fig. 45

$$u'(\xi) = \frac{u_1'(x)}{1 + \frac{k+1}{2} u_1'(x)(t - t_1)} \tag{100}$$

Let $x = x_m$ be the abscissa of the point at which the curve $u_1(x)$ has the slope $u_1'(x_m)$, the absolute magnitude of which is maximum. Then, assuming that at the time instant $t = t_\infty$, the left side of Equation 100 will turn to infinity at $\xi = \xi_m$, we obtain the following equation from the condition of the finite character of the quantity $u_1'(x_m)$:

$$1 + \frac{k+1}{2} u_1'(x_m)(t_\infty - t_1) = 0$$

From this it follows that

$$t_\infty - t_1 = -\frac{2}{(k+1)\, u_1'(x_m)} = \frac{2}{(k+1)\, |u_1'(x_m)|}$$

This expression can serve as an estimate of the time interval after the lapse of which the distribution of velocities with the maximum (in absolute magnitude) slope $|u_1'(x_m)|$ will go over into a jumpwise distribution of velocities which corresponds to the formation of the shock wave. This time interval is finite and the smaller it is, the greater the maximum slope of the curve of the initial disturbance of velocities.

In the case of piston motion considered above, rarefaction is formed to the left of the piston, which propagates with the aid of the waves of the second family (rarefaction waves), and which also leads to a certain continuous distribution of disturbance velocities directed to the side of positive x. In this case, however, the distribution of velocities will be represented not by a descending curve, as shown in Figure 44, but by an ascending curve (Figure 45), because the points closer to the left side of the piston will have higher velocities than those away from this side. The conclusion derived from Equations 96 concerning the rise in the absolute velocity of propagation of the second family waves shows that in this case the slope of the curves becomes smaller and there is no tendency to the formation of discontinuities.

11. Rarefaction Waves behind a Moving Piston. Centered Waves. Self-Similar and General Problems

Let us consider in more detail[12] the problem of rarefaction waves formed in a one-dimensional gas flow behind a moving piston by beginning with the simplest case of the so-called centered rarefaction waves. First, it should be noted that according to Equations 86 or 87 the preservation condition of the gas motion velocity u and of the speed of sound a along the characteristics of the given family gives their rectilinearity in the plane (x, t). In fact, if, for example, it is assumed that the invariant r is constant, i.e., if it is adopted in the isentropic case that

$$u + \frac{2a}{k-1} = \text{const} = u_1 + \frac{2a}{k-1} \tag{101}$$

then along the characteristics C_2 of the second family (rarefaction waves) the quantities u and a should be preserved, so that the equation

$$\frac{dx}{dt} = u - a = \text{const} = u_1 - a_1$$

will yield

$$\frac{dx}{dt} = \frac{x - x_1}{t - t_1} = u_1 - a_1$$

where (x_1, t_1) is a certain point of the plane (x, t) located on the characteristic C_2 and u_1, and a_1 are the gas velocity and the speed of sound at this point, respectively. This equation yields the equation for the family of the characteristics C_2 in the plane (x, t)

$$x - x_1 = (u_1 - a_1)(t - t_1) \tag{102}$$

These are straight lines on each of which the quantities u_1 and a_1 preserve their constant values.

When all the characteristics of the given family intersect at the same point, the waves that correspond to them are called centered.

As an example of the centered waves we shall consider the problem of gas flow behind a moving piston (Figure 46) and determine what should be the law of piston motion for the rarefaction waves induced by it to be called centered.

Denoting by $(\bar{x}, \bar{t})$ the coordinates of the common point (poles) at which the characteristics of the second family (rarefaction waves) intersect, we shall assume that at the initial instant ($t = 0$) the piston (its velocity will be designated by w) and simultaneously the gas were at rest ($w = 0$, $u = u_0 = 0$, $a = a_0$), and that from the instant $t = t_1$ the piston, having reached the velocity w_1, began to move uniformly with this velocity. Then, according to Equations 101 and 102, we have the equations of characteristics in the planes (x, t) and (u, a) in the closed form

$$\frac{x - \bar{x}}{t - \bar{t}} = u - a \qquad u + \frac{2a}{k-1} = \frac{2a_0}{k-1} \tag{103}$$

It should be noted that when the characteristics of the family C_2 in the plane (x, t) represent the family of straight lines, which depends on the value of the difference $u - a$, all the characteristics in the plane (u, a) merge into one straight line.

---

[12] See Zeldovich, Ya. B. and Raizer, Yu. P., *Fizika udarnykh voln i vysokotemperaturnykh gidrodinamicheskikh yavleniy* (*The Physics of Shock Waves and of High Temperature Hydrodynamic Phenomena*), Fizmatgiz Press, Moscow, 1963, pp. 25 to 43 and Patterson, G. N., *Molekulyarnoye techeniye gazov* (*Molecular Flow of Gases*), Fizmatgiz Press, Moscow, 1960, pp. 63 to 72.

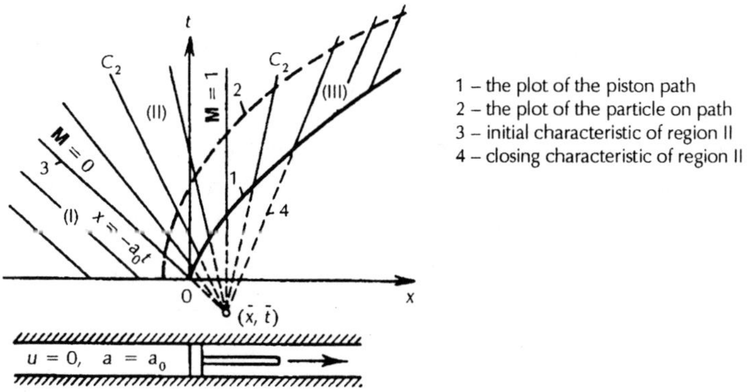

Fig. 46

We find from Equation 103 that

$$u = \frac{2}{k+1}\left(\frac{x-\bar{x}}{t-\bar{t}} + a_0\right) \tag{104}$$

$$a = \frac{2a_0}{k+1} - \frac{k-1}{k+1}\frac{x-\bar{x}}{t-\bar{t}} \tag{105}$$

As seen in Figure 46, initial characteristic 3 of the sector of centered waves II has the equation

$$x = -a_0 t$$

which directly follows from Equation 87 at $u_0 = 0$, $a = a_0 = \text{const}$, and from the condition of the passage of this characteristic through the coordinate origin (the initial position of the piston at the initial time instant). Region I located to the left from this characteristic corresponds to the quiescent gas, and here the characteristics will be represented by the straight lines parallel to the initial characteristic. Region II is closed by characteristic 4, which passes through that point of the plot giving the piston path (1). Starting here, this plot becomes rectilinear and the piston velocity becomes constant and equal to w_1. The closing characteristic will have the equation

$$x - \bar{x} = (u_1 - a_1)(t - \bar{t})$$

where $u_1 = w_1$ and a_1 is the speed of sound to the left from the piston already moving uniformly with the velocity w_1. To the right from closing characteristic 4, in region III, the characteristics will be straight lines parallel to this characteristic.

In the considered motion of the piston, the equation of motion of the gas particles is determined by the integration of the equation

$$u = \frac{dx}{dt} = \frac{2}{k+1}\left(\frac{x-\bar{x}}{t-\bar{t}} + a_0\right) \tag{106}$$

and has the form

$$x - \bar{x} = (t-\bar{t})\left[\alpha(t-\bar{t})^{-\frac{k-1}{k+1}} + \frac{2a_0}{k-1}\right] \qquad t > \bar{t} \tag{107}$$

Different values of the integration constant α correspond to the selection of particles, and the motions are considered. Assuming $x = \xi \le 0$ at $t = -x/a_0$, we shall find the value of $\alpha(\xi)$ and, substituting it into Equation 107, we obtain the equation of the motion of the particle which at the initial instant (at the start of piston motion) had the abscissa ξ. In particular, this is the way in which it is possible to find the law of piston motion in which we are interested, and which leads to centered rarefaction waves. For this, it is sufficient to form the equation of particle motion (107) by having selected the constant α from the condition that $\xi = 0$ at $t = 0$. Then,

$$\alpha = (-t)^{-\frac{k-1}{k+1}}\left(\frac{\bar{x}}{\bar{t}} - \frac{2a_0}{k-1}\right)$$

According to Equations 104 and 105, the local Mach number $\mathbf{M} = u/a$ is equal to

$$\mathbf{M} = \frac{1 + \dfrac{x-\bar{x}}{a_0(t-\bar{t})}}{1 - \dfrac{k-1}{2}\dfrac{x-\bar{x}}{a_0(t-\bar{t})}}$$

and it preserves a constant value along each characteristic C_2 and therefore serves as a natural "mark" of the characteristics of the family (Figure 46).

Using the second of Equations 103, it is possible to obtain the relation

$$a_0 = a\left(1 + \frac{k-1}{2}\mathbf{M}\right)$$

and the group of equations

$$T_0 = T\left(1 + \frac{k-1}{2}\mathbf{M}\right)^2 \qquad \rho_0 = \rho\left(1 + \frac{k-1}{2}\mathbf{M}\right)^{\frac{2}{k-1}} \qquad p_0 = p\left(1 + \frac{k-1}{2}\mathbf{M}\right)^{\frac{2k}{k-1}}$$

which differ markedly from the isentropic formulas derived in Chapter 5 for the steady gas motion. First, in contrast to the steady motion, the number $\mathbf{M}$ in the brackets is to the first, not the second, power; moreover, the exponents of the round brackets on the right exceed by a factor of 2 the corresponding values in the steady case. To assess this difference, we assume, for example, that the piston moves with an increasing velocity so that $\mathbf{M} \to \infty$, $a \to 0$ and, according to the second of Equation 103,

$$u \to u_{max} = \frac{2}{k-1}a_0$$

If the piston velocity exceeds this velocity, then a vacuum forms between the piston and the gas, where the velocity of discharge into the vacuum u_{max} in this unsteady case differs from the steady maximum velocity (19) by the fact that the latter has the factor $\sqrt{2/(k-1)}$ before the quantity a_0 and not $2/(k-1)$. Thus, for example, for air the velocity of discharge into the vacuum in the considered unsteady motion turns to be higher by a factor of $\sqrt{5}$ than in the case of steady motion. In steady motion of the gas the specific kinetic energy of discharge into the vacuum is exactly equal, according to Equation 20, to the total enthalpy in the tank from which the discharge takes place; on the other hand, in the case of unsteady discharge the kinetic energy exceeds the total enthalpy by a factor of $2/(k-1)$.

The simplicity of the above solution for the centered waves is due to the fact that the solution is represented by the function of the ratio $(x - \bar{x})/(t - \bar{t})$ and not of each separate variable x and t. Using polar coordinates, it is possible to say that the solution depends only on the polar angle $\theta = \operatorname{arctg} \dfrac{x - \bar{x}}{t - \bar{t}}$ and does not depend on the distance $r = \sqrt{(x - \bar{x})^2 + (t - \bar{t})^2}$ of the instantaneous point of the spatial-time plane up to the point of the intersection of the characteristics $(\bar{x}, \bar{t})$. It is for this reason that the solution was reduced to an ordinary differential equation, which also turned out to be rather simple and made it possible to represent the integral in a closed form.

These kinds of cases are frequently encountered when it is possible to reduce partial to ordinary differential equations. The solution is represented as a function of one argument which involves a certain combination of basic arguments of the problem. To the constant values of this complex argument correspond varieties of solutions with separate arguments which can be considered to be being similar. For this reason, in general, the solution of a partial differential equation expressed as a function of one complex argument, which represents a one-term combination of the arguments contained in the statement of the problem, and which satisfies an ordinary differential equation (to which the basic partial differential equation is reduced), is called the self-similar solution and the problem itself is called self-similar.

An example of a self-similar solution is the solution for the problem concerning the centered rarefaction waves, in which the complex argument was represented by the ratio $(x - \bar{x})/(t - \bar{t})$. In complete analogy with this one-dimensional nonstationary problem, we consider later a plane stationary problem of centered rarefaction waves in a supersonic gas flow around an external obtuse angle. A great number of self-similar problems are also considered in subsequent chapters dealing with the theory of motion of a viscous fluid in a boundary layer.[13]

The self-similar solutions are sure to represent only certain of the simplest particular solutions of the general problem posed, but at the same time in the majority of cases they turn to be useful in as allowing judgment about the basic aspects of the phenomenon considered. It should be noted (later confirmed by numerous examples) that the possibility of the existence of self-similar solutions is attributable to the absence in the statement of the problem (equations, boundary, and initial conditions) of certain characteristic scales: time, length, mass, etc. In other words, it is due to a certain limitation in the statement of the problem itself, in departing from generality. Thus, for example, in the previous problem of the centered rarefaction waves behind the moving piston, an arbitrary *a priori* prescribed law of piston motion was out of the question, but instead a particular law of piston motion was determined in the course of the problem solution which made the existence of centered waves possible.

Now, consider the general statement of the problem concerning the rarefaction waves behind an arbitrarily moving piston.

As before, assume that Riemann's first invariant r preserves a constant value throughout the entire flow region, i.e., that Equation 101 is fulfilled. Then, noting that the quantities u and a preserve their values along rectilinear, but generally not centered characteristics of the second family C_2 (rarefaction waves), we obtain, after having performed integration of differential Equation 87, that

$$x = (u - a)\,t + \varphi(u) \tag{108}$$

[13] Various examples of self-similar problems can be found in the books by Sedov, L. I., *Metody podobiya i razmernosti v mekhanike (The Similarity and Dimensionality Methods in Mechanics)*, Fizmatgiz Press, Moscow, 1981; and *Similarity and Dimensional Methods in Mechanics*, Academic Press, New York.

where $\varphi(u)$ is the integration constant that behaves different on different characteristics. To determine the function $\varphi(u)$, we prescribe the law of piston motion $x_p = X(t_p)$, and thereby the piston velocity $w = X'(t_p)$. Then, the set of equations

$$\varphi(w) = X(t_p) - [w - a(w)]\, t_p \qquad w = X'(t_p) \tag{109}$$

after the elimination of t_p, will determine the form of the function $\varphi(w)$; the function $a(w)$ is found from Equation 101. The solution of the problem is given by the equation

$$x = [u - a(u)]\, t + X(t_p) - [w - a(w)]\, t_p$$

which should be considered to be an equation for determining u as a function of x and t. In contrast to the problem of centered waves, u is generally a function of x and t taken separately and not of any of their combinations.

In the specific case of centered waves with the pole at the point (x_0, t_0), Equation 108 will give the relation

$$\varphi(u) = x_0 - (u - a)\, t_0$$

and also the following relation to determine $u(x, t)$:

$$u - a(u) = \frac{x - x_0}{t - t_0}$$

In other words, we see again that in centered waves the velocities of particles, just as the speed of sound and other thermodynamic parameters of the gas, will be the functions of the ratio $(x - x_0)/(t - t_0)$ and not of x and t taken separately. An illustrative example and some details of the problem considered can be found in the book by Zeldovich and Raizer (cited earlier).

12. An Elementary Theory of the Shock Tube

The scheme of the shock tube for producing supersonic flows of short duration in laboratory conditions and not requiring large quantities of energy for the purpose is shown in Figure 47a.

A cylindrical tube with closed entrance and exit is partitioned into two compartments by a diaphragm. On both sides of the diaphragm are gases with different physical properties. The gas in the left compartment is compressed to a substantial pressure, as against the gas in the other compartment. At a certain time the diaphragm bursts and the gas from the left compartment rushes to the right compartment. The difference of pressures that existed before the rupture of the diaphragm propagates as a shock wave to the right entraining the accompanying gas flow. On its way the flow encounters the model studied, which is located in the right compartment of the tube. A special optical facility makes it possible to obtain an instantaneous photographic record of the flow spectrum around the model as well as to make other measurements needed by the researcher.

An elementary calculation of the shock tube follows. Let the physical properties of the gas be characterized by two constants: molecular weight μ and specific heat ratio k. For the gas, which initially was in the right compartment, corresponds the constants μ_1 and k_1, and for the gas in the left compartment, μ_4 and k_4. The initial pressures are denoted by p_1 and p_4, respectively, where $p_4 \gg p_1$. The tube is considered to be thermally insulated and the motion of gas to be adiabatic.

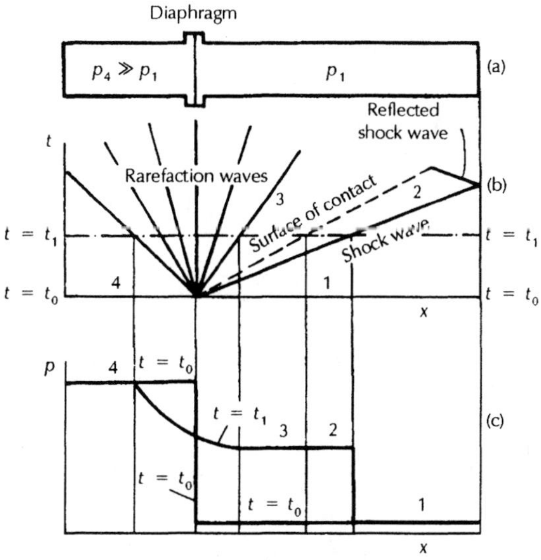

Fig. 47

Figure 47b presents the plots of the propagation of a shock wave, contact surfaces of the gases,[14] and of rarefaction waves. It is seen from the plots that the greatest velocity, expressed in terms of the derivative dx/dt, is exhibited by the shock wave, then by the contact surface of the gases and, finally, by the rarefaction waves which propagate through the gas from right to left, but which are swept to the right by the accompanying flow. The region of the tube occupied by the rarefaction waves (this region expands with time) represents the region of continuous isentropic motion, so that using Equations 48 of Chapter 5 we obtain the following equation on the boundaries of this region

$$\frac{p_4}{p_3} = \left(\frac{a_4}{a_3}\right)^{\frac{2k_4}{k_4 - 1}} \tag{110}$$

Considering the rarefaction waves as simple waves of finite intensity, we apply to them the theory given in Section 10. According to this theory, in the entire region filled with rarefaction waves which move with the absolute velocity $u - a$, the sum $\tilde{\mathcal{P}} + u$ (Equation 94) equal

$$\tilde{\mathcal{P}} + u = \frac{2}{k-1} a + u + \text{const}$$

preserves the same magnitude so that it is possible to write (on the left boundary of region 3 the gas is motionless):

$$\frac{2}{k_4 - 1} a_4 = \frac{2}{k_4 - 1} a_3 + u_3 \tag{111}$$

Noting that near the contact surface of the gases the pressure and velocity are continuous, we have (Figure 47c):

[14] It is assumed that for the small period of time during which the tube operates, the gases have no time to noticeably mix.

$$p_3 = p_2 \qquad u_3 = u_2 = V$$

where V is the velocity of the accompanying flow. Thus, using Equation 111 instead of Equation 110, we obtain the following equation:

$$p_4 = p_2\left(\frac{a_4}{a_4 - \frac{k_4 - 1}{2}V}\right)^{\frac{2k_4}{k_4 - 1}} \tag{112}$$

Recalling Equation 53 for gas compression in passage through a shock wave ($\mathbf{M}_1 = \theta/a_1$ is the Mach number of the shock wave)

$$\frac{p_2}{p_1} = \frac{2k_1}{k_1 + 1}\mathbf{M}_1^2 - \frac{k_1 - 1}{k_1 + 1} \tag{113}$$

and Equation 69 for the accompanying flow velocity behind the shock wave, which in the present case has the form

$$V = \frac{2}{k_1 + 1}\left(\mathbf{M}_1 - \frac{1}{\mathbf{M}_1}\right)a_1 \tag{114}$$

we substitute these expressions into Equation 112. This will give

$$\frac{p_4}{p_1} = \left(\frac{2k_1}{k_1 + 1}\mathbf{M}_1^2 - \frac{k_1 - 1}{k_1 + 1}\right)\left[1 - \frac{k_4 - 1}{k_1 + 1}\left(\mathbf{M}_1 - \frac{1}{\mathbf{M}_1}\right)\frac{a_1}{a_4}\right]^{-\frac{2k_4}{k_4 - 1}} \tag{115}$$

Here, p_4/p_1 is the given initial ratio of pressures in the compartments of the tube, whereas the ratio a_1/a_4 at identical initial temperatures of gases can be calculated, according to Section 4 of Chapter 5, as

$$\frac{a_1}{a_4} = \sqrt{\frac{k_1\mu_4}{k_4\mu_1}} \tag{116}$$

Substituting this ratio into Equation 115, we find $\mathbf{M}_1$ and, consequently, from Equation 114 we also find the velocity of the accompanying flow V in the tube.

In Figure 48[15] the plots of (Mach number $\mathbf{M}_1$ vs. $\lg(p_4/p_1)$ at different values of the parameter a_4/a_1 are given for $k_1 = 5/3$ and $k_4 = 7/5$.

Let us determine approximately the maximum value of $\mathbf{M}_1$ corresponding to an infinite value of the ratio p_4/p_1. For this, suppose that the expression in the square brackets on the right side of Equation 115 is equal to zero and neglect the term containing $\mathbf{M}_1$ in the denominator. This will give

$$(\mathbf{M}_1)_{p_4/p_1 = \infty} = \frac{k_1 + 1}{k_4 - 1}\frac{a_4}{a_1} \tag{117}$$

[15] Resler, E. L., Lin, C. C., and Kantrowitz, A., The production of high-temperature gases in shock tubes, *J. Appl. Phys.*, 23, 1390, 1952.

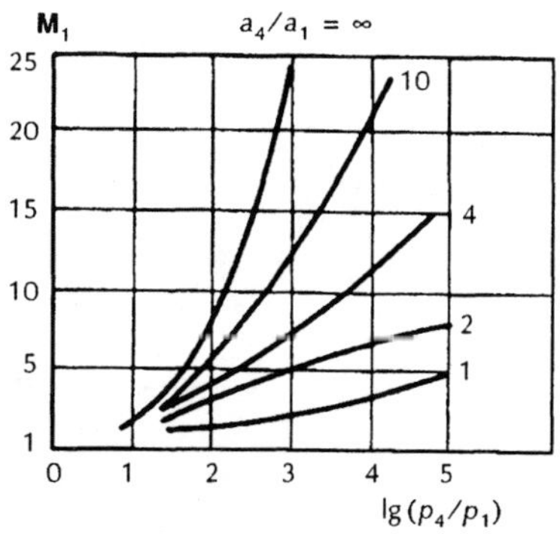

Fig. 48

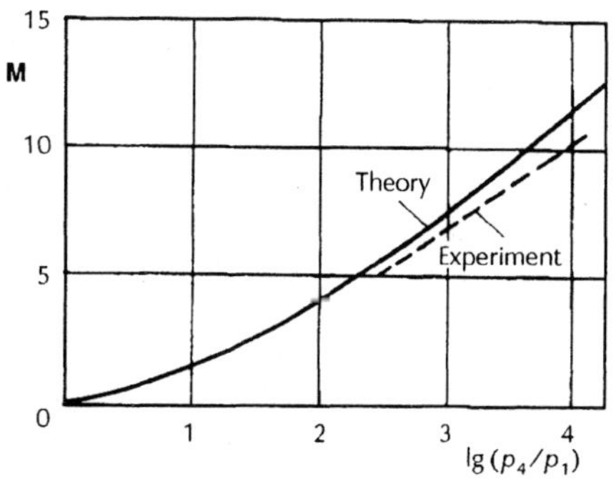

Fig. 49

From this it is seen that an as great as possible ratio a_4/a_1 should be selected and therefore it is expedient that the gas selected for the left high-pressure compartment of the tube have a small molecular weight, for example, hydrogen ($\mu_4 = 2.016$). Similarly, the gas for the right compartment should be heavy, for example, argon ($\mu_1 = 39.94$).

An elementary theory of a shock tube in its simplest form satisfactorily agrees with the experiment, as testified to by the plot (Figure 49) borrowed from the work cited in footnote 15, and corresponding to the use of hydrogen and argon.

7

Irrotational Motions of an Ideal Medium. Plane Irrotational Motion of an Ideal Incompressible Fluid

1. The Theorems of Kelvin and Lagrange: Conditions for the Existence of Irrotational Flows

The study of ideal fluid flows other than one-dimensional – plane, axisymmetric, and more general three-dimensional motions – presents mathematical difficulties. The Euler equations of the dynamics of an ideal liquid and gas were used in Chapter 6 to derive two basic equations, Bernoulli and Helmholtz-Fridman, with the theorem on the conservation of vortex lines which follows the latter equation. Direct integration of Euler equations for the general case of rotational motions is not considered in this book, although the advent of electronic computers has considerably facilitated the solution of such problems. Planning to devote the present chapter to the case of irrotational motions which are more accessible for solution, we refer the readers interested in rotational motions to the special book written by Goldshtik.[1] The basic assumption which has played an historical role in bringing the theoretical hydrodynamics to specific applications was the supposition about the absence of vorticity in a moving ideal fluid. The possibility for the existence of such an irrotational motion is justified by the following two theorems.

The theorem of Kelvin: In the case of barotropic motion of an ideal fluid under the action of a field of volumetric forces with single-valued potential, the circulation of velocity along a closed fluid contour does not change.

This theorem is easily proved with the aid of the kinematic theorem of Kelvin, which concerns the change in time of velocity circulation and which was presented in Chapter 2, Section 7. According to this theorem

[1] Goldshtik, M. A. *Vikhrevye potoki* (*Vortical Streams*), Nauka Press, Novosibirsk, 1981.

$$\frac{d}{dt}\oint \boldsymbol{V}\cdot\delta\boldsymbol{r} = \oint \dot{\boldsymbol{V}}\cdot\delta\boldsymbol{r}$$

Let us substitute into the right side of this equation the expression of acceleration by the Euler basic equation (3) of Chapter 5, which in the case of potential volumetric forces and barotropic motion can be rewritten in the form

$$\dot{\boldsymbol{V}} = -\text{grad}\ (\Pi + \mathcal{P})$$

then,

$$\frac{d}{dt}\oint \boldsymbol{V}\cdot\delta\boldsymbol{r} = -\oint \text{grad}\ (\Pi + \mathcal{P})\cdot\delta\boldsymbol{r} = -\oint \delta(\Pi + \mathcal{P})$$

because the scalar product of the gradient of a certain function by the space-oriented element of the arc of the curve is nothing but the differential of this function taken along the arc of the curve.

When the function Π is single-valued (the single-valuedness of $\mathcal{P}$ is obvious), the contour integral along the closed contour of the differential is equal to zero, so that

$$\frac{d}{dt}\oint \boldsymbol{V}\cdot\delta\boldsymbol{r} = 0$$

and, consequently,

$$\oint \boldsymbol{V}\cdot\delta\boldsymbol{r} = \text{const}$$

which proves the Kelvin theorem. Recalling that the velocity circulation along the closed contour is equal to the overall intensity of vortex tubes enclosed by this contour, we may conclude on the basis of the Kelvin theorem that with the specified provisos about the barotropicity of motion and the presence of single-valued potential of volumetric forces the intensities of vortex tubes are also preserved

$$\int_{\sigma} (\text{rot}\ \boldsymbol{V})_n d\sigma = \text{const} \tag{1}$$

Now, let us suppose that at a certain initial time instant there is no vorticity at any point in the region filled with fluid (rot $\boldsymbol{V} = 0$), i.e., the fluid volume elements move without rotation, performing only translational and deformational motions; then, the constant on the right side of Equation 1 will be equal to zero and the following equation will be preserved at any time instant

$$\int_{\sigma} (\text{rot}\ \boldsymbol{V})_n d\sigma = 0 \tag{2}$$

The arbitrariness of the choice of the surface σ in Equation 2 will make it possible to conclude that at any time instant and at any point of the region

$$(\text{rot}\ \boldsymbol{V})_n = 0$$

Finally, due to the arbitrariness of the choice of the normal direction we find

$$\operatorname{rot} V = 0 \tag{3}$$

From this the *Lagrange theorem* follows: If at all the points of an ideal fluid moving barotropically under the action of volumetric forces with a single-valued potential the vorticity was equal to zero at a certain initial time instant, the motion will remain irrotational at any subsequent time instant. It also follows from analogous arguments that if the motion was rotational at the beginning, it will remain such afterward.

Let us assume that a solid body immersed in a quiescent fluid starts to move. There was no vorticity in the fluid at rest; consequently, under the conditions of the validity of the Lagrange theorem the vortices could not be formed, and the motion will remain irrotational during the rest of the time. If at a certain time instant, the vorticity was created in the ideal fluid due to the violation of the conditions for the Lagrange theorem, then if these conditions are preserved, the motion will be rotational. In reality one can observe both the formation and disappearance of rotational motions. The main reason for the violation of the validity of the Kelvin and Lagrange theorems is the presence in a real fluid of internal friction (viscosity) which is especially important in a thin boundary layer on the surface of a body in a flow and in the aerodynamic make behind the body.

The internal friction is not the sole reason for the origin of vortices. Thus, in the free atmosphere the factors responsible for vortex formations are deviations from the barotropicity during the motion of air; the density of air layers depends not only on pressure, but also on temperature, which is determined by solar radiation, on the quantity of water vapors, and other reasons.

The model of an ideal fluid admits still another possibility for the violation of the validity of the Kelvin and Lagrange theorems. This is the formation of the surfaces of discontinuity in the continuous fluid which are unstable and roll up into discrete vortices; e.g., the Kármán streets observed in the wake behind a body in flow (see the spectra in the atlas by van Dyke cited in an earlier chapter).

Despite the presence of these factors that upset the existence of irrotational motion, the scheme of the irrotational motion gives a picture that is close to the real one in many practical cases. Thus, let us postulate the absence of vorticity, and direct our attention to the basic properties of such irrotational flows.

2. Velocity Potential and Its Definition from the Given Velocity Field

When the fluid motion is irrotational, then the condition that the curl of the velocity rot V is equal to zero presupposes that the function φ of the coordinates and time exists, which is connected to velocity V by the equality

$$V = \operatorname{grad} \varphi \tag{4}$$

or, in projections onto the axes of rectangular Cartesian coordinates,

$$u = \frac{\partial \varphi}{\partial x} \qquad v = \frac{\partial \varphi}{\partial y} \qquad \omega = \frac{\partial \varphi}{\partial z} \tag{5}$$

and which is called the potential of the velocity field (or the potential of velocities).[2]

[2]The function φ is better known in the West as the "velocity potential".

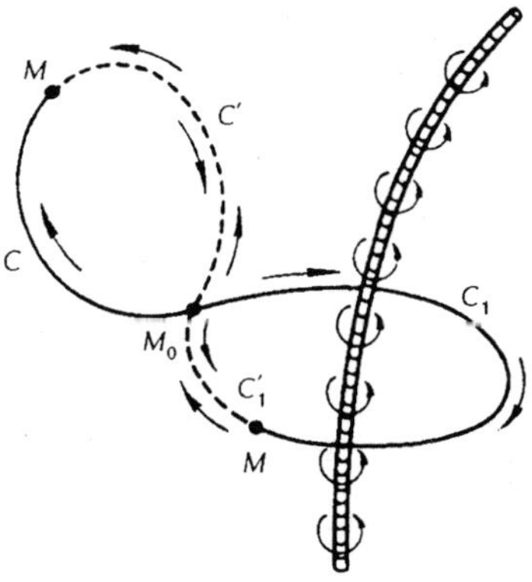

Fig. 50

According to Equations 4 or 5, the velocity potential is determined to within a constant. The equation of the level surface for the potential of velocities is

$$\varphi(x, y, z; t) = \text{const}$$

and in the case of the stationary velocity field the time t as an argument of φ is absent. The potential φ can be found from the given velocity field with the aid of Equation 4. For this it is sufficient to perform the scalar multiplication of both parts of Equation 4 by the directed element $\delta \boldsymbol{r}$ of any curve C originating from the point M_0 (Figure 50) and integrate the resulting expression along the portion M_0M of this curve. Then to determine the velocity potential $\varphi(M)$ at point M, we obtain the equation

$$\int_{\substack{M_0 \\ (C)}}^{M} \boldsymbol{V} \cdot \delta \boldsymbol{r} = \int_{\substack{M_0 \\ (C)}}^{M} \text{grad}\, \varphi \cdot \delta \boldsymbol{r} = \int_{\substack{M_0 \\ (C)}}^{M} \delta\varphi = \varphi(M) - \varphi(M_0) \tag{6}$$

where it follows that

$$\varphi(M) = \varphi(M_0) + \int_{\substack{M_0 \\ (C)}}^{M} \boldsymbol{V} \cdot \delta \boldsymbol{r} = \varphi(M_0) + \int_{\substack{M_0 \\ (C)}}^{M} (u\delta x + v\delta y + w\delta z) \tag{7}$$

If the flow in a simply connected region[3] is irrotational, then having closed (in the figure, by means of a dashed curve) the curve C with the aid of the curve C' so that the point M would coincide with M_0, and noting that in this case the velocity circulation, which takes place along the closed contour $(C + C')$ and which is equal to the sum of intensities of the vortex tubes enclosed by this contour, reduces to zero in the considered velocity field without vortices, we obtain (Equation 7), the following equation

$$\varphi(M) = \varphi(M_0) \quad \text{when} \quad M \to M_0 \tag{8}$$

i.e., the velocity potential represents a single-valued function of coordinates. From this it also follows that the integral in Equation 7 is independent of the shape of integration contour C,

[3]The region is called simple connected of any closed contour in it can be continuously shrunk to a point.

since, by virtue of the fact that the integral along the closed contour consisting of the segment M_0CM (represented in Figure 50 by the solid curve) and of the segment $MC'M_0$ (represented by the dashed curve) is equal to zero, it is possible to conclude that

$$\underset{(C)}{\int_{M_0}^{M}} + \underset{(C')}{\int_{M}^{M_0}} = 0 \qquad \underset{(C)}{\int_{M_0}^{M}} = \underset{(C')}{\int_{M_0}^{M}}$$

The situation is different if there is an isolated vortex tube in irrotational motion (Figure 50). Then, having made integration over the contour C, we again obtain Equation 7; but the result will be quite different if instead of the contour C the contour C_1 is taken, which encloses the vortex tube. The integral occurring on the right side of Equation 7 and calculated from the closed contour ($C_1 + C_1'$) (the closure is shown in the figure by dashes) will be equal, as it follows from the Stokes, theorem (Chapter 2, Section 5), to the intensity of the vortex tube

$$\oint_{(C_1 + C_1')} \boldsymbol{V} \cdot \delta \boldsymbol{r} = \Gamma$$

According to Equation 7, the potential at the point M_0, after traversing the vortex tube, will turn out to be equal to $\varphi(M_0) + \Gamma$. Moving out of the point M_0 and taking, as the contour of integration, the loop-like curve (not shown in the figure) which k times encircles the vortex tube, we return to the point M_0, with the value of the potential differing from the initial one by the value multiple of the intensity Γ

$$\varphi(M_0) \pm k\Gamma$$

Thus, if in a simply connected region of irrotational fluid motion there is a vortex tube, the velocity potential expressed in terms of the velocities by Equation 7 will be a many-valued function of the field points. The value of the velocity potential at the point will be dependent (in this case) on the shape of the curve along which the integration is carried out.

Bearing in mind the subsequent hydrodynamic applications, we employ a different approach to the problem of the many-valuedness of the potential in irrotational motion. From the fluid flow region we isolate a purely irrotational part and consider the stream surfaces that bound the vortex tubes as solid walls. It should be clarified that near the vortex lines there are always closed streamlines located on the stream surfaces that separate the vortex lines from the surrounding fluid.

As there is no friction in an ideal medium, it is possible to replace (mentally, not disturbing at all the occurring motion) the stream surfaces by solid surfaces impermeable to the moving medium.

In such an approach there are no vortex tubes in the moving fluid, but instead the flow region itself becomes multiply connected.[4] In fact, according to the second theorem of Helmholtz, the vortex tubes cannot terminate in the fluid itself: they form either closed tubes (vortex rings) or rest on boundary surfaces (solid walls, free interfaces). In all of these cases, the closed contour that encircles the tube cannot be reduced to a point by continuous conversion while it remains in the region of irrotational flow (Figure 51); this proves that the region of a purely irrotational motion, in the presence of the above-indicated isolation of vortex tubes, is

[4] The domain is said to be k-ply connected when there are $k - 1$ contours not reduced to each other by continuous conversion and not shrunk to a point.

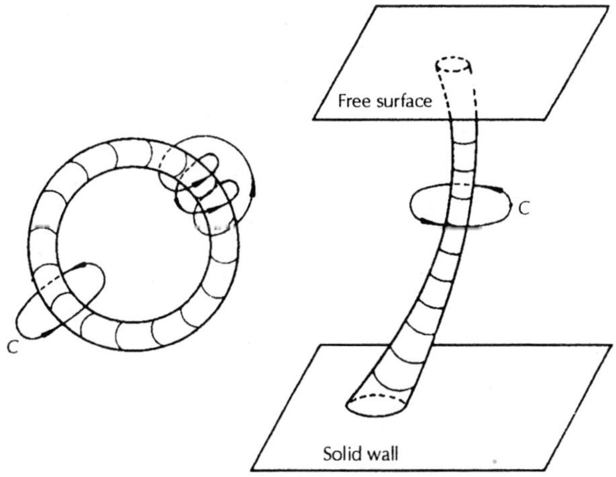

Fig. 51

generally not simply connected. The values of velocity circulations Γ along the contours that singly en circle the tubular surface, which breaks the simple connectedness of the space with the fluid, are called the cyclic constants of multiply connected regions in general cases of any physical fields.

If vortex tubes are found in a multiply connected flow region, the Stokes, theorem should be formulated as follows: velocity circulation along a closed contour constructed arbitrarily in a multiply connected region differs from the sum of the intensities of vortex tubes encircled by the contour by the sum of the integral multiples of the cyclic constants of the multiply connected region.

Constructing additional surfaces in the multiply connected region, it is possible to decrease the degree of its connectedness and, if so desired, to convert the multiply connected region into a simply connected region. Thus, for example (Figure 52), a doubly connected region outside the ring (torus) can be made simply connected by constructing the additional surface σ which closes the hole of the ring. In the presence of the surface σ it becomes impossible to construct the closed contour C which would encircle the ring. If the cyclic constant of the doubly connected region, considered prior to the construction of σ, differs from zero, the value of the velocity potential $\varphi_+(M)$ on one, say frontal, side of the surface σ will differ from the value of $\varphi_-(M)$ on the rear side of the surface σ by the value of the cyclic constant, although the value of the potential was taken at the same point M (Figure 52b). In this case the velocity potential $\varphi(M)$ undergoes, when passing through the surface σ, change by the finite amount $\varphi_+ - \varphi_-$, and the surface σ represents the surface of the potential discontinuity. Considering the surface σ together with the surface S as the boundary of the region, it is possible to consider the potential φ to be continuous throughout the entire region.

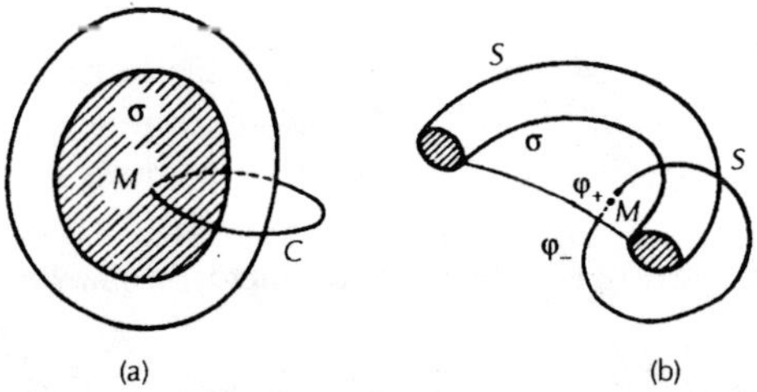

Fig. 52

3. Lagrange-Cauchy Integral: Some General Properties of Irrotational Motion

In the case of irrotational motion of an ideal fluid, the first integral of the equation of motion can be easily indicated. For this purpose we take the Euler equation in the form of the Gromeka-Lamb equation (Equation 8, of Chapter 5)

$$\frac{\partial \boldsymbol{V}}{\partial t} + \text{grad}\left(\frac{V^2}{2} + \mathcal{P} + \Pi\right) + \text{rot}\,\boldsymbol{V} \times \boldsymbol{V} = 0 \tag{9}$$

and, according to Equation 4, we assume in it that

$$\boldsymbol{V} = \text{grad}\,\varphi \qquad \text{rot}\,\boldsymbol{V} = 0 \tag{10}$$

Then, noting that the independent character of the operations of partial, or local, differentiation with respect to time $\partial/\partial t$ and of spatial differentiation expressed by the operation grad allows the change in the order of differentiation,

$$\frac{\partial \boldsymbol{V}}{\partial t} = \frac{\partial}{\partial t}\text{grad}\,\varphi = \text{grad}\left(\frac{\partial \varphi}{\partial t}\right)$$

we have, instead of Equation 9,

$$\text{grad}\left(\frac{\partial \varphi}{\partial t} + \frac{V^2}{2} + \mathcal{P} + \Pi\right) = 0 \tag{11}$$

the integration of which leads to the expression of the first integral of the motion equation

$$\frac{\partial \varphi}{\partial t} + \frac{V^2}{2} + \mathcal{P} + \Pi = f(t) \tag{12}$$

called the Lagrange-Cauchy integral; here, $f(t)$ is an arbitrary time function which is the same for the entire flow region and which is found from the boundary conditions.

In the theory of unsteady motion of an ideal fluid the Lagrange-Cauchy integral plays the same part as the Bernoulli integral in steady motion.[5] In the latter case

$$\frac{\partial \varphi}{\partial t} = 0 \qquad f(t) = \text{const}$$

and Equation 12 is converted into the ordinary Bernoulli integral

$$\frac{V^2}{2} + \mathcal{P} + \Pi = \text{const} \tag{13}$$

In irrotational motion the constant on the right side has the same value at all points of the moving fluid and not only along the streamlines. In irrotational motion the Lagrange-Cauchy integral, just as the Bernoulli integral, serves to express the pressure p in terms of the kinematic elements φ, V, and the coordinates on which Π depends. Expressing V^2 in terms of the projections of grad φ onto the axes of Cartesian coordinates, we have

[5] Readers are reminded that the terms "steady" and "stationary" may be used interchangeably (likewise, "unsteady" and "nonstationary").

$$\frac{\partial \varphi}{\partial t} + \frac{1}{2}\left[\left(\frac{\partial \varphi}{\partial x}\right)^2 + \left(\frac{\partial \varphi}{\partial y}\right)^2 + \left(\frac{\partial \varphi}{\partial z}\right)^2\right] + \mathcal{P} + \Pi = f(t) \tag{14}$$

For the incompressible fluid motion in the absence of volumetric forces ($\mathcal{P} = p/\rho$, $\Pi = 0$) we find

$$\frac{\partial \varphi}{\partial t} + \frac{V^2}{2} + \frac{p}{\rho} = f(t) \tag{15}$$

In the particular case of unsteady motion of a solid body through an incompressible fluid being at rest at an infinite distance from this body, assuming at infinity that

$$p = p_\infty \qquad V = 0 \qquad \frac{\partial \varphi}{\partial t} = 0$$

we obtain

$$f(t) = \text{const} = \frac{p_\infty}{\rho} \tag{16}$$

In the case of irrotational motion of liquid or gas, the three unknown quantities – the velocity projections u, v, w – are expressed as one unknown function, the velocity potential $\varphi(x, y, z, t)$. The adopted assumption about the absence of vorticity, together with the assumption concerning the barotropic character of motion $\rho = \rho(p)$, reduces the solution of the problem about the motion of liquid or gas to the search for two unknown quantities φ and p. This can be done with the aid of Equation 14 and the mass conservation equation

$$\frac{\partial \rho}{\partial t} + \operatorname{div}(\rho \mathbf{V}) = 0$$

Dwelling in detail on the case of an incompressible fluid, we use two basic conditions: incompressibility

$$\operatorname{div} \mathbf{V} = 0$$

and the presence of the velocity potential, Equation 4. Then, to determine the unknown quantity φ, we obtain the Laplace equation

$$\nabla^2 \varphi = 0 \tag{17}$$

specifically, in the Cartesian coordinate system

$$\frac{\partial^2 \varphi}{\partial x^2} + \frac{\partial^2 \varphi}{\partial y^2} + \frac{\partial^2 \varphi}{\partial z^2} = 0$$

This equation must be integrated at the prescribed boundary conditions that depend on the type of the problem set. Thus, in the problem of the motion of a body through a fluid being quiescent at infinity, the following boundary conditions should be fulfilled:[6]

1. The impermeability condition on the body surface, i.e., the requirement that the normal velocity components $V_n = \partial\varphi/\partial n$ of liquid particles on the solid body surface should be equal to normal velocity components V_n^* of the solid body points contacting these particles

[6] The solution to this problem, named after Kirchhoff, is given in Chapter 9.

$$\frac{\partial \varphi}{\partial n} = V_n^* \tag{18}$$

2. The velocities of the liquid particles should be equal to zero at infinity from the body

$$\text{grad } \varphi = 0 \tag{19}$$

in the Cartesian system

$$\frac{\partial \varphi}{\partial x} = \frac{\partial \varphi}{\partial y} = \frac{\partial \varphi}{\partial z} = 0$$

The problem of a fixed solid body being immersed in the flow of fluid which has the prescribed velocity $\boldsymbol{V}_\infty$ at infinity the following boundary conditions:

On the body surface

$$\frac{\partial \varphi}{\partial n} = 0 \tag{20}$$

At infinity

$$\frac{\partial \varphi}{\partial x} = V_\infty \cos\,(\widehat{\boldsymbol{V}_\infty, x}) \qquad \frac{\partial \varphi}{\partial y} = V_\infty \cos\,(\widehat{\boldsymbol{V}_\infty, y}) \qquad \frac{\partial \varphi}{\partial z} = V_\infty \cos\,(\widehat{\boldsymbol{V}_\infty, z}) \tag{21}$$

As is known, the problem of integrating the Laplace equation (17) or of searching for a harmonic function that would satisfy conditions (18), (19), or (20), (21) affords an example of the external problem in the theory of potential. Later, different examples of the solution of such problems are considered for both the flow of fluid around bodies (external problem) and inner flow of fluid through channels (internal problem).

Having determined the velocity potential $\varphi(x, y, z, t)$, we find the pressure p with the aid of the Lagrange-Cauchy integral, (Equation 15); we have

$$p = \rho\left\{f(t) - \frac{\partial \varphi}{\partial t} - \frac{1}{2}\left[\left(\frac{\partial \varphi}{\partial x}\right)^2 + \left(\frac{\partial \varphi}{\partial y}\right)^2 + \left(\frac{\partial \varphi}{\partial z}\right)^2\right]\right\} \tag{22}$$

The irrotational motion of an ideal incompressible fluid has many interesting properties. We shall prove the following theorem of Kelvin: if on the boundary of a certain simply connected region the rotational motion coincides with the irrotational one, the kinetic energy of the irrotational motion in the region considered is smaller than the kinetic energy of the rotational motion.

The Kelvin theorem can be proved based on the fact that velocities in irrotational motion are represented by the velocity potential gradient and that the divergence of the incompressible fluid velocity is equal to zero for both irrotational and rotational motion. In fact, let us agree to use the symbol Δ for denoting the difference between the corresponding elements of the rotational and irrotational motion. The following is the expression for the difference of kinetic energies

$$\Delta T = \frac{\rho}{2}\int_\tau [(V + \Delta V)^2 - V^2]\, d\tau = \rho \int_\tau \boldsymbol{V} \cdot \Delta \boldsymbol{V} d\tau + \frac{\rho}{2}\int_\tau |\Delta V|^2 d\tau \tag{23}$$

The first integral on the right is equal to

$$\int_\tau V \cdot \Delta V d\tau = \int_\tau \operatorname{grad} \varphi \cdot \Delta V d\tau$$

and according to the well-known equation

$$\operatorname{div}(\varphi \boldsymbol{a}) = \varphi \operatorname{div} \boldsymbol{a} + \operatorname{grad} \varphi \cdot \boldsymbol{a}$$

it can be transformed thus:

$$\int_\tau V \cdot \Delta V d\tau = \int_\tau \operatorname{grad} \varphi \cdot \Delta V d\tau = \int_\tau \operatorname{div}(\varphi \Delta V)\, d\tau - \int_\tau \varphi \operatorname{div}(\Delta V)\, d\tau =$$

$$\int_\sigma \varphi (\Delta V)_n d\sigma - \int_\tau \varphi \Delta(\operatorname{div} V)\, d\tau$$

where σ is the surface that confines the simply connected volume, whereas the divergence of the difference between two vector functions is replaced by the difference between the divergences of these functions. By hypothesis, the irrotational and rotational motions on the surface σ coincide, i.e., $\Delta V = 0$ on σ; moreover, according to the incompressibility condition, div $V = 0$. Thus, the first integral in Equation 23 turns out to be equal to zero and it remains that

$$\Delta T = \frac{\rho}{2} \int_\tau |\Delta V|^2 d\tau > 0$$

from which the Kelvin theorem follows. The Kelvin theorem can be interpreted from the variational point of view as the assertion about the minimum quantity of kinetic energy in irrotational flow as compared with any other, rotational, motion, if only these motions coincide on the boundary of the region.

It follows from the Kelvin theorem that if the velocities on the boundary of the simply connected region are equal to zero, then the only possible irrotational motion of an incompressible fluid inside this region is the state of rest. Actually, one can always imagine an arbitrary (rotational) motion, as slow as possible, in which the velocities on the boundary of the region are equal to zero. The kinetic energy of such rotational motion will be as small as possible, whereas the kinetic energy of the corresponding (by the Kelvin theorem) irrotational motion, while being a positive quantity smaller than the other (as small as desired) quantity, should be identically equal to zero in the entire region.

The same result can also be arrived at directly without resorting to the Kelvin theorem. For this purpose, we derive a useful (for what follows) general formula of the kinetic energy of a simply connected volume of an incompressible fluid moving irrotationally. We have

$$T = \frac{\rho}{2} \int_\tau V^2 d\tau = \frac{\rho}{2} \int_\tau \operatorname{grad} \varphi \cdot \operatorname{grad} \varphi\, d\tau$$

Let us again apply the formula for the divergence of the product of a scalar by a vector assuming in it that $\boldsymbol{a} = \operatorname{grad} \varphi$; this will give

$$T = \frac{\rho}{2}\int\limits_{\tau} \text{div}(\varphi \text{ grad } \varphi)\, d\tau - \frac{\rho}{2}\int\limits_{\tau} \varphi \text{ div grad } \varphi\, d\tau =$$

$$= -\frac{\rho}{2}\int\limits_{\sigma} \varphi (\text{grad } \varphi)_n\, d\sigma - \frac{\rho}{2}\int\limits_{\tau} \varphi \nabla^2 \varphi\, d\tau$$

In the surface integral, which is obtained from the volumetric one by the Gauss-Ostrogradskiy formula, the quantity $\boldsymbol{n}$ is understood to be the ort of the internal normal directed inside the liquid volume. (Due to this the minus sign is ahead of the integral.[7]) Noting that according to Equation 17 the second integral is equal to zero, we finally have

$$T = -\frac{\rho}{2}\int\limits_{\sigma} \varphi \frac{\partial \varphi}{\partial n}\, d\sigma \tag{24}$$

If on the surface σ, which confines the simply connected volume of fluid, the velocity is equal to zero, then

$$V_n = \frac{\partial \varphi}{\partial n} = 0$$

from which, according to Equation 24, it also follows that $T = 0$. Thus, we again arrive at the same result that followed from the Kelvin theorem.

4. A Plane Irrotational Motion of an Incompressible Fluid: Application of the Functions of a Complex Variable

We begin the study of irrotational motions from the simplest class of such kind of motions – a plane steady motion of an incompressible fluid. The definition of the plane motion in hydrodynamics in no way differs from the corresponding definition for a solid body. In plane motion all the liquid particles undergo displacements parallel to a certain plane which will be taken to be the $x0y$ plane, with the motions being identical in all the parallel planes. Therefore, we consider only the motion in the plane $x0y$. Each line in such a plane motion, constructed in the plane $x0y$, is actually the guide of the cylindrical surface with generatrices perpendicular to the plane $x0y$. The contour of the body immersed in the flow will be represented by some line in the plane, although there is actually an infinite cylindrical body in the flow. All the values of the fluid flow rates and of the forces applied to the bodies immersed in flows, etc., will be related to the unit length in the direction of the perpendicular to the $x0y$ plane, i.e., in the direction of the $0z$ axis, which is later omitted from the figures.

As mentioned in the previous section, in the case considered the problem is reduced to solving the problem of integration, at these or other boundary conditions, of the Laplace equation. For the plane case it has the form

$$\nabla^2 \varphi = \frac{\partial^2 \varphi}{\partial x^2} + \frac{\partial^2 \varphi}{\partial y^2} = 0 \tag{25}$$

---

[7] In other words, $\boldsymbol{n}$ is the unit vector for the direction of the inward-facing normal at the volume surface.

In the case of plane motion this problem can be successfully solved with the aid of the method of complex variable, the application of which is basic to the hydrodynamics of the plane irrotational motion of an incompressible fluid.

From the continuity (incompressibility) equation

$$\operatorname{div} \boldsymbol{V} = \frac{\partial u}{\partial x} + \frac{\partial v}{\partial y} = 0$$

it follows that it is always possible to find the function $\psi(x, y)$ which identically satisfies Equation 25, and which is connected to the velocity projections u and v by the equations

$$u = \frac{\partial \psi}{\partial y} \qquad v = -\frac{\partial \psi}{\partial x} \tag{26}$$

The function $\psi(x, y)$ has a simple hydrodynamic meaning. In fact, write the differential equation of streamlines (Equation 7, Chapter 2) which has the following form in the case of plane motion

$$\frac{dx}{u} = \frac{dy}{v}$$

and substitute into it the values of velocity projections by Equation 26; this will yield

$$\frac{dx}{\partial \psi / \partial y} = \frac{dy}{-\partial \psi / \partial x}$$

or

$$\frac{\partial \psi}{\partial x} dx + \frac{\partial \psi}{\partial y} dy = d\psi = 0$$

It follows from the last equation that the function ψ preserves constant values along streamlines; in other words, the family of the level lines of the function

$$\psi(x, y) = C \tag{27}$$

that correspond to different values of C, represents the set of streamlines. In view of this, the function $\psi(x, y)$ is called the stream function. To elucidate the kinematic sense of the stream function, consider two infinitely close streamlines in the plane of the flow: one which passes through the point M_0 with the corresponding value of the stream function ψ_0 and the other which passes through the point M_1 with the value of the stream function $\psi = \psi_0 + d\psi$ (Figure 53).

A portion of the fluid between the cylindrical surfaces, which are perpendicular to the plane of the figure and which have the streamlines as guides, forms an elementary stream tube, which will be bounded by the plane of the figure and by the plane that is parallel to it and is located at a distance of unit length from the plane of the figure. In the case of plane motion, the above-described image of the stream tube is represented by the region confined between two streamlines in the plane of fluid motion. This explanation should be kept in mind for what follows.

Noting (Figure 53) that the elementary volumetric flow rate of fluid dQ through any section of the stream tube is independent of the shape of this section, we select it in the form of the set of two segments, $M_0 M_0' = dy$ and $M_0' M_1 = -dx$, parallel to the coordinate axes. Then, as it follows directly from the figure, the expression for dQ will acquire the form

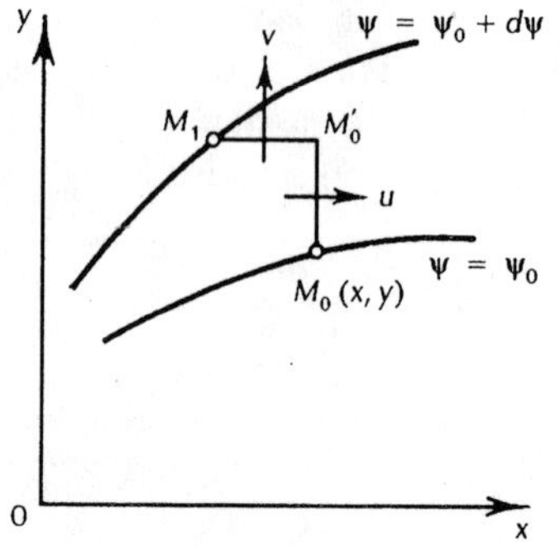

Fig. 53

$$dQ = udy - vdx = \frac{\partial \psi}{\partial y} dy + \frac{\partial \psi}{\partial x} dx = d\psi$$

from which follows the expression for the flow rate through the tube having the finite dimensions $(M_0 M)$

$$Q = \int_{M_0}^{M} dQ = \psi(M) - \psi(M_0) \tag{28}$$

Let us agree for the purpose of further discussion that any one streamline is arbitrarily regarded as the null one, assuming that along this line

$$\psi(x, y) = 0$$

This can be always done because, according to the system of Equations 26, the stream function is determined accurate to the additive constant. With this kind of condition taken into account, the value of the constant in Equation 27 on some streamline will be equal to the volumetric fluid flow rate through the section of the stream tube formed by this streamline and by the null line chosen arbitrarily.

Let us compare expressions (5) of the velocity projections in terms of the velocity potential, which in the case of plane motion are reduced to the system of two equations

$$u = \frac{\partial \varphi}{\partial x} \qquad v = \frac{\partial \varphi}{\partial y} \tag{29}$$

and expressions (26) of the same projections in terms of the stream function ψ; we have the following system of relations:

$$\frac{\partial \varphi}{\partial x} = \frac{\partial \psi}{\partial y} \qquad \frac{\partial \varphi}{\partial y} = -\frac{\partial \psi}{\partial x} \tag{30}$$

These equations express the well-known Cauchy-Riemann conditions in compliance with which the complex quantity

$$\chi = \varphi + i\psi = \varphi(x, y) + i\psi(x, y) = \chi(z) \tag{31}$$

will be not merely the function of two variables (coordinates x, y), but rather the function of one complex variable $z = x + iy$.[8] In fact, if the quantity χ is only the function of the position

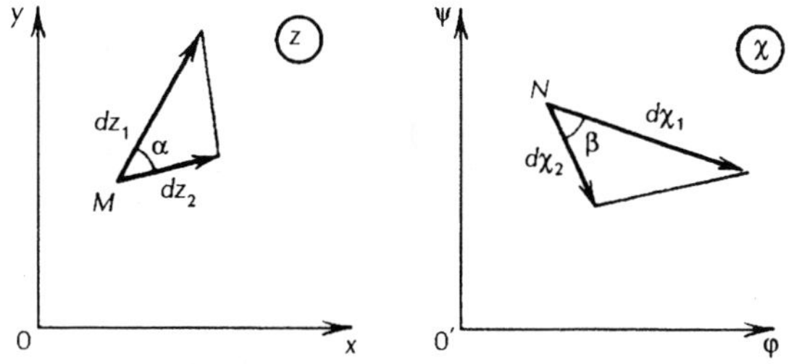

Fig. 54

of the point M with the coordinate z, then the derivative of it at this point must be only the function of the position of the point, i.e., of the coordinate z. It must not depend on the direction of differentiation in the plane. In other words, the derivatives $d\chi/dz$ and the derivatives with respect to the directions of the real and imaginary axes must be equal among themselves

$$\frac{d\chi}{dz} = \frac{\partial\chi}{\partial x} = \frac{\partial\chi}{\partial(iy)} \tag{32}$$

Noting that

$$\frac{\partial\chi}{\partial x} = \frac{\partial(\varphi + i\psi)}{\partial x} = \frac{\partial\varphi}{\partial x} + i\frac{\partial\psi}{\partial x}$$

$$\frac{\partial\chi}{\partial(iy)} = -i\frac{\partial(\varphi + i\psi)}{\partial y} = \frac{\partial\psi}{\partial y} - i\frac{\partial\varphi}{\partial y}$$

and taking into account the Cauchy-Riemann conditions (30), we convince ourselves of the validity of relations (32).

It is useful to recall for what follows that the function of the complex variable $\chi(z)$ uniquely maps the points of the plane of the complex variable $z = x + iy$ onto the plane of the complex variable $\chi = \varphi + i\psi$. In this case the closed curves and the parts of the plane z bounded by these curves, map into the corresponding figures or parts of the plane χ. Such mapping is called conformal. Many examples of such conformal mappings are given later.

The basic property of the conformal transformation (mapping), which will be dealt with in what follows, consists of the similarity of infinitesimal figures: the given and the mapped. Figure 54 shows an example of an infinitesimal triangle. At the point M of the plane of the complex variable z let us prescribe two infinitesimal increments, dz_1 and dz_2, which form the finite angle α. Correspondingly, at the point N of the plane of the complex variable χ appears the increments $d\chi_1$, $d\chi_2$ which form a certain angle β.

The transition from the plane z into the plane χ takes place according to Equation 31

$$\chi = \chi(z)$$

From this it follows that the increments $d\chi$ are connected to the increments dz by the relations (it is assumed that the derivative $\chi'(z)$ exists and does not convert either into zero or infinity at the point M)

[8] The coordinate z, which corresponds to the axis $0z$, perpendicular to the plane $0xy$, is not encountered in plane motion; this allows the use of the letter z to denote the complex quantity $x + iy$.

$$d\chi_1 = \chi'(z)\,dz_1 \qquad d\chi_2 = \chi'(z)\,dz_2$$

so that

$$\frac{d\chi_1}{d\chi_2} = \frac{dz_1}{dz_2}$$

Let us assume that

$$dz_1 = ds_1 e^{i\alpha_1} \qquad dz_2 = ds_2 e^{i\alpha_2}$$

$$d\chi_1 = d\sigma_1 e^{i\beta_1} \qquad d\chi_2 = d\sigma_2 e^{i\beta_2}$$

Then, it follows from the previous equation that

$$\frac{d\sigma_1}{d\sigma_2} e^{i(\beta_1 - \beta_2)} = \frac{ds_1}{ds_2} e^{i(\alpha_1 - \alpha_2)}$$

Comparing the moduli and arguments, we obtain

$$\frac{d\sigma_1}{d\sigma_2} = \frac{ds_1}{ds_2} \qquad \beta_1 - \beta_2 = \alpha_1 - \alpha_2 \quad \text{i.e.} \quad \beta = \alpha$$

which proves the similarity of the infinitesimal figures (triangles) depicted in Figure 54.

It is important for what follows to emphasize that in the case of the conformal mapping of the figures the corresponding angles are preserved, except for probably one or more points which are called singular.

Separating the real, Re, and imaginary, Im, parts in the arbitrary function of the complex variable $\chi(z)$, we obtain the velocity potential $\varphi(x, y)$ and the stream function $\psi(x, y)$ of a certain plane irrotational motion

$$\varphi(x, y) = \operatorname{Re}\chi(z) \qquad \psi(x, y) = \operatorname{Im}\chi(z) \tag{33}$$

Equating the function $\varphi(x, y)$ to different constants

$$\varphi(x, y) = C \tag{34}$$

we obtain the family of isopotential lines; analogously, the set of equations

$$\psi(x, y) = C' \tag{35}$$

will represent, according to Equation 27, the family of streamlines.

It can be easily seen that the isopotential lines and the streamlines are mutually orthogonal at any point of the plane of the flow. For this it is sufficient to show that the vectors – the gradients of the functions φ and ψ – are mutually perpendicular. We have

$$\operatorname{grad}\varphi \cdot \operatorname{grad}\psi = \frac{\partial\varphi}{\partial x}\frac{\partial\psi}{\partial x} + \frac{\partial\varphi}{\partial y}\frac{\partial\psi}{\partial y} = \frac{\partial\varphi}{\partial x}\left(-\frac{\partial\varphi}{\partial y}\right) + \frac{\partial\varphi}{\partial y}\frac{\partial\varphi}{\partial x} \equiv 0$$

and it is this that proves the mutual orthogonality of the isopotential lines and the streamlines.

If, instead of the function $\chi(z)$ we consider the function $i\chi(z)$, in the new motion the velocity potential will change places with the stream function and the isopotential lines will change places with the streamlines; on must often use this technique when constructing flows past bodies. From this it follows that the stream function $\psi(x, y)$ always plays a conjugate part with the function $\varphi(x, y)$ which is the velocity potential; each of these functions may be both the stream function and the velocity potential in two conjugate irrotational plane motions of an ideal fluid. The function $\chi(z) = \varphi + i\psi$ is called the complex potential.

Let us show the way in which the velocity vector V or its projections u and v can be determined when the complex potential $\chi(z)$ is known. Corresponding to each complex number in the plane there is a vector with projections that are respectively equal to the real and imaginary parts of this complex number. When studying the plane motion, let us agree that the letter V will denote the complex velocity $V = u + iv$; the conventional designation of the complex number modulus will be preserved for the velocity magnitude:

$$|V| = \sqrt{u^2 + v^2}$$

Along with the complex velocity V we introduce into consideration the conjugate velocity $\bar{V}$ equal to

$$\bar{V} = u - iv \tag{36}$$

If θ is the angle between $\bar{V}$ and the axis $0x$, then

$$V = u + iv = |V|(\cos\theta + i\sin\theta) = |V|e^{i\theta}$$

$$\bar{V} = u - iv = |V|(\cos\theta - i\sin\theta) = |V|e^{-i\theta} \tag{37}$$

It can be noted that the conjugate velocity $\bar{V}$ is the mirror image of V relative to the axis $0u$.

The plane $0xy$ of the values of the complex coordinate z is called the physical plane or the plane of the flow. The set of the values of the complex velocity V forms the plane of the hodograph of velocity or simply the hodograph plane $u0v$; this plane will incorporate the velocity hodographs, i.e., the geometric loci of the ends of the fluid particle velocity vectors drawn from the origin 0.

Let us consider the derivative $d\chi/dz$ of the complex potential with respect to the complex argument. According to the earlier noted property of the functions of the complex variable

$$\frac{d\chi}{dz} = \frac{\partial\chi}{\partial x} = \frac{\partial(\varphi + i\psi)}{\partial x} = \frac{\partial\varphi}{\partial x} + i\frac{\partial\psi}{\partial x}$$

from this

$$\frac{d\chi}{dz} = u - iv = \bar{V} = |V|e^{-i\theta} \tag{38}$$

follows, i.e., the derivative of the complex potential with respect to the complex coordinate is equal to the conjugate velocity. The velocity projections u and v will be defined as

$$u = \mathrm{Re}\left(\frac{d\chi}{dz}\right) \qquad v = -\mathrm{Im}\left(\frac{d\chi}{dz}\right) \tag{39}$$

It is also useful to consider the contour integral of the conjugate velocity $\bar{V}$ along the closed contour C in the flow plane. This integral is equal to

$$\oint \bar{V} dz = \oint \frac{d\chi}{dz} dz = \oint d\chi = \oint (d\varphi + i d\psi) \tag{40}$$

Computing the real and imaginary parts of this integral, we find that the real part determines the circulation of the velocity Γ along the closed contour and the imaginary part, the volumetric fluid flow rate Q through the closed contour. In fact,

$$\mathrm{Re} \oint \bar{V} dz = \oint (u dx + v dy) = \oint d\varphi = \Gamma$$

$$\mathrm{Im} \oint \bar{V} dz = \oint (u dy - v dx) = \oint d\psi = Q \tag{41}$$

5. Complex Potentials of Simple Flows

Making use of the technique (33) that separates the real and imaginary parts in the expression for the complex potential, we may determine the velocity potentials and the stream functions. Using Equation 39 we can also determine the velocity distribution for several simple plane flows of an ideal incompressible fluid.

1. A uniform flow with the velocity vector V_∞ (u_∞, v_∞) directed to the axis $0x$ at the angle θ_∞ (Figure 55):

$$\chi(z) = \bar{V}_\infty z \qquad \bar{V}_\infty = u_\infty - i v_\infty = |V_\infty| e^{-i\theta_\infty} = |V_\infty| (\cos\theta_\infty - i \sin\theta_\infty)$$

$$\varphi = u_\infty x + v_\infty y \qquad \psi = -v_\infty x + u_\infty y \qquad V = V_\infty = u_\infty + i v_\infty$$

2. The source (sink) at the coordinate origin (Figure 56) with the volumetric fluid flow rate (discharge) Q, which is the real quantity that is positive in the case of the source and negative in the case of the sink:

$$\chi(z) = \frac{Q}{2\pi} \ln z \qquad \varphi = \frac{Q}{2\pi} \ln r \qquad \psi = \frac{Q}{2\pi} \varepsilon \quad (r, \varepsilon \text{ are polar coordinates})$$

$$\bar{V} = \frac{Q}{2\pi z} \qquad |V| = \frac{|Q|}{2\pi r}$$

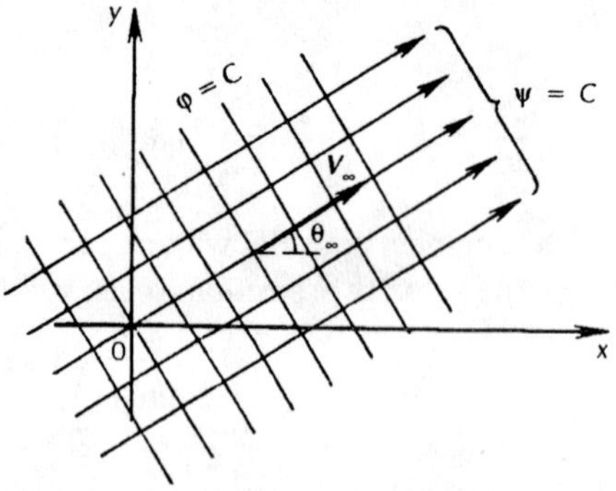

Fig. 55

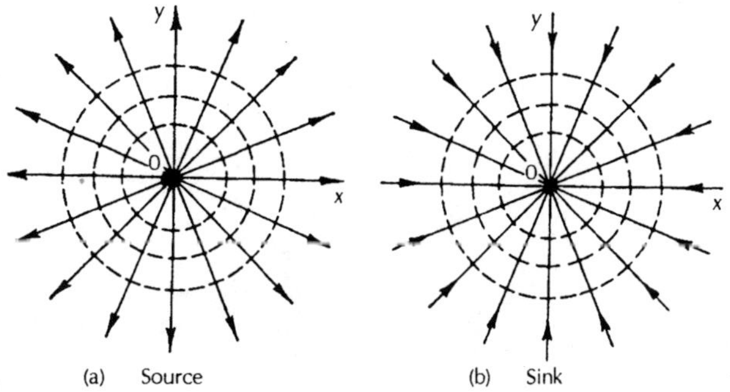

Fig. 56

3. A vortex, isolated at the coordinate origin (Figure 57), with the circulation Γ (real quantity):

$$\chi(z) = \frac{\Gamma}{2\pi i}\ln z \qquad \varphi = \frac{\Gamma}{2\pi}\varepsilon \qquad \psi = -\frac{\Gamma}{2\pi}\ln r \qquad \bar{V} = \frac{\Gamma}{2\pi i z} \qquad |V| = \frac{|\Gamma|}{2\pi r}$$

4. A dipole at the coordinate origin with the moment m (Figure 58) (real quantity) and the axis directed along the axis $0x$:

$$\chi(z) = \frac{m}{2\pi z} \qquad \varphi = \frac{mx}{2\pi(x^2+y^2)} = \frac{mx}{2\pi r^2} \qquad \psi = -\frac{my}{2\pi(x^2+y^2)} = -\frac{my}{2\pi r^2}$$

$$\bar{V} = \frac{m}{2\pi z^2} \qquad |V| = \frac{|m|}{2\pi r^2}$$

The technique of the superposition of flows, justifiable because of the linearity of the equations for φ and ψ, makes it possible to obtain new flows. Thus, for example, the flow of the dipole (item 4) can be obtained by summing up the flows of the source and sink of identical powers located along the axis $0x$ symmetrically relative to the coordinate origin 0 at the points with the abscissas $\pm h$ in the case of the limiting transition: $h \to 0$, $|Q| \to \infty$; the quantity $Q \cdot 2h$ tends to the finite value of m, which is the moment of the dipole

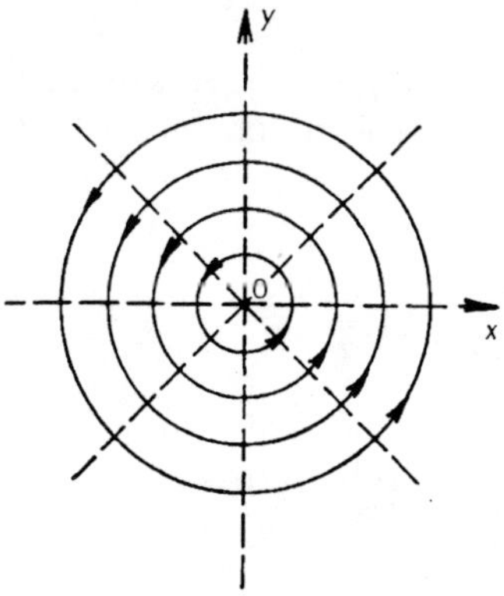

Fig. 57

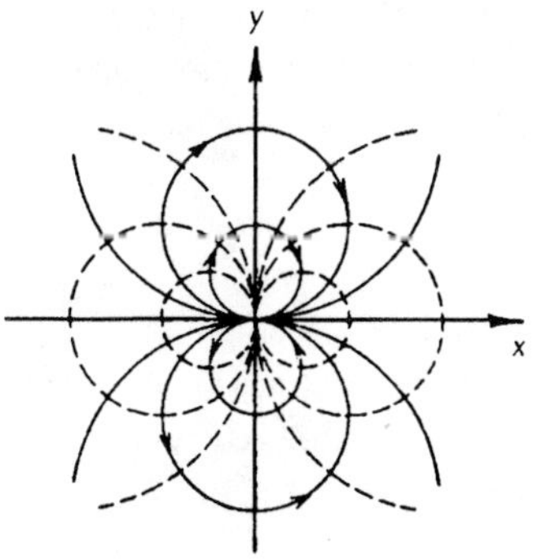

Fig. 58

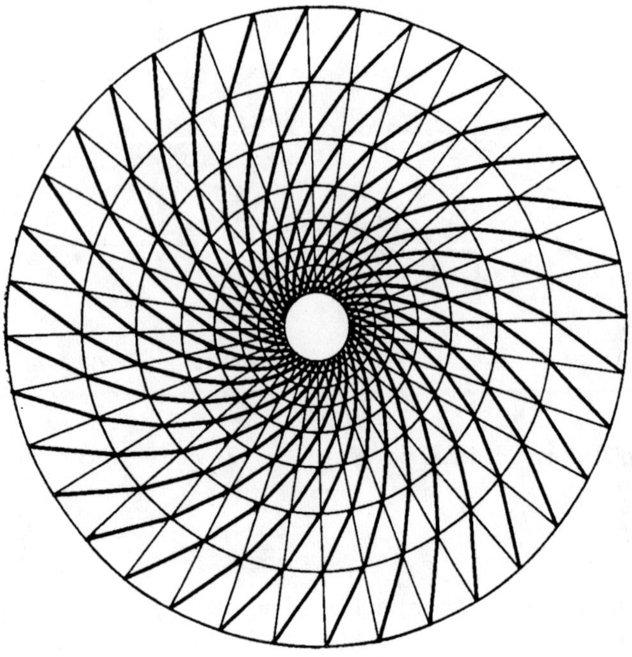

Fig. 59

$$\chi(z) = \lim_{\substack{h \to 0,\, |Q| \to \infty,\\ Q \cdot 2h \to m}} \frac{Q \cdot 2h}{2\pi} \frac{\ln(z+h) - \ln(z-h)}{2h} = \frac{m}{2\pi}\frac{d\ln z}{dz} = \frac{m}{2\pi z}$$

The value $m > 0$ corresponds to the position of the sink on the positive side of the axis $0x$, $m < 0$, to the opposite case.

Such superposition makes it possible to obtain the flows in Figures 57 to 59.

5. The vortex source (sink) (Figure 59) can be obtained by summing up the complex potentials of the vortex and of the source (sink):

$$\chi = \frac{Q - \Gamma i}{2\pi} \ln z$$

6. The noncirculation flow past a circular cylinder of radius a along the axis $0x$ (Figure 60) can be obtained by superimposing the uniform flow (item 1) onto the flow of the dipole (item 4) at $m = 2\pi a^2 V_\infty$ (V_∞ is a real positive quantity):

$$\chi = V_\infty z + \frac{2\pi a^2 V_\infty}{2\pi} \frac{1}{z} = V_\infty \left(z + \frac{a^2}{z}\right) \tag{42}$$

The null streamline ($\psi = 0$) is split in this case into two curves: the circle ($x^2 + y^2 = a^2$) of radius a and the axis $0x$ ($y = 0$). When $V_\infty = |V_\infty| e^{i\theta_\infty}$, the same flow is determined by the complex potential

$$\chi(z) = V_\infty z + V_\infty \frac{a^2}{z} \tag{43}$$

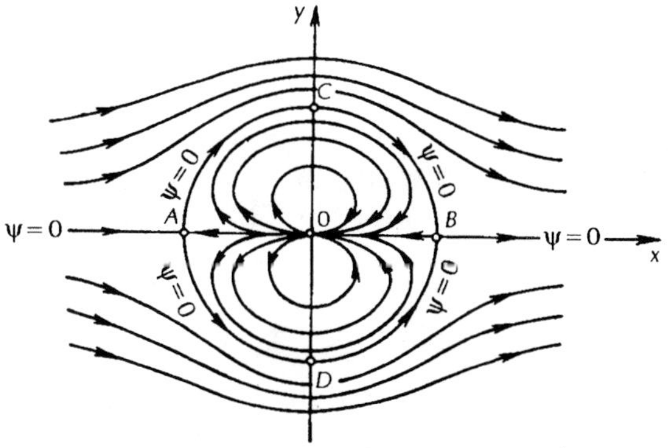

Fig. 60

the expression of which is easily derived from Equation 42 by applying it in the plane of the complex variable z' with the axis $0x'$ directed along the velocity vector $\boldsymbol{V}_\infty$, so that via Equation 42

$$\chi(z') = |V_\infty|\left(z' + \frac{a^2}{z'}\right)$$

and returning to the plane z by replacing z' by $ze^{-i\theta_\infty}$, we obtain

$$\chi(z) = V_\infty z = V_\infty \frac{a^2}{z} \tag{43}$$

The velocity $\bar{V}$ at an arbitrary point of the flow (42) is equal to

$$\bar{V} = \frac{d\chi}{dz} = V_\infty\left(1 - \frac{a^2}{z^2}\right) \tag{44}$$

The velocity distribution along the contour of the circle $z = ae^{i\varepsilon}$ is defined by the formula

$$|V| = 2V_\infty|\sin\varepsilon| \tag{45}$$

where ε is the polar angle between the circle radius and the axis $0x$. By taking, the local angle of attack θ, instead of the polar angle ε (which brings the polar angle to $\pi/2$ and which is the angle between the velocity vector at the point on the circle contour and the direction of the flow impinging against the body), Equation 45 will acquire the form

$$|V| = 2V_\infty \cos\theta$$

which expresses the proportionality of the velocity modulus at the points of the circle contour to the cosine of the local angle of attack at these points. This trend is valid for the flow around an ellipse and for a number of other flows (see Section 7). At the points A ($\varepsilon = \pi$) and B ($\varepsilon = 0$) the velocities are equal to zero; these points are called the stagnation points of the flow. In the case of the flow direction, as shown in Figure 60, the point A is called the front stagnation point and the point B is called the rear stagnation point.

The velocity on the cylinder surface acquires its maximum value at $\varepsilon = \pm\pi/2$ at the points C and D of the frontal area of the cylinder; this maximum velocity is equal to

$$|V_{max}| = 2V_\infty$$

i.e., to the doubled free stream velocity or to the doubled velocity at infinity.

The pressure distribution in the flow and, in particular, along the cylinder contour, can be represented in the form of the pressure coefficient $c_p = (p - p_\infty)/(1/2\rho V_\infty^2)$. Applying the Bernoulli theorem in the form of Equation 26 (Chapter 5) and defining the constant as $p_\infty + 1/2\rho V_\infty^2$, we find in the case considered that

$$c_p = \frac{p - p_\infty}{1/2\rho V_\infty^2} = 1 - \frac{|V|^2}{V_\infty^2} = 1 - 4\sin^2\varepsilon \tag{46}$$

The resulting distribution of pressure along the circle contour does not give the net force, as it follows directly from the symmetry of flow with respect to the axes 0x and 0y. This is a case particular to d'Alembert's paradox which is established in Chapter 9 for a body of any shape during its translational, rectilinear, and uniform motion through an ideal incompressible fluid being at rest at a distance from this body.

7. Circulational flow around a circular cylinder can be obtained by superposing a vortex with the circulation Γ (item 3) onto the circulation-free flow around a circular cylinder (item 6). According to Equation 42 and item 3, the complex potential of the composite motion is equal to

$$\chi(z) = |V_\infty|\left(z + \frac{a^2}{z}\right) + \frac{\Gamma}{2\pi i}\ln z \tag{47}$$

which, when $\Gamma > 0$, corresponds to the counterclockwise circulation motion. Dwelling on this case, we determine the conjugate velocity

$$\bar{V} = \frac{d\chi}{dz} = |V_\infty|\left(1 - \frac{a^2}{z^2}\right) + \frac{\Gamma}{2\pi i z} \tag{48}$$

and find the position of the stagnation points by solving the equation

$$|V_\infty|\left(1 - \frac{a^2}{z^2}\right) + \frac{\Gamma}{2\pi i z} = 0$$

or the quadratic equation

$$z^2 + \frac{\Gamma}{2\pi i|V_\infty|}z - a^2 = 0$$

Its roots are

$$z = \frac{\Gamma i}{4\pi|V_\infty|} \pm \sqrt{a^2 - \frac{\Gamma^2}{16\pi^2|V_\infty|^2}} \tag{49}$$

Depending on the magnitude of circulation, three types of the flow around bodies are possible:

1. The circulation is great: $\Gamma > 4\pi a|V_\infty|$. In this case, expression (49) will have the negative quantity under the radical sign and it is possible to write

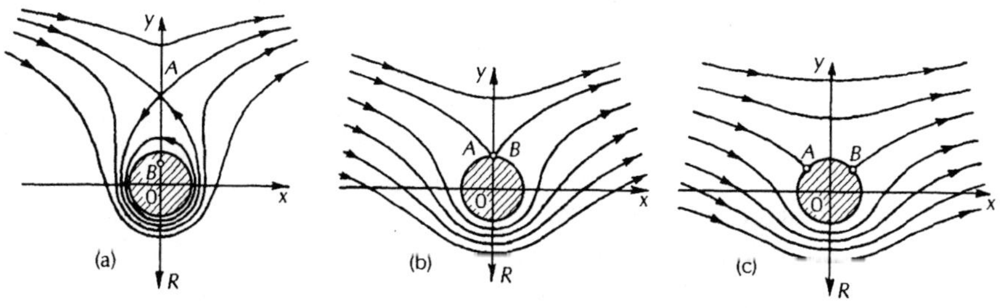

Fig. 61

$$z = \left(\frac{\Gamma}{4\pi|V_\infty|} \pm \sqrt{\frac{\Gamma^2}{16\pi^2|V_\infty|^2} - a^2}\right)i$$

Both roots of the quadratic equation are imaginary, with the modulus of one being greater than the cylinder radius; the other is smaller. In fact, the root

$$z_1 = \left(\frac{\Gamma}{4\pi|V_\infty|} + \sqrt{\frac{\Gamma^2}{16\pi^2|V_\infty|^2} - a^2}\right)i$$

has the modulus ($\Gamma > 0$)

$$|z_1| = \frac{\Gamma}{4\pi|V_\infty|} + \sqrt{\frac{\Gamma^2}{16\pi^2|V_\infty|^2} - a^2} > \frac{\Gamma}{4\pi|V_\infty|} > a$$

the second root

$$z_2 = \left(\frac{\Gamma}{4\pi|V_\infty|} - \sqrt{\frac{\Gamma^2}{16\pi^2|V_\infty|^2} - a^2}\right)i$$

has the modulus

$$|z_2| = \frac{\Gamma}{4\pi|V_\infty|} - \sqrt{\frac{\Gamma^2}{16\pi^2|V_\infty|^2} - a^2} = \frac{a^2}{\dfrac{\Gamma}{4\pi|V_\infty|} + \sqrt{\dfrac{\Gamma^2}{16\pi^2|V_\infty|^2} - a^2}}$$

Let us replace in the denominator of the latter expression $\Gamma/(4\pi|V_\infty|)$ by the smaller quantity a; this yields

$$|z_2| < \frac{a^2}{a} = a$$

The first root gives the critical point A (Figure 61a), which lies outside the circle on the positive side of the imaginary axis; the second root gives the critical point B on the same axis but inside of the circle.

2. The intermediate case: $\Gamma = 4\pi a|V_\infty|$. The roots z_1 and z_2 are equal; the stagnation points coincide (Figure 61b) and lie on the imaginary axis at the point

$$z_1 = z_2 = ai$$

3. The circulation is small: $\Gamma < 4\pi a|V_\infty|$. In this case the roots (49) are complex

$$z = \pm\sqrt{a^2 - \frac{\Gamma^2}{16\pi^2|V_\infty|^2}} + \frac{\Gamma}{4\pi|V_\infty|}i$$

they have the common ordinate $\Gamma/(4\pi a|V_\infty|)$ and differ only by the signs of the abscissas, the absolute values of which are smaller than a. The modulus of each of the roots is equal to a, i.e., they are located on the circle of radius a. The position of the stagnation points is shown in Figure 61c. When Γ decreases to zero, the stagnation points shift and strive to occupy the position at the intersection of the circle with the axis $0x$, as should be the case when $\Gamma = 0$.

In the case of circulation flow around a circular cylinder the symmetry about the axis $0y$ persists, but the symmetry about the axis $0x$ is invalidated. In view of this, the resultant vector of the forces of fluid pressure on the cylinder surface will be different from zero and will be directed along the axis $0y$. It should be noted that in the layers of the fluid under the cylinder the velocities of the noncirculational flow around the cylinder and of the purely circulational flow around the cylinder add up; those above the cylinder are subtracted. Under the cylinder the velocities are greater and the pressures are smaller, according to the Bernoulli equation. Conversely, the velocities above the cylinder are smaller and the pressures are greater. This leads to the situation that in the flow considered the resultant vector $\boldsymbol{R}$ of the forces of fluid pressure on the cylinder will be directed along the axis $0y$ in the negative direction (downward).

In the case of clockwise circulational flow ($\Gamma < 0$) the pattern of flow, with the same position of the coordinate axes, will change to that turned over about the axis $0x$ by 180° and the resultant vector will be directed along the axis $0y$ into the positive direction (upward). It is possible to give a simple rule for determining the direction of the resultant vector of the forces of fluid pressure on the cylinder surface: by placing the beginning of the velocity vector $\boldsymbol{V}_\infty$ at the cylinder center 0, we shall rotate it through 90° in a sense opposite to the direction of circulational motion; this will give the direction of the resultant vector $\boldsymbol{R}$.

The quantity R_y remains to calculated. We have

$$\boldsymbol{R} = -\oint p\boldsymbol{n}\,ds \qquad R_y = -\oint pn_y ds = -a\oint_0^{2\pi} p\sin\varepsilon\, d\varepsilon \tag{50}$$

By Equation 48, on the circle contour

$$\bar{V} = |V_\infty|(1 - e^{-2i\varepsilon}) + \frac{\Gamma}{2\pi i a}e^{-i\varepsilon} =$$

$$= 2|V_\infty|ie^{-i\varepsilon}\frac{e^{i\varepsilon} - e^{-i\varepsilon}}{2i} - \frac{\Gamma}{2\pi a}ie^{-i\varepsilon} = ie^{-i\varepsilon}\left(2|V_\infty|\sin\varepsilon - \frac{\Gamma}{2\pi a}\right)$$

so that

$$|V| = |V_\infty|\left|2\sin\varepsilon - \frac{\Gamma}{2\pi|V_\infty|a}\right|$$

$$c_p = 1 - \left(2\sin\varepsilon - \frac{\Gamma}{2\pi|V_\infty|a}\right)^2 \tag{51}$$

In contrast to the noncirculation flow around a cylinder, in the case of circulation flow considered above the pressure coefficient depends on the parameter $\Gamma/(|V_\infty|a)$ which contains an arbitrary value of the circulation imposed. As follows from Equation 51, for the circulation flows around bodies to be similar it is necessary to impose the requirement that the parameter $\Gamma/(|V_\infty|a)$ (the parameter of dimensionless circulation) be identical in the flows compared.

Substituting the value of p into expression (50) for R_y according to Bernoulli's equation and taking into account Equation 51, we obtain

$$R_y = \frac{\rho a}{2}\oint_0^{2\pi}\left(2|V_\infty|\sin\varepsilon - \frac{\Gamma}{2\pi a}\right)^2 \sin\varepsilon\, d\varepsilon = -\rho|V_\infty|\Gamma \tag{52}$$

Making analogous computations it could be shown that $R_x = 0$, but this is obvious from the symmetry considerations.

Just as in the case of noncirculation flow around a cylinder there is no resistance, but in the case of circulation flow the transverse force is equal to the product of the fluid density by the free stream velocity and by the circulation. The resulting expression (52) for R_y is a particular case of the general theorem of Zhukovskiy relating to any contour immersed in a flow; the proof of this theorem is given later.

In the case of rotational motion of bodies in a real fluid which has internal friction (viscosity), it is possible to observe the appearance of circulational motions qualitatively resembling those just studied. The effect of the transverse force formation (the Magnus effect) helps explain many interesting phenomena, such as the origin of the aerodynamic moment in the action of air flow onto a spinning shell which, in conjuction with the gyroscopic moment, causes the rotation of the shell in the plane of shooting and the approach of its axis to the tangent to the trajectory. This kind of flow problem also includes an historical attempt (which even now attracts the attention of shipbuilders) to create a ship propulsive device in the form of vertical rotating cylindrical towers (the so-called Fletner rotors) sited on the ship deck and creating the motive force perpendicular to the wind direction. An analogous effect is observed during the flight of spinning soccer and tennis balls. This or other intensity of spinning and the direction of spinning create ball trajectories quite unexpected by the opponent.

It should be noted that the complex potential (47) is a particular case of a more general potential:

$$\chi(z) = \bar{V}_\infty z + V_\infty\frac{a^2}{z} + \frac{\Gamma}{2\pi i}\ln z \tag{53}$$

corresponding, as follows from Equation 43, to the presence of a certain angle θ_∞ between the vector V_∞ and the axis $0x$; the complex potential (53) may form the basis for the method of searching for complex potentials for plane flows past bodies of other shapes.

6. Solution of the Problem of Flow Past Bodies by the Method of Conformal Mappings: Zhukovskiy-Chaplygin Hypothesis – The Circulation Formula

Consider the application of the method of conformal mappings to the solution of the direct problem of determining the flow past wing profiles. The wing profile is understood to be a smooth, elongated (in the direction of oncoming flow), closed, and nonself-intersecting geometric contour with a rounded leading edge (the "head" of the profile) and the rounded rear

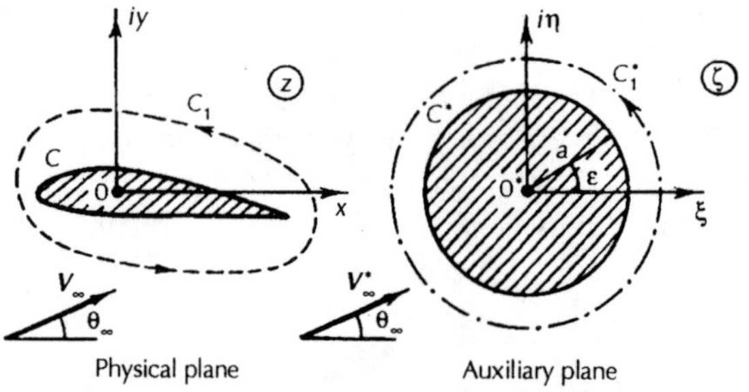

Fig. 62

edge (the "tail" of the profile). A portion of the straight line that connects a certain point at the leading edge with the vertex of the angle at the rear edge is called the chord of the wing profile (very different chords can be selected) and the length of the chord is called the length of the profile; the maximum thickness of the profile in the direction perpendicular to the chord is called the thickness of the profile and the thickness-to-length ratio is called the relative thickness of the wing profile. The angle formed by the oncoming flow velocity vector far from the profile (velocity vector "at infinity") and the direction of the chord is called the angle of attack. Having made the convention regarding the terminology, we move to the statement of the basic problem of flow past a wing profile by a plane, uniform at infinity, irrotational stream of an ideal incompressible fluid.

The oncoming flow will be prescribed by the complex velocity vector V_∞, which forms the angle θ_∞ with the axis $0x$. The physical plane z has a cut, hatched in Figure 62, which makes this plane doubly connected; for the problem to be definite it is necessary to prescribe *a priori* the velocity circulation Γ along the arbitrary contour C_1 which encircles the profile.

We consider that the first, purely geometric, and essentially most difficult problem about the mapping of the external (with respect to the area shaded in Figure 62) region C in the physical plane onto the external (with respect to the shaded circle C^*) region of the auxiliary plane as previously resolved.

Let the function of the complex variable

$$z = f(\zeta) \tag{54}$$

represent the sought-after transforming, or mapping, function which makes the conformal mapping of the external (with respect to the area encircled by the contour C and shaded in Figure 62) region of the plane of the complex variable $z = x + iy$ onto the external (with respect to the circle C^* shaded in the same figure and having the radius a and the center at the coordinate origin of the system $0^*\xi\eta$) part of the auxiliary plane of the complex variable $\zeta = \xi + i\eta$. Let us impose additional conditions onto the transforming function (54): (1) the infinitely removed point $z = \infty$ should convert on mapping into an infinitely removed point $\zeta = \infty$, and (2) the velocity direction at infinity V_∞ should be preserved on transition from the plane z into the plane ζ. As is proved in the theory of the functions of the complex variable (Riemann's theorem), the transformation (54) is unique if these conditions are fulfilled.

Let $\chi(z)$ be the sought-after complex potential of the flow in the physical plane and $\chi^*(\zeta)$ be the complex potential of the flow in the auxiliary plane, i.e., the complex potential of circulation flow past a circular cylinder which was defined in the previous section. According to Equation 53, in the present case

$$\chi^*(\zeta) = \bar{V}^*_\infty \zeta + \frac{V^*_\infty a^2}{\zeta} + \frac{\Gamma^*}{2\pi i} \ln \zeta \tag{55}$$

where V_∞^* and Γ^* are, respectively, the velocity at infinity and velocity circulation along the arbitrary contour C_1^* which encircles C^* in the auxiliary plane ζ. Making use of the connection (54) between z and ζ, we conclude that ($\chi = \varphi + i\psi$ is the invariant)

$$\chi(z) = \chi[f(\zeta)] = \chi^*(\zeta)$$

Having taken the derivative of the both parts of this equation with respect to ζ, we obtain

$$\frac{d\chi^*}{d\zeta} = \frac{d\chi}{dz}\frac{dz}{d\zeta} = \frac{d\chi}{dz}f'(\zeta)$$

or, by Equation 38,

$$\bar{V}^* = \bar{V}f'(\zeta) \tag{56}$$

and at infinitely removed points

$$\bar{V}_\infty^* = m_\infty \bar{V}_\infty \qquad m_\infty = f'(\infty) \tag{57}$$

In conformity with the earlier-adopted condition, the direction of the velocity vector at infinity V_∞ is preserved on conformal mapping, i.e., the vectors V_∞^* and V_∞ are parallel. From this follows the parallelism of the conjugate vectors $\bar{V}_\infty^*$ and $\bar{V}_\infty$. The real nature of the quantity m_∞ follows from Equation 57,

$$V_\infty^* = m_\infty V_\infty$$

For definiteness we regard m_∞ to be a positive quantity.

The mapping function (54) can be represented by the Laurent series

$$z = f(\zeta) = m_\infty \zeta + \sum_{n=0}^{\infty} \frac{m_n}{\zeta^n} \tag{58}$$

converging in the entire region $|\zeta| > a$, which is external with respect to the circle C^*. The coefficients of the principal part of this series, represented by the sum of the terms with the negative powers of ζ, can be determined with the aid of the contour integrals ($n = 0, 1, 2, \cdots$)

$$m_n = \frac{1}{2\pi i}\oint_{C_1^*} f(\zeta)\,\zeta^{n-1}d\zeta \tag{59}$$

calculated along the circle C^* or along any other circle which contains C^* inside.

The mapping function can be also expressed with the aid of the Cauchy integral

$$z = f(\zeta) = \frac{1}{2\pi i}\oint_{C_1^*} \frac{f(\zeta')\,d\zeta'}{\zeta - \zeta'} \tag{60}$$

where ζ' is the complex variable of integration.

Consider now the circulation Γ^*. Having represented it, according to the first of Equations 41, as the real part of the integrals

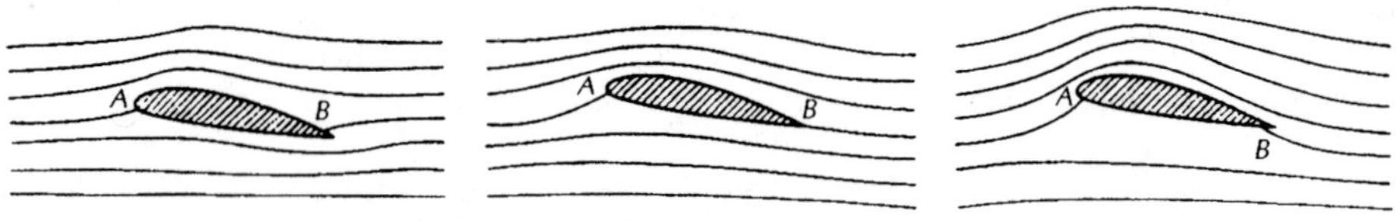

Fig. 63

$$\Gamma^* = \mathrm{Re}\oint_{C_1^*} \bar{V}^* d\zeta = \mathrm{Re}\oint_{C_1^*} \bar{V}\frac{dz}{d\zeta}d\zeta = \mathrm{Re}\oint_{C_1} \bar{V}dz = \Gamma$$

we conclude that the circulation of velocity along any closed contour, which encircles the profile in the flow, preserves its value during conformal mapping.

The above arguments have made it possible to express the unknown quantities V_∞^* and Γ^* in terms of the specified quantities V_∞, Γ, and the coefficient m_∞, which is determined (Equation 57), by the familiar mapping function $f(\zeta)$. Consequently, we have the final expression for the complex potential χ in the plane of flow in the form of the parametric dependence on the parameter ζ

$$\chi(z) = \chi^*(\zeta) = m_\infty\left(\bar{V}_\infty\zeta + \frac{V_\infty a^2}{\zeta}\right) + \frac{\Gamma}{2\pi i}\ln\zeta \quad z = f(\zeta) \tag{61}$$

Thus, if the solution is known for the geometric problem about the conformal mapping of the external (with respect to the contour C in a flow) region of the physical plane z onto the external (with respect to the circle C^* of arbitrary radius a) region of the auxiliary plane ζ, then the solution of the hydrodynamic problem about the determination of the complex potential $\chi(z)$ may be obtained without difficulty.

It follows from the system of Equations 61 that the problem about the flow past the profile C by the stream with the velocity specified in magnitude and direction at infinity has an infinite number of solutions that depend on the arbitrary choice of the value of circulation Γ. From the viewpoint of the theory of an ideal fluid this arbitrariness corresponds to the essence of the problem. As shown earlier for the case of flow around the circle, by imposing this or other circulation, it is possible to obtain an infinite variety of the shapes of flow past the circular cylinder with different positions of stagnation points.

In exactly the same way, three types of flow depicted in Figure 63 are theoretically possible for the same wing profile with the angular point on the trailing edge and for the same (in magnitude and direction) velocity at infinity. The letters A and B indicate the stagnation points in flow. In case (a), just as in case (c), the fluid should flow from one side of the wing surface to the other; from the upper to the lower one in case (c) and from the lower to the upper one in case (a). Either infinitely great velocities should originate on the sharp edge, which lead to physically unfeasible great negative pressures, or separation of flow from the profile surface and vortex formation should take place. Among the three indicated possible forms of flow only one form, (b), with the rear stagnation point B coinciding with the angular point on the trailing edge of the profile, leads to smooth running of fluid streams from the trailing edge of the wing with a finite velocity.

As evidenced by contemporaries, at the end of 1909, Chaplygin, in discussing the report made by Zhukovskiy, advanced the following hypothesis as a generalization of the familiar experimental fact: Among an infinite number of theoretically possible flows past a profile

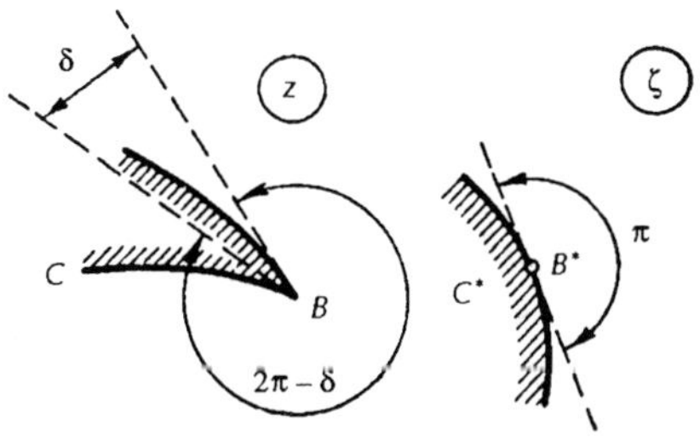

Fig. 64

with an angular point at the trailing edge, smooth flow with a finite velocity takes place actually at this point.

The stated priority of Chaplygin, who gave the general interpretation of this postulate, is subject to question in view of the use of an analogous hypothesis by Zhukovskiy in his earlier works dealing with wing profiles of specific shapes. We shall keep the name "the Zhukovskiy-Chaplygin hypothesis" as is generally accepted in the literature.

Experience shows that for each wing profile there is the range of attack angles, within which there is no separation of fluid from the profile surface and there is a smooth flow past the trailing edge. The wing profiles corresponding to the Zhukovskiy-Chaplygin hypothesis are usually called well-streamlined; the remainder are badly streamlined. Streamlining is not a purely geometric property of profiles, but depends not only on the shape of the profile, but also on the flow velocity, angle of attack, physical properties of the fluid, presence of other bodies near the profile, etc. (see later material).

The Zhukovskiy-Chaplygin hypothesis makes it possible to unequivocally determine the value of circulation Γ, the superposition of which leads to the separation-free form of flow past a wing profile with a finite velocity at its trailing edge.

To the angular point B (Figure 64) on the profile C corresponds a certain point B^* on the circle C^*. These points are the singular points of transformation, because here the basic property of the conformal mapping is violated; i.e., the conservation of angles between the tangents to the contours transformed. In fact, the external angle with the vertex at the point B at the trailing edge (equal to $2\pi - \delta$, where δ is the internal acute angle at the trailing edge), passes in the plane ζ into the angle π, differing from the former one, with the vertex at the point B^*.

Let us consider the conformal mapping of the external (with respect to the profile C) small region in the vicinity of the vertex of the angle B onto the small, external (with respect to the circle C^*) region near the point B^* in the plane ζ. This conformal mapping can be presented by the formula

$$z - z_B = M(\zeta - \zeta_{B^*})^{\frac{2\pi - \delta}{\pi}} \tag{62}$$

where z_B and ζ_{B^*} are complex coordinates of the points B and B^*, corresponding to each other, in the planes z and ζ, and M is a certain real number. In fact, having assumed in the vicinity of the points B and B^* that

$$z - z_B = re^{i\beta} \qquad \zeta - \zeta_{B^*} = r^* e^{i\beta^*}$$

and substituted these expressions into Equation 62, we find

$$re^{i\beta} = Mr^{*\frac{2\pi-\delta}{\pi}} e^{i\frac{2\pi-\delta}{\pi}\beta^*}$$

Equating the arguments we obtain

$$\beta = \frac{2\pi-\delta}{\pi}\beta^* \pm 2k\pi$$

and convince ourselves that to the change of β^* by π corresponds the change of β by $2\pi - \delta$.

Using the mapping function (54), we may establish the connection between the velocities at the points B and B^*; we obtain

$$\bar{V}_{B^*} = \left(\frac{d\chi^*}{d\zeta}\right)_{\zeta=\zeta_{B^*}} = \left(\frac{d\chi}{dz}\right)_{z=z_B}\left(\frac{dz}{d\zeta}\right)_{\zeta=\zeta_{B^*}} = \bar{V}_B\left(\frac{dz}{d\zeta}\right)_{\zeta=\zeta_{B^*}}$$

or, computing the derivative by Equation 62,

$$\bar{V}_{B^*} = \bar{V}_B\frac{2\pi-\delta}{\pi}M(\zeta-\zeta_{B^*})^{\frac{\pi-\delta}{\pi}}_{\zeta=\zeta_{B^*}}$$

According to the Zhukovskiy-Chaplygin hypothesis, the velocity V_B must be finite and, since $\delta < \pi$, the last factor must be equal to zero; consequently, the entire product is equal to zero. From this it follows that the point of the circle, which corresponds to the trailing edge of the profile and which is located in the auxiliary plane, should be the stagnation point.

From this condition we find the circulation Γ if, using Equation 61, we write that the velocity at the point B^* is equal to zero

$$\bar{V}_{B^*} = \left(\frac{d\chi^*}{d\zeta}\right)_{\zeta=\zeta_{B^*}} = m_\infty\bar{V}_\infty - \frac{m_\infty V_\infty a^2}{\zeta_{B^*}^2} + \frac{\Gamma}{2\pi i}\frac{1}{\zeta_{B^*}} = 0$$

Assuming here that

$$\zeta_{B^*} = ae^{i\varepsilon_0} \qquad V_\infty = |V_\infty|e^{i\theta_\infty}$$

where ε_0 is the polar angle of the point B^* on the circle C^*, θ_∞ is the angle formed by the velocity at infinity with the axes $0x$ or $0^*\xi$, we obtain

$$m_\infty|V_\infty|e^{-i\theta_\infty} - m_\infty|V_\infty|e^{i(\theta_\infty-2\varepsilon_0)} + \frac{\Gamma}{2\pi ai}e^{-i\varepsilon_0} = 0$$

where

$$\Gamma = -4\pi am_\infty|V_\infty|\frac{e^{i(\theta_\infty-\varepsilon_0)} - e^{-i(\theta_\infty-\varepsilon_0)}}{2i}$$

or, passing from exponential to trigonometric functions,

$$\Gamma = -4\pi am_\infty|V_\infty|\sin(\theta_\infty-\varepsilon_0) \tag{63}$$

Let us introduce the notation $\theta_\infty - \varepsilon_0 = \alpha$ and rewrite Equation 63 in the form

$$\Gamma = -4\pi am_\infty|V_\infty|\sin\alpha \tag{64}$$

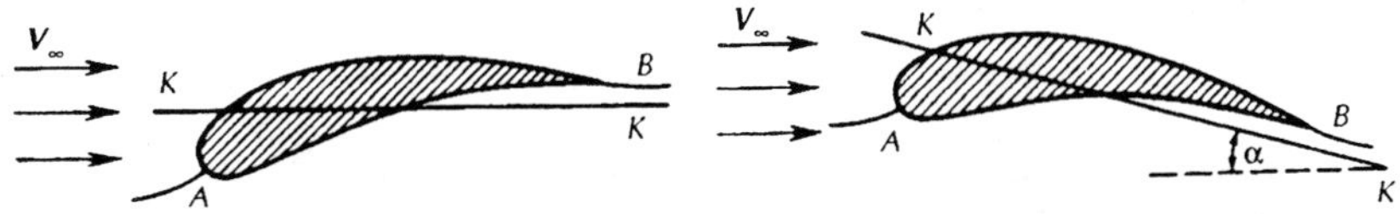

(a) Noncirculatory flow past a profile (b) Circulatory flow past a profile

Fig. 65

Let us rotate the profile with respect to the given flow in such a way that without the imposition of circulation ($\Gamma = 0$), the trailing edge could coincide with the stagnation point B. Let us denote on the profile the straight line KK (Figure 65a) which determines the direction of velocity at infinity corresponding to this noncirculatory flow. The straight line KK, rigidly connected with the profile, is called the direction of the noncirculatory flow and the corresponding value of the angle $\theta_\infty = \varepsilon_0$ will be referred to as the angle of noncirculatory flow past the profile.

Having rotated the profile through the angle α (Figure 65b), we obtain the circulatory flow with the stagnation point at the trailing edge B, with the necessary circulation being determined by Equation 64.

The acute angle α between the direction of the free stream velocity and the direction of noncirculatory flow will be called the theoretical angle of attack, in contrast to the practical angles of attack, defined as the angles between the direction of velocity at infinity, and the chords of the wing which are specified in different ways.

Equation 64 defines the so-called theoretical circulation. As experiments indicate, the theoretical circulation somewhat exceeds the actual one; the explanation is associated with the presence of internal friction in actual fluids.

Excluding circulation from the parametric system (61) with the aid of Equation 64, we obtain a single-valued solution of the problem about external flow past the wing profile.

The derivation of Equation 64 is based on the fact that the wing profile has a sharp trailing edge. In the case of flow past a profile of smooth shape without the angular point at the trailing edge, the Zhukovskiy-Chaplygin hypothesis does not hold and the circulation remains indeterminate. Theoretical calculation of flow past such profiles requires either special assumptions or the assignment of the position for the rear stagnation point.

7. Examples of the Use of Conformal Mappings: Flow Around an Ellipse and a Plate

Consider some simple conformal transformations of the exterior of a circle in the auxiliary plane on the exterior of the closed profile in the plane of the flow.

The first transformation of this kind, indicated Zhukovskiy and employed by Chaplygin in 1910, has the form

$$z = \frac{1}{2}\left(\zeta + \frac{c^2}{\zeta}\right) \tag{65}$$

or, in a more symmetric form,

$$\frac{z-c}{z+c} = \left(\frac{\zeta - c}{\zeta + c}\right)^2 \tag{66}$$

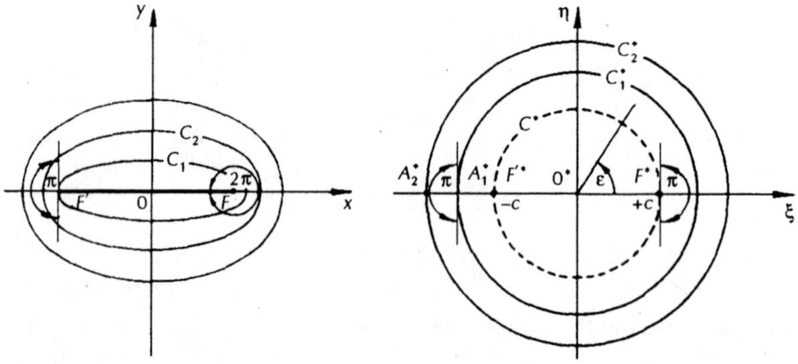

Fig. 66

Zhukovskiy's transformation (65) is the simplest specific case of the general transformation (58) corresponding to the values of the coefficients

$$m_\infty = \frac{1}{2} \qquad m_1 = \frac{1}{2}c^2 \qquad m_0 = m_2 = \cdots = 0$$

It satisfies all the conditions posed in Section 6.

The circle C^* of radius c in the plane ζ, which is called the basic circle, is transformed in the plane z into the segment $F'F$ (Figure 66) on the axis $0x$ with the ends at the points $(-c, 0)$ and $(+c, 0)$. In fact, on the circle $\zeta = ce^{i\varepsilon}$, and after transformation we shall find

$$z = \frac{c}{2}(e^{i\varepsilon} + e^{-i\varepsilon}) = c\cos\varepsilon \tag{67}$$

so that to the complete circuit round the circle ($0 \le \varepsilon \le 2\pi$) corresponds to the double traverse of the segment FF', from right to left along the upper part of the cut FF', and from left to right along the lower part of the cut

The circles C_1^*, $C_2^*, \cdots$, which are concentric with the basic circle C^*, are transformed into the confocal ellipses C_1, $C_2, \cdots$ with the foci F, F'. In fact, denoting by r the radius of any one the circles and assuming in Equation 65 that

$$\zeta = re^{i\varepsilon} \quad (r > c)$$

we obtain

$$z = \frac{1}{2}\left(re^{i\varepsilon} + \frac{c^2}{r}e^{-i\varepsilon}\right)$$

where it follows that

$$x = \frac{1}{2}\left(r + \frac{c^2}{r}\right)\cos\varepsilon$$

$$y = \frac{1}{2}\left(r - \frac{c^2}{r}\right)\sin\varepsilon \qquad \frac{x^2}{\frac{1}{4}\left(r + \frac{c^2}{r}\right)^2} + \frac{y^2}{\frac{1}{4}\left(r - \frac{c^2}{r}\right)^2} = 1$$

Let us denote the semi-axes of the ellipses by a and b. From the equations

$$a = \frac{1}{2}\left(r + \frac{c^2}{r}\right) \qquad b = \frac{1}{2}\left(r - \frac{c^2}{r}\right) \tag{68}$$

we find

$$a^2 - b^2 = c^2$$

All the ellipses considered have the common focal distance $2c$ and foci at the points F and F'. It follows from Equation 68 that

$$r = a + b \tag{69}$$

The complex potential $\chi(z)$ of flow around any of the ellipses C_1, C_2, also including in the limit the segment FF', with the velocity at infinity V_∞ (which forms the angle θ_∞ with the axis $0x$) and circulation Γ can, as before, be composed in the parametric form (61):

$$\chi = \frac{1}{2}\left[\bar{V}_\infty \zeta + \frac{V_\infty (a+b)^2}{\zeta}\right] + \frac{\Gamma}{2\pi i}\ln\zeta \qquad z = \frac{1}{2}\left(\zeta + \frac{c^2}{\zeta}\right) \tag{70}$$

in the case of flow past the segment FF' it is necessary to assume that $b = 0$, $c = a$.

The complex parameter ζ can be easily eliminated from system (70). Resolving the second equation in ζ, we obtain

$$\zeta = z \overset{+}{\underset{(-)}{}} \sqrt{z^2 - c^2} \tag{71}$$

with the upper sign being taken in front of the root in correspondence with the mapping of the exterior of the ellipse onto the exterior of the circle; in fact, taking the lower sign, we could have

$$\zeta = z - \sqrt{z^2 - c^2} = \frac{c^2}{z + \sqrt{z^2 - c^2}}$$

and $\zeta = 0$ at $z = \infty$, i.e., the interior of the circle would have corresponded to the exterior of the ellipse.

Thus, instead of the parametric representation (70) of the complex potential of flow around the ellipse, we obtain the explicit expression of this potential

$$\chi(z) = \frac{1}{2}\left[\bar{V}_\infty (z + \sqrt{z^2 - c^2}) + \frac{V_\infty (a+b)^2}{z + \sqrt{z^2 - c^2}}\right] + \frac{\Gamma}{2\pi i}\ln(z + \sqrt{z^2 - c^2})$$

which can be easily simplified if the irrationality in the denominator of the second term in square brackets is eliminated by the usual technique. We have

$$\chi(z) = \frac{1}{2}\bar{V}_\infty (z + \sqrt{z^2 - c^2}) + \frac{1}{2}\frac{(a+b)^2}{c^2} V_\infty (z - \sqrt{z^2 - c^2}) +$$

$$+ \frac{\Gamma}{2\pi i}\ln(z + \sqrt{z^2 - c^2}) \tag{72}$$

The noncirculatory ($\Gamma = 0$) flow around an ellipse is shown in Figure 67. The null streamline passing through the stagnation points A and B consists of the ellipse itself immersed in

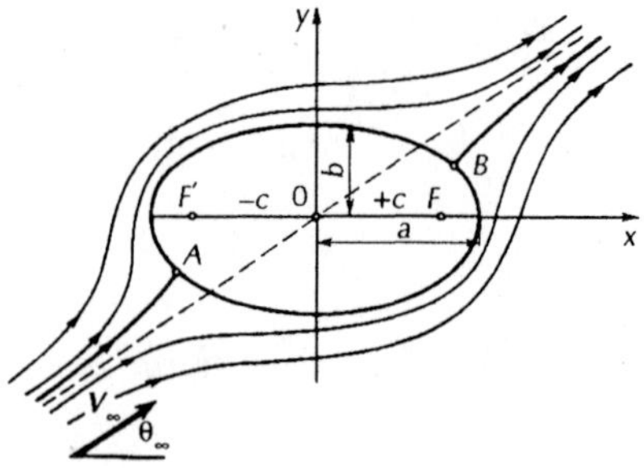

Fig. 67

flow and two segments of confocal hyperbola, the parameters of which depend on the angle of attack θ_∞.

Having the expression of the complex velocity potential (72) and having taken the derivative of it with respect to z, we find the conjugate velocity and, consequently, the distribution of velocities and pressures along the contour of the ellipse in flow.

Gonor[9] put these distributions in an elegant form by having proved the theorem: in the case of irrotational noncirculatory flow around an ellipse by an incompressible fluid in the direction of its axes the velocity at the points of the contour is proportional to the cosine of the local angle of attack at these points.

Let us prove this theorem for the case of the stream oncoming parallel to the greater axis of the ellipse. The proof is analogous in the case of the stream being parallel to the small axis; and by superposing these streams, the proof is obtained for the flow at any angle of attack.

The expression of the velocity potential φ can be found by assuming in Equation 72 that $\Gamma = 0$, $\bar{V}_\infty = V_\infty = U_\infty$ and having separated from it the real part. For this, let us go over from the Cartesian, x, y, to the elliptical, ξ, η, coordinates:

$$x = c\,\mathrm{ch}\,\xi\cos\eta \qquad y = c\,\mathrm{ch}\,\xi\sin\eta$$

The coordinate lines will be represented by the family of confocal ellipses and hyperbolas (c is the distance from the coordinate origin to the foci):

$$\frac{x^2}{c^2\,\mathrm{ch}^2\,\xi} + \frac{y^2}{c^2\,\mathrm{sh}^2\,\xi} = 1 \qquad \frac{x^2}{c^2\cos^2\eta} - \frac{y^2}{c^2\sin^2\eta} = 1$$

Considering the ellipse with the semi-axes a and b as one of the ellipses of the first family ($\xi = \text{const}$), we obtain the following relations valid on the contour of the ellipse:

$$\mathrm{ch}\,\xi = a/c \qquad \mathrm{sh}\,\xi = b/c \qquad c = \sqrt{a^2 - b^2}$$

$$\cos\eta = x/a \qquad \sin\eta = y/b$$

The transition from the Cartesian to the elliptical coordinates corresponds the transformation $z = c\,\mathrm{ch}\,\zeta$ where $z = x + iy$, $\zeta = \xi + i\eta$.

In this case the complex potential of velocities χ acquires the form

[9]Gonor, A. L., Determination of the flow field on the surface of some bodies in the flow of an incompressible fluid, *Mekh. Zhidk. Gaza*, No. 2, pp. 187 to 190, 1976.

$$\chi = U_\infty \sqrt{(a+b)/(a-b)}\,(a\,\mathrm{ch}\,\xi - b\,\mathrm{sh}\,\xi)$$

and its real part and the derivatives of it with respect to ξ and η are

$$\varphi = U_\infty \sqrt{(a+b)/(a-b)}\,(a\,\mathrm{ch}\,\xi - b\,\mathrm{sh}\,\xi)\cos\eta$$

$$\partial\varphi/\partial\xi = U_\infty \sqrt{(a+b)/(a-b)}\,(a\,\mathrm{sh}\,\xi - b\,\mathrm{ch}\,\xi)\cos\eta$$

$$\partial\varphi/\partial\eta = -U_\infty \sqrt{(a+b)/(a-b)}\,(a\,\mathrm{ch}\,\xi - b\,\mathrm{sh}\,\xi)\sin\eta$$

To determine the derivatives $\partial\varphi/\partial x$ and $\partial\varphi/\partial y$, we use the system of equalities

$$\frac{\partial\varphi}{\partial\xi} = \frac{\partial\varphi}{\partial x}\frac{\partial x}{\partial\xi} + \frac{\partial\varphi}{\partial y}\frac{\partial y}{\partial\xi} \qquad \frac{\partial\varphi}{\partial\eta} = \frac{\partial\varphi}{\partial x}\frac{\partial x}{\partial\eta} + \frac{\partial\varphi}{\partial y}\frac{\partial y}{\partial\eta}$$

which, according to the just-obtained relations, are reduced on the contour of the ellipse to the following

$$\frac{b}{a}x\frac{\partial\varphi}{\partial x} + \frac{a}{b}y\frac{\partial\varphi}{\partial y} = 0 \qquad \frac{a}{b}y\frac{\partial\varphi}{\partial x} - \frac{b}{a}x\frac{\partial\varphi}{\partial y} = U_\infty\frac{a+b}{b}y$$

The solutions of this system of equations will be expressed in the form

$$\frac{\partial\varphi}{\partial x} = U_\infty\frac{a+b}{ab^4}y^2\left(\frac{x^2}{a^4}+\frac{y^2}{b^4}\right)^{-1}$$

$$\frac{\partial\varphi}{\partial y} = -U_\infty\frac{a+b}{a^3b^2}xy\left(\frac{x^2}{a^4}+\frac{y^2}{b^4}\right)^{-1}$$

and the square of the velocity modulus will be equal to

$$|V|^2 = \left(\frac{\partial\varphi}{\partial x}\right)^2 + \left(\frac{\partial\varphi}{\partial y}\right)^2 = U_\infty^2\frac{(a+b)^2}{a^2b^4}y^2\left(\frac{x^2}{a^4}+\frac{y^2}{b^4}\right)^{-1}$$

The squared cosine of the local angle of attack θ on the contour of the ellipse will be equal to

$$\cos^2\theta = \left(\frac{dx}{ds}\right)^2 = \left[1+\left(\frac{dy}{dx}\right)^2\right]^{-1}$$

or, according to the equation of an ellipse in Cartesian coordinates,

$$\cos^2\theta = (y^2/b^4)\,(x^2/a^4 + y^2/b^4)^{-1}$$

Comparing the expressions for $|V|^2$ and $\cos^2\theta$, we see the validity of the following relation on the contour of the ellipse

$$|V| = (1 + b/a)\,U_\infty\cos\theta$$

which proves Gonor's theorem. The pressure coefficient $c_p = (p - p_\infty)/(1/2\rho U_\infty^2)$ will be defined by the equation

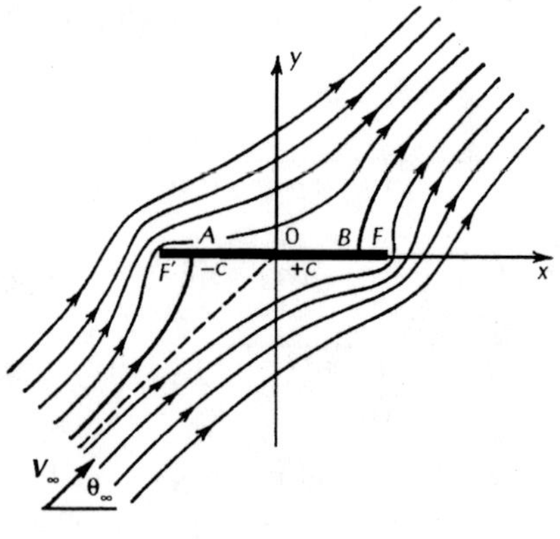

Fig. 68

$$c_p = 1 - (1 + b/a)^2 \cos^2\theta$$

When $b = a$, we again find the formulas of noncirculatory flow around a circular cylinder which were given in Section 5.

Gonor's theorem holds not only for a plane noncirculatory irrotational flow around an elliptical cylinder, but also in the case of three-dimensional flow past a slender and flattened ellipsoid of revolution as well as a three-axes ellipsoid. The proof of these results can be found in Gonor's paper (cited earlier).

Assuming that $c = 0$, $b = a$ in Equation 72, we obtain the complex potential of circulatory flow around a circle. For this it is sufficient only to calculate the limiting expression

$$\left.\frac{z - \sqrt{z^2 - c^2}}{c^2}\right|_{c \to 0} = \frac{1}{2z}$$

which enters into the second term. We have

$$\chi(z) = \bar{V}_\infty z + \frac{V_\infty a^2}{z} + \frac{\Gamma}{2\pi i} \ln z$$

in complete agreement with expression 53.

Assuming in the same formula (72) that $b = 0$, $a = c$, we find the complex potential of the circulatory flow past a plate of length $2c$ (segment FF' in Figure 68)

$$\chi = \frac{1}{2}\bar{V}_\infty (z + \sqrt{z^2 - c^2}) + \frac{1}{2} V_\infty (z - \sqrt{z^2 - c^2}) + \frac{\Gamma}{2\pi i} \ln (z + \sqrt{z^2 - c^2}) =$$

$$= \frac{1}{2} (V + \bar{V}_\infty) z - \frac{1}{2} (V_\infty - \bar{V}_\infty) \sqrt{z^2 - c^2} + \frac{\Gamma}{2\pi i} \ln (z + \sqrt{z^2 - c^2}) =$$

$$= u_\infty z - i v_\infty \sqrt{z^2 - c^2} + \frac{\Gamma}{2\pi i} \ln (z + \sqrt{z^2 - c^2}) \quad (73)$$

where u_∞, v_∞ are the projections of V_∞ on the coordinate axes.

The streamlines of the flow past a plate with $\Gamma = 0$ are depicted in Figure 68. The null streamline consists of the plate surface and two segments of the confocal hyperbola the parameters of which, as in the case of flow around an ellipse, depend on the angle of attack θ_∞.

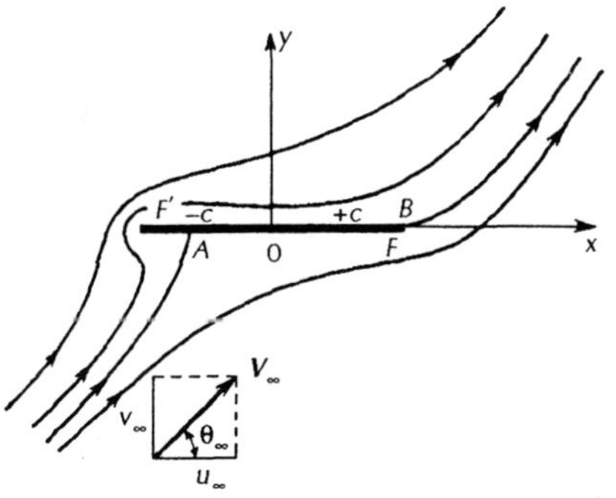

Fig. 69

Excluding the analysis of the flow determined in the general case by formula (72), we consider in more detail expression (73) for the complex potential of flow past a plate.

Calculating the velocity, we obtain

$$\bar{V} = \frac{d\chi}{dz} = u_\infty - iv_\infty \frac{z}{\sqrt{z^2-c^2}} + \frac{\Gamma}{2\pi i} \frac{1+\frac{z}{\sqrt{z^2-c^2}}}{z+\sqrt{z^2-c^2}} = u_\infty - \frac{iv_\infty z - \frac{\Gamma}{2\pi i}}{\sqrt{z^2-c^2}} \tag{74}$$

When the magnitude of the circulation Γ is arbitrary and $z = \pm c$, the velocity has infinite values, thus corresponding to the flow past sharp frontal and rear edges.

Now, let us impose on Γ the condition of the finite velocity at the rear edge ($z = c$) in conformity with the Zhukovskiy-Chaplygin hypothesis. For this, the following equation should hold

$$iv_\infty c - \frac{\Gamma}{2\pi i} = 0$$

i.e.,

$$\Gamma = -2\pi c v_\infty \tag{75}$$

In the case of the plate considered above, the noncirculatory direction coincides with the direction of the plate itself, so that the angle ε_0 in the circulation equation (63) is equal to zero. Determining the circulation by this equation and replacing $m_\infty = 1/2$ by Equation 65, we obtain

$$\Gamma = -4\pi\frac{1}{2}c|V_\infty|\sin\theta_\infty = -2\pi c|V_\infty|\sin\theta_\infty \tag{76}$$

in complete conformity with Equation 75.

Substituting the expression of circulation (75) into Equation 74, we obtain the following formula for the distribution of velocities

$$\bar{V} = u_\infty - iv_\infty\frac{z-c}{\sqrt{z^2-c^2}} = u_\infty - iv_\infty\sqrt{\frac{z-c}{z+c}} \tag{77}$$

At the rear edge ($z = c$) the velocity is finite and equal to u_∞, at the frontal edge ($z = -c$) the velocity remains equal to infinity. By selecting the circulation, it is impossible to make the velocity finite at both sharp edges. The position of streamlines in the case of smooth flow past the rear edge is shown in Figure 69.

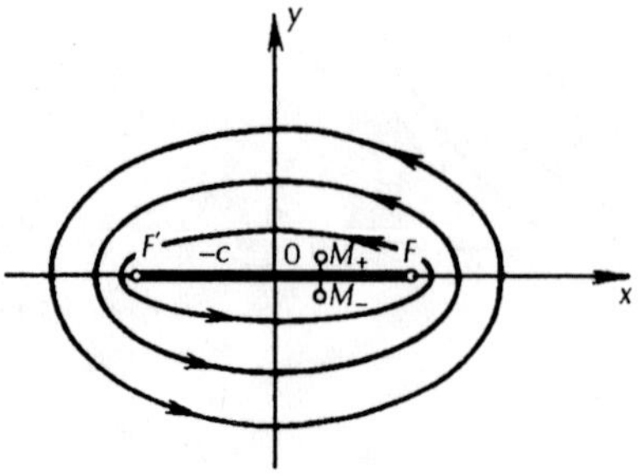

Fig. 70

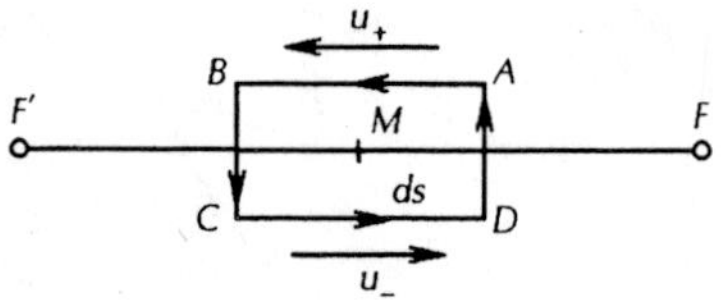

Fig. 71

The last term in Equations 72 and 73 is represented by the complex potential of purely circulatory flow around an ellipse or a plate

$$\chi_0(z) = \frac{\Gamma}{2\pi i}\ln\left(z + \sqrt{z^2 - c^2}\right) \tag{78}$$

The streamlines (ellipses) corresponding to this flow are shown in Figure 70. The conjugate velocity is

$$\bar{V} = \frac{d\chi_0}{dz} = -\frac{\Gamma}{2\pi}\frac{1}{\sqrt{c^2 - z^2}}$$

On the upper surface of the plate ($y = 0$, $-c < x < c$), the plus sign corresponds to the square root, so that the conjugate velocity is real and is equal to

$$u_+ = -\frac{\Gamma}{2\pi}\frac{1}{\sqrt{c^2 - x^2}} \qquad (u_+ < 0 \text{ when } \Gamma > 0)$$

On the lower surface, the root should be taken with the minus sign, so that

$$u_- = \frac{\Gamma}{2\pi}\frac{1}{\sqrt{c^2 - x^2}} \qquad (u_- > 0 \text{ when } \Gamma > 0)$$

Let us ignore the fact that the segment $F'F$ represents a solid wall and imagine that the entire plane $x0y$ is occupied by a fluid. Then the line $F'F$ will be the line of the discontinuity of velocities in the flow. In fact, according to what has just been proved, when crossing the line $F'F$ (Figure 70) the velocity u undergoes the finite jump

$$u_- - u_+ = \frac{\Gamma}{\pi\sqrt{c^2 - x^2}}$$

Let us construct on the infinitesimal segment ds of the line $F'F$ (Figure 71), a rectangular contour which will enclose the point M. The velocity circulation along this closed contour

$$(u_- - u_+)\,ds = \frac{\Gamma\,ds}{\pi\sqrt{c^2 - x^2}}$$

differs from zero; consequently, along the segment ds of the velocity discontinuity line there are vortices with the common intensity equal to this circulation.

Let γ denote the density of the distribution of vortices, i.e., the intensity of their continuous distribution per unit length of the segment $F'F$; then,

$$\gamma ds = (u_- - u_+)\, ds$$

and, consequently,

$$\gamma = u_- - u_+ = \frac{\Gamma}{\pi\sqrt{c^2 - x^2}} \tag{79}$$

Continuous distribution of vortices on the surface (in the case of plane motion, the distribution along a certain line) forms a vortex layer.

It follows from the foregoing that a purely circulatory motion (78) around an elliptical cylinder (in particular, a plate) is equivalent to the flow formed by a vortical layer located along the line which connects the foci of the ellipse. The density of the distribution of vortices in the layer are determined by Equation 79.

8. The Zhukovskiy-Chaplygin Wing Profiles

The examples considered in the previous section and dealing with the flow past an elliptical contour (in particular, past a plate) do not furnish a complete understanding of the flow past wing profiles.

Forming the expression for the derivative of the mapping function (65)

$$\frac{dz}{d\zeta} = \frac{1}{2}\left(1 - \frac{c^2}{\zeta^2}\right) \tag{80}$$

we see that the points F^* and F'^* (Figure 66) of the plane ζ with the coordinates $\zeta = \pm c$ are singular points of the conformal mapping (65) because at these points the derivative is equal to zero. At the points F^* and F'^* the conformity of the mapping is violated: the angles 2π at the points F and FF' of the plane z correspond to the angles π.

The circle C^* drawn through both singular points was transformed into the rectilinear segment FF' or, more precisely, into the cut of the plane z with two angular points, whereas the circles C_1^* and C_2^*, which do not pass through the singular points, were converted into the ellipses, smooth curves which did not have angular points. Drawing in the plane ζ circles or any other closed curves so that they could pass through only one singular point, we construct in the plane z the profiles with one angular point from which it is possible to select those that are suited to wing profiles.

The examples of such theoretical wing profiles can be represented by the Zhukovskiy-Chaplygin profiles formed by the conformal mapping (65) of the circles K^* drawn in the auxiliary plane (Figure 72) through the singular point F^* and enclosing the second singular point F'^*. The specific feature of these profiles is the zero angle at the rear edge.

If the center of the circle K_1^* is located at the point N_1 of the axis $0^*\xi$, then in the plane z we obtain the symmetrical profile K_1 – Zhukovskiy's rudder (shown by the dashed line in the figure). The circle C^* passes into the segment $F'F$ which serves as the skeleton of Zhukovskiy's rudder in the sense that as the relative thickness of the rudder decreases, its contour K_1 will shrink to the segment $F'F$. To obtain the rudder of small thickness (as compared with its length), we displace the point N_1 slightly to the left from the point 0^* with the absolute value of the displacement being equal to λc where $\lambda \ll 1$. Then, the equation of the circle K_1^* can be presented in the form

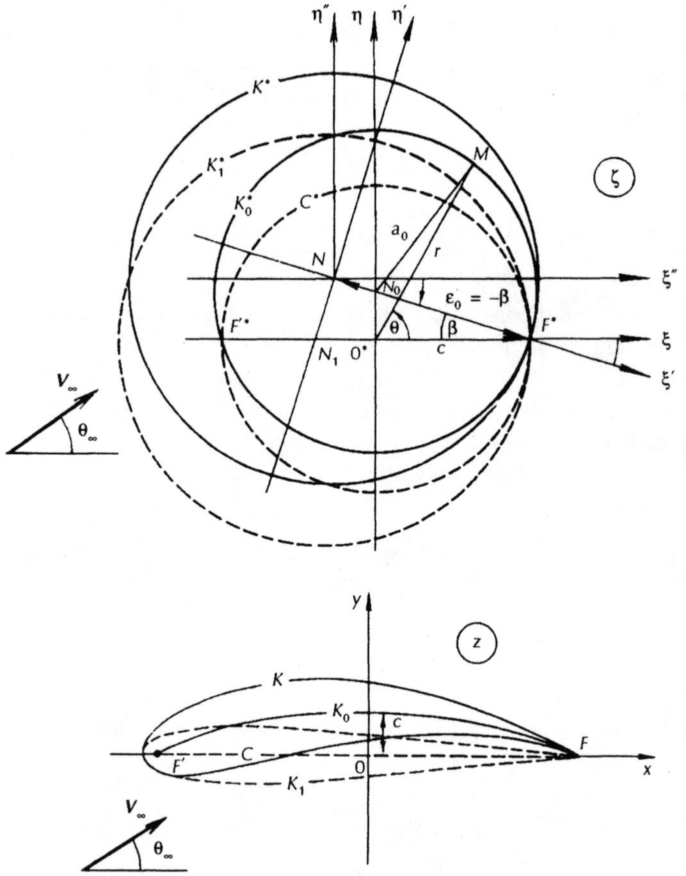

Fig. 72

$$\zeta = -\lambda c + (1+\lambda)\, c e^{i\theta}$$

where θ is the polar angle of the points of the circle K_1^* (not shown in Figure 72). Substituting this expression into Equation 65, we obtain the equation for Zhukovskiy's rudder

$$z = \frac{1}{2}\left[-\lambda c + (1+\lambda)\, c e^{i\theta} + \frac{c}{-\lambda + (1+\lambda)\, e^{i\theta}}\right]$$

Because of the smallness of the parameter λ it will not be difficult to represent the right side in the form of series expansion in the powers of λ. Restricting ourselves to the terms with the first power of λ, we obtain

$$z = c\cos\theta + \frac{1}{2}\lambda c(\cos 2\theta - 1) + i\lambda c\left(\sin\theta - \frac{1}{2}\sin 2\theta\right)$$

or

$$x = c\cos\theta + \frac{1}{2}\lambda c(\cos 2\theta - 1) \qquad y = \lambda c(1 - \cos\theta)\sin\theta$$

It follows from the second equation that the values θ = 0 and θ = π correspond to the extreme abscissas of the rudder, so that the length, or the chord, of the profile b is equal to

$$b = x(0) - x(\pi) = 2c$$

i.e., in the approximation adopted it does not differ from the length of the skeleton, i.e., of the segment FF'. Let us find the maximum thickness of the profile. To the maximum value of the ordinate of the profile there corresponds the root of the equation

$$\frac{dy}{d\theta} - \lambda c(\cos\theta - \cos 2\theta) = 0$$

equal to $\theta_m = 120°$. Consequently, the maximum ordinate will be

$$y_{max} = \lambda c \sin 60°(1 + \cos 60°) = \frac{3\sqrt{3}}{4}\lambda c$$

and the thickness

$$t = 2y_{max} = \frac{3\sqrt{3}}{2}\lambda c$$

The circle K^* having the center at any point N of the plane ζ and passing through the singular point F^* is transformed into the Zhukovskiy-Chaplygin cambered profile K. The arc K_0 serves as the skeleton for this profile just as the segment FF' does for the rudder K_1. The camber of the arc K_0 represents at the same time the camber of the profile K. If the camber of the profile K is preserved but its thickness is decreased, the profile will shrink to its skeleton, the arc K_0.

To solve the problem about the flow of the stream past the profile K at the velocity V_∞ directed at the angle θ_∞ to the axis $0x$, it will be simpler to do the following.

Let us draw in the auxiliary plane ζ the axes $N\xi'$ and $N\eta'$ with the origin at the center of the shifted circle N (Figure 72). The plane of the complex variable $\zeta' = \xi' + i\eta'$ is rotated with respect to the plane ζ through the angle $(-\beta)$ so that by adopting

$$\zeta'' = e^{-i\beta}\zeta'$$

we come to the correspondence between the planes ζ and ζ'' with the parallel axes of coordinates (a is the radius of the circle K^*)

$$\zeta = c - ae^{-i\beta} + \zeta''$$

Then, regarding the plane ζ'' to be auxiliary, we pass from transformation (65) to the more general transformation

$$z = \frac{1}{2}\left(\zeta + \frac{c^2}{\zeta}\right) = \frac{1}{2}\left(c - ae^{-i\beta} + \zeta'' + \frac{c^2}{c - ae^{-i\beta} + \zeta''}\right)$$

which can be rewritten in the form of the Laurent series

$$z = \frac{1}{2}\zeta'' + \frac{1}{2}(c - ae^{-i\beta}) + \frac{1}{2}\frac{c^2}{\zeta''} - \frac{1}{2}\frac{c^2(c - ae^{-i\beta})}{\zeta''^2} + \cdots$$

The complex potential in the plane ζ'' will preserve the previous form

$$\chi^*(\zeta'') = m_\infty\left(\bar{V}_\infty\zeta'' + \frac{V_\infty a^2}{\zeta''}\right) + \frac{\Gamma}{2\pi i}\ln\zeta''$$

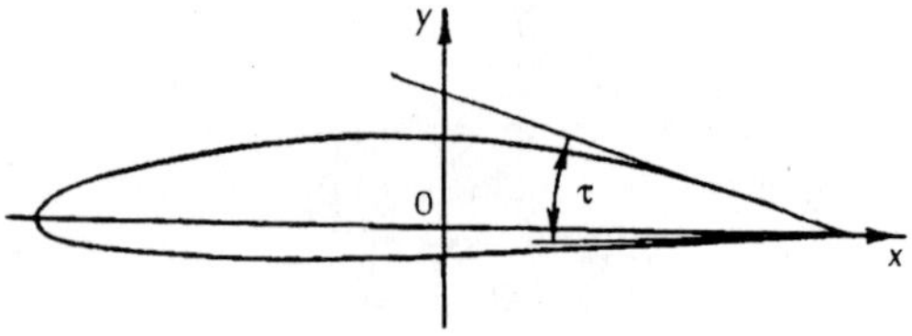

Fig. 73

The set of the latter two equations represents the sought-after solution to the problem of flow past the Zhukovskiy-Chaplygin theoretical profiles.

The generalized Zhukovskiy-Chaplygin profiles which correspond to the transformation

$$\frac{z-\sigma c}{z+\sigma c} = \left(\frac{\zeta-c}{\zeta+c}\right)^{\sigma} \qquad \sigma = 2-\frac{\tau}{\pi}$$

lead to the wing profiles with the acute angle τ at the rear edge (Figure 73).

Approximate methods for calculating the flow past a wing profile of arbitrary shape were presented in all previous editions (see 4th edition) of this book and also in Section 10. The present capabilities of computers make such approximate methods virtually redundant.

9. Resultant Vector and Resultant Moment of Forces of Flow Pressure on the Closed Contour. Chaplygin Formulas. Zhukovskiy Theorem. Coefficients of the Lifting Force and of the Moment of the Plate

Let us determine the dynamic effect of flow around the profile immersed in it. Let us form the expression for the resultant vector $\boldsymbol{R}$ and for the resultant moment L_0 of the forces of flow pressure on the profile C (Figure 74) relative to the coordinate origin 0. Using the Bernoulli equation

$$p = \text{const} - \frac{\rho|V|^2}{2}$$

we obtain the expression of the resultant vector

$$\boldsymbol{R} = -\oint_C p\boldsymbol{n}\,ds = \frac{\rho}{2}\oint_C |V|^2\boldsymbol{n}\,ds$$

and of the resultant moment

$$L_0 = -\oint_C (xn_y - yn_x)\,p\,ds = \frac{\rho}{2}\oint_C (xn_y - yn_x)\,|V|^2 ds$$

Passing in these formulas to complex quantities, we note that

$$n = -ie^{i\theta} \qquad ds = dze^{-i\theta} \qquad xn_y - yn_x = \operatorname{Re}(iz\bar{n}) \qquad V = \pm|V|e^{i\theta}$$

Then the previous formulas for the force and for the moment will be reduced to the form

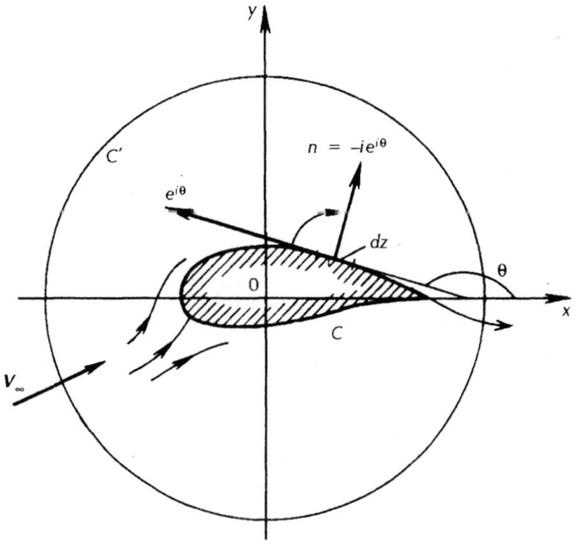

Fig. 74

$$R = R_x + iR_y = \frac{\rho i}{2}\oint_C |V|^2 dz \qquad L_0 = -\frac{\rho}{2}\mathrm{Re}\oint_C |V|^2 e^{-2i\theta} z\,dz$$

Let the following replacement be made in these formulas;

$$|V| = \pm V e^{-i\theta} = \pm \bar{V} e^{i\theta}$$

we find

$$\bar{R} = R_x - iR_y = \frac{\rho i}{2}\oint_C |V|^2 dz = \frac{\rho i}{2}\oint_C \bar{V}^2 dz \qquad L_0 = -\frac{\rho}{2}\mathrm{Re}\oint_C \bar{V}^2 z\,dz \tag{81}$$

Recalling that as before $\bar{V} = d\chi/dz$, we rewrite these expression in the following form

$$\bar{R} = \frac{\rho i}{2}\oint_C \left(\frac{d\chi}{dz}\right)^2 dz \qquad L_0 = -\frac{\rho}{2}\mathrm{Re}\oint_C \left(\frac{d\chi}{dz}\right)^2 z\,dz \tag{82}$$

The above formulas for the resultant vector and resultant moment of the forces of flow pressure on the profile were suggested by Chaplygin in 1910.[10]

We shall show that in order to calculate the contour integrals entering into Chaplygin's formulas (81) or (82), it is not obligatory to know the entire complex potential, but it is sufficient to have only the first three coefficients in the expansion of the derivative of the complex potential $d\chi/dz$ or of the conjugate complex velocity $\bar{V}$, in the Laurent series.

Assuming that there are no singular points in the entire external (with respect to the profile immersed in a flow) region of the physical plane, including the contour C itself, we

[10] Chaplygin, S. A., About the pressure of a plane-parallel flow on obstructing bodies (towards the theory of an airplane), *Mat. Sb.*, Vol. 28, 1910.

may represent the conjugate velocity $\bar{V}$ by its expansion into the Laurent series in the physical plane

$$\bar{V} = a_0 + \frac{a_1}{z} + \frac{a_2}{z^2} + \cdots \tag{83}$$

It can be easily verified that the first two coefficients are expressed in terms of the free stream velocity V_∞ and circulation Γ, *viz.*,

$$a_0 = (\bar{V})_{z=\infty} = \bar{V}_\infty \qquad a_1 = \frac{1}{2\pi i}\oint_C \bar{V}dz = \frac{\Gamma}{2\pi i} \tag{84}$$

Let us substitute into the Equations 81 the expansion of the conjugate velocity into the Laurent series (83), preserving under the integral sign only those terms that will give the results of integration differing from zero; we have (C' is the arbitrary circular contour in the physical plane that encircles the profile C)

$$\bar{R} = \frac{\rho i}{2}\oint_{C'}\left(\cdots + \frac{2a_0a_1}{z} + \cdots\right)dz = -2\pi\rho a_0 a_1$$

$$L_0 = -\frac{\rho}{2}\operatorname{Re}\oint_{C'}\left(\cdots + \frac{a_1^2 + 2a_0a_2}{z} + \cdots\right)dz = -\pi\rho\operatorname{Re}[i(a_1^2 + 2a_0a_2)]$$

Using expressions (84) for the first two coefficients a_0 and a_1, we obtain

$$R = -i\rho V_\infty\Gamma \qquad L_0 = -2\pi\rho\operatorname{Re}(i\bar{V}_\infty a_2) \tag{85}$$

The first of Equations 85 represents the well-known theorem of Zhukovskiy about the lifting force of the wing in a plane-parallel irrotational flow of an ideal incompressible fluid. This theorem was published in 1906 in the classical work "On Bound Vortices",[11] in which Zhukovskiy was first to establish the vortex origin of the forces exerted on the wing by the flow and to indicate the simple dependence between this force and velocity circulation along the contour which encircles the streamlined wing.

In order that within the framework of the ideal irrotational flow one could determine the magnitude of the effect of flow on the body immersed in it, Zhukovskiy replaced the wing by a certain imaginary liquid wing bounded by a closed streamline and assumed that inside of this liquid wing occurred motion with a singularity, i.e., with a vortex.[12]

This kind of a vortex was identified by Zhukovskiy as a bound vortex. The intensity of the bound vortex can be calculated only with the aid of a certain additional assumption. As we already know, such an assumption is the Zhukovskiy-Chaplygin hypothesis about the finite nature of the velocity at the rear sharp edge of the wing profile. By using this hypothesis it has become possible to theoretically determine the magnitude of the imposed circulation or the intensity of the bound vortex. This magnitude is specified by Equations 63 or 64 of the present chapter.

---

[11] Zhukovskiy, N. Ye., *Selected Papers*, Vol. 2, Gostekhizdat, Moscow, 1948, p. 97.

[12] See Stepanov, G. Yu., About some inaccuracies in the explanations of the theory of the wing, *Mekh. Zhidk. Gaza*, No. 3, pp. 188, 189; 1975, Stepanov, G. Yu., Concerning the comments on the selected papers by S. A. Chaplygin, *Mekh. Zhidk. Gaza*, No.4, pp. 180 to 183, 1977.

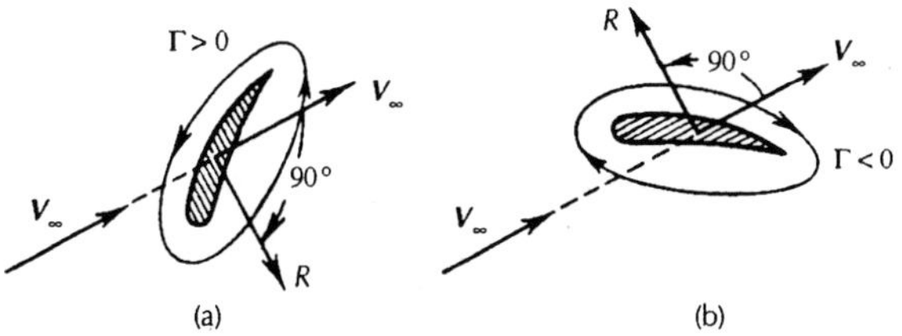

Fig. 75

The idea of the bound vortex and the hypothesis about the finite nature of the velocity at the rear sharp edge of the wing profile form the basis of the entire theory of the wing in a plane-parallel flow. These ideas have been further developed and applied in a more general, three-dimensional theory of the wing of finite span, system of wings, and also in the theory of the blade apparatus of turbines, compressors, and pumps.

Returning to the first formula of Equation 85 and taking into account its vector character, we conclude that the magnitude of the resultant vector $\boldsymbol{R}$ is equal to the product of the fluid density by the stream velocity and by the imposed circulation (Γ is an algebraic quantity which can be both positive and negative)

$$|R| = \rho|V_\infty||\Gamma| \tag{86}$$

and its direction is determined (Figure 75a and b) by the rotation through 90° of the free stream velocity vector $\boldsymbol{V}_\infty$ in a clockwise sense when $\Gamma > 0$ and in a counterclockwise sense when $\Gamma < 0$. Thus, as has already been noted in Section 5, the resultant vector $\boldsymbol{R}$ is the force which is transverse to the free stream direction or to the direction of translational motion of a body moving in the reverse direction. According to Zhukovskiy's formulation, in order to obtain the direction of the resultant vector of pressure forces $\boldsymbol{R}$ it is necessary to rotate the velocity vector of the free stream $\boldsymbol{V}_\infty$ through 90° to the side opposite of the direction of circulation.

Note that the direction of circulation in the general case of the wing profile of arbitrary shape and any direction of inflow is not liable to direct understanding, as was, for example, the case with circulatory flow around a circular cylinder which admitted the simple physical interpretation (Section 5). In order to determine the "direction of circulation", it is necessary in each separate case to find the sign of Γ using Equation 63 for this purpose.

Very essential is the fact that the only force acting on the profile in a plane-parallel irrotational flow of an ideal incompressible fluid is the force which is perpendicular to the free stream direction, or, in the reversed flow, transverse to the direction of profile motion and which can be called the lifting or sustaining force because it is this force that ensures the lift of the aircraft into air and supports its wing in horizontal flight. Emphasis is placed on the absence of the force component directed along the fluid motion or of the direction of body motion with respect to the fluid, the resistance force. This is the specific case of d'Alembert's paradox.

The Zhukovskiy theorem confirms d'Alembert's paradox for any plane irrotational motion of an ideal incompressible fluid, both in the presence of bound vortices and in their absence. The general proof of the paradox for a three-dimensional flow will be given in Chapter 9.

Making use of the Zhukovskiy theorem and of the Zhukovskiy-Chaplygin hypothesis it is possible to obtain from Equations 63, 64, and 86 the expression for the lifting force in the form

$$|R| = 4\pi a m_\infty \rho|V_\infty|^2\left|\sin(\varepsilon_0 - \theta_\infty)\right| = 4\pi a m_\infty \rho|V_\infty|^2|\sin\alpha| \tag{87}$$

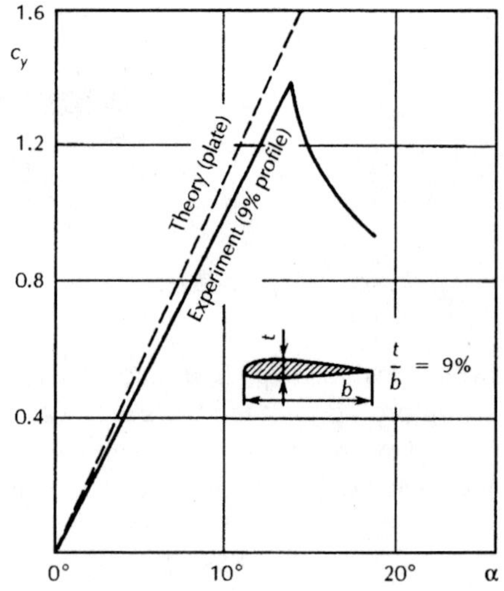

Fig. 76

for the first time given by Chaplygin. The product am_∞ entering this equation depends on the shape of the streamlined contour.

To illustrate the application of Equations 85, first consider the simplest example (given at the end of Section 7), i.e., the flow past a plate with smooth streaming of jets from the trailing edge. In this case $am_\infty = 1/2c$ and the lifting force is equal to

$$|R| = 2\pi\rho c|V_\infty|^2 \sin\alpha \tag{88}$$

The coefficient of the lifting force is introduced as the ratio of the lifting force $|R|$ to the velocity head of the free stream $1/2\rho|V_\infty|^2$ and the length of the chord b. Usually the axis $0x$ is directed along the velocity V_∞; the lifting force will be directed along the axis $0y$ and can be designated by Y or R_y. That is why the coefficient of the lifting force is conventionally designated in the literature a c_y and the resistance coefficient as c_x.

With this notation, we have

$$c_y = \frac{|R|}{1/2\rho|V_\infty|^2 b} = 8\pi\frac{am_\infty}{b}\sin\alpha \tag{89}$$

and in the case of the plate now considered ($b = 2c$)

$$c_y = 2\pi\sin\alpha \tag{90}$$

As numerous experiments show, at comparatively small angles of attack under which the condition of smooth streaming of jets from the trailing edge is really fulfilled, Equation 90, rewritten in the form ($\sin\alpha \approx \alpha$)

$$c_y \approx 2\pi\alpha \approx 6.28\alpha \tag{91}$$

satisfactorily reflects the regularity revealed in the experiments: at small angles of attack the coefficient of the lifting force of the plate is directly proportional to the angle of attack. However, the theoretical value of the proportionality coefficient $2\pi = 6.28$ is somewhat overestimated. Figure 76 presents for comparison the theoretical straight line and experimental curve

$c_y(\alpha)$ for the symmetrical profile with the ratio of the maximum thickness to the chord equal to 9%. As is seen from the figure, within the range of the angles of attack $-13° < \alpha < 13°$ the difference between the theoretical coefficient of the lifting force of the plate and the experimental is not great for a thin profile. (The region of negative values of α is not given in the figure, but by symmetry of the profile it differs from the region of positive value of α only by the sign).

The possibility that the plate would be considered as the limiting case of a thin symmetric profile with its relative thickness tending to zero allows the justification of the validity of Equation 91. Physically it is possible to imagine that no matter how thin the plate may be, its frontal edge is yet rounded. At it, due to a very great (for a theoretically infinite plate) velocity, considerable rarefaction is formed which creates the suction force directed opposite to the flow along the plate surface. Together with the pressure forces perpendicular to the plate, it is this suction force that produces the lifting force transverse to the direction of the flow coming onto the plate.

In order to calculate the resultant moment L_0 of the forces of flow pressure onto the plate, we expand the conjugate velocity $\bar{V}$ of flow past the plate, determined in the now-considered case of smooth streaming of fluid from the trailing edge of the plate by Equation 77, into the series in the negative powers of z:

$$\bar{V}(z) = u_\infty - iv_\infty\sqrt{\frac{z-c}{z+c}} = u_\infty - iv_\infty + \frac{cv_\infty i}{z} - \frac{1}{2}\frac{c^2 v_\infty i}{z^2} + \cdots$$

Comparing this expansion with series (83), we obtain

$$a_0 = u_\infty - iv_\infty = \bar{V}_\infty \qquad a_1 = civ_\infty = ci|V_\infty|\sin\alpha$$

$$a_2 = -\frac{1}{2}c^2 iv_\infty = -\frac{1}{2}c^2 i|V_\infty|\sin\alpha$$

We find from the first of Equation 85 that

$$R = R_x + iR_y = -i\rho(u_\infty + iv_\infty)\Gamma = \rho v_\infty\Gamma - i\rho u_\infty\Gamma$$

or, using the formula of circulation (75), that

$$R_x = -2\pi\rho c v_\infty^2 \qquad R_y = 2\pi\rho c u_\infty v_\infty$$

According to the second formula of Equation 85, the moment L_0 is equal to

$$L_0 = -2\pi\rho\,\mathrm{Re}\left[-\frac{1}{2}c^2 iv_\infty i(u_\infty - iv_\infty)\right] = -\pi\rho c^2 v_\infty \mathrm{Re}(u_\infty - iv_\infty) = -\pi\rho c^2 v_\infty u_\infty$$

Going over from the velocity projections u_∞, v_∞ to their expressions in terms of the velocity modulus and the angle of attack $\alpha = \theta_\infty$, we obtain

$$R_x = -2\pi\rho c|V_\infty|^2\sin^2\alpha \qquad R_x = 2\pi\rho c|V_\infty|^2\sin\alpha\cos\alpha$$

$$L_0 = -\pi\rho c^2|V_\infty|^2\sin\alpha\cos\alpha$$

Having the expression for the projections of the lifting force and of the moment relative to the point 0, we may find the equation for the line of action of the resultant force. Let x and

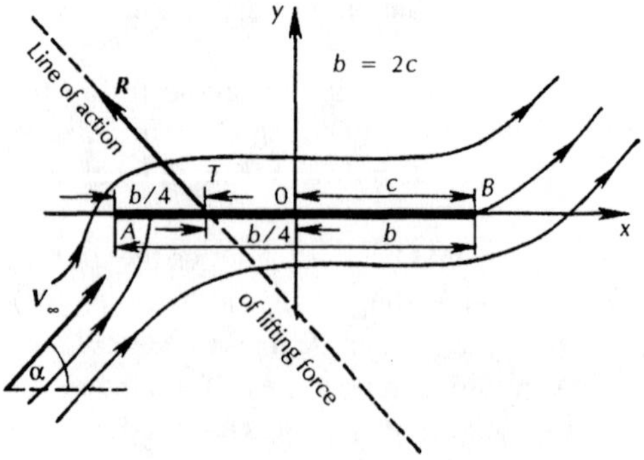

Fig. 77

y denote the current coordinates of the point at the line of action of the resultant force; the equation for this line is obtained in the form

$$xR_y - yR_x = L_0$$

or, using the previous expressions and making obvious simplifications,

$$x \cos \alpha + y \sin \alpha = -\frac{1}{2} c \cos \alpha$$

The point T (Figure 77) at the intersection of the line of action of the lifting force with the plate is called the center of thrust. If all the forces of flow pressure onto the plate are reduced to one force R, this force will be applied at the center of thrust T. Assuming in the latter equation that $y = 0$, we find the abscissa of the position of the center of thrust T on the plate

$$x = -\frac{c}{2}$$

The center of flow pressure onto the plate is located at one fourth of its length from the frontal edge, with the position of the center of thrust being independent either of the free stream velocity or of the angle of attack, as is clear from the latter equation.

By introducing into consideration the coefficient of the moment

$$c_m = \frac{L_0}{1/2\rho|V_\infty|^2 b^2}$$

we have at small angles of attack ($\sin \alpha \approx \alpha$, $\cos \alpha \approx 1$) that

$$c_m = \frac{\pi}{2}\alpha$$

Comparing this to the formula for the coefficient of the lifting force $c_y = 2\pi\alpha$, we see that

$$c_m : c_y = 1:4$$

Note that this relation, usually expressed in terms of the coefficients $dc_m/d\alpha$ and $dc_y/d\alpha$ in the form

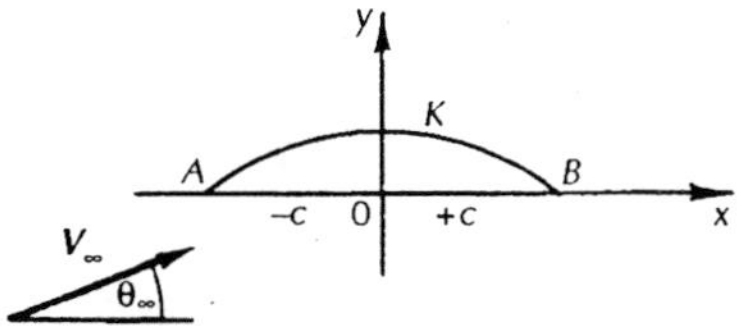

Fig. 78

$$\frac{dc_m}{dc_y} = \frac{dc_m}{d\alpha} : \frac{dc_y}{d\alpha} = \frac{1}{4}$$

satisfactorily corresponds to the experimental data, not only for the flow past a plate, but also for thin symmetric profiles.

By adopting the point $T(-c/2, 0)$ to be the point relative to which the resultant moment of pressure forces is taken, then the moment L_T will be equal to zero This property of the plate and also of thin symmetric profiles is used in designing the fastening of rudders. The axis of rudder rotation is taken to be the line passing through the centers of thrust of the rudder sections normal to the axis of rotation; this reduces to minimum the moment required to rotate the rudder and facilitates its control.

10. The Theory of a Thin Profile (Arc)

A thin, slightly cambered wing profile can be considered to be arc, the equation of which in the coordinate system shown in Figure 78, is represented in the form

$$y = h(x) = \frac{1}{2}[h_1(x) + h_2(x)] \tag{92}$$

where $y_1 = h_1(x)$ and $y_2 = h_2(x)$ are the equations for the upper and lower surfaces of the thin profile approximately replaced by the arc.

Let us place the arc into a plane-parallel irrotational flow of an ideal incompressible fluid having the velocity at infinity V_∞ which forms the angle of attack θ_∞ with the chord of the profile $AB = 2c$. To calculate the flow past a thin profile, we apply the method of small perturbations, having accepted that the perturbation of the velocity $\boldsymbol{V}^*$ caused by the arc in the oncoming flow is small in magnitude. Thus, the velocity vector $\boldsymbol{V}$ at any point of the flow can be represented by the sum

$$\boldsymbol{V} = \boldsymbol{V}_\infty + \boldsymbol{V}^* \tag{93}$$

The condition of the arc impermeability will be written in the form of the projection of the velocity $\boldsymbol{V}$ onto the normal to the contour being equal to zero on the arc contour:

$$V_n = V_{\infty n} + V_n^* = u_\infty \cos(\widehat{\boldsymbol{n}, x}) + v_\infty \cos(\widehat{\boldsymbol{n}, y}) + V_n^* = 0$$

From this follows the boundary condition

$$V_n^* = -V_{\infty n} = -[u_\infty \cos(\widehat{\boldsymbol{n}, x}) + v_\infty \cos(\widehat{\boldsymbol{n}, y})] = u_\infty \sin\theta - v_\infty \cos\theta \tag{94}$$

which must be fulfilled on the both sides of the arc (θ is the angle of the tangent to the arc with the axis $0x$; u_∞ and v_∞ are the projections of the velocity $\boldsymbol{V}_\infty$ onto the axes $0x$ and $0y$). From the condition of small camber it follows that the angle θ is small and it is possible to assume that

$$\sin\theta \approx \operatorname{tg}\theta = h'(x) \qquad \cos\theta \approx 1$$

Using the smallness of the deviation of the arc from the chord AB, we replace in boundary condition (94) the normal component of the perturbation velocity V_n^* by the projection v^* of this velocity onto the axis $0y$, an approximation that is accurate to the small higher-order values. Moreover, boundary condition (94) will be satisfied not on the arc K, but on the chord AB. Such transport of boundary conditions from one curve onto the other, deviating little from the former, introduces an error of second order smallness. Thus, we have the boundary condition

$$v^*(x, y) = u_\infty h'(x) - v_\infty \quad \text{when} \quad -c \le x \le c \quad y = \pm 0 \tag{95}$$

The double sign before the zero expresses the applicability of boundary condition (95) on both the upper and lower boundaries of the cut AB along the axis $0x$ in the plane of the flow, i.e., the continuity of the change in the function $v^*(x, y)$ which presents the imaginary part of $\bar{V}^*$ taken with the reverse sign, on transition from one side of the cut AB to the other.

At an infinite distance from the arc the perturbation velocity vanishes; moreover, the Zhukovskiy-Chaplygin condition of the finite character of velocity at the point B should also be fulfilled.

Thus, the problem is reduced to the search for the holomorphic, vanishing at infinity, function $\bar{V}^*(z)$ the imaginary part of which along the segment of the real axis ($-c \le x \le c$) satisfies the condition

$$\operatorname{Im} \bar{V}^*(z) = v_\infty - u_\infty h'(x) \tag{96}$$

directly following from Equation 95; the function $\bar{V}^*(z)$ must be finite at $z = c$.

Let us compare the problem posed with the earlier considered problem of flow past a plate with the finite velocity at the rear edge (Section 7). From the distribution of velocities (77) obtained in Section 7, we find the conjugate velocity of perturbation

$$\bar{V}^* = \bar{V} - \bar{V}_\infty = u_\infty - iv_\infty\sqrt{\frac{z-c}{z+c}} - u_\infty + iv_\infty = iv_\infty\left(1 - \sqrt{\frac{z-c}{z+c}}\right)$$

In this case condition (96) seems to be satisfied because on the plate

$$\bar{V}^* = iv_\infty\left(1 - i\sqrt{\frac{c-x}{c+x}}\right) = u_\infty\sqrt{\frac{c-x}{c+x}} + iv_\infty,$$

and $h'(x) = 0$; moreover, $\bar{V}^* \to 0$ when $z \to \infty$ and $u^* = 0$, $v^* = -v_\infty$ at $z = c$, i.e., $\bar{V}^*$ is finite. On the upper side of the cut ($-c \le x \le c$, $y = +0$) the root is given the positive value, and on the lower side ($-c \le x \le c$, $y = -0$) the root is given the negative value. Thus equating the real and imaginary parts in the previous formula, we obtain

$$u^*(x, +0) = v_\infty\sqrt{\frac{c-x}{c+x}} \qquad u^*(x, -0) = -v_\infty\sqrt{\frac{c-x}{c+x}}$$

$$v^*(x, +0) = v^*(x, -0) = -v_\infty$$

where

$$u^*(x, +0) = -u^*(x, -0) \qquad v^*(x, +0) = v^*(x, -0) \tag{97}$$

The finite jump in the value of the perturbation velocity component u^* on transition Across the cut from the positive to negative y is equal to

$$u^*(x, +0) - u^*(x, -0) = 2v_\infty \sqrt{\frac{c-x}{c+x}}$$

As explained at the end of Section 7, this discontinuity in the component u^* with the component v^* being continuous can be interpreted as the presence of a vortex layer with the total intensity equal to the circulation imposed on the plate.

Returning to the problem of flow past an arc, imagine[13] the sought-after conjugate velocity of perturbed motion as the product (the radical plays the same part as in Equation 77 for the plate)

$$\bar{V}^* = \sqrt{\frac{z-c}{z+c}}\, f(z) \tag{98}$$

where $f(z)$ is the limited, holomorphic function vanishing at infinity; with the function $\bar{V}^*(z)$ thus selected, the conditions $\bar{V}^*(c) = 0$, $\bar{V}^*(c) = u_\infty$, necessary for separation-free flow past the rear edge ($z = c$), will be fulfilled. At the frontal edge ($z = -c$) the velocity in the general case becomes infinite.[14]

According to the familiar Cauchy formula, the following integral representation of the function $f(z)$ in terms of its values on the contour is

$$f(z) = \frac{1}{2\pi i} \oint_L \frac{f(\zeta)}{\zeta - z}\, d\zeta \tag{99}$$

where L is the contour of the cut AB with two infinitely small circles that isolate the branching points A and B of the integrand

$$f(\zeta) = \sqrt{\frac{\zeta + c}{\zeta - c}}\, \bar{V}^*(\zeta)$$

in the upper part of the cut AB the root should be taken with the plus sign and it should be regarded that

$$f(\xi) = \sqrt{\frac{\xi + c}{\xi - c}}\, [u^*(\xi, +0) - iv^*(\xi, +0)] =$$

$$= -i\sqrt{\frac{c + \xi}{c - \xi}}\, [u^*(\xi, +0) - iv^*(\xi, +0)]$$

and on the lower part, the minus sign, so that

[13] Sedov, L. I., *Plane Problems of the Hydrodynamics and Aerodynamics*, Gostekhizdat, Moscow, 1950, pp. 46 to 64; Kibel, I. A., Kochin, N. Ye., and Roze, N. V., *Theoretical Hydromechanics*, Part 1, Fizmatgiz, Moscow, 1963, pp. 297 to 321; Sedov, L. I., *Two-dimensional Problems in Hydrodynamics and Aerodynamics*, John Wiley & Sons, New York, 1965; Kochin, N. E., Kibel, J. A., and Rose, N. V., *Theoretische Hydromechanik*, Volo 1 and 2, Berlin, 1954.

[14] Regarding this, see remark in Section 7.

$$f(\xi) = -\sqrt{\frac{\xi + c}{\xi - c}}\,[u^*(\xi, -0) - iv^*(\xi, -0)] =$$

$$= i\sqrt{\frac{c + \xi}{c - \xi}}\,[u^*(\xi, -0) - iv^*(\xi, -0)]$$

Forming the integral (99) we see that as the radii of small circles tend to zero, the integrals along these small circles with the centers at the points A and B also tend to zero, whereas the Cauchy integral (99) is reduced to the difference of the integrals found within in the limits ($-c$, c) calculated in the positive direction of the axis $0x$ along the upper and lower sides of the cut AB. As regards the basic property of the vortex layer located, in this general case, along the arc or, with the error of the second order of smallness, along the cut AB, conditions (97) will be fulfilled. Thus, terms containing $u^*(\xi, +0)$ and $u^*(\xi, -0)$ in the difference of integrals will be eliminated and the terms with $v^*(\xi, +0) = v^*(\xi, -0) = v^*(\xi, 0)$ will add up; we obtain

$$\bar{V}^*(z) = -\frac{1}{2\pi i}\sqrt{\frac{z - c}{z + c}}\int_{-c}^{+c}\frac{2v^*(\xi, 0)}{\xi - z}\sqrt{\frac{c + \xi}{c - \xi}}\,d\xi$$

and, according to boundary condition (95),

$$\bar{V}^*(z) = -\frac{1}{\pi i}\sqrt{\frac{z - c}{z + c}}\int_{-c}^{+c}\frac{u_\infty h'(\xi) - v_\infty}{\xi - z}\sqrt{\frac{c + \xi}{c - \xi}}\,d\xi \qquad (100)$$

Returning to the full velocity $\bar{V}$ and noting that at small θ_∞ it is possible to assume that

$$u_\infty = |V_\infty| \qquad v_\infty = |V_\infty|\theta_\infty$$

we finally obtain

$$\bar{V}(z) = \bar{V}_\infty + \bar{V}^*(z) = \bar{V}_\infty - \frac{|V_\infty|}{\pi i}\sqrt{\frac{z - c}{z + c}}\int_{-c}^{+c}\frac{h'(\xi) - \theta_\infty}{\xi - z}\sqrt{\frac{c + \xi}{c - \xi}}\,d\xi \qquad (101)$$

Let us calculate the resultant vector and the resultant moment of the forces of flow pressure on the arc. For this, we expand into a series in negative powers of z the following expressions

$$\sqrt{\frac{z - c}{z + c}} = \sqrt{\frac{1 - c/z}{1 + c/z}} = 1 - \frac{c}{z} + \frac{1}{2}\frac{c^2}{z^2} - \cdots$$

$$\frac{1}{\xi - z} = -\frac{1}{z}\frac{1}{1 - \xi/z} = -\frac{1}{z} - \frac{\xi}{z^2} - \cdots$$

and substitute them into Equation 101; comparing the result of the substitution with the expansion of the conjugate velocity (83), we find the values of the expansion coefficients

$$a_0 = \bar{V}_\infty$$

$$a_1 = \frac{\Gamma}{2\pi i} = \frac{|V_\infty|}{\pi i}\int_{-c}^{+c}[h'(\xi) - \theta_\infty]\sqrt{\frac{c+\xi}{c-\xi}}\,d\xi$$

$$a_2 = -\frac{|V_\infty|}{\pi i}\int_{-c}^{+c}[h'(\xi) - \theta_\infty]\sqrt{c^2-\xi^2}\,d\xi$$

and, consequently, according to Equation 85, the general expressions of the resultant vector and of the resultant moment

$$R_x = -2\pi\rho c|V_\infty|^2\theta_\infty^2 + 2\rho|V_\infty|^2\theta_\infty\int_{-c}^{+c}h'(\xi)\sqrt{\frac{c+\xi}{c-\xi}}\,d\xi$$

$$R_y = 2\pi\rho c|V_\infty|^2\theta_\infty - 2\rho|V_\infty|^2\int_{-c}^{+c}h'(\xi)\sqrt{\frac{c+\xi}{c-\xi}}\,d\xi$$

$$L_0 = -\pi\rho c^2|V_\infty|^2\theta_\infty + 2\rho|V_\infty|^2\int_{-c}^{+c}h'(\xi)\sqrt{c^2-\xi^2}\,d\xi \qquad (102)$$

In the approximation adopted $R_x = 0$ because both terms on the right side of the equation of system (102) represent second-order small quantities, the disregard of which in the theory of a thin wing is obligatory; this result corresponds to d'Alembert's paradox about the absence of resistance.

In the case of a plate, $h'(\xi) \equiv 0$ and Equations 102 lead to the familiar formulas (accurate to θ_∞ in the first degree)

$$R_y = 2\pi\rho c|V_\infty|^2\theta_\infty \qquad L_0 = -\pi\rho c^2|V_\infty|^2\theta_\infty$$

In the case of an arbitrary shape of the arc $y = h(x)$ the quantities R_y and L_0 can be easily calculated from Equations 102, and it should be emphasized that all terms on the right sides of the second and third equations of system (102) have the first order of smallness. The integral terms on the right sides of expressions 102 for R_y and L_0 are supplementary to the corresponding expressions for the plate; these terms determine the influence of the arc camber. Thus, for example, for the arc of the parabola

$$h(x) = \delta\left(1 - \frac{x^2}{c^2}\right) \qquad (103)$$

where δ is the height of the camber line,

$$R_y = 2\pi\rho c|V_\infty|^2\left(\theta_\infty + \frac{\delta}{c}\right) \qquad L_0 = -\pi\rho c^2|V_\infty|^2\theta_\infty \qquad (104)$$

As is seen from these equations, in the approximation adopted, the relative camber $\delta/(2c)$ increases the lifting force, but does not influence the moment with respect to the coordinate origin 0 located in the middle of the segment AB.

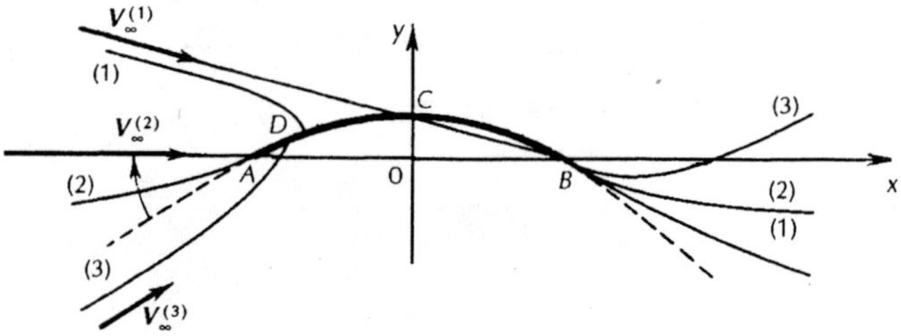

Fig. 79

According to the first of Equations 104, the direction of the zero lifting force ($\theta_\infty = -\delta/c$) or the noncirculatory direction coincides with the direction of the straight line drawn through the vertex C of the arc and through the rear edge B (Figure 79).

The calculation of velocity on the surface of the arc by Equation 101 offers some difficulties because the integral on the right side is improper and must be calculated in the sense of its principal value.[15]

Omitting computational details, the formula of the distribution of velocities over the surface of the parabolic arc specified by Equation 103 is

$$\bar{V} = |V_\infty|\left(1 - i\theta_\infty\sqrt{\frac{z-c}{z+c}}\right) + \frac{2i|V_\infty|\delta}{z + \sqrt{z^2 - c^2}} \tag{105}$$

The first term gives the distribution of velocity over the plate (for small values of θ_∞); the second term presents the influence of the arc camber $\delta/(2c)$.

At $\theta_\infty = 0$ the previous formula is reduced to the form

$$\bar{V} = |V_\infty|\left(1 + \frac{2i\delta}{z + \sqrt{z^2 - c^2}}\right)$$

assuming here that $z = \pm c$, we obtain

$$\bar{V}(\pm c) = |V_\infty|\left(1 \pm i\frac{2\delta}{c}\right) \tag{106}$$

Noting that the tangent or sine of the small slope of the tangent to the arc at the points ($x = \pm c$, $y = 0$) is equal to $\mp 2\delta/c$ and the cosine is equal to unity, we conclude that in the case considered the velocity vectors at the edges are directed along the tangents to the arc at these points. In other words, at $\theta_\infty = 0$, the frontal edge is the point of smooth or, in everyday language, shock-free approach of fluid to the arc, and the trailing edge is the point at which the stream flows smoothly past this edge (Figure 79).

In this case the lifting force, equal to

$$(R_y)_{\theta_\infty = 0} = 2\pi\rho|V_\infty|^2\delta = 4\pi\rho c|V_\infty|^2\frac{\delta}{2c} \tag{107}$$

is proportional to the relative camber of the arc.

---

[15] See, for example, Smirnov, V. I., *Kurs vysshei matematiki* (*The Course of High Mathematics*), Vol. 4, Gostekhizdat, Moscow, 1941, pp. 252 to 253.

In Figure 79 the patterns of flow past a parabolic arc are superimposed for three different and most characteristic directions of the oncoming flow, with only null streamlines being shown for each of the three flows in order not to make the figure look awkward. The dashes show the directions of the tangents to the arc at the frontal and rear edges. The directions of oncoming flows far from the arc are represented by the velocity vectors at infinity $V_\infty^{(1)}$, $V_\infty^{(2)}$, and $V_\infty^{(3)}$.

According to the condition adopted when deriving Equation 105, the imposed circulations are selected for all three flows in such a way that the rear edge B could be the point of finite velocity and smooth flow discharge. As is seen from the figure, at this B point all three null streamlines, (1), (2), and (3), have the common point B of flow discharge and the common tangent, which is shown by dashes and which coincides with the tangent to the arc at the rear edge B.

On the other hand, the locations of the points of flow branching near the frontal edge A differ drastically. Smooth, shock-free approach of the flow to the frontal edge with the finite velocity determined by Equation 106 is formed only when the oncoming flow is directed along the velocity vector at infinite $V_\infty^{(2)}$, i.e., parallel to the chord of the arc AB along the axis $0x$. It is natural to call this direction the shock-free direction of flow.

In the case of noncirculatory impingement of the flow corresponding to the velocity vector at infinity $V_\infty^{(1)}$, just as in the case of the impingement in the direction of the tangent to the arc at the frontal edge with the velocity at infinity $V_\infty^{(3)}$, the velocity at point A will be infinitely great; the flow onto the arc at the frontal edge will not be shock-free.

The above-described situations, which are (strictly speaking) relevant only to the considered parabolic arcs symmetric with respect to the axis $0y$, also hold to some degree for the arcs of other shapes. Moreover, accurate calculations and experimental results show that these properties of flow past an arc are qualitatively valid also for the profiles of small finite thicknesses and curvatures. In this case the shock-free flow should be conventionally understood to be a flow in which there are no sharp rarefaction peaks that form near the tip of the profile and that correspond to great local velocities on the frontal surface of the profile. To determine the direction of the oncoming flow corresponding to this conventional shock-free flow, it is necessary in each separate case to carry out a number of analytical calculations of flow past the profile considered. In these calculations the angles of attack can be selected on the grounds of their small difference from the angle of attack corresponding to shock-free flow past the "skeleton" of the profile.

11. Zhukovskiy Theorem of the Lifting Force of the Profile in a Cascade

The plane cascade of profiles (Figure 80) is understood to be an infinite set of periodically located (in the plane) identical wing profiles, each of which is obtained from the adjacent one by a parallel displacement through a certain length t, called the gap, in a direction that determines the axis of the cascade. The angle β between the chord of a profile and perpendicular to the axis of the cascade is sometimes called the angle of stagger. The additional angle β' is called the setting angle of the profile. The vector $\boldsymbol{t}$, equal in length to the gap and directed perpendicularly to the axis of the cascade to the side of flow, is called the gap vector.

In contrast to a single profile, velocities at an infinite distance upstream and downstream of the cascade are generally different in both magnitude and direction. The cascade not only alters the velocity of the impinging flow, but also rotates the flow.

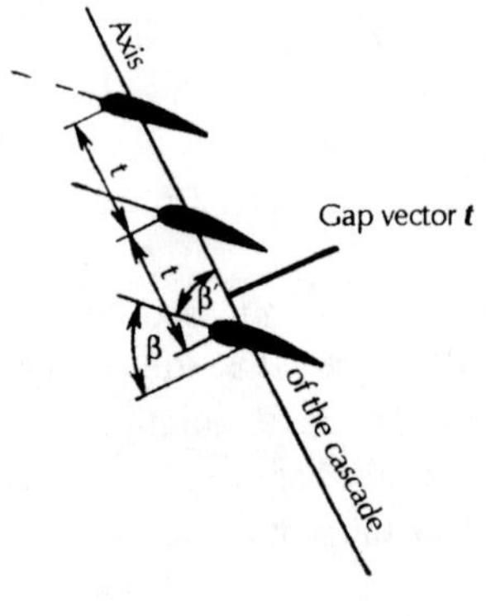

Fig. 80

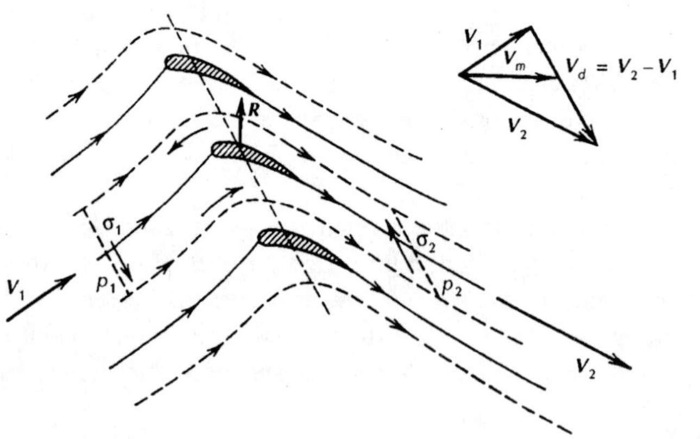

Fig. 81

Let us denote (Figure 81) the velocity vector of the flow at infinity before the cascade by $\boldsymbol{V}_1$, the pressure by p_1, and the velocity vector and pressure at infinity distance behind the cascade by $\boldsymbol{V}_2$ and p_2, respectively; the density is identical throughout and is equal to ρ.

Let us consider in the plane of the figure, the stream tube formed by any two streamlines displaced from each other in the direction of the axis of the cascade by a distance equal to the gap. The entire flow apparently can be divided into such equal stream tubes because the flow has the property of periodicity, with the period equal to the gap.

Let us apply the theorem of momentum in the form of Euler, having taken as the control surface the side surface of the above-described stream tube and two infinitely spaced sections of the tube σ_1, σ_2, parallel to the axis of the cascade and equal in length to the gap. Then, designating the resultant vector of the forces of flow pressure on the profile by $\boldsymbol{R}$, we have

$$(p_1 - p_2)\,\boldsymbol{t} + \rho\,(\boldsymbol{t}\cdot\boldsymbol{V}_1)\,\boldsymbol{V}_1 - \rho\,(\boldsymbol{t}\cdot\boldsymbol{V}_2)\,\boldsymbol{V}_2 - \boldsymbol{R} = 0 \tag{108}$$

The quantities $\boldsymbol{t}\cdot\boldsymbol{V}_1 = \boldsymbol{t}\cdot\boldsymbol{V}_2$ represent the equal volumetric liquid flow rates through the sections of the stream tube and $-\boldsymbol{R}$ is the resultant vector of the forces of profile pressure on the flow.

Assuming the flow to be irrotational and applying the Bernoulli equation, we obtain

$$p_1 - p_2 = \frac{1}{2}\rho\,(V_2^2 - V_1^2)$$

or

$$p_1 - p_2 = \frac{1}{2}\rho\,(\boldsymbol{V}_1 + \boldsymbol{V}_2)\cdot(\boldsymbol{V}_2 - \boldsymbol{V}_1)$$

if the difference of the squared velocities is represented as the scalar product of the sum of velocity vectors by their difference.

Let us introduce two velocities characteristic for the flow past the cascade: the mean vector velocity

$$\boldsymbol{V}_m = \frac{1}{2}(\boldsymbol{V}_1 + \boldsymbol{V}_2)$$

and the velocity of flow deviation

$$\boldsymbol{V}_d = \boldsymbol{V}_2 - \boldsymbol{V}_1$$

which characterizes the deviation of the flow by the cascade.

Then,

$$p_1 - p_2 = \rho \boldsymbol{V}_m \cdot \boldsymbol{V}_d$$

$$\boldsymbol{t} \cdot \boldsymbol{V}_1 = \boldsymbol{t} \cdot \boldsymbol{V}_2 = \boldsymbol{t} \cdot \boldsymbol{V}_m \qquad \boldsymbol{t} \cdot \boldsymbol{V}_d = \boldsymbol{t} \cdot (\boldsymbol{V}_2 - \boldsymbol{V}_1) = 0 \tag{109}$$

and Equation 108 will be rewritten in the form

$$\boldsymbol{R} = \rho (\boldsymbol{V}_m \cdot \boldsymbol{V}_d) \boldsymbol{t} + \rho (\boldsymbol{t} \cdot \boldsymbol{V}_1)(\boldsymbol{V}_1 - \boldsymbol{V}_2) = \rho (\boldsymbol{V}_m \cdot \boldsymbol{V}_d) \boldsymbol{t} - \rho (\boldsymbol{t} \cdot \boldsymbol{V}_m) \boldsymbol{V}_d$$

The expression on the right side represents the expansion of the double vector product, so that finally

$$\boldsymbol{R} = \rho \boldsymbol{V}_m \times (\boldsymbol{t} \times \boldsymbol{V}_d) \tag{110}$$

Noting that according to Equation 109 the gap $\boldsymbol{t}$ is perpendicular to the flow deviation velocity $\boldsymbol{V}_d$ and the vector $\boldsymbol{t} \times \boldsymbol{V}_d$ is perpendicular to the flow plane (i.e., also to $\boldsymbol{V}_m$), we find the value of the resultant vector in the form

$$R = \rho t V_m V_d \tag{111}$$

Vector equation (110) gives in explicit form the dependence of the resultant vector $\boldsymbol{R}$ on the liquid density, gap $\boldsymbol{t}$ of the cascade, and on two characteristic velocities: mean $\boldsymbol{V}_m$ and flow deviation $\boldsymbol{V}_d$. Scalar equation (111) determines the magnitude of the resultant vector of the forces of flow pressure on a profile in the cascade as the product of the liquid density, cascade gap, mean, and deviation velocities. It follows from vector equation (110) that the resultant vector $\boldsymbol{R}$ lies in the flow plane and is directed along the perpendicular to the mean velocity $\boldsymbol{V}_m$ to the side determined by vector product (110).

We now introduce the unit vectors of the cascade axis $\boldsymbol{a}$ and of the normal to the plane of the figure in the direction of the reader $\boldsymbol{k}$, having directed them so that the set of the vectors $\boldsymbol{t}$, $\boldsymbol{a}$, and $\boldsymbol{k}$ could form the trihedral aligned with the adopted right side system of coordinates. (The vector $\boldsymbol{t}$ is not a unit vector; its magnitude is equal to the gap.) Noting that, according to Equation 109, the vector $\boldsymbol{V}_d = \boldsymbol{V}_2 - \boldsymbol{V}_1$ is directed parallel to the cascade axis, we obtain

$$\boldsymbol{V}_d = (V_{2a} - V_{1a}) \boldsymbol{a}$$

and, consequently,

$$\boldsymbol{t} \times \boldsymbol{V}_d = \boldsymbol{t} \times \boldsymbol{a} (V_{2a} - V_{1a}) = t (V_{2a} - V_{1a}) \boldsymbol{k}$$

Let us determine the velocity circulation Γ along the contour composed of the directing sections σ_1, σ_2 and two streamlines displaced by the gap. Forming the integral $\oint V_s ds$ along the positive direction of the contor surface, shown in Figure 81, we obtain ($\sigma_1 = \sigma_2 = t \cdot 1$)

$$\Gamma = (V_{2a} - V_{1a}) t$$

because the integrals along the streamlines displaced by the gap are equal in magnitude and opposite in sign. Thus, $\boldsymbol{t} \times \boldsymbol{V}_d = \boldsymbol{k}\Gamma$ and, consequently, according to Equation 110:

$$\boldsymbol{R} = \rho \boldsymbol{V}_m \times \boldsymbol{k}\Gamma \tag{112}$$

This formula corresponds to the formula for the unit profile, with the mean vector velocity $\boldsymbol{V}_m$ replacing the velocity at infinity $\boldsymbol{V}_\infty$.

We now let the gap $\boldsymbol{t}$ go to infinity, and at the same time preserve the magnitude of circulation. In this case the difference $V_{2a} - V_{1a}$ should tend to zero, which in the presence of Equation 109 establishes the common velocity $\boldsymbol{V}_1 \to \boldsymbol{V}_2 \to \boldsymbol{V}_\infty$ instead of $\boldsymbol{V}_m$, and Equation 112 goes over into Zhukovskiy's equation for the unit profile

$$\boldsymbol{R} = \rho \boldsymbol{V}_\infty \times \boldsymbol{k}\Gamma$$

The law for determining the direction of the vector $\boldsymbol{R}$ from Equation 112 remains the same as that for the unit profile. The vector $\boldsymbol{R}$ has the direction of the vector $\boldsymbol{V}_m$ which was rotated in the plane of the flow through 90° to the side opposite to the direction of circulation, i.e., to that direction of circulation around the contour of integration for which the circulation is positive.[16]

12. Application of the Method of Conformal Mappings to the Theory of Jet Flows

The method of conformal mappings presented earlier was long ago found successful, not only for solving the problems of plane flow past closed contours, but particularly for wing profiles. One of the most important regions of application of this method was the theory of flows of an ideal incompressible fluid with a discontinuous potential known as jet flows. Due to the absence of internal friction in ideal fluid flows the disruption of the continuity of flow becomes possible as does the formation of "dead" zones of quiescent fluid.

Examples of flows with the disruption of continuity can be furnished by cavitational "cavities" filled with the vapor of liquid and air, stalling zones behind blunt bodies, jets of dense medium surrounded by a liquid (gas) of small density, water discharge through a barrier and from under a panel, and flows relating to the problems of transport on "air cushions". Some of the above problems, especially hydrotechnical problems associated with the motion of water in the gravity field and often having essentially spatial character, present considerable mathematical difficulties and cannot be presented in this textbook.[17]

The direct method for solving the problems of plane discontinuous flows is the method of conformal mappings.[18] "Free" streamlines (Figure 82) separated from the body immersed

---

[16] Theoretical problems in this field were analyzed in Kochin, N. Ye., *Gidrodinamicheskaya teoriya reshyotok. Sovremennye problemy mekhaniki* (*The Hydrodynamic Theory of Cascades. The Present-day Problems of Mechanics*), Gostekhizdat, Moscow, 1949, and Stepanov, G. Yu., *Gidrodinamika reshyotok turbomashin* (*The Hydrodynamics of the Cascades of Turbines*), Fizmatgiz, Moscow, 1962. Concerning a wider subject matter, Sedov, L. I., *Ploskiye zadachi gidrodinamiki i aerodinamiki* (*The Plane Problems of Hydrodynamics and Aerodynamics*), Nauka Press, Moscow, 1980, the theory of cascades is discussed in Chapter 3. The practical problems of numerical calculation of the cascades of profiles are covered in detail in Zhukovskiy, M. I., *Raschyot obtekaniya reshyotok profiley turbomashin* (*Calculation of Flow Past the Cascades of the Profiles of Turbines*), Mashgiz, Moscow – Leningrad, 1960, and in Sedov, L. I., *Two-Dimensional Problem in Hydrodynamics and Aerodynamics*, John Wiley & Sons, New York, 1965.

[17] The overview of these difficulties and the means of approximate solutions of this kind of problems can be found in Chapters 11 and 12 of the book by M. I. Gurevich, *Teoriya struy idealnoy zhidkosti* (*The Theory of Jets of an Ideal Fluid*), Nauka Press, Moscow, 1979.

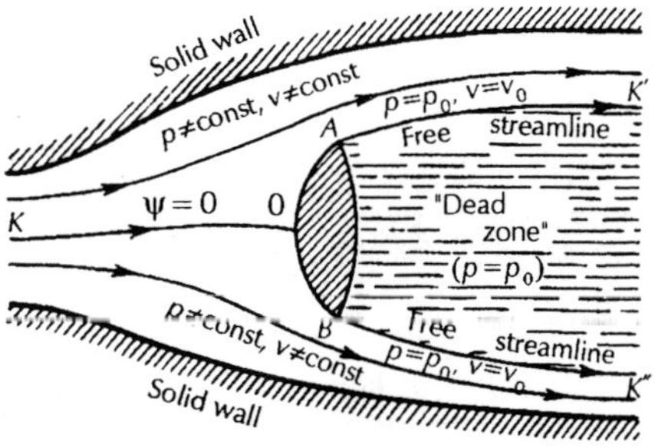

Fig. 82

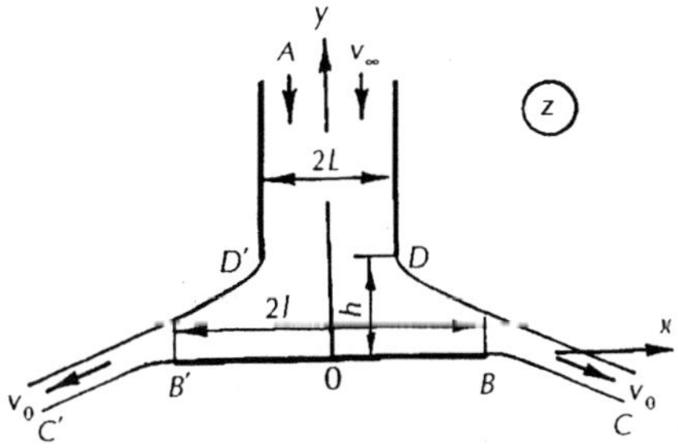

Fig. 83

in a flow are the flow boundaries of unknown shape. At the same time, in the hodograph plane of the conjugate velocity $\bar{V}$ or of the inverse quantity – the Kirchhoff function $\zeta = 1/\bar{V}$ – just as in the plane of the complex potential χ, the boundaries are represented by the segments of the simplest lines: straight lines and circles if the streamlined bodies consist of rectilinear segments (plates, wedges) located in channels with rectilinear boundaries, etc. In cases such as these, it is not difficult to establish a one-to-one correspondence between the regions in the planes of complex variables $\bar{V}$ (or ζ) and χ, and the method of conformal mappings gives the mapping function $\bar{V}(\chi)$ or $\zeta(\chi)$. The problem is reduced to the examination of the coupling between these functions and the basic complex argument z in the physical plane. This can be done with the aid of the obvious relation

$$z = \int \frac{d\chi}{\bar{V}(\chi)} = \int \zeta(\chi)\, d\chi$$

Let us illustrate the method using as an example the impact of the jet issuing from a channel of finite width on the plate located normally to it (Figure 83). Let us consider the right side of the physical plane z. Designate the half-width of the supplying channel by L, the half-width of the plate by l, and the distance between the plate and the exit section of the channel by h. The velocity constant in magnitude along the boundaries of free streamlines BC and DC is denoted by v_0, the velocity in the channel far from the exit by v_∞; then

$$\bar{V}_A = v_\infty i \qquad \bar{V}_0 = 0 \qquad \bar{V}_B = v_0 \qquad \bar{V}_D = v_0 i \qquad |V_C| = v_0$$

Let us introduce the complex variable (the Kirchhoff function)

$$\zeta = \frac{1}{\bar{V}} = \frac{1}{|V|}\, e^{i\theta}$$

and establish the correspondence between the points of the physical plane z and the hodograph plane of the inverse conjugate velocity vector ζ.

To the boundaries of the flow $A0BC$ and ADC in the plane z corresponds the boundaries in the plane ζ indicated in Figure 84 by the same letters; the flow region is hatched. Moreover, let us denote the flow region (Figure 85) in the plane of the complex potential χ. This will be the band between two straight lines parallel to the real axis and corresponding to the constancy of the stream function ψ along the streamlines. The streamline $A0BC$ will be taken

[18] Historically, the first and simplest scheme is that of Helmholtz and Kirchhoff (Helmholtz, H., Über discontinuierliche Flüssigkeitsbewegungen, *Monatsber. Berl. Akad. Wiss.*, April, Bd. 23, 1868, Kirchhoff, G., Zur Theorie freier Flüssigkeitsstrahlen, *Grelle's J. Mathe.*, 70, 1869).

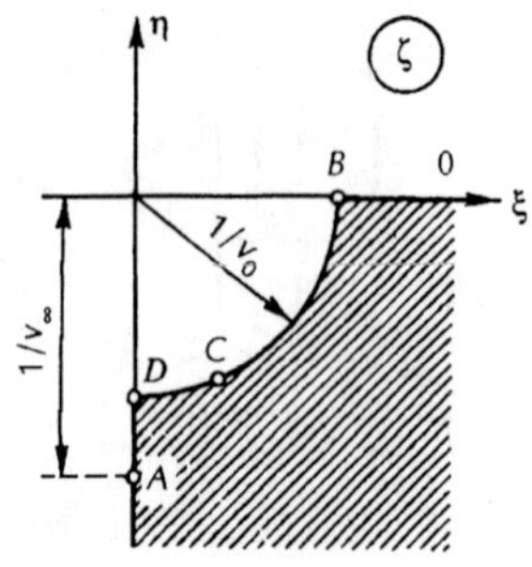

Fig. 84

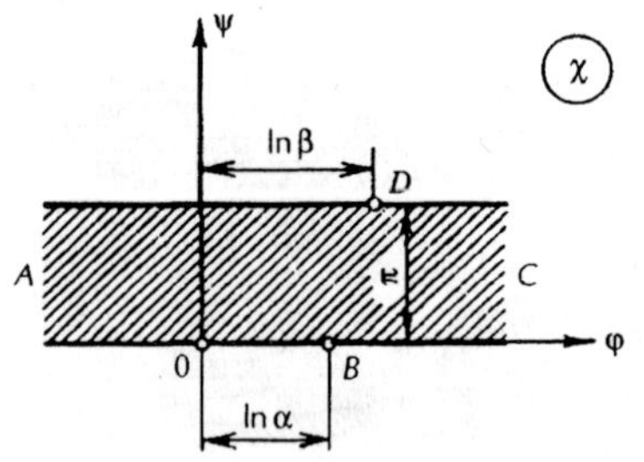

Fig. 85

as the null, the streamline *ADC* will be given the value $\psi = \pi$. Then the band width will be equal to π.

The one-to-one correspondence between the hatched regions of the planes ζ and χ can be established by conformally mapping both regions onto this or an other half-plane, and then, employing the well-known theorem of the functions of the complex variable, connecting the mapping functions by the fractional-linear substitution. Thus, the hatched region of the plane ζ is transformed into the lower half-plane with the aid of the relation

$$Z_1 = \frac{(\zeta^2 - 1/v_0^2)^2}{(\zeta^2 + 1/v_0^2)^2}$$

and the band of the plane χ into the upper half-plane with the aid of the transformation

$$Z_2 = e^{\chi}$$

For the one-to-one correspondence between the half-planes Z_1 and Z_2 we should have

$$Z_1 = \frac{aZ_2 + b}{Z_2 + c}$$

or

$$\left(\frac{\zeta^2 - 1/v_0^2}{\zeta^2 + 1/v_0^2}\right)^2 = \frac{ae^{\chi} + b}{e^{\chi} + c}$$

The constants a, b, c will be determined from the conditions of correspondence

$$\chi = 0 \text{ at } \zeta = \infty \quad \text{(at the point 0)}$$

$$\chi = \ln\alpha \text{ at } \zeta = 1/v_0 \quad \text{(at the point } B\text{)}$$

$$\chi = \pi i + \ln\beta \text{ at } \zeta = -i/v_0 \quad \text{(at the point } D\text{)}$$

where $\ln\alpha$ and $\ln\beta$ are presently unknown constants equal to the abscissas of the points *B* and *D* in the plane χ. Substituting these values into the previous equation, we obtain

$$a = -\frac{\beta+1}{\alpha-1} \qquad b = \frac{\beta+1}{\alpha-1}\alpha \qquad c = \beta$$

so that

$$\left(\frac{\zeta^2 - 1/v_0^2}{\zeta^2 + 1/v_0^2}\right)^2 = \frac{\beta + 1}{\alpha - 1}\frac{\alpha - e^\chi}{e^\chi + \beta}$$

or, after simple transformations,

$$\zeta = \frac{1}{v_0}\frac{\sqrt{(\alpha - 1)(e^\chi + \beta)} + \sqrt{(\beta + 1)(\alpha - e^\chi)}}{\sqrt{(\alpha + \beta)(e^\chi - 1)}} \tag{113}$$

This equation is supplemented with the obvious relation

$$z = \int_0^\chi \zeta d\chi \tag{114}$$

where the constant is determined from the condition of correspondence at the point 0. Equations 113 and 114 give the sought-after solution of the problem; it is only necessary to compare the unknown constants α, β with the geometric parameters of the problem l, L, and h and the given velocity v_∞, having taken into account that the flow rate is adopted to be equal to π.

For this let us first determine the half-width of the plate l. We have

$$l = \int_0^{\ln\alpha} \zeta d\chi = \frac{1}{v_0}\sqrt{\frac{\alpha - 1}{\alpha + \beta}}\int_0^{\ln\alpha}\sqrt{\frac{e^\chi + \beta}{e^\chi - 1}}d\chi + \frac{1}{v_0}\sqrt{\frac{\beta + 1}{\alpha + \beta}}\int_0^{\ln\alpha}\sqrt{\frac{\alpha - e^\chi}{e^\chi - 1}}d\chi$$

Integration can be accomplished with the aid of the substitutions

$$\frac{e^\chi + \beta}{e^\chi - 1} = \frac{1}{t^2}$$

for the first integral and

$$\frac{\alpha - e^\chi}{e^\chi - 1} = \frac{1}{\tau^2}$$

for the second. We have

$$l = \frac{1}{v_0}\left[2\sqrt{\beta\frac{\alpha - 1}{\alpha + \beta}}\operatorname{arctg}\left(\sqrt{\beta\frac{\alpha - 1}{\alpha + \beta}}\right) + \right.$$

$$\left. + \sqrt{\frac{\alpha - 1}{\alpha + \beta}}\ln\frac{1 + \sqrt{\dfrac{\alpha - 1}{\alpha + \beta}}}{1 - \sqrt{\dfrac{\alpha - 1}{\alpha + \beta}}} + \pi\sqrt{\frac{\beta + 1}{\alpha + \beta}}(\sqrt{\alpha} - 1)\right] \tag{115}$$

The half-width of the channel L will be expressed in terms of the flow rate π and velocity v_∞:

$$L = \frac{\pi}{v_\infty}$$

The ratio

$$\frac{l}{L} = \frac{1}{\pi}\frac{v_\infty}{v_0}\left[2\sqrt{\beta\frac{\alpha-1}{\alpha+\beta}}\operatorname{arctg}\left(\sqrt{\beta\frac{\alpha-1}{\alpha+\beta}}\right) + \right.$$

$$\left. + \sqrt{\frac{\alpha-1}{\alpha+\beta}}\ln\frac{1+\sqrt{\dfrac{\alpha-1}{\alpha+\beta}}}{1-\sqrt{\dfrac{\alpha-1}{\alpha+\beta}}} + \pi\sqrt{\frac{\beta+1}{\alpha+\beta}}(\sqrt{\alpha}-1)\right]$$

can be expressed only in terms of α and β by noting that at the point A

$$|V| = v_\infty \qquad \zeta = -\frac{i}{v_\infty} \qquad \chi = -\infty$$

According to Equation 113, corresponding to the above is the equation

$$-\frac{i}{v_\infty} = \frac{1}{v_0}\frac{\sqrt{\beta(\alpha-1)} + \sqrt{\alpha(\beta+1)}}{i\sqrt{\alpha+\beta}}$$

where

$$\frac{v_\infty}{v_0} = \frac{\sqrt{\alpha+\beta}}{\sqrt{\alpha(\beta+1)} + \sqrt{\beta(\alpha-1)}} \tag{116}$$

Thus, we have

$$\frac{l}{L} = \frac{2\sqrt{\beta(\alpha-1)}\operatorname{arctg}\left(\sqrt{\beta\dfrac{\alpha-1}{\alpha+\beta}}\right)}{\pi[\sqrt{\alpha(\beta+1)} + \sqrt{\beta(\alpha-1)}]} +$$

$$+ \frac{\sqrt{\alpha-1}\ln\dfrac{1+\sqrt{(\alpha-1)/(\alpha+\beta)}}{1-\sqrt{(\alpha-1)/(\alpha+\beta)}} + \pi\sqrt{\beta+1}(\sqrt{\alpha}-1)}{\pi[\sqrt{\alpha(\beta+1)} + \sqrt{\beta(\alpha-1)}]} \tag{117}$$

The force of the jet pressure on the entire plate will be found from the Bernoulli theorem

$$P = 2\int_0^l (p-p_0)\,dx = \rho\int_0^l [v_0^2 - |V|^2]\,dx$$

where p_0 is the pressure in the dead zone behind the plate. Performing the replacement on the surface of the plate

$$dx = dz \qquad |V| = \frac{1}{\zeta} \qquad dz = \zeta d\chi$$

we obtain

$$P = \rho \int_0^{\ln \alpha} \left(v_0^2 - \frac{1}{\zeta^2}\right)\zeta d\chi = \rho v_0 \int_0^{l} \left(v_0\zeta - \frac{1}{v_0\zeta}\right) d\chi$$

To compute the integral, it should be noted that according to Equation 113:

$$v_0\zeta = \frac{\sqrt{(\alpha-1)(e^\chi+\beta)} + \sqrt{(\beta+1)(\alpha-e^\chi)}}{\sqrt{(\alpha+\beta)(e^\chi-1)}}$$

$$\frac{1}{v_0\zeta} = \frac{\sqrt{(\alpha-1)(e^\chi+\beta)} - \sqrt{(\beta+1)(\alpha-e^\chi)}}{\sqrt{(\alpha+\beta)(e^\chi-1)}}$$

so that

$$P = 2\rho v_0 \sqrt{\frac{\beta+1}{\alpha+\beta}} \int_0^{\ln \alpha} \sqrt{\frac{\alpha-e^\chi}{e^\chi-1}} d\chi = 2\pi\rho v_0 \sqrt{\frac{\beta+1}{\alpha+\beta}} (\sqrt{\alpha}-1)$$

Let us introduce explicitly the length of the plate $2l$; for this purpose from the previous equation we find separately

$$\pi\sqrt{\frac{\beta+1}{\alpha+\beta}}(\sqrt{\alpha}-1) = \frac{P}{2\rho v_0} \qquad \frac{1}{\sqrt{\alpha+\beta}} = \frac{P}{2\rho v_0 \pi \sqrt{\beta+1}(\sqrt{\alpha}-1)}$$

and substitute into Equation 115 the half-lengths of the plate; then simple transformations will give

$$P = \frac{1}{2}\rho v_0^2 \cdot 2l \frac{2\pi(\sqrt{\alpha}-1)}{\pi(\sqrt{\alpha}-1) + 2\sqrt{\beta\frac{\alpha-1}{\beta+1}}\operatorname{arctg}\left(\sqrt{\beta\frac{\alpha-1}{\beta+1}}\right) + \sqrt{\frac{\alpha-1}{\beta+1}}\ln\frac{1+\sqrt{(\alpha-1)/(\alpha+\beta)}}{1-\sqrt{(\alpha-1)/(\alpha+\beta)}}} =$$

$$= \frac{1}{2}\rho v_0^2 \cdot 2l \times$$

$$\times \frac{2\pi}{\pi + 2\sqrt{\frac{\beta(\sqrt{\alpha}+1)}{(\beta+1)(\sqrt{\alpha}-1)}}\operatorname{arctg}\left(\sqrt{\beta\frac{\alpha-1}{\beta+1}}\right) + \sqrt{\frac{(\sqrt{\alpha}+1)}{(\beta+1)(\sqrt{\alpha}-1)}}\ln\frac{1+\sqrt{(\alpha-1)/(\alpha+\beta)}}{1-\sqrt{(\alpha-1)/(\alpha+\beta)}}} \quad (118)$$

To determine the constants α and β we have still only one equation (117). Forming the second equation requires the introduction of the constant h, which determines the distance of the plate from the exit section of the duct. It can be easily done having written that according to Equation 114

$$h = \operatorname{Im} z_D = \operatorname{Im} \int_0^{\chi_D} \zeta d\chi = \operatorname{Im} \int_0^{\pi i + \ln \beta} \zeta d\chi$$

Then, having composed the ratio between h and l or L, we obtain the sought-after second equation that determines α and β. Omitting the calculations which are analogous to those

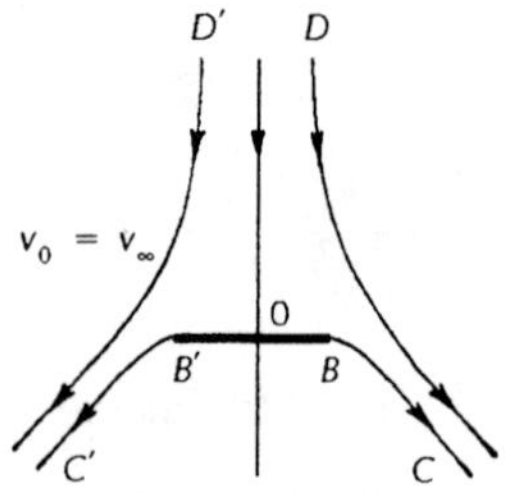

Fig. 86

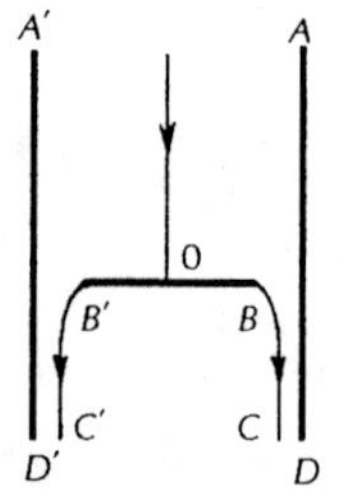

Fig. 87

made when composing Equation 115 for l, we satisfy ourselves with the consideration of several particular cases not requiring the assignment of h:

1. A plate immersed in an infinite jet flow. In this limiting case ($L \to \infty$), it directly follows from Equation 117 that the quantity α is close to unity, and it is clear from physical considerations that the value of h and, consequently, the value of the parameter β, are unnecessary for calculating P. This can be proved by computations. Let us pass in Equation 118 for the force P to the limit at $\alpha = 1$ and arbitrary β. This can be done most simply by expanding the functions arctg and ln in the denominator into the series in the powers of small arguments $\sqrt{\beta\frac{\alpha-1}{\alpha+\beta}}$ and $\sqrt{\frac{\alpha-1}{\alpha+\beta}}$ and retaining only the first terms of these expansions; thus,

$$P = \frac{1}{2}\rho v_0^2 \cdot 2l\left[\frac{2\pi}{\pi + 2\sqrt{\beta\frac{\sqrt{\alpha}+1}{(\beta+1)(\sqrt{\alpha}-1)}}\sqrt{\beta\frac{\alpha-1}{\beta+1}} + \sqrt{\frac{\sqrt{\alpha}+1}{(\beta+1)(\sqrt{\alpha}-1)}}\,2\sqrt{\frac{\alpha-1}{\beta+1}}}\right]_{\alpha=1} =$$

$$= \frac{1}{2}\rho v_0^2 \cdot 2l\,\frac{2\pi}{\pi + 2\frac{2\beta}{\beta+1} + 2\frac{2}{\beta+1}} = \frac{1}{2}\rho v_0^2 \cdot 2l\,\frac{2\pi}{\pi+4}$$

which is the well-known Rayleigh equation; it follows from Equation 116 that $v_0 = v_\infty$ when $\alpha = 1$.

2. A plate immersed in a finite-width jet flow restricted by free surfaces (Figure 86). Let us move the exit of the channel to infinity in the positive direction of the axis $0y$ ($h = \infty$). Recalling that $\ln\beta$ is the abscissa of the point D on the straight line $\psi = \pi$ (Figure 86) i.e., the value of the velocity potential φ which varies from $-\infty$ far upstream to $+\infty$ downstream), we assume that $\ln\beta = -\infty$, and, consequently, $\beta = 0$. Then, Equation 118 will give the expression for the force of the pressure on the plate

$$P = \frac{1}{2}\rho v_0^2 \cdot 2l\,\frac{2\pi}{\pi + \sqrt{\frac{\sqrt{\alpha}+1}{\sqrt{\alpha}-1}}\ln\frac{1+\sqrt{(\alpha-1)/\alpha}}{1-\sqrt{(\alpha-1)/\alpha}}} \tag{119}$$

where, according to Equation 116, $v_0 = v_\infty$. Equation 119, together with Equation 117 rewritten as

$$\frac{l}{L} = \frac{1}{\pi}\sqrt{\frac{\alpha-1}{\alpha}}\ln\frac{1+\sqrt{(\alpha-1)/\alpha}}{1-\sqrt{(\alpha-1)/\alpha}} + 1 - \frac{1}{\sqrt{\alpha}} \tag{120}$$

determines P at the given ρ, v_∞, and l, and at the ratio l/L of the plate width to the free stream width. When $L \to \infty$ and l is fixed; i.e., according to Equation 120 when $\alpha \to 1$, Equation 119 will again yield the Rayleigh equation.

3. Jet flow past a plate located in a channel of finite width (Figure 87). This case can be considered as the limiting one, when $\beta \to \infty$. From Equation 116:

$$v_0 = v_\infty(\sqrt{\alpha} + \sqrt{\alpha - 1})$$

The force will be determined from Equation 118 by the parametric expression

$$P = \frac{1}{2}\rho v_\infty^2 \cdot 2l \frac{2\pi(\sqrt{\alpha} + \sqrt{\alpha - 1})^2}{\pi + 2\sqrt{\frac{\sqrt{\alpha} + 1}{\sqrt{\alpha} - 1}} \operatorname{arctg} \sqrt{\alpha - 1}} \qquad (121)$$

which should be considered together with the expression for the plateto-channel width ratio

$$\frac{l}{L} = \frac{2\sqrt{\alpha - 1} \operatorname{arctg} \sqrt{\alpha - 1} + \pi(\sqrt{\alpha} - 1)}{\pi(\sqrt{\alpha} + \sqrt{\alpha - 1})} \qquad (122)$$

It can be easily verified from Equation 121 that when $L \to \infty$ and l is fixed, i.e., when $\alpha \to 1$, according to Equation 122, we again obtain the Rayleigh equation for the case of transverse flow past a plate.

Cases 2 and 3 give a general idea about the necessity of changes in the flow boundaries which are introduced when blowing against plates in wind tunnels with open (free jet) and closed working sections.

The example above quite clearly illustrates the simple classical method of Kirchhoff, which in the past was employed in practice, but which does not characterize the present state-of-the-art in the theory of jet flows. Development of this theory has been due to the works of many outstanding scientists at the end of the 19th and first half of the 20th centuries. The literature devoted to the theory of jet flows of an ideal fluid is voluminous, therefore, we restrict ourselves to listing the most notable scientific sources. First, we mention the relevant fundamental works of Russia's outstanding scientists: Zhukovskiy,[19] Chaplygin,[20] Nekrasov,[21] Lavrentiyev,[22] and Sedov.[23] Also cited are the historically important investigations of the problem

[19] Zhukovskiy, N. Ye., The modification of the Kirchhoff method for determining the motion of fluid in two dimensions at a constant velocity given on the unknown streamline, *Mat. Sb.*, 15, 1890.

[20] Chaplygin, S. A., Concerning the jets in an incompressible fluid, *Tr. Otd. Fiz. Nauk Obshchestva Lyubitelei Estestv. Nauk*, 10(1), 1889; Chaplygin, S. A., About the forces acting on the cylinder in a flow with the formation of discontinuous surfaces, *Sb. Obshcheteoreticheskoi Gruppy TsAGI, Tr. TsAGI*, No. 240, 1935; Chaplygin, S. A., and Lavrentiyev, A. L., About the lifting force and resistance of a long plane wing on the assumption of stalling from its upper surface, *Tr. TsAGI*, No. 240, 1935.

[21] Nekrasov, A. I., About the discontinuous flow of fluid in two dimensions past a barrier in the form of the arc of a circle, *Izv. Ivanovo-Voznesenskogo Politekh. Inst.*, No. 5, 1922 (*Vyp. Mat.*). This work is closely related to the studies Sekerzh-Zenkovich, Ya. I., Concerning the theory of flow past a curvilinear arc with separation of jets, *Tr. TaAGI*, No. 299, 1937 and other works by the same author.

[22] Lavrentiyev, M. A., Concerning certain properties of one-sheet functions with applications to the theory of jets, Mat. Sb. (*Novaya Seriya*), 4(46) (No. 3), 1938; The cumulative charge and the principle of its operation. *Uspekhi Mat. Nauk*, 12(4)(76), 1957.

[23] Sedov, L. I., *Ploskiye zadachi gidrodinamiki i aerodinamiki* (*The Plane Problems of the Hydrodynamics and Aerodynamics*), Nauka Press, Moscow, 1980; a number of works by the same author published in *Tr. TsAGI* and *Teoreticheskiye Zametki TsAGI* between 1934 and 1938; *Two-Dimensional Problems in Hydrodynamics and Aerodynamics*, John Wiley & Sons, New York, 1965.

about a plane jet flow past a wedge, which were carried out in the last quarter of the 19th century by Bobylyov[24] and Meshcherskiy.[25]

Having no possibility to dwell on numerous foreign publications concerning the theory of jet flows (the reader is related to the book by M. I. Gurevich[28] and to the list of literature in it), we shall mention only two basic works which played an essential role in for the development of the modern theory of jets: Levi-Civitta[26] and Villat.[27]

In Russian literature a very notable book is that of M. I. Gurevich,[28] which contains a wealth of information and presents rigorously and with clarity the theoretical foundations of the methods for calculating jet flows and the solutions for many corresponding problems, including plane and axisymmetric problems, as well as the problems of jets of "heavy" liquid which have great practical importance. Of considerable interest is the book by Gogish and Stepanov[29] and the work solely by Stepanov.[30] A new approach to jet flows based on the solution of the Euler equation in the general case of rotational flows is indicated in the book by M. A. Goldshtik.[31]

---

[24] Bobylyov, D. K., A note on the pressure produced by a stream of infinite width on two walls located at any angle to each other, *Zh. Russkogo Fiz.-Khim.* Obshch., 13, 1881.

[25] Meshcherskiy, I. V., Concerning the problem of the resistance of fluids, *Zh. Russkogo Fiz.-Khim. Obshch.*, 18, 1886.

[26] Levi-Civitta, T., Scie e leggi di resistenza, *Rendiconti Circolo Mathem. Palermo*, 23 (1), 1907.

[27] Villat, H., Aperçus theoriques de la resistance des fluides, *Phys. Math.*, 38, Gauthier-Villars, Paris, 1920, p. 101.

[28] Gurevich, M. I., *Teoriya struy idealnoy zhidkosti (The Theory of Ideal Fluid Jets)*, Nauka Press, Moscow, 1979. The book contains an extensive literature list.

[29] Gogish, L. V., and Stepanov, G. Yu., *Turbulentnye otryvnye techeniya (Turbulent Separation Flows)*, Nauka Press, Moscow, 1979, pp. 286 to 313.

[30] Stepanov, G. Yu., *Gidrodinamicheskaya teoriya apparatov na vozdushnoi podushke (The Hydrodynamic Theory of Apparatus on Air Cushion)*, Mashgiz Press, Moscow, 1963.

[31] Goldshtik, M. A., *Vikhrevye potoki (Rotational Flows)*, Nauka Press, Novosibirsk, 1981.

8

Plane Irrotational Motion of an Ideal Gas

1. Basic Equations of Motion and their Linearization

The continuity equation (9) of Chapter 4 in the case of plane steady motion of gas will acquire the form

$$\frac{\partial(\rho u)}{\partial x} + \frac{\partial(\rho v)}{\partial y} \equiv \rho\left(\frac{\partial u}{\partial x} + \frac{\partial v}{\partial y}\right) + u\frac{\partial \rho}{\partial x} + v\frac{\partial \rho}{\partial y} = 0 \qquad (1)$$

Provided the motion of gas is adiabatic

$$\frac{\partial \rho}{\partial x} = \frac{d\rho}{dp}\frac{\partial p}{\partial x} = \frac{1}{a^2}\frac{\partial p}{\partial x} \qquad \frac{\partial \rho}{\partial y} = \frac{d\rho}{dp}\frac{\partial p}{\partial y} = \frac{1}{a^2}\frac{\partial p}{\partial y}$$

The Euler equations (4) of Chapter 5 which, in the case of plane steady motion in the absence of volumetric forces, can be rewritten as

$$\rho\left(u\frac{\partial u}{\partial x} + v\frac{\partial u}{\partial y}\right) = -\frac{\partial p}{\partial x} \qquad \rho\left(u\frac{\partial v}{\partial x} + v\frac{\partial v}{\partial y}\right) = -\frac{\partial p}{\partial y}$$

and we obtain

$$\frac{\partial \rho}{\partial x} = -\frac{\rho}{a^2}\left(u\frac{\partial u}{\partial x} + v\frac{\partial u}{\partial y}\right) \qquad \frac{\partial \rho}{\partial y} = -\frac{\rho}{a^2}\left(u\frac{\partial v}{\partial x} + v\frac{\partial v}{\partial y}\right)$$

Substituting these values of $\partial\rho/\partial x$ and $\partial\rho/\partial y$ into Equation 1, we find the equation

$$(a^2 - u^2)\frac{\partial u}{\partial x} - uv\left(\frac{\partial u}{\partial y} + \frac{\partial v}{\partial x}\right) + (a^2 - v^2)\frac{\partial v}{\partial y} = 0 \qquad (2)$$

which is valid for both irrotational and rotational motion. In the case of irrotational motion this equation is supplemented also with the condition of the absence of a vortex

$$\frac{\partial u}{\partial y} - \frac{\partial v}{\partial x} = 0 \tag{3}$$

In the case of irrotational steady adiabatic motion of an ideal gas the following Bernoulli equation is valid throughout the entire region (plane) of motion

$$\frac{1}{2}V^2 + \frac{a^2}{k-1} = \frac{1}{2}(u^2 + v^2) + \frac{a^2}{k-1} = \frac{1}{2}V_\infty^2 + \frac{a_\infty^2}{k-1} \tag{4}$$

with the aid of which the speed of sound can be expressed in terms of the gas velocity.

Condition (3) of the absence of vorticity in the flow allows one to introduce the velocity potential $\varphi(x, y)$ which is connected with velocity projections by the equations

$$u = \frac{\partial \varphi}{\partial x} \qquad v = \frac{\partial \varphi}{\partial y} \tag{5}$$

As regards the stream function, its existence follows from the left side of Equation 1, according to which it is possible to assume (by introducing the constant factor ρ_∞ into the right sides of the equations) that

$$\rho u = \rho_\infty \frac{\partial \psi}{\partial y} \qquad \rho v = -\rho_\infty \frac{\partial \psi}{\partial x} \tag{6}$$

After the substitution of the expressions of velocity projections (5), Equation 2 acquires the final form

$$\left[a^2 - \left(\frac{\partial \varphi}{\partial x}\right)^2\right]\frac{\partial^2 \varphi}{\partial x^2} - 2\frac{\partial \varphi}{\partial x}\frac{\partial \varphi}{\partial y}\frac{\partial^2 \varphi}{\partial x \partial y} + \left[a^2 - \left(\frac{\partial \varphi}{\partial y}\right)^2\right]\frac{\partial^2 \varphi}{\partial y^2} = 0 \tag{7}$$

the squared speed of sound a^2 entering into this equation must be eliminated with the aid of the following equation (resulting from Equation 4)

$$a^2 = \frac{k-1}{2}V_\infty^2 + a_\infty^2 - \frac{k-1}{2}\left[\left(\frac{\partial \varphi}{\partial x}\right)^2 + \left(\frac{\partial \varphi}{\partial y}\right)^2\right] \tag{8}$$

The integration of Equation 7, with the squared sound speed having been eliminated by Equation 8, under the usual conditions of impermeability of the solid walls of bodies in flow and at the prescribed velocities at infinity presents great mathematical difficulties associated with the nonlinearity of the equation. Let us consider the simplest case of a plane flow past thin, slightly curved bodies located in a uniform gas flow at a small angle of attack. In this case the disturbances produced by the body in the uniform flow will be small and Equations 7 and 8 can be linearized.

Let us select the direction of the uniform flow to be coincident with the direction of the axis $0x$ and use the notation U_∞, p_∞, ρ_∞, a_∞, etc., for the velocity, pressure, density, and speed of sound in the uniform flow, respectively. The disturbances introduced into this flow by the thin body will be designated by a tilde (~), so

$$u = U_\infty + \tilde{u} \qquad v = \tilde{v} \qquad p = p_\infty + \tilde{p} \qquad \rho = \rho_\infty + \tilde{\rho} \qquad a = a_\infty + \tilde{a} \tag{9}$$

Let us first consider in detail the two basic cases in which the uniform flow is subsonic or supersonic, i.e. when the Mach number of the uniform flow $\mathbf{M}_\infty = U_\infty / a_\infty$ is smaller or greater than unity ($\mathbf{M}_\infty \neq 1$). The case of the sonic flow ($\mathbf{M}_\infty = 1$) possesses certain specific features and will be considered separately in what follows.

Let us substitute into Equations 2 and 3 the values of velocity projections and of the speed of sound, according to the set of equations (9), and neglect the products of disturbances and their spatial derivatives as small quantities of second and higher orders. In this approximation we have

$$(a_\infty^2 - U_\infty^2)\frac{\partial \tilde{u}}{\partial x} + a_\infty^2 \frac{\partial \tilde{v}}{\partial y} = 0 \qquad \frac{\partial \tilde{u}}{\partial y} - \frac{\partial \tilde{v}}{\partial x} = 0$$

or, dividing both parts of the first equation by a_∞^2,

$$(1 - \mathbf{M}_\infty^2)\frac{\partial \tilde{u}}{\partial x} + \frac{\partial \tilde{v}}{\partial y} = 0 \qquad \frac{\partial \tilde{u}}{\partial y} - \frac{\partial \tilde{v}}{\partial x} = 0 \tag{10}$$

Let us represent the velocity potential φ and the stream function ψ of the disturbed motion as the sum of the velocity potential φ_∞ and stream function ψ_∞ of the uniform motion and corresponding potential $\tilde{\varphi}$ and stream function $\tilde{\psi}$ of small disturbances produced by the thin body in the uniform flow

$$\varphi = \varphi_\infty + \tilde{\varphi} \qquad \psi = \psi_\infty + \tilde{\psi} \tag{11}$$

Substituting these expressions into Equations 5 and 6, we shall make a linearization of them by discarding small quantities of higher orders:

$$U_\infty + \tilde{u} = \frac{\partial \varphi_\infty}{\partial x} + \frac{\partial \tilde{\varphi}}{\partial x} \qquad \tilde{v} = \frac{\partial \varphi_\infty}{\partial y} + \frac{\partial \tilde{\varphi}}{\partial y}$$

$$(\rho_\infty + \tilde{\rho})(U_\infty + \tilde{u}) = \rho_\infty\left(\frac{\partial \psi_\infty}{\partial y} + \frac{\partial \tilde{\psi}}{\partial y}\right)$$

$$(\rho_\infty + \tilde{\rho})\tilde{v} = -\rho_\infty\left(\frac{\partial \psi_\infty}{\partial x} + \frac{\partial \tilde{\psi}}{\partial x}\right)$$

Comparison of the finite (small quantities of zero order) quantities gives

$$\varphi_\infty = U_\infty x \qquad \psi_\infty = U_\infty y \tag{12}$$

whereas comparison of small quantities of the first order leads to the system of equations

$$\tilde{u} = \frac{\partial \tilde{\varphi}}{\partial x} \qquad \tilde{v} = \frac{\partial \tilde{\varphi}}{\partial y} \tag{13}$$

$$\tilde{\rho} U_\infty + \rho_\infty \tilde{u} = \rho_\infty \frac{\partial \tilde{\psi}}{\partial y} \qquad \tilde{v} = -\frac{\partial \tilde{\psi}}{\partial x} \tag{14}$$

On the left side of the first of Equations 14, the disturbance of the density $\tilde{\rho}$ will be expressed in terms of the velocity disturbance $\tilde{u}$ using for this purpose Equation 29 of Chapter 5, rewritten in the form ($p_0 = p_\infty$, $\rho_0 = \rho_\infty$, $kp_\infty/\rho_\infty = a_\infty^2$, const $= U_\infty^2/2$)

$$\frac{V^2}{2} + \frac{a_\infty^2}{k-1}\left[\left(\frac{\rho}{\rho_\infty}\right)^{k-1} - 1\right] = \frac{U_\infty^2}{2}$$

Linearize this equation as made earlier. First, rewrite it in the form

$$\frac{(U_\infty + \tilde{u})^2 + \tilde{v}^2}{2} + \frac{a_\infty^2}{k-1}\left[\left(1 + \frac{\tilde{\rho}}{\rho_\infty}\right)^{k-1} - 1\right] = \frac{U_\infty^2}{2}$$

and then, after simple transformations and disregard of small quantities of higher order, we obtain

$$U_\infty \tilde{u} + \frac{a_\infty^2}{\rho_\infty}\tilde{\rho} = 0$$

Returning to Equation 14, we find, having excluded $\tilde{\rho}$, that

$$\tilde{u} = \frac{1}{1 - \mathbf{M}_\infty^2}\frac{\partial\tilde{\psi}}{\partial y} \qquad \tilde{v} = -\frac{\partial\tilde{\psi}}{\partial x} \tag{15}$$

By substituting the values of $\tilde{u}$, $\tilde{v}$ from Equation 13 into the first of Equations 10, we obtain the basic linearized equation for determining the potential of the velocities of disturbances $\tilde{\varphi}$

$$(1 - \mathbf{M}_\infty^2)\frac{\partial^2\tilde{\varphi}}{\partial x^2} + \frac{\partial^2\tilde{\varphi}}{\partial y^2} = 0 \tag{16}$$

Having taken the values of $\tilde{u}$, $\tilde{v}$ from Equation 15 and substituted them into the second equation of system (10), we find an analogous linearized equation for determining the stream function of the disturbances $\tilde{\psi}$

$$(1 - \mathbf{M}_\infty^2)\frac{\partial^2\tilde{\psi}}{\partial x^2} + \frac{\partial^2\tilde{\psi}}{\partial y^2} = 0 \tag{17}$$

For solving the problems of flow past a thin wing, either Equations 16 or 17 may be employed. Only the boundary conditions on the contour of the profile in the gas flow will differ.

The condition of the impenetrability of the contour can be assigned in two different forms: either zero flow velocity component normal to the surface is required or the surface of the profile is considered to be the null streamline.

Let us assign the contour of the profile located in the vicinity of the segment *ab* of the axis $0x$ by the equation (a and b are the abscissas of the frontal and rear edges of the profile)

$$y = h(x) \qquad a \le x \le b$$

so that the angle θ of the tangent with the axis x is determined by equalities

$$\mathrm{tg}\,\theta \approx \theta \approx h'(x)$$

The condition of the impenetrability of the profile contour in the first of the above forms is

$$V_n = U_{\infty n} + \tilde{V}_n = 0 \quad \text{when} \quad y = h(x) \quad a \le x \le b$$

or, with the same degree of approximation,

$$\tilde{v} = \frac{\partial \tilde{\varphi}}{\partial y} = U_\infty h'(x) \quad \text{when} \quad y = h(x) \quad a \le x \le b$$

According to Equations 11 and 12, the second form of the impenetrability condition will be

$$\psi = \psi_\infty + \tilde{\psi} = U_\infty y + \psi = 0 \quad \text{when} \quad y = h(x) \quad a \le x \le b$$

or

$$\tilde{\psi} = -U_\infty h(x) \quad \text{when} \quad y = h(x) \quad a \le x \le b$$

As it was done in Section 10 of Chapter 7, in the adopted degree of approximation we require fulfillment of the just-derived conditions of impenetrability, not on the contour itself, but on the segment of the axis $0x$ between $x = a$ and $x = b$, with the upper and lower sides of this segment being distinguished conventionally, assuming for this purpose that $y = \pm 0$.

Thus, for Equation 16 we have the boundary condition

$$\frac{\partial \tilde{\varphi}}{\partial y} = U_\infty h'(x) \quad \text{when} \quad y = \pm 0 \quad a \le x \le b \tag{18}$$

and for Equation 17

$$\tilde{\psi} = -U_\infty h(x) \quad \text{when} \quad y = \pm 0 \quad a \le x \le b \tag{19}$$

As regards the boundary condition at infinity from the profile, it is formulated very simply for subsonic motion ($\mathbf{M}_\infty < 1$) and is reduced to disturbances vanishing with distance from the profile contour

$$\tilde{\varphi} \to 0 \quad \tilde{\psi} \to 0 \quad \text{when} \quad \sqrt{x^2 + y^2} \to \infty$$

For a supersonic flow ($\mathbf{M}_\infty > 1$) this condition is not fulfilled, as will be soon demonstrated.

In concluding this section, the expression of the pressure coefficient c_p is formed, which is required for calculating the flow pressure on the body surface, having retained for it the same definition (Equation 46 of Chapter 7) as that in the case of an incompressible fluid. However, the choice of $\rho = \rho_\infty$ is emphasized as the characteristic value of density at infinity, so that

$$c_p = \frac{p - p_\infty}{\frac{1}{2}\rho_\infty V_\infty^2} \tag{20}$$

Using isentropic equations (87) in Chapter 5, which relate to pressures, and assuming in them that $p_1 = p$, $p_2 = p_\infty$, $\mathbf{M}_1 = \mathbf{M}$, $\mathbf{M}_2 = \mathbf{M}_\infty$, we transform Equation 20 for the pressure coefficient c_p

$$c_p = \frac{2p_\infty}{\rho_\infty V_\infty^2}\left(\frac{p}{p_\infty} - 1\right) = \frac{2}{k}\frac{k(p_\infty/\rho_\infty)}{V_\infty^2}\left[\left(\frac{1 + \frac{k-1}{2}\mathbf{M}_\infty^2}{1 + \frac{k-1}{2}\mathbf{M}^2}\right)^{\frac{k}{k-1}} - 1\right]$$

or, making the substitution of $k\,(p_\infty/\rho_\infty)$ by a_∞^2 and V_∞/a_∞ by $\mathbf{M}_\infty$,

$$c_p = \frac{2}{k\mathbf{M}_\infty^2} = \left[\left(\frac{1+\dfrac{k-1}{2}\mathbf{M}_\infty^2}{1+\dfrac{k-1}{2}\mathbf{M}^2}\right)^{\frac{k}{k-1}} - 1\right] \tag{21}$$

This expression for the pressure coefficient is in general not associated with the assumption about the smallness of disturbances. In the case of small disturbances we shall linearize Equation 28 of Chapter 5 in the fashion analogous to that made with Equation 29 of the same chapter. Then,

$$\tilde{p} = -\rho_\infty U_\infty \tilde{u}$$

after which we obtain for the pressure coefficient c_p the sought-after approximate expression valid only in the case of small disturbances

$$c_p = \frac{2\tilde{p}}{\rho_\infty U_\infty^2} = -\frac{2\tilde{u}}{U_\infty} \tag{22}$$

The resulting equations for the potential and the stream function of disturbances represent linear partial differential equations of second order with constant coefficients. In such an approximate linearized formulation the problems of gas dynamics can be solved by relatively simple means. Depending on whether the gas motion is subsonic ($\mathbf{M}_\infty < 1$) or supersonic ($\mathbf{M}_\infty > 1$), Equations 16 and 17 will belong to the elliptic or hyperbolic type. In the first case ($\mathbf{M}_\infty < 1$) the equation can be retained in the previous form; in the second case ($\mathbf{M}_\infty > 1$) it can be rewritten as

$$(\mathbf{M}_\infty^2 - 1)\frac{\partial^2\tilde{\varphi}}{\partial x^2} - \frac{\partial^2\tilde{\varphi}}{\partial y^2} = 0 \qquad (\mathbf{M}_\infty^2 - 1)\frac{\partial^2\tilde{\psi}}{\partial x^2} - \frac{\partial^2\tilde{\psi}}{\partial y^2} = 0 \tag{23}$$

The presence of the negative sign in the hyperbolic equations corresponds to the specific (as shown in Chapter 5) wave character of the processes of supersonic gas flow.

2. Subsonic Flow Past a Thin Profile: Prandtl-Glauert Rule

The contour of the profile placed in a flow will be specified by the equations for the upper (subscript 1) and lower (subscript 2) parts of the contour

$$y_1 = h_1(x) \qquad y_2 = h_2(x) \qquad a \le x \le b$$

which will be designated for simplicity as

$$y = h_{1,2}(x) \qquad a \le x \le b$$

Use the equation of the stream function (17) and, correspondingly, the boundary conditions (19). Introducing the notation $\omega^2 = 1 - \mathbf{M}_\infty^2$, we come to the following boundary-value problem

$$\frac{\partial^2 \tilde{\psi}}{\partial x^2} + \frac{1}{\omega^2}\frac{\partial^2 \tilde{\psi}}{\partial y^2} = 0$$

$$\tilde{\psi} = -U_\infty h_{1,2}(x) \quad \text{when} \quad y = \pm 0 \quad a \le x \le b$$

$$\tilde{\psi} \to 0 \text{ at infinity} \tag{24}$$

Let us transform from the coordinates x and y to the new coordinates ξ and η, having performed the affine transformation (the deformation of the ordinates)[1]

$$\xi = x \qquad \eta = \omega y \tag{25}$$

Then, because the boundary conditions use only zero and infinitely great values of y, the boundary-value problem (24) will be reduced to

$$\frac{\partial^2 \tilde{\psi}}{\partial \xi^2} + \frac{\partial^2 \tilde{\psi}}{\partial \eta^2} = 0$$

$$\tilde{\psi} = -U_\infty h_{1,2}(\xi) \quad \text{when} \quad \eta = \pm 0 \quad a \le \xi \le b$$

$$\tilde{\psi} \to 0 \text{ at infinity} \tag{26}$$

Let us compare this problem with problem (24), in which we set *a priori* that $\mathbf{M}_\infty = 0, \omega = 1$, which will correspond to the incompressible fluid flow past the same thin profile (this will be denoted by subscript 0 at $\tilde{\psi}$). We have

$$\frac{\partial^2 \tilde{\psi}_0}{\partial x^2} + \frac{\partial^2 \tilde{\psi}_0}{\partial y^2} = 0$$

$$\tilde{\psi}_0 = -U_\infty h_{1,2}(x) \quad \text{when} \quad y = \pm 0 \quad a \le x \le b$$

$$\tilde{\psi}_0 \to 0 \text{ at infinity} \tag{27}$$

Comparing Equations 26 and 27 we arrive at the identity of the solutions to these equations expressed respectively in terms of the variables ξ, η and x, y, so that

$$\tilde{\psi}(\xi, \eta) \equiv \tilde{\psi}_0(x, y) \qquad \frac{\partial \tilde{\psi}}{\partial \xi} \equiv \frac{\partial \tilde{\psi}_0}{\partial x} \qquad \frac{\partial \tilde{\psi}}{\partial \eta} \equiv \frac{\partial \tilde{\psi}_0}{\partial y} \tag{28}$$

Using these identities, equalities (15) and transformations (25), we obtain

$$\tilde{u} = \frac{1}{1 - \mathbf{M}_\infty^2}\frac{\partial \tilde{\psi}}{\partial y} = \frac{1}{1 - \mathbf{M}_\infty^2}\frac{\partial \tilde{\psi}}{\partial \eta}\frac{d\eta}{dy} = \frac{1}{\sqrt{1 - \mathbf{M}_\infty^2}}\frac{\partial \tilde{\psi}_0}{\partial y} = \frac{\tilde{u}_0}{\sqrt{1 - \mathbf{M}_\infty^2}}$$

[1]Prandtl, L., Über Strömungen deren Gerschwindigkeiten mit der Schallgeschwindigkeit vergleichbar sind, *J. Aeron. Res. Inst. Tokyo*, 65, 14, 1930; Glauert, H., The effect of compressibility on the lift on airfoil, *Proc. R. Soc. A*, 118, 1928; Sedov, L. I., *Ploskiye zadachi gidrodinamiki* (*Plane Problems of Hydrodynamics*), Nauka Press, Moscow, 1980; Sedov, L. I., *Two-Dimensional Problems in Hydrodynamics and Aerodynamics*, John Wiley & Sons, New York, 1965.

$$\tilde{v} = -\frac{\partial\tilde{\psi}}{\partial x} = -\frac{\partial\tilde{\psi}}{\partial\xi} = -\frac{\partial\tilde{\psi}_0}{\partial x} = \tilde{v}_0 \tag{29}$$

after which we obtain, on the basis of Equation 22, the following connection between the pressure coefficients c_p in the linearized subsonic gas flow ($\mathbf{M}_\infty < 1$) and c_{p0} in the incompressible liquid ($\mathbf{M}_\infty = 0$)

$$c_p = \frac{c_{p0}}{\sqrt{1-\mathbf{M}_\infty^2}} \tag{30}$$

The above relation expresses the following rule of Prandtl-Glauert: the distribution of the pressure coefficient in a plane irrotational linearized subsonic gas flow at the given value of $\mathbf{M}_\infty < 1$ can be obtained from the corresponding distribution in an incompressible fluid flow if all the ordinates of this distribution are increased by a factor of $1/\sqrt{1-\mathbf{M}_\infty^2}$.

Recalling (Section 9 of Chapter 7) that the lifting force of the profile R_y and the lifting force coefficient c_y are determined from the equations

$$R_y = -\oint pn_y ds = \frac{1}{2}\rho_\infty U_\infty^2 \oint c_p dx$$

$$c_y = \frac{R_y}{\frac{1}{2}\rho_\infty U_\infty^2 b \cdot 1} = \oint c_p d\bar{x} \qquad \bar{x} = x/b$$

where b is the chord of the profile, we conclude that for the profile with the same contour in the flows of a gas and of an incompressible liquid the following relation is valid for the lifting force coefficients c_y and c_{y0}

$$c_y = \frac{c_{y0}}{\sqrt{1-\mathbf{M}_\infty^2}} \tag{31}$$

which is analogous to Equation 30.

The results of experiments show that integral equation 31 is valid within a somewhat wider range of the numbers $\mathbf{M}_\infty < 1$ than the local equation (30). Figure 88 presents for comparison the results of an experimental determination of c_y for a thin (the relative thickness is 6.5%), slightly curved screw profile at the two angles of attack: 2° and 4°. It is possible to note that at the angle of attack of 4° the experimental curve (dashes with circles) deviates from the theoretical curve (solid line) (31) earlier than at the angle of attack of 2° (dashes with crosses).

Of interest is the fact that the greater the perturbation of the flow (in the present case the greater the angle of attack), the smaller the range of the numbers $\mathbf{M}_\infty$ within which the linearized theory is valid. The theory presented in this section does not give quantitative explanation of this important fact. Later, more general ideas are given about the connection between linearized subsonic gas and incompressible liquid flows past bodies which will confirm theoretically the just-noted fact.

The Prandtl-Glauert rule serves only to convert the values of c_{p0} and c_{y0} determined previously in an incompressible liquid (for example, by the method of a thin wing given in Section 10 of Chapter 7) into their values at the given number $\mathbf{M}_\infty < 1$ in a subsonic gas flow.

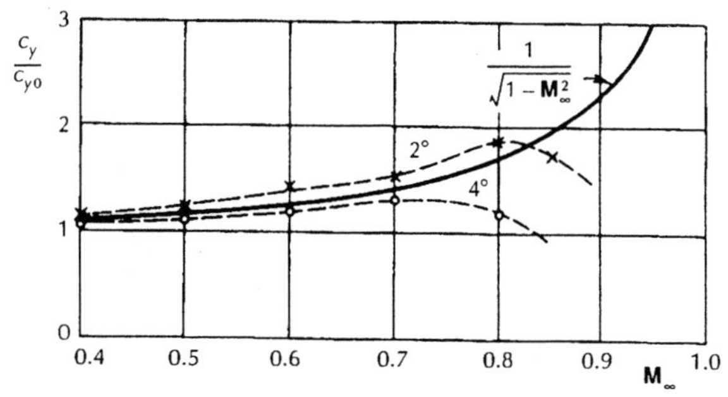

Fig. 88

3. Supersonic Flow Past a Thin Profile: Ackeret's Formulas

In the case of supersonic gas flow ($\mathbf{M}_\infty > 1$) the second of linearized equations (23) acquires the form

$$\frac{\partial^2\tilde{\psi}}{\partial x^2} - \frac{1}{\omega^2}\frac{\partial^2\tilde{\psi}}{\partial y^2} = 0 \tag{32}$$

where, in contrast to the previous section, it is assumed that

$$\omega^2 = \mathbf{M}_\infty^2 - 1 \tag{33}$$

The wave equation (32) is already familiar from Chapter 5. The difference in the mathematical structure reflects the physical aspects of the phenomena described by Equation 32 as compared to the phenomena of subsonic flow.

It can be easily verified by simple substitution that the general solution of Equation 32 can be expressed by the formula

$$\tilde{\psi}(x, y) = \Omega_1(x - \omega y) + \Omega_2(x + \omega y)$$

where Ω_1 and Ω_2 are the arbitrary functions of the arguments $x - \omega y$ and $x + \omega y$; the form of these functions is determined with the aid of the boundary conditions of the problem.

Let us consider the specific solution

$$\tilde{\psi}_1 = \Omega_1(x - \omega y) \tag{34}$$

It has the following meaning: in the flow plane (x, y) exists the family of straight lines

$$x - \omega y = \text{const}$$

along which the stream function of disturbances and, consequently in general, of the disturbance of the parameters of the gas motion and state will retain constant values. These straight lines represent the first family (C_1) of the characteristics of wave equation (32) and play the role of the lines of disturbance in the supersonic flow considered. They are called the Mach lines or waves.

In the same wave, to the particular solution

$$\tilde{\psi}_2 = \Omega_2(x + \omega y) \tag{35}$$

there corresponds the second family (C_2) the characteristics or lines of disturbance,

$$x + \omega y = \text{const}$$

along which the disturbances of the parameters of gas motion and state also retain constant values.

Considering the angular coefficients of these families of straight lines

$$\text{tg}\,\alpha = \pm\frac{1}{\omega} = \pm\frac{1}{\sqrt{\mathbf{M}_\infty^2 - 1}} \tag{36}$$

we conclude that the angles α formed by the disturbance lines with the direction of the nonperturbed motion (axis $0x$), are equal to

$$\alpha = \pm\arcsin\frac{1}{\mathbf{M}_\infty} \tag{37}$$

Figure 89 shows two lines of disturbance from the disturbance point source S located on the axis $0x$. Only along the two rays originating from the point S is it possible to observe the disturbances of the uniform flow oncoming on the point S; at all the rest of the points of the flow plane the flow is not disturbed and retains its uniformity. We recall that the "point" source in a plane flow is actually represented by the source of disturbances in the form of a straight line perpendicular to the plane of the figure (for example, a thin wire located transverse to the flow). From the actually point source in space the lines of disturbances will be located on the conical surface with the vertex at the point S and the semivertex angle α determined from Equation 37. This cone of disturbances is also called the Mach cone. From the slope of the disturbance lines it is possible to determine the value of $\mathbf{M}_\infty$ for a uniform flow.

Thus, by introducing small disturbances into the flow in aerodynamic supersonic tunnels by means of thin needles or probes, the lines of disturbances are observed and the number $\mathbf{M}_\infty$ in the tunnel is determined from the angles of their inclination to the nonperturbed flow direction. The disturbance lines are made visible because along them the density of the gas (air) and, consequently, the refractive index, differ from the corresponding values in a nonperturbed flow.

Let us consider the problem of supersonic flow past a thin profile. Just as in a subsonic flow, its contour will be prescribed by the ordinates of the upper (subscript 1) and lower (subscript 2) surfaces: $y = h_{1,2}(x)$.

As before, the boundary condition will be presented in the form

$$\tilde{\psi} = -U_\infty h_{1,2}(x) \quad \text{when} \quad y = \pm 0 \quad x_A \le x \le x_B \tag{38}$$

Fill the flow region from above and from below of the profile contour (Figure 90) respectively with the characteristics of the first (C_1) and second (C_2) families. The above-indicated properties of the characteristics (Equation 34 for the first family and Equation 35 for the second family) make it possible to conclude directly that the solution of Equation 32 under boundary condition (38) can be presented in the form

$$\tilde{\psi} = -U_\infty h_{1,2}(x \mp \omega y) \tag{39}$$

where, we recall, to subscript 1 at h there corresponds the upper sign in round brackets and to subscript 2, the lower.

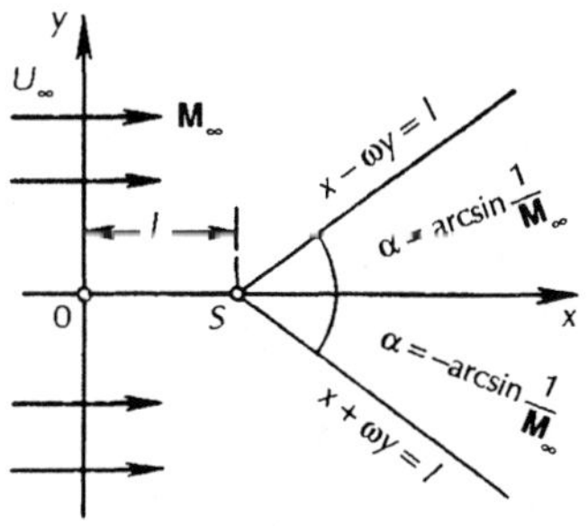

Fig. 89

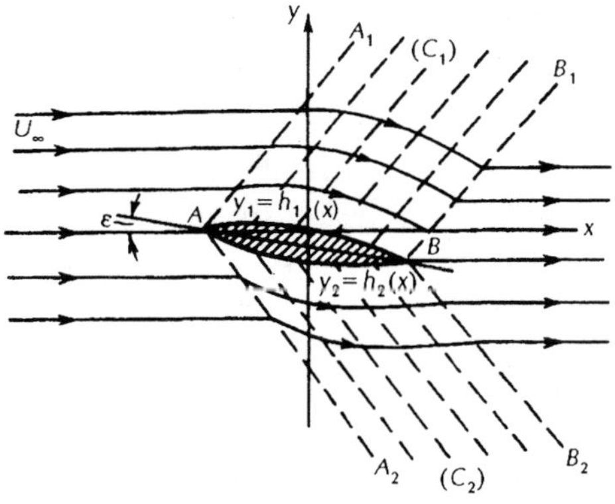

Fig. 90

In contrast to a subsonic flow, the stream function of disturbances $\tilde{\psi}(x, y)$ does not vanish at any distance, as great as we wish, from the profile contour, but inside of the upper and lower bands bounded by the extreme characteristics AA_1, BB_1 and AA_2, BB_2 for $y \to \pm\infty$ it retains the same distribution over x just as on upper and lower surfaces of the profile.[2] Outside these bands the flow remains uniform and its velocity is equal to U_∞. As it is directly seen from solution (39) and shown in Figure 90, the streamlines of the perturbed motion ($\psi = U_\infty y + \tilde{\psi} = \text{const}$) are the curves that can be obtained by parallel displacement of the upper and lower contours of the profile respectively along the characteristics of the first and second families.

Having solution (39), we find from Equation 15 the distribution of the disturbances of the velocity components (the prime means the derivative with respect to the entire argument in parentheses

$$\tilde{u} = -\frac{1}{M_\infty^2 - 1}\frac{\partial \tilde{\psi}}{\partial y} = \mp \frac{U_\infty}{\sqrt{M_\infty^2 - 1}} h'_{1,2}(x \mp \omega y)$$

$$\tilde{v} = -\frac{\partial \tilde{\psi}}{\partial x} = U_\infty h'_{1,2}(x \mp \omega y) \tag{40}$$

which is valid throughout the entire region of disturbed motion. The second equation of system (40) can give the angle $\theta_{1,2}$ of deviation of the tangent to the streamline in the disturbed region from the streamline of the nonperturbed flow. We have the streamlines by definition and due to the smallness of the deviation $\theta_{1,2}$

$$\theta_{1,2} \approx \operatorname{tg} \theta_{1,2} = \frac{\tilde{v}}{(U_\infty + \tilde{u})} \approx \frac{\tilde{v}}{U_\infty} \approx h'_{1,2}(x \mp \omega y)$$

and then Equations 40 give

---

[2] The asymptotic methods of the theory of small disturbances show that at great distances from the profile the effect of small second-order quantities becomes essential in the first approximation and distort the flow pattern depicted in Figure 3. The characteristics become distorted and cease to be parallel. See van Dyke, M., *The Perturbation Methods in the Mechanics of a Fluid* (Russian translation), Mir Press, Moscow, 1967, pp. 147 to 161.

$$\tilde{u} = \mp \frac{U_\infty}{\sqrt{\mathbf{M}_\infty^2 - 1}} \theta_{1,2} \qquad \tilde{v} = U_\infty \theta_{1,2} \tag{41}$$

These equations express the basic property of the linearized supersonic flow. The longitudinal and transverse velocity components of perturbation at the given velocity and Mach number of the nondisturbed flow are Proportional to the local inclination angle of the streamline of the disturbed flow with respect to the direction of the nondisturbed flow and have the local character. The same property is typical of the pressure, density, and other quantities characteristic for the flow, thus distinguishing in principle the supersonic linearized flow from the subsonic one in which the values of the flow parameters at the given point depend on the distribution of these parameters throughout the entire flow.

Recalling the expression of the pressure coefficient (22), identical for a subsonic and supersonic linearized flows, we find, according to Equation 41, the expressions for the pressure coefficient at any point of the disturbed flow

$$c_p = \pm \frac{2}{\sqrt{\mathbf{M}_\infty^2 - 1}} h'_{1,2}(x \mp \omega y) = \pm \frac{2\theta_{1,2}(x \mp \omega y)}{\sqrt{\mathbf{M}_\infty^2 - 1}}$$

and then also on the profile surface (contour), where it is possible to assume approximately that $y = \pm 0$,

$$c_p(x) = \frac{\pm 2h'_{1,2}(x)}{\sqrt{\mathbf{M}_\infty^2 - 1}} = \frac{\pm 2\theta_{1,2}(x)}{\sqrt{\mathbf{M}_\infty^2 - 1}} \tag{42}$$

This equation, which expresses the proportionality of the pressure coefficient in a linearized supersonic flow to the local angle of the tangent to the thin profile contour with the direction of the nondisturbed flow (this angle is commonly called the local angle of attack) resembles the well-known "shock" theory of Newton, the application of which in the theory of incompressible liquid flow past bodies was opposed by Euler. It will be soon shown that Newton's "shock" theory will be employed as quite a satisfactory, though only approximate, method for calculating supersonic flows past bodies at very great Mach numbers.

The lifting force coefficient c_y will be found by performing the integration ($b = AB$ is the chord of the profile equal, in the approximation adopted, to the difference $x_B - x_A$ of the abscissas of the points B and A)

$$c_y = \oint c_p d\left(\frac{x}{b}\right) = \frac{1}{b}\int_{x_A}^{x_B} (c_{p2} - c_{p1})\, dx \tag{43}$$

Substituting into this equation the values of c_{p1} and c_{p2} from Equation 42, we obtain

$$c_y = -\frac{2}{b\sqrt{\mathbf{M}_\infty^2 - 1}} \int_{x_A}^{x_B} [h'_2(x) + h'_1(x)]\, dx = -\frac{2}{\sqrt{\mathbf{M}_\infty^2 - 1}} \frac{2(y_B - y_A)}{x_B - x_A}$$

Let us introduce the angle of attack of the profile ε as an acute angle between the direction of the chord AB and the oncoming flow

$$\operatorname{tg} \varepsilon \approx \varepsilon \approx \frac{y_A - y_B}{x_B - x_A}$$

Then the previous equation will acquire the final form

$$c_y = \frac{4\varepsilon}{\sqrt{\mathbf{M}_\infty^2 - 1}} \tag{44}$$

This is Ackeret's formula.[3]

In the linearized theory of supersonic flow past a thin profile the lifting force coefficient is independent of the profile shape and depends only on the angle of attack and on the Mach number $\mathbf{M}_\infty > 1$ of the oncoming flow.

In contrast to the linearized subsonic flow in which there is no resistance of the profile, as it follows directly from the Prandtl-Glauert rule (Section 2), in a supersonic flow the resistance differs from zero; it is called the wave resistance. The appearance of this resistance can be physically explained by that longitudinal flow asymmetry which distinguishes supersonic from subsonic flow. While in a subsonic flow the pressure that is at the rear portion of the profile recovers and creates the force that opposes the resultant vector of the pressure forces at the front of the profile, there is no such balancing in a supersonic flow. In the rear, diverging region of the flow the phenomenon takes place, which is similar to that observed in the Laval nozzle: when diverging the supersonic flow is accelerated, the pressure in the rear portion does not recover but continues to decrease, and this leads to the appearance of an additional suction force directed downstream. Thus, in contrast to a subsonic flow the resultant vectors of pressure forces over the front and rear portions of the profile surface do not cancel each other but, conversely, sum up forming the overall force of wave resistance.

The wave resistance coefficient c_x will be found from the expression formed for the resistance force

$$R_x = -\oint p n_x ds = -\oint p dy = -\frac{1}{2}\rho_\infty U_\infty^2 \oint c_p \frac{dy}{dx} dx =$$

$$= \frac{1}{2}\rho_\infty U_\infty^2 \int_{x_A}^{x_B} \left(c_{p1}\frac{dh_1}{dx} - c_{p2}\frac{dh_2}{dx} \right) dx$$

According to Equation 42 we shall have the following Ackeret formula for the wave resistance coefficient

$$c_x = \frac{R_x}{\frac{1}{2}\rho_\infty U_\infty^2 b} = \frac{2}{b\sqrt{\mathbf{M}_\infty^2 - 1}} \int_{x_A}^{x_B} \{ [h_1'(x)]^2 + [h_2'(x)]^2 \} dx \tag{45}$$

As is seen from this formula, the wave resistance coefficient represents a small second-order quantity; in contrast to the lifting force coefficient, it depends on the shape of the profile immersed in a flow. Thus, for a plate

[3] Ackeret, J., Über Luftkräfte auf Flügel, die mit grosserer als Schallgeschwindigkeit bewegt werden, *Z. Fluegtechn.*, 16, 1925.

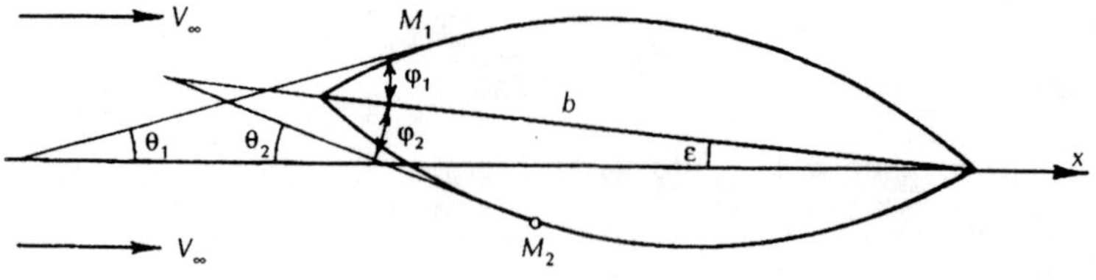

Fig. 91

$$h_1'(x) = h_2'(x) = -\frac{y_A - y_B}{x_B - x_A} = -\operatorname{tg}\varepsilon \approx -\varepsilon$$

and Equation 45 yields

$$c_x = \frac{4\varepsilon^2}{\sqrt{\mathbf{M}_\infty^2 - 1}} \tag{46}$$

To calculate c_x for the profiles of more complex shapes, it is useful to assume in Equation 45 that

$$h_1'(x) \approx \theta_1 = \varphi_1 - \varepsilon \qquad h_2'(x) \approx \theta_2 = \varphi_2 + \varepsilon$$

where ε is the angle of attack (Figure 91), and φ_1 and φ_2 are the angles between the tangents at the points $\mathbf{M}_1$ and $\mathbf{M}_2$ of the upper and lower surfaces and the chord of the profile. Then, according to Equation 45, c_x will be expressed as

$$c_x = \frac{2}{b\sqrt{\mathbf{M}_\infty^2 - 1}} \int_{x_A}^{x_B} [\varphi_1^2 + \varphi_2^2 + 2\varepsilon^2 + 2\varepsilon(\varphi_2 - \varphi_1)]\, dx$$

In the case of a symmetric wing ($\varphi_1 = \varphi_2 = \varphi$) we obtain

$$c_x = \frac{4\varepsilon^2}{\sqrt{\mathbf{M}_\infty^2 - 1}} + \frac{4}{\sqrt{\mathbf{M}_\infty^2 - 1}} \frac{1}{b} \int_{x_A}^{x_B} \varphi^2 dx = \frac{4(\varepsilon^2 + \overline{\varphi}^2)}{\sqrt{\mathbf{M}_\infty^2 - 1}} \tag{47}$$

where $\overline{\varphi}^2$ denotes the mean integral value of the squared angle of inclination of the tangent to the wing contour with respect to its chord. It follows from Equation 47 that according to the linearized theory, of all the profiles located at a small angle of attack to the flow, the plate has the smallest wave resistance.

For a diamond-shaped profile of maximum thickness (small diagonal) t and chord (great diagonal) b we have $\varphi = t/b$, so that

$$c_x = \frac{4}{\sqrt{\mathbf{M}_\infty^2 - 1}} \left(\varepsilon^2 + \frac{t^2}{b^2}\right) \tag{48}$$

In the case of a biconvex profile (Figure 91) composed of two identical arcs of circles, it is possible to write in the approximation adopted:

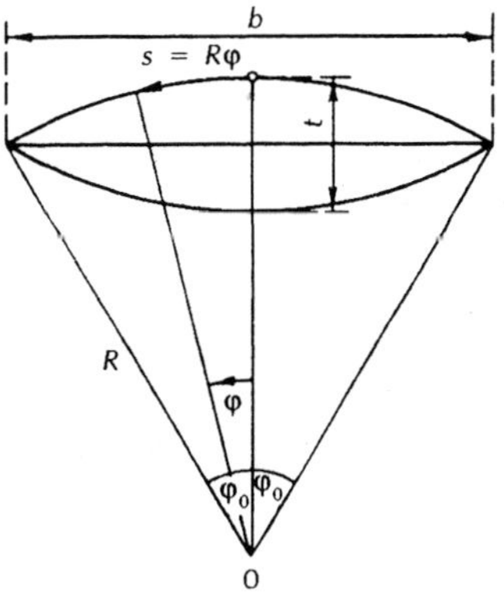

Fig. 92

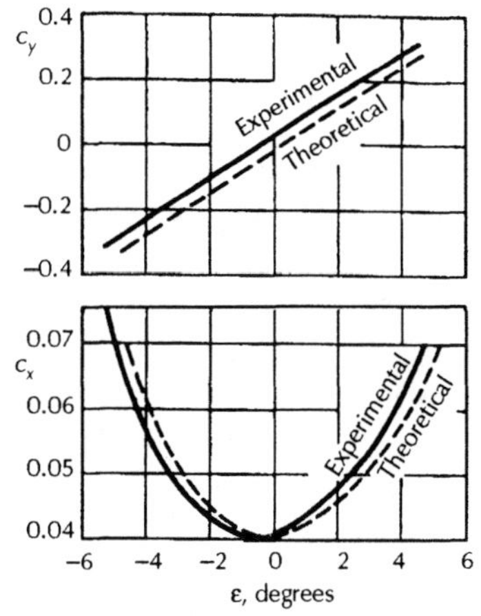

Fig. 93

$$\bar{\varphi}^2 = \frac{2}{b}\int_0^{b/2} \varphi^2 dx = \frac{2R}{b}\int_0^{\varphi_0} \varphi^2 d\varphi = \frac{2}{3}\frac{R}{b}\varphi_0^3$$

and to note that

$$b = 2R\sin\varphi_0 \approx 2R\varphi_0 \qquad \frac{t}{2} = R - R\cos\varphi_0 \approx \frac{1}{2}R\varphi_0^2 \qquad \frac{R}{b} = \frac{1}{2\varphi_0} \qquad \varphi_0 = 2\frac{t}{b}$$

Finally, from Equation 47 we find

$$c_x = \frac{4}{\sqrt{\mathbf{M}_\infty^2 - 1}}\left(\varepsilon^2 + \frac{4}{3}\frac{t^2}{b^2}\right) \tag{49}$$

The theory of Ackeret, as the theory of first approximation, gives the results which satisfactorily coincide with the experiment if the profile is thin enough, the angles of attack are small and the Mach number $\mathbf{M}_\infty$ is not very close to unity. As an example (Figure 93), we compare the theoretical curves $c_y(\varepsilon)$ and $c_x(\varepsilon)$, calculated from Ackeret's formulas (44) and (47), with the experimental curves of Hilton and Pruden[4] for a diamond-shaped profile of relative thickness 8.7% with the rounded angles in the middle at $\mathbf{M}_\infty = 1.45$. The experimental curve is given by a solid line; the theoretical curve is given by a dashed line. The ordinates of the theoretical curve c_x are increased by the factor 0.008, which corresponds to the friction resistance coefficient and which is not taken into account by the Ackeret theory.[5] A more comprehensive treatment of the problem about the limits of the applicability of Ackeret's formulae as well as additional experimental information can be found in the special literature.[6]

[4]Hilton, W. F. and Pruden, F. W., Subsonic and Supersonic High-Speed Tunnel Tests of a Faired Double Wedge Airfoil, Rep. Memo. No. 2057, 1943.

[5]The *L* curves for experimental and theoretical values of c_x for negative angles of a attack in Figure 93 appear to reversed.

[6]See, for example, Hilton, W. F., *The Aerodynamics of High Speeds* (in Russian translation), IL Press, Moscow, 1955, pp. 165 to 231.

4. The Similarity Laws of Plane Sub- and Supersonic Flows Past a Thin Profile: Case of Transonic Flow

In the previous two sections dealing with the problems of plane sub- and supersonic flows past a thin profile, formulas were derived which made it possible to convert the pressure and lifting force coefficients and also the wave resistance coefficient for a supersonic flow past the given profile from one Mach number of the oncoming flow into the other number.

In the case of a subsonic flow the connection was established between c_p or c_y of a thin profile of the given shape and location in a uniform flow at an arbitrary, smaller than unity, Mach number $\mathbf{M}_\infty$ and the corresponding values of these coefficients for the same profile, positioned in the same fashion in a uniform flow of incompressible liquid ($\mathbf{M}_\infty = 0$). From equations 30 and 31 directly follow somewhat more general formulas

$$\frac{(c_p)_{\mathbf{M}_\infty = \mathbf{M}_1}}{(c_p)_{\mathbf{M}_\infty = \mathbf{M}_2}} = \frac{(c_y)_{\mathbf{M}_\infty = \mathbf{M}_1}}{(c_y)_{\mathbf{M}_\infty = \mathbf{M}_2}} = \sqrt{\frac{1-\mathbf{M}_2^2}{1-\mathbf{M}_1^2}} \tag{50}$$

for converting the coefficients c_p and c_y from one value, $\mathbf{M}_\infty = \mathbf{M}_1 < 1$, into the other, $\mathbf{M}_\infty = \mathbf{M}_2 < 1$. Of course, and this should be again emphasized, Equation 50 is applicable only for the same (as to the shape and location) thin profile.

Judging from Equations 42, 44, and 45, for two thin profiles identical in shape and identically located in two uniform supersonic flows with $\mathbf{M}_\infty$ numbers equal to $\mathbf{M}_1 > 1$ and $\mathbf{M}_2 > 1$, there hold the following conversion formulas:

$$\frac{(c_p)_{\mathbf{M}_\infty = \mathbf{M}_1}}{(c_p)_{\mathbf{M}_\infty = \mathbf{M}_2}} = \frac{(c_y)_{\mathbf{M}_\infty = \mathbf{M}_1}}{(c_y)_{\mathbf{M}_\infty = \mathbf{M}_2}} = \frac{(c_x)_{\mathbf{M}_\infty = \mathbf{M}_1}}{(c_x)_{\mathbf{M}_\infty = \mathbf{M}_2}} = \sqrt{\frac{\mathbf{M}_2^2-1}{\mathbf{M}_1^2-1}} \tag{51}$$

which can be rewritten in the form identical to that of Equation 50 if the terms in the numerator and denominator under the root sign are replaced.

The above formulas represent the specific case of more general relations of dynamic similarity which can be derived for different in shape but affine-similar thin profiles located at small angles of attack in two uniform flows with different Mach numbers (but, of course, such that the both flows compared be either subsonic or the both supersonic).

Let us elucidate the conditions for the dynamic similarity of two linearized gas flows and determine the relationships between the coefficient c_p and c_y, and, in the case of a supersonic flow, also c_x for such similar flows. Consider two linearized flows with the velocities at infinity U_i and Mach numbers $\mathbf{M}_i$ prescribed by their stream functions of small perturbations $\tilde{\psi}_i$ in the physical planes of motion (x_i, y_i). Here and hereafter, the subscript i acquires the values $i = 1, 2$ corresponding to the two flows compared.

According to equation 17, the differential equations of motion can be written in the form

$$\frac{\partial^2 \tilde{\psi}_i}{\partial x_i^2} \pm \frac{1}{|1-\mathbf{M}_i^2|}\frac{\partial^2 \tilde{\psi}_i}{\partial y_i^2} = 0 \quad (i = 1, 2) \tag{52}$$

where the upper sign relates to the subsonic, and the lower to the supersonic, flow. Without loss of generality, it is assumed that the chords of both the profiles are identical and equal to $b = x_B - x_A$ and that the equations

$$y_i = h_i(x_i) \qquad (i = 1, 2)$$

correspond to the entire contour of the profile, which, in contrast to (Section 2 and 3) is not divided into the upper and lower parts.

According to Equation 19, the boundary conditions on the surfaces of thin profiles immersed in the flows will be equal to

$$\tilde{\psi}_i = -U_i h_i(x_i) \quad \text{when} \quad y_i = \pm 0 \qquad x_A \le x_i \le x_B \qquad (i = 1, 2) \tag{53}$$

Let us perform the transition to dimensionless variables in Equations 52 and boundary conditions (53), having assumed that

$$x_i = b\xi_i \qquad y_i = \frac{b}{\sqrt{|1 - \mathbf{M}_i^2|}}\eta_i$$

$$\tilde{\psi}_i = bU_i\sqrt{|1 - \mathbf{M}_i^2|}\,\Psi_i\theta_i(\xi_i, \eta_i) \qquad h_i = b\tau_i H_i(\xi_i) \tag{54}$$

In these transition formulas the quantities Ψ_i are certain arbitrary constants, due to the homogeneity of Equations 52, thus reflecting the arbitrary nature of the scales of $\tilde{\psi}_i$. The scales of the ordinates of the profiles are taken to be the products of the length of the chord b by the small relative thickness τ_i, which is understood here to be not only the usual ratio of the maximum thickness of the profile $h_{i\,\max}$ to the length of the chord b, but also (more generally), the relative camber of the profile equal to the ratio of the height of the profile camber line to the chord length, or, simply, the angle of attack. ξ_i, η_i, θ_i, and H_i are the dimensionless coordinates and their functions.

Expressing all the variables in Equations 52 and boundary conditions (53) in terms of their dimensionless values, we shall obtain after simple reductions ($i = 1, 2$)

$$\frac{\partial^2\theta_i}{\partial\xi_i^2} \pm \frac{\partial^2\theta_i}{\partial\eta_i^2} = 0 \tag{55}$$

$$\Psi_i\sqrt{|1 - \mathbf{M}_i^2|}\,\theta_i = -\tau_i H_i \quad \text{when} \quad \eta_i = \pm 0 \qquad \frac{x_A}{b} \le \xi_i \le \frac{x_B}{b} \tag{56}$$

In the case of a subsonic flow conditions (56) are also supplemented with the condition: $\theta_i \to 0$ when $\sqrt{\xi_i^2 + \eta_i^2} \to \infty$.

If at the identical points of two flows which are compared and determined by the equalities of the dimensionless coordinates of these points

$$\xi_1 = \xi_2 = \xi \qquad \eta_1 = \eta_2 = \eta$$

under the conditions of the immersed in flow

$$H_1(\xi_1) = H_2(\xi_2) = H(\xi)$$

the following equalities are fulfilled

$$\theta_1(\xi_1, \eta_1) = \theta_2(\xi_2, \eta_2) = \theta(\xi, \eta)$$

then such flows are kinematically similar to each other.

Resorting to Equation 55 and boundary conditions (56), we see that for the above requirements to be fulfilled, the following equalities should be satisfied (λ is a certain constant)

$$\frac{\tau_1}{\Psi_1\sqrt{|1-\mathbf{M}_1^2|}} = \frac{\tau_2}{\Psi_2\sqrt{|1-\mathbf{M}_2^2|}} = \lambda \tag{57}$$

The fulfillment of the additional condition at infinity in a subsonic flow will not lead to a new similarity condition.

If the conditions of the affinity of the profiles and of Equations 57 are fulfilled, the set of equations (55) and boundary conditions (56) will be reduced to one equation and corresponding boundary conditions

$$\frac{\partial^2\theta}{\partial\xi^2} \pm \frac{\partial^2\theta}{\partial\eta^2} = 0$$

$$\theta = -\lambda H(\xi) \quad \text{when} \quad \eta = \pm 0 \qquad \frac{x_A}{b} \le \xi \le \frac{x_B}{b}$$

$$\theta \to 0 \quad \text{when} \quad \sqrt{\xi^2+\eta^2} \to \infty \quad \text{(subsonic flow)}$$

At the given value of λ the above equation has the unique solution $\theta = \theta(\xi, \eta)$, thus testifying to the existence of the kinematic similarity between the flows compared.

As now elucidated, the different values of the constant λ depend on the choice of the expression for Ψ in terms of the given constants τ and $\mathbf{M}_\infty$. This choice remains arbitrary, and to each value of λ corresponds its own condition (57) and the resulting specific condition of kinematic similarity. Some of these conditions are considered in what follows.

To establish the conditions of dynamic similarity, let us consider the pressure coefficient c_p in two kinematically similar flows. According to Equation 22, passing to dimensionless quantities at identical points, we have

$$c_{pi} = -\frac{2\tilde{u}_i}{U_i} = -\frac{2}{U_i(1-\mathbf{M}_i^2)}\frac{\partial\tilde{\psi}_i}{\partial y_i} = \mp 2\Psi_i\frac{\partial\theta_i}{\partial\eta_i}$$

where it follows that the dynamic similarity condition is

$$\frac{c_{p1}}{\Psi_1} = \frac{c_{p2}}{\Psi_2} = \mp 2\frac{\partial\theta}{\partial\eta} = f(\lambda) \tag{58}$$

Applying the symbol "idem" (meaning "same" in Latin) to designate identical quantities in two dynamically similar flows being compared, we may write conditions (57) and (58) in the following more concise form

$$\lambda = \frac{\tau}{\Psi\sqrt{|1-\mathbf{M}_\infty^2|}} = \text{idem} \qquad \frac{c_p}{\Psi} = \text{idem} \tag{59}$$

The set of equations (59) can be presented in the form of the following functional relation:[7]

[7]Liepmann, H. W. and Roshko, A., *Elementy gazovoy dinamiki* (*The Elements of Gas Dynamics*) (translated from English), IL Press, Moscow, 1960, pp. 302 to 305.

$$\frac{c_p}{\Psi} = f\left(\frac{\tau}{\Psi\sqrt{|1-\mathbf{M}_\infty^2|}}\right) = f(\lambda) \tag{60}$$

which expresses the fact that at $\tau/(\Psi\sqrt{|1-\mathbf{M}_\infty^2|})$ = idem, we shall also have

$$\frac{c_p}{\Psi} = \text{idem}$$

As is conventional, the right sides of Equations 58 to 60 have no arguments ξ and η, which are identical at the identical points of the flows compared. It must not be forgotten, of course, that at $\tau/(\Psi\sqrt{|1-\mathbf{M}_\infty^2|}) = \lambda =$ idem, the ratio c_p/Ψ is the function of ξ and η, which depends on the choice of λ.

Assigning different expressions for the arbitrary constant Ψ in terms of the constants τ and $\mathbf{M}_\infty$, we obtain different values of λ and, correspondingly, different specific laws of similarity. Thus, assuming $\Psi = |1-\mathbf{M}_\infty^2|^{-1/2}$, we have, according to Equations 57 and 58, the following formula ($\lambda = \tau$)

$$c_p = \frac{f(\tau)}{\sqrt{|1-\mathbf{M}_\infty^2|}}$$

which, in the case of a subsonic flow, coincides with formula (30) showing the familiar (the Prandtl-Glauert rule) fact that if the compared profiles have identical relative thicknesses (in the generalized sense, see above), the pressure coefficients at the identical points of both flows will be inversely proportional to the square roots of the difference $1 - \mathbf{M}^2$ or $1 - \mathbf{M}_2^2$.

By assuming that $\Psi = 1$, the similarity relation (60) will acquire the form

$$\lambda = \frac{\tau}{\sqrt{|1-\mathbf{M}_\infty^2|}} \qquad c_p = f\left(\frac{\tau}{\sqrt{|1-\mathbf{M}_\infty^2|}}\right)$$

and will express another aspect of the same Prandtl-Glauert rule: if in two subsonic flows being compared the relative thicknesses of the thin profiles relate as the square roots of the differences $(1 - \mathbf{M}_1^2)$ and $(1 - \mathbf{M}_2^2)$, the pressure coefficients at the identical points of such two flows will be identical.

This rule clarifies the earlier-noted property of the curves $c_y(\mathbf{M}_\infty)$ given in Figure 88. The condition of the identity of the ratio $\tau\sqrt{|1-\mathbf{M}_\infty^2|}$ leads to a situation in which the greater the angle of attack, the smaller values of $\mathbf{M}_\infty$ are needed for the coincidence between the predicted and experimental lifting force coefficients.

It is interesting also to assume that $\Psi = \tau$; then,

$$\lambda = \frac{1}{\sqrt{|1-\mathbf{M}_\infty^2|}} \qquad c_p = \tau f(\sqrt{|1-\mathbf{M}_\infty^2|})$$

The validity of such an equation stating that at the given value of $\mathbf{M}_\infty$ the pressure coefficient is proportional to the relative thickness of the profile was verified many times for the case $\mathbf{M}_\infty = 0$ in Section 10 of Chapter 7 (see the first of Equations 104 and Equation 107).

If, conversely, the relative thickness of the profile is varied in inverse proportion to $\sqrt{|1-\mathbf{M}_\infty^2|}$, i.e., to retain the quantity $\tau\sqrt{|1-\mathbf{M}_\infty^2|}$ as intact, then the pressure coefficient will be inversely proportional to $|1-\mathbf{M}_\infty^2|$ according to the formula

$$c_p = \frac{1}{|1-\mathbf{M}_\infty^2|} f(\tau\sqrt{|1-\mathbf{M}_\infty^2|})$$

which results from Equation 60 on assuming that $\Psi = 1/|1-\mathbf{M}_\infty^2|$, $\lambda = \tau\sqrt{|1-\mathbf{M}_\infty^2|}$.

Finally, assume in Equation 60 that

$$\Psi = \tau/\sqrt{|1-\mathbf{M}_\infty^2|}$$

then we have

$$\lambda = 1 \qquad c_p = \frac{\tau}{\sqrt{|1-\mathbf{M}_\infty^2|}} f(1)$$

where, as already noted, $f(1)$ is the function of the coordinates ξ, η, the form of which cannot, of course, be determined from the similarity considerations. The equation obtained for c_p is fully consistent with Equations 30 and 42.

The noted variety of the "similarity rules", which at different Mach numbers of the oncoming flow allow comparison of past profiles of flows with different laws of connection between relative thicknesses and Mach numbers, is due to the linearity and uniformity of the theory of sub- and supersonic flow past a thin profile.

Now, let us consider the case of transonic flow past a thin profile characterized by the Mach number $\mathbf{M}_\infty$ of the oncoming flow being close to unity. In this special case the simplest means of linearization, presented in Section 1, is no longer applicable and must be supplemented with some more refined considerations concerning the relative smallness of the terms of basic equation (7) and of the expression for the speed of sound (8) applied to the problem of flow past a thin profile. For this, it proves most convenient to consider the general equation (2) and the expression for the speed of sound (4). In these equations replace u, v, and a by Equations 9. In contrast to the previous linearization (Section 1), we shall retain not only the finite, but also the small first-order quantities in the factors in front of the perturbation velocity derivatives $\partial\tilde{u}/\partial x$, $\partial\tilde{u}/\partial y$, $\partial\tilde{v}/\partial x$, $\partial\tilde{v}/\partial y$, which are the small first-order quantities. For this purpose, we shall avail ourselves of the following approximate formula for the squared speed of sound

$$a^2 = a_\infty^2 + \frac{k-1}{2} U_\infty^2 - \frac{k-1}{2} (U_\infty^2 + 2U_\infty\tilde{u} + \tilde{u}^2 + \tilde{v}^2) \approx a_\infty^2 - (k-1)\, U_\infty\tilde{u}$$

In this approximation we shall obtain

$$\left[a_\infty^2 - U_\infty^2 - (k+1)\, U_\infty\tilde{u}\right] \frac{\partial\tilde{u}}{\partial x} - U_\infty\tilde{v}\left(\frac{\partial\tilde{u}}{\partial y} + \frac{\partial\tilde{v}}{\partial x}\right) +$$

$$+ \left[a_\infty^2 - (k-1)\, U_\infty\tilde{u}\right] \frac{\partial\tilde{v}}{\partial x} = 0$$

or, separating the small first-order quantities on the left-hand side and dividing the both parts by a_∞^2,

$$(1-\mathbf{M}_\infty^2)\frac{\partial\tilde{u}}{\partial x} + \frac{\partial\tilde{v}}{\partial y} = (k+1)\,\mathbf{M}_\infty^2 \frac{\tilde{u}}{U_\infty}\frac{\partial\tilde{u}}{\partial x} +$$

$$+ (k-1)\,\mathbf{M}_\infty^2 \frac{\tilde{u}}{U_\infty}\frac{\partial \tilde{v}}{\partial y} + \mathbf{M}_\infty^2 \frac{\tilde{v}}{U_\infty}\left(\frac{\partial \tilde{u}}{\partial y} + \frac{\partial \tilde{v}}{\partial x}\right)$$

The linearization requires the omission of the right side resulting in Equation 10 for the cases of sub- and supersonic motions. In the event of transonic flow now considered the value $\mathbf{M}_\infty = 1$ is special because in this case the coefficient at the derivative $\partial\tilde{u}/\partial x$ vanishes. The derivative $\partial\tilde{u}/\partial x$ at the value $\mathbf{M}_\infty$ close to unity cannot already be considered to be a small quantity whereas the rest of the derivatives $\partial\tilde{u}/\partial y$, $\partial\tilde{v}/\partial x$, $\partial\tilde{v}/\partial y$ retain their smallness. That is why the first term on the right side should be retained and the remainder may be omitted as before. Thus, in the case of transonic flow past a thin profile, the linearization is violated and we have the following, already nonlinear, system of motion equations

$$(1-\mathbf{M}_\infty^2)\frac{\partial \tilde{u}}{\partial x} + \frac{\partial \tilde{v}}{\partial y} = (k+1)\,\mathbf{M}_\infty^2 \frac{\tilde{u}}{U_\infty}\frac{\partial \tilde{u}}{\partial x}$$

$$\frac{\partial \tilde{u}}{\partial y} - \frac{\partial \tilde{v}}{\partial x} = 0$$

In the case considered it is simpler to deal with the perturbation velocity potential $\tilde{\varphi}$ rather than with the stream function of perturbations $\tilde{\psi}$. Using the presence of the velocity potential of small perturbations $\tilde{\varphi}(x, y)$, we reduce the system obtained to one equation:

$$(1-\mathbf{M}_\infty^2)\frac{\partial^2 \tilde{\varphi}}{\partial x^2} + \frac{\partial^2 \tilde{\varphi}}{\partial y^2} = \frac{(k+1)\,\mathbf{M}_\infty^2}{U_\infty}\frac{\partial \tilde{\varphi}}{\partial x}\frac{\partial^2 \tilde{\varphi}}{\partial x^2} \tag{61}$$

with corresponding boundary condition (18)

$$\frac{\partial \tilde{\varphi}}{\partial x} = U_\infty h'(x) \quad \text{when} \quad y = \pm 0 \qquad x_A \le x \le x_B$$

To establish the relationship of similarity between two plane transonic flows past thin profiles with the prescribed relative thicknesses τ_1, τ_2, oncoming uniform flow velocities U_1, U_2, Mach numbers $\mathbf{M}_1$ and $\mathbf{M}_2$, and specific heat ratios k_1 and k_2, we set up the following equations and boundary conditions:

For the first flow

$$\frac{\partial^2 \tilde{\varphi}_1}{\partial x^2} + \frac{1}{1-\mathbf{M}_1^2}\frac{\partial^2 \tilde{\varphi}_1}{\partial y_1^2} = \frac{(k_1+1)\,\mathbf{M}_1^2}{U_1(1-\mathbf{M}_1^2)}\frac{\partial \tilde{\varphi}_1}{\partial x}\frac{\partial^2 \tilde{\varphi}_1}{\partial x^2}$$

$$\frac{\partial \tilde{\varphi}_1}{\partial y_1} = U_1 h_1'(x) \quad \text{when} \quad y_1 = \pm 0 \qquad x_A \le x \le x_B \tag{62}$$

and for the second flow

$$\frac{\partial^2 \tilde{\varphi}_2}{\partial x^2} + \frac{1}{1-\mathbf{M}_2^2}\frac{\partial^2 \tilde{\varphi}_2}{\partial y_2^2} = \frac{(k_2+1)\,\mathbf{M}_2^2}{U_2(1-\mathbf{M}_2^2)}\frac{\partial \tilde{\varphi}_2}{\partial x}\frac{\partial^2 \tilde{\varphi}_2}{\partial x^2}$$

$$\frac{\partial \tilde{\varphi}_2}{\partial y_2} = U_2 h_2'(x) \quad \text{when} \quad y_2 = \pm 0 \qquad x_A \le x \le x_B \tag{63}$$

Just as at the beginning of the section, we introduce the transformations

$$\tilde{\varphi}_1 = \Phi_1 \tilde{\varphi} \qquad \tilde{\varphi}_2 = \Phi_2 \tilde{\varphi} \qquad x = \xi \qquad \sqrt{|1-\mathbf{M}_1^2|}\, y_1 = \sqrt{|1-\mathbf{M}_2^2|}\, y_2 = \eta$$

$$h_1(x) = b\tau_1 H(\xi) \qquad h_2(x) = b\tau_2 H(\xi) \tag{64}$$

By similarity condition the functions $\tilde{\varphi}(\xi, \eta)$ and $H(\xi)$ are identical in the systems compared. With the aid of these functions, equations and boundary conditions (62) and (63) are rewritten as

$$\frac{\partial^2 \tilde{\varphi}}{\partial \xi^2} \pm \frac{\partial^2 \tilde{\varphi}}{\partial \eta^2} = \frac{(k_1+1)\,\mathbf{M}_1^2 \Phi_1}{U_1(1-\mathbf{M}_1^2)} \frac{\partial \tilde{\varphi}}{\partial \xi} \frac{\partial^2 \tilde{\varphi}}{\partial \xi^2}$$

$$\frac{\partial \tilde{\varphi}}{\partial \eta} = \frac{U_1 b \tau_1}{\Phi_1 \sqrt{|1-\mathbf{M}_1^2|}} H'(\xi) \quad \text{when} \quad \eta = \pm 0 \qquad x_A \le \xi \le x_B$$

$$\frac{\partial^2 \tilde{\varphi}}{\partial \xi^2} \pm \frac{\partial^2 \tilde{\varphi}}{\partial \eta^2} = \frac{(k_2+1)\,\mathbf{M}_2^2 \Phi_2}{U_2(1-\mathbf{M}_2^2)} \frac{\partial \tilde{\varphi}}{\partial \xi} \frac{\partial^2 \tilde{\varphi}}{\partial \xi^2}$$

$$\frac{\partial \tilde{\varphi}}{\partial \eta} = \frac{U_2 b \tau_2}{\Phi_2 \sqrt{|1-\mathbf{M}_2^2|}} H'(\xi) \quad \text{when} \quad \eta = \pm 0 \qquad x_A \le \xi \le x_B \tag{65}$$

The requirement of the coincidence of these two systems consisting of the equations and boundary conditions for the first and the second flows, is reduced to two equalities

$$\frac{(k_1+1)\,\mathbf{M}_1^2 \Phi_1}{U_1(1-\mathbf{M}_1^2)} = \frac{(k_2+1)\,\mathbf{M}_2^2 \Phi_2}{U_2(1-\mathbf{M}_2^2)} \qquad \frac{\Phi_1 \sqrt{|1-\mathbf{M}_1^2|}}{U_1 \tau_1} = \frac{\Phi_2 \sqrt{|1-\mathbf{M}_2^2|}}{U_2 \tau_2} \tag{66}$$

According to the general formula

$$c_p = -\frac{2\tilde{u}}{U_\infty} = -\frac{2}{U_\infty}\frac{\partial \tilde{\varphi}}{\partial x}$$

the expressions of the pressure coefficients c_{p1} and c_{p2} in the flows compared will be

$$c_{p1} = -\frac{2\Phi_1}{U_1}\frac{\partial \tilde{\varphi}}{\partial \xi} \qquad c_{p2} = -\frac{2\Phi_2}{U_2}\frac{\partial \tilde{\varphi}}{\partial \xi}$$

where it follows that

$$\frac{U_1 c_{p1}}{\Phi_1} = \frac{U_2 c_{p2}}{\Phi_2} \tag{67}$$

Eliminating Φ_1/U_1 and Φ_2/U_2 from the system of Equations 66 and 67, we obtain the sought-after similarity conditions for two plane transonic flows past thin bodies

$$\frac{c_{p1}\sqrt{|1-\mathbf{M}_1^2|}}{\tau_1}=\frac{c_{p2}\sqrt{|1-\mathbf{M}_2^2|}}{\tau_2}\qquad\frac{(k_1+1)\,\mathbf{M}_1^2\tau_1}{|1-\mathbf{M}_1^2|^{3/2}}=\frac{(k_2+1)\,\mathbf{M}_2^2\tau_2}{|1-\mathbf{M}_2^2|^{3/2}}\tag{68}$$

which, with the same arbitrariness as before, can be expressed in the form of the general functional relation

$$\frac{c_p\sqrt{|1-\mathbf{M}_\infty^2|}}{\tau}=f\left(\frac{(k+1)\,\mathbf{M}_\infty^2\tau}{|1-\mathbf{M}_\infty^2|^{3/2}}\right)\tag{69}$$

Since the number $\mathbf{M}_\infty$ is rather close to unity (the flow is close to the sonic one), it can be omitted in the numerator on the right side of Equation 69, and this equation can be rewritten as

$$\frac{c_p\sqrt{|1-\mathbf{M}_\infty^2|}}{\tau}=f\left(\frac{(k+1)\,\tau}{|1-\mathbf{M}_\infty^2|^{3/2}}\right)\tag{70}$$

Moreover, if the comparison is made in the flows with identical physical properties (the same value of k), the similarity relation (70) becomes still more simple and acquires the form

$$\frac{c_p\sqrt{|1-\mathbf{M}_\infty^2|}}{\tau}=f\left(\frac{\tau}{|1-\mathbf{M}_\infty^2|^{3/2}}\right)\tag{71}$$

As distinct from the earlier considered similarity conditions for sub- and supersonic flows past thin bodies there is only one similarity relation in the case of transonic flow past a thin body. This fact follows from the nonlinear character of Equation 62 with respect to the potential $\tilde{\varphi}$; $\tilde{\varphi}$ enters linearly on the left side and nonlinearly on the right side in the form of the product of two derivatives.

Equation 69 also can be represented in other forms. Thus, for example, multiplying both parts by the argument from the right side, the relative thickness τ can be discarded from the left side. Retaining the same notation for the function f:

$$\frac{(k+1)\,c_p\mathbf{M}_\infty^2}{|1-\mathbf{M}_\infty^2|}=f\left(\frac{(k+1)\,\mathbf{M}_\infty^2\tau}{|1-\mathbf{M}_\infty^2|^{3/2}}\right)\tag{72}$$

Denoting by χ the argument that occurs on the right side and which is raised to the power $(-2/3)$, i.e., by assuming[8] that

$$\chi=\frac{|1-\mathbf{M}_\infty^2|}{[(k+1)\,\tau\mathbf{M}_\infty^2]^{2/3}}\tag{73}$$

then, having multiplied both parts of Equation 72 by χ, it can be rewritten in the form

[8]Liepmann, H. W. and Roshko, A., *Elementy gazovoy dinamiki (The Elements of Gas Dynamics)*, IL Press, Moscow, 1960, p. 307.

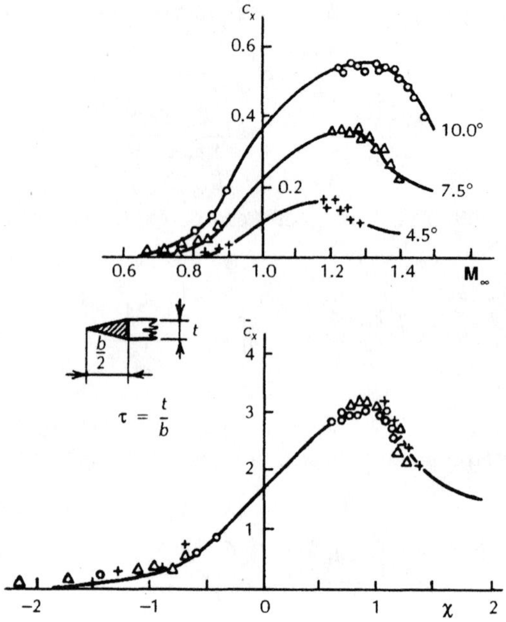

Fig. 94

$$\frac{c_p\,[(k+1)\,\mathbf{M}_\infty^2]^{1/3}}{\tau^{2/3}} = f\left(\frac{|1-\mathbf{M}_\infty^2|}{[(k+1)\,\tau\mathbf{M}_\infty^2]^{2/3}}\right) = f(\chi) \tag{74}$$

or in simplified forms that correspond to Equations 70 or 71.

It follows from Equation 74 that in contrast to the sub- and supersonic flows the transonic flows can be compared only when the relative thicknesses of thin profiles, Mach numbers, and specific heat ratios of these flows are connected by the condition of the identity of the similarity parameter χ in both flows.

According to the formulas

$$c_x = -\frac{1}{b}\oint c_p dy = -\frac{1}{b}\oint c_p \frac{dy}{dx}dx \qquad c_y = \frac{1}{b}\oint c_p dx$$

we have for the transonic flow

$$c_y = \frac{\tau^{2/3}}{[(k+1)\,\mathbf{M}_\infty^2]^{1/3}}\,f\left(\frac{|1-\mathbf{M}_\infty^2|}{[(k+1)\,\tau\mathbf{M}_\infty^2]^{2/3}}\right)$$

$$c_x = \frac{\tau^{5/3}}{[(k+1)\,\mathbf{M}_\infty^2]^{1/3}}\,f_1\left(\frac{|1-\mathbf{M}_\infty^2|}{[(k+1)\,\tau\mathbf{M}_\infty^2]^{2/3}}\right) \tag{75}$$

A convincing illustration of the similarity law for the transonic flow past thin bodies can be provided by the curve[9] presented in Figure 94 and relating to the longitudinal flow past thin wedges with the semivertex angles equal to 4.5°, 7.5°, and 10°. In the upper half of the figure three experimental curves are shown for the resistance coefficient c_x vs. the Mach

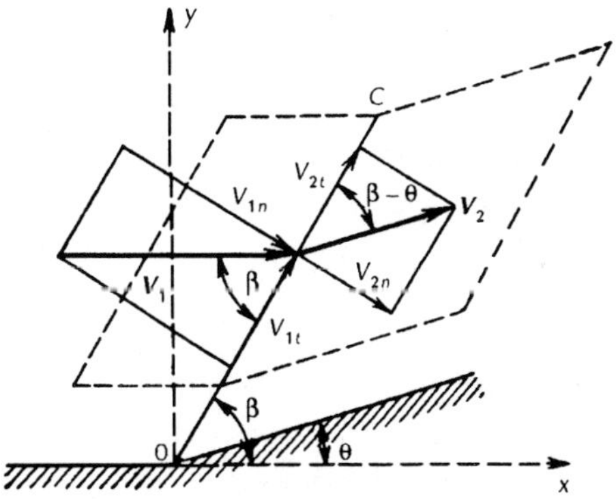

Fig. 95

number $\mathbf{M}_\infty$ in the transonic region for the above three semivertex angles. In the lower half of the same figure the same curves are plotted in the similarity parameters ($\tau = t/b$)

$$\bar{c}_x = \frac{c_x [(k+1)\mathbf{M}_\infty^2]^{1/3}}{\tau^{5/3}} \qquad \chi = \frac{|\mathbf{M}_\infty^2 - 1|}{[(k+1)\mathbf{M}_\infty^2 \tau]^{2/3}}$$

Quite obvious is the fact that the three curves of the upper half of the figure converge into one common curve in the lower half, thus confirming the correctness of the similarity law derived for the transonic flow past thin bodies.

5. A Converging Supersonic Flow. An Oblique Shock Wave

Turning our attention to the discussion of nonlinearized supersonic flows that correspond to the general case of finite disturbances of the flow by bodies or flows through nozzles, we first consider two basic phenomena: retardation occurring under the conditions of supersonic flow convergence and, conversely, acceleration the flow on its divergence.

Let us start with the phenomenon of the supersonic flow retardation, originating, for example, during flow incidence on a solid body immersed in it. The simplest case that admits an elementary analysis is a symmetric supersonic flow past an infinite wedge with the vertex angle of 2θ having a certain finite value. According to the well-known property of ideal flows, it is possible to replace the null streamline belonging to the flow incident on the wedge and directed into the wedge vertex 0 (Figure 95) by a solid wall and to consider only the upper part of the flow which will represent a plane flow inside of an obtuse angle equal to $\pi - \theta$.

To elucidate the general character of the flow we invert the motion and consider the flow induced in a quiescent gas by a symmetric infinite wedge gas by a symmetric infinite wedge moving with a supersonic velocity along its symmetry line. This flow might be likened to that considered in Chapter 6, Section 5, and caused by a pushing piston in a gas.

The sides of the moving wedge also play the role of pistons pushing the gas in front of them and causing the formation of shock waves in it. These waves, driving one another, form the discontinuity front of the moving gas the parameters of which, in contrast to the case of the shock wave parallel to the piston plane and perpendicular to the direction of its motion

[9] Spreiter, J. R., *The Aerodynamics of Wings and Bodies at Transonic Velocities*, collected Volume of Translations Mechanics, No. 3, 1960.

(Chapter 6, Section 5), will be somewhat inclined to the direction of wedge motion (to the line of its symmetry).

Let us elucidate under what conditions the formation of the plane bow shock wave $0C$ emanating from the wedge vertex 0 (Figure 95) and otherwise called oblique shock wave is possible. For this purpose, proceeding in a way analogous to that used in Chapter 6, Section 5 for calculating a normal shock wave, we apply three basic conservation laws, mass, total momentum, and total enthalpy, to an arbitrary stream tube that intersects the oblique shock wave.

Let us agree that hereafter subscript 1 will denote the values before the shock and subscript 2 denote the values behind the shock; moreover, the subscript t will designate the velocity component in the shock plane $0C$ and the subscript n will designate the normal velocity component.

Selecting the control surface in the manner shown in Figure 95, we have:

1. The mass conservation law

$$\rho_1 V_{1n} = \rho_2 V_{2n}$$

2. The law of total momentum conservation in the direction tangent to the discontinuity line

$$\rho_1 V_{1n} V_{1t} = \rho_2 V_{2n} V_{2t}$$

3. Same in the direction normal to the discontinuity line

$$p_1 + \rho_1 V_{1n}^2 = p_2 + \rho_2 V_{2n}^2$$

4. The law of total enthalpy conservation

$$h_1 + \frac{1}{2}(V_{1t}^2 + V_{1n}^2) = h_2 + \frac{1}{2}(V_{2t}^2 + V_{2n}^2)$$

The equations under items (1) and (2) directly yield the following equality which is basic to the theory of the oblique shock

$$V_{1t} = V_{2t} = V_t$$

and which states that for a gas passing through an oblique shock the tangent velocity component is retained; only the normal velocity component varies stepwise.

Taking into account this property of the oblique shock when rewriting the equation of the total enthalpy [item (4)] in the form

$$h_1 + \frac{1}{2}V_{1n}^2 = h_2 + \frac{1}{2}V_{2n}^2$$

and comparing it, and also the equations under items (1) and (3), with corresponding equations (39), (40), and (41) of the theory of the normal shock wave (Chapter 6, Section 5), we see that the three basic equations for calculating the elements of the oblique shock wave

$$\rho_1 V_{1n} = \rho_2 V_{2n} \qquad p_1 + \rho_1 V_{1n}^2 = p_2 + \rho_2 V_{2n}^2 \qquad h_1 + \frac{1}{2}V_{1n}^2 = h_2 + \frac{1}{2}V_{2n}^2$$

coincide completely with the corresponding equations of the theory of the normal shock wave provided that the velocity before and after the shock is the normal velocity component. This relieves us of the necessity to repeat the conclusions of Chapter 6, Section 5.

From this it follows that the relationship between the pressure and density resulting from the previous equations on the elimination of velocity, i.e., the Hugoniot shock adiabatic expressed by Equation 43 of Chapter 6 and by the plot of Figure 36 of Chapter 6, in the case of an oblique shock must remain the same as in the case of the normal shock.

Just as in the case of the normal shock, the following equations based on the conservation law of the total enthalpy h_0 will remain the same

$$h_{10} = h_{20} = h_0 \qquad T_{10} = T_{20} = T_0 \qquad a_{10} = a_{20} = a_0$$

and, consequently, also the following equations

$$T_1^* = T_2^* = T^* \qquad a_1^* = a_2^* = a^*$$

Thus, when the gas passes through an oblique shock, the enthalpy, temperature, and the speed of sound in a gas in adiabatic and isentropic stagnation state, as well as the critical gas temperature and also its critical and maximum velocities remain invariant.

Together with the equation for the Hugoniot shock adiabatic (Equation 43 of Chapter 6) also preserved for the normal velocities in the theory of the shock is the Prandtl equation (49, Chapter 6) but with the altered right side. There is no need to fully repeat the derivation of this equation. The difference in the derivations lies in the fact that in the theory of the oblique shock the Bernoulli equation written in the form (the second formula in Equation 77 of Chapter 5)

$$\frac{V^2}{2} + \frac{a^2}{k-1} = \frac{k+1}{2\,(k-1)}\,a^{*2}$$

can be rewritten, after the replacement $V^2 = V_t^2 + V_n^2$,

$$\frac{V_n^2}{2} + \frac{a^2}{k-1} = \frac{k+1}{2\,(k-1)}\,a^{*2} - \frac{V_t^2}{2} = \frac{k+1}{2\,(k-1)}\,\bar{a}^{*2}$$

where the "reduced" critical velocity $\bar{a}^*$, which has appeared on the right side and which is equal to

$$\bar{a}^* = \sqrt{a^{*2} - \frac{k-1}{k+1}V_t^2}$$

will also be retained, by conditions of the preservation of a^* and V_t^*, when the gas passes through an oblique shock. Repeating the conclusion presented in the theory of the normal shock (Chapter 6, Section 5), but using the Bernoulli equation, which has been formed for the normal velocity V_n and which contains the "reduced" critical velocity $\bar{a}^*$ on its right, we obtain the Prandtl equation for an oblique shock in the following two forms:

$$V_{1n}V_{2n} = \bar{a}^{*2} \qquad \bar{\lambda}_{1n}\bar{\lambda}_{2n} = 1 \tag{76}$$

understanding that by $\bar{\lambda}_n$ is meant the ratio $V_n/\bar{a}^*$.

Directing our attention to the velocity triangles shown in Figure 95 we conclude about the validity of the following system of equations (θ is the angle of flow deviation by the shock, β is the angle formed by the shock line $0C$ with the incident flow direction):

$$V_{1n} = V_1 \sin\beta \qquad V_{2n} = V_2 \sin(\beta - \theta)$$

$$V_{1t} = V_1 \cos\beta = V_{2t} = V_2 \cos(\beta - \theta) = V_t \tag{77}$$

Using the previous designations for the velocity projections on the axis of the rectangular Cartesian system of coordinates $0xy$

$$V_{1x} = V_1 \qquad V_{1y} = 0 \qquad V_{2x} = V_2 \cos\theta = u_2 \qquad V_{2y} = V_2 \sin\theta = v_2$$

we may rewrite system (77) also in the form

$$V_{1n} = V_1 \sin\beta \qquad V_{2n} = u_2 \sin\beta - v_2 \cos\beta$$

$$V_t = V_1 \cos\beta = u_2 \cos\beta + v_2 \sin\beta \tag{78}$$

The last equation of system (78) easily yields the expressions of the trigonometric functions of the shock angle β in terms of the Cartesian velocity projections before and after the shock

$$\sin\beta = \frac{V_1 - u_2}{N} \qquad \cos\beta = \frac{v_2}{N} \qquad N = \sqrt{(V_1 - u_2)^2 + v_2^2} \tag{79}$$

These relations and Equations 78 make it possible to find the expressions for the tangent and normal velocity components in terms of its Cartesian projections

$$V_t = \frac{V_1 v_2}{N} \qquad V_{1n} = \frac{1}{N} V_1 (V_1 - u_2) \qquad V_{2n} = \frac{1}{N}[u_2(V_1 - u_2) - v_2^2]$$

substituting them into the first of Equations 76 and making simple reductions, we obtain

$$v_2^2 = \frac{(V_1 - u_2)^2 (V_1 u_2 - a^{*2})}{a^{*2} + \dfrac{2}{k+1} V_1^2 - V_1 u_2} \tag{80}$$

Dividing both parts of this equation by a^{*2} or by a_1^2, we may rewrite it in the following two forms:

$$\left(\frac{v_2}{a^*}\right)^2 = \frac{\left(\lambda_1 - \dfrac{u_2}{a^*}\right)^2 \left(\lambda_1 \dfrac{u_2}{a^*} - 1\right)}{1 + \dfrac{2}{k+1}\lambda_1^2 - \lambda_1 \dfrac{u_2}{a^*}} \qquad \left(\frac{v_2}{a_1}\right)^2 = \frac{\left(\mathbf{M}_1 - \dfrac{u_2}{a_1}\right)^2 \left(\mathbf{M}_1 \dfrac{u_2}{a_1} - \left(\dfrac{a^*}{a_1}\right)^2\right)}{\left(\dfrac{a^*}{a_1}\right)^2 + \dfrac{2}{k+1}\mathbf{M}_1^2 - \mathbf{M}_1 \dfrac{u_2}{a_1}} \tag{81}$$

which represent the equations of the family of curves respectively in the planes $(u_2/a^*, v_2/a^*)$ or $(u_2/a_1, v_2/a_1)$ with the parameters λ_1 or $\mathbf{M}_1$. It is assumed in the latter of Equations 81 that the quantity a^*/a_1 entering the right side is expressed in terms of $\mathbf{M}_1$ by the familiar isentropic formula. The resulting families represent the geometric loci for the termini of the velocity vectors $\mathbf{V}_2$ behind an oblique shock wave related in the first case to a^* and in the second case to a_1, with the value of the velocity $\mathbf{V}_1$ before the jump, related to a^* or a_1, being used as the parameters of the families.

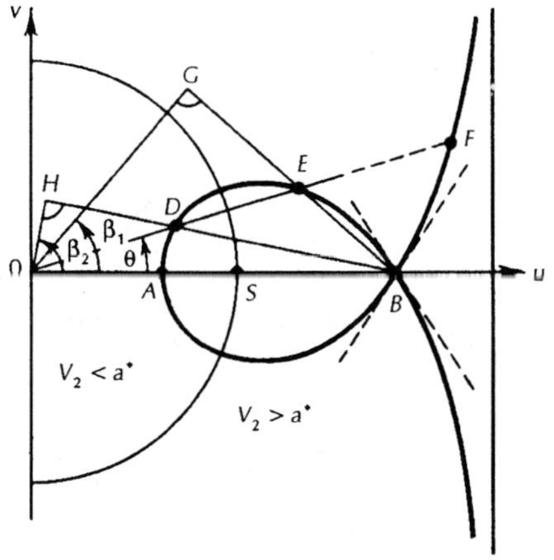

Fig. 96

The curves of the families (81) are the strophoids (also called hypercissoids or leaves of Descartes) the graphing of which is not difficult.

One of such strophoids is shown in Figure 96 in dimensional coordinates u, v. The ray $0F$ drawn from the coordinate origin at an angle θ, equal to the rotation of the flow, or for example, to the semivertex angle of the wedge, intersects the strophoid at three points: D, E, and F and thus determines three values of the velocity vector $\mathbf{V}_2$ behind the shock wave.

As is seen from Equation 80, to the double point B (two tangents shown by dashes pass through this point) corresponds the value $u = V_1$ of the velocity before the shock.

Because $0F > 0B$ and because we refer to the retardation of the flow behind the shock, the point F and, generally, both the infinite branches of the strophoid located to the right from the point B and approaching the asymptote, are idle and can be omitted.

Only two values of the velocity vector $\mathbf{V}_2$ behind the shock have physical meaning, $0D$ and $0E$. As follows directly from Equations 77, the segments $0G$ and $0H$ express two possible values of the common tangent of the velocity component $\mathbf{V}_1$ and $\mathbf{V}_2$, whereas (BG, EG) and (BH, DH) are the normal components of these velocities for two possible directions of the shock which correspond to two values β_1 and β_2 of the angle β.

The free stream is supersonic by condition; consequently, $V_1 = 0B > a^*$. On the other hand, it can be easily concluded from Equation 80 that the point A of the intersection of the strophoid with the axis $0u$ will have the abscissa $0A = a^{*2}/V_1 = a^*/\lambda_1 < a^*$, because $\lambda_1 > 1$. From this it follows that the point S on the axis $0u$, which corresponds to the critical velocity $0S = a^*$, should be located between the points A and B and in such a manner that the inversion condition $0A \cdot 0B = 0S^2$ could be fulfilled. The circle of radius $0S = a^*$ divides the regions of the sub- and supersonic flows. The velocities behind the shock can be both sub- and supersonic. This will be considered in more detail later.

It should be also noted that in each given case, i.e., with the number λ_1 or $\mathbf{M}_1$ being specified, there exists such a value of $\theta = \theta_{max}$ at which the points D and E will merge into one and, consequently, to these values of θ, λ_1, or $\mathbf{M}_1$ there will correspond only one value of the angle β and only one location of the oblique shock. When at the given λ_1 or $\mathbf{M}_1$ the angle of flow rotation is prescribed to be greater than θ_{max}, the solution will become impossible. This means that the considered scheme (Figure 95) of the rectilinear shock $0C$ emanating from the vertex of the angle (vertex of the wedge) cannot be realized in this case and must be replaced by another scheme, *viz.*, by the scheme of the bow shock wave which "receded" from the vertex 0; this will be treated later.

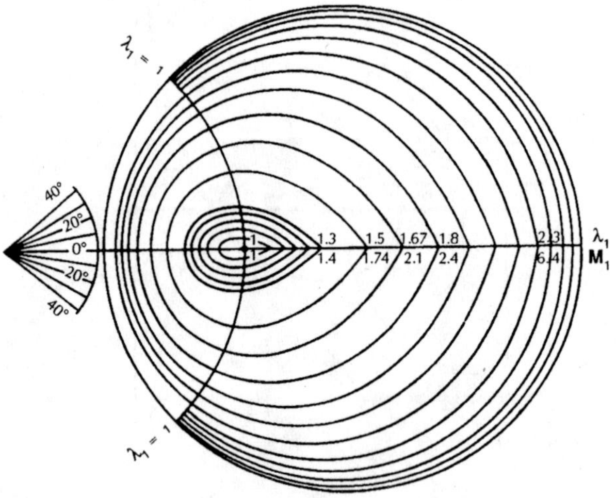

Fig. 97

In Figure 97 the diagram of a family of shock polars, strophoids, is given plotted for $k = 1.4$ according to the first of Equations 81. The infinite branches are omitted as being idle. To each strophoid correspond their own values of the parameters λ_1 or $\mathbf{M}_1$ indicated at the double point of the curve (λ_1 from above, $\mathbf{M}_1$ from below).

Let us go over to a more detailed analysis of the phenomena of gas passage through an oblique shock wave. First, use the technique indicated at the beginning of the section for deriving the formulas of an oblique shock from the corresponding formulas for a normal shock by means of the replacement of V_1 and V_2 by $V_{1n} = V_1 \sin\beta$, $V_{2n} = V_2 \sin(\beta - \theta)$ and, consequently, of $\mathbf{M}_1$ and $\mathbf{M}_2$ by $\mathbf{M}_{1n} = \mathbf{M}_1 \sin\beta$ and $\mathbf{M}_{2n} = \mathbf{M}_2 \sin(\beta - \theta)$.

The existence of this simple rule allows the easy set up of the ratios of pressures, densities, and stagnation pressures after and before an oblique shock and makes it easier to memorize them.

Thus, instead of Equations 53 and 55 of Chapter 6, we have in the case of an oblique shock, the following expressions for the pressure and density ratios before and after the shock

$$\frac{p_2}{p_1} = \frac{2k}{k+1}\mathbf{M}_1^2 \sin^2\beta - \frac{k-1}{k+1} \qquad \frac{\rho_2}{\rho_1} = \frac{\dfrac{k+1}{2}\mathbf{M}_1^2 \sin^2\beta}{1 + \dfrac{k-1}{2}\mathbf{M}_1^2 \sin^2\beta} \tag{82}$$

An analogous transformation is made with the pressure ratio of the gas in adiabatic and isentropic stagnation states after the shock p_{20} and before the shock p_{10}, which characterizes the losses of the mechanical energy of the gas when the latter passes through an oblique shock (Chapter 6, Section 6). For an oblique shock wave, Equation 59 of Chapter 6 converts into the following formula

$$æ = \frac{p_{20}}{p_{10}} = \left(\frac{k+1}{2}\right)^{\frac{k+1}{k-1}} \frac{(\mathbf{M}_1 \sin\beta)^{\frac{2k}{k-1}}}{\left(1 + \dfrac{k-1}{2}\mathbf{M}_1^2 \sin^2\beta\right)^{\frac{k}{k-1}} \left(k\mathbf{M}_1^2 \sin^2\beta - \dfrac{k-1}{2}\right)^{\frac{1}{k-1}}} \tag{83}$$

Just as in the theory of the normal shock, we shall have the following asymptotic expressions for this ratio:

1. For $\mathbf{M}_1 \sin\beta \gg 1$ (~ is the symbol of asymptotic equality)

$$æ = \frac{p_{20}}{p_{10}} \sim \left[\frac{(k+1)^{k+1}}{2k(k-1)^k}\right]^{\frac{1}{k-1}} (\mathbf{M}_1 \sin\beta)^{-\frac{2}{k-1}} \tag{84}$$

and for the change in entropy

$$\frac{s_2 - s_1}{R} = -\ln\frac{p_{20}}{p_{10}} \sim \frac{2}{k-1}\ln(\mathbf{M}_1 \sin\beta) \tag{85}$$

Note that the right side of this equation is always positive because ($\beta > \alpha$)

$$\mathbf{M}_1 \sin\beta = \frac{\sin\beta}{1/\mathbf{M}_1} = \frac{\sin\beta}{\sin\alpha} > 1$$

According to asymptotic equation 84 the quantity æ in the stream of air ($k = 1.4$) for $\mathbf{M}_1 \sin\beta \gg 1$ is inversely proportional to the fifth power of $\mathbf{M}_1 \sin\beta$.

2. When the numbers $\mathbf{M}_1 \sin\beta$ are close to unity

$$æ = \frac{p_{20}}{p_{10}} \approx 1 - \frac{2k}{3(k+1)^2}[(\mathbf{M}_1 \sin\beta)^2 - 1]^3$$

$$\frac{s_2 - s_1}{R} \approx \frac{2k}{3(k+1)^2}[(\mathbf{M}_1 \sin\beta)^2 - 1]^3 \tag{86}$$

As noted, the expression in the square brackets is always positive. Equations 86 are the extensions of Equations 61 and 64 of Chapter 6 to the case of an oblique shock.

Let us determine the connection between the angles β and θ at the given value of the number $\mathbf{M}_1$ of the free stream. It seems to be most simple to use for the purpose basic relation (76), which, taking into account relations (77), can be rewritten in the form

$$V_{1n}V_{2n} = V_1 \sin\beta \cdot V_2 \sin(\beta - \theta) = V_1^2 \sin\beta\cos\beta\,\mathrm{tg}(\beta - \theta) = \bar{a}^{*2} =$$

$$= a^{*2} - \frac{k-1}{k+1}V_1^2\cos^2\beta$$

This equation directly yields the relationship between the angles β, θ, and the velocity coefficient λ_1

$$\left[\mathrm{tg}\,\beta\,\mathrm{tg}(\beta - \theta) + \frac{k-1}{k+1}\right]\cos^2\beta = \frac{1}{\lambda_1^2} \tag{87}$$

Replacing here λ_1 by the Mach number $\mathbf{M}_1$ via formula (78) of Chapter 5, we obtain

$$\frac{\mathrm{tg}(\beta - \theta)}{\mathrm{tg}\,\beta} = \frac{k-1}{k+1} + \frac{2}{(k+1)\mathbf{M}_1^2\sin^2\beta} = \frac{1 + \frac{k-1}{2}\mathbf{M}_1^2\sin^2\beta}{\frac{k+1}{2}\mathbf{M}_1^2\sin^2\beta} \tag{88}$$

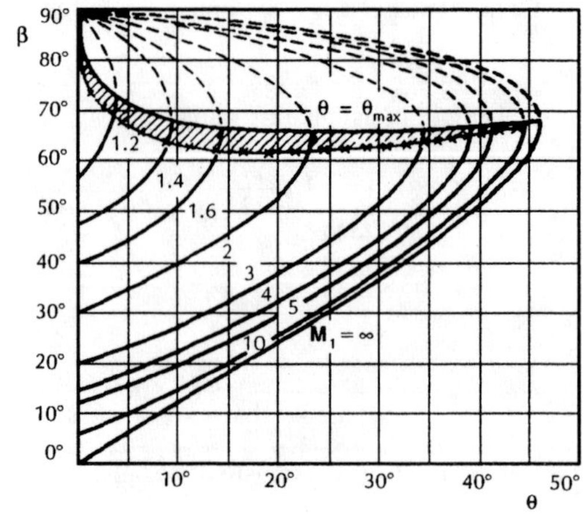

Fig. 98

This equation can be solved in tg θ to give

$$\operatorname{tg}\theta = \frac{\sin^2\beta - \dfrac{1}{\mathbf{M}_1^2}}{\dfrac{k+1}{2} - \sin^2\beta + \dfrac{1}{\mathbf{M}_1^2}}\operatorname{ctg}\beta \tag{89}$$

In Figure 98 the plot of θ vs. β in the form of a family of curves with the parameter $\mathbf{M}_1$ is given. As noted earlier, to each specified value $\theta < \theta_{max}$ correspond two values of β. Denoting by β_m the value of β which corresponds to the greatest θ at the given $\mathbf{M}_1$, it is possible to conclude from the curves of Figure 98 that one of the values of β obtained will lie within the range (α, β_m) where α is the Mach angle at the given $\mathbf{M}_1$; the other value of β will lie within the range $(\beta_m, \pi/2)$. To determine the first value of β, it is necessary to use the solid curves, whereas the dashed curves should be used to determine the other value of β. The solid curves are separated from the dashed curves by the line drawn through the points of the greatest values of $\theta = \theta_{max}$.

This duplicity in the determination of β from the given value of θ corresponds to the essence of the phenomenon of gas passage through an oblique shock. Just as in the theory of the Laval nozzle, the choice of the mode of flow behind an oblique shock depends on the pressure there. As follows from the first equation of system (82), to the greater β corresponds the greater value of the ratio p_2/p_1 between the pressures after and before the shock and to the smaller value of β, the smaller ratio of these pressures, respectively.

Recalling that the pressure ratio considered serves as the measure of the intensity (strength) of the shock, we call strong those oblique shocks that are determined by the upper region of the diagram (Figure 98) from the values $\pi/2 > \beta > \beta_m$ given by the dashed curves; the shocks that correspond to the range of angles $\alpha < \beta < \beta_m$ calculated from the lower solid curves are called weak. The front of a strong shock is the surface (a line in a two-dimensional flow) of a strong variation in the kinematic, dynamic, and thermodynamic characteristics of gas flow; the front of a weak shock is the surface (line) of a weak variation in these quantities. The both types of variations are observed, for example, in "detached" waves (Figure 99).

Let us elucidate the conditions under which the flow behind an oblique shock will be sub- or supersonic. Let us find the connection between the Mach numbers $\mathbf{M}_1$ and $\mathbf{M}_2$ before

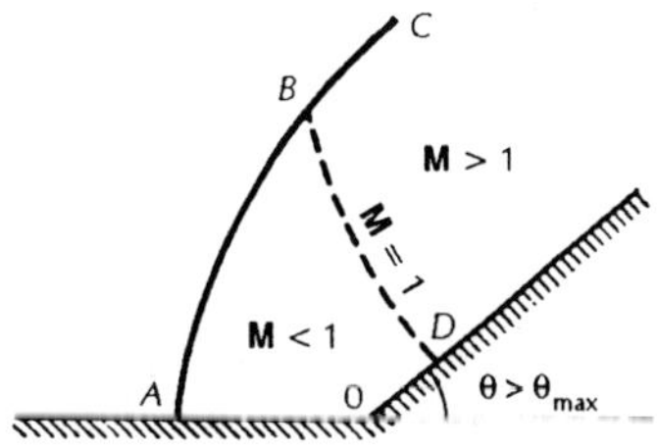

Fig. 99

and after the shock. For this purpose, we shall use Equation 51 of Chapter 6 for a normal shock and, following the general technique (frequently used earlier), we shall replace in this equation $\mathbf{M}_1$ by $\mathbf{M}_1 \sin\beta$ and of $\mathbf{M}_2$ by $\mathbf{M}_2 \sin(\beta - \theta)$. This will yield the sought-after connection formula

$$\mathbf{M}_2^2 \sin(\beta - \theta) = \frac{1 + \dfrac{k-1}{2}\mathbf{M}_1^2 \sin^2\beta}{k\mathbf{M}_1^2 \sin^2\beta - \dfrac{k-1}{2}} \tag{90}$$

Using this expression and relation (89), it is possible to express the Mach number behind the shock $\mathbf{M}_2$ in terms of the Mach number before the shock $\mathbf{M}_1$ and the angle β. In this case, at the same value of $\mathbf{M}_1$ to two different values of β for a strong and a weak shock, correspond two different values of $\mathbf{M}_2$, with the strong shock (just like the normal shock) converting the supersonic flow into the subsonic, and the weak shock almost always retaining the supersonic flow. The only exception is a small region depicted in Figure 98 and adjacent to $\theta = \theta_{max}$. The curve with crosses shows the geometric loci where $\mathbf{M}_2 = 1$. Above this curve, $\mathbf{M}_2 < 1$ and below, $\mathbf{M}_2 > 1$.

It follows from the diagram that for θ close to θ_{max} and at any $\mathbf{M}_1$ both values of β may correspond to the transition from the supersonic to the subsonic velocity; i.e., in this region of the values of θ, both the strong and weak oblique shocks behave as normal, transforming the supersonic flow into the subsonic one. It should be noted, however, that in this case the direction of the strong shock differs from that of the weak shock by $< 10°$, so that the difference between the intensities of the shocks also becomes insignificant.

It is difficult to use computational equations because they are cumbersome. There are detailed tables for air ($k = 1.41$) which connect the quantities $\mathbf{M}_1$, β, θ, $\mathbf{M}_2$, p_2/p_1, ρ_2/ρ_1 with the entropy increment $\Delta s = s_2 - s_1$,[10] but for the range of the numbers $\mathbf{M}_1$ from 1.05 to 4.00 at 0.05 intervals and relatively rough division of the angles θ and corresponding values of β, they occupy 27 pages in the Ferri book.

In view of this we advise that calculations not requiring great accuracy be made using a simple nomogram[11] also set up for the air ($k = 1.41$). In contrast to the diagram in Figure 98, this nomogram is based on the family of curves for $\beta(\mathbf{M}_1; 0)$ with the parameter θ. These curves are shown on the nomogram by solid bell-like curves (when looking along the $\mathbf{M}_1$ axis) which represent the double-value connection between β and $\mathbf{M}_1$. It should be noted that the family of the $\beta(\mathbf{M}_1; \theta)$ curves are constructed from Equation 89 at several values of θ

[10] Ferri, A., *Aerodinamika sverkhzvukovykh techeniy* (*Aerodynamics of Supersonic Flows*) (translation from English), Gostekhizdat, Moscow, 1953, Table 5, pp. 431 to 458.

[11] Liepmann, H. W. and Pucket, A. E., *Vvedeniye v aerodinamiku szhimayemoi zhidkosti* (*Introduction into the Compressible Fluid Aerodynamics*), IL Press, Moscow, 1949, pp. 244 to 245.

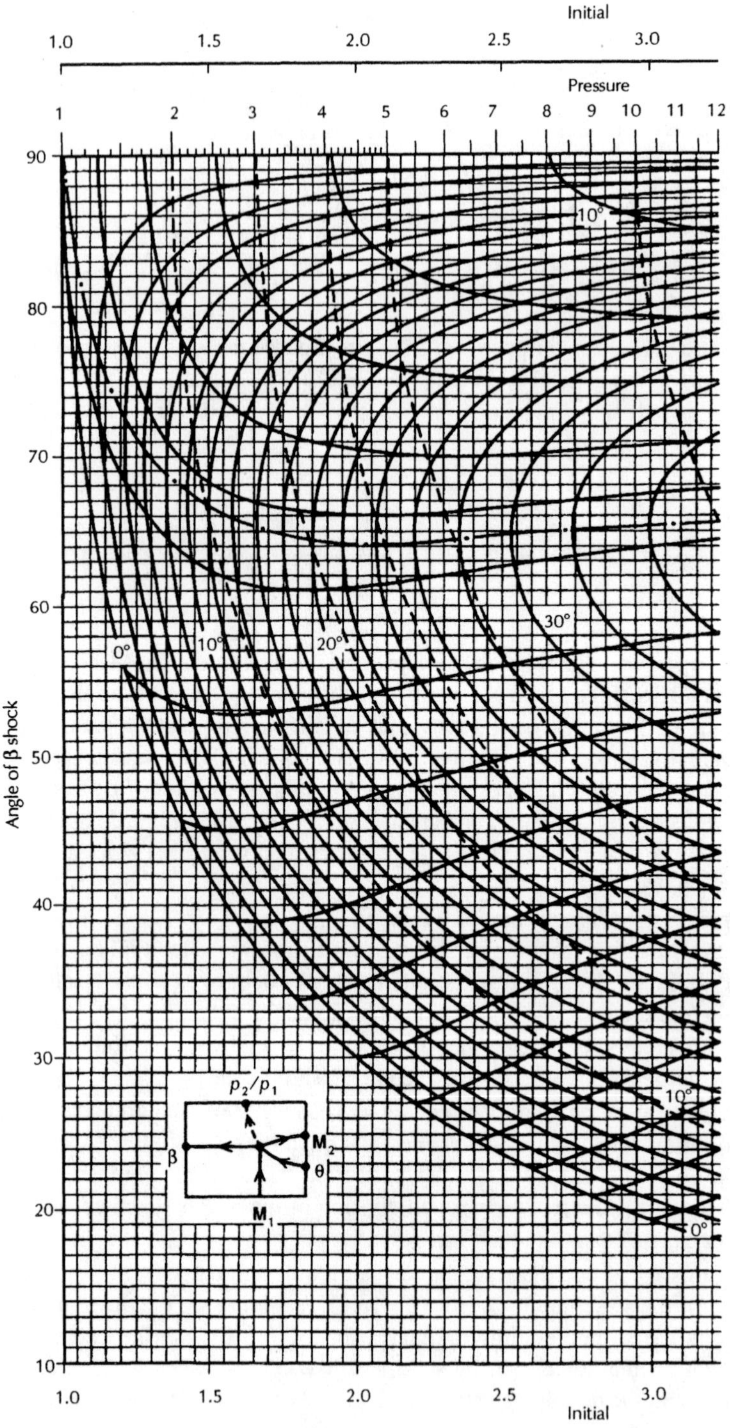
Initial
1.0
1.5
2.0
2.5
3.0
Pressure
1
2
3
4
5
6
7
8
9
10
11
12
90
80
70
60
50
40
30
20
10
Angle of β shock
10°
0°
10°
20°
30°
10°
0°
p_2/p_1
β
M_2
θ
M_1
1.0
1.5
2.0
2.5
3.0
Initial

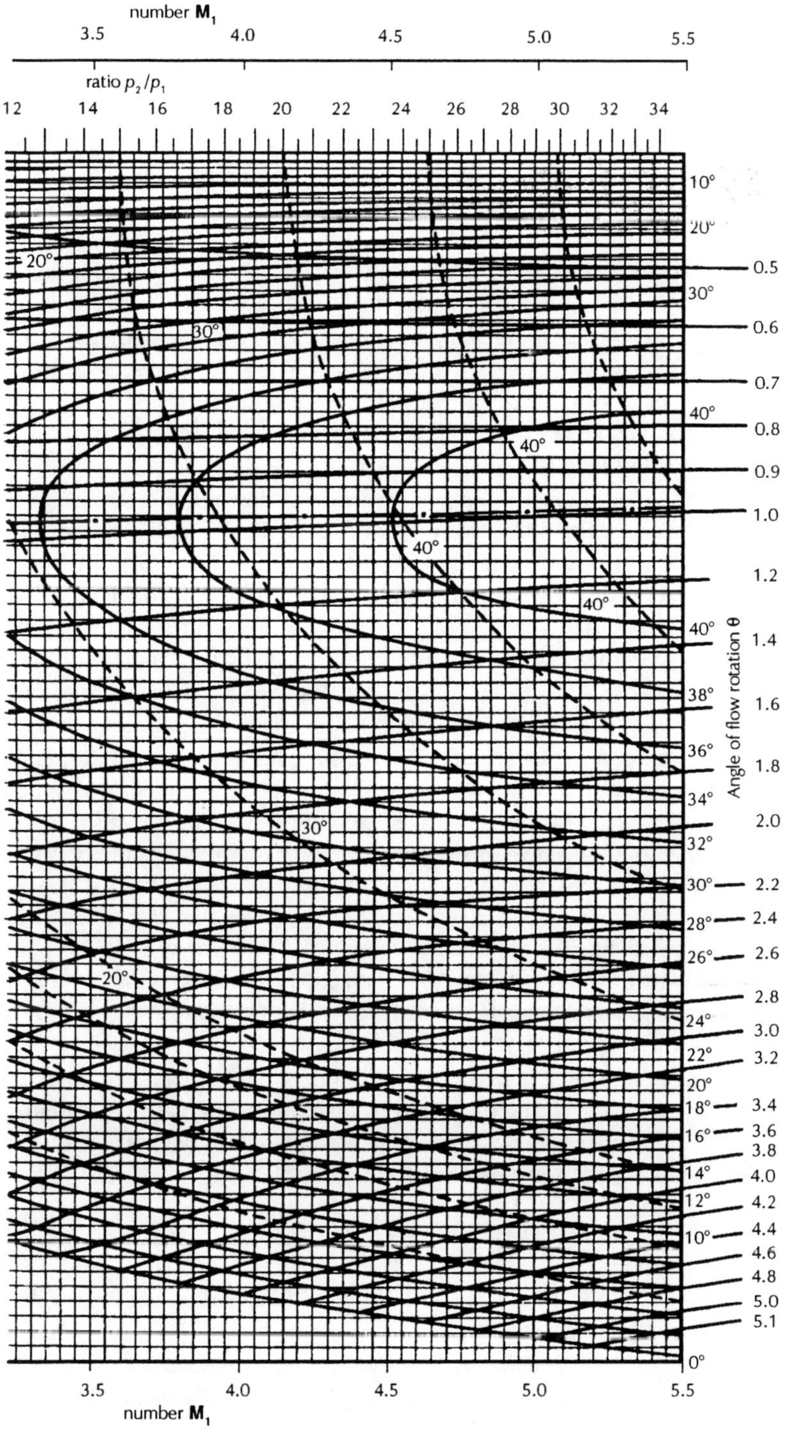
number $\mathbf{M}_1$
ratio p_2/p_1
Angle of flow rotation θ
Finite number $\mathbf{M}_2$
3.5
4.0
4.5
5.0
5.5
12
14
16
18
20
22
24
26
28
30
32
34
10°
20°
30°
40°
38°
36°
34°
32°
28°
26°
24°
22°
18°
16°
14°
12°
0°
0.5
0.6
0.7
0.8
0.9
1.0
1.2
1.4
1.6
1.8
2.0
2.2
2.4
2.6
2.8
3.0
3.2
3.4
3.6
3.8
4.0
4.2
4.4
4.6
4.8
5.0
5.1

considered to be the parameter, and this family should not be confused with the family of curves in Figure 98 in which θ is the argument and $\mathbf{M}_1$ is the parameter.

The nomogram is used as follows. For the required value of θ on the vertical scale on the right we isolate that curve of the family $\beta(\mathbf{M}_1; \theta)$ which corresponds to the selected value of θ. Then, having prescribed the number $\mathbf{M}_1$, which is among the known quantities, we find two values of β corresponding to this number. Two curves pass through each of these two points of the intersection of the selected curve $\beta(\mathbf{M}_1)$ with the vertical line $\mathbf{M}_1$ = const; one (solid) leads to the right extreme vertical scale of the numbers $\mathbf{M}_2$, the second (dashed) leads to the upper horizontal scale of the shock intensity p_2/p_1. To the higher value of β corresponds greater intensity of the shock; to the smaller corresponds the smaller intensity p_2/p_1.

In the lower left corner of the nomogram there is its key. The entrance is denoted by the arrows directed from the scales of θ and $\mathbf{M}_1$ to the point of the intersection of the curves; the exit is shown by the arrows directed from this point to the scales of β, p_2/p_1, and $\mathbf{M}_2$ on which the answers can be read off. The key is given for one value of β; it also remains the same for the second value of β.

The foregoing is valid only so long as the angle of flow rotation (the semivertex angle of the wedge) θ does not exceed the greatest possible (at the given $\mathbf{M}_1$) value of θ_{max}. Within the range $1 \leq \mathbf{M}_1 \leq 4$ the quantity θ_{max} varies from 0 to 38°47'. When $\mathbf{M}_1 \to \infty$, $\theta_{max} \to 45°22'$. As is directly seen from Figure 98, at $\theta = \theta_{max}$ the flow after the shock is always subsonic (the line $\theta = \theta_{max}$ lies above the line $\mathbf{M}_2 = 1$ shown by crosses).

If $\theta > \theta_{max}$, then the presence of a rectilinear oblique shock attached to the angle vertex (wedge) 0 is impossible. A curvilinear "bow" shock wave, or a detached shock wave *AC* originate upstream ahead of the point 0 (Figure 99).

Close to the point *A* the detached shock *AC* behaves as a normal one, and with distance from the point *A* it behaves first as a strong oblique shock and then, with a decrease in the local angle β, it gradually becomes "weaker" and goes over into a rectilinear oblique shock. Downstream after the detached shock there are both the sub- and supersonic gas flows. Downstream of the segment *AB* of the shock, the subsonic zone of the gas flow is formed and downstream of the segment *BC*, the supersonic zone. These two characteristic zones of flow after the shock are divided by the line *BD* along which the gas velocity is equal to the local speed of sound. The flow downstream of the detached curvilinear wave is "mixed", transonic. Up to now the analytical methods for investigating such flows present great mathematical difficulties, which are obviated only with the aid of electronic computers.[12]

To contrast the rectilinear attached shock the flow downstream of the detached curvilinear shock is rotational and the entropy field is nonuniform. Actually, by definition according to Equation 62 of Chapter 6, the change in the entropy during the passage of gas through the shock is determined by the equation (*R* is the gas constant)

$$s_2 - s_1 = -R \ln (p_{20}/p_{10})$$

according to Equation 83, the change in entropy will depend on the local value of β which varies along the curvilinear wave *AC*.

The presence of the segments of strong waves in the bow detached wave (including the central portion close to the normal shock) naturally leads to the idea that it is necessary to create such a "wave breaker" which would break the detached wave and replace it by a system of oblique shocks accompanied by smaller losses of mechanical energy.

---

[12] ***Obtekaniye zatuplennykh tel potokom realnogo gaza (The Real Gas Flow Past Blunt Bodies)***, Belotserkovskiy, O. M., Ed., VTs AN SSSR Press, Moscow, 1966; Belotserkovskiy, O. M., ***Chislennoye modelirovaniye v mekhanike sploshnykh sred (Numerical Simulation in the Mechanics of Continuous Media)***, Nauka Press, Moscow, 1980.

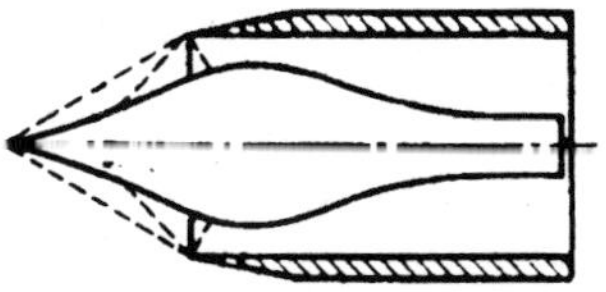

Fig. 100

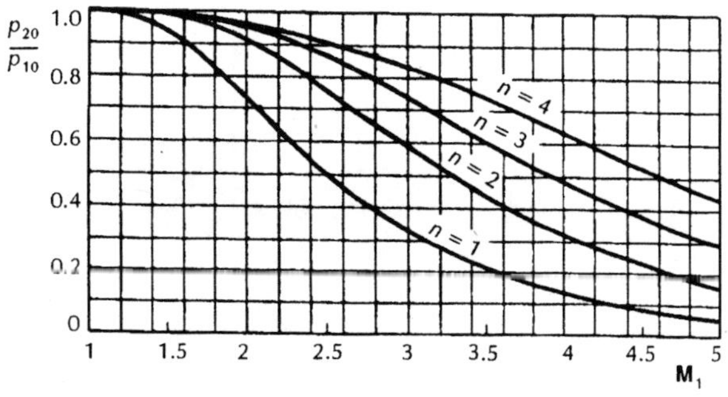

Fig. 101

To this effect, the obtuse nose of the body, or the vertex angle of the wedge or cone which exceeds the admissible one, is replaced by an acute, gradually expanding "needle" on the surface of which weak shocks with relatively small angles β originate in a supersonic flow. As is seen from Equation 83, the losses of mechanical energy due to the presence of the factor $\sin\beta$ at the Mach number $\mathbf{M}_1$ will decrease.

In Figure 100 the inlet to the jet engine chamber is shown with a "needle" which breaks the bow wave into oblique conical (see Chapter 9) shocks. It should be recalled that the supersonic diffusor, the theory of which was presented in Chapter 6 served the same purpose.

Such a "needle", intended to decrease the entry losses of mechanical energy, is called the Oswatitsch needle after the German scientist who was first to publish[13] (1944) a detailed investigation concerning the dependence of the ratio of total pressures in the free stream and at the inlet to the jet engine chamber after the passage of the flow through a system of oblique shocks (two are shown in Figure 100) ended with a normal shock. He also found the possible optimum value of this ratio at the given number n of shocks: $n - 1$ oblique shocks plus one normal shock.

In Figure 101 this optimum ratio is given as a function of the number $\mathbf{M}$ of the free stream for $n = 1, 2, 3, 4$. The lower curve ($n = 1$) corresponds to the location of an ordinary shock wave at the entrance. The plot of Figure 101 shows that the "needle" allows one, even at great Mach numbers, to improve the recovery of the total pressure, whereas the supersonic diffusor described in Chapter 6 gives real results only at flight Mach numbers not above 1.6 to 1.8. The summary of practical information about air intakes in high-speed airplanes can be found in manuals on applied aerodynamics.[14]

6. An Expanding Supersonic Flow. Gas Motion in the Sector of Rarefaction

Let us consider the problem of Prandtl-Meyer,[15] which differs in principle from the previous problem and concerns the supersonic flow rotation around the edge of the convex angle 0 (Figure 102). As becomes clear later, it is possible, not violating the generality at all, to

[13] The theory of K. Oswatitsch is presented in detail in the book by R. Hermann, *Sverkhzvukovye vkhodnye diffuzory (Supersonic Entry Diffusors)*, Fizmatgiz, Moscow, 1960, pp. 161 to 180. Also see Petrov, G. I., and Ukhov, Ye. P., *Raschyot vosstanovleniya davleniya pri perekhode ot sverkhzvukovogo potoka k dozvukovomu pri razlichnykh sistemakh ploskikh skachkov uplotneniya (Calculation of Pressure Recovery on Transition from a Supersonic to a Subsonic Flow with Different Systems of Plane Shocks)*, Moscow, 1974 (the citation is borrowed from the book by Abramovich, G. N., *Prikladnaya gazovaya dinamika (Applied Gas Dynamics)*, Nauka Press, Moscow, 1976, p. 457.

[14] Martynov, A. K., *Prikladnaya aerodinamika (Applied Aerodynamics)*, Mashinostroyeniye Press, Moscow, 1972, pp. 388 to 405.

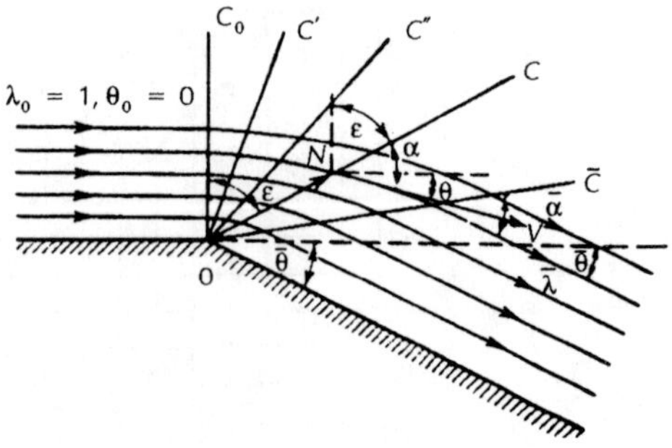

Fig. 102

assume that the initial flow to the left from the straight line $0C_0$ ($0C_0$ is perpendicular to the initial flow) is sonic ($\lambda_0 = \mathbf{M}_0 = 1$). To the flow after the full rotation through the angle $\bar{\theta}$ corresponds a uniform flow to the right from the straight line $0\bar{C}$ with the dimensionless velocity $\bar{\lambda}$ and number $\bar{\mathbf{M}}$, both exceeding unity, because the flow expands (we shall assume that the pressure $\bar{p}$ on the line $0\bar{C}$ satisfies the condition of transition to a supersonic flow). The rotation through the finite angle $\bar{\theta}$ can be considered to be the result of the occasion of successive small rotations in the region $C_0 0C'$ and then in the region $C'0C''$ and so on. The nonlinearized problem is thus split into a number of linearized problems. From this it directly follows that the rays $0C_0$, $0C'$, $0C''$, $\cdots$ are the perturbation lines and at the some time they are characteristics. The disturbances of the flow are constant along these lines; the transition downstream from one to the next very close perturbation line we shall fill with such perturbation lines the entire sector $C_0 0\bar{C}$ called the rarefaction sector.

In the theory of linearized supersonic flows the following formula was obtained

$$\frac{\tilde{u}}{U_\infty} = \frac{\mp\theta_{1,2}}{\sqrt{\mathbf{M}_\infty^2 - 1}} \tag{91}$$

which gives the magnitude of the small increment of the gas velocity $\tilde{u}$ on flow rotation through small angle θ. Applying this formula here to any instantaneous rarefaction sector element (the flow as a whole is not linearized) and assuming that $d\theta > 0$, we rewrite it in the form

$$\frac{dV}{V} = \frac{d\theta}{\sqrt{\mathbf{M}^2 - 1}} \tag{92}$$

The quantities V, θ, and $\mathbf{M}$ figuring in this equation represent the instantaneous values of these quantities for a certain intermediate line $0C$ of the rarefaction sector.

In order to integrate Equation 92, let us go from the number $\mathbf{M}$ to the velocity coefficient λ. Then, using the familiar formula of transition (80) of Chapter 5 we have (in the motion considered $d\lambda > 0$; θ, $d\theta > 0$ when calculating clockwise)

$$\frac{d\lambda}{\lambda} = \frac{d\theta}{\sqrt{\dfrac{\lambda^2 - 1}{1 - \dfrac{k-1}{k+1}\lambda^2}}} \tag{93}$$

[15] Prandtl, L., Neue Untersuchungen über die strömende Bewegung der Gase und DämpferVorträge und Diskussionen von der 78 Natur forschersammlung zu Stuffgart, *Phys. Z.*, 1907; Bewegungs vorgänge in einem Gas, das mit Überschallgeschwindigkeit strömt, *Forschung. VDI*, Vol. 62. 1908.

The variables are separated, and the integration yields

$$\theta = \theta(\lambda) =$$

$$= \sqrt{\frac{k+1}{k-1}}\operatorname{arctg}\left(\sqrt{\frac{k-1}{k+1}}\sqrt{\frac{\lambda^2-1}{1-\frac{k-1}{k+1}\lambda^2}}\right) - \operatorname{arctg}\sqrt{\frac{\lambda^2-1}{1-\frac{k-1}{k+1}\lambda^2}} \tag{94}$$

Here, the initial condition $\theta = 0$ at $\lambda = 1$ is used, which was given earlier.

By the definition of the perturbation line, the perturbation angle α, formed by it with the flow direction at this point, is equal to

$$\alpha = \arcsin\frac{1}{\mathbf{M}} = \operatorname{arctg}\frac{1}{\sqrt{\mathbf{M}^2-1}} = \operatorname{arctg}\sqrt{\frac{1-\frac{k-1}{k+1}\lambda^2}{\lambda^2-1}} \tag{95}$$

Let us prescribe the position of the perturbation line $0C$ by the angle ε formed by this line with the initial line $0C_0$. Then, from Figure 102 it can be easily concluded that

$$\varepsilon = \frac{\pi}{2} + \theta - \alpha = \frac{\pi}{2} + \sqrt{\frac{k+1}{k-1}}\operatorname{arctg}\left(\sqrt{\frac{k-1}{k+1}}\sqrt{\frac{\lambda^2-1}{1-\frac{k-1}{k+1}\lambda^2}}\right) -$$

$$- \operatorname{arctg}\sqrt{\frac{\lambda^2-1}{1-\frac{k-1}{k+1}\lambda^2}} - \operatorname{arctg}\sqrt{\frac{1-\frac{k-1}{k+1}\lambda^2}{\lambda^2-1}}$$

Noting that the arguments of the latter two arctangents are the mutually inverse quantities and that the sum of such arctangents is equal to $\pi/2$, we obtain

$$\varepsilon = \sqrt{\frac{k+1}{k-1}}\operatorname{arctg}\left(\sqrt{\frac{k-1}{k+1}}\sqrt{\frac{\lambda^2-1}{1-\frac{k-1}{k+1}\lambda^2}}\right) \tag{96}$$

The set of equations (94 to 96) give the parametric solution of the problem posed, with λ playing the part of a parameter.

Using the above-established relationship between λ and **M**

$$\frac{\lambda^2-1}{1-\frac{k-1}{k+1}\lambda^2} = \mathbf{M}^2 - 1$$

we replace the set of the previous equations that connect θ, α, and ε with λ by a corresponding group of equations that couple with the **M** number

$$\theta = \sqrt{\frac{k+1}{k-1}}\operatorname{arctg}\left(\sqrt{\frac{k-1}{k+1}}\sqrt{\mathbf{M}^2-1}\right) - \operatorname{arctg}(\sqrt{\mathbf{M}^2-1})$$

$$\alpha = \operatorname{arctg}\frac{1}{\sqrt{\mathbf{M}^2 - 1}} \qquad \varepsilon = \sqrt{\frac{k+1}{k-1}}\operatorname{arctg}\left(\sqrt{\frac{k-1}{k+1}}\sqrt{\mathbf{M}^2 - 1}\right) \tag{97}$$

One could also arrive at these relations by direct integration of Equation 92 rewritten as

$$d\theta = \frac{\sqrt{\mathbf{M}^2 - 1}\,d\mathbf{M}}{\mathbf{M}\left(1 + \frac{k-1}{2}\mathbf{M}^2\right)}$$

which can be easily derived from isentropic equation (72) of Chapter 5 and from the obvious equality $V = \mathbf{M}a$.

The pressure and density in the rarefaction sector are determined from the ordinary isentropic formulas

$$\frac{p}{p_0} = \left(1 + \frac{k-1}{2}\mathbf{M}^2\right)^{-\frac{k}{k-1}} = \left(1 - \frac{k-1}{k+1}\lambda^2\right)^{\frac{k}{k-1}}$$

$$\frac{\rho}{\rho_0} = \left(1 + \frac{k-1}{2}\mathbf{M}^2\right)^{-\frac{1}{k-1}} = \left(1 - \frac{k-1}{k+1}\lambda^2\right)^{\frac{1}{k-1}} \tag{98}$$

Here, the subscript 0 denotes the condition for the gas in the adiabatic and isentropic stagnation states.

The formulas being somewhat cumbersome, use is made of tables and plots which can be set up once and for all for any values of k. Table 7 which compares the characteristic values of θ, α, ε, p/p_0, $\mathbf{M}$, λ, and also the plot (Figure 103) of the same quantities, except ε, vs. the angle θ for air ($k = 1.41$).

Let us analyze the results obtained. First, it should be noted that all the quantities, which characterize the flow, are independent of the radius-vector r of the points in the flow region relative to the pole 0, i.e., on the distance to the angle vertex, and depend only on the polar angle ε. Here, we again meet with a specific case in which the partial differential equation, which describes a plane, two-dimensional motion, is reduced to an ordinary differential equation, i.e., to the equation involving one independent variable.

Similar to the case of centered rarefaction waves behind a piston (Chapter 6), the investigation of which resulted in the simple similarity problem in the plane (x, t), in the case of the

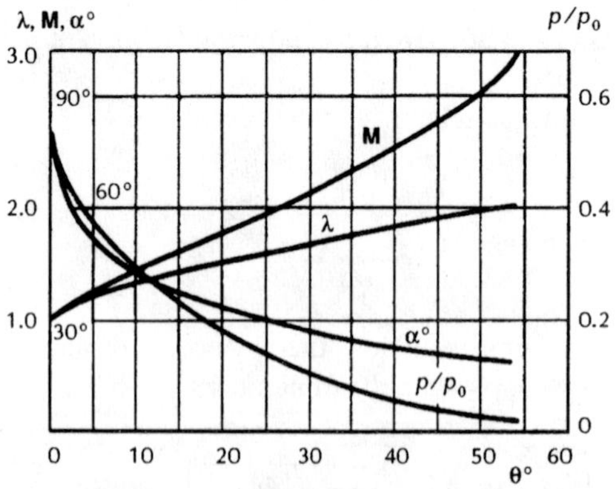

Fig. 103

Table 7

$\theta°$	$\varepsilon°$	$\alpha°$	M	λ	p/p_0
0	0.00	90.00	1.000	1.000	0.527
1	23.72	67.28	1.084	1.068	0.477
2	30.04	61.96	1.133	1.107	0.449
3	34.82	58.18	1.178	1.141	0.424
4	38.88	55.12	1.220	1.173	0.401
5	42.34	52.66	1.258	1.201	0.381
6	45.42	50.58	1.295	1.227	0.363
7	48.30	48.70	1.332	1.253	0.345
8	50.93	47.07	1.366	1.276	0.329
9	53.46	45.54	1.401	1.299	0.313
10	55.84	44.16	1.435	1.322	0.298
11	58.16	42.84	1.470	1.344	0.284
12	60.38	41.62	1.505	1.366	0.270
13	62.49	40.51	1.539	1.387	0.257
14	64.52	39.48	1.572	1.407	0.245
15	66.53	38.47	1.608	1.428	0.233
16	68.47	37.53	1.641	1.448	0.221
17	70.33	36.67	1.675	1.467	0.210
18	72.18	35.82	1.710	1.486	0.199
19	73.98	35.02	1.744	1.504	0.189
20	75.74	34.26	1.779	1.523	0.179
21	77.49	33.51	1.815	1.541	0.170
22	79.20	32.80	1.850	1.559	0.161
23	80.90	32.10	1.884	1.576	0.153
24	82.55	31.45	1.918	1.592	0.145
25	84.20	30.80	1.954	1.609	0.137
26	85.81	30.19	1.989	1.625	0.130
27	87.42	29.58	2.025	1.641	0.123
28	89.02	28.98	2.062	1.657	0.116
29	90.58	28.42	2.098	1.673	0.110
30	92.12	27.88	2.135	1.688	0.104
31	93.66	27.34	2.174	1.704	0.097
32	95.18	26.82	2.214	1.720	0.092
33	96.68	26.32	2.251	1.735	0.086
34	98.20	25.80	2.296	1.752	0.080
35	99.67	25.33	2.339	1.767	0.075
36	101.13	24.87	2.378	1.781	0.071
37	102.58	24.42	2.422	1.795	0.066
38	104.02	23.98	2.466	1.810	0.062
39	105.46	23.54	2.508	1.824	0.058
40	106.88	23.12	2.550	1.837	0.054
41	108.30	22.70	2.595	1.851	0.051
42	109.71	22.29	2.640	1.864	0.047
43	111.11	21.89	2.689	1.878	0.044
44	112.51	21.49	2.734	1.891	0.041
45	113.89	21.11	2.778	1.903	0.038
46	115.27	20.73	2.826	1.917	0.036
47	116.63	20.37	2.873	1.928	0.033
48	118.00	20.00	2.920	1.939	0.031
49	119.36	19.64	2.968	1.951	0.029
50	120.71	19.29	3.021	1.963	0.027
51	122.07	18.93	3.074	1.975	0.025
52	123.41	18.59	3.131	1.987	0.023
53	124.74	18.26	3.188	1.999	0.021
54	126.03	17.97	3.250	2.012	0.019
219.32	219.32	0.000	∞	2.437	0.000

plane motion in the rarefaction sector we also have a similarity problem, but in the plane (x, y) or (r, ε). The rectilinear characteristics, which fill the rarefaction sector, are also centered because all of them pass through the angle vertex 0. The independence of the radius-vector r, and thus of the self-similarity of the problem, are attributable to the condition of flow uniformity to the left from the first perturbation line.

Further, it follows directly from the solution obtained that as λ increases up to the maximum possible value

$$\lambda_{max} = \sqrt{\frac{k+1}{k-1}}$$

or as **M** increases to **M** = ∞, i.e., when the fluid escapes with the maximum velocity V_{max} into a perfect vacuum, the angle θ increases to its maximum value

$$\theta_{max} = \left(\sqrt{\frac{k+1}{k-1}} - 1\right)\frac{\pi}{2}$$

In this case, according to Equations 95 and 96,

$$\alpha = 0 \qquad \varepsilon_{max} = \frac{\pi}{2} + \theta_{max} = \frac{\pi}{2}\sqrt{\frac{k+1}{k-1}}$$

Thus, there exists the limiting angle of flow rotation θ_{max} for air ($k = 1.41$) equal to 129.32°, which can be realized only in the ideal case of gas expansion up to a perfect vacuum. In practice, the flow may rotate only through angles smaller than θ_{max}.

Let us find the equation for the streamlines in the rarefaction sector in the polar coordinates ($r = 0N$, ε). For this purpose, it should be noted that according to the well-known formula of differential geometry $\frac{1}{r}\frac{dr}{d\varepsilon} = \text{ctg}\,\alpha$. On the other hand, one can easily eliminate λ from Equations 95 and 96 and find the direct coupling between a and ε; we have

$$\frac{1}{r}\frac{dr}{d\varepsilon} = \sqrt{\frac{k+1}{k-1}}\,\text{tg}\left(\sqrt{\frac{k-1}{k+1}}\,\varepsilon\right)$$

Integration leads to the following equation for the family of streamlines with the parameter r_0 equal to the initial radius-vector of the streamline on the straight line $0C_0$ ($\varepsilon = 0$):

$$r = r_0\left[\cos\left(\sqrt{\frac{k-1}{k+1}}\,\varepsilon\right)\right]^{-\frac{k+1}{k-1}}$$

The streamlines form the family of identical curves with the similarity center at the point 0.

The solution above of the problem concerning the flow about the exterior of the angle is general and independent of the earlier introduced limitation that $\lambda_0 = \mathbf{M}_0 = 1$ at $\theta = 0$. This condition was advanced only to simplify Equation 94 and the associated relations which will be given further on. By the well-known property of supersonic flows the solution obtained can be used for an initial flow with any values of the numbers $\lambda > 1$ or $\mathbf{M} > 1$. In this case one should only begin from the characteristic (perturbation line), which corresponds to the given initial value of λ or **M** and to supply a uniform rectilinear flow to it.

In the same way, the end of the flow rotation is determined by specifying θ or λ and **M** at the exit and by constructing the exit uniform rectilinear flow with the velocity and angles calculated by the theory given.

Let us reject the condition $\lambda_0 = \mathbf{M}_0 = 1$ at $\theta = 0$ and adopt the initial characteristic to be any characteristic $0C_1$ corresponding to $\mathbf{M} = \mathbf{M}_1 > 1$. Let the subscript 1 denote the flow parameters on the entry characteristic $0C_1$; the quantities without the subscripts will be retained for the instantaneous values of the parameters inside the rarefaction sector. Let us agree to reckon the angle of flow rotation θ from the flow direction at the entrance and ε, from the initial characteristic; the definition of α remains the same. Then, Equations 97 will acquire the form

$$\theta = \sqrt{\frac{k+1}{k-1}}\left[\operatorname{arctg}\left(\sqrt{\frac{k-1}{k+1}}\sqrt{\mathbf{M}^2-1}\right) - \operatorname{arctg}\left(\sqrt{\frac{k-1}{k+1}}\sqrt{\mathbf{M}_1^2-1}\right)\right] -$$

$$- \operatorname{arctg}\sqrt{\mathbf{M}^2-1} + \operatorname{arctg}\sqrt{\mathbf{M}_1^2-1}$$

$$\alpha = \operatorname{arctg}\frac{1}{\sqrt{\mathbf{M}^2-1}}$$

$$\varepsilon = \sqrt{\frac{k+1}{k-1}}\left[\operatorname{arctg}\left(\sqrt{\frac{k-1}{k+1}}\sqrt{\mathbf{M}^2-1}\right) - \operatorname{arctg}\left(\sqrt{\frac{k-1}{k+1}}\sqrt{\mathbf{M}_1^2-1}\right)\right] \tag{99}$$

These equations can be supplemented with the familiar isentropic equations for pressure and density

$$\frac{p}{p_1} = \left(\frac{1+\frac{k-1}{2}\mathbf{M}_1^2}{1+\frac{k-1}{2}\mathbf{M}^2}\right)^{\frac{k}{k-1}} \qquad \frac{\rho}{\rho_1} = \left(\frac{1+\frac{k-1}{2}\mathbf{M}_1^2}{1+\frac{k-1}{2}\mathbf{M}^2}\right)^{\frac{1}{k-1}} \tag{100}$$

These quantities can be calculated from Table 7 or from the curves of Figure 103.

7. The Case of Great Mach Numbers. The Law of Similarity of Hypersonic Flows

Let us consider the case of flow past a thin profile with very great Mach numbers ($\mathbf{M} \gg 1$); such a flow is called hypersonic. We continue to regard the gas to be homogeneous and neglect those complex processes that actually originate in hypersonic flows due to high temperatures occurring during gas stagnation on the body surface and when passing through the surfaces of strong discontinuities. Let us agree that in this section we shall neglect the dissociation phenomena and subsequent possible recombination of gas molecules, gas ionization, and other physical and chemical processes typical of hypersonic motions of gas. Later, we will return to some of these essential phenomena in order to study the gas model that is closer to reality and which involves internal friction (viscosity) and thermal conductivity.

A strict mathematical statement of the problem of hypersonic flow past bodies presents great difficulties because under these conditions there are complex interactions of the flow with strong discontinuities which cause the flow nonisentropicity in a thin layer near the body surface. The presence in this region of the considerable effect of such processes as viscous friction, heat conduction, mechanical energy dissipation, radiation and, it should be repeated, dissociationrecombination of molecules, and gas ionization make this problem very complex.

Let us additionally consider the simplest phenomena, which were given in the two previous sections and which concerned the retardation and acceleration of a uniform gas flow, from the view point of those specific features which appear at great Mach numbers of the free stream ($\mathbf{M}_1 \gg 1$) and at small angles of flow rotation.

Taking first the passage of the gas through an oblique shock, we shall consider Equations 88 and 82. By the body thinness condition we shall consider the body to be tapered with a small vertex angle θ. Then, in the case of weak shocks and rather high values of $\mathbf{M}_1$ the angle between the oblique shock and the oncoming flow β will have the same order of smallness as the angle θ. This will make it possible in the equations considered to replace the sines of the angles by the angles themselves and the cosines – by unities. Rejecting small quantities of higher orders, we shall have, according to equation (88), the quadratic equation in $\mathbf{M}_1\beta = K_c$:

$$K_c^2 - \frac{k+1}{2}(\mathbf{M}_1\theta)K_c - 1 = 0 \qquad (101)$$

the solution of which under the obvious condition

$$K_c = \mathbf{M}_1\beta > 0$$

can be represented as

$$K_c = \mathbf{M}_1\beta = \frac{k+1}{4}K + \sqrt{\left(\frac{k+1}{4}\right)^2 K^2 + 1} \qquad (102)$$

where the following notation is introduced

$$K_c = \mathbf{M}_1\theta \qquad (103)$$

for the quantity which plays the part of the similarity criterion for gas flow past thin, tapered bodies at great numbers $\mathbf{M}_1$.

Equations 82 will go over into the following:

$$\frac{p_2}{p_1} = \frac{2k}{k+1}K_c^2 - \frac{k-1}{k+1} \qquad \frac{\rho_2}{\rho_1} = \frac{\frac{k+1}{2}K_c^2}{1 + \frac{k-1}{2}K_c^2} \qquad (104)$$

On retention of the number K_c, the ratios (104) between the pressures and densities after and before the shock also will be conserved presumably.

According to the first of Equations 104, we shall find the pressure coefficient in the flow after the shock

$$c_p = \frac{p_2 - p_1}{\frac{1}{2}\rho_1 V_1^2} = \frac{2p_1}{\rho_1 V_1^2}\left(\frac{p_2}{p_1} - 1\right) = \frac{2}{k\mathbf{M}_1^2}\frac{2k}{k+1}[(\mathbf{M}_1\beta)^2 - 1] = \frac{4}{k+1}\frac{(\mathbf{M}_1\beta)^2 - 1}{K^2}\theta^2 \qquad (105)$$

which is proportional, at constant K, to the squared angle θ of the flow rotation on its retardation with the proportionality coefficient representing the function of the number K. Using directly quadratic equation (101) and its solution (102), we can rewrite Equation 105 in the form of explicit dependence on K

$$c_p = \left[\frac{k+1}{2} + \sqrt{\left(\frac{k+1}{2}\right)^2 + \frac{4}{K^2}}\right]\theta^2 \qquad (106)$$

Equation 106 expresses the law of similarity of hypersonic, slightly converging flows, in particular, the flow past wedge-like profiles with small semivertex angles θ. According to this law, if the values of K are identical in two such flows, the pressure coefficients will relate to each other as the squared semivertex angles of the wedges.

We see that an analogous similarity law also takes place in the case of diverging flows if the divergence angle is small and the initial Mach number is great. Let us apply to the first equation of system (99) the following asymptotic equations, which are valid for sufficiently great values of x ($\sim$ is the sign of asymptotic equality),

$$\operatorname{arctg} x \sim \frac{\pi}{2} - \frac{1}{x} \qquad \sqrt{x^2 - 1} \sim x$$

Then,

$$\theta \sim \frac{2}{k-1}\left(\frac{1}{\mathbf{M}_1} - \frac{1}{\mathbf{M}}\right)$$

where

$$\frac{\mathbf{M}_1}{\mathbf{M}} \sim 1 - \frac{k-1}{2}\mathbf{M}_1\theta = 1 - \frac{k-1}{2}K \tag{107}$$

The maximum possible angle of rotation θ_{max} will correspond to the escape into vacuum ($T = 0$, $a = 0$, $\mathbf{M} = \infty$)

$$\theta_{max} = \frac{2}{(k-1)\,\mathbf{M}_1} \tag{108}$$

The pressure coefficient c_p is equal to

$$c_p = \frac{p - p_1}{\frac{1}{2}\rho_1 V_1^2} = \frac{2}{k\mathbf{M}_1^2}\left(\frac{p}{p_1} - 1\right)$$

and according to Equations 100 and 107

$$c_p \sim \frac{2}{k\mathbf{M}_1^2}\left[\left(\frac{\mathbf{M}_1}{\mathbf{M}}\right)^{\frac{2k}{k-1}} - 1\right] = \frac{2}{k\mathbf{M}_1^2}\left[\left(1 - \frac{k-1}{2}\mathbf{M}_1\theta\right)^{\frac{2k}{k-1}} - 1\right] \tag{109}$$

Just as in the case of gas retardation in passage through an oblique shock, Equation 109 can be rewritten, in complete analogy with Equation 106:

$$c_p \sim \frac{2}{kK^2}\left[\left(1 - \frac{k-1}{2}K\right)^{\frac{2k}{k-1}} - 1\right]\theta^2 \tag{110}$$

This leads to the following general law of similarity of hypersonic flows near thin bodies: if in two similar plane hypersonic flows the values of K are identical, the pressure coefficients are related as the squared deviation angles.

Introducing, as before, the generalized concept of the "relative thickness" τ, which can be equal either to the relative thickness of the profile in the usual sense of the word, or to the relative camber of the arc, or finally, to the angle of attack, we shall consider the quantity K, which is equal to the product of the characteristic Mach number (for example, the Mach number $\mathbf{M}_\infty$ of a uniform free stream) by the relative thickness τ, to be the criterion of similarity of two hypersonic flows weakly deviating from the given uniform hypersonic flow. We write the similarity law for this case in the general form

$$c_p = \tau^2 f(K) \tag{111}$$

Going from the pressure coefficient c_p to the lifting force, c_y, and wave resistance, c_x, coefficients, which are connected to c_p by the integral relations

$$c_y = \oint c_p d\bar{x} \qquad c_x = -\oint c_p d\bar{y} = -\oint c_p \frac{dy}{dx} d\bar{x}$$

we see that (the mean value of dy/dx is proportional to τ) in the flows with identical values of the number K the lifting force coefficients, as with the pressure coefficients, will relate as the squares of the relative thickness. On the other hand, the wave resistance coefficients will relate as the cubes of the relative thickness. The indicated similarity law seems to be involved as a particular case in the general law (60), where it is necessary only to assume that $\Psi = \tau^2$ and to neglect unity under the root sign as compared with $\mathbf{M}_\infty^2$.

Using Equations 106 and 110, one can easily obtain the expressions for the lifting force and wave resistance coefficients for the plate AB (Figure 104), located at a small angle of attack $\bar{\theta}$ (in Figure 104 this angle is greatly exaggerated) to the incident uniform flow with the Mach number $\mathbf{M}_\infty$ greatly exceeding unity. Designating, as before, by subscripts 1 and 2 the upper and lower surfaces of the plate, respectively

$$c_y = \frac{p_2 - p_1}{\frac{1}{2}\rho_\infty U_\infty^2} = c_{p_2} - c_{p_1} = \frac{c_x}{\bar{\theta}}$$

and substituting into this equation the values of c_{p2} and c_{p1}, specified by Equations 106 and 110, we finally obtain[16] (the angle of attack $\bar{\theta}$ corresponds to the angle of full flow rotation)

$$c_y = \frac{c_x}{\bar{\theta}} = \left\{\frac{k+1}{2} + \sqrt{\left(\frac{k+1}{2}\right)^2 + \frac{4}{K^2} + \frac{2}{kK^2}\left[1 - \left(1 - \frac{k-1}{2}K\right)^{\frac{2k}{k-1}}\right]}\right\}\bar{\theta}^2 \tag{112}$$

In this equation, when $K \geq \frac{2}{k-1}$, and according to Equation 108 this corresponds to $\bar{\theta} \geq \bar{\theta}_{max}$, the square bracket should be regarded equal to unity. This means that on the upper surface of the plate at $\bar{\theta} = \bar{\theta}_{max}$ the zero pressure (a perfect vacuum) has formed which is also retained for $\bar{\theta} > \bar{\theta}_{max}$.

In Figure 105, from the book by Chyornyi[16] the solid curves represent the functions $c_y(\bar{\theta})$ at different values of $\mathbf{M}_\infty$ within the range $3 < \mathbf{M}_\infty < \infty$, which were calculated by Equation 112 for air ($k = 1.4$); the dashed straight lines show the corresponding values of $c_y(\bar{\theta})$ obtained by Equation 44 of Ackeret's linear theory (Section 3). Finally, the upper straight line ($\mathbf{M}_\infty = 0$), shown by dashes and dots, corresponds to the well-known equation $c_y = 2\pi\bar{\theta}$ for an incompressible fluid.

---

[16] Chyornyi, G., *Techeniye gaza s bolshoi sverkhzvukovoi skorostiyu (The Gas Flow with a High Supersonic Velocity)*, Fizmatgiz Press, Moscow, 1959, p. 47; *Gas Dynamics*, Nauka Press, Moscow, 1988, p. 406.

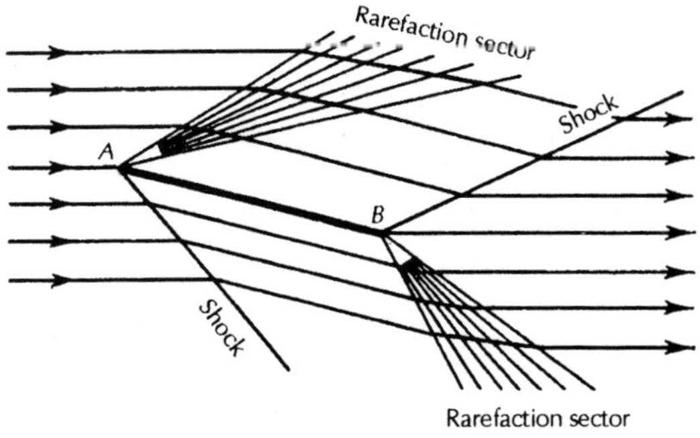

Fig. 104

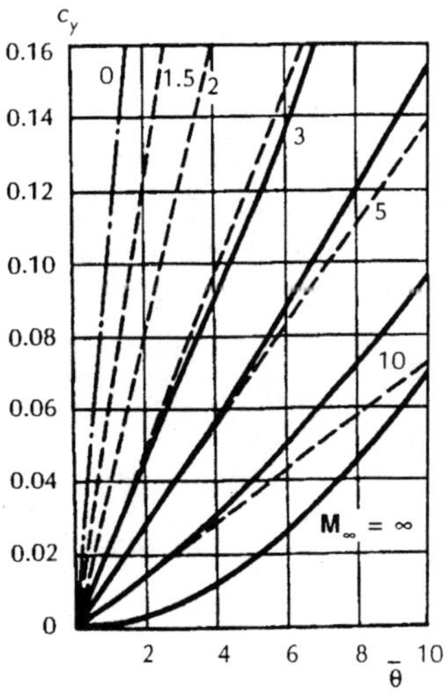

Fig. 105

The analysis of the curves given in Figure 105 shows that when $\mathbf{M}_\infty \geq 5$ Ackeret's formula underestimates the lifting force coefficient.

It can also be noted that as the $\mathbf{M}_\infty$ number rises, the curves $c_y(\bar{\theta})$ deviate increasingly from the linear law and approach the quadratic law. Thus, in the limiting case $\mathbf{M}_\infty = \infty$ Equation 112 acquires a purely quadratic form

$$c_y = (k+1)\bar{\theta}^2 \tag{113}$$

and only by the numerical coefficient does it differ from the corresponding formula of the Newton "shock" theory, which briefly consists of the following. At the values of $\mathbf{M}_\infty$ close to infinity, the oblique shock at the point A (Figure 104) of the plate approaches very close to the lower surface so that without great error it is possible to consider that the streamlines of the oncoming flow closely approach the lower surface of the plate and that the pressure on it is equal to the full momentum loss corresponding to an absolutely nonelastic impact of gas particles on the lower surface of the plate. The velocity is equal to the normal free stream velocity component (the tangent velocity component is conserved):

$$p_2 = \rho_\infty (U_\infty \sin\bar{\theta})^2 \approx \rho_\infty U_\infty^2 \bar{\theta}^2$$

On the other hand, a complete vacuum is attained on the upper surface of the plate at $\mathbf{M}_\infty = \infty$, i.e., in this case $p_1 = 0$. Thus, the lifting force coefficient will be equal to (l is the plate length)

$$c_y = \frac{(p_2 - p_1)\,l}{\frac{1}{2}\rho_\infty U_\infty^2 l} = 2\bar{\theta}^2$$

which, at $k = 1$, coincides with Equation (113) and at $k = 1.4$ (air) gives a smaller result, by about 17%.

8. Equations of Gas Dynamics in the Hodograph Plane of Velocity

Of fundamental importance for the development of contemporary gas dynamics was the transformation of general equations to independent variables in the hodograph plane, which was established by Chaplygin[17] in his doctoral thesis, defended in 1904. This transition from the physical plane to the plane of hodograph leads to a remarkable result: the nonlinear equations of gas dynamics become linear.

From the condition of the absence of vorticity

$$\frac{\partial u}{\partial y} - \frac{\partial v}{\partial x} = 0$$

and from the continuity equation,

$$\frac{\partial(\rho u)}{\partial x} + \frac{\partial(\rho v)}{\partial y} = 0$$

it is possible to make use of the existence of the velocity potential φ and of the stream function ψ, so that

$$u = \frac{\partial\varphi}{\partial x} \qquad v = \frac{\partial\varphi}{\partial y} \qquad \frac{\rho}{\rho_0}u = \frac{\partial\psi}{\partial y} \qquad \frac{\rho}{\rho_0}v = -\frac{\partial\psi}{\partial x} \tag{114}$$

where ρ_0 is the density of the gas at rest. From this it follows that

$$udx + vdy = d\varphi \qquad -vdx + udy = \frac{\rho_0}{\rho}d\psi$$

Multiply the second of these equations by $i = \sqrt{-1}$ and sum using the first equation. This will yield

$$(u - iv)\,d(x + iy) = d\varphi + i\frac{\rho_0}{\rho}d\psi$$

Making the replacement (V is the magnitude of the velocity vector, θ is its angle with the axis $0x$)

$$u - iv = Ve^{-i\theta} \qquad x + iy = z$$

we shall find the well-known connection between the conjugate velocity and the derivative of the complex potential with respect to the coordinate, extended to the case of a compressible gas:

$$dz = \left(d\varphi + i\frac{\rho_0}{\rho}d\psi\right)\frac{1}{V}e^{i\theta} \tag{115}$$

Let us make the transition to the hodograph plane (V, θ). With this purpose in mind, adopt in Equation 115 the variables x, y (and consequently, z) and also φ and ψ to act as the functions of the new variables V and θ; then, Equation (115) becomes:

---

[17] Chaplygin, S. A., About gas jets, *Uchyon. Zapiski Moskov Univ., Otd. Fiz.-Mat. Nauk*, No. 21, 1904.

$$\frac{\partial z}{\partial V}dV + \frac{\partial z}{\partial \theta}d\theta = \left[\frac{\partial \varphi}{\partial V}dV + \frac{\partial \varphi}{\partial \theta}d\theta + i\frac{\rho_0}{\rho}\left(\frac{\partial \psi}{\partial V}dV + \frac{\partial \psi}{\partial \theta}d\theta\right)\right]\frac{1}{V}e^{i\theta} =$$

$$= \frac{1}{V}\left(\frac{\partial \varphi}{\partial V} + i\frac{\rho_0}{\rho}\frac{\partial \psi}{\partial V}\right)e^{i\theta}dV + \frac{1}{V}\left(\frac{\partial \varphi}{\partial \theta} + i\frac{\rho_0}{\rho}\frac{\partial \psi}{\partial \theta}\right)e^{i\theta}d\theta$$

Comparing in this equation the coefficients at the differentials of independent variables, we obtain

$$\frac{\partial z}{\partial V} = \frac{1}{V}\left(\frac{\partial \varphi}{\partial V} + i\frac{\rho_0}{\rho}\frac{d\psi}{\partial V}\right)e^{i\theta} \qquad \frac{\partial z}{\partial \theta} = \frac{1}{V}\left(\frac{\partial \varphi}{\partial \theta} + i\frac{\rho_0}{\rho}\frac{d\psi}{\partial \theta}\right)e^{i\theta} \tag{116}$$

The involved ratio ρ_0/ρ is the function of the velocity V or of the velocity coefficient λ. Actually, from the familiar isentropic equation

$$\frac{\rho_0}{\rho} = \left(1 - \frac{k-1}{k+1}\lambda^2\right)^{-\frac{1}{k-1}} \tag{117}$$

In order to eliminate the independent variable z from the system of equations (116), we differentiate the first equation of the system with respect to θ and the second with respect to V and subtract the results from each other. Then, by virtue of the obvious relation

$$\frac{\partial^2 z}{\partial \theta\, \partial V} = \frac{\partial^2 z}{\partial V\, \partial \theta}$$

we obtain the equation

$$\frac{1}{V}\left(\frac{\partial^2 \varphi}{\partial \theta\, \partial V} + i\frac{\rho_0}{\rho}\frac{\partial^2 \psi}{\partial \theta\, \partial V}\right)e^{i\theta} + \frac{1}{V}ie^{i\theta}\frac{\partial \varphi}{\partial V} - \frac{\rho_0}{\rho}\frac{1}{V}e^{i\theta}\frac{\partial \psi}{\partial V} =$$

$$= \frac{1}{V}\left(\frac{\partial^2 \varphi}{\partial V\, \partial \theta} + i\frac{\rho_0}{\rho}\frac{\partial^2 \psi}{\partial V\, \partial \theta}\right)e^{i\theta} - \frac{1}{V^2}e^{i\theta}\frac{\partial \varphi}{\partial \theta} + ie^{i\theta}\frac{d}{dV}\left(\frac{1}{V}\frac{\rho_0}{\rho}\right)\frac{\partial \psi}{\partial \theta}$$

which, after cancellations and comparison of the real and imaginary parts, will lead to the system of equations

$$\frac{1}{V}\frac{\partial \varphi}{\partial V} = \frac{d}{dV}\left(\frac{1}{V}\frac{\rho_0}{\rho}\right)\frac{\partial \psi}{\partial \theta} \qquad \frac{1}{V}\frac{\partial \varphi}{\partial \theta} = \frac{\rho_0}{\rho}\frac{\partial \psi}{\partial V} \tag{118}$$

Let us now go from the quantity V to the velocity coefficient λ and calculate the derivative on the right side of the first equation of system (118)

$$\frac{d}{d\lambda}\left(\frac{1}{\lambda}\frac{\rho_0}{\rho}\right) = \frac{d}{d\lambda}\left[\frac{\left(1 - \frac{k-1}{k+1}\lambda^2\right)^{-\frac{1}{k-1}}}{\lambda}\right]$$

then, system (118) will acquire the form

$$\frac{\partial\varphi}{\partial\theta} = \frac{\lambda}{\left(1 - \frac{k-1}{k+1}\lambda^2\right)^{\frac{1}{k-1}}}\frac{\partial\psi}{\partial\lambda} \qquad \frac{\partial\varphi}{\partial\lambda} = -\frac{1-\lambda^2}{\lambda\left(1 - \frac{k-1}{k+1}\lambda^2\right)^{\frac{k}{k-1}}}\frac{\partial\psi}{\partial\theta} \tag{119}$$

Replacing λ by the variable τ:

$$\tau = \frac{k-1}{k+1}\lambda^2 = \left(\frac{V}{V_{\max}}\right)^2 \qquad V_{\max} = \sqrt{2h_0} = \sqrt{\frac{2}{k-1}}\,a_0 = \sqrt{\frac{k+1}{k-1}}\,a^*$$

we obtain the system of Chaplygin's equations

$$\frac{\partial\varphi}{\partial\tau} = -\frac{1}{2\tau}\frac{1 - \frac{k+1}{k-1}\tau}{(1-\tau)^{\frac{k}{k-1}}}\frac{\partial\psi}{\partial\theta} \qquad \frac{\partial\varphi}{\partial\theta} = \frac{2\tau}{(1-\tau)^{\frac{1}{k-1}}}\frac{\partial\psi}{\partial\tau} \tag{120}$$

By cross-differentiation and subtraction it is possible to obtain separate equations for φ and ψ, with these equations being linear partial differential second-order equations. Thus, for example, the equation for the stream function ψ will have the form

$$\frac{\partial}{\partial\tau}\left[2\tau(1-\tau)^{-\frac{1}{k-1}}\frac{\partial\psi}{\partial\tau}\right] + \frac{1 - \frac{k+1}{k-1}\tau}{2\tau(1-\tau)^{\frac{k}{k-1}}}\frac{\partial^2\psi}{\partial\theta^2} = 0 \tag{121}$$

Let us dwell on the case of subsonic flow ($\lambda < 1$). Replacing λ in Equation 119 by the new variable s connected in the case of subsonic flow with λ by the differential relation

$$ds = \sqrt{\frac{1-\lambda^2}{1 - \frac{k-1}{k+1}\lambda^2}}\,\frac{d\lambda}{\lambda} \tag{122}$$

we reduce system (119) to the form

$$\frac{\partial\varphi}{\partial\theta} = \sqrt{K}\frac{\partial\psi}{\partial s} \qquad \frac{\partial\varphi}{\partial s} = -\sqrt{K}\frac{\partial\psi}{\partial\theta} \tag{123}$$

where the quantity K represents the following function of λ:

$$K = \frac{1-\lambda^2}{\left(1 - \frac{k-1}{k+1}\lambda^2\right)^{\frac{k+1}{k-1}}} \tag{124}$$

Equations 120 and 121, derived by Chaplygin, were reduced to the different, so-called canonical forms by Leibenzon,[18] Slyozkin,[19] and Khristianovich.[20]

---

[18] Leibenzon, L. S., About the theory of the motion of gases, *Dokl. Akad. Nauk SSSR.*, No. 9, 1935.

[19] Slyozkin, N. A., Concerning the problem of plane gas motion, Tr. Moscow. Gos. Univ., 1935; *Dokl. Akad. Nauk SSSR* 3(9), 1936 (new series).

[20] Khristianovich, S. A., Gas flow past a body at high subsonic velocities, *Tr. TsAGI*, 481, 1940; Khristianovich, S. A. and Yuriyev, I. M., Flow past a wing profile at the subcritical flow velocity, *Prikl. Mat. Mekh.*, 11(1), 1947.

9. Effect of Compressibility on Velocity and Pressure Distributions in a Plane Subsonic Flow

A significant practical importance of the problem of including the effect of compressibility or, in other words, of the number $\mathbf{M}_\infty$ of the oncoming subsonic flow on the pressure distribution over the wing profile surface and on its lifting force spurred the appearance of a number of approximate techniques.

As indicated earlier, the Prandtl-Glauert rule is applicable only in the case of thin, slightly cambered profiles located at a small angle of attack in the flow with comparatively small $\mathbf{M}_\infty$ numbers. The techniques presented in this section make it possible to introduce the correction for compressibility for thicker and cambered profiles at greater angles of attack and greater ranges of $\mathbf{M}_\infty$ numbers. The function $\sqrt{K(\lambda)}$ occurring in the system of equations (123) at the values of λ not very close to unity, differ little from unity, as can be seen from Table 8 compiled for air (k = 1.41). This table also gives for comparison the dependence of $\sqrt{K}$ on $\mathbf{M}$ expressed analytically by the following formula, easily derived from equation 124,

$$\sqrt{K} = (1-\mathbf{M}^2)^{1/2}\left(1+\frac{k-1}{2}\mathbf{M}^2\right)^{\frac{1}{k-1}} \tag{125}$$

Thus, for example, at $\lambda = 0.65$ ($\mathbf{M} = 0.615$) the quantity $\sqrt{K}$ differs from unity by only 5%. Having replaced in system (123) the quantity $\sqrt{K}$ approximately by unity, we obtain in the plane (s, θ) the system of equations

$$\frac{\partial \varphi}{\partial \theta} = \frac{\partial \psi}{\partial s} \qquad \frac{\partial \varphi}{\partial s} = -\frac{\partial \psi}{\partial \theta} \tag{126}$$

Table 8

λ	$\mathbf{M}$	$\sqrt{K}$
0	0	1.0000
0.05	0.0457	1.0000
0.10	0.0913	1.0000
0.15	0.1372	0.9999
0.20	0.1832	0.9996
0.25	0.2294	0.9991
0.30	0.2759	0.9982
0.35	0.3228	0.9965
0.40	0.3701	0.9940
0.45	0.4179	0.9899
0.50	0.4663	0.9840
0.55	0.5152	0.9754
0.60	0.5649	0.9632
0.65	0.6154	0.9461
0.70	0.6668	0.9221
0.75	0.7192	0.8925
0.80	0.7727	0.8416
0.85	0.8274	0.7740
0.90	0.8834	0.6788
0.95	0.9409	0.5092
1.00	1.0000	0

which is analogous to the exact conditions of Cauchy-Rieman (in what follows, the subscript 0 at the quantity means that it relates to the incompressible fluid flow)

$$\frac{\partial \varphi_0}{\partial \theta_0} = \frac{\partial \psi_0}{\partial s_0} \qquad \frac{\partial \varphi_0}{\partial s_0} = -\frac{\partial \psi_0}{\partial \theta_0} \tag{127}$$

in the plane (s_0, θ_0) and, according to the definition (122) adopted for the introduction of the function s, in the case of very small values of λ_0, we have

$$ds_0 = \frac{d\lambda_0}{\lambda_0} \tag{128}$$

The coincidence of approximate Equations 126 in the plane (s, θ) with exact conditions (127) in the plane (s_0, θ_0) does not indicate the connection between the compressible and incompressible flows because the boundary conditions in the planes s_0, θ_0, and s, θ differ from one another.

In view of this it becomes impossible to directly obtain universal (i.e., applicable for all possible types of subsonic flows) "conversion" formulas. To explain the above statement, let us consider two simplest flows – from a vortex and from a source – for which exact solutions are available.

It can be shown that in the case of a vortex (at the same circulation Γ) the velocity distributions in the flows of gas and of incompressible fluid do not differ by anything and, consequently, $\lambda_0 = \lambda$.

In the case of the source (at the same mass flow rate Q) in the gas flow $Q = 2\pi r \lambda a^* \rho(\lambda)$ and in the incompressible fluid $Q = 2\pi r \lambda_0 a^* \rho_0$, where ρ_0 is the constant density of the incompressible fluid (it can be adopted to be equal, for example, to the density of the gas in

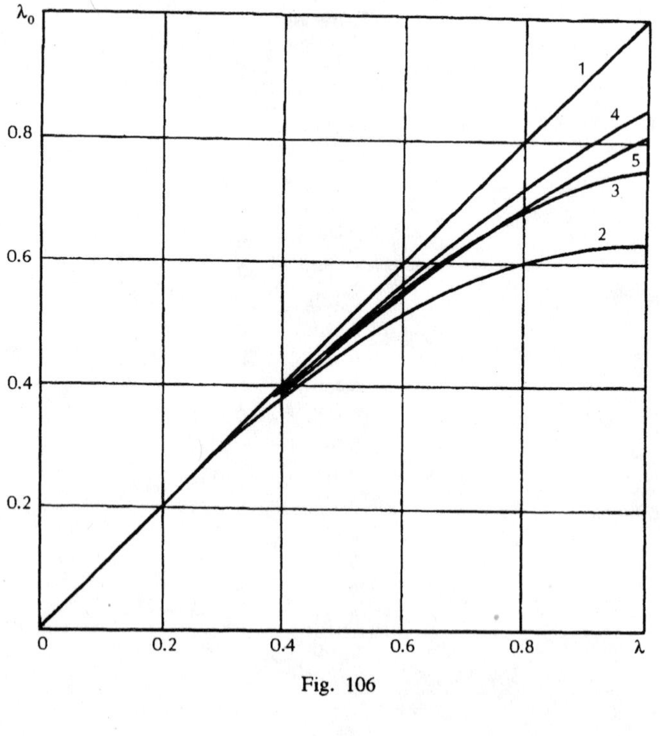

Fig. 106

Table 9

λ	M	λ_0
0	0	0
0.05	0.0457	0.0500
0.10	0.0913	0.0998
0.15	0.1372	0.1493
0.20	0.1832	0.1983
0.25	0.2294	0.2467
0.30	0.2759	0.2943
0.35	0.3228	0.3410
0.40	0.3701	0.3862
0.45	0.4179	0.4307
0.50	0.4663	0.4734
0.55	0.5152	0.5144
0.60	0.5649	0.5535
0.65	0.6154	0.5904
0.70	0.6668	0.6251
0.75	0.7192	0.6568
0.80	0.7727	0.6857
0.85	0.8274	0.7110
0.90	0.8834	0.7324
0.95	0.9409	0.7483
1.00	1.0000	0.7577

isentropical stagnation state $\rho(0)$). Thus, for this flow we have already another relationship between λ and λ_0:

$$\lambda_0 = \lambda\rho(\lambda)/\rho_0$$

which probably corresponds to the greatest of the possible corrections for the compressibility.[21] Both functions are given in Figure 106 (straight line 1 for the vortex and curve 2 for the source).

There are different approximate methods for taking into account the compressibility, the precision, and the limits of applicability, which can be assessed only by comparing with experiment. We consider some of these techniques below.

Let us ignore the difference between ds (Equation 122) and ds_0 (Equation 128) and assume that

$$ds = \sqrt{\frac{1-\lambda^2}{1-\frac{k-1}{k+1}\lambda^2}}\frac{d\lambda}{\lambda} = ds_0 = \frac{d\lambda_0}{\lambda_0} \tag{129}$$

Let us integrate this ordinary first-order equation which connects λ and λ_0; the constant of integration will be found from the condition of the coincidence of velocity fields in the limiting transition to an incompressible fluid

$$\lim_{\lambda \to 0} \frac{\lambda}{\lambda_0} = 1 \tag{130}$$

The integration will lead to a certain approximate connection between λ and λ_0, independent either of the shape of the profiles or of the character of flow past them. This, due

[21] These examples, giving the upper and lower limits of corrections for the compressibility, were kindly indicated to the author by G. Yu. Stepanov.

to the cumbersome form of the analytical expression, will be represented by curve 3 (Figure 106) and Table 9.

Having this connection, it is not difficult to obtain the dependence of the pressure coefficient c_p in a gas on the coefficient c_{p0} in an incompressible fluid.

From the familiar isentropic relations

$$\frac{p}{p_\infty} = \left(\frac{1-\frac{k-1}{k+1}\lambda^2}{1-\frac{k-1}{k+1}\lambda_\infty^2}\right)^{\frac{k}{k-1}} \qquad \frac{a_\infty^2}{a^{*2}} = \frac{1-\frac{k-1}{k+1}\lambda_\infty^2}{1-\frac{k-1}{k+1}\cdot 1} = \frac{k+1}{2}\left(1-\frac{k-1}{k+1}\lambda_\infty^2\right)$$

we find, in the presence of compressibility, that

$$c_p = \frac{p-p_\infty}{\frac{1}{2}\rho_\infty V_\infty^2} = \frac{2p_\infty}{\rho_\infty V_\infty^2}\left[\left(\frac{1-\frac{k-1}{k+1}\lambda^2}{1-\frac{k-1}{k+1}\lambda_\infty^2}\right)^{\frac{k}{k-1}} - 1\right] =$$

$$= \frac{k+1}{k}\,\frac{1-\frac{k-1}{k+1}\lambda_\infty^2}{\lambda_\infty^2}\left[\left(\frac{1-\frac{k-1}{k+1}\lambda^2}{1-\frac{k-1}{k+1}\lambda_\infty^2}\right)^{\frac{k}{k-1}} - 1\right] \tag{131}$$

Adding the equation corresponding to the incompressible fluid

$$c_{p_0} = 1 - \left(\frac{\lambda_0}{\lambda_{0\infty}}\right)^2 \tag{132}$$

and the relationship between λ and λ_0 given by Table 9, we consider the set of Equations 131 and 132 as the parametric connection between c_p and c_{p0} via the parameter λ. This is the essence of the approximate technique suggested by Khristianovich for taking the compressibility into account. With the number $\mathbf{M}_\infty$ being selected, we find λ_∞ and, consequently, also $\lambda_{0\infty}$ from Table 9, after which, having prescribed different values of λ_0 and determined the corresponding values of λ, we obtain the sought-after connection between c_p and c_{p0} from Equations 131 and 132.

Figures 107 and 108 present the nets of curves[22] expressing this connection at different $\mathbf{M}_\infty$ numbers separately for positive and negative c_p. The dashed line $\lambda = 0.85$ in Figure 108 shows the applicability limit of the technique; at greater λ the results are unsatisfactory; the segments of curves to the right from this line should be used.

Among the drawbacks of the above approximate technique is the absence of the explicit dependence between c_p and c_{p0} and the resulting necessity for using the indicated nets of curves. This drawback can be eliminated by applying the method first suggested by Chaplygin.[23] Let us turn again to Equations 118 having adopted in them $\lambda = V/a^*$ and θ as arguments:

---

[22] Polyadskiy, V. F., *Rashyot raspredeleniya davleniya pri bolshikh skorostyakh polyota* (*Calculation of Pressure Distribution at High Velocities of Flight*), NKAP Press, 1943; ibid., the conversion tables and cirrections by a more accurate approximation.

[23] Chaplygin, S. A., About gas jets, *Uchyon. Zapiski Moskov. Univ., Otd. Fiz. Mat. Nauk*, 21, 1904.

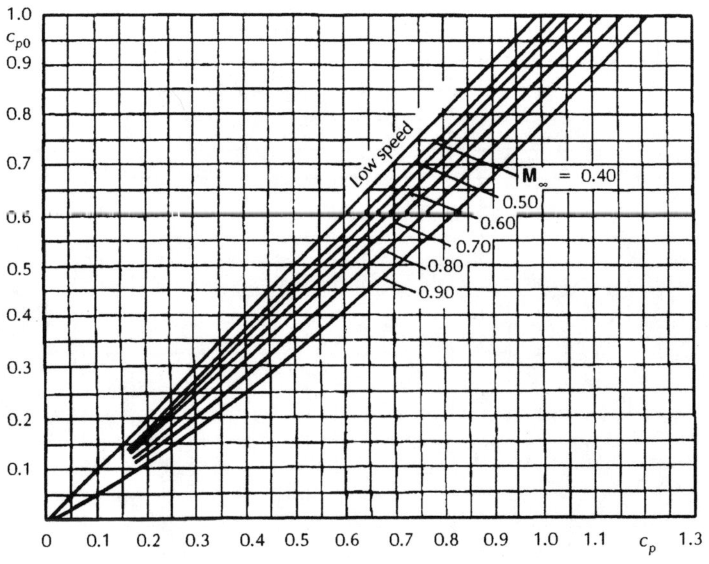

Fig. 107

$$\frac{\partial\varphi}{\partial\lambda} = \lambda\frac{d}{d\lambda}\left(\frac{1}{\lambda}\frac{\rho_0}{\rho}\right)\frac{\partial\psi}{\partial\theta} \qquad \frac{\partial\varphi}{\partial\theta} = \lambda\frac{\rho_0}{\rho}\frac{\partial\psi}{\partial\lambda}$$

Analogous equations for an incompressible ($\rho = \rho_0$) fluid in the previous notation will have the form:

$$\frac{\partial\varphi_0}{\partial\lambda_0} = \lambda_0\frac{d}{d\lambda_0}\left(\frac{1}{\lambda_0}\right)\frac{\partial\psi_0}{\partial\theta_0} \qquad \frac{\partial\varphi_0}{\partial\theta_0} = \lambda_0\frac{\partial\psi_0}{\partial\lambda_0}$$

The approximate method of Chaplygin consists of the reduction of the compressible gas motion equations to the equations of motion of an incompressible fluid by replacing the variable λ by its certain function $\tilde{\lambda}(\lambda)$. As will be clear from what follows, such a transformation is equivalent to the use of a certain approximate function instead of the actual function $\rho(\lambda)/\rho_0$. After this replacement the equation of compressible gas motion will be reduced to the form:

$$\frac{\partial\varphi}{\partial\tilde{\lambda}} = \lambda\frac{d}{d\tilde{\lambda}}\left(\frac{1}{\lambda}\frac{\rho_0}{\rho}\right)\frac{\partial\psi}{\partial\theta} \qquad \frac{\partial\varphi}{\partial\theta} = \lambda\frac{\rho_0}{\rho}\frac{d\tilde{\lambda}}{d\lambda}\frac{\partial\psi}{\partial\tilde{\lambda}}$$

We shall assume that $\varphi = \varphi_0$, $\psi = \psi_0$ at the same ($\tilde{\lambda} = \lambda_0$, $\theta = \theta_0$) values of the arguments by having required the coincidence of corresponding differential equations. For this purpose the following equations would be fulfilled:

$$\lambda\frac{d}{d\lambda_0}\left(\frac{1}{\lambda}\frac{\rho_0}{\rho}\right) = \lambda_0\frac{d}{d\lambda_0}\left(\frac{1}{\lambda_0}\right) = -\frac{1}{\lambda_0} \qquad \lambda\frac{\rho_0}{\rho}\frac{d\lambda_0}{d\lambda} = \lambda_0$$

This system of ordinary differential equations determines the functions $\lambda(\lambda_0)$ and $\rho(\lambda_0)/\rho$. Taking into consideration the coincidence condition for λ and λ_0 at small velocities [$\lim\limits_{\lambda\to 0}(\lambda/\lambda_0) = 1$], we obtain the solution

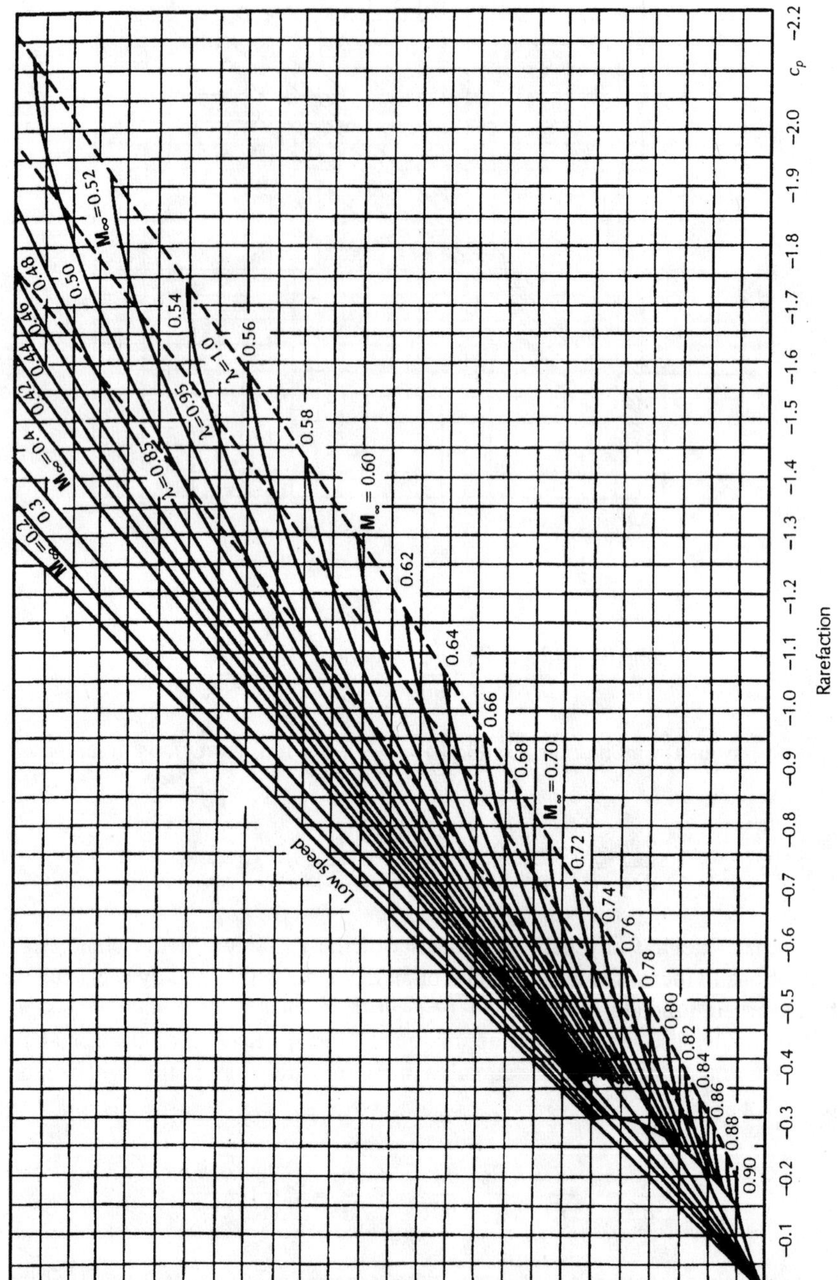

c_{p0}
c_p
Rarefaction
Low speed
$M_\infty=0.2$
0.3
$M_\infty=0.4$
0.42
0.44
0.46
0.48
0.50
$M_\infty=0.52$
0.54
0.56
0.58
$M_\infty=0.60$
0.62
0.64
0.66
0.68
$M_\infty=0.70$
0.72
0.74
0.76
0.78
0.80
0.82
0.84
0.86
0.88
0.90
$\lambda=0.85$
$\lambda=0.95$
$\lambda=1.0$

Fig. 108

$$\lambda = \frac{\lambda_0}{1 - C\lambda_0^2} \qquad \frac{\rho}{\rho_0} = \frac{1 - C\lambda_0^2}{1 + C\lambda_0^2} = \frac{1}{\sqrt{1 + 4C\lambda^2}}$$

where C is an integration constant.

The resulting dependence of density on the velocity coefficient for Chaplygin's hypothetical gas

$$\frac{\rho}{\rho_0} = (1 + 4C\lambda^2)^{-1/2} = 1 - 2C\lambda^2 + O(\lambda^4)$$

differs from the familiar relation for adiabatic motion

$$\frac{\rho}{\rho_0} = \left(1 - \frac{k-1}{k+1}\lambda^2\right)^{\frac{1}{k-1}} = 1 - \frac{1}{k+1}\lambda^2 + O(\lambda^4)$$

The selection of the constant C can be used for obtaining the best approximation to the actual function $\rho(\lambda)/\rho_0$. Thus, assuming $2C = 1/(k+1)$, $C = 0.2083$ ($k = 1.4$; curve 4 in Figure 106) we have the error of approximation of order 14. The calculations show that for $C = 0.296$ (curve 5 in the same figure) the maximum relative error does not exceed 2% when $\lambda < 0.95$.[24]

The use of these or other functions $\rho(\lambda)/\rho_0$ corresponds to the replacement of the Poisson actual adiabatic by different simple curves or straight lines that approximate it.[25] Thus, Kármán and Tsien,[26] using a rectilinear adiabatic, suggested the simple approximate analytical formula of coupling between c_p and c_{p0}:

$$c_p = \frac{c_{p0}}{\sqrt{1 - \mathbf{M}_\infty^2} + \frac{1}{2}\frac{\mathbf{M}_\infty^2}{1 + \sqrt{1 - \mathbf{M}_\infty^2}}c_{p0}} = \frac{c_{p0}}{\sqrt{1 - \mathbf{M}_\infty^2} + \frac{1}{2}\left(1 - \sqrt{1 - \mathbf{M}_\infty^2}\right)c_{p0}} \tag{133}$$

not involving the specific heat ratio k.

At small values of $\mathbf{M}_\infty$ Equation 133 becomes the Prandtl-Glauert formula

$$c_p = \frac{c_{p0}}{\sqrt{1 - \mathbf{M}_\infty^2}}$$

In contrast to the latter formula, Equation 133 (containing c_{p0} in the denominator) does not make it possible to establish simple relationship between c_y and c_{y0}; the quantity c_y is computed by integrating the converted pressure distribution in each separate case.

The Kármán-Tsien equation is convenient for computations and, as comparison with experiments shows, gives a satisfactory estimate of the effect of compressibility ($\mathbf{M}_\infty$ value on the pressure coefficient c_{p0} for the incompressible fluid flow past the same profile, even at relatively large (but smaller than unity, of course) Mach numbers.

---

[24] Stepanov, G. Yu., *Gidrodinamika reshyotok turbomashin* (*The Hydrodynamics of the Cascades of Turbines*), Fizmatgiz Press, Moscow, 1962, p. 197.

[25] A detailed coverage of this kind of technique can be found in the book by G. A. Dombrovskiy, *Metod approksimatsii adiabaty v teorii ploskikh techeniy gaza* (*The Method of Approximation of Adiabatic in the Theory of Plane Gas Flows*), Nauka Press, Moscow, 1964.

[26] Kármán, Th., Compressibility effects in aerodynamics, *J. Aeron. Sci.*, Vol. 8, July 1941; Tsien, H. S., Two-dimensional subsonic flow of compressible fluids, *J. Aeron. Sci.*, Vol. 6, August 1939; see the 4th edition of the present book, pp. 303 to 307.

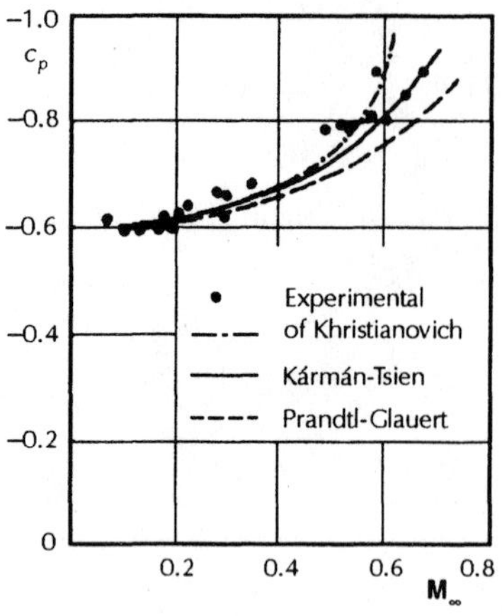

Fig. 109

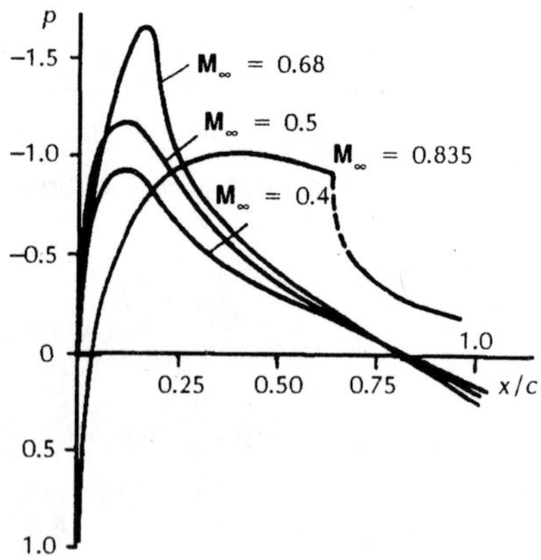

Fig. 110

As an illustration Figure 109 compares with the experiment the calculated pressure coefficient c_p at the point located at a distance of 30% of the chord from the body nose at an angle of attack of $\alpha = -2°$ and different $\mathbf{M}_\infty$ numbers for the upper surface of the wing profile NACA 4412.

By relating the upper experimental point at $\mathbf{M}_\infty = 0.6$ to the number of those not involved, it is possible to note that the curve plotted by the Kármán-Tsien equation corresponds well to the experimental points.

The effect of compressibility in subsonic flow past profiles shows up as an increase in rarefactions on the upper surface of the profile. This fact has already been noted in the presentation of the Prandtl-Glauert theory of small perturbations. Figure 110 shows the experimental pressure distributions over the upper surface of a certain wing profile at different $\mathbf{M}_\infty$ numbers of the oncoming flow. It can be noted that the increase in the $\mathbf{M}_\infty$ number from 0.4 to 0.68 resulted in almost double the increase of the rarefaction peak.

A formula similar in structure to that of Kármán-Tsien can also be derived from simple considerations, not connected with the Chaplygin theory, by suggesting, according to Laitone,[27] that the pressure coefficient at the given point of the compressible subsonic flow can be expressed in terms of the corresponding pressure coefficient in an incompressible flow by the Prandtl-Glauert equation in which the number $\mathbf{M}_\infty$ should be interpreted as the local $\mathbf{M}$ number varying from point to point. In fact, in a subsonic flow described by elliptic-type partial differential equations the effect of compressibility is not reduced to the local effect characterized by the number $\mathbf{M}$ at the given point, but requires taking into account the effect of compressibility on the flow as a whole. This is the principle of the approximate nature of Laitone's assumption.

According to this assumption, the pressure coefficients at the corresponding points are connected by the equation

$$c_p = \frac{c_{p0}}{\sqrt{1 - \mathbf{M}^2}} \tag{134}$$

Let us eliminate $\mathbf{M}$ by using the isentropic relation

[27] Laitone, E. V, New compressibility correction for two-dimensional subsonic flow, *J. Aeron. Sci.*, 18(5), 350, 1951.

$$c_p = \frac{p - p_\infty}{\frac{1}{2}\rho_\infty V_\infty^2} = \frac{2p_\infty}{\rho_\infty V_\infty^2}\frac{p - p_\infty}{p_\infty} = \frac{2}{k\mathbf{M}_\infty^2}\left[\left(\frac{1 + \frac{k-1}{2}\mathbf{M}_\infty^2}{1 + \frac{k-1}{2}\mathbf{M}^2}\right)^{\frac{k}{k-1}} - 1\right] \tag{135}$$

here, finding $\mathbf{M}^2$ and expanding the expression obtained for it into the series in the powers of the small quantity $\frac{k\mathbf{M}_\infty^2}{2}c_p$,

$$\mathbf{M}^2 = \left(\mathbf{M}_\infty^2 + \frac{2}{k-1}\right)\left(1 + \frac{k\mathbf{M}_\infty^2}{2}c_p\right)^{-\frac{k-1}{k}} - \frac{2}{k-1} = \mathbf{M}_\infty^2 - \left(1 + \frac{k-1}{2}\mathbf{M}_\infty^2\right)\mathbf{M}_\infty^2 c_p + \cdots$$

Substituting this equation into Equation 134, we obtain Laitone's formula

$$c_p = \frac{c_{p0}}{\sqrt{1 - \mathbf{M}_\infty^2} + \dfrac{\mathbf{M}_\infty^2\left(1 + \frac{k-1}{2}\mathbf{M}_\infty^2\right)}{2\sqrt{1 - \mathbf{M}_\infty^2}}c_{p0}} \tag{136}$$

which involves, as does the Kármán-Tsien formula, the correction terms in the denominator in the Prandtl-Glauert formula. Comparison of the Laitone and Kármán-Tsien formulas shows that with $\mathbf{M}_\infty$ approaching unity, the Laitone correction increases much more quickly than the correction in the Kármán-Tsien formula. At relatively small values of c_p and $\mathbf{M}_\infty$ numbers, not very close to unity, the Laitone formula gives better agreement with the experiment than the Prandtl-Glauert formula and is close to the Kármán-Tsien formula.

All the simplified methods of calculation given above did not entail the determination of the full solution of the subsonic flow problem, in particular, the effect of compressibility on the attached circulation which ensures smooth flow from the trailing sharp edge of the wing profile.[28]

It should be noted that at present there is the possibility for an accurate numerical calculation of velocity and pressure distributions in flows past bodies at any velocity.[29]

10. Subcritical Flow Past a Wing Profile

As the $\mathbf{M}_\infty$ number increases, originates complex phenomena hardly susceptible to theoretical analysis originates near the surface of the wing profile immersed in the flow. Let us return to the curves of Figure 110. Attention is drawn to the phenomenon of the displacement of the rarefaction peak downstream starting from about $\mathbf{M}_\infty = 0.5$.

The last curve in Figure 110, relating to the number $\mathbf{M}_\infty - 0.835$, drops out distinctly from the general behavior of the pressure curves with the growth of $\mathbf{M}_\infty$. Very apparent at first is a substantial decrease (in absolute magnitude) and smoothing of the rarefaction peak;

[28] The exact (in this sense) solutions were simultaneously obtained by Sedov and Ts. Ts. Lin. See Sedov, L. I., *Ploskiye zadachi gidrodinamiki i aerodinamiki* (*Plane Problems of Hydrodynamics and Aerodynamics*), Nauka Press, Moscow, 1980; Sedov, L. I., *Two-dimensional Problems in Hydrodynamics and Aerodynamics*, John Wiley & Sons, New York, 1965.

[29] Davydov, Yu. M., *Tablitsy raschyota vneshnikh okolozvukovykh techeniy gaza* (*Tables of Calculation of External Transonic Gas Flows*), Tsents AN SSSR Press, Moscow, 1983.

then a jumpwise pressure recovery is clearly seen, which is shown in the figure by dashes. These phenomena can be explained by the formation of the critical section in the stream tube which converges to the point of maximum velocity in the subsonic flow. Further expansion of the stream tube creates motion analogous to the motion in the Laval nozzle. The velocity becomes supersonic and then returns to the subsonic value in the shock wave. The presence of shock waves leads to considerable losses of mechanical energy and adversely affects the aerodynamic characteristics of the wing profile. One of the measures for eliminating this phenomenon was the creation of profiles with the possibly later formation of the critical velocity on their surface.

Let the critical number $\mathbf{M}_{cr}$ be such a value of the number $\mathbf{M}_\infty$ of the flow incident on the wing, at which the local number $\mathbf{M}$ somewhere on the wing surface becomes equal to unity, i.e., the velocity originates which is equal to the local speed of sound.

The value $\mathbf{M}_{cr}$ can be easily connected with the minimal value of the pressure coefficient c_{pm}. For this purpose, using expression (135) for c_p, we assume that somewhere on the profile the velocity attained the local speed of sound and the local number $\mathbf{M}$ became equal to unity; then c_p attains the minimum value c_{pm} as compared with other points of the flow and the number $\mathbf{M}_\infty$ becomes equal to $\mathbf{M}_{cr}$. Consequently, by assuming in Equation 135 that

$$c_p = c_{pm} \qquad \mathbf{M}_\infty = \mathbf{M}_{cr} \qquad \mathbf{M} = 1$$

we thereby determine the sought-after connection between $\mathbf{M}_{cr}$ and c_{pm}

$$c_{pm} = \frac{2}{k\mathbf{M}_{cr}^2}\left[\left(\frac{2}{k+1}\right)^{\frac{k}{k-1}}\left(1 + \frac{k-1}{2}\mathbf{M}_{cr}^2\right)^{\frac{k}{k-1}} - 1\right] \tag{137}$$

where, it should be emphasized, the quantity c_{pm} denotes the pressure coefficient at the given $\mathbf{M}_\infty = \mathbf{M}_{cr}$, i.e., in an actual gas flow with the compressibility taken into account.

In practical calculations it is more useful to have the connection between $\mathbf{M}_{cr}$ and the pressure coefficient c_{p0m} for the flow of an incompressible fluid past the same profile, because the latter quantity can be easily calculated by the techniques described in Chapter 7, or is determined by blowing in low velocity tunnels. It is hardly realistic to use the Prandtl-Glauert rule or approximate Khristianovich technique to convert c_{pm} into c_{p0m}, because c_{pm} relates to the number $\mathbf{M}_\infty = \mathbf{M}_{cr}$ with the approach to which the indicated calculation techniques become unjustifiable.

Recalling a purely experimental fact that the Kármán-Tsien formula is comparatively satisfactory, we may use it at the values of $\mathbf{M}_\infty$ close to $\mathbf{M}_{cr}$ and determine $\mathbf{M}_{cr}$ as a function of c_{p0m} from the set of Equations 133 and 137 rewritten in the form

$$c_{p0m} = \frac{c_{pm}\sqrt{1 - \mathbf{M}_{cr}^2}}{1 - \frac{1}{2}\left(1 - \sqrt{1 - \mathbf{M}_{cr}^2}\right)c_{pm}}$$

$$c_{pm} = \frac{2}{k\mathbf{M}_{cr}^2}\left[\left(\frac{2}{k+1}\right)^{\frac{k}{k-1}}\left(1 + \frac{k-1}{2}\mathbf{M}_{cr}^2\right)^{\frac{k}{k-1}} - 1\right] \tag{138}$$

Substituting the value of c_{pm} from the second equation into the first, we obtain the sought-after connection between c_{p0m} and $\mathbf{M}_{cr}$; however, in view of the awkward form of the final result, we do not give it here.

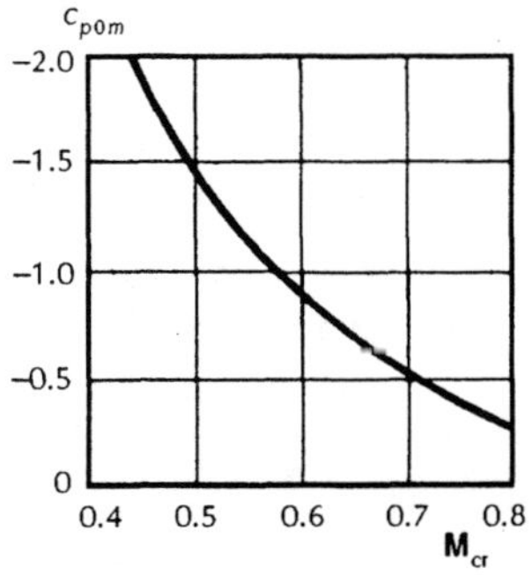

Fig. 111

Different approximate formulas have been suggested for determining $\mathbf{M}_{cr}$ from the calculated c_{p0m}. Here is one set up by Temple[30] for air ($k = 1.4$)

$$c_{p0m} = 1 - 0.522 \frac{(1 + 0.2\mathbf{M}_{cr}^2)^2}{\mathbf{M}_{cr}^2 (1 - 0.05\mathbf{M}_{cr}^2)^2} \tag{139}$$

which is virtually equivalent to the set of Equations 138 and which leads to a good agreement with the experiment. The corresponding curve is shown in Figure 111.

It is possible to conclude from the definition of the concept of the critical $\mathbf{M}_{cr}$ number that in subsonic flows one should prefer such profiles, which at the same value of the required lifting force (coefficient c_y) have the highest possible value of $\mathbf{M}_{cr}$ and, consequently, the smaller value of c_{pm}. In other words, it is necessary to strive for such a situation when the same area confined between the pressure distribution curves over the upper and lower wing surfaces could be achieved at flattened pressure distribution curves, rather than at the expense of sharp rarefaction peaks. Such profiles with increased values $\mathbf{M}_{cr}$ have been designed in Russia and abroad and have been widely applied in aviation and in the manufacturing of turbines.

11. A Plane Supersonic Flow. General Properties of Characteristics. The Graphic Method for Calculating Supersonic Flows

The passage to the hodograph plane led to the development of analytical methods for integrating the differential equations of supersonic ideal gas flow. However, in practical calculations one prefers the use of numerical methods or simple graphic techniques of integration of these equations on the basis of the method of characteristics.

Let us return to the basic system of equations of a plane irrotational gas motion (2) and (3) of the present chapter

$$(a^2 - u^2)\frac{\partial u}{\partial x} - uv\left(\frac{\partial u}{\partial y} + \frac{\partial v}{\partial x}\right) + (a^2 - v^2)\frac{\partial v}{\partial y} = 0 \qquad \frac{\partial u}{\partial y} - \frac{\partial v}{\partial x} = 0 \tag{140}$$

Generalizing the technique presented in Chapter 6 for solving the Riemann problem of the propagation of finite perturbations, we shall set up the linear combination of Equations 140;

[30] Hilton, W. F., *Aerodinamika, bolshikh skorostei (The Aerodynamics of High Velocities)*, IL Press, Moscow, 1955, p. 29.

multiply the first of these equations, respectively, by the indeterminate factor Λ_2 and the second by Λ_1 and sum them. Then,

$$\Lambda_2(a^2-u^2)\frac{\partial u}{\partial x}+(\Lambda_1-\Lambda_2 uv)\frac{\partial u}{\partial y}-(\Lambda_1+\Lambda_2 uv)\frac{\partial v}{\partial x}+\Lambda_2(a^2-v^2)\frac{\partial v}{\partial y}=0$$

or, separately collecting the terms with the derivatives of u and v,

$$\Lambda_2(a^2-u^2)\left[\frac{\partial u}{\partial x}+\frac{\Lambda_1-\Lambda_2 uv}{\Lambda_2(a^2-u^2)}\frac{\partial u}{\partial y}\right]-(\Lambda_1+\Lambda_2 uv)\left[\frac{\partial v}{\partial x}-\frac{\Lambda_2(a^2-v^2)}{\Lambda_1+\Lambda_2 uv}\frac{\partial v}{\partial y}\right]=0 \qquad (141)$$

Now, using the quantities Λ_1 and Λ_2, try to determine at each point of the plane (x, y) such direction with the angular coefficient $m = dy/dx$ so that the expressions in square brackets of Equation 141 could represent the derivatives of u and v with respect to x corresponding to this direction

$$\frac{\partial u}{\partial x}+\frac{\Lambda_1-\Lambda_2 uv}{\Lambda_2(a^2-u^2)}\frac{\partial u}{\partial y}=\frac{\partial u}{\partial x}+\frac{\partial u}{\partial y}\frac{dy}{dx}=\frac{\partial u}{\partial x}+\frac{\partial u}{\partial y}m=\frac{du}{dx}$$

$$\frac{\partial v}{\partial x}-\frac{\Lambda_2(a^2-v^2)}{\Lambda_1+\Lambda_2 uv}\frac{\partial v}{\partial y}=\frac{\partial v}{\partial x}+\frac{\partial v}{\partial y}\frac{dy}{dx}=\frac{\partial v}{\partial x}+\frac{\partial v}{\partial y}m=\frac{dv}{dx}$$

and, consequently, that Equation 141 could transform into a differential equation in the hodograph plane (u, v)

$$\frac{dv}{du}=\frac{\Lambda_2(a^2-u^2)}{\Lambda_1+\Lambda_2 uv} \qquad (142)$$

For these conditions to be fulfilled, it is necessary to subject the quantities Λ_1 and Λ_2 to the proportion

$$\frac{\Lambda_1-\Lambda_2 uv}{\Lambda_2(a^2-u^2)}=\frac{\Lambda_2(a^2-v^2)}{-(\Lambda_1+\Lambda_2 uv)}=m \qquad (143)$$

or to a system of homogeneous equations

$$\Lambda_1-\Lambda_2(m(a^2-u^2)+uv)=0 \qquad m\Lambda_1+\Lambda_2[a^2-v^2+muv]=0$$

which will have the nonzero solutions only when the determinant of the system is equal to zero, i.e., at

$$(u^2-a^2)m^2-2uvm+v^2-a^2=0 \qquad (144)$$

Forming the discriminant of this quadratic equation

$$u^2v^2-(u^2-a^2)(v^2-a^2)=a^2(u^2+v^2-a^2)=a^2(V^2-a^2)$$

we see that the real solutions will take place when the condition $V \geq a$ or $\mathbf{M} \geq 1$ is fulfilled.

Thus, at each point of the supersonic flow it is possible to determine two characteristic directions with the angular coefficients m_1 and m_2 which are equal to the roots of quadratic equation (144)

$$m_{1,2} = \left(\frac{dy}{dx}\right)_{1,2} = \frac{uv + a\sqrt{V^2 - a^2}}{u^2 - a^2} \tag{145}$$

so that the differentials *dx*, *dy*, which correspond to these directions in the physical plane of flow, were correlated with the differentials *du*, *dv* in the hodograph plane, which are connected by Equation 142.

Of fundamental importance is the fact that differential equation (142) can be integrated in the final form for an arbitrary supersonic flow. Rewriting Equation 142 in the form of identity

$$\frac{dv}{du} = \frac{\Lambda_2 (a^2 - u^2)}{\Lambda_1 + \Lambda_2 uv} = \frac{\Lambda_2 (a^2 - v^2)}{\Lambda_1 + \Lambda_2 uv} \frac{a^2 - u^2}{a^2 - v^2}$$

and comparing it with Equations 143 and 144, we obtain

$$\frac{dv}{du} = -m\frac{a^2 - u^2}{a^2 - v^2} = \frac{uv \pm a\sqrt{V^2 - a^2}}{a^2 - v^2} \tag{146}$$

Now, go from the projections of the velocity u, v to the velocity vector V and angle θ formed by this vector with the axis $0x$ of the physical plane, having assumed that

$$u = V\cos\theta \qquad v = V\sin\theta$$

$$du = dV\cos\theta - V\sin\theta\, d\theta \qquad dv = dV\sin\theta + V\cos\theta\, d\theta$$

Then, Equation 146 will convert to

$$\frac{\sin\theta\, dV + V\cos\theta\, d\theta}{\cos\theta\, dV - V\sin\theta\, d\theta} = \frac{V^2\sin\theta\cos\theta \pm a^2\sqrt{\mathbf{M}^2 - 1}}{a^2 - V^2\sin^2\theta}$$

Let us collect the terms with dV and $d\theta$; then, after simple transformations and cancellations we obtain

$$d\theta = \pm\sqrt{\mathbf{M}^2 - 1}\,\frac{dV}{V} \tag{147}$$

In other words, Equation 92, which is familiar from Section 6, has an integral that is represented by Equation 94, and in the present case can be written in the general form as

$$\theta = \pm\sigma(\lambda) + \text{const}$$

$$\sigma(\lambda) = \sqrt{\frac{k+1}{k-1}}\,\text{arctg}\left(\sqrt{\frac{k-1}{k+1}}\sqrt{\frac{\lambda^2 - 1}{1 - \frac{k-1}{k+1}\lambda^2}}\right) - \text{arctg}\left(\sqrt{\frac{\lambda^2 - 1}{1 - \frac{k-1}{k+1}\lambda^2}}\right) \tag{148}$$

The family of the integral curves of Equation 145 which correspond to the presence of different signs before the radical, form the characteristics in the physical plane (x, y), whereas the quantities m_1 and m_2 represent the angular coefficients of the tangents to the characteristics or the characteristic directions in the physical plane.

For definiteness, the integral curves that correspond to differential equation (145) with the positive sign before the radical, i.e.,

$$\frac{dy}{dx} = \frac{uv + a\sqrt{V^2 - a^2}}{u^2 - a^2}$$

will be called the characteristics of the first family, and the integral curves of the equation

$$\frac{dy}{dx} = \frac{uv - a\sqrt{V^2 - a^2}}{u^2 - a^2}$$

will be called the characteristics of the second family.

In the same way, differential equations (146) or their integrals (148) determine two families of curves at each point of the hodograph plane of velocity. These curves are the characteristics in the hodograph plane. To the plus sign corresponds the characteristics of the first family, to the minus sign corresponds the characteristics of the second family. Denoting by n the angular coefficient of the characteristic directions in the hodograph plane, we have from Equation 146

$$n_{1,2} = \left(\frac{dv}{du}\right)_{1,2} = \frac{uv \pm a\sqrt{V^2 - a^2}}{a^2 - v^2} \tag{149}$$

The characteristic directions in the physical plane and in the hodograph plane are interconnected, as it follows directly from Equations 145 and 149, by the obvious relations

$$n_1 m_2 + 1 = 0 \qquad n_2 m_1 + 1 = 0$$

Thus, it follows that when selecting the axes Ox and Oy parallel to the axes $O'u$ and $O'v$, the characteristic directions of the first family in the physical plane will be perpendicular to the characteristic directions of the second family at the corresponding point of the hodograph plane and, conversely, the characteristic directions of the second family in the physical plane are perpendicular to the characteristic directions of the first family in the hodograph plane.

Thus, the characteristic directions in the physical plane are rigidly conjugated with the characteristic directions in the hodograph plane. However, the differential equations of the characteristics in the hodograph plane (147) were integrated and led to the final formulas (148) of the characteristics that represent two certain distinct families of curves identical for all plane supersonic flows.

Noting that in supersonic flows the dimensionless velocity λ varies within the limits $1 \le \lambda \le \sqrt{\frac{k+1}{k-1}}$, with the left boundary corresponding to the critical velocity $V = a^*$ and the right boundary corresponding to the limiting maximum velocity $V = V_{max}$, we draw the concentric circles $\lambda = 1$ and $\lambda = \sqrt{\frac{k+1}{k-1}}$ and fill the space between them with the net of curves (148). These curves represent two families of epicycloids (Figure 112) which are described by the points of the circle of radius $\frac{1}{2}\left(\sqrt{\frac{k+1}{k-1}} - 1\right)$ rolling along the circle $\lambda = 1$.

Having such a net of epicycloids drawn once and for all for the given k (in Figure 112 for air $k = 1.4$), it is not difficult, with the aid of simple graphic means, to construct characteristic directions at the points of the physical plane, having drawn through these points perpendicu-

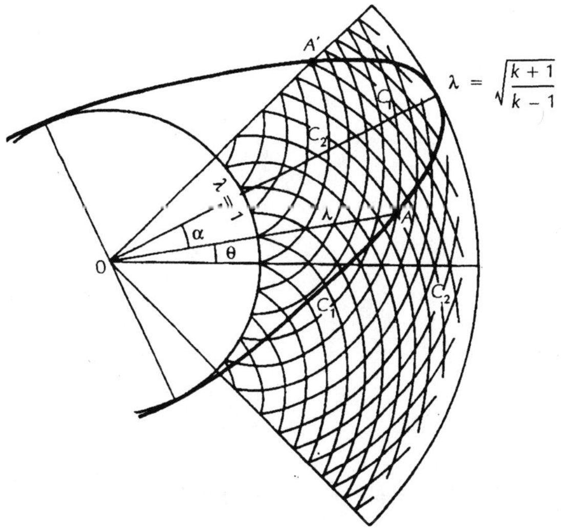

Fig. 112

lars to the corresponding characteristic directions in the hodograph plane which are given for the families (148).

Before moving to the presentation of these graphic means, we first consider some general properties of characteristics in the physical plane.

1. The characteristics of the equations for a plane supersonic flow form the angles with the gas velocity vector at the given point which are equal to the perturbation angles (Mach angles)

$$\alpha = \pm \arcsin \frac{1}{\mathbf{M}}$$

i.e., they coincide with the perturbation lines (Mach lines).

For the above to be proved, we avail ourselves of the well-known formula of the analytical geometry for the tangent of the angle δ between the directions with the specified angular coefficients m and v/u; according to Equation 145

$$\operatorname{tg} \delta = \frac{m - \frac{v}{u}}{1 + m\frac{v}{u}} = \frac{\frac{uv \pm a\sqrt{V^2 - a^2}}{u^2 - a^2} - \frac{v}{u}}{1 + \frac{uv \pm a\sqrt{V^2 - a^2}}{u^2 - a^2}\frac{v}{u}} =$$

$$= \frac{a^2 v \pm au\sqrt{V^2 - a^2}}{u(V^2 - a^2) \pm av\sqrt{V^2 - a^2}} = \pm\frac{a}{\sqrt{V^2 - a^2}} = \pm\frac{1}{\sqrt{\mathbf{M}^2 - 1}}$$

so that actually

$$\sin \delta = \pm\frac{1}{\mathbf{M}} = \sin \alpha$$

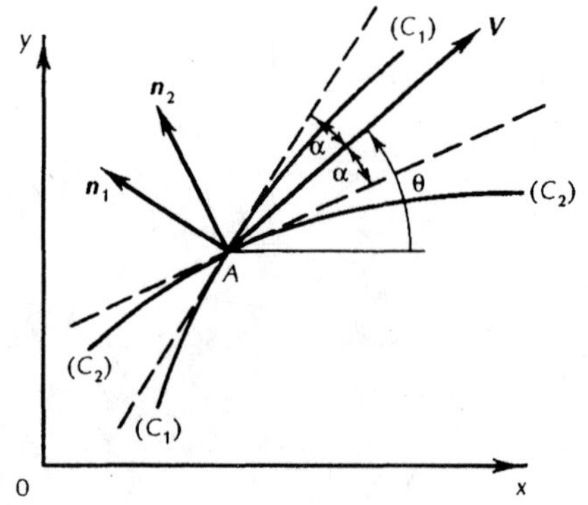

Fig. 113

From this directly follows the second property of the characteristics in a physical plane (Figure 113).

2. The velocity vector is directed along the bisectrix of the angle between the characteristics in the physical plane.

3. The gas velocity projections onto the normals to the characteristics at the given point of the physical plane are equal in absolute magnitude to the local speed of sound.

Actually (Figure 113),

$$V_{n1} = V\cos(90° + \alpha) = -V\sin\alpha = -V\cdot\frac{1}{\mathbf{M}} = -a$$

$$V_{n2} = V\cos(90° - \alpha) = V\sin\alpha = V\cdot\frac{1}{\mathbf{M}} = a$$

Using the last property of the characteristics in the physical plane, it can be easily shown that the curve of the function $\lambda(\alpha)$ for any supersonic flow represents an ellipse, the shape of which depends only on the specific heat ratio k which expresses the physical properties of the gas. For this purpose we use the well-known isentropic relation (V_{max} is the velocity of discharge into vacuum at $a = 0$)

$$\frac{V^2}{2} + \frac{a^2}{k-1} = \frac{V_{max}^2}{2} = \frac{1}{2}\frac{k+1}{k-1}a^{*2}$$

Let us use u and v to denote the projections of the velocity vector $\boldsymbol{V}$ onto the coordinate axis, the first of which is directed along the tangent to the characteristic and the second of which is directed along the normal to it. Then, according to property 3, v will be equal to $\pm a$, and the previous equation can be rewritten in the form

$$\frac{u^2 + v^2}{2} + \frac{v^2}{k-1} = \frac{1}{2}\frac{k+1}{k-1}a^{*2}$$

or, assuming that $\lambda_u = u/a^*$, $\lambda_v = v/a^*$,

$$\frac{k-1}{k+1}\lambda_u^2 + \lambda_v^2 = 1 \tag{150}$$

which in the plane (λ_u, λ_v) represents an ellipse with the major semiaxis $\sqrt{\frac{k+1}{k-1}}$ and minor semiaxis equal to unity. The ellipse (150) – the Busemann ellipse – is located in the region between the circles $\lambda = 1$ and $\lambda = \sqrt{\frac{k+1}{k-1}}$. In conformity with the foregoing, its major semiaxis is directed along the tangent to the characteristic at the given point of the physical plane and the minor semiaxis is directed along the perpendicular to it, i.e., along the tangent to the conjugated characteristic in the hodograph plane of velocities.

The possible location of the Busemann ellipse with respect to the net of epicycloids is shown in Figure 112. If the center 0 of the ellipse drawn on a piece of transparent material is made coincident with the center of the circles that confine the family of epicycloids and the ellipse is rotated so that its certain point A could coincide with the prescribed point (λ, θ) of the hodograph plane, then the major semiaxis of the ellipse, which forms the angle α with the velocity vector, will show the direction of one of the characteristics (perturbation lines) in the physical plane. The direction of the other characteristic can be obtained if the velocity vector end (i.e., the point (λ, θ)), is made coincident with the point A' of the ellipse which is the mirror reflection of the point A of the ellipse with respect to its major axis.

Thus, knowing the magnitude and direction of velocity at a certain point of the physical plane, we may, using the Busemann ellipse (without additional computations and only by graphic means), draw through it two characteristic directions in this (physical) plane. The minor semiaxis of the ellipse will indicate the conjugated characteristic direction in the hodograph plane.

The graphic method of calculation of plane supersonic flows is based on the use of the properties (listed above) of the families of characteristics in the physical plane of the flow and in the hodograph plane of velocity.[31]

Generally, this method is analogous to the graphic method for calculating the propagation of finite-amplitude waves presented in Chapter 6. Its certain specific feature only consists in the convenient use of: (1) the nets of characteristics drawn once and for all in the hodograph plane, familiar epycycloids (148), and (2) the Busemann ellipse (150) manufactured in the form of a transparent gauge.

We shall rewrite the equations of characteristics in the hodograph plane (148) in the form of two separate equations

$$\theta = C_1 - \sigma(\lambda) \qquad \theta = \sigma(\lambda) - C_2$$

the first of which represents the family of epicycloids running from below upward, and the second running from above downward.

Adding and substracting these equations term-by-term, we obtain

$$\theta = \frac{1}{2}(C_1 - C_2) \qquad \sigma(\lambda) = \frac{1}{2}(C_1 + C_2) \tag{151}$$

so that the rays radially emanating from the point 0 (Figure 112) (θ = const) will be determined by the condition $C_1 - C_2$ = const, whereas the isotachs (λ = const) and, consequently, the isobars (p/p_0 = const) by the condition $C_1 + C_2$ = const.

---

[31] Prandtl, L. and Busemann, A., Näherungsverfahren zur zeichnerischen Ermittlung von ebenen Stromungen mit Überschallgeschwindigkeit-Zürich: Stodola Festschrift, 1929. The details of the method and numerous illustrations of its application can be found in the survey taken by Busemann and published in *Handbuch der Experimental Physik*, Vol. 4, Part 1, Leipzig, 1931, Section 26, p. 421.

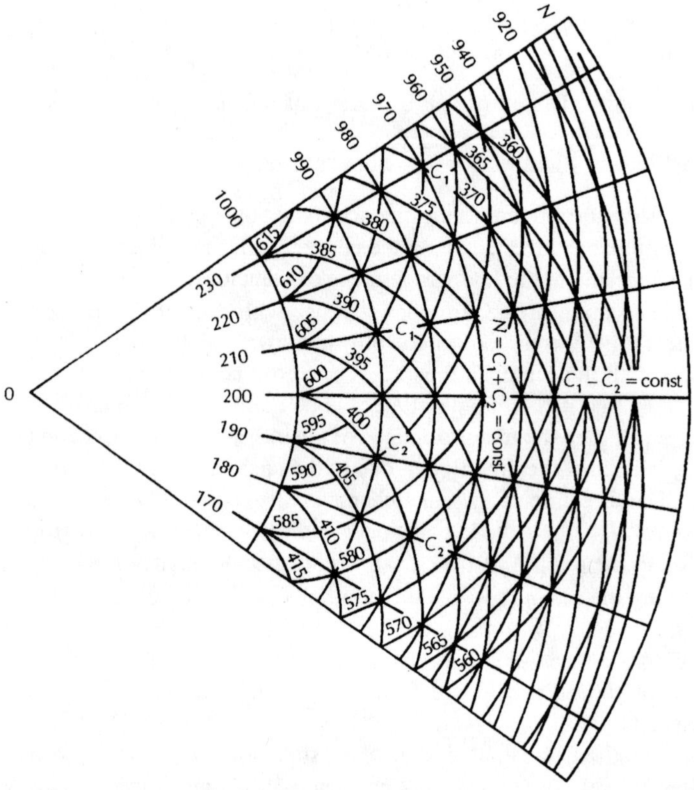

Fig. 114

Figure 114 the generally adopted marking of the family of epicycloids.

The net of epicycloids applied in graphic operations is finer than that corresponding to the conventional range of variation of C_1 and C_2 equal to unity. The family of the epicycloids C_1 have figures from 560 to 615 and the epicycloids C_2 have figures from 360 to 415. Instead of the marking of the angles θ indicated in Figure 112, in Figure 114 the marking of the radii is made by the differences between the constants $C_1 - C_2$ from 165 to 235 (for example, mark 210 stands on the intersection of epicycloids 605 and 395). The circles concentric with the basic ($\lambda = 1$ and $\lambda = \sqrt{\frac{k+1}{k-1}}$) are marked by the sums of the constants $N = C_1 + C_2$. According to Equation 151, this sum is a characteristic number adopted in the present marking which determines the dimensionless values: λ or **M**, pressure p/p_0, temperature T/T_0, density ρ/ρ_0.

Detailed tables of these quantities exist which were calculated for air depending on the number $N = C_1 + C_2$. Table 10 is such an example.[32] The rest of the characteristic quantities can be determined from the familiar isentropic relations or tables.

Specifying the flow velocities on a certain initial curve which does not coincide with the characteristics of the physical plane, for example, at the entrance into the nozzle being calculated and using the Busemann ellipse, we shall draw in the physical plane the characteristic directions at sufficiently close points of the initial curve. The points of their intersections

---

[32] See the survey by Busemann cited earlier and also the book by A. Ferri, *Aerodinamika sverkhzvukovykh techenii* (*The Aerodynamics of Supersonic Flows*), Gostekhizdat Press, Moscow, 1953, p. 424 (Table 2).

Table 10

N	λ	M	p/p_0
1000	1.000	1.000	0.527
999	1.073	1.090	0.476
998	1.110	1.142	0.449
997	1.141	1.186	0.424
996	1.172	1.228	0.402
995	1.200	1.265	0.382
994	1.227	1.305	0.363
993	1.253	1.342	0.345
992	1.278	1.376	0.329
991	1.300	1.413	0.313
990	1.322	1.443	0.298
989	1.343	1.474	0.284
988	1.365	1.506	0.270
987	1.387	1.542	0.257
986	1.409	1.575	0.245
985	1.426	1.608	0.233
984	1.447	1.643	0.221
983	1.466	1.680	0.210
982	1.486	1.718	0.200
981	1.503	1.750	0.190
980	1.520	1.780	0.180
979	1.539	1.815	0.171
978	1.556	1.850	0.162
977	1.575	1.885	0.153
976	1.590	1.923	0.145
975	1.608	1.958	0.137
974	1.625	1.995	0.130
973	1.640	2.028	0.123
972	1.656	2.065	0.116
971	1.671	2.101	0.109
970	1.686	2.138	0.103
969	1.700	2.178	0.097
968	1.718	2.215	0.091
967	1.732	2.258	0.086
966	1.748	2.298	0.081
965	1.763	2.338	0.076
964	1.776	2.378	0.071
963	1.791	2.421	0.067
962	1.805	2.460	0.062
961	1.819	2.506	0.057
960	1.832	2.548	0.055
959	1.845	2.592	0.051
958	1.858	2.636	0.048
957	1.872	2.680	0.044
956	1.884	2.730	0.041
955	1.898	2.778	0.039
954	1.910	2.825	0.036
953	1.923	2.875	0.033
952	1.936	2.920	0.031
951	1.948	2.978	0.029
950	1.960	3.028	0.027
949	1.972	3.078	0.025
948	1.984	3.135	0.023
947	1.995	3.180	0.021

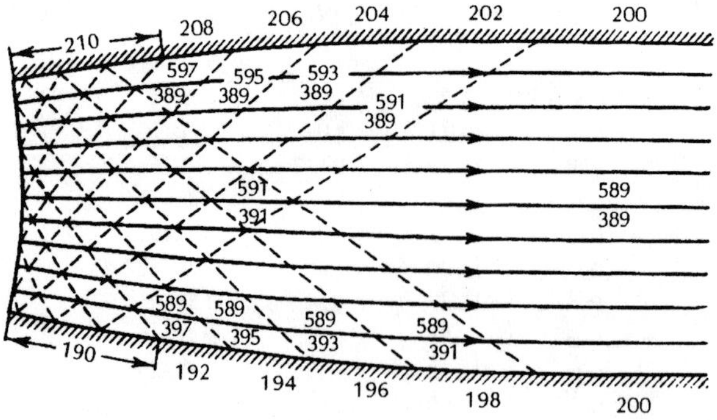

Fig. 115

determine the new curve split into small intervals. The net of characteristics in the hodograph plane will make it possible to find epicycloids conjugated with the characteristics in the physical plane and thus to obtain the numbers C_1, C_2 and their sum. Using Table 10 and the latter numbers, we shall find the thermodynamic quantities at the points of the new curve, which again can be adopted as the initial one and so on.

Thus, a portion of the physical plane in which the supersonic gas flow takes place will be filled with the characteristics that split it into small "rhombs" along the diagonals of which the sought-after segments of streamlines will be directed. Inside each rhomb the values of the marks for the epicycloids are placed one above the other (the characteristics in the hodograph plane) relating to the given rhomb of the physical plane. The upper number (greater) relates to the characteristic running from below upward, the lower (smaller) to the characteristic running from above downward. The presence of solid boundaries or free surfaces in the flow slightly complicates the graphic construction. The process of mapping is reduced to the transition from one family of characteristics in the physical plane to the other. With the aid of the available tables the sums of indices are easily converted into the mean values of thermodynamic parameters relating to the small region of the given rhomb. An example of the graphic construction of the nozzle is shown in Figure 115; the flow represents the transition from the flow expanding radially with the full vertex angle of 20° to the plane-parallel flow. The corner spacing of the net is equal to 2° (see the upper and lower figures).

In the presence of flow convergences that cause the occurrence of oblique shocks, it is convenient in graphic construction to use the diagram of shock polars (strophoids) (see Figure 97).

In addition to this general graphic method, different approximate analytical methods exist, especially for such an important case as the plane irrotational gas motion at very great Mach numbers. The Newton method was mentionin an earlier chapter which, despite its approximate nature, allows one, in some cases, to obtain a satisfactory accuracy when determining overall dynamic characteristics.

The problem of approximate methods for calculating supersonic flows past thin bodies will be dealt with at the end of Chapter 9 in a consideration of three-dimensional supersonic flows.

9

Three-Dimensional Irrotational Motion of Liquids and Gases

1. Velocity Potentials of the Simplest Three-Dimensional Flows

Based on the general considerations given in the beginning of Chapter 7, the problem of flow around a body can be substantially simplified by having assumed the irrotational nature of the flow. With this assumption, for the entire region of motion we have

$$\text{rot}\, V = 0$$

and, consequently, the velocity vector has the potential φ which is governed by the equality $V = \text{grad}\,\varphi$. In the rectangular Cartesian system

$$u = \frac{\partial\varphi}{\partial x} \qquad v = \frac{\partial\varphi}{\partial y} \qquad w = \frac{\partial\varphi}{\partial z}$$

In projections on the axes of curvilinear coordinates (see Equation 84 of Chapter 1)

$$V_{q_i} = (\text{grad}\,\varphi)_{q_i} = \frac{1}{H_i}\frac{\partial\varphi}{\partial q_i} \qquad (i = 1, 2, 3) \tag{1}$$

In cylindrical and spherical curvilinear coordinates, according to Equations 85 and 86 of Chapter 1, we have cylindrical coordinates

$$V_r = \frac{\partial\varphi}{\partial r} \qquad V_\varepsilon = \frac{1}{r}\frac{\partial\varphi}{\partial\varepsilon} \qquad V_z = \frac{\partial\varphi}{\partial z} \tag{2}$$

spherical coordinates

$$V_R = \frac{\partial\varphi}{\partial R} \qquad V_\theta = \frac{1}{R}\frac{\partial\varphi}{\partial\theta} \qquad V_\varepsilon = \frac{1}{R\sin\theta}\frac{\partial\varphi}{\partial\varepsilon} \tag{3}$$

Moreover, assuming the liquid to be incompressible, we establish the equality

$$\operatorname{div} V = \operatorname{div} \operatorname{grad} \varphi = \nabla^2 \varphi = 0 \tag{4}$$

which is the well-known Laplace equation. The unknown velocity potential φ is the solution for the Laplace equation which satisfies certain boundary conditions. Let us consider the problem of an external infinite flow around a fixed solid body with the surface σ and with the external unit normal $\boldsymbol{n}$; the flow at infinity is considered uniform and has the velocity $\boldsymbol{V}_\infty$. Then, the boundary conditions will be the condition of the body surface impermeability

$$V_n = \operatorname{grad}_n \varphi = \frac{\partial \varphi}{\partial n} = 0 \quad \text{on the surface } \sigma$$

condition at infinity

$$\boldsymbol{V} = \operatorname{grad} \varphi = \boldsymbol{V}_\infty \quad \text{when} \quad r \to \infty$$

where r is the radius-vector of the points of the flow region with respect to the coordinate origin located near the body considered.

As is proven in the theory of the potential, under very wide assumptions about the form of the surface σ and under the boundary conditions (see above) the Laplace equation has the single solution (the Neumann problem). We shall not consider the general theory, but some simplest specific hydrodynamic problems and thus more general three-dimensional flows.

Let us begin, just as in the case of two-dimensional motion, by finding the potentials of simple motions.

1. The infinite uniform rectilinear flow having the prescribed velocity $\boldsymbol{V}_\infty$ with the projections u_∞, v_∞, w_∞, satisfies the obvious system of equations

$$\frac{\partial \varphi}{\partial x} = u_\infty = \text{const} \qquad \frac{\partial \varphi}{\partial y} = v_\infty = \text{const} \qquad \frac{\partial \varphi}{\partial z} = w_\infty = \text{const}$$

Consequently, the velocity potential in this case is equal to

$$\varphi = u_\infty x + v_\infty y + w_\infty z = V_\infty (x \cos \alpha + y \cos \beta + z \cos \gamma) \tag{5}$$

where α, β, and γ are the angles of the prescribed flow direction with the coordinate axes $0x$, $0y$, and $0z$.

2. The flow from a source (sink) of strength Q placed at the coordinate origin 0 in an infinite fluid is symmetrical and gives the velocity field corresponding to the obvious condition of flow rate conservation

$$V \cdot 4\pi R^2 = Q$$

where R is the distance of the flow point from the source; thus,

$$V = \frac{Q}{4\pi R^2}$$

Noting that in the spherical coordinate system

$$V_R = \frac{\partial \varphi}{\partial R} = V = \frac{Q}{4\pi R^2} \qquad V_\theta = \frac{1}{R}\frac{\partial \varphi}{\partial \theta} = 0 \qquad V_\varepsilon = \frac{1}{R \sin \theta}\frac{\partial \varphi}{\partial \varepsilon} = 0$$

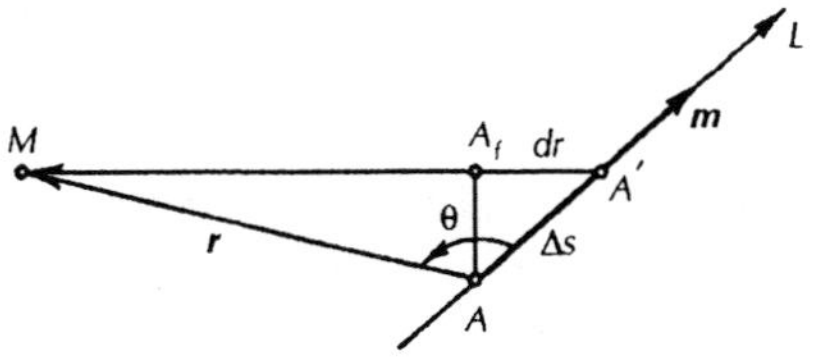

Fig. 116

we find the sought after velocity potential

$$\varphi = -\frac{Q}{4\pi R} \tag{6}$$

in the case of the source $Q > 0$, in the case of the sink $Q < 0$. It is not difficult to recognize that Equation 6 is the simplest case of the Newtonian potential from the theory of attraction, electrostatics, etc.

3. The flow due to a dipole in an infinite fluid can be obtained by employing the technique of the superposition of flows. First, determine the velocity potential of the field created by the set of the source and sink with the strength $\pm Q$ equal in absolute value. Let us place the sink (Figure 116) at the point A of the straight line AL and the source at the point A' located from the point A at the distance $AA' = \underset{\to}{\Delta s}$. Let us determine the velocity potential φ at a certain point M with the vector-radius $\overrightarrow{AM} = \mathbf{r}$, which forms the angle θ with the direction of the straight line AL; we find $(r' = A'M)$

$$\theta = -\frac{Q}{4\pi r'} + \frac{Q}{4\pi r}$$

Now, assume that the source comes closer to the sink and the strength increases to infinity and that the following equality is fulfilled

$$\lim_{\substack{A' \to A \\ Q \to \infty}} Q \cdot AA' = m$$

Then, rewriting the velocity potential φ in the form

$$\varphi = -\frac{1}{4\pi} Q \cdot AA' \cdot \frac{1/r' - 1/r}{AA'}$$

and proceeding to the limit, we obtain

$$\varphi = -\frac{m}{4\pi} \frac{d}{ds} \frac{1}{r} \tag{7}$$

or, calculating the derivative and noting that according to Figure 116

$$\frac{d}{ds}\left(\frac{1}{r}\right) = -\frac{1}{r^2}\frac{dr}{ds} = \frac{\cos\theta}{r^2}$$

we obtain

$$\varphi = -\frac{m\cos\theta}{4\pi r^2} \tag{8}$$

The resulting limiting flow with the velocity potential φ, determined by Equations 7 or 8, is called the flow due to a dipole at the point A with the axis AL and moment m. Sometimes the dipole moment is considered to be the vector $\boldsymbol{m}$, which has the magnitude m and direction along the dipole axis AL. The dipole potential can be represented in this case with the aid of the scalar product of the moment by the vector-radius as:

$$\varphi = -\frac{\boldsymbol{m} \cdot \boldsymbol{r}}{4\pi r^3}$$

4. The continuous distribution of sources in an infinite fluid. Let sources (sinks) inside a certain volume τ be continuously distributed so that the unit volume has the strength q. The quantity q, which is the function of the position of points in the volume τ, plays the part of volumetric density of the distribution of sources ($q > 0$) or sinks ($q < 0$). To the volume element $d\tau$ at a certain point A of the volume τ corresponds the source of strength $qd\tau$, and the velocity potential of this source element at any point M of the space filled with the fluid both inside and outside the volume τ is equal to

$$d\varphi = -\frac{qd\tau}{4\pi r}$$

where r is the length of the vector-radius $\overrightarrow{AM} = \boldsymbol{r}$ which connects the elementary source at the point A with the representative point of the space M. Using the technique of the superposition of flows, determine the velocity potential at the point M from the sources continuously distributed in the volume τ:

$$\varphi = -\frac{1}{4\pi}\int_{\tau} \frac{qd\tau}{r} \tag{9}$$

It should be emphasized that integration is made over all the volume elements that form the volume τ, i.e., over the variable coordinates of the point A, whereas the point M, at which the velocity potential is determined, is fixed. By denoting by a, b, c the Cartesian coordinates of the point A and by x, y, z the coordinates of the point M, formula (9) can be explicitly rewritten as

$$\varphi(x, y, z) = -\frac{1}{4\pi}\iiint_{(\tau)} \frac{q(a, b, c)\, dadbdc}{\sqrt{(x-a)^2 + (y-b)^2 + (z-c)^2}} \tag{10}$$

If the region of fluid flow is infinite, the function φ will tend to zero when the point M recedes to infinity. Let us designate by R the mean distance between the point M and the particles of the finite volume τ. Then, with the point M at a sufficiently far distance, it is possible to say that the velocity potential φ will tend to zero, as does $1/R$ for $R \to \infty$; i.e., the function φ has the order of the small quantity $1/R$:

$$\varphi = O(1/R)$$

The velocity potential (9) coincides in form with the general expression for the Newtonian potential. If q is understood to be the density of mass distribution in the volume τ, then expression (9) will give the potential of the forces of gravitation of the unit mass at the point M to a certain, in the general case inhomogeneous, mass confined in the volume τ; if q is understood to be the density of distribution of electrical charges, then φ will be the potential of the electrostatic field.

Recalling the definition of the velocity vector divergence as the fluid flow rate from continuously distributed sources per unit volume, we may write the following expression for any point of the volume

$$\operatorname{div} \boldsymbol{V} = q$$

where, using the equalities $\boldsymbol{V} = \operatorname{grad} \varphi$, $\operatorname{div} \boldsymbol{V} = \nabla^2\varphi$, we obtain the following equation of Poisson

$$\nabla^2\varphi = q \tag{11}$$

From this it follows that the function φ defined by Equation 9 in a certain infinite region which confines the finite volume τ filled up with sources, is the solution of Poisson equation (11) inside the volume; in the remaining region, where $q = 0$, the function φ represents the solution of the Laplace equation

$$\nabla^2\varphi = 0$$

where $\varphi = O(1/R)$.

It is proved in the theory of the potential that Newtonian potential (9) gives the single, finite, continuous, and single-value solution to Poisson equation (11), which transforms into zero at infinity.

It should, of course, be kept in mind that the above solution is only the simplest specific solution to the Poisson equation corresponding to the infinite region and not obeying the boundary conditions that occur in the problems concerning the determination of the potential in finite regions.

Along with the volumetric distribution of sources the field of hydrodynamics, just as other branches of physics, also considers the surface and linear distributions of sources. Retaining the same notation q for the surface and linear density of the distribution of sources, we have the corresponding velocity potentials in the form of the surface and linear integrals

$$\varphi = -\frac{1}{4\pi}\int_\sigma \frac{q\,d\sigma}{r} \qquad \varphi = -\frac{1}{4\pi}\int_L \frac{q\,ds}{r} \tag{12}$$

The first of these expressions, which represents the velocity potential of the continuous distribution of sources over a certain surface σ, yields the hydrodynamic interpretation of the simple layer potential known in the theory of gravity and electrostatic attraction. The simple layer potential is the solution of the Laplace equation and, as proved in the theory of the potential, the simple layer potential is finite and continuous throughout the entire region, including the surface σ.[1] The derivative of the simple layer potential along the normal to the surface σ undergoes discontinuity – finite shock – when representative point M passes through the surface σ.

Just as the velocity potentials of continuous distributions of sources were considered, it is possible to introduce analogous concepts also for the continuous distribution of dipoles. Let us consider in detail one of the most important distribution of dipoles which forms the so-called double layer. Take a certain surface and cover it with continuously distributed dipoles so that their moments (or axes) may coincide in direction with the external normals $\boldsymbol{n}$ to the surface σ. Having designated the density of the distribution of dipoles by m, we obtain the dipole

[1] At the points of the surface σ the simple layer potential is expressed, according to Equation 12, in terms of the improper integral which is taken in the sense of its principal value.

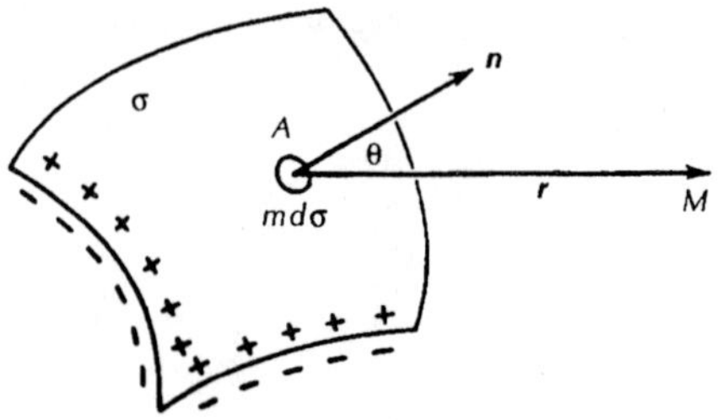

Fig. 117

moment vector per area element $d\sigma$ with the unit external normal $\boldsymbol{n}$ in the form of $md\sigma\boldsymbol{n}$, and the elementary velocity potential $d\varphi$, according to Equations 7 or 8, will be equal to

$$d\varphi = -\frac{1}{4\pi} m \frac{\partial}{\partial n}\left(\frac{1}{r}\right) d\sigma = -\frac{1}{4\pi}\frac{m \cos\theta}{r^2} d\sigma$$

where θ (Figure 117) is the angle between the external normal to the surface σ and the vector-radius $\boldsymbol{r} = \overrightarrow{AM}$ of the representative point M with respect to the point A taken on the surface.

The total velocity potential due to the surface σ covered with dipoles

$$\varphi = -\frac{1}{4\pi}\int_{\sigma} m \frac{\partial}{\partial n}\left(\frac{1}{r}\right) d\sigma = -\frac{1}{4\pi}\int_{\sigma} \frac{m\cos\theta}{r^2} d\sigma \tag{13}$$

serves as the hydrodynamic analogy of the double layer potential known in the theory of electricity and magnetism. While the simple layer potential represents, for example, the electrostatic potential of the charged surface, the double layer potential gives the magnetic potential of the magnetized surface (magnetic shell).

It should be mentioned that the double layer potential (13) is also the solution of the Laplace equation, but, in contrast to the simple layer, the double layer potential undergoes breaks of continuity when the representative point M passes through the surface σ.

Combining the potentials of the simple and double layers, it is possible to solve different problems of flows past bodies.

2. Velocity Field Induced by the Given System of Vortices in an Infinite Fluid. Biot-Savart Formula

Along with the basic aspects of the velocity fields, sources, sinks, and dipoles, we also consider the vortex tubes and lines. Assume that in a certain volume τ (finite or infinite, as for example, in the case of an infinitely long vortex tube) a continuous distribution of the vorticity $\boldsymbol{\Omega}$ is given and the distribution of velocities must be found throughout the entire flow region. The simplest problem of this kind is the determination of the velocity field in an infinite region from the given field of vortices. In this case the problem is reduced to the composition of such a solution, in $\boldsymbol{V}$, of the equation

$$\text{rot}\, \boldsymbol{V} = \boldsymbol{\Omega} \tag{14}$$

which would tend to zero in receding to infinity from the region occupied by the vortices.

Let us introduce into consideration the vector potential $\boldsymbol{A}$ connected with the velocity vector $\boldsymbol{V}$ by the relation

$$\boldsymbol{V} = \operatorname{rot} \boldsymbol{A} \tag{15}$$

and subject the vector potential to the additional condition

$$\operatorname{div} \boldsymbol{A} = 0$$

Then, Equation 14, if we recall Equation 88, Chapter 1,

$$\operatorname{rot} \operatorname{rot} \boldsymbol{A} = \operatorname{grad} \operatorname{div} \boldsymbol{A} - \nabla^2 \boldsymbol{A}$$

will convert to

$$\nabla^2 \boldsymbol{A} = -\boldsymbol{\Omega} \tag{16}$$

Considering this equation as the vector analog of Poisson equation (11), we may set up the solution of Equation 16 in the form of the generalization of Newtonian potential (9)

$$\boldsymbol{A} = \frac{1}{4\pi} \int_{\tau} \frac{\boldsymbol{\Omega}\, d\tau}{r} \tag{17}$$

where r is the radius-vector of the representative point of the field M with respect to the volume element $d\tau$.

According to Equation 15, we obtain the sought-after value of the velocity vector

$$\boldsymbol{V} = \frac{1}{4\pi} \operatorname{rot} \int_{\tau} \frac{\boldsymbol{\Omega}\, d\tau}{r} \tag{18}$$

Let us consider in more detail the case of the elementary vortex tube (with the finite circulation Γ) which encircles the vortex filament L in the field of the vector $\operatorname{rot} \boldsymbol{V} = \boldsymbol{\Omega}$ (Figure 118). Let us denote by $d\boldsymbol{r}$ the element of the filament which is oriented in the same direction as $\boldsymbol{\Omega}$; then, using the well-known Stokes' theorem of the connection between the

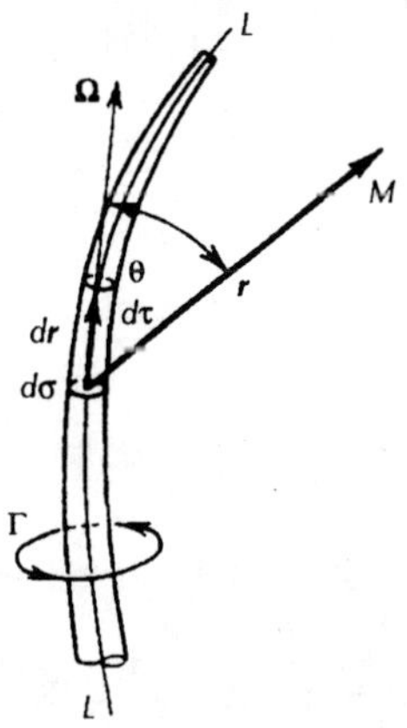

Fig. 118

vortex tube intensity and velocity circulation around the contour encircling the tube, we replace, under the integral sign of Equation 18,

$$\mathbf{\Omega} d\tau = \mathbf{\Omega} d\sigma ds = \mathbf{\Omega} d\sigma d\mathbf{r} = \Gamma d\mathbf{r}$$

and, instead of Equation 18, we obtain

$$\mathbf{V} = \frac{\Gamma}{4\pi} \operatorname{rot} \int_L \frac{1}{r} d\mathbf{r} = \frac{\Gamma}{4\pi} \int_L \operatorname{rot}\left(\frac{1}{r} d\mathbf{r}\right)$$

Using the third formula in Equation 88, Chapter 1:

$$\operatorname{rot}\left(\frac{1}{r} d\mathbf{r}\right) = \frac{1}{r} \operatorname{rot}(d\mathbf{r}) + \operatorname{grad}\left(\frac{1}{r}\right) \times d\mathbf{r}$$

and noting that $d\mathbf{r}$ is the potential vector so that $\operatorname{rot}(d\mathbf{r}) \equiv 0$, we can rewrite the previous expression for $\mathbf{V}$ as

$$\mathbf{V} = \frac{\Gamma}{4\pi} \int_L \operatorname{grad}\left(\frac{1}{r}\right) \times d\mathbf{r} \tag{19}$$

The resulting solution of the problem of the velocity field around the given vortex filament L with the circulation Γ can be simplified further by directly computing the gradient under the integral sign

$$\operatorname{grad}\left(\frac{1}{r}\right) = -\frac{1}{r^2} \operatorname{grad} r = -\frac{1}{r^2} \frac{\mathbf{r}}{r} = -\frac{\mathbf{r}}{r^3}$$

This leads to the hydrodynamic analog of the Biot-Savart formula well known in the theory of electromagnetism

$$\mathbf{V} = \frac{\Gamma}{4\pi} \int_L \frac{d\mathbf{r} \times \mathbf{r}}{r^3} \tag{20}$$

By considering the elementary velocity $d\mathbf{V}$, formed (induced) at the point M by the vortex filament element $d\mathbf{r}$, it is possible (instead of Equation 20) to write

$$d\mathbf{V} = \frac{\Gamma}{4\pi} \frac{d\mathbf{r} \times \mathbf{r}}{r^3}$$

or, going over to the value of the elementary velocity

$$|d\mathbf{V}| = \frac{\Gamma}{4\pi} \frac{|d\mathbf{r} \times \mathbf{r}|}{r^3} = \frac{\Gamma}{4\pi} \frac{ds \sin\theta}{r^2} \tag{21}$$

The analogous Biot-Savart formula is used for determining the magnetic field from the electric current element.

To illustrate the application of formula (20), let us determine the velocity induced at the points of the space by the rectilinear segment AB of the vortex filament with the circulation Γ (Figure 119).

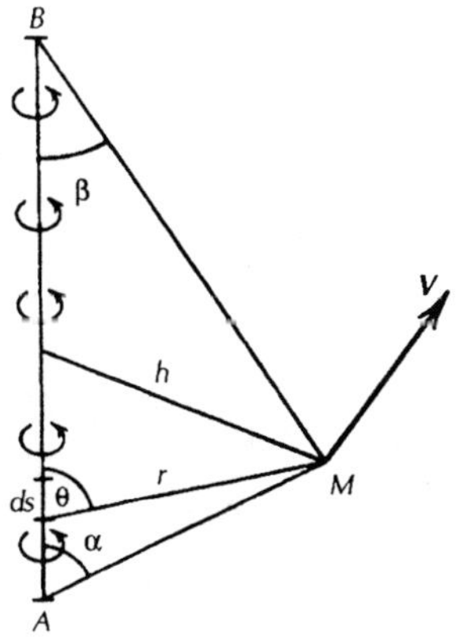

Fig. 119

Noting that all the elements of the rectilinear vortex will give identically directed elementary velocities dV at the given point M (along the perpendicular to the plane drawn through the segment AB and the point M to the side of rotation produced by the vortex), and using the obvious equalities (h is the shortest distance of the point M from the segment AB)

$$h = r\sin\theta \qquad ds = -d(h\operatorname{ctg}\theta) = h\frac{d\theta}{\sin^2\theta}$$

we find from Equation 21 the expression for $|dV|$

$$|dV| = \frac{\Gamma}{4\pi}\,\frac{\sin\theta\sin^2\theta}{h^2}\,\frac{h\,d\theta}{\sin^2\theta} = \frac{\Gamma}{4\pi h}\sin\theta\,d\theta$$

The integration over θ from $\theta = \alpha$ to $\theta = \pi - \beta$ gives the sought after value of the velocity V induced by the vortex segment AB:

$$V = \frac{\Gamma}{4\pi h}\int_{\alpha}^{\pi-\beta}\sin\theta\,d\theta = \frac{\Gamma}{4\pi h}(\cos\alpha + \cos\beta) \tag{22}$$

Formula (22) plays the major part in the calculations of the velocity field around vortex lines and will be used in what follows in the theory of the wing of finite span.

Assuming in formula (22) that $\alpha = \beta = 0$, we again obtain the formula, known from the theory of plane motion, for the velocity induced by an infinitely long rectilinear vortex filament

$$V = \frac{\Gamma}{2\pi h}$$

3. Potential of the Velocity Field Induced by a Closed Vortex Filament

In the case of a closed vortex filament, it is easy to obtain a simple expression for the velocity potential. Consider a closed vortex filament L of strength Γ in the vector field $\operatorname{rot}\boldsymbol{a}$ (Figure 120) around a certain nonclosed surface σ and compose the integral

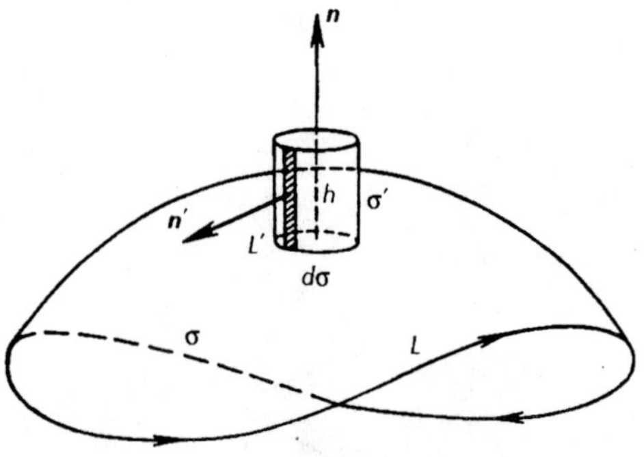

Fig. 120

$$\int_L \boldsymbol{a} \times d\boldsymbol{r}$$

Construct an elementary cylinder with the generatrices parallel to the unit vector of the normal $\boldsymbol{n}$ to the surface σ and with the directing line L' which confines the elementary area $d\sigma$; then

$$\int_{L'} \boldsymbol{a} \times d\boldsymbol{r}' = \frac{1}{h} \int_{\sigma'} \boldsymbol{a} \times (\boldsymbol{n} \times \boldsymbol{n}')\, d\sigma'$$

where σ' is the total surface of the elementary cylinder consisting of the side surface and two bases $d\sigma$, and $d\boldsymbol{r}'$ and $d\sigma'$ are, respectively, the elements of the contour L' and of the surface σ' of the elementary cylinder ($d\sigma'$ in Figure 120 is represented by the hatched strip). Applying the formula of the triple vector product, we find

$$\int_{L'} \boldsymbol{a} \times d\boldsymbol{r}' = \frac{1}{h} \int_{\sigma'} \boldsymbol{n} a_{n'}\, d\sigma' - \frac{1}{h} \int_{\sigma'} \boldsymbol{n}' a_n d\sigma' = \boldsymbol{n} \frac{d\tau}{h} \operatorname{div} \boldsymbol{a} - \operatorname{grad}\left(a_n \frac{d\tau}{h} \right) =$$

$$= \boldsymbol{n} \operatorname{div} \boldsymbol{a} d\sigma - \operatorname{grad}(a_n d\sigma)$$

By summing both parts of the above equation over all the elementary contours L' on the left and over all the elementary areas $d\sigma$ on the right, we obtain

$$\int_L \boldsymbol{a} \times d\boldsymbol{r} = \int_\sigma \boldsymbol{n} \operatorname{div} \boldsymbol{a} d\sigma - \operatorname{grad} \int_\sigma a_n d\sigma \tag{23}$$

Let us assume in this formula that $\boldsymbol{a} = \operatorname{grad}(1/r)$; then, instead of Equation 19,

$$\boldsymbol{V} = \frac{\Gamma}{4\pi} \int_\sigma \boldsymbol{n} \nabla^2 \left(\frac{1}{r} \right) d\sigma - \frac{\Gamma}{4\pi} \operatorname{grad} \int_\sigma \frac{\partial}{\partial n} \left(\frac{1}{r} \right) d\sigma$$

However, again, the function $1/r$ satisfies the Laplace equation

$$\nabla^2 \left(\frac{1}{r} \right) = 0$$

so that finally

$$V = -\frac{\Gamma}{4\pi}\operatorname{grad}\int_{\sigma}\frac{\partial}{\partial n}\left(\frac{1}{r}\right)d\sigma \tag{24}$$

Consequently, the sought-after potential of the velocity field of the closed vortex filament is equal to

$$\varphi = -\frac{\Gamma}{4\pi}\int_{\sigma}\frac{\partial}{\partial n}\left(\frac{1}{r}\right)d\sigma \tag{25}$$

Comparing Equation 25 with the expression of the double layer potential (13), we conclude that the velocity potential of the closed vortex filament L with the circulation Γ coincides with the double layer potential of dipoles that are located over the surface σ resting on the contour L, and which have the moment distribution density identical over the entire surface and equal to the circulation of the vortex filament.

The hydrodynamic theorem above represents the analogue of the well-known electrodynamics theorem about the equivalency of the circular electric current to the magnetic shell field.

4. Stream Function in Three-Dimensional Motions

In three-dimensional motions one cannot introduce the stream function in the general case of motion as was done when studying two-dimensional motions; the stream function exists only in separate specific cases; some are considered below.

According to the second equation of system (84) of Chapter 1, the incompressibility equation (4) has the form

$$\frac{\partial}{\partial q_1}\left(H_2H_3V_{q_1}\right)+\frac{\partial}{\partial q_2}\left(H_3H_1V_{q_2}\right)+\frac{\partial}{\partial q_3}\left(H_1H_2V_{q_3}\right) = 0$$

Let us suppose that one of the velocity components, for example, V_{q3}, is everywhere equal to zero or that it retains the value independent of q_3; in the latter case the Lamé coefficients H_1 and H_2 are also independent of q_3. Then, the previous equation is reduced to the simple

$$\frac{\partial}{\partial q_1}\left(H_2H_3V_{q_1}\right)+\frac{\partial}{\partial q_2}\left(H_3H_1V_{q_2}\right) = 0$$

and it is possible to state that the function ψ exists which satisfies the system of equations

$$H_2H_3V_{q_1} = \pm\frac{\partial\psi}{\partial q_2} \qquad H_3H_1V_{q_2} = \mp\frac{\partial\psi}{\partial q_1} \tag{26}$$

where it follows that

$$V_{q_1} = \pm\frac{1}{H_2H_3}\frac{\partial\psi}{\partial q_2} \qquad V_{q_2} = \mp\frac{1}{H_3H_1}\frac{\partial\psi}{\partial q_1} \tag{27}$$

By analogy with the case of the two-dimensional motion, function ψ will be called the stream function in curvilinear coordinates. The choice of the superscripts or subscripts is arbitrary and is governed by additional considerations.

Let us emphasize that the presence of the stream function depends not only on the character of motion, but also on the choice of the curvilinear coordinate system with the aid of which the motion is described.

Let us consider, for example, the axisymmetric (with respect to the axis $0z$) motion of incompressible fluid in meridian planes passing through the axis $0z$. In such motion all three Cartesian velocity projections u, v, and w exist and all depend on the three coordinates x, y, z, so that from the incompressibility equation $\frac{\partial u}{\partial x} + \frac{\partial v}{\partial y} + \frac{\partial w}{\partial z} = 0$, formed in the Cartesian coordinates, it is possible to postulate the absence of the stream function.[2] At the same time, when using the cylindrical system of coordinates r, ε, z with the meridian nature of motion ($V_\varepsilon = 0$), the incompressibility equation has the form

$$\frac{\partial(rV_r)}{\partial r} + \frac{\partial(rV_z)}{\partial z} = 0 \tag{28}$$

and it makes it possible to find the stream function $\psi(r, z)$ connected with the velocity projections on the axis of the cylindrical coordinates by the relations (the choice of the signs will be explained)

$$rV_r = -\frac{\partial\psi}{\partial z} \qquad rV_z = -\frac{\partial\psi}{\partial r}$$

where

$$V_r = -\frac{1}{r}\frac{\partial\psi}{\partial z} \qquad V_z = \frac{1}{r}\frac{\partial\psi}{\partial r} \tag{29}$$

Similarly, in the spherical system of coordinates (R, θ, ε) at $V_\varepsilon = 0$ the incompressibility equation has the form

$$\frac{\partial(R^2 V_R \sin\theta)}{\partial R} + \frac{\partial(RV_\theta \sin\theta)}{\partial\theta} = 0 \tag{30}$$

and the velocity projections on the axis of the spherical coordinate system are expressed in terms of the corresponding stream function ψ as

$$V_R = \frac{1}{R^2\sin\theta}\frac{\partial\psi}{\partial\theta} \qquad V_\theta = -\frac{1}{R\sin\theta}\frac{\partial\psi}{\partial R} \tag{31}$$

The stream function introduced by Equations 27 has properties analogous to the stream function in a two-dimensional motion. Noting that in the orthogonal curvilinear coordinates the streamline equation at $dq_3 = 0$ has the form

---

[2]The dependence upon the three coordinate directions makes the stream function undefinable in the usual sense. However, in the flow is symmetrical about same axis, then a dimensional dependency is removed and a useful stream function can be defined. This is illustrated in subsequent paragraphs.

$$\frac{H_1 dq_1}{V_{q_1}} = \frac{H_2 dq_2}{V_{q_2}} \qquad V_{q_1} = H_1\frac{dq_1}{dt} \qquad V_{q_2} = H_2\frac{dq_2}{dt} \qquad V_{q_3} = H_3\frac{dq_3}{dt} = 0$$

we find from Equation 27 that

$$d\psi = \frac{\partial\psi}{\partial q_1}dq_1 + \frac{\partial\psi}{\partial q_2}dq_2 \qquad dq_3 = 0$$

Consequently, along the streamline

$$\psi = \text{const} \qquad q_3 = \text{const}$$

In the case of earlier considered axisymmetric motion of fluid over meridian planes ($\varepsilon = \text{const}$) the equations $\psi = \text{const}$ will represent the surfaces formed by the rotation of streamlines around the axis $0z$. The surfaces $\psi = \text{const}$ will be called the stream surfaces. In the specific case of axisymmetric motion considered above it is possible to assume that $\psi = 0$ on the axis $0z$; the values of ψ will be proportional to the volumetric liquid flow rates through the orthogonal (to the axis) cross-section of the stream tube confined by the given stream surface.

In fact, the volumetric flow rate through the orthogonal (to the axis $0z$) section confined by the circle of the given radius r will be equal, according to Equation 29, and condition $\psi(0) = 0$, to

$$Q = \int_0^r V_z \cdot 2\pi r dr = 2\pi\int_0^r \frac{1}{r}\frac{\partial\psi}{\partial r} r dr = 2\pi\psi(r)$$

It is clear that the choice of the sign on the right sides of Equation 27 is made so that $\psi > 0$ when $Q > 0$.

The stream function can be considered one of the components of the vector velocity potential $\boldsymbol{A}$ connected with the velocity vector by Equation 15. In fact, according to this equation and to Equations 84 of Chapter 1, we have

$$V_{q_1} = \text{rot}_{q_1}\boldsymbol{A} = \frac{1}{H_2H_3}\left[\frac{\partial(H_3A_{q_3})}{\partial q_2} - \frac{\partial(H_2A_{q_2})}{\partial q_3}\right]$$

$$V_{q_2} = \text{rot}_{q_2}\boldsymbol{A} = \frac{1}{H_3H_1}\left[\frac{\partial(H_1A_{q_1})}{\partial q_3} - \frac{\partial(H_3A_{q_3})}{\partial q_1}\right]$$

$$V_{q_3} = \text{rot}_{q_3}\boldsymbol{A} = \frac{1}{H_1H_2}\left[\frac{\partial(H_2A_{q_2})}{\partial q_1} - \frac{\partial(H_1A_{q_1})}{\partial q_2}\right]$$

By selecting the vector $\boldsymbol{A}$ to be orthogonal to the coordinate surfaces $q_3 = \text{const}$ throughout the entire space, we find

$$A_{q_1} = A_{q_2} = 0 \qquad V_{q_3} = 0 \qquad V_{q_1} = \frac{1}{H_2H_3}\frac{\partial(H_3A_{q_3})}{\partial q_2} \qquad V_{q_2} = -\frac{1}{H_3H_1}\frac{\partial(H_3A_{q_3})}{\partial q_1}$$

having assumed that $H_3A_{q_3} = \pm\psi(q_1, q_3)$ and that the Lamé coefficients and the quantity A_{q_3} are independent of q_3, we obtain Equations 27. Thus, for example, in the spherical or cylindrical coordinate systems the vector $\boldsymbol{A}$ should be directed along the tangent to parallel circles, corresponding to the change of ε alone, and should be independent of ε.

The following are examples of stream functions for the simplest motions:

1. The uniform rectilinear flow with the velocity $\boldsymbol{V}_\infty$ parallel to the axis $0z$. In the cylindrical coordinate system we have

$$V_r = 0 = -\frac{1}{r}\frac{\partial\psi}{\partial z} \qquad V_z = V_\infty = \frac{1}{r}\frac{\partial\psi}{\partial r}$$

consequently,

$$\psi = \frac{1}{2}V_\infty r^2 \tag{32}$$

In the spherical coordinate system

$$V_R = V_\infty \cos\theta = \frac{1}{R^2 \sin\theta}\frac{\partial\psi}{\partial\theta} \qquad V_\theta = -V_\infty \sin\theta = -\frac{1}{R\sin\theta}\frac{\partial\psi}{\partial R}$$

Integration of this system of equations in total differentials gives

$$\psi = \frac{1}{2}V_\infty R^2 \sin^2\theta \tag{33}$$

2. The source (sink). The expression for the stream function in the spherical coordinate system can be found by integrating the system of equations

$$V_R = \frac{Q}{4\pi R^2} = \frac{1}{R^2\sin\theta}\frac{\partial\psi}{\partial\theta} \qquad V_\theta = 0 = -\frac{1}{R\sin\theta}\frac{\partial\psi}{\partial R}$$

We obtain

$$\psi = -\frac{Q\cos\theta}{4\pi} + \text{const}$$

or, selecting the constant from the condition $\psi = 0$ at $\theta = 0$,

$$\psi = \frac{Q}{4\pi}(1 - \cos\theta) \tag{34}$$

3. The dipole. Using the expression for the velocity potential (8), we have the following system of equations in spherical coordinates

$$V_R = \frac{m\cos\theta}{2\pi R^3} = \frac{1}{R^2\sin\theta}\frac{\partial\psi}{\partial\theta} \qquad V_\theta = \frac{m\sin\theta}{4\pi R^3} = -\frac{1}{R\sin\theta}\frac{\partial\psi}{\partial R}$$

where

$$\frac{\partial\psi}{\partial\theta} = \frac{m}{2\pi R}\sin\theta\cos\theta \qquad \frac{\partial\psi}{\partial R} = -\frac{m}{4\pi R^2}\sin^2\theta$$

It is easy to find the integral of this system:

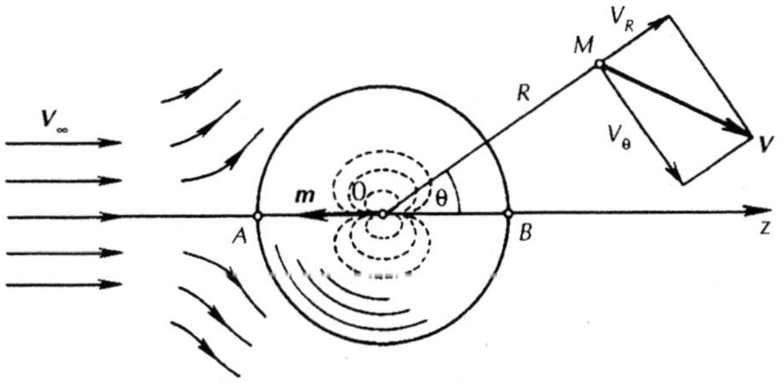

Fig. 121

$$\psi = \frac{m \sin^2\theta}{4\pi R} \tag{35}$$

which vanishes at $\theta = 0$.

5. Flow Past a Sphere. D'Alembert's Paradox

Just as in the case of a two-dimensional flow around a circular cylinder, it is possible to find the three-dimensional flow around a sphere by imposing a uniform flow parallel, for example, to the axis $0z$, with the velocity V_∞, onto the flow due to a dipole with the moment oriented along this axis but to the side opposite of the incident flow (Figure 121). Adding up the stream functions (33) and (35), we find the stream function of the combined flow.[3]

$$\psi = \frac{1}{2} V_\infty R^2 \sin^2\theta - \frac{m}{4\pi R} \sin^2\theta = \left(\frac{1}{2} V_\infty R^2 - \frac{m}{4\pi R}\right) \sin^2\theta \tag{36}$$

The equation of the zero stream surface

$$\psi = \left(\frac{1}{2} V_\infty R^2 - \frac{m}{4\pi R}\right) \sin^2\theta = 0$$

is divided into the equation for the sphere surface

$$R = \sqrt[3]{\frac{m}{2\pi V_\infty}} = a$$

where a is the radius of the sphere, and into the equation for the axis $0z$

$$\theta = 0, \pi$$

From this it follows that, wishing to find the stream function of the flow past a sphere of radius a in a stream with the velocity V_∞ at infinity directed along the axis $0z$, it is necessary to require in the expression for the stream function that

$$m = 2\pi a^3 V_\infty$$

[3] In conformity with the choice of the direction, for the moment, m in Equation 35 is replaced by $-m$, where $m > 0$.

then we obtain

$$\psi = \frac{1}{2} V_\infty R^2 \left[1 - \left(\frac{a}{R}\right)^3\right] \sin^2\theta \tag{37}$$

It is also not difficult to find the velocity potential by having integrated the system of equations connecting the potential φ with the stream function ψ, or more simply, by directly setting up the sum of the potentials of the flow (5) and (8) being added:

$$\varphi = V_\infty z + \frac{m \cos\theta}{4\pi R^2} = V_\infty R\left[1 + \frac{1}{2}\left(\frac{a}{R}\right)^3\right] \cos\theta \tag{38}$$

Let us study the flow obtained. First, we find the distribution of velocities

$$V_R = \frac{\partial\varphi}{\partial R} = V_\infty\left[1 - \left(\frac{a}{R}\right)^3\right] \cos\theta$$

$$V_\theta = \frac{1}{R}\frac{\partial\varphi}{\partial\theta} = -V_\infty\left[1 + \frac{1}{2}\left(\frac{a}{R}\right)^3\right] \sin\theta \tag{39}$$

It is directly seen that on the surface of the sphere ($R = a$) the basic boundary condition of the impermeability of the solid wall is fulfilled:

$$V_n = V_R = 0$$

and at infinity ($R \to \infty$)

$$V_R = V_\infty \cos\theta \qquad V_\theta = -V_\infty \sin\theta$$

i.e., the uniform flow velocity at infinity is equal in magnitude to V_∞ and is directed along the axis $0z$ to the positive side.

The velocity distribution over the sphere surface is characterized by the equation

$$V_\theta = -\frac{3}{2} V_\infty \sin\theta$$

The points A and B (Figure 121) are critical because the velocity vanishes at these points. The maximum velocity takes place in the middle plane at $\theta = \pi/2$; by absolute value it is equal to

$$|V_\theta|_{\max} = \frac{3}{2} V_\infty$$

Comparing this result with the case of flow past a circular cylinder (Chapter 7), we see that in the case of a three-dimensional flow past a sphere the maximum velocity on its surface attains only $3/2$ of the free stream velocity, whereas in the case of a two-dimensional flow past a circular cylinder the maximum velocity exceeds the incident flow velocity by a factor of 2 (i.e., the cylinder produces a much more considerable perturbation of the uniform flow than the sphere). This is natural because the cylinder section normal to the flow is infinite, whereas the section of the sphere is finite. It should be noted that (as in the case of a plane flow) in reality the maximum velocity does not attain such a great value. The sphere is a badly streamlined body. The flow of a real fluid is shed from the sphere surface, not attaining

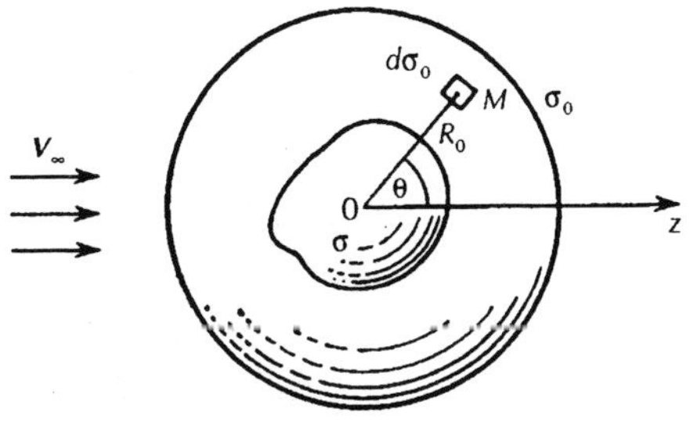

Fig. 122

even the middle plane under some conditions, and going somewhat farther than this plane under other conditions.

The pressure distribution over the sphere surface can be obtained from the Bernoulli equation

$$p + \rho\frac{V^2}{2} = p_\infty + \rho\frac{V_\infty^2}{2}$$

from which the expression for the pressure coefficient follows:

$$c_p = \frac{p - p_\infty}{\frac{1}{2}\rho V_\infty^2} = 1 - \left(\frac{V}{V_\infty}\right)^2 = 1 - \frac{9}{4}\sin^2\theta$$

As is directly seen from the latter formula, the resultant vector of the forces of the ideal liquid flow pressure on the sphere surface is equal to zero. The sphere does not offer resistance to the incident flow which is uniform at infinity. In other words, when the sphere uniformly moves in an ideal fluid, it does not experience resistance. This is the principle of the particular case of the well-known d'Alembert paradox, which was discussed in an earlier chapter. In the case of the sphere just considered this paradox follows from the considerations of symmetry in the distribution of pressure over the sphere surface; however, the paradox is also valid in the general case.

Let us prove the validity of d'Alembert's paradox for a three-dimensional irrotational flow past a body of finite size and arbitrary shape. For this, we first determine the order of the decrease in the velocities of a uniform flow disturbed by a certain body bounded by the closed surface σ (Figure 122) on moving away from this body.

Divide the potential φ of the flow past the body into the potential of the uniform flow with the velocity V_∞, parallel, for example, to the axis 0z, and into the perturbation velocity potential φ′ caused by the perturbing action of the body on the uniform flow into which it is placed.

Let us first show that on moving away to infinity ($R \to \infty$) the perturbation potential φ′ decreases as $1/R^2$. Using the flow around a sphere as an example, this decrease can be easily verified by turning to Equation 38 and separating in it the perturbation potential:

$$\varphi' = \frac{1}{2}V_\infty\frac{a^3}{R^2}\cos\theta = O\left(\frac{1}{R^2}\right) \qquad V_R', V_\theta' = O\left(\frac{1}{R^3}\right)$$

Remembering the general property of the solutions to the Laplace equation – that the perturbations of the flow decrease on moving away from the source of these perturbations – it may be assumed that at great mean distances from the body (which disturbs the flow) the details of its shape cannot influence the law of the decrease in the perturbation velocity potential. Thus, it is possible to conclude that for any body of finite dimensions the law of the decrease of φ' will be

$$\varphi' = O\left(\frac{1}{R^2}\right)$$

The same result can be attained on the basis of more rigorous considerations by noting that the perturbation potential φ' caused by an arbitrarily shaped body must satisfy the Laplace equation, which can be written in the spherical coordinates in the form (Equation 86, Chapter 1)

$$\frac{\partial}{\partial R}\left(R^2 \frac{\partial \varphi'}{\partial R}\right) + \frac{1}{\sin\theta}\frac{\partial}{\partial \theta}\left(\sin\theta \frac{\partial \varphi'}{\partial R}\right) + \frac{1}{\sin^2\theta}\frac{\partial^2 \varphi'}{\partial \varepsilon^2} = 0 \tag{40}$$

Wishing to find the general form of the solution of this equation, we assume

$$\varphi'(R, \theta, \varepsilon) = X(R)\, Y(\theta, \varepsilon) \tag{41}$$

Substituting this product into the previous equation,

$$Y\frac{d}{dR}\left(R^2 \frac{dX}{dR}\right) + \frac{X}{\sin\theta}\frac{\partial}{\partial \theta}\left(\sin\theta \frac{\partial Y}{\partial \theta}\right) + \frac{X}{\sin^2\theta}\frac{\partial^2 Y}{\partial \varepsilon^2} = 0$$

or, separating the function R from the rest of the variables,

$$\frac{1}{X}\frac{d}{dR}\left(R^2 \frac{dX}{dR}\right) = -\frac{1}{Y}\left[\frac{1}{\sin\theta}\frac{\partial}{\partial \theta}\left(\sin\theta \frac{\partial Y}{\partial \theta}\right) + \frac{1}{\sin^2\theta}\frac{\partial^2 Y}{\partial \varepsilon^2}\right] \tag{42}$$

On the left side is the function only of R and on the right side only of ε and θ. Because the variables R, θ, and ε are independent of each other, the previous equation yields

$$\frac{1}{X}\frac{d}{dR}\left(R^2 \frac{dX}{dR}\right) = \text{const}$$

It is easy to see that the number of the solutions $X(R)$ of this equation will include the integer positive or negative exponents of the variable R if only the arbitrary constant is taken to be equal to $n(n+1)$. Dealing only with the integer negative values of the numbers $n = -k$ $(k > 0)$, because the perturbation potential φ' should decrease with an increasing R, we obtain from Equation 41 the system of partial solutions of the Laplace equation (40) in the form

$$\frac{Y_k(\theta, \varepsilon)}{R^k} \qquad (k = 1, 2, 3, \cdots, \infty)$$

Among the functions $Y_k(\theta, \varepsilon)$ (the spherical functions), which according to Equation 42 and the condition of the choice of const $= k(k-1)$ satisfy the partial differential equation,

$$\frac{1}{\sin\theta}\frac{\partial}{\partial \theta}\left(\sin\theta \frac{\partial Y_k}{\partial \theta}\right) + \frac{1}{\sin^2\theta}\frac{\partial^2 Y_k}{\partial \varepsilon^2} + k(k-1)\,Y_k = 0$$

we shall select only those solutions which are limited at all values of θ and ε.

At $k = 1$, the solution of this equation, which is limited for all the values $0 \leq \theta \leq \pi$, will be Y_1 = const in conformity with the simplest particular solution const$/R$, which represents the familiar Newtonian potential of the single source (sink). When $k = 2$, the equation has the solution const $\cdot \cos\theta$, which leads to the velocity potential of the dipole.

Due to the linear character of the Laplace equation, the sought-after potential φ' can be represented as the sum of particular solutions:

$$\varphi' = \sum_{k=1}^{\infty} \frac{Y_k(\theta, \varepsilon)}{R^k} = \frac{C}{R} + \sum_{k=2}^{\infty} \frac{Y_k(\theta, \varepsilon)}{R^k} \tag{43}$$

Let us prove that the constant C is equal to zero. For this purpose, let us enclose the body into the sphere C_0 of greater radius R_0. Noting that there are no sources or sinks between the body surface σ and the sphere surface σ_0, we shall write the condition of the zero overall fluid flow rate through the surface σ_0

$$\int_{\sigma_0} V'_R d\sigma_0 = \int_{\sigma_0} \frac{\partial \varphi'}{\partial R} d\sigma_0 = -\frac{C}{R_0^2} \int_{\sigma_0} d\sigma_0 - \sum_{k=2}^{\infty} \frac{k}{R_0^{k+1}} \int_{\sigma_0} Y_k(\theta, \varepsilon)\, d\sigma_0 = 0$$

Using the equalities

$$d\sigma_0 = R_0^2 \sin\theta d\theta d\varepsilon \qquad \int_{\sigma_0} d\sigma_0 = 4\pi R_0^2$$

we obtain

$$-4\pi C - \sum_{k=2}^{\infty} \frac{k}{R_0^{k-1}} \int_0^{2\pi} d\varepsilon \int_0^{\pi} Y_k(\theta, \varepsilon) \sin\theta d\theta$$

where for $R_0 \to \infty$, with the functions $Y_k(\theta, \varepsilon)$ being limited, it follows that $C = 0$.

Thus, the general form of the perturbation velocity potential will be

$$\varphi' = \sum_{k=2}^{\infty} \frac{Y_k(\theta, \varepsilon)}{R^k}$$

consequently, at great values of R the potential φ' and, according to Equation 3, also the perturbation velocity V' will respectively have the orders

$$\varphi' = O\left(\frac{1}{R^2}\right) \qquad V' = O\left(\frac{1}{R^3}\right)$$

With this result, it is now not difficult to prove d'Alembert's paradox. We shall apply the momentum theorem in Euler's form to the fluid volume confined between the control surfaces σ and σ_0, assuming that there are no sources (sinks) between them ($\boldsymbol{F}$ is the resultant vector of forces of pressure on the body),

$$-\rho\int_{\sigma_0} V_{n_0} \boldsymbol{V} d\sigma_0 - \int_{\sigma_0} p\boldsymbol{n}_0 d\sigma_0 - \boldsymbol{F} = 0$$

because the momentum transfer through the solid body surface σ is equal to zero.

In the case of the irrotational character of flow in the considered region the Bernoulli theorem is valid

$$p = \text{const} - \frac{\rho V^2}{2}$$

Substituting this expression into the previous relation,

$$\boldsymbol{F} = -\rho\int_{\sigma_0} V_{n_0} \boldsymbol{V} d\sigma_0 + \frac{\rho}{2}\int_{\sigma_0} V^2 \boldsymbol{n}_0 d\sigma_0$$

since the surface integral $\int_{\sigma_0} \boldsymbol{n}_0 d\sigma_0$ is equal to zero.

Dividing, as before, the flow velocity into the basic incidence velocity $\boldsymbol{V}_\infty$ and the perturbation velocity $\boldsymbol{V}'$,

$$\boldsymbol{F} = -\rho \boldsymbol{V}_\infty \int_{\sigma_0} V_{n_0} d\sigma_0 - \rho\int_{\sigma_0} V_{n_0} \boldsymbol{V}' d\sigma_0 + \frac{\rho}{2}\int_{\sigma_0} |\boldsymbol{V}_\infty + \boldsymbol{V}'|^2 \boldsymbol{n}_0 d\sigma_0$$

noting that

$$\int_{\sigma_0} V_{n_0} d\sigma_0 = 0 \qquad \int_{\sigma_0} \boldsymbol{n}_0 d\sigma_0 = 0$$

we find

$$\boldsymbol{F} = -\rho\int_{\sigma_0} V_{n_0} \boldsymbol{V}' d\sigma_0 + \rho\int_{\sigma_0} (\boldsymbol{V}_\infty \cdot \boldsymbol{V}')\, \boldsymbol{n}_0 d\sigma_0 + \frac{\rho}{2}\int_{\sigma_0} V'^2 \boldsymbol{n}_0 d\sigma_0$$

However, according to what has just been proved, the perturbation velocity V' at great values of R_0 is on the order of $1/R_0^3$, whereas the integration element $d\sigma_0$ is on the order of R_0^2. Letting R_0 go to infinity, we see that the resultant vector $\boldsymbol{F}$ of the forces of flow pressure on the body tends to zero. $\boldsymbol{F}$ cannot depend on the arbitrary radius R_0 of the sphere drawn according to our imagination; consequently, the resultant vector $\boldsymbol{F}$ is equal to zero, and this proves d'Alembert's paradox: In irrotational steady flow of an ideal incompressible fluid past a finite body in the absence of sources or sinks around the body, the resultant vector of the forces of flow pressure on the body is equal to zero.

D'Alembert's paradox is valid for finite-size bodies bounded by the closed surface. The resultant vector, which propagates to infinity (for example on the half-body Figure 123), depends on the law of the increase in the width d of this body section with an increase in the distance z to infinity. Thus, the resistance of the half-body, which has the transverse dimen-

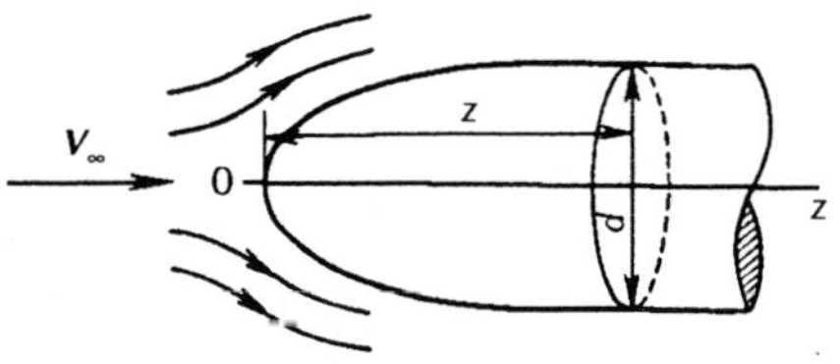

Fig. 123

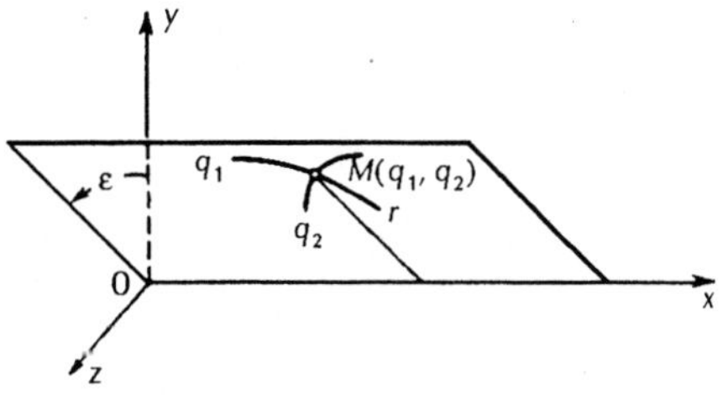

Fig. 124

sion of the cross-section tending to a finite value on moving away to infinity (for example, for the half-body formed by the superposition of a uniform flow on the source), is equal to zero.

The paraboloid of revolution presents an example of the half-body of infinite resistance. Among the half-bodies, the width of which increases more slowly than that of the paraboloid, can be the bodies of finite resistance.[4]

6. Equation of the Longitudinal Axisymmetric Motion. Flow through Channels

One of the most abundant forms of three-dimensional flows is the motion symmetric with respect to a certain axis (for example, the axis $0x$), called axisymmetric. These forms include the motion in nozzles of circular section, in convergent and divergent nozzles, axial flow around bodies of rotation, airship, and other shapes.

Establish the general equation of longitudinal axisymmetric motion occurring in the meridian planes (Figure 124) which form the angle ε with the plane $x0y$ and select in them a certain system of the orthogonal curvilinear coordinates q_1, q_2 independent of the angle ε. Then, in each meridian plane

$$r = r(q_1, q_2) \qquad x = x(q_1, q_2)$$

and in general for any point M

$$y = r(q_1, q_2)\cos\varepsilon \qquad z = r(q_1, q_2)\sin\varepsilon \qquad x = x(q_1, q_2)$$

from this, using Equations 83 of Chapter 1, it is easy to find the Lamé coefficients

$$H_1 = \sqrt{\left(\frac{\partial y}{\partial q_1}\right)^2 + \left(\frac{\partial z}{\partial q_1}\right)^2 + \left(\frac{\partial x}{\partial q_1}\right)^2} = \sqrt{\left(\frac{\partial r}{\partial q_1}\right)^2 + \left(\frac{\partial x}{\partial q_1}\right)^2}$$

$$H_2 = \sqrt{\left(\frac{\partial y}{\partial q_2}\right)^2 + \left(\frac{\partial z}{\partial q_2}\right)^2 + \left(\frac{\partial x}{\partial q_2}\right)^2} = \sqrt{\left(\frac{\partial r}{\partial q_2}\right)^2 + \left(\frac{\partial x}{\partial q_2}\right)^2}$$

$$H_3 = H_\varepsilon = \sqrt{\left(\frac{\partial y}{\partial \varepsilon}\right)^2 + \left(\frac{\partial z}{\partial \varepsilon}\right)^2 + \left(\frac{\partial x}{\partial \varepsilon}\right)^2} = r(q_1, q_2) \tag{44}$$

[4] Special investigation of the problem about the effect of the half-body shape on its resistance by M. I. Gurevich, The flow around an axisymmetric half-body of finite resistance, *Prikl. Mat. Mekh.* 11(1), 1947.

According to the sixth formula of Equation 84 from Chapter 1, the Laplace equation for determining the velocity potential will have the form

$$\frac{\partial}{\partial q_1}\left(\frac{H_2}{H_1} r \frac{\partial \varphi}{\partial q_1}\right) + \frac{\partial}{\partial q_2}\left(\frac{H_1}{H_2} r \frac{\partial \varphi}{\partial q_2}\right) = 0 \tag{45}$$

because the third term of this equation, which involves the derivative with respect to the coordinate ε, vanishes due to the adopted symmetry of motion.

It should be emphasized that the equation of axisymmetric motion (45), established in the coordinates q_1 and q_2, does not coincide with the equation of plane motion in the same coordinates.

Let us select the rectangular coordinates x, r in the meridian planes as the curvilinear coordinates; we have $H_x = 1$, $H_r = 1$, and, consequently, the motion equation will be reduced to the form

$$\frac{\partial}{\partial x}\left(r \frac{\partial \varphi}{\partial x}\right) + \frac{\partial}{\partial r}\left(r \frac{\partial \varphi}{\partial r}\right) = 0 \tag{46}$$

which corresponds to the Laplace equation in cylindrical coordinates in the absence of the dependence of motion on ε.

Let us dwell on the solution of the problem concerning the axisymmetric flow of an incompressible fluid through a channel, the surface of which is the surface of revolution. Assume that the rotational motion of the fluid around the channel axis is absent. Consider only a comparatively simple, "inverse" problem of determing the channel surface shape and velocity field in the channel from the given law of velocity variation along the channel axis. This problem is of practical interest for designing the shape of channels on the basis of the specified general dimensions.

Such problems arise, for example, when designing convergent and divergent nozzles of wind tunnels, ventilation, and other ducts limited in their dimensions by the volume of the rooms in which they are installed. An analogous method also can be employed to advantage for calculating the shortest possible pipes that connect two cylindrical tubes of different diameters and in other problems.

Let us first prove that the solution of Equation 46, which converts into the function $\varphi_0(x)$ prescribed on the symmetry axis, can be presented in the form of a certain integral

$$\varphi(x, r) = \frac{1}{\pi}\int_0^{\pi} \varphi_0(x + ir\cos\omega)\, d\omega \tag{47}$$

where $\varphi_0(x + ir\cos\omega)$ is the analytical function of the complex variable

$$t = x + ir\cos\omega \tag{48}$$

in a certain region of the plane t enclosing the coordinate origin $t = 0$.[5] Forming the derivatives

$$\frac{\partial \varphi}{\partial r} = \frac{i}{\pi}\int_0^{\pi} \varphi_0'(t)\cos\omega\, d\omega \qquad \frac{\partial^2 \varphi}{\partial r^2} = -\frac{1}{\pi}\int_0^{\pi} \varphi_0''(t)\cos^2\omega\, d\omega$$

[5]E. T. Whittaker and G. N. Watson, *Modern Analysis*, Part 2, Fizmatgiz, Moscow, 1963, pp. 250, 251.

and applying the integration by parts, we find the second term in Equation 46

$$\frac{\partial}{\partial r}\left(r\frac{\partial\varphi}{\partial r}\right) = r\frac{\partial^2\varphi}{\partial r^2} + \frac{\partial\varphi}{\partial r} = -\frac{r}{\pi}\int_0^\pi \varphi_0''(t)\cos^2\omega\, d\omega + \frac{i}{\pi}\int_0^\pi \varphi_0'(t)\cos\omega\, d\omega =$$

$$= -\frac{r}{\pi}\int_0^\pi \varphi_0''(t)\cos^2\omega\, d\omega + \frac{i}{\pi}\left[\varphi_0'(t)\sin\omega\right]_{\omega=0}^{\omega=\pi} - \frac{r}{\pi}\int_0^\pi \varphi_0''(t)\sin^2\omega\, d\omega = -\frac{r}{\pi}\int_0^\pi \varphi_0''(t)\, d\omega$$

The first term of Equation 46, equal to

$$\frac{\partial}{\partial x}\left(r\frac{\partial\varphi}{\partial x}\right) = r\frac{\partial^2\varphi}{\partial x^2} = r\frac{1}{\pi}\int_0^\pi \varphi_0''(t)\, d\omega$$

differs from the second term only by the sign. Thus, we see that Equation 47 actually gives the integral representation of the solution to Equation 46. Let us investigate this solution. Substituting $r = 0$ into the right side of Equation 47, we directly see that the function $\varphi_0(x)$ represents the distribution of the velocity potential $\varphi(x, r)$ along the motion symmetry axis $0x$. Let us go over to velocities: formula (47) yields

$$V_x = \frac{\partial\varphi}{\partial x} = \frac{1}{\pi}\int_0^\pi \varphi_0'(t)\, d\omega \qquad V_r = \frac{\partial\varphi}{\partial r} = \frac{i}{\pi}\int_0^\pi \varphi_0'(t)\cos\omega\, d\omega \tag{49}$$

When $r = 0$, we have

$$V_x = \varphi_0'(x) \qquad V_r = \frac{i}{\pi}\varphi_0'(x)\int_0^x \cos\omega\, d\omega = 0$$

Instead of the function $\varphi_0(t)$, we introduce into consideration the function $f_0(t) = \varphi_0'(t)$. The form of this function depends on the distribution of the velocity V_x along the axis $0x$, because by definition,

$$f_0(x) = \varphi_0'(x) = (V_x)_{r=0}$$

In what follows we consider this distribution to be given. Then, formulas (49) will be represented as

$$V_x = \frac{1}{\pi}\int_0^\pi f_0(t)\, d\omega \qquad V_r = \frac{i}{\pi}\int_0^\pi f_0(t)\cos\omega\, d\omega \tag{50}$$

Let us also introduce the stream function $\psi(x, r)$. According to Equations 29 and 50 and taking into account that the axial coordinate in Equation 29 now corresponds to x, we have

$$\frac{d\psi}{\partial r} = rV_x = \frac{r}{\pi}\int_0^\pi f_0(t)\,d\omega$$

where the integration over r yields

$$\psi = \frac{1}{\pi}\int_0^r r\,dr\int_0^r f_0(x + ir\cos\omega)\,d\omega \tag{51}$$

the function of the coordinate x, which appears in this partial integration over r, is equal to zero, because the axis $0x$ ($r = 0$) can be taken as the zero streamline by virtue of the flow symmetry ($\psi = 0$ at $r = 0$ and any x).

The resulting expressions (47), (50), and (51) can be considered to be the solution of the problem about an irrotational axisymmetric flow of an ideal incompressible liquid through a channel, the boundary of which is represented by the surfaces of revolution

$$\psi(x, r) = \text{const} \tag{52}$$

and the law of variation of the longitudinal velocity V_x along the channel axis $0x$ is specified by the function $f_0(x)$. This type of solution does not make it possible to directly find the fluid flow in a channel with an *a priori* prescribed profile;[6] however, it is possible to design diverse channels and to select those which would best satisfy the imposed conditions; for example, the required degree of uniformity of the velocity field at the given section of the channel.

For this purpose, establish the series expansion

$$f_0(x + ir\cos\omega) = f_0(x) + ir\cos\omega f_0'(x) + \cdots$$

and substitute it into formulas (50) and (51). Noting that

$$\frac{1}{\pi}\int_0^\pi \cos^{2n}\omega\,d\omega = \frac{(2n)!}{2^{2n}\cdot(n!)^2} \qquad \frac{1}{\pi}\int_0^\pi \cos^{2n-1}\omega\,d\omega = 0$$

we obtain

$$V_x(x, r) = \sum_{n=0}^{\infty}\frac{(-1)^n}{2^{2n}(n!)^2}r^{2n}f_0^{(2n)}(x)$$

$$V_r(x, r) = \sum_{n=1}^{\infty}\frac{(-1)^n 2n}{2^{2n}(n!)^2}r^{2n-1}f_0^{(2n-1)}(x)$$

$$\psi(x, r) = -\sum_{n=1}^{\infty}\frac{(-1)^n 2n}{2^{2n}(n!)^2}r^{2n}f_0^{(2n-2)}(x) \tag{53}$$

[6]See Sanoyan, V. G., The motion of a fluid in an axisymmetric channel of given profile and the calculation of actual pressures, *Tr. Leningr. Politekh. Inst.*, No. 176, 160 to 174, 1955.

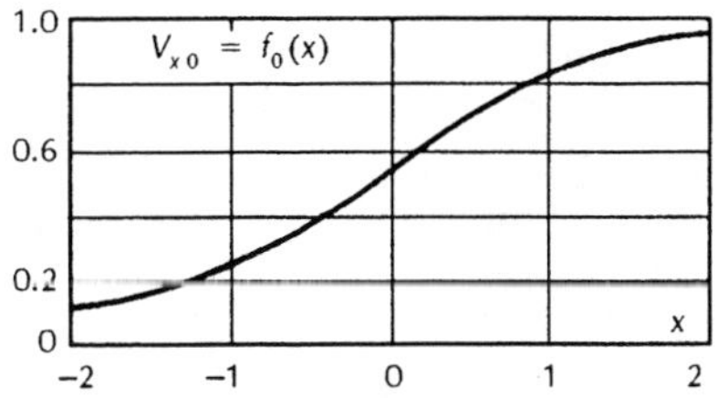

Fig. 125

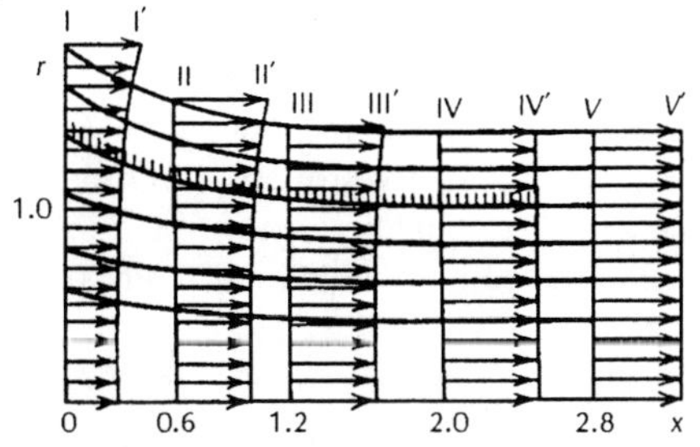

Fig. 126

Using expansions (53), it is possible to design different shapes of convergent and divergent nozzles and other channels. Thus, for example, we assume[7]

$$f_0(x) = 0.55 + 0.90\int_0^x \Phi(x)\,dx \qquad \Phi(x) = \frac{1}{\sqrt{2\pi}}e^{-\frac{1}{2}x^2}$$

$$\int_0^x \Phi(x)\,dx = \frac{1}{2}\operatorname{Erf}\left(\frac{x}{\sqrt{2}}\right) \qquad \operatorname{Erf}(\pm\infty) = \pm 1$$

which gives the change of V_x along the axis $0x$ from 0.1 at $x = -\infty$ to 1 at $x = \infty$, as shown in Figure 125. The successive derivatives of the function $f_0(x)$ are determined by the explicit equation

$$f_0^{(n+1)}(x) = 0.90\Phi^{(n)}(x)$$

where

$$\Phi^{(n)}(x) = \frac{1}{\sqrt{2\pi}}\frac{d^n}{dx^n}(e^{-\frac{1}{2}x^2})$$

Recalling the definition of the Hermite polynomials[8]

$$H_n(x) = (-1)^n e^{\frac{1}{2}x^2}\frac{d^n}{dx^n}(e^{-\frac{1}{2}x^2})$$

we obtain the following expression for the successive derivatives of the given function $f_0(x)$:

$$f_0^{(n+1)}(x) = (-1)^n 0.90\frac{1}{\sqrt{2\pi}}e^{-\frac{1}{2}x^2}H_n(x)$$

Using the tables of the Hermite polynomials and carrying out the summation indicated in system (53), we find the sought-after values of the longitudinal and radial velocities, whereas equating the expression of the stream function to different constants, we determine the possible shapes of the channels. In Figure 126 the streamlines and the distributions of longitudinal

[7] Tsien, H. S., On the design of the contraction cone for a wind tunnel, *J. Aeron. Sci.*, 10(2), 68 to 70, 1943.
[8] Janke, E., Emde, F., and Lösch, F., *Special Functions: Formulae, Plots, Tables*, Nauka Press, Moscow, 1977.

velocities are presented which correspond to the considered axisymmetrical flow in a convergent nozzle.

The Roman digits mark the sections of stream tubes, and the primed Roman digits denote the velocity profiles corresponding to these sections. Assuming the streamline to be a solid wall, we obtain the profile of the convergent nozzle, and the profiles will show to what extent the velocity field in different sections of the confusor is uniform. Thus, for example, it is seen that the profile of the convergent nozzle (Figure 126: dashes) is of rather good shape; the inevitable increase in the velocity at the convergent nozzle walls is not harmful because the small retardation of the fluid due to viscosity near the walls must straighten the field.

7. Axisymmetric Longitudinal Flow Past Bodies of Revolution

When investigating three-dimensional flows, one must use different curvilinear systems of coordinates: cylindrical, spherical, elliptical, etc. This approach not only simplifies the description of the pattern of motion, but sometimes it is inevitable: the possibility for the separation of variables in differential equations and the simplicity of the techniques used to satisfy the boundary conditions depend on the successful selection of the coordinate system. In a plane irrotational motion the transition from the physical plane $z = x + iy$ to the auxiliary plane $\zeta = \xi + i\eta$ was equivalent to the use of the curvilinear coordinates ξ, η in the physical plane instead of the rectilinear coordinates x, y. In the space of three dimensions there is no such convenient analytical apparatus as the functions of complex variables, and one must directly use the formulas of transition from rectilinear coordinates to curvilinear ones, expressing in them the differential equations themselves and the corresponding boundary conditions.

For calculating the external axisymmetric flow around bodies of revolution (Figure 127a), we take in the meridian planes r, x the elliptic coordinate system ξ, η connected to r, x by the relations

$$x = c \operatorname{ch} \xi \cos \eta \qquad 0 \leq \xi < \infty$$

$$x = c \operatorname{sh} \xi \sin \eta \qquad 0 \leq \eta \leq \pi$$

where the quantity c is the distance of the foci of the family of coordinate lines – confocal ellipses and hyperbolas – from the coordinate origin.

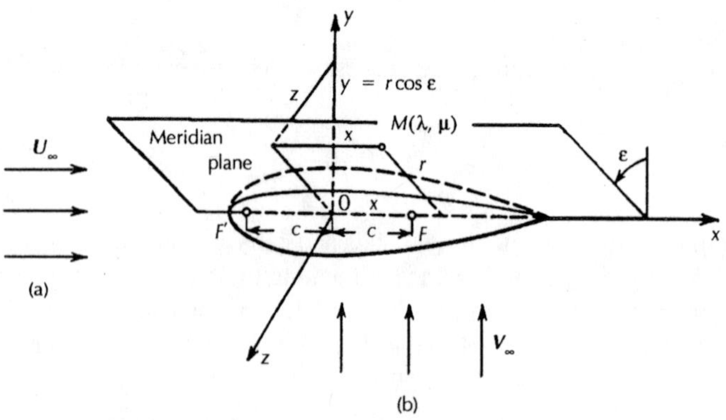

Fig. 127

We shall assume that

$$\operatorname{ch} \xi = \lambda \qquad \cos \eta = \mu \qquad 1 \le \lambda < \infty \qquad -1 \le \mu \le 1$$

Then the connection between the coordinates r, x and λ, μ will have the form

$$x = c\lambda\mu \qquad r = c\sqrt{\lambda^2 - 1}\sqrt{1 - \mu^2} \tag{54}$$

Having determined the derivatives

$$\frac{\partial r}{\partial \lambda} = c\lambda\sqrt{\frac{1-\mu^2}{\lambda^2-1}} \qquad \frac{\partial r}{\partial \mu} = -c\mu\sqrt{\frac{\lambda^2-1}{1-\mu^2}} \qquad \frac{\partial x}{\partial \lambda} = c\mu \qquad \frac{\partial x}{\partial \mu} = c\lambda$$

we find the Lamé coefficients (Equation 83, of Chapter 1)

$$H_\lambda = \sqrt{\left(\frac{\partial r}{\partial \lambda}\right)^2 + \left(\frac{\partial x}{\partial \lambda}\right)} = c\sqrt{\frac{\lambda^2-\mu^2}{\lambda^2-1}}$$

$$H_\mu = \sqrt{\left(\frac{\partial r}{\partial \mu}\right)^2 + \left(\frac{\partial x}{\partial \mu}\right)^2} = c\sqrt{\frac{\lambda^2-\mu^2}{1-\mu^2}}$$

$$H_\varepsilon = r = c\sqrt{\lambda^2-1}\sqrt{1-\mu^2} \tag{55}$$

After this, it is not difficult to form the Laplace basic differential equation for the velocity potential. According to the sixth formula of Equation 84, Chapter 1, we obtain

$$\frac{\partial}{\partial \lambda}\left[(\lambda^2 - 1)\frac{\partial \varphi}{\partial \lambda}\right] + \frac{\partial}{\partial \mu}\left[(1-\mu^2)\frac{\partial \varphi}{\partial \mu}\right] = 0 \tag{56}$$

We seek the particular solution of this equation in the form of the product of two functions of the variables λ and μ separately

$$\varphi = L(\lambda)\, M(\mu) \tag{57}$$

The variables in Equation 56 will then separate, and it will follow from the equation

$$\frac{1}{L(\lambda)}\frac{d}{d\lambda}\left[(\lambda^2-1)\frac{dL}{d\lambda}\right] = -\frac{1}{M(\mu)}\frac{d}{d\mu}\left[(1-\mu^2)\frac{dM}{d\mu}\right]$$

due to the independence of λ and μ, that each of the sides of the equation must be constant. Assuming this constant to be equal to $n(n+1)$, where n is the positive integer, we shall obtain two ordinary linear second-order Legendre-type equations for determining $L(\lambda)$ and $M(\mu)$

$$\frac{d}{d\lambda}\left[(1-\lambda^2)\frac{dL}{d\lambda}\right] + n(n+1)L = 0$$

$$\frac{d}{d\mu}\left[(1-\mu^2)\frac{dM}{d\mu}\right] + n(n+1)M = 0 \tag{58}$$

Two classes of independent solutions satisfy these equations:[9]

---

[9]E. T. Whittaker and G. N. Watson, *Modern Analysis*, Part 2, Fizmatgiz, Moscow, 1963, p. 109 and further.

1. The Legendre first-kind functions – the Legendre polynomials $P_n(x)$ defined by the equalities

$$P_0(x) = 1 \qquad P_1(x) = x \qquad P_2(x) = \frac{1}{2}(3x^2 - 1) \qquad P_3(x) = \frac{1}{2}(5x^3 - 3x) \cdots$$

and by the recurrent relation for calculating the subsequent polynomials

$$(n+1)P_{n+1}(x) = (2n+1)xP_n(x) - nP_{n-1}(x)$$

2. The Legendre second-kind functions $Q_n(x)$ defined by the equalities

$$Q_0(x) = \frac{1}{2}\ln\frac{x+1}{x-1} \qquad Q_1(x) = \frac{1}{2}x\ln\frac{x+1}{x-1} - 1$$

$$Q_2(x) = \frac{1}{4}(3x^2 - 1)\ln\frac{x+1}{x-1} - \frac{3}{2}x$$

$$Q_3(x) = \frac{1}{4}(5x^3 - 3x)\ln\frac{x+1}{x-1} - \frac{5}{2}x^2 + \frac{2}{3} \cdots$$

and by the recurrent relation

$$(n+1)Q_{n+1}(x) = (2n+1)xQ_n(x) - nQ_{n-1}(x)$$

which coincides with the previous one for the Legendre polynomials.

Let us represent the solution of Equation 56 as a sum of two potentials:

1. The potential φ_∞ of a uniform flow incident on the body with the velocity U_∞; according to the first formula of Equation 54 this potential is equal to

$$\varphi_\infty = U_\infty x = U_\infty c\lambda\mu$$

2. The potential φ' of perturbation velocities which will be represented by the sum of particular solutions (57).

As the nth power polynomial, the function $P_n(x)$ goes to infinity when the argument increases infinitely; at the same time the function $Q_n(x)$ tends to zero, but it is logarithmically infinite at $x = \pm 1$. In the case of external flow around a body, the coordinate $\lambda = \operatorname{ch}\xi$ may attain infinite values, whereas the coordinate μ is limited. We take into consideration that the velocity potential of perturbation motion (i.e., the flow minus a uniform flow with the velocity equal to that at infinity) must tend to zero on moving away from the body surface, and as before (Section 5), we must have

$$\varphi' = O\left(\frac{1}{R^2}\right)$$

It follows from the above considerations that the sought-after particular solutions must have the form of the products

$$Q_n(\lambda)P_n(\mu) \qquad (n = 1, 2, \cdots)$$

where it should be emphasized that in the summation the reckoning of n should begin from unity rather than zero. This is confirmed by the presence of the following obvious asymptotic equalities, which are valid at great values of λ, and, consequently, according to Equation 54, at great values of the quantity $R = \sqrt{x^2 + r^2}$ having the same order as λ:

$$Q_0(\lambda) = \frac{1}{2}\ln\frac{\lambda+1}{\lambda-1} \sim \frac{1}{\lambda} + \frac{1}{3\lambda^3} + \cdots = O\left(\frac{1}{\lambda}\right) = O\left(\frac{1}{R}\right)$$

$$Q_1(\lambda) = \frac{1}{2}\lambda\ln\frac{\lambda+1}{\lambda-1} - 1 = \lambda\left(\frac{1}{\lambda} + \frac{1}{3\lambda^3} + \cdots\right) - 1 \sim \frac{1}{3\lambda^3} + \cdots = O\left(\frac{1}{\lambda^2}\right) = O\left(\frac{1}{R^2}\right)$$

Thus, we have the needed order of the decrease of φ' at infinity if we assume that

$$\varphi' = cU_\infty \sum_{n=1}^{\infty} A_n Q_n(\lambda) P_n(\mu) \tag{59}$$

where A_n are the constant coefficients that depend on the shape of the body surface.

Adding up the potentials φ_∞ and φ', we obtain the sought-after velocity potential of the longitudinal flow around the body of revolution with the velocity at infinity equal to U_∞,

$$\varphi = cU_\infty \left(\sum_{n=1}^{\infty} A_n Q_n(\lambda) P_n(\mu) + \lambda\mu \right) \tag{60}$$

In order to determine the coefficients A_n, we find the expression for the stream function ψ. According to Equations 1, 26, and 55,

$$\frac{\partial\psi}{\partial\lambda} = -\frac{H_\lambda H_\varepsilon}{H_\mu}\frac{\partial\varphi}{\partial\mu} = -c(1-\mu^2)\frac{\partial\varphi}{\partial\mu} \qquad \frac{\partial\psi}{\partial\mu} = \frac{H_\mu H_\varepsilon}{H_\lambda}\frac{\partial\varphi}{\partial\lambda} = c(\lambda^2-1)\frac{\partial\varphi}{\partial\lambda}$$

or, after the substitution of expression (60),

$$\frac{\partial\psi}{\partial\lambda} = -c^2 U_\infty \left[(1-\mu^2) \sum_{n=1}^{\infty} A_n Q_n \frac{dP_n}{d\mu} + \lambda(1-\mu^2) \right]$$

$$\frac{\partial\psi}{\partial\mu} = c^2 U_\infty \left[(\lambda^2-1) \sum_{n=1}^{\infty} A_n \frac{dQ_n}{d\lambda} P_n + \mu(\lambda^2-1) \right]$$

Rewriting the second equation in the form

$$\frac{\partial\psi}{\partial\mu} = c^2 U_\infty (\lambda^2-1) \left(\sum_{n=1}^{\infty} A_n \frac{dQ_n}{d\lambda} P_n + \mu \right)$$

substitute under the summation the expression for P_n from the basic differential equation of the Legendre functions (58)

$$P_n = -\frac{1}{n(n+1)} \frac{d}{d\mu}\left[(1-\mu^2)\frac{dP_n}{d\mu}\right]$$

This will yield

$$\frac{\partial \psi}{\partial \mu} = -c^2 U_\infty (\lambda^2 - 1)\left\{\sum_{n=1}^{\infty} \frac{A_n}{n(n+1)} \frac{dQ_n}{d\lambda} \frac{d}{d\mu}\left[(1-\mu^2)\frac{dP_n}{d\mu}\right] - \mu\right\}$$

Integrating over μ and adding the needed function of λ, we obtain the final expression for the stream function

$$\psi = -\frac{1}{2} c^2 U_\infty (\lambda^2 - 1)(1-\mu^2)\left[\sum_{n=1}^{\infty} \frac{2A_n}{n(n+1)} \frac{dQ_n}{d\lambda}\frac{dP_n}{d\mu} + 1\right] \tag{61}$$

The equation of the null stream surface will be

$$\sum_{n=1}^{\infty} \frac{2A_n}{n(n+1)} \frac{dQ_n}{d\lambda}\frac{dP_n}{d\mu} + 1 = 0 \tag{62}$$

Comparing this equation with the prescribed equation for the profile of the body of revolution in elliptic coordinates, we can determine the coefficients A_n.[10] In Kaplan's paper the determination of the coefficients A_n is reduced to the solution of the algebraic system of equations of the first power, and this presents a simple problem with the modern state of computational technique.

Having the expression for the velocity potential, we find the velocity from the formula

$$V^2 = V_\lambda^2 + V_\mu^2 = \frac{1}{H_\lambda^2}\left(\frac{\partial \varphi}{\partial \lambda}\right)^2 + \frac{1}{H_\mu^2}\left(\frac{\partial \varphi}{\partial \mu}\right)^2 =$$

$$= \frac{U_\infty^2}{\lambda^2 - \mu^2}\left[(\lambda^2 - 1)\left(\sum_{n=1}^{\infty} A_n \frac{dQ_n}{d\lambda} P_n + \mu\right)^2 + (1-\mu^2)\left(\sum_{n=1}^{\infty} A_n Q_n \frac{dP_n}{d\mu} + \lambda\right)^2\right] \tag{63}$$

The method is illustrated with the aid of the simplest example. Consider the flow about an ellipsoid of revolution, the meridian section of which has the equation $\lambda = \lambda_0$. Assuming in Equation 62 that $A_n = 0$ when $n > 1$ and $\lambda = \lambda_0$, we obtain

$$A_1 = -\frac{1}{\left(\dfrac{dQ_1}{d\lambda}\right)_{\lambda=\lambda_0}} = -\frac{1}{\dfrac{1}{2}\ln\dfrac{\lambda_0 + 1}{\lambda_0 - 1} - \dfrac{\lambda_0}{\lambda_0^2 - 1}}$$

According to Equation 60, the velocity potential is equal to

[10] Kaplan, C., Potential Flow About Elongated Bodies of Revolution, NACA Rep. No. 516, 1935.

$$\varphi(\lambda, \mu) = -cU_\infty \left(\frac{\frac{1}{2}\lambda \ln \frac{\lambda+1}{\lambda-1} - 1}{\frac{1}{2}\ln \frac{\lambda_0+1}{\lambda_0-1} - \frac{\lambda_0}{\lambda_0^2-1}} - \lambda \right) \mu \tag{64}$$

This expression can be represented in a somewhat different form by introducing explicitly the ellipsoid half-axes a and $b < a$ located on the axes $0x$ and $0r$, respectively. According to Equation 54, we have the equation of the ellipse $\lambda = \lambda_0$ in the form

$$\frac{x^2}{c^2\lambda_0^2} + \frac{r^2}{c^2(\lambda_0^2-1)} = 1$$

where it follows that

$$c\lambda_0 = a \qquad c\sqrt{\lambda_0^2-1} = b$$

or, introducing the eccentricity $e = c/a$,

$$\lambda_0 = \frac{1}{e} \qquad \sqrt{\lambda_0^2-1} = \frac{b}{c}$$

In these notations we obtain

$$\varphi(\lambda, \mu) = -U_\infty a \left(\frac{\frac{1}{2}\lambda \ln \frac{\lambda+1}{\lambda-1} - 1}{\frac{1}{2e}\ln \frac{1+e}{1-e} - \frac{1}{1-e^2}} - e\lambda \right) \mu \tag{65}$$

It is possible, by using this expression, to obtain the potential of the flow around a sphere of radius a by noting that by definition of the elliptic coordinates, in this case $e \to 0$, $c\lambda \to R$, $\mu \to \cos\theta$ when $c \to 0$, where R and θ are the spherical coordinates. Carrying out the expansion ($\lambda > 1$, $e < 1$)

$$\ln\frac{\lambda+1}{\lambda-1} = \ln\frac{1+\frac{1}{\lambda}}{1-\frac{1}{\lambda}} = 2\left(\frac{1}{\lambda} + \frac{1}{3\lambda^3} + \cdots\right) \qquad \ln\frac{1+e}{1-e} = 2\left(e + \frac{1}{3}e^3 + \cdots\right)$$

and replacing e by c/a, we see that for $c \to 0$ the quantity φ tends to

$$U_\infty R\left[1 + \frac{1}{2}\left(\frac{a}{R}\right)^3\right]\cos\theta$$

i.e., to Equation 38.

The velocity projections on the axes of the elliptic coordinates will be

$$V_\lambda = \frac{1}{H_\lambda}\frac{\partial\varphi}{\partial\lambda} = -U_\infty\sqrt{\frac{\lambda^2-1}{\lambda^2-\mu^2}}\left(\frac{\frac{1}{2}\ln\frac{\lambda+1}{\lambda-1} - \frac{\lambda}{\lambda^2-1}}{\frac{1}{2}\ln\frac{1+e}{1-e} - \frac{e}{1-e^2}} - 1 \right)\mu$$

$$V_\mu = \frac{1}{H_\mu}\frac{\partial\varphi}{\partial\mu} = -U_\infty\sqrt{\frac{1-\mu^2}{\lambda^2-\mu^2}}\left(\frac{\frac{1}{2}\lambda\ln\frac{\lambda+1}{\lambda-1}-1}{\frac{1}{2}\ln\frac{1+e}{1-e}-\frac{e}{1-e^2}}-\lambda\right)$$

Assuming here that $\lambda = \lambda_0$, we see that on the ellipsoid surface $V_\lambda = 0$; this is natural because the coordinate lines (λ) are perpendicular to the ellipsoid surface and the condition $V_\lambda = 0$ is equivalent to the condition of zero velocity component normal to the surface. The distribution of velocities over the ellipsoid surface will be determined by the equation

$$V = V_\mu = \frac{e^3 U_\infty}{e - \frac{1}{2}(1-e^2)\ln\left(\frac{1+e}{1-e}\right)}\sqrt{\frac{1-\mu^2}{1-e^2\mu^2}}$$

The above solution relates to the flow about an ellipsoid of revolution aligned with the flow. Similarly, it is possible to investigate the case of flow about a prolate ellipsoid, the foci of the meridian section of which lie not on the axis $0z$, but rather in meridian planes.[11] The books cited in footnote 11 also present the solution for a more general problem of flow around an ellipsoid, of which all the axes are different.

8. Transverse Flow about the Bodies of Revolution

Of interest along with the longitudinal flow about the bodies of revolution is the transverse flow perpendicular (Figure 127b) to the symmetry axis of a body. Adding these two flows gives the flow about a body of revolution at any angle of attack. The solution of the problem concerning the transverse flow about the bodies of revolution is given here.

The axisymmetric motion does not already exist. According to the sixth formula of Equation 84, Chapter 1, the Laplace equation, which determines the velocity potential, has the following form in orthogonal curvilinear coordinate system:

$$\frac{\partial}{\partial q_1}\left(\frac{H_2H_3}{H_1}\frac{\partial\varphi}{\partial q_1}\right)+\frac{\partial}{\partial q_2}\left(\frac{H_3H_1}{H_2}\frac{\partial\varphi}{\partial q_2}\right)+\frac{\partial}{\partial q_3}\left(\frac{H_1H_2}{H_3}\frac{\partial\varphi}{\partial q_3}\right) = 0$$

Retaining the same system of coordinates (λ, μ, ε) as in the case of axisymmetric flow around a body of revolution, and recalling the expression for the Lamé coefficients (55), we rewrite the previous expression in the form

$$\frac{\partial}{\partial\lambda}\left[(\lambda^2-1)\frac{\partial\varphi}{\partial\lambda}\right]+\frac{\partial}{\partial\mu}\left[(1-\mu^2)\frac{\partial\varphi}{\partial\mu}\right]+\left[\frac{1}{\lambda^2-1}+\frac{1}{1-\mu^2}\right]\frac{\partial^2\varphi}{\partial\varepsilon^2} = 0 \qquad (66)$$

The solution of this equation is sought as the product of two functions

$$\varphi = N(\lambda, \mu)\,E(\varepsilon)$$

---

[11] See, for example, Kibel, I. A., Kochin, N. Ye., and Roze, N. V., *Theoretical Hydromechanics*, Part 1, Fizmatgiz, Moscow, 1963, p. 365; Lamb, G., *Hydrodynamics*, Gostekhizdat, Moscow, 1947, pp. 175 to 181; Kochin, N. E., Kibel, J. A., and Rose, N. V., *Theoretische Hydromechanik*, Vol. 1 and 2, Berlin, 1954.

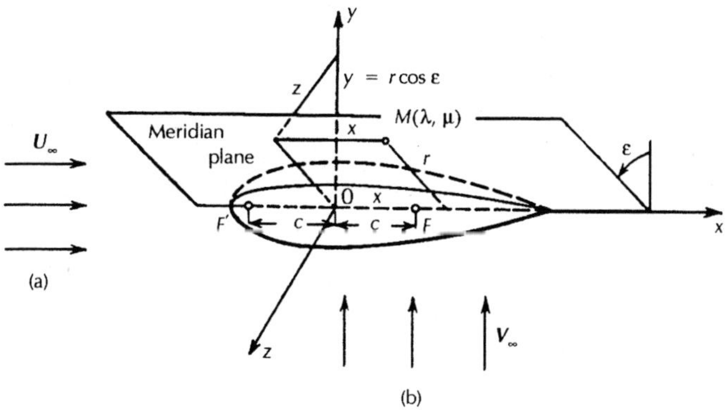

Fig. 127

Then, substituting this expression into Equation 66 and separating the functions of the independent variables, we obtain a system of equations (k is an arbitrary number which will be considered positive and an integer)

$$\frac{d^2E}{d\varepsilon^2} + k^2E = 0 \qquad \frac{\partial}{\partial\lambda}\left[(\lambda^2-1)\frac{\partial N}{\partial\lambda}\right] + \frac{\partial}{\partial\mu}\left[(1-\mu^2)\frac{\partial N}{\partial\mu}\right] - k^2\frac{\lambda^2-\mu^2}{(\lambda^2-1)(1-\mu^2)}N = 0$$

The first equation has the solution

$$E = A\cos k\varepsilon + B\sin k\varepsilon$$

The second equation (if it is assumed that $N = L(\lambda)\,M(\mu)$) and the variables are separated, can be reduced to the following system of equations:

$$\frac{d}{d\lambda}\left[(1-\lambda^2)\frac{dL}{d\lambda}\right] + \left[n(n+1) - \frac{k^2}{1-\lambda^2}\right]L = 0$$

$$\frac{d}{d\mu}\left[(1-\mu^2)\frac{dM}{d\lambda}\right] + \left[n(n+1) - \frac{k^2}{1-\mu^2}\right]M = 0$$

whose partial solutions are the so-called associated Legendre functions[12]

$$P_n^k(\mu) = (1-\mu^2)^{k/2}\frac{d^kP_n(\mu)}{d\mu^k} \qquad Q_n^k(\lambda) = (\lambda^2-1)^{k/2}\frac{d^kQ_n(\lambda)}{d\lambda^k} \tag{67}$$

Combining these functions so that the expression for the velocity potential of perturbed motion could be limited when $\lambda \to \infty$, we obtain the general expression for the velocity potential

$$\varphi(\lambda,\mu) = \sum_{n=1}^{\infty}\sum_{k=1}^{\infty} Q_n^k(\lambda)\,P_n^k(\mu)\,(A_{nk}\cos k\varepsilon + B_{nk}\sin k\varepsilon) + V_\infty y$$

Here, the last term represents the velocity potential of the uniform flow incident on the body with the velocity at infinity V_∞ directed parallel to the axis Oy (Figure 127b).

---

[12] E. T. Whittaker and G. N. Watson, *Modern Analysis*, Part 2, Fizmatgiz, Moscow, 1963, pp. 140 to 144.

Assuming in the above general equation of the potential that

$$A_{n1} = cV_{\infty}C_n \qquad A_{n2} = A_{n3} = \cdots = 0 \qquad B_{n1} = B_{n2} = \cdots = 0$$

i.e., being satisfied with the solution containing $\cos \varepsilon$ and, moreover, representing y as a function of λ, μ, and ε, according to the equations given in the beginning of the previous section

$$y = r\cos \varepsilon = c\,\mathrm{sh}\,\xi \sin \eta \cos \varepsilon = c\sqrt{\lambda^2 - 1}\sqrt{1 - \mu^2}\cos \varepsilon$$

we obtain the following expression for the velocity potential of the flow incident transversely with the velocity V_{∞} along the axis $0y$:

$$\varphi = cV_{\infty}\cos \varepsilon \sum_{n=1}^{\infty} C_n Q_n^1(\lambda) P_n^1(\lambda) + cV_{\infty}\sqrt{\lambda^2 - 1}\sqrt{1 - \mu^2}\cos \varepsilon$$

or, using the definition of the associated Legendre functions (67),

$$\varphi = cV_{\infty}\sqrt{\lambda^2 - 1}\sqrt{1 - \mu^2}\left(\sum_{n=1}^{\infty} C_n \frac{dQ_n}{d\lambda}\frac{dP_n}{d\mu} + 1\right)\cos \varepsilon \tag{68}$$

To determine the constants C_n, it is necessary to establish the boundary condition on the given surface of the body immersed in a flow. In this case of axisymmetric motion the stream function is absent and it is necessary to calculate the normal velocity $V_n = \partial\varphi/\partial n$ and equate it to zero.

Making the computations somewhat easier, we write out, in the selected coordinate system (λ, μ), the condition stating that for the impermeable surface of a body immersed in a flow the element of the arc of its meridian section is parallel to the velocity component in the meridian plane (the condition of fluid sliding along the solid surface):

$$\frac{ds_{\lambda}}{V_{\lambda}} = \frac{ds_{\mu}}{V_{\mu}}$$

or, recalling the expressions for the elements of the arcs of coordinate lines and projections of the potential gradient onto the directions of these lines:

$$H_{\lambda}d\lambda : \frac{1}{H_{\lambda}}\frac{\partial\varphi}{\partial\lambda} = H_{\mu}d\mu : \frac{1}{H_{\mu}}\frac{\partial\varphi}{\partial\mu}$$

From this follows the sought-after boundary condition

$$H_{\mu}^2\frac{\partial\varphi}{\partial\lambda} = H_{\lambda}^2\frac{\partial\varphi}{\partial\mu}\frac{d\lambda}{d\mu} \tag{69}$$

in which λ is the given function of μ according to the equation of the streamlined body contour in the meridian plane. Establishing the partial derivatives $\partial\varphi/\partial\lambda$, $\partial\varphi/\partial\mu$, we obtain from Equation 68

$$\frac{1}{cV_{\infty}\cos \varepsilon}\frac{\partial\varphi}{\partial\lambda} =$$

$$= \lambda\sqrt{\frac{1-\mu^2}{\lambda^2-1}}\left(\sum_{n=1}^{\infty} C_n \frac{dQ_n}{d\lambda}\frac{dP_n}{d\mu} + 1\right) + \sqrt{(\lambda^2-1)(1-\mu^2)}\sum_{n=1}^{\infty} C_n \frac{d^2Q_n}{d\lambda^2}\frac{dP_n}{d\mu}$$

$$\frac{1}{cV_\infty \cos\varepsilon}\frac{\partial\varphi}{\partial\mu} =$$

$$= -\mu\sqrt{\frac{\lambda^2-1}{1-\mu^2}}\left(\sum_{n=1}^{\infty} C_n \frac{dQ_n}{d\lambda}\frac{dP_n}{d\mu} + 1\right) + \sqrt{(\lambda^2-1)(1-\mu^2)}\sum_{n=1}^{\infty} C_n \frac{dQ_n}{d\lambda}\frac{d^2P_n}{d\mu^2}$$

Replacing the expressions of the second derivatives in the above equations on the basis of the differential equations for the functions P_n and Q_n

$$(1-\lambda^2)\frac{d^2Q_n}{d\lambda^2} = 2\lambda\frac{dQ_n}{d\lambda} - n(n+1)Q_n \qquad (1-\mu^2)\frac{d^2P_n}{d\mu^2} = 2\mu\frac{dP_n}{d\mu} - n(n+1)P_n$$

and making simple transformations, we obtain

$$\frac{1}{cV_\infty \cos\varepsilon}\frac{\partial\varphi}{\partial\lambda} = -\lambda\sqrt{\frac{1-\mu^2}{\lambda^2-1}}\sum_{n=1}^{\infty} C_n \frac{dQ_n}{d\lambda}\frac{dP_n}{d\mu} +$$

$$+ \sqrt{\frac{1-\mu^2}{\lambda^2-1}}\sum_{n=1}^{\infty} n(n+1)C_n Q_n \frac{dP_n}{d\mu} + \lambda\sqrt{\frac{1-\mu^2}{\lambda^2-1}}$$

$$\frac{1}{cV_\infty \cos\varepsilon}\frac{\partial\varphi}{\partial\mu} = \mu\sqrt{\frac{\lambda^2-1}{1-\mu^2}}\sum_{n=1}^{\infty} C_n \frac{dQ_n}{d\lambda}\frac{dP_n}{d\mu} -$$

$$- \sqrt{\frac{\lambda^2-1}{1-\mu^2}}\sum_{n=1}^{\infty} n(n+1)C_n \frac{dQ_n}{d\lambda}P_n - \mu\sqrt{\frac{\lambda^2-1}{1-\mu^2}}$$

Substituting these expressions of the derivatives into Equation 69 and using the Lamé coefficients derived earlier

$$H_\lambda^2 = c^2\frac{\lambda^2-\mu^2}{\lambda^2-1} \qquad H_\mu^2 = c^2\frac{\lambda^2-\mu^2}{1-\mu^2}$$

we obtain, after an obvious simplification,

$$\sum_{n=1}^{\infty} C_n\left[\left(\lambda + \mu\frac{d\lambda}{d\mu}\right)\frac{dQ_n}{d\lambda}\frac{dP_n}{d\mu} - n(n+1)\left(Q_n\frac{dP_n}{d\mu} + \frac{dQ_n}{d\lambda}P_n\frac{d\lambda}{d\mu}\right)\right] = \lambda + \mu\frac{d\lambda}{d\mu}$$

Bearing in mind that on the body surface λ is the given function of μ, we rewrite the boundary condition in the final form as

$$\sum_{n=1}^{\infty} C_n \left[\frac{d(\lambda\mu)}{d\mu} \frac{dQ_n}{d\lambda} \frac{dP_n}{d\mu} - n(n+1) \frac{d}{d\mu}(Q_n P_n) \right] = \frac{d(\lambda\mu)}{d\mu} \tag{70}$$

Let us consider the transverse flow about an ellipsoid of revolution $\lambda = \lambda_0$, the longitudinal axisymmetric flow about which was studied in the previous section. In this case, boundary condition (70) can be fulfilled having assumed $C_n = 0$ when $n > 1$; then we shall have $(P_1 = \mu)$

$$C_1 \left[\lambda_0 \left(\frac{dQ_1}{d\lambda} \right)_{\lambda = \lambda_0} - 2Q_1(\lambda_0) \right] = \lambda_0$$

where, according to the expression obtained earlier for $Q_1(\lambda)$, it follows that

$$C_1 = \frac{\lambda_0}{2 - \dfrac{\lambda_0^2}{\lambda_0^2 - 1} - \dfrac{1}{2}\lambda_0 \ln \dfrac{\lambda_0 + 1}{\lambda_0 - 1}}$$

Recall that here $\lambda_0 = 1/e$, where e is the eccentricity of the ellipse that represents the meridian section of an ellipsoid. According to Equation 68, the velocity potential of the considered transverse flow about the ellipsoid of revolution is equal to

$$\varphi = cV_\infty \sqrt{\lambda^2 - 1}\sqrt{1 - \mu^2} \left\{ C_1 \left[Q_0(\lambda) + \frac{1}{1 - \lambda^2} \right] + 1 \right\} \cos \varepsilon \tag{71}$$

and the velocities are determined by simple differentiation of Equation 71

$$V_\lambda = \frac{1}{H_\lambda} \frac{\partial \varphi}{\partial \lambda} \qquad V_\mu = \frac{1}{H_\mu} \frac{\partial \varphi}{\partial \mu} \qquad V_\varepsilon = \frac{1}{H_\varepsilon} \frac{\partial \varphi}{\partial \varepsilon}$$

As is seen from this and previous sections, the solution of the problems concerning the longitudinal and transverse flow about the body of revolution leads to the necessity to carry out laborious calculations in each case. These calculations can be made easier by applying approximate methods, the use of which is limited only to the case of flow past slender bodies with the ratio of the length to the maximum thickness on the of order 8 to 12 (see Section 76 of the fourth edition of this book). With the present state of computational technique, such approximate methods are not of great significance.

9. Application of the Method of Singularities for Calculating Longitudinal and Lateral Flows about the Bodies of Revolution. Slender Bodies

The method of investigation of the longitudinal and lateral flow past the bodies of revolution which was presented in the previous sections and which is based on direct solution of the Laplace equation in elliptical coordinates, is not the only method of solving this problem.

Originally, the shapes of the streamlined bodies of revolution for airships were determined by the superposition of a uniform flow, parallel to a certain axis, onto the flow from a system of sources (sinks) distributed along the same axis. For this purpose, first the discrete singularities of the flow were used, i.e., the systems of sources (sinks) or dipoles, and afterward, their continuous distributions.

Let us assume for the sake of definiteness that over the segment $(-c, +c)$ of the axis $0x$ a continuous distribution of sources (sinks) of the intensity $q(x)$ is prescribed. Then, according to the second formula of Equation 12 (the minus sign will be introduced into the definition of the intensity q), the potential φ' of the disturbed motion produced by this system of singularities will be equal to

$$\varphi'(r, x) = \frac{1}{4\pi} \int_{-c}^{c} \frac{q(x')\,dx'}{\sqrt{r^2 + (x - x')^2}} \qquad (72)$$

By specifying the form of the function $q(x')$ and computing integral (72), we obtain the perturbation velocity potential, whereas the differentiation with respect to r and x will make it possible to calculate also the velocity projections V_r' and V_x'. Conversely, by prescribing the shape of the streamlined body, converting from the velocity potential of disturbed motion to the total potential of the longitudinal uniform flow past a body, with the prescribed velocity at infinity, and having written the impermeability condition for the body surface, it is possible to obtain the integral equation in which $q(x')$ will be the unknown function. Replacing the velocity potential by the stream function, von Kármán[13] developed the method of approximate integration of the corresponding integral equation based on the replacement of the integral by the finite sum. However, the von Kármán method was not general enough, and moreover, in each case it required the solution of a system of a great number of linear algebraic equations, thus making it very laborious for that time.

Analogously, using the expression for the dipole potential (8), it is also possible to form the potential φ_1 of the transverse flow past a body of revolution by adding the potential of a uniform flow with the given velocity at infinity and the velocity potential of the disturbed flow of liquid from continuously distributed (over the segment $-c < x < c$) dipoles of intensity $m(x')$ with the axes directed along $0y$

$$\varphi'(r, x, \varepsilon) = \frac{r\cos\varepsilon}{4\pi} \int_{-c}^{c} \frac{m(x')\,dx'}{[r^2 + (x - x')^2]^{3/2}} \qquad (73)$$

Here, it is also possible to specify the distribution of the intensity $m(x')$ or, conversely, to determine this intensity from the integral equation which represents the impermeability condition of the given body surface with respect to the flow composed of the perturbed and uniform flows at infinity.

Not dwelling on the details of the methods which are rarely used at present, we only indi cate that they are simply connected
with the methods presented in the previous sections. It is shown that for the given shape of the surfaces of streamlined bodies of revolution the unknown functions $q(x')$ and $m(x')$ can be expressed in terms of the coefficients A_n and C_n[10] derived earlier.

---

[13] Von Kármán, Th., Bereshnung der Druckverteilung an Lüftschiffkörpern, *Abh. Aerodyn. Inst. Aachen*, 6, 1927. For a more detailed description of this and other methods and also their application to calculations, see Fabrikant, N. Ya., *Kurs aerodinamiki (The Course of Aerodynamics)*, Part 1, Gostekhizdat, Moscow, 1938, chap. 3.

Let us divide the symmetry axis of the body of revolution $0x$ into two regions: one determined by the interval $-c \le x \le c$ filled with singularities, and the other which represents the remaining part of the axis $0x$, where $|x| > c$. In the elliptical coordinates λ, μ, introduced at the beginning of Section 7, the segment over which the singularities are located will be represented, according to the second of Equations 54, as

$$\lambda = 1 \qquad -1 \le \mu \le 1$$

and the remaining portion of the axis $0x$ as

$$\mu = \pm 1 \qquad 1 < \lambda < \infty$$

Then, comparing outside the segment $(-c < x' < c)$ the expressions of the perturbation potentials (72) and (73) with the corresponding expressions of the same potentials taken from Equations 60 and 68 and taking into account that $P_n(1) = 1$, $dP_n/d\mu|_{\mu=1} = n(n+1)/2$, we obtain the following two equations:

$$\frac{1}{4\pi}\int_{-1}^{+1}\frac{q(c\mu')\,d\mu'}{\lambda-\mu'} = cU_\infty\sum_{n=1}^{\infty}A_nQ_n(\lambda) \tag{74}$$

$$\frac{1}{4\pi c^2}\int_{-1}^{+1}\frac{m(c\mu')\,d\mu'}{(\lambda-\mu')^3} = V_\infty\sum_{n=1}^{\infty}\frac{n(n+1)}{2}C_n\frac{dQ_n}{d\lambda} \tag{75}$$

which at the given coefficients A_n and C_n can be considered to be two integral equations for determining the unknown functions q and m.

The integral equation (74) can be solved if the solution is sought in the form of the series

$$q(c\mu') = \sum_{n=1}^{\infty}a_nP_n(\mu') \qquad -1 \le \mu' \le 1$$

Substituting this expansion into Equation 74,

$$\frac{1}{4\pi}\sum_{n=1}^{\infty}a_n\int_{-1}^{+1}\frac{P_n(\mu')\,d\mu'}{\lambda-\mu'} = cU_\infty\sum_{n=1}^{\infty}A_nQ_n(\lambda)$$

Noting that by the well-known formula of the Legendre theory of functions[14]

$$\int_{-1}^{+1}\frac{P_n(\mu')\,d\mu'}{\lambda-\mu'} = 2Q_n(\lambda)$$

the previous integral equation is rewritten in the form

[14] E. T. Whittaker and G. N. Watson, *Kurs sovremennogo analiza* (*Modern Analysis*), Part 2, Fizmatgiz, Moscow, 1963, p. 135 (138 for the expansion).

$$\frac{1}{2\pi}\sum_{n=1}^{\infty} a_n Q_n(\lambda) = cU_\infty \sum_{n=1}^{\infty} A_n Q_n(\lambda)$$

from which the sought-after solution follows directly

$$a_n = 2\pi c U_\infty A_n \qquad q(x') = 2\pi c U_\infty \sum_{n=1}^{\infty} A_n P_n(x'/c) \tag{76}$$

In order to find the second unknown function $m(x')$, integrate the well-known expansion[14] once with respect to λ and the second time with respect to μ':

$$\frac{1}{\lambda - \mu'} = \sum_{n=1}^{\infty} (2n+1)\, Q_n(\lambda)\, P_n(\mu')$$

then

$$\frac{1}{(\lambda - \mu')^3} = -\frac{1}{2}\sum_{n=1}^{\infty} (2n+1) \frac{dQ_n}{d\lambda}\frac{dP_n}{d\mu'}$$

Substituting this expansion into integral equation (75), we obtain it in the form

$$-\frac{1}{8\pi c^2}\sum_{n=1}^{\infty} (2n+1)\frac{dQ_n}{d\lambda}\int_{-1}^{1} m(c\mu')\frac{dP_n}{d\mu'}d\mu' = V_\infty \sum_{n=1}^{\infty} \frac{n(n+1)}{2} C_n \frac{dQ_n}{d\lambda}$$

Further, using the expansion of the unknown function $m(c\mu')$ in the form

$$m(c\mu') = -2\pi c^2 V_\infty (1-\mu'^2)\sum_{n=1}^{\infty} c_k \frac{dP_k}{d\mu'}$$

and noting that, by virtue of the orthogonality of the Legendre polynomials,

$$\int_{-1}^{1} (1-\mu'^2)\frac{dP_n}{d\mu'}\frac{dP_k}{d\mu'}d\mu' = \begin{cases} 0 & \text{at } k \neq n, \\ \dfrac{2n(n+1)}{2n+1} & \text{at } k = n \end{cases}$$

we see that the following equation is valid:

$$c_n = C_n$$

Thus, we have

$$m(c\mu') = m(x') = -2\pi c^2 V_\infty \left[1 - \left(\frac{x'}{c}\right)^2\right]\sum_{n=1}^{\infty} C_k \frac{dP_k}{d\left(\frac{x'}{c}\right)} \tag{77}$$

The sets of formulas (72) and (76) and (73) and (77) allow, if desired, the use of the perturbation velocity potential in cylindrical coordinates if the coefficients A_n and C_n have been calculated in advance. Note that these coefficients can be more simply calculated by expanding the equation for the meridian contour into series in the functions of elliptical coordinates, and then the calculations in elliptical or cylindrical coordinates can be carried out.

It should also be noted in conclusion that for very slender bodies it is possible to determine $q(x)$ and $m(x)$ from the following two simplest assumptions:

1. In the case of longitudinal flow we shall regard the perturbation velocity V_n', which is normal to the body surface, to be equal to the velocity of plane motion from the source located at the nearest point of the axis. Then the impermeability condition for the surface will give

$$V_n' = \frac{q(x)}{2\pi r} = U_\infty \frac{dr}{dx}$$

where

$$q(x) = 2\pi U_\infty r \frac{dr}{dx} = U_\infty \frac{dA}{dx} \tag{78}$$

with the prescribed function $r(x)$ describing the meridian section contour, A is the cross-sectional area.

2. In the case of transverse flow about a body of revolution, we select $m(x)$ from the requirement that the flow past the body element, cut by the planes x and $x + dx$, be the same as the flow past the element of an infinite cylinder in plane motion. This will lead to the equation

$$m(x) = 2\pi V_\infty r^2(x) = 2V_\infty A(x) \tag{79}$$

10. Elementary Theory of the Wing of Finite Span

When considering the plane flow past a cylindrical wing of infinite span, the vortices formed from the interaction of the wing with the flow can be replaced by one bound vortex responsible for the lifting force of the wing. According to the Helmholtz classical theory, this bound vortex cannot begin or terminate in the fluid. Coinciding in direction with the axis of the wing of infinite span, the bound vortex arrives from infinity and recedes to infinity. The bound vortex intensity is constant along the cylindrical wing axis, as is the velocity circula-

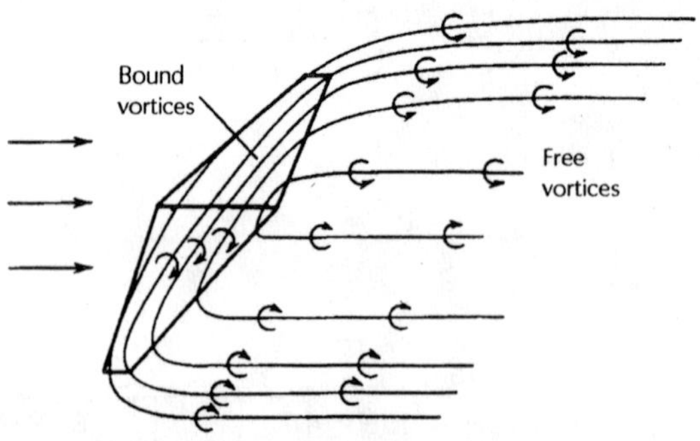

Fig. 128

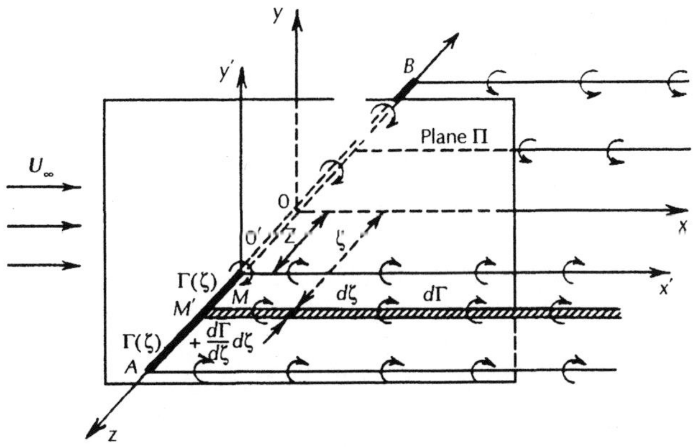

Fig. 129

tion along the contour which encircles any section of the wing and also the lift force of the wing per unit length.

Experience shows that on the wing of finite span, for example, on the wing of an aircraft, the circulation does not persist along the axis, but attains its maximum value in the middle of the wing and vanishes at its ends. This is attributable to the possibility for the equalization of pressures on the lower and upper surfaces of the wing due to the arising overflows of air at the ends of the wing from the region of elevated pressure on the lower surface into the region of rarefaction on the upper surface.

The equalization of pressures leads to the disappearance of the lift force and, consequently, of the bound vortex circulation on the wing ends. The overflow of air from the lower to the upper surface forms transverse streams on the wing which are washed away from the surface by the incident flow and which form vortices when they run off from the trailing wing edge.

The first steps in the creation of the theory of the wing of finite span were made in Russia by Chaplygin[15] and in Germany by Finsterwalder[16] in 1910. However, Prandtl's theory of the lifting line,[17] which was developed in the period from 1913 to 1918, and whose essentials are given in this section, found wide application due to its exceptional simplicity and clarity.

The essence of this scheme of the wing of finite span follows. The main bound vortical filament on the wing sheds the so-called free vortices, the axes of which at a certain distance from the wing coincide with the streamlines of the fluid that carries them away. In the case of translational uniform motion of the wing of finite span in the direction perpendicular to the wing axis or with the incidence of a uniform flow on the wing, the wing can be replaced by a certain imaginary stationary system of motionless vortices which consists of the bound vortices of the wing and of the free vortices shed from the wing; this scheme is shown in Figure 128.

Somewhat idealizing the scheme, we replace the bound vortex of the wing by a lifting vortex line represented by the segment $-l \leq z \leq l$ of the axis $0z$, and locate the free vortices in the plane $x0z$ in the form of rays receding to infinity parallel to the axis $0x$ along which the flow approaches the wing (Figure 129). Downstream of the lifting line, the free vortices form a vortex sheet which, similar to the vortex layer (Chapter 7), represents the surface of the discontinuity of velocity components parallel to the axis $0z$.

---

[15] Chaplygin, S. A., The Results of Theoretical Investigations of the Motion of Airplanes, the paper read on 9/XI, 1910, *Collected Papers*, Vol. 2, Gostekhizdat, Moscow, 1948, pp. 230 to 245.

[16] Finsterwalder, Die Aerodynamik als Grundlage der Luftschiffahrt, *ZFM*, No. 1 and 2, 1910.

[17] Prandtl, L., Ergebnisse und Ziele der Göttinger Modellversuchanstalt, *ZFM*, No. 3, 1913 and also *Tragflügeltheorie* I, II, Göttinger, Nachrichten, 1918.

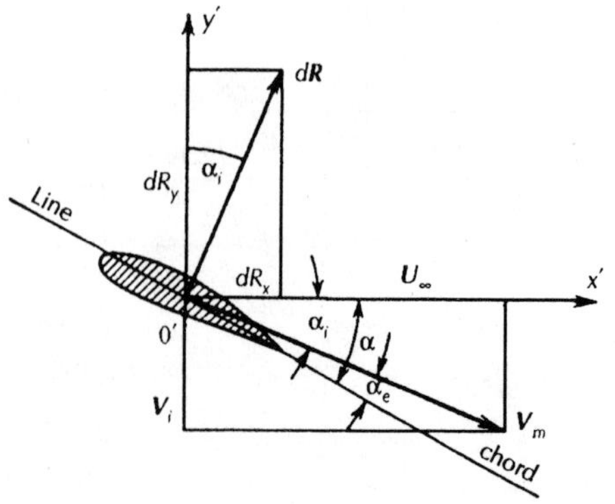

Fig. 130

Let the continuous and differentiable function $\Gamma(z)$ characterize the distribution of circulation along the lifting line ($-l \le z \le l$). To the change in the circulation of the bound vortex from the value of $\Gamma(\zeta)$ at the point M ($z = \zeta$) to $\Gamma(\zeta) + \frac{d\Gamma}{d\zeta}\, d\zeta$ at the point M' ($z = \zeta + d\zeta$) by $d\Gamma = \frac{d\Gamma}{d\zeta}\, d\zeta$ corresponds the shedding of the vortex strip (hatched in Figure 129) which forms the vortex sheet element whose circulation is also equal to $d\Gamma$.

The vortex system of the wing of finite span induces the velocity field which is compounded with the uniform incident flow. This superposition produces the nonuniform velocity field which admits approximate consideration.

Draw through the points of the lifting line the planes perpendicular to it one of which Π ($x'0'y'$) is shown in Figure 130. Consider the projection of the real velocity field at the points of the plane Π onto this plane and call the corresponding flow (not having transverse velocities w) as the section of the actual flow by the plane Π or, for brevity, the plane section of the flow.

If the wing had an infinite span, the flow would have been plane; then, having removed the wing, we would have obtained a uniform field of the incident flow with a certain velocity at infinity $\boldsymbol{U}_\infty$. In the case of the wing of finite span this is not so. If in the plane section the field of disturbances from the lifting line element (located in this plane) is subtracted from the total field, the remaining field of the plane section of the flow will contain both the uniform portion $\boldsymbol{U}_\infty$ of the incident flow and the additional nonuniform portion $\boldsymbol{V}_i$ induced by the free vortices of the sheet located in the plane $x0z$. The nonuniformity of the field of this induced velocities $\boldsymbol{V}_i$ results from different distances of separate points of the plane from the elements of the free vortices of the sheet.

Analyzing the quantitative difference between the induced velocities at the points of the plane Π near the point $0'$, it would be possible to prove[18] that in all the plane sections of the flow at a distance from the ends A and B of the lifting line the differences between the fields of the induced velocities near the point of the intersection of the lifting line with the close planes of the section are smaller, and thus the greater the aspect ratio of the wing, i.e., the greater is the ratio of its span to the mean chord.

---

[18] Dorodnitsyn, A. A., Extension of the lifting line theory to the case of a wing with a curved axis not perpendicular to the flow, *Prikl. Mat. Mekh.*, 8, 1944.

It seems admissible for each plane section of the flow to introduce the concept about its local velocity of approach $\boldsymbol{V}_m$ (Figure 130) equal to the sum of the flow velocity vectors at infinity in front of the wing $\boldsymbol{U}_\infty$ and the induced velocity $\boldsymbol{V}_i$ from the free vortices of the sheet

$$\boldsymbol{V}_m = \boldsymbol{U}_\infty + \boldsymbol{V}_i \tag{80}$$

As seen in Figure 130, the vector of the induced velocity $\boldsymbol{V}_i$ at the point $0'$ of the lifting line must be directed along the axis $0'y'$. Its orientation toward the negative side of the axis $0'y'$ corresponds to the directions of the rotation of liquid particles around free vortex lines, as shown in the figure.

Adopt the following hypothesis of plane sections: at rather great aspect ratios of the wing of finite span the flow past each plane section of stream at a distance from the ends of the wing can be considered as a plane flow with the local approach velocity equal to the sum of the flow velocity vectors at infinity in front of the wing and of the velocity induced by the free vortices of the sheet at the corresponding point of the lifting line.

This hypothesis, which endows the conditional plane sections of the flow with the meaning of actual plane motions, reduces the calculation of the wing of finite span to the solution of the problem (given in Chapter 7) of a plane flow past wing profiles formed at the intersection of the wing of finite span by the planes normal to the wing axis and to the subsequent summation of the results over all the plane sections of the wing. Such an assumption has meaning only for the wings of great aspect ratios.

Let us denote by α (Figure 130) the angle of attack of the incident flow at infinity in front of the wing, i.e., the angle between the vector $\boldsymbol{U}_\infty$ and the chord of the wing cross-section and call this angle the geometric angle of attack. Let us also introduce into consideration the actual (or effective) angle of attack α_e as the angle between the local velocity at infinity $\boldsymbol{V}_m$ and the same chord. The angle between the velocities $\boldsymbol{U}_\infty$ and $\boldsymbol{V}_m$ will be designated by α_i and it will be called the angle of flow skewness or the induced angle. As is seen from Figure 130.

$$\alpha_e = \alpha - \alpha_i \tag{81}$$

Find the projection dv_i onto the axis $0y$ of the elementary induced velocity at the point $0'$ (Figure 129) from the vortex strip bounded by the vortex rays emanating from the points M and M'. Considering this strip as a vortex filament with the circulation $d\Gamma$ and applying formula (22) at $\alpha = 90°$, $\beta = 0°$, we obtain

$$dv_i = -\frac{1}{4\pi}\frac{d\Gamma}{z-\zeta} = -\frac{1}{4\pi}\frac{d\Gamma}{d\zeta}\frac{d\zeta}{z-\zeta} \tag{82}$$

Here (Figure 129) $d\Gamma < 0$, $d\zeta > 0$, $z < \zeta$ and, according to Equation 82, $dv_i < 0$ which corresponds to the location of the point $0'$ (shown in the figure) where the elementary induced velocity is determined with respect to the vortex ray emanating from the point M. From the elementary induced velocity dv_i we go to the total induced velocity v_i at the point $0'$, having summed up the values of dv_i over all the elementary strips of the vortex sheet except for that which emanates from the lifting line segment $(z-\varepsilon, z+\varepsilon)$ (contained within the point $0'$ with the coordinate z). This is explained by the fact that the vortex filament does not induce certain velocities at its points that are the singular points of the velocity field around the vortex filament. The value of ε can be taken as small as possible, and the above summation will give the following expression for the induced velocity

$$v_i = -\frac{1}{4\pi}\lim_{\varepsilon\to 0}\left(\int_{-l}^{z-\varepsilon}\frac{d\Gamma}{d\zeta}\frac{d\zeta}{z-\zeta} + \int_{z+\varepsilon}^{l}\frac{d\Gamma}{d\zeta}\frac{d\zeta}{z-\zeta}\right) \tag{83}$$

The limit involved in this formula is called the principal value (in Cauchy's sense) of the improper integral[19]

$$\int_{-l}^{l} \frac{d\Gamma}{d\zeta} \frac{d\zeta}{z-\zeta} \tag{84}$$

This means in what follows that integral (84) must be calculated in the sense of its principal value determined on the right side of Equation 83. We have in brief notation

$$v_i = -\frac{1}{4\pi} \int_{-l}^{l} \frac{d\Gamma}{d\zeta} \frac{d\zeta}{z-\zeta} \tag{85}$$

Assume that the induced velocities v_i are small as compared with the free stream velocity U_∞. As is clear from the following, this corresponds to the cases of small angles of attack for which the linear law of coupling between the lifting force coefficient and the angle of attack is valid. Then, noting that (Figure 130)

$$\alpha_i \approx \operatorname{tg} \alpha_i = -\frac{v_i}{U_\infty}$$

we shall obtain, according to Equation 85, the formula for the induced angle or flow skewness

$$\alpha_i = \frac{1}{4\pi U_\infty} \int_{-l}^{l} \frac{d\Gamma}{d\zeta} \frac{d\zeta}{z-\zeta} \tag{86}$$

For calculating the induced velocity and angle, we assume the distribution of circulation over the span $\Gamma(z)$ to be given. Let us represent it in the form of the trigonometric series

$$\Gamma(\theta) = 4U_\infty l \sum_{n=1}^{\infty} A_n \sin n\theta \tag{87}$$

where the angle θ is connected with the coordinate z, which is variable over the span, by the equation

$$z = -l \cos \theta \tag{88}$$

To the ends of the lifting line segment ($z = -l$, $z = l$) correspond the values $\theta = 0$ and $\theta = \pi$; in this case the circulation Γ vanishes.

If the distribution of circulation is symmetric with respect to the coordinate origin ($z = 0$, $\theta = \pi/2$), then it should be that $\Gamma(\theta) = \Gamma(\pi - \theta)$ and, consequently,

$$A_2 = A_4 = \cdots = A_{2n} = \cdots = 0$$

[19] See, e.g., Smirnov, V. I., *Kurs vysshey matematiki (The Course of Higher Mathematics)*, Vol. 3, GTTI Press, Moscow, 1933, p. 415; Vol. 4, Gostekhizdat Press, Moscow, 1941, p. 240.

Let us calculate by Equation 87 the derivative $d\Gamma/d\zeta$ assuming, along with Equation 88, that $\zeta = -\lambda \cos \theta'$; we have

$$\frac{d\Gamma}{d\zeta} = \frac{d\Gamma}{d\theta'} : \frac{d\zeta}{d\theta'} = 4U_\infty l \sum_{n=1}^{\infty} nA_n \cos n\theta' \frac{1}{l \sin \theta'} = 4U_\infty \sum_{n=1}^{\infty} nA_n \frac{\cos n\theta'}{\sin \theta'}$$

Substituting this expression into Equation 85, we obtain the expression for the induced velocity

$$v_i(\theta) = -\frac{U_\infty}{\pi} \int_0^\pi \sum_{n=1}^{\infty} \frac{nA_n \cos n\theta'}{\cos \theta' - \cos \theta} d\theta' = -\frac{U_\infty}{\pi} \sum_{n=1}^{\infty} nA_n \int_0^\pi \frac{\cos n\theta'}{\cos \theta' - \cos \theta} d\theta'$$

The integral under the summation sign is calculated in the sense of its principal value and is equal to[20]

$$\int_0^\pi \frac{\cos n\theta'}{\cos \theta' - \cos \theta} d\theta' = \frac{\pi \sin n\theta}{\sin \theta}$$

so that finally the following expressions can be obtained: for the induced velocity

$$v_i(\theta) = -U_\infty \sum_{n=1}^{\infty} nA_n \frac{\sin n\theta}{\sin \theta} \tag{89}$$

and, by Equation 86, also for the angle of skewness

$$\alpha_i(\theta) = \sum_{n=1}^{\infty} nA_n \frac{\sin n\theta}{\sin \theta} \tag{90}$$

Examine the determination of the forces that act on the wing of finite span. According to the Zhukovskiy theorem, for the lifting line segment of length dz we have the elementary force which is perpendicular to the local approach velocity $\boldsymbol{V}_m$ (Figure 130) and equal in magnitude to

$$dR = \rho V_m \Gamma dz$$

accurate to the small quantities of second order with respect to α_i, it is possible to assume that

$$V_m = \sqrt{U_\infty^2 + v_i^2} \approx U_\infty \sqrt{1 + \alpha_i^2} \approx U_\infty$$

Thus, instead of the previous formula we obtain

$$dR = \rho U_\infty \Gamma dz \tag{91}$$

[20] The calculation of this integral can be found, e.g., in the book Golubev, V. V., *Lektsii po teorii kryla* (*Lectures on the Theory of the Wing*), Gostekhizdat Press, Moscow, 1949, pp. 215 to 216.

Calculating the components over the coordinate axes, we obtain in the approximation adopted (Figure 130) that

$$dR_x = dR \sin \alpha_i \approx dR \cdot \alpha_i = \rho U_\infty \Gamma \alpha_i dz$$

$$dR_y = dR \cos \alpha_i \approx dR = \rho U_\infty \Gamma dz$$

and, integrating over the span,

$$R_x = \rho U_\infty \int_{-l}^{l} \Gamma(z)\, \alpha_i(z)\, dz \qquad R_y = \rho U_\infty \int_{-l}^{l} \Gamma(z)\, dz \tag{92}$$

To calculate the integrals, we avail ourselves again of the trigonometric representations of circulation (87) and of induced angle (90):

$$R_x = 4\rho U_\infty^2 l^2 \int_0^\pi \sum_{n=1}^{\infty} A_n \sin n\theta \sum_{m=1}^{\infty} mA_m \frac{\sin m\theta}{\sin\theta} \sin\theta\, d\theta =$$

$$= \rho U_\infty^2 (2l)^2 \sum_{n,\, m=1}^{\infty} mA_n A_m \int_0^\pi \sin n\theta \sin m\theta\, d\theta$$

Noting that by the property of the orthogonality of the functions of the sines of multiple arcs

$$\int_0^\pi \sin n\theta \sin m\theta\, d\theta = \begin{cases} \pi/2 & \text{if } n = m \\ 0 & \text{if } n \neq m \end{cases}$$

we obtain

$$R_x = \pi \frac{\rho U_\infty^2}{2} (2l)^2 \sum_{n=1}^{\infty} nA_n^2 \tag{93}$$

Analogously,

$$R_y = \rho U_\infty \int_{-l}^{l} \Gamma(z)\, dz = 4\rho U_\infty^2 l^2 \int_0^\pi \sum_{n=1}^{\infty} A_n \sin n\theta \sin\theta\, d\theta =$$

$$= \rho U_\infty^2 (2l)^2 \sum_{n=1}^{\infty} A_n \int_0^\pi \sin n\theta \sin\theta\, d\theta$$

However, according to the above property of the orthogonality of the sines of multiple arcs

$$\int_0^{\pi} \sin n\theta \sin \theta \, d\theta = \begin{cases} \frac{\pi}{2} & \text{when} \quad n = 1 \\ 0 & \text{when} \quad n > 1 \end{cases}$$

The sum involved in the previous equation will preserve only one term, so that

$$R_y = \pi \frac{\rho U_\infty^2}{2} (2l)^2 A_1 \tag{94}$$

Considering the flow past a wing of finite span the same as its uniform, translational and rectilinear motion with the velocity U_∞ in the fluid being at rest at infinity, it is natural to call the component R_x directed to the side opposite to the motion of the body as the drag of the wing and the component R_y perpendicular to the direction of motion and to the lifting line as the lift force. At the same time, noting the vortical nature of the drag, which represents a part of the lift force in the flow skewed near the lifting line due to the induced action of the vortex sheet, this drag is called the induced drag.

Denote by S the wing plan area, i.e., its projection on the plane $x0z$ containing the incident flow velocity and the wing axis (lifting line). Introduce into consideration the coefficients of the induced drag c_{xi} and of the lift force c_y, having assumed that

$$c_{xi} = \frac{R_x}{\frac{1}{2}\rho U_\infty^2 \cdot S} \qquad c_y = \frac{R_y}{\frac{1}{2}\rho U_\infty^2 \cdot S}$$

Then, using Equations 93 and 94

$$c_{xi} = \pi \frac{(2l)^2}{S} \sum_{n=1}^{\infty} n A_n^2 \qquad c_y = \pi \frac{(2l)^2}{S} A_1 \tag{95}$$

The quantity

$$\lambda = \frac{(2l)^2}{S} \tag{96}$$

which in the case of the wing with rectangular plan is equal to the ratio of the wing span $2l$ to its chord b, is called the aspect ratio of the wing. Introducing this quantity into Equation 95, we finally obtain the following formulas for the coefficients of the induced drag and lift force:

$$c_{xi} = \pi\lambda \sum_{n=1}^{\infty} n A_n^2 \qquad c_y = \pi\lambda A_1 \tag{97}$$

As follows directly from the first equation of system (95), the induced drag represents the sum of essentially positive quantities. From this it follows that the induced drag of the wing of finite span at the lift force other than zero ($A_1 \neq 0$) will be at minimum if all the coefficients in the expansion of circulation, except for the first one, are equal to zero. According to Equation 87, this corresponds to the distribution of circulation

$$\Gamma = 4U_\infty l A_1 \sin\theta \tag{98}$$

or, returning to the variable z of Equation 88, to

$$\Gamma = 4U_\infty l A_1 \sqrt{1 - \left(\frac{z}{l}\right)^2} \tag{99}$$

Rewriting the latter equation in the form

$$\frac{\Gamma^2}{(4U_\infty l A_1)^2} + \frac{z^2}{l^2} = 1$$

we see that the distribution of circulation over the wing span (lifting line) will be represented by the ellipse with the semiaxes: along the axis z, equal to the semispan of the wing l, along the axis Γ, to the circulation Γ_m maximum along the span; the coefficient A_1 can be expressed in terms of this maximum circulation Γ_m as

$$4U_\infty l A_1 = \Gamma_m \qquad A_1 = \frac{\Gamma_m}{4U_\infty l} \tag{100}$$

The equation of the ellipse is

$$\frac{\Gamma^2}{\Gamma_m^2} + \frac{z^2}{l^2} = 1 \tag{101}$$

The resulting distribution of circulation is called elliptical. According to what has just been proved, in the case of elliptical distribution of circulation the induced drag is at a minimum. In connection with this, the wing with the elliptical distribution of circulation is of fundamental importance in the theory of the wing; we shall consider its main properties. First, it directly follows from formulas (89) and (90) that in the case of the elliptical distribution of circulation the induced velocity and the induced angle (skewness) are constant along the entire span. In fact, substituting into formulas (89) and (90) the values of the coefficients A_n

$$A_1 = \frac{\Gamma_m}{4U_\infty l} \qquad A_2 = A_3 = \cdots = 0$$

we obtain

$$v_i = -\frac{\Gamma_m}{4l} \qquad \alpha_i = \frac{\Gamma_m}{4U_\infty l} \tag{102}$$

The wing with the geometric angle of attack constant over the span is called the geometrically straight wing; the wing with the actual angle of attack constant over the span is called the aerodynamically straight wing.

From the above property that the angle of skewness along the wing span is constant with the elliptical distribution of circulation, it follows that the geometrically straight wing with the elliptical distribution of circulation will also be aerodynamically straight.

We conclude from formulas (102) that with an increase in the span at the given maximum circulation the induced velocity and the skewness angle tend to zero as it should be in transition to the wing of infinite span.

Let us prove that the geometrically straight wing with the elliptical distribution of circulation and with the aerodynamic characteristics of sections being identical over the entire span has elliptical shape in the plan.

To prove the above, connect the lift force coefficient c_y' of the wing section with the corresponding value of circulation $\Gamma(z)$. According to the Zhukovskiy theorem, for the wing unit length (b is the chord)

$$\rho U_\infty \Gamma = c_y' \frac{\rho U_\infty^2}{2} b$$

or, recalling also that for small angles of attack α_e', reckoned from the direction of the zero lift force,

$$c_y' = \left(\frac{dc_y'}{d\alpha}\right)_{c_y'=0} \cdot \alpha_e' = a_0 \alpha_e'$$

where a_0 is the slope of the curve of c_y' vs. α and α_e' is the actual angle of attack, we find the general formula of the sought-after coupling in the form

$$\Gamma = \frac{1}{2} a_0 b U_\infty \alpha_e' \tag{103}$$

From this it follows directly that the geometrically straight wing ($\alpha = \text{const}$, $\alpha_e' = \text{const}$), with the elliptical distribution of circulation and constant (along the span) aerodynamic characteristic of the wing sections a_0, has the law of variation of the chord b along the span identical with the law of variation of the circulation Γ; i.e., it will also be elliptical. According to Equations 101 and 103 and the obvious relation $\Gamma_m = 1/2 a_0 b_m U_\infty \alpha_e'$, the shape of the wing in the plan will be represented by the equation (b_m is the maximum chord of the section corresponding to $z = 0$)

$$\frac{b^2}{b_m^2} + \frac{z^2}{l^2} = 1 \tag{104}$$

Thus, under the adopted conditions of geometric straightness and of identical aerodynamic characteristics along the span, the wing with the elliptical distribution of circulation will also have the elliptical shape in the plan; such a wing can be called elliptical.

Let us determine the coupling between the coefficients of the induced drag and the lift force of the elliptical wing. According to Equation 97, at $A_2 = A_3 = \cdots = 0$,

$$c_{xi} = \pi\lambda A_1^2 \qquad c_y = \pi\lambda A_1$$

Eliminating A_1, we obtain

$$c_{xi} = \frac{1}{\pi\lambda} c_y^2 \tag{105}$$

An analogous formula can be also derived for the wing of any shape in the plan. We assume that

$$\frac{1}{A_1^2} \sum_{n=1}^{\infty} n A_n^2 = 1 + \frac{1}{A_1^2} \sum_{n=2}^{\infty} n A_n^2 = 1 + \delta \tag{106}$$

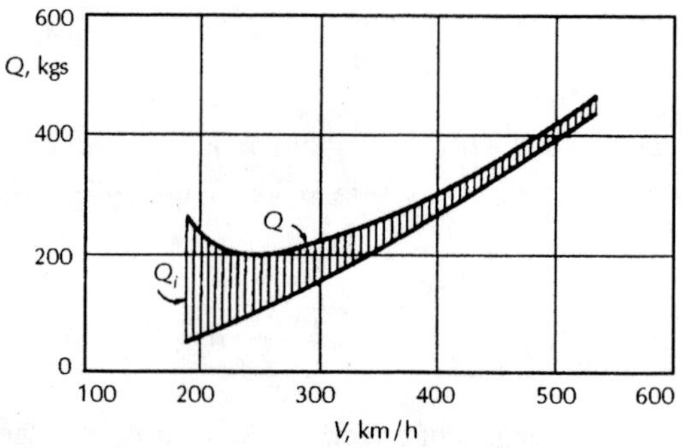

Fig. 131

where the smaller δ is, the closer the wing considered is to be elliptical. Then, from Equation 97 we shall obtain in the general case that

$$c_{xi} = \frac{1+\delta}{\pi\lambda} c_y^2 \qquad (107)$$

Let us assume that during the flight at maximum velocities the values of c_y required for sustaining an aircraft are small ($c_y \approx 0.15$ to 0.20). In this case the induced drag coefficients c_{xi} will be small as compared with the coefficients of profile resistance c_{xp} caused by the friction and pressure resistances originating due to the nonideality of air. Conversely, at small flight velocities the induced drag becomes of major importance.

In Figure 131 a typical curve of the total drag Q of an aircraft is presented for illustration; the role of induced drag (hatched strip) at different flight velocities is emphasized.[21] In flight with relatively great values of c_y (for example, longrange transport airplanes) it is advantageous to increase the aspect ratio with the limits determined by the strength of the wing and by other design considerations.

These problems, as with the problems of application of formulas (97) to particular wings, are considered in special books dealing with the theory of the wing and aerodynamics of the aircraft.

Let us consider the most complex problem of the theory of the wing, *viz.*, the problem of determining the circulation formed on the wing of specified shape in the plan with the prescribed aerodynamic characteristics of sections. We shall retain the designations $b(z)$, $\alpha(z)$, and $a_0(z)$ for the prescribed variables along the span of the chord, geometrical angle of attack, and the derivative of the lift the angle of attack. Then, from the formulas (103) and (81) for the circulation $\Gamma(z)$:

$$\Gamma(z) = \frac{1}{2} a_0(z)\, b(z)\, U_\infty \alpha_e' = \frac{1}{2} a_0(z)\, b(z)\, U_\infty [\alpha(z) - \alpha_i(z)] \qquad (108)$$

If in this equation the induced angle $a_i(z)$ is replaced according to Equation 86, then, to determine the unknown circulation $\Gamma(z)$, we find the following integro-differential equation of Prandtl

$$\Gamma(z) = \frac{1}{2} a_0(z)\, b(z)\, U_\infty \left[\alpha(z) - \frac{1}{4\pi U_\infty} \int_{-l}^{l} \frac{d\Gamma}{d\zeta} \frac{d\zeta}{z-\zeta} \right] \qquad (109)$$

[21] See Goroshchenko, B. T., *Aerodinamika skorostnogo samolyota (The Aerodynamics of a High-Speed Aircraft)*, Oborongiz Press, Moscow, 1948, p. 25.

It should be reemphasized that in this equation the geometric angle of attack $\alpha(z)$, just as the actual angle in the previous equation, is meant to be the angle reckoned from the direction of the zero lift force.

The present-day special courses in the aerodynamics of aircraft suggest the methods of solution to the Prandtl equation (109), including the methods that employ machine computation. As previously mentioned, the above theory of the "lifting line" is applicable only for calculating the aircraft wings with a large relative aspect ratio. The theory of the wings of small aspect ratios is based on the replacement of the wing by the vortex surface instead of the vortex "lifting line". There is an extensive literature on the problem both in the former Soviet Union and abroad. The reader is referred to the *Collected Papers on the Aerodynamics* (Oborongiz Press, 1957). In the papers by P. I. Chushkin and G. A. Kolesnikov, the methods for designing the wings with small aspect ratios are presented and the main bibliography on the problem is given.

In large part due to the studies carried out by Soviet republics scientists, the theory of unsteady motion of a wing in an irrotational flow of incompressible liquid and gas has been advanced greatly.[22]

11. General Case of Solid Body Motion in an Infinite Ideal Incompressible Fluid – Kirchhoff's Problem

When considering the external flow around a solid body, it has been assumed that the body is fixed and that the incident flow is uniform and steady, or that the fluid far from the body is quiescent and the body moves through it translationally, rectilinearly, and uniformly. It is under this assumption that d'Alembert's paradox about the vanishing resultant vector of the forces of liquid pressure on the surface of the body of finite dimensions was proved. Consider the general case of a nonuniform and nontranslational motion of a solid body in an infinite, incompressible ideal fluid which is quiescent at infinity.

Let us agree that in what follows all the quantities relating to a solid body will be designated by an asterisk, and the same quantities in the fluid surrounding the body denoted by the standard notation. Thus, the velocity vector of the points of the solid body will be designated by $\boldsymbol{V}^*$ (u^*, v^*, w^*) and will be regarded as equal to

$$\boldsymbol{V}^* = \boldsymbol{V}_0^* + \boldsymbol{\omega}^* \times \boldsymbol{r}^* \tag{110}$$

where $\boldsymbol{V}_0^*$ (u_0^*, v_0^*, w_0^*) is the velocity of the arbitrary point 0^* of the solid body taken to be a pole; $\boldsymbol{\omega}^*$ is the angular velocity of body rotation around an instantaneous axis passing through the pole 0^*; $\boldsymbol{r}^*$ is the vector-radius of the solid body point relative to the pole 0^*.

We introduce into consideration two coordinate systems: (1) the absolute fixed system $0xyz$ and (2) the relative mobile system $0^*x^*y^*z^*$ connected with the solid body. If there is no differentiation with respect to time in the course of derivation, the time loses its meaning as an independent variable and simply becomes a parameter which notes the succeeding spatial pictures of the phenomenon. Without loss of generality, it is possible at any fixed instant of time to consider both coordinate systems coinciding, and to use either the coordinates x^*, y^*, z^*, or x, y, z to describe the phenomenon.

---

[22] Belotserkovskiy, S. M., Skripach, B. K., and Tabachnikov, V. G., *Krylo v nestatsionarnom potoke gaza* (*A Wing in an Unsteady Gas Flow*), Nauka Press, Moscow, 1971 (with extensive bibliography).

In cases in which time is an argument with respect to which the differentiation is carried out, it is already nonpermissible to ignore the mutual motion of the coordinate systems. It therefore becomes necessary to distinguish two kinds of derivatives: the absolute, d/dt calculated in the fixed coordinate system $0xyz$ and the relative, d^*/dt determined in the mobile system $0^*x^*y^*z^*$ connected with the solid body. Recall that the absolute and relativetime derivatives of a certain vector-function $\boldsymbol{a}(t)$ are connected by the relation[23]

$$\frac{d\boldsymbol{a}}{dt} = \frac{d^*\boldsymbol{a}}{dt} + \boldsymbol{\omega}^* \times \boldsymbol{a} \tag{111}$$

According to the Legendre theorem proved at the beginning of Chapter 7, it is possible to assume that the motion in the fluid caused by a body moving in it will be irrotational. In contrast to the previous steady motions, the velocity potential φ is the function not only of the coordinates, but also of time.

From the equation $\boldsymbol{V} = \operatorname{grad} \varphi$ and from the liquid incompressibility equation $\operatorname{div} \boldsymbol{V} = 0$, it follows that the sought-after potential $\varphi(x, y, z, t)$ should satisfy, at each given time instant, the Laplace equation

$$\nabla^2\varphi = 0 \tag{112}$$

The boundary condition of impermeability for the solid body surface σ which requires that the projections of the velocity $\boldsymbol{V}$ of liquid particles that slide along the surface onto the normal $\boldsymbol{n}$ to the surface σ should coincide with the corresponding projections of the velocity $\boldsymbol{V}^*$ of the solid surface points. The condition of the decrease of the potential φ on receding from the body can be written in the form

$$V_n = \frac{\partial\varphi}{\partial n} = V_n^* = V_{0n}^* + (\boldsymbol{\omega}^* \times \boldsymbol{r}^*) \cdot \boldsymbol{n} = u_0^* n_x + v_0^* n_y + w_0^* n_z +$$

$$+ \omega_x^* (yn_z - zn_y) + \omega_y^* (zn_x - xn_z) + \omega_z^* (xn_y - yn_x) \quad (\text{at } \sigma)$$

$$\varphi = O(1/R^2) \to 0 \quad \text{when} \quad R \to \infty \tag{113}$$

In these equations the following notations are adopted: $V_{0x}^* = u_0^*$, $V_{0y}^* = v_0^*$, $V_{0z}^* = w_0^*$, with the projections which are taken on the axis $0xyz$ and which at the given time instant coincide with $0^*x^*y^*z^*$.

It should be emphasized that to determine the nonstationary field of the velocity potential $\varphi(x, y, z, t)$, no initial conditions are required because Equation 112 does not involve the partial time derivative. At any time instant t the solution is independent of the flow prehistory. In this problem the time serves only as a parameter, the influence of which is exhibited in the specific form of the right side of boundary condition (113) involving the solid body motion characteristics.

Using the linear character of Equation 112, we seek, following Kirchhoff,[24] the solution of this equation in the form

$$\varphi = u_0^*\varphi_1 + v_0^*\varphi_2 + w_0^*\varphi_3 + \omega_x^*\varphi_4 + \omega_y^*\varphi_5 + \omega_z^*\varphi_6 \tag{114}$$

---

[23] Loitsyanskiy, L. G. and Luriye, A. I., *Kurs teoreticheskoy mekhaniki* (*The Course of Theoretical Mechanics*), Vol. 1, Nauka Press, Moscow, 1982, p. 303 (Equation 12).

[24] Kirchhoff, G. R., *Mekhanika. Lektsii po matematicheskoy fizike* (*Mechanics. Lectures on Mathematical Physics*), AN SSSR Press, Moscow, 1962; Lecture 18th, pp. 122 to 197.

Then, Equation 112 decomposes into a system of the following six Laplace equations for each of the functions φ_i:

$$\nabla^2\varphi_i = 0 \quad (i = 1, 2, \cdots, 6) \tag{115}$$

with the boundary conditions

$$\frac{\partial\varphi_1}{\partial n} = n_x \qquad \frac{\partial\varphi_2}{\partial n} = n_g \qquad \frac{\partial\varphi_3}{\partial n} = n_z$$

$$\frac{\partial\varphi_4}{\partial n} = yn_z - zn_y \qquad \frac{\partial\varphi_5}{\partial n} = zn_x - xn_z \qquad \frac{\partial\varphi_6}{\partial n} = xn_y - yn_x$$

$$\varphi_i = O(1/R^2) \to 0 \text{ when } R \to \infty \tag{116}$$

Equations 115 do not involve time derivatives. Noting that the right sides of boundary conditions (116) in the coordinate system $0xyz$ are identical to the corresponding expressions in the coordinate system $0^*x^*y^*z^*$, coinciding with the former system and connected with the body, we see that the solutions φ_i of Equations 115 under boundary conditions (116) are also independent of time. From this it follows that Equation 114 represents the sought-after solution for the potential $\varphi(x, y, z, t)$ as a sum of the products of the *a priori* given time functions $u_0^*(t)$, $v_0^*(t)$, $w_0^*(t)$; $\omega_x^*(t)$, $\omega_y^*(t)$, $\omega_z^*(t)$, which determine the motion of the body in a fluid by the unknown functions $\varphi_i(x, y, z)$ of the coordinates alone.

The functions $\varphi_i(x, y, z)$ can be interpreted as the velocity potentials of the following fluid motions relative to the coordinate system connected with the solid body. The first three potentials, φ_1, φ_2, φ_3, correspond to the flows past the body of interest during its translational and uniform motions, with the velocities equal to unity along the coordinate axes; the last three potentials, φ_4, φ_5, φ_5, correspond to the rotational motions of the body with unit angular velocities around the coordinate axes. In view of this the functions φ_i are called the unit potentials.

Assuming the functions φ_i to be determined, we seek the resultant vector and the resultant moment of the forces of liquid pressure on the solid body moving in it. Enclose the moving body inside a certain liquid sphere of great radius R_0 with the surface σ_0 and apply the momentum theorem to the liquid mass located in the time-variable volume τ between the surfaces σ and σ_0. Let us designate by $\boldsymbol{Q}$ the resultant vector of the momenta of liquid in the volume τ, and by $\boldsymbol{F}$ the sought-after resultant vector of the pressure forces of liquid on the body surface σ, and by $\boldsymbol{F}'$ the resultant vector of pressure forces applied from outside to the surface σ_0. Then,

$$\frac{d\boldsymbol{Q}}{dt} = -\boldsymbol{F} + \boldsymbol{F}'$$

where it follows that

$$\boldsymbol{F} = \boldsymbol{F}' - \frac{d\boldsymbol{Q}}{dt} \tag{117}$$

It should be noted that in Equation 117 and in the equation above it the time derivative is the absolute derivative; i.e., it expresses the rapidity of variation in time of the resultant vector of momenta of fluid relative to the fixed coordinate system $0xyz$.

The vector $\boldsymbol{F}'$ will be found from the formula

$$F' = -\int_{\sigma_0} p\boldsymbol{n}_0 d\sigma_0$$

where, according to the Lagrange-Cauchy integral (Chapter 7), the pressure p should be replaced by the expression

$$p = \rho f(t) - \frac{\rho V^2}{2} - \rho\frac{\partial\varphi}{\partial t}$$

and according to the condition of the fluid being at rest at infinity

$$p \to p_\infty \qquad V \to 0 \qquad \varphi \to 0 \quad \text{when} \quad R \to \infty$$

the function $f(t)$ can be replaced by the constant value p_∞/ρ. Discarding the integral from the constant term p_∞, we obtain

$$F' = \rho\int_{\sigma_0} \frac{\partial\varphi}{\partial t}\boldsymbol{n}_0 d\sigma_0 + \frac{\rho}{2}\int_{\sigma_0} V^2\boldsymbol{n}_0 d\sigma_0 \tag{118}$$

The resultant vector of the momenta of fluid in the volume enclosed between the surfaces σ and σ_0 will be expressed in terms of the surface integrals as

$$\boldsymbol{Q} = \int_\tau \rho\boldsymbol{V}d\tau = \rho\int_\tau \text{grad}\ \varphi\, d\tau = -\rho\int_\sigma \varphi\boldsymbol{n}d\sigma + \rho\int_{\sigma_0} \varphi\boldsymbol{n}_0 d\sigma_0 \tag{119}$$

the minus sign before the first integral on the right side is due to the fact that the outward normal to the body surface σ is the inward normal relative to the fluid volume τ enclosed between the surfaces σ and σ_0.

Let us now return to the calculation of the resultant vector $\boldsymbol{F}$ of the forces of flow pressure on the body moving in it. According to Equation 117, in order to determine the vector $\boldsymbol{F}$, it is necessary to calculate the ordinary derivative of the resultant vector of momenta $\boldsymbol{Q}$ represented by the right side of Equation 119. Forming the expression for the derivative

$$\frac{d\boldsymbol{Q}}{dt} = -\frac{d}{dt}\int_\sigma \rho\varphi\boldsymbol{n}d\sigma + \frac{d}{dt}\int_{\sigma_0} \rho\varphi\boldsymbol{n}_0 d\sigma_0 \tag{120}$$

the term on the right side remains intact and we express the second term as the sum of the local time derivative, easily calculated for the immovable (independent of time) surface σ_0 in the form

$$\frac{\partial}{\partial t}\int_{\sigma_0} \rho\varphi\boldsymbol{n}_0 d\sigma_0 = \int_{\sigma_0} \rho\frac{\partial\varphi}{\partial t}\boldsymbol{n}_0 d\sigma_0$$

and the convective derivative, which requires for its calculation directly finding the limit for the ratio of the difference between the increased and initial values of the differentiable integral to the increment of time. According to Chapter 4, this limit is reduced to momentum transfer through the surface, i.e., in the given case to the integral

$$\int_{\sigma_0} \rho V_{n_0} \boldsymbol{V} d\sigma_0$$

Substituting the resulting expressions for the local and convective derivatives into the right side of Equation 120, we obtain

$$\frac{d\boldsymbol{Q}}{dt} = -\frac{d}{dt}\int_{\sigma} \rho\varphi \boldsymbol{n}\, d\sigma + \int_{\sigma_0} \rho \frac{\partial\varphi}{\partial t} \boldsymbol{n}_0 d\sigma_0 + \int_{\sigma_0} \rho V_{n_0} \boldsymbol{V} d\sigma_0$$

and, returning to Equations 117 and 118, we find the following expression for the resultant vector of pressure forces applied from the liquid side to the surface of the body moving in it:

$$\boldsymbol{F} = \frac{d}{dt}\int_{\sigma} \rho\varphi \boldsymbol{n}\, d\sigma + \rho \int_{\sigma_0} \left(\frac{1}{2} V^2 \boldsymbol{n}_0 - V_{n_0} \boldsymbol{V}\right) d\sigma_0 \tag{121}$$

Now, let the radius R_0 of the surface σ_0 go to infinity. Then, using the same arguments as those employed to prove d'Alembert's paradox (Section 5), we see that the expression under the integral sign in the last term on the right side of Equation 121 is on the order of $1/R_0^6$, whereas the integration surface is on the order of R_0^2; consequently, when $R_0 \to \infty$, this term tends to zero, Finally, we find the sought-after formula for the resultant vector of the forces of fluid pressure on the body surface

$$\boldsymbol{F} = \frac{d}{dt}\int_{\sigma} \rho\varphi \boldsymbol{n}\, d\sigma \tag{122}$$

Using analogous arguments, we find the expression for the resultant moment of the forces of fluid pressure on the body surface

$$\boldsymbol{M} = \frac{d}{dt}\int_{\sigma} \rho\varphi \boldsymbol{r} \times \boldsymbol{n}\, d\sigma \tag{123}$$

Following up on the earlier adopted notation, we prescribe the resultant vector $\boldsymbol{Q}^*$ and the resultant moment $\boldsymbol{K}^*$ of the momenta of the solid body moving in a fluid. Then, according to the theorems of momenta and moment of momenta of motion as applied to a solid body, we obtain the equation for the solid body motion

$$\frac{d\boldsymbol{Q}^*}{dt} = \boldsymbol{F}^* + \boldsymbol{F} \qquad \frac{d\boldsymbol{K}^*}{dt} = \boldsymbol{M}^* + \boldsymbol{M}$$

where $\boldsymbol{F}^*$ and $\boldsymbol{M}^*$ designate the resultant vector and the resultant moment of external forces applied to the solid body, respectively, besides the fluid responses $\boldsymbol{F}$ and $\boldsymbol{M}$. Taking into account Equations 122 and 123, the previous equations are rewritten in the form

$$\frac{d}{dt}\left(\boldsymbol{Q}^* - \int_{\sigma} \rho\varphi \boldsymbol{n}\, d\sigma\right) = \boldsymbol{F}^* \qquad \frac{d}{dt}\left(\boldsymbol{K}^* - \int_{\sigma} \rho\varphi \boldsymbol{r} \times \boldsymbol{n}\, d\sigma\right) = \boldsymbol{M}^* \tag{124}$$

Introducing the notation

$$-\int_{\sigma} \rho\varphi \boldsymbol{n}\, d\sigma = \boldsymbol{B} \qquad -\int_{\sigma} \rho\varphi \boldsymbol{r} \times \boldsymbol{n}\, d\sigma = \boldsymbol{I} \tag{125}$$

we shorten Equations 124 to

$$\frac{d}{dt}(\boldsymbol{Q}^* + \boldsymbol{B}) = \boldsymbol{F}^* \qquad \frac{d}{dt}(\boldsymbol{K}^* + \boldsymbol{I}) = \boldsymbol{M}^* \tag{126}$$

The following interpretation is assigned to the equations obtained: the equations of motion of a solid body in a fluid can be considered as the equations of body motion in vacuum if the resultant vectors of momenta and moments of momenta of the solid body motion are supplemented with the additional vectors $\boldsymbol{B}$ and $\boldsymbol{I}$, respectively, determined by Equations 125. They are called the vectors of momenta and moments of momenta of fluid motion which are added to the solid body.

Now, consider in more detail expressions (122) obtained for the resultant vector $\boldsymbol{F}$ and (123) for the resultant moment $\boldsymbol{M}$ of the forces of fluid pressure on the surface. With this purpose in mind, making use of Equations 125, they are rewritten for brevity in the form

$$\boldsymbol{F} = -\frac{d\boldsymbol{B}}{dt} \qquad \boldsymbol{M} = -\frac{d\boldsymbol{I}}{dt} \tag{127}$$

and go from the variables in these equations of absolute derivatives to relative derivatives, according to Equation 111.

For the vector $\boldsymbol{F}$,

$$\boldsymbol{F} = -\frac{d^2\boldsymbol{B}}{dt} - \boldsymbol{\omega}^* \times \boldsymbol{B} \tag{128}$$

As regards Equation 123 for the resultant moment $\boldsymbol{M}$, here, it is necessary to take into account the difference (which is very essential for the absolute time derivative to be correctly calculated) between the vector-radius $\boldsymbol{r}$ in the absolute coordinate system and the relative radius-vector $\boldsymbol{r}^*$ coinciding with the former radius-vector at the given time instant. In the case of the adopted instantaneous coincidence of the mobile $0^*x^*y^*z^*$ and fixed $0xyz$ coordinate systems the absolute radius-vector $\boldsymbol{r}$ should be considered to be the limiting value of the sum of the radius-vector of the point 0^* and of the relative radius-vector $\boldsymbol{r}^*$

$$\boldsymbol{r} = (\boldsymbol{r}_{0^*} + \boldsymbol{r})_{r_{0^*} = 0}$$

which should not be ignored when calculating the absolute derivative. According to the second equations of systems (125) and (127),

$$\frac{d\boldsymbol{I}}{dt} = -\frac{d}{dt}\int_{\sigma} \rho\varphi \boldsymbol{r} \times \boldsymbol{n}\, d\sigma = \left[-\frac{d}{dt}\int_{\sigma} \rho\varphi(\boldsymbol{r}_{0^*} + \boldsymbol{r}) \times \boldsymbol{n}\, d\sigma\right]_{r_{0^*} = 0} =$$

$$= -\left[\frac{d\boldsymbol{r}_{0^*}}{dt} \times \int_{\sigma} \rho\varphi \boldsymbol{n}\, d\sigma + \boldsymbol{r}_{0^*} \times \frac{d}{dt}\int_{\sigma} \rho\varphi \boldsymbol{n}\, d\sigma + \frac{d}{dt}\int_{\sigma} \rho\varphi \boldsymbol{r}^* \times \boldsymbol{n}\, d\sigma\right]_{r_{0^*} = 0} =$$

$$= \boldsymbol{V}_0^* \times \boldsymbol{B} + \frac{d}{dt}(\boldsymbol{I})_{\boldsymbol{r} = \boldsymbol{r}^*}$$

After this, in going from the absolute to the relative derivatives, we finally obtain the sought-after formula for the resultant moment of the forces of flow pressure on the body surface

$$M = -\frac{d^*I}{dt} - \omega^* \times I - V_0^* \times B \tag{129}$$

where I is meant to be its expression in the relative coordinate system

$$I = -\int_\sigma \rho\varphi r^* \times n \, d\sigma \tag{130}$$

Let us consider in detail two particular cases of the general motion of a solid body in a fluid.

1. The body moves in the fluid translationally and uniformly. In this case $\omega^* = 0$, B is independent of time and Equation 128 leads to the d'Alembert paradox (see Section 5). As it follows from Equation 129, in this case the resultant moment of the pressure forces is not equal to zero

$$M = -V_0^* \times B$$

To the body equilibrium corresponds the fulfillment of the requirement for the vectors V_0^* and B to be parallel.
2. The motion of the body is composed of the uniform motions of the pole and rotation around the pole. The d'Alembert paradox is not valid in this case; according to Equation 128,

$$F = -\omega^* \times B$$

The resultant moment consists of two terms

$$M = -\omega^* \times I - V_0^* \times B$$

For what follows it is useful to change the designations of the velocity vector projections V_0^* of the body pole and angular velocity ω^* of its rotation having assumed that

$$u_0^* = q_1 \qquad v_0^* = q_2 \qquad w_0^* = q_3 \qquad \omega_x^* = q_4 \qquad \omega_y^* = q_5 \qquad \omega_z^* = q_6$$

Analogously we adopt the designations

$$B_x = B_1 \qquad B_y = B_2 \qquad B_z = B_3 \qquad I_x = B_4 \qquad I_y = B_5 \qquad I_z = B_6$$

In new notation, the velocity potential (114) is expressed as

$$\varphi = \sum_{k=1}^{6} q_k \varphi_k \tag{131}$$

Let us now avail ourselves of the system of Equations 116, which is fulfilled at any time instant on the surface of the body σ; then, according to Equation 131, in new notation:

$$B_i = -\rho\int_\sigma \varphi \frac{\partial\varphi_i}{\partial n} d\sigma = -\rho\int_\sigma \sum_{k=1}^{6} q_k\varphi_k \frac{\partial\varphi_i}{\partial n} d\sigma = \sum_{k=1}^{6} \lambda_{ik} q_k \quad (i = 1, 2, \cdots, 6) \tag{132}$$

where the following designation is introduced

$$\lambda_{ik} = -\rho \int_{\sigma} \frac{\partial \varphi_i}{\partial n} \varphi_k d\sigma \tag{133}$$

The quantities λ_{ik} calculated in the coordinate system, which is connected with the solid body, are certain constants that depend only on the fluid density and on the body surface shape because, as proved earlier, φ_i is independent of time.

As viewed in Equation 132, the coefficients of the momentum and moment of momentum of motion associated with the velocities q_k, the quantities λ_{ik} play the part of inertia coefficients added to the inertia coefficients that enter into the similar expressions for the momentum and moment of momentum of the solid body itself.

Thus, for example, the projection (onto the axis $0x$) of the momentum of the solid body, the mass of which will be denoted by m^*, is equal to

$$Q_x^* = \int_{m^*} V_x^* dm^* = \int_{m^*} (u_0^* + \omega_y^* z - \omega_z^* y)\, dm^* =$$

$$= m^* u_0^* + \omega_y^* \int_{m^*} z dm^* - \omega_z^* \int_{m^*} y dm^* = m^* u_0^* + m^* z_c \omega_y^* - m^* y_c \omega_z^*$$

where y_c and z_c are the coordinates of the body gravity center; from this it follows in new notation

$$Q_1^* = m^* q_1 + m^* z_c q_5 - m^* y_c q_6$$

The projection, onto the $0x$ axis, of the sum of the momentum $\boldsymbol{Q}^*$ and of the "added" momentum $\boldsymbol{B}$, will be equal to

$$Q_1^* + B_1 = (m^* + \lambda_{11})\, q_1 + \lambda_{12} q_2 + \lambda_{13} q_3 + \lambda_{14} q_4 + (m^* z_c + \lambda_{15})\, q_5 + (-m^* y_c + \lambda_{16})\, q_6$$

As is seen from the structure of this expression, the inertia coefficients λ_{ik} are added to the inertia coefficients in the expression for the solid body momentum projection: λ_{11} to the mass, λ_{15} and λ_{16} to the static moments of masses; the remaining coefficients are represented in the general case by the terms that are absent in the expression for the projection of the resultant vector of the solid body momentum. The inertia coefficients λ_{ik} are called the coefficients of added masses.

Thirty-six coefficients of the added masses

$$\lambda_{ik} \begin{pmatrix} i = 1, 2, \cdots, 6 \\ k = 1, 2, \cdots, 6 \end{pmatrix}$$

possess the property of symmetry, i.e., they are independent of the order of indices. In order to prove this, let us form in application to the volume τ, bounded by the surfaces of the body σ and of the sphere σ_0 of great radius R_0, the following familiar relation (the second relation of Equation 88, Chapter 1 at a = grad φ_k is applied):

$$\int_{\tau} \varphi_i \nabla^2 \varphi_k d\tau = \int_{\tau} \varphi_i \operatorname{div}(\operatorname{grad} \varphi_k)\, d\tau = \int_{\tau} \operatorname{div}(\varphi_i \operatorname{grad} \varphi_k) d\tau - \int_{\tau} \operatorname{grad} \varphi_i \cdot \operatorname{grad} \varphi_k\, d\tau$$

and subtract from it the analogous relation with the altered order of indices; this will yield the general formula

$$\int_\tau (\varphi_i \nabla^2 \varphi_k - \varphi_k \nabla^2 \varphi_i)\, d\tau = \int_\tau \operatorname{div}(\varphi_i \operatorname{grad} \varphi_k) d\tau - \int_\tau \operatorname{div}(\varphi_k \operatorname{grad} \varphi_i) d\tau$$

Noting that the unit potentials φ_i and φ_k satisfy the Laplace equation, and applying the Gauss-Ostrogradskiy formula to the right side, we obtain the equation

$$\int_\sigma \left(\varphi_i \frac{\partial \varphi_k}{\partial n} - \varphi_k \frac{\partial \varphi_i}{\partial n}\right) d\sigma = \int_{\sigma_0} \left(\varphi_i \frac{\partial \varphi_k}{\partial n_0} - \varphi_k \frac{\partial \varphi_i}{\partial n_0}\right) d\sigma_0$$

As before, take into account that the integral on the right side tends to zero when the sphere surface σ_0 recedes to infinity (φ_i is on the order of $1/R_0^2$, $\partial \varphi_i / \partial n_0$ is on the order of $1/R_0^3$). Then,

$$\int_\sigma \varphi_i \frac{\partial \varphi_k}{\partial n} d\sigma = \int_\sigma \varphi_k \frac{\partial \varphi_i}{\partial n} d\sigma$$

or, by the definition of the coefficients of added masses,

$$\lambda_{ik} = \lambda_{ki}$$

which proves the symmetry property of these quantities.

The added masses λ_{ik} enter as coefficients into the expression for the quadratic dependence of the kinetic energy T of the perturbed fluid motion on the solid body velocities.

Calculating the kinetic energy of the fluid in the volume τ between the surfaces σ and σ_0, and noting that the outward normal to σ coincides with the inward with respect to the volume τ, we obtain

$$T = \frac{\rho}{2} \int_\tau V^2 d\tau = \frac{\rho}{2} \int_\tau \operatorname{grad} \varphi \cdot \operatorname{grad} \varphi d\tau =$$

$$= \frac{\rho}{2} \int_\tau \operatorname{div}(\varphi \operatorname{grad} \varphi)\, d\tau - \frac{\rho}{2} \int_\tau \varphi \nabla^2 \varphi d\tau = -\frac{\rho}{2} \int_\sigma \varphi \frac{\partial \varphi}{\partial n} d\sigma + \frac{\rho}{2} \int_{\sigma_0} \varphi \frac{\partial \varphi}{\partial n_0} d\sigma_0$$

and noting again that as the surface σ_0 recedes to infinity, the second integral on the right vanishes, we obtain

$$T = -\frac{\rho}{2} \int_\sigma \varphi \frac{\partial \varphi}{\partial n} d\sigma \qquad (134)$$

Let us substitute the expansion of the velocity potential φ into unit potentials of the composite motions φ_i. Then, multiplying the sums, we find the sought-after expression for the kinetic energy of perturbed fluid motion in terms of the body velocities and added masses

$$T = \frac{1}{2}\sum_{i=1}^{6}\sum_{k=1}^{6}\lambda_{ik}q_i q_k \tag{135}$$

Comparing this expression to Equation 132, we obtain the connection between the added kinetic energy of the perturbed motion T and the added momentum B_i:

$$B_i = \frac{\partial T}{\partial q_i} \quad (i = 1, 2, \cdots, 6) \tag{136}$$

By writing in expanded form the expression for the kinetic energy of the moving body itself

$$T^* = \frac{1}{2}\int_{m^*} V^{*2}dm^* = \frac{1}{2}\,[m^*(u_0^{*2} + v_0^{*2} + \omega_0^{*2}) + 2m^*x_c(v_0^*\omega_z^* - \omega_0^*\omega_y^*) +$$

$$+ 2m^*y_c(\omega_0^*\omega_x^* - u_0^*\omega_z^*) + 2m^*z_c(u_0^*\omega_y^* - v_0^*\omega_x^*) + J_x\omega_x^{*2} + J_y\omega_y^{*2} + J_z\omega_z^{*2} -$$

$$- 2J_{yz}\omega_y^*\omega_z^* - 2J_{zx}\omega_z^*\omega_x^* - 2J_{xy}\omega_x^*\omega_y^*\,]$$

we easily see that on adding the kinetic energy T of the fluid perturbed by the body to the energy of the moving body itself T^*, the coefficients λ_{ik}, as in the case of the vectors of momenta and moments of momenta, are added to the corresponding inertia coefficients in the expression for T^* (to the mass, static moments, moments of inertia, and centrifugal moments). This once again explains the sense of the coefficients λ_{ik} and the origin of the term "added masses".

We present calculations of several of the simplest "added masses". Let a circular cylinder of radius a surrounded by an infinite ideal incompressible fluid of density r perform translational motion along the axis $0x$, perpendicular to the cylinder axis, with the velocity u_0, which is the given function of time t. In this case the velocity potential of the perturbed motion will be

$$\varphi = -\text{Re}\left(u_0\frac{a^2}{z}\right) = -u_0(t)\,a^2\text{Re}\left(\frac{1}{z}\right) = -u_0(t)\,a^2\frac{x}{x^2+y^2} =$$

$$= -u_0(t)\,a^2\frac{x}{r^2} = -u_0(t)\,a^2\frac{\cos\varepsilon}{r}$$

The coefficient at $u_0(t)$

$$\varphi_1 = -\frac{a^2\cos\varepsilon}{r}$$

will play the part of the unit potential, whereas the coefficient of the added mass will be equal, according to Equation 133, to

$$\lambda_{11} = -\rho\int_0^{2\pi}\varphi_1\left(\left.\frac{\partial\varphi_1}{\partial r}\right|_{z=a}\right)a\,d\varepsilon = \rho a^2\int_0^{2\pi}\cos^2\varepsilon\,d\varepsilon = \pi\rho a^2 = m$$

where m is the mass of the fluid in the volume of the cylinder per its unit length.

The fluid response to the cylinder will be determined from the formula

$$F_x = -\frac{dB_x}{dt} = -\lambda_{11}\frac{du_0}{dt} = -m\frac{du_0}{dt}$$

In the case of a uniform motion of the cylinder ($du_0/dt = 0$) this force vanishes and d'Alembert's paradox takes place. In the case of an accelerated motion of the cylinder, there is the response of the fluid. The greater the response, the greater the cylinder acceleration.

Setting up the differential equation of cylinder motion (m^* is the mass of the cylinder unit length, $\boldsymbol{F}^*$ is the external force, besides the fluid response)

$$m^*\frac{du_0}{dt} = F_x^* + F_x$$

we see that it can be rewritten as

$$(m^* + m)\frac{du_0}{dt} = F_x^*$$

Under the action of the applied force $\boldsymbol{F}^*$ the cylinder will move in the fluid in the same way as in vacuum if only its mass is increased by the value of the added mass of the fluid in the volume of the cylinder.

In the same simple way the problem of rectilinear motion of a sphere is solved. In this case,

$$\varphi = -\frac{a^3\cos\theta}{2R^2}u_0(t) \qquad \varphi_1 = -\frac{a^3}{2}\frac{\cos\theta}{R^2}$$

$$\lambda_{11} = -\rho\int_0^{2\pi} d\varepsilon\int_0^{\pi}\left(-\frac{a^3}{2}\frac{\cos\theta}{a^2}\right)\left(\frac{a^3}{2}\frac{2\cos\theta}{a^3}\right)a^2\sin\theta\, d\theta = \frac{2}{3}\pi\rho a^3 = \frac{1}{2}m$$

where m is the mass of the fluid in the volume of the sphere.

The differential equation of the sphere motion will be

$$m^*\frac{du_0}{dt} = F_x^* + F_x = -\frac{1}{2}m\frac{du_0}{dt} + F_x^*$$

or

$$\left(m^* + \frac{1}{2}m\right)\frac{du_0}{dt} = F_x^*$$

Comparing this equation with the equation of rectilinear motion of the sphere under the action of the same force $\boldsymbol{F}^*$ in vacuum

$$m^*\frac{du_0}{dt} = F_x^*$$

we conclude that the motion of the sphere in the fluid can be considered similar to that occurring in vacuum if only the mass of the sphere is supplemented with the additional mass equal to the half the mass of the fluid in the volume of the sphere. If the mass of the fluid in the volume of a moving body is small as against the mass of the moving body itself (for example, a shell or an airplane in air), the added mass can be neglected. In other cases (an airship in air, a ship or a torpedo in water, etc.), conversely, the role of the added masses is of primary importance.

It should, of course, be borne in mind that the added masses in the case of actual fluid flow past bodies differ from those calculated here in the model of nonseparated flow of an inviscid incompressible fluid.

Bearing in mind an especially great applied importance of the concept of the added mass for the bodies of revolution (airship and torpedo shapes), we derive the general formulas of added masses for the longitudinal (with respect to the symmetry axis) and transverse (with respect to it) motion of the body of revolution.

In the case of the longitudinal motion along the axis $0x$,

$$\lambda_{11} = \lambda_{xx} = -\rho \int_\sigma \varphi_1 \frac{\partial \varphi_1}{\partial n} d\sigma$$

or, by virtue of boundary condition (116) on the body surface and the obvious equality $d\sigma = 2\pi r ds$,

$$\lambda_{11} = -\rho \int_\sigma \varphi_1 n_x d\sigma = -2\pi\rho \int \varphi_1 r n_x ds = -2\pi\rho \int \varphi_1 r dr$$

Using Equation 54, we obtain

$$\lambda_{11} = 2\pi\rho c^2 \int_{-1}^{1} \varphi_1 \left[(1-\mu^2)\lambda \frac{d\lambda}{d\mu} - (\lambda^2 - 1)\mu \right] d\mu$$

According to Equation 59, for the potential of perturbed motion with the unit velocity,

$$\varphi_1 = -c \sum_{n=1}^{\infty} A_n Q_n(\lambda) P_n(\mu)$$

so that for the added mass in longitudinal motion or (briefly) longitudinal added mass, we obtain the following general expression:

$$\lambda_{11} = -2\pi\rho c^2 \int_{-1}^{1} \left\{ \left[(1-\mu^2)\lambda \frac{d\lambda}{d\mu} - (\lambda^2 - 1)\mu \right] \sum_{n=1}^{\infty} A_n Q_n(\lambda) P_n(\mu) \right\} d\mu$$

where it is presumed that the coordinate λ is the given function of μ according to the equation for the path around the meridian section of the body.

In the case of the ellipsoid of revolution with the major axis a directed along the axis $0x$, for which the equation of the path around its surface is $\lambda = \lambda_0 = 1/e$ (e is the eccentricity), the previous integral is easily calculated. As in Section 7,

$$\lambda_{11} = \frac{4}{3}\pi\rho c^3 (\lambda_0^2 - 1) \frac{\frac{1}{2}\lambda_0 \ln \frac{\lambda_0 + 1}{\lambda_0 - 1} - 1}{\frac{\lambda_0}{\lambda_0^2 - 1} - \frac{1}{2} \ln \frac{\lambda_0 + 1}{\lambda_0 - 1}} = \frac{4}{3}\pi\rho a b^2 \frac{\frac{1}{2e} \ln \frac{1+e}{1-e} - 1}{\frac{1}{1-e^2} - \frac{1}{2e} \ln \frac{1+e}{1-e}}$$

where, it should be recalled, a and b are the major and minor semiaxes, e is the eccentricity. Assuming in the latter formula that $e = 0$ and expanding the indeterminacy, we again obtain the added mass of the sphere

$$(\lambda_{11})_{e=0} = \frac{4}{3}\pi\rho a^3 \left[\frac{\left(1 + \frac{1}{3}e^2 + \cdots\right) - 1}{1 + e^2 + \cdots - \left(1 + \frac{1}{3}e^2 + \cdots\right)} \right]_{e=0} = \frac{2}{3}\pi\rho a^3$$

Analogously, we also determine the added mass of the body of revolution in the case of its transverse translation motion along the axis Oy, or the transverse added mass.

Retaining the notation of Section 8, we find

$$\lambda_{22} = \lambda_{yy} = \pi\rho c^3 \int_{-1}^{1} (1 - \mu^2)(\lambda^2 - 1)\left(\mu \frac{d\lambda}{d\mu} + \lambda\right) \sum_{n=1}^{\infty} C_n \frac{dP_n}{d\mu}\frac{dQ_n}{d\lambda}\, d\mu$$

and in the specific case of the transverse motion of the ellipsoid of revolution (in the direction normal to the major axis)

$$\lambda_{22} = \frac{4}{3}\pi\rho ab^2 \frac{\frac{1}{e^2} - \frac{1 - e^2}{2e^3}\ln\frac{1+e}{1-e}}{2 - \frac{1}{e^2} + \frac{1 - e^2}{2e^3}\ln\frac{1+e}{1-e}}$$

At $e = 0$, the latter formula also goes into the formula for the added mass of the sphere.

The added masses are very important for calculating the unsteady motions of bodies in fluids, phenomena related to the impact of a body against the free surface of the fluid, etc.[25]

The problem of a plane unsteady motion of a fluid caused by a nonuniformly moving profile is a particular case of the above general theory if the circulation around the profile is taken to be constant. The classical investigation of this case of profile motion and the derivation of formulas for the force and moment belongs to Chaplygin[26] and dates back to 1926, and further development of this problem was done by Sedov.[27] The main difficulty in studying the unsteady motions of the wing profile consists of the time-variable nature of circulation and of the origination of the vortex sheet which sheds from the profile and exerts the inductive influence on the flow past it.

Presentation of the classical theories of the unsteady motion of a wing in a fluid can be found in the special monograph by Nekrasov[28] and in the monograph by Sedov;[27] the latter also contains the investigation of wing vibrations in a gas.

---

[25] Sedov, L. I., *Ploskiye zadachi gidrodinamiki i aerodinamiki* (*Plane Problems of Hydrodynamics and Aerodynamics*), Nauka Press, Moscow, 1980. This book contains tables of added masses for cylindrical bodies. Also see the book by A. M. Basin, *Teoriya ustoychivosti na kurse i povorotlivosti sudna, Ser. Sovremennye problemy mekhaniki* (*Theory of Stability and Control of Ships, in the series Modern Problems of Mechanics*), Gostekhizdat Press, Moscow, 1949, where one can find the graphs of the added masses of ellipsoids and other bodies; the monographs by Kochin, N. Ye., Kibel, I. A., and Roze, N. V., *Teoreticheskaya Gidromekhanika* (*Theoretical Hydromechanics*), Part 1, Fizmatgiz Press, Moscow, 1963, chap. 7; Fabrikant, N. Ya., *Kurs aerodinamiki* (*The Course of Aerodynamics*), Part 1, GONTI Press, Moscow, 1938.

[26] Chaplygin, S. A., On the effect of a plane-parallel flow of air on a cylindrical wing moving through it, *Tr. TsAGI*, No. 19, 1926; see also *Collected Works*, Nauka Press, Moscow, 1964.

[27] Sedov, L. I., Concerning the theory of the unsteady motion of a wing in a fluid, *Tr. TsAGI*, No. 229, 1935; On the unsteady motion of bodies of revolution inside a fluid, *Tr. TsAGI*, No. 515, 1940.

[28] Nekrasov, A. I., *The Theory of a Wing in an Unsteady Flow*, AN SSSR Press, Moscow, 1947.

The theory of the unsteady motion of a wing of finite span and various shapes in the plan is presented in special monographs.[29]

12. Axisymmetric Sub- and Supersonic Flow Past a Thin Body of Revolution

As in the case of a two-dimensional flow, the equations of the three-dimensional irrotational motion of an ideal perfect gas can be obtained using the conditions of (1) flow continuity, (2) absence of vorticity in the flow, and (3) adiabatic and isentropic nature of the process. The continuity equation in cylindrical coordinates can be easily obtained with the aid of Equation 85 of Chapter 1 for the divergence. Assuming the circumferential (transversal) velocity V_ε to be equal to zero and the motion to be independent of the azimuth ε,

$$\mathrm{div}\,(\rho V) = \frac{1}{r}\frac{\partial(\rho r V_r)}{\partial r} + \frac{\partial(\rho V_z)}{\partial z} = 0$$

Let us agree that the longitudinal coordinate and velocity, which are parallel to the symmetry axis of motion, are designated by x and u and the transverse ones are denoted by r and v. Then, the continuity equation acquires the form

$$\frac{\partial(\rho r u)}{\partial x} + \frac{\partial(\rho r v)}{\partial r} = 0 \qquad (137)$$

In the case of the flow symmetry, the only vortex component, which is not identically equal to zero, will be the azimuthal component, so that the condition of the absence of vorticity in the adopted notation will be reduced to one equation:

$$\frac{\partial u}{\partial r} - \frac{\partial v}{\partial x} = 0 \qquad (138)$$

which coincides with Equation 3 for the two-dimensional motion of a gas (Section 1, Chapter 8). Equation 137 differs from Equation 1 of Section 1, Chapter 8 by the presence of the factor r under the signs of the derivatives.

The adopted condition of the adiabatic and isentropic nature of the process of gas motion will be used in what follows in different forms most convenient to the present stage of consideration, just as was done for the two-dimensional case in Chapter 8, Section 1.

Using Equations 137 and 138, it is possible to introduce two functions: the velocity potential $\varphi(x, r)$ and the stream function $\psi(x, r)$, having assumed, according to Equation 138, that

$$u = \frac{\partial \varphi}{\partial x} \qquad v = \frac{\partial \varphi}{\partial x} \qquad (139)$$

and having satisfied Equation 137 with the aid of the equations (ρ_∞ is the density in the nonperturbed uniform flow; the choice of the sign was explained in Section 4)

[29] Polyakhov, N. N., *Teoriya nestatsionarnykh dvizheniy nesushchey poverkhnosti (The Theory of Unsteady Motions of the Lifting Surface)*, LGU Press, Leningrad, 1960; Belotserkovskiy, S. M., Skripach, B. K., and Tabachnikov, V. G., *Krylo v nestatsionarnom potoke gaza (A Wing in the Unsteady Gas Flow)*, Nauka Press, Moscow, 1971; Belotserkovskiy, S. M., and Nisht, M. I., *Otryvnoye i bezotryvnoye obtekaniye tonkikh kryliyev neszhimayemoy zhidkostyu (Separation and Non-separation Flow Past Thin Wings in an Incompressible Fluid)*, Nauka Press, Moscow, 1978.

$$\frac{\rho}{\rho_\infty} ru = \frac{\partial \psi}{\partial r} \qquad \frac{\rho}{\rho_\infty} rv = -\frac{\partial \psi}{\partial x} \tag{140}$$

Substituting the values of u and v from Equation 139 into 137, we obtain the equation to determine the velocity potential $\varphi(x, r)$:

$$\frac{\partial}{\partial x}\left(\rho r \frac{\partial \varphi}{\partial x}\right) + \frac{\partial}{\partial r}\left(\rho r \frac{\partial \varphi}{\partial x}\right) = 0 \tag{141}$$

and carrying out an analogous operation with the expressions for the velocity projections (140) and with Equation 138, we find the equation to determine the stream function $\psi(x, r)$

$$\frac{\partial}{\partial x}\left(\frac{1}{\rho r}\frac{\partial \psi}{\partial x}\right) + \frac{\partial}{\partial r}\left(\frac{1}{\rho r}\frac{\partial \psi}{\partial r}\right) = 0 \tag{142}$$

Equations 141 and 142 are nonlinear because the density ρ accompanies, in accordance with the Bernoulli equation, the velocity function V. Recall the derivation of this and another equation for the pressure p, which will be needed for what follows. Using Saint-Venant and Wantzel's formula (Equation 30 of Chapter 5) and determining the constant from the condition at infinity, we shall have

$$\frac{V^2}{2} - \frac{k}{k-1}\frac{p_\infty}{\rho_\infty}\left[1 - \left(\frac{p}{p_\infty}\right)^{\frac{k-1}{k}}\right] = \frac{U_\infty^2}{2}$$

from which it follows that ($kp_\infty/\rho_\infty = a_\infty^2$, $U_\infty/a_\infty = \mathbf{M}_\infty$)

$$\frac{p}{p_\infty} = \left[1 + \frac{k-1}{2}\mathbf{M}_\infty^2\left(1 - \frac{V^2}{U_\infty^2}\right)\right]^{\frac{k}{k-1}} \tag{143}$$

and from the isentropic relation $p/p_\infty = (\rho/\rho_\infty)^k$ we find the sought-after expression for ρ/ρ_∞

$$\frac{\rho}{\rho_\infty} = \left[1 + \frac{k-1}{2}\mathbf{M}_\infty^2\left(1 - \frac{V^2}{U_\infty^2}\right)\right]^{\frac{1}{k-1}} \tag{144}$$

The linearization of Equations 141 and 142 in application to the problem of a three-dimensional flow around a thin body of revolution, which induces small disturbances in the incident flow, leads to some difficulties.[30]

As occurred in the two-dimensional case, we divide the velocity potential and the stream function into the parts that correspond to the nonperturbed uniform flow φ_∞, ψ_∞ and to small disturbances $\tilde{\varphi}$, $\tilde{\psi}$, having assumed that

$$\varphi = \varphi_\infty + \tilde{\varphi} \qquad \psi = \psi_\infty + \tilde{\psi}$$

Analogously, regarding the uniform flow as parallel to the symmetry axis Ox and its velocity to be equal to U_∞, we represent the velocity projections u, v, and the density ρ as

[30] Van Dyke, M., *Metody vozmushcheniy v mekhanike zhidkosti (The Perturbation Methods in the Mechanics of a Fluid)*, Mir Press, Moscow, 1967, pp. 236 to 242.

also being split into the quantities corresponding to the nonperturbed flow U_∞, 0, ρ_∞ and small disturbances $\tilde{u}$, $\tilde{v}$, $\tilde{\rho}$

$$u = U_\infty + \tilde{u} \qquad v = \tilde{v} \qquad \rho = \rho_\infty + \tilde{\rho}$$

Substituting these expressions into Equations 139 and 140, we obtain

$$U_\infty + \tilde{u} = \frac{\partial \varphi_\infty}{\partial x} + \frac{\partial \tilde{\varphi}}{\partial x} \qquad \tilde{v} = \frac{\partial \varphi_\infty}{\partial r} + \frac{\partial \tilde{\varphi}}{\partial r}$$

$$\left(1 + \frac{\tilde{\rho}}{\rho_\infty}\right) r (U_\infty + \tilde{u}) = \frac{\partial \psi_\infty}{\partial r} + \frac{\partial \tilde{\psi}}{\partial r} \qquad \left(1 + \frac{\tilde{\rho}}{\rho_\infty}\right) r\tilde{v} = -\frac{\partial \psi_\infty}{\partial x} - \frac{\partial \tilde{\psi}}{\partial x} \tag{145}$$

From this, in the zero approximation ($\tilde{\varphi} = \tilde{\psi} = \tilde{u} = \tilde{v} = \tilde{\rho} = 0$),

$$U_\infty = \frac{\partial \varphi_\infty}{\partial x} \qquad 0 = \frac{\partial \varphi_\infty}{\partial r} \qquad rU_\infty = \frac{\partial \psi_\infty}{\partial r} \qquad 0 = \frac{\partial \psi_\infty}{\partial x}$$

which will lead to ordinary expressions for the velocity potential and stream function of the nonperturbed flow φ_∞ and ψ_∞:

$$\varphi_\infty = U_\infty x \qquad \psi_\infty = \frac{1}{2} U_\infty r^2 \tag{146}$$

From the first line of the system of Equations 145 it follows directly that

$$\tilde{u} = \frac{\partial \tilde{\varphi}}{\partial x} \qquad \tilde{v} = \frac{\partial \tilde{\varphi}}{\partial r} \tag{147}$$

In order to derive analogous connections of the perturbation velocity projections with the stream function, corresponding to them, it is necessary in the last two lines of system (145) to eliminate the density disturbance $\tilde{\rho}$. For this purpose, we note first that

$$V^2 = (U_\infty + \tilde{u})^2 + \tilde{v}^2 = U_\infty^2 + 2U_\infty \tilde{u} + \tilde{u}^2 + \tilde{v}^2$$

$$1 - \frac{V^2}{U_\infty^2} = -2\frac{\tilde{u}}{U_\infty} - \left(\frac{\tilde{u}}{U_\infty}\right)^2 - \left(\frac{\tilde{v}}{U_\infty}\right)^2$$

and rewrite Equations 143 and 144 as

$$\frac{\tilde{p}}{p_\infty} = -\frac{k}{2}\mathbf{M}_\infty^2 \left[2\frac{\tilde{u}}{U_\infty} + \left(\frac{\tilde{u}}{U_\infty}\right)^2 + \left(\frac{\tilde{v}}{U_\infty}\right)^2 + \cdots\right]$$

$$\frac{\tilde{\rho}}{\rho_\infty} = -\frac{1}{2}\mathbf{M}_\infty^2 \left[2\frac{\tilde{u}}{U_\infty} + \left(\frac{\tilde{u}}{U_\infty}\right)^2 + \left(\frac{\tilde{v}}{U_\infty}\right)^2 + \cdots\right] \tag{148}$$

It should be recalled that in Chapter 8, when the linearized equations of the two-dimensional gas flow were considered, we used analogous formulas in which only the first terms in square brackets were retained and the quadratic terms were neglected.

It will be shown that in the case of a three-dimensional axisymmetric flow around thin bodies of revolution such discarding is admissible only for the term $(\tilde{u}/U_\infty)^2$ which, as compared with the small quantity $\tilde{u}/U_\infty$, represents the small quantity of the higher order. As to the quantity $(\tilde{v}/U_\infty)^2$, it will become clear from further estimates that it does not have the second order of smallness, as compared with $\tilde{u}/U_\infty$. Therefore discarding the term $(\tilde{v}/U_\infty)^2$ in the square bracket, and retaining the first term $2\tilde{u}/U_\infty$, is not justifiable.

To consider this not very simple problem, we employ the following procedure. First, we carry out the ordinary, analogous to the plane case, rough linearization, having retained only the first term $2\tilde{u}/U_\infty$ in the square brackets of Equation 148; i.e., having assumed that

$$\frac{\tilde{p}}{p_\infty} \approx -k\mathbf{M}_\infty^2 \frac{\tilde{u}}{U_\infty} \qquad \frac{\tilde{\rho}}{\rho_\infty} \approx -\mathbf{M}_\infty^2 \frac{\tilde{u}}{U_\infty} \tag{149}$$

and then, using the thus obtained (we shall not call it the "first approximation") approximate solution, we estimate *a posteriori* the magnitude of the discarded term $(\tilde{v}/U_\infty)^2$.

Substituting the second of relations (149) into the last two equations (145) and equating the coefficients at small quantities of the first order, we find the expressions for the perturbation velocity projections in terms of the perturbation stream function

$$\tilde{u} = \frac{1}{1-\mathbf{M}_\infty^2}\frac{1}{r}\frac{\partial\tilde{\psi}}{\partial r} \qquad \tilde{v} = -\frac{1}{r}\frac{\partial\tilde{\psi}}{\partial x} \tag{150}$$

Formulas (147) coincide with corresponding formulas (13) for the plane linearized motion (Chapter 8), and formulas (150) differ from formulas (15) of the same chapter by the factor $1/r$ before the derivatives on the right sides.

Substituting the resulting density perturbation $\tilde{\rho}$ from Equation 149 into Equation 141, and equating the small quantities of the first order, we establish the linearized equation for the perturbation velocity potential

$$(1-\mathbf{M}_\infty^2)\frac{\partial^2\tilde{\varphi}}{\partial x^2} + \frac{\partial^2\tilde{\varphi}}{\partial r^2} + \frac{1}{r}\frac{\partial\tilde{\varphi}}{\partial r} = 0 \tag{151}$$

and, analogously, the linearized equation for the stream function of perturbations

$$(1-\mathbf{M}_\infty^2)\frac{\partial^2\tilde{\psi}}{\partial x^2} + \frac{\partial^2\tilde{\psi}}{\partial r^2} - \frac{1}{r}\frac{\partial\tilde{\psi}}{\partial r} = 0 \tag{152}$$

which differs from the previous one by the sign at the last term on the left side.

Let us now consider the boundary conditions on the surface of the body immersed in a flow. For Equation 151 the boundary condition can be obtained as the impermeability condition on the body surface. Let the equation of the body surface in cylindrical coordinates be $r = r_0(x)$. Then the condition of the body surface impermeability can be written as the equality of the angles at which the tangent to the body contour and the streamline are inclined to the longitudinal axis:

$$\frac{dr_0}{dx} = r_0'(x) = \left(\frac{\tilde{v}}{U_\infty + \tilde{u}}\right)_{r=r_0(x)} \approx \frac{\tilde{v}_{r=r_0(x)}}{U_\infty} \tag{153}$$

This boundary condition cannot be "shifted" to the body axis, as in the two-dimensional case (i.e., cannot be applied at $r = 0$), because in the case of the axisymmetric motion the $0x$ axis is a singular line – on it $\tilde{v} = \infty$ – which follows directly from the second equation of (150). This singularity can be obviated approximately, having replaced the previous condition by the following:

$$r_0(x)\, r_0'(x) = \frac{(r\tilde{v})_{r=0}}{U_\infty} \tag{154}$$

It will be quite different with the use of Equation 152 for the stream function. In this case the boundary condition expresses the fact that the body surface is the null stream surface

$$\psi = (\psi_\infty + \tilde{\psi})_{r=r_0(x)} = \frac{1}{2} U_\infty r_0^2 + \tilde{\psi}_{r=r_0(x)} = 0$$

or

$$\tilde{\psi} = -\frac{1}{2} U_\infty r_0^2 \quad \text{when} \quad r = r_0(x)$$

This condition can be "shifted" to the $0x$ axis and it is reduced to the following approximate condition

$$\tilde{\psi} = -\frac{1}{2} U_\infty r_0^2(x) \quad \text{when} \quad r = 0 \tag{155}$$

Let us consider, in the above-mentioned simplest statement, the von Kármán problem[31] of the longitudinal supersonic flow past a thin body of revolution.

The basic differential equation of this flow in cylindrical coordinates can be presented, according to Equation 151, in the form

$$\frac{\partial^2\tilde{\varphi}}{\partial x^2} - \frac{1}{\omega^2}\left(\frac{\partial^2\tilde{\varphi}}{\partial r^2} + \frac{1}{r}\frac{\partial\tilde{\varphi}}{\partial r}\right) = 0 \qquad \omega^2 = \mathbf{M}_\infty^2 - 1 \tag{156}$$

Following von Kármán, we shall prove that Equation 156 has the integral

$$\tilde{\varphi} = \int_0^{x-\omega r} \frac{f(\xi)\,d\xi}{\sqrt{(x-\xi)^2 - \omega^2 r^2}} \tag{157}$$

where $x = \xi = 0$ corresponds to the frontal point of the body and $f(\xi)$ is a certain continuous, twice differentiable function which is identically equal to zero when $\xi \le 0$. Noting that differentiation is hindered by the fact that the function to be integrated becomes infinite in the upper limit, the integration variable is replaced, assuming that

$$\xi = x - \omega r \operatorname{ch} t \qquad d\xi = -\omega r \operatorname{sh} t\, dt \tag{158}$$

Integral (157) has a simpler form

$$\tilde{\varphi}(x, r) = \int_0^{\operatorname{arch}\frac{x}{\omega r}} f(x - \omega r \operatorname{ch} t)\, dt \tag{159}$$

[31] Kármán, Th., *Problemy soprotivleniya c szhimayemoy zhidkosti (The problem of resistance in a compressible fluid)* (translation from English), in *Gazovaya dinamika (Gas Dynamics)*, GONTI Press, Moscow, 1939, pp. 81 to 90.

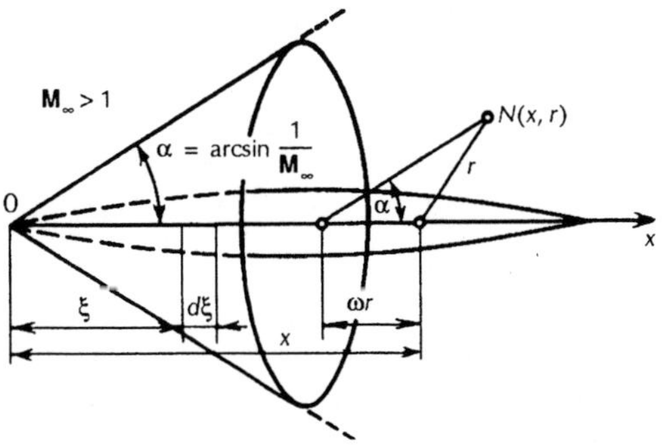

Fig. 132

which can be made still more simple if it is noted that to the change of the variable t in the region

$$\operatorname{arch}\frac{x}{\omega r} \leq t < \infty$$

corresponds the region of the variation of ξ

$$0 \geq \xi > -\infty$$

where, according to the assumption made, $f(\xi) \equiv 0$. This allows one to replace the upper integration limit in Equation 159 by ∞ and to obtain

$$\tilde{\varphi}(x, r) = \int_0^\infty f(x - \omega r \operatorname{ch} t)\,dt \tag{160}$$

The integral is determined in the entire space (x, r), but in the region outside the cone (Figure 132)

$$x = \omega r \tag{161}$$

with the vertex semiangle α equal to

$$\alpha = \operatorname{arctg}\frac{r}{x} = \operatorname{arctg}\frac{1}{\omega} = \arcsin\frac{1}{\mathbf{M}_\infty}$$

This integral is identically equal to zero, because in this region the argument of the function f in front of the integral becomes negative; in fact,

$$x - \omega r \operatorname{ch} t < x - \omega r < 0 \quad \text{when} \quad x < \omega r$$

This shows that the considered disturbances of the uniform flow are concentrated inside the cone, Equation 161, which is called the cone of perturbation or the Mach cone. The vertex angle of this cone 2α is equal to the double angle of perturbation (Mach angle), just as in the case in the two-dimensional supersonic flow.

It is not difficult to prove that the integral (160) and, consequently, the integral (159) equal to it, satisfy the differential equation of small perturbations in a supersonic flow (156).

Let us calculate directly from Equation 160 the derivatives[32]

$$\frac{\partial\tilde{\varphi}}{\partial x} = \int_0^\infty f'(x - \omega r \operatorname{ch} t)dt \qquad \frac{\partial^2\tilde{\varphi}}{\partial x^2} = \int_0^\infty f''(x - \omega r \operatorname{ch} t)dt$$

$$\frac{\partial\tilde{\varphi}}{\partial r} = -\omega\int_0^\infty f'(x - \omega r \operatorname{ch} t)\operatorname{ch} t\, dt \qquad \frac{\partial^2\tilde{\varphi}}{\partial r^2} = \omega^2\int_0^\infty f''(x - \omega r \operatorname{ch} t)\operatorname{ch}^2 t\, dt$$

and note that the integration by parts gives

$$\frac{\partial\tilde{\varphi}}{\partial r} = -\omega\int_0^\infty f'(x - \omega r \operatorname{ch} t)d(\operatorname{sh} t) =$$

$$= -\,\omega\,[f'(x - \omega r \operatorname{ch} t)\operatorname{sh} t]_0^\infty - \omega^2 r\int_0^\infty f''(x - \omega r \operatorname{ch} t)\operatorname{sh}^2 t\, dt$$

From the condition $f(\xi) \equiv 0$, when $\xi < 0$, it follows that

$$\lim_{t \to \infty} [f'(x - \omega r \operatorname{ch} t)\operatorname{sh} t] = 0$$

and the substitution vanishes; the calculation of the left-hand side of equation (156) yields

$$\frac{\partial^2\tilde{\varphi}}{\partial x^2} - \frac{1}{\omega^2}\left(\frac{\partial^2\tilde{\varphi}}{\partial r^2} + \frac{1}{r}\frac{\partial\tilde{\varphi}}{\partial r}\right) = \int_0^\infty [f'' - (\operatorname{ch}^2 t - \operatorname{sh}^2 t)f''] \, dt \equiv 0$$

The integral (157) can be considered to be the supersonic analog of the velocity potential of perturbations from the sources distributed along the positive portion of the axis $0x$. In the case of an incompressible fluid ($\mathbf{M}_\infty = 0$, $\omega^2 = -1$), this integral is represented by Equation 72. There is a fundamental difference between these two hydrodynamic images, mathematically expressed in the fact that the corresponding velocity potential in the incompressible fluid satisfies the elliptic-type equation (Laplace equation), whereas the potential (157) satisfies the hyperbolic-type equation, which is the wave equation (156). Physically, this difference is localized in the fact that in the first case, the effect of perturbations extends over the entire region filled with the fluid, and in the second case it extends over only that part of it located inside the cone of perturbations with the vertex at the point 0 from which the distribution starts and propagates along the axis ($0x$ axis) of the cone of the supersonic sources.

The perturbation velocity potential (160) can be used for calculating the supersonic flow past slender bodies of revolution in a uniform stream parallel to their axis of symmetry. For this purpose, let us subject the unknown function $f(\xi)$ to the condition of the body surface impermeability. In the approximation adopted this condition can be written by having

---

[32] The primes on f under the integral sign designate the derivatives over the entire argument in brackets.

expressed the equality of the slopes, with respect to the axis Ox, of the tangents to the streamline and to the contour of the meridian section of the immersed body at the points of its surface

$$\left(\frac{\tilde{v}}{U_\infty + \tilde{u}}\right)_{r=r_0} \approx \left(\frac{\tilde{v}}{U_\infty}\right)_{r=r_0} = \frac{dr_0}{dx} \tag{162}$$

where $r = r_0(x)$ represents the equation of the meridian section of the body. To calculate the perturbation velocities $\tilde{u}$, $\tilde{v}$, we use Equations 147 and 160; this will give

$$\tilde{u} = \int_0^\infty f'(x - \omega r \operatorname{ch} t)\,dt \qquad \tilde{v} = -\omega\int_0^\infty f'(x - \omega r \operatorname{ch} t) \operatorname{ch} t\,dt \tag{163}$$

or, returning to the variable ξ,

$$\tilde{u} = \int_0^{x-\omega r} \frac{f'(\xi)}{\sqrt{(x-\xi)^2 - \omega^2 r^2}}\,d\xi \qquad \tilde{v} = -\frac{1}{r}\int_0^{x-\omega r} \frac{(x-\xi)f'(\xi)}{\sqrt{(x-\xi)^2 - \omega^2 r^2}}\,d\xi \tag{164}$$

Substituting the last expression into Equation 162, the following integral equation is found for determining the unknown function $f(\xi)$:

$$\int_0^{x-\omega r_0(x)} \frac{(x-\xi)f'(\xi)}{\sqrt{(x-\xi)^2 - \omega^2 r_0^2(x)}}\,d\xi = -U_\infty r_0(x)\, r_0'(x) \tag{165}$$

In rough approximation this equation is easily integrable, if for very slender bodies it is assumed under the integral sign and in its upper limit that

$$\omega r_0 \ll x - \xi \qquad \omega r_0 \ll x$$

Then (let us recall that $f(0) = 0$),

$$\int_0^x f'(\xi)\,d\xi = f(x) = -U_\infty r_0(x)\, r_0'(x) \tag{166}$$

or, characterizing the body by the law of change in the area of its cross-section $A(x) = \pi r_0^2(x)$,

$$f(x) = -\frac{U_\infty}{2\pi} A'(x) \tag{167}$$

This approximation is equivalent to expression (78) given at the end of Section 9 for an incompressible fluid.

Substituting the resulting value of $f(x)$ into Equation 157, we find the sought-after expression of the potential

$$\tilde{\varphi}(x, r) = -\frac{U_\infty}{2\pi}\int_0^{x-\omega r} A'(\xi)\frac{d\xi}{\sqrt{(x-\xi)^2 - \omega^2 r^2}} \tag{168}$$

This makes it less difficult to determine the perturbation velocity projections onto the axes of the cylindrical coordinate systems

$$\tilde{u} = -\frac{U_\infty}{2\pi} \int_0^{x-\omega r} A''(\xi) \frac{d\xi}{\sqrt{(x-\xi)^2 - \omega^2 r^2}}$$

$$\tilde{v} = \frac{U_\infty}{2\pi r} \int_0^{x-\omega r} A''(\xi) \frac{(x-\xi)\, d\xi}{\sqrt{(x-\xi)^2 - \omega^2 r^2}} \qquad (169)$$

and the expression for the pressure coefficient

$$c_p = -\frac{2\tilde{u}}{U_\infty} = \frac{1}{\pi} \int_0^{x-\omega r} A''(\xi) \frac{d\xi}{\sqrt{(x-\xi)^2 - \omega^2 r^2}} \qquad (170)$$

Summing up the pressures over the surface of the body of revolution of the given length l and considering that $r_0(0) = r_0(l) = 0$, it is possible to obtain the wave resistance of the body. Omitting the calculations, the final formula belonging to von Kármán is

$$R_x = -\pi\rho_\infty \int_0^l \int_0^l f'(\xi) f'(x) \ln|x-\xi| dx d\xi \qquad (171)$$

The calculation of the resistance by this formula becomes easier if the formal analogy between Equation 171 and the first of Equations 92 for the induced drag of the wing of finite span is noted. In fact, replacing $\alpha_i(z)$ in Equation 92 by formula (86),

$$R_x = \frac{\rho_\infty}{4\pi} \int_{-l}^{l} \Gamma(z) \left[\int_{-l}^{l} \Gamma'(\zeta) \frac{d\zeta}{z-\zeta} \right] dz$$

and after the integration by parts

$$R_x = -\frac{\rho_\infty}{4\pi} \int_{-l}^{l} \int_{-l}^{l} \Gamma'(z) \Gamma'(\zeta) \ln|z-\zeta|\, d\zeta dz$$

Thus, it is possible to formulate the following result: the wave resistance of the body of revolution in a longitudinal flow can be calculated from the formulas of the induced drag of the wing of finite span if, instead of the distribution of circulation over the span, the distribution of the strength of the sources is given along the axis of the body of revolution.

When calculating the wave resistance of the body in a supersonic flow, this allows one to apply the same method of expansion into trigonometric series as that used for calculating the induced drag of the wing of finite span by the theory of the lifting line.

In particular, in complete analogy to the theory of the wing of finite span it is possible to conclude that at the given aspect ratio of the body of revolution, the coefficient of wave resis-

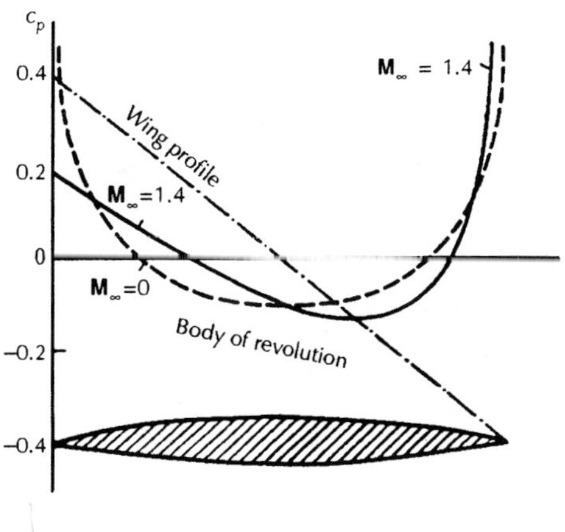

Fig. 133

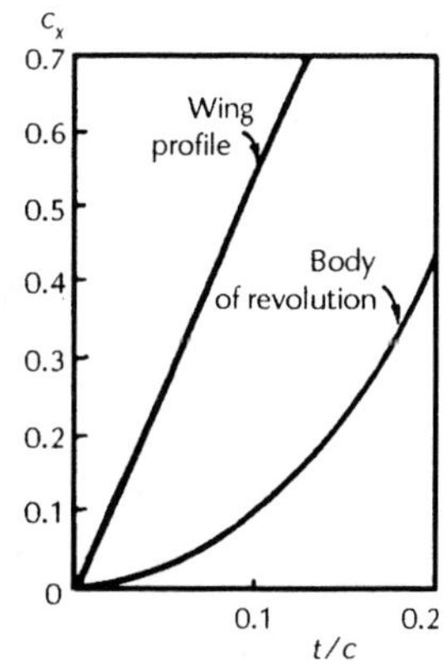

Fig. 134

tance will be minimal if it is assumed that the strength of the sources is distributed according to the elliptical law.

In order to obtain idea about the difference between the pressure distributions over the surface of the slender body of revolution and over the analogous (in shape) cross-section of the wing profile, we give comparative plots of the distribution of the pressure coefficient for the biconvex 10% symmetric profile formed by the arcs of the parabola and having the same meridian section of the body of revolution at $\mathbf{M}_\infty = 1.4$ (Figure 133). The plot clearly reveals a sharp fall of pressure on the wing profile as compared with the corresponding body of revolution. This leads to a considerable difference (Figure 134) in the laws of increase in the wave resistance with the rise of the relative thickness of the wave profile and body of revolution. It is natural that in the case of the body of revolution this increase is much weaker.

The origin of the wave resistance (explained in Section 2, Chapter 8) is proved by the plots presented in Figure 133. Comparing the distribution of pressure over the body of revolution immersed in an incompressible fluid flow ($\mathbf{M}_\infty = 0$) with the corresponding distribution at $\mathbf{M}_\infty = 1.4$, we find the appearance of asymmetry in the distribution of pressures.

It is because of the partial delay in the pressure recovery in the rear portion of the body of revolution in a supersonic flow (the downstream shift of the solid curve with respect to the dashed curve) that the wave resistance originates. The absence of pressure recovery observed in the case of a plane wing leads to a sharp difference between the wave resistances of the wing and of the body of revolution the meridian section of which coincides with the wing profile.

Figure 135 compares the pressure coefficient distribution predicted by the formulas of this section with the experiment[33] for a shell, having an ogive nose and the stern and cylindrical middle portion, at the number $\mathbf{M}_\infty = 1.87$.

In order to solve the problem of flow past a thin body of revolution located in the incident flow at a certain small angle of attack, one has, along with the longitudinal flow, also to consider the transverse flow past the body of revolution in complete conformity with the theory of incompressible fluid flow past bodies (see Section 8).

The differential equation of small perturbations in a uniform supersonic flow directed perpendicularly to the axis of the body of revolution (transverse flow) will contain the polar angle ε in the plane Oxy. In this case one cannot discard the derivative with respect to the

[33] Figures 133 to 135 are derived from the book by A. Shapiro, *The Dynamics and Thermodynamics of Compressible Fluid Flow*, Vol. 1 and 2, New York, 1953.

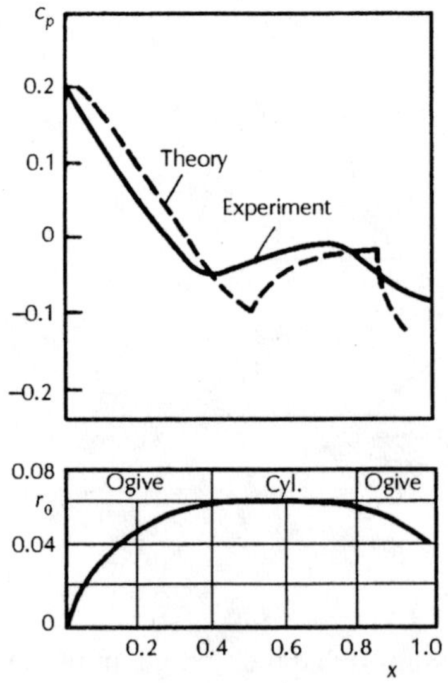

Fig. 135

angle ε, and the equation for determining the perturbation velocity potential $\tilde{\varphi}_1$ in the case of transverse flow will be

$$\frac{\partial^2 \tilde{\varphi}_1}{\partial x^2} - \frac{1}{\omega^2}\left(\frac{\partial^2 \tilde{\varphi}_1}{\partial r^2} + \frac{1}{r}\frac{\partial \tilde{\varphi}_1}{\partial r} + \frac{1}{r^2}\frac{\partial^2 \tilde{\varphi}_1}{\partial \varepsilon^2}\right) = 0$$

and its integral has the form

$$\tilde{\varphi}_1 = \frac{1}{\omega}\frac{\partial}{\partial r}\left[\int_0^{x-\omega r} \frac{f_1(\xi)\,d\xi}{\sqrt{(x-\xi)^2 - \omega^2 r^2}}\right]\cos\varepsilon = \frac{\cos\varepsilon}{r}\int_0^{x-\omega r} \frac{(x-\xi)\,m(\xi)\,d\xi}{\sqrt{(x-\xi)^2 - \omega^2 r^2}}$$

$m(\xi)$ determines the density of distribution of the intensity of supersonic dipoles on the symmetry axis of the body of revolution.

In order to calculate the pressure coefficient in the case of very thin bodies, the same simplified technique is used as that in the case of longitudinal flow, but based on approximate formula (79), which connects the density of distribution of the moments of dipoles with the law of change in the cross-sectional area of the body of revolution.

The potential of transverse flow in this rough approximation which is valid only for very thin bodies, can be presented in the final form as

$$\tilde{\varphi}_1 = \frac{1}{\pi} V_{1\infty} \frac{A(x)}{r} \cos\varepsilon$$

The presence of the potentials of the longitudinal and transverse flows makes it possible, by means of the summation of the solutions, to obtain the flow past a thin body at any angle of attack α and then to calculate the lift force and resistance coefficients. Omitting the computations,[34] we shall only indicate that the lift force coefficient is equal to $c_y = 2\alpha$ and

[34] Liepmann, H. W. and Roshko, A., *Elementy gazovoy dinamiki* (*The Elements of Gas Dynamics*) (translated from English), IL Press, Moscow, 1960, pp. 290 to 294.

the resistance coefficient in a longitudinal flow, which can be calculated from Equation 171, is supplemented in addition (from the transverse flow) to the term $c_{xi} = \alpha^2$ called the coefficient of induced resistance. These results, which express the independence of the coefficients c_y and c_{xi} of the body shape, are very approximate and cannot compete with more accurate theories. Such theories differ from the theory of von Kármán (given above) first by the fact that they take into account the presence of the bow shock wave on the nose portion of the body and, in the case of a body of revolution with an acute nose, the presence of the conical added density shock (see Section 9).

Now, let us return to the problem raised at the beginning of the section, dealing with the admissibility of the linearization just employed. For estimating the order of the values of perturbation velocities $\tilde{u}$, $\tilde{v}$ depending on the basic small parameter which characterizes the axisymmetric flow past a thin body of revolution (its relative thickness τ), we shall avail ourselves of Equations 164 and apply them to the simplest case $f'(x) = \text{const}$. According to Equation 167, this case will correspond to the flow past a cone with a small semivertex angle θ_0 defined by the equation

$$\frac{dr_0}{dx} = \frac{r_0}{x} = \operatorname{tg}\theta_0 \approx \theta_0 \tag{172}$$

Simple calculations show that in this case the following expressions for velocity projections on the cone surface will be valid $(r = r_0(x))$

$$\tilde{u}_0 = -\text{const} \cdot \operatorname{arch}\frac{x}{\omega r_0} \qquad \tilde{v}_0 = \text{const} \cdot \omega\sqrt{\left(\frac{x}{\omega r_0}\right)^2 - 1}$$

or, taking into account Equation 172,

$$\tilde{u}_0 = -\text{const} \cdot \operatorname{arch}\frac{1}{\omega\theta_0} \qquad \tilde{v}_0 = \text{const} \cdot \omega\sqrt{\frac{1}{(\omega\theta_0)^2} - 1}$$

Noting that θ_0 is small and assuming that the Mach number is not so great for the product

$$\omega\theta_0 = \sqrt{\mathbf{M}_\infty^2 - 1}\theta \approx \mathbf{M}_\infty\theta_0$$

to cease to be small, we obtain the following simple asymptotic estimates[35] which are valid for $\omega\theta_0 \ll 1$:

$$\tilde{u}_0 \sim -\text{const} \cdot \ln\frac{2}{\omega\theta_0} \qquad \tilde{v}_0 \sim \text{const} \cdot \frac{1}{\theta_0} \tag{173}$$

But from boundary condition (162) and Equation 172 it follows directly that $\tilde{v}_0 = O(\theta_0)$. Thus, according to the second of Equations 173, we see that the constant introduced is on the order of θ_0^2, whereas the perturbation velocity projection $\tilde{u}_0$ at the given Mach number is on the order of $\theta_0^2 \ln\theta_0$. Returning to the adopted designation τ for the generalized concept of the relative thickness of the body of revolution $[\tau = O(\theta_0)]$, we finally obtain the following general estimate of the order of smallness of the perturbation velocity projections

[35] See, for example, Dwight, H. B., *Tablitsy integralov i drugiye matematicheskiye formuly* (*Tables of Integrals and Other Mathematical Formulae*) (translated from English), Nauka Press, Moscow, 1983.

$$\tilde{u} = O(\tau^2 \ln \tau) \qquad \tilde{v} = O(\tau) \tag{174}$$

From this it follows that the terms $\tilde{u}/U_\infty$ and $(\tilde{v}/U_\infty)^2$ in the square brackets of Equations 148 have the same order, because at not very small values of τ the factor $\ln \tau$ exerts a little effect on the order of the product $\tau^2 \ln \tau$ Thus, for example, if $\tau = 10^{-1}$, then $\tilde{u}/U_\infty$ will be on the order of $2 \cdot 10^{-2}$, and $(\tilde{v}/V_\infty)^2$ on the order of 10^{-2}, i.e., the same order.

Having retained in the equation for the relative density perturbation $\tilde{\rho}/\rho_\infty$ (148) the quadratic term $(\tilde{v}/U_\infty)^2$, which has as we have just seen the same order as the first term $2\tilde{u}/U_\infty$, we thus reject the admissibility of the linearization of the equation for small perturbations. However, as Lighthill's[36] detailed investigation shows, it is possible to establish the correct first approximation based on the use of linearized equations (151) and (152) with boundary conditions (154) and (155). To calculate the pressure coefficient

$$c_p = \frac{p - p_\infty}{\frac{1}{2}\rho_\infty U_\infty^2} = \frac{2\tilde{p}}{\rho_\infty U_\infty^2} = 2\frac{p_\infty}{\rho_\infty U_\infty^2}\frac{\tilde{p}}{p_\infty} = \frac{2}{k\mathbf{M}_\infty^2}\frac{\tilde{p}}{p_\infty}$$

instead of the incorrect equation (149), one applies the expansion (148), having retained in the square bracket the two terms that have identical orders; thus,

$$c_p = -2\frac{\tilde{u}}{U_\infty} - \left(\frac{\tilde{v}}{U_\infty}\right)^2 \tag{175}$$

This approach is usually used when the first approximation is sought in the problems of supersonic flow past thin bodies of revolution.

Let us give the equations which without proof were obtained in this first approximation for the pressure coefficient c_p and resistance c_x of the front portion of the body (i.e., without taking into account the so-called bottom pressure at the rear section; l is the body length, $A(x)$ is the area of the section normal to the axis):

$$c_p = \frac{A''(x)}{\pi}\ln\frac{2}{\omega r_0} + \frac{1}{\pi}\frac{d}{dx}\int_0^x A''(\xi)\ln(x-\xi)\,d\xi - \left(\frac{dr_0}{dx}\right)^2$$

$$c_x = \frac{[A'(l)]^2}{2\pi}\ln\frac{2}{\omega r_0(l)} + \frac{A'(l)}{\pi}\int_0^l A''(\xi)\ln(l-\xi)\,d\xi -$$

$$-\frac{1}{\pi}\int_0^l\int_0^x A''(x)A''(\xi)\ln(x-\xi)\,d\xi dx$$

If $A'(l) = 0$, the last formula is reduced to the earlier derived formula of von Kármán (171). This occurs in two cases: (1) when $r_0(l) = 0$, i.e., the body has no rear cutting, (2) $r_0'(l) = 0$, i.e., the slope of the body contour to the free stream direction at the boundary of the rear cutting is equal to zero, which follows directly from the obvious equation $A'(x) = 2\pi r_0(x) r_0'(l)$.

---

[36] Lighthill, M., *Supersonic Flow Past Bodies of Revolution*, Rep. 8, Mem. Aero. Res. Council No. 2003, 1945.

The first approximation, determined by Equation 152, boundary condition (155), and Equation 175, is valid for both the supersonic and subsonic motions. This approximation can be used as the basis for the theory of similarity of axisymmetric three-dimensional flows past thin bodies of revolution.

13. Similarity Laws for the Flows Past Thin Bodies of Revolution and Thin Wings of Finite Span

Let us reapply, in complete analogy with Chapter 8 for a two-dimensional motion, the similarity arguments that allow conversion from one sub- or supersonic flow into the other one similar to it. We shall not repeat *in toto* the arguments given in Chapter 8. Note that in the axisymmetric case, just as in the two-dimensional one, the conversion similar to Equation 54 of Chapter 8, retains its force and reduces differential equation (152), formed for two flows being compared, to the same equation

$$\frac{\partial^2\theta}{\partial\xi^2} \pm \frac{\partial^2\theta}{\partial\eta^2} \mp \frac{1}{\eta}\frac{\partial\theta}{\partial\eta} = 0$$

with transformation (54), giving no similarity conditions at all.

The difference is already felt in the similarity of boundary conditions (155). Assuming, just as in the two-dimensional case, for the two flows compared that

$$r_{01}(x) = b\tau_1 h(\xi) \qquad r_{02}(x) = b\tau_2 h(\xi)$$

we have two boundary conditions

$$\tilde{\psi}_1(0,\xi) = \Psi_1\theta = -\frac{1}{2}U_1 b^2\tau_1^2 h^2(\xi) \qquad \tilde{\psi}_2(0,\xi) = \Psi_2\theta = -\frac{1}{2}U_2 b^2\tau_2^2 h^2(\xi)$$

from which the first similarity condition follows

$$\frac{U_1\tau_1^2}{\Psi_1} = \frac{U_2\tau_2^2}{\Psi_2} \tag{176}$$

The pressure coefficient c_p will be determined from formula (175) into which the values of $\tilde{u}$ and $\tilde{v}$, obtained from Equation 150, will be substituted. Again, the transformation will be used which is analogous to Equation 54 of Chapter 8. For the two motions compared,

$$c_{p_1} = \mp\frac{2\Psi_1}{U_1}\frac{1}{\eta}\frac{\partial\theta}{\partial\eta} - \frac{|1-\mathbf{M}_1^2|\Psi_1^2}{U_1^2}\frac{1}{\eta^2}\left(\frac{\partial\theta}{\partial\xi}\right)^2$$

$$c_{p_2} = \mp\frac{2\Psi_2}{U_2}\frac{1}{\eta}\frac{\partial\theta}{\partial\eta} - \frac{|1-\mathbf{M}_2^2|\Psi_2^2}{U_2^2}\frac{1}{\eta^2}\left(\frac{\partial\theta}{\partial\xi}\right)^2$$

or

$$\frac{U_1 c_{p_1}}{\Psi_1} = \mp\frac{2}{\eta}\frac{\partial\theta}{\partial\eta} - \frac{|1-\mathbf{M}_1^2|\Psi_1}{U_1}\frac{1}{\eta^2}\left(\frac{\partial\theta}{\partial\xi}\right)^2$$

$$\frac{U_2 c_{p_2}}{\Psi_2} = \mp \frac{2}{\eta}\frac{\partial \theta}{\partial \eta} - \frac{|1-\mathbf{M}_2^2|\Psi_2}{U_2}\frac{1}{\eta^2}\left(\frac{\partial \theta}{\partial \xi}\right)^2 .$$

From this still another two similarity conditions follow for the flows compared

$$\frac{U_1 c_{p_1}}{\Psi_1} = \frac{U_2 c_{p_2}}{\Psi_2} \qquad \frac{|1-\mathbf{M}_1^2|\Psi_1}{U_1} = \frac{|1-\mathbf{M}_2^2|\Psi_2}{U_2} \tag{177}$$

which are added to condition (176). Eliminating the ratios U_1/Ψ_1 and U_2/Ψ_2, we obtain two similarity conditions

$$\frac{c_{p_1}}{\tau_1^2} = \frac{c_{p_2}}{\tau_2^2} \qquad (1-\mathbf{M}_1^2)\,\tau_1^2 = (1-\mathbf{M}_2^2)\,\tau_2^2$$

which can be interpreted as one similarity relation general for axisymmetric sub- and supersonic flows:

$$c_p = \tau^2 f\left(\tau\sqrt{|1-\mathbf{M}_\infty^2|}\right) \tag{178}$$

It can be easily noted that Equation 178 enters as a particular case ($\Psi = \tau^2$) into the general law (60) derived in Chapter 8 for two-dimensional sub- and supersonic flows.

It should be emphasized that in the cases of axisymmetric sub- and supersonic flows past thin bodies there is no such vast variety of similarity laws as in the case of the two-dimensional flow. In particular, it is impossible to obtain the analog of the Prandtl-Glauert relation (30), which was given in Chapter 8, and which allowed one (on the basis of the pressure coefficient c_{p0} distribution over the surface of the given body in the flow of an incompressible fluid, to judge directly about the distribution of the same coefficient c_p in a subsonic flow of gas. As it follows from formula (178), it is only possible to establish the ratio of the pressure coefficients c_p in a gas to this coefficient c_{p0} in an incompressible liquid

$$\frac{c_p}{c_{p_0}} = \frac{f\left(\tau\sqrt{1-\mathbf{M}_\infty^2}\right)}{f(\tau)} \tag{179}$$

Knowing the pressure coefficient distribution over the surface of a certain thin body of revolution in a flow of incompressible liquid, it is possible from formula (179) to obtain the distribution of the same coefficient in a gas flow, but only on a thin body altered affinely by the factor $\sqrt{1-\mathbf{M}_\infty^2}$ in the direction transverse to the flow.

Similarity law (178) expressed for supersonic flows in the form

$$c_p = \tau^2 f\left(\tau\sqrt{\mathbf{M}_\infty^2 - 1}\right)$$

at very great values of $M_\infty \gg 1$ goes into the relation

$$c_p = \tau^2 f(\mathbf{M}_\infty \tau) = \tau^2 f(K)$$

coinciding with Equation 111, Chapter 8.

Relatively simple is the solution of the problem of the similarity law for thin wings of finite span in the flows of sub- and supersonic velocities. Employing the same coordinates as in Section 10, we form the equation of small perturbations in Cartesian coordinates

$$\frac{\partial^2 \tilde{\varphi}}{\partial x^2} + \frac{1}{1 - \mathbf{M}_\infty^2}\left(\frac{\partial^2 \tilde{\varphi}}{\partial y^2} + \frac{\partial^2 \tilde{\varphi}}{\partial z^2}\right) = 0 \tag{180}$$

In this three-dimensional case the perturbation stream function is absent and one should use the perturbation velocity potential $\tilde{\varphi}(x, y, z)$. The boundary condition of the wing surface impermeability can be shifted to the $0xz$ plane

$$\tilde{v} = \left(\frac{\partial \tilde{\varphi}}{\partial y}\right)_{y=0} = U_\infty \tau = \frac{\partial h}{\partial (x/b)} \tag{181}$$

where the shape of the wing surface is assumed to be specified by the equation

$$y = b\tau h\left(\frac{x}{b}, \frac{z}{l}\right) \tag{182}$$

Here, b is the mean chord of the wing, l is its span.

Just as in the two-dimensional case, we designed the scales of the velocity potentials in the two motions compared by Φ_1 and Φ_2, and, having replaced the variables

$$x_1 = x_2 = \xi \qquad y_1\sqrt{|1 - \mathbf{M}_1^2|} = y_2\sqrt{|1 - \mathbf{M}_2^2|} = \eta$$

$$z_1\sqrt{|1 - \mathbf{M}_1^2|} = z_2\sqrt{|1 - \mathbf{M}_2^2|} = \xi$$

$$\tilde{\varphi}_1 = \Phi_1\theta(\xi, \eta, \zeta) \qquad \tilde{\varphi}_2 = \Phi_2\theta(\xi, \eta, \zeta) \tag{183}$$

analogously to Equation 64 of Chapter 8, instead of Equations 180, we have the following equation for each flow

$$\frac{\partial^2 \theta}{\partial \xi^2} \pm \left(\frac{\partial^2 \theta}{\partial \eta^2} + \frac{\partial^2 \theta}{\partial \zeta^2}\right) = 0$$

In the two flows compared, boundary condition (181) will be reduced to:

$$\Phi_1\sqrt{|1 - \mathbf{M}_1^2|}\left(\frac{\partial \theta}{\partial \eta}\right)_{\eta=0} = U_1\tau_1 b\frac{\partial h}{\partial \xi} \qquad \Phi_2\sqrt{|1 - \mathbf{M}_2^2|}\left(\frac{\partial \theta}{\partial \eta}\right)_{\eta=0} = U_2\tau_2 b\frac{\partial h}{\partial \xi}$$

That they are identical in similar flows, it is necessary to assume that

$$\frac{U_1\tau_1}{\Phi_1\sqrt{|1 - \mathbf{M}_1^2|}} = \frac{U_2\tau_2}{\Phi_2\sqrt{|1 - \mathbf{M}_2^2|}} \tag{184}$$

which gives the first similarity condition.

According to Equations 182 and 183 the condition of geometric similarity will involve the relation

$$\frac{l_1}{b}\sqrt{|1 - \mathbf{M}_1^2|} = \frac{l_2}{b}\sqrt{|1 - \mathbf{M}_2^2|}$$

where, noting that the relative aspect ratio of the wing $\lambda = l^2/S$ (where S is the wing area in the plan) is proportional in similar wings to the product lb, we shall find the second similarity condition

$$\lambda_1\sqrt{|1-\mathbf{M}_1^2|} = \lambda_2\sqrt{|1-\mathbf{M}_2^2|} \tag{185}$$

The pressure coefficients in similar systems are defined by the equations[37]

$$c_{p_1} = -\frac{2\tilde{u}_1}{U_1} = -\frac{2}{U_1}\frac{\partial\tilde{\varphi}_1}{\partial x} = -\frac{2\Phi_1}{U_1}\frac{\partial\theta}{\partial\xi}$$

$$c_{p_2} = -\frac{2\tilde{u}_2}{U_2} = -\frac{2}{U_2}\frac{\partial\tilde{\varphi}_2}{\partial x} = -\frac{2\Phi_2}{U_2}\frac{\partial\theta}{\partial\xi}$$

from the coincidence of which for similar flows the third similarity condition follows

$$\frac{c_{p_1}U_1}{\Phi_1} = \frac{c_{p_2}U_2}{\Phi_2} \tag{186}$$

In the system of Equations 184 and 186 the quantities Φ_1 and Φ_2 are completely arbitrary. Introducing the designations for new arbitrary quantities

$$\frac{\Phi_1}{U_1} = A_1 \qquad \frac{\Phi_2}{U_2} = A_2$$

we finally obtain the following set of similarity conditions (the symbol "idem" means "the same", in the last condition this means "at identical points in the motions compared")

$$\frac{\tau}{A\sqrt{|1-\mathbf{M}_\infty^2|}} = \text{idem} \qquad \lambda\sqrt{|1-\mathbf{M}_\infty^2|} = \text{idem} \qquad \frac{c_p}{A} = \text{idem} \tag{187}$$

which can be replaced by the conventional functional dependence

$$c_p = Af\left(\frac{\tau}{A\sqrt{|1-\mathbf{M}_\infty^2|}};\ \lambda\sqrt{|1-\mathbf{M}_\infty^2|}\right) \tag{188}$$

Just as in the two-dimensional case, the quantity A is perfectly arbitrary here; employing this arbitrariness, it is possible to obtain diverse similarity laws. For example, assuming in the case of a subsonic flow that $A = \dfrac{1}{\sqrt{1-\mathbf{M}_\infty^2}}$ we obtain the extension (to the case of a thin wing of finite span) of the Prandtl-Glauert rule (Equation 30, Chapter 8),

$$c_p = \frac{f(\tau;\ \lambda\sqrt{|1-\mathbf{M}_\infty^2|})}{\sqrt{|1-\mathbf{M}_\infty^2|}} \tag{189}$$

---

[37] In this case the singularity typical of the axisymmetric motion is absent and it is possible to use the ordinary linear coupling between the pressure coefficient c_p and the perturbation velocity $\tilde{u}$.

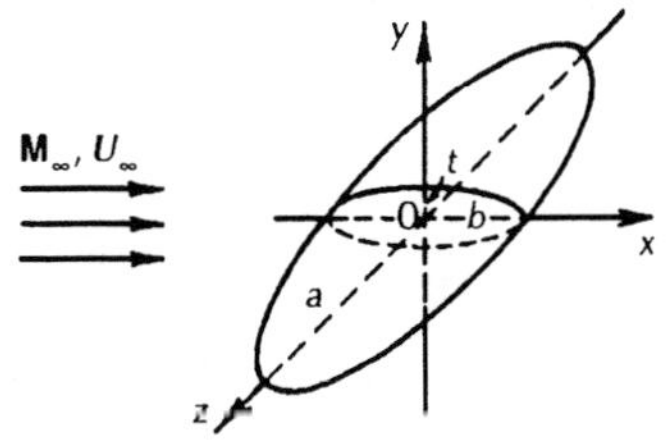

Fig. 136

After changing the relative aspect ratio of the wing λ by a factor of $\sqrt{1-\mathbf{M}_\infty^2}$, this rule makes it possible to use the usual Prandtl-Glauert rule for plane sections of a wing in a subsonic flow.

On the assumption that $A = |1-\mathbf{M}_\infty^2|^{-1}$, relation (188) acquires the following form, which is symmetric with respect to both directions transverse to the flow,

$$c_p = \frac{1}{|1-\mathbf{M}_\infty^2|} f(\tau\sqrt{|1-\mathbf{M}_\infty^2|};\ \lambda\sqrt{|1-\mathbf{M}_\infty^2|}) \tag{190}$$

which represents the following similarity law: if the chords of the plane sections remain intact, the dimensions of the wing, which are transverse to the flow, are changed by the factor $\sqrt{|1-\mathbf{M}_\infty^2|}$, i.e., the relative thickness and aspect ratio, the pressure coefficient will change by the factor of $|1-\mathbf{M}_\infty^2|^{-1}$.

Finally, assuming that $A = \tau/\sqrt{|1-\mathbf{M}_\infty^2|}$, we obtain the simplest similarity condition

$$c_p = \frac{\tau}{\sqrt{|1-\mathbf{M}_\infty^2|}} f(\lambda\sqrt{|1-\mathbf{M}_\infty^2|}) \tag{191}$$

The case of gas flow past a thin wing (considered above) was based on the same hypothesis about plane sections as that mentioned in Section 10. The smallness of the perturbation velocity in the direction of the $0z$ axis (wing span) presupposes the presence of the great wing aspect ratio, which makes the flow past the wing almost plane. This is the explanation for the noticeable likeness between the similarity laws obtained and those for the two-dimensional motion (Section 3, Chapter 8). The difference is only in the presence of the additional argument $\lambda\sqrt{|1-\mathbf{M}_\infty^2|}$ which determines the effect of the finite nature of the wing span.

It is of interest to assess (though only approximately) the effect of this factor on the pressure coefficient distribution over the wing. For this purpose, it is possible to use the theoretical calculations of Hess and Gardner[38] for the family of three-axes ellipsoids for which the equations of velocity distribution over the surface in a flow of incompressible fluid are known.[39] Let us consider the ellipsoids (Figure 136) with the fixed small relative thickness $\tau = t/b$ and different relative aspect ratios $\lambda = \frac{(2a)^2}{S} = \frac{(2a)^2}{\pi ab} = \frac{4}{\pi}\frac{a}{b}$. The value $\lambda = \infty$ corresponds to the flow past a thin elliptical cylinder of infinite span; when $a = t$ and, conse-

[38] Hess, R. V. and Gardner, C. S., Study by the Prandtl-Glauert Method of Compressibility Effects and Critical Mach-Number for Ellipsoids of Various Aspect Ratios and Thickness ratios, NACA Rep. NL7B03a, 1947.

[39] See, e.g., Kochin, N. Ye., Kibel, I. A., and Roze, N. V., *Teoreticheskaya gidromekhanika* (*Theoretical Hydromechanics*), Part 1, Fizmatgiz Press, Moscow, 1963, pp. 387 to 396; *Aerodinamika* (*Aerodynamics*), W. F. Durand, Ed., ONTI Press, Moscow, 1937, pp. 317 to 327; Kochin, N. E., Kibel, J. A., and Rose, N. V., *Theoretische Hydromechanik*, Vol. 1 and 2, Berlin, 1954.

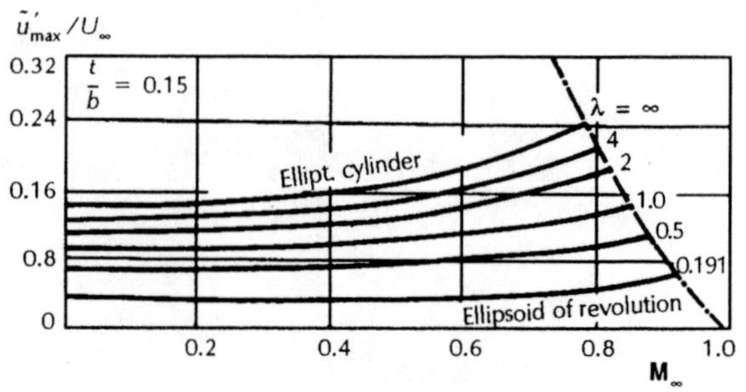

Fig. 137

quently, $\lambda = \frac{4}{\pi}\tau$, we shall have the longitudinal flow past an ellipsoid of revolution with the relative thickness τ and with the axis of revolution coinciding with the free stream direction.

Hess and Gardner calculated the dependence of the ratio $\tilde{u}_{max}/U_\infty$ (which differs from the quantity $c_{p\,min}$ only by a factor of 2) on the number $\mathbf{M}_\infty$ in a subsonic gas flow at different aspect ratios of the wing λ for the given value of its relative thickness $t/b = 15\%$. The ratio $\tilde{u}_{max}/U_\infty$ vividly estimates the perturbation introduced by the wing into a uniform flow.

The results of calculations are given in Figure 137. The upper curve corresponds to the elliptical cylinder ($\lambda = \infty$), the lower corresponds to the ellipsoid of revolution ($t = a$); in the latter case the aspect ratio is equal to

$$\lambda = \frac{4}{\pi}\frac{a}{b} = \frac{4}{\pi}\frac{t}{b} = \frac{4}{\pi}0.15 = 0.191$$

The difference between two extreme curves for the variation in the maximum dimensionless perturbation velocity as a function of the number $\mathbf{M}_\infty$ is very distinct. The upper curve follows the Prandtl-Glauert rule

$$\frac{\tilde{u}_{max}}{U_\infty} = \frac{\tilde{u}_{0\,max}}{U_\infty}\frac{1}{\sqrt{1-\mathbf{M}_\infty^2}}$$

and gives a relatively appreciable change in the ratio considered. The lower curve represents a weaker change in the same ratio with the number $\mathbf{M}_\infty$, which is characteristic for the body of revolution. It is possible to obtain an analytical formula for this change by making use of the last equation of Section 7, which represents the velocity distribution over the surface of the ellipsoid of revolution in an incompressible fluid flow. The maximum value of this velocity is obtained at $\mu = 0$:

$$\frac{\tilde{u}_{0\,max}}{U_\infty} = \frac{U_{0\,max}}{U_\infty} - 1 = \frac{e^3}{e - \frac{1}{2}(1-e^2)\ln\frac{1+e}{1-e}} - 1 = \frac{(1-e^2)\left(\frac{1}{2}\ln\frac{1+e}{1-e} - e\right)}{e - \frac{1}{2}(1-e^2)\ln\frac{1+e}{1-e}}$$

Note that the eccentricity of the ellipse will be equal to

$$e = \frac{\sqrt{b^2-t^2}}{b} = \sqrt{1-\left(\frac{t}{b}\right)^2} = \sqrt{1-\tau^2}$$

so that

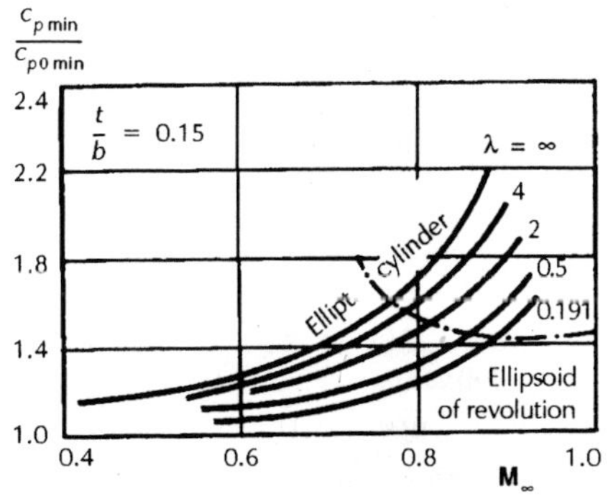

Fig. 138

$$\frac{\tilde{u}_{0\,\max}}{U_\infty} = \frac{\tau^2\left(\frac{1}{2}\ln\frac{1+\sqrt{1-\tau^2}}{1-\sqrt{1-\tau^2}} - \sqrt{1-\tau^2}\right)}{\sqrt{1-\tau^2} - \frac{1}{2}\tau^2\ln\frac{1+\sqrt{1-\tau^2}}{1-\sqrt{1-\tau^2}}}$$

Taking into account the smallness of the quantity τ, which is indispensable when using the linearized theory, and expanding the radicals into a series, we find the following approximate relation for the incompressible liquid flow

$$\frac{\tilde{u}_{0\,\max}}{U_\infty} = \tau^2\left(\ln\frac{2}{\tau} - 1\right) + O(\tau^4 \ln^2 \tau)$$

Returning to rule (178) (previously formulated for calculating the pressure coefficient c_p) or to the perturbation velocity $\tilde{u}/U_\infty$ (proportional to this coefficient), from the given c_{p0} or $\tilde{u}_0/U_\infty$ in the flow of an incompressible liquid for a body with reduced dimensions transverse to the flow (i.e., with the relative thickness $\tau\sqrt{|1-\mathbf{M}_\infty^2|}$) we shall obtain

$$\frac{\tilde{u}_{\max}}{U_\infty} = \tau^2\left(\ln\frac{2}{\tau\sqrt{|1-\mathbf{M}_\infty^2|}} - 1\right) \tag{192}$$

This formula can be further converted to the form

$$\frac{\tilde{u}_{\max}}{\tilde{u}_{0\,\max}} = \frac{c_{p\,\min}}{c_{p0\,\min}} = 1 + \frac{1}{2}\,\frac{\ln(1-\mathbf{M}_\infty^2)}{1-\ln 2+\ln\tau} \tag{193}$$

Figure 138 presents the curves that correspond to the dependence of this relation on $\mathbf{M}_\infty$ at different aspect ratios of the wing. The upper curve corresponds to the Prandtl-Glauert formula (Equation 30, Chapter 8), the lower to formula (193).

It is seen from the plot given in Figure 137 that at $\mathbf{M}_\infty = 0.7$ the relative increase in the pressure coefficient due to compressibility attains 40% for an elliptical cylinder, while at the same $\mathbf{M}_\infty$ for an ellipsoid of revolution this increase does not reach even 20%. This fact depicts the well-known property of flows; the elliptical cylinder, as the body of an infinite span, leads to a more significant perturbation of the flow than the body of revolution having the same meridian section.

The intermediate curves in Figure 137 and 138 make it possible to conclude that the smaller the effect of the number $\mathbf{M}_\infty$ in a three-dimensional subsonic flow, the smaller the relative aspect ratio of the body. The dashed-dotted curves on the right side of the plots that intersect the grids of solid curves bound the regions of their applicability. The abscissas of the points of intersection of the dashed-dotted and solid curves correspond to the critical values of $\mathbf{M}_\infty = \mathbf{M}_{cr}$, i.e., to such values at which the sonic velocity originates at the point of maximum perturbation on the body surface.

Attention should be paid to the following essential fact: The critical number $\mathbf{M}_{cr}$ increases with a decrease in the relative aspect ratio λ. While for an elliptical cylinder of 15% relative thickness $\mathbf{M}_{cr} \approx 0.78$, for an ellipsoid of revolution of the same relative thickness this number attains $\mathbf{M}_{cr} \approx 0.93$, thus reconfirming the comparatively weak effect of compressibility on the three-dimensional subsonic flow around bodies.

The above arguments relate, of course, only to the above-considered flow past a family of ellipsoids. However, they qualitatively reflect the specific features of flows around other bodies that differ from ellipsoids.

The above-noted likeness between the flow past wings of finite span and two-dimensional flows makes it possible, using the same arguments as those given at the end of Section 4 of Chapter 8, to establish the following similarity law for the transonic "quasi-plane" flows:

$$\frac{c_p\,[(k+1)\,\mathbf{M}_\infty^2]^{1/3}}{\tau^{2/3}} = f\left\{\frac{1-\mathbf{M}_\infty^2}{\tau^{2/3}\,[(k+1)\,\mathbf{M}_\infty^2]^{2/3}};\ \lambda\sqrt{|1-\mathbf{M}_\infty^2|}\right\}$$

which is the analog of similarity law (74) of Chapter 8 for plane motion.

14. Longitudinal Supersonic Flow Past a Circular Cone. Conical Shock Wave

The major feature of supersonic flow past sharpened bodies of revolution is the formation of the discontinuity surface near the frontal portion of the body. Under certain conditions this surface has the shape of the attached conical shock wave. As can be concluded from Figure 139, which depicts the patterns of a plane (on the left) flow past a wedge and spatial (on the right) flow after the conical shock, differs in principle from that after a plane shock by curvilinear streamlines are present in the spatial gas flow. The Mach lines are also curvilinear; in the figure they are shown by dashes.

In contrast to a plane oblique shock, in which the velocity vector direction directly after the shock was known – it coincided with the direction of the wedge side – in the case of the conical shock such a condition simplifying the solution of the problem does not take place. It should be replaced by the solution of the problem of curvilinear motion of gas particles downstream of the conical shock, which satisfies the boundary conditions on the discontinuity surface and on the cone surface, the flow past which is being sought.

To consider this problem, we shall avail ourselves of the spherical coordinate system (R, θ, ε) and, being satisfied with the particular case of purely meridian flow, omit the azimuthal (along the axis ε) components and derivatives with respect to the coordinate ε. Establish the continuity equation in spherical coordinates (see Equation 86, Chapter 1)

$$\operatorname{div}\,(\rho V) = \frac{1}{R^2}\frac{\partial(\rho R^2 V_R)}{\partial R} + \frac{1}{R\sin\theta}\frac{\partial(\rho V_\theta \sin\theta)}{\partial\theta} = 0$$

Making differentiation to eliminate the density ρ, we shall convert it to the form

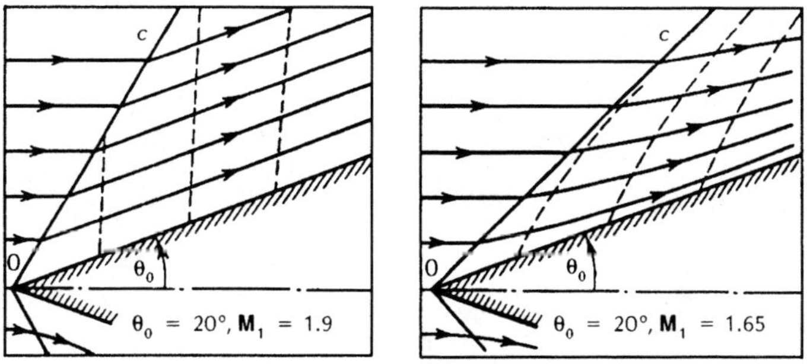

Fig. 139

$$\sin\theta\frac{\partial(R^2V_R)}{\partial R}+R\frac{\partial(V_\theta\sin\theta)}{\partial\theta}+\frac{1}{\rho}\frac{\partial\rho}{\partial R}R^2V_R\sin\theta+\frac{1}{\rho}\frac{\partial\rho}{\partial\theta}RV_\theta\sin\theta=0 \tag{194}$$

Noting that (a is the local speed of sound)

$$\frac{\partial\rho}{\partial R}=\frac{d\rho}{dp}\frac{\partial p}{\partial R}=\frac{1}{a^2}\frac{\partial p}{\partial R}\qquad\frac{d\rho}{\partial\theta}=\frac{d\rho}{dp}\frac{\partial p}{\partial\theta}=\frac{1}{a^2}\frac{\partial p}{\partial\theta}$$

we shall eliminate the pressure derivatives by using the Euler equation in spherical coordinates (Equation 46, Chapter 5)

$$V_R\frac{\partial V_R}{\partial R}+\frac{V_\theta}{R}\frac{\partial V_R}{\partial\theta}-\frac{V_\theta^2}{R}=-\frac{1}{\rho}\frac{\partial p}{\partial R}$$

$$V_R\frac{\partial V_\theta}{\partial R}+\frac{V_\theta}{R}\frac{\partial V_\theta}{\partial\theta}+\frac{V_RV_\theta}{R}=-\frac{1}{\rho}\frac{1}{R}\frac{\partial p}{\partial\theta}$$

We have

$$\frac{1}{\rho}\frac{\partial\rho}{\partial R}=-\frac{1}{a^2}\left(V_R\frac{\partial V_R}{\partial R}+\frac{V_\theta}{R}\frac{\partial V_R}{\partial\theta}-\frac{V_\theta^2}{R}\right)$$

$$\frac{1}{\rho}\frac{\partial\rho}{\partial\theta}=-\frac{1}{a^2}\left(RV_R\frac{\partial V_\theta}{\partial R}+V_\theta\frac{\partial V_\theta}{\partial\theta}+V_RV_\theta\right)$$

and Equation 194, after the substitution of the above equations into it, will become

$$(a^2-V_R^2)\frac{\partial V_R}{\partial R}-V_RV_\theta\left(\frac{1}{R}\frac{\partial V_R}{\partial\theta}+\frac{\partial V_\theta}{\partial R}\right)+$$

$$+(a^2-V_\theta^2)\frac{1}{R}\frac{\partial V_\theta}{\partial\theta}+\frac{2V_R+V_\theta\operatorname{ctg}\theta}{R}a^2=0 \tag{195}$$

As the generatrices of the conical discontinuity surface are rectilinear, the flow downstream of this surface will be irrotational. This allows Equation 195 to be supplemented in addition to the condition of the absence of vorticity. In the considered specific case of axisymmetric meridian motion this equation, according to Equation 86 of Chapter 1, will have the form

$$\operatorname{rot}_\varepsilon V = \frac{1}{R}\left[\frac{\partial (RV_\theta)}{\partial R} - \frac{\partial V_R}{\partial \theta}\right] = 0$$

or

$$R\frac{\partial V_\theta}{\partial R} - \frac{\partial V_R}{\partial \theta} + V_\theta = 0 \tag{196}$$

In the absence of vorticity in the entire flow region the Bernoulli equation holds which can be written as (C is a constant)

$$a^2 = C - \frac{k-1}{2}(V_R^2 + V_\theta^2) \tag{197}$$

The set of equations (195), (196), and (197) represents a closed system of equations which should be used as the basis for solving the specified problem of axisymmetric supersonic flow past a circular cone. The boundary conditions are

1. The condition of the impenetrability of the surface of the cone immersed in a flow ($\theta = \theta_0$)

$$V_\theta = 0 \quad \text{when} \quad \theta = \theta_0 \tag{198}$$

2. The conditions on the discontinuity surface (β is the angle between the conical shock generating line and the free stream direction)

$$V_{R_2} = V_1 \cos\beta \qquad V_{\theta_2} = -\frac{\rho_1}{\rho_2} V_1 \sin\beta \quad \text{when} \quad \theta = \beta \tag{199}$$

which are identical to the conditions on a plane oblique shock (Chapter 8); the first condition expresses the condition of the constant velocity projection onto the direction of the conical shock generating line, the second expresses the constancy of the gas mass per second during the passage of the gas through a shock.

The solution of the problem set will be self-similar, i.e., one which allows, instead of the system of partial differential equations (195) and (196), the use of a system of ordinary differential equations. We have already dealt with such similarity problems (centered waves in unsteady one-dimensional and steady plane two-dimensional flows). We shall use the conical symmetry of the boundary conditions for the problem and shall seek the solution of the equations from the condition that all motion and gas state parameters are functions only of the polar angle θ and are independent of the radius-vector R.

This solution is a particular case of a more general class of spatial conical motions of a gas which can be nonmeridian; i.e., may also involve the azimuthal velocity component.[40]

Under the assumption adopted, Equations 195 and 196 will be reduced to a system of two ordinary differential equations:

[40] See, e.g., Ferri, A., *Aerodinamika sverkhzvukovykh techeniy* (*Aerodynamics of Supersonic Flows*) (translated from English), Gostekhizdat Press, Moscow, 1953, chap. 12.

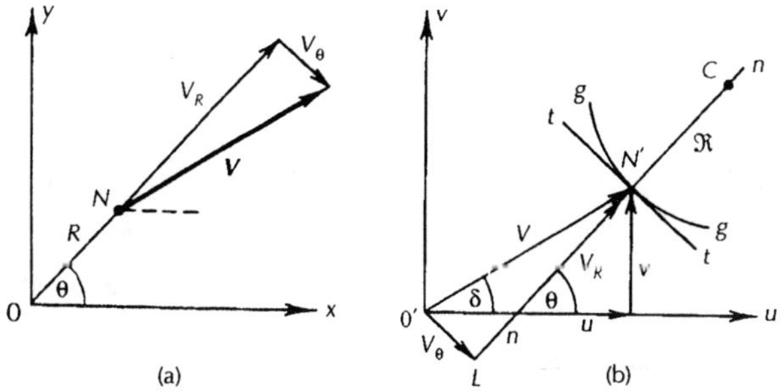

Fig. 140

$$V_R V_\theta \frac{dV_R}{d\theta} + (V_\theta^2 - a^2)\frac{dV_\theta}{d\theta} - (2V_R + V_\theta \text{ctg } \theta)\, a^2 = 0 \qquad \frac{dV_R}{d\theta} - V_\theta = 0$$

which, by virtue of the second equation, will be further simplified and will finally acquire the form

$$\frac{dV_\theta}{d\theta} + V_R = \frac{a^2 (V_R + V_\theta \text{ctg } \theta)}{V_\theta^2 - a^2} \qquad \frac{dV_R}{d\theta} - V_\theta = 0 \tag{200}$$

The classical interpretation of the first of these equations and the simple graphic method based on this interpretation and used for integrating system (200) was suggested by Busemann.[41]

Let us designate by u and v the projections of vector velocity $\boldsymbol{V}$ onto the symmetry axis $0x$ (Figure 140) and the perpendicular direction $0y$.

The projections u, v are connected with the spherical components V_R and V_θ by the relations

$$u = V_R \cos\theta - V_\theta \sin\theta \qquad v = V_R \sin\theta + V_\theta \cos\theta \tag{201}$$

so that

$$\frac{du}{d\theta} = \frac{dV_R}{d\theta}\cos\theta - \frac{dV_\theta}{d\theta}\sin\theta - V_R \sin\theta - V_\theta \cos\theta$$

$$\frac{dv}{d\theta} = \frac{dV_R}{d\theta}\sin\theta + \frac{dV_\theta}{d\theta}\cos\theta + V_R \cos\theta - V_\theta \sin\theta$$

Taking into account the second of equations 200, the first and the last terms on the right side are mutually cancelled to give

$$\frac{du}{d\theta} = -\left(V_R + \frac{dV_\theta}{d\theta}\right)\sin\theta \qquad \frac{dv}{d\theta} = \left(V_R + \frac{dV_\theta}{d\theta}\right)\cos\theta \tag{202}$$

Let us designate by ds the differential of the arc of the hodograph gg (Figure 140b). Then, Equation 202 will give

[41] Busemann, A., Drücke auf Kegelförmige Spitzen bei Bewegung mit Überschallgeschwindigkeit, *Z. Angew. Math. Mech.*, 9(6), 1929.

$$ds = \sqrt{du^2 + dv^2} = \left(V_R + \frac{dV_\theta}{d\theta}\right) d\theta \qquad (203)$$

Moreover, the cosines of the angles between the tangent tt (Figure 140b) to the hodograph at the point N' and the axes $0'u$ and $0'v$, proportional to $du/d\theta$ and $dv/d\theta$, are connected with the cosines $\cos\theta$ and $\cos[(\pi/2) - \theta] = \sin\theta$ (which determine the direction of the radius-vector $0N = R$ with respect to the axes $0x$, $0y$ or $0'u$, $0'v$) by the equation

$$\frac{du}{d\theta}\cos\theta + \frac{dv}{d\theta}\sin\theta = 0$$

From this it follows that the tangent tt is perpendicular to $0N$, whereas the normal nn forms with the $0'u$ axis the angle θ which can be considered additional to the adjacent angle (angle of curvature).

Thus, accurate to the sign, Equation 203 yields a formula for the curvature radius $\mathfrak{R}$ of the hodograph gg at the point N':

$$\mathfrak{R} = \frac{ds}{d\theta} = V_R + \frac{dV_\theta}{d\theta} \qquad (204)$$

In this case the first of Equations 200 is reduced the expression for the curvature radius of the hodograph in terms of the spherical components of the velocity vector, the radius-vector angle between the point of flow and the abscissa axis in the physical plane and of the local speed of sound

$$\mathfrak{R} = \frac{V_R + V_\theta \operatorname{ctg}\theta}{(V_\theta/a)^2 - 1} \qquad (205)$$

The curvature radius of the hodograph naturally has the dimension of velocity. Equation 205 can be brought to dimensionless form by dividing, for example, both parts of the last relation by the maximum velocity $V_{max} = \sqrt{\frac{k+1}{k-1}}\,a^*$. Then, using the designations of Equation 22 from Chapter 6, we obtain

$$\frac{\mathfrak{R}}{V_{max}} = \frac{V_R/V_{max} + (V_\theta/V_{max})\operatorname{ctg}\theta}{\frac{2}{k-1}(V_\theta/V_{max})^2/(1 - V^2/V_{max}^2) - 1} = \frac{\sqrt{\tau_R} - \sqrt{\tau_\theta}\operatorname{ctg}\theta}{[2/(k-1)]\,\tau_\theta/(1-\tau) - 1}$$

$$\tau_R = (V_R/V_{max})^2 \qquad \tau_\theta = (V_\theta/V_{max})^2 \qquad \tau = (V/V_{max})^2 \qquad (206)$$

Using Equation 205 or its dimensionless form (206), it is possible to use simple graphic means to construct the hodographs of the velocities of gas particles.

Having the given velocity V_1 or the quantity $\sqrt{\tau_1}$ and the angle θ_1 at the point N_1 of the physical plane, we find the position of the point N_1' in the hodograph plane (Figure 141). Projecting the vector V_1 onto the direction of the radius-vector and onto the direction perpendicular to it, we find V_{1R} and $V_{1\theta}$ or $\sqrt{\tau_{1R}}$ and $\sqrt{\tau_{1\theta}}$, which in the hodograph plane are, respectively, equal to the segments L_1N_1' and $0'L_1$. With the aid of Equation 206 we find the curvature radius of the hodograph at the point N_1' equal in nondimensional form to

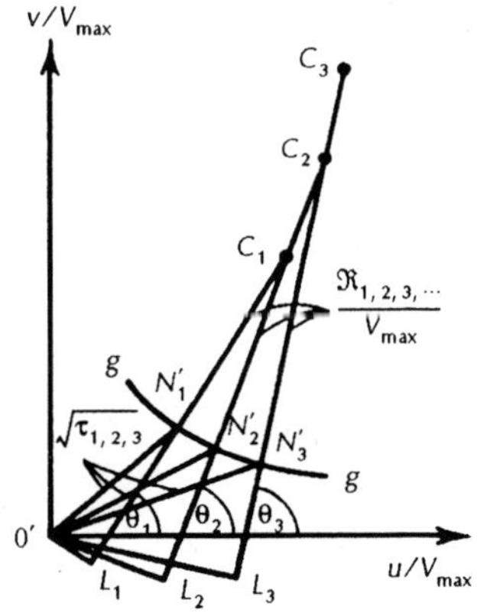

Fig. 141

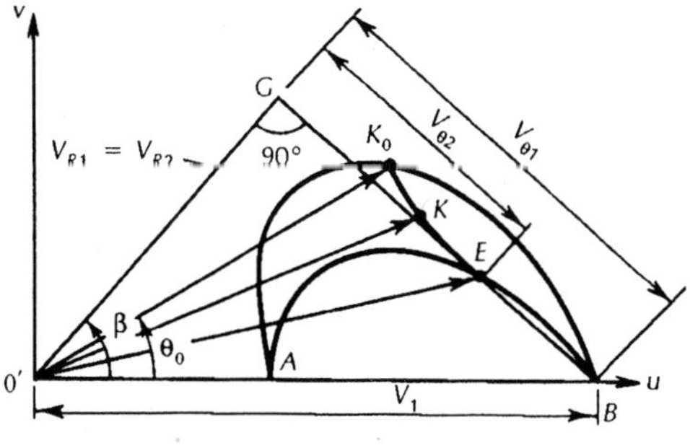

Fig. 142

$$\frac{\mathfrak{R}_1}{V_{max}} = \frac{\sqrt{\tau_{1R}} - \sqrt{\tau_{1\theta}}\,\mathrm{ctg}\,\theta_1}{[2/(k-1)]\,\tau_{1\theta}/(1-\tau_1) - 1}$$

Drawing the circular arc $N_1'N_2'$ of radius $\mathfrak{R}_1$, we adopt it to be approximately the sought-after arc of the hodograph gg. After this, we again find the values of $\sqrt{\tau_2} = 0'N_2'$, $\sqrt{\tau_{2R}} = L_2N_2'$, $\sqrt{\tau_{2\theta}} = 0'L_2$ and the new value of the angle $\theta = \theta_2$ determined by the direction of $N_2'C_1$. Then, from Equation 206, we find the new value of $\mathfrak{R}_2/V_{max}$, and, using this radius, we draw the arc $N_2'N_3'$ and so on. Thus, the sought-after hodograph is approximately represented by the set of the contacting circular arcs.

Now, it remains to show how in the presented graphic method the boundary conditions are used. Let us prescribe *a priori* a certain value of the angle β between the generatrices of the conical shock, the direction of the free stream, and the numbers $\mathbf{M}_1$ and λ_1 of the free stream. Then, selecting the strofoid corresponding to these numbers (Figure 97 of Chapter 8), we shall determine by the given β, the point E (Figure 142) on the strofoid. For this it is possible, for example, to draw the ray $0'G$ at the given angle β to the axis $0'u$ and then, from the double point B of the strofoid, to drop the perpendicular BG to the ray $0'G$. The point of intersection of this perpendicular with the strofoid will give the sought-after point E, whereas the segments $0'G$ and EG will give the radial V_{R2} and transverse $V_{\theta 2}$ components of the gas velocity vector on the surface of the conical shock directly after passing through this shock. The segments $0'G$ and BG are equal to the same components before the shock, i.e., V_{R1} and $V_{\theta 1}$.

Having these data, it is possible to employ the graphic method of Busemann and, calculating successively the centers of curvature of the hodograph from Equations 205 or 206, to construct the sought-after hodograph with the aid of the small circular arcs of the curvature. The construction is to be made until the angle of the radius-vector of the representative point K of the hodograph with the axis $0'u$ (Figure 142) becomes equal to the semivertex angle θ_0 of the circular cone immersed in a gas flow. This point K_0 will become the finite point of graphic construction and the segment $0'K_0$ will determine the limiting velocity on the surface of the cone in a flow.

Specifying different values of the angle β (but retaining the value of $\mathbf{M}_1$) and repeating the indicated construction, we obtain the geometric locus K_0 which represents a curve usually called "apple-like" due to its specific form.

Thus, it is possible to supplement in advance the grid of strofoids, constructed for different values of $\mathbf{M}_1$ and λ_1 (Figure 97 of Chapter 8), with a grid of apple-like curves and hodo-

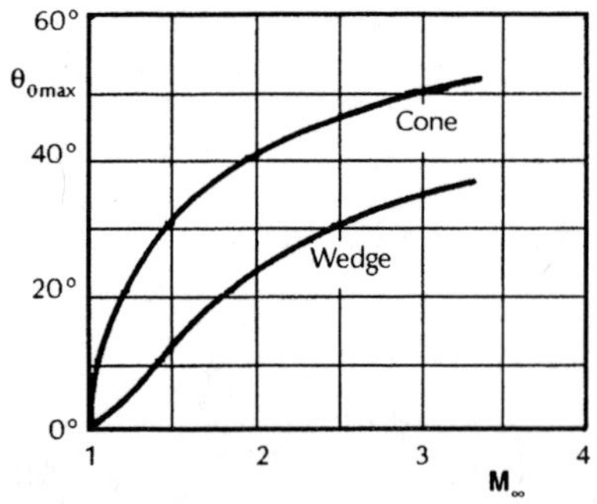

Fig. 143

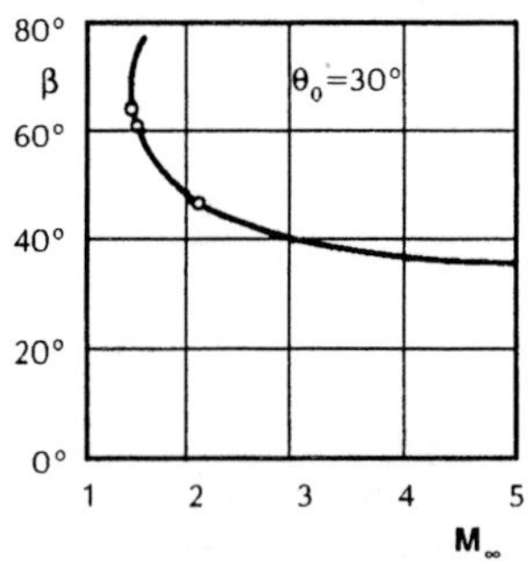

Fig. 144

graphs. This will allow one to comparatively simply solve the problems of longitudinal flow past circular cones, the vertex angles of which correspond to the condition of the presence of a shock wave attached to the cone vertex.

The practical means of constructing the conical shock front is very simple (Figure 142). Having selected the corresponding shock polar and apple-like curve by the value of the dimensionless velocity of the flow incident on the cone and having constructed the angle θ_0 equal to the semivertex angle of the cone, we find the position of the point K_0, which determines the magnitude and direction of velocity on the cone surface. Going down from this point along the portion of the hodograph curve to the point *E*, we determine the velocity vector directly behind the shock front. Then (just as with the problem of a plane shock), dropping the perpendicular $0'G$ to the straight line *BG*, which passes through the point *E*, we shall determine the angle β of the direction of the conical shock front.

The plot contained in Figure 143 illustrates the difference in the maximum semivertex angles $\theta_{0\,max}$ ($\mathbf{M}_\infty$) which admit the existence of the attached oblique shock in the case of a wedge and of the attached conical shock in the case of a flow past a cone. As is clearly seen from the plot given, the specific feature of the three-dimensional flow is the much greater value of $\theta_{0\,max}$ for the cone at the same free stream Mach number than in the case of a flow past a wedge. As always, we deal here with the softening effect of the symmetrical spreading of the gas near the cone vertex.

In practice, the plots are used which, in addition to strofoids, apple-like curves, and hodographs, also have the grids of curves that allow direct reading of pressures and other thermodynamic parameters.[42]

In conclusion, the plot of Figure 144[43] makes it possible to judge to what extent the theoretical calculation of the semivertex angle β of the conical shock (the cone semivertex angle $\theta_0 = 30°$) correctly predicts its actual values at different Mach numbers; unfortunately, the range of experimental values of Mach numbers is not great on the plot. As Figure 144 shows the left branch of the curve corresponding to a weak conical shock deviates upward and to the right in order to form an extra branch corresponding to a strong, detached shock. The same duplicity will naturally take place in the distribution of the pressure coefficient over the cone surface. The explanation for the possibilities of originating a weak or strong conical shock is the same as in the case of a plane oblique shock.

[42] See the book by A. Ferri (see citation 43); review by A. Busemann in *Handbuch der Experimental Physik*, Vol. 4, Part 1, Leipzig, 1931, Sect. 421; Krasnov, N. F., *Aerodinamika tel vrashcheniya (The Aerodynamics of the Bodies of Revolution)*, Oborongiz Press, Moscow, 1958.

[43] Ferri, A., *Elements of Aerodynamics of Supersonic Flow*, Macmillan, New York, 1949, p. 254, Figure 174.

15. Supersonic Flow Past a Thin Body of Revolution at Great Mach Numbers

In the case of a longitudinal supersonic flow past a thin body of revolution ($\mathbf{M}_\infty \gg 1$) the differential equations derived at the beginning of Section 14 have simple asymptotic solutions.

Let us return to the consideration of the system of Equations 200 and boundary conditions (198) and (199). At very great Mach numbers and with the fulfillment of the condition that the bow wave is a conical shock attached to the cone vertex, it is possible to assume that the difference $\theta - \theta_0$ will be a small quantity within the entire flow region between the surfaces of the cone and the shock.

Then, to approximately determine the functions $V_R(\theta)$ and $V_\theta(\theta)$, it is possible to avail ourselves of their expansions into series in the powers of the small difference $\theta - \theta_0$.[44]

Let us designate by $V_k = V_R(\theta_0)$ the constant value of the gas velocity on the cone surface, and note that by the condition of the cone surface impermeability, the normal velocity component $V_\theta(\theta_0)$ is equal to zero. Moreover, it directly follows from the system of Equations 200 that the following equations are valid on the cone surface:

$$\frac{dV_R}{d\theta} = 0 \qquad \frac{dV_\theta}{d\theta} = \frac{d^2V_R}{d\theta^2} = -2V_k \tag{207}$$

Thus, the sought-after first terms of expansions can be presented in the form

$$V_R/V_k = 1-(\theta-\theta_0)^2 \qquad V_\theta/V_k = -2(\theta-\theta_0) \tag{208}$$

Substituting these expressions into boundary conditions (199) on the conical shock ($\theta = \beta$), we have two equations:

$$2(\beta-\theta_0)V_k = \frac{\rho_1}{\rho_2}V_1\sin\beta \qquad [1-(\beta-\theta_0)^2]V_k = V_1\cos\beta$$

from which at small quantities β (which correspond to the condition of the thinness of the body of revolution) the simple approximate relation follows directly

$$\beta-\theta_0 = \frac{1}{2}\frac{\rho_1}{\rho_2}\beta \tag{209}$$

The involved density ratio ρ_1/ρ_2 is expressed by the following formula

$$\frac{\rho_1}{\rho_2} = \frac{k-1}{k+1}+\frac{2}{k+1}\frac{1}{\mathbf{M}_1^2\sin^2\beta}$$

which, in the case of a conical shock, coincides with the second of Equations 82 of Chapter 8 for a plane oblique shock.

Substituting this value of the density ratio into Equation 209 and using the same designations (i.e., $K = \mathbf{M}_1\theta_0$, $K_c = \mathbf{M}_1\beta$) as in Section 7 of Chapter 8 for a plane hypersonic flow past a wedge, we establish the quadratic equation

[44] Chyornyi, G. G., *Techeniye gaza s bolshoi sverkhzvukovoy skorostiyu* (*Gas Flow at a Great Supersonic Velocity*), Fizmatgiz Press, Moscow, 1959, pp. 107, 108.

$$\frac{k+3}{2(k+1)}K_c^2 - KK_c - \frac{1}{k+1} = 0$$

the positive root of which is

$$\frac{K_c}{K} = \frac{k+1}{k+3} + \sqrt{\left(\frac{k+1}{k+3}\right)^2 + \frac{2}{k+3}\frac{1}{K^2}} \tag{210}$$

As the comparison of prediction by this equation with the exact theory shows, formula (210) can be successfully used for not very great semivertex cone angles, approximately up to $\theta_0 = 15°$, for all $K < 3$, and at somewhat greater angles, also for $K > 3$.

Let us determine, with the same degree of approximation, the pressure on the cone surface which will be designated by p_k. For this purpose, the expression

$$\frac{p_k}{p_1} = \frac{p_k}{p_2}\frac{p_2}{p_1} = \left(\frac{a_k}{a_2}\right)^{\frac{2k}{k-1}}\frac{p_2}{p_1} \tag{211}$$

if formed and we note that the first factor on the right can be calculated by the Bernoulli theorem because the gas motion after the conical shock is irrotational. The second factor is calculated from the well-known relationship of pressures before and after the shock.

We have

$$\frac{a_k^2}{k-1} + \frac{V_k^2}{2} = \frac{a_2^2}{k-1} + \frac{V_{R2}^2 + V_{\theta 2}^2}{2}$$

where, after the division of both parts by the first term on the right, it follows that

$$\frac{a_k^2}{a_2^2} = 1 + \frac{k-1}{2}\frac{V_{R2}^2}{a_2^2}\left(1 - \frac{V_k^2}{V_{R2}^2} + \frac{V_{\theta 2}^2}{V_{R2}^2}\right)$$

Using the approximate relations (208), we rewrite the previous equation in the form

$$\frac{a_k^2}{a_2^2} = 1 + \frac{k-1}{2}\frac{V_1^2}{a_1^2}\frac{a_1^2}{a_2^2}\left\{1 - \left[\frac{1}{1-(\beta-\theta_0)^2}\right]^2 + \left[\frac{2(\beta-\theta_0)}{1-(\beta-\theta_0)^2}\right]^2\right\} =$$

$$= 1 + \frac{k+1}{2}\mathbf{M}_1^2\frac{a_1^2}{a_2^2}\{1 - [1 + 2(\beta-\theta_0)^2] + 4(\beta-\theta_0)^2\} =$$

$$= 1 + (k-1)\,\mathbf{M}_1^2\frac{a_1^2}{a_2^2}(\beta-\theta_0)^2 = 1 + (k-1)\frac{a_1^2}{a_2^2}(K_c - K)^2$$

Let us substitute this expression into Equation 211 and also apply formulas 104 of Chapter 8 according to which

$$\frac{p_2}{p_1} = \frac{2k}{k+1}K_c^2 - \frac{k-1}{k+1}$$

$$\frac{a_1^2}{a_2^2} = \frac{T_1}{T_2} = \frac{p_1}{p_2}\frac{\rho_2}{\rho_1} = \left(\frac{2k}{k+1}K_c^2 - \frac{k-1}{k+1}\right)^{-1}\frac{k+1}{2}K_c^2\left(1 + \frac{k-1}{2}K_c^2\right)^{-1}$$

Then, after simple transformations, we obtain the following formula for the pressure coefficient c_p on the cone surface:

$$\frac{c_p}{\theta_0^2}K^2 = \frac{2}{k}\frac{p_k - p_1}{p_1} = \frac{4}{k+1}(K_c^2 - 1) + 2(K_c - K)^2\frac{k+1}{k-1+2/K_c^2} \tag{212}$$

in which, if desired, K_c can be expressed in terms of K, according to Equation 210.

Formula (212) also coincides very well with exact calculations in the above-mentioned range of the semivertex angles of the cone in a flow.

Using the general idea of the Newtonian method, presented in Chapter 8 for a plane hypersonic motion, it is possible, with a certain degree of approximation, to also apply the formulas above to thin bodies of revolution that differ from a circular cone. For this it is sufficient to compare the cone semivertex angles θ_0 with the local angles of attack on the given surface of the body of revolution; i.e., to assume that

$$K = \mathbf{M}_1\frac{dy}{dx} = \mathbf{M}_1\tau\frac{d\bar{y}}{d\bar{x}} = K_0\frac{d\bar{y}}{d\bar{x}} \tag{213}$$

where $y = y(x)$ is the equation for the meridian section of the body of revolution and τ is the relative thickness of the body in the generalized sense (described earlier).

This approximate method is based on the Newtonian method, suggested in 1947 by Vallander, and was called the method of contacting cones (in the plane case, of contacting wedges).

Let us consider, for example, the body of ogive shape with a parabolic contour of the meridian section

$$\bar{y} = \bar{x} - \frac{1}{2}\bar{x}^2$$

for which $K = K_0(1 - \bar{x})$ according to Equation 213. In this case,[45] as calculations in formulas (210) and (212) show, the pressure distribution over the surface of the body of revolution of the selected ogive shape will be determined for air ($k = 1.4$) by the formula

$$\frac{2}{k}\left(\frac{p}{p_1} - 1\right) = 1.041K_0^2(1-\bar{x})^2 - 0.454 +$$

$$+\sqrt{1.084K_0^2(1-\bar{x})^4 + 1.656K_0^2(1-\bar{x})^2}$$

In this case the resistance formula will have the form

$$c_x\mathbf{M}_1^2 = 0.347K_0^2 - 0.454 - \frac{1}{K_0^4}[4.278N^3 - (1.382K_0^2 + 1.055)^2N +$$

$$+ (0.608K_0^2 + 0.464)\ln(2.400N + 1.309K_0^2 + 1)]$$

where

[45] Probstein, R. F. and Bray, K., Hypersonic similarity and the tangent-cone approximation for unyawed bodies of revolution, *J. Aeron. Sci.*, 22(1), 1955.

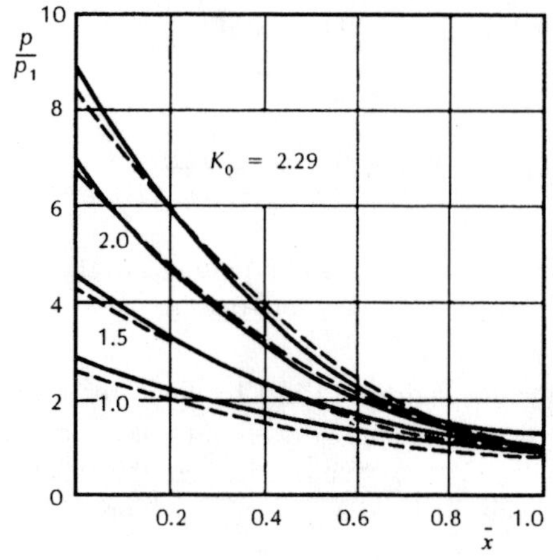

Fig. 145

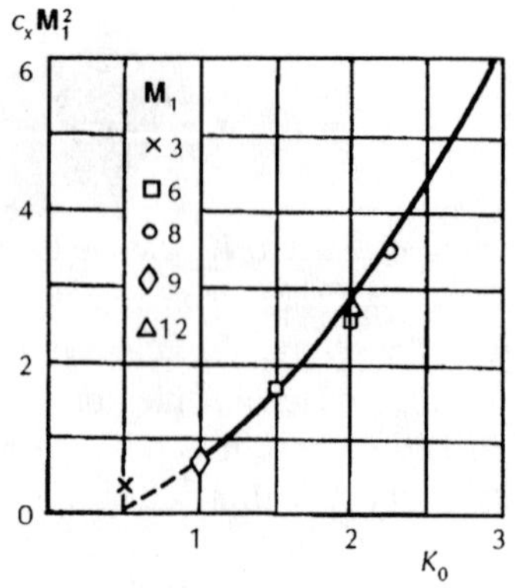

Fig. 146

$$N = \sqrt{0.297K_0^4 + 0.454K_0^2}$$

Comparison of the above approximate relations established by the method of contacting cones for the selected ogive shape with the exact solution is presented in Figure 145 and 146 and may be regarded as quite satisfactory.

In special books, which have been repeatedly cited, and which deal with supersonic flows at very great Mach numbers, other approximate calculation methods are also given; for example, the method of plane sections, the method of a shock layer, etc.

The problem of supersonic flow past thin bodies of revolution at very great Mach numbers in the case in which the bow wave "recedes" from the sharp nose of the body due to a very great vertex angle or due to the nose bluntness, presents great difficulties. Just as in the plane case, the separated shock near the symmetry axis of the flow has a nearly plane segment that corresponds to the normal shock and the adjacent segments of "strong" discontinuity downstream, the flow of which is subsonic. The motion in the region between the bow wave and the surface of the body in the flow has a mixed (sub-, super-, and trans-) sonic character.

The presence of the finite curvature near the surface of the bow wave leads to the vortex character of a gas flow behind the wave surface and to the variable nature of entropy. All these increase the difficulties of consideration of flows past bodies with the detached wave. However, in practice one has to deal with blunt noses. This is due to the processes of destruction of the sharpened side of the body during flight via great supersonic velocities (fusing, sometimes transpiration-sublimation). The destruction occurs because of high heating, a result of mechanical energy dissipation.

Along with this, during gas heating to comparatively high temperatures (on the order of 1000 K), physicochemical conversions occur in the gas which vary its original composition. Thus, in air, on attaining 2000 K, a great portion of the molecular oxygen dissociates and converts into the atomic; at 4000 K the dissociation of nitrogen begins, and at higher temperatures (on the order of 7000 to 10,000 K), a noticeable air ionization starts which is accompanied by the formation of free electrons (electronic gas). Under these conditions the gas displays a sharp increase in the thermal and electrical conductivities; the Coulomb interaction forces originate between its molecules. All this makes it possible to ascribe to the gas a specific state of aggregation called plasma (more precisely, low-temperature plasma).

The study of such gas flows presents great difficulties and cannot be incorporated into this text. It constitutes the subject of the special course of hypersonic aerodynamics. Those who wish to expand and deepen their knowledge in the field of the aerothermodynamics of hypersonic motions of an inviscid gas are referred to the book by V. V. Lunyov.[46]

At the present time, the study of supersonic gas flows is increasingly associated with the application of numerical methods. Widely used are the finite-difference methods with different versions and also the method of "large particles", which was suggested by Harlow[47] in 1957 and was developed in Russia in the works of Belotserkovskiy and co-workers.[48]

The numerical methods made it possible to solve complex problems of supersonic flow in an ideal gas past bodies of different geometric shapes.[49] These methods are dealt with in Chapter 16.

---

[46] Lunyov, V. V., *Giperzvukovaya aerodinamika (The Hypersonic Aerodynamics)*, Mashinostroyeniye Press, Moscow, 1975.

[47] See the Russian translation Harlow, F., *Chislennyi metod chastits v yacheykakh (The numerical method of particles in cells)*, in *Vychislitelnye metody v gidrodinamike (Computational Methods in Hydrodynamics)*, Mir Press, Moscow, 1967, pp. 316 to 342.

[48] Belotserkovskiy, O. M. and Davydov, Yu. M., *Metod krupnykh chastits v gazovoy dinamike: vychislitelnyi eksperiment (The Method of Large Particles in Gas Dynamics: Computational Experiments)*, Nauka Press, Moscow, 1982.

[49] Godunov, S. K., Zabrodin, A. V., Ivanov, M. Ya., and Kraiko, A. N., *Chislennoye resheniye mnogomernykh zadach gazovoy dinamiki (Numerical Solution of the Multi-Dimensional Problems of Gas Dynamics)*, Nauka Press, Moscow, 1976.

10

The Dynamics of an Incompressible Viscous Fluid (Linear Problems)

1. Newtonian Viscous Fluid and Its Rheological Equation

The differential equations of motion, energy, and mass balance in liquid and gas flows, which were derived in Chapter 4, referred to fully arbitrary media only if these media had two rather general properties – continuity and fluidity. The equations were derived with the use of the second law of dynamics as applied to the continuous system of material particles and the general thermodynamic law of the total energy conservation of the system.

The chapters following Chapter 4 considered the simplest models of a continuous medium: an ideal (devoid of internal friction) incompressible liquid or gas under the conditions of motion with small values of the Mach number, which characterizes the gas compressibility, and a more general model of an ideal gas at great sub- and supersonic velocities when the property of the gas compressibility acquires the utmost importance. In the latter case, for the model adopted to be definite, it was necessary to additionally impose the condition of gas "perfection" expressed by the gas state equation or to specify *a priori* the thermodynamic character of gas motion (adiabaticity, isothermicity).

Let us recall that the property of the liquid or gas ideality was expressed by the absence of shear stresses in them and the resulting condition of the stress tensor sphericity (E is the tensor unity)

$$P = -pE \tag{1}$$

in the presence of which all the normal stresses at the given point of the medium can be expressed in terms of one scalar quantity, i.e., pressure. Equation 1 represents the simplest example of the rheological equation of the medium. The rheological equations (laws) of media are meant to be the equations that relate the components of the tensors of stresses to deformations and their time derivatives. Such equations can be identical for different motions

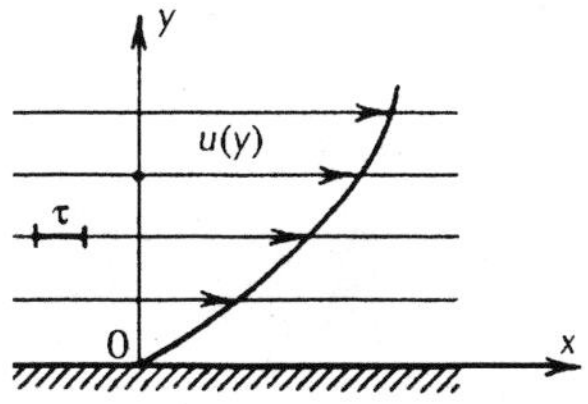

Fig. 147

of the medium considered or may depend on the character of its different possible motions, specifically, on the construction of the apparatus in which the motions take place, on the pre-history of flows, etc. As mentioned in the introduction to the volume, the most general science of the fluidity of continuous media is rheology.

The next most complex rheological equation after Equation 1, is the equation of fluidity of an ordinary viscous liquid. In the simplest case of a rectilinear laminar "shear" motion this equation is reduced to Newton's well-known rheological law[1] of the proportionality between the shear component of the stress tensor (friction) τ and the shear velocity derivative transverse to the flow direction – the tangent component of the rate of strain tensor $\partial u/\partial y$ (Figure 147).

$$\tau = \mu \frac{\partial u}{\partial y} \tag{2}$$

The proportionality factor μ which can depend only on the liquid temperature, but not on its pressure (discussed in more detail later; in fact, in real fluids exposed to very great pressures the quantity μ also depends on pressure), is called the dynamic coefficient of viscosity in contrast to the kinematic coefficient of viscosity ν, which is equal to

$$\nu = \mu/\rho \tag{3}$$

i.e., to the ratio between the dynamic coefficient of viscosity and liquid density.

According to Equation 2, the dimensionality of the dynamic coefficient of viscosity μ will be

$$\frac{\text{Force} \cdot \text{Length}}{\text{Length}^2 \cdot \text{Velocity}} = \frac{\text{Force}}{\text{Velocity} \cdot \text{Length}}$$

In the physical system of units (CGS) the dynamic coefficient of viscosity (or, simply "viscosity", as it is sometimes called for brevity) is expressed in terms of poises (P) (after the French scientist, Poiseuille) which are equal to

$$1\ \text{P} = 1\,\frac{\text{dyne} \cdot \text{s}}{\text{cm}^2} = 1\,\frac{\text{g}}{\text{cm} \cdot \text{s}}$$

Usually the unity is used which is smaller by two orders, i.e., centipoise, to which the dynamic viscosity of water corresponds at 20.5 °C.

In the technical system the viscosity unit is taken to be the quantity

[1]Newton, I., *Principia Mathematica Philosophiae Naturalis*, Sections 7 and 9, Assumptions (translated by A. N. Krylov, publication by the Naval Academy, 1915, p. 436. See also the collection of works by academician A. N. Krylov, Vol. 7, USSR Academy of Sciences Press, Moscow–Leningrad, 1936.

Table 11 Temperature Dependence of the Viscosity Coefficient of Water

Temperature (°C)	$\mu \cdot 10^2$ (P)	$\nu \cdot 10^2 \left(\frac{cm^2}{s}\right)$
0	1.792	1.792
5	1.519	1.519
10	1.308	1.308
15	1.140	1.141
20	1.005	1.007
25	0.894	0.897
30	0.801	0.804
35	0.723	0.727
40	0.656	0.661
45	0.599	0.605
50	0.549	0.556
60	0.469	0.477
70	0.406	0.415
80	0.357	0.367
90	0.317	0.328
100	0.284	0.296

Table 12 Temperature Dependence of the Viscosity Coefficient of Air

Temperature (°C)	$\mu \cdot 10^2$ (P)	$\nu \cdot 10^2 \left(\frac{cm^2}{s}\right)$
0	1.709	0.132
20	1.808	0.150
40	1.904	0.169
60	1.997	0.188
80	2.088	0.209
100	2.175	0.230
120	2.260	0.252
140	2.344	0.274
160	2.425	0.298
180	2.505	0.322
200	2.582	0.346
220	2.658	0.371
240	2.733	0.397
260	2.806	0.424
280	2.877	0.451
300	2.946	0.481
320	3.014	0.507
340	3.080	0.535
360	3.146	0.565
380	3.212	0.595
400	3.277	0.625
420	3.340	0.656
440	3.402	0.688
460	3.463	0.720
480	3.523	0.752
500	3.583	0.785

Table 13 Glycerine Viscosity

t (°C)	3	18	21
μ (P)	42.20	10.69	7.78
$\nu\ \left(\frac{\text{cm}^2}{\text{s}}\right)$	33.40	8.48	6.18

$$\frac{\text{kg} \cdot \text{s}}{\text{m}^2}$$

In the International System of Units (SI) the viscosity unity is represented by the pascal-second

$$1\ \text{P} \cdot \text{s} = 10\ \text{P} = 1\ \text{N} \cdot \text{s/m}^2 = 1\ \text{kg/(m} \cdot \text{s)}$$

equal to 10^3 centipoise.

The coefficient of kinematic viscosity is expressed in square centimeters per second (cm^2/s), square meters per second (m^2/s); the quantity equal to 1 cm^2/s is called the Stokes; the quantity which is smaller by two orders is called the centistokes.

The dynamic and kinematic coefficients of viscosity of liquids and gases depend considerably on temperature; these dependences are presented in Tables 11 and 12. It should be noted that, as is seen from these tables, both coefficients of water viscosity, dynamic and kinematic, decrease with increasing temperature, whereas the viscosity coefficients of air and also of other gases increase.There are very viscous fluids, for example, glycerine at 3°C μ = 42.20 P, ν = 33.40 cm^2/s; machine oil at 10°C μ = 6.755 P, ν = 7.34 cm^2/s.

As a rule, the viscosity of these fluids decreases quickly with an increasing temperature (see Table 13). Different empirical formulas[2] for the dependence of liquid viscosity on temperature are used.

The temperature dependence of the viscosity coefficient of a gas is given by Sutherland's formula derived theoretically in the kinetic theory of gases (the constants depend on the kind of gas)

$$\mu = \frac{\text{const} \cdot T^{3/2}}{T + C} \tag{4}$$

where $C \approx 122$ (for air). In practice, the use of the approximate power-law formula

$$\frac{\mu}{\mu_0} = \left(\frac{T}{T_0}\right)^n \tag{5}$$

is preferred, where the exponent n differs for different gases and depends on the temperature range. As the temperature increases, the exponent n in Equation 5 decreases. For approximate estimations it is adopted that $n = 1$ for comparatively low temperatures, and $n = 0.76$ for high temperatures.

In this chapter primary attention is given to the simplest cases of isothermal motion of a viscous incompressible liquid with constant values of density and viscosity coefficients. More complex motions of the gas with variable physical properties are considered in Chapter 16.

---

[2]P. A. Rebinder, Viscosity, Physical Dictionary, *Great Soviet Encyclopedia*, Vol. 2, 1960.

Rheological equation (2) represents a specific case of a more general (corresponding to any spatial motion of a viscous fluid) law of the linear connection between the stress tensor and the rate of strain tensor. This law, called the generalized Newton law, is valid for not very rarefied gases and for the majority of liquids called "Newtonian". The dynamics of "non-Newtonian" fluids is presented in special courses in rheology. Some of the simplest rheological laws of motion of non-Newtonian fluids are given in the next section; moreover, Section 7 gives the solution to the problem of motion of the most abundant "viscoelastic" (Bingham-Shvedov) fluid through a cylindrical channel of circular cross-section.

Among diverse fluid media, as to the internal nature of viscosity, the most abundant and attractive by their simplicity are isotropic viscous liquids and gases that completely fall into this category. Isotropy is defined as the identical mechanism of viscous interaction in all directions and, as a consequence, the presence of scalar (invariant) viscosity coefficients: one (the above-mentioned dynamic coefficient of viscosity) in noncompressible media, and two coefficients for the motion of gases with great velocities (see the beginning of Chapter 15).

Newton's generalized law provides the linear connection between the stress tensor and the rate of strain tensor expressed in the case of an isotropic medium by the tensor relation

$$P = a\dot{S} + bE \tag{6}$$

where a and b are scalars, E is the tensor unity

$$E_{ij} = \begin{cases} 0 & \text{when } j \neq i \\ 1 & \text{when } j = i \end{cases}$$

In order that Equation 6 may give the linear connection between the tensors, the scalar a should not depend on the tensor components P and $\dot{S}$. This scalar represents a physical constant which should be taken equal to 2μ based on the condition of Equation 6 being coincident with its specific case (2). The scalar b can be connected linearly with the tensors P and $\dot{S}$, and, due to the isotropic nature of viscosity, this scalar should represent the linear combination of the first invariants of these tensors (Chapter 1, Section 6). The first invariant of the stress tensor P is the sum of three normal stresses

$$p_{11} + p_{22} + p_{33}$$

The first invariant of the strain rate tensor $\dot{S}$ is the sum

$$\dot{S}_{11} + \dot{S}_{22} + \dot{S}_{33} = \frac{\partial V_1}{\partial x_1} + \frac{\partial V_2}{\partial x_2} + \frac{\partial V_3}{\partial x_3} = \operatorname{div} \mathbf{V}$$

which is equal to zero in the case of incompressible fluid considered in this chapter. The first invariant of the tensor unity E is equal to 3.

In order to find the scalar b, we equate the linear invariants of the tensors on the left and right sides of the basic equation (6). This will yield

$$p_{11} + p_{22} + p_{33} = 3b$$

where

$$b = \frac{1}{3}(p_{11} + p_{22} + p_{33}) \tag{7}$$

Generalizing the concept of pressure introduced into the dynamics of an ideal liquid according to the system of equations $p_{11} = p_{22} = p_{33} = -p$, we make the simplest assumption that in a Newtonian incompressible viscous fluid the mean arithmetic (with the reverse sign) of the three normal stresses applied to mutually perpendicular areas at the given point of the medium represents the pressure at this point

$$-\frac{1}{3}(p_{11} + p_{22} + p_{33}) = p \tag{8}$$

The above assumption is an additional hypothesis to the Newtonian generalized law. Proceeding from the general hydrodynamic considerations, it is impossible to prove that the thus-defined invariant scalar quantity p actually will be the very thermodynamic characteristic of a liquid or gas, which, for example, in the case of a perfect gas will be associated with other thermodynamic characteristics of the gas such as density and temperature Clapeyron's formula. The validity of the adopted hypothesis (8) is justified by its practical application for calculating the motions of a Newtonian viscous incompressible fluid.

According to Equations 6 through 8, the following expression is obtained for the Newtonian generalized law for an incompressible viscous fluid

$$P = 2\mu\dot{S} - pE = -pE + 2\mu \operatorname{def} \boldsymbol{V} \tag{9}$$

or, in a component (analytical) form,

$$p_{ij} = \begin{cases} \mu\left(\dfrac{\partial V_i}{\partial x_j} + \dfrac{\partial V_j}{\partial x_i}\right) & \text{when } j \neq i \\ -p + 2\mu\dfrac{\partial V_i}{\partial x_i} & \text{when } j = i \text{ (no summation over } i) \end{cases} \tag{10}$$

Equation 9 represents the rheological equation of a Newtonian incompressible viscous fluid. Write out in expanded form the equations of the adopted connection (9) in three basic coordinate systems: rectangular Cartesian (x_1, x_2, x_3), cylindrical (r, ε, z), and spherical (R, θ, ε). In the rectangular Cartesian system (x_1, x_2, x_3):

$$p_{11} = -p + 2\mu\frac{\partial V_1}{\partial x_1} \qquad p_{22} = -p + 2\mu\frac{\partial V_2}{\partial x_2} \qquad p_{33} = -p + 2\mu\frac{\partial V_3}{\partial x_3}$$

$$p_{12} = p_{21} = \mu\left(\frac{\partial V_1}{\partial x_2} + \frac{\partial V_2}{\partial x_1}\right) \qquad p_{23} = p_{32} = \mu\left(\frac{\partial V_2}{\partial x_3} + \frac{\partial V_3}{\partial x_2}\right)$$

$$p_{13} = p_{31} = \mu\left(\frac{\partial V_3}{\partial x_1} + \frac{\partial V_1}{\partial x_3}\right) \tag{11}$$

In the cylindrical (polar if $V_z = 0$) system (r, ε, z), recalling the components def V in this system (Equation 100 of Chapter 1), we obtain

$$p_{rr} = -p + 2\mu\frac{\partial V_r}{\partial r} \qquad p_{\varepsilon\varepsilon} = -p + 2\mu\left(\frac{1}{r}\frac{\partial V_\varepsilon}{\partial \varepsilon} + \frac{V_r}{r}\right) \qquad p_{zz} = -p + 2\mu\frac{\partial V_z}{\partial z}$$

$$p_{r\varepsilon} = p_{\varepsilon r} = \mu\left(\frac{1}{r}\frac{\partial V_r}{\partial \varepsilon} + \frac{\partial V_\varepsilon}{\partial r} - \frac{V_\varepsilon}{r}\right) \qquad p_{\varepsilon z} = p_{z\varepsilon} = \mu\left(\frac{\partial V_\varepsilon}{\partial z} + \frac{1}{r}\frac{\partial V_z}{\partial \varepsilon}\right)$$

$$p_{zr} = p_{rz} = \mu\left(\frac{\partial V_z}{\partial r} + \frac{\partial V_r}{\partial z}\right) \tag{12}$$

In the spherical system (R, θ, ε), see the components def V in this system (Equation 100 of Chapter 1)

$$p_{RR} = -p + 2\mu\frac{\partial V_R}{\partial R} \qquad p_{\varepsilon\varepsilon} = -p + 2\mu\left(\frac{1}{R\sin\theta}\frac{\partial V_\varepsilon}{\partial \varepsilon} + \frac{V_R}{R} + \frac{V_\theta \operatorname{ctg}\theta}{R}\right)$$

$$p_{\theta\theta} = -p + 2\mu\left(\frac{1}{R}\frac{\partial V_\theta}{\partial \theta} + \frac{V_R}{R}\right) \qquad p_{R\theta} = p_{\theta R} = \mu\left(\frac{1}{R}\frac{\partial V_R}{\partial \theta} + \frac{\partial V_\theta}{\partial R} - \frac{V_\theta}{R}\right)$$

$$p_{\theta\varepsilon} = p_{\varepsilon\theta} = \mu\left(\frac{1}{R\sin\theta}\frac{\partial V_\theta}{\partial \varepsilon} + \frac{1}{R}\frac{\partial V_\varepsilon}{\partial \theta} - \frac{V_\varepsilon \operatorname{ctg}\theta}{R}\right)$$

$$p_{\varepsilon R} = p_{R\varepsilon} = \mu\left(\frac{\partial V_\varepsilon}{\partial R} + \frac{1}{R\sin\theta}\frac{\partial V_R}{\partial \varepsilon} - \frac{V_\varepsilon}{R}\right) \tag{13}$$

The above expressions for the Newtonian generalized law, as with similar expressions of it in any orthogonal curvilinear system of coordinates that can be easily established with the aid of Equation 99 of Chapter 1, should be fundamental for the three-dimensional motions of a viscous incompressible fluid. It should be emphasized that the use of Newton's classical equation (2) for curvilinear streamlines by replacing in it the derivative d/dy by d/dn, where n is the direction normal to the streamlines, is inadmissible as is easily illustrated in the following simple example. In quasi-solid rotation of liquid around a fixed axis, with the velocity distribution according to the law $V = \omega r$ (r is the distance to the axis of rotation), the shears and, consequently, the rates of shears are absent. According to the definition of fluidity given in Chapter 4, tangential stress should also be absent in the quasi-solid motion. However, arbitrarily applied equation (2) will give $\tau = \mu dV/dr = \mu\omega \neq 0$. The correct result for $p_{r\varepsilon}$ is given in this case ($V_r = 0$) by formula (12)

$$p_{r\varepsilon} = \mu\left(\frac{\partial V_\varepsilon}{\partial r} - \frac{V_\varepsilon}{r}\right) = \mu(\omega - \omega) = 0$$

It should be recalled that in anisotropic viscous fluids the Newtonian generalized law becomes much complicated because the scalar coefficients of viscosity are replaced by tensors which express the three-dimensional anisotropy of the fluid viscosity and are not connected with the stress tensors, strain rate tensors, and their time derivatives.

2. Rheological Laws of Non-Newtonian Viscous Incompressible Fluids

The majority of liquids and all gases possess the properties of Newtonian fluids, which are described by rheological equation (9). Fine suspensions, clay solutions, and oil paints furnish examples of fluids that differ from the Newtonian fluids by their properties. The fact that non-

Newtonian fluids have diverse properties distinguishing them from the Newtonian is attributable to the specific features of molecular structures and internal molecular motions.

Of special interest, due to their abundance, are "viscoplastic" fluids which, along with viscosity, also maintain plastic properties consisting of the presence of a certain limiting shear stress after the attainment of which the "fluidity" of the medium sets in. The rheological laws of viscoplastic fluids are usually ascribed to Bingham (1916), though they were known to F. N. Shvedov long before 1889.

Limiting ourselves to the simplest case of plane shear rectilinear motion along the axis $0x$ with the shear rate $\dot{\varepsilon} = du/dy$ (Figure 147), we give the rheological equation for such a viscoplastic fluid in the form

$$\tau = \tau_0 + \mu'\dot{\varepsilon} \quad \text{when} \quad \tau > \tau_0 \tag{14}$$

where τ_0 is the limiting shear stress, and μ' is the dynamic coefficient of structural viscosity (the dot above the letter means a time derivative). When $\tau < \tau_0$, the fluidity is absent, i.e., the medium behaves similar to a solid body. An example of viscoplastic fluid flow along a cylindrical tube of circular profile is considered in Section 7.

The viscoplastic model (described above) complies, for example, with the motions of media occurring in practice such as clay and cement solutions[3] employed in the oil industry for washing boreholes, oil paints, slurries, and also some pastes. The physical explanation of the special features of all these fluids is based on the assumption that at rest they have a certain spatial rigid structure that can resist any external load as long as the shear stress induced by it does not exceed the limiting stress corresponding to this structure. After this the structure breaks down completely and the fluid starts to behave as an ordinary Newtonian viscous fluid at the apparent stress equal to the excess $\tau - \tau_0$ of the actual stress over the limiting one. As this apparent stress decreases until zero, i.e., when the actual stress regains its limiting value, the spatial rigid structure recovers.[4]

Other, so-called pseudoplastic fluids are devoid of the limiting yield stress, but their apparent viscosity is determined by the coefficient which depends on the shear rate. Such "nonlinear" fluids (suspensions of asymmetric particles, solutions of nigh polymers) obey rheological equations of the type (Ostwald, Reiner)

$$\tau = k\dot{\varepsilon}^n \tag{15}$$

where k and $n < 1$ are almost constant within wide ranges of stress and strain rates, whereas the apparent viscosity coefficient $\tau/\dot{\varepsilon} = k\dot{\varepsilon}^{n-1}$ decreases with increasing $\dot{\varepsilon}$.

The absence of the limiting stress makes the pseudoplastic fluids akin to the so-called dilatant fluids, the apparent viscosity of which (in contrast to pseudoplastic fluids) increases with the stress ($n > 1$). This tendency is characteristic for suspensions of solid particles at their high concentrations and also for starch glues which cannot be related to concentrated suspensions of solid particles.

Great interest is directed at present to viscoelastic media possessing both properties of viscosity and elasticity. Such media include very viscous synthetic materials and weak solutions of polymers in Newtonian fluids. It is remarkable that sometimes even small amounts (by weight) of additives of polymers convert Newtonian fluids into non-Newtonian fluids, imparting specific viscoelastic properties to them.

---

[3] Mirzadzhanzade, A. Kh., Mirzoyan, A. A., Gevinyan, G. M., and Seid-rza M. K., *Gidravlika glinistykh i tsementnykh rastvorov* (*Hydraulics of Clay and Cement Solutions*), Nedra Press, Moscow, 1966, pp. 31 to 43.

[4] Wilkinson, W. L., *Non-Newtonian Fluids* (translated from English), Mir Press, Moscow, 1964, pp. 20, 21.

Depending on the approach to the determination of the simultaneous effect of elasticity and viscosity, two rheological models are distinguished:

1. The Voigt model based on the superposition of elastic and viscous stresses:

$$\tau = G\varepsilon + \mu\dot{\varepsilon} \tag{16}$$

where G is the rigidity modulus, ε is the shear deformation, $\dot{\varepsilon} = du/dy$ is the shear rate, and μ is the dynamic coefficient of viscosity, with plane shear motion being considered similar to that depicted in Figure 147 ($v = 0$, $u = u(y)$).

2. The Maxwell model based on the superposition of the rates of elastic and viscous deformations

$$\dot{\varepsilon} = \frac{\dot{\tau}}{G} + \frac{\tau}{\mu} \tag{17}$$

where the same designations are preserved and the same plane shear motion is considered.

The difference between these models can be clearly seen if it is assumed that in the first model (16) $\tau = \text{const} = \tau_0$ and in the second model $\dot{\varepsilon} = 0$. Integrating Equation 16, we obtain the following equation for the Voigt model at $\tau = \tau_0$:

$$\varepsilon = \frac{\tau_0}{G}\left[1 - \exp\left(-\frac{Gt}{\mu}\right)\right] \tag{18}$$

which shows that with an increase of the time t retardation occurs in the establishment of elastic deformation τ_0/G, which corresponds to the constant strain $\tau = \tau_0$. The constant μ/G is the characteristic time of retardation.

When integrating Equation 17, the Maxwell model, with the assumption that $\dot{\varepsilon} = 0$, it leads to the solution

$$\tau = \tau_0 \exp\left(-\frac{Gt}{\mu}\right) \tag{19}$$

which shows that with an increase of the time t the stress τ regains that zero value which would have corresponded to the equilibrium state of the medium in the absence of the rate of strain $\dot{\varepsilon}$.

The process of the recovery of the equilibrium state from which the medium was disturbed by some perturbation is called relaxation and the characteristic time of the development of this process is called the time of relaxation.

The decrease in the stress in Maxwell's model as given by Equation 19 furnishes an example of stress relaxation in the flow of a viscoelastic fluid. The quantity

$$\lambda = \frac{\mu}{G} \tag{20}$$

occurring in both Voigt's and Maxwell's models can be interpreted in the first case as the "time of retardation" for the establishment of elastic deformation and in the second case as "the time of relaxation" of the stress.

The Maxwell equation (17), with regard for the character of the rectilinear shear motion ($v = 0$, $\dot{\varepsilon} = du/dy$), can be rewritten in the form

$$\lambda\dot{\tau} + \tau = \mu\dot{\varepsilon} = \mu\frac{du}{dy} \tag{21}$$

At the zero relaxation time ($\lambda = 0$), Equation (21) leads to Newton's law (2) which corresponds to the instantaneous following of the stress upon the shear rate (in other words, to the infinite velocity of the propagation of disturbances). The content of Section 4 of Chapter 5, which showed that there was an infinitely great velocity of the propagation of disturbances in an incompressible fluid and the final velocity of this process in a compressible medium (e.g., should be recalled gas). The presence of the property of elasticity in a viscoelastic medium leads, just as in a gas, to the final velocity of the propagation of disturbances.

To determine the magnitude of this velocity note that in the plane rectilinear shear motion considered the basic equation of the dynamics "in terms of stresses" (Equation 8, Chapter 4) in the absence of volumetric forces can be put in the form

$$\rho\frac{\partial u}{\partial t} = \frac{\partial \tau}{\partial y} \tag{22}$$

Having taken the time derivative of both parts of Equation 21,

$$\lambda\frac{\partial^2 \tau}{\partial t^2} + \frac{\partial \tau}{\partial t} = \mu\frac{\partial^2 u}{\partial y \partial t}$$

or, according to Equation 22,

$$\frac{\partial^2 \tau}{\partial t^2} + \frac{1}{\lambda}\frac{\partial \tau}{\partial t} = \frac{\nu}{\lambda}\frac{\partial^2 \tau}{\partial y^2} \tag{23}$$

This hyperbolic-type partial differential ("telegraph") equation describes the wave propagation of the stress, with the velocity of this propagation being determined by the square root of the coefficient of $\partial^2\tau/\partial y^2$; it is independent of the medium viscosity and is equal to

$$w = \sqrt{\frac{\nu}{\lambda}} = \sqrt{\frac{\nu G}{\mu}} = \sqrt{\frac{G}{\rho}} \tag{24}$$

The fact that viscoelastic media have final velocity of the propagation of disturbances results in the appearance, in the moving viscoelastic medium, of the dependence of motion at the given point in the flow on the previous motions at the points located upstream. This phenomenon is called the "effect of the flow prehistory"; the fluids that possess this property are called "hereditary". The viscoelastic fluid furnishes an example of such kind of "hereditary" fluids.

The essential difference between Voigt's and Maxwell's models is as follows. Characteristic of the Voigt model is the fact that under the action of constant stress the shear rate $\dot{\varepsilon}$, which can be obtained from Equation 18 by differention in time, tends rapidly to zero, when $t \to \infty$; i.e., under the action of constant loading the "Voigt body" does not possess the property of infinite fluidity. In contrast to the Voigt "body", the Maxwell "body" for which (see Equation 17, the following relation holds when $\tau = \tau_0$ and $\dot{\tau} = 0$:

$$\dot{\varepsilon} \to \frac{\tau_0}{\mu} \neq 0 \quad \text{at any } t$$

will flow, under the action of constant loading with the constant shear rate $\dot{\varepsilon}_0 = \tau_0/\mu$. This is the reason that the medium obeying the Voigt rheological law is often called the viscoelastic, "hereditary" solid body in contrast to the Maxwellian "body" which represents the viscoelastic "hereditary" fluid.

The above examples do not exhaust the diversity of the specific properties of non-Newtonian fluids. The mechanical properties of many fluids depend substantially not only on the deformation rate, but also on the duration of deformation and on the flow prehistory. Such fluids are called thixotropic. Some of them, the rheopectic fluids, are capable of increasing the stiffness of their structures in the presence of shear motion. Others, conversely, are capable of weakening their structures. To the first type will belong, for example, cement mortars in mode, "setting". the melted metals which in the liquid state are purely Newtonian fluids, and at the initial stage of solidification are being filled up with the tiniest crystalline formations which bring them closer to being dilatant fluids.

The thixotropy may show up in the reverse, also associated with time, effect of the breakdown of a rigid structure under the action of a shear deformation motion, as is the case, for example, in a kefir-type liquid (kefir is a kind of yogurt). When shaken, the kefir, which represents a jelly-like body, pours out freely from a bottle, but restores its structure after a certain rest period.

Modern synthetic materials used in machine building, textile, food, and other industries, furnish many examples of diverse non-Newtonian (sometimes called rheological) fluids, the mechanical laws of motion of which are very complicated. In addition, with a certain degree of approximation, they can be represented by a combination of simplest laws briefly described in this section.

The rheology of non-Newtonian fluids borders on the dynamics of inhomogeneous fluids which are understood to be multiphase, multicomponent media. In such media, for example, fine suspensions of particles, it is generally very difficult to differentiate an inhomogeneous medium from a non-Newtonian fluid. Some think that in certain cases non-Newtonian fluids can lose their initial molecular microstructure in transition from laminar to turbulent motion and can become non-Newtonian, involving admixtures, i.e., inhomogeneous fluids.

Of special complexity is the problem of finding shear tensors for separate phases in a moving mixture and general laws for "interphase" shear tensors in which the very concept of deformation, as of the rate of deformation, becomes especially complex.[5]

The basic method of overcoming these difficulties is the approximate approach, consisting of the preservation (the mixture as a whole) of the rheological equation of a homogeneous, Newtonian, or non-Newtonian medium, the physical constants of which take into account on average the specific features of separate components of the nonhomogeneous mixture.

For media with small volumetric concentrations of admixtures the application of Einstein's correction to the dynamic coefficient of viscosity of the carrying phase is very popular. The corrected dynamic viscosity coefficient of the mixture μ^* is expressed in terms of the corresponding coefficients: μ for a "pure" carrying phase and $\bar{\mu}$ for a liquid or gaseous admixture with spherically shaped particles, according to the equation[6]

$$\frac{\mu^*}{\mu} = 1 + \alpha \frac{\mu + \frac{5}{2}\bar{\mu}}{\mu + \bar{\mu}} \tag{25}$$

where α is the volumetric fraction of the admixture.

---

[5]Nigmatulin, R. I., The methods of mechanics of a continuous medium for describing multiphase mixtures, *Prikl. Mat. Mekh.*, 34(6), 1097–1112, 1970; *Osnovy mekhaniki geterogennykh sred* (*The Fundamentals of the Mechanics of Heterogeneous Media*), Nauka Press, Moscow, 1978, chap. 1.

[6]The substantiation of this equation can be found in Batchelor, J., *Vvedeniye v dinamiku zhidkosti* (*Introduction Into the Dynamics of a Fluid*) (translated from English), Mir Press, Moscow, 1973, pp. 313 to 321.

In the case of a solid admixture one should assume that $\bar{\mu} = \infty$; then, the previous formula can be reduced to a simpler formula which is valid for solid suspensions of spherical particles:

$$\mu^* = \mu\left(1 + \frac{5}{2}\alpha\right) \tag{26}$$

In the case of admixtures of nonspherical particles, the viscosity of the mixture increases. Thus, for solid particles having the shape of the ellipsoids of revolution with the semiaxis ratio 6:1,

$$\mu^* = \mu(1 + 5\alpha) \tag{27}$$

It should be noted that Einstein's correction is essential. For example, for blood consisting of the Newtonian carrying phase – plasma with the dynamic viscosity coefficient $\mu = 0.015$ P – and blood cells (transported by the plasma) with the volumetric concentration $\alpha \approx 40\%$, these cells are considered to be solid ellipsoids of revolution,[7] with the relative thickness 1:6 (close to reality). Thus, we have $\mu^* = 3\mu$.

At present, the rheology of non-Newtonian fluids represents a great, independent area of the mechanics of continuous media.[8] Its modern development owes a great deal to the well-known Soviet school headed by Z. P. Shulman at the Heat and Mass Transfer Institute of the BSSR Academy of Sciences in Minsk.[9]

3. The Navier-Stokes Equations of the Dynamics of a Newtonian Incompressible Medium

To derive the basic differential equations of the dynamics of a viscous incompressible fluid, use will be made of the equation of the dynamics of continuous media in terms of "stresses" (Equations 6 and 8 of Chapter 4) and augment them with the equation of incompressibility (Equation 10, Chapter 3).

Let us substitute the stresses p_{ij} from Equation 11 (this chapter) into the right sides of the system of equations in terms of "stresses" (Equation 8, Chapter 4), written with the use of numerical indexation,

$$\rho \frac{dV_1}{dt} = \rho F_1 + \frac{\partial p_{11}}{\partial x_1} + \frac{\partial p_{21}}{\partial x_2} + \frac{\partial p_{31}}{\partial x_3}$$

$$\rho \frac{dV_2}{dt} = \rho F_2 + \frac{\partial p_{12}}{\partial x_1} + \frac{\partial p_{22}}{\partial x_2} + \frac{\partial p_{32}}{\partial x_3}$$

---

[7] Jones, R. T., Blood flow, *Annu. Rev. Fluid Mech.*, 1, 223–224, 1969.

[8] Among the general works on rheology, the author recommends: Astarita and Marruci, *Osnovy gidromeknaniki nenyutonovskikh zhidkostei (The Foundation of the Hydromechanics of Non-Newtonian Fluids)* (translated from English), Mir Press, Moscow, 1973; Truesdell, G., *Pervonachalnyi kurs ratsionalnoi meknaniki sploshnykh sred (The Primary Course of the Rational Mechanics of Continuous Media)* (translated from English), Mir Press, Moscow, 1975; Reiner, M., *Deformatsiya i techeniye. Vvedeniye v reologiyu (Deformation and Flow. Introduction into Rheology)* (translated from English), GNTI Press, Moscow, 1963; Noll, W., A mathematical theory of the mechanical behavior of continuous media, *Arch. Rat. Mech. Anal.*, 2(3), 197–226, 1958 and also the book cited in Chapter 9 by Wilkinson.

[9] Shulman, Z. P. and Khusid, B. M., *Nestatsionarnye protsessy konvektivnogo perenosa v nasledstvennykh sredakh (Unsteady-State Processes of Convective Transfer in Hereditary Media)*, Nauka i Tekhnika Press, Minsk, 1983 (extensive bibliography).

$$\rho\frac{dV_3}{dt} = \rho F_3 + \frac{\partial p_{13}}{\partial x_1} + \frac{\partial p_{23}}{\partial x_2} + \frac{\partial p_{33}}{\partial x_3} \tag{28}$$

or

$$\rho\frac{dV_i}{dt} = \rho F_i + \frac{\partial p_{ik}}{\partial x_k} \qquad (i, k = 1, 2, 3)$$

Considering in more detail the case of isothermal motion when μ = const, ρ = const, and ν = const, we obtain for the first equation of system (28)

$$\frac{\partial p_{11}}{\partial x_1} + \frac{\partial p_{21}}{\partial x_2} + \frac{\partial p_{31}}{\partial x_3} = -\frac{\partial p}{\partial x_1} + 2\mu\frac{\partial^2 V_1}{\partial x_1^2} + \mu\frac{\partial}{\partial x_2}\left(\frac{\partial V_1}{\partial x_2} + \frac{\partial V_2}{\partial x_1}\right) +$$

$$+\mu\frac{\partial}{\partial x_3}\left(\frac{\partial V_1}{\partial x_3} + \frac{\partial V_3}{\partial x_1}\right) = -\frac{\partial p}{\partial x_1} + \mu\left(\frac{\partial^2 V_1}{\partial x_1^2} + \frac{\partial^2 V_1}{\partial x_2^2} + \frac{\partial^2 V_1}{\partial x_3^2}\right) +$$

$$+\mu\frac{\partial}{\partial x_2}\left(\frac{\partial V_1}{\partial x_1} + \frac{\partial V_2}{\partial x_2} + \frac{\partial V_3}{\partial x_3}\right) = -\frac{\partial p}{\partial x_1} + \mu\nabla^2 V_1 + \mu\frac{\partial}{\partial x_1}\operatorname{div}\boldsymbol{V}$$

In an incompressible fluid div $\boldsymbol{V} = 0$, and, consequently, the last term on the right is equal to zero.

Repeating similar operations with the remaining two formulas of Equation 28, dividing both sides by ρ, expanding the expression for the derivative dV_i/dt and adding the equation of incompressibility (Equation 10, Chapter 3), we obtain the following system of the Navier-Stokes equations

$$\frac{\partial V_1}{\partial t} + V_1\frac{\partial V_1}{\partial x_1} + V_2\frac{\partial V_1}{\partial x_2} + V_3\frac{\partial V_1}{\partial x_3} = F_1 - \frac{1}{\rho}\frac{\partial p}{\partial x_1} + \nu\nabla^2 V_1$$

$$\frac{\partial V_2}{\partial t} + V_1\frac{\partial V_2}{\partial x_1} + V_2\frac{\partial V_2}{\partial x_2} + V_3\frac{\partial V_2}{\partial x_3} = F_2 - \frac{1}{\rho}\frac{\partial p}{\partial x_2} + \nu\nabla^2 V_2$$

$$\frac{\partial V_3}{\partial t} + V_1\frac{\partial V_3}{\partial x_1} + V_2\frac{\partial V_3}{\partial x_2} + V_3\frac{\partial V_3}{\partial x_3} = F_3 - \frac{1}{\rho}\frac{\partial p}{\partial x_3} + \nu\nabla^2 V_3$$

$$\frac{\partial V_1}{\partial x_1} + \frac{\partial V_2}{\partial x_2} + \frac{\partial V_3}{\partial x_3} = 0 \tag{29}$$

or briefly

$$\frac{\partial V_i}{\partial t} + V_k\frac{\partial V_i}{\partial x_k} = F_i - \frac{1}{\rho}\frac{\partial p}{\partial x_i} + \nu\nabla^2 V_i \qquad \frac{\partial V_k}{\partial x_k} = 0 \qquad (i, k = 1, 2, 3)$$

Based on the considerations of the interaction of molecules, these equations were obtained first by Navier in 1822 and by Poisson in 1829 and then by Saint-Venant in 1843 and Stokes[10] in 1845 by a technique very close to that described earlier.

In vector representation, Equations 29 has the form

$$\frac{\partial \boldsymbol{V}}{\partial t} + (\boldsymbol{V} \cdot \nabla)\, \boldsymbol{V} = \boldsymbol{F} - \frac{1}{\rho}\, \mathrm{grad}\, p + \nu \nabla^2 \boldsymbol{V} \qquad \mathrm{div}\, \boldsymbol{V} = 0 \tag{30}$$

with the symbol $\nabla^2 \boldsymbol{V}$ (Laplacian $\boldsymbol{V}$) designating the vector with the projections $\nabla^2 V_i$ $(i = 1, 2, 3)$.

In applications one has to deal with the Navier-Stokes equations in projections on the axis of orthogonal curvilinear coordinates. For setting up the equations of such a form, it is useful to take into account identity (50) of Chapter 2

$$(\boldsymbol{V} \cdot \nabla)\, \boldsymbol{V} = \mathrm{grad}\left(\frac{V^2}{2}\right) + \mathrm{rot}\, \boldsymbol{V} \times \boldsymbol{V}$$

and also the equality which follows the last line of Equation 88 of Chapter 1, provided there is incompressibility (div $\boldsymbol{V} = 0$):

$$\nabla^2 \boldsymbol{V} = -\mathrm{rot}\ \mathrm{rot}\, \boldsymbol{V} \tag{31}$$

Due to which the Navier-Stokes equations, under the condition of potentiality of volumetric forces, can be reduced to the form analogous to the Gromeka-Lamb equations (Equations 8 of Chapter 5) for an ideal fluid

$$\frac{\partial \boldsymbol{V}}{\partial t} + \mathrm{rot}\, \boldsymbol{V} \times \boldsymbol{V} = -\,\mathrm{grad}\left(\frac{V^2}{2} + \Pi + \frac{p}{\rho}\right) - \nu\, \mathrm{rot}\ \mathrm{rot}\, \boldsymbol{V}$$

$$\mathrm{div}\, \boldsymbol{V} = 0 \tag{32}$$

Using the projections onto the curvilinear axes of the operators rot, grad, and div, we obtain the following expressions for the Navier-Stokes equations in cylindrical and spherical coordinates (the volumetric forces are omitted):

- Cylindrical coordinates (r, ε, z):

$$\frac{\partial V_r}{\partial t} + V_r \frac{\partial V_r}{\partial r} + \frac{V_\varepsilon}{r} \frac{\partial V_r}{\partial \varepsilon} + V_z \frac{\partial V_r}{\partial z} - \frac{V_\varepsilon^2}{r} = -\frac{1}{\rho} \frac{\partial p}{\partial r} + \nu\left(\nabla^2 V_r - \frac{V_r}{r^2} - \frac{2}{r^2} \frac{\partial V_\varepsilon}{\partial \varepsilon}\right)$$

$$\frac{\partial V_\varepsilon}{\partial t} + V_r \frac{\partial V_\varepsilon}{\partial r} + \frac{V_\varepsilon}{r} \frac{\partial V_\varepsilon}{\partial \varepsilon} + V_z \frac{\partial V_\varepsilon}{\partial z} + \frac{V_r V_\varepsilon}{r} = -\frac{1}{\rho} \frac{1}{r} \frac{\partial p}{\partial \varepsilon} + \nu\left(\nabla^2 V_\varepsilon + \frac{2}{r^2} \frac{\partial V_r}{\partial \varepsilon} - \frac{V_\varepsilon}{r^2}\right)$$

$$\frac{\partial V_z}{\partial t} + V_r \frac{\partial V_z}{\partial r} + \frac{V_\varepsilon}{r} \frac{\partial V_z}{\partial \varepsilon} + V_z \frac{\partial V_z}{\partial z} = -\frac{1}{\rho} \frac{\partial p}{\partial z} + \nu \nabla^2 V_z$$

$$\frac{\partial (r V_r)}{\partial r} + \frac{\partial V_\varepsilon}{\partial \varepsilon} + \frac{\partial (r V_z)}{\partial z} = 0$$

$$\nabla^2 \equiv \frac{\partial^2}{\partial r^2} + \frac{1}{r} \frac{\partial}{\partial r} + \frac{1}{r^2} \frac{\partial^2}{\partial \varepsilon^2} + \frac{\partial^2}{\partial z^2} \tag{33}$$

---

[10] Citation according to the fundamental book by Lamb, G., *Gidrodinamika* (*Hydrodynamics*) (translated from English), Gostekhizdat Press, Moscow, 1947, p. 723.

- Spherical coordinates (R, θ, ε):

$$\frac{\partial V_R}{\partial t} + V_R \frac{\partial V_R}{\partial R} + \frac{V_\theta}{R}\frac{\partial V_R}{\partial \theta} + \frac{V_\varepsilon}{R \sin\theta}\frac{\partial V_R}{\partial \varepsilon} - \frac{V_\theta^2 + V_\varepsilon^2}{R} =$$

$$= -\frac{1}{\rho}\frac{\partial p}{\partial R} + \nu\left(\nabla^2 V_R - \frac{2V_R}{R^2} - \frac{2}{R^2}\frac{\partial V_\theta}{\partial \theta} - \frac{2V_\theta \operatorname{ctg}\theta}{R^2} - \frac{2}{R^2 \sin\theta}\frac{\partial V_\varepsilon}{\partial \varepsilon}\right)$$

$$\frac{\partial V_\theta}{\partial t} + V_R \frac{\partial V_\theta}{\partial R} + \frac{V_\theta}{R}\frac{\partial V_\theta}{\partial \theta} + \frac{V_\varepsilon}{R \sin\theta}\frac{\partial V_\theta}{\partial \varepsilon} + \frac{V_R V_\theta}{R} - \frac{V_\varepsilon^2 \operatorname{ctg}\theta}{R} =$$

$$= -\frac{1}{\rho}\frac{1}{R}\frac{\partial p}{\partial \theta} + \nu\left(\nabla^2 V_\theta + \frac{2}{R^2}\frac{\partial V_R}{\partial \theta} - \frac{V_\theta}{R^2 \sin^2\theta} - \frac{2\cos\theta}{R^2 \sin^2\theta}\frac{\partial V_\varepsilon}{\partial \varepsilon}\right)$$

$$\frac{\partial V_\varepsilon}{\partial t} + V_R \frac{\partial V_\varepsilon}{\partial R} + \frac{V_\theta}{R}\frac{\partial V_\varepsilon}{\partial \theta} + \frac{V_\varepsilon}{R \sin\theta}\frac{\partial V_\varepsilon}{\partial \varepsilon} + \frac{V_\varepsilon V_R}{R} + \frac{V_\theta V_\varepsilon \operatorname{ctg}\theta}{R} =$$

$$= -\frac{1}{\rho}\frac{1}{R \sin\theta}\frac{\partial p}{\partial \varepsilon} + \nu\left(\nabla^2 V_\varepsilon - \frac{V_\varepsilon}{R^2 \sin^2\theta} + \frac{2}{R^2 \sin\theta}\frac{\partial V_R}{\partial \varepsilon} + \frac{2\cos\theta}{R^2 \sin^2\theta}\frac{\partial V_\theta}{\partial \varepsilon}\right)$$

$$\frac{\partial}{\partial R}(R^2 V_R \sin\theta) + \frac{\partial}{\partial \theta}(R V_\theta \sin\theta) + \frac{\partial (R V_\varepsilon)}{\partial \varepsilon} = 0$$

$$\nabla^2 = \frac{1}{R^2}\frac{\partial}{\partial R}\left(R^2 \frac{\partial}{\partial R}\right) + \frac{1}{R^2 \sin\theta}\frac{\partial}{\partial \theta}\left(\sin\theta \frac{\partial}{\partial \theta}\right) + \frac{1}{R^2 \sin^2\theta}\frac{\partial^2}{\partial \varepsilon^2} \tag{34}$$

Equations (29) represent a nonlinear system of four second-order partial equations with four unknown functions V_1, V_2, V_3, and p; the quantities ρ and ν are prescribed constants, whereas the projections of the volumetric force F_1, F_2, F_3 (gravity forces, inertial, centrifugal, or Coriolis forces) are the specified functions of coordinates and velocities. The nonlinearity of the system is due to the presence of the convective component of acceleration on the left side of Equations 29.

To obtain specific solutions when integrating Equations 29, the boundary conditions should be used, whereas in the case of unsteady-state motion the initial conditions should also be used. Let us recall that in an ideal liquid the basic boundary condition on the solid surface immersed in a flow consists of an impermeable surface and, as a result, in the coincidence of the velocity components of liquid particles normal to the surface and points of the surface itself. In the case of a viscous fluid this boundary condition is replaced by the condition of the "adherence" of liquid particles to the solid wall. This means the absence of both the normal (to the solid surface) relative velocity between liquid particles and adjacent points of the surface, and the tangent components of the relative velocity, i.e., the absence of the fluid slip velocity on the surface.

The above assumption concerning the absence of the slip of fluid particles on the solid surface ("adherence") should not be confused with the phenomenon of "wetting" by a liquid (or the absence of wetting) of a solid surface which characterizes the so-called contact effect (the formation of a meniscus) on the boundary of three phases (solid, liquid, gaseous) or a solid surface and two fluids of different densities. Mercury does not wet the inner surface of a glass pipe through which it flows, but adheres to it.

Thus, in contrast to the ideal liquid, when a viscous fluid flows past solid surfaces the boundary condition should be fulfilled which says that the fluid velocity on a stationary streamlined surface should be equal to zero or that the velocities of liquid particles should coincide with the velocities of the moving solid surface points with which the liquid particles contact. Even at the end of the 19th century this boundary condition was questioned, but it has since been completely justified.[11] An exception to this general behavior is represented by the boundary conditions in highly diluted gases, where the slip of a gas on a solid surface is possible which is proportional to the derivative of the tangential velocity component over the normal to the surface.

The boundary conditions also include the specification of velocity far from the body in cases of external flow or of the discharge in cases of liquid flow through a channel, and also the specification of pressure at any one point of the flow, particularly at an infinite distance from the body in the flow.

The initial conditions occur in the problems of unsteady-state motions and represent the specifications of velocity distributions in the flow region at a certain initial instant, of the variation of pressure in time at the given point of the space, etc.

For specific classes of problems of the motion of a viscous fluid there are rigorous proofs of the theorems about the existence and uniqueness of the solution. In addition to their general mathematical content, these theorems are also important because they indicate which should be the boundary and initial conditions to be supplemented to differential equations, and also other additional requirements without the fulfillment of which the solution of the problem will not be unique. Sometimes the solution may not exist at all.

Such are the requirements of the smoothness of the surface in flow or the presence on it of the finite number of crisps or curvature discontinuities, the conditions imposed on the distribution of physical quantities, continuity, the existence of derivatives, etc.

Among the conditions of uniqueness there are some integral equations that are analogous to the case of an ideal fluid, where the "bound" circulation was used when calculating the lift force of the wing profile (Chapter 7). In the dynamics of a viscous fluid an analogous role is played by the assignment of the magnitude of jet momentum when calculating the jet propagation in space flooded by the same fluid, the assignment of body resistance for determining the flow in the aerodynamic wake behind it, etc.

4. Similarity of Flows of a Viscous Incompressible Fluid

The similarity method is very fruitful when studying not only hydrodynamic, but also many other physical and technical problems. Worthy of note is the direct use of this method to scientifically substantiate the techniques of simulation of actual, "natural" processes in laboratory conditions. The similarity method makes it possible to establish the requirements that should be imposed on the laboratory model and for the carrying out on it the process studied in order that the results of simulation could be used for designing the actual objects. Moreover, processing laboratory measurements and the generalizing the results of these measurements in the form of empirical equations are also carried out according to the specifications of the similarity method.

This purely applied aspect of the similarity method is far from exhausting its importance. For many years the similarity method has been used for theoretically studying the phenomena as a means of predicting the internal structure of variables and parameters occurring in ana-

[11] See the survey "Concerning the conditions on the surface of liquid contact with a solid body, in *Modern Developments in Fluid Dynamics*, Vol. 2, Goldstein, S., Ed., IL Press, Moscow, 1948, p. 356.

lytical relations derived from the theory and sometimes even the very form of these relations. One should recall, for example, the similarity relations derived in Chapters 8 and 9 for sub- and supersonic flows around slender bodies as well as the construction of "self-similar" solutions presented in the same chapters. In this and subsequent chapters many examples show the use of the ideas of the similarity method.

Two physical phenomena are called "similar" if the quantities that characterize one phenomenon can be obtained from the respective quantities of other phenomena, taken at identical space-time points, by simply multiplying by the identical (at all points) factors (similarity coefficients).

Let $\varphi(\boldsymbol{r}; t)$, $\boldsymbol{a}(\boldsymbol{r}; t)$, $Q(\boldsymbol{r}; t)$ represent some generally nonstationary fields of the distribution of physical scalar, vector or tensor quantities in the space-time region $(\boldsymbol{r}; t)$, here, $\boldsymbol{r}$ is the vector-radius of the point, whereas its projections (x, y, z) are the coordinates of this point. Compare this phenomenon with some other phenomena characterized by the scalar, vector, or tensor functions $\bar{\varphi}(\bar{\boldsymbol{r}}; \bar{t})$, $\bar{\boldsymbol{a}}(\bar{\boldsymbol{r}}; \bar{t})$, $\bar{Q}(\bar{\boldsymbol{r}}; \bar{t})$, respectively, in the region $(\bar{\boldsymbol{r}}; \bar{t})$. The space-time point $\bar{M}(\bar{\boldsymbol{r}}; \bar{t})$ is called "identical" with respect to the point $M(\boldsymbol{r}; t)$ if the vector-radii of these points (or their coordinates) and the corresponding time instants can be obtained (one out of the other) by simple linear transformations

$$\bar{\boldsymbol{r}} = k_l \boldsymbol{r} \quad (\text{or } \bar{x} = k_l x, \bar{y} = k_l y, \bar{z} = k_l z) \quad \bar{t} = k_t t \tag{35}$$

Here, the similarity coefficients k_l and k_t are the same for all points of comparable regions. Moreover, *the similarity coefficient k_l is the same for all the coordinates, i.e., it is independent of the direction of coordinate axes in space.*

The two physical phenomena are considered to be similar (in the primary meaning of the word) if the functions φ, $\boldsymbol{a}$, Q and $\bar{\varphi}$, $\bar{\boldsymbol{a}}$, $\bar{Q}$ that characterize them and are determined at identical points of the regions $(\boldsymbol{r}; t)$ and $(\bar{\boldsymbol{r}}; \bar{t})$ also can be obtained from each other by simple linear transformations,

$$\bar{\varphi} = k_\varphi \varphi \qquad \bar{\boldsymbol{a}} = k_a \boldsymbol{a} \qquad \bar{Q} = k_Q Q \tag{36}$$

This equation has the same values of the similarity coefficients k_φ, k_a, k_Q for different identical points and *with the same similarity coefficients*: k_a for all the projections of the vector a and k_Q for all the components of the tensor Q.

Expanding this definition of similarity, we also introduce the affine similarity (referred to in Chapters 8 and 9). In the case of affine similarity the set of transformations (35) and (36) is replaced by the following more general transformations:

$$\bar{\varphi} = k_\varphi \varphi \qquad \bar{a}_x = k_{a_x} a_x \qquad \bar{a}_y = k_{a_y} a_y \qquad \bar{a}_z = k_{a_z} a_z$$

$$\bar{Q}_{xx} = k_{Q_{xx}} Q_{xx} \qquad \bar{Q}_{xy} = k_{Q_{xy}} Q_{xy} \tag{37}$$

which are valid at "identical" points defined by the transition formulas

$$\bar{x} = k_x x \qquad \bar{y} = k_y y \qquad \bar{z} = k_z z \qquad \bar{t} = k_t t \tag{38}$$

The similarity coefficients $k_\varphi, k_{a_x}, \cdots, k_{Q_{xx}}, k_{Q_{xy}}, \cdots, k_x, k_y, k_z$ do not vary in transition from one point to the other, but the isotropy is already absent and $k_{a_x} \neq k_{a_y} \neq k_{a_z}$, $k_{Q_{xx}} \neq k_{Q_{xy}} \neq \cdots$, $k_x \neq k_y \neq k_z$.

Transformations (35) and (36), which characterize the ordinary similarity, or (37) and (38) which characterize the affine similarity, can be interpreted in another fashion by introducing, for each of the phenomena considered, certain constant quantities that characterize the quantitative order (scale) of variable physical quantities describing the phenomena. These constant quantities will be called the scales of the corresponding variable quantities (lengths, time, velocities, pressures, etc.). In the field of one of the phenomena compared, say, the first phenomenon, in which the designations do not have overbars, we designate some characteristic length and time by L and T and take them to be the scales of these quantities. In the field of the other phenomenon we analogously isolate the corresponding scales $\bar{L}$ and $\bar{T}$. Restricting ourselves for simplicity to the case of simple similarity, according to Equation 35,

$$\bar{L} = k_l L \qquad \bar{T} = k_l T \tag{39}$$

In the same way, we also determine the scales Φ, A, Q^* and $\bar{\Phi}$, $\bar{A}$, $\bar{Q}^*$ for the quantities φ, $\boldsymbol{a}$, Q, $\bar{\varphi}$, $\bar{\boldsymbol{a}}$, $\bar{Q}$; then

$$\bar{\Phi} = k_\varphi \Phi \qquad \bar{A} = k_a A \qquad \bar{Q}^* = k_Q Q^* \tag{40}$$

Eliminating from Equations 39 and 40 the similarity coefficients k_l, k_t, k_φ, etc., the transformations (36), which are valid for identical points, at which, according to Equation 35,

$$\frac{\bar{x}}{\bar{L}} = \frac{x}{L} \qquad \frac{\bar{y}}{\bar{L}} = \frac{y}{L} \qquad \frac{\bar{z}}{\bar{L}} = \frac{z}{L} \qquad \frac{\bar{t}}{\bar{T}} = \frac{t}{T} \tag{41}$$

can be rewritten in the form of dimensionless relations

$$\frac{\bar{\varphi}}{\bar{\Phi}} = \frac{\varphi}{\Phi} \qquad \frac{\bar{\boldsymbol{a}}}{\bar{A}} = \frac{\boldsymbol{a}}{A} \qquad \frac{\bar{Q}}{\bar{Q}^*} = \frac{Q}{Q^*} \tag{42}$$

Using the designation "idem" to denote the identity of the comparable dimensionless quantities at identical points of the regions where the phenomena studied take place, Equations 41 and 42 can be replaced by the following definition of the similarity of phenomena:

$$\frac{\varphi}{\Phi} = \text{idem} \qquad \frac{\boldsymbol{a}}{A} = \text{idem} \qquad \frac{Q}{Q^*} = \text{idem} \tag{43}$$

when

$$\frac{x}{L} = \text{idem} \qquad \frac{y}{L} = \text{idem} \qquad \frac{z}{L} = \text{idem} \qquad \frac{t}{T} = \text{idem}$$

In other words, two similar phenomena at identical space-time points of the regions of their occurrence differ from one another only by the scales of the quantities describing, the phenomena.

From this it follows directly that if in differential equations, boundary, initial, and other conditions of the uniqueness of solutions to these differential equations, one goes over from ordinary dimensional variables to nondimensional, obtainable by relating dimensional quantities to their scales, then both the non-dimensional differential equations themselves and the corresponding non-dimensional boundary, initial, and other conditions of uniqueness will become identical for both phenomena being compared.

What we have outlined regarding the similar phenomena in the conventional sense of the term "similarity" also fully relates to the case of affine similarity, with the only difference being that in the case of affine similarity different length scales should be selected for different coordinates, i.e., X, Y, Z. In the same way for different projections of the vectors a_x, a_y, a_z, different scales must be taken, say, A_x, A_y, A_z, etc. Recall that the method of affine similarity was applied in Chapters 8 and 9 in just this manner.

The similarity of flows past bodies by an ideal incompressible fluid (or by an ideal gas at small Mach numbers) was ensured by a simple geometric similarity of the bodies immersed in the flow and by their similar location with respect to incident streams in the flows compared (by the equality of the angles of attack and other angles that determine the location of the body with respect to the incident uniform flow).

Thus, two-dimensional flows around two circular cylinders in an ideal incompressible fluid at $\Gamma/(V_\infty a)$ = idem (see Chapter 7, Section 5) were identical irrespective of the radii of the cylinders, incident flow velocities, and fluid densities in the flows compared. The pressure coefficients c_p and, consequently, the coefficients of the lift force c_y were also the same at the identical points.

For two geometrically similar wing profiles the hydrodynamic similarity would have also required the identity of the angles of attack and, moreover, the fulfillment of the Zhukovskiy-Chaplygin hypothesis about the finite nature of velocity at the trailing sharp edge. The spatial flows past geometrically identical bodies located identically in uniform flows of ideal incompressible fluids with different velocities are similar.

Let us now consider the similarity conditions for two isothermal flows of Newtonian viscous incompressible fluids with different but constant densities and viscosities. Following the approach (described above) of comparison between dimensionless differential equations and corresponding boundary and initial conditions, we reduce the Navier-Stokes equations (29) to dimensionless form by selecting some, characteristic for the flow, constant quantities: T, L, V, P, F as the scales of time, length (specifically, coordinates), velocities, pressures, and volumetric forces, respectively.

Priming the nondimensional values of time, coordinates, velocities, pressures, and forces, we assume (here, it is more convenient to use the letter symbols x, y, z, u, v, w) that

$$t = Tt' \qquad x = Lx' \qquad y = Ly' \qquad z = Lz'$$

$$u = Vu' \qquad v = Vv' \qquad w = Vw' \qquad p = Pp'$$

$$F_x = FF'_x \qquad F_y = FF'_y \qquad F_z = FF'_z \tag{44}$$

Substituting these values of $t, x, \cdots, u, \cdots, p, F_x, \cdots$ into Equations 29,

$$\frac{V}{T}\frac{\partial u'}{\partial t'} + \frac{V^2}{T}\left(u'\frac{\partial u'}{\partial x'} + v'\frac{\partial u'}{\partial y'} + w'\frac{\partial u'}{\partial z'}\right) = FF'_x - \frac{P}{\rho L}\frac{\partial p'}{\partial x'} + \frac{\nu V}{L^2}\left(\frac{\partial^2 u'}{\partial x'^2} + \frac{\partial^2 u'}{\partial y'^2} + \frac{\partial^2 u'}{\partial z'^2}\right)$$

$$\frac{V}{T}\frac{\partial v'}{\partial t'} + \frac{V^2}{T}\left(u'\frac{\partial v'}{\partial x'} + v'\frac{\partial v'}{\partial y'} + w'\frac{\partial v'}{\partial z'}\right) = FF'_y - \frac{P}{\rho L}\frac{\partial p'}{\partial y'} + \frac{\nu V}{L^2}\left(\frac{\partial^2 v'}{\partial x'^2} + \frac{\partial^2 v'}{\partial y'^2} + \frac{\partial^2 v'}{\partial z'^2}\right)$$

$$\frac{V}{T}\frac{\partial w'}{\partial t'} + \frac{V^2}{T}\left(u'\frac{\partial w'}{\partial x'} + v'\frac{\partial w'}{\partial y'} + w'\frac{\partial w'}{\partial z'}\right) = FF'_z - \frac{P}{\rho L}\frac{\partial p'}{\partial z'} + \frac{\nu V}{L^2}\left(\frac{\partial^2 w'}{\partial x'^2} + \frac{\partial^2 w'}{\partial y'^2} + \frac{\partial^2 w'}{\partial z'^2}\right)$$

$$\frac{V}{L}\left(\frac{\partial u'}{\partial x'} + \frac{\partial v'}{\partial y'} + \frac{\partial w'}{\partial z'}\right) = 0 \tag{45}$$

after which, having canceled both parts of the first three equations by the appropriately selected combination of the scales T, L, V, P and physical constants, we reduce by unity the number of their complexes composed in equations. Thus, assuming in the general case that convective accelerations are not omitted, we divide both sides of the first three equations by V^2/L; this will give

$$\mathbf{Sh}\frac{\partial u'}{\partial t'} + u'\frac{\partial u'}{\partial x'} + v'\frac{\partial u'}{\partial y'} + w'\frac{\partial u'}{\partial z'} = \frac{1}{\mathbf{Fr}}F'_x - \mathbf{Eu}\frac{\partial p'}{\partial x'} + \frac{1}{\mathbf{Re}}\left(\frac{\partial^2 u'}{\partial x'^2} + \frac{\partial^2 u'}{\partial y'^2} + \frac{\partial^2 u'}{\partial z'^2}\right)$$

$$\mathbf{Sh}\frac{\partial v'}{\partial t'} + u'\frac{\partial v'}{\partial x'} + v'\frac{\partial v'}{\partial y'} + w'\frac{\partial v'}{\partial z'} = \frac{1}{\mathbf{Fr}}F'_y - \mathbf{Eu}\frac{\partial p'}{\partial y'} + \frac{1}{\mathbf{Re}}\left(\frac{\partial^2 v'}{\partial x'^2} + \frac{\partial^2 v'}{\partial y'^2} + \frac{\partial^2 v'}{\partial z'^2}\right)$$

$$\mathbf{Sh}\frac{\partial w'}{\partial t'} + u'\frac{\partial w'}{\partial x'} + v'\frac{\partial w'}{\partial y'} + w'\frac{\partial w'}{\partial z'} = \frac{1}{\mathbf{Fr}}F'_z - \mathbf{Eu}\frac{\partial p'}{\partial z'} + \frac{1}{\mathbf{Re}}\left(\frac{\partial^2 w'}{\partial x'^2} + \frac{\partial^2 w'}{\partial y'^2} + \frac{\partial^2 w'}{\partial z'^2}\right)$$

$$\frac{\partial u'}{\partial x'} + \frac{\partial v'}{\partial y'} + \frac{\partial w'}{\partial z'} = 0 \tag{46}$$

Equations 46 involve the following dimensionless one-term complexes called the "similarity numbers":

$$\frac{L}{VT} = \mathbf{Sh} \quad \text{(the Struhal number)} \qquad \frac{P}{\rho V^2} = \mathbf{Eu} \quad \text{(the Euler number)}$$

$$\frac{VL}{\nu} = \mathbf{Re} \quad \text{(the Reynolds number)} \qquad \frac{V^2}{FL} = \mathbf{Fr} \quad \text{(the Froude number)} \tag{47}$$

Equations 46 represent the Navier-Stokes dimensionless equations of the dynamics of a viscous incompressible fluid. These equations are supplemented with the dimensionless initial and boundary conditions, which correspond to the given specific problem, and (in a number of cases) with other conditions of the uniqueness of solutions to the Navier-Stokes equations.

Let us suppose that two generally unsteady-state flows of a Newtonian viscous incompressible fluid are similar. Then, as before, the dimensionless initial, boundary, and other conditions of uniqueness, as with the Navier-Stokes dimensionless equations (46) themselves, should be identical for both motions compared. However, according to the assumption made about the existence of similarity, all the dimensionless, primed variables are identical at identical points of the flows. Consequently, for the differential equations to be coincident, it remains only to require that the similarity numbers be identical, i.e.,

$$\mathbf{Sh} = \text{idem} \qquad \mathbf{Eu} = \text{idem} \qquad \mathbf{Fr} = \text{idem} \qquad \mathbf{Re} = \text{idem} \tag{48}$$

The above similarity conditions, including the last system of equations, are the necessary similarity conditions. Difficulties are encountered on the path of elucidation of the sufficient similarity conditions. These difficulties arise due to the fact that for the sufficient similarity conditions to be rigorously determined it is necessary that the theorem of the existence and uniqueness of the solutions to the Navier-Stokes equations[12] be proved within a wide range

[12] To become acquainted with this problem, we recommend Ladyzhenskaya, O. A., *Matematicheskiye voprosy dinamiki vyazkoy neszhimayemoy zhidkosti. Sovremennye problemy matematiki* (*Mathematical Problems of the Dynamics of a Viscous Incompressible Fluid. The Current Problems of Mathematics*), Fizmatgiz Press, Moscow, 1961. The proofs of the existence and uniqueness of solutions to the Navier-Stokes equations for certain problems can be found in Goldshtik, M. A., *Vikhrevye potoki* (*Rotational Flows*), Nauka Press, Novosibirsk, 1981.

of boundary conditions. A great variety of problems encountered in practice (along with the ordinary problems of flow around bodies and through tubes and channels, there are also the problems of free convection, propagation of jets, formation of wakes behind bodies, development of boundary layers, etc.) do not allow one to believe that the problem of finding sufficient similarity conditions for motion of a viscous incompressible fluid has been resolved.

Let us agree that from among the similarity numbers (47) we separate those that are composed only of scales of compared flows and physical constants of the medium that are contained in the statement of the problem about the definition of motion, i.e., given *a priori*. The identity of such similarity numbers is responsible for the similarity of the two flows compared, and therefore the numbers themselves can be called the similarity criteria. A corresponding class of flows has a smaller quantity of similarity criteria than of similarity numbers because not all of the scale quantities, introduced when establishing dimensionless equations and boundary and initial conditions, actually can be given beforehand. The values of some of them are determined only after obtaining the single solution to the given specific problem. From this it follows that the number of sufficient conditions represented by the system of Equations 48 will be smaller than the general number of the necessary conditions, whereas the similarity numbers left after the isolation of the applicable similarity criteria will be expressed in terms of those criteria. This result is confirmed by the dimensionality theory in Section 5.

Let us illustrate the foregoing by several examples; many other examples will accompany this presentation. Place a cylinder of diameter d into a uniform flow of a viscous incompressible fluid with the kinematic viscosity coefficient ν, density ρ, and constant velocity V_∞ and set the problem to determine the resistance of the cylinder to the incident flow under the assumption that the motion is steady and no volumetric forces exist. Then, out of the necessary similarity conditions (48), only two conditions remain: **Eu** = idem and **Re** = idem. The Reynolds number, equal in the given case to **Re** = $V_\infty d/\nu$, is the similarity criterion because it contains the *a priori* given scales of velocities V_∞ and lengths d and the given physical constant ν. The resistance force, designated by W, can be determined only after the solution of the problem of flow. This is because it is computed by summing, over the cylinder surface, the forces of flow pressure on the surface and the forces of liquid friction against the cylinder surface, which in turn depend on the solution of the problem of flow. The Euler number, involving the scale of the *a priori* unknown pressure, cannot be a similarity criterion in this case. Rather it is a function of the criterion, i.e., of the Reynolds number.

The resistance coefficient c_w of the unit length of the cylinder determined by the ratio (σ is the frontal area of the cylinder)

$$c_w = \frac{W}{(1/2)\,\rho V_\infty^2 \sigma}$$

plays the role of the Euler number (because W/σ has the dimensionality of the pressure drop) and depends on the Reynolds number. The experiment fully justifies this conclusion. The results of experiments carried out with cylinders of different diameters placed in liquid flows with different velocities and physical properties lay quite satisfactorily on the same curve c_w (**Re**) shown in Figure 148.

By taking a closer look at the experimental pattern of flow around the cylinder, it is possible to note that the flow is not steady; in the rear portion of the cylinder rotating masses of the fluid retarded by the cylinder stall from side or the other of its surface. The masses produce flow oscillations, the frequency of which depends on the flow velocity, its viscosity, and diameter of the cylinder. More precisely, the frequency depends on the Reynolds number.

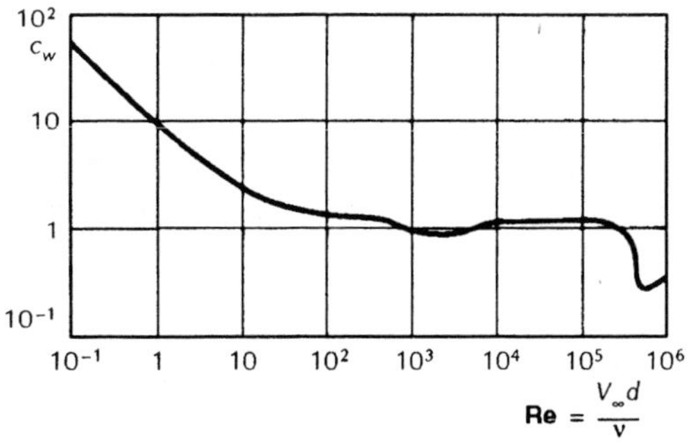

Fig. 148

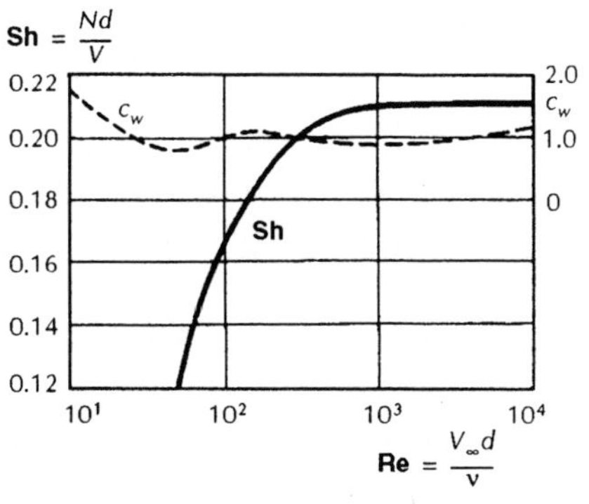

Fig. 149

Such oscillations of the cylinder in a constant velocity flow occurring in the boundary layer on the cylinder surface due to the internal phenomena that lead to separations of fluid masses from the cylinder surface (mentioned above), relate to self-oscillations. They can be observed on all kinds of badly streamlined bodies. Originating in the fluid, these periodic processes cause vibrations of bodies immersed in the fluid. The self-oscillations of stacks and high buildings during wind are well known; the frequency of these oscillations are not associated with the frequency of wind bursts as had been the case with forced oscillations. The periscope of a submarine, the pipes of the steam turbine condenser, etc., perform analogous self-oscillations.

As far back as the end of the 19th century, the Czech physicist Strouhal[13] studied the vibrations of strings in a uniform air flow (analogous to vibrations of telegraph wires in windy weather) and determined the frequency of vibrations $N = 1/T$ (T is the period of vibrations) by the pitch of the sound. He was the first to note that the dimensionless one-term combination of the frequency of vibrations N, wire diameter d, and incident flow velocity V_∞, currently called the Strouhal number

$$\mathbf{Sh} = \frac{Nd}{V_\infty}$$

preserves a certain constant value close to 0.2.

Consider this phenomenon from the viewpoint of the similarity method. Neglecting the action of volumetric forces, we obtain three similarity numbers, **Sh**, **Eu** (or c_w), and **Re**. Among these numbers, only the **Re** number consists of the *a priori* given quantities V_∞, d, and ν, and consequently represents the similarity criterion of the self-oscillatory motions of a viscous incompressible fluid that originate in the flow past the cylinder. The similarity numbers **Sh** and c_w contain the *a priori* unknown frequency of self-oscillations $N = 1/T$ and the resistance W and are therefore the functions of the similarity criterion **Re**. The experiments confirm this statement. Figure 149 presents the curves **Sh** (**Re**) and c_w (**Re**) composed on the basis of the experiments of Roshko.[14] The experiments with cylinders were carried out in a wide range of diameters, from 0.235 to 6.35 mm. Judging from the curve, at large **Re** values the value **Sh** = 0.21 is established. This is well justified at not very great numbers **Re**, but it is known that when the Reynolds number attains values on the order of $5 \cdot 10^5$, the resistance falls steeply (see Figure 148), the character of flow past the cylinder alters, and the ten-

[13] Strouhal, V., *Ann. Phys. Chem.* (*Wiedemann's Ann.*), 5, 216–251, 1878.

[14] Roshko, A., On the Development of Turbulent Wakes from Vortex Streets, NACA Rep. 1954, Vol. 1191.

dency of the Strouhal number toward the constant value 0.21 is violated. This problem is discussed in Chapter 13 in a consideration of the phenomenon of the "crisis of flow" past a circular cylinder.

Now, we shall alter the statement of the problem. Having placed the cylinder in a uniform flow of constant velocity V_∞, we bring it into the oscillatory regime given at will with the frequency N. In this case the Strouhal number **Sh** will become, along with the Reynolds number **Re**, the similarity criterion, whereas the resistance coefficient c_w will be the function of two similarity criteria **Sh** and **Re**

$$c_w = f(\mathbf{Sh}, \mathbf{Re})$$

A curious example of the difference in the selection of the similarity criteria from among the similarity numbers relates to the case considered in Section 6 dealing with the motion of a viscous incompressible fluid through a cylindrical pipe of circular profile. The fluid with density ρ and with viscosity coefficient μ flows through a horizontal cylindrical circular tube of diameter d under the action of a constant pressure difference over the tube section l equal to Δp; the rate of volume flow through the tube is equal to Q. Let us elucidate what can be suggested by the similarity method regarding the general form of the dependence between the pressure difference in the tube Δp (provided by the pump or by the head of the liquid column between the reservoir and the tube) and the rate of the fluid volume flow through the tube Q. Two statements of this problem are possible:

1. The required flow rate Q is given; it is necessary to calculate the appropriate pressure difference Δp over the given section of the tube of length l.
2. The pressure difference Δp over the section of the tube of length l is given; it is necessary to determine the fluid volume flow rate Q through the tube.

Let us ignore the action of volumetric forces, in the given case the weight, because of the horizontal orientation of the tube, and take into account the stationary character of the problem.

Then, among the similarity numbers remain only $\mathbf{Eu} = P/\rho V^2$ and $\mathbf{Re} = Vd/\nu$. As the scale of pressures P we take the pressure difference over the tube length characteristic for this tube, i.e., its diameter d; this difference can be expressed in terms of the given difference Δp over a certain length l by the formula

$$P = \frac{\Delta p}{l} d$$

The velocity scale will be selected to be the tube section-average velocity of fluid V_{av} determined by the obvious ratio

$$V_{\text{av}} = \frac{Q}{\frac{1}{4}\pi d^2}$$

With the scales thus selected, the similarity numbers will be

$$\mathbf{Eu}_Q = \frac{\Delta p}{\rho V_{\text{av}}^2}\frac{d}{l} \qquad \mathbf{Re}_Q = \frac{V_{\text{av}} d}{\nu}$$

In the first statement (given the quantity Q, Δp should be determined over the given section of the tube of length l) the similarity criterion will be represented by the Reynolds number $\mathbf{Re}_Q$, whereas the Euler number $\mathbf{Eu}_Q$ will be its function. Then, introducing the coefficient of tube resistance λ, according to the resistance formula ($\mathbf{Eu}_Q = \lambda/2$),

$$\Delta p = \lambda \frac{l}{d} \frac{\rho V_{av}^2}{2}$$

we have

$$\lambda = f(\mathbf{Re}_Q)$$

The experiments carried out for many dozens of years fully confirmed the validity of this relation for the flow of very dissimilar fluids in smooth tubes of different diameters in both laminar and turbulent regimes of flow within a wide range of the rate of volume flow or of tube section-average velocities. The problem of the form of the function $f(\mathbf{Re}_Q)$ is considered in detail in what follows.

In the second statement, the pressure difference $P = \frac{\Delta p}{l} d$ is given, but the rate of volume flow (i.e., the tube section-average velocity V_{av}) remains unknown. In this case, there is no single criterion among the similarity numbers **Eu** and **Re**, because both quantities involve the *a priori* unknown quantity V_{av}. In this case, it is easy to construct the criterion that contains the given quantity $P = \frac{\Delta p}{l} d$ and does not contain $V_{av} = Q/(\pi d^2/4)$; in order to eliminate the quantity V_{av}^2 from **Eu**, we compose the dimensionless group

$$\mathbf{Eu} \cdot \mathbf{Re}^2 = \frac{\rho \Delta p d^3}{\mu^2 l} = \frac{\rho d^2 \Delta p}{\mu^2} \frac{d}{l}$$

In the second statement of the problem of the similarity it is this group that will play the role of the similarity criterion, i.e., of the Reynolds number $\mathbf{Re}_P$, whereas the Reynolds number $\mathbf{Re}_Q = V_{av} d/\nu$ will remain simply the similarity number. In this case we have[15]

$$\mathbf{Re}_Q = \frac{V_{av} d}{\nu} = \frac{4Q}{\pi \nu d} = f_1\left(\frac{\rho d^2 \Delta p}{\mu^2} \frac{d}{l}\right) \tag{49}$$

where f_1 is the symbol of the new functional dependence. By agreeing that in both systems compared the quantity Δp is understood to be the pressure difference over the identical section $l = d$ or multiple of d, the previous formula can be also rewritten as:

$$Q = \frac{\pi \nu d}{4} f_1\left[\frac{\Delta p}{\rho}\left(\frac{d}{\nu}\right)^2\right]$$

As another example, we consider the case of unsteady-state motion of a viscous incompressible fluid, the physical properties of which are characterized by the constants ρ and μ, through an infinitely long, circular cylindrical tube of diameter d under the action of the pressure difference Δp, which is a certain harmonic function with the period T (or frequency $N = 1/T$) and amplitude P. In this case (the action of volumetric forces is omitted) no characteristic velocity is prescribed, and thus none of the similarity numbers (**Sh**, **Eu**, and **Re**) can be

[15] The designations $\mathbf{Re}_Q$ and $\mathbf{Re}_P$ are borrowed from the paper by Rozhdestvenskiy, B. L., and Simakin, I. N., The modelling of turbulent flows in a plane channel, *Zh. Vych. Mat. Mat. Fiz.*, 97, 98, 1985.

a criterion. Just as in the previous case, because the pressure difference is prescribed (the pressure scale can be taken to be, for example, the amplitude of the fluctuations of the pressure P) and the frequency N is of unsteady motion (for simplicity, we consider only the developed forced oscillations of the fluid), the similarity criteria will be composed by combining the numbers **Sh** and **Eu** with the Reynolds number **Re** so that the velocity V can be eliminated. We have the following similarity criteria:

$$\mathbf{Eu} \cdot \mathbf{Re}^2 = \frac{P}{\rho V^2}\left(\frac{Vd}{\nu}\right)^2 = \frac{P}{\rho}\left(\frac{d}{\nu}\right)^2$$

$$\mathbf{Sh} \cdot \mathbf{Re} = \frac{Nd}{V} \cdot \frac{Vd}{\nu} = \frac{Nd^2}{\nu}$$

In the case considered the Reynolds number $\mathbf{Re} = Vd/\nu$ will not be a similarity criterion, but will be defined as a function of the two similarity criteria above:

$$\mathbf{Re} = \frac{Vd}{\nu} = f_2\left[\frac{Nd^2}{\nu}, \frac{P}{\rho}\left(\frac{d}{\nu}\right)^2\right]$$

Not dwelling on the details, we note in conclusion that as with the Reynolds number, the Froude number, which in the specific case of the weight force ($F = g$) is equal to $\mathbf{Fr} = V^2/(gl)$, will be the similarity criterion in the majority of practical problems of ship motion. If, when testing ship models in basins, the liquid is water, then it is impossible to carry out rigorous modeling which requires the identity of the similarity criteria **Fr** and **Re** in the model and in the real case. In fact, the conditions

$$\frac{V^2}{gl} = \text{idem} \qquad \frac{Vl}{\nu} = \text{idem} \quad \text{at} \quad g = \text{idem} \quad \text{and} \quad \nu = \text{idem}$$

lead to the following two conditions:

$$\frac{V^2}{l} = \text{idem} \qquad Vl = \text{idem}$$

i.e., V = idem, l = idem. This means that in the model and in the real case the velocities and dimensions of the ship should coincide; thus, the modeling turns out to be impossible. In the practice of shipbuilding the modeling is done separately. When based only on the Reynolds number, of primary importance are the resistance of water friction against the ship's skin and the resistance of pressures due to the shape of the ship regardless of its position with respect to the free surface where the waves originate and the wave resistance is induced. On the other hard, modeling is based only on the Froude number when the wave resistance is of primary importance.

Cases are possible in which characteristic velocity of liquid motion can be indicated in the phenomenon studied. In this case, just as in the previous examples, it is possible to be rid of the velocity when forming the dimensionless group

$$\frac{\mathbf{Fr}}{\mathbf{Re}^2} = \frac{V^2}{gl}\frac{\nu^2}{V^2 l^2} = \frac{\nu^2}{gl^3}$$

This group plays the part of the similarity criterion in the problems of free convection of fluid; the inverse quantity is called the Galilean criterion. In the present section, only a very narrow class of flows was considered deliberately; the fluid was considered to be incompressible, the flow to be isothermal and physically uniform.

The same method will be applied in what follows to a wider range of phenomena (nonisothermal motions of fluid in the presence of admixtures, gas motion with great sub- and supersonic velocities). All of these cases are based on the mathematical model of the phenomenon of differential equations of the dynamics and thermodynamics with the appropriate initial and boundary conditions and sometimes with additional conditions without which the problem cannot be stated rigorously. Examples are the conditions of the nontrivial character of solutions we shall meet in the problems of the development of jets, wakes, and other motions "free" from the effect of solid walls.

5. Foundations of the Dimensionality Theory: The Π Theorem

In parallel with the similarity method, widely applied in practice is the method of dimensionalities usually presented in the same books in which the similarity method is treated.[16] The fundamental difference between the dimensionality and the similarity theory consists of the fact that it does not require the assignment of the mathematical model of the phenomenon, but is satisfied with the establishment of the list of basic physical quantities that determine the phenomenon and with the consideration of their dimensionalities.

The dimensionality theory is indispensable in those cases in which the mathematical simulation, i.e., the formation of basic equations, was not yet made. In these especially complex cases the dimensionality theory makes it possible to predict the structure of generalized (dimensionless) empirical formulas, which for the time being successfully (for practice) replace the laws established theoretically.

For the relatively narrow field (the mechanics of liquids and gases) of physics considered in this book, the theories of similarity and dimensionality identically satisfy the practice and they can be applied in parallel. Thus, in the similarity theory presented in Section 4 the system of Navier-Stokes equations (45) was used, the coefficients of which are represented by constant dimensional complexes of the scales: V/T, V^2/L, etc. After the division of both sides by the complex V^2/L, this system of equations went into the system of dimensionless equations (46), which can be conventionally represented, as explained in Section 4, Chapter 8 by the functional relation

$$G(\mathbf{Sh}, \mathbf{Fr}, \mathbf{Eu}, \mathbf{Re}) = 0$$

This equation, derived from the similarity theory considerations for the particular case of unsteady viscous incompressible fluid motion in the presence of volumetric forces, can furnish an example of the general relation which is proved in the dimensionality theory and which constitutes the subject matter of the so-called Π theorem.

[16] The literature on similarity theory is extensive. Only a few sources are included that interpret this problem thoroughly: Sedov, L. I., *Metody podobiya i razmernosti v mekhanike (The Similarity and Dimensionality Methods in Mechanics)*, Nauka Press, Moscow, 1981; Gukhman, A. A., *Vvedeniye v teoriyu podobiya (Introduction Into the Similarity Theory)*, Vysshaya Shkola Press, Moscow, 1973; Gukhman, A. A., *Primeneniye teorii podobiya k issledovaniyu protsessov teplomassoperenosa (Application of the Similarity Theory to the Study of Heat and Mass Transfer Processes)*, Vysshaya Shkola Press, Moscow, 1974; Ovsyannikov, L. V., *Vvedeniye v mekhaniku sploshnykh sred (Introduction Into the Mechanics of Continua)*, Part 2, Gosuniversitet Press, Novosibirsk, 1976; Loitsyanskiy, L. G., *Metody podobiya i razmernostey v mekhanike zhidkosti i gaza (The Similarity and Dimensionality Methods in the Mechanics of Liquids and Gases). Collected Papers on Theoretical Mechanics*, No. 11, Vysshaya Shkola Press, Moscow, 1981; Sedov, L. I., *Similarity and Dimensional Methods in Mechanics*, Academic Press, New York, 1959.

In presenting the fundamentals of the dimensionality theory, we first note that instead of selecting the scales of the physical quantities typical for the similarity theory on the basis of the so-called relationship (i.e., of the velocity for velocities) of pressure for pressures, etc., the scales in the dimensionality theory are represented by the units of measure adopted in the study and the combinations of which define the dimensions of the involved physical (both variable and constant) quantities. Thus, in purely dynamic phenomena, which are not associated with thermodynamics or other branches of physics, the basic scales are mass, length, and time, the MLT system; in the SI system these scales are kilogram, meter, second.

Referring to the case of the most general physical phenomenon in the arguments below, x_i $(i = 1, 2, \cdots, n)$ designates the numerical values of the physical quantities entering the problem statement, $[x_i]$ denotes their dimensions, and M_j $(j = 1, 2, \cdots, m)$ denotes the measurement units adopted. Then, the dimensions of x_i will be represented in the form of the power monomial

$$[x_i] = M_1^{a_{i1}} M_2^{a_{i2}} \cdots M_m^{a_{im}} \tag{50}$$

where the exponents a_{ij} are called the dimensions of the quantities x_i with respect to the basic units M_j $(j = 1, 2, \cdots, m)$ selected.

Following Görtler,[17] the matrix

$$A = \|a_{ij}\| \qquad (i = 1, 2, \cdots, n \quad j = 1, 2, \cdots, m \quad m \leq n)$$

the matrix of dimensions is introduced, and its rank, equal to the maximum order of the nonzero minor, will be designated by the letter r. Let us consider the set of power monomials:

$$\Pi_s = x_1^{k_{1s}} x_2^{k_{2s}} \cdots x_n^{k_{ns}} \quad (s = 1, 2 \cdots, p) \tag{51}$$

the dimensions of which in the system of units M_j $(j = 1, 2, \cdots, m)$ will be found by substituting into (51) the dimensions of x_i found from Equation 50 and will be equal to $(s = 1, 2, \cdots, p)$

$$[\Pi_s] = M_1^{a_{i1}k_{is}} M_2^{a_{i2}k_{is}} \cdots M_m^{a_{im}k_{is}} \tag{52}$$

where in the exponents the summation over the repeated index i from $i = 1$ to $i = n$ is meant. If each of the exponents in Equation 52 is equal to zero, i.e., if k_{is} satisfies a homogeneous system of linear algebraic equations (summation over i!)

$$a_{i1}k_{is} = 0 \qquad a_{i2}k_{is} = 0, \cdots, a_{im}k_{is} = 0 \tag{53}$$

then all Π_s $(s = 1, 2, \cdots, p)$ will become nondimensional. Then, according to the terminology of the similarity theory, Π_s will be able to express the similarity numbers, but on the condition that all Π_s are independent; i.e., none of them can be expressed in terms of the product of the others.

If a system of exponential monomials $\Pi_1, \Pi_2, \cdots, \Pi_p$ satisfies the conditions of independence and nondimensionality and, moreover, is such that any exponential one-term dependence of Π on $x_1, x_2, \cdots, x_n$ can be expressed in terms of a power monomial composed of Π_s, then such a system of Π_s $(s = 1, 2, \cdots, p)$ forms the fundamental system of dimensionless power monomials.

---

[17] Görtler, H., Zur Geschichte des Π-Theorems, Z. *Angew. Math. Mech.*, 55, 3–8, 1975.

It can be easily verified that the number p of fundamental groups Π_s $(s = 1, 2, \cdots, p)$ is equal to the difference of the total number of the physical quantities n considered and the rank r of the matrix of dimensions A, i.e.,

$$p = n - r \tag{54}$$

In fact, the system of linear algebraic equations (53), with the rank of the matrix of coefficients A equal to $r < m$, can be replaced by a system of r inhomogeneous equations with the determinant not equal to zero. These equations will contain as unknowns the r number of quantities $k_{1s}, k_{2s}, \cdots, k_{rs}$ on the left side and the $n - r$ number of quantities $k_{r+1,s}, k_{r+2,s}, \cdots, k_{ns}$ on the right side. The latter remain arbitrary and form the set of exponents at the fundamental groups $\Pi_1, \Pi_2, \cdots, \Pi_p$, where according to what has just been proved, $p = n - r$.

The function $f(x_1, x_2, \cdots, x_n)$ of the real variables $x_1, x_2, \cdots, x_n$, determined in a certain domain D of the values of these variables, is called the dimensionally homogeneous function of these variables. If such a set of real numbers $b_k = (k = 1, 2, \cdots, m)$ exists that the following equality holds

$$f(\alpha_1^{a_{11}}\alpha_2^{a_{12}}\cdots\alpha_m^{a_{1m}}x_1, \cdots, \alpha_1^{a_{n1}}\cdots\alpha_m^{a_{nm}}x_n) = \alpha_1^{b_1}\alpha_2^{b_2}\cdots\alpha_m^{b_m}f(x_1, x_2, \cdots, x_n)$$

which is valid for all α_k $(k = 1, 2, \cdots, m)$ and x_i $(i = 1, 2, \cdots, n)$ in the domain D.

Going from the system of units M_j $(j = 1, 2, \cdots, m)$ to the units which are α_j times smaller, we replace the system of variables x_i by the system of variables

$$\bar{x}_i = \alpha_1^{a_{i1}}\cdots\alpha_m^{a_{nm}}x_i$$

If the function

$$y = f(x_1, x_2, \cdots, x_n)$$

represents a certain physical connection between the variables x_i, it must possess the property of dimensionally homogeneous functions, so that the quantity y has the dimensions

$$[y] = M_1^{b_1}M_2^{b_2}\cdots M_m^{b_m}$$

For any dimensionally homogeneous function of the variables $x_1, x_2, \cdots, x_n$, not converting into zero, there is at least one power monomial of these variables

$$x_1^{k_1}x_2^{k_2}\cdots x_n^{k_n}$$

such that

$$[x_1^{k_1}x_2^{k_2}\cdots x_n^{k_n}] = [y]$$

and, consequently, the dimensions of the ratio

$$\frac{f(x_1, x_2, \cdots, x_n)}{x_1^{k_1}x_2^{k_2}\cdots x_n^{k_n}}$$

will be zero. From this follows the fundamental theorem of the dimensionality theory called the Π theorem, usually accredited to Buckingham.[18] As the investigations by Görtler (cited earlier) showed this theorem was proved 4 years before the appearance of Buckingham's work by Federman,[19] a teacher at the St. Peters-burgh Politechnical Institute, whose work was not widely recognized.

The Π theorem is formulated as follows: if $x_1, x_2, \cdots, x_n$ are the numerical values of n physical quantities, $A = \|a_{ij}\|$ $(i = 1, \cdots, n;\ j = 1, \cdots, m)$ is the matrix of their dimensions with respect to the units of measurements $M_1, M_2, \cdots, M_m$, f is an arbitrary dimensionally homogeneous function of the variables $x_1, x_2, \cdots, x_n$, and $\Pi_1, \cdots, \Pi_p$ $(p = n - r$, r is the rank of the matrix $A)$ is the fundamental system of power monomials of the variables $x_1, x_2, \cdots, x_n$. Then at the arbitrary real numbers $k_1, k_2, \cdots, k_n$ the following equation holds

$$f(x_1, x_2, \cdots, x_n) = x_1^{k_1} x_2^{k_2} \cdots x_n^{k_n} G(\Pi_1, \Pi_2, \cdots, \Pi_p) \tag{55}$$

If the equality $f(x_1, x_2, \cdots, x_n) = 0$ represents a certain physical connection between the variables $x_1, x_2, \cdots, x_n$, then the Π theorem states that this connection can be expressed in the form of the dependence

$$G(\Pi_1, \Pi_2, \ldots, \Pi_p) = 0 \quad (p = n - r) \tag{56}$$

between the fundamental nondimensional power monomials in terms of the physical variables $x_1, x_2, \cdots, x_n$.

Thus, not having the mathematical models of the phenomenon, it is possible, based on the dimensions of physical quantities, to find the number and composition of similarity numbers which are represented by the complex $\Pi_1, \Pi_2, \cdots, \Pi_p$.

Let us illustrate this with the example of an incompressible viscous heavy liquid. We introduce the basic measures (units of measurement) of all the variable and constant physical quantities that determine the motion of liquid in the form of the matrix of their dimensions. Noting that

$$[x, y, z] = L \qquad [t] = T \qquad [u, v, w] = LT^{-1} \qquad [g] = LT^{-2}$$

$$[p] = ML^{-1}T^{-2} \qquad [\rho] = ML^{-3} \qquad [\mu] = ML^{-1}T^{-1}$$

we have the following matrix of dimensions (the lines are the dimensions of the quantities, the columns are the units of measures):

	M	L	T
x, y, z	0	1	0
t	0	0	1
u, v, w	0	1	−1
p	1	−1	−2
g	0	1	−2
ρ	1	−3	0
μ	1	−1	−1

[18] Buckingham, E., On physically similar systems; illustration of the use of dimensional equations, *Phys. Rev.*, 4, 345–376, 1914; Model experiments and the forms of empirical equations, *Trans. Am. Soc. Mech. Eng.*, 37, 263–288, 1915; Bridgeman, P. V., *Analiz razmernostey* (*The Dimensionality Analysis*), ONTI Press, Moscow, 1934.

[19] Federman, A., About the general methods of integrating first-order partial differential equations, *Izv. St. Petersburg. Politekhn. Inst.*, 16, 97–154, 1911.

The rank of the matrix is $r = 3$; the number of dimensional quantities that determine the motion is $n = 7$. The fundamental system of the parameters Π contains $p = n - r = 7 - 3 = 4$ parameters. These are the similarity numbers **Sh**, **Fr**, **Eu**, and **Re**.

As seen from the foregoing, the dimensionality theory, as with the similarity theory, establishes the presence of implicit connections (56) between the parameters – the similarity numbers – but says nothing regarding which of the similarity numbers are the similarity criteria that enter into the sufficient conditions of similarity. As seen in the previous section, the solution of the latter problem depends on the statement of the problem involving the mathematical model (equations – initial and boundary conditions) just as in the similarity theory, but it is less definite when the dimensionality theory is used.

6. Examples of the Solution of the Navier-Stokes Equations. Simplest Linear Problems

First, let us examine some relatively easily solvable linear problems. Such problems either follow exactly from their very statement (steady motion with rectilinear trajectories when nonlinear terms are absent), or result from approximate linearization of the Navier-Stokes equations by means of discarding small convective accelerations, and this corresponds to small Reynolds numbers (the "slow" or "slug" motions).

One of the simplest cases of motion of a viscous incompressible fluid is the so-called laminar (laminated) motion through a cylindrical tube of arbitrary cross-section in which the streamlines are straight lines parallel to the tube axis.

Experiments show such motion is realized in cylindrical tubes with different cross-sectional shapes if only the Reynolds number does not exceed its certain definite critical value after which the motion ceases to be laminar. Thus, the fluid particles acquire complex curvilinear trajectories and the solution presented in this section no longer holds. The results presented here have practical importance only in motions with very small velocities, in thin capillaries, or of highly viscous fluids. A more detailed presentation of the conditions of the existence of laminar mode of flow and the phenomena of its transition into a more complex, turbulent mode is considered in more detail at the beginning of Chapter 13.

The axis Oz is directed along the tube axis and we assume the tube to be infinitely long (Figure 150). The flow is directed along the tube axis so that of the three velocity components, u, v, and w, only one component (w) remains and the other two are equal to zero. Ignoring the action of volumetric forces and considering the flow to be isothermal and consequently the density ρ and the viscosity coefficient μ to be constant, we have (according to the Navier-Stokes equations (29)), the system of equations

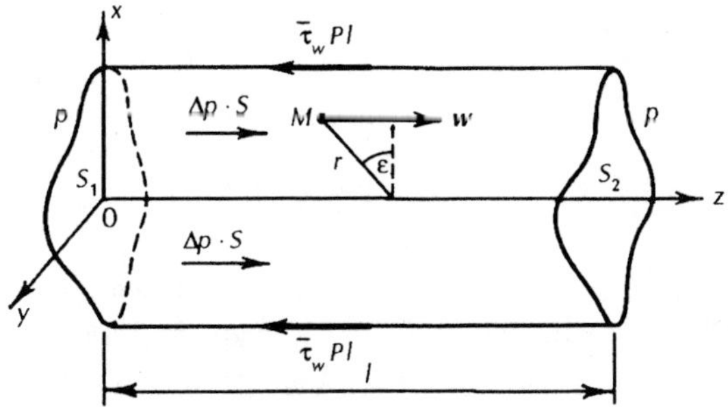

Fig. 150

$$0 = -\frac{1}{\rho}\frac{\partial p}{\partial x} \qquad 0 = -\frac{1}{\rho}\frac{\partial p}{\partial y}$$

$$w\frac{\partial w}{\partial z} = -\frac{1}{\rho}\frac{\partial p}{\partial z} + \nu\left(\frac{\partial^2 w}{\partial x^2} + \frac{\partial^2 w}{\partial y^2} + \frac{\partial^2 w}{\partial z^2}\right) \qquad \frac{\partial w}{\partial z} = 0$$

It follows from the last equation of this system that w represents only the function of x and y and from the first two equations that p is the function of only z. In other words, by passing the sections normal to the tube axis, in all of such sections the velocity distributions are identical, whereas the pressure varies only from section to section, preserving identical value in the given section. Such motions are called fully developed.

The previous system of equations is reduced to one equation

$$\mu\left(\frac{\partial^2 w}{\partial x^2} + \frac{\partial^2 w}{\partial y^2}\right) = \frac{dp}{dz} \tag{57}$$

The left side of Equation 57 represents the function only of x and y, the right side, only of z. When the coordinates are independent of each other, this can occur only in cases in which the left and right sides of the equation are separately constant.

We introduce the notation convenient for what follows

$$\frac{dp}{dz} = \text{const} = -\frac{\Delta p}{l} \tag{58}$$

where Δp is the pressure drop constant along the tube over an arbitrarily selected portion of length l.

In fully developed motion of a viscous fluid through a cylindrical tube the pressure drop Δp, being multiplied by the cross-sectional area $S_1 = S_2 = S$ (Figure 150), plays the part of the moving force $\Delta p \cdot S$ counterbalanced by the forces of the resistance caused by the friction of the fluid on the tube surface. The result is equal to $\int \tau_w ds \cdot l$, where $\tau_w(s)$ is the skin friction variable around the perimeter.

From this it follows directly that the pressure in the cylindrical tube should decrease downstream and, consequently, $\Delta p > 0$. For a tube of variable cross-section where the motion can be either accelerated or retarded this conclusion cannot be made.

In specific calculations the pressure drop Δp over the tube portion of length l is either prescribed directly or can be expressed in terms of other assigned quantities: the fluid flow rate through the tube and the section-averaged or maximum velocities.

Equation 57 is reduced to a linear second-order partial differential equation in the plane Oxy (the Poisson equation)

$$\nabla^2 w = \frac{\partial^2 w}{\partial x^2} + \frac{\partial^2 w}{\partial y^2} = -\frac{\Delta p}{\mu l} \tag{59}$$

which must be solved under the boundary condition stating that the velocity w converts into zero on the contour C of the section normal to the cylinder axis and under the additional condition which determines the prescribed pressure drop over the tube length selected arbitrarily, the rate of volume fluid flow through the tube section (or the average velocity based on it), or the maximum velocity on the tube axis.

Mathematically, the problem set is analogous to the problem of prismatic rod twisting well known in elasticity theory.[20]

As the first, most simple example of integration of Equation 59, we analyze a plane laminar motion of a viscous incompressible fluid between two planes $y = \pm h$ which can be visualized as the limiting case of flow through a prismatic tube of rectangular cross-section if one side of the rectangle is preserved as equal to $2h$ and the other is allowed to tend toward infinity. In this sense, the motion considered can be called the flow of fluid through a "plane" tube. In this case the coordinate x vanishes and Equation 59 is reduced to an ordinary second-order differential equation

$$\frac{\partial^2 w}{\partial y^2} = -\frac{\Delta p}{\mu l} \tag{60}$$

and its integral under the boundary conditions $w = 0$ at $y = \pm h$ is

$$w = \frac{\Delta p \cdot h^2}{2\mu l}\left[1 - \left(\frac{y}{h}\right)^2\right] \tag{61}$$

The distribution of velocities is represented by the second-order parabola. The maximum velocity w_{max} is achieved in the plane of flow symmetry ($y = 0$) and is equal to

$$w_{max} = \frac{1}{2}\frac{\Delta p \cdot h^2}{\mu l} \tag{62}$$

Calculating the rate of volume fluid flow based on the unit length in the direction $0x$ perpendicular to the plane parallel to which the motion of fluid takes place, we find

$$Q = \int_{-h}^{h} w dy = \frac{2}{3}\frac{\Delta p \cdot h^3}{\mu l} \tag{63}$$

from which the section-averaged velocity is found

$$w_{max} = \frac{Q}{2h} = \frac{1}{3}\frac{\Delta p \cdot h^2}{\mu l} = \frac{2}{3} w_{max} \tag{64}$$

An essential circumstance is noted: At the given pressure drop (determined by the power and design of the pump) over the section of a plane tube of selected length, the flow rate is proportional to the cube of the distance between the planes; i.e., it decreases sharply with a decrease of this distance and, conversely, increases sharply with its increase. By assigning the necessary flow rate, the pressure drop, which is required for its provision and which sets the fluid into motion, will vary in inverse proportion to the cube of the width of the slit between the planes. As shown further on, this fact assumes added strength when going to a circular tube where the flow rate is proportional to the fourth power of tube diameter.

Introducing the coefficient of resistance λ to the fluid motion through the slit between the planes with the aid of the formula

$$\Delta p = \lambda \frac{l}{2h}\frac{\rho w_{av}^2}{2} \tag{65}$$

---

[20] See, for example, Arutyunyan, N. Kh. and Abramyan, B. L., *Krucheniye uprugikh tel* (*Twisting of Elastic Bodies*), Moscow, Fizmatgiz Press, 1963.

and using in its left side the expression of $\Delta p/l$ in terms of w_{av} according to Equation 64, we obtain the resistance law for a plane tube

$$\lambda = \frac{24}{\mathbf{Re}} \quad \text{where} \quad \mathbf{Re} = \frac{w_{av}2h}{\nu} \tag{66}$$

An analogous motion will take place in a "plane" long tray, the width of the bottom of which greatly exceeds its height (flow depth) when the tray is inclined. Due to the presence of free surface, along which the pressure is constant (it is equal to atmospheric pressure in an open tray), there will not be a longitudinal pressure drop in the flow, i.e., $dp/dz = 0$; the transverse pressure drop will be hydrostatic, identical in all sections.

If the tray is inclined to the horizon at a certain angle α, the role of the volumetric force will be played by the gravity force acceleration vector. Its projection onto the axis $0z$ directed, as before, along the flow (in the given case, at the angle α to the horizon) will apparently be equal to $F_z = g \sin \alpha$, so that the equation of fluid motion in the direction of the $0z$ axis will take the form

$$\rho g \sin \alpha + \mu \frac{d^2w}{dy^2} = 0$$

or ($\rho g = \gamma$ is the specific weight of the fluid)

$$\frac{d^2w}{dy^2} = -\frac{\gamma}{\mu} \sin \alpha \tag{67}$$

The boundary conditions are determined by the fluid "adherence" to the tray bottom and by the absence of friction on the free surface. Designating the depth of the flow by h, we obtain the boundary conditions (the axis $0z$ is located along the bottom)

$$w = 0 \quad \text{at} \quad y = 0 \qquad \frac{dw}{dy} = 0 \quad \text{at} \quad y = h \tag{68}$$

Comparing Equation 67 to Equation 60, we see that motion in the tray is determined by the same equation and the forced motion in a plane tube if it is assumed that

$$\frac{\Delta p}{l} = \rho g \sin \alpha = \gamma \sin \alpha$$

Integration of Equations 67 under boundary conditions (68) yields

$$w = \frac{\gamma \sin \alpha}{2\mu} y(2h - y)$$

from which it follows that in this case the velocity profile is also given by the second-order parabola. The formulas of the rate of volume flow, mean, and maximum velocities are

$$Q = \frac{\gamma h^3 \sin \alpha}{3\mu} \qquad w_{av} = \frac{Q}{h} = \frac{\gamma h^2 \sin \alpha}{3\mu} \qquad w_{max} = \frac{\gamma h^2 \sin \alpha}{2\mu} = \frac{3}{2} w_{av}$$

Consider the flow through a cylindrical tube of elliptical cross-section. If the tube section represents an ellipse with half-axes a and b, the equation of which in the plane $0xy$ is

$$\frac{x^2}{a^2} + \frac{y^2}{b^2} = 1$$

the solution of Equation 59 satisfying the boundary condition of the vanishing of velocity on the section contour is

$$w = A\left(1 - \frac{x^2}{a^2} - \frac{y^2}{b^2}\right) \tag{69}$$

where, according to Equation 59, the constant A is determined from the condition

$$-2A\left(\frac{1}{a^2} + \frac{1}{b^2}\right) = -\frac{\Delta p}{\mu l}$$

and is equal to

$$A = \frac{\Delta p}{2\mu l} \frac{a^2 b^2}{a^2 + b^2}$$

Thus, we obtain velocity distribution in the section of an elliptic tube in the form

$$w = \frac{\Delta p}{2\mu l} \frac{a^2 b^2}{a^2 + b^2}\left(1 - \frac{x^2}{a^2} - \frac{y^2}{b^2}\right) \tag{70}$$

The velocity diagram of the vectors drawn from the points of the tube cross-section is represented by the surface of an elliptical paraboloid; the curves of identical (in magnitude) velocities, isotaches, is represented by similar ellipses (with the same ratios of half-axes).

The velocity distribution (70) will yield the section-maximum velocity on the axis of the elliptical tube

$$w_{max} = \frac{\Delta p}{2\mu l} \frac{a^2 b^2}{a^2 + b^2} \tag{71}$$

after which the velocity distribution can be rewritten in the form

$$w = w_{max} = \left(1 - \frac{x^2}{a^2} - \frac{y^2}{b^2}\right) \tag{72}$$

Let us determine the rate of volume fluid flow Q through the elliptic tube section:

$$Q = \iint_S w\,dx\,dy = w_{max} \iint_S \left(1 - \frac{x^2}{a^2} - \frac{y^2}{b^2}\right) dx\,dy$$

For simplicity of calculations, we assume that

$$x = ax' \qquad y = by' \qquad r' = \sqrt{x'^2 + y'^2}$$

Then, the integral on the ellipse area is reduced to the integral on the area S' of a single circle and is equal to

$$Q = w_{max}\, ab \iint_{S'} (1 - x'^2 - y'^2)\, dx'dy' = w_{max}\, ab \int_0^1 (1 - r'^2)\, 2\pi r' dr' = \frac{\pi}{2} ab w_{max}$$

According to Equation 71, we obtain

$$Q = \frac{1}{2}\pi ab w_{\max} = \frac{\pi a^3 b^3 \Delta p}{4\mu l (a^2 + b^2)} \tag{73}$$

The average velocity w_{av} will be determined as the ratio of the rate of volume fluid flow Q to the tube sectional area $S = \pi ab$; according to Equation 73, this gives

$$w_{\max} = \frac{Q}{\pi ab} = \frac{1}{2} w_{\max} = \frac{\Delta p}{4\mu l} \frac{a^2 b^2}{a^2 + b^2} \tag{74}$$

It should be noted that in contrast to a plane tube, in an elliptical tube the average velocity is equal to half the maximum velocity. This trend is also preserved in the specific case of a cylindrical tube of circular cross-section.

Assuming in the previous formulas that $b = a$, we obtain the basic formulas of laminar flow through a cylindrical tube of circular cross-section.

The velocity distribution (72) will take the form (a is the tube radius, $x^2 + y^2 = r^2$)

$$w = w_{\max}\left(1 - \frac{r^2}{a^2}\right) \tag{75}$$

where, according to Equation 71,

$$w_{\max} = \frac{a^2 \Delta p}{4\mu l} = 2w_{av} \tag{76}$$

In this case the velocity profile is given by the paraboloid of revolution. The meridian section in the form of a parabola (75) is usually called the Poiseuille parabola, after the French physician and physiologist who investigated the laws of blood motion through capillary vessels and who published the results of his work in the *Reports of the Paris Academy of Sciences* in 1840.

According to Equation 73, the rate of volume fluid flow is equal to

$$Q = \frac{Qa^4 \Delta p}{8\mu l} \tag{77}$$

It expresses the well-known Poiseuille law: In developed laminar viscous incompressible fluid flow through a cylindrical tube of circular cross-section the rate of volume flow is proportional to the pressure drop per unit length of the tube and to the fourth power of its radius (or diameter). It is emphasized in the formulation of the Poiseuille law that the motion is fully developed. This means that there is identical velocity distribution in all tube sections. This character of motion is attained in the portion of the tube that is rather far away from its entrance. Conversely, over the initial tube section located after the inlet the motion is unsteady and the Poiseuille law is not valid.

Note once again an essential feature of the flow. The pressure drop required to obtain the prescribed flow rate over the given tube section, called the resistance of this section, is inversely proportional to the fourth power of the tube diameter. (It should be recalled that in the case of a plane tube this drop is inversely proportional to the third power of the width of gap between the planes). This circumstance is of great importance in the problems of pumping fluids through tubes of small diameters (for example, capillary tubes, capillary blood vessels, etc.) and also in cases of the motion of very viscous fluids.

Just as was done earlier for the plane tube, let us determine the coefficient of resistance λ of the circular tube from the formula

$$\Delta p = \lambda \frac{l}{d} \frac{\rho w_{av}^2}{2} \tag{78}$$

Then, expressing Δp in terms of w_{av} by Equation 76, we obtain the following law of resistance for cylindrical circular tubes

$$\lambda = \frac{64}{\mathbf{Re}} \quad \text{where} \quad \mathbf{Re} = \frac{w_{av} d}{\nu} \tag{79}$$

Substituting this expression of λ in terms of **Re** into Equation 78, we see that in the case of laminar flow the resistance of a circular tube, just as with a plane tube, is proportional to the first degree of the average velocity of fluid motion through a tube. The resistance formula (78) only in appearance has the form of the quadratic dependence on the average velocity. The actual dependence on velocity is determined only on the basis of the resistance law (79) derived from the equation of fluid motion.

Equation 79 remains valid also for an elliptic tube if its effective diameter is understood to be the quantity d, the square of which represents the mean harmonic quantity of the squared large and small axes of the ellipse

$$\frac{1}{a^2} = \frac{1}{2}\left[\frac{1}{(2a)^2} + \frac{1}{(2b)^2}\right] = \frac{1}{8}\frac{a^2 + b^2}{a^2 b^2}$$

which can be easily verified by substituting the expression from Equation 74 into Equation 78.

The velocity distribution (75) in a cylindrical tube of circular cross-section can be also obtained directly by replacing the Laplacian in the left side of Equation 59 by its expression in polar coordinates. According to Equation 85 of Chapter 1, in the case of cylindrical symmetry and absence of the azimuthal component,

$$\frac{1}{r}\frac{d}{dr}\left(r\frac{dw}{dr}\right) = -\frac{\Delta p}{\mu l} \tag{80}$$

In integrating, we find the general solution

$$w = -\frac{\Delta p}{4\mu l} r^2 + C_1 \ln r + C_2 \tag{81}$$

It follows from the condition of the limited velocity on the tube axis at $r = 0$ that $C_1 = 0$; the second constant is given by the condition that

$$w = 0 \quad \text{at} \quad r = a$$

which leads to the parabolic velocity profile (75).

Solution (81) represents a greater generality than that presented by Equation 75. Thus, for example, using Equation 81, we obtain velocity distribution in the region between two coaxial circular cylinders of radii a and $b > a$. Subjecting solution (81) to the boundary conditions

$$w = 0 \quad \text{at} \quad r = a \quad \text{and} \quad r = b$$

we obtain the distribution of velocities in the section

$$w = \frac{\Delta p}{4\mu l}\left[a^2 - r^2 + \frac{b^2 - a^2}{\ln(b/a)}\ln\left(\frac{r}{a}\right)\right] \tag{82}$$

as well as the formulas for the flow rate and average velocity

$$Q = \frac{\pi\Delta p}{8\mu l}\left[b^4 - a^4 - \frac{(b^2 - a^2)^2}{\ln(b/a)}\right] \qquad w_{av} = \frac{\Delta p}{8\mu l}\left[b^4 + a^2 + \frac{b^2 - a^2}{\ln(b/a)}\right] \tag{83}$$

Employing the technique analogous to that used for composing solution (69) for a cylindrical tube of elliptical cross-section, it becomes possible to form the solution for a prismatic tube, the cross-section of which is given by the equilateral triangle. Directing the axis $0x$ along the base of the triangle and the axis $0y$ along its height, we obtain the equations of the straight lines that form the sides of the triangle (a is the side): $y = 0$ (base), $y = \frac{a\sqrt{3}}{2} \pm \sqrt{3}x$ (lateral sides).

Composing the following product, which converts into zero on the contour of the triangular tube section,

$$y\left(y - \sqrt{3}x - \frac{a\sqrt{3}}{2}\right)\left(y + \sqrt{3}x - \frac{a\sqrt{3}}{2}\right)$$

we see that the Laplacian in the plane (x, y) of this product is equal to a constant value $(-2a\sqrt{3})$; consequently, the sought-after solution is

$$w = \frac{\Delta p}{2\sqrt{3}\mu a l}y\left(y - \sqrt{3}x - \frac{a\sqrt{3}}{2}\right)\left(y + \sqrt{3}x - \frac{a\sqrt{3}}{2}\right)$$

The rate of volume flow through the triangular tube section and the section-average velocity are

$$Q = \frac{a^4\sqrt{3}}{320}\frac{\Delta p}{\mu l} \qquad w_{av} = \frac{a^2}{80}\frac{\Delta p}{\mu l}$$

As in the previous examples, the resistance coefficient λ in the formula

$$\Delta p = \lambda\frac{l}{a}\frac{\rho w_{av}^2}{2}$$

is inversely proportional to the Reynolds number

$$\lambda = \frac{160}{\mathbf{Re}} \qquad \mathbf{Re} = \frac{w_{av}a}{\nu}$$

Using expansions into infinite series, it is possible to solve the problem of incompressible viscous fluid flow through a tube of rectangular cross-section. Let us designate the rectangle height, parallel to the axis $0y$, by $2h$ and the base, parallel to the axis $0x$, by $2æh$, where $æ$ is any positive constant. As before, the axis $0z$ is drawn through the center of the rectangle and is directed downstream.

Let us dimensionalize Equation 59, taking the height h to be the length scale and the quantity $\frac{h^2}{\mu}\cdot\frac{\Delta p}{l}$, which has the velocity dimensions, to be the velocity scale. Let us introduce the following notation for the new dimensionless variables ξ, η, and w^*:

$$\xi = \frac{x}{h} \qquad \eta = \frac{y}{h} \qquad w^* = \frac{w\mu l}{h^2 \Delta p} \tag{84}$$

Substituting them into Equation 59, we establish the dimensionless equations and boundary conditions

$$\frac{\partial^2 w^*}{\partial \xi^2} + \frac{\partial^2 w^*}{\partial \eta^2} = -1$$

$$w^* = 0 \quad \text{when} \quad \xi = \pm æ \quad |\eta| < 1 \quad \text{and when} \quad \eta = \pm 1 \quad |\xi| < æ \tag{85}$$

To solve this equation, we use the familiar Fourier series

$$\sum_{n=0}^{\infty} \frac{(-1)^n}{2n+1} \cos \frac{2n+1}{2} \pi t = \begin{cases} \pi/4 & \text{when} \quad |t| < 1 \\ 0 & \text{when} \quad |t| = 1 \end{cases} \tag{86}$$

Then, Equation 85 can be presented in the form ($|\xi| < æ$)

$$\frac{\partial^2 w^*}{\partial \xi^2} + \frac{\partial^2 w^*}{\partial \eta^2} = -\frac{4}{\pi} \sum_{n=0}^{\infty} \frac{(-1)^n}{2n+1} \cos\left(\frac{2n+1}{2} \frac{\pi}{æ} \xi\right) \tag{87}$$

It is natural that the solution of this equation is sought in the form of the series

$$w^* = \sum_{n=0}^{\infty} Y_n(\eta) \cos\left(\frac{2n+1}{2} \frac{\pi}{æ} \xi\right) \tag{88}$$

in which, according to Equation 86, the first boundary condition of system (85) is already fulfilled.

Substituting expansion (88) into Equation 87 and equating the coefficients at the cosines with identical arguments, we obtain a system of ordinary differential equations (the prime means the derivative with respect to η)

$$Y_n'' - \left(\frac{2n+1}{2} \frac{\pi}{æ}\right)^2 Y_n = -\frac{4}{\pi} \frac{(-1)^n}{2n+1}$$

$$Y_n = 0 \quad \text{at} \quad \eta = \pm 1 \quad (n = 0, 1, \cdots) \tag{89}$$

The general solutions of these equations can be presented in the form

$$Y_n = A_n + B_n \operatorname{ch}\left(\frac{2n+1}{2} \frac{\pi}{æ} \eta\right) + C_n \operatorname{sh}\left(\frac{2n+1}{2} \frac{\pi}{æ} \eta\right) \tag{90}$$

where the constants A_n are determined by directly substituting Equation 90 into Equation 89. They are equal to

$$A_n = \frac{\dfrac{(-1)^n 2}{æ}}{\left(\dfrac{2n+1}{2} \dfrac{\pi}{æ}\right)^3} = \frac{16 æ^2}{\pi^3} \frac{(-1)^n}{(2n+1)^3} \tag{91}$$

The constants B_n and C_n are found from boundary conditions (89), i.e., from the system of equations

$$B_n \operatorname{ch}\left(\frac{2n+1}{2}\frac{\pi}{æ}\right) + C_n \operatorname{sh}\left(\frac{2n+1}{2}\frac{\pi}{æ}\right) = -A_n$$

$$B_n \operatorname{ch}\left(\frac{2n+1}{2}\frac{\pi}{æ}\right) - C_n \operatorname{sh}\left(\frac{2n+1}{2}\frac{\pi}{æ}\right) = -A_n$$

Thus, we find

$$B_n = -\frac{A_n}{\operatorname{ch}\left(\frac{2n+1}{2}\frac{\pi}{æ}\right)} \qquad C_n = 0 \quad (n = 0, 1, \cdots) \tag{92}$$

where the numbers A_n are already determined by Equations 91. Returning to Equation 88, we find the sought-after solution in dimensionless form

$$w^* = \frac{16æ^2}{\pi^3}\sum_{n=0}^{\infty}\frac{(-1)^n}{(2n+1)^3}\left[1 - \frac{\operatorname{ch}\left(\frac{2n+1}{2}\frac{\pi}{æ}\eta\right)}{\operatorname{ch}\left(\frac{2n+1}{2}\frac{\pi}{æ}\right)}\right]\cos\left(\frac{2n+1}{2}\frac{\pi}{æ}\xi\right) \tag{93}$$

and in dimensional form

$$w = \frac{16æ^2}{\pi^3}\frac{h^2\Delta p}{\mu l}\sum_{n=0}^{\infty}\frac{(-1)^n}{(2n+1)^3}\left[1 - \frac{\operatorname{ch}\left(\frac{2n+1}{2}\frac{\pi y}{æh}\right)}{\operatorname{ch}\left(\frac{2n+1}{2}\frac{\pi}{æ}\right)}\right]\cos\left(\frac{2n+1}{2}\frac{\pi x}{æh}\right) \tag{94}$$

Omitting simple computations, we note that the rate of volume flow Q and section-average velocity w_{av} are determined by the equations (the parameter $æ > 1$ represents the ratio of the width of the rectangular cross-section $2æh$ to its height $2h$)

$$Q = \frac{\Delta p}{\mu l}æh^4 f(x) \qquad w_{av} = \frac{Q}{4æh^2} = \frac{\Delta p h^2}{16\mu l}f(æ)$$

$$f(æ) = \frac{16}{3} - \frac{1024}{\pi^5 æ}\left(\operatorname{th}\frac{\pi æ}{2} + \frac{1}{3^3}\operatorname{th}\frac{3\pi æ}{2} + \cdots\right) \tag{95}$$

and the function $f(æ)$ has the following values:

æ	1	2	3	5	10	12	100	∞
$f(æ)$	2.253	3.664	4.203	4.665	5.000	5.059	5.299	5.333

Relating the flow rate Q to the tube width $2æh$, i.e., introducing the quantity $Q' = Q/(2æh)$ and then in the expressions of Q' and w_{av} going to the limit for $æ \to \infty$, we shall find from the last column of the table above that

$$Q' = \frac{\Delta p h^3}{8\mu l}f(æ) \to \frac{\Delta p h^3}{8\mu l}\frac{16}{3} = \frac{2}{3}\frac{\Delta p h^3}{\mu l} \qquad w_{av} \to \frac{\Delta p h^2}{16\mu l}\frac{16}{3} \to \frac{1}{3}\frac{\Delta p h^2}{\mu l}$$

in full conformity with formulas (63) and (64) for the plane tube. Assuming in Equations 94 and 95 that æ = 1, we obtain the solution of the problem about the flow of an incompressible viscous fluid through a prismatic tube of quadratic (square) cross-section.

Introducing, as before, the resistance coefficient λ by the formula

$$\Delta p = \lambda \frac{l}{2h} \frac{\rho w_{av}^2}{2}$$

and expressing Δp on the left side in terms of w_{av} according to the second equation of system (95), and also making simple transformations, we find the law of the resistance of prismatic tubes of rectangular cross-section

$$\lambda = \frac{128}{\mathbf{Re}} \frac{1}{f(æ)} \qquad \mathbf{Re} = \frac{w_{av} 2h}{\nu} \tag{96}$$

Proceeding to the limit $æ \to \infty$, we again obtain the law of resistance of a plane tube (66), whereas at æ = 1, we also obtain the law of resistance of the square tube from the foregoing table.

To approximately estimate the resistance of cylindrical or prismatic complex shaped tubes, a technique is employed when the resistance of these tubes is compared with the tube of circular cross-section, the resistance of which is equivalent to the resistance of the former and the radius (or diameter) of which is taken to be the so-called "hydraulic" radius r_h (or the diameter $d_h = 2r_h$) equal to the ratio of the area of the normal section S of the tube (Figure 150) to the perimeter P of the cross-section

$$r_h = \frac{1}{2} d_h = \frac{S}{P}$$

This technique is very rough and makes sense only if the cross-sections of the tubes compared are geometrically close to each other.

To clarify the sense of this technique, let us first establish the connection between the pressure drop Δp over some arbitrarily selected length of the tube l and the overall friction over the inner, "wetted" surface of this tube. Let us take into consideration that as was pointed out at the beginning of this section, the motion of fluid is identical in all sections. This corresponds to the equilibrium of the fluid volume bounded by two sections (Figure 150) $S_1 = S_2 = S$ of the tube, a distance l apart, and the lateral surface of the tube equal to the product of the perimeter of the cross-section P by the length l. The condition for equilibrium is represented by the straightforward equation (τ_w is the skin friction variable over the section perimeter, ds is the differential of the perimeter arc)

$$S\Delta p = \int_P \tau_w ds \cdot l = P \cdot \frac{1}{P} \int_P \tau_w ds \cdot l = P\bar{\tau}_w l$$

where $\bar{\tau}_w$ means the perimeter-average skin friction τ_w. From this, using the concept of the hydraulic radius, we obtain

$$\bar{\tau}_w = \frac{\Delta p}{l} r_h \tag{97}$$

i.e., the skin friction average over the perimeter of a cylindrical (prismatic) tube is equal to the pressure drop over the portion of the tube, the length of which is equal to the hydraulic radius.

In the particular case of a plane tube with the distance between the parallel planes equal to $2h$, the following relations will hold per unit length in the direction transverse to the flow:

$$S = 2h \cdot 1 \qquad P = 2 \cdot 1 \qquad r_{\text{h}} = S/P = h \qquad d_{\text{h}} = 2h$$

in a circular tube of radius a:

$$S = \pi a^2 \qquad P = 2\pi a \qquad r_{\text{h}} = a/2 \qquad d_{\text{h}} = a$$

The skin friction τ_w in a circular tube coincides in magnitude with the pressure drop over the portion of the tube of length equal to half the tube radius, i.e.,

$$\tau_w = \frac{\Delta p}{l}\frac{a}{2} = \frac{1}{4}\frac{\Delta p}{l}d \tag{98}$$

Analogously, for the skin friction τ on the coaxial cylindrical surface of arbitrary radius r drawn inside the round tube,

$$\tau = \frac{\Delta p}{l}\frac{r}{2}$$

Consequently, dividing both sides of the latter equations by each other, we obtain the relation (y is the distance of the surface from the tube wall)

$$\tau = \tau_w \frac{r}{a} = \tau_w\left(1 - \frac{y}{a}\right) \tag{99}$$

which expresses the linear nature of the connection between the skin friction, applied to the area that is perpendicular to the vector-radius of any point of some section relative to the section center, and the magnitude of this vector-radius. Let us emphasize that in deriving Equation 97 the laminar nature of motion was used in no way as well as the velocity profile typical of this mode of flow. Relation (97) can also be applied to turbulent motion of fluid through a cylindrical (prismatic) tube of arbitrary cross-section, just as Equations 98 and 99 to laminar and turbulent motions in a cylindrical tube of circular cross-section and in a plane tube.

Moreover, these equations, which express the balance between the pressure drop, moving the fluid, and the skin friction, retarding the motion, can be applied to the motions of any continuous media through cylindrical tubes and, in particular, to the motions of non-Newtonian fluids. The simplest example of such a motion constitutes the content of the next section.

7. Developed Motion of a Viscoplastic Fluid through a Cylindrical Tube of Circular Profile

In a cylindrical tube of circular profile let there be a developed laminar flow of a viscoplastic fluid, the rheological equation for the flow of which, according to Equation 14, is presented in the form (r is the instantaneous radius of the point, a is the tube radius)

$$\dot{\varepsilon} = -\frac{dw}{dr} = \begin{cases} \dfrac{\tau - \tau_0}{\mu'} & \text{when } \tau_0 < \tau < \tau_w \\ 0 & \text{when } \tau \le \tau_0 \end{cases} \tag{100}$$

Let us recall that in this equation τ is the skin friction at any point of the tube section applied to the area which is perpendicular to the tube radius, τ_0 is the limiting stress characteristic of the given viscoplastic fluid and after the attainment of which the fluid only starts to flow; μ' is the dynamic coefficient of the structural (plastic) viscosity.

In view of the fact that the limits of applicability of the rheological law (100) are indicated in the quantities τ, it is useful to adopt the stress τ itself as the argument to ease the integration of this equation. To this end, using Equation 99, rewrite Equation 100 in the form

$$\frac{dw}{dr} = \frac{dw}{d\tau}\frac{d\tau}{dr} = \frac{\tau_w}{a}\frac{dw}{d\tau} = \begin{cases} -\dfrac{\tau-\tau_0}{\mu'} & \tau_0<\tau<\tau_w \\ 0 & \tau=\tau_0 \end{cases} \tag{101}$$

Simple integration with the fulfillment of the obvious boundary condition $w = 0$ at $\tau = \tau_w$ will lead to the following connection between the velocity and the skin friction at the same point:

$$w = \begin{cases} \dfrac{\tau_w a}{2\mu'}\left(1-2\dfrac{\tau_0}{\tau_w}+\dfrac{\tau}{\tau_w}\right)\left(1-\dfrac{\tau}{\tau_w}\right) & \tau_0<\tau<\tau_w \\ \dfrac{\tau_w a}{2\mu'}\left(1-\dfrac{\tau_0}{\tau_w}\right)^2 = \text{const} = w_0 & \tau=\tau_0 \end{cases} \tag{102}$$

The stress τ_w can be expressed by Equation 98 in terms of the given pressure drop Δp over the length l. Equation 102 gives the following velocity profile ($r_0 = \dfrac{2l\tau_0}{\Delta p}$):

$$w = \begin{cases} \dfrac{a^2\Delta p}{4\mu' l}\left[1-\dfrac{r^2}{a^2}-\dfrac{4l\tau_0}{a\Delta p}\left(1-\dfrac{r}{a}\right)\right] & r_0<r<a \\ \dfrac{a^2\Delta p}{4\mu' l}\left(1-\dfrac{2l\tau_0}{a\Delta p}\right)^2 = w_0 & 0<r\le r_0 \end{cases} \tag{103}$$

The velocity profile consists partly of the surface of the paraboloid of revolution (from the tube wall to the cylindrical surface of radius ρ_0) and partly of a plane area perpendicular to the tube axis (at the central part of the tube inside the indicated cylindrical surface) (Figure 151). In the central part of the tube the viscoplastic fluid moves as a solid rod. When $\Delta p/l \le 2\tau_0/a$,

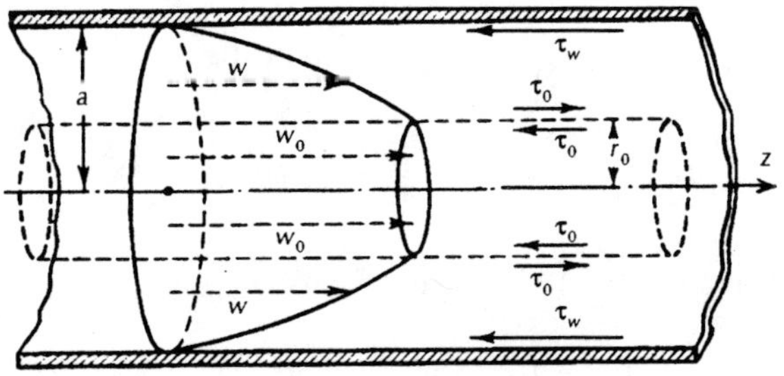

Fig. 151

this "solid rod" fills the tube section and, due to the property of viscous fluid "adherence" to a solid surface, the fluid remains motionless.

By assuming in Equations 103 that $r_0 = 0$, which corresponds to $\tau_0 = 0$ (i.e., by going to an ordinary Newtonian viscous fluid), the velocity profile is reduced to the well-known paraboloid of revolution and the plane area disappears.

Having the velocity profile one can easily determine the rate of volume flow Q of a viscoplastic fluid through the tube section. The quantity Q can be most simply calculated by going over in the flow rate integral

$$Q = \int_0^a 2\pi w r dr = \pi \int_0^a w d(r^2) = \pi [r^2 w]_0^a - \pi \int_0^a r^2 \frac{dw}{dr} dr = -\pi \int_0^a r^2 \frac{dw}{dr} dr$$

from the independent variable r to τ by formula (99). Using rheological law (100),

$$Q = \frac{\pi a^3}{\mu' \tau_w^3} \int_{\tau_0}^{\tau_w} \tau^2 (\tau - \tau_0)\, d\tau = \frac{\pi a^3}{\mu' \tau_w^3} \left[\frac{1}{4} (\tau_w^4 - \tau_0^4) - \frac{\tau_0}{3} (\tau_w^3 - \tau_0^3) \right] =$$

$$= \frac{\pi a^3 \tau_w}{4\mu'} \left[1 - \frac{4}{3} \frac{\tau_0}{\tau_w} + \frac{1}{3} \left(\frac{\tau_0}{\tau_w} \right)^4 \right] = \frac{\pi a^3 \tau_w}{4\mu'} f\left(\frac{\tau_0}{\tau_w} \right) \tag{104}$$

Replacing here τ_w by its expression in terms of Δp, according to Equation 98, we obtain Buckingham's formula

$$Q = \frac{\pi a^4 \Delta p}{8\mu' l} \left[1 - \frac{4}{3} \left(\frac{2\tau_0 l}{a \Delta p} \right) + \frac{1}{3} \left(\frac{2\tau_0 l}{a \Delta p} \right)^4 \right] = \frac{\pi a^4 \Delta p}{8\mu' l} f\left(\frac{2\tau_0 l}{a \Delta p} \right) \tag{105}$$

At $\tau_0 = 0$ and $\mu' = \mu$ we return to the Poiseuille formula (77).

According to Equation 105, the section-average velocity w_{av} is equal to

$$w_{av} = \frac{Q}{\pi a^2} = \frac{a^2 \Delta p}{8\mu' l} \left[1 - \frac{4}{3} \left(\frac{2\tau_0 l}{a \Delta p} \right) + \frac{1}{3} \left(\frac{2\tau_0 l}{a \Delta p} \right)^4 \right] =$$

$$= \frac{d^2 \Delta p}{32\mu' l} f\left(\frac{4\tau_0 l}{d \Delta p} \right) = \frac{d^2 \Delta p}{32\mu' l} f\left(\frac{\tau_0}{\tau_w} \right) \tag{106}$$

Let us use the same resistance formula (78) as was employed for the Newtonian viscous fluid. Unfortunately, in the given case Equation 106 cannot be solved at all in Δp. Therefore, we avail ourselves of the following technique. Equation 106 yields

$$\lambda = \frac{64}{\mathbf{Re}} \left[f\left(\frac{4\tau_0 l}{d \Delta p} \right) \right]^{-1} \qquad \mathbf{Re} = \frac{\rho w_{av} d}{\mu'} \tag{107}$$

On the other hand, from Equation 106 it is possible to determine the Reynolds number

$$\mathbf{Re} = \frac{\rho d^3 \Delta p}{32\mu'^2 l} f\left(\frac{4\tau_0 l}{d \Delta p} \right) = \frac{1}{8} \frac{\rho d^2 \tau_0}{\mu'^2} f_1\left(\frac{4\tau_0 l}{d \Delta p} \right)$$

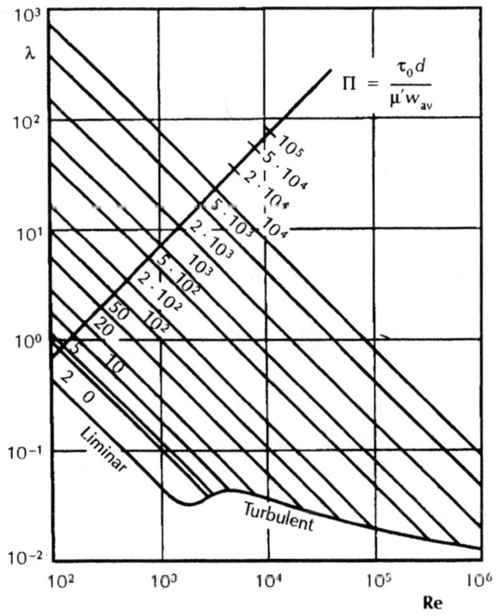

Fig. 152

$$f_1\left(\frac{4\tau_0 l}{d\Delta p}\right) = f\left(\frac{4\tau_0 l}{d\Delta p}\right)\Big/\left(\frac{4\tau_0 l}{d\Delta p}\right) \tag{108}$$

The dimensionless number $\rho d^2\tau_0/\mu'^2$ does not contain Δp. It characterizes the visco-plastic properties of the fluid. The set of equations

$$\lambda = \frac{64}{\mathbf{Re}}\left[f\left(\frac{4\tau_0 l}{d\Delta p}\right)\right]^{-1} \qquad \mathbf{Re} = \frac{1}{8}\frac{\rho d^2\tau_0}{\mu'^2} f_1\left(\frac{4\tau_0 l}{d\Delta p}\right)$$

can be considered to be the parametric (the role of the parameter is played by the quantity $4\tau_0 l/(d\Delta p)$) expression for the law of resistance of viscoplastic fluids to the motion in tubes. Excluding the parameter $4\tau_0 l/(d\Delta p)$ from the last system of equations,

$$\frac{\lambda \mathbf{Re}}{64} = \Omega_1\left(\frac{\mu'^2}{\rho d^2\tau_0}\frac{\rho w_{av} d}{\mu'}\right)$$

where the general law of resistance to the motion of viscoplastic fluids through tubes of circular cross-section follows:

$$\lambda = \frac{64}{\mathbf{Re}}\Omega\left(\frac{\tau_0 d}{\mu' w}\right) \qquad \mathbf{Re} = \frac{\rho w_{av} d}{\mu'} \tag{109}$$

Thus, in the theory of similarity for the flows of viscoplastic fluids through circular cylindrical tubes there are two similarity criteria:

1. The Reynolds number

$$\mathbf{Re} = \frac{\rho w_{av} d}{\mu'}$$

characterizes the effect of structural viscosity.

2. The "plasticity parameter"

$$\Pi = \frac{\tau_0 d}{\mu' w}$$

determines the effect of the fluid plasticity.

The above criteria could be replaced by others consisting of different combinations.[21]

Figure 152 depicts the resistance law (109) in the logarithmic scale. The law is presented in the coordinates (lg **Re**, lg λ) in the form of a family of parallel straight lines with the parameter Π. To the value $\Pi = 0$ corresponds the lower straight line, which expresses the usual trend $\lambda = 64/\mathbf{Re}$ for an ordinary viscous fluid. The lower ends of the straight lines rest on the curve corresponding to the transition of laminar to turbulent flow of an ordinary viscous fluid.

8. Developed Flow of an Electrically Conducting Viscous Fluid through Tubes in the Presence of Transverse Magnetic Fields

Present-day metallurgical processes make wide use of the control of flow of liquid metals through tubes and channels with the aid of external, constant, or variable magnetic fields. The resulting mixed hydrodynamic and electricomagnetic problems constitute a relatively new field of the mechanics of liquids and gases called magnetic hydrodynamics (MHD).[22]

Our discussion is restricted to the case of the steady motion of an incompressible fluid having the constant electrical conductivity coefficient and being subjected to the action of an external steady uniform magnetic field. We neglect the presence of free electrical charges in the fluid. The magnetic permeability (the commonly adopted symbol μ, which should be preserved in this section, should not be confused with the designation of the dynamic viscosity coefficient; for the latter the product ρν of the liquid density ρ by the kinematic viscosity coefficient ν must be used) is considered identical for all fluid and solid boundaries by equating it to the value of μ_0 in vacuum. Finally, we neglect the action of all volumetric forces except for the ponderomotive force (the Lorentz force) $\boldsymbol{j} \times \boldsymbol{B}$, where $\boldsymbol{j}$ is the density of the electric current originating in the electrically conducting fluid, which has the electrical conductivity coefficient σ and moves with the velocity $\boldsymbol{V}$, due to the local electrical field of strength $\boldsymbol{E}$ and to the magnetic field with the magnetic induction $\boldsymbol{B}$. This force is defined by the generalized Ohm's law

$$\boldsymbol{j} = \sigma(\boldsymbol{E} + \boldsymbol{V} \times \boldsymbol{B}) \tag{110}$$

---

[21] Wilkinson, W. L., *Nenyutonovskiye zhidkosti* (*Non-Newtonian Fluids*) (translation from English), Mir Press, Moscow, 1964.

[22] The foundations of electrical dynamics, the knowledge of which is a prerequisite for understanding the content of the present section, as well as a more extensive coverage of these problems can be found in Sedov, L. I., *Mekhanika sploshnoy sredy* (*Mechanics of a Continuum*), Vol. 1, Nauka Press, Moscow, 1983; Vatazhin, A. B., Lyubimov, G. A., and Regirer., S. A., *Magnitogidrodinamicheskiye techeniya v kanalakh* (*Magnetohydrodynamic Channel Flows*), Nauka Press, Moscow, 1970; Gogosov, V. V. and Polyanskiy, V. A., Electrodynamics: problems and applications, basic Equations, different solutions, Advances in Science and Technology, *Mekh. Zhidk. Gaza*, 10, 5, 1976; Gogosov, V. V., Nalyotova, V. A., and Shaposhnikova, G. A., The hydrodynamics of Magnetizable fluids, *Ibid.*, 16, 75–208, 1981; Sedov, L. I., *A Course in Continuum Mechanics*, Vol. 1 to 4, Wolters-Northoff, 1971–1972.

With the above simplifications, the system of MHD equations to be considered is reduced to the following:

$$\rho\,(V \cdot \nabla)\,V = -\operatorname{grad} p + \rho\nu\nabla^2 V + j \times B \qquad \operatorname{div} V = 0$$

$$\operatorname{rot} B = \mu_0 j \qquad \operatorname{div} B = 0 \qquad \operatorname{rot} E = 0 \qquad \operatorname{div} E = 0 \tag{111}$$

with Ohm's law (110) supplemented to it.

The main specific feature of magnetohydrodynamic investigations is the fact that according to the very essence of the phenomena, it is quite insufficient to use ordinary equations of liquid motion, only supplementing mechanical volumetric forces with the Lorentz ponderomotive force, which expresses the action of the external magnetic field on the moving electrically conducting fluid. In fact, a much more complex phenomenon of the interaction between the magnetic field and a fluid flow is studied under conditions in which the solid boundaries of the flow, depending on their electrical conductivity, themselves exert their influence on the magnetic field in the region of fluid flow.

Mathematically, this means that one cannot consider the equations of fluid motion (the Navier-Stokes and continuity equations) separately from the electromagnetic field equations (Maxwell equations). Only in a very simplified statement, can the motion equations be considered "autonomous" and admitting the self-sustained integration separately from the general equations of the electrical dynamics of continuous media. As regards the problem of boundary conditions that correspond to the possible stationary problems, it should be noted that they consist of the hydrodynamic conditions previously discussed (adherence of fluid to the surface of bodies in a flow, the conditions at infinity, etc.) and specific electromagnetic conditions on the interfaces between liquid and solid phases (for example, tube wall), and also between the solid phase and the outer region (gas, vacuum). Consider the magnetohydrodynamic extension of the problem of incompressible viscous fluid flow through cylindrical (prismatic) tubes to the case of an electrically conducting fluid in the presence of a uniform magnetic field being transverse to the direction of the fluid flow (see Section 6). There is much theoretical and experimental work along these lines,[23] beginning in 1937 with the well-known work by Hartmann.[24]

We shall make use of the same location of the coordinate axes with respect to the tube as in Figure 149. The unknown quantities in this problem are the velocity field V, electrical voltage E, and magnetic induction B. We will show that in the considered case of an infinitely long tube of constant cross-section with the walls made from homogeneous material of constant thickness δ, which are considered small as compared to the dimensions of the tube

---

[23] Regirer, S. A., About the flow of an electrically conducting fluid in the presence of a magnetic field through tubes of arbitrary profile, *Prikl. Mat. Mekh.*, 24(3), 541–542, 1960; Uflyand, Ya. S., The developed flow of an electrically conducting fluid in a rectangular duct in the presence of a transverse magnetic field, *Zh. Tekh. Fiz.*, 30(10), 1960; Grinberg, G. A.,

[24] Concerning the developed flow of a conducting fluid in a rectangular tube with two non-conducting and two conducting walls parallel to the external magnetic field, *Prikl. Mat. Mekh.*, 25, 1024–1034, 1961 and Concerning certain gases of flow of conducting fluid in tubes of rectangular cross-section situated in the magnetic fields, *Ibid.*, 26, 80–87, 1962; Kulikovskiy, A. G. and Lyubimov, G. A., *Magnitnaya gidrodinamika* (*Magnetic Hydrodynamics*), Fizmatgiz Press, Moscow, 1962, pp. 56–58; Shercliff, J. A., Steady motion of conducting fluids in Pipes under transverse magnetic fields, *Proc. Cambr. Phil. Soc.*, 49(1), 136–144, 1953; Chang Chieh, C. and Lundgren, T. S., Duct flow in magnetohydrodynamics, *Z. Angew. Math. Phys.*, 12, 100–114, 1961; Sloan, D. M. and Smith, P., Magnetohydrodynamic flow in a rectangular pipe between conducting plates, *Z. Angew. Math. Mech.*, 7, 439–443, 1966; Hunt, J. C. R., Magnetohydrodynamic flow in rectangular ducts, *J. Fluid Mech.*, 21, 577–590, 1965; Branover, H., *Magnetohydrodynamic Flow in Ducts*, John Wiley & Sons, New York, 1978; Hartmann, J., Hg-Dynamics. I. Theory of the laminar flow of an electrically conductive liquid in a homogeneous magnetic field, *Mat.-Fys. Medd. Kgl. Dauske Vissenkab.*, 15(6), 1937.

cross-section, the solution of the problem can be reduced to the determination of two functions $V_z = w(x, y)$ and $B_z = B(x, y)$ and to exclude the electric field E from the equations.

Just as in the purely hydrodynamic problem considered in Section 6, it follows from the condition of the equality of cross-sections in an infinitely long tube with constant geometric parameters of the cross-sections that all the distributions of mechanical and physical quantities will depend only on coordinates x and y in the plane of the pipe cross-section. The exception is the pressure which decreases linearly along the tube, but the pressure drop over the portion of fixed length also preserves its value along the tube. Under these conditions, the existence of electric current along the tube is impossible, so that according to Equation 110, at $u = 0$ and $v = 0$,

$$j_z = \sigma E_z = 0 \qquad j_z = 0 \qquad E_z = 0$$

i.e., the induced currents are located in the planes of the normal sections of the tube, whereas the induced component of the magnetic field is directed along the axis $0z$ according to the Biot-Savart law. Taking into account the presence of the homogeneous external transverse magnetic field with the components $(0,\ B_0,\ 0)$ and the condition of the solenoidality of the magnetic field, expressed in this case by the equality $\operatorname{div}\boldsymbol{B} = \partial B_z/\partial z = 0$, we conclude[25] that the overall components of the magnetic field will be equal to

$$B_x = 0 \qquad B_y = B_0 = \text{const} \qquad B_z = B(x, y)$$

As can be easily concluded from the first electrodynamic equation of system (111), the quantity B/μ_0 can be considered a "stream function" of the plane field of the vectors $\boldsymbol{j}(j_x,\ j_y)$, so that

$$\mu_0 j_x = \frac{\partial B}{\partial y} \qquad \mu_0 j_y = -\frac{\partial B}{\partial x} \tag{112}$$

Returning to the first dynamic equation of system (111), the Navier-Stokes equation, we rewrite it in terms of the projections onto the coordinate axes in the form

$$\frac{\partial p}{\partial x} = j_y B_z = -\frac{1}{\mu_0} B \frac{\partial B}{\partial x}$$

$$\frac{\partial p}{\partial y} = -j_x B_z = -\frac{1}{\mu_0} B \frac{\partial B}{\partial y}$$

$$\frac{\partial p}{\partial z} = \rho\nu\nabla^2 w + j_x B_y - j_y B_x = \rho\nu\nabla^2 w + \frac{B_0}{\mu_0}\frac{\partial B}{\partial y} \tag{113}$$

Note that the first two of these equations are reduced to the equalities

$$\frac{\partial}{\partial x}\left(p + \frac{B^2}{2\mu_0}\right) = 0 \qquad \frac{\partial}{\partial y}\left(p + \frac{B^2}{2\mu_0}\right) = 0$$

---

[25] In the general presentation of the problem of electrically conducting fluid motion through a cylindrical (prismatic) tube of any cross-section and a plane tube, the works of Sherclift, Chang, and Lundgren (see earlier footnotes) were used.

which express the fact that the sum $p + B^2/(2\mu_0)$ is the function of only z. Recalling that, according to the foregoing, B depends only on x and y, we conclude that $\partial p/\partial z$ represents the function of only z. Then, from the last equality of system (113), where the left side is the function of only z and the right side of only x and y, it follows that each of these sides separately is equal to a constant; let us represent, as before, this constant in the form

$$\frac{\partial p}{\partial z} = \text{const} = -\frac{\Delta p}{l}$$

Thus, we obtain $\nabla^2 = \dfrac{\partial^2}{\partial x^2} + \dfrac{\partial^2}{\partial y^2}$ which is the equation of liquid motion in a tube

$$\rho\nu\nabla^2 w + \frac{B_0}{\mu_0}\frac{\partial B}{\partial y} = -\frac{\Delta p}{l} \tag{114}$$

The second equation can be found by applying the operation rot to the left and right sides of Equation 110, which makes it possible to get rid of the electric field $\boldsymbol{E}$, and then by setting up the projection of both sides of the resulting equation onto the axis $0z$. Using the first equation of the second line of system (111) and of (88) of Chapter 1,

$$\operatorname{rot}\boldsymbol{j} = \frac{1}{\mu_0}\operatorname{rot}\operatorname{rot}\boldsymbol{B} = \frac{1}{\mu_0}\operatorname{grad}\operatorname{div}\boldsymbol{B} - \frac{1}{\mu_0}\nabla^2\boldsymbol{B} = -\frac{1}{\mu_0}\nabla^2\boldsymbol{B} \qquad \operatorname{rot}\boldsymbol{E} = 0$$

$$\operatorname{rot}\,(\boldsymbol{V}\times\boldsymbol{B}) = (\boldsymbol{B}\cdot\nabla)\,\boldsymbol{V} - (\boldsymbol{V}\cdot\nabla)\,\boldsymbol{B} + \boldsymbol{V}\operatorname{div}\boldsymbol{B} - \boldsymbol{B}\operatorname{div}\boldsymbol{V} = (\boldsymbol{B}\cdot\nabla)\,\boldsymbol{V} - (\boldsymbol{V}\cdot\nabla)\,\boldsymbol{B}$$

and projecting onto the axis $0z$, we obtain the second

$$\nabla^2 B + \sigma\mu_0 B_0\frac{\partial w}{\partial y} = 0 \tag{115}$$

which expresses the structure of the magnetic field in the moving fluid.

Let us isolate the following three regions in the plane $0xy$ (Figure 153):

1. The region S_1 inside the tube in which integration of the system of Equations 114 and 115 should be made
2. The thin (the thickness will be designated by δ) region S_2 occupied by the solid walls of the tube; in this region $w = 0$, and for determining the function B, it is sufficient to integrate the Laplace equation

$$\nabla^2 B = 0 \tag{116}$$

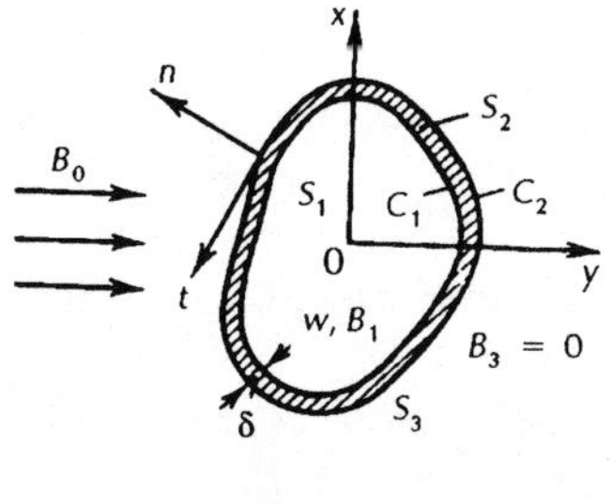

Fig. 153

3. The external (with respect to the tube) region S_3 filled with an electrical nonconducting medium ($\sigma = 0$) in which $j_x = j_y = 0$ and, consequently, according to the first of the electrodynamic equations in this region

$$\text{rot}\,\boldsymbol{B} = 0 \qquad \frac{\partial B_z}{\partial x} = \frac{\partial B_z}{\partial y} = 0 \qquad B_z = \text{const}$$

At an infinite distance from the tube, the magnetic field is directed parallel to the axis 0y and, consequently, $B_z = B = 0$ in the entire region S_3.

Let us designate by C_1 and C_2 the boundaries that separate S_1 from S_2 and S_2 from S_3, and agree to use the indices 1, 2, and 3 for designating the quantities B and σ which correspond to the regions S_1, S_2, and S_3. In the equations formed, we move to dimensionless quantities having assumed (a is a certain characteristic length, l is the length over which the pressure drop Δp is prescribed)

$$x = a\xi \qquad y = a\eta \qquad w = \frac{a^2}{\nu\rho}\frac{\Delta p}{l}w^* \qquad B = \frac{a^2\Delta p}{l}\mu_0\left(\frac{\sigma_1}{\nu\rho}\right)^{1/2}B^* \tag{117}$$

Then, we obtain the dimensionless system of equations in the region S_1

$$\frac{\partial^2 w^*}{\partial \xi^2} + \frac{\partial^2 w^*}{\partial \eta^2} + \mathbf{Ha}\frac{\partial B_1^*}{\partial \eta} = -1 \qquad \frac{\partial^2 B_1^*}{\partial \xi^2} + \frac{\partial^2 B_1^*}{\partial \eta^2} + \mathbf{Ha}\frac{\partial w^*}{\partial \eta} = 0 \tag{118}$$

where the dimensionless number

$$\mathbf{Ha} = B_0 a\left(\frac{\sigma_1}{\nu\rho}\right)^{1/2} \tag{119}$$

is called the Hartmann number.

In the second region (S_2), we have the Laplace equation

$$\frac{\partial^2 B_2^*}{\partial \xi^2} + \frac{\partial^2 B_2^*}{\partial \eta^2} = 0 \tag{120}$$

As has already been indicated, the third region (S_3) will have

$$w = 0 \qquad B_3^* = 0$$

Let us now compose the boundary conditions on the boundaries C_1 and C_2 between the regions. In addition to the obvious hydrodynamic condition

$$w = 0 \quad (\text{on } C_1)$$

which coincides with the conventional condition of the hydrodynamic problem, and to the prescribed pressure drop Δp over the length l or the rate of volume flow Q (average velocity w_{av}), the electrodynamic boundary conditions that are derived from the conditions of conjugation (continuity) of the tangent component E_t on the contour and of the value B also should be composed.

On the contour C_1, which is the common boundary of the regions S_1 and S_2, the velocity $w = 0$; therefore, besides the condition $B_1 = B_2$, we also have, according to the continuity condition $(E_t)_1 = (E_t)_2$ and Ohm's law (110), the condition $(j_t/\sigma)_1 = (j_t/\sigma)_2$. This condition can be easily expressed in terms of the condition of conjugation of the normal derivatives of the function B when, by using Equation 112, the replacement $j_t = (1/\mu_0)(\partial B/\partial n)$ is made in the previous condition of conjugation. Then, we have the following final form of the condition of conjugation on the boundary C_1 (according to an earlier assumption, the magnetic permeability μ_0 is identical in all the regions)

$$\frac{1}{\sigma_1}\frac{\partial B_1}{\partial n} = \frac{1}{\sigma_2}\frac{\partial B_2}{\partial n} \tag{121}$$

Due to the continuity of B, we have $B_2 = B_3 = 0$ on the contour C_2.

In order to simplify the problem with the conditions of conjugation (above) on the boundaries of the regions, use is made of the adopted smallness of the ratio between the thickness of the region S_2 the tube wall thickness δ and the characteristic dimension a of the region S_1. Neglecting the curvature of the tube walls in the region S_2, we adopt the linear function as an approximate solution of Laplace equation (120) for B_2; i.e., to assume that in the region S_2

$$\frac{\partial B_2}{\partial n} = -B_2\frac{a}{\delta}$$

and on the boundary C_1, where $B_2 = B_1$, that

$$\frac{\partial B_2}{\partial n} = -B_1\frac{a}{\delta}$$

This assumption will be rigorous for the case of a plane tube with any thickness of the walls, whereas in the approximation adopted it is valid only locally.

Thus, according to Equation 121, the following approximate system of boundary conditions in dimensionless coordinates is adopted:

$$w^* = 0 \qquad \frac{\partial B_1^*}{\partial n} + \frac{B_1^*}{\varphi} = 0 \quad (\text{on } C_1) \qquad B_2^* = 0 \quad (\text{on } C_2) \tag{122}$$

where the notation

$$\varphi = \frac{\sigma_2\delta}{\sigma_1 a} \tag{123}$$

is used.

If $\sigma_2 = 0$, which corresponds to the case in which an insulator serves as the tube wall, then $\varphi = 0$ and, consequently, according to Equation 122, we have $B_1^* = 0$ on the contour C_1. In this case, the problem is simplified and reduced to the solution of the set of differential equations (118) under the boundary conditions

$$w^* = 0 \qquad B_1^* = 0 \quad (\text{on } C_1)$$

In the opposite case ($\sigma_2 = \infty$), in which the tube walls have absolute conductivity, the coefficient $\varphi = \infty$ and the boundary conditions for the system of equations (118) are

$$w^* = 0 \qquad \partial B_1^*/\partial n = 0 \quad (\text{on } C_1)$$

The above statement of the problem of the motion of a viscous incompressible and electrically conductive fluid through a cylindrical (prismatic) tube with the arbitrary cross-section is rather general, since, along with the hydrodynamic generality, it also contains the possibility for arbitrary assignment of the quantity φ, which characterizes the relative electric conductivity of a fluid and tube walls.

To illustrate consider the most simple problem for which, the previous statement is rather rigorous; i.e., the problem of the motion of an electrically conducting fluid in a plane tube.

Just as in Section 6, we locate the infinite planes, which represent the tube walls, perpendicularly to the axis $0y$ and, thus, perpendicularly to the external magnetic field at distances $\pm h$ from the symmetry plane of the tube $0xz$. Then, the sought-after quantities will not depend on x or, in dimensionless variables, on ξ. It will be possible to conclude from Equation 120 that the function B_2^* is strictly linear with respect to the variable η.

Equations 118 will be reduced to a system of two ordinary second-order equations with constant coefficients

$$\frac{\partial^2 w^*}{\partial \eta^2} + \mathbf{Ha}\frac{dB^*}{d\eta} = -1 \qquad \frac{\partial^2 B^*}{\partial \eta^2} + \mathbf{Ha}\frac{dw^*}{d\eta} = 0 \tag{124}$$

which must be solved under the boundary conditions (the characteristic length a will be taken to be half the distance h between the plane walls of the tube)

$$w^* = 0 \qquad \frac{dB^*}{d\eta} \pm \frac{B^*}{\varphi} = 0 \quad \text{at} \quad \eta = \pm 1 \tag{125}$$

The solution is

$$w^* = \frac{1}{\mathbf{Ha}}\frac{\varphi + 1}{\mathbf{Ha}\,\varphi + \operatorname{th}\mathbf{Ha}}\left[1 - \frac{\operatorname{ch}(\mathbf{Ha}\,\eta)}{\operatorname{ch}\mathbf{Ha}}\right]$$

$$B^* = -\frac{\eta}{\mathbf{Ha}} + \frac{1}{\mathbf{Ha}}\frac{\varphi + 1}{\mathbf{Ha}\,\varphi + \operatorname{th}\mathbf{Ha}}\frac{\operatorname{sh}(\mathbf{Ha}\,\eta)}{\operatorname{ch}\mathbf{Ha}} \tag{126}$$

The first solution of this problem relating to the specific case of a nonconducting wall ($\varphi = 0$) was suggested in 1937 by Hartmann. He was also the first to note the main effect of the presence of magnetic field transverse to the flow. With an increasing magnetic induction B_0, more precisely the Hartmann number (defined by Equation 119), the velocity profiles in the sections of a plane tube become increasingly steeper or, as is sometimes said, the profiles become "filled".

Let us calculate the section-average velocity w_{av}. First, determine the rate of volume flow

$$Q = \int_{-h}^{h} w\,dy = \frac{h^3}{\nu\rho}\frac{\Delta p}{l}\int_{-1}^{1} w^* d\eta$$

Substituting the value of w^* found from the first of Equations 126,

$$Q = \frac{2h^3}{\nu\rho}\frac{\Delta p}{l}\frac{\varphi + 1}{\mathbf{Ha}^2}\frac{\mathbf{Ha} - \operatorname{th}\mathbf{Ha}}{\mathbf{Ha}\,\varphi + \operatorname{th}\mathbf{Ha}} \tag{127}$$

Compare this expression with the equation corresponding to it when $\varphi = 0$ and $\mathbf{Ha} = 0$, i.e., Equation 63. Noting that

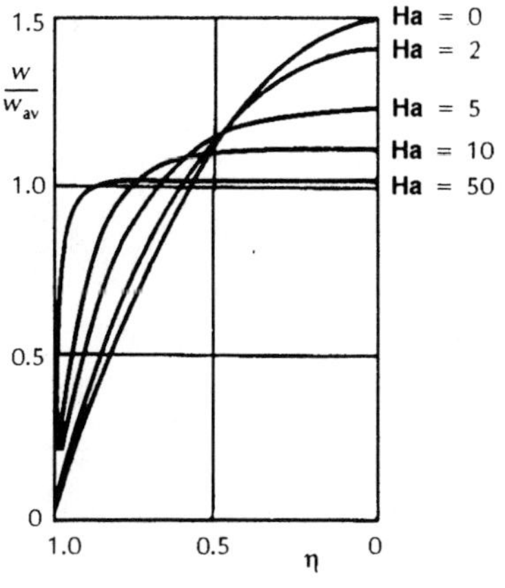

Fig. 154

$$\text{th}\,\mathbf{Ha} = \mathbf{Ha} - \frac{1}{3}\mathbf{Ha}^3 + \cdots$$

we see that evaluating of this limit leads to Equation 63. From formula (127), we determine the average velocity

$$w_{av} = \frac{Q}{2h} = \frac{h^2}{\nu\rho}\frac{\Delta p}{l}\frac{\varphi + 1}{\mathbf{Ha}^2}\frac{\mathbf{Ha} - \text{th}\,\mathbf{Ha}}{\mathbf{Ha}\,\varphi + \text{th}\,\mathbf{Ha}} \tag{128}$$

From this, we find the dimensionless velocity

$$\frac{w}{w_{av}} = \frac{\mathbf{Ha}}{\mathbf{Ha} - \text{th}\,\mathbf{Ha}}\left[1 - \frac{\text{ch}\,(\mathbf{Ha}\,\eta)}{\text{ch}\,\mathbf{Ha}}\right] \tag{129}$$

Note that w/w_{av} is independent of φ, i.e., of the conductivity of the walls. Figure 154, where the ratio of the dimensional velocity w to its average value over the section w_{av} is plotted on the ordinate axis and the dimensionless distance η from the pipe axis is laid off on the abscissa axis, shows the effect (noted earlier) of the influence of the transverse magnetic field on the shape of velocity profiles in the sections of a plane tube. This effect, independent, of the conductivity of the walls, is called the Hartmann effect.

It can be easily verified that when $\mathbf{Ha} \to 0$, velocity profiles (129) convert into an ordinary second-order parabola (Equation 61); the ratio w_{av}/w_{max} which characterizes the "fullness" of the profiles, is equal to $2/3$ at $\mathbf{Ha} = 0$, and tends to unity when $\mathbf{Ha} \to \infty$, thus corresponding to a uniform profile over the entire section excluding the near-wall region thickness $1/\mathbf{Ha}$.

Let us introduce, as in the absence of magnetic field, the formula of resistance in the form

$$\Delta p = \lambda\frac{l}{2h}\frac{\rho w_{av}^2}{2} \qquad \left(\mathbf{Re} = \frac{w_{av}2h}{\nu}\right)$$

Substituting the value of Δp from Equation 128 into the left side of this equation and making simple cancellations, we find the law of resistance for the motion of an electrically conducting fluid in a plane tube located in a transverse magnetic field

$$\lambda = \frac{8}{\mathbf{Re}} \frac{\mathbf{Ha}^2}{\varphi + 1} \frac{\mathbf{Ha}\,\varphi + \text{th}\,\mathbf{Ha}}{\mathbf{Ha} - \text{th}\,\mathbf{Ha}} \tag{130}$$

The limiting transition for $\mathbf{Ha} \to 0$ leads to the law of resistance (66) in the absence of a magnetic field

$$(\lambda)_{\mathbf{Ha}=0} = \lambda_0 = \frac{24}{\mathbf{Re}}$$

The ratio

$$\frac{\lambda}{\lambda_0} = \frac{1}{3} \frac{\mathbf{Ha}^2}{\varphi + 1} \frac{\mathbf{Ha}\,\varphi + \text{th}\,\mathbf{Ha}}{\mathbf{Ha} - \text{th}\,\mathbf{Ha}} \tag{131}$$

represents the function of φ and **Ha**, the diagram of which is shown in Figure 155. It can be noted that the ratio λ/λ_0 coincides with the inverse ratio Q/Q_0 where $Q_0 = (Q)_{\mathbf{Ha}=0}$.

Some details of the analysis of electric current density distribution can be found in the work by Chang and Lundgren (cited earlier).

The consideration of the flow of a conducting fluid in a prismatic tube of rectangular cross-section presents no fundamental difficulties as compared to the case of a nonconducting fluid (Section 6).

The work by Sloan and Smith[26] considers (Figure 156) the motion of an electrically conducting fluid (the coefficient of electric conductivity σ_1 in the direction perpendicular to the plane of the figure and through a channel of rectangular cross-section S_1 between two plane conducting walls (the coefficient of electric conductivity σ_2) perpendicular to the axis $0y$ along which the external homogeneous magnetic field with the induction B_0 is directed. The thickness of the conducting walls can be considered finite and have the same order as the channel height $2h$ and equal to $(q-1)h$. The two other walls are nonconducting.

With the dimensionless quantities (117) being preserved, the problem will be reduced to the necessity of integrating the system of differential equations (118) and (120) under the boundary conditions

$$w^* = 0 \quad \text{when} \quad \xi = \pm æ \quad |\eta| < 1 \quad \text{and} \quad \eta = \pm 1 \quad |\xi| < æ$$

$$B_1^* = 0 \quad \text{when} \quad \xi = \pm æ \quad |\eta| < 1$$

$$B_2^* = 0 \quad \text{when} \quad \xi = \pm æ \quad 1 < |\eta| < q \quad \text{and} \quad \eta = \pm 1 \quad |\xi| < æ$$

$$\sigma_2 \frac{\partial B_1^*}{\partial \eta} = \sigma_1 \frac{\partial B_2^*}{\partial \eta} \quad B_1^* = B_2^* \quad \text{when} \quad \eta = \pm 1 \quad |\xi| < æ \tag{132}$$

Following the method applied in Section 6 for the flow of a nonconducting fluid in a prismatic tube of rectangular profile in the absence of magnetic field, we seek the solution of Equations 183 and 120 in the form of the Fourier series

$$w^* = \sum_{n=0}^{\infty} f_n(\eta) \cos(a_n \xi)$$

---

[26] Sloan, D. M. and Smith, P., Magnetohydrodynamic flow in a rectangular pipe between conducting plates, *Z. Angew. Math. Mech.*, 46, 439–443, 1966.

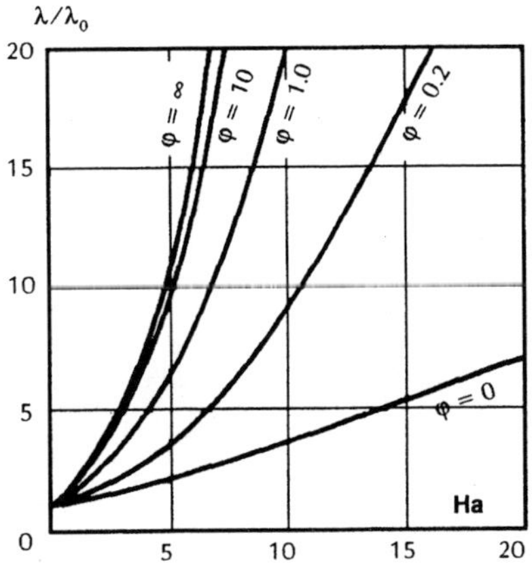

Fig. 155

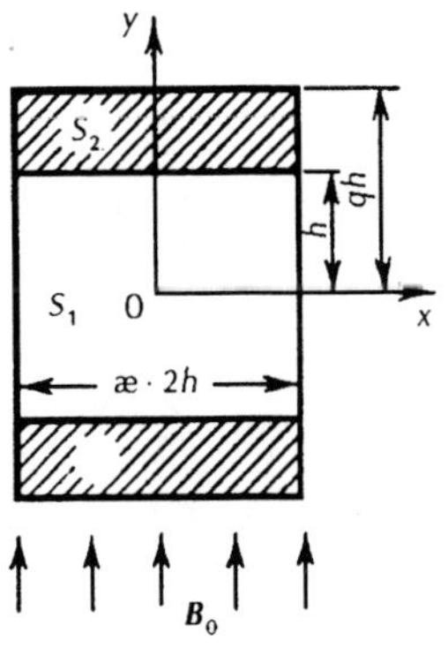

Fig. 156

$$B_1^* = \sum_{n=0}^{\infty} h_n(\eta)\cos(a_n\xi)$$

$$B_2^* = \begin{cases} \sum_{n=0}^{\infty} h_n(\eta)\cos(a_n\xi) & 1 < \eta < q \\ \sum_{n=0}^{\infty} s_n(\eta)\cos(a_n\xi) & -q < \eta < -1 \end{cases} \tag{133}$$

where it is assumed that

$$a_n = \left(n + \frac{1}{2}\right)\frac{\pi}{æ} \tag{134}$$

the boundary conditions for ξ have already been fulfilled.

Using also, as in Section 6, the expansion

$$1 = \frac{4}{\pi}\sum_{n=0}^{\infty}\frac{(-1)^n}{2n+1}\cos(a_n\xi) \quad \text{when} \quad |\xi| < æ$$

and substituting it and series (133) into Equations 118 and 120, we obtain, after equating the coefficients at the cosines with identical arguments, the following system of ordinary differential second-order equations with constant coefficients (the prime is the derivative with respect to η):

$$f_n'' - a_n^2 f_n + \mathbf{Ha}\, g_n' = -\frac{4}{\pi}\frac{(-1)^n}{2n+1}$$

$$g_n'' - a_n^2 g_n + \mathbf{Ha} f_n' = 0$$

$$h_n'' - a_n^2 h = 0 \qquad s_n'' - a_n^2 s_n = 0 \tag{135}$$

with the boundary conditions

$$h_n = 0 \quad \text{when} \quad \eta = q$$

$$f_n = 0 \quad g_n = h_n \quad \sigma_2 g_n' = \sigma_1 h_n' \quad \text{when} \quad \eta = 1$$

$$f_n = 0 \quad g_n = s_n \quad \sigma_2 g_n' = \sigma_1 s_n' \quad \text{when} \quad \eta = -1$$

$$s_n = 0 \quad \text{when} \quad \eta = -q \tag{136}$$

Not dwelling on the details of the solution of this mathematically relatively simple problem, we present in the final form the expressions for the dimensionless velocity w^* and magnetic induction of the field B_1^* in the region S_1 of fluid flow

$$w^* = \sum_{n=0}^{\infty} \frac{(-1)^n 2}{æ a_n^3}\left(1 - \frac{D_n \text{ch } \alpha_n \eta - E_n \text{ch } \beta_n \eta}{D_n \text{ch } \alpha_n - E_n \text{ch } \beta_n}\right) \cos a_n \xi$$

$$B_1^* = \sum_{n=0}^{\infty} \frac{(-1)^n 2}{æ a_n^3} \frac{E_n \text{sh } \beta_n \eta - D_n \text{sh } \alpha_n \eta}{D_n \text{ch } \alpha_n - E_n \text{ch } \beta_n} \cos a_n \xi$$

It is assumed here that

$$D_n = \sigma_1 a_n \text{sh } \beta_n - \sigma_2 \beta_n \text{th } [a_n (1-q)] \text{ ch } \beta_n$$

$$E_n = \sigma_1 a_n \text{sh } \alpha_n - \sigma_2 \alpha_n \text{th } [a_n (1-q)] \text{ ch } \alpha_n$$

$$\alpha_n, \beta_n = \frac{1}{2}\left(-\mathbf{Ha} \pm \sqrt{\mathbf{Ha}^2 + 4a_n^2}\right)$$

We shall satisfy ourselves with the statement of the general form of the solution, referring those interested in the details to the paper by Sloan and Smith.[26]

The efficiency of the Fourier method in constructing the above-mentioned solution depends on the rapidity with which the series converge. To obtain numerical results it is necessary that there be a rather rapid convergence of these series in the practically interesting ranges of Hartmann number and other physical parameters characteristic for separate specific problems. At very great Hartmann numbers, special asymptotic solutions can be constructed.

It should be noted that the Fourier method is not the unique method for solving this kind of problem. Grinberg,[27] by applying Green's functions, derives integral equations, the numerical solutions of which can be made with the aid of successive approximations. The problem

[27] See the two works cited earlier of Grinberg.

of the efficiency of the method is certainly solved by considering the rapidity with which the approximations converge.

9. Pulsating Laminar Motion of a Viscous Fluid through a Cylindrical Tube of Circular Profile

The problem of unsteady motion of a viscous fluid through a cylindrical pipe of circular profile has long attracted the attention of investigators. The simplest case was considered by Helmholtz as early as 1879.[28] For any initial conditions and prescribed law of the time dependence of pressure drop in a tube, the problem was systematically investigated in the work of I. S. Gromeka in 1882.[29] Specific cases of the same problem were developed by other authors.[30]

This section presents the solution of the problem of developed pulsating motion of a viscous fluid in a circular tube under the action of the pressure drop harmonically varying with time.

The basic differential equation of laminar unsteady motion in a cylindrical tube of circular cross-section is obtained as prescribed in Section 6, the only difference being that the local acceleration $\partial w/\partial t$ is preserved on the left.

We have in cylindrical coordinates[31]

$$\frac{\partial w}{\partial t} - \nu\left(\frac{\partial^2 w}{\partial r^2} + \frac{1}{r}\frac{\partial w}{\partial r}\right) = \frac{1}{\rho} f(t) \tag{137}$$

where it is assumed that

$$-\frac{\partial p}{\partial z} = f(t)$$

This equation should be integrated under the boundary condition

$$w = 0 \quad \text{when} \quad r = a \tag{138}$$

and under the initial condition (in the case of axisymmetric motion)

$$w = w_0(r) \quad \text{when} \quad t = 0 \tag{139}$$

As an example, consider the developed pulsating motion corresponding to the harmonic law of pressure drop variation in a tube

$$f(t) = \rho A \cos \omega t$$

In this case the initial condition loses importance and only boundary condition (138) is preserved. Assuming that

---

[28] Helmholtz, H., Über electrische Grenzschichten, *Ann. Phys. Chem.*, 7, 337–382, 1879.

[29] Gromeka, I. S., Concerning the theory of fluid motion in narrow cylindrical tubes, *Uchyon. Zapiski Kazan. Univ.*, 1882, and *Collected Works*, AN USSR Press, Moscow, 1952, pp. 149–171.

[30] Szymanski, P., Quelques solutions exactes des equations de l'hydrodynamique de fluide visqueux dans un tube cylindrique, *J. Math.*, 11, 67–107, 1932; Luriye, A. I., *Operatsionnoye ischisleniye* (*Operational Calculus*), ONTI Press, Moscow, 1935, 205–209; Lambosi, P., Forced vibrations of an incompressible viscous fluid in a rigid horizontal tube, in *Mechanics*, No. 3, IL Press, Moscow, 1953, 67–77.

[31] As in the steady-state case, the velocity field depends not on z, but only on r and t; the pressure p is a linear function of z.

$$w(r,t) = \varphi(r,t) + \frac{A}{\omega}\sin \omega t \tag{140}$$

and introducing the new variable $\tau = \nu\tau$ instead of t, we rewrite Equation 137 in the form

$$\frac{\partial \varphi}{\partial \tau} = \frac{\partial^2 \varphi}{\partial r^2} + \frac{1}{r}\frac{\partial \varphi}{\partial r} \tag{141}$$

We correspondingly transform boundary condition (138) into

$$\varphi(a,\tau) = -\frac{A}{\omega}\sin\frac{\omega}{\nu}\tau \tag{142}$$

Equation 141 relates to the parabolic type and coincides with the heat conduction equation. The problem posed is equivalent to the problem of heat propagation in an infinite cylinder on the surface of which the temperature pulsates with time according to the law (142).

Let us set up the particular solution of Equation 141 in the form of the product of the functions of r and τ

$$\varphi = R(r)\,e^{-i\lambda\tau} \tag{143}$$

where λ is still undetermined but real quantity. Substituting Equation 143 into Equation 141, we obtain, for determining $R(r)$, the ordinary second-order linear equation

$$\frac{d^2R}{dr^2} + \frac{1}{r}\frac{dR}{dr} + i\lambda R = 0 \tag{144}$$

Its particular solution, of course at $r = 0$ (on the tube axis), is[32]

$$R(r) = J_0(r\sqrt{\lambda i}) = \operatorname{ber}(r\sqrt{\lambda}) - i\operatorname{bei}(r\sqrt{\lambda}) \tag{145}$$

Here, $J_0(x\sqrt{i})$ is the zero-order Bessel function of the complex argument, whereas the Kelvin real functions $\operatorname{ber}(x)$ and $\operatorname{bei}(x)$ represent the real and imaginary (taken with the inverse sign) parts of $J_0(x\sqrt{i})$ (Re is the real part, Im is the imaginary part)

$$\begin{aligned} \operatorname{ber}(x) &= \operatorname{Re} J_0(x\sqrt{i}) = 1 - \frac{x^4}{2^2 4^2} + \frac{x^8}{2^2 4^2 6^2 8^2} - \cdots \\ \operatorname{bei}(x) &= -\operatorname{Im} J_0(x\sqrt{i}) = \frac{x^2}{2^2} - \frac{x^6}{2^2 4^2 6^2} + \frac{x^{10}}{2^2 4^2 6^2 8^2 10^2} - \cdots \end{aligned} \tag{146}$$

Introducing the complex constant of integration $B - iC$, we compose the general solution of Equation 141 in the form of the real part of the expression

$$\varphi(r,\tau) = \operatorname{Re}\{(B - iC)[\operatorname{ber}(r\sqrt{\lambda}) - i\operatorname{bei}(r\sqrt{\lambda})](\cos\lambda\tau - i\sin\lambda\tau)\} =$$

[32] Gray, A. and Mathews, G. B., *Funktsii Besselya i ikh prilozheniya v fizike i mekhanike (The Bessel Functions and Their Applications in Physics and Mechanics)*, IL Press, Moscow, 1949, 39; our formula (145) can be easily derived from formula (64) by transition from the function I_0 to J_0 of the corresponding argument. The tables of the functions $\operatorname{ber}(x)$ and $\operatorname{bei}(x)$ can be found in Janke, E. and Emde, F. *Tablitsy funktsiy (Tables of Functions)*, Nauka Press, Moscow, 1977.

$$= B\,[\mathrm{ber}\,(r\sqrt{\lambda})\cos\lambda\tau - \mathrm{bei}\,(r\sqrt{\lambda})\sin\lambda\tau] -$$

$$- C\,[\mathrm{bei}\,(r\sqrt{\lambda})\cos\lambda\tau + \mathrm{ber}\,(r\sqrt{\lambda})\sin\lambda\tau] \qquad (147)$$

Substituting this expression into boundary condition (142) and equating the arguments of trigonometric functions and the coefficients at the identical trigonometric functions, we obtain a system of equations for determining λ and the constants B and C

$$\lambda = \frac{\omega}{\nu}$$

$$B\,\mathrm{ber}\left(a\sqrt{\frac{\omega}{\nu}}\right) - C\,\mathrm{bei}\left(a\sqrt{\frac{\omega}{\nu}}\right) = 0 \qquad B\,\mathrm{bei}\left(a\sqrt{\frac{\omega}{\nu}}\right) + C\,\mathrm{ber}\left(a\sqrt{\frac{\omega}{\nu}}\right) = \frac{A}{\omega}$$

Calculating the constants, we find

$$B = \frac{A}{\omega}\,\frac{\mathrm{bei}\left(a\sqrt{\frac{\omega}{\nu}}\right)}{\mathrm{ber}^2\left(a\sqrt{\frac{\omega}{\nu}}\right) + \mathrm{bei}^2\left(a\sqrt{\frac{\omega}{\nu}}\right)}$$

$$C = \frac{A}{\omega}\,\frac{\mathrm{ber}\left(a\sqrt{\frac{\omega}{\nu}}\right)}{\mathrm{ber}^2\left(a\sqrt{\frac{\omega}{\nu}}\right) + \mathrm{bei}^2\left(a\sqrt{\frac{\omega}{\nu}}\right)} \qquad (148)$$

Returning to the variable $t = \tau/\nu$ and substituting the resulting value of $\varphi(r, \tau)$ from Equation 147 into Equation 140, we finally obtain the following expression for the distribution of velocities:

$$w(r, t) = \frac{A}{\omega}\left\{\left[1 - \frac{\mathrm{bei}\left(a\sqrt{\frac{\omega}{\nu}}\right)\mathrm{bei}\left(r\sqrt{\frac{\omega}{\nu}}\right)}{\mathrm{ber}^2\left(a\sqrt{\frac{\omega}{\nu}}\right) + \mathrm{bei}^2\left(a\sqrt{\frac{\omega}{\nu}}\right)} - \right.\right.$$

$$\left. - \frac{\mathrm{ber}\left(a\sqrt{\frac{\omega}{\nu}}\right)\mathrm{ber}\left(r\sqrt{\frac{\omega}{\nu}}\right)}{\mathrm{ber}^2\left(a\sqrt{\frac{\omega}{\nu}}\right) + \mathrm{bei}^2\left(a\sqrt{\frac{\omega}{\nu}}\right)}\right]\sin\omega t +$$

$$\left. + \frac{\mathrm{bei}\left(a\sqrt{\frac{\omega}{\nu}}\right)\mathrm{ber}\left(r\sqrt{\frac{\omega}{\nu}}\right) - \mathrm{ber}\left(a\sqrt{\frac{\omega}{\nu}}\right)\mathrm{bei}\left(r\sqrt{\frac{\omega}{\nu}}\right)}{\mathrm{ber}^2\left(a\sqrt{\frac{\omega}{\nu}}\right) + \mathrm{bei}^2\left(a\sqrt{\frac{\omega}{\nu}}\right)}\cos\omega t\right\} \qquad (149)$$

The involved functions were tabulated so that the computation of the velocity profiles at different time instants presents no difficulty.

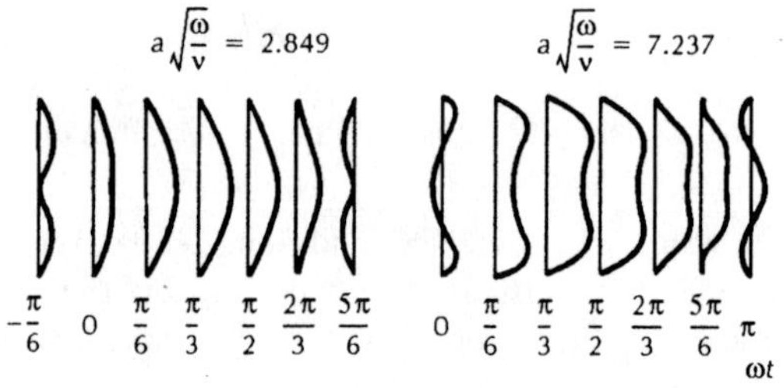

Fig. 157

In order to obtain a certain idea about the character of variation in the velocity profile with time, Figure 157 gives the velocity profiles borrowed from the paper by Lambosi (cited earlier) at the time instants indicated on the abscissas axis for the values of the parameter

$$a\sqrt{\frac{\omega}{\nu}} = 2.849 \quad \text{and} \quad 7.237$$

which are the roots of the function $\operatorname{ber}(a\sqrt{\omega/\nu})$ to ease calculations by formula (149). Judging from these curves, the reverse currents appear under the considered pressure fluctuations in the tube. Also observed is the outstripping of the layers located near the tube axis by the wall layers.

The solution of the problem of setting a quiescent viscous fluid in motion in a circular cylindrical tube under the action of an instantly applied prescribed constant pressure drop can be found in the monograph by Slyozkin.[33]

On the right side of 137, it is necessary to adopt in this case that

$$f(t) = \text{const} = \frac{\Delta p}{l}$$

and the initial and boundary conditions are

$$w = 0 \quad \text{when} \quad t = 0 \qquad w = 0 \quad \text{when} \quad r = a$$

The solution of this problem is reduced to the series that contain Bessel functions. Relating the reader for conclusions to the Slyozkin monograph, we give the formula of velocity distributions

$$w(r, t) = a^2\frac{\Delta p}{4\mu l}\left[1 - \frac{r^2}{a^2} - 8\sum_{k=1}^{\infty}\exp\left(-\frac{\nu\lambda_k^2 t}{a^2}\right)\frac{J_0(\lambda_k r/a)}{\lambda_k^3 J_1(\lambda_k)}\right]$$

In this formula λ_k are the roots of the equation

$$J_0(\lambda_k) = 0$$

and J_0 and J_1 are the Bessel functions of the zero and first orders. The first two terms in the square brackets express the developed-for $(t \to \infty)$ motion and correspond to the earlier-obtained Poiseuille parabola.

---

[33] Slyozkin, N. A., *Dinamika vyazkoy neszhimayemoy zhidkosti (The Dynamics of a Viscous Incompressible Fluid)*, Gostekhizdat Press, Moscow, 1955, 322–326.

We also give the formula of the rate of volume flow

$$Q = \frac{\pi a^4 \Delta p}{8\mu l}\left[1 - 32\sum_{k=1}^{\infty}\frac{\exp\left(-\nu\lambda_k^2 t/a^2\right)}{\lambda_k^4}\right]$$

All the examples given in this chapter for motion of a viscous incompressible fluid were based on the precise solution of the Navier-Stokes equations, which were linear in view of the absence of convective acceleration in a rectilinear developed flow. Approximate solutions obtained due to the neglect of convective terms and thus to linearization of the Navier-Stokes equations, just as some precise self-similar solutions of nonlinearized equations and numerical solutions of nonsimilar problems, are considered in Chapter 11.

10. Dissipation of Mechanical Energy in the Flow of a Viscous Fluid: Variational Principle of Helmholtz

In contrast to the case of the steady motion of an ideal fluid, the total mechanical energy in the flow of a viscous fluid is not preserved, but dissipates in space. This phenomenon is called the dissipation of mechanical energy. The dissipation occurs at the expense of that part of work (power) of internal forces which is determined by viscosity.

To elucidate the quantitative aspect of this phenomenon, recall that according to the second law of thermodynamics the irreversible losses of mechanical energy in the form of its conversion into heat are associated with the growth of specific entropy s of the medium. In turn, the increment of entropy is determined in accordance with the basic (first) law of thermodynamics by the equality $T ds = c_v dT + p d\left(\frac{1}{\rho}\right) = dU + p d\left(\frac{1}{\rho}\right)$, from which it follows for an incompressible fluid that

$$T\frac{ds}{dt} = \frac{dU}{dt} \tag{150}$$

Now, let us consider the equation of the internal energy balance, expressed in the general form by Equation 22 of Chapter 4. Then,

$$\rho T\frac{ds}{dt} = \rho q - N_{\text{in}} \tag{151}$$

According to this equation, the change in entropy in the flow of a viscous fluid occurs for two reasons: either at the expense of the inflow of heat q, or of the lost power of the internal forces $(-N_{\text{in}})$. Neglecting the inflow of heat from outside, i.e., considering the motion to be adiabatic ($q = 0$), we shall prove that during the flow of an incompressible viscous fluid, which satisfies the generalized Newton law (9), the quantity $(-N_{\text{in}})$ is non-negative and, according to Equation 151, it generally corresponds to the entropy increase.

According to Equation 19 of Chapter 4 and Equation 9 of this Chapter,

$$-N_{\text{in}} = P \cdot \dot{S} = 2\mu\dot{S} \cdot \dot{S} - pE \cdot \dot{S}$$

or, noting that for an incompressible fluid (in letter indexation)

$$E \cdot \dot{S} = \frac{\partial u}{\partial x} + \frac{\partial v}{\partial y} + \frac{\partial w}{\partial z} = \operatorname{div} V = 0$$

$$-N_{\text{in}} = 2\mu \dot{S}^2 \geq 0 \tag{152}$$

where when composing this sum it was taken into account that the squares of identical nondiagonal terms are united in pairs

$$\dot{S}^2 = \dot{S} \cdot \dot{S} = \left(\frac{\partial u}{\partial x}\right)^2 + \left(\frac{\partial v}{\partial y}\right)^2 + \left(\frac{\partial w}{\partial z}\right)^2 +$$

$$+ \frac{1}{2}\left(\frac{\partial u}{\partial y} + \frac{\partial v}{\partial x}\right)^2 + \frac{1}{2}\left(\frac{\partial v}{\partial z} + \frac{\partial w}{\partial y}\right)^2 + \frac{1}{2}\left(\frac{\partial w}{\partial x} + \frac{\partial u}{\partial z}\right)^2 \tag{153}$$

Equation 151 acquires the form

$$\rho T \frac{ds}{dt} = -N_{\text{in}} = 2\mu \dot{S}^2 \tag{154}$$

from which it follows directly that the power of the internal forces of viscosity causes an increase of entropy, so that to the work of viscosity forces corresponds the irreversible transition of mechanical energy into heat – its dissipation. For the power (taken with the minus sign) of the internal forces of viscosity ($-N_{\text{in}}$) we introduce the notation N_{dis} having assumed that

$$N_{\text{dis}} = 2\mu \dot{S}^2 \tag{155}$$

where $\dot{S}^2$ is defined by the right side of Equation 153. The energy dissipated per unit time in the finite volume τ is defined by the integral

$$\bar{N}_{\text{dis}} = \int_\tau N_{\text{dis}}\, d\tau = 2\mu \int_\tau \dot{S}^2 d\tau \tag{156}$$

Formula 155 makes it possible to answer the question whether such motions of a viscous fluid exist, in which the mechanical energy is not scattered (is not dissipated) throughout the flow.

It should be emphasized that the dissipated energy, as the sum of the squares, is the nonnegative quantity that corresponds to the positive nature of the entropy increment which expresses the irreversibility of the viscous fluid flow mechanical energy transition into heat. From expressions (153) and (155) it follows that the only motion of a viscous incompressible fluid not accompanied by the dissipation of mechanical energy is its quasisolid motion in which all the terms in square brackets (the squared deformation) convert into zero individually.

The concept of dissipated energy constituted the basis of the principle of "minimum dissipated energy" established by Helmholtz[34] and valid for every "slow" steady motion which admits the discarding of inertia terms in the equations of motion of an incompressible viscous fluid under the action of the conservative field of volumetric forces.

Let us compare the actual "slow" motion in a certain region bounded by the closed surface σ with an arbitrary other motion of the same (incompressible, viscous, Newtonian) fluid

---

[34] Helmholtz, H., Zur Theorie der stationären Ströma in reibenden Flussigkeiten, *Verhandl. Naturalist. Med. Vereines*, Oktober 30, 1868.

coinciding with the former as to velocities on the surface σ. Let us designate by V, P, and $\dot{S}$ the velocity vector, the tensors of stresses, and tensors of deformation rates in actual motion, and by primed letters the differences between these quantities for the arbitrary and actual motions, so that for the arbitrary motion the velocity vector and the tensors of stresses and of deformation rates will be equal to

$$V + V' \qquad P + P' \qquad \dot{S} + \dot{S}'$$

Let us compose, according to Equation 156, an expression for the power $\overline{N}^*_{\text{dis}}$ dissipated in arbitrary motion of a fluid in a volume bounded by the surface σ; we find

$$\overline{N}^*_{\text{dis}} = \int_\tau (P + P') \cdot (\dot{S} + \dot{S}')\, d\tau =$$

$$= \int_\tau P \cdot \dot{S} d\tau + \int_\tau P \cdot \dot{S}' d\tau + \int_\tau P' \cdot \dot{S} d\tau + \int_\tau P' \cdot \dot{S}' d\tau \tag{157}$$

The first integral represents the power $\overline{N}_{\text{dis}}$, which was dissipated in actual motion. The second and third integrals are equal between themselves. In fact, according to Equation 9:

$$P \cdot \dot{S}' = 2\mu \dot{S} \cdot \dot{S}' - pE \cdot \dot{S}' \qquad P' \cdot \dot{S} = 2\mu \dot{S}' \cdot \dot{S} - p' E \cdot \dot{S}$$

for an incompressible fluid $E \cdot \dot{S}' = \text{div}\, V' = 0$, $E \cdot \dot{S} = \text{div}\, V = 0$. Consequently

$$P \cdot \dot{S}' = P' \cdot \dot{S}$$

We shall prove that under the assumptions made these two integrals are equal to zero. For this, calculate the expression (summation over the repeated index!)

$$V' \cdot \text{Div}\, P - \text{div}\,(PV') = V_i' \frac{\partial p_{ij}}{\partial x_j} - \frac{\partial}{\partial x_j}(p_{ij} V_i') = -p_{ij} \frac{\partial V_i'}{\partial x_j} =$$

$$= -\frac{1}{2} p_{ij} \left(\frac{\partial V_i'}{\partial x_j} + \frac{\partial V_j'}{\partial x_i} \right) - \frac{1}{2} p_{ij} \left(\frac{\partial V_i'}{\partial x_j} - \frac{\partial V_j'}{\partial x_i} \right) = -p_{ij} \dot{S}_{ij}' = -P \cdot \dot{S}'$$

after which

$$\int_\tau P \cdot \dot{S}' d\tau = \int_\tau \text{div}\,(PV') d\tau - \int_\tau V' \cdot \text{Div}\, P d\tau$$

According to the Gauss-Ostrogradskiy formula

$$\int_\tau \text{div}\,(PV') d\tau = \int_\sigma (PV')_n d\sigma = 0$$

because by the condition $V' = 0$ on the surface σ. Then, taking into consideration that the actual slow steady motion in the presence of the conservative field of volumetric forces with the potential Π satisfies the equation in terms of stresses

$$-\operatorname{grad}\Pi + \frac{1}{\rho}\operatorname{Div}P = 0$$

and that due to this (div $V' = 0$ according to the condition of incompressibility)

$$\frac{1}{\rho}V'\cdot\operatorname{Div}P = V'\cdot\operatorname{grad}\Pi = \operatorname{div}(\Pi V') - \Pi\operatorname{div}V' = \operatorname{div}(\Pi V')$$

and again applying the Gauss-Ostrogradskiy formula and using the condition $V' = 0$ on σ,

$$\frac{1}{\rho}\int_\tau V'\cdot\operatorname{Div}P\,d\tau = \int_\tau \operatorname{Div}(\Pi V')\,d\tau = \int_\sigma \Pi V'_n\,d\sigma$$

Finally, the fourth integral can be presented in the form of the non-negative quantity

$$2\mu\int_\tau \dot{S}'^2 d\tau$$

Now, let us return to Equation 157 and rewrite it in the form

$$\bar{N}^* - \bar{N} = 2\mu\int_\tau \dot{S}'^2 d\tau \geq 0 \tag{158}$$

From this, the following principle of Helmholtz follows: the mechanical energy dissipated in actual "slow" steady motion of a viscous incompressible fluid in a certain volume is not greater than that in an analogous arbitrary motion of an incompressible fluid with the same distribution of velocities on the surface bounding this volume.

The principle of Helmholtz can be interpreted as the variational principle of the minimum dissipated energy

$$\delta\int_\tau \dot{S}^2 d\tau = 0 \tag{159}$$

To a certain extent, this principle is an analogue of the principle of the minimum potential energy of deformations that is widely used in the elasticity theory. The Helmholtz principle in the hydrodynamics of a viscous fluid, just as with the principle of the minimum potential energy in the elasticity theory, can be used as the basis for applying direct straight methods of variational calculus for solving the problems of "slow" motion.

The irreversibility of the process of mechanical energy dissipation is responsible for the fact that the viscous fluid set into motion and left to itself scatters (dissipates) the imparted mechanical energy until it comes to the state of rest. The mechanical energy imparted to a certain volume of fluid dissipates not only in the same volume of fluid, but gradually starts to propagate over the entire region occupied by the flow and redistribute in it. This propagation, or dispersion, of mechanical energy is carried out by two different processes. The first consists in simple transport of energy by the fluid flow and is called convection. The second, diffusion, results from the presence of intramolecular transfer in the fluid. The nature of this process is the same as in viscous friction, which represents the macroscopic manifestation of the microscopic (molecular) transfer of momentum.

The molecular transfer (diffusion of mechanical energy) and the analogous transfer of momentum (viscous friction) have a common carrier and a common coefficient of transfer (diffusion) – the kinematic coefficient of viscosity ν. Section 11 describes the processes of transfer of other quantities: vorticity, heat (heat transfer), and mass (mass transfer). The carriers also will be molecules in their thermal motion, but along with the coefficient ν for vorticity, other coefficients of transfer will appear. Thus, in the case of heat transfer, the coefficient of thermal diffusivity will be used; in the case of mass transfer, the coefficient of mass conductivity is used.

11. Diffusion of Vorticity

Let us write out the Navier-Stokes equation in a form analogous to the Gromeka-Lamb equation (32), but use the initial form for the viscous term

$$-\nu \operatorname{rot} \operatorname{rot} \boldsymbol{V} = \nu \nabla^2 \boldsymbol{V}$$

Apply the operation with curl(rot) on the left and right sides of the thus-obtained equation:

$$\frac{\partial \boldsymbol{V}}{\partial t} + \operatorname{rot} \boldsymbol{V} \times \boldsymbol{V} = -\operatorname{grad}\left(\frac{V^2}{2} + \Pi + \mathcal{P}\right) + \nu \nabla^2 \boldsymbol{V} \tag{160}$$

Then, not repeating the manipulations made at the end of Section 1 of Chapter 5 when deriving the Helmholtz equation (16) and also noting that

$$\operatorname{rot}(\nabla^2 \boldsymbol{V}) = \nabla^2 (\operatorname{rot} \boldsymbol{V}) = \nabla^2 \boldsymbol{\Omega}$$

we obtain the equation of dispersion (propagation) of vorticity in a viscous incompressible fluid

$$\frac{\partial \boldsymbol{\Omega}}{\partial t} + (\boldsymbol{V} \cdot \nabla) \boldsymbol{\Omega} = (\boldsymbol{\Omega} \cdot \nabla) \boldsymbol{V} + \nu \nabla^2 \boldsymbol{\Omega} \tag{161}$$

or, in a more brief form by introducing the ordinary derivative of $\boldsymbol{\Omega}$ with respect to t into the left side:

$$\frac{d\boldsymbol{\Omega}}{dt} = (\boldsymbol{\Omega} \cdot \nabla) \boldsymbol{V} + \nu \nabla^2 \boldsymbol{\Omega} \tag{162}$$

Equations 161 and 162 can be considered to be the extension of the Helmholtz equation (16) of Chapter 5 to the case of an incompressible viscous fluid.

The last term on the right represents the diffusion of vorticity proper $\boldsymbol{\Omega}$, the role of diffusion coefficient being played by the kinematic coefficient of viscosity ν.

In the case of a plane flow, the vector $\boldsymbol{\Omega}$ is perpendicular to the plane of the flow, and the quantity $(\boldsymbol{\Omega} \cdot \nabla) \boldsymbol{V}$, proportional to the derivative of $\boldsymbol{V}$ in the direction of the perpendicular to this plane, is equal to zero, so that Equation 162 in the plane flow acquires the form

$$\frac{d\boldsymbol{\Omega}}{dt} = \frac{\partial \boldsymbol{\Omega}}{\partial t} + (\boldsymbol{V} \cdot \nabla) \boldsymbol{\Omega} = \nu \nabla^2 \boldsymbol{\Omega} \tag{163}$$

Let us consider the simplest process of diffusion of a separate rectilinear vortical filament in a liquid quiescent at an infinite distance from it. Let at a certain time instant $t = 0$, an infinite rectilinear vortical filament, be introduced into a viscous incompressible fluid. To this filament, in the plane perpendicular to it, corresponds a vortex with the circulation Γ and velocity field

$$V = \frac{\Gamma}{2\pi r}$$

Such a field can exist both in a viscous and in an ideal fluid because the plane irrotational velocity field with the potential $\varphi = \mathrm{Re}\left(\frac{\Gamma}{2\pi i}\ln z\right)$ satisfies the condition $\nabla^2\varphi = 0$, and the single boundary condition $V \to 0$ for $r \to \infty$ seems to be fulfilled. The only difference is that in an ideal fluid no scattering (dissipation) of energy occurs and the vortex introduced into the ideal fluid will exist infinitely long, whereas in a viscous fluid, in order to sustain motion with the velocity field $V = \Gamma/(2\pi r)$, it is necessary to supply energy continually, for example, by rotating in the fluid a thin cylinder which is the source of vorticity.

Consider the unsteady process that originates if in the initial moment $t = 0$ the source of vorticity is expelled from the motion of a viscous fluid induced by the vortex.

Assuming the motion to be plane and, due to the symmetry, with circular streamlines, we omit in Equation 163 the convective term $\boldsymbol{V} \cdot \mathrm{grad}\,\Omega$ equal to zero, because the vorticity along the streamlines preserves a constant value. Then, Equation 163 projected onto the axis $0z$ acquires the form

$$\frac{\partial\Omega}{\partial t} = \nu\nabla^2\Omega$$

or in polar coordinates at $\partial\Omega/\partial\varepsilon = 0$, according to Equation 85 of Chapter 1,

$$\frac{\partial\Omega}{\partial t} = \frac{\nu}{r}\frac{\partial}{\partial r}\left(r\frac{\partial\Omega}{\partial r}\right)$$

This equation can be rewritten in the form of the well-known equation of the theory of heat propagation

$$\frac{\partial\Omega}{\partial t} = \nu\left(\frac{\partial^2\Omega}{\partial r^2} + \frac{1}{r}\frac{\partial\Omega}{\partial r}\right) \tag{164}$$

Its integral, which satisfies the initial condition $\Omega = 0$ for $t = 0$ and $r > 0$ and the boundary condition $\Omega \to 0$ for $r \to \infty$, is represented by the well-known solution of the "heat source" type

$$\Omega = \frac{A}{t}e^{-\frac{r^2}{4\nu t}} \tag{165}$$

which can be easily verified by substituting this expression into the equation, initial and boundary conditions. In order to find the value of A, we shall avail ourselves of the Stokes theorem and write that at any instant the intensity of the vortex tube of radius r

$$\int_0^r \Omega \cdot 2\pi r\,dr$$

is equal to velocity circulation around the circle of radius r equal to $V \cdot 2\pi r$. Thus,

$$V = \frac{1}{2\pi r}\int_0^r \frac{A}{t} e^{-\frac{r^2}{4\nu t}} 2\pi r\, dr = \frac{2A\nu}{r}\left(1 - e^{-\frac{r^2}{4\nu t}}\right)$$

Comparing with the initial distribution of velocities

$$V = \frac{\Gamma}{2\pi r} \quad \text{when} \quad t = 0, r > 0$$

we shall find

$$A = \frac{\Gamma}{4\pi\nu}$$

Thus, we obtain the final formulas of vortex distribution

$$\Omega = \frac{\Gamma}{4\pi\nu t} e^{-\frac{r^2}{4\nu t}} \tag{166}$$

and of velocity distribution

$$V = \frac{\Gamma}{2\pi r}\left(1 - e^{-\frac{r^2}{4\nu t}}\right) \tag{167}$$

It can be seen easily that the results obtained correspond to the self-similar solution of Equation 164, which exists due to the absence of length and velocity scales in the statement of the problem. The trivial solution of Equation 164, $\Omega \equiv 0$ and $V \equiv 0$, does not satisfy the condition of the specified intensity of the vortex or of the circulation Γ.

Let us analyze the results obtained. At the initial time instant $t = 0$, the motion is irrotational everywhere ($r > 0$). After the removal of the vorticity source (i.e., when $t > 0$), in the entire space there instantly originated a certain distribution of the propagating vorticity which is represented by function 166, rapidly decreasing with an increase of the distance r. The vorticity at the center ($r = 0$) decreases monotonically with an increase of time, whereas at the point at a certain distance from the center, it first increases and then decreases to zero at $t = \infty$.

Let us consider any circle of radius $r = a$. The change of vorticity with time at the points of this circle is presented in the form

$$(\Omega)_{r=a} = \frac{\Gamma}{4\pi\nu t} e^{-\frac{a^2}{4\nu t}}$$

Investigating the maximum-minimum of this function, we conclude that at the time $t_m = a^2/(4\nu)$ the vorticity attains its maximum value

$$\Omega_m = \frac{\Gamma}{4\pi\nu e t_m} = \frac{\Gamma}{\pi a^2 e}$$

With a further increase of time, the vorticity decreases.

The general character of the time-dependence of vorticity at the points located at different distances from the center can be judged from the curves given in Figure 158. The curves of velocity distributions at different successive time instants are presented in Figure 159.

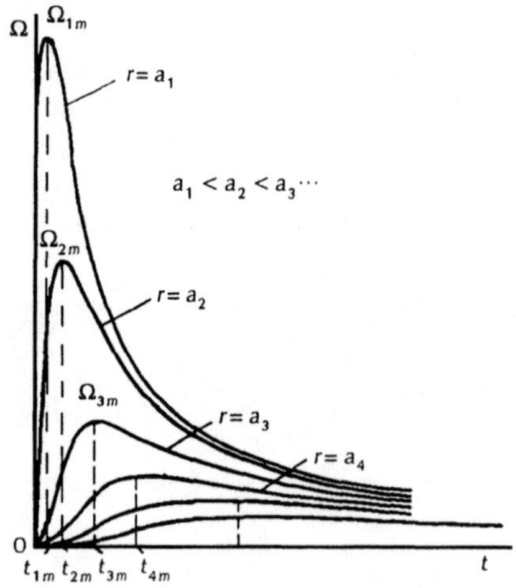

Fig. 158

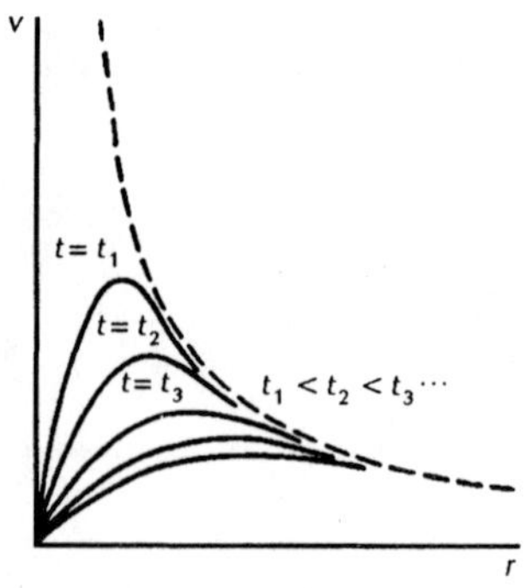

Fig. 159

Using the formula and plots obtained, it is possible to form a general idea about the phenomenon of diffusion of a single vortex in an infinite viscous fluid. Mathematically more complex is the solution of the problem of the diffusion of the finite vortex tube in an infinite viscous fluid and also of plane and cylindrical rotational layers.[35]

In this case Equation 164 remains valid but the initial conditions become more complex and acquire the form

$$\Omega(r,0) = \text{const} = \Omega_0 \quad \text{when} \quad 0 \le r \le a$$
$$\Omega(r,0) = 0 \quad \text{when} \quad r > a$$

Relating those interested to the first Kochin et al. book, we give the final result

$$\Omega(r,t) = \begin{cases} \Omega_0 e^{-\frac{r^2+a^2}{4\nu t}} \sum\limits_{k=1}^{\infty} \left(\dfrac{a}{r}\right)^k \mathrm{I}_k\left(\dfrac{ar}{2\nu t}\right) & \text{when} \quad r \ge a \\ \Omega_0 - \Omega_0 e^{-\frac{r^2+a^2}{4\nu t}} \sum\limits_{k=0}^{\infty} \left(\dfrac{r}{a}\right)^k \mathrm{I}_k\left(\dfrac{ar}{2\nu t}\right) & \text{when} \quad r < a \end{cases} \tag{168}$$

where I_k is the Bessel function defined in terms of the Bessel function $J_k(z)$ with the imaginary argument by the equality

$$\mathrm{I}_k(z) = \frac{1}{i^k} J_k(iz) \tag{169}$$

Wishing to determine the rapidity with which the vorticity is scattered, depending on the initial radius of the vortex tube a, we note that according to the second formula of system (168)

[35] Kochin, N. Ye., Kibel, I. A., and Roze, N. V., *Teoreticheskaya gidromekhanika* (*Theoretical Hydromechanics*), Part 2, Fizmatgiz Press, Moscow, 1963, 453–460. Kochin, N. E., Kibel, J. A., and Rose, N. V., *Theoretische Hydromechanik*, Vol. 1 and 2, Berlin, 1954.

$$\Omega(0,t) = \Omega_0\left(1 - e^{-\frac{a^2}{4\nu t}}\right) \tag{170}$$

In order to estimate the relaxation time τ of the vortex tube of finite radius a, i.e., the time interval after the elapse of which Ω at the center of the tube will attain half its initial value, we find

$$1 - e^{-\frac{a^2}{4\nu\tau}} = \frac{1}{2} \qquad \tau = \frac{a^2}{4\nu \ln 2} \approx 0.36\frac{a^2}{\nu}$$

If the liquid is water ($\nu = 0.01$ cm^2/s), then $\tau = 36\ a^2$, so that at $a = 1$ cm the relaxation time will be equal to 36 s. For air ($\nu = 0.133$ cm^2/s) $\tau = 2.7\ a^2$, i.e., at $a = 1$ cm, $\tau = 2.7$ s. When $a = 1$ m, τ will become equal to 450 min (!). These considerations, borrowed from Kochin,[35] will be useful later when considering relaxational phenomena in fine- and large-scale turbulence.

It should be noted that the more vigorous the diffusion of vortices, the smaller are their dimensions; fine vortices relax most rapidly.

12. Diffusion of Heat and Mass in Incompressible Viscous Fluid Flows

Assuming a viscous fluid to be incompressible, i.e., that which in motion does not vary its density, we shall renounce its isothermicity and homogeneity. At the same time, we shall satisfy ourselves using the simplest case, in which the variability of the temperature and concentration of admixture do not influence the viscosity, thermal conductivity, heat capacity, and other thermodynamic properties of the fluid, and in particular, the diffusion coefficient of the admixture. In the case of weak heatings of the medium we ignore the radiant heat transfer and consider the transfer of heat to be fully effected by heat conduction.

Taking into account the fact that sometimes it is important to include free convection originating in the medium due to the difference of its densities in a nonhomogeneous temperature field, we consider the volumetric Archimedes force

$$\boldsymbol{F} = \frac{\Delta\rho}{\rho}\boldsymbol{g}$$

The concept of the Archimedes force acting on a solid body immersed in a fluid is extended directly to the case of a liquid body of density ρ which differs from the density ρ_0 of the surrounding fluid. The result of the gravity force $\rho \boldsymbol{g}\delta\tau$ of the volume element $\delta\tau$ and of the Archimedes force applied to it ($-\rho_0 \boldsymbol{g}\delta\tau$), being related to the mass element $\rho\delta\tau$, will give that volumetric force

$$\boldsymbol{F} = \frac{\rho \boldsymbol{g}\delta\tau - \rho_0 \boldsymbol{g}\delta\tau}{\rho\delta\tau} = \frac{\rho - \rho_0}{\rho}\boldsymbol{g} = \frac{\Delta\rho}{\rho}\boldsymbol{g}$$

which should be taken into account when setting up the differential equation of free convection of the fluid. Introducing the thermal coefficient of fluid expansion β, defined by the equality

$$\frac{\Delta\rho}{\rho}\boldsymbol{g} = \beta\Delta T$$

where $\Delta\rho$ and ΔT are the "excess" density and temperature, the quantity $\boldsymbol{F}$ can be presented in the form

$$\boldsymbol{F} = \beta \boldsymbol{g} \Delta T \tag{171}$$

The Navier-Stokes equations (30) that characterize the dynamic aspect of the phenomenon have the form

$$\frac{\partial \boldsymbol{V}}{\partial t} + (\boldsymbol{V} \cdot \nabla)\boldsymbol{V} = -\frac{1}{\rho}\operatorname{grad} p + \beta \boldsymbol{g}\Delta T + \nu\nabla^2 \boldsymbol{V}$$

$$\operatorname{div} \boldsymbol{V} = 0 \tag{172}$$

In order to derive the equation of heat propagation in an incompressible medium (the case of the gas was considered in Section 149) we use the general equation of energy balance (22) of Chapter 4, having assumed in it that $U = cT$ where c is the heat capacity of the medium and T is the absolute temperature. For the heat flux vector q related to the unit q_e area and unit time, we use the Fourier law

$$q = -\lambda \operatorname{grad} T \tag{173}$$

where λ is the thermal conductivity of the medium. The specific inflow of heat ρq (per unit volume) from outside to the given point of the medium is determined by the limit

$$\rho q = -\lim_{\tau \to 0} \frac{1}{\tau}\int_\sigma q_n d\sigma = \lim_{\tau \to 0} \frac{1}{\tau}\int_\sigma (\lambda \operatorname{grad} T)_n d\sigma = \operatorname{div}(\lambda \operatorname{grad} T)$$

In the case of a homogeneous medium with the constant thermal conductivity coefficient λ, this inflow of heat is determined by the equality

$$\rho q = \lambda \operatorname{div} \operatorname{grad} T = \lambda \nabla^2 T$$

The power of internal forces in Equation 22 of Chapter 4 in the case of an incompressible viscous fluid is reduced, as shown in Section 10, to the mechanical energy dissipating into heat (Equation 155); at small velocities, more precisely, at small Mach numbers, it is negligible and can be omitted. This follows directly from the last but one dimensionless equation (33) of Chapter 15, in which the terms involving the mechanical energy dissipation are proportional to the squared Mach number.

The equation of heat conduction, sometimes called the heat balance equation, acquires the form

$$\frac{\partial T}{\partial t} + \boldsymbol{V} \cdot \operatorname{grad} T = \frac{\lambda}{\rho c}\nabla^2 T = a\nabla^2 T \tag{174}$$

where the constant a is called the thermal diffusivity coefficient. Equation 174 is combined with Equation 172, forming the general system of equations of the process of heat transfer in an incompressible medium.

Let us apply to this system of equations a common technique of the similarity theory which serves for finding the similarity numbers and criteria. Let us form the system of dimensionless equations

$$\mathbf{Sh}\frac{\partial V'}{\partial t'} + (V' \cdot \nabla')V' = -\mathbf{Eu}\,\mathrm{grad}'p' + \frac{1}{\mathbf{Fr}}g' + \frac{1}{\mathbf{Re}}\nabla'^2V'$$

$$\mathbf{Sh}\frac{\partial T'}{\partial t'} + V' \cdot \mathrm{grad}'T' = \frac{1}{\mathbf{Pe}}\nabla'^2T' \tag{175}$$

Here, in addition to the Struhal, Euler, and Reynolds numbers

$$\mathbf{Sh} = \frac{NL}{V} \qquad \mathbf{Eu} = \frac{P}{\rho V^2} \qquad \mathbf{Re} = \frac{VL}{\nu} \tag{176}$$

the Peclet number appears

$$\mathbf{Pe} = \frac{VL}{a} \tag{177}$$

Moreover, the Froude number has the following value specific for the nonisothermal motion of fluid (ΔT is the temperature drop scale)

$$\mathbf{Fr} = \frac{V^2}{gL}\frac{1}{\beta\Delta T} \tag{178}$$

Usually, instead of the Peclet and Froude numbers, use is made of their combination with the rest of the similarity numbers (176): the Prandtl number, **Pr**, equal to the ratio of the Peclet number, **Pe**, to the Reynolds number, **Re**,

$$\mathbf{Pr} = \frac{\mathbf{Pe}}{\mathbf{Re}} = \frac{\nu}{a} = \frac{\mu c}{\lambda} \tag{179}$$

and the Grashof number

$$\mathbf{Gr} = \frac{\mathbf{Re}^2}{\mathbf{Fr}} = \frac{\beta g L^3 \Delta T}{\nu^2} \tag{180}$$

The Grashof number is used when studying the phenomena of free convection because it does not involve the velocity scale, which is absent in the specification of free convection.

Consider the allied phenomenon of admixture propagation (diffusion) in an homogeneous medium – "carrier". Referring the reader for the detailed description of diffusion to Chapter 15, we note here that the phenomena of heat and mass transfer are closely interconnected and admit the same mathematical model. The analogue of the heat flux vector q (Equation 173), in the processes of substance diffusion is represented by the admixture mass flux density vector $\rho C V^*$ which contains, other then the density, the rate of diffusion V^* and the admixture concentration C. (Preserving the notation close to the generally accepted one, we introduce the capital letter C for concentration so that it is not confused with the heat capacity c). The rate of admixture diffusion in the "carrier" is determined by the Fick law:

$$V^* = -\frac{D}{C}\,\mathrm{grad}\,C \tag{181}$$

which is analogous to the Fourier law of heat propagation (Equation 173), the only difference being that the thermal conductivity coefficient λ is replaced by the mass conductivity coefficient D, usually called the coefficient of admixture diffusion.

The differential equation of admixture diffusion is derived in the same fashion as the heat balance equation (174) and has the same form

$$\frac{\partial C}{\partial t} + \boldsymbol{V} \cdot \text{grad } C = D\nabla^2 C \tag{182}$$

Resulting Equation 182, as with Equation 174, is combined with Equations 172. It is usually assumed that there is no effect of admixture on the dynamic portion of the phenomenon (the condition of the passive state of admixture). A more general approach relating to multicomponent mixtures of gases is presented in Chapter 15 in application to the theory of laminar boundary layer in the gas flow of sub- and supersonic velocities. Going over in Equation 182 to dimensionless variables,

$$\mathbf{Sh}\frac{\partial C'}{\partial t'} + \boldsymbol{V}' \cdot \text{grad}' C' = \frac{1}{\mathbf{Pe}_d}\nabla'^2 C' \tag{183}$$

where $\mathbf{Pe}_d$ is the diffusion Peclet number, which differs from the number **Pe** (Equation 177) by the coefficient of substance (admixture) diffusion D. Replaces the thermal diffusivity coefficient a:

$$\mathbf{Pe}_d = \frac{VL}{D} \tag{184}$$

The ratio

$$\mathbf{Sc} = \mathbf{Pr}_d = \frac{\mathbf{Pe}_d}{\mathbf{Re}} = \frac{\nu}{D} = \frac{\mu}{\rho D} \tag{185}$$

is called the Schmidt number or the diffusion Prandtl number.

The numerical values of the transfer coefficient D for different liquids and gases can be found in various books.[36]

The solution of linear equations of heat and admixture propagation often requires complex computations.[37]

The equations of heat transfer that take into account the heat inflow at the expense of mechanical energy dissipation are considered in Chapter 16.

---

[36] Belashchenko, D. K. and Zhukhovitskiy, A. A., Diffusion, in *Physical Encyclopedic Dictionary*, Vol. 1, Soviet Encyclopedea Moscow, 1960, 622, 623.

[37] The examples showing the application of the equations of heat and mass transfer (above) would have required too much attention in the book, and our concern is not these self-sustained problems. Those interested may consult the following books: Spalding, D. B., *Konvektivnyi massoperenos* (*Convective Mass Transfer*), Energiya Press, Moscow–Leningrad, 1965: Groeber, H., Erk, S., and Grigull, U., *Osnovy uchyeniya o teploobmene* (*Fundamentals of Heat Transfer*), IL Press, Moscow, 1958; Luikov, A. V. and Mikhailov, Yu. A., *Teoriya teplo i massoperenosa* (*The Theory of Heat and Mass Transfer*), Gosenergoizdat Press, Moscow, 1963; Zhukauskas, A. A., *Konvektivnyi perenos v teploobmennikakh* (*Convective Transfer in Heat Exchangers*), Nauka Press, Moscow, 1982.

11

Integration of the Navier-Stokes Equations: Linearized and Self-Similar Solutions

1. Flow Past a Sphere at Small Reynolds Numbers; The Stokes Equation and Its Generalizations

The derivation of exact solutions to the Navier-Stokes equations by comparatively simple mathematical means was stipulated by linearity of basic equations, which follows from the assumption on a rectilinear character of streamlines in cylindrical (prismatic) tubes. Solution to the problems of an external flow of a viscous fluid past bodies requires solution of nonlinear equations, with nonlinearity being contained in the inertia term on the left side of the equation that expresses a convective portion of acceleration. Discarding this term or substitution by an approximate linear expression leads to linearization of equations.

The classical Stokes problem of a slow stationary flow past a sphere when the main emphasis is attached to forces of friction and pressure and inertia terms are discarded can serve as the simplest example of this linearization.

Denote the uniform flow velocity at the infinity as V_∞ and the sphere radius as a. Direct the main axis $0x$ (Figure 160) of the spherical system of coordinates (R, θ, ε) parallel to the vector V_∞. Neglecting the volumetric forces and inertia terms we bring the Navier-Stokes equation (30) (Chapter 10) to the form

$$\operatorname{grad} p + \mu \operatorname{rot} \mathbf{\Omega} = 0$$

$$\mathbf{\Omega} = \operatorname{rot} \mathbf{V} \qquad \operatorname{div} \mathbf{V} = 0 \tag{1}$$

Eliminate the pressure p by applying the curl procedure to both sides of the first equation of system (1). We obtain

$$\operatorname{rot} \operatorname{rot} \mathbf{\Omega} = 0 \tag{2}$$

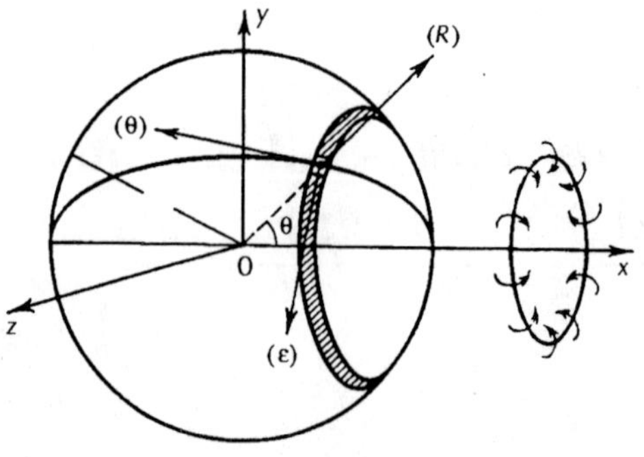

Fig. 160

Taking notice of the fact that thanks to the flow axial symmetry vortex lines are circles in the planes normal to the axis $0x$ with their centers on this axis, we conclude that the vorticity vector has the only component Ω_ε, which for brevity will be denoted as Ω, with the sign $\pm$ being included in this notation; the components Ω_R and Ω_θ are apparently equal to zero because the vector vortex is directed along the tangent to the vortex line. By virtue of the same symmetry,

$$\frac{\partial \Omega}{\partial \varepsilon} = 0 \qquad \Omega = \Omega(R, \theta)$$

Recalling Equation 86 of Chapter 1, we have the expressions for vector components rot $\mathbf{\Omega}$ in a spherical system of coordinates

$$\text{rot}_R \mathbf{\Omega} = \frac{1}{R \sin\theta} \frac{\partial}{\partial \theta}(\Omega \sin\theta) \qquad \text{rot}_\theta \mathbf{\Omega} = -\frac{1}{R}\frac{\partial(R\Omega)}{\partial R} \qquad \text{rot}_\varepsilon \mathbf{\Omega} = 0$$

and repeating the same procedure yields

$$\text{rot}_R(\text{rot}\,\mathbf{\Omega}) = 0 \qquad \text{rot}_\theta(\text{rot}\,\mathbf{\Omega}) = 0$$

$$\text{rot}_\varepsilon(\text{rot}\,\mathbf{\Omega}) = \frac{1}{R}\frac{\partial}{\partial R}(R\,\text{rot}_\theta \mathbf{\Omega}) - \frac{1}{R}\frac{\partial}{\partial \theta}(\text{rot}_R \mathbf{\Omega}) =$$

$$= \frac{1}{R}\frac{\partial}{\partial R}\left[-R\frac{1}{R}\frac{\partial(R\Omega)}{\partial R}\right] - \frac{1}{R}\frac{\partial}{\partial \theta}\left[\frac{1}{R\sin\theta}\frac{\partial}{\partial\theta}(\Omega\sin\theta)\right] =$$

$$= -\frac{1}{R}\frac{\partial^2(R\Omega)}{\partial R^2} - \frac{1}{R^2}\frac{\partial}{\partial\theta}\left[\frac{1}{\sin\theta}\frac{\partial}{\partial\theta}(\Omega\sin\theta)\right]$$

Thus, Equation 2 in spherical coordinates is reduced to the simpler

$$R\frac{\partial^2(R\Omega)}{\partial R^2} + \frac{\partial}{\partial\theta}\left[\frac{1}{\sin\theta}\frac{\partial}{\partial\theta}(\Omega\sin\theta)\right] = 0 \tag{3}$$

The solution of which $\Omega(R, \theta)$ in the meantime can be subjected to one boundary condition only

$$\Omega \to 0 \quad \text{when} \quad R \to \infty \tag{4}$$

Searching for the solution of Equation 3 in the form of the product of the two functions $P(R)$ and $\Theta(\theta)$, each of which depends on only one variable, we substitute $\Omega = P(R)\,\Theta(\theta)$ into Equation 3; this yields

$$\frac{R}{P(R)}\frac{d^2}{dR^2}[RP(R)] = -\frac{1}{\Theta(\theta)}\frac{d}{d\theta}\left\{\frac{1}{\sin\theta}\frac{d}{d\theta}[\Theta(\theta)\sin\theta]\right\}$$

By virtue of the independence of the coordinates R and θ the left and right sides of this equation should be constant independently; hence, it follows (α is the arbitrary constant)

$$\frac{R}{P(R)}\frac{d^2}{dR^2}[RP(R)] = \alpha \qquad \frac{1}{\Theta(\theta)}\frac{d}{d\theta}\left\{\frac{1}{\sin\theta}\frac{d}{d\theta}[\Theta(\theta)\sin\theta]\right\} = -\alpha$$

Making use of an arbitrary choice of the constant, we select it so that the second of the above equations could have the periodical solution corresponding to the problem sense. Note that at $\alpha = 2$ this equation has the obvious solution

$$\Theta(\theta) = \sin\theta$$

and the first equation of the system is converted to

$$\frac{d^2}{dR^2}[RP(R)] = \frac{2}{R}P(R)$$

The only solution of the last equation that satisfies the condition of converting to zero when $R \to \infty$ is const/R^2. Designating the constant as A we obtain the solution for the vorticity Ω in the form

$$\Omega = \frac{A\sin\theta}{R^2} \tag{5}$$

Apply now to the problem of seeking after the velocity components V_R and V_θ ($V_\varepsilon = 0$ due to the symmetry of the flow); for their determination two equations are at our disposal: (1) Equation 5, which, expressing the velocity vortex $\mathbf{\Omega}$ in terms of the velocity components in spherical coordinates V_R and V_θ, can be rewritten in the form

$$\frac{1}{R}\frac{\partial(RV_\theta)}{\partial R} - \frac{1}{R}\frac{\partial V_R}{\partial\theta} = \frac{A\sin\theta}{R^2} \tag{6}$$

and (2) the equation of incompressibility in spherical coordinates (at $V_\varepsilon = 0$)

$$\frac{1}{R^2}\frac{\partial(R^2V_R)}{\partial R} + \frac{1}{R\sin\theta}\frac{\partial(V_0\sin\theta)}{\partial\theta} = 0 \tag{7}$$

The system of Equations 6 and 7 should be solved under the boundary conditions

$$\begin{aligned} V_R &= 0 & V_\theta &= 0 & &\text{at} \quad R = a \\ V_R &= U_\infty\cos\theta & V_\theta &= -U_\infty\sin\theta & &\text{at} \quad R = \infty \end{aligned} \tag{8}$$

Taking into account these conditions, we seek the solution in the form

$$V_R = \left(U_\infty + \sum_{k=1}^{n} \frac{\lambda_k}{R^k}\right)\cos\theta \quad V_\theta = \left(-U_\infty + \sum_{k=1}^{n} \frac{\lambda'_k}{R^k}\right)\sin\theta \tag{9}$$

where the number n is considered to be indeterminate. Substituting expressions (9) into Equations 6 and 7 and equating the coefficients at similar trigonometric functions we obtain

$$\sum_{k=1}^{n} [\lambda_k + (1-k)\lambda'_k] R^{1-k} = A \quad \sum_{k=1}^{n} [(2-k)\lambda_k + 2\lambda'_k] R^{1-k} = 0$$

By virtue of an arbitrariness of the quantity R, for $k = 1$,

$$\lambda_1 = A \quad \lambda_1 + 2\lambda'_1 = 0 \quad \lambda'_1 = -\frac{1}{2}\lambda_1 = -\frac{1}{2}A$$

and for $k > 1$

$$\lambda_k + (1-k)\lambda'_k = 0 \quad (2-k)\lambda'_k + 2\lambda'_k = 0$$

The last uniform system has solutions different from zero only when the determinant of the system is equal to zero

$$2 - (1-k)(2-k) = 0$$

The roots of this equation are $k = 0$ and $k = 3$, with the first being discarded because $k \geq 1$. Hence, the equality follows

$$\lambda'_3 = \frac{1}{2}\lambda_3$$

the remainder λ_k and λ'_k are identically equal to zero.

Returning to Equation 9, formulate the general expressions for velocities

$$V_R = \left(U_\infty + \frac{A}{R} + \frac{\lambda_3}{R^3}\right)\cos\theta \quad V_\theta = \left(-U_\infty - \frac{A}{2R} + \frac{\lambda_3}{2R^3}\right)\sin\theta$$

Subjecting these velocities to the boundary conditions (8) we obtain two equations for the determination of the coefficients A and λ_3

$$\frac{A}{a} + \frac{\lambda_3}{a^3} = -U_\infty \quad -\frac{A}{2a} + \frac{\lambda_3}{2a^3} = U_\infty$$

Let us find

$$A = -\frac{3}{2}aU_\infty \quad \lambda_3 = \frac{1}{2}a^3U_\infty$$

Finally, we obtain the following expressions for the velocity components in the spherical system of coordinates:

$$V_R = U_\infty\left[1 - \frac{3}{2}\frac{a}{R} + \frac{1}{2}\left(\frac{a}{R}\right)^3\right]\cos\theta$$

$$V_\theta = -U_\infty\left[1 - \frac{3}{4}\frac{a}{R} - \frac{1}{4}\left(\frac{a}{R}\right)^3\right]\sin\theta \qquad (10)$$

Subtracting from these expressions the velocity components at infinity: $U_\infty \cos\theta$ and $(-U_\infty \sin\theta)$, we obtain the components of the velocity of infinite viscous fluid perturbation by a sphere

$$V'_R = -U_\infty\left[\frac{3}{2}\frac{a}{R} - \frac{1}{2}\left(\frac{a}{R}\right)^3\right]\cos\theta \qquad V'_\theta = U_\infty\left[\frac{3}{4}\frac{a}{R} + \frac{1}{4}\left(\frac{a}{R}\right)^3\right]\sin\theta$$

Note that contrary to the case of a flow of an ideal fluid past a sphere, where the order of perturbation velocities was $1/R^3$, a much stronger perturbation takes place in a viscous fluid that decreases with the distance from the sphere only as $1/R$.

The vorticity distribution is determined by Equation 5 as

$$\Omega = -\frac{3}{2}aU_\infty\frac{\sin\theta}{R^2}$$

The distribution of pressure in the flow, the friction on a sphere surface, and the total resistance of a sphere remains to be found. We have from the first equation of system (1)

$$\text{grad}\, p = -\mu \,\text{rot}\, \mathbf{\Omega}$$

or in the spherical coordinates

$$\frac{\partial p}{\partial R} = -\mu\frac{1}{R\sin\theta}\frac{\partial}{\partial\theta}(\Omega\sin\theta) = 3\mu a U_\infty\frac{\cos\theta}{R^3}$$

$$\frac{1}{R}\frac{\partial p}{\partial\theta} = \mu\frac{1}{R}\frac{\partial}{\partial R}(R\Omega) = \frac{3}{2}\mu a U_\infty\frac{\sin\theta}{R^3}$$

This system of equations in full differential is easily integrated and gives the expression for pressure

$$p = -\frac{3}{2}\mu a U_\infty\frac{\cos\theta}{R^2} + p_\infty$$

Then, formulating the pressure coefficient by the foregoing, we obtain

$$c_p = \frac{p - p_\infty}{\frac{1}{2}\rho U_\infty^2} = -\frac{3\mu}{\rho U_\infty a}\frac{\cos\theta}{(R/a)^2} = -\frac{6}{\mathbf{Re}}\frac{\cos\theta}{(R/a)^2} \qquad (11)$$

where **Re** means the Reynolds number characteristic for a flow past a sphere ($d = 2a$ is the sphere diameter)

$$\mathbf{Re} = \frac{\rho U_\infty d}{\mu} = \frac{U_\infty d}{\nu}$$

Note the characteristic differences of pressure distributions in a "slow" flow of a viscous fluid past a sphere from that of an ideal fluid:

1. In an ideal fluid the pressure coefficient on the surface is dependent on the relative position of the point only (angle θ), where the pressure is determined and does not depend on body dimensions, velocity, and density of fluid; in a viscous fluid the pressure coefficient is the function of the flow Reynolds number, i.e., it depends on the absolute body dimensions, velocity, density; and viscosity of fluid.
2. Pressure distribution over the sphere surface is not symmetrical relative to the frontal plane so that the resultant vector of pressure forces in the case of a viscous fluid flow past a sphere is not equal to zero (the d'Alembert paradox does not take place).
3. The coefficient of pressure at the stagnation points is not equal to unity; it depends on the Reynolds number and has different signs at the front and rear stagnation points; in a frontal area ($\theta = \pi/2$) the pressure on the sphere surface is equal to the pressure in an unperturbed flow, and maximum rarefaction is attained at the rear stagnation point.

The tangent component of the skin friction stress of a sphere $\tau_{R\theta}$ is equal, according to Equation 13 of Chapter 10, to

$$\tau_{R\theta} = p_{R\theta} = \mu\left(\frac{\partial V_\theta}{\partial R} + \frac{1}{R}\frac{\partial V_R}{\partial \theta} - \frac{V_\theta}{R}\right)_{R=a} = \mu\left(\frac{\partial V_\theta}{\partial R}\right)_{R=a} = -\frac{3}{2}\mu\frac{U_\infty}{a}\sin\theta$$

Having taken a band on the sphere surface (it is cross-hatched in Figure 160) with the area $2\pi a \sin\theta \cdot a\,d\theta = 2\pi a^2 \sin\theta\,d\theta$, we multiply the skin friction $\tau_{R\theta}$ and pressure p by this area; we project the elementary forces onto the axis $0x$ and perform the summation through the sphere entire surface (from $\theta = 0$ to $\theta = \pi$). Then we obtain the projection in the direction of the free stream velocity of the resistance to body motion in a viscous flow that is applied to the body from the fluid side, in the form

$$W_x = \int_0^\pi (-\tau_{R\theta}\sin\theta - p\cos\theta)\cdot 2\pi a^2 \sin\theta\,d\theta = 3\pi\mu a U_\infty \int_0^\pi \sin\theta\,d\theta = 6\pi\mu a U_\infty \quad (12)$$

This expression of the resistance force is called the Stokes equation.

The solution obtained is suitable only at very small values of the number $\mathbf{Re} = U_\infty d/\nu$. This follows from the fact that the ratio of the discarded inertia terms to the terms characterizing friction forces has the order of the Reynolds number. In fact,

$$\frac{\rho|(V\cdot\nabla)V|}{\mu|\text{rot}\,\mathbf{\Omega}|} \sim \frac{\rho U_\infty^2}{d} : \frac{\mu U_\infty}{d^2} = \mathbf{Re}$$

It can be stated that **Re** serves as the measure *f* a comparative role of inertia and viscous terms in the equations of motion. The smaller is **Re**; the greater is the role of viscosity forces in the motion considered.

Passing in Equation 12 from the resistance force to the coefficient of resistance c_x,

$$c_x = \frac{W_x}{\frac{1}{2}\rho U_\infty^2 \pi a^2} = \frac{6\pi\mu a U_\infty}{\frac{1}{2}\rho U_\infty^2 \pi a^2} = \frac{24}{\mathbf{Re}} \quad (13)$$

The more precise Oseen and Oseen-Goldstein theories give for the coefficient of resistance, instead of Equation 13, the expansion into series by the powers of the number **Re**, which is taken to be small:

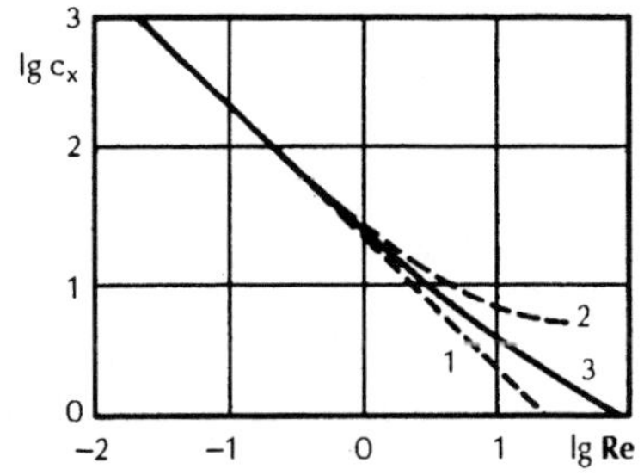

Fig. 161

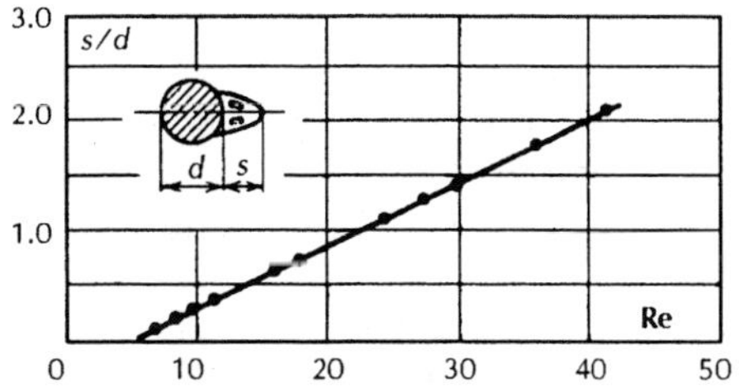

Fig. 162

$$c_x = \frac{24}{\mathbf{Re}}\left(1 + \frac{3}{16}\mathbf{Re} - \frac{19}{1280}\mathbf{Re}^2 + \cdots\right)$$

Retaining only the first term of the series, we obtain the Stokes equation; the two terms give the Oseen formula

$$c_x = \frac{24}{\mathbf{Re}}\left(1 + \frac{3}{16}\mathbf{Re}\right) = \frac{24}{\mathbf{Re}} + 4.5 \tag{14}$$

Figure 161[1] shows the curves calculated by Stokes (1), by Oseen (2), and found experimentally (3). Passing the value of the Reynolds number which has the order of unity, curves 1 and 2 noticeably deviate from curve 3 in different directions. As can be judged by these results, the Stokes equation (Equations 12 or 13) may be used only in cases of very small Reynolds numbers, for example, for small rain droplets or dust in atmospheric air, falling down of steel pellets in viscous fluids, etc. At large Reynolds numbers the formulas become unsatisfactory. This is explained by the fact that with an increase of **Re** in the rear region of a sphere (as well as of other "blunt" bodies), separation and complex nonstationary phenomena of the self-oscillation type are formed (discussed in Section 4 of Chapter 10). These phenomena are developed past the rear end of the body, with effective dimensions of the body increasing with the Reynolds number, as can be judged by Figure 162[2] which refers to a plane flow past a circular cylinder.

For a wide range of Reynolds numbers one has to use empirical data (see Chapter 13). As to approximate calculations of the resistance of a sphere or a circular cylinder of an infinite size, they become quite realizable due to the development of numerical methods.

An example of the numerical solution of the problem of a flow past a circular cylinder is considered in Section 9. The literature on the methods of linearization of the Navier-Stokes equations in the case of a flow past a sphere or a circular cylinder is voluminous.[3]

Along with the stationary problem considered, its nonstationary analogue is also studied thoroughly. We present without derivation[4] the Boussinesq equation of the resistance $\mathbf{W}$ of a

---

[1] Batchelor, J., *Vvedeniye v dinamiku zhidkosti* (*Introduction into Fluid Dynamics*), Mir Press, Moscow, 1973, 298.

[2] *Ibid*, p. 365.

[3] The book by Oseen, S. W., *Hydrodynamik*, Akademie Verlag, Leipzig, 1927, at one time was a classic monograph containing linearization methods. Among the courses most widely used at present, where great attention is paid to this problem, are Kochin, N. Ye., Kibel, I. A., and Roze, N. V., *Teoreticheskaya gidrodinamika* (*Theoretical Hydrodynamics*), Part 2, Fizmatgiz, Moscow, 1963, 504–534; Slyozkin, N. A., *Dinamika vyazkoi neszhimaemoi zhidkosti* (*Dynamics of Viscous Incompressible Fluid*), Gostekhizdat, Moscow, 1965 and the just-cited monograph by Batchelor. Criticism of the linearization methods is available in van Dyke, M., *Metody vozmushchenii v mekhanike zhidkosti* (*Perturbation Methods in Fluid Mechanics*), Mir Press, Moscow, 1967.

sphere with the radius a moving translatory with the prescribed variable velocity $V(t)$ in an infinite region filled in with a viscous fluid $[V(0) = 0]$

$$W = -6\pi\mu a V(t) - \frac{2}{3}\pi\rho a^3 V'(t) - 6\pi\mu a^2 \frac{1}{\sqrt{\pi\nu}} \int_0^t \frac{V'(\tau)\, d\tau}{\sqrt{t-\tau}}$$

Here, the first term represents the Stokes equation (12); the second is the inertia component of resistance corresponding to the presence of sphere virtual mass.

In the case of a sphere impulsively put from rest to translational rectilinear uniform motion with the velocity V_0, we have the value of the resistance[5]

$$W = 6\pi\mu a V_0 \left(1 + \sqrt{\frac{a^2}{\pi\nu t}}\right)$$

From this equation at $t = \infty$ we again obtain the Stokes equation (12).

2. Spatial Viscous Fluid Flow between Close Parallel Planes. Hele-Shaw "Spectra". The Darcy Law

Linearization, which corresponds to the notion of a "slow" viscous fluid flow (i.e., of the flow with negligible inertia forces as compared with viscous and pressure forces), is used also in problems of viscous fluid flow through thin slots. To these refer such practically important problems as filtration of viscous fluids (water, oil) through porous media (sand, soil, jointing rocks) and motion of liquid lubricating oils in a thin clearance between a rotating shaft and a bearing pad, among others.

We shall begin from the consideration of spatial incompressible viscous fluid flow between two infinite parallel planes positioned at small distance $2h$ from each other, precisely at such a distance that $\mathbf{Re} = u \cdot 2h/\nu$ is small. Between the planes, perpendicular to them, cylindrical obstacles of the same height $2h$ can be placed.

We position the coordinate plane $0xy$ in a midplane and direct the axis $0z$ perpendicularly to the planes bounding the flow. We assume that the flow takes place in planes parallel to the flow limits, thus taking $w = 0$. Moreover, to linearize the equations we discard convective terms (acceleration). Neglecting the effect of volumetric forces and assuming the flow to be stationary, we formulate the following linearized system of the Navier-Stokes equations

$$\mu\left(\frac{\partial^2 u}{\partial x^2} + \frac{\partial^2 u}{\partial y^2} + \frac{\partial^2 u}{\partial z^2}\right) = \frac{\partial p}{\partial x}$$

$$\mu\left(\frac{\partial^2 v}{\partial x^2} + \frac{\partial^2 v}{\partial y^2} + \frac{\partial^2 v}{\partial z^2}\right) = \frac{\partial p}{\partial y}$$

$$0 = \frac{\partial p}{\partial z}$$

---

[4]Luriye, A. I., *Operatsionnoe ischisleniye i ego prilozheniya k zadacham mekhaniki* (*Operational Calculus and its Applications to Problem of Mechanics*), GONTI, Moscow, 1938, 209–216.

[5]Slyozkin, N. A., *Dinamika vyaskoi neszhimaemoi zhidkosti* (*Dynamics of Viscous Noncompressible Fluid*), Gostekhizdat, Moscow, 1955, 341–349.

$$\frac{\partial u}{\partial x} + \frac{\partial v}{\partial y} = 0 \tag{15}$$

Subject the velocity components u, v to the boundary conditions

$$u = v = 0 \quad \text{at} \quad z = \pm h \tag{16}$$

According to the third equation of the system the pressure p is the function of x and y only. On the other hand, taking the derivative with respect to x of both sides of the first equation of system (15), with respect to y of both sides of the second equation of the same system, and summing up the results, to determine $p(x, y)$, we obtain the Laplace equation

$$\frac{\partial^2 p}{\partial x^2} + \frac{\partial^2 p}{\partial y^2} = 0 \tag{17}$$

Making use of Equation 17 and recalling equality (61) of Chapter 10 it is easy to see from simple substitution that Equations 15 and boundary conditions (16) can be satisfied, assuming that

$$u = -\frac{h^2}{2\mu}\frac{\partial p}{\partial x}\left[1 - \left(\frac{z}{h}\right)^2\right] \qquad v = -\frac{h^2}{2\mu}\frac{\partial p}{\partial y}\left[1 - \left(\frac{z}{h}\right)^2\right] \tag{18}$$

Formulating the velocities $\bar{u}$ and $\bar{v}$, which are mean along the normal to the planes bounding the flow, by the formulas

$$\bar{u} = \frac{1}{2h}\int_{-h}^{h} u\,dz = -\frac{h^2}{3\mu}\frac{\partial p}{\partial x} \qquad \bar{v} = \frac{1}{2h}\int_{-h}^{h} v\,dz = -\frac{h^2}{3\mu}\frac{\partial p}{\partial y} \tag{19}$$

and determining maximum velocities by Equations 18

$$u_{\max} = -\frac{h^2}{2\mu}\frac{\partial p}{\partial x} \qquad v_{\max} = -\frac{h^2}{2\mu}\frac{\partial p}{\partial y} \tag{20}$$

we see that the fields of these local mean or maximum velocities in the plane $0xy$ correspond to some imaginary plane[6] irrotational flows of an ideal fluid with one of the following velocity potentials

$$\bar{\varphi} = -\frac{h^2}{3\mu}p \qquad \varphi_m = -\frac{h^2}{2\mu}p \tag{21}$$

These are proportional to the real pressure p at the given point of the viscous fluid flow. In this case isopotential curves will coincide with isobars.

Placing the models with a small height of the given profile between two parallel planes which are very close to each other and letting an incompressible viscous fluid move through this slot, one can obtain flows similar to plane irrotational floes past the bodies of the same profile but infinitely long in a span (Section 6 of Chapter 7). In this sense the considered fictitious flow can be called "viscous analogy" of a plane irrotational flow of an ideal fluid. How-

[6]We emphasize that, as directly follows from Equations 18, there is no plane motion. Only the fields of mean and maximum velocities are plane, as determined by Equations 19 and 20.

ever, an interesting specific feature of such a flow should be noted with implies that to determine the field of pressures, use cannot be made of the Bernoulli equation which is simply absent both in this case and in other cases of viscous flows. It should be specified that previous considerations used in derivation of solutions (18) and consequences (19 to 21) become invalid in the vicinity of the surface of a short cylindrical body placed into the flow, although this region is small as compared to the body size and its influence on a potential flow can be neglected. As is shown by observations this effect is slightly noticeable even in the rear region of the streamlined body and in the wake behind it.

The analogy mentioned is used to demonstrate plane irrotational flows of an ideal fluid past the profiles of the given shape[7] with the aid of a real viscous fluid (Hele-Shaw "spectra"). For this purpose the profile tested is cut out of a thin sheet and fastened between transparent plates bounding the flow. To visualize streamlines, thin jets of pigmented fluid are injected into the flow between the planes. In this case one can obtain distinct "spectra" of plane flows.

The mechanism of filtration of viscous fluids through porous media is close to that of the viscous fluid flow (considered above) through a thin slot between parallel planes. The law making the foundation of the theory of these motions was formulated in the 19th century by the famous French hydraulicist Darcy on the basis of his experiments.[8] This law is virtually a simple and natural generalization of linear dependences (19) of mean velocity on the pressure gradient.

In a general case of a spatial filtration flow[9] penetrating through porous medium due to gravity the Darcy law is expressed as

$$\mathfrak{b} = -k\,\mathrm{grad}\left(\frac{p}{\gamma} + z\right) \tag{22}$$

Here, the vector $\mathfrak{b}$ is the vector of filtration rate at the given point determined as the limit of the ratio of flow rate per second through a plate normal to the direction of maximum rate to the plate size when this quantity tends to zero. In brackets the known Bernoulli trinomial $V^2/(2g) + p/\gamma + z$, which in this case degenerated into binomial because the rate of flow through pores has, as a rule, the order of several millimeters per second and, sometimes, even less. In this case the square of velocity can be neglected as compared with other terms: the piezometric head p/γ and the gravity head z. At the same time a small velocity or, more precisely, a small Reynolds number of viscous fluid flow through pores allows one to disregard convective accelerations caused by curvature of pores and variability of their cross-sectional areas.

These specific features of a porous medium at small Reynolds numbers slightly affect mean resistance of pores and thus a flow rate component of the filtration velocity. This is the reason for the resemblance of the Darcy law (22) (deduced on the basis of experimental data and being virtually the result of spatial averaging of a viscous fluid flow over randomly oriented and differently shaped pores of a filtrating medium with the laws of strictly defined flows of the same fluid in a thin slot between parallel planes.

The coefficient of proportionality k entering into the Darcy equation 22 is called the coefficient of filtration. It is a constant value if the filtrating medium is uniform and the fluid

[7] Hele-Shaw, H. S., Investigation of the nature of surface resistance of water and of stream motion under certain experimental conditions, *Trans. Inst. Nav. Arch.*, 25, 1898; Riegels, F., Zur Kritik des Hele-Shaw, *Versuches. Z. Angew. Math. Mekh.*, 18, 95–106, 1938.

[8] Darcy, H., Les fontanies publiques de la ville de Dijon, Paris, 1856.

[9] Aravin, V. I. and Numerov, S. N., *Teoriya dvizheniya zhidkostei i gazov v nedeformiruemoi poristoi srede* (*Theory of Fluid Flow in Non-Deformable Porous Media*), Nauka Press, Moscow, 1973; Polubarinova-Kochina, P. Ya., *Teoriya dvizheniya gruntovykh vod* (*Theory of Motion of Underground Waters*), 2nd ed., Nauka Press, Moscow, 1977.

possesses constant physical properties. To reveal the influence of viscosity of a moving fluid on the coefficient of filtration, it is expressed in the form

$$k = \frac{\gamma c}{\mu} = \frac{gc}{\nu}$$

(μ and ν are the dynamic and kinematic coefficients of viscosity, respectively; γ is the fluid specific gravity; and g is the gravity acceleration).

The quantity c is called the coefficient of medium porosity; it depends only on the geometry of pores, soil graininess, etc.

In flowing of viscous fluids through porous media with comparatively large mean dimensions of pores (large-grain rocks, pebble, ore, bituminous coal), the linear Darcy law is not yet justified and should be replaced by a more complex nonlinear one. Physically this is explained first by the effect of convective accelerations in a flow and then by the loss in the stability of laminar fluid flow in pores and in the transition to the regime of turbulent filtration. The latter can be judged by the change in the filtration Reynolds number which is equal to $\mathfrak{b}d/\nu$, where d is the mean diameter of pores.

The methods of solving the problems of filtrational flow of water through water development works as well as of oil in its infiltration through soil are well developed. If we leave aside the complex combined problems of filtration of multiphase media (for example, oil-gas, water-solid suspension) through nonhomogeneous anisotropic soils, motions with physicochemical conversions (evaporation, condensation, chemical reactions in "backfills"), then these methods are close to those used in hydrodynamics of plane irrotational flows of an ideal fluid.[9]

3. Viscous Fluid Flow between Rotating Coaxial Cylinders. The Problem of Slider Motion Along the Plane Coated with Viscous Fluid

As the following section (4) is a detailed analysis of the hydrodynamic theory of bearing lubrication, this section presents the solution to two problems, each of which in a simple manner reflects a viscous incompressible flow in a thin film between a shaft and a bearing. The first refers to the viscous fluid flow between the surfaces of two coaxial cylinders rotating relative to each other; i.e., it does not take into account their eccentricity which is typical of real bearings. The second problem deals with slider translational motion with a supporting plane inclined at a small angle with respect to a sliding plane coated with a viscous fluid. The region of fluid flow has a form of a wedge into which the fluid is entrained due to sticking to the wall. This problem is sometimes called the problem of a "viscous wedge". The "viscous wedge" effect, which allows for a curvilinear nature of walls, serves as the basis of the hydrodynamic theory of bearings.

Consider a plane flow of a viscous fluid between two coaxial cylinders with the radii R and R', respectively, rotating at different angular velocities ω and ω' (the prime refers to the outer cylinder). Taking the flow to be stationary and occurring around the concentric circles situated in a plane normal to the common axis of cylinders $0z$, we conclude from the considerations of symmetry that (in this section the conventional notation of the azimuthal angle ε is replaced by φ)

$$V_r \equiv 0 \qquad V_z \equiv 0 \qquad \frac{\partial}{\partial \varphi} \equiv 0 \qquad \frac{\partial}{\partial z} \equiv 0$$

Then the Navier-Stokes equations (33 of Chapter 10) in polar coordinates will be reduced to the system of two ordinary differential equations

$$\frac{d^2V_\varphi}{dr^2} + \frac{1}{r}\frac{dV\varphi}{dr} - \frac{V_\varphi}{r^2} = 0 \qquad \frac{V_\varphi^2}{r} = \frac{1}{\rho}\frac{dp}{dr} \tag{23}$$

with the boundary conditions

$$V_\varphi = R\omega \quad \text{at} \quad r = R \qquad V_\varphi = R'\omega' \quad \text{at} \quad r = R' \tag{24}$$

The common solution of the first of Equations 23 can be presented as

$$V_\varphi = \frac{C_1}{r} + C_2 r \tag{25}$$

and the constants C_1 and C_2 can be found from the boundary conditions (24). Omitting simple calculations, we have

$$V_\varphi = \frac{1}{R'^2 - R^2}\left[(\omega' R^2 - \omega R'^2)\, r + \frac{(\omega - \omega')\, R^2 R'^2}{r}\right]$$

$$\frac{p}{\rho} = \frac{1}{(R'^2 - R^2)^2}\left[(\omega' R^2 - \omega R'^2)^2 \frac{r^2}{2} + 2R^2R'^2(\omega - \omega') \times \right.$$

$$\left. \times\, (\omega' R^2 - \omega R'^2) \ln\left(\frac{r}{R}\right) - \frac{(\omega - \omega')^2 R^4 R'^4}{2r^2}\right] + \text{const} \tag{26}$$

The tangential skin friction between annular layers is equal to

$$\tau_{r\varphi} = -2\mu \frac{(\omega - \omega')\, R^2 R'^2}{(R'^2 - R^2)\, r^2} \tag{27}$$

and the total moment relative to the axis of rotation of friction forces around some circle of the radius r

$$L = \int_0^{2\pi} \tau_{r\varphi} r^2 d\varphi = -4\pi\mu \frac{(\omega - \omega')\, R^2 R'^2}{R'^2 - R^2} \tag{28}$$

will be independent of the radius of this circle. Assuming in the last equation $\omega' = 0$, $R' - R = \varepsilon$ and taking the clearance between the cylinders ε to be small as compared with the radii of cylinders, we obtain the formula, which historically is one of the first and belongs to Petrov,[10] for the moment of resistance to rotation of a shaft in a coaxial bearing

$$L = -\frac{4\pi\mu\omega R^2 (R + \varepsilon)^2}{\varepsilon (R' + R)} \approx -\frac{2\pi\mu\omega R^3}{\varepsilon} \tag{29}$$

---

[10] Petrov, N. P., Friction in machines and influence of lubricant on it, Inz. Zh., 1883; *Collected Papers: Gidrodinamicheskaya teoriya smazki (Hydrodynamic Theory of Lubrication)*, L. S. Leibenzon, Ed., ONII Press, Moscow, 1934.

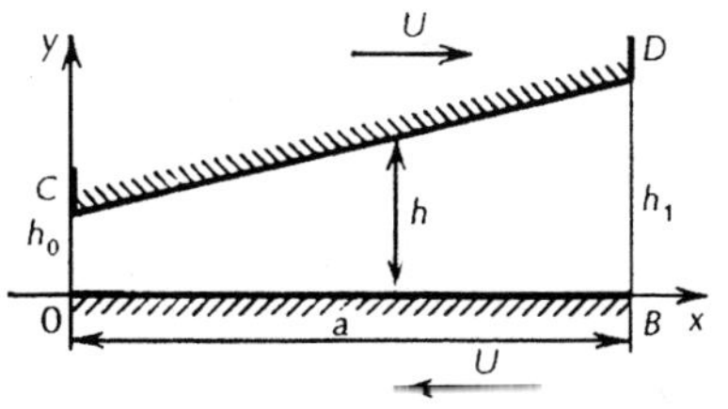

Fig. 163

This moment is directly proportional to the dynamic coefficient of viscosity and inversely proportional to the width of the clearance between the cylinders. Now tackle the second problem solved by Reynolds and Rayleigh.[11]

Restricting ourselves to the plane case ($V_z = 0$) we prescribe (Figure 163) by the segment CD a supporting plane of a slider and place the segment $0B$ in a plane along which the slider moves (the segment $0B$ is equal to the projection of CD onto this plane). The angle between CD and $0B$ is considered to be small as well as the mean distance between these planes within $0B$.

Assuming slider motion to be translational, rectilinear, and uniform with the velocity U parallel to $0B$, we reverse the motion, imparting to the drawing the same velocity in the opposite direction. In this case the slider appears to be motionless and the segment $0B$ will move in the plane with the velocity U from the wide portion of a slot BD to the narrow $0C$. According to the property of viscous fluid sticking to the solid surface, it will be entrained in the same direction.

Designate the heights of outlet and inlet sections of the slot by h_0 and h_1, respectively, the current ordinate of the segment CD by h, the projection of CD onto the axis $0x$ directed along $0B$ by a, as shown in Figure 163. The equation of the segment CD is prescribed by the formula

$$h = h_0\left(1 + k\frac{x}{a}\right) \tag{30}$$

where k is the constant coefficient equal to

$$k = \frac{h_1}{h_0} - 1 > 0 \tag{31}$$

Note that (the prime denotes the derivative with respect to x)

$$h' = \frac{kh_0}{a} > 0 \tag{32}$$

In constructing the equations for a flow of a viscous fluid entrained into a wedge-like slot by a moving segment $0B$, we shall take notice of slot thinness and impermeability of walls bounding the slot. This makes it possible to consider the transverse velocity V_y negligibly small compared with V_x, and the derivative $\partial^2 V_x/\partial y^2$ considerably greater than $\partial^2 V_x/\partial x^2$.

[11] We shall cite the two most available books: Lamb, H., *Gidrodinamika* (*Hydrodynamics*) (translated from English), Gostekhizdat, Moscow–Leningrad, 1947, 730, 731; Targ, S. M., *Osnovnye zadachi teorii laminarnykh techenii* (*Basic Problems in the Theory of Laminar Flows*), Gostekhizdat, Moscow–Leningrad, 1951, 319–324.

Moreover, neglecting the convective terms, we reduce the system of the Navier-Stokes equations to the form

$$\mu \frac{\partial^2 V_x}{\partial y^2} = \frac{\partial p}{\partial x} \qquad 0 = \frac{\partial p}{\partial y} \qquad \frac{\partial V_x}{\partial x} + \frac{\partial V_y}{\partial y} = 0 \tag{33}$$

The boundary conditions will be

$$V_x = -U \qquad V_y = 0 \quad \text{at} \quad y = 0$$

$$V_x = 0 \qquad V_y = 0 \quad \text{at} \quad y = h$$

$$p = p_0 \quad \text{at} \quad x = 0 \quad \text{and} \quad x = a \tag{34}$$

with the last condition corresponding to leveling the pressure outside the viscous wedge.

Integrating both sides of the first equation of system (33) with respect to y and allowing for the boundary conditions (34), we obtain

$$V_x = \frac{1}{2\mu}\frac{dp}{dx}(y^2 - hy) - \frac{U}{h}(h - y) \tag{35}$$

The condition of the constancy of fluid flow rate through any flow section yields

$$\int_0^h V_x dy = \text{const}$$

or, after substitution of V_x from (35) and integration

$$\frac{h^3}{6\mu}\frac{dp}{dx} + Uh = \text{const}$$

Judging by the last boundary condition of system (34), at an intermediate point with the abscissa x_m within the range $0 < x_m < a$ the condition $dp/dx = 0$ should be valid. Introducing the notation $h = h_m$ at $x = x_m$, we find from the previous equality

$$Uh_m = \text{const}$$

Eliminating the constant, we have from the last two equations

$$\frac{h^3}{6\mu}\frac{dp}{dx} + Uh = Uh_m$$

where it follows that

$$\frac{dp}{dx} = -\frac{6\mu U}{h^3}(h - h_m) \tag{36}$$

Having calculated at the point $x = x_m$ the derivative $(d^2p/dx^2)_{x_m} = -6\mu U h'(x_m)/h_m^3$, we see that due to condition (32), it has a negative value. Hence it follows that at the point $x = x_m$ pressure is maximum.

We change over from the argument x to h using for this purpose the basic geometric relation (30), according to which

$$\frac{dp}{dx} = \frac{dp}{dh}\frac{dh}{dx} = k\frac{h_0}{a}\frac{dp}{dh}$$

Rewriting equality (36) in the form

$$\frac{dp}{dh} = -\frac{6\mu Ua}{kh_0}\frac{h-h_m}{h^3} \tag{37}$$

and integrating it with respect to h we have

$$p = \frac{6\mu Ua}{kh_0}\left(\frac{1}{h}-\frac{h_m}{2h^2}\right)+b \tag{38}$$

Making use of the boundary conditions (34) and equality (30), we find the integration constants h_m and b

$$h_m = 2h_0\frac{1+k}{2+k} \qquad b = p_0-\frac{6\mu Ua}{kh_0^2}\frac{1}{2+k} \tag{39}$$

Substituting this value of the constants into (38) and reverting to the variable x, we obtain the following final expression for $p(x)$

$$p(x) = p_0+\frac{6\mu aU}{h_0^2k}\left[\frac{a}{a+kx}-\frac{1}{2+k}-\frac{1+k}{2+k}\frac{a^2}{(a+kx)^2}\right] \tag{40}$$

with the abscissa x_m, as it directly follows from (30) and the first equality of system (39), being equal to

$$x_m = \frac{a}{2+k} \tag{41}$$

Solve the main problem on the magnitude of the force $\boldsymbol{R}$ directed along 0y and supporting the slide. We have, according to Equation 40 (over CD pressure is equal to p_0)

$$R = \int_0^a (p-p_0)\,dx = \frac{6\mu Ua^2}{k^2h_0^2}\left[\ln(1+k)-\frac{2k}{2+k}\right] \tag{42}$$

The force R attains its maximum when $k \approx 1.2$, i.e., when $h_1/h_0 \approx 2.2$ and is equal to

$$R_{\max} \approx 0.16\mu\frac{Ua^2}{h_0^2} \tag{43}$$

To determine the resistance F to slider motion, we first find the skin friction τ on its supporting surface or due to the smallness of the wedge-like slot opening, on the segment 0B

$$\tau = \mu\left(\frac{\partial V_x}{\partial y}\right)_{y=0}$$

where, according to Equation 35,

$$\tau = \mu \frac{U}{h} - \frac{h}{2} \frac{dp}{dx}$$

and after substitution of dp/dx by (36) and h by (30)

$$\tau = \mu \frac{U}{h_0} \left[\frac{4a}{a + kx} - \frac{6(1+k)}{2+k} \frac{a^2}{(a+kx)^2} \right] \tag{44}$$

Integrating with respect to x from $x = 0$ to $x = a$, we determine the total force of air resistance to slider motion

$$F = \frac{\mu a U}{h_0} \left[\frac{4}{k} \ln(1+k) - \frac{6}{2+k} \right] \tag{45}$$

At $k = 1.2$, corresponding to maximum supporting force,

$$F = F_{\max} \approx 0.753 \frac{\mu a U}{h_0} \tag{46}$$

The coefficient of friction, conditionally determined by the resistance to supporting force ratio (seen in Equations 45 and 42), is proportional to h_0/a, i.e., it is very small.

Referring to the above-cited literature for additional data, we illustrate the main fact of attaining large supporting forces of the slider due to smallness of the gap between the slider supporting surface and the plane of sliding. At optimal value of the relation $h_1/h_0 = 2.2$, we consider the case of square slider in plane, $a = 0.1$ m, $h_0 = 10^{-4}$ m, $U = 1$ m/s, $\mu = 0.706 \cdot 10^{-1}$ N $\cdot$ s/m^2.

Calculation by Equation 43 shows that under these conditions $R_{\max}$ attains the value 1130 N. It explains the existence of large supporting forces in plentifully lubricated sliding bearings and indicates the main cause of formation of these forces the "viscous wedge" effect.

We find the abscissa x_R of the application point of the resultant $\boldsymbol{R}$ of the elementary supporting forces distributed along the segment *CD*. For this purpose we compose the moment of the resultant L_C as an integral (the smallness of the angle of "wedge" opening makes it possible to take the cosine of this angle equal to unity)

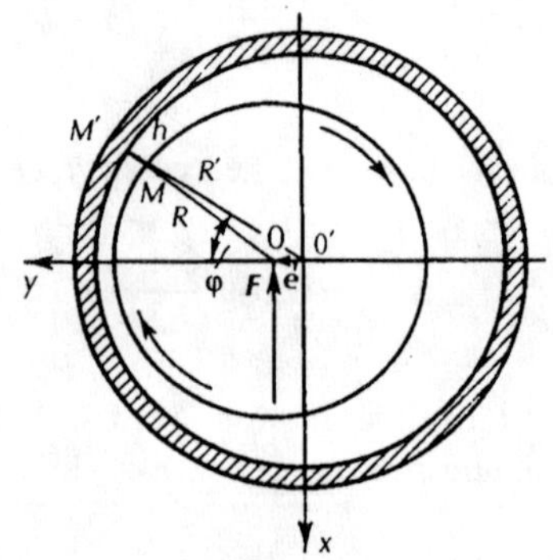

Fig. 164

$$L_C = \int_0^a (p - p_0)\, x dx = \frac{6\mu U a^3}{2h_0^2 k^3 (2+k)} [6k + k^2 - 2(3+2k) \ln(1+k)] \tag{47}$$

The abscissa x_R will be determined as $x_R = L_C/R$; i.e., according to Equations 42 and 47 it will be equal to

$$x_R = \frac{6k + k^2 - 2(3+2k) \ln(1+k)}{2k[(2+k)\ln(1+k) - 2k]} a \tag{48}$$

At $k = 1.2$ $x_R \approx 0.42a$. In other words, the application point of the supporting force $\boldsymbol{R}$ lies not in the center of the carrying surface (the segment CD), but somewhat closer to the point C. This fact is attributed to considerable shift of the point of maximum pressure ($x_m = 0.31a$ at $k = 1.2$) to the left side of the slider.

4. Hydrodynamic Theory of Bearing Lubrication. Plane Problem

Moving to the case of a shaft eccentric position relative to a bearing, which is closer to the hydrodynamic theory of bearing, we consider the Sommerfeld[12] problem. We neglect the "end" effects in a bearing; i.e., we assume that the bearing is infinitely long in the direction of the rotation axis and the flow in the gap between the shaft and bearing insert is plane.

In this schematic formulation the problem is reduced to the study of viscous incompressible fluid flow between two eccentric circles (Figure 164). One has its center at the point $0'$ (outer) and is motionless, while the other (inner) has its center at the point 0 rotating at the given angular velocity ω, with the eccentricity $e = 0'0$ being very small compared to the radii of circles R and $R' > R$

$$e \ll R, R'$$

The range of the change of the radius-vector r in the gap between the circles is $R \le r \le R + h$.

Making use of the last condition and assuming the fluid flow in the gap between the circles is slow in the sense that inertia terms may be neglected compared with the terms which take into account viscous forces and pressure variation, we shall reduce the Navier-Stokes equation in the polar coordinates (Equation 33, Chapter 10) to a simplified form

$$\mu \frac{\partial^2 V_\varphi}{\partial r^2} = \frac{1}{r}\frac{\partial p}{\partial \varphi} \qquad \mu \frac{\partial^2 V_r}{\partial r^2} = \frac{\partial p}{\partial r} \qquad \frac{\partial V_r}{\partial r} + \frac{1}{r}\frac{\partial V_\varphi}{\partial \varphi} = 0 \tag{49}$$

When constructing these equations account is taken of the relative smallness of the gap that allows one to assume that

$$V_\varphi \gg V_r \qquad \frac{\partial^2 V_\varphi}{\partial r^2} \gg \frac{1}{r}\frac{\partial V_\varphi}{\partial r} \quad \frac{1}{r^2}\frac{\partial^2 V_\varphi}{\partial \varphi^2} \quad \frac{1}{r^2}\frac{\partial V_r}{\partial \varphi} \quad \frac{V_\varphi}{r^2}$$

[12] See the earlier cited book of collected papers: *Gidrodinamicheskaya teoriya smazki* (*Hydrodynamic Theory of Lubrication*), L. S. Leibenzon, Ed., ONTI Press, Moscow, 1934; Targ, S. M., *Osnovnye zadachi teorii laminarnykh techenii* (*Basic Problems of the Theory of Laminar Flows*), Gostekhizdat, Moscow, 1951; monograph by N. A. Slyozkin, which was cited earlier.

$$\frac{\partial^2 V_r}{\partial r^2} \gg \frac{1}{r}\frac{\partial V_r}{\partial r} \quad \frac{1}{r^2}\frac{\partial^2 V_r}{\partial \varphi^2} \quad \frac{1}{r^2}\frac{\partial V_\varphi}{\partial \varphi} \quad \frac{V_r}{r^2} \qquad \frac{\partial V_r}{\partial r} \gg \frac{\partial V_r}{r}$$

By the same taken,

$$\frac{\partial^2 V_r}{\partial r^2} \ll \frac{\partial^2 V_\varphi}{\partial r^2}$$

and from the first two equalities of system (49) it follows

$$r\frac{\partial p}{\partial r} \ll \frac{\partial p}{\partial \varphi}$$

thus making it possible to take

$$\frac{\partial p}{\partial r} = 0 \qquad p = p(\varphi)$$

Moreover, in the last equation of system (49), preceding the derivative $1/r$ can be substituted by $1/R$, and from the variable $r(R \le r \le R+h)$ one can move to the variable $\zeta = r - R$ which changes within the range $0 \le \zeta \le h$, where $h = h(\varphi) = MM'$ (Figure 164) means the local thickness of the gap between the cylinders. It can be easily found from the triangle $0'0M'$, taking approximately,

$$0'0 \cos\varphi + 0M' = 0'M'$$

or

$$e \cos\varphi + R + h = R'$$

Introducing further the notation $R' - R = \varepsilon$ and relative eccentricity $\lambda = e/\varepsilon$, we have with the adopted accuracy

$$h(\varphi) = \varepsilon - e\cos\varphi = \varepsilon(1 - \lambda\cos\varphi)$$

In the figure the axis $0y$ is directed through the line of centers $0'0$, so that at the given point of the origin of angles φ

$$h_{\min} = \varepsilon - e \qquad \text{at } \varphi = 0$$

$$h_{\max} = \varepsilon + e \qquad \text{at } \varphi = \pi$$

Equations 49 can be rewritten as

$$\mu\frac{\partial^2 V_\varphi}{\partial \zeta^2} = \frac{1}{R}\frac{dp}{d\varphi} \qquad \frac{\partial V_r}{\partial \zeta} + \frac{1}{R}\frac{\partial V_\varphi}{\partial \varphi} = 0 \tag{50}$$

The first of these equations allows reintegration with respect to the variable ζ and gives

$$V_\varphi = \frac{1}{2\mu R}\frac{dp}{d\varphi}\zeta^2 + C_1(\varphi)\zeta + C_2(\varphi)$$

Satisfying the boundary conditions

$$V_\varphi = \omega R \quad \text{at} \quad \zeta = 0 \qquad V_\varphi = 0 \quad \text{at} \quad \zeta = h$$

we have

$$V_\varphi - \frac{1}{2\mu R}\frac{dp}{d\varphi}(\zeta - h)\zeta + \frac{\omega R}{h}(h - \zeta) \tag{51}$$

Change over to the velocity $\bar{V}_\varphi$ mean over the gap cross-section h, determining it as

$$\bar{V}_\varphi = \frac{1}{h}\int_0^h V_\varphi d\zeta$$

then from the previous equality we find

$$\bar{V}_\varphi = -\frac{h^2}{12\mu R}\frac{dp}{d\varphi} + \frac{1}{2}\omega R \tag{52}$$

Integrate for ζ from 0 to h both sides of the second equation of system (50). Then, noting that by the stationary condition in a radial direction and impermeability of both solid walls $V_r = 0$ at $\zeta = 0$ and $\zeta = h$, we obtain

$$\frac{d}{d\varphi}(h\bar{V}_\varphi) = 0 \qquad h\bar{V}_\varphi = Q = \text{const}$$

This equality expresses the condition of conservation of the fluid flow rate per second through any gap cross-section. Comparing it with formula (52) we have the equation for pressure determination

$$\frac{dp}{d\varphi} = \frac{6\mu\omega R^2}{h^2} - \frac{12\mu RQ}{h^3} \tag{53}$$

The integral of this, if a new additive constant of integration is introduced into the determination of p, assuming (e.g.) $p = 0$ at $\varphi = 0$, will have the form

$$p(\varphi) = 6\mu\omega R^2\int_0^\varphi \frac{d\varphi}{h^2(\varphi)} - 12\mu RQ\int_0^\varphi \frac{d\varphi}{h^3(\varphi)} \tag{54}$$

The unknown *a priori* constant Q can be eliminated if use is made of the condition of the periodicity of pressure distribution

$$p(\varphi + 2\pi) = p(\varphi)$$

or, in particular, of the relation

$$p(2\pi) = p(0) = 0$$

This yields

$$Q = \frac{1}{2}\omega R \frac{\int_0^{2\pi} \frac{d\varphi}{h^2}}{\int_0^{2\pi} \frac{d\varphi}{h^3}} \tag{55}$$

Thus, pressure distribution in the bearing gap is determined completely. The level of pressure at the point of the minimum gap or at any other point can be given arbitrarily and does not enter into the expression of the supporting force.

Determine the skin friction τ_w using the expression for the velocity V_φ by Equation 51. According to Equation 53 we have

$$\tau_w = \mu\left(\frac{\partial V_\varphi}{\partial \zeta}\right)_{\zeta=0} = -\frac{h}{2R}\frac{dp}{d\varphi} - \mu\frac{\omega R}{h} = \frac{6\mu Q}{h^2} - \frac{4\mu\omega R}{h} \tag{56}$$

Introduce the integrals

$$S_k(\varphi;\lambda) = \int_0^{\varphi} \frac{d\varphi}{(1-\lambda\cos\varphi)^k} \qquad (k = 1, 2, 3)$$

Having integrated, we obtain

$$S_1(\varphi;\lambda) = \frac{2}{(1-\lambda^2)^{1/2}}\operatorname{arctg}\left(\sqrt{\frac{1+\lambda}{1-\lambda}}\operatorname{tg}\frac{\varphi}{2}\right)$$

$$S_2(\varphi;\lambda) = \frac{2}{(1-\lambda^2)^{3/2}}\operatorname{arctg}\left(\sqrt{\frac{1+\lambda}{1-\lambda}}\operatorname{tg}\frac{\varphi}{2}\right) + \frac{\lambda}{1-\lambda^2}\frac{\sin\varphi}{1-\lambda\cos\varphi}$$

$$S_3(\varphi;\lambda) = \frac{2+\lambda^2}{(1-\lambda^2)^{5/2}}\operatorname{arctg}\left(\sqrt{\frac{1+\lambda}{1-\lambda}}\operatorname{tg}\frac{\varphi}{2}\right) +$$

$$+\frac{3}{2}\frac{\lambda}{(1-\lambda^2)^2}\frac{\sin\varphi}{1-\lambda\cos\varphi} + \frac{1}{2(1-\lambda^2)}\frac{\sin\varphi}{(1-\lambda\cos\varphi)^2}$$

Noting that at $\varphi = 2\pi$ the value of π corresponds to the arc tangent, we obtain the values of these functions at $\varphi = 2\pi$ which are necessary for the future

$$S_1(2\pi;\lambda) = \frac{2\pi}{(1-\lambda^2)^{1/2}} \qquad S_2(2\pi;\lambda) = \frac{2\pi}{(1-\lambda^2)^{3/2}} \qquad S_3(2\pi;\lambda) = \frac{\pi(2+\lambda^2)}{(1-\lambda^2)^{5/2}}$$

Using these quantities, we find by Equation 55

$$Q = \omega R\varepsilon\frac{1-\lambda^2}{2+\lambda^2} \tag{57}$$

Reverting to Equations 53 and 54, we obtain

$$p(\varphi) = \frac{6\mu\omega R^2}{\varepsilon^2}\left[S_2(\varphi;\lambda) - \frac{2(1-\lambda^2)}{2+\lambda^2}S_3(\varphi;\lambda)\right] \tag{58}$$

and by Equation 56 the skin friction

$$\tau_w(\varphi) = \frac{2\mu\omega R}{\varepsilon(1-\lambda\cos\varphi)}\left[\frac{3(1-\lambda^2)}{(2+\lambda^2)(1-\lambda\cos\varphi)} - 2\right] \tag{59}$$

Comparing the last two expressions, we draw the conclusion that

$$p(\varphi) = O\left(\frac{R^2}{\varepsilon^2}\right) \qquad \tau_w(\varphi) = O\left(\frac{R}{\varepsilon}\right)$$

this makes it possible, when calculating the resultant $\boldsymbol{F}$ of fluid reactions, to sum up only pressure forces and to neglect friction forces. We have, relating the main vector $\boldsymbol{F}$ to the unit length along the bearing axis,

$$F_x = \int_0^{2\pi} p\cos\left(\frac{\pi}{2}-\varphi\right)R\,d\varphi - R\int_0^{2\pi} p\sin\varphi\,d\varphi \qquad F_y = -R\int_0^{2\pi} p\cos\varphi\,d\varphi$$

Note that due to periodicity and oddness of pressure and evenness of cos φ the second integral is identically equal to zero. The first integral is most easily calculated by parts; this yields

$$F_x = -R\int_0^{2\pi} p\,d(\cos\varphi) = -Rp\cos\varphi\Big|_0^{2\pi} + R\int_0^{2\pi}\frac{dp}{d\varphi}\cos\varphi\,d\varphi$$

By the condition of pressure periodicity the first term vanishes; placing instead of $dp/d\varphi$ its expression (53), we have

$$F_x = \frac{6\mu\omega R^3}{\varepsilon^2}\left[\int_0^{2\pi}\frac{\cos\varphi\,d\varphi}{(1-\lambda\cos\varphi)^2} - \frac{2(1-\lambda^2)}{2+\lambda^2}\int_0^{2\pi}\frac{\cos\varphi\,d\varphi}{(1-\lambda\cos\varphi)^3}\right]$$

Noting that

$$\int_0^{2\pi}\frac{\cos\varphi\,d\varphi}{(1-\lambda\cos\varphi)^2} = \frac{1}{\lambda}[S_2(2\pi;\lambda) - S_1(2\pi;\lambda)] = \frac{2\pi\lambda}{(1-\lambda^2)^{3/2}}$$

$$\int_0^{2\pi}\frac{\cos\varphi\,d\varphi}{(1-\lambda\cos\varphi)^3} = \frac{1}{\lambda}[S_3(2\pi;\lambda) - S_2(2\pi;\lambda)] = \frac{3\pi\lambda}{(1-\lambda^2)^{5/2}}$$

we find after simple transformations

$$F_x = -\frac{12\pi\mu\omega R^3\lambda}{\varepsilon^2(2+\lambda^2)\sqrt{1-\lambda^2}} \tag{60}$$

The vector $\boldsymbol{F}$ is shown in Figure 164 to be applied about the center of the inner circle. We can give a vector equation (the vector $\boldsymbol{\omega}$ of the angular velocity of shaft rotation is directed into the plane of figure, and the vector $\boldsymbol{e}$ of the eccentricity is directed from the point 0′ to the point 0)

$$\boldsymbol{F} = \frac{12\pi\mu R^3}{\varepsilon^3 (2+\lambda^2)\sqrt{1-\lambda^2}} \boldsymbol{\omega} \times \boldsymbol{e} \tag{61}$$

The force $\boldsymbol{F}$ can counterbalance vertical load (the weight of a shaft with the rotor, etc.) if at the horizontal axis of the shaft its displacement $\boldsymbol{e}$, as shown in Figure 164, will also lie in a horizontal plane. In this case the resultant $\boldsymbol{F}$ will play the role of supporting force. We resort to the calculation of the moment of fluid resistance to shaft rotation per unit length. We have (the axis $0z$ is directed in toward the figure)

$$L_z = \int_0^{2\pi} \tau_w R^2 d\varphi = \frac{2\mu\omega R^3}{\varepsilon}\left[\frac{3(1-\lambda^2)}{2+\lambda^2} S_2(2\pi;\lambda) - 2S_1(2\pi;\lambda)\right] =$$

$$= -\frac{4\pi\mu\omega R^3 (1+2\lambda^2)}{\varepsilon(2+\lambda^2)\sqrt{1-\lambda^2}} \tag{62}$$

Petrov[10] considers the case of coaxial cylinders or of concentric circles is considered. This corresponds to $e = 0$ and $\lambda = 0$. As is seen from Equation 60, in this case there is no supporting force (the fact follows directly from the symmetry reasons) and Equation 62 changes over to Petrov's equation (29).

In the modern state of the hydrodynamic theory of lubrication of bearings much more complex problems are solved which are concerned with the allowance for finiteness of the bearing width, that disturbs a plane character of lubrication motion in a gap between a shaft and a bearing pad, with incomplete following of a gap by lubricant, with the effect of convective accelerations, etc.

5. Spatial Problem of the Hydrodynamic Theory of Lubrication of Bearings. Ball Bearings and Suspensions

We give an example of spatial lubricant motion in a ball bearing, making the same assumptions on the "slowness" of lubricant motion in the cavity between a rotating inner sphere and a motionless outer sphere. This makes it possible to linearize the problem discarding nonlinear terms in the Navier-Stokes equations, and on a relatively small size of the cavity[13] lateral to flow.

Consider the motion of a viscous incompressible fluid in a cavity between two eccentric spheres with the centers 0 and 0′ (Figure 165a) and the radii R and R' with $R' > R$. The difference of radii $\varepsilon = R' - R$ will be taken to be small compared with the radii of the spheres and we shall adopt the ratio ε/R as the first-order small quantity and agree to compare in future all the quantities with it.

[13] Loitsyanskiy, L. G., Hydrodynamic theory of a ball bearing, *Prikl. Mat. Mekh.*, 19(5), 531–540, 1955; *ibid*, 1(1), 133–135, 1956; Wannier, G. H., A contribution to the hydrodynamics of lubrication, *Q. Appl. Math.*, 8(1), 1, 1–32, 1950.

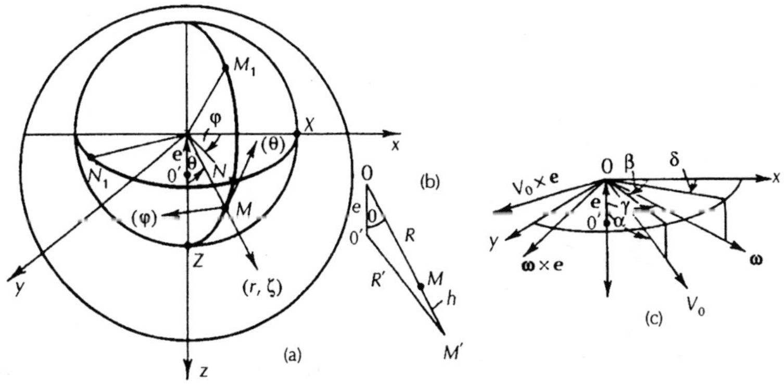

Fig. 165

We shall use a spherical coordinate system r, θ, φ, taking the axis $0z$, drawn through the centers of the spheres, to be the polar axis, with the plane $0xz$ of the origin of longitudes φ remaining arbitrary. To the latitude θ and the longitude φ of any point M of the inner sphere surface there correspond the arcs ZM and XN. Write $h(0, \varphi)$ for the cross-sectional dimension of the cavity at the point M and determine it as the distance between the points M and M' on the inner and outer spheres that lie on the same radius drawn from the center 0 of the inner sphere. With the order of $(\varepsilon R)^2$ the equality

$$h = \varepsilon + e \cos \varphi \tag{63}$$

will be valid (Figure 165b), where the eccentricity e is determined as the distance $00'$ between the centers of the spheres.

Assuming the motion to be stationary, neglecting the inertia terms, and taking into account (as was done with a prior case of a plane bearing) the dominant effect of the derivatives across the cavity (with respect to r) as compared with the derivatives with respect to θ and φ, we shall reduce the Navier-Stokes equations in spherical coordinates (Equation 34 of Chapter 10) to the form (r also stands for the current radius-vector, and φ for the azimuth)

$$\frac{\partial p}{\partial r} = \mu \frac{\partial^2 V_r}{\partial r^2} \qquad \frac{\partial p}{\partial \theta} = \mu r \frac{\partial^2 V_\theta}{\partial r^2} \qquad \frac{\partial p}{\partial \varphi} = \mu r \sin\theta \frac{\partial^2 V_\varphi}{\partial r^2}$$

$$\frac{\partial V_r}{\partial r} + \frac{1}{r\sin\theta}\frac{\partial}{\partial\theta}(V_\theta \sin\theta) + \frac{1}{r\sin\theta}\frac{\partial V_\varphi}{\partial\varphi} = 0 \tag{64}$$

It follows from the continuity equation (the last equation) that the derivative $\partial V_r/\partial r$ has the same order as the values V_θ/R or V_φ/R. On this basis, reverting to the first three equations of system (64), we conclude that

$$\frac{r\,\partial p/\partial r}{\partial p/\partial\theta} = O\left(\frac{\varepsilon}{R}\right) \qquad \frac{r\,\partial p/\partial r}{\partial p/\partial\varphi} = O\left(\frac{\varepsilon}{R}\right)$$

so that in the adopted approximation we can neglect the quantity $\partial p/\partial r$ compared with $\partial p/\partial\theta$ and $\partial p/\partial\varphi$ and consider the pressure p to be the function of θ and φ only. Moreover, with the same error we can replace in the second, third, and fourth formulas of Equations 64 the quantity r, which does not enter under the differentiation sign, by R.

Introducing the variable $\zeta = r - R$ instead of r, which changes across the flow in the cavity filled in with fluid within the limits $0 \le \zeta \le h$, based on the simplifications mentioned, we reduce system (64) to the form

$$\frac{\partial^2 V_\theta}{\partial \zeta^2} = \frac{1}{\mu R}\frac{\partial p}{\partial \theta} \qquad \frac{\partial^2 V_\varphi}{\partial \zeta^2} = \frac{1}{\mu R \sin\theta}\frac{\partial p}{\partial \varphi}$$

$$\frac{\partial V_r}{\partial \zeta} + \frac{1}{R\sin\theta}\frac{\partial}{\partial\theta}(V_\theta \sin\theta) + \frac{1}{R\sin\theta}\frac{\partial V_\varphi}{\partial\varphi} = 0 \qquad (65)$$

We write out the boundary conditions on the inner and outer spheres, putting the superscript "zero" at the velocity components of the points of the moving inner sphere surface, in this case, in contrast to that with a plane bearing. We assume that at the given moment the inner sphere not only carries out a rotatory motion, but also some small (of the order of gap thickness) translational motion. We have

$$V_r = V_r^0 \qquad V_\theta = V_\theta^0 \qquad V_\varphi = V_\varphi^0 \quad \text{at} \quad \zeta = 0$$

$$V_r = V_\theta = V_\varphi = 0 \quad \text{at} \quad \zeta = h \qquad (66)$$

Integration of the first two equations of system (65) under the boundary conditions (66) yields

$$V_\theta = -\frac{h^2}{2\mu R}\frac{\partial p}{\partial\theta}\frac{\zeta}{h}\left(1 - \frac{\zeta}{h}\right) + V_\theta^0\left(1 - \frac{\zeta}{h}\right)$$

$$V_\varphi = -\frac{h^2}{2\mu R \sin\varphi}\frac{\partial p}{\partial\varphi}\frac{\zeta}{h}\left(1 - \frac{\zeta}{h}\right) + V_\varphi^0\left(1 - \frac{\zeta}{h}\right) \qquad (67)$$

We shall change over, in complete analogy with the previous example, to mean velocities

$$\bar{V}_\theta = \frac{1}{h}\int_0^h V_\theta d\zeta = -\frac{h^2}{12\mu R}\frac{\partial p}{\partial\theta} + \frac{1}{2}V_\theta^0$$

$$\bar{V}_\varphi = \frac{1}{h}\int_0^h V_\varphi d\zeta = -\frac{h^2}{12\mu R\sin\theta}\frac{\partial p}{\partial\varphi} + \frac{1}{2}V_\varphi^0 \qquad (68)$$

Let us now integrate both sides of the last equation of system (65), the continuity equations, for ζ from 0 to h. Then, in the above notation, we have

$$\frac{\partial}{\partial\theta}(h\bar{V}_\theta \sin\theta) + \frac{\partial}{\partial\varphi}(h\bar{V}_\varphi) = V_r^0 R\sin\theta \qquad (69)$$

Substituting the expressions $\bar{V}_\theta$ and $\bar{V}_\varphi$ from Equation 68 into 69, we obtain the following equation for pressure distribution over the inner sphere surface. (In transformations it is taken into account that according to Equation 63, h is independent of φ and therefore it can be taken outside the derivatives with respect to this variable.)

$$\sin\theta\frac{\partial}{\partial\theta}\left(h^3\frac{\partial p}{\partial\theta}\sin\theta\right)+h^3\frac{\partial^2 p}{\partial\varphi^2} =$$

$$= 6\mu R\left[\frac{\partial}{\partial\theta}(hV_\theta^0\sin\theta)+h\frac{\partial V_\varphi^0}{\partial\varphi}\right]\sin\theta - 12\mu R^2V_r^0\sin^2\theta \tag{70}$$

To calculate the values of V_r^0, V_θ^0 and V_φ^0, note that by the familiar equation of the solid body kinetics,

$$\boldsymbol{V}^0 = \boldsymbol{V}_0+\boldsymbol{\omega}\times\boldsymbol{r}^0 \tag{71}$$

where the vector $\boldsymbol{V}_0$ defines the small velocity of the inner sphere center 0, $\boldsymbol{\omega}$ is its final angular velocity, and $\boldsymbol{r}^0$ is the radius-vector of the point on the sphere. Then, prescribing the vector $\boldsymbol{V}_0$ by its magnitude V_0, by the angle α with the axis $0z$ and by the angle β of its projection onto the plane $0xy$ with the axis $0x$ and, similarly, the vector $\boldsymbol{\omega}$ by its magnitude ω and by the angles γ and δ, we shall have, applying the formulas of spherical trigonometry ($r_r^0 = R$, $r_\theta^0 = V_\varphi^0 = 0$) (see Figure 165*a* and *c*),

$$V_r^0 = V_{0r} \qquad V_\theta^0 = V_{0\theta}+R\omega_\varphi \qquad V_\varphi^0 = V_{0\varphi}-R\omega_\theta$$

$$V_{0r} = V_0\cos(\boldsymbol{V}_0, 0M) = V_0[\cos\alpha\cos\theta+\sin\alpha\sin\theta\cos(\varphi-\beta)]$$

$$V_{0\theta} = V_0\cos(\boldsymbol{V}_0, 0M_1) = V_0[-\cos\alpha\sin\theta+\sin\alpha\cos\theta\cos(\varphi-\beta)]$$

$$V_{0\varphi} = V_0\cos(\boldsymbol{V}_0, 0N_1) = -V_0\sin\alpha\sin(\varphi-\beta)$$

$$\omega_\theta = \omega\cos(\boldsymbol{\omega}, 0M_1) = \omega[-\cos\gamma\sin\theta+\sin\gamma\cos\theta\cos(\varphi-\delta)]$$

$$\omega_\varphi = \omega\cos(\boldsymbol{\omega}, 0N_1) = -\omega\sin\gamma\sin(\varphi-\delta)$$

and, consequently,

$$V_\theta^0 = V_0[-\cos\alpha\sin\theta+\sin\alpha\cos\theta\cos(\varphi-\beta)]-\omega R\sin\gamma\sin(\varphi-\delta)$$

$$V_\varphi^0 = -V_0\sin\alpha\sin(\varphi-\beta)+\omega R[\cos\gamma\sin\theta-\sin\gamma\cos\theta\cos(\varphi-\delta)] \tag{72}$$

Substituting these expressions into the right side of Equation 70, we note that the last term on the right side of this equation has the order R^2V_0. After substitution and use of Equation 63, other terms appear involving V_0eR as a multiplier, and, consequently, they can be omitted. Thus, constructing the final form of the approximate equation which serves to find the pressure distribution $p(\theta, \varphi)$ over the moving sphere surface,

$$\sin\theta\frac{\partial}{\partial\theta}\left(h^3\frac{\partial p}{\partial\theta}\sin\theta\right)+h^3\frac{\partial^2 p}{\partial\varphi^2} = 6\mu R^2\omega e\sin\gamma\sin^3\theta\sin(\varphi-\delta)-$$

$$-12\mu V_0R^2\sin^2\theta[\cos\alpha\cos\theta+\sin\alpha\sin\theta\cos(\varphi-\beta)] \tag{73}$$

We seek the solution to this equation in the form

$$p(\theta, \varphi) = \Theta_0(\theta) + \Theta_1(\theta)\cos(\varphi - \beta) + \Theta_2(\theta)\sin(\varphi - \delta) \tag{74}$$

Substituting the expression of p into Equation 73 and equating the coefficients at $\cos(\varphi - \beta)$ and $\sin(\varphi - \delta)$ on both sides of the equation obtained, we come to the system of ordinary differential equations of the second order

$$\sin\theta \frac{d}{d\theta}\left(h^3 \sin\theta \frac{d\Theta_0}{d\theta}\right) = -12\mu V_0 R^2 \cos\alpha \sin^2\theta \cos\theta$$

$$\sin\theta \frac{d}{d\theta}\left(h^3 \sin\theta \frac{d\Theta_1}{d\theta}\right) - h^3\Theta_1 = -12\mu V_0 R^2 \sin\alpha \sin^3\theta$$

$$\sin\theta \frac{d}{d\theta}\left(h^3 \sin\theta \frac{d\Theta_2}{d\theta}\right) - h^3\Theta_2 = 6\mu R^2 \omega e \sin\gamma \sin^3\theta \tag{75}$$

the last two of which differ only by the constant multiplier on the right side.

The first of Equations 75 is easily integrated. In fact, reducing both sides of this equation by $\sin\theta$ and making one integration, we obtain

$$h^3 \frac{d\Theta_0}{d\theta} = -6\mu R^2 V_0 \cos\alpha \sin\theta + \frac{C_1}{\sin\theta}$$

From the condition of finiteness of the quantity $\partial p/\partial\theta$, according to Equation 74, there follows finiteness of $\partial\theta_0/\partial\theta$, which in turn requires the integration constant C_1 to be equal to zero. Reintegration gives

$$\Theta_0(\theta) = -6\mu V_0 R^2 \cos\alpha \int_0^\theta \frac{\sin\theta \, d\theta}{h^3}$$

In this case the integration constant is chosen so that $\Theta_0(0) = 0$; this determines the choice of constant pressure in the cavity between the spheres. The integral on the right side, h being replaced by its expression by (63), is easily calculated and gives

$$\Theta_0(\theta) = -\frac{3\mu V_0 R^2 \cos\alpha}{\varepsilon^3 (1+\lambda)^2} \frac{(2 + \lambda + \lambda\cos\theta)(1 - \cos\theta)}{(1 + \lambda\cos\theta)^2} \tag{76}$$

where the notation $\lambda = e/\varepsilon$ $(0 \le \lambda \le 1)$ is introduced. This also holds true in the case of a plane bearing.

The second and third equations of system (75) should be integrated

$$\Theta_1 = \Theta_2 = 0 \quad \text{at} \quad \theta = 0 \quad \text{and} \quad \theta = \pi$$

under the boundary conditions which follow directly from the condition of finiteness of the quantity $\bar{V}_\varphi$ determined by the second equation of system (68) and, consequently, from the necessity for the derivative $\partial p/\partial\varphi$ at $\theta = 0$ and $\theta = \pi$ to vanish. This, in the presence of Equation 74, leads to the boundary conditions (see above).

The solutions of the second and third equations of system (75) under the mentioned boundary conditions can be also found in a closed form, *viz.*:

$$\Theta_1(\theta) = \frac{12\mu V_0 R^2 \cos\alpha}{\varepsilon^3} \frac{2+\lambda\cos\theta}{(4+\lambda^2)(1+\lambda\cos\theta)^2}\sin\theta$$

$$\Theta_2(\theta) = -\frac{6\mu R^2 \omega e \sin\gamma}{\varepsilon^3} \frac{2+\lambda\cos\theta}{(4+\lambda^2)(1+\lambda\cos\theta)^2}\sin\theta \tag{77}$$

The aggregate of Equations 74, 76, and 77 fully determines the distribution of pressures $p(\theta, \varphi)$ over the surface of a spherical tenon in a bearing, thus making it possible to easily calculate the "supporting" force, distribution of velocities, and skin frictions required for calculation of the moment of resistance to tenon rotation.

When calculating the resultant $\boldsymbol{F}$ of the fluid reactions, which is applied to the inner sphere, we can neglect viscous normal and tangential stresses having, with the accuracy to the square of the characteristic length, the order of $\mu R^2 V_0/\varepsilon$ and $\mu R^2 V_\theta/\varepsilon$, respectively, compared with pressure which has the order of $\mu R^2 V_\theta/\varepsilon^2$. We obtain ($\sigma$ is the inner sphere surface, $\boldsymbol{n}$ is the unit vector of the outer normal to the surface)

$$\boldsymbol{F} = -\int_\sigma p\boldsymbol{n}\,d\sigma = -R^2\int_0^{2\pi} d\varphi \int_0^{\pi} p\boldsymbol{n}\sin\theta\,d\theta$$

or in the projections onto the coordinate axis (Figure 165*a*)

$$F_x = -R^2\int_0^{2\pi} d\varphi\int_0^{\pi} p\sin\theta\cos(\widehat{M0x})\,d\theta = -R^2\int_0^{2\pi}\cos\varphi\,d\varphi\int_0^{\pi} p\sin^2\theta\,d\theta$$

$$F_y = -R^2\int_0^{2\pi} d\varphi\int_0^{\pi} p\sin\theta\cos(\widehat{M0y})\,d\theta = -R^2\int_0^{2\pi}\sin\varphi\,d\varphi\int_0^{\pi} p\sin^2\theta\,d\theta$$

$$F_z = -R^2\int_0^{2\pi} d\varphi\int_0^{\pi} p\sin\theta\cos(\widehat{M0z})\,d\theta = -R^2\int_0^{2\pi} d\varphi\int_0^{\pi} p\sin\theta\cos\theta\,d\theta$$

Substituting into these equations $p(\theta, \varphi)$ according to Equation 74 and integrating with respect to φ, construct the following formulas for the projections F_x, F_y, and F_z:

$$F_x = -\pi R^2\cos\beta\int_0^{\pi}\Theta_1(\theta)\sin^2\theta\,d\theta + \pi R^2\sin\delta\int_0^{\pi}\Theta_2(\theta)\sin^2\theta\,d\theta$$

$$F_x = -\pi R^2\sin\beta\int_0^{\pi}\Theta_1(\theta)\sin^2\theta\,d\theta - \pi R^2\cos\delta\int_0^{\pi}\Theta_2(\theta)\sin^2\theta\,d\theta$$

$$F_z = -\pi R^2\int_0^{\pi}\Theta_0(\theta)\sin 2\theta\,d\theta \tag{78}$$

It remains only to substitute the values of the functions $\Theta_0(\theta)$ by Equation 76 and of the functions $\Theta_1(\theta)$, $\Theta_2(\theta)$ by Equation 77. Calculating the integrals we finally obtain

$$F_x = -\frac{4\pi\mu R^4(2V_0 \sin\alpha\cos\beta + \omega e\sin\gamma\sin\delta)}{\varepsilon^3}K_1(\lambda)$$

$$F_y = -\frac{4\pi\mu R^4(2V_0 \sin\alpha\sin\beta - \omega e\sin\gamma\cos\delta)}{\varepsilon^3}K_1(\lambda)$$

$$F_z = -\frac{8\pi\mu R^4 V_0\cos\alpha}{\varepsilon^3}K_2(\lambda) \tag{79}$$

where for brevity the notation

$$K_1(\lambda) = \frac{3}{2(4+\lambda^2)}\left[\left(\frac{1}{\lambda}+\frac{1}{\lambda^3}\right)\ln\frac{1+\lambda}{1-\lambda}-\frac{2}{\lambda^2}\right]$$

$$K_2(\lambda) = \frac{3}{4\lambda^3}\left(\frac{2\lambda}{1-\lambda^2}-\ln\frac{1+\lambda}{1-\lambda}\right)$$

$$K_1(0) = K_2(0) = 1 \tag{80}$$

Is introduced the set of Equations 79 is equivalent to one vector formula

$$\boldsymbol{F} = -\frac{8\pi\mu R^4}{\varepsilon^3}K_1(\lambda)\boldsymbol{V}_0 + \frac{4\pi\mu R^4}{\varepsilon^3}K_1(\lambda)\boldsymbol{\omega}\times\boldsymbol{e} + $$

$$+\frac{8\pi\mu R^4}{\varepsilon^5}\frac{K_1(\lambda)-K_2(\lambda)}{\lambda^2}(\boldsymbol{V}_0\cdot\boldsymbol{e})\boldsymbol{e} \tag{81}$$

where the vector $\boldsymbol{e} = \overrightarrow{O'O}$ of the inner sphere center displacement relative to the outer sphere is introduced.

In the absence of translational motion of the inner sphere ($\boldsymbol{V}_0 = 0$) Equation (81) is substantially simplified and reduced to

$$\boldsymbol{F} = \frac{4\pi\mu R^4}{\varepsilon^3}K_1(\lambda)\boldsymbol{\omega}\times\boldsymbol{e} \tag{82}$$

According to this vector equation, in the case of inner sphere rotation about some axis that is not accompanied by translational motion of the sphere, the result of the reactions of a viscous incompressible fluid is perpendicular to the plane drawn through this axis and the line of the centers of the inner and motionless outer spheres. By its structure Equation 82 is similar to the previously derived Equation 61 for a plane motion of a cylindrical shaft in a bearing.

In the particular case ($\boldsymbol{V}_0 = 0$, $\boldsymbol{\omega}\perp\boldsymbol{e}$) we obtain the Wannier equation[14]

$$F = \frac{6\pi\mu R^4\omega}{e^2(4\varepsilon^2+e^2)}\left[(\varepsilon^2+e^2)\ln\frac{\varepsilon+e}{\varepsilon-e}-2\varepsilon e\right]$$

---

[14] Wannier, G. H., A contribution to the hydrodynamics of lubrication, *Q. J. Appl. Math.*, 8(1), 1–32, 1950.

At the moment of the coincidence of the centers of the spheres ($\lambda = 0$) the resultant $\boldsymbol{F}$ will be determined only by the translational motion of the inner sphere and will be equal to

$$\boldsymbol{F} = -\frac{8\pi\mu R^4}{\varepsilon^3}\boldsymbol{V}_0$$

A sharply pronounced dependence of this vector on the sphere radius R and on the cavity lateral dimension ε attracts attention. Comparing using the Stokes formula (which is concerned with sphere motion in an infinite viscous incompressible fluid), the resistance of this fluid to sphere motion inside the envelope (which is close to its size by the order of $(R/\varepsilon)^3$ exceeds the resistance to the motion in an infinite fluid.

Calculate the resultant moment $\boldsymbol{L}$ of the viscous friction forces applied to the inner sphere surface. Making use of the set of equations (13) in Chapter 10, we have (adopting the notation τ instead of p_{ij}, r instead of R, φ instead of ε, and $\partial\zeta$ instead of ∂R)

$$\tau_{r\theta} = \mu\left(\frac{1}{r}\frac{\partial V_r}{\partial\theta} + \frac{\partial V_\theta}{\partial\zeta} - \frac{V_\theta}{r}\right)_{\zeta=0} = \mu\left(\frac{\partial V_\theta}{\partial\zeta}\right)_{\zeta=0}$$

$$\tau_{r\varphi} = \mu\left(\frac{\partial V_\varphi}{\partial\zeta} + \frac{1}{r\sin\theta}\frac{\partial V_r}{\partial\varphi} - \frac{V_\varphi}{r}\right)_{\zeta=0} = \mu\left(\frac{\partial V_\varphi}{\partial\zeta}\right)_{\zeta=0}$$

The simplifications made are obvious; the quantities

$$V_\theta \quad V_\varphi \quad \frac{\partial V_r}{\partial\theta} \quad \frac{\partial V_r}{\partial\varphi}$$

are small as compared to the terms retained.

Substituting the values of V_θ and V_φ from Equation 67 into the expressions above for $\tau_{r\theta}$ and $\tau_{r\varphi}$, we obtain

$$\tau_{r\theta} = -\frac{h}{2R}\frac{\partial p}{\partial\theta} - \frac{\mu V_\theta^0}{h} \qquad \tau_{r\varphi} = -\frac{h}{2R\sin\theta}\frac{\partial p}{\partial\varphi} - \frac{\mu V_\varphi^0}{h}$$

Then, using the previously derived expressions (72) for V_θ^0 and V_φ^0 and the three-term presentation of the pressure p (Equation 74), we find

$$\tau_{\tau\theta} = -\frac{h}{2R}\left[\Theta_0'(\theta) + \Theta_1'(\theta)\cos(\varphi-\beta) + \Theta_2'(\theta)\sin(\varphi-\delta)\right] + $$
$$+\frac{\mu\omega R}{h}\sin\gamma\sin(\varphi-\delta)$$

$$\tau_{r\varphi} = \frac{h}{2R\sin\theta}\left[\Theta_1(\theta)\sin(\varphi-\beta) - \Theta_2(\theta)\cos(\varphi-\delta)\right] -$$
$$-\frac{\mu\omega R}{h}\left[\cos\gamma\sin\theta - \sin\gamma\cos\theta\cos(\varphi-\delta)\right] \qquad (83)$$

The elementary moments of friction forces dL_x, dL_y, and dL_z relative to rectangular Cartesian coordinate axes are determined as the sums of the moments of forces relative to these axes tangential to the sphere surface that have spherical coordinates

$$dF_\theta = \tau_{r\theta} R^2 \sin\theta\, d\theta\, d\varphi \qquad dF_\varphi = \tau_{r\varphi} R^2 \sin\theta\, d\theta\, d\varphi$$

and the corresponding Cartesian components

$$dF_\theta \cos\theta \cos\varphi \qquad dF_\theta \cos\theta \sin\varphi \qquad -dF_\theta \sin\theta \qquad -dF_\varphi \sin\varphi \qquad dF_\varphi \cos\varphi \qquad 0$$

Noting that the application point M of these forces has the coordinates ($R\sin\theta \cos\varphi$, $R\sin\theta\sin\varphi$, $R\cos\theta$), according to the conventional rule we construct the equations of moments relative to the coordinate axes

$$dL_x = -R^3 (\tau_{r\theta} \sin\varphi + \tau_{r\varphi} \cos\theta \cos\varphi) \sin\theta\, d\theta\, d\varphi$$

$$dL_y = R^3 (\tau_{r\theta} \cos\varphi - \tau_{r\varphi} \cos\theta \sin\varphi) \sin\theta\, d\theta\, d\varphi$$

$$dL_z = R^3 \tau_{r\varphi} \sin^2\theta\, d\theta\, d\varphi$$

Substituting the values of $\tau_{t\theta}$ and $\tau_{r\varphi}$ into these equations by Equation 83, collecting the terms with the trigonometrical functions of the angle φ and making first integration for φ from 0 to 2π and then also for θ from 0 to π, we find

$$L_x = -\frac{4\pi\mu R^4}{\varepsilon}\left[\frac{2}{3}(1+\lambda^2)\,\omega \sin\gamma \cos\delta - \frac{1}{\varepsilon^2} V_0 e \sin\alpha \sin\beta\right] K_1(\lambda)$$

$$L_y = -\frac{4\pi\mu R^4}{\varepsilon}\left[\frac{2}{3}(1+\lambda^2)\,\omega \sin\gamma \sin\delta + \frac{1}{\varepsilon^2} V_0 e \sin\alpha \cos\beta\right] K_1(\lambda)$$

$$L_y = -\frac{8\pi\mu R^4}{3\varepsilon}\,\omega\,(1-\lambda^2) \cos\gamma\, K_2(\lambda) \tag{84}$$

where the functions $K_1(\lambda)$ and $K_2(\lambda)$ are the same.

The obtained set of equations for the moments relative to the coordinate axes is equivalent to one vector equation for the moment relative to the coordinate origin

$$\boldsymbol{L} = -\frac{8\pi\mu R^4}{3\varepsilon}(1+\lambda^2) K_1(\lambda)\,\boldsymbol{\omega} - \frac{4\pi\mu R^4}{\varepsilon^3} K_1(\lambda)\, \boldsymbol{V}_0 \times \boldsymbol{e} +$$
$$+\frac{8\pi\mu R^4}{3\varepsilon^3}\,\frac{(1+\lambda^2) K_1(\lambda) - (1-\lambda^2) K_2(\lambda)}{\lambda^2}\,(\boldsymbol{\omega}\cdot\boldsymbol{e})\,\boldsymbol{e} \tag{85}$$

In the case of a pure rotational motion of the inner sphere ($\boldsymbol{V}_0 = 0$) we have

$$\boldsymbol{L} = -\frac{8\pi\mu R^4}{3\varepsilon}(1+\lambda^2) K_1(\lambda)\,\boldsymbol{\omega} +$$
$$+\frac{8\pi\mu R^4}{3\varepsilon^3}\,\frac{(1+\lambda^2) K_1(\lambda) - (1-\lambda^2) K_2(\lambda)}{\lambda^2}\,(\boldsymbol{\omega}\cdot\boldsymbol{e})\,\boldsymbol{e} \tag{86}$$

and in the particular case of the perpendicularity of the axis of rotation and the center line of the spheres ($\boldsymbol{\omega}\cdot\boldsymbol{e} = 0$), studied by Wannier, we obtain the simpler equation

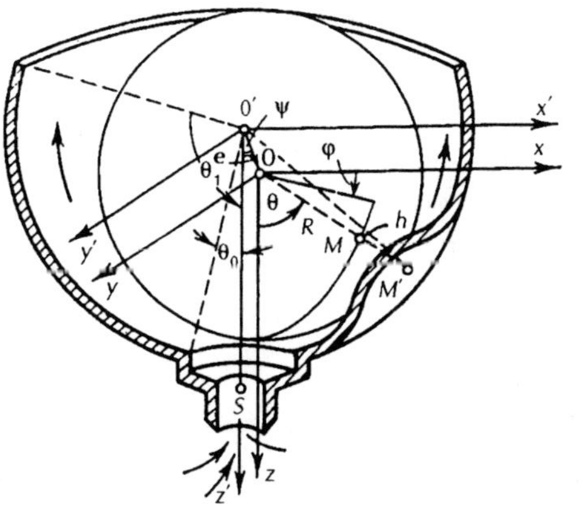

Fig. 166

$$L = \frac{4\pi\mu R^4\omega}{e^3}\frac{\varepsilon^2+e^2}{4\varepsilon^2+e^2}\left[(\varepsilon^2+e^2)\ln\frac{\varepsilon+e}{\varepsilon-e}-2\varepsilon e\right]$$

In the case of concentric spheres ($e = 0$) Equation 86 is reduced to the form

$$L = -\frac{8\pi\mu R^4}{3\varepsilon}\boldsymbol{\omega} \tag{87}$$

which is the generalization of the Petrov equation (29) for a cylindrical bearing to the case of a ball bearing.

Equations of lubricant motion in a ball bearing refer to the type of equations used for calculating a spherical suspension or, as it is sometimes referred to, a hydrostatic bearing.

A schematic diagram of this suspension is depicted in Figure 166. A heavy sphere is motionlessly suspended in an incompressible viscous fluid flow produced due to the supply of this fluid through the opening S on the bottom of a motionless spherical cup which embraces the sphere suspended in a flow. In this case the gap between the surfaces of the sphere and the cup is assumed to be very small as compared with their radii R and R'.

Drawing the coordinate axes as shown in Figure 166, we had, in order to simplify calculations, to choose the direction of the axis $0'z'$ so that it passes through the center of the opening for fluid supply S. In this connection equation (63), which determines the gap width h between the spheres in the case of a ball bearing, becomes complicated.

As is easily seen from the figure

$$h = MM' \approx 0'M' - 0'0\cos(\overrightarrow{0'0}, \overrightarrow{0'M'}) - 0M$$

$$0'M' - 0M \approx R' - R = \varepsilon \qquad 0'0 = e$$

$$\cos(\overrightarrow{0'0}, \overrightarrow{0'M'}) = \cos\psi\cos\theta + \sin\psi\sin\theta\cos\varphi$$

so that (ψ is the constant with the given position of the sphere relative to the cup angle between the displacement vector $\boldsymbol{e}$ of the sphere center 0 and the axis $0z$)

$$h(\theta, \varphi) = \varepsilon - e(\cos\psi\cos\theta + \sin\psi\sin\theta\cos\theta) \tag{88}$$

Thus, in the case of suspension the quantity h is the function of both θ and φ thus making it impossible, when deriving Equation 70, to take it outside the sign of the derivative with respect to φ. Moreover $V^0 = 0$, so that instead of (70) the following equation should be integrated:

$$\sin\theta\frac{\partial}{\partial\theta}\left(h^3\frac{\partial p}{\partial\theta}\sin\theta\right)+\frac{\partial}{\partial\varphi}\left(h^3\frac{\partial p}{\partial\varphi}\right) = 0 \tag{89}$$

where h is prescribed by Equation 89.

The development of the methods for integrating equations similar to Equation 89 is the problem of the lubrication theory. We restrict ourselves to the consideration of the simplest case of a suspended vertically displaced sphere ($\psi = 0, \pi$), when the gap width h is determined by Equation 63. The more general problem, when along with vertical displacement of the sphere center a horizontal (lateral) displacement takes place, also can be easily considered in an approximate formulation (small $\lambda = e/\varepsilon$).[15]

We present the fluid supply through an opening in the cup bottom, schematically placing the source with volumetric flow rate per second Q at the point S ($\theta = 0$), and assuming that efflux occurs through a circular slot ($\theta = \theta_1$) on the upper edge of the cup into a space with the absolute pressure p_1. Equation 89 is rewritten in the form

$$\frac{d}{d\theta}\left(h^3\frac{dp}{d\theta}\sin\theta\right) = 0$$

and has the first integral

$$h^3\frac{dp}{d\theta}\sin\theta = \text{const}$$

To determine the constants we formulate the condition of the constancy of flow rate through the horizontal cross-section of a gap

$$2\pi Rh\sin\theta\cdot\bar{V}_\theta = \text{const} = Q$$

or by the first of Equations 68 at $V_\theta^0 = 0$

$$-\frac{\pi h^3}{6\mu}\frac{dp}{d\theta}\sin\theta = Q$$

Thus, we have

$$\frac{dp}{d\theta} = -\frac{6\mu Q}{\pi}\frac{1}{(\varepsilon - e\cos\theta)^3\sin\theta} \tag{90}$$

Integrating once more and using the boundary condition

$$p = p_1 \quad \text{at} \quad \theta = \theta_1$$

we obtain the unknown pressure distribution

---

[15] Loitsyanskiy, L. G. and Stepanyants, L. G., Hydrodynamic theory of a spherical suspension, *Tr. LPI*, No. 198, 89–98, 1958.

$$p - p_1 = \frac{6\mu Q}{\pi\varepsilon^3}[\Phi(\theta_1) - \Phi(\theta)] \tag{91}$$

where it is assumed that

$$\Phi(\theta) = \frac{1}{2}\frac{\lambda}{1-\lambda^2}\frac{1}{(1-\lambda\cos\theta)^2} + \frac{2\lambda}{(1-\lambda^2)^2}\frac{1}{1-\lambda\cos\theta} - \frac{\lambda(3+\lambda^2)}{(1-\lambda^2)^3}\ln(1-\lambda\cos\theta) -$$

$$-\frac{1}{2(1+\lambda)^3}\ln(1+\cos\theta) + \frac{1}{2(1-\lambda)^3}\ln(1-\cos\theta) \tag{92}$$

At $\theta = 0$ (i.e., directly in the source S itself) the pressure is equal to infinity. To make the solution more physical, we assume that close to the point S in the cup body there is a cavity with the constant pressure p_0; the size of this cavity can be prescribed by the angle θ_0. Then the working difference of pressures $p_0 - p_1$ in a suspension is, according to Equation 91, related to the flow rate by the equation

$$p_0 - p_1 = \frac{6\mu Q}{\pi\varepsilon^3}[\Phi(\theta_1) - \Phi(\theta_0)] \tag{93}$$

In spite of a comparatively complicated character of the equation for pressure distribution (91), the force $\boldsymbol{F}$, supporting the sphere, is calculated simply. Note that the orders of skin frictions and pressure are

$$\tau = \mu\left(\frac{\partial V_\theta}{\partial\zeta}\right)_{\zeta=0} \sim \frac{\mu Q}{R\varepsilon^2} \qquad p_0 - p_1 = \frac{\mu Q}{\varepsilon^3}$$

so that

$$\tau/(p_0 - p_1) = O(\varepsilon)$$

This proves the possibility of neglecting friction when determining the supporting force $\boldsymbol{F}$ and finding the force to be the integral of the elementary pressure forces only. We have, integrating by parts and reckoning the pressure from the value p_1,

$$F_z = -2\pi R^2 p_0 \int_0^{\theta_0} \sin\theta\cos\theta\, d\theta - 2\pi R^2 \int_0^{\theta_1} p\sin\theta\cos\theta\, d\theta =$$

$$= \pi R^2 \int_{\theta_0}^{\theta_1} \frac{dp}{d\theta}\sin^2\theta\, d\theta = \left(-\frac{3\mu R^2 Q}{\varepsilon^3}\right)\frac{1}{f(\lambda;\theta_1)} \tag{94}$$

where it is assumed that

$$f(\lambda;\theta_1) = \frac{(1-\lambda\cos\theta_0)^2(1-\lambda\cos\theta_1)^2}{\left(1-\lambda\cos\frac{\theta_1+\theta_0}{2}\cos\frac{\theta_1-\theta_0}{2}\right)\sin\frac{\theta_1+\theta_0}{2}\sin\frac{\theta_1-\theta_0}{2}} \tag{95}$$

The dependence of the supporting force component F_z on the dimensionless vertical displacement of the inner sphere λ is very complex because at the given working pressure difference $p_0 - p_1$, Q is in turn the function of the dimensionless displacement λ. In fact the problem is even more complex because the pressure difference $p_P - p_1$ is usually maintained, where p_P is the absolute pressure in the plenum chamber from which gas through some channel is supplied to the opening S.

The problem of the stability of the heavy sphere equilibrium state in a somewhat simplified presentation of a blowing device (a point source at the point S) is considered in the fourth edition of this book.[16] It is found that when $\theta_1 < 90°$ only one equilibrium position is possible, while when $\theta_1 > 90°$ two are possible.

The theory related to the use of an incompressible fluid or gas under the conditions when its compressibility can be neglected. Under great loads on a suspension or on a rotating shaft in a bearing the problem becomes more complex due to the necessity to take into account medium compressibility in the Reynolds equation. This equation, even in the isothermal approach, is nonlinear and to be integrated requires application of approximate numerical methods.

6. Application of the Dimensionality Theory to Determination of the Structure of the Solution of the Navier-Stokes Equations. Self-Similar Solutions

Formation of the dimensionality matrix (Section 5 of Chapter 10) and of the main relation of the Π theorem (Equation 56 of Chapter 10) makes it possible in a number of cases to simplify seeking the solutions to the nonlinear Navier-Stokes equations by reducing these partial differential equations to ordinary differential equations. This simplification is always possible if, by the essence of the problem formulation itself, some scale quantities in the problem, e.g., length and velocity, do not appear. The problems allowing the simplification mentioned refer to the class of self-similar ones (compare to the content of Chapter 6, Section 11). We shall illustrate this fact on the classical von Kármán problem on a stationary viscous fluid flow in a half-space over the disk of an infine of an radius uniformly rotating in its plane.

A cylindrical coordinate system z, r, ε is selected, placing the origin in the disk center, and directing the axis z normally to the plane of its rotation. The following nine quantities serve as the variables and constants describing the phenomenon of the coordinates r and z (ε vanishes due to the symmetry), the angular velocity of disk rotation ω, the velocity components of fluid particles V_r, V_ε, V_z, the pressure difference $\Delta p = p - p_\infty$, the fluid density p, and the coefficient of kinematic viscosity ν. The equations of motion are a complex nonlinear system of partial differential equations (derivatives with respect to ε are omitted due to symmetry), which according to Equation 33 of Chapter 10 has the form

$$V_r\frac{\partial V_r}{\partial r} + V_z\frac{\partial V_r}{\partial z} - \frac{V_\varepsilon^2}{r} = -\frac{1}{\rho}\frac{\partial p}{\partial r} + \nu\left(\frac{\partial^2 V_r}{\partial r^2} + \frac{1}{r}\frac{\partial V_r}{\partial r} + \frac{\partial^2 V_r}{\partial z^2} - \frac{V_r}{r^2}\right)$$

$$V_r\frac{\partial V_\varepsilon}{\partial r} + V_z\frac{\partial V_\varepsilon}{\partial z} - \frac{V_r V_\varepsilon}{r} = \nu\left(\frac{\partial^2 V_\varepsilon}{\partial r^2} + \frac{1}{r}\frac{\partial V_\varepsilon}{\partial r} + \frac{\partial^2 V_\varepsilon}{\partial z^2} - \frac{V_\varepsilon}{r^2}\right)$$

$$V_r\frac{\partial V_z}{\partial r} + V_z\frac{\partial V_z}{\partial z} = -\frac{1}{\rho}\frac{\partial p}{\partial z} + \nu\left(\frac{\partial^2 V_z}{\partial r^2} + \frac{1}{r}\frac{\partial V_z}{\partial r} + \frac{\partial^2 V_z}{\partial z^2}\right)$$

---

[16] Loitsyanskiy, L. G., *Mekhanika zhidkosti i gaza* (*Fluid Mechanics*), 4th ed., Izd. Nauka, Moscow, 1973, 483.

$$\frac{\partial}{\partial r}(rV_r) + \frac{\partial}{\partial z}(rV_z) = 0 \tag{96}$$

under the boundary conditions

$$V_r = 0 \qquad V_\varepsilon = \omega r \qquad V_z = 0 \quad \text{at} \quad z = 0$$

$$V_r \to 0 \qquad V_\varepsilon \to 0 \quad \text{when} \quad z \to \infty \tag{97}$$

The solution of system (96) with the boundary conditions (97) does not contain the scales of lengths and velocities. They can be composed from dimensionality considerations. We shall take the quantity $\sqrt{\nu/\omega}$, which has the dimensional representation of length, as the length scale and the quantities $\sqrt{\nu\omega}$ and $\rho\nu\omega$ as the scales of velocities and pressures, respectively. With the fact that the boundary conditions in the first line of Equation 97 are satisfied at any r and V_ε at $z = 0$ is equal to $V_\varepsilon = \omega r$ taken into account, we seek the solution in the form ($V_z < 0$)

$$V_r = \omega r F(\zeta) \qquad V_\varepsilon = \omega r G(\zeta) \qquad V_z = \sqrt{\nu\omega}\, H(\zeta)$$

$$p - p_0 = -\rho\nu\omega P(\zeta) \tag{98}$$

where it is assumed that $\zeta = z\sqrt{\omega/\nu}$. It can be easily shown that the result of substitution of velocity and pressure components (98) into the system of differential equations (96) will lead to the system of ordinary differential equations (the prime denotes the derivative with respect to ζ)[17]

$$F^2 - G^2 + F'H = F'' \qquad 2FG + G'H = G'' \qquad HH' = P' + H''$$

$$2F + H' = 0 \tag{99}$$

with the boundary conditions

$$F = 0 \qquad G = 1 \qquad H = 0 \qquad P = 0 \quad \text{when} \quad \zeta = 0$$

$$F \to 0 \qquad G \to 0 \quad \text{when} \quad \zeta \to \infty \tag{100}$$

This proves self-similarity of the solution to the von Kármán problem. From the first, second, and fourth formulas of Equation 99, F, G, and H are determined, and from the third, P. An approximate solution of the system of ordinary differential equations (99) with boundary conditions (100) was made by Cochran[18] by the method of expansion of the functions $F(\zeta)$, $G(\zeta)$, $H(\zeta)$ into a series

$$F = a_0\zeta - \frac{1}{2}\zeta^2 - \frac{1}{3}b_0\zeta^3 - \frac{1}{12}b_0^2\zeta^4 - \cdots$$

$$G = 1 + b_0\zeta + \frac{1}{3}a_0\zeta^3 + \frac{1}{12}(a_0b_0 - 1)\zeta^4 + \cdots$$

---

[17] Equations 99 and the boundary conditions to them (100) were found by von Kármán (von Kármán, Th., Über laminare und turbulente Reibung, *Z. Angew. Math. Mech.*, 1, 233–252, 1921.

[18] Cochran, W. G., The flow due to a rotating disk, *Proc. Cambr. Phil. Soc.*, 30, 365–375, 1934.

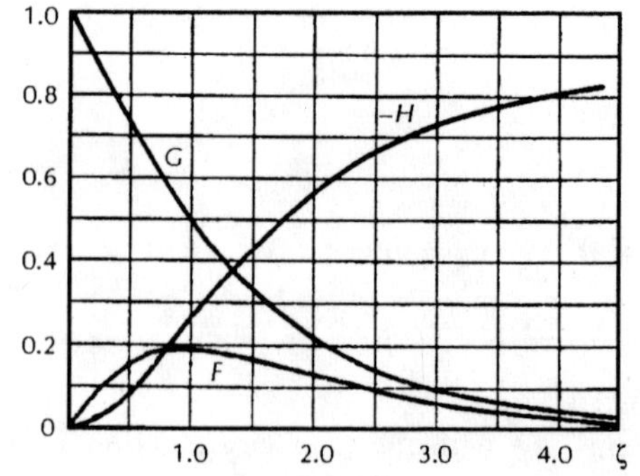

Fig. 167

$$H = -a_0\zeta^2 + \frac{1}{3}\zeta^3 + \frac{1}{6}b_0\zeta^4 + \cdots \tag{101}$$

at small ζ and

$$F = Ae^{-c\zeta} - \frac{A^2 + B^2}{2c^2}e^{-2c\zeta} + \frac{A(A^2 + B^2)}{4c^2}e^{-3c\zeta} + \cdots$$

$$G = Be^{-c\zeta} - \frac{B(A^2 + B^2)}{12c^4}e^{-3c\zeta} + \cdots$$

$$H = -c + \frac{2A}{c}e^{-c\zeta} - \frac{A^2 + B^2}{2c^3}e^{-2c\zeta} + \frac{A(A^2 + B^2)}{6c^5}e^{-3c\zeta} + \cdots \tag{102}$$

at large ζ where the constants a_0, b_0, A, B, and c were determined from the conditions of continuous transition of the functions F, G, H, F', and G' from small values of ζ to large ones. By this method the following values of the constants were found

$$a_0 = 0.510 \quad b_0 = -0.616 \quad c = 0.866 \quad A = 0.934 \quad B = 1.208 \tag{103}$$

and the tables of the dependences of F, G, H, and P, on ζ available in the cited paper by Cochran were compiled.

We restrict ourselves here to the presentation of the graphs of the first three of these functions (Figure 167). It follows from the graphs that the functions F and G exponentially tend to zero when $\zeta \to \infty$. In this case, when ν/ω is small, F and G are virtually different from zero only within the limits of a thin layer on the disk surface. This case presents the first example of the boundary layer, the theory of which will be given in detail in Chapter 12. At this point we only emphasize that the ratio of the order of this layer thickness $\sqrt{\nu/\omega}$ to the disk radius a, artificially introduced for the von Kármán problem conditions, will have the order

$$\sqrt{\frac{\nu}{\omega}} : a = \frac{1}{\sqrt{\mathbf{Re}}} \qquad \mathbf{Re} = \frac{a^2\omega}{\nu} \tag{104}$$

Note the phenomenon typical of the flow considered. At a distance from the disk surface ($\zeta \to \infty$), according to the expansion of the function H (102), we have $(H)_{\zeta=\infty} = -c$, and, consequently, from Equation 98 $(V_z)_{\zeta=\infty} = -c\sqrt{\nu/\omega}$, or by Equation 103

$$(V_z)_{\zeta=\infty} = -0.866\pi a^2\sqrt{\nu\omega} \tag{105}$$

With this velocity the rotating disk of an infinitely large radius will draw in (note the minus sign on the right side of Equation 105) the fluid surrounding it and will throw this fluid away thanks to centrifugal volumetric forces to the disk periphery (in the von Kármán problem, to infinity).

Neglecting the end effect at $r = a$, where a is the disk radius, we can apply the obtained solution of the von Kármán problem to the disk of a finite radius a. Calculating the fluid flow rate per second Q, thrown from the disk surface by a centrifugal force

$$Q = 2\pi a\int_0^\infty V_r dz$$

we obtain

$$Q = 0.866\pi a^2\sqrt{\nu\omega} \tag{106}$$

In complete accordance with Equation 105:

$$(V_z)_{\zeta=\infty}\pi a^2 = Q$$

In the same approximation we determine the moment of fluid resistance to rotation of disk with a finite radius a

$$M = -2\pi\int_0^a r^2 p_{z\varepsilon}dr$$

where it is assumed that

$$p_{z\varepsilon} = \mu\left(\frac{\partial V_\varepsilon}{\partial z}\right)_{z=0} = \rho\sqrt{\nu\omega^3}rG'(0)$$

The moment of the resistance of a disk wetted by fluid from both sides will be equal to

$$2M = -\pi\rho a^4\sqrt{\nu\omega^3}G'(0) = 0.616\pi\rho a^4\sqrt{\nu\omega^3} \tag{107}$$

and introducing the coefficient of resistance

$$c_M = \frac{2M}{\frac{1}{2}\rho\omega^2 a^5} = -\frac{2\pi\sqrt{\nu}G'(0)}{a\sqrt{\omega}}$$

we can obtain the resistance equation

$$c_M = \frac{3.87}{\sqrt{\mathbf{Re}}} \qquad \mathbf{Re} = \frac{a^2\omega}{\nu} \tag{108}$$

This equation is in good agreement with the test data on the resistance of disks whose radii are large compared with the thickness of a boundary layer on their surface, i.e., by Equation 104 at large Reynolds numbers, but limited by critical values (see Chapter 13), so that the regime remains laminar.

The detailed example of the exact solution to the Navier-Stokes equations in the case of the infinite viscous fluid flow during uniform rotation of a disk in it is one of the simplest and most evident. This example of a self-similar solution is not the only one. Indeed, a class of self-similar solutions corresponding to plane radial flows of an incompressible viscous fluid in a convergent (divergent) duct with rectilinear walls intersecting at the source of the power Q[19] was considered long ago.

Noting that the dimensions of Q (in a plane motion $[Q] = VL$) coincides with the dimensions of ν, we can easily see that it is impossible to express V and L separately in terms of Q and of the combination of physical constants μ and ρ. This indicates the fact of the existence of self-similar solutions of the type (in the polar coordinates r, ε)

$$V_r = \frac{\nu}{L} f_r\left(\frac{r}{L}, \varepsilon\right) \qquad V_\varepsilon = \frac{\nu}{L} f_\varepsilon\left(\frac{r}{L}, \varepsilon\right)$$

$$\frac{p - p_0}{\rho} = \left(\frac{\nu}{L}\right)^2 P\left(\frac{r}{L}, \varepsilon\right) \tag{109}$$

The absence of the characteristic length L in the problem formulation leads one to take the functions f_r, f_ε, and P in the form

$$f_r\left(\frac{r}{L}, \varepsilon\right) = \frac{L}{r} F(\varepsilon) \qquad f_\varepsilon\left(\frac{r}{L}, \varepsilon\right) = \frac{L}{r} G(\varepsilon)$$

$$P\left(\frac{r}{L}, \varepsilon\right) = \left(\frac{L}{r}\right)^2 E(\varepsilon) \tag{110}$$

thus leading to the structure of solutions

$$V_r = \frac{\nu}{r} F(\varepsilon) \qquad V_\theta = \frac{\nu}{r} G(\varepsilon) \qquad \frac{p - p_0}{\rho} = \frac{\nu^2}{r^2} E(\varepsilon) \tag{111}$$

We will not reproduce the calculations but only note that in substituting the values of these velocities and pressures into the system of stationary Navier-Stokes equations (Equation 33 of Chapter 10) for a plane flow (the prime denotes the derivative with respect to ε), we have

$$G(\varepsilon) = \text{const} \qquad F'' + F^2 = -2P \qquad P' = 2F' \tag{112}$$

Hence, it follows (C is the integration constant)

$$P = 2F + C$$

and the only unknown function $F(\varepsilon)$ is determined from the ordinary differential equation

$$F'' + F^2 + 4F + 2C = 0 \tag{113}$$

on the details of whose integration[20] we will not pause. We only emphasize that the considered class of flows, as well as the previous one, allows exact self-similar solutions. Note that

---

[19] Hamel, G., Spiralförmige Bewegungen zähen Flüssigkeiten, *Jahr. Ber. Dtsch. Math. Ver.*, 25, 34–60, 1917.

[20] Rosenhead, L., Ed., *Laminar Boundary Layers*, Clarendon Press, Oxford, 1963, 144–150; Loitsyanskiy, L. G., *Mekhanika Zhidkosti i gaza* (*Fluid Mechanics*), 4th edition, Nauka Press, Moscow, 1973, 491–494.

though a similar spatial problem with the prescribed intensity of the source will not be self-similar yet (see the fourth edition (1973) of this book, pp. 432, 433), the dimensional analysis, having retained the necessity to solve partial differential equations, reduces the number of independent variables by one.

To the exact self-similar solutions of the Navier-Stokes equations pertains the class of motions considered previously purely mathematically by Slyozkin[21] and later specified by Landau.[22] This is the case of spatial round jets of a viscous fluid spreading in the medium of the same physical properties (submerged straight jets). We use the system of spherical coordinates (R, θ, ε) with the axis directed along the jet symmetry axis. The condition of jet unswirled nature will be $V_\varepsilon = 0$; the derivatives with respect to ε vanish.

The self-similar solution of the system can be obtained when it is assumed that

$$V_R = \frac{\nu}{R}F(\eta) \qquad V_\theta = \frac{\nu}{R}G(\eta) \qquad \frac{p-p_0}{\rho} = \left(\frac{\nu}{R}\right)^2 P(\eta) \tag{114}$$

where it is expedient to introduce the argument $\eta = \cos\theta$ instead of θ.

Substituting expressions (114) of the components of velocities V_R and V_θ and the pressure $p - p_0$ into the system of stationary Navier-Stokes equations (Equation 34 of Chapter 10) and introducing the function $f(\eta)$, which is related to F and G by the equations (the prime denotes the derivative with respect to η)

$$F(\eta) = -f'(\eta) \qquad G(\eta) = -\frac{f(\eta)}{\sqrt{1-\eta^2}} \tag{115}$$

and which is within a factor νR the stream function, we come to the system of ordinary differential equations

$$\frac{d}{d\eta}[ff' - (1-\eta^2)f''] + \frac{f^2}{1-\eta^2} + 2P = 0$$

$$\frac{d}{d\eta}\left[\frac{f^2}{2(1-\eta^2)}\right] + f'' + P' = 0 \tag{116}$$

It follows directly from the second equation of system (116) that (C_1 is the arbitrary integration constant)

$$P = -\frac{f^2}{2(1-\eta^2)} - f' - \frac{1}{2}C_1 \tag{117}$$

and then the independent equation for f can be easily obtained from the first (C_2 is a new integration constant)

$$(1-\eta^2)f'' - ff' + 2f + C_1\eta + \frac{1}{2}C_2 = 0$$

which is directly integrated and gives (C_3 is one more arbitrary integration constant)

---

[21] Slyozkin, N. A., On one case of integrability of complete differential equations of a viscous fluid flow, *Uchycn. Zap. MGU*, 2, 89, 90, 1934.

[22] Landau, L. D., New exact solution of the Navier-Stokes equations, *Dokl. AN SSSR*, 44, 311–314, 1944.

$$2(1-\eta^2)f' - f^2 + 4\eta f + C_1\eta^2 + C_2\eta + C_3 = 0 \tag{118}$$

Equation 118 belongs to the Riccati-type equations and is not integrated in the familiar functions. By substitution

$$f = -\frac{2(1-\eta^2)\,g'}{g} \tag{119}$$

it is reduced to the hypergeometrical equation

$$-4(1-\eta^2)^2 g'' + (C_1\eta^2 + C_2\eta + C_3)\,g = 0 \tag{120}$$

which in the case

$$C_1\eta^2 + C_2\eta + C_3 = A(1-\eta^2) \tag{121}$$

has simple exponential solutions

$$g = (1+\eta)^{\lambda} \qquad \lambda = \frac{1}{2}(1 \pm \sqrt{1+A}) \tag{122}$$

To the case of an unbounded round jet issuing from a point source, as is shown by Landau,[22] correspond the values of the constants $C_1 = C_2 = C_3 = 0$, and consequently, $A = 0$. Here, the function f is equal to

$$f = \frac{2(1-\eta^2)}{a+1-\eta} \tag{123}$$

Prescribing the jet by its projection onto the axis of the full impulse J (which includes the momentum flow, pressure, and viscous stresses) – the only physical quantity characterizing a certain submerged round jet issuing from a point source – we find the relation between a and J

$$\frac{J}{2\pi\rho\nu^2} = \frac{32(1+a)}{3a(2+a)} + 8(1+a) - 4(1+a)^2 \log\left(\frac{2+a}{a}\right) \tag{124}$$

In the considered case of the finiteness of the quantity J the point source power Q should be equal to zero because at the infinite fluid velocity in the point source its power Q is proportional to the first degree of the velocity, and J the to the second, thus leading, in the case of J finiteness, to the vanishing Q.

The case of the finiteness of both these quantities will result in two dimensional scales of lengths and velocities

$$L = \frac{\mu Q}{J} \qquad V = \frac{J}{\rho Q}$$

As explained earlier, in the presence of the scales of lengths and velocities the flow will not be determined by a self-similar solution to the Navier-Stokes equations. A nonself-similar solution corresponding to the assignment of Q, J, μ, and ρ, was mathematically studied by Yatseyev.[23] The physical interpretation of this solution as a viscous fluid flow ejected by a jet

[23] Yatseyev, V. I., On one class of exact solutions of a viscous fluid flow, *Zh. Eksp. Teor. Fiz.*, 20(11), 1950.

issuing from the opening of a finite width, for the region at a distance from the source, was given by Rumer.[24]

The self-similar solutions of the Navier-Stokes equations considered in the given section and those not mentioned were reduced to integration of the systems of ordinary differential equations that can be performed by different approximate or classical numerical methods. Seeking common nonself-similar solutions, which requires integration of partial differential equations, offered many difficulties. However, due to the relatively recent appearance of computers, computer numerical methods have been greatly advanced. It should be recognized that the attraction of attention to these methods, which make it possible to approach exact solutions very closely, is quite deserved. We should not understate the significance of different approximate analytical methods, which as a rule give the solutions in an easy form, vividly expressing the main tendencies of the phenomena studied. Further examples of both numerical and approximate analytical solutions to the problems of fluid mechanics are given elsewhere.[25]

---

[24] Rumer, Yu. B., The problem of a submerged jet, *Prikl. Mat. Mekh.*, 16(2), 1952.

[25] A voluminous review of the solutions to the Navier-Stokes equations is undertaken in Berker, R., Integration des equations du mouvement d'un fluid visqueux incompressible, *Handb. Phys.*, 7(2), 1–384, 1963.

12

Laminar Boundary Layer in Incompressible Fluid

1. Interaction between Convection and Diffusion in an Incompressible Viscous Fluid Flow. Laminar Boundary Layer

The insight into a laminar boundary layer on a solid body surface in a viscous fluid flow or in the case of mixing of flows and formation of the region of interaction between them forms the basis for explanation and utilization of a great number of practically important phenomena.

By a boundary layer we mean a flow region which under certain conditions (see below) is thin in a direction lateral to the flow. In this region, in contrast to an irrotational flow surrounding it, the motion is vortical and it is characterized by sharp variations of velocity (a velocity boundary layer), temperature (temperature boundary layer), and admixture concentration (a concentration or diffusion boundary layer) that are concentrated in this region.

There can be laminar or turbulent boundary layers, depending on the laminar or turbulent flow conditions in them. This chapter deals with laminar boundary layers only, so in view of this the epithet "laminar" is often omitted.

Small fluid viscosity, or more precisely, a great value of the flow Reynolds number **Re** that (though it does not reach the critical value at which the flow mode in a boundary layer becomes turbulent) is the main condition for the formation of velocity laminar boundary layers. Analogously, the attainment of great values of the Peclet number **Pe** and great values of $\mathbf{Pe}_d$ (the diffusional Peclet number) becomes a similar condition for temperature and concentration layers, correspondingly.

A physical picture of the formation of boundary layers on solid surfaces is as follows. A homogeneous irrotational flow, having reached a solid body surface, "sticks" to it by the particles that directly contact the body surface, whereas the neighboring layers continue to move with the velocities, sharply increasing at a distance from the surface, thus leading to flow twisting. Vortices formed in the vicinity of the surface are carried by a free stream, thus participating in convection, while they diffuse into the surrounding fluid.

In case convection is large as compared with diffusion (this, as shown later, corresponds to large Reynolds numbers), a very thin layer of noticeably vortical fluid, the so-called near-wall boundary layer, remains on and near the surface. Similarly, with some initial vorticity which appeared at a reservoir outlet or in the flow past the body surface "free" boundary layers are formed: "submerged jets" and "wakes" beyond the body trailing edge.

The balance of convection and diffusion, produced in the process of flow in a boundary layer region, and also of local variation of vorticity in the case of a non stationary boundary layer is quantitatively determined by the Helmholtz equations Equation 162 and Equation 163 of Chapter 10. This balance may give qualitative estimation of the order of the boundary layer "thickness" at large Reynolds numbers composed by the velocity of a flow approaching the body, a linear size of the body L, and the kinematic coefficient of fluid viscosity.

The growth of the Reynolds number indicates the increase in the role of convective acceleration and thus, as follows from the derivation of the Helmholtz equation, of the vorticity convection. The growth of the role of this process leads to two consequences:

1. As the Reynolds number grows, the dimension of the vorticity region, transverse to the flow, decreases
2. The processes of convection and diffusion concentrate in this thin region due to the balance between them

The first of these consequences serves as an explanation of the existence of the boundary layer at large Reynolds numbers; the second speaks of the relative intensity of diffusion in the boundary layer region and of the determining role of the diffusion of vorticity in the direction lateral to the body surface.

Before discuss the analysis of a viscous fluid flow in the boundary layer region we focus on some considerations which allow one to give a quantitative estimation of the law of the boundary layer thickness reduction depending on the growth of the flow Reynolds number.

First, we restrict ourselves to the case of plane laminar flow, in which the equation of balance between the processes of convection and vorticity diffusion Equation 163 of Chapter 10 can be written in the simplest form

$$V \cdot \mathrm{grad}\Omega = \nu\nabla^2\Omega \tag{1}$$

To estimate the order of the change in the quantities on the left (vorticity convection) and on the right sides (vorticity diffusion) of this equation with the growth of the Reynolds number, we apply the technique which is used at the beginning of Chapter 10 to derive the conditions of similarity between two flows of viscous fluid and which is the expression of variable quantities entering into the equation in forms that relate them to constant scales typical of them. When considering the processes of vorticity convection and diffusion in the boundary layer region, we agree to distinguish between the scales of longitudinal lengths and velocities L_0 and U_0 and the corresponding scales of lateral lengths and velocities δ_0 and V_0. We also introduce the scale Ω_0 for vorticity.

As to the first two scales, their specification does not require special explanation. The quantity L_0 can designate, for example, the diameter of a round contour in a fluid flow, of a chord or the maximum thickness of a wing profile, the inner radius of a tube at whose inlet a boundary layer is formed, etc. It is natural to take the velocity U_∞ of the flow approaching the body or the velocity at the tube inlet as the scale of transverse velocities U_0. The vorticity scale Ω_0 falls out from further consideration and it should not be specified.

The problem of the choice of the transverse scale of lengths δ_0 requires special explanation. It is natural to connect this scale with the characteristic distance to which vorticity diffusion propagates in the direction lateral to the streamlined body surface, which is the source of vorticity. This finite-invalue distance does not exist in the problems of viscous fluid dynamics stated in previous chapters.

Recall, for instance, the Stokes problem of a viscous fluid flow past a sphere (Chapter 11, Section 1) or calculation of vorticity diffusion formed by a vortex filament (Chapter 10, Section 11). In all of these cases the viscous effect distributes instantly and, in unbounded infinite flows, at infinitely large distances. This principal fact is a direct consequence of the generalized Newton law which expresses a linear relation between the tensors of stresses and of deformation velocities and causes an elliptical character of the Navier-Stokes differential equations. Further presentation is based on these concepts of infinite velocity of vorticity diffusion and of infinite range of its distribution in an unbounded viscous fluid flow.

According to the generalized Newton law (taken as a basis for viscous fluid dynamics, typical for which is an infinite velocity of propagation of disturbances, the change in the velocity value would be correspond to an asymptotic approach to limiting values. This asymptotic formulation is usually retained in forming the boundary conditions for differential equations of a boundary layer on its outer edge and is designated by the expression $y \to \infty$, where y is the distance of the given point from the body surface.

By the boundary layer thickness, as some finite quantity δ, is meant the distance from the surface of a streamlined body to such a point in the flow ($y = \delta$). Here, with the prescribed degree of approximation, the longitudinal velocity in a boundary layer can be assumed virtually equal to its value at the same point of the outer irrotational flow. Geometrical location of these points gives an approximate, "finite", representation of the boundary layer outer edge.

The presence of the above-mentioned conditions justifies an historically established notion the finite boundary layer thickness, though the arbitrariness in the value of the permissible error makes this definition somewhat obscure. Further, it appears more favorable to deal with less obvious but more unambiguous determinations of the boundary layer thickness that are based on integral characteristics of distributions of longitudinal velocities in the boundary layer cross-sections normal to the body surface.

Each of these characteristics, being variable along the body surface, can serve as a local scale for a boundary layer cross-section normal to the flow. The previously introduced scale of transverse lengths in a flow δ_0 will be treated as the order of the boundary layer thickness general for all of the sections. In order to find the law of the dependence of the thickness δ_0 on the flow Reynolds number $\mathbf{Re}_0 = U_0L_0/\nu$, we shall equate the orders of the left and right sides of the equation of vorticity balance (1).

Introducing the capital letter O as a symbol of the order of the quantity variation with the Reynolds number growth, we note that the left and right sides can be estimated as

$$V \cdot \mathrm{grad}\Omega = O\left(\frac{U_0\Omega_0}{L_0}\right) \qquad \nu\nabla^2\Omega = O\left(\frac{\nu\Omega_0}{\delta_0^2}\right)$$

Having equated these orders, we obtain

$$\frac{U_0\Omega_0}{L_0} = O\left(\frac{\nu\Omega_0}{\delta_0^2}\right)$$

where

$$\frac{\delta_0}{L_0} = \left(\sqrt{\frac{\nu}{U_0L_0}}\right) = O\left(\frac{1}{\sqrt{\mathbf{Re}_0}}\right) \tag{2}$$

The obtained equation expresses the law, which is general for all plane stationary laminar boundary layers, of the variation of their relative thicknesses in inverse proportion to the square root of the flow Reynolds number. This is the first basic property of the laminar boundary layer.

We reemphasize that the question is of the dependency of the order of boundary layer thicknesses on the flow Reynolds number, common for all boundary layer cross-sections, rather than of their variation along the body surface at the fixed Reynolds number $\mathbf{Re}_0$.

Along with the velocity or vorticity boundary layer, we have to deal with boundary layers that differ in physical properties of substance transferred such as temperature and diffusion (concentration). These are regions thin in the direction transverse with respect to the body surface where intense heat (temperature) or substance (concentration) diffusion is concentrated. The thinness of these regions is stipulated by great values of the Peclet number ($\mathbf{Pe}$) and the diffusion Peclet number ($\mathbf{Pe}_d$) (see Chapter 10), rather than by the Reynolds number.

There is no need to repeat all the considerations made for the velocity boundary layer, as applied to these two layers. We shall only note that in the case of a plane stationary heat and substance propagation the equations of convection and diffusion balance can be presented in the form of equalities

$$V \cdot \operatorname{grad} T = a\nabla^2 T \qquad V \cdot \operatorname{grad} c = D\nabla^2 c$$

which coincide with the previously mentioned equation for vorticity Ω, if Ω is replaced by T or c, and ν is substituted by a or D. Estimating the orders or the quantities that enter the latter equations and introducing along with previous scales also additional δ_{0T} and δ_{0d} for transverse lengths in temperature and diffusion (concentration) layers, respectively, we obtain the expressions for these scales and write them together with the relation obtained for the order of velocity boundary layer thickness:

$$\frac{\delta_0}{L} = O\left(\frac{1}{\sqrt{\mathbf{Re}_0}}\right) \qquad \frac{\delta_{0T}}{L} = O\left(\frac{1}{\sqrt{\mathbf{Pe}_0}}\right) \qquad \frac{\delta_{0d}}{L} = O\left(\frac{1}{\sqrt{\mathbf{Pe}_{d0}}}\right) \tag{3}$$

Introducing Chapter 10, section 12 the Prandtl and Schmidt (diffusion Prandtl number) numbers

$$\mathbf{Pr}_0 = \frac{\mathbf{Pe}_0}{\mathbf{Re}_0} \qquad \mathbf{Sc}_0 = \mathbf{Pr}_{d0} = \frac{\mathbf{Pe}_{d0}}{\mathbf{Re}_0}$$

we obtain the relationships between the orders of the thicknesses of boundary layers

$$\frac{\delta_0}{\delta_{0T}} = O(\sqrt{\mathbf{Pr}_0}) \qquad \frac{\delta_0}{\delta_{0d}} = O(\sqrt{\mathbf{Sc}_0}) = O(\sqrt{\mathbf{Pr}_{d0}}) \tag{4}$$

which express the fact that at large values of $\mathbf{Pr}_0$ and $\mathbf{Sc}_0$ ($\mathbf{Pr}_{d0}$) temperature and concentration boundary layers are much thinner than velocity ones, and conversely, at small $\mathbf{Pr}_0$ and $\mathbf{Sc}_0$ ($\mathbf{Pr}_{d0}$), they are much thicker.

At present the theory of laminar and turbulent boundary layers still attracts researchers, forming an independent branch of fluid mechanics. Along with voluminous literature dealing

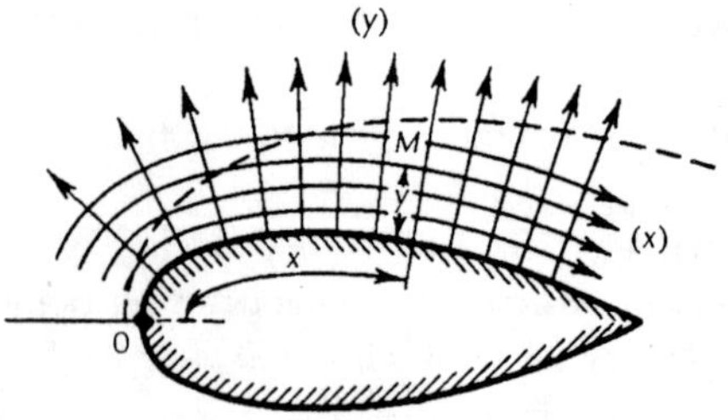

Fig. 168

with this branch there exist special monographs are available on the boundary layer theory.[1]

2. Derivation of the Prandtl Equation for Viscous Fluid Flow in a Laminar Boundary Layer. Separation Phenomenon

Moving to the derivation of the basic differential equation of viscous medium flow in a laminar boundary layer region, this section concentrates only on the case of a plane near-wall stationary velocity boundary layer. In subsequent sections more complex cases of both nonstationary and spatial flows, not only in near-wall but also in free boundary layers, are considered. At the end of the chapter some information on temperature and concentration boundary layers is given. The concept of a turbulent boundary layer and empirical and semiempirical methods of their calculation can be found in Chapter 13; laminar and turbulent boundary layers in a compressible nonisothermal medium – gas at supersonic velocities – comprise Chapters 15 and 16.

The smallness of the order of the boundary layer thicknesses δ_0 at large flow Reynolds numbers makes it possible to consider the orthogonal grid of the curves parallel to the body surface and normal to them (Figure 168) as a rectilinear Cartesian system (x, y) in the boundary layer region and to preserve the Navier-Stokes equations in their ordinary form

$$u\frac{\partial u}{\partial x} + v\frac{\partial u}{\partial y} = -\frac{1}{\rho}\frac{\partial p}{\partial x} + \nu\left(\frac{\partial^2 u}{\partial x^2} + \frac{\partial^2 u}{\partial y^2}\right)$$

$$u\frac{\partial v}{\partial x} + v\frac{\partial v}{\partial y} = -\frac{1}{\rho}\frac{\partial p}{\partial y} + \nu\left(\frac{\partial^2 v}{\partial x^2} + \frac{\partial^2 v}{\partial y^2}\right)$$

$$\frac{\partial u}{\partial x} + \frac{\partial v}{\partial y} = 0 \qquad (5)$$

The above simplification in the choice of the coordinate system is not central. The derivation can be made without assuming the grid of coordinate lines given in Figure 168 of

[1] Schlichting, G., *Boundary Layer Theory*, Nauka Press, Moscow, 1974; Loitsyanskiy, L. G., *Aerodinamika pogranichnogo sloya* (*Aerodynamics of Boundary Layer*), Gostekhizdat, Moscow, 1941; Loitsyanskiy, L. G., *Laminarnyi pogranichnyi sloi* (*Laminar Boundary Layer*), Fizmatgiz, Moscow, 1962; Rosenhead, L., Ed., *Laminar Boundary Layers*, Clarendon press, Oxford, 1963; Evans H.L., *Laminar Boundary Layer Theory*, Addison-Wesley, Reading, MA, 1968.

Chapter 11 to be rectilinear.[2]

Making use of the previously adopted method for estimating the orders of separate terms of the system of Equations 5 with the aid of constant scales (longitudinal velocities U_0, longitudinal lengths L_0, transverse δ_0), we first determine the scale of transverse velocities V_0. For this, note that according to the last formula of Equation 5 the incompressibility equation, there is ($y = 0$ corresponds to the body contour)

$$v = -\int_0^y \frac{\partial u}{\partial x} dy$$

so that

$$V_0 = O(v) = \frac{U_0 \delta_0}{L_0}$$

and consequently,

$$\frac{V_0}{U_0} = \frac{\delta_0}{L_0} = \frac{1}{\sqrt{\mathbf{Re}_0}} \tag{6}$$

i.e., the transverse velocity in the boundary layer region has the same relative order as the layer thickness. This is the second basic property of the boundary layer.

Estimating similarly the terms on the left and right sides of the second equation of the system, we find

$$u\frac{\partial v}{\partial x} = O\left(\frac{U_0 V_0}{L_0}\right) = O\left(\frac{U_0 U_0 \delta_0}{L_0^2}\right) = O\left(\frac{U_0^2/L_0}{\sqrt{\mathbf{Re}_0}}\right)$$

$$v\frac{\partial v}{\partial y} = O\left(U_0 \frac{\delta_0}{L_0} \frac{U_0 \delta_0}{L_0 \delta_0}\right) = O\left(\frac{U_0^2/L_0}{\sqrt{\mathbf{Re}_0}}\right)$$

$$\nu\frac{\partial^2 v}{\partial x^2} = O\left(\nu \frac{U_0 \delta_0}{L_0} \frac{1}{L_0^2}\right) = O\left(\frac{\nu}{U_0 L_0} \frac{\delta_0}{L_0} \frac{U_0^2}{L_0}\right) = O\left(\frac{U_0^2/L_0}{\mathbf{Re}_0 \sqrt{\mathbf{Re}_0}}\right)$$

$$\nu\frac{\partial^2 v}{\partial y^2} = O\left(\nu \frac{U_0 \delta_0}{L_0} \frac{1}{\delta_0^2}\right) = O\left(\frac{\nu}{U_0 L_0} \frac{U_0^2 L_0}{L_0} \frac{\sqrt{\mathbf{Re}_0}}{L_0}\right) = O\left(\frac{U_0^2/L_0}{\sqrt{\mathbf{Re}_0}}\right)$$

According to the estimations given, at large Reynolds numbers $\mathbf{Re}_0$ these terms have the order $1/\sqrt{\mathbf{Re}_0}$ or even greater order of smallness $1/(\mathbf{Re}_0\sqrt{\mathbf{Re}_0})$. As to the term involving the derivative $\partial p/\partial y$, it is a measure of the difference of other terms with values of the order $1/\sqrt{\mathbf{Re}_0}$, and therefore must have the same order. Thus, in the adopted approximation of the boundary layer theory, i.e., under the specified great values of $\mathbf{Re}_0$, we can

[2]See Kochin, N. E., Kibel, I. A. and Roze, N. B., *Teoreticheskaya gidromekhanika (Theoretical Hydromechanics)*, Part 2, Fizmatgiz, Moscow, 1963, 550–552; *Laminar Boundary Layers*, Rosenhead, L., Ed., Clarendon Press, Oxford, 1963, 201–203.

neglect small values of this term and assume $\partial p/\partial y = 0$, where the third basic property of the boundary layer follows: at all the points of the given boundary layer cross-section normal to the body surface pressure has one and the same value.

Recall that this property is typical of all uniform rectilinear viscous fluid flows in cylindrical or prismatic tubes (considered in Chapters 10 and 11), and also between parallel planes placed in close proximity to each other and in small bearing clearances. In a boundary layer this property is provided by its small thickness and, as a consequence, by the parallelness of streamlines with the body surface.

In the considered case of a boundary layer this property makes it possible to convert the partial derivative on the right side of the first equation of system (5) $\partial p/\partial x$ into the full derivative dp/dx. According to the same property the pressure distribution $p(x)$ along the boundary layer coincides with pressure distribution in an outer irrotational flow. By the Bernoulli theorem, which is valid for a irrotational flow of ideal fluid approaching the body, this distribution can be connected with the velocity in the outer flow. Thanks to boundary layer thinness this velocity can be "transferred" to the body surface, imagining it to be equal to velocity $U(x)$, dependent on x only, of fluid sliding over the body surface, which would occur in ideal fluid (in the absence of the boundary layer). The velocity $U(x)$ will be called the velocity at the boundary layer outer edge or the outer velocity.

According to the Bernoulli theorem we have

$$\frac{1}{\rho}\frac{dp}{dx} = -U\frac{dU}{dx} \tag{7}$$

Making use of the previously adopted method of estimating the order of separate terms in the system of Equations 5 and starting with equality (7), we obtain the estimation of the first term on the right side of the first equation of system (5)

$$\frac{1}{\rho}\frac{dp}{dx} = O\left(\frac{U_0^2}{L_0}\right) \tag{8}$$

which will be added to the following evident estimation of the remaining terms of this equation

$$u\frac{\partial u}{\partial x} = O\left(\frac{U_0^2}{L_0}\right) \qquad v\frac{\partial u}{\partial y} = O\left(V_0\frac{U_0}{\delta_0}\right) = O\left(\frac{U_0^2}{L_0}\right)$$

$$\nu\frac{\partial^2 u}{\partial x^2} = O\left(\frac{\nu U_0}{L_0^2}\right) = O\left(\frac{U_0^2/L_0}{\mathbf{Re}_0}\right) \qquad \nu\frac{\partial^2 u}{\partial y^2} = O\left(\frac{\nu U_0}{\delta_0^2}\right) = O\left(\frac{U_0^2}{L_0}\right) \tag{9}$$

Taking into account these estimations, we omit the first term within the parentheses on the right side of the first equation of system (5) because it is small as compared to the second formula. We also omit the entire second equation of system (5), which has been used to substantiate the third property of a boundary layer.

Now, rewrite system (5) in the form

$$u\frac{\partial u}{\partial x} + v\frac{\partial u}{\partial y} = -\frac{1}{\rho}\frac{dp}{dx} + \nu\frac{\partial^2 u}{\partial y^2}$$

$$\frac{\partial u}{\partial x} + \frac{\partial v}{\partial y} = 0 \tag{10}$$

or in form equivalent to it by Equation 7

$$u\frac{\partial u}{\partial x} + v\frac{\partial u}{\partial y} = U\frac{dU}{dx} + \nu\frac{\partial^2 u}{\partial y^2}$$

$$\frac{\partial u}{\partial x} + \frac{\partial v}{\partial y} = 0 \qquad (11)$$

which is more convenient for further presentation.

These equations are a system of nonlinear equations in partial derivatives of the parabolic type of second order which were presented by Prandtl in 1904.[3]

According to Prandtl the outer velocity $U(x)$ entering the first equation of system (11) is considered to be prescribed, calculated previously by the theory of a plane irrotational flow of ideal incompressible fluid past a body (Chapter 7). In this formulation of the problem it is assumed that a boundary layer around the entire contour of a streamlined body is so thin that its distorting influence on the outer flow is negligibly small. It can be said that in this case the inverse influence of the boundary layer on an irrotational flow is disregarded. In some cases (a smooth flow past slender bodies) such disregard for the boundary layer inverse influence on the outer flow is quite permissible, in other cases this influence is so large that one must calculate the outer velocity $U(x)$ by Equation 7 using the experimentally measured pressure distribution over the body contour.

Equations 11 are subjected to integration under the following boundary conditions

$$u = 0 \qquad v = 0 \quad \text{at} \quad y = 0$$

$$u \to U(x) \quad \text{for} \quad y \to \infty$$

$$u = u_0(y) \quad \text{at} \quad x = x_0 \qquad (12)$$

The first two express the condition of viscous fluid adhesion to a solid wall ($y = 0$) – the streamlined body contour. The third condition is the requirement of the longitudinal velocity u asymptotic approach in the boundary layer region to the velocity $U(x)$ at the interface between a boundary layer and an irrotational flow. This boundary condition can be interpreted as the operation of "matching" the solution of the Prandtl equations for viscous fluid flow in a boundary layer with the solution of the Euler equations for irrotational flow of ideal incompressible fluid past a body.

The last boundary condition, which expresses the assignment of the velocity profile $u_0(y)$ in some initial boundary layer cross-section ($x = x_0$), is typical of parabolic equations. According to the general theory of partial differential equations the solution of parabolic equations under the given boundary conditions is determined only in the flow region downstream from this initial section.

In the region of the flow described by parabolic equations the initial velocity profile does not have any effect upstream. However, because of the existence of the outer irrotational flow described by elliptic equations, and the previously mentioned inverse influence of the boundary layer on the outer flow in this region, there is a combined effect of parabolic and elliptic equations. We also had to observe the influence of the initial profile on the boundary layer regions located upstream relative to the initial cross-section.

If we leave aside a comparatively narrow class of special problems concerned with the

---

[3]Prandtl, L., Über Flüssigkeitsbewegung bei sehr kleiner Reibung, *Verhandl. 3rd. Kongr., Heidelberg*, 1904.

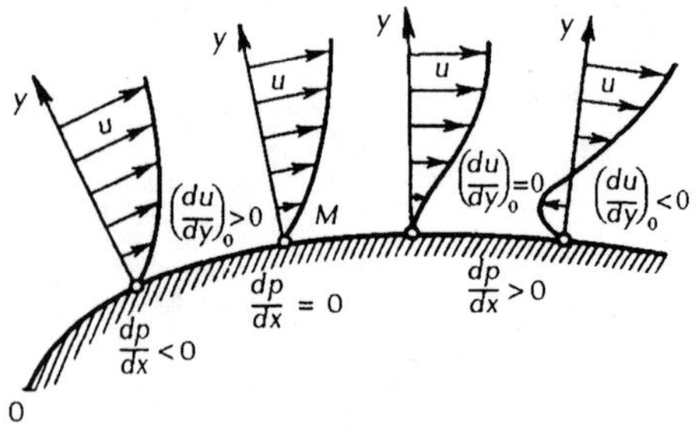

Fig. 169

necessity of strict extension of the prescribed initial velocity profile downstream in the boundary layer, then we can see that the initial boundary condition is not especially substantial because its downstream effect rapidly decreases with the relative (expressed in multiples of the initial layer thickness) distance from this cross-section. As compared to longitudinal distances the boundary layer transverse dimension is so small that starting from some cross-section all the remaining boundary layer flow will occur many boundary layer thicknesses from the initial cross-section. Unfortunately, the problem of the existence and the uniqueness of the solution of Equations 11 under the boundary conditions (12) deviates from the theme of this book.[4]

The boundary layer theory makes it possible to explain the nature of flow separation from a smooth surface. This phenomenon is closely connected with the property of viscous fluid adhesion to a solid surface of the body in this fluid flow and with boundary layer formation on the surface.

The mechanism of viscous separation differs from the mechanism described in Section 12 of Chapter 7 of inertial stalling of an ideal fluid irrotational flow from the protruding sharp body edges. In the presence of viscous separation complex nonstationary reverse motions with closed streamlines arise directly behind viscous separation in the so-called near wake.

Stationary separation is the result of the interaction of three factors: flow inertia, viscous interaction between adjacent fluid layers and a solid surface, and the reverse pressure drop directed opposite to the flow. The process of origination and development of separation under nonstationary conditions will be considered later in the Chapter.

In the profile rear edge downstream from the point of pressure minimum M (Figure 169) pressure grows and $dp/dx > 0$ (this region is called diverging); in this case the fluid in the boundary layer moves from the region of smaller pressure into the region with higher pressure with the effect of an opposite retarding pressure drop. If the fluid was ideal and the velocity on the wing surface was not equal to zero, then the content of fluid kinetic energy would appear to be sufficient to overcome the retarding pressure field. In a boundary layer the pressure field does not differ from the pressure field in an ideal fluid, but in direct vicinity to the body surface velocities are very small and, consequently, kinetic energy of fluid particles is negligible. Under these conditions retardation can cause stoppage and then reverse (Figure 169) motion under the effect of pressure drop opposite to the flow. The meeting of the free stream with the fluid reversely moving in a boundary layer leads to sharp displacement of streamlines away from the body surface, to boundary layer thickening and its separation from

---

[4]Among the works concentrated on this approach, the following deserve attention: Oleinik, O. A., On the system of boundary layer equations for non-stationary incompressible fluid flow, *Dokl. AN SSSR*, 168(4), 751–754, 1966 and references therein; Nickel, K., Ein Eindeutigkeitssatz für instantionäre Grenzchichten, *Math. Leitschr.*, 74, 1960, and his review Prandtl 's boundary layer theory from the viewpoint of a mathematician, *Annu. Rev. Fluid Mech.*, 5, 405–428, 1973.

the body surface. Before the separation point S, as is seen in Figure 169, $(\partial u/\partial x)_{y=0} > 0$, behind the separation point $(\partial u/\partial x)_{y=0} < 0$, at the point itself we have the separation condition

$$\left(\frac{\partial u}{\partial y}\right)_{y=0} = 0 \tag{13}$$

The above explanation of the phenomenon of viscous separation shows that separation of this nature can appear only in a diverging region of a boundary layer where viscous interactions in fluid exist with the pressure drop opposite to the flow direction. Thus, the separation point S always locates more downstream as compared with the point M of pressure minimum (the outer velocity maximum).

The separation condition (13) also preserves its form after passing over to the dimensionless quantities $u' = u/U$, $y' = y/\delta$. Noting that the equation, expressed in terms of these new variables, for the determination of the separation point position $(\partial u'/\partial y')_{y'=0} = 0$ evidently will not involve the Reynolds number, we draw the conclusion that the dimensionless abscissa of the separation point $x_s' = x_s/L$, which is the root of this equation, will also be independent of the Reynolds number.

This conclusion on the separation point independence of the Reynolds number is valid only on the assumption of the applicability of the Prandtl equation in a preseparation region. In fact in the separation region (its dimensions require special evaluation by the Reynolds number) the Prandtl equations in the considered form are invalid. When approaching the separation point the retarding effect of the wall sharply decreases up to zero and the prevailing effect of the derivatives along the normal to the wall vanishes as compared to the derivatives in the direction parallel to the wall. In this case there are no grounds to neglect the quantity $\partial^2 u/\partial x^2$ as compared to $\partial^2 u/\partial y^2$ within brackets on the right side of the first equation in system (5). The transverse size of the boundary layer, as well as the transverse velocity cease to be small, and the transverse pressure variation also becomes substantial.

Formal determination of the separation point position from relation (13), in which u is substituted into the solution of the Prandtl equation (11), cannot give an accurate result because Equations 11 are invalid in the vicinity of the separation point. The Prandtl equations should be replaced either by the exact Navier-Stokes equations, thus offering great difficulties, or by the available approximate asymptotic techniques.[5]

Of great importance is also the fact of the distortion of the outer irrotational flow due to the displacement of its streamlines from the body surface, which is noticeable under these conditions. In this case the possibility of directly solving the boundary layer equations is lost. It is impossible to a priori prescribe the distributions of velocities in the outer irrotational flow that would take place in an ideal fluid in the absence of the boundary layer. In this case the inverse effect of the boundary layer on a potential flow should be taken into account or an experimentally obtained pressure distribution over the profile surface should be used.

Disregard of the "inverse effect" of the boundary layer on the outer irrotational flow, which is determined by the theory of ideal fluid, can lead to substantial distortion of the separation phenomenon. For example, Blasius[6] calculated the position of the point of flow separation from a round cylinder surface, neglecting the "inverse effect", (i.e., considering the distribution of the outer velocity by the "sine law"; see Chapter 7) and obtained the value of the angular coordinate of the separation point $\theta_s^0 \approx 110°$, i.e., behind the cylinder frontal

[5]Sychyov, V. V., On laminar separation, *Izv. AN SSSR Mekh. Zhidk. Gaza*, N3, 1972; Ruban, A. A. and Sychyov, V. V. Asymptotic theory of boundary layer separation in incompressible fluid, *Usp. Mekh. Izd. VINITI*, 2(4), 57–95, 1979; Brown, S. N. and Stewartson, K., Laminar separation, *Annu. Rev. Fluid Mech.*, 1, 45–72, 1969.

[6]Blasius, H., Grenzschichten in Flüssigkeiten mit kleiner Reibung, *Z. Math. Phys.*, 56, 1908.

area, whereas experiments give $\theta_s^0 \approx 82°$.

Hiemenz[7] showed the cause of the Blasius error. Having solved the boundary layer equations (11) using the experimental distribution of the outer velocity that was approximated by a power-law relation, he obtained full coincidence of the computational result with the experimental value $\theta_s^0 \approx 82°$.This result was not distorted by the application of the Prandtl equations (11) up to the separation point. The fact presented may serve as some justification that we can satisfied with an approximate determination of the separation point based on the Prandtl equations and relation (13) (or equivalent to it, equality to zero of the coefficient of friction c_f at the separation point; see Section 4), not overlooking the tentativeness of this determination of the separation point.

The retarding effect of the reverse pressure drop is a necessary condition for the boundary layer separation from the body surface. Therefore, when pressure is constant along the boundary layer separation cannot take place. The condition of pressure constancy arises, for example, in a jet (slender compared with the body size) past a body. The outer edge of this jet is a free surface because it has an interface with an immovable medium through which pressure is constant. Boundary layer separation from the body surface does not take place in this jet; slender jets adhere to the body surface along which they spread. This curious, frequent phenomenon is sometimes called the Coanda effect after a Romanian engineer who studied this phenomenon as early as 1910.[8]

Boundary layer separation is usually classified as a deleterious phenomenon which causes a sharp increase of the resistance of bodies in fluid flows, their dangerous vibrations, and in cases of flows inside tubes and channels to the reduction of useful fluid flow rate, growth of energy losses, and decreased of efficiency.

3. Different Forms of the Prandtl Equation. Mises and Crocco Equations

In the application of the theory of a plane laminar boundary layer in an incompressible fluid it is preferable to deal with one equation obtained from system (11) by introduction of the stream function $\psi(x, y)$ related to u and v by the equations

$$u = \partial\psi/\partial y \qquad v = -\partial\psi/\partial x \tag{14}$$

In this case the equation of incompressibility (the second formula in Equation 11) is fulfilled identically, and the Prandtl equations lead to one partial differential equation of the third order with the corresponding boundary conditions

$$\frac{\partial\psi}{\partial y}\frac{\partial^2\psi}{\partial x\partial y} - \frac{\partial\psi}{\partial x}\frac{\partial^2\psi}{\partial y^2} = U\frac{dU}{dx} + \nu\frac{\partial^3\psi}{\partial y^3}$$

$$\psi = 0 \qquad \partial\psi/\partial y = 0 \quad \text{when} \quad y = 0$$

[7] Hiemenz, K., Die Grenzschicht an eimen in der gleichformigen Flüssigkeitstrom eingetauchten geraden Kreiszylinder, *Dingl. Polytec. J.*, 326, 321, 1911. A detailed presentation of this problem is found in *Laminar Boundary Layer*, Rosenhead, L. Ed., Clarendon Press, Oxford, 1963, 156, 231, 260, 264, 285, 329; Schlichting, H., *Teoriya pogranichnogo sloya* (*The Boundary Layer Theory*), Izd. Nauka, Moscow, 1969, 160–166; Loitsyanskiy, L. G.,. *Laminarnyi pogranichnyi sloy* (*Laminar Boundary Layer*), Fizmatgiz, Moscow, 1962, 82.

[8] A detailed description of this phenomenon, the history of its development, and subsequent use can be found in Reba, J. Applications of the Coanda – effect, *Sci. Am.*, June, 84–92, 1966.

$$\partial\psi/\partial y \to U(x) \quad \text{when} \quad y \to \infty$$

$$\partial\psi/\partial y = u_0(y) \quad \text{when} \quad x = x_0 \tag{15}$$

The mentioned forms of the Prandtl equations (11) and (15) are not the only applicable ones. We shall touch on two common forms: the Prandtl-Mises equation[9] and the Crocco equation.[10]

The Prandtl-Mises equation is based on the use, along with x, of the stream function ψ as the second independent variable.

The transition from the variables x, y in Equations 15 to the Prandtl-Mises variables $\xi = x$, $\eta = \psi$ can be made by the formulas (by definition of the stream function: $\partial\eta/\partial x = \partial\psi/\partial x = -v$, $\partial\eta/\partial y = \partial\psi/\partial y = u$)

$$\frac{\partial}{\partial x} = \frac{\partial}{\partial \xi} + \frac{\partial\eta}{\partial x}\frac{\partial}{\partial\eta} = \frac{\partial}{\partial\xi} - v\frac{\partial}{\partial\eta} \qquad \frac{\partial}{\partial y} = \frac{\partial\eta}{\partial y}\frac{\partial}{\partial\eta} = u\frac{\partial}{\partial\eta}$$

$$\frac{\partial^2}{\partial y^2} = u\frac{\partial}{\partial\eta}\left(u\frac{\partial}{\partial\eta}\right) \tag{16}$$

Using this transformation to the first equation of system (11), we have (U is dependent on ξ only)

$$u\frac{\partial u}{\partial\xi} = U\frac{dU}{d\xi} + \nu u\frac{\partial}{\partial\eta}\left(u\frac{\partial u}{\partial\eta}\right)$$

Introducing into consideration "the defect" of kinetic energy in a boundary layer compared with the outer flow

$$z = \frac{1}{2}(U^2 - u^2) \tag{17}$$

or, according to the main property of the boundary layer, pressure constancy over the layer cross-section, the defect of total pressure

$$z = \left(\frac{p}{\rho} + \frac{1}{2}U^2\right) - \left(\frac{p}{\rho} + \frac{1}{2}u^2\right) \tag{18}$$

we obtain the laminar boundary layer equation in the Mises form

$$\frac{\partial z}{\partial\xi} = \nu u\frac{\partial^2 z}{\partial\eta^2} = \sqrt{2}\nu\sqrt{Z - z}\frac{\partial^2 z}{\partial\eta^2} \tag{19}$$

where it is assumed $Z = (z)_{\eta=0} = \frac{1}{2}U^2$

The boundary conditions for this partial differential equation of the second order is

[9]Mises, R., Bemerkungen zur Hydrodynamik, Z. *Angew. Math. Mech.*, 7, 425–431, 1927. Via some information these equations were derived by Prandtl as early as 1914.

[10]Crocco, L., A Characteristic Transformation of the Equations of the Boundary Layer in Gases, ARC Rep. No. 4582, London, 1969.

$$z = Z(\xi) \quad \text{at} \quad \eta = 0 \qquad z \to 0 \quad \text{at} \quad \eta \to \infty$$

$$z = z_0(\eta) \quad \text{at} \quad \xi = \xi_0 \tag{20}$$

Having determined after integration $z(\xi, \eta)$, we can revert to the ordinary variable y, inverting the integral

$$y = \int_0^\psi \frac{d\psi}{u} = \int_0^\psi \frac{d\psi}{\sqrt{2(Z-z)}} \tag{21}$$

The Prandtl-Mises boundary layer equation (19) resembles in form the heat conduction equation except for the nonlinear case, in which the coefficient of thermal diffusivity (the coefficient at the second derivative on the right side equal to $\sqrt{2}\nu\sqrt{Z-z}$) is temperature dependent. (In this case the role of temperature is played by the defect of kinetic energy.)

Crocco suggested taking the new independent variables $\xi = x$ and $\eta = u$ instead of x and y and to use the skin friction $\tau = \mu \partial u/\partial y$ as the dependent variable.

To derive the equations of transition from x and y to $\xi = x$ and $\eta = u$ we have two obvious equalities

$$\frac{\partial \eta}{\partial \xi} = 0 = \frac{\partial u}{\partial x}\frac{\partial x}{\partial \xi} + \frac{\partial u}{\partial y}\frac{\partial y}{\partial \xi} = \frac{\partial u}{\partial x} + \frac{\partial u}{\partial y}\frac{\partial y}{\partial \xi}$$

$$\frac{\partial \eta}{\partial \eta} = 1 = \frac{\partial u}{\partial x}\frac{\partial x}{\partial \eta} + \frac{\partial u}{\partial y}\frac{\partial y}{\partial \eta} = \frac{\partial u}{\partial y}\frac{\partial y}{\partial \eta}$$

from which it follows

$$\frac{\partial \eta}{\partial y} = \frac{\partial u}{\partial y} = \frac{\tau}{\mu} = \frac{1}{\partial y/\partial \eta} \quad \text{or} \quad \frac{\partial y}{\partial \eta} = \frac{\mu}{\tau}$$

$$\frac{\partial \eta}{\partial x} = \frac{\partial u}{\partial x} = -\frac{\partial u}{\partial y}\frac{\partial y}{\partial \xi} = -\frac{\tau}{\mu}\frac{\partial y}{\partial \xi} \tag{22}$$

Making use of these expressions we find the transition equations to the Crocco variables

$$\frac{\partial}{\partial x} = \frac{\partial \xi}{\partial x}\frac{\partial}{\partial \xi} + \frac{\partial \eta}{\partial x}\frac{\partial}{\partial \eta} = \frac{\partial}{\partial \xi} - \frac{\tau}{\mu}\frac{\partial y}{\partial \xi}\frac{\partial}{\partial \eta}$$

$$\frac{\partial}{\partial y} = \frac{\partial \xi}{\partial y}\frac{\partial}{\partial \xi} + \frac{\partial \eta}{\partial y}\frac{\partial}{\partial \eta} = \frac{\tau}{\mu}\frac{\partial}{\partial \eta} \tag{23}$$

Applying the transition equations (23) to the first equation of system (11) and dividing both sides of the obtained equation by τ/μ, we have

$$-\rho\eta\frac{\partial y}{\partial \xi} + \rho v = -\frac{dp}{d\xi}\frac{\partial y}{\partial \eta} + \frac{\partial \tau}{\partial \eta} = -\frac{dp}{d\xi}\frac{\mu}{\tau} + \frac{\partial \tau}{\partial \eta} \tag{24}$$

The second equation of system (11), the continuity equation, is transformed to the form $(\partial \eta/\partial \xi = 0)$

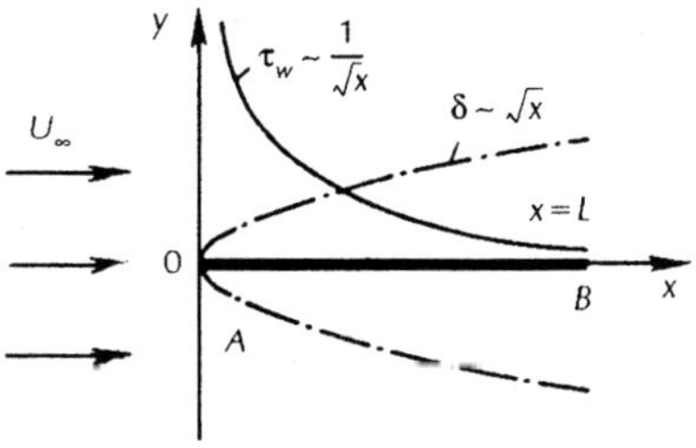

Fig. 170

$$-\frac{\partial y}{\partial \xi} + \frac{\partial v}{\partial \eta} = 0 \tag{25}$$

Differentiating both sides of Equation 24 with respect to η, we find

$$-\rho\eta \frac{\partial}{\partial \xi}\left(\frac{\partial y}{\partial \eta}\right) - \rho \frac{\partial y}{\partial \xi} + \rho \frac{\partial v}{\partial \eta} = -\frac{dp}{d\xi}\frac{\partial}{\partial \eta}\left(\frac{\mu}{\tau}\right) + \frac{\partial^2 \tau}{\partial \eta^2}$$

To eliminate v, we subtract termwise both sides of this equation from both sides of Equation 25, which are multiplied by ρ; then, noting also that $\partial y/\partial\eta = \mu/\tau$ we obtain the following boundary layer equation in the Crocco form

$$\mu\rho\eta \frac{\partial}{\partial \xi}\left(\frac{1}{\tau}\right) + \frac{\partial^2 \tau}{\partial \eta^2} - \mu \frac{dp}{d\xi}\frac{\partial}{\partial \eta}\left(\frac{1}{\tau}\right) = 0 \tag{26}$$

According to Equation 24 the boundary conditions to it are

$$\frac{\partial(\tau^2)}{\partial \eta} = 2\mu \frac{dp}{d\xi} \quad \text{at} \quad \eta = 0 \qquad \tau = 0 \quad \text{at} \quad \eta = U$$

$$\tau = \tau_0(\eta) \quad \text{at} \quad \xi = \xi_0 \tag{27}$$

4. Simplest Self-Similar Solutions of Laminar Boundary Layer Equations. Boundary Layer on a Plate in a Longitudinal Flow

To the problem admitting the simplest self-similar solutions of the Prandtl equations (11) or (15) refers the case of a near-wall boundary layer on a plate in a longitudinal flow. Directing the axis Ox (Figure 170) along the surface of the plate (the thickness of which is negligibly small), the axis Oy along the perpendicular to the plate surface, and taking the coordinate origin in the leading edge O of the plate, we note that in this case the outer flow with respect to the boundary layer will be homogeneous; i.e., it will have constant velocity U_∞ over the entire flow region.

We shall use the Prandtl equation (15), assuming that in this case $U = \text{const} = U_\infty$ and $dU/dx = 0$. We have the equation

$$\frac{\partial \psi}{\partial y}\frac{\partial^2 \psi}{\partial x \partial y} - \frac{\partial \psi}{\partial x}\frac{\partial^2 \psi}{\partial y^2} = \nu \frac{\partial^3 \psi}{\partial y^3} \tag{28}$$

The boundary conditions of this parabolic equation will be taken in the form simplified by Blasius[11]

$$\psi = 0 \qquad \frac{\partial \psi}{\partial y} = 0 \quad \text{when} \quad y = 0, x > 0$$

$$\frac{\partial \psi}{\partial y} \to U_\infty \quad \text{when} \quad y \to \infty$$

$$\frac{\partial \psi}{\partial y} = U_\infty \quad \text{when} \quad x = 0 \quad y > 0 \tag{29}$$

The Blasius simplification of the problem (28) and (29) formulation is first in the substitution of the velocity profile determined by the inverse effect (retardation) of the plate on a free stream into the unperturbed velocity profile $u = U_\infty$ when $x = 0$, $y > 0$; this is expressed by the last equation in the system of boundary conditions (29). Then, the condition of fluid adhesion to the plate surface ($0 \le x \le L$, $y = 0$) is replaced by the first condition in system (29) satisfied on the ray ($x > 0$, $y = 0$).

These two simplifications follow the conventional formulation of parabolic problems and have a simple physical interpretation – disregard of "end effects" on the leading and rear edges of the plate. The fact that the velocity at the point $x = 0$, $y = 0$ undergoes discontinuity, $u = U_\infty$ at $x = -0$, $u = 0$ at $x = +0$, makes the point $x = 0$, $y = 0$ singular, but does not complicate the solution given below.

The typical property of the independence of its solution at the given point of subsequent conditions of flow makes the solution of the parabolic equation (15) independent of the plate length L. This fact will be then used to elucidate a solution general structure.

We shall move in Equation 28 and boundary conditions (29) to dimensionless variables assuming (prime corresponds to the dimensionless variables; $\mathbf{Re} = U_\infty L/\nu$)

$$x = Lx' \qquad y = \frac{1}{\sqrt{\mathbf{Re}}}y' = \sqrt{\frac{\nu L}{U_\infty}}y' \qquad \psi = \frac{U_\infty L}{\sqrt{\mathbf{Re}}}\psi' = \sqrt{\nu U_\infty L}\,\psi'$$

Equations 28 and boundary conditions (29) in this case are written in the form

$$\frac{\partial \psi'}{\partial y'}\frac{\partial^2 \psi'}{\partial x' \partial y'} - \frac{\partial \psi'}{\partial x'}\frac{\partial^2 \psi'}{\partial y'^2} = \frac{\partial^3 \psi'}{\partial y'^3}$$

$$\psi' = 0, \quad \frac{\partial \psi'}{\partial y'} = 0 \quad \text{when} \quad y' = 0, \quad x' > 0$$

$$\frac{\partial \psi'}{\partial y'} \to 1 \quad \text{when} \quad x' > 0, y' \to \infty$$

$$\frac{\partial \psi'}{\partial y'} = 1 \quad \text{when} \quad x' = 0, y' > 0 \tag{30}$$

which does not involve the characteristic constants of the problem.

[11] Blasius, H., Grenzschichten in Flüssigkeiten mit kleiner Reibung, *Z. Math. Phys.*, 56, 1908.

The unknown integral of this equation satisfying the boundary conditions is

$$\psi' = \psi'(x', y')$$

and in the dimensional variables

$$\psi(x, y) - \sqrt{\nu U_\infty x}\, \psi'\left(\frac{x}{L}, y\sqrt{\frac{U_\infty}{\nu L}}\right)$$

The solution of the problem $\psi(x, y)$ at any fixed point (x, y) is independent of L. Consequently, the dependence of ψ' on its arguments x', y' should provide this property. We can obtain this result, having assumed

$$\psi'(x', y') = \sqrt{x'}\, f\left(\frac{y'}{\sqrt{x'}}\right)$$

In fact, the problem is reduced to seeking solution in the form that does not involve L

$$\psi = \sqrt{\nu U_\infty x}\, f\left(y\sqrt{\frac{U_\infty}{\nu x}}\right) \tag{31}$$

or, in case

$$\eta = \frac{1}{2} y \sqrt{\frac{U_\infty}{\nu x}} \tag{32}$$

is introduced as an argument, in the following form

$$\psi = \sqrt{\nu U_\infty x}\, \varphi(\eta) \tag{33}$$

Calculating the derivatives in Equation 28 (the prime stands for derivatives with respect to η) and noting also that

$$\frac{\partial \eta}{\partial x} = -\frac{1}{4} y \sqrt{\frac{U_\infty}{\nu x^3}} \qquad \frac{\partial \eta}{\partial y} = \frac{1}{2}\sqrt{\frac{U_\infty}{\nu x}}$$

we find

$$\frac{\partial \psi}{\partial y} = \frac{1}{2} U_\infty \varphi'(\eta) \qquad \frac{\partial^2 \psi}{\partial y^2} = \frac{1}{4} U_\infty \sqrt{\frac{U_\infty}{\nu x}}\, \varphi''(\eta)$$

$$\frac{\partial^3 \psi}{\partial y^3} = \frac{1}{8} \frac{U_\infty^2}{\nu x} \varphi'''(\eta) \qquad \frac{\partial \psi}{\partial x} = \frac{1}{2}\sqrt{\frac{\nu U_\infty}{x}}\, [\varphi(\eta) - \eta \varphi'(\eta)]$$

$$\frac{\partial^2 \psi}{\partial x \partial y} = -\frac{1}{4} \frac{U_\infty}{x} \eta \varphi''(\eta)$$

Substituting the obtained derivatives into Equation 28, we obtain, after simple transformations, an ordinary differential equation of the third order

$$\varphi''' + \varphi \varphi'' = 0$$

$$\varphi = 0 \qquad \varphi' = 0 \quad \text{when} \quad \eta = 0$$

$$\varphi' \to 2 \text{ when } \eta \to \infty \tag{34}$$

The Blasius problem has a self-similar solution.

Note that the initial condition for $x = 0$, $y > 0$ is satisfied automatically and falls out of the system of boundary conditions. Equation 34 cannot be integrated in elementary functions. Blasius integrated it by the method of series matching at small and large η similar to the procedure made later by Cochran (Chapter 10, Section 6); then Equation 34 was integrated numerically.

We present the results of the numerical solution of Equation 34 in the form of the dependence of the dimensionless velocity u/U_∞ on η (Table 13).

The skin friction on the plate surface is given by:

$$\tau_\omega = \mu\left(\frac{\partial u}{\partial y}\right)_{y=0} = \frac{1}{4}\mu\sqrt{\frac{U_\infty^3}{\nu x}}\,\varphi''(0)$$

Table 13

η	$\frac{u}{U_\infty} = \frac{1}{2}\varphi'(\eta)$
0	0
0.1	0.0664
0.2	0.1328
0.3	0.1989
0.4	0.2647
0.5	0.3298
0.6	0.3938
0.7	0.4563
0.8	0.5168
0.9	0.5748
1.0	0.6298
1.1	0.6813
1.2	0.7290
1.3	0.7725
1.4	0.8115
1.5	0.8460
1.6	0.8761
1.7	0.9018
1.8	0.9233
1.9	0.9411
2.0	0.9555
2.1	0.9670
2.2	0.9759
2.3	0.9827
2.4	0.9878
2.5	0.9915
2.6	0.9942
2.7	0.9962
2.8	0.9975
2.9	0.9984

To do this, making use of Table 13, we determine the magnitude of $\varphi''(0)$

$$\varphi''(0) = \frac{\varphi'(0.1) - \varphi'(0)}{0.1} = 1.328$$

which follows from the linearity of the relation u/U_∞ at small η, and obtain the Blasius formula

$$\tau_\omega = 0.332\sqrt{\frac{\mu\rho U_\infty^3}{x}} \tag{35}$$

Note two important regularities expressed by the last equation: the local skin friction is proportional to one and a half power of the velocity of the flow approaching the plate and changes along the plate in inverse proportion to the square root of the distance from the leading edge.

The Blasius equation is coincidental with the test data at all points of the plate, with the exception of those close to the plate ends.

Introducing the local coefficient of friction c_f we obtain

$$c_f = \frac{\tau_\omega}{\frac{1}{2}\rho U_\infty^2} = \frac{0.664}{\sqrt{\mathbf{Re}_x}} \qquad \mathbf{Re}_x = \frac{U_\infty x}{\nu} \tag{36}$$

Summing up the skin friction on both sides of the plate we find the total friction drag of the plate

$$W_f = 2\int_0^L \tau_\omega dx = 1.328\sqrt{\mu\rho L U_\infty^3} \tag{37}$$

and the corresponding coefficient of the total friction drag of the plate

$$C_f = \frac{W_f}{\frac{1}{2}\rho U_\infty^2 S} = \frac{1.328}{\sqrt{\mathbf{Re}}} \qquad \mathbf{Re} = \frac{U_\infty L}{\nu} \tag{38}$$

where S is the area of the plate "wetted" surface equal to $2L \cdot 1$.

The distribution of dimensional longitudinal velocities over the plate boundary layer cross-sections

$$\frac{u}{U_\infty} = \frac{1}{2}\varphi'\left(\frac{1}{2}y\sqrt{\frac{U_\infty}{\nu x}}\right) \tag{39}$$

is shown in Figure 171 in the coordinates u/U_∞, $y\sqrt{U_\infty/(\nu x)}$. In the same figure are the experimental points from the Hansen[12] experiments for boundary layer cross-sections with abscissas $x = 3$, 10, and 15 cm. With general good agreement between the theory and

[12] Hansen, M., Die Geschwindigkeitsverteilung in der Grenzschicht an einer eingetauchten Platte, Z. *Angew. Math. Mech.*, 8, 185–199, 1928.

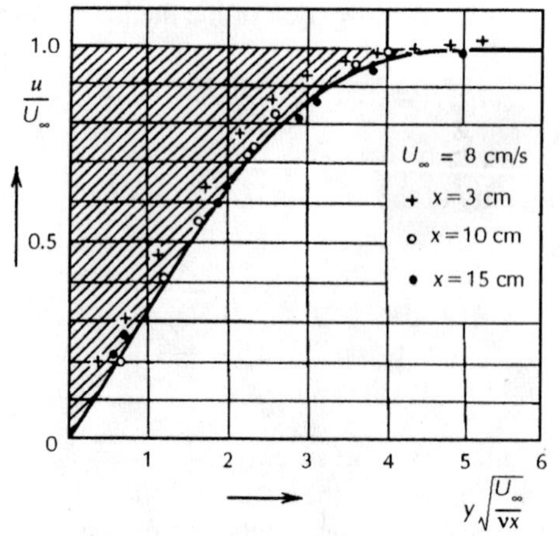

Fig. 171

experiments, some deviation of points at x = 3 cm can be observed that is explained by the effect of the plate wedge-like leading edge.

Making use of Table 13, we find

$$\delta_{99} \approx 5\sqrt{\frac{\nu x}{U_\infty}} \tag{40}$$

where δ_{99} is introduced to denote the ordinate y at the point of the boundary layer cross-section where the velocity u is equal to $0.99U_\infty$, i.e., it differs from U_∞ by 1%; similarly we could introduce δ_{999} for the ordinate in which the velocity differed by 0.1% from U_∞ and so on. Such boundary layer thicknesses will be then called nominal.

Integrating velocity profiles in boundary layer cross-sections numerically, we find the integral boundary layer thicknesses: the displacement thickness

$$\delta^* = \int_0^\infty \left(1 - \frac{u}{U_\infty}\right) dy = 1.721\sqrt{\frac{\nu x}{U_\infty}} \tag{41}$$

and the momentum thickness

$$\delta^{**} = \int_0^\infty \frac{u}{U_\infty}\left(1 - \frac{u}{U_\infty}\right) dy = 0.664\sqrt{\frac{\nu x}{U_\infty}} \tag{42}$$

The last two "thicknesses" of the boundary layer do not give quantitative presentation of the nominal boundary layer thickness ($\delta_{99} \approx 3\delta^* \approx 7.5\delta^{**}$, but they have the advantage of being only slightly dependent on the inaccuracy in the agreement between u and U_∞ on the layer edge (at large y). For example, δ^* is determined by the cross-hatched region in Figure 171 the value of which is only slightly dependent on whether integration will be made up to finite abscissas 5 or 6.

It can be concluded from Equations 40 to 42 that the layer thicknesses grow from the leading edge to the rear edge in direct proportion to the square root of the abscissa of the

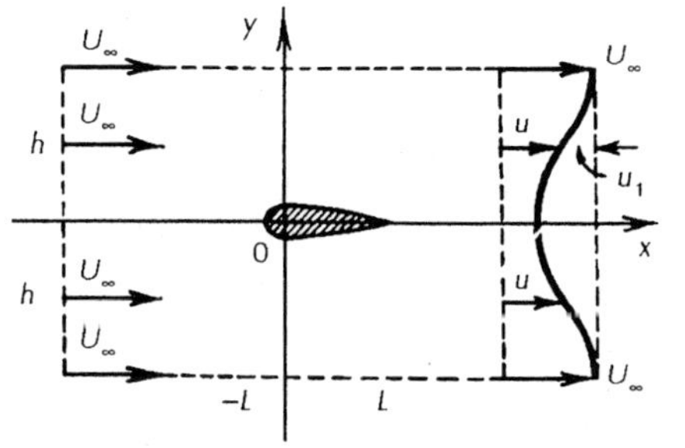

Fig. 172

point on the plate surface.

Comparing Equation 39 with Equations 40 to 42 we note that velocity distribution in boundary layer cross-sections can be considered in one of the following, forms of affine similarity

$$\frac{u}{U_\infty} = f_1\left(\frac{y}{\delta_{99}}\right) = f_2\left(\frac{y}{\delta^*}\right) = f_3\left(\frac{y}{\delta^{**}}\right)$$

thus being equivalent to the self-similarity condition of the Blasius problem solution.

5. Two-dimensional "Free" Boundary Layers: Far Laminar Wake, "Submerged" Jet Issuing from the Point Source

In the previous section we considered a "near-wall" boundary layer. Now we give two examples of "free" boundaries layers that are not connected with the presence of a solid wall in a flow.

The viscous fluid rotating due to adhesion to the solid body surface, leaves the body surface, thus forming a "wake" past its rear edge that consists of fluid with conserved vorticity. "Near" and "far" wakes are distinguished depending on the distance of the considered·wake region from the body rear edge. Motion in the "near" wake depends on the boundary layer on the solid body surface and therefore its calculation offers great difficulties. In a simpler way the problem on a "far" wake is solved in which motion is independent of the shape of the body in the flow, but it depends only on the resistance W the body experiences, on the free stream velocity U_∞, and, of course, on the physical constants of fluid μ and ρ. Direct the axis $0x$ (Figure 172) along the free stream velocity. Disturbance made by the body in a free stream causes a "gap" on the velocity profiles in the wake cross-sections (Figure 172). This gap will be smaller the more downstream the chosen cross-section is situated.

We denote the longitudinal velocity disturbance as u_1, having assumed

$$u_1 = U_\infty - u$$

and take it for large values of x as a small quantity. The square, the highest powers, and its derivatives can be neglected. Transverse velocities will also be small. One can judge this by the incompressibility equation. The equation of motion in a wake will be reduced in this case to

$$U_\infty \frac{\partial u_1}{\partial x} = \nu \frac{\partial^2 u_1}{\partial y^2}$$

$$\frac{\partial u_1}{\partial y} = 0 \quad \text{when} \quad y = 0$$

$$u_1 \to 0 \quad \text{when} \quad y \to \pm\infty \tag{43}$$

From this equation and the assumption on a smooth transition to zero at infinity ($\partial u_1/\partial y \to 0$ for $y \to \pm\infty$) the integral condition

$$\frac{d}{dx}\int_{-\infty}^{\infty} u_1 dy = 0 \quad \text{when} \quad \int_{-\infty}^{\infty} u_1 dy = \text{const} \tag{44}$$

follows.

To understand the sense of the constant and to calculate it we derive an integral expression of body resistance in terms of velocities in the wake cross-section at a large distance from the body. We shall apply the theorem of momentum in the Euler form to the control surface, embracing the body and having the shape of a rectangle with the sides parallel to the coordinate axes. In Figure 172 this rectangle is cross-hatched. Taking its sides $2h$ and $2L$ rather large, we have the smaller error the larger these sides are, and in particular, the larger the h,

$$\rho U_\infty^2 \cdot 2h - W - \int_{-h}^{h} \rho u^2 dy = 0$$

where W is the body resistance. In the same approximation we obtain the equation for flow rate

$$\rho U_\infty \cdot 2h = \int_{-h}^{h} \rho u dy$$

Hence, we find the approximate formula of the body resistance W

$$W = \rho U_\infty^2 \cdot 2h - \int_{-h}^{h} \rho u^2 dy = U_\infty \int_{-h}^{h} \rho u dy - \int_{-h}^{h} \rho u^2 dy = \int_{-h}^{h} \rho u (U_\infty - u)\, dy$$

and considering the cross-section of the wake rather remote ($L \to \infty$) and its width rather large ($h \to \infty$), we obtain

$$W = \int_{-\infty}^{\infty} \rho u (U_\infty - u)\, dy \tag{45}$$

Making substitution $u = U_\infty - u_1$, $U_\infty - u = u_1$ and discarding the squares of small quantities, we have

$$W = \rho U_\infty \int_{-\infty}^{\infty} u_1 dy$$

so that the integral condition (44) can be rewritten in the form

$$\int_{-\infty}^{\infty} u_1 dy = \frac{W}{\rho U_\infty} \tag{46}$$

where W is the resistance of the body, the wake of which is studied. Prove now that at a rather large distance from the body the motion in the wake behind the body is determined not by the shape of the body surface and its dimensions, but by the constant only – by the body resistance.

Equation 43 is similar to the equation of heat propagation and has a simple integral of the source type

$$u_1 = \frac{C}{\sqrt{x}} \exp\left(-\frac{U_\infty y^2}{4\nu x} \right) \tag{47}$$

satisfying the adopted boundary conditions. To determine the constant C substitute this integral into Equation 46, then, having calculated the quadrature, we obtain

$$C = \frac{W}{2\sqrt{\pi\rho\mu U_\infty}}$$

Finally, we have

$$u_1 = U_\infty - u = \frac{W}{2\sqrt{\pi\rho\mu U_\infty x}} \exp\left(-\frac{U_\infty y^2}{4\nu x} \right)$$

The law of decrease in the maximum disturbance on the wake axis ($y = 0$) is presented by the formula

$$u_{1m} = \frac{W}{2\sqrt{\pi\mu\rho U_\infty}} \frac{1}{\sqrt{x}} \tag{48}$$

Applying the obtained formula to the wake at a distance from the longitudinally streamlines plate of the length L, we obtain according to Equation 37

$$\frac{u}{U_\infty} = 1 - \frac{0.664}{\sqrt{\pi}} \sqrt{\frac{L}{x}} \exp\left(-\frac{U_\infty}{4\nu} \frac{y^2}{x} \right)$$

$$\frac{u_m}{U_\infty} = 1 - \frac{0.664}{\sqrt{\pi}} \sqrt{\frac{L}{x}} \tag{49}$$

We introduce the concept of the wake conditional width $2\delta(x)$ as a twice ordinate of such a point at which the ratio of $1 - u/U_\infty$ to $1 - u_m/U_\infty$ takes a small nearly constant value; we obtain from the first equality of system (49)

$$\delta(x) = \text{const} \cdot \sqrt{x} \tag{50}$$

Then the profiles of relative velocity disturbances

$$\frac{u_1}{u_{1m}} = \frac{U_\infty - u}{U_\infty - u_m} = \frac{1 - u/U_\infty}{1 - u_m/U_\infty}$$

can be presented in the form

$$\frac{u_1}{u_{1m}} = f\left(\frac{y}{\delta}\right) \tag{51}$$

speaking of the presence of the affine similarity of the distribution of longitudinal velocity disturbances over the wake normal cross-sections at a distance from the source of disturbances, the body, the wake of which is considered. The problem of the "far" wake is self-similar.

Another characteristic type of free boundary layers are the so-called submerged jets propagating in a motionless medium with the same physical constants that the wake itself has, or, in a general case, different from them.

In their simplest formulation these problems refer to the case of the jet source ultimately thin over the cross-section, but finite, thanks to very high velocity of issuing, initial pulse (momentum per second) J_0. To this simplified formulation one can assign the problem of a submerged jet, issuing from a nozzle of finite diameter and having the finite initial velocity, in case the flow region at a rather large distance from the jet source is considered. In this remote region the flow will also be determined by the pulse which can be treated as an initial pulse of the equivalent jet from a thin source.

The problem on a two-dimensional laminar submerged jet was solved by Schlichting.[13] The Prandtl-Mises equation was applied to the problem on a two-dimensional jet by Loitsyanskiy.[14]

We shall direct the axis $0x$ along the axis of two-dimensional jet symmetry (Figure 173). The equality to zero of the outer velocity U makes it possible to write the equation of viscous incompressible fluid flow in the two-dimensional jet region in one of the two following forms

$$u\frac{\partial u}{\partial x} + v\frac{\partial u}{\partial y} = \nu\frac{\partial^2 u}{\partial y^2}$$

$$\frac{\partial u}{\partial x} + \frac{\partial v}{\partial y} = 0$$

$$\frac{\partial \psi}{\partial y}\frac{\partial^2 \psi}{\partial x \partial y} - \frac{\partial \psi}{\partial x}\frac{\partial^2 \psi}{\partial y^2} = \nu\frac{\partial^3 \psi}{\partial y^3} \tag{52}$$

[13] Schlichting, H., Laminare Strahlausbreitung, Z. *Angew. Math. Mech.*, 13 (4), 260, 1933.

[14] Loitsyanskiy, L. G., On the theory of two-dimensional laminar and turbulent jets, *Tr. LPI*, No. 176, 101–114, 1955.

In view of symmetry we have on the jet axis

$$\frac{\partial u}{\partial y} = 0, v = 0 \quad \text{or} \quad \frac{\partial^2 \psi}{\partial y^2} = 0, \psi = 0 \quad \text{at} \quad y = 0 \tag{53}$$

and on the jet outer edge with the transition to the motionless fluid surrounding the jet there will be

$$u \to 0 \quad \text{or} \quad \frac{\partial \psi}{\partial y} \to 0 \quad \text{for} \quad y \to \pm\infty \tag{54}$$

The presence of these zero boundary condition for y values and the absence of the boundary condition for x lead to a trivial solution $u = 0$, $v = 0$ over the entire region that is in contradiction with the fluid flow presence in a jet. To eliminate this contradiction we rewrite the first equation of systém (52), making use of the second one, as

$$\frac{\partial (u^2)}{\partial x} + \frac{\partial (uv)}{\partial y} = \nu \frac{\partial^2 u}{\partial y^2}$$

and integrate both sides with respect to y within the limits $(-\infty, \infty)$. Then, on the assumption of integral existence within infinite limits from the first summand and in the feasibility of the change in the order of operations of differentiation and integration, we obtain

$$\frac{d}{dx}\int_{-\infty}^{\infty} u^2 dy + uv\Big|_{-\infty}^{\infty} = \nu \frac{\partial u}{\partial y}\Big|_{-\infty}^{\infty}$$

The substitution on the left side of this equation vanishes due to evaluation of the limits and, on the right side due to the assumption of a smooth transition to these limit which requires $\partial u/\partial y \to 0$ when $y \to \pm\infty$. Thus, we obtain

$$\frac{d}{dx}\int_{-\infty}^{\infty} u^2 dy = 0 \qquad \int_{-\infty}^{\infty} u^2 dy = \text{const}$$

Finishing multiplication of both sides of the last equality by the constant density ρ, we rewrite it in the form

$$J = \int_{-\infty}^{\infty} \rho u^2 dy = \text{const} = J_0 \tag{55}$$

The physical sense of this equality is that the momentum transferred through the jet transverse cross-section per second is equal for all cross-sections. The constant J_0 serves as the same characteristic constant of the jet as the intensity for a point source or sink, the moment for a point dipole, etc. Assignment of the quantity J_0 makes the problem of jet propagation quite definite. The solution $u \equiv v \equiv 0$ does not satisfy the integral condition (55) which can be considered the condition of the nontriviality of system (52) solution under the boundary conditions (53) and (54).

Express the longitudinal coordinates x as the length scale L, the transverse coordinates y

as the scale $Y = L/\sqrt{\mathbf{Re}} = \sqrt{\nu L/V}$. Then, the quantity $\Psi = \sqrt{\nu VL}$ (V is some velocity scale) will serve as the scale Ψ of the stream function ψ. On the other hand, according to Equation 55,

$$\rho\int_{-\infty}^{\infty}\left(\frac{\partial\psi}{\partial y}\right)^2 dy = J_0 \quad \text{and} \quad \frac{\rho\Psi^2}{Y}\int_{-\infty}^{\infty}\left(\frac{\partial\psi'}{\partial y'}\right)^2 dy' = J_0$$

(the prime denotes dimensionless quantities), so that assuming

$$\Psi^2 = \frac{J_0}{\rho}\sqrt{\frac{\nu L}{V}}$$

we compose condition (55) in the dimensionless form

$$\int_{-\infty}^{\infty}\left(\frac{\partial\psi'}{\partial y'}\right)^2 dy' = 1 \tag{56}$$

Comparing the obtained values of Ψ, we have the expressions of the scales L and Ψ in terms of V

$$L = \frac{J_0^2}{\mu\rho V^3} \qquad \Psi = \frac{J_0}{\rho V}$$

by the character of the problem itself, the velocity V as a scale should not enter the solution. Write the general form of the solution in terms of dimensionless coordinates

$$\psi' = \psi'(x', y')$$

and dimensional coordinates

$$\psi = \Psi\psi'\left(\frac{x}{L}, y\sqrt{\frac{V}{\nu L}}\right) = \frac{J_0}{\rho V}\psi'\left(\frac{\mu\rho V^3}{J_0^2}x, \frac{\rho V^2}{J_0}y\right)$$

For the right side of this expression to be independent of V, the general form of the solution should be

$$\psi' = \sqrt[3]{x'}\,\varphi(y'/x'^{2/3}) \tag{57}$$

In fact, here

$$\psi = \frac{J_0}{\rho V}\frac{\sqrt[3]{\mu\rho V^3 x}}{J_0^{2/3}}\varphi\left(\frac{\rho V^2}{J_0}\frac{J_0^{4/3}}{\mu^{2/3}\rho^{2/3}V^2}\frac{y}{x^{2/3}}\right) = \sqrt[3]{\frac{\nu J_0 x}{\rho}}\,\varphi\left(\sqrt[3]{\frac{J_0}{\rho\nu^2}}\frac{y}{x^{2/3}}\right) \tag{58}$$

Assuming

$$\frac{y'}{x'^{2/3}} = \sqrt[3]{\frac{J_0}{\rho\nu^2}}\frac{y}{x^{2/3}} = \eta \tag{59}$$

substitute the expression of ψ' from Equation 57, rewritten in the form

$$\psi' = x'^{1/3}\varphi(\eta)$$

into the third equation of system (52) transform.ed to the dimensionless coordinates

$$\frac{\partial\psi'}{\partial y'}\frac{\partial^2\psi'}{\partial x'\partial y'} - \frac{\partial\psi'}{\partial x'}\frac{\partial^2\psi'}{\partial y'^2} = \nu\frac{\partial^3\psi'}{\partial y'^3}$$

Beforehand calculate (the primes of φ denote the corresponding derivatives with respect to η)

$$\frac{\partial\psi'}{\partial y'} = x'^{-1/3}\varphi'(\eta) \qquad \frac{\partial^2\psi'}{\partial y'^2} = x'^{-1}\varphi''(\eta)$$

$$\frac{\partial^3\psi'}{\partial y'^3} = x'^{-5/3}\varphi'''(\eta) \qquad \frac{\partial\psi'}{\partial x'} = \frac{1}{3}x'^{-2/3}\left[\varphi(\eta) - 2\eta\varphi'(\eta)\right]$$

$$\frac{\partial^2\psi'}{\partial x'\partial y'} = -\frac{1}{3}x'^{-4/3}\left[\varphi'(\eta) + 2\eta\varphi''(\eta)\right]$$

Making the substitution mentioned we come to integration of the ordinary third order equation

$$\varphi''' + \frac{1}{3}(\varphi'^2 + \varphi\varphi'') = 0 \tag{60}$$

under the boundary conditions

$$\varphi = 0, \quad \varphi'' = 0 \quad \text{at} \quad \eta = 0$$

$$\varphi' \to 0 \quad \text{when} \quad \eta \to \pm\infty \tag{61}$$

and to the integral condition (56) which can be written in the form

$$\int_{-\infty}^{\infty} \varphi'^2(\eta)\,d\eta = 1 \tag{62}$$

Integrating both sides of Equation 60 termwise, we obtain

$$\varphi'' + \frac{1}{3}\varphi\varphi' = 0$$

with the first two conditions of (61) being satisfied. Making one more integration and introducing the notation

$$\varphi = \varphi_0 \quad \text{at} \quad \varphi' = 0$$

we have

$$\varphi' = \frac{1}{6}\left(\varphi_0^2 - \varphi^2\right)$$

Further integration gives

$$\int_0^{\varphi} \frac{d\varphi}{\varphi_0^2 - \varphi^2} = \frac{1}{2\,\varphi_0}\ln\frac{\varphi_0 + \varphi}{\varphi_0 - \varphi} = \frac{1}{6}\eta$$

or

$$\frac{\varphi_0 + \varphi}{\varphi_0 - \varphi} = e^{\frac{1}{3}\varphi_0\eta}$$

Hence, we find

$$\varphi = \varphi_0 \frac{e^{\frac{1}{3}\varphi_0\eta} - 1}{e^{\frac{1}{3}\varphi_0\eta} + 1} = \varphi_0 \frac{e^{\frac{1}{6}\varphi_0\eta} - e^{-\frac{1}{6}\varphi_0\eta}}{e^{\frac{1}{6}\varphi_0\eta} + e^{-\frac{1}{6}\varphi_0\eta}} = \varphi_0 \,\mathrm{th}\left(\frac{1}{6}\varphi_0\eta\right) \tag{63}$$

Calculating the derivative with respect to η

$$\varphi'(\eta) = \frac{1}{6}\varphi_0^2 \cdot \frac{1}{\mathrm{ch}^2\left(\frac{1}{6}\varphi_0\eta\right)}$$

and substituting it into (62), we obtain the following equation for determining φ_0

$$\frac{1}{36}\varphi_0^4 \int_{-\infty}^{\infty} \frac{d\eta}{\mathrm{ch}^4\left(\frac{1}{6}\varphi_0\eta\right)} = 1$$

Assume for integral calculation

$$\frac{1}{6}\varphi_0\eta = \xi$$

Then, the previous equation changes over to

$$\frac{\varphi_0^3}{6}\int_{-\infty}^{\infty} \frac{d\xi}{\mathrm{ch}^4\xi} = 1$$

and, noting that the determined integral is equal to 4/3, we find the unknown value of φ_0

$$\varphi_0 = \sqrt[3]{\frac{9}{2}} = 1.6510$$

Reverting to Equation 63 and then to (58), we obtain

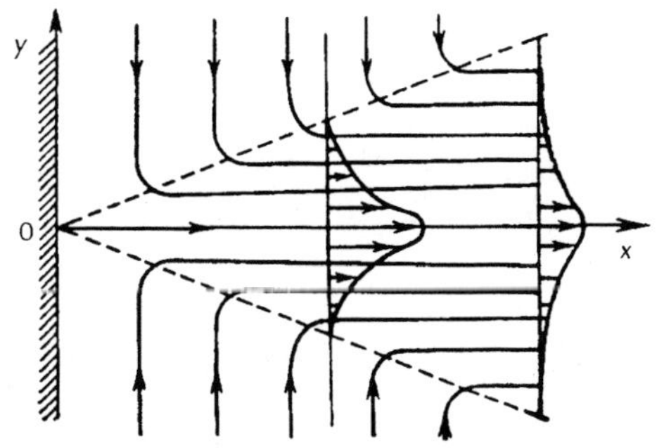

Fig. 173

$$\psi = 1.6510 \sqrt[3]{\frac{\nu J_0 x}{\rho}} \, \mathrm{th}\left(0.2752 \sqrt[3]{\frac{J_0}{\rho \nu^2}} \frac{y}{\sqrt[3]{x^2}} \right) \qquad (64)$$

It follows from the sense itself of the stream function that the rate of volume discharge Q through the jet cross-section will be equal to

$$Q = 2(\psi)_{y=\infty} = 3.3020 \sqrt[3]{\frac{\nu J_0 x}{\rho}}$$

The rate grows from cross-section to cross-section proportionally to the cubic root of the cross-section distance from the jet source. At the slot outlet ($x = 0$) the flow rate is equal to zero. At the same time, the momentum per second is everywhere equal and finite; consequently, it is also finite at the outlet. With an infinitely large velocity of jet issuing from an infinitely thin slot, the result obtained is not contradictory. From the qualitative viewpoint this result expresses the familiar property of ejection:[15] fluid flow rate in a jet is small compared with the flow rate objected due to the substantial quantity of jet momentum. As is seen from Figure 173 in the considered case of an infinitely thin slot the jet fully consists of particles of fluid filling the space into which the jet ejects. In the case of a finite jet, this will not be so.

Having found the expression of the stream function (64), we find the distribution of the longitudinal velocity u over jet cross-sections and of the maximum value of this velocity u_m along the jet axis

$$u = 0.4543 \sqrt[3]{\frac{J_0^2}{\rho^2 \nu x}} \frac{1}{\mathrm{ch}^2\left(0.2752 \sqrt[3]{\frac{J_0}{\rho \nu^2}} \frac{y}{\sqrt[3]{x}} \right)} \qquad u_m = 0.4543 \sqrt[3]{\frac{J_0^2}{\rho^2 \nu x}} \qquad (65)$$

6. Problem on a Two-dimensional "Near-Wall" Jet

Consider the example of the mixed boundary layer, a near-wall jet[16], which joints a near-wall layer on a plate with a free jet (Figure 174).

---

[15] Also referred to as entrainment.

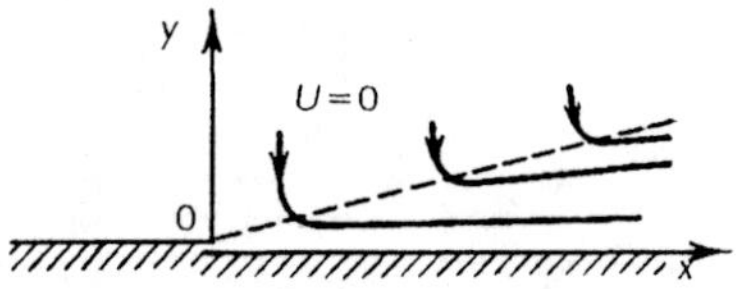

Fig. 174

It is obvious that with nonlinearity of the motion equations, the superposition of flows is out of the question. However, some similarity is still observed of the profile of longitudinal velocities near the plane bounding the jet with the corresponding profile near the plate and the velocity profile at a distance from the plane with the profile in the jet.

The motion equation is

$$\frac{\partial\psi}{\partial y}\frac{\partial^2\psi}{\partial x\partial y}-\frac{\partial\psi}{\partial x}\frac{\partial^2\psi}{\partial y^2}=\nu\frac{\partial^3\psi}{\partial y^3}$$

$$\psi=0 \qquad \frac{\partial\psi}{\partial y}=0 \quad \text{at} \quad y=0$$

$$\frac{\partial\psi}{\partial y}\to 0 \quad \text{when} \quad y\to\infty \tag{66}$$

In order that a nontrivial solution may exist it is necessary to specify such an integral characteristic of motion that, similar to the jet pulse, could correspond to some conservation law and thus could serve as a quantitative measure of the given specific semibounded near-wall jet. Note that in the given case a pulse cannot be preserved along the jet because, in contrast to an infinite jet, an outer force of fluid friction exists against a semiplane boundary.

We derive the unknown nontriviality equation having rewritten the boundary layer equation in the form

$$\frac{\partial}{\partial x}(u^2)+\frac{\partial}{\partial y}(uv)=\nu\frac{\partial^2 u}{\partial y^2} \tag{67}$$

and, having multiplied both sides termwise by ψ and integrated them across the layer going from 0 to ∞, wee have successively

$$\psi\frac{\partial}{\partial x}(u^2)+\psi\frac{\partial}{\partial y}(uv)=\nu\psi\frac{\partial^2 u}{\partial y^2}$$

$$\frac{\partial}{\partial x}(\psi u^2)+\frac{\partial}{\partial y}(\psi uv)-u^2\frac{\partial\psi}{\partial x}-uv\frac{\partial\psi}{\partial y}=\nu\frac{\partial}{\partial y}\left(\psi\frac{\partial u}{\partial y}\right)-\nu\frac{\partial\psi}{\partial y}\frac{\partial u}{\partial y}$$

After obvious reduction

$$\frac{\partial}{\partial x}(\psi u^2)+\frac{\partial}{\partial y}(\psi uv)=\nu\frac{\partial}{\partial y}\left(\psi\frac{\partial u}{\partial y}\right)-\nu\frac{\partial}{\partial y}\left(\frac{u^2}{2}\right)$$

---

[16] Akatnov, N. I., Propagation of a two-dimensional laminar fluid jet along a solid wall, *Tr. LPI* (*Energomashinostroeniye, Tekhnicheskaya gidromekhanika*), No. 5, Mashgiz, Moscow, 1953, 24–31; Glauert, M. B., Wall jet, *J. Fluid Mech.*, 1, 1956.

and termwise integration yields

$$\frac{d}{dx}\int_0^\infty \psi u^2 dy + \psi u v\Big|_0^\infty = \nu\psi\frac{\partial u}{\partial y}\Big|_0^\infty - \nu\frac{u^2}{2}\Big|_0^\infty \tag{68}$$

Noting that all substitutions are equal to zero we find

$$\frac{d}{dx}\int_0^\infty \psi u^2 dy = 0$$

were it follows that the quantity (α is the stream function-value on the boundary layer outer edge)

$$\int_0^\infty \psi u^2 dy = \int_0^\alpha u\psi d\psi = \text{const} = E \tag{69}$$

retains equal values along the entire boundary layer and its assignment can serve as a condition of solution nontriviality.

Further, it will be found in view of the analysis of the solution obtained that the quantity E differs from the product of the rate of volume flow Q and the momentum K related to mass unit

$$Q = \int_0^\infty u dy \qquad K = \frac{J}{\rho} = \int_0^\infty u^2 dy$$

only by the numerical constant.

The dimension of the quantity E will be determined as $[U^3\delta^2] = [U^3L^2/(UL/\nu)]$ and is equal to

$$[E] = [\nu U^2 L]$$

Dimensional considerations will lead in this case to a general solution $[L = E/(\nu U^2)]$

$$\psi = \sqrt{\frac{E}{U}} f\left(\frac{\nu x U^2}{E}, y\sqrt{\frac{U^3}{E}}\right)$$

which, due to the absence of the characteristic velocity in the problem conditions, should become

$$\psi = \sqrt{\frac{E}{U}}\sqrt[4]{\frac{\nu x U^2}{E}}\, F\left[\frac{y\sqrt{U^3/E}}{\sqrt[4]{(\nu x U^2/E)^3}}\right]$$

i.e., in the final form

$$\psi = \sqrt[4]{\nu E x}\, F(\eta) \qquad \eta = \sqrt[4]{\frac{E}{\nu^3}}\frac{y}{\sqrt[4]{x^3}} \tag{70}$$

Having substituted this expression of ψ into Equation 66 we convince ourselves of problem self-similarity; the equation will be reduced to the ordinary differential equation (the prime indicates derivative with respect to η)

$$4F''' + FF'' + 2F'^2 = 0 \tag{71}$$

The boundary conditions, according to Equation 66, will take the form

$$F = F' = 0 \quad \text{at} \quad \eta = 0$$

$$F' \to 0 \quad \text{when} \quad \eta \to \infty \tag{72}$$

with the integral condition

$$\int_0^\infty F'^2(\eta)\, F(\eta)\, d\eta = 1 \tag{73}$$

easily derived from the system of Equations 69 and 70, being added.

Integrating Equation 71 allows determination of the multiplier F. Integrating once and determining the integration constant from the boundary conditions (72), we find

$$4FF'' - 2F'^2 + F^2F' = 0 \tag{74}$$

Replacing the variables $F' = \Phi$ and taking F as the independent variable, we obtain the linear equation of the first order

$$\frac{d\Phi}{dF} - \frac{1}{2F}\Phi = -\frac{F}{4} \tag{75}$$

the solution of which will be

$$\Phi = F' = C\sqrt{F} - \frac{1}{6}F^2 \tag{76}$$

Designate the quantity F at $\eta = \infty$ and $\Phi = 0$ as F_∞. Then the constant C is equal to

$$C = \frac{1}{6}F_\infty\sqrt{F_\infty}$$

and Equation 76 may be transformed to

$$\frac{d\theta}{d\eta} = \frac{F_\infty}{6}(\sqrt{\theta} - \theta^2) \qquad \theta = \frac{F}{F_\infty} \tag{77}$$

Determine the constant F_∞ substituting the values

$$F' = \frac{1}{6}\left(F_\infty\sqrt{F_\infty}\sqrt{F} - F^2\right) \qquad d\eta = \frac{6dF}{F_\infty\sqrt{F_\infty}\sqrt{F} - F^2}$$

into the integral condition (73). We obtain

$$F_\infty = 2.515 \tag{78}$$

Equation 77 is integrated and leads to the relation

$$\eta = 0.7952\left[\ln\frac{1+\sqrt{\theta}+\theta}{(1-\sqrt{\theta})^2} + 2\sqrt{3}\,\mathrm{arctg}\frac{\sqrt{3\theta}}{\sqrt{\theta}+2}\right] \tag{79}$$

Combining it with Equation 77 rewritten in the form

$$F' = \frac{F_\infty^2}{6}(\sqrt{\theta}-\theta^2) = 1.054\,(\sqrt{\theta}-\theta^2) \tag{80}$$

we obtain F' in a parametric form. It is not difficult to determine the quantities Q and K. We have

$$Q = \int_0^\infty u\,dy = 2.515\,\sqrt[4]{\nu E x} \qquad K = \int_0^\infty u^2 dy = 0.884\,\sqrt[4]{\frac{E^3}{\nu x}} \tag{81}$$

Each of these quantities is a function of the longitudinal coordinate x; i.e., it changes with the transition from one jet cross-section to another, but their product retains a constant value and is simply related to the main constant E

$$QK = \frac{20}{9}E \tag{82}$$

We also present the formula of shearing skin friction on a solid body along which a jet propagates

$$\tau_\omega = \mu\left(\frac{\partial u}{\partial y}\right)_{y=0} = 0.221\,\mu\,\sqrt[4]{\frac{E^3}{\nu^5 x^5}} \tag{83}$$

In Figure 175 and 176 compare the distributions of velocities in the inner (from the wall to the velocity maximum $u = u_m$ in the cross-section with the ordinate y_m) and the outer

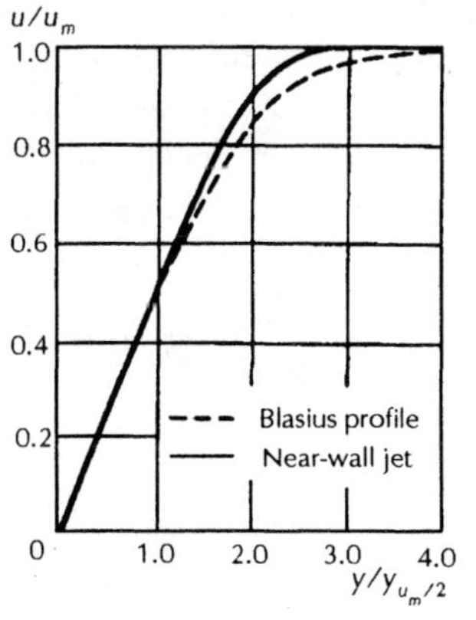

Fig. 175

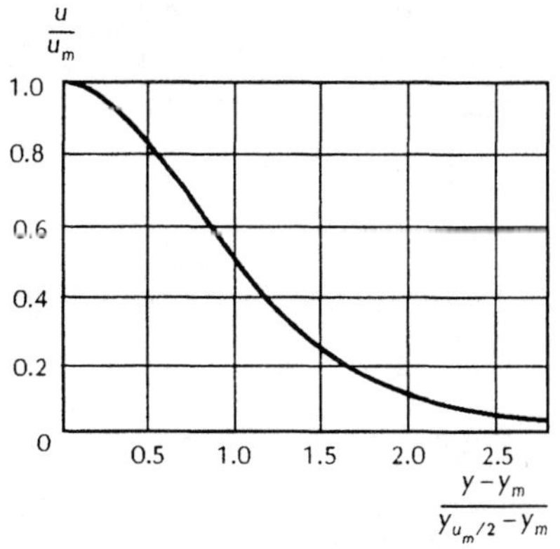

Fig. 176

(beyond the velocity maximum) jet regions with distributions of velocities on a plate in a longitudinal flow and in an infinite jet. In both figures the same ratio u/u_m is laid off on the ordinate, while on the abscissa the ratio $y/y_{u_{m/2}}$ of the current ordinate to the ordinate corresponding to the cross-section point with $u = u_m/2$ is laid off for the inner region and $(y - y_m)/\left(y_{u_{m/2}} - y_m\right)$ for the outer one.

The graphs justify the earlier statement of the similarity between the velocity profiles for the considered case of a complex flow and for the previously considered simpler cases. Virtually full coincidence is observed in the outer region (Figure 176).

7. General Case of Exact Asymptotic Solutions to the Equations of a Stationary Two-Dimensional Near-Wall Boundary Layer

Among general solutions of the equations for a stationary two-dimensional near-wall boundary layer, which correspond to arbitrary velocity assignment on the boundary layer outer edge, the class of exact self-similar problems with power-law assignment of the outer velocity is distinguished by its simplicity and clarity. In these cases, the solution is reduced to integration of the ordinary differential equation of the third order.

Now we determine the conditions of the existence of self-similar solutions of Equation 15. Introduce the dimensionless variables

$$\xi = \frac{y}{\delta} \qquad \Phi = \frac{\psi}{U\delta} \tag{84}$$

where $U(x)$ is the velocity on the boundary layer outer edge in the given cross-section x and also the scale of longitudinal velocities u; $\delta(x)$ is the boundary layer thickness (in one of its possible definitions), which is taken as the scale of ordinates in the corresponding boundary layer cross-section; and the product $U\delta$ is the stream function scale. The quantity Φ is called the reduced stream function.

From Equations 84 the dimensional stream function in terms of its "reduced" value is

$$\psi = U\delta\Phi \tag{85}$$

Upon substitution into Equation 15 with the accompanying boundary conditions we find what should be the functions $U(x)$ and $\delta(x)$ needed to the partial differential Equation 15, as a result of this substitution, to the ordinary differential equation relative to the reduced stream function Φ which should be a function of ξ only (i.e., it should be independent of x). First, we find (the prime denotes the derivative with respect to x and the dot above the symbol stands for the derivative with respect to ξ)

$$\frac{\partial\psi}{\partial x} = (U'\delta + U\delta')\Phi - U\delta'\xi\dot{\Phi} = \frac{\nu}{\delta}\left[\left(\beta + \frac{1}{2}\bar{\beta}\right)\Phi - \frac{1}{2}\bar{\beta}\xi\dot{\Phi}\right],$$

$$\frac{\partial^2\psi}{\partial x\partial y} = \frac{\nu}{\delta^2}\left(\beta\dot{\Phi} - \frac{1}{2}\bar{\beta}\xi\ddot{\Phi}\right),$$

$$\frac{\partial\psi}{\partial y} = U\dot{\Phi} \qquad \frac{\partial^2\psi}{\partial y^2} = \frac{U}{\delta}\ddot{\Phi} \qquad \frac{\partial^3\psi}{\partial y^3} = \frac{U}{\delta^2}\dddot{\Phi} \tag{86}$$

Here, the notations for "similarity parameters" are introduced

$$U'z = \beta \qquad Uz' = \bar{\beta} \qquad z = \frac{\delta^2}{\nu} \tag{87}$$

The quantities β and $\bar{\beta}$ are the unknown functions of x, and designation of one of them by the overbar is explained by the presence of the offbeat conjugation between them which resides in the fact that in the case of U substitution into z one of them changes to the other

$$\beta \rightleftarrows \bar{\beta} \quad \text{when} \quad U \rightleftarrows z\,. \tag{88}$$

This property is retained for a wider class of motions which satisfy the conditions of generalized similarity.

Substitution of the values of derivatives (86) into the equations and boundary conditions (15) leads to the differential equation of the third order

$$\dddot{\Phi} + \left(\beta + \frac{1}{2}\bar{\beta}\right)\Phi\ddot{\Phi} + \beta\left(1 - \dot{\Phi}^2\right) = 0$$

$$\Phi = \dot{\Phi} = 0 \quad \text{at} \quad \xi = 0 \qquad \dot{\Phi} \to 1 \quad \text{for} \quad \xi \to \infty\,,$$

$$\dot{\Phi} = \dot{\Phi}_0(\xi) \quad \text{for} \quad x = x_0 \tag{89}$$

where Φ is a function of ξ, whereas the quantities β and $\bar{\beta}$, entering into the equation, are functions of x.

Thus, by definition of self-similarity Equation 89 was an ordinary differential equation, we should assume that

$$\beta = U'z = \text{const} \qquad \bar{\beta} = Uz' = \text{const} \tag{90}$$

and omit the "initial" condition given in the last line of Equation 89.

Determination of $U(x)$ and $\delta(x)$ requires integration of the system of ordinary differential equations (90). Imposing the condition of finiteness $z = \delta^2/\nu$ at $U = 0$ and introducing the integration constant c, we obtain the following expressions for U and along with them also for z and δ

$$U = cx^m \qquad z = \frac{\beta}{cm}x^{1-m} \qquad \delta = \left(\frac{\nu\beta}{cm}\right)^{1/2} x^{(1-m)/2}\,, \tag{91}$$

where the notation is introduced

$$m = \frac{\beta}{\beta + \bar{\beta}} \qquad \beta = \frac{m}{1-m}\beta \qquad \beta = \frac{1-m}{m}\beta\,. \tag{92}$$

As follows from the first equality of Equation 91, Equation 89 will have a self-similar solution only in the case of power-law distribution of velocity $U(x)$. A specific case of the sum $\beta + \frac{1}{2}\bar{\beta}$ being equal to zero that leads to a revealing law of U(x) distribution is then eliminated. We will show that integral thicknesses of the boundary layer δ^* and δ^{**} determined by Equations 41 and 42 as well as the skin friction τ_w on the body surface ($\xi = 0$) also will be expressed by power-law formulas. We have according to Equation 85

$$\frac{u}{U} = \frac{1}{U}\frac{\partial[\psi/(U\delta)]}{\partial(y/\delta)}\frac{U\delta}{\delta} = \dot{\Phi}\left(\xi;\beta,\bar{\beta}\right)$$

$$\tau_w = \mu\left(\frac{\partial u}{\partial y}\right)_{y=0} = \frac{\mu U}{\delta}\left[\frac{\partial(u/U)}{\partial(y/\delta)}\right]_{y=0} = \frac{\mu U}{\delta}\ddot{\Phi}\left(0;\beta,\bar{\beta}\right), \tag{93}$$

and, consequently, it may be assumed that

$$\delta^* = \int_0^\infty\left(1-\frac{u}{U}\right)dy = \int_0^\infty\left(1-\dot{\Phi}\right)d\xi\cdot\delta = A\left(\beta,\bar{\beta}\right)\delta$$

$$\delta^{**} = \int_0^\infty\frac{u}{U}\left(1-\frac{u}{U}\right)dy = \int_0^\infty\dot{\Phi}\left(1-\dot{\Phi}\right)d\xi\cdot\delta = B\left(\beta,\bar{\beta}\right)\delta$$

$$\tau_w = \frac{\mu U}{\delta}\zeta\left(\beta,\bar{\beta}\right), \tag{94}$$

where the following notation is taken

$$\int_0^\infty\left(1-\dot{\Phi}\right)d\xi = A\left(\beta,\bar{\beta}\right) \qquad \int_0^\infty\dot{\Phi}\left(1-\dot{\Phi}\right)d\xi = B\left(\beta,\bar{\beta}\right)$$

$$\ddot{\Phi}\left(0;\beta,\bar{\beta}\right) = \zeta\left(\beta,\bar{\beta}\right). \tag{95}$$

Retaining the parameters β and $\bar{\beta}$ independent we could come to a two-term solution $\Phi\left(\xi;\beta,\bar{\beta}\right)$. We restrict ourselves to such a ratio of parameters when the coefficient of the second term of the left side of Equation 89 is equal to unity, i.e., we assume

$$\beta + \frac{1}{2}\bar{\beta} = 1 \quad \bar{\beta} = 2(1-\beta)\,. \tag{96}$$

Note that with the choice of $\bar{\beta}$ the case of $\beta + \frac{1}{2}\bar{\beta} = 0$ is eliminated that leads to the exponential form of $U(x)$.

Equation 89, under conditions (96), takes on a simpler one-term form

$$\dddot{\Phi} + \Phi\ddot{\Phi} + \beta\left(1+\dot{\Phi}^2\right) = 0$$

$$\Phi = \dot{\Phi} = 0 \quad \text{at} \quad \xi = 0 \qquad \dot{\Phi}\to 1 \quad \text{when} \quad \xi\to\infty \tag{97}$$

In this case the parameter β will be related, according to Equation 92, to m by the equations

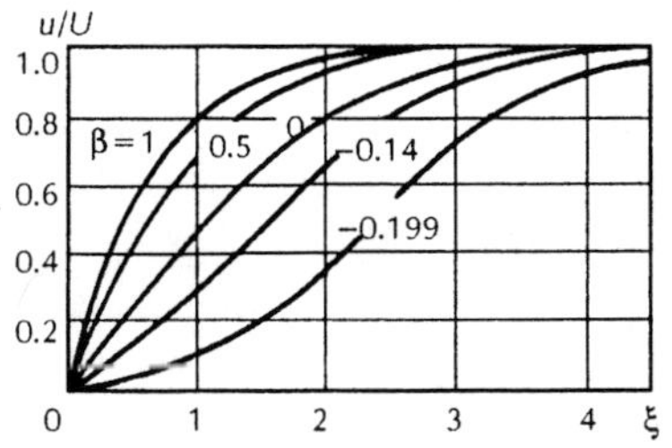

Fig. 177

$$\beta = \frac{2m}{m+1} \qquad m = \frac{\beta}{2-\beta} \tag{98}$$

and formulas (91) and (94), which serve for determination of the thicknesses δ, δ^*, δ^{**} and the skin friction τ_w will take on the form

$$z(x) = \frac{\delta^2}{\nu} = \frac{2}{(m+1)c} x^{1-m} \qquad \delta(x) = \left[\frac{2\nu}{(m+1)c}\right]^{1/2} x^{(1-m)/2}$$

$$\delta^*(x) = A(\beta)\,\delta(x) \qquad \delta^{**}(x) = B(\beta)\,\delta(x)$$

$$\tau_w = \frac{\mu U}{\delta}\zeta(\beta) \tag{99}$$

In this one-term approximation the relation between the physical variables x, y, ψ, and the "similarity variables" ξ, Φ, according to Equations 84 and 91 will be

$$\xi = \frac{y}{\delta} = \sqrt{\frac{m+1}{2}\cdot\frac{c}{\nu}}\, yx^{(m-1)/2} = \sqrt{\frac{c}{(2-\beta)\nu}}\, yx^{(\beta-1)/(2-\beta)}$$

$$\Phi = \sqrt{\frac{m+1}{2\nu c}}\, x^{-(m+1)/2}\psi = \frac{1}{\sqrt{(2-\beta)\nu c}}\, x^{-1/(2-\beta)}\psi \tag{100}$$

Equation 97 was deduced in some other form by Falkner and Skan[17] and integrated numerically by Hartree[18], who indicated the fact that solution of this equation for $m < 0$, $\beta < 0$ ceases to be the only possibility and allows bifurcation (two different branches on the integral curve). In numerical integration of Equation 97 he chose that branch of the solution for which asymptotic tendency $u/U \to 1$ when $\xi \to \infty$ was the quickest.

Soviet authors[19] further sustantiated this demand that was based on the consideration of asymptotics of a disturbed, nonself-similar solution which is close to the considered self-similar one.

Table 14 presents the values of the dimensionless velocity $u/U = \dot{\Phi}(\xi, \beta)$ (Figure 177) in the boundary layer region within the range from $\beta = -0.1988$ corresponding to the

[17] Falkner V. M., and Skan S. W., Some approximate solutions of the boundary layer equations, *Phil. Mag.*, 12, 1931.

[18] Hartree D. R., On an equation occurring in Falkner-Skan's approximate treatment of the equations of boundary layer. II., *Proc. Cambr. Phil. Soc.*, 33, 1937.

[19] Kulikovskiy, A. G. and Slobodkina, F. A., On the choice of self-similar solution in the boundary layer theory, *Mekh. Zhidk. Gaza*, No. 4, 42–46, 1974.

Table 14

ξ \ β	−0.1988	−0.19	−0.18	−0.16	−0.14	−0.10	0	0.1	0.2	0.3
0.0	0.0000	0.0000	0.0000	0.0000	0.0000	0.0000	0.0000	0.0000	0.0000	0.0000
0.1	0.0010	0.0095	0.0137	0.0198	0.0246	0.0324	0.0469	0.0582	0.0677	0.0760
0.2	0.0040	0.0209	0.0293	0.0413	0.0507	0.0659	0.0939	0.1154	0.1334	0.1490
0.3	0.0089	0.0343	0.0467	0.0643	0.0781	0.1003	0.1408	0.1715	0.1970	0.2189
0.4	0.0158	0.0495	0.0659	0.0889	0.1069	0.1356	0.1876	0.2265	0.2584	0.2858
0.5	0.0248	0.0665	0.0868	0.1151	0.1370	0.1718	0.2342	0.2803	0.3177	0.3495
0.6	0.0358	0.0855	0.1094	0.1427	0.1684	0.2088	0.2806	0.3328	0.3747	0.4100
0.7	0.0487	0.1063	0.1338	0.1719	0.2010	0.2466	0.3266	0.3839	0.4294	0.4672
0.8	0.0636	0.1289	0.1598	0.2023	0.2347	0.2849	0.3720	0.4335	0.4816	0.5212
0.9	0.0803	0.1533	0.1874	0.2341	0.2694	0.3237	0.4167	0.4815	0.5312	0.5718
1.0	0.0991	0.1794	0.2166	0.2671	0.3050	0.3628	0.4606	0.5274	0.5782	0.6190
1.2	0.1423	0.2364	0.2791	0.3362	0.3784	0.4415	0.5453	0.6135	0.6640	0.7033
1.4	0.1927	0.2991	0.3463	0.4083	0.4534	0.5194	0.6244	0.6907	0.7383	0.7743
1.6	0.2498	0.3665	0.4170	0.4820	0.5284	0.5948	0.6967	0.7583	0.8011	0.8326
1.8	0.3126	0.4372	0.4896	0.5555	0.6016	0.6660	0.7610	0.8160	0.8528	0.8791
2.0	0.3802	0.5095	0.5621	0.6269	0.6712	0.7314	0.8167	0.8637	0.8940	0.9151
2.2	0.4509	0.5814	0.6327	0.6944	0.7354	0.7896	0.8633	0.9019	0.9260	0.9421
2.4	0.5230	0.6509	0.6995	0.7561	0.7927	0.8398	0.9011	0.9315	0.9500	0.9617
2.6	0.5946	0.7162	0.7605	0.8107	0.8422	0.8817	0.9306	0.9537	0.9612	0.9754
2.8	0.6635	0.7754	0.8146	0.8574	0.8836	0.9153	0.9529	0.9697	0.9792	0.9847
3.0	0.7278	0.8273	0.8607	0.8959	0.9168	0.9413	0.9691	0.9808	0.9873	0.9908
3.2	0.8158	0.8713	0.8986	0.9265	0.9425	0.9607	0.9804	0.9883	0.9924	0.9946
3.4	0.8364	0.9071	0.9286	0.9499	0.9616	0.9746	0.9880	0.9931	0.9957	0.9970
3.6	0.8789	0.9352	0.9515	0.9669	0.9752	0.9841	0.9929	0.9961	0.9976	0.9984
3.8	0.9132	0.9563	0.9681	0.9789	0.9845	0.9904	0.9959	0.9978	0.9987	0.9991
4.0	0.9399	0.9716	0.9798	0.9871	0.9907	0.9944	0.9978	0.9988	0.9993	0.9995
4.2	0.9598	0.9822	0.9876	0.9924	0.9946	0.9969	0.9988	0.9994	0.9996	0.9997
4.4	0.9741	0.9893	0.9927	0.9957	0.9970	0.9983	0.9994	0.9997	0.9998	0.9999
4.6	0.9839	0.9938	0.9959	0.9977	0.9984	0.9991	0.9997	0.9998	0.9999	0.9999
4.8	0.9904	0.9965	0.9978	0.9988	0.9992	0.9996	0.9999	0.9999		
5.0	0.9945	0.9981	0.9988	0.9994	0.9996	0.9998	0.9999			
5.2	0.9969	0.9990	0.9994	0.9997	0.9998	0.9999				
5.4	0.9984	0.9995	0.9997	0.9999	0.9999					
5.6	0.9992	0.9997	0.9999	0.9999						
5.8	0.9996	0.9999	0.9999							
6.0	0.9998	0.9999								
6.2	0.9999									
6.4	1.0000									

Table 14 **(Continued)**

0.4	0.5	0.6	0.8	1.0	1.2	1.6	2.0	2.4	β \ ξ
0.0000	0.0000	0.0000	0.0000	0.0000	0.0000	0.0000	0.0000	0.0000	0.0
0.0834	0.0903	0.0966	0.1080	0.1183	0.1276	0.1441	0.1588	0.1720	0.1
0.1628	0.1756	0.1872	0.2081	0.2266	0.2433	0.2726	0.2980	0.3206	0.2
0.2382	0.2558	0.2719	0.3003	0.3252	0.3475	0.3859	0.4186	0.4472	0.3
0.3097	0.3311	0.3506	0.3848	0.4144	0.4405	0.4849	0.5219	0.5537	0.4
0.3771	0.4015	0.4235	0.4619	0.4946	0.5231	0.5708	0.6096	0.6424	0.5
0.4403	0.4670	0.4907	0.5317	0.5662	0.5959	0.6446	0.6834	0.7155	0.6
0.4994	0.5276	0.5524	0.5947	0.6298	0.6596	0.7076	0.7449	0.7752	0.7
0.5545	0.5834	0.6086	0.6512	0.6859	0.7150	0.7610	0.7858	0.8235	0.8
0.6055	0.6344	0.6596	0.7015	0.7350	0.7629	0.8058	0.8376	0.8624	0.9
0.6526	0.6811	0.7056	0.7460	0.7778	0.8037	0.8432	0.8717	0.8934	1.0
0.7351	0.7615	0.7837	0.8194	0.8467	0.8682	0.8997	0.9214	0.9373	1.2
0.8027	0.8258	0.8449	0.8748	0.8968	0.9137	0.9375	0.9530	0.9640	1.4
0.8568	0.8860	0.8917	0.9154	0.9324	0.9450	0.9620	0.9726	0.9726	1.6
0.8988	0.9141	0.9264	0.9443	0.9569	0.9658	0.9775	0.9845	0.9845	1.8
0.9305	0.9421	0.9514	0.9644	0.9732	0.9793	0.9871	0.9914	0.9944	2.0
0.9537	0.9621	0.9689	0.9779	0.9841	0.9879	0.9928	0.9954	0.9970	2.2
0.9700	0.9760	0.9807	0.9867	0.9905	0.9931	0.9961	0.9976	0.9985	2.4
0.9812	0.9852	0.9884	0.9922	0.9946	0.9962	0.9980	0.9989	0.9993	2.6
0.9886	0.9913	0.9933	0.9956	0.9971	0.9980	0.9990	0.9994	0.9996	2.8
0.9933	0.9952	0.9962	0.9976	0.9985	0.9989	0.9995	0.9997	0.9998	3.0
0.9962	0.9974	0.9979	0.9987	0.9992	0.9995	0.9998	0.9999	0.9999	3.2
0.9979	0.9986	0.9989	0.9993	0.9996	0.9997	0.9999			3.4
0.9989	0.9993	0.9995	0.9997	0.9998	0.9999				3.6
0.9994	0.9994	0.9997	0.9998	0.9999					3.8
0.9997	0.9999	0.9999	0.9999						4.0
0.9999									4.2
0.9999									4.4
									4.6
									4.8
									5.0
									5.2
									5.4
									5.6
									5.8
									6.0
									6.2
									6.4

Table 15

β	$A(\beta)$	$B(\beta)$	$\xi(\beta)$
−0.1988	2.3588	0.5854	0.0000
−0.1950	2.1170	0.5814	0.0552
−0.1900	2.0068	0.5765	0.0857
−0.1800	1.8716	0.5677	0.1286
−0.1700	1.7789	0.5600	0.1621
−0.1600	1.7067	0.5522	0.1908
−0.1500	1.6470	0.5452	0.2164
-0.1400	1.5959	0.5386	0.2397
−0.1200	1.5113	0.5263	0.2818
−0.1000	1.4427	0.5150	0.3193
−0.0500	1.3124	0.4905	0.4003
0.0000	1.2168	0.4696	0.4696
0.0500	1.1417	0.4515	0.5311
0.1000	1.0803	0.4355	0.5870
0.2000	0.9842	0.4082	0.6867
0.3000	0.9110	0.3857	0.7748
0.4000	0.8526	0.3667	0.8544
0.5000	0.8046	0.3503	0.9277
0.6000	0.7640	0.3360	0.9958
0.8000	0.6987	0.3118	1.1203
1.00	0.6479	0.2923	1.2326
1.20	0.607	0.276	1.336
1.60	0.544	0.250	1.521
2.00	0.498	0.231	1.687

boundary layer cross-section at the separation point to $\beta = 2.4$. According to this one-term solution the conversion factors $A(\beta)$ and $B(\beta)$ (Equation 94) were calculated. Their values are given in Table 15. The last column of this table presents the values of the function $\zeta(\beta)$ necessary for the determination of the skin friction τ_w on the body surface

$$\tau_w = \frac{\mu U}{\delta}\zeta(\beta) = \sqrt{\frac{(m+1)\,c^3\mu\rho}{2}}\,x^{(3m-1)/2}\zeta(\beta) =$$

$$= \sqrt{\frac{c^3\mu\rho}{2-\beta}}\,x^{(2\beta-1)/(2-\beta)}\zeta(\beta) \qquad (101)$$

We do not consider the general character of the obtained one-term solution depending on the change in the parameter β. The reader is referred to special monographs on the laminar boundary-layer theory[20] where the properties of one- and two-term solutions, depending on the choice of the values of the parameters, are thoroughly described. In these monographs one can also find the solution for the case of a permeable wall through which fluid with the same physical constants that possesses the fluid flowing around the body surface is either injected of suctioned off.

---

[20] *Laminar Boundary Layers*, Rosenhead, L., Clarendon Press, Oxford, 1963, 243-252; Evans, H. L., *Laminar Boundary – Layer Theory*, Addison-Wesley, London, 1969.

Power-law distribution of velocities on the outer edge of the boundary layer corresponds to the outer plane irrotational flow past an infinite wedge placed symmetrically with respect to the incident homogeneous flow[21]. The apex angle of the wedge is equal to $[2m/(m+1)]\pi = \beta\pi$. The values $m = 0$, $\beta = 0$ refer to the wedge with the zero apex angle, i.e., to the case of the flow past an infinitely thin and long plate in full conformity with the conditions to the Blasius problem considered earlier in the Chapter. The values of the parameters $m = 1$, $\beta = 1$ make the apex angle equal to π, thus leading to the boundary layer on the body surface near a front critical point. The thicknesses of the boundary layer in this region will be, as follows from Equations 99 for δ, δ^*, δ^{**}, constant and equal to

$$\delta = \sqrt{\frac{\nu}{c}} \qquad \delta^* = A(1)\sqrt{\frac{\nu}{c}} = 0.648\sqrt{\frac{\nu}{c}}$$

$$\delta^{**} = B(1)\sqrt{\frac{\nu}{c}} = 0.292\sqrt{\frac{\nu}{c}} \tag{102}$$

and the skin friction τ_w (see Equation 101) changes along the flow according to the linear law

$$\tau_w = \zeta(1)\sqrt{\mu\rho c^3}\,x = 1.2326\,(\sqrt{\mu\rho c^3}\,x) \tag{103}$$

Note also that to the converging flow in the inner portion of the wedge having the velocity $U = -c/x$ (a plane converging duct) corresponds the value $m = -1$, and consequently $\beta = \infty$. The case of large β was thoroughly considered in the Evans monograph,[22] which is concerned with the description of self-similar plane and spatial isothermal and nonisothermal flows in boundary layers. This monograph involves tables referring r to the case of $\beta \gg 1$.

A diverging flow inside a wedge (a plane diverging duct) is possible only when $\beta > -0.1988$; $m > -0.0904$, because at these limiting values of β (see Table 15) $\zeta = 0$, thus corresponding to flow separation from the diverging duct walls and subsequent back flow of fluid.

A general self-similar solution of the laminar near-wall boundary layer equation with power-law distribution of velocity on its outer edge was then extended to the region of "free" boundary layers – jets and aerodynamic wakes[23, 24] and the solution stated in this section formed the basis of an approximate method suitable for an arbitrary distribution of velocity on the outer edge of the boundary layer.[25]

The latter method opened the way to the development of other approximate methods and is still used in engineering calculations. It is presented in the following section which considers approximate methods for boundary layer calculation.

8. Approximate Methods for Calculating Laminar Boundary Layers. von Kármán Integral Relation. Pohlhausen's and Kochin-Loitsyanskiy's Methods

Wide application of electronic computers to numerical calculations somewhat lessens the

[21] See previously cited monographs edited by L. Rosenhead and H.L. Evans.

[22] Evans, K. L., *Laminar Boundary-Layer Theory*, Addison-Wesley, London, 1969.

[23] Stewartson, K. Further solutions of the Falkner – Skan equation, Proc. Cambr. Phil. Soc., 50, 454–465, 1954; 2 (7), Stewartson, K., Falkner – Skan equation for wakes, *AIAA J.*, 2 (7), 1964.

[24] Kennedy, E. D., Wake-like solutions of the laminar boundary layer equations, *AIAA J.*, 2, 225–231, 1964.

[25] Kochin, N. E. and Loitsyanskiy, L. G., On one approximate method for calculating laminar boundary layer, *Dokl AN SSSR*, 36(9), 1942.

interest in approximate computational methods. However, due to some circumstances, viz., the absence of rather powerful computers, the form of the computational results that is not easy to grasp, etc., approximate methods did not lose their practical value.

Approximate methods of laminar boundary layer calculation that appeared in the 1920s and 1930s are tightly connected with the Ritz and Galerkin methods. Both deal with semi-intuitive velocity profiles, which are not related to accurate solutions of the Prandtl equation, but only satisfy boundary conditions fixed for this equation and with determination of the parameters of the families of the so-called "competing" profiles on the basis of "integral relations" derived from the Prandtl equation.

We deal with one of these, the von Kármán relation; the others may be found in special monographs[26] and separate papers.[27] In accordance with the Ritz method integral relations were also used that proceed from the variational principles of mechanics.[28]

At the end of the 1930s these methods had improved dramatically. Exact solutions of the Prandtl equations for particular specifications of velocity on the outer edge of the boundary layer were used as "competing" profiles. The first attempt was made by Howarth,[29] who used the exact velocity profiles corresponding to linear "one-slope" distribution of velocities on the outer edge of the layer. Only in 1940s did this idea attain complete development both in Russia and abroad independently due to World War II. For the same reasons publication of the first Soviet paper dealing with these problems, the concern of which was the Kochin-Loitsyanskiy method,[30] was suppressed. In this work the "competing" profiles of velocities were taken from an exact self-similar solution of the boundary layer equation in the case of power-law velocity distribution on its outer edge (see Section 7).

The method of "generalized similarity",[31] published in 1965, signaled the completion of the development of the methods of this approach. This method acquired wide application both in problems of different laminar boundary layers in an incompressible liquid and in gas flows of high sub- and supersonic velocities (see Sections 10 and Chapter 15), and recently in turbulent boundary layers (Chapter 14).

The algorithm of constructing consecutive approximate solutions principally different from both the Ritz-Galerkin and methods mentioned earlier formed the basis of the method of "generalized similarity". The prescribed arbitrary distribution of velocities on the boundary layer outer edge in the first approximation is substituted by linear, in the second approximation by quadratic, etc. This is done so that at each point coincidence of the approximate velocity distribution on the outer edge is maintained with the tangent, curvature, and subsequent "curvatures" assigned by the slope. It is well known this requiring coincidence between the tangent and the first curvature by the slope, and the relatively small subsequent curvatures, leads to the second approximation being rather satisfactory, as confirmed by calculations. This indicates fast convergence of the method of "generalized similarity". Each approximation leads to its own equation independent on the form of velocity distribution $U(x)$ on the boundary layer outer edge for particular problems and is thus called "universal". The universal equation for each approximation may be integrated once and permanently having

[26] Kochin, N. E., Kibel, I. A., and Roze, N. N., *Teoreticheskaya gidromekhanika* (*Theoretical Hydromechanics*), Part II, Fizmatgiz, Moscow, 1963, 556-566; see also Loitsyanskiy, L. G., *Laminarnyi pogranichnyi sloi* (*Laminar Boundary Layer*), Fizmatgiz, Moscow, 1962, 92–113.

[27] Loitsyanskiy, L. G. Approximate method of integrating the equations of laminar boundary layer in an incompressible gas, *Prikl. Mat. Mekh.*, 13(5), 1949.

[28] Loitsyanskiy, L. G., Integral methods in the boundary layer theory, *Prikl. Mat. Mekh.*, 5(3), 453–470, 1941.

[29] Howarth, L., On the solution of the laminar boundary layer equations, *Proc. R. Soc. A*, 164(919), 547, 1938.

[30] Kochin, N. E. and Loitsyanskiy, L. G., On one approximate method of calculating laminar boundary layer, *Dokl. AN SSSR*, 36(9), 1942.

[31] Loitsyanskiy, L. G., Universal equations and parametric approximations in the theory of laminar boundary layer, *Prikl. Mat. Mekh.*, 29(4), 1965.

composed corresponding tables. To calculate a separate specific problem requires the integration of an ordinary differential equation of the first order or to make use of simple quadrature.

Approximate methods of the laminar boundary layer theory are based on the use of this or other forms of the von Kármán integral relation, which expresses the theorem of momentum as applied to a liquid flow in the boundary layer region.

Making use of the second of equations of system (11), this system may be identically transformed to

$$\frac{\partial}{\partial x}(u^2) + \frac{\partial}{\partial y}(uv) = U\frac{dU}{dx} + \nu\frac{\partial^2 u}{\partial y^2}$$

$$\frac{\partial}{\partial x}(Uu) + \frac{\partial}{\partial y}(Uv) = u\frac{dU}{dx}$$

Then substracting separately the right and left sides of these equations yields

$$\frac{\partial}{\partial x}[u(U-u)] + \frac{\partial}{\partial y}[v(U-u)] + (U-u)\frac{dU}{dx} = -\nu\frac{\partial^2 u}{\partial y^2}.$$

Term-by-term integration with respect to y from $y = 0$ to $y = \infty$ or δ, thus corresponding to the notion of asymptotic approach $u \to U$ when $y \to \infty$ or to the notion of the boundary layer thickness finiteness ($u = U$ at $y = \delta$), gives

$$\int_0^{\infty,\delta} \frac{\partial}{\partial x}[u(U-u)]\,dy + v(U-u)\Big|_{y=0}^{y=\infty} + \frac{dU}{dx}\int_0^{\infty,\delta}(U-u)\,dy = -\nu\frac{\partial u}{\partial y}\Big|_{y=0}^{y=\infty,\delta}.$$

We shall use the boundary conditions (12) or the conditions $u = U$, $\partial u/\partial y = 0$ at $y = \delta$, which correspond to the assumption on the boundary layer thickness finiteness.

Assuming the existence of the integrals

$$\int_0^{\infty,\delta} u(U-u)\,dy \qquad \int_0^{\infty,\delta}(U-u)\,dy$$

and that reversing the order of differentiation and integration is permissible in the first integral of the left side of the previous equation, we obtain

$$\frac{d}{dx}\int_0^{\infty,\delta} u(U-u)\,dy + \frac{dU}{dx}\int_0^{\infty,\delta}(U-u)\,dy = \nu\left(\frac{\partial u}{\partial y}\right)_{y=0}. \qquad (104)$$

Introducing the following generalization of integral thicknesses of a boundary layer (41) and (42) to the case of an arbitrary velocity $U(x)$ on its outer edge

$$\delta^* = \int_0^{\infty,\delta}\left(1-\frac{u}{U}\right)dy \qquad \delta^{**} = \int_0^{\infty,\delta}\frac{u}{U}\left(1-\frac{u}{U}\right)dy, \qquad (105)$$

we first obtain from Equation 104

$$\frac{d}{dx}(U^2\delta^{**}) + U\frac{dU}{dx}\delta^* = \frac{\tau_w}{\rho U^2},$$

and then evaluating the derivative on the left side we come to the agreed-upon expression of the von Kármán integral relation ($U' = dU/dx$)[32]

$$\frac{d\delta^{**}}{dx} + \frac{U'}{U}(2\delta^{**} + \delta^*) = \frac{\tau_w}{\rho U^2}, \tag{106}$$

or

$$\frac{d\delta^{**}}{dx} + \frac{U'\delta^{**}}{U}(2 + H) = \frac{\tau_w}{\rho U^2} \qquad H = \frac{\delta^*}{\delta^{**}}. \tag{107}$$

On the right sides of Equations 104, 106, and 107 use is made of the notation of the skin friction on the solid surface of a streamlined body

$$\tau_w = \mu\left(\frac{\partial u}{\partial y}\right)_{y=0}. \tag{108}$$

The most substantiated of the approximate methods of calculation of a laminar boundary layer known in the 1940s was the Kochin-Loitsyanskiy method cited at the end of the previous section. In this method the general self-similar solutions of the near-wall stationary boundary layer equations are used applied to nonself-similar problems, assuming that the parameters β and $\underline{\beta}$ change from section to section. We restrict ourselves to the one-term approach relating β and β by Equation 96, which leads Equation 89 to the form of Equation 97. Thus, having at hand only the parameter β we can use the available tables of the dimensionless velocity $u/U = \dot{\Phi}(\xi, \beta)$ and of the functions $\zeta(\beta)$, $A(\beta)$, and $B(\beta)$ presented in the previous section.

The essence of the Kochin-Loitsyanskiy method is in seeking the unknown function $\beta(x) = U'(x)z(x) = U'\delta^2/\nu$ and thus $\delta(x)$, because possessing tabular determination of the dimensionless velocity $u/U = \dot{\Phi}(\xi, \beta)$ (see Table 14) and having found $\delta(x)$ one can obtain the solution of the specific problem with the assigned distribution of $U(x)$ in the form

$$\frac{u(x, y)}{U(x)} = \dot{\Phi}\left[\frac{y}{\delta(x)}, \frac{U'(x)\delta^2(x)}{\nu}\right].$$

To determine $\delta(x)$ use can be made of the arbitrarily adopted condition (96) and according to Equation 9) the equation

$$U'z + \frac{1}{2}Uz' = 1,$$

may be obtained, the integral of which is

$$z = \frac{\delta^2}{\nu} = \frac{2}{U^2}\int_0^x U dx. \tag{109}$$

[32] von Kármán, Th., Über laminare und turbulente Reibung, *Z. Angew. Math. Mech.*, 1, 1921.

However, this solution, which is connected only with the assignment of the velocity $U(x)$ on the boundary layer outer edge and does not reflect the effect of velocity profiles in the boundary layer cross-sections, should be discarded and substituted by the following one based on the application of the von Kármán integral relation in any form (Equations 104, 106, or 107).

We further agree to use the integral "thickness of momentum loss" δ^{**} (Equation 42) as a characteristic boundary layer thickness. This simplifies a computational aspect of both the method under consideration and all the rest. In such a choice of the boundary layer thickness the role of the parameters β and $\bar{\beta}$ is executed by

$$f = U'z^{**} = \frac{U'\delta^{**2}}{\nu} \qquad \bar{f} = Uz^{**\prime} = \frac{2U\delta^{**}\delta^{**\prime}}{\nu} \tag{110}$$

Multiply both sides of Equation 107 into $U\delta^{**}/\nu$. Then, after simple transformations it is rewritten in one of the following forms

$$\frac{df}{dx} = \frac{U'}{U}F(f) + \frac{U''}{U'}f \tag{111}$$

$$\bar{f} = U\frac{dz^{**}}{dx} = F(f) \tag{112}$$

where, besides Equation 110, the notations

$$F(f) = 2\{\zeta(f) - [2 + H(f)]f\}$$

$$\zeta(f) = \left[\frac{\partial(u/U)}{\partial(y/\delta^{**})}\right]_{y=0} = \ddot{\Phi}(0, f) \tag{113}$$

are also introduced.

Expressions for $\zeta(f)$, $H(f)$ and $F(f)$ are obtained with the use of equalities (99), (100), and (113). Now, compose the following parametric system of relations (parameter β):

$$f = \frac{U'\delta^{**2}}{\nu} = \frac{U'\delta^2}{\nu}\left(\frac{\delta^{**}}{\delta}\right)^2 = \beta B(\beta) \qquad H = \frac{\delta^*}{\delta^{**}} = \frac{A(\beta)}{B(\beta)}$$

$$\frac{\tau_w\delta^{**}}{\mu U} = \ddot{\Phi}(0, \beta)B(\beta) = \zeta(\beta)B(\beta)$$

$$F(\beta) = -4\beta B^2(\beta) - 2\beta A(\beta)B(\beta) + 2\ddot{\Phi}(0, \beta)B(\beta). \tag{114}$$

Specifying different β, we determine, using Table 15, the values of $A(\beta)$, $B(\beta)$, $\ddot{\Phi}(0, \beta)$. Then, using Equation 114 we obtain $f(\beta)$, $H(\beta)$, $\zeta(\beta)$, and $F(\beta)$. After this, discarding β, we find the unknowns $\zeta(f) = \tau_w\delta^{**}/(\mu U)$, $H(f)$, and $F(f)$. The numerical values of these quantities are given in Table 16 which the graphs of $\zeta(f)$, $H(f)$, and $F(f)$ are constructed (Figure 178).

The graph of the function $F(f)$ may be replaced by the linear presentation

$$F(f) \approx a - bf \qquad a \approx 0.45 \qquad b \approx 5.35 \tag{115}$$

Table 16

f	$\zeta(f)$	$H(f)$	$F(f)$
−0.0681	0.0000	4.03	0.821
−0.06	0.064	3.35	0.772
−0.05	0.098	3.12	0.715
−0.04	0.130	2.96	0.658
−0.03	0.155	2.84	0.602
−0.02	0.178	2.74	0.548
−0.01	0.200	2.66	0.495
0.00	0.221	2.59	0.441
0.01	0.240	2.53	0.388
0.02	0.257	2.48	0.336
0.03	0.274	2.43	0.283
0.04	0.291	2.38	0.232
0.05	0.307	2.34	0.180
0.06	0.323	2.30	0.130
0.07	0.338	2.26	0.078
0.08	0.352	2.23	0.028
0.09	0.366	2.20	−0.023
0.10	0.380	2.18	−0.074

with the deviation from the linear law (115) not exceeding, the value of 0.03.

Equation 111 in this case became the linear ordinary differential equation

$$\frac{df}{dx} = a\frac{U'}{U} + \left(\frac{U''}{U'} - b\frac{U'}{U}\right)f$$

whose integral is

$$f = \frac{aU'}{U^b}\int_0^x U^{b-1}(x)\,dx + C\frac{U'}{U^b}$$

It follows from the condition of finiteness of f at $x = 0$, where $U = 0$, that $C = 0$ and the

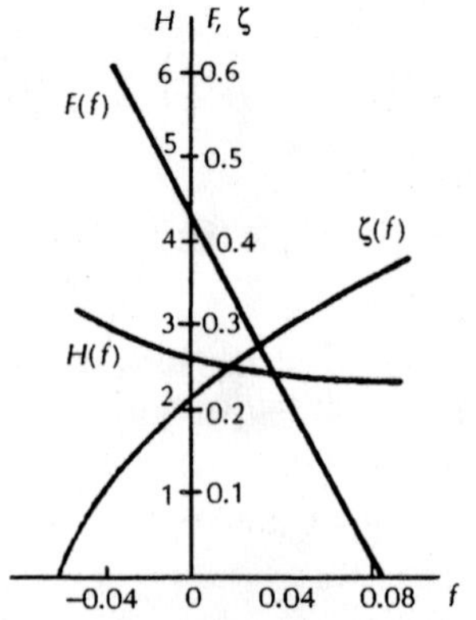

Fig. 178

solution will take on the finite form

$$f = \frac{aU'}{U^b}\int_0^x U^{b-1}(x)\,dx \tag{116}$$

with the parameter f at the initial point ($x = 0$) being equal to

$$f(0) = \left[\frac{aU'(x)\,U^{b-1}(x)}{bU^{b-1}(x)\,U'(x)}\right]_{x=0} = \frac{a}{b}$$

Table 16 is common for all problems, with the values of $f(x)$ (according to Equation 116) relating the distribution $U(x)$ to each specific problem. Table 16 completely solves the matter of the approximate calculation of the boundary layer by the method stated. At the separation point S, where following Equation 13, the condition

$$\left(\frac{\partial u}{\partial y}\right)_{y=0} = \frac{U}{\delta^{**}}\left[\frac{\partial(u/U)}{\partial(y/\delta^{**})}\right]_{y=0} = \frac{U}{\delta^{**}}\zeta(f) = 0 \qquad \zeta(f) = 0$$

is fulfilled, the parameter f_s as it directly follows from Table 16, is $f_s = -0.0681$

Judging by the calculations made and from the comparison with exact solutions, the given method leads to underestimated values of skin friction on the surface of a streamlined body, and, in conformity with this, to smaller (compared to other methods) values of the separation point abscissa. We defer treatment of this problem in to following section, in which results are obtained based on more perfect approximate methods.

The formulated approximate method of laminar boundary layer calculation is based on the use of the one-term family of velocity profiles, representing exact self-similar solutions of the Prandtl equation (Equation 11). This approach or a somewhat more general one consisting of the selection of "competing" one-term families of velocity profiles among the other familiar exact solutions, appeared only at the end of the 1930s. Earlier, artificially formed analytical families of profiles were used for this purpose, which were similar in form to real profiles and which coincided with them on the outer ($y = \delta$) and inner ($y = 0$) edges of the boundary layer. Arbitrariness in the selection of this kind of competing sets of velocity profiles generated a great number of various approximate methods.

The basic idea of the method is illustrated with the example of the Pohlhausen method[33] which historically was the first and which evoked numerous initiations. Being published simultaneously and in the same journal as the (earlier cited) paper written by von Kármán, the Pohlhausen paper was aimed at the illustration of the application of the von Kármán integral relation.

Pohlhausen suggested taking the family of fourth-order polynomials

$$\frac{u}{U} = \frac{12+\lambda}{6}\frac{y}{\delta} - \frac{\lambda}{2}\left(\frac{y}{\delta}\right)^2 - \frac{4-\lambda}{2}\left(\frac{y}{\delta}\right)^3 + \frac{6-\lambda}{6}\left(\frac{y}{\delta}\right)^4 \equiv$$

$$\equiv 2\xi - 2\xi^3 + \xi^4 + \frac{1}{6}\lambda\xi(1-\xi)^3 \tag{117}$$

---

[33] Pohlhausen, K., Zur näherungsweisen Integration der Differenzialgleichungen der laminaren Grenzschichten, *Z. Angew. Math. Mech.*, 1, 1921.

with the family parameter equal to $\lambda = U'\delta^2/\nu$, where δ is the boundary layer thickness adopted to be finite and $\xi = y/\delta$, as the one-term family of profiles of longitudinal velocities in the boundary layer cross-sections. The coefficients of the polynomial (117) were chosen according to the boundary conditions

$$u = 0 \qquad \frac{\partial^2 u}{\partial y^2} = -\frac{UU'}{\nu} \quad \text{at} \quad y = 0$$

$$u = U \qquad \frac{\partial u}{\partial y} = 0 \qquad \frac{\partial^2 u}{\partial y^2} = 0 \quad \text{at} \quad y = \delta \tag{118}$$

The first of these conditions coincided with the similar condition Equation 12, and the second is obtained from Equation 11 by its application to a solid surface ($y = 0$, $u = v = 0$). The third expresses the equality of longitudinal velocities in a boundary layer and in an outer flow at the point of their match at the finite distance $y = \delta$ from the solid surface instead of the corresponding asymptotic condition (12). Finally, the fourth and the fifth express smoothness of the transition from u to U at the matching point ($y = \delta$).

The latter conditions and their quantity incorporate arbitrariness which, as was shown by subsequent studies,[34] greatly affects the method. Thus, the additional condition $(\partial^3 u/\partial y^3)_{y=0} = 0$, proceeding from Equation 11 and introduced instead of the latter condition in the system of Equations 118 greatly improves the accuracy of the Pohlhausen method, emphasizing once more the intuitive character of the method.

From a routine viewpoint the Pohlhausen method is very similar to the Kochin-Loitsyanskiy method. In the same way, but based on the velocity profiles in the layer sections of Equation 117, the functions

$$A(\lambda) = \frac{\delta^*}{\delta} = \int_0^1 \left(1 - \frac{u}{U}\right) d\xi = \frac{3}{10} - \frac{\lambda}{120}$$

$$B(\lambda) = \frac{\delta^{**}}{\delta} = \int_0^1 \frac{u}{U}\left(1 - \frac{u}{U}\right) d\xi = \frac{37}{315} - \frac{\lambda}{945} - \frac{\lambda^2}{9072}$$

$$\zeta(\lambda) = \left[\frac{\partial(u/U)}{\partial(y/\delta)}\right]_{y=0} = 2 + \frac{\lambda}{6} \tag{119}$$

are constructed.

Their substitution into the von Kármán integral relation leads to a complex nonlinear equation of the first order which Pohlhausen solved graphically by the method of isoclinic lines. $\delta(x)$ determined from this equation was then substituted into previous equalities, thus providing the solution of the problem.

The transition from $\lambda = U'\delta^2/\nu$ to that used in the given section and to the adopted later parameter $f = U'\delta^{**2}/\nu$ makes it possible to reduce the solution to integration of Equation 111 with the adopted notation (113) for the function $F(f)$. In Table 17 the recalculated

---

[34] Loitsyanskiy, L. G. *Aerodinamika pogranichnogo sloya (Boundary Layer Aerodynamics)*, Gostekhizdat, Moscow-Leningrad, 1941, 170–189; *Laminarnyi pogranichnyi sloy (Laminar Boundary Layer)*, Fizmatgiz, Moscow, 1962, 96–98.

Table 17

λ	f	$\zeta(f)$	$H(f)$	$F(f)$
8	0.0831	0.340	2.29	−0.033
7.052	0.0770	0.332	2.31	0
7	0.0767	0.331	2.31	0.002
6	6.0689	0.321	2.33	0.046
5	0.0599	0.310	2.36	0.098
4	0.0497	0.297	2.39	0.158
3	0.0385	0.283	2.43	0.226
2	0.0264	0.268	2.47	0.300
1	0.0135	0.252	2.51	0.382
0	0	0.235	2.55	0.470
−1	−0.0140	0.217	2.60	0.563
−2	−0.0284	0.199	2.66	0.662
−3	−0.0429	0.179	2.72	0.764
−4	−0.0575	0.160	2.78	0.870
−5	−0.0720	0.140	2.85	0.978
−6	−0.0862	0.119	2.92	1.085
−7	−0.0999	0.100	3.00	1.198
−8	−0.1130	0.079	3.08	1.308
−9	−0.1254	0.059	3.18	1.417
−10	−0.1369	0.039	3.28	1.523
−11	−0.1474	0.019	3.38	1.625
−12	−0.1567	0	3.50	1.724

values of the functions $\zeta(f)$, $H(f)$, and $F(f)$ corresponding to the Pohlhausen method are presented. As before, we can, although not so accurately, substitute $F(f)$ into its linear approximation (115) and obtain the solution in the form of Equation 116. The values of the constants a and b differ from those presented earlier for the Kochin-Loitsyanskiy method. Their approximate values are: $a = 0.47$, $b = 6.10$.

As shown by calculations, the Pohlhausen method, in contrast to the Kochin-Loitsyanskiy method, leads to overestimated values of skin friction on the body surface (τ_w). Correspondingly, it leads to overestimated value of the modulus of the separation point parameter, i.e., to the delayed, as compared to the real, separation point position.

Descriptions of other approximate methods of laminar boundary layers calculation may be found in the monograph edited by Rosenhead (footnote 1).

9. Generalization of Affine Similarity to the Case of Nonself-Similar Flows in Boundary Layers. Method of Generalized Similarity

Relations of affine similarity have been mentioned in previous sections. In particular, in the case of the boundary layer on a plate streamlined by a longitudinal flow, the following expression of the affine similarity of velocity profiles in boundary layer cross-section ($\xi = y/\delta^{**}$) was used

$$\frac{u}{U_\infty} = \varphi\left(\frac{y}{\delta^{**}}\right) = \varphi(\xi)$$

where U_∞ and δ^{**} represented the scales of velocities and ordinates of the points in the boundary layer region. Now, considering the approximate method for calculating any non-self-similar boundary layer with arbitrary, and not only with power-law velocity distribution (Section 7) on its outer edge, we extend the definition of the affine similarity to its possible dependence on the arbitrary number of parameters. We call them the parameters of the generalized similarity. In other words, we consider the previous relation of an ordinary affine similarity in the form of the "affine similarity with the parameters"

$$\frac{u}{U} = \varphi(\xi, \chi_1, \chi_2, \ldots)$$

where χ_k ($k = 1, 2, \ldots$) are some functions of x, and U is an arbitrary function of x.

If it is assumed that the values of χ_k entering into the previous relation may take any independent value, then this expression will represent a multiparametric family of affine transformations in their ordinary sense. Taking χ_k to be the continuous functions of x, we include them, along with φ and ξ, in the total number of "similarity variables". Omitting the word "affine" we apply the term "generalized" to this similarity, and the name "generalized similarity variables" to the variables $\varphi, \xi, \chi_1, \chi_2, \ldots$.

To make use of the adopted definition of "generalized similarity" and "generalized similarity variables" for the construction of the approximate method of the Prandtl equation (15) integration, we suggest the following heuristic technique.

Take an ordinary affine similarity expression of the stream function ψ

$$\psi = U\delta^{**}\Phi(\xi) \tag{120}$$

where U and δ^{**} are the arbitrary values of the velocity on the boundary layer outer edge and of the momentum loss thickness, and $\xi = y/\delta^{**}$ is the "zero" approximation. Substituting the expression of ψ by Equation 120 into the Prandtl equation (15), the result is

$$\frac{\partial^3\Phi}{\partial\xi^3} + \left(U'z + \frac{1}{2}Uz'\right)\Phi\frac{\partial^2\Phi}{\partial\xi^2} + U'z\left[1 - \left(\frac{\partial\Phi}{\partial\xi}\right)^2\right] = 0$$

(prime stands for the derivative with respect to x, z is equal to δ^{**2}/ν). Here, taking into account that in a general case expression (120) is not the solution of Equation 15, we obtain the "residuals" of the adopted assumption (120) in the form of the functions of x

$$f_1(x) = U'z \qquad \bar{f}_1(x) = Uz' \tag{121}$$

The presence of the variable x by Equation 121 along with ξ in the previous differential equation induces the symbols of partial derivatives to express the derivatives of Φ.

Having included the "residual" f_1 and $\bar{f}_1$ among the "generalized similarity variables", we seek the solution of the Prandtl equation (15) in the form

$$\psi = U\delta^{**}\Phi(\xi, f_1, \bar{f}_1)$$

Substitute once again this expression into Equation 15. Note that the derivatives entering into Equation 15 will take the form

$$\frac{\partial\psi}{\partial x} = \frac{\nu}{\delta^{**}}\left[\left(f_1 + \frac{1}{2}\bar{f}\right)\Phi - \frac{1}{2}\bar{f}_1\xi\frac{\partial\Phi}{\partial\xi} + Uz\left(\frac{\partial\Phi}{\partial f_1}f_1' + \frac{\partial\Phi}{\partial\bar{f}_1}\bar{f}_1'\right)\right]$$

$$\frac{\partial^2\psi}{\partial x\partial y} = \frac{\nu}{\delta^{**2}}\left[f_1\frac{\partial\Phi}{\partial\xi} - \frac{1}{2}\bar{f}_1\xi\frac{\partial^2\Phi}{\partial\xi^2} + Uz\left(\frac{\partial^2\Phi}{\partial\xi\partial f_1}f_1' + \frac{\partial^2\Phi}{\partial\xi\partial\bar{f}_1}\bar{f}_1'\right)\right]$$

$$\frac{\partial\psi}{\partial y} = U\frac{\partial\Phi}{\partial\xi} \qquad \frac{\partial^2\psi}{\partial y^2} = \frac{U}{\delta^{**}}\frac{\partial^2\Phi}{\partial\xi^2} \qquad \frac{\partial^3\psi}{\partial y^3} = \frac{U}{\delta^{**2}}\frac{\partial^3\Phi}{\partial\xi^3}$$

Eliminate the expressions Uzf_1' and $Uz\bar{f}_1'$ obtained by direct differentiation of the parameters f_1 and $\bar{f}_1$ (Equation 121) from the right sides of the first two equalities. We find the relations

$$Uzf_1' = U'zUz' + UU''z^2 = f_1\bar{f}_1 + UU''z^2$$

$$Uz\bar{f}_1' = U'zUz' + U^2zz'' = f_1\bar{f}_1 + U^2zz''$$

Introducing the new variables of the generalized similarity that follow f_1 and $\bar{f}_1$

$$f_2 = UU''z^2 \qquad \bar{f}_2 = U^2zz''$$

rewrite the previous equalities in the form

$$Uzf_1' = f_1\bar{f}_1 + f_2 = \theta_1$$

$$Uz\bar{f}_1' = f_1\bar{f}_1 + \bar{f}_2 = \bar{\theta}_1 \tag{122}$$

Substituting the expressions obtained into Equation 15, we come to

$$\frac{\partial^3\Phi}{\partial\xi^3} + \left(f_1 + \frac{1}{2}\bar{f}_1\right)\Phi\frac{\partial^2\Phi}{\partial\xi^2} + f_1\left[1 - \left(\frac{\partial\Phi}{\partial\xi}\right)^2\right] =$$

$$= \theta_1\frac{D(\Phi, \partial\Phi/\partial\xi)}{D(\xi, f_1)} + \bar{\theta}_1\frac{D(\Phi, \partial\Phi/\partial\xi)}{D(\xi, \bar{f}_1)}$$

where the ordinary Jacobians are used:

$$\frac{D(\Phi, \partial\Phi/\partial\xi)}{D(\xi, f_1)} = \frac{\partial\Phi}{\partial\xi}\frac{\partial^2\Phi}{\partial\xi\partial f_1} - \frac{\partial\Phi}{\partial f_1}\frac{\partial^2\Phi}{\partial\xi^2}$$

$$\frac{D(\Phi, \partial\Phi/\partial\xi)}{D(\xi, f_1)} = \frac{\partial\Phi}{\partial\xi}\frac{\partial^2\Phi}{\partial\xi\partial f_1} - \frac{\partial\Phi}{\partial f_1}\frac{\partial^2\Phi}{\partial\xi^2}$$

Extending this technique we seek the solution of the Prandtl equation (15) in the form

$$\psi = U\delta^{**}\Phi(\xi, f_1, \bar{f}_1, f_2, \bar{f}_2, \ldots) \tag{123}$$

with the independent variables ξ, f_k, $\bar{f}_k$ ($k = 1, 2, \ldots$) determined by the equalities ($z = \delta^{**2}/\nu$)

$$\xi = \frac{y}{\delta^{**}} \qquad f_k = U^{k-1}\frac{d^kU}{dx^k}z^k \qquad \bar{f}_k = z^{k-1}\frac{d^kz}{dx^k}U^k \tag{124}$$

and by the relations

$$Uzf_k' = [(k-1)f_1 + k\bar{f}_1]f_k + f_{k+1} = \theta_k$$

$$Uz\bar{f}_k{}' = [(k-1)\bar{f}_1 + kf_1]\bar{f}_k + \bar{f}_{k+1} = \bar{\theta}_k \tag{125}$$

The introduced variables f_k, $\bar{f}_k$ (k = 1, 2, ...) and their functions θ_k and $\bar{\theta}_k$ posses the following properties:

1. We shall convince ourselves directly that the symmetry condition

$$f_k \rightleftarrows \bar{f}_k \qquad \theta_k \rightleftarrows \bar{\theta}_k \quad \text{of} \quad U \rightleftarrows z$$

is satisfied.

2. According to the estimate Equation 6 of the scale of transverse lengths, we obtain the following expression for the order of magnitude and dimensions of z

$$z = \frac{\delta^{**2}}{\nu} = O\left(\frac{L^2}{\mathbf{Re}\,\nu}\right) = O\left(\frac{L}{U}\right)$$

$$[z] = \left[\frac{L}{U}\right] = [T]$$

Thus, the quantity z has the dimensions of time. Hence, it follows that the dimensions of the parameters f_k and $\bar{f}_k$ are determined by the equalities

$$[f_k] = \left[\frac{U^{k-1}U}{L^k}T^k\right] = \left[\left(\frac{UT}{L}\right)^k\right] = [1]$$

$$[\bar{f}_k] = \left[\frac{T^{k-1}T}{L^k}\left(\frac{L}{T}\right)^k\right] = [1] \tag{126}$$

thus indicating dimensionlessness of the variables f_k and $\bar{f}_k$; their independence of the Reynolds number follows from the z independence of it.

Moving on to the formulation of Equation 15 transformed to the generalized similarity variables Φ, ξ, f_k, $\bar{f}_k$ (k = 1, 2, ...), we substitute the values of ψ and of its derivatives in the generalized similarity variables into this equation. The same is done with the boundary conditions.

As a result of simple calculations we find that the left side of this equation will retain the form obtained in the initial stages of the technique formulated, and to the right side the sum of Jacobians will be added that allow for the effect of the variables f_k and $\bar{f}_k$ following f_1 and $\bar{f}_1$. As a result of subsequent assumptions the unknown equation in its the most general unsimplified form is

$$\frac{\partial^3\Phi}{\partial\xi^3} + \left(f_1 + \frac{1}{2}\bar{f}_1\right)\Phi\frac{\partial^2\Phi}{\partial\xi^2} + f_1\left[1 - \left(\frac{\partial\Phi}{\partial\xi}\right)^2\right] =$$

$$= \sum_{k=1}^{\infty} \left[\theta_k \frac{D(\Phi, \partial\Phi/\partial\xi)}{D(\xi, f_k)} + \bar{\theta}_k \frac{D(\Phi, \partial\Phi/\partial\xi)}{D(\xi, \bar{f}_k)} \right] \tag{127}$$

To this equation the boundary conditions

$$\Phi = \dot{\Phi} = 0 \quad \text{at} \quad \xi = 0 \qquad \dot{\Phi} \to \infty \quad \text{when} \quad \xi \to \infty$$

$$\Phi = \Phi_0(\eta) \quad \text{at} \quad f_k = f_{k0}, \quad \bar{f}_k = \bar{f}_{k0} \quad (k = 1, 2, ...) \tag{128}$$

The last condition marked by the subscript "0" expresses the fact that the integral curve should emerge corresponding to some self-similar solution. Note that the following values of the parameters f_{k0} and $\bar{f}_{k0}$ dependent on the exponent of power m in the velocity law $U = cx^m$

$$f_{k0} = \frac{2^k}{(m+1)^k} m(m-1) \ldots (m-k+1) B^{2k}\left(\frac{2m}{m+1}\right)$$

$$\bar{f}_{k0} = (-1)^k \frac{2^k}{(m+1)^k} (m-1) \ldots (m+k-2) B^{2k}\left(\frac{2m}{m+1}\right) \tag{129}$$

where $B\left(\frac{2m}{m+1}\right) = B(\beta)$, the function given in Table 15, will correspond to the common self-similar solution stated in Section 7. Common to a self-similar solution, the parameters f_{k0} and $\bar{f}_{k0}$ are independent of x.

Equation 127 is substantially more complicated than the Prandtl equation (15), although it possesses great advantages. In contrast to Equation 15,which should be solved separately for each specific assignment of $U(x)$, Equation 127 does not involve $U(x)$ explicitly, and consequently, it may be integrated permanently, disregarding whatever problem is suggested as solved. This makes it possible to call Equation 127 "universal". As a result of numerical integration of the universal equation (127) with the "universal" boundary conditions (128) all the characteristics of boundary layers are tabulated that are used directly at the first stage of calculation of any boundary layer. The solution of problems (127) and (128) is also valuable because it gives a multiparametric family of profiles of dimensionless velocity that are obtained rationally from the boundary layer equations (15), rather than advanced from them or the intuitive considerations.

Also worthy of attention is that the only solution of this universal equation, which combines a great number of problems of the boundary layer theory, expresses some general regularities relating the quantities characteristic for the boundary layer to such values as, e.g., inclination of the outer velocity curve $U(x)$ and its curvature. Like regularities may be revealed on the basis of the similar approach to temperature on concentration boundary layers.

The method presented is based on the application of the infinite process of subsequent constructions of solutions of the universal equation (127) at the boundary conditions (128). It requires mathematical proof of the convergence of this process that has not been found as yet.[35]

It should be qualified that the universal equation (127) and the Prandtl equation (15) may be assumed to be equal because the first of them was obtained from the second by the transformation of physical variables x, y, ψ (we shall restrict ourselves to the stationary case) to the variables of generalized similarity ξ, f_1, $\bar{f}_1$, f_2, $\bar{f}_2$, ..., Φ.

An infinite number of variables of generalized similarity in the "universal" formulation of the problem of the laminar boundary layer theory is virtually reduced to a small number of them in the interest of fast convergence of the calculation process. The resulting approximation of the method of generalized similarity is implemented with the present-day technique of computer calculation, by the selection of a small number of variables. This leads one to simplify the universal equation, applying different assumptions. To them belongs "truncation" of the number of variables, followed by their "localization" by the given variable by means of truncation of derivatives with respect to this variable.

Passing on to specific simplification of the universal equation (127), we abandon, as we did it the general case of self-similar solutions (Section 7), the search for the variables $\bar{f}_k$ ($k = 1, 2, ...$) and $\bar{\theta}_k$ ($k = 1, 2, ...$), having only retained the variable $\bar{f}_1$ that explicitly enters into the left side of Equation 127 and expressions θ_k (k = 1, 2) of Equation 125 on its right side. The sense of this simplification will soon become clear. The retained variable $\bar{f}_1$ is expressed in terms of the basic variables f_k ($k = 1, 2, ...$), having used the von Kármán integral relation (Section 8). We have the following expression for $\bar{f}_1$ (see Equations 112 and 113) (the dot above the latter indicates the derivative with respect to ξ):

$$\bar{f}_1 = 2\{\zeta(f_1, \ldots) - [2 + H(f_1, \ldots)]f_1\} \tag{130}$$

where it is assumed that

$$\zeta(f_1, \ldots) = \left[\frac{\partial(u/U)}{\partial(y/\delta^{**})}\right]_{y=0} = \ddot{\Phi}(0, f_1, \ldots)$$

$$H(f_1, \ldots) = \int_0^\infty \left(1 - \frac{u}{U}\right) d\xi = \int_0^\infty \left[1 - \dot{\Phi}(\xi, f_1, \ldots)\right] d\xi \tag{131}$$

Equation 130 may be obtained by integration of both sides of Equation 127 from $\xi = 0$ to $\xi = \infty$, similar to the procedure of Section 8 for Equation 11. For this it should be noted that the resulting integrals are equal to (the layer thickness is taken as δ^{**}; $\xi = y/\delta^{**}$)

$$\int_0^\infty \dddot{\Phi} d\xi = \left(\ddot{\Phi}\right)_{\xi=0}^{\xi=\infty} = -\ddot{\Phi}[0, (f_k)] = -\zeta[(f_k)]$$

$$\int_0^\infty \Phi\ddot{\Phi} d\xi = \int_0^\infty \Phi \frac{\partial}{\partial\xi}\left(\dot{\Phi} - 1\right) d\xi = \Phi\left(\dot{\Phi} - 1\right)\Big|_{\xi=0}^{\xi=\infty} + \int_0^\infty \dot{\Phi}\left(1 - \dot{\Phi}\right) d\xi =$$

---

[35] Presentation of the method of generalized similarity in its different aspects has been the concern of a number of works, among which the followings should be noted: Loitsyanskiy, L. G., Universal equations and parametric approximations in the laminar boundary layer theory, *Prikl. Mat. Mekh.*, 29 (1), 1965; Universal equations of the laminar boundary layer theory and parametric methods of their integration, *Tr. LPI*, No.280, 1967; *Generalized-Similar Solution of Boundary Layer Equation, Collected Papers Dedicated to the 60th Anniversary of L. I. Sedov*, Nauka Press, Moscow, 1969, 301; Similarity methods in the theory of integration of laminar boundary layer equations, Collected papers "*Voprosy matematicheskoy fiziki*" (*Problems of Mathematical Physics*) dedicated to the 75th anniversary of G. A. Grinberg, Nauka Press, Leningrad, 1976, 237–254. References to the papers, including various applications of the method of generalized similarity, are given later.

$$= \int_0^\infty \frac{u}{U}\left(1 - \frac{u}{U}\right)\frac{dy}{\delta^{**}} = 1$$

$$\int_0^\infty \left(1 - \dot{\Phi}^2\right)d\xi = \int_0^\infty \left(1 - \dot{\Phi}\right)d\xi + \int_0^\infty \dot{\Phi}\left(1 - \dot{\Phi}\right)d\xi = \int_0^\infty \left(1 - \frac{u}{U}\right)\frac{dy}{\delta^{**}} + \int_0^\infty \frac{u}{U}\left(1 - \frac{u}{U}\right)\frac{dy}{\delta^{**}} =$$

$$= \frac{\delta^*}{\delta^{**}} + 1 = H + 1$$

The integrals of Jacobians, as follows from asymptotic (when $\xi \to \infty$) equations (~ is the sign of asymptotic equality)

$$\Phi \equiv \xi - \int_0^\xi \left(1 - \dot{\Phi}\right)d\xi = \xi - \int_0^\infty \left(1 - \dot{\Phi}\right)d\xi + \int_\xi^\infty \left(1 - \dot{\Phi}\right)d\xi \sim \xi - \frac{\delta^*}{\delta^{**}}$$

$$\frac{\partial \Phi}{\partial f_k} \sim -\frac{\partial}{\partial f_k}\left(\frac{\delta^*}{\delta^{**}}\right)$$

are equal to zero; in fact,

$$\int_0^\infty \frac{D(\Phi, \partial\Phi/\partial\xi)}{D(\xi, f_k)}d\xi = \int_0^\infty \left(\frac{\partial\Phi}{\partial\xi}\frac{\partial^2\Phi}{\partial\xi\partial f_k} - \frac{\partial\Phi}{\partial f_k}\frac{\partial^2\Phi}{\partial\xi^2}\right)d\xi =$$

$$= -\frac{\partial}{\partial f_k}\int_0^\infty \left[1 - \left(\frac{\partial\Phi}{\partial\xi}\right)^2\right]d\xi - \int_0^\infty \frac{\partial}{\partial\xi}\left(\frac{\partial\Phi}{\partial\xi}\frac{\partial\Phi}{\partial f_k}\right)d\xi =$$

$$= -\frac{\partial}{\partial f_k}\left(\frac{\delta^*}{\delta^{**}} + 1\right) + \frac{\partial}{\partial f_k}\left(\frac{\delta^*}{\delta^{**}}\right) = 0$$

thus leading to Equation 130.

With such a "truncation" of the variables $\bar{f}_k$ (k = 1, 2,...) and assignment of $\bar{f}_1$ by Equation 130, the universal equation (127) and the boundary conditions to it take the form

$$\frac{\partial^3\Phi}{\partial\xi^3} + \left(f_1 + \frac{1}{2}\bar{f}_1\right)\Phi\frac{\partial^2\Phi}{\partial\xi^2} + f_1\left[1 - \left(\frac{\partial\Phi}{\partial\xi}\right)^2\right] = \sum_{k=1}^{\infty}\theta_k\frac{D(\Phi, \partial\Phi/\partial\xi)}{D(\xi, f_k)} \tag{132}$$

$$\Phi = 0 \qquad \dot{\Phi} = 0 \quad \text{at} \quad \xi = 0$$

$$\dot{\Phi} \to 1 \quad \text{when} \quad \xi \to \infty$$

$$\Phi = \Phi_*(\xi) \quad \text{at} \quad f_1 = f_2 = \ldots = 0$$

where it is assumed for brevity that $\bar{f}_1$ is substituted everywhere into its expression (130).

The boundary condition on the last line is related for the sake of simplicity to the case of a longitudinal flow past a plate (Section 4) but with the argument $\xi = y/\delta^{**}$, which corresponds to the equation

$$\dddot{\Phi}_* + \zeta_* \Phi_* \ddot{\Phi}_* = \dddot{\Phi}_* + \ddot{\Phi}_*(0)\, \Phi_* \ddot{\Phi}_* = 0 \tag{133}$$

that by simple affine transformation (substitution of Φ_* into Φ_*/ζ_*, ξ into ξ/ζ_* is reduced to Equation 34.

It may be noted that Equation 132 in replacement of $\bar{f}_1$ by Equation 130 containing $\ddot{\Phi}(0, f_1, \ldots)$ and the integral $\int_0^\infty \left[1 - \dot{\Phi}(\xi, f_1, \ldots)\right] d\xi$ takes on the integro-differential form which only slightly complicates numerical integration on a computer.

Making the two simplifications mentioned ("truncation" and "localization") we come to have the following approximations:

1. Local one-term approximation with one variable f_1 and localization over this variable amounting to the truncation of the right side of Equation 132. The approximation is based on the integration of the equation (the superscripts at Φ and $\bar{f}_1$ stand for the number of approximation)

$$\frac{\partial^3 \Phi^{(0)}}{\partial \xi^3} + \left(f_1 + \frac{1}{2}\bar{f}_1^{(0)}\right) \Phi^{(0)} \frac{\partial^2 \Phi^{(0)}}{\partial \xi^2} + f_1\left[1 - \left(\frac{\partial \Phi^{(0)}}{\partial \xi}\right)^2\right] = 0$$

$$\Phi^{(0)} = 0 \qquad \dot{\Phi}^{(0)} = 0 \quad \text{at} \quad \xi = 0$$

$$\dot{\Phi}^{(0)} \to 1 \quad \text{when} \quad \xi \to \infty$$

$$\Phi^{(0)} = \Phi_*(\xi) \quad \text{at} \quad f_1 = 0 \tag{134}$$

it is assumed in this case that

$$\bar{f}_1^{(0)} = 2\{\zeta^{(0)}(f_1) - [2 + H^{(0)}(f_1)] f_1\}$$

$$\zeta^{(0)}(f_1) = \ddot{\Phi}^{(0)}(0, f_1) \qquad H^{(0)}(f_1) = \int_0^\infty \left[1 - \dot{\Phi}^{(0)}(\xi, f_1)\right] d\xi \tag{135}$$

This local one-term approximation corresponds to the Kochin-Loitsyanskiy method.

2. Full one-term approximation (ξ, f_1). The equation

$$\frac{\partial^3 \Phi^{(1)}}{\partial \xi^3} + \left(f_1 + \frac{1}{2}\bar{f}_1^{(1)}\right) \Phi^{(1)} \frac{\partial^2 \Phi^{(1)}}{\partial \xi^2} + f_1\left[1 - \left(\frac{\partial \Phi^{(1)}}{\partial \xi}\right)^2\right] =$$

$$= f_1 \bar{f}_1^{(1)} \left(\frac{\partial \Phi^{(1)}}{\partial \xi} \frac{\partial^2 \Phi^{(1)}}{\partial \xi \partial f_1} - \frac{\partial \Phi^{(1)}}{\partial f_1} \frac{\partial^2 \Phi^{(1)}}{\partial \xi^2} \right)$$

$$\Phi^{(1)} = 0 \qquad \dot{\Phi}^{(1)} = 0 \quad \text{at} \quad \xi = 0$$

$$\dot{\Phi}^{(1)} \to 1 \quad \text{when} \quad \xi \to \infty$$

$$\Phi^{(1)} = \Phi_*(\xi) \quad \text{at} \quad f_1 = 0$$

$$\bar{f}_1^{(1)} = 2\{\zeta^{(1)}(f_1) - [2 + H^{(1)}(f_1)] f_1\}$$

$$\zeta^{(1)}(f_1) = \ddot{\Phi}^{(1)}(0, f_1) \qquad H^{(1)}(f_1) = \int_0^\infty \left[1 - \dot{\Phi}^{(1)}(\xi, f_1)\right] d\xi \tag{136}$$

corresponds to it.

3. Full two-term approximation with the equation

$$\frac{\partial^3 \Phi^{(2)}}{\partial \xi^3} + \left(f_1 + \frac{1}{2}\bar{f}_1^{(2)}\right) \Phi^{(2)} \frac{\partial^2 \Phi^{(2)}}{\partial \xi^2} + f_1 \left[1 - \left(\frac{\partial \Phi^{(2)}}{\partial \xi}\right)^2\right] =$$

$$= (f_1 \bar{f}_1^{(2)} + f_2) \left(\frac{\partial \Phi^{(2)}}{\partial \xi} \frac{\partial^2 \Phi^{(2)}}{\partial \xi \partial f_1} - \frac{\partial \Phi^{(2)}}{\partial f_1} \frac{\partial^2 \Phi^{(2)}}{\partial \xi^2} \right) +$$

$$+ (f_1 + 2\bar{f}_1^{(2)}) f_2 \left(\frac{\partial \Phi^{(2)}}{\partial \xi} \frac{\partial^2 \Phi^{(2)}}{\partial \xi \partial f_2} - \frac{\partial \Phi^{(2)}}{\partial f_2} \frac{\partial^2 \Phi^{(2)}}{\partial \xi^2} \right)$$

$$\Phi^{(2)} = 0 \qquad \dot{\Phi}^{(2)} = 0 \quad \text{at} \quad \xi = 0$$

$$\dot{\Phi}^{(2)} \to 1 \quad \text{at} \quad \xi \to \infty$$

$$\Phi^{(2)} = \Phi_0(\xi) \quad \text{when} \quad f_1 = 0 \qquad f_2 = 0$$

$$\bar{f}_1^{(2)} = 2\{\zeta^{(2)}(f_1, f_2) - [2 + H^{(2)}(f_1, f_2)] f_1\}$$

$$\zeta^{(2)}(f_1, f_2) = \ddot{\Phi}^{(2)}(0, f_1, f_2) \qquad H^{(2)}(f_1, f_2) = \int_0^\infty \left[1 - \dot{\Phi}^{(2)}(\xi, f_1, f_2)\right] d\xi \tag{137}$$

The approximations mentioned may be explained as follows. Full one-term approximation would strictly correspond to the assumption of a linear character of $U(x)$ variation. Numerical integration of the Prandtl equation (15) for "one-slope" profiles of $U(x)$ was made by Howarth.[36]

One-term approximation in the solution of the universal equation (127) supposes the account of one variable f_1, which, by its definition allows for the effect of tangent slope to the profile of outer velocities $U(x)$. The sign at f_1 determines either converging ($f_1 > 0$) or diverging ($f_1 < 0$) character of the boundary layer region in which calculation is made.

Two-term approximation, which along with f1 includes the variable $f_2 = UU''z^2$ makes it possible to allow for convexity or concavity of the curve $U(x)$ and its local curvature. Moreover the value of the momentum loss thickness entering into both approximations allows one to integrally evaluate the effect of the flow previous history on the boundary layer characteristics.

Numerical integration of Equation 137 was made by Ozerova and Simuni,[37] in whose paper the numerical method is presented in detail.

In Tables 18 and 19 the two functions $\zeta^{(2)}(f_1,f_2)$ and $\bar{f}_1^{(2)}(f_1,f_2)$ are given that are the most important for boundary layer calculation. The first of them occurs in the definition of skin friction τ_w on the body surface according to

$$\tau_w^{(2)} = \frac{\mu U}{\delta^{**}}\zeta^{(2)}(f_1,f_2) \tag{138}$$

The second serves for determining the relation $\delta^{**}(x)$ and thus, $f_1(x)$ and $f_2(x)$. This second final stage of boundary layer calculation is based on the integration of ordinary differential equations of the first order (111) and (112), which in the case of generalized similarity in the second approximation will be rewritten ($f = f_1^{(2)}$, $F = \bar{f}_1^{(2)}$) in the form

$$\frac{df_1^{(2)}}{dx} = \frac{U'}{U}\bar{f}_1^{(2)}[f_1^{(2)},f_2^{(2)}] + \frac{U''}{U'}f_1^{(2)} \tag{139}$$

$$U\frac{dz^{(2)}}{dx} = \bar{f}_1^{(2)}[f_1^{(2)},f_2^{(2)}] \tag{140}$$

where

$$f_1^{(2)} = U'z^{(2)} = \frac{U'[\delta^{**(2)}]^2}{\nu} \qquad f_2^{(2)} = UU''[z^{(2)}]^2 = UU''\left\{\frac{[\delta^{**(2)}]^2}{\nu}\right\}^2$$

The methods of solving the nonlinear first older equation (140) with respect to the quantity $z^{(2)} = [\delta^{**(2)}]^2/\nu$ are stated in the following section. The integral of this equation involves an arbitrary constant for those determination the assignment of some characteristic of the velocity profile is necessary in the initial section ($x = x_0$); that is naturally $z_0^{(2)} = [\delta_0^{**(2)}]^2/\nu$. This quantity contains scanty information on the initial velocity profile. We could enlage the volume of this information having increased the older of the equation. This requires only to retain the variables $\bar{f}_2$, $\bar{f}_3$, ... on the right side of Equation 140 which include the derivatives with respect to $z^{(2)}$ of the order higher than 2. However, this would extremely complicate practical calculations, thus explaining our rejection of the account for the variables $\bar{f}_2$, $\bar{f}_3$, ...

Consideration of the graphs $\zeta^{(2)}$ and $\bar{f}_1^{(2)}$ (Figures 179 and 180) composed in accordance with Tables 18 and 19 in the form of the family of curves $\zeta^{(2)}(f_1)$ and $\bar{f}_1^{(2)}(f_1)$

---

[36] Howarth, L., On the solution of the boundary layer equations, *Proc. R. Soc. London*, A919, 164, 1938.

[37] Ozerova, Ye. F. and Simuni, L. M., Numerical solution of the equations of two-term boundary layer theory, *Tr. LPI, Aerotermodinamika*, No. 313, 44–54, 1970.

Table 18. The Values of the Function $\zeta^{(2)}(f_1, f_2)$

f_2 \ f_1	−0.13	−0.1225	−0.115	−0.11	−0.105	−0.1	−0.095	−0.09	−0.08	−0.06	−0.04	−0.02	0	0.02	0.04	0.06	0.08
−0.25	−0.0012	0.0354	0.0608	0.0751	0.0882	0.1004	0.1118	0.1227	0.1430	0.1787	0.2101	0.2380	0.2639	0.2806	–	–	–
−0.20	–	0.0049	0.0396	0.0570	0.0721	0.0858	0.0985	0.1104	0.1323	0.1704	0.2037	0.2333	0.2608	0.2787	–	–	–
−0.15	–	–	0.0037	0.0293	0.0488	0.0654	0.0802	0.0937	0.1183	0.1598	0.1956	0.2274	0.2569	0.2764	–	–	–
−0.12	–	–	–	0.0035	0.0296	0.0489	0.0655	0.0804	0.1069	0.1514	0.1895	0.2230	0.2538	0.2745	–	–	–
−0.10	–	–	–	–	0.0143	0.0368	0.0548	0.0705	0.0982	0.1444	0.1842	0.2194	0.2514	0.2734	–	–	–
−0.08	–	–	–	–	−0.0090	0.0234	0.0437	0.0606	0.0894	0.1364	0.1776	0.2147	0.2487	0.2721	–	–	–
−0.06	–	–	–	–	–	0.0056	0.0314	0.0501	0.0806	0.1284	0.1699	0.2083	0.2448	0.2704	–	–	–
−0.04	–	–	–	–	–	–	0.0172	0.0391	0.0714	0.1204	0.1620	0.2002	0.2386	0.2687	–	–	–
−0.02	–	–	–	–	–	–	−0.0093	0.0243	0.0614	0.1129	0.1548	0.1926	0.2303	0.2616	0.2923	–	–
0	–	–	–	–	–	–	–	−0.0051	0.0597	0.1036	0.1471	0.1855	0.2208	0.2545	0.2870	0.3190	0.3511
0.02	–	–	–	–	–	–	–	–	0.0572	0.1031	0.1457	0.1836	0.2187	0.2521	0.2853	0.3131	0.3429
0.04	–	–	–	–	–	–	–	–	0.0471	0.1026	0.1442	0.1817	0.2166	0.2498	0.2822	0.3127	0.3415
0.06	–	–	–	–	–	–	–	–	0.0360	0.0934	0.1371	0.1754	0.2103	0.2431	0.2744	0.3045	–
0.08	–	–	–	–	–	–	–	–	0.0166	0.0849	0.1306	0.1696	0.2046	0.2375	0.2687	0.2986	–

Table 19. The Values of Function $\bar{f}_1^{(2)}(f_1, f_2)$

f_2 \ f_1	−0.13	−0.1225	−0.115	−0.11	−0.105	−0.1	−0.095	−0.09	−0.08	−0.06	−0.04	−0.02	0	0.02	0.04	0.06	0.08
−0.25	1.396	1.337	1.2801	1.2436	1.2080	1.1730	1.1387	1.1049	1.038	0.9083	0.7805	0.6534	0.5277	0.3896	–	–	–
−0.20	–	1.329	1.2704	1.2329	1.1965	1.1611	1.1265	1.0925	1.026	0.8966	0.7702	0.6450	0.5217	0.3826	–	–	–
−0.15	–	–	1.2615	1.2219	1.1838	1.1471	1.1117	1.0772	1.010	0.8818	0.7571	0.6342	0.5139	0.3775	–	–	–
−0.12	–	–	–	1.2156	1.1766	1.1385	1.1020	1.0767	0.9990	0.8703	0.7471	0.6262	0.5075	0.3733	–	–	–
−0.10	–	–	–	–	1.1722	1.1331	1.0957	1.0596	0.9908	0.8611	0.7387	0.6196	0.5028	0.3701	–	–	–
−0.08	–	–	–	–	1.1650	1.1281	1.0897	1.0529	0.9828	0.8511	0.7284	0.6111	0.4974	0.3680	–	–	–
−0.06	–	–	–	–	–	1.1224	1.0840	1.0463	0.9748	0.8411	0.7165	0.5997	0.4895	0.3642	–	–	–
−0.04	–	–	–	–	–	–	1.0783	1.0401	0.9669	0.8308	0.7042	0.5853	0.4771	0.3574	0.2794	–	–
−0.02	–	–	–	–	–	–	1.0711	1.0333	0.9591	0.8216	0.6932	0.5717	0.4606	0.3451	0.2363	–	–
0	–	–	–	–	–	–	–	1.0266	0.9500	0.8110	0.6817	0.5591	0.4416	0.3295	0.2220	0.1196	0.0235
0.02	–	–	–	–	–	–	-	1.0023	0.9485	0.8073	0.6774	0.5547	0.4374	0.3251	0.2210	0.1064	0.0024
0.04	–	–	–	–	–	–	–	–	0.9420	0.8037	0.6732	0.5503	0.4332	0.3211	0.2152	0.0974	–
0.06	–	–	–	–	–	–	–	–	0.9356	0.7932	0.6627	0.5391	0.4207	0.3064	0.1957	0.0884	–
0.08	–	–	–	–	–	–	–	–	0.9282	0.7839	0.6530	0.5287	0.4093	0.2939	0.1820	0.0732	–

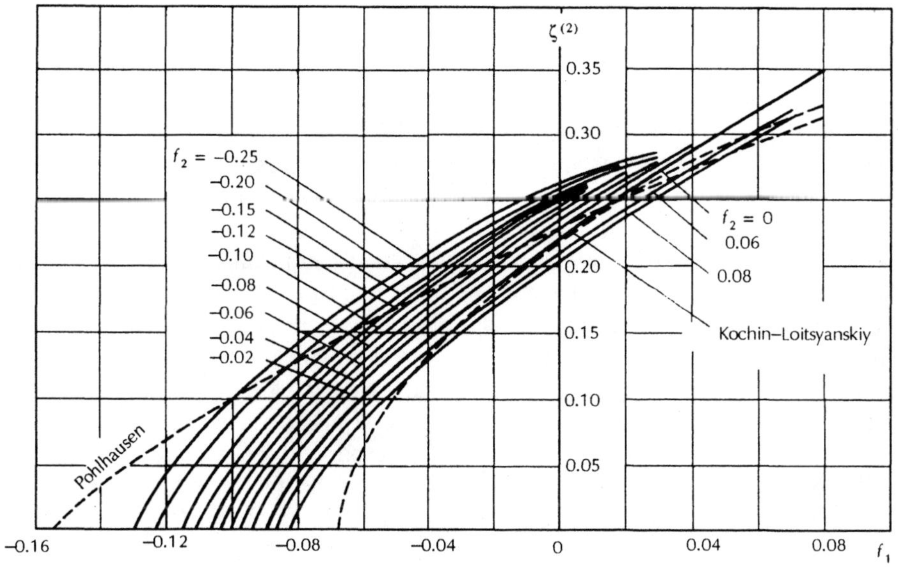

Fig. 179

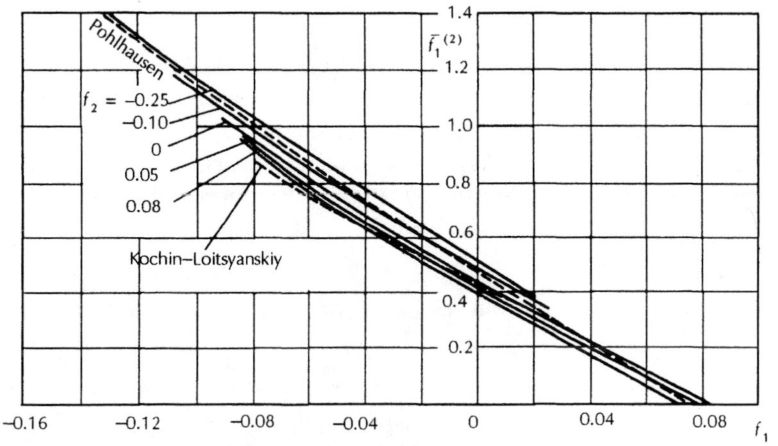

Fig. 180

dependent on the parameter f_2 allows one to make some general conclusions:

1. Distributions of the "reduced" friction $\zeta^{(2)}(f_1,f_2)$ are strongly dependent on the parameter f_2. In particular, the values of f_{1s} at the separation point ($\zeta^{(2)} = 0$) really differ from each other depending on the value of f_2.

2. The curves of the distribution $\bar{f}_1^{(2)}(f_1,f_2)$ combine into a narrow band thus making it possible to approximately assume

$$\bar{f}_1^{(2)}(f_1,f_2) = a - bf_1$$

taking both quantities a and b as constants or considering a as a function of f_2 and b as the constant which is equal to about 6. Application of the linearity of $\bar{f}_1^{(2)}(f_1,f_2)$ will lead, as is already known, to quadrature (116). Equations 139 or 140 may be easily integrated by tabular values of $\bar{f}_1^{(2)}(f_1, f_2)$ on small computers or even on calculators.

3. The results of the first approximation are given in the presented tables at $f_2 = 0$. The curves $\zeta^{(1)}(f_1)$ and $\bar{f}_1^{(1)}(f_1)$ are shown in Figures 179 and 180 as the curves of the second approximation between which they are located.

4. The curves $\zeta(f)$ and $F(f) = \bar{f}_1(f)$ corresponding to the Kochin-Loitsyanskiy method are shown on the same figures by dashed lines. This method gives very early separation ($\zeta = 0$) at $f_s = -0.0681$.

5. The Pohlhausen method stands apart from the content of the method of generalized similarity. The upper dashed curve corresponding to it demonstrates delay of the separation condition ($f_s = -0.1567$).

6. In converging regions of boundary layers ($f_1 > 0$) the curves $\zeta^{(2)}(f_1, f_2)$ converge, thus indicating the possibility of using the first approximation ($f_2 = 0$) in this region. In diverging regions ($f_1 < 0$) the curves $\zeta^{(2)}(f_1, f_2)$ diverge; it may be noted here that the Kochin-Loitsyanskiy method everywhere gives, with the exception of the vicinity of the separation point ($\zeta = 0$) of the boundary layer, the results close to the curves of the second approximation of the method of generalized similarity when $f_2 > 0$ [with curve $U(x)$ convex facing the axis 0x], and the curve of the Pohlhausen method conversely, under the same conditions is closer to the curves of the method of generalized similarity when $f_2 < 0$ [with curve $U(x)$ concave facing the axis $0x$].

The method of generalized similarity, in the presence of Tables 18 and 19, is easily applied to practical calculations of boundary layers. It should be noted that large scientific institutions with powerful computers possess standard programs for laminar boundary layer calculation which make it possible to deal without approximate methods in general and without the method of generalized similarity in particular. Of course, in this case its becomes difficult to determine general laws governing the influence of the curve $U(x)$ shape (the slope of the tangent, the curve curvature) on the boundary layer characteristics.

10. Examples of the Application of the Method of Generalized Similarity

Determination of the convergence of subsequent approximations when calculating boundary layer by the method of generalized similarity, and in the presence of convergence also estimation of its speed, is a very complicated mathematical problem which has not been solved. It remains only to assume that a two-term approximation allowing for the effect not only of the slope of tangential lines to the curve of velocity distribution on the outer edge of the boundary layer $U(x)$, but also of the curvature of this curve, should lead to rather accurate results. An example stated illustrates the validity of this assumption.

The absence of rigorous estimation of the speed of approximation convergence leads one to take the course of comparing the solutions obtained in the first and second approximations by the method of generalized similarity with the available, rather accurate numerical solutions of the Prandtl equation. The results of numerical calculation of a plane stationary boundary layer on the surface of a circular cylinder in a transverse flow with the theoretical law of velocity distribution over the cylinder contour made by. Terrill[38] may serve as an example of this numerical solution. Referring one for a detailed description of the technique of numerical integration to the paper mentioned, we note that the solution was calculated in a

[38] Terrill, R. M., Laminar boundary layer flow near separation with and without suction, *Phil. Trans.*, A253, 55–100, 1960.

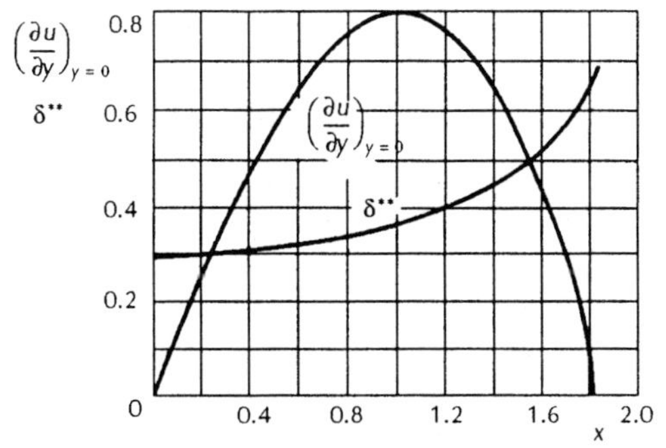

Fig. 181

dimensionless form with the longitudinal lengths being related to the cylinder radius a, longitudinal velocities in the boundary layer sections to the free stream velocity U_∞, and the transverse lengths and velocities to the quantities $a/\sqrt{\mathbf{Re}_\infty}$, $U_\infty/\sqrt{\mathbf{Re}_\infty}$, respectively, where $\mathbf{Re}_\infty = U_\infty a/\nu$. The angle x expressed in radians served as the dimensionless abscissa reckoned from the front critical point.

We present the master graph of the two quantities $\delta^{**}(x)$ and $(\partial u/\partial y)_{y=0}$ which are greatly interesting (Figure 181). As is seen in the figure the dimensionless thickness of the momentum loss $\delta^{**}(x)$ grows monotonously from some initial value at the front critical point equal to about 0.29. This coincides with the value of $B(\beta)$ determined from Table 15 at $m = \beta = 1$ and $c = 2$, thus corresponding to the law of velocity distribution on the outer edge of the boundary layer in the vicinity of the front critical point $U = cx$.

Dimensionless skin friction grows from zero at $x = 0$ and attains its maximum at $x = 1$, thus corresponding to $58°18'$. Then, skin friction diminishes to zero at $x_s = 1.82$, or in degrees $x_s^{\circ} = 104°30'$. This point is the separation point S of the boundary layer from the circular cylinder surface. In this calculation the reverse effect of the boundary layer on the outer flow is disregarded, i.e., that substantial distortion which is introduced by separation into a theoretical potential flow past a body.

The Terrill example, as an exact solution, is taken as a standard (test calculation) with which the values of the quantities calculated by two-term approximation of the method of generalized similarity are compared. The approximate curve $\delta^{**}(x)$ virtually coincides with the accurate one; maximum deviation was observed near the separation point and did not exceed 3% from the accurate one. The curve of one-term approximation for the value of the reduced friction $(\partial u/\partial y)_{y=0}$ noticeably deviates from the Terrill exact curve (Figure 182)

Calculation by two-term approximation leads to the value of friction nearly fully coinciding with an accurate value, with the exception of a small region in close vicinity to the separation point (hardly noticeable deviation is shown by dashes; in Figure 183 this region is given in a large scale).

One may lift the restrictions on the number of the parameters applied using expansion of the function $\Phi[\zeta, (f_k)]$ in a power series of (f_k) instead of direct integration of Equation 132; however, convergence of these series is limited by small values of the parameters. As can be judged by Figure 184, smallness of the parameters f_2 and f_3 calculated for the Terrill example is provided only at a distance from the separation point, which is a singular point of Equation 132. The zero value of $f_3(x)$ nearly within the entire range of x, with the nearest vicinity of the separation point excluded, is an indication that the second approximation may be used as rather accurate solution.

As shown by calculations, to determine $\delta^{**}(x)$ and then f_1, f_2, f_3 one may be restricted to the first approximation, since a relative error on finding δ^{**}, which quickly

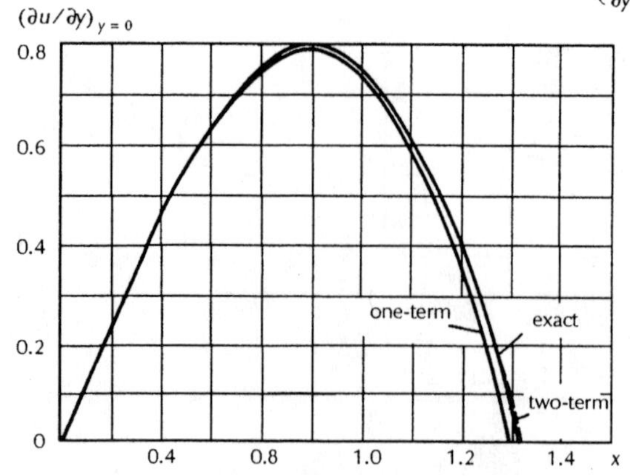

Fig. 182

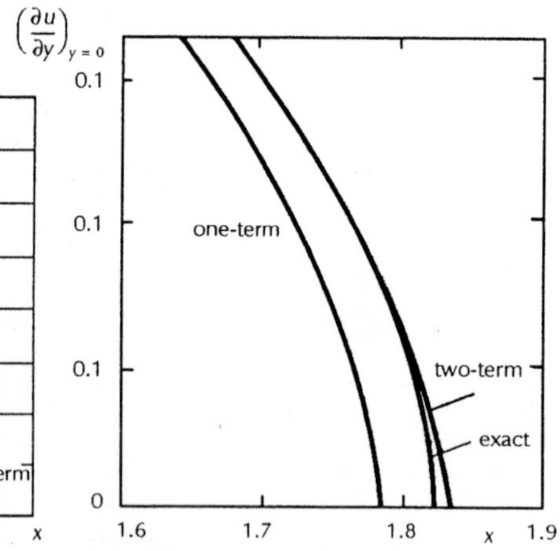

Fig. 183

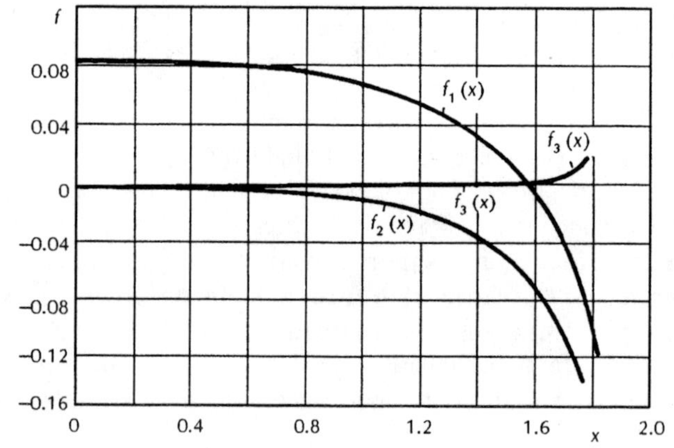

Fig. 184

grows in a diverging portion, is small. The effect of the second parameter $f_2(x)$ should be taken into account only when calculating the skin friction τ_w in terms of $\zeta^{(2)}(f_1, f_2)$ near the separation point. In this case it may be assumed that $a = 0.4408$, $b = 5.714$.

With a more strict approach the error ε may be taken into account, which is introduced by the application of a simplified representation

$$\overline{f_1} = a(f_2) - bf_1 + \varepsilon$$

Integrating formally under this assumption equality (140), we obtain

$$U^b z(x) = \int_{x_0}^{x} U^{b-1}(t)\,\{a[f_2(t)] + \varepsilon[f_1(t), f_2(t)]\}\,dt + U^b(x_0)\,z(x_0) \qquad (141)$$

where $z(x_0)$ is determined either by the *a priori* prescribed value of $\delta^{**}(x_0) = \delta_0^{**}$ or by the value of z_0 at a front critical point of a streamlined body ($x = 0$, $U = 0$).

The value of z_0 may be determined from the condition of the finiteness of the derivative dz/dx at $x = 0$, i.e., from the equation

$$\bar{f}_1^{(2)}(f_{10}, f_{20}) = 0$$

which, with the prescribed behavior of $U(x) = cx$ near the point $x = 0$ in one-term approximation, is reduced to the equation with one variable and which has the root

$$f_{10} = 0.0854 \qquad z_0 = \frac{\delta_0^{**2}}{\nu} = \frac{f_{10}}{U_0'} = \frac{0.0854}{c}$$

where c depends on the form of the tip of the profile.

Equation 141 may be solved by consecutive approximations of the form

$$U^b(x_i)\,z(x_i) = U^b(x_{i-1})\,z(x_{i-1}) +$$

$$+ \{a[f_2(x_{i-1})] + \varepsilon[f_1(x_{i-1}), f_2(x_{i-1})]\}\int_{x_{i-1}}^{x_i} U^{b\ 1}(t)\,dt$$

if the entire range (x_0, x) is divided into small ranges with the nodes at the points x_i, and $f_2(x)$ and $\varepsilon(x)$ are substituted by stepwise representations. This "step by step" calculation can be easily performed.

The quantities ζ, H, and $\bar{f}_1$ in the function from f_1, f_2, f_3, ... may be presented[39] by the power series

$$\zeta = 0.2204 + 1.7350 f_1 - 2.4188 f_1^2 - 0.2992 f_2 +$$

$$+ 18.234 f_1^3 - 0.1653 f_1 f_2 + 0.0937 f_3 - \ldots$$

$$H = 2.5919 - 5.4282 f_1 + 21.914 f_1^2 + 1.4741 f_2 -$$

$$- 163.06 f_1^3 - 4.8076 f_1 f_2 - 0.5061 f_3 + \ldots$$

$$\bar{f}_1 = 0.4408 - 5.7139 f_1 + 6.0189 f_1^2 - 0.5984 f_2 -$$

$$- 7.3611 f_1^3 - 3.2753 f_1 f_2 - 1.0123 f_3 + \ldots$$

the application of which in practice is limited by their convergence radii which are determined by direct calculations. The separation point is the singular point on approaching to which the speed of convergence of these series sharply decreases.

In this section, among various applications of the method of generalized similarity in the laminar boundary layer theory we consider two near-wall layers: a permeable surface (suction or injection of liquid from a solid surface) and a magnetohydrodynamic (MHD) boundary layer in an electrically conducting liquid.

---

[39] Loitsyanskiy, L. G., Universal equations and parametric approximations in the laminar boundary layer theory, *Prikl. Mat. Mekh.*, 29 (1), 1965.

A detailed presentation of the laminar boundary layer theory on permeable surfaces (injection and suction of the same liquid as that in a free stream, or injection of liquid with different properties) is given in special monographs dealing with the boundary layer theory.[40]

It was Chan who initially applied the method of generalized similarity to the problems of a laminar boundary layer on a permeable surface, who composed a universal equation, and who used the method of expanding the solution into the series in powers of the parameters for solving the universal equation. A numerical solution of the universal equation in the simplest approximations on an electronic computer for the case of a permeable surface was performed by Lesnikov[41] and Shishkina[42] in the Department of Hydroaerodynamics of the M. I. Kalinin Leningrad Polytechnical Institute.

Equations of a laminar boundary layer on a permeable surface have the form of (11) and (15) similar to that on an impermeable one. The difference tells only on the first boundary condition on the body surface. If $v_0(x)$ denotes the prescribed velocity with which the liquid having the same physical constants ρ, μ, ν, that the liquid in a free stream has, passes through a solid surface ($v_0 > 0$ in the case of liquid injection, $v_0 < 0$ in its suction) normally to it, then the first line of the boundary conditions for Equation 11 will, in contrast to Equation 12, have the form

$$u = 0 \qquad v = v_0 \quad \text{at} \quad y = 0$$

Introducing in this case the notation for the stream function $\psi(x, 0) = \psi_0(x)$ variable along the body surface, we substitute the boundary conditions on the body surface ($y = 0$) for Equation 15 by the following (a prime denotes the derivative with respect to x)

$$\frac{\partial \psi}{\partial x} = \psi_0' = -v_0 \qquad \frac{\partial \psi}{\partial y} = 0 \quad \text{at} \quad y = 0 \tag{142}$$

We may retain the old boundary conditions of Equation 15 if we introduce a new stream function $\psi^*(x, y)$ related to $\psi(x, y)$ as

$$\psi(x, y) = \psi_0(x) + \psi^*(x, y) \qquad \psi^*(x, 0) = 0 \tag{143}$$

Thus, the problem is reduced to the integration of the equation

$$\frac{\partial \psi^*}{\partial y}\frac{\partial^2 \psi^*}{\partial x \partial y} - \frac{\partial \psi^*}{\partial x}\frac{\partial^2 \psi^*}{\partial y^2} - \psi_0'\frac{\partial^2 \psi^*}{\partial y^2} = UU' + \nu\frac{\partial^3 \psi^*}{\partial y^3}$$

$$\psi^* = 0 \qquad \frac{\partial \psi^*}{\partial y} = 0 \quad \text{when} \quad y = 0$$

$$\frac{\partial \psi^*}{\partial y} \to U(x) \quad \text{when} \quad y \to \infty$$

$$\frac{\partial \psi^*}{\partial y} = u_0(y) \quad \text{when} \quad x = x_0 \tag{144}$$

---

[40] Schlichting, G., *Teoriya pogranichnogo sloya* (*Boundary Layer Theory*) (translated from German) Nauka Press, Moscow, 1969, 359–386; *Laminar Boundary Layers* Rosenhead, L., Ed., Clarendon Press, Oxford, 1963, 339–348.

[41] Lesnikov, A. L., Laminar boundary layer on a permeable surface, *J. Eng. Phys.*, No. 5, 1972.

[42] Shishkina, L. G., Two-term solution of the equations of a laminar boundary layer on a permeable surface, *Iz. AN SSSR, Mekh. Zhidk. Gaza*, No. 6, 1973.

with the additional term (compare with Equation 15),

$$-\psi_0' \frac{\partial^2 \psi^*}{\partial y^2} = v_0 \frac{\partial^2 \psi^*}{\partial y^2}$$

on the left side of Equation 144.

An attempt to solve this equation in an affine-like form

$$\psi^*(x, y) = U(x)\,\delta^{**}(x)\,\Phi^*\left(\frac{y}{\delta^{**}}\right) = U(x)\,\delta^{**}(x)\,\Phi^*(\xi)$$

will lead, similar to that described in Section 7, to the equation

$$\dddot{\Phi}^* + \left(f + \frac{1}{2}\bar{f}\right)\Phi^*\ddot{\Phi}^* + f\left(1 - \dot{\Phi}^{*2}\right) + \lambda\ddot{\Phi}^* = 0 \tag{145}$$

where, along with the previous parameters $f = U'z = U'\delta^{**2}/\nu$, $\bar{f} = Uz'$, a new "injection parameter" appears that is equal to

$$\lambda(x) = -\frac{v_0\delta^{**}}{\nu} = \frac{\psi_0'\delta^{**}}{\nu} \tag{146}$$

Repeating the algorithm adopted in Section 9 for deriving the boundary layer equation in the variables of the generalized similarity for a permeable surface, we introduce the transformation

$$\psi^*(x, y) = U(x)\,\delta^{**}(x)\,\Phi^*[\xi; (f_k), (\bar{f}_k), (\lambda_k)]$$

$$f_k = U^{k-1}\frac{d^kU}{dx^k}z^k \qquad \bar{f}_k = z^{k-1}\frac{d^kz}{dx^k}U^k$$

$$\lambda_k = -U^{k-1}\frac{d^{k-1}v_0}{dx^{k-1}}\frac{z^{k-1/2}}{\nu^{1/2}} = U^{k-1}\frac{d^k\psi_0}{dx^k}\frac{z^{k-1/2}}{\nu^{1/2}} \tag{147}$$

Then, we come to the following universal equation of viscous flow in a boundary layer on a permeable surface

$$\frac{\partial^3\Phi^*}{\partial\xi^3} + \left(f_1 + \frac{1}{2}\bar{f}_1\right)\Phi^*\frac{\partial^2\Phi^*}{\partial\xi^2} + f_1\left[1 - \left(\frac{\partial\Phi^*}{\partial\xi}\right)^2\right] + \lambda_1\frac{\partial^2\Phi^*}{\partial\xi^2} =$$

$$= \sum_{k=1}^{\infty}\left[\theta_k\left(\frac{\partial\Phi^*}{\partial\xi}\frac{\partial^2\Phi^*}{\partial\xi\partial f_k} - \frac{\partial\Phi^*}{\partial f_k}\frac{\partial^2\Phi^*}{\partial\xi^2}\right) + \bar{\theta}_k\left(\frac{\partial\Phi^*}{\partial\xi}\frac{\partial^2\Phi^*}{\partial\xi\partial\bar{f}_k} - \frac{\partial\Phi^*}{\partial\bar{f}_k}\frac{\partial^2\Phi^*}{\partial\xi^2}\right) +\right.$$

$$\left. + \Lambda_k\left(\frac{\partial\Phi^*}{\partial\xi}\frac{\partial^2\Phi^*}{\partial\xi\partial\lambda_k} - \frac{\partial\Phi^*}{\partial\lambda_k}\frac{\partial^2\Phi^*}{\partial\xi^2}\right)\right] \tag{148}$$

where the old notation (125) is introduced for θ_k and $\bar{\theta}_k$ and a new one for Λ_k

$$\Lambda_k = \left[(k-1)f_1 + \left(\frac{2k-1}{2}\right)\bar{f_1}\right]\lambda_k + \lambda_{k+1} \tag{149}$$

with the relations

$$Uzf_k' = [(k-1)f_1 + k\bar{f_1}]f_k + f_{k+1} = \theta_k$$

$$Uz\bar{f_k}' = [(k-1)\bar{f_1} + kf_1]\bar{f_k} + \bar{f}_{k+1} = \bar{\theta}_k$$

$$Uz\lambda_k' = \left[(k-1)f_1 + \left(\frac{2k-1}{2}\right)\bar{f_1}\right]\lambda_k + \lambda_{k+1} = \Lambda_k \tag{150}$$

being used to derive Equation 148.

The integral relation derived similarly to that stated in Section 9 by the integration of both sides of Equation 148 with respect to ξ from $\xi = 0$ to $\xi = \infty$ will be reduced to the form

$$\bar{f_1} = 2[\zeta - (2+H)f_1 - \lambda_1] \tag{151}$$

which differs from Equation 130 in the presence of the last term within square brackets on the right side of Equation 151.

Restricting ourselves to the localization over all the parameters, with $f_1 = f$ and $\lambda_1 = \lambda$ excepted, and to the truncation over the parameters by which localization was made, we obtain the universal two-term equation

$$\frac{\partial^3\Phi^*}{\partial\xi^3} + \left(f + \frac{1}{2}\bar{f}\right)\Phi^*\frac{\partial^2\Phi^*}{\partial\xi^2} + f\left[1 - \left(\frac{\partial\Phi^*}{\partial\xi}\right)^2\right] + \lambda\frac{\partial^2\Phi^*}{\partial\xi^2} =$$

$$= f\bar{f}\left(\frac{\partial\Phi^*}{\partial\xi}\frac{\partial^2\Phi^*}{\partial f\partial\xi} - \frac{\partial\Phi^*}{\partial f}\frac{\partial^2\Phi^*}{\partial\xi^2}\right) + \bar{f}\lambda\left(\frac{\partial\Phi^*}{\partial\xi}\frac{\partial^2\Phi^*}{\partial\xi\partial\lambda} - \frac{\partial\Phi^*}{\partial\lambda}\frac{\partial^2\Phi^*}{\partial\xi^2}\right)$$

$$\Phi^* = \frac{\partial\Phi^*}{\partial\xi} = 0 \quad \text{at} \quad \xi = 0 \qquad \frac{\partial\Phi^*}{\partial\xi} \to 1 \quad \text{when} \quad \xi \to \infty$$

$$\Phi^* = \Phi_0^*(\xi) \quad \text{at} \quad f = \lambda = 0 \tag{152}$$

This equation was integrated on an computer by Shishkina. The graphs of the relations $\zeta(f, \lambda)$ and $\bar{f}\,(f, \lambda)$ presented by Shishkina are given in Figures 185 and 186.

The effect of suction ($\lambda > 0$) on the boundary layer separation delay is vividly seen. Injection ($\lambda < 0$) with the velocity normal to the surface, on the contrary, precedes separation, but facilitates laminar-to-turbulent flow transition, thus also leading to the delay of the separation point position.

The boundary layer thickness may be determined by the equation

$$\frac{dz}{dx} = \frac{\bar{f}\,(f, \lambda)}{U(x)} = \frac{2\{\zeta(f, \lambda) - [2 + H(f, \lambda)]f - \lambda\}}{U(x)} \tag{153}$$

with the numerator on the right (similar to the case of an impermeable surface) being approximately replaced by the linear function

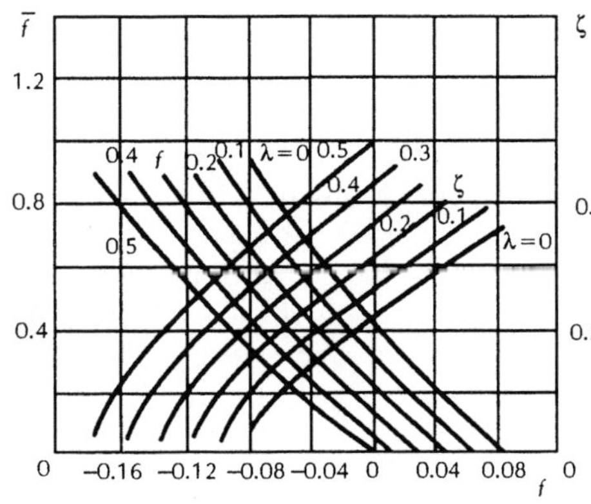

Fig. 185

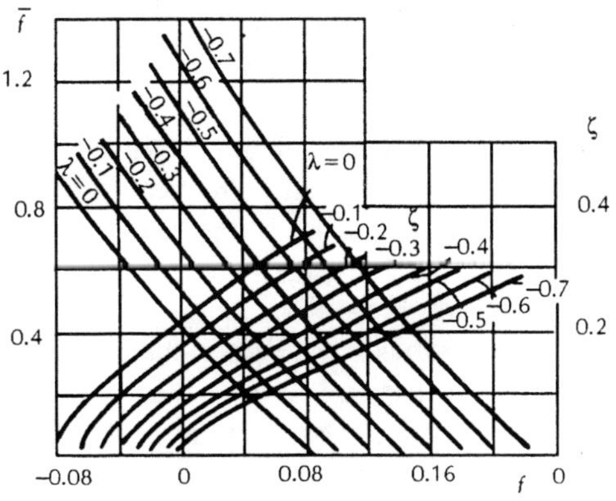

Fig. 186

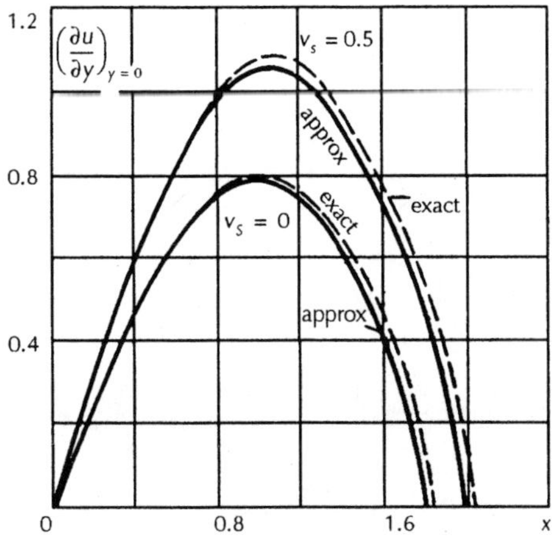

Fig. 187

$$\bar{f}\,(f, \lambda) = a - bf - 2a\lambda \tag{154}$$

thus making integration of Equation 153 easier. The constant a was adopted to be equal to 0.44, and b to 5.15. The difference from the earlier $b = 5.75$ is insignificant. The Terrill solution for two values of the constant of the suction dimensionless velocity $v_s = 0$ and $v_s = 0.5$ cited at the end of the previous section was taken as a standard of the comparison with the exact solution. As is seen (Figure 187), the approximate solutions obtained by Shishkina (the lower curves) slightly differ from the exact ones (the upper curves). At $v_s = 0$ an approximate solution coincides with the one-term solution, the comparison of which with a two-term and the Terrill exact solutions was shown in Figure 182.

At the end of the 1950s and the beginning of the 1960s studies of MHD-boundary layers were conducted by many scientists (Likoudis, Rosso, Wu, and others) who used different methods of the boundary layer theory (self-similar solutions, one-tern methods, expansions in terms of a small parameter).

The method of generalized similarity was first applied to the MHD-boundary layer by

Yuferev[43] in the Department of Hydroaerodynamics of the Kalinin Polytechnical Institute in Leningrad. Calculation of the following approximations was then performed by the Yugoslavian scientist Z. Boričič.[44]

Assuming that the outer magnetic field is normal to the body surface ($B_x = 0$, $B_y = B$; notation is same as in Section 8 of chapter 10), that there is no electric field ($E = 0$), and that fluid is neutral, we shall use the general equations (110) and (111) (Chapter 1) of the conducting viscous incompressible fluid flow. We shall pause on the case of small magnetic Reynolds numbers $\mathbf{Re}_m = U_\infty L/\nu_m$, where ν_m is the kinematic coefficient of magnetic viscosity when the dependence of magnetic induction B on y may be neglected and $B = B(x)$.

Having made simplifications typical of the boundary layer theory, we have (σ is the coefficient of fluid electric conductivity)

$$u\frac{\partial u}{\partial x} + v\frac{\partial u}{\partial y} = -\frac{1}{\rho}\frac{dp}{dx} + \nu\frac{\partial^2 u}{\partial y^2} - \frac{\sigma B^2}{\rho}u$$

and on the outer edge of the boundary layer

$$U\frac{dU}{dx} = -\frac{1}{\rho}\frac{dp}{dx} - \frac{\sigma B^2}{\rho}U$$

where, substracting term-by-term the left and right sides of these equations, we obtain under the adopted conditions the MHD-boundary layer equation in the form

$$u\frac{\partial u}{\partial x} + v\frac{\partial u}{\partial y} = UU' + \nu\frac{\partial^2 u}{\partial y^2} + \frac{\sigma B^2}{\rho}(U - u)$$

$$\frac{\partial u}{\partial x} + \frac{\partial v}{\partial y} = 0 \qquad (155)$$

which differs from Equation 11 by the presence of the additional term on the right that expresses the effect of the volumetric ponderomotive force.

Introducing the stream function $\psi(x, y)$ and assuming for brevity $\sigma B^2/\rho = N$, we compose the equation

$$\frac{\partial\psi}{\partial y}\frac{\partial^2\psi}{\partial x\partial y} - \frac{\partial\psi}{\partial x}\frac{\partial^2\psi}{\partial y^2} = UU' + \nu\frac{\partial^3\psi}{\partial y^3} + N\left(U - \frac{\partial\psi}{\partial y}\right)$$

$$\psi = 0 \qquad \frac{\partial\psi}{\partial y} = 0 \quad \text{when} \quad y = 0$$

$$\frac{\partial\psi}{\partial y} \to U \quad \text{at} \quad y \to \infty$$

$$\frac{\partial\psi}{\partial y} = u_0(y) \quad \text{when} \quad x = x_0 \qquad (156)$$

---

[43] Yuferev, V. S., Parametric method for calculating laminar boundary layer in magnetohydrodynamics, *Magnitnaya Gidrodinamika*, No. 4, 1966; Yuferev V. S., On one approximate method of laminar boundary layer calculation in magnetohydrodynamics, *Mekh. Zhidk. Gaza*, No. 1, 124–127, 1967.

[44] Boričič, Z., Locally two-term equations of a plane flow of conducting incompressible fluid, *Magnitnaya Gidrodinamika*, No 1, 5–10, No. 1.

which subsequently will be basic.

Having made the transition to the earlier introduced variables of the generalized similarity

$$\xi = \frac{y}{\delta^{**}} \qquad \Phi = \frac{\psi}{U\delta^{**}}$$

$$f_k = U^{k-1}\frac{d^k U}{dx^k} z^k \qquad \bar{f}_k = z^{k-1}\frac{d^k z}{dx^k} U^k \tag{157}$$

and using also a new succession of the variables

$$g_k = U^{k-1}\frac{d^{k-1}N}{dx^{k-1}} z^k \tag{158}$$

we obtain the universal MHD-boundary layer equation in the form

$$\frac{\partial^3\Phi}{\partial\xi^3} + \left(f_1 + \frac{1}{2}\bar{f}_1\right)\Phi\frac{\partial^2\Phi}{\partial\xi^2} + f_1\left[1 - \left(\frac{\partial\Phi}{\partial\xi}\right)^2\right] + g_1\left(1 - \frac{\partial\Phi}{\partial\xi}\right) =$$

$$= \sum_{k=1}^{\infty}\left[\theta_k\left(\frac{\partial\Phi}{\partial\xi}\frac{\partial^2\Phi}{\partial\xi\partial f_k} - \frac{\partial\Phi}{\partial f_k}\frac{\partial^2\Phi}{\partial\xi^2}\right) + \bar{\theta}_k\left(\frac{\partial\Phi}{\partial\xi}\frac{\partial^2\Phi}{\partial\xi\partial\bar{f}_k} - \frac{\partial\Phi}{\partial\bar{f}_k}\frac{\partial^2\Phi}{\partial\xi^2}\right) + \right.$$

$$\left. + G_k\left(\frac{\partial\Phi}{\partial\xi}\frac{\partial^2\Phi}{\partial\xi\partial g_k} - \frac{\partial\Phi}{\partial g_k}\frac{\partial^2\Phi}{\partial\xi^2}\right)\right]$$

$$\Phi = \frac{\partial\Phi}{\partial\xi} = 0 \quad \text{at} \quad \xi = 0 \qquad \frac{\partial\Phi}{\partial\xi} \to 1 \quad \text{when} \quad \xi \to \infty$$

$$\Phi = \Phi_0(\xi) \quad \text{at} \quad f_1 = f_2 = \ldots = 0, \quad \bar{f}_1 = \bar{f}_{10}, \quad \bar{f}_2 = \bar{f}_{20}, \quad \ldots,$$

$$g_1 = g_{10}, \quad g_2 = g_{20}, \quad \ldots, \tag{159}$$

where along with θ_k and $\bar{\theta}_k$ appear the coefficients

$$G_k = [(k-1)f_1 + k\bar{f}_1]g_k + g_{k+1} \tag{160}$$

and the relations

$$Uzf_k' = \theta_k \qquad Uz\bar{f}_k' = \bar{\theta}_k \qquad Uzg_k' = G_k \tag{161}$$

are used on derivation of Equations 159.

The integral relation leads to the following expression for $\bar{f}_1$

$$\bar{f}_1 = 2[\zeta - (2 + H)f_1 - Hg_1] \tag{162}$$

which forms the basis when determining the momentum loss thickness $\delta^{**}(x)$ according to the equation

$$\frac{dz}{dx} = \frac{\bar{f}_1[(f_k), (\bar{f}_k), (g_k)]}{U(x)} \tag{163}$$

The generality of the method in different fields of application and analogy with the preceding are rather obvious.

In the Boričič paper Equation 159, truncated by the variables f_k, $\bar{f}_k$, g_k when $k > 1$ and localized over $\bar{f}_1 = \bar{f}$ and also alternately over $f_1 = f$ or $g_1 = g$, was integrated numerically on a computer.

A general form of this equation is

$$\frac{\partial^3 \Phi}{\partial \xi^3} + \left(f + \frac{1}{2}\bar{f}\right)\Phi \frac{\partial^2 \Phi}{\partial \xi^2} + f\left[1 - \left(\frac{\partial \Phi}{\partial \xi}\right)^2\right] + g\left(1 - \frac{\partial \Phi}{\partial \xi}\right) =$$

$$= f\bar{f}\left(\frac{\partial \Phi}{\partial \xi}\frac{\partial^2 \Phi}{\partial \xi \partial f} - \frac{\partial \Phi}{\partial f}\frac{\partial^2 \Phi}{\partial \xi^2}\right) + \bar{f}\, g\left(\frac{\partial \Phi}{\partial \xi}\frac{\partial^2 \Phi}{\partial \xi \partial g} - \frac{\partial \Phi}{\partial g}\frac{\partial^2 \Phi}{\partial \xi^2}\right)$$

$$\Phi = \frac{\partial \Phi}{\partial \xi} = 0 \quad \text{at} \quad \xi = 0 \qquad \frac{\partial \Phi}{\partial \xi} \to 1 \quad \text{when} \quad \xi \to \infty$$

$$\Phi = \Phi_0(\xi, g) \quad \text{at} \quad f = 0 \qquad \Phi = \Phi_1(\xi, f) \quad \text{at} \quad g = 0 \tag{164}$$

We shall bear in mind that in the case of additional (besides general over $\bar{f}$) localization over g there is no second bracket on the right side of Equation 164 and it that the first boundary condition in the last line, that relates to a longitudinal flow of electrically conducting fluid past a plate (i.e. to the equation),

$$\dddot{\Phi}_{00} + \frac{1}{2}\bar{f}_{00}\Phi_{00}\ddot{\Phi}_{00} + g\left(1 - \dot{\Phi}_{00}\right) = 0$$

is taken into account and with additional localization over f there is no first bracket on the right side of Equation 164 and the second boundary condition is used for the equation

$$\dddot{\Phi}_{10} + \left(f_{00} + \frac{1}{2}\bar{f}_{10}\right)\Phi_{10}\ddot{\Phi}_{10} + f_{00}\left(1 - \dot{\Phi}_{10}^2\right) = 0$$

which expresses a locallysimilar solution with arbitrary distribution of velocities on the outer edge of the boundary layer $U(x)$, but in a nonconducting fluid. In conformity with that or this localization the functions $\bar{f}_{00}(g)$ or $\bar{f}_{10}(f)$ are determined.

The tables and graphs presented by Boričič vividly show that the magnetic factor g, coinciding with the square of the Hartman number, where the momentum loss thickness δ^{**} is taken as the characteristic length, considerably affects the flow of electrically conducting fluid in a MHD-boundary layer. The reduced coefficient of friction ζ grows with the parameter g and the boundary layer separation is delayed. This is vividly seen in Figure 188 where the numerals 1, 2,...,5 on the curves denote the following values of g_1: 1, $g_1 = 0$; 2, $g_1 =$ = 0.05 3, $g_1 = 0.10$; 4, $g_1 = 0.15$; 5, $g_1 = 0.20$.

The main purpose of this section is the demonstration of the possibilities of applying the method of the generalized similarity to the calculation of laminar boundary layers under different physical conditions (surface permeability, the presence of electric conductivity of fluid moving in a magnetic field). This trend left no place for detailed quantitative description of separation phenomena; in view of this we had to restrict ourselves to the composition of universal equations and to the analysis of some results of their numerical solution, in particular, of the problem of the delay of boundary layer separation.

The method of generalized similarity is successfully applied not only in the theory of laminar boundary layers, but also in many other problems.

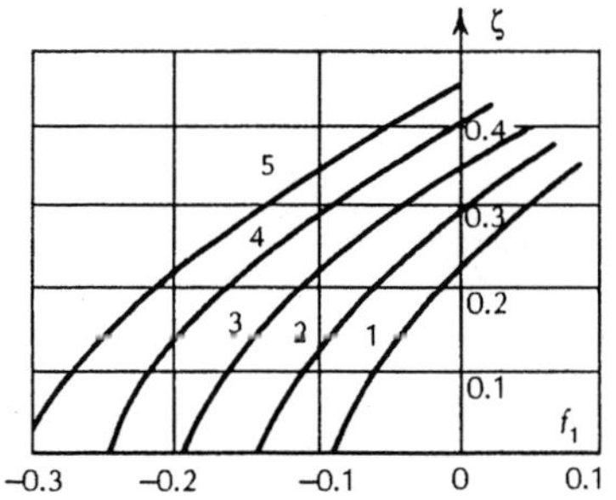

Fig. 188

Stepanyants used the method of generalized similarity to the inverse problem of the calculation of velocity and temperature distributions by the specified layer thickness distribution between two solid surfaces and to the problems of film flow in the boundary layer approximation. He considered the following problems: (1) laminar nonisothermal viscous gas flow through thin tubes of an alternative cross-sections,[45] (2) hydrodynamic theory of lubrication at large Reynolds numbers,[46] (3) viscous gas flow in a thin layer,[47] (4) heat transfer in a film flow.[48] Under all these conditions the corresponding universal equations were formulated and new results were obtained that extended the ranges of the applicability of the method of generalized similarity.

11. Spatial Near-Wall Boundary Layers. Free Spatial Jets

This book presents no opportunity to go into the detail of multiple problems of the laminar boundary layer theory that are combined in certain special guides (already cited). We present only some examples of the simplest spatial problems and give the notion of nonstationary boundary layer flows. In Section 12 we shall discuss temperature boundary layers in nonisothermal motions of fluid with homogeneous physical composition and the notion of boundary concentration layers in the flows with admixtures.

The simplest example of a spatial boundary layer gives a longitudinal axisymmetric flow past a body of revolution. Similar to a plane case, x may be reckoned along the body contour and y along the normal to it (Figure 189). These coordinates may be considered as rectilinear, and the radius-vector r if the point M relative to the body axis may be considered, with sufficient approximation, coinciding with the radius of transverse curvature of the body $r_0(x)$ in the corresponding cross-section of the body normal to its axis.[49] With such an approach the basic boundary layer equation will remain the form similar to that in a plane case, and the continuity equation will take on the form typical of the longitudinal axisymmetric motion in cylindrical coordinates

[45] Smirnov, B. I. and Stepanyants, K. G., Generalized similarity of laminar non-isothermal viscous gas flow in the tubes of alternative cross-sections, *J. Eng. Phys.* 42 (2), 285–289, 1982.

[46] Smaga, T. I., Smirnov,B. I., and Stepanyants, L. G., On one method of the solution of the hydrodynamic theory of lubrication at large Reynolds numbers, *Izv. AN BSSR, Ser. Fiz-Energ. Nauk*, No. 1, 119–124, 1978.

[47] Smirnov, B. I. and Stepanyants, L. G., Method of generalized similarity in the problems of viscous gas flow in a thin layer, *J. Eng. Phys*, 40 (2), 204–213, 1981.

[48] Kotelnikova, O. P. and Stepanyants, L. G., Parametric method for solving heat transfer problems in liquid film flow, *J. Eng. Phys.*, 44 (4), 632–636, 1983.

[49] Denial of this simplifying assumption would make calculation substantially difficult. [See Loitsyanskiy, L. G. *Laminarnyy pogranichnyy sloy* (*Laminar Boundary Layer*), Fizmatgiz, Moscow, 1962, 154–167

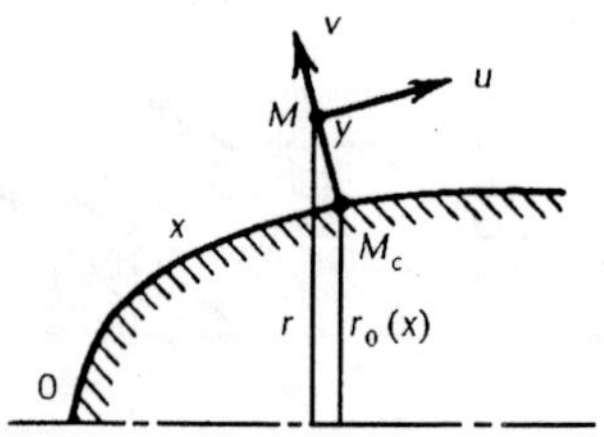

Fig. 189

$$\frac{\partial(ru)}{\partial x}+\frac{\partial(rv)}{\partial y}=0$$

but with the substitution of the coordinate r into $r_0(x)$. The basic system of equations in this case is

$$u\frac{\partial u}{\partial x}+v\frac{\partial u}{\partial y}=U\frac{dU}{dx}+\nu\frac{\partial^2 u}{\partial y^2}$$

$$\frac{\partial(r_0 u)}{\partial x}+\frac{\partial(r_0 v)}{\partial y}=0$$

$$u=v=0 \quad \text{at} \quad y=0 \qquad u\to U(x) \quad \text{when} \quad y\to\infty \tag{165}$$

By the transformation[50] equations (165) (the prime denotes the derivative with respect to x)

$$\bar{x}=\int_0^x r_0^2(\xi)\,d\xi \qquad \bar{y}=r_0(x)\,y$$

$$\bar{u}=u \qquad \bar{U}=U \qquad \bar{v}=\frac{v}{r_0}+\frac{r_0'(x)\,yu}{r_0^2} \tag{166}$$

with the transition equations

$$\frac{\partial}{\partial x}=r_0^2\frac{\partial}{\partial\bar{x}}+r_0'y\frac{\partial}{\partial\bar{y}} \qquad \frac{\partial}{\partial y}=r_0\frac{\partial}{\partial\bar{y}}$$

are converted to the equations

$$\bar{u}\frac{\partial\bar{u}}{\partial\bar{x}}+\bar{v}\frac{\partial\bar{u}}{\partial\bar{y}}=\bar{U}\frac{d\bar{U}}{d\bar{x}}+\nu\frac{\partial^2\bar{u}}{\partial\bar{y}^2}$$

[50] Stepanov, E. I., On the integration of laminar boundary layer equations for motion with axias symmetry, Prikl. Mat. Mekh., 2 (1), 1947; Magler, W., *Ber. Aerodyn. Vers., Anst. Göettingen*, 45/A/17, 1945, and *Math. Mech.*, 28, 97–103, 1948.

$$\frac{\partial \bar{u}}{\partial \bar{x}} + \frac{\partial \bar{v}}{\partial \bar{y}} = 0$$

$$\bar{u} = \bar{v} = 0 \quad \text{at} \quad \bar{y} = 0 \qquad \bar{u} \to \bar{U}(\bar{x}) \quad \text{when} \quad \bar{y} \to \infty \tag{167}$$

which do not differ from the plane boundary layer equations.Thus, in the calculations of boundary layers on bodies of revolution, with axisymmetric longitudinal flow past them, the problem is reduced to the determination of the prescribed velocity distribution $U(x)$ on the outer edge of the boundary layer that is transformed to new coordinates (166) and subsequent consideration of the plane boundary layer.

So, for instance, we shall apply transformation (166) to the case of a boundary layer forming near a front critical point of the body of revolution with a longitudinal flow past it. The velocity distribution in the outer flow near the critical point on the body of revolution will be linear ($U = cx$); it also may be considered in this region that $r_0(x) = x$. Then transformation (166) will be reduced to the following

$$\bar{x} = \frac{1}{3}x^3 \qquad \bar{y} = xy \qquad \bar{u} = u \qquad \bar{v} = \frac{v}{x} + \frac{yu}{x^2} \qquad \bar{U} = c\sqrt[3]{3\bar{x}} = c_1\bar{x}^{1/3}$$

As can be judged by the last equality the considered axisymmetric problem is reduced to the plane self-similar Falkner-Skan problem with the power index $m = 1/3$ ($\beta = 1/2$) given earlier. The tables of this problem solution were given in Section 7. Making use of Table 15 we may determine all the values of interest in this spatial problem

$$\delta^* = \frac{\bar{\delta}^*}{r_0} \qquad \delta^{**} = \frac{\bar{\delta}^{**}}{r_0} \qquad \tau_w = r_0\bar{\tau}_w$$

Applying transformation (166), we introduce the analogue of the integral von Kármán relation to the case of axisymmetric boundary layer on a body of revolution in a longitudinal flow. If use is made of the new variables (166), then, as follows from Equation 167, the von Kármán equation for an axisymmetric flow should have the form of Equation 106, similar to the plane flow

$$\frac{d\bar{\delta}^{**}}{d\bar{x}} + \frac{\bar{U}'_{\bar{x}}\bar{\delta}^{**}}{\bar{U}}(2 + \bar{H}) = \frac{\bar{\tau}_w}{\rho\bar{U}^2} \tag{168}$$

where

$$\bar{\delta}^{**} = \int_0^\infty \frac{\bar{u}}{\bar{U}}\left(1 - \frac{\bar{u}}{\bar{U}}\right)d\bar{y} = r_0\int_0^\infty \frac{u}{U}\left(1 - \frac{u}{U}\right)dy = r_0\delta^{**}$$

and similarly

$$\bar{\delta}^* = \int_0^\infty \left(1 - \frac{\bar{u}}{\bar{U}}\right)d\bar{y} = r_0\int_0^\infty \left(1 - \frac{u}{U}\right)dy = r_0\delta^*$$

so that

$$\bar{H} = \frac{\bar{\delta}^*}{\bar{\delta}^{**}} = \frac{\delta^*}{\delta^{**}} = H$$

The skin friction $\bar{\tau}_w$ will be equal to

$$\bar{\tau}_w = \mu\left(\frac{\partial\bar{u}}{\partial\bar{y}}\right)_{\bar{y}=0} = \frac{\mu}{r_0}\left(\frac{\partial u}{\partial y}\right)_{y=0} = \frac{\tau_w}{r_0}$$

Substituting the obtained expressions of $\bar{\delta}^{**}$, $\bar{\delta}^*$, $\bar{H}$, and $\bar{\tau}_w$ into Equation 168, we find

$$\frac{1}{r_0^2(x)}\frac{d}{dx}(r_0\delta^{**}) + \frac{U_x' r_0\delta^{**}}{Ur_0^2(x)}(2+H) = \frac{\tau_w}{\rho r_0 U^2}$$

and expanding the differentiation in the first term of the left side, we obtain the unknown expression of the integral von Kármán relation for an axisymmetric boundary layer on the body of revolution

$$\frac{d\delta^{**}}{dx} + \frac{U'\delta^{**}}{U}(2+H) + \frac{r_0'}{r_0}\delta^{**} = \frac{\tau_w}{\rho U^2} \tag{169}$$

which differs from the similar equation for a plane case by the last term $(r_0'\delta^{**}/r_0)$ on the left side. The same result may also be obtained by rewriting the first equation of the system (165) on the basis of the second one in the form

$$\frac{\partial(r_0u^2)}{\partial x} + \frac{\partial(r_0uv)}{\partial y} = r_0UU' + \nu r_0\frac{\partial^2 u}{\partial y^2}$$

and the second equation after the multiplication of both sides by U as

$$\frac{\partial(r_0Uu)}{\partial x} + \frac{\partial(r_0Uv)}{\partial y} = r_0uU'$$

Substracting term-by-term these two equations one from the other, we obtain

$$\frac{\partial}{\partial x}[r_0u(U-u)] + \frac{\partial}{\partial y}[r_0v(U-u)] = r_0U'(u-U) - \nu r_0\frac{\partial^2 u}{\partial y^2}$$

Integrating both sides with respect to y from $y = 0$ to $y = \infty$, we find

$$\frac{d}{dx}(r_0U^2\delta^{**}) + r_0U\frac{dU}{dx}\delta^* = r_0\frac{\tau_w}{\rho}$$

and expanding the expression of the derivative in the first term on the left, we again come to Equation 169. These results are valid under the assumption that $r = r_0$ or under the equivalent assumption regarding the smallness of the ratio δ/r_0. On approaching the rear part of the body of revolution, when δ grows and r_0 decreases, the considerations mentioned loose their force.

We give an example of the simplest one-term calculation of a boundary layer on a slender body of revolution. Retaining the previous notation, we shall multiply, as was done in the case of a plane boundary layer, both sides of Equation 169 into $U\delta^{**}/\nu$. Having then transformed its left side, we obtain the analogue of the integral von Kármán relation

$$\frac{df}{dx} = \frac{U'}{U}F(f) + \left(\frac{U''}{U'} - \frac{2r_0'}{r_0}\right)f$$

that differs from the plane case only by the presence of the term $(-2r_0'/r_0)$ within the round brackets on the right side of the equation. Here, the function $F(f) = \bar{f}$ is similar to that in the plane case. Substituting it into the linear representation

$$F(f) = a - bf$$

we find the following formula for the parameter f:

$$f = \frac{aU'(x)}{U^b(x)\, r_0^2(x)} \int_0^x U^{b-1}(t)\, r_0^2(t)\, dt$$

thus solving the problem in the simplest one-term approximation.[51]

The presence of transformation (166) of an axisymmetric flow in a boundary layer on slender bodies of revolution into a plane one makes the special presentation of the method of generalized similarity with reference to this case redundant.

To the class of the problems on spatial "near-wall" laminar boundary layer belongs the problems on viscous fluid flow around a rotating disk in infinite or bounded volumes of fluid. The first of them has an exact solution based on the Navier-Stokes equations (see the von Kármán problem in Chapter 11). The second problem, as with many others dealing with boundary layers on rotating bodies (round cylinder, sphere), were the concern of the monograph by Dorfman.[52] Of great interest is a boundary layer flow that is induced around a plane by a vortex filament normal to this plane (a water-spout model). This problem is detailed in the monograph by Goldshtik.[53]

Literature on spatial (three-dimensional) boundary layers is voluminous.[54]

Propagation of an axisymmetric non swirled jet issuing from an infinitely thin hole into infinite space filled in with the same fluid may serve as a simple example of a self-similar flow in a "free" axisymmetric boundary layer.[55]

General equations of a stationary axisymmetric viscous fluid flow in cylindrical coordinates r, ε, x will be obtained from formulas (33) (Chapter 10) discarding the derivatives with

---

[51] Loitsyanskiy, L. G., Laminar boundary layer on the body of revolution, *Dokl. AN SSSR*, 36 (6), 1942.

[52] Dorfman, L. A., *Gidrodinamicheskoe soprotivleniye i teplootdacha (Hydrodynamic Resistance and Heat Transfer)*, Fizmatgiz Press, Moscow, 1960.

[53] Goldshtik, M. A., *Vikhrevye potoki (Vortical Flows)*, Nauka Press, Novosibirsk, 1981, 190–235.

[54] Bogdanova, V. V., Universal equations in the theory of a spatial boundary layer, *Mekh. Zhidk Gaza*, No. 6, 1968; Karyakin, Yu. E., Magnetohydrodynamic laminar boundary layer in an axisymmetric swirled flow, *Magnitnaya Gidrodinamika*, No. 2, 1968; Struminskiy, V. V., Wing sliding in viscous fluid, *Dokl. AN SSSR*, 54(7), 575–578, 1946; Sliding of wing in viscous and compressible gas, *Dokl. AN SSSR*, 54(9), 769–772; Struminskiy, V. V., *Theory of a spatial boundary layer on a sliding wing, in Sbornik teoreticheskhikh rabot po aerodinamike (Collection of Theoretical Papers on Aerodynamics)* (TsAGI), Oborongiz Press, Moscow, 1957, 174–205; Sedney, R. Some aspects of three-dimensional boundary layers, *Q. Appl. Math.*, 15(2), 1957; Rott, N. and Crabtree, L., Simplified laminar boundary layers calculations for bodies of revolution and for yawed wing, *J. Aeron. Sci.*, 19(8), 556–560, 1952; Loos, H. G., A simple laminar boundary layer with secondary flow, *J. Aeron. Sci.*, 22, 35–40, 1955; Shevelev, Yu. D., *Trekhmernye zadachi teorii laminarnogo pogranichnogo sloya (Three-Dimensional Problems in the Laminar Boundary Layer Theory)*, Nauka Press, Moscow, 1977.

[55] Schlichting, H., Laminare Strahlausbreitung, *Z. Angew. Math. Mech.*, 13, 260, 1933.

respect to t and ε.

We shall have, denoting the axial V_x, radial V_r, and transverse V_ε velocity components in terms of u, v, and w, respectively

$$u\frac{\partial u}{\partial x}+v\frac{\partial u}{\partial r} = -\frac{1}{\rho}\frac{\partial p}{\partial x}+\nu\left(\frac{\partial^2 u}{\partial r^2}+\frac{1}{r}\frac{\partial u}{\partial r}+\frac{\partial^2 u}{\partial x^2}\right)$$

$$u\frac{\partial v}{\partial x}+v\frac{\partial v}{\partial r}-\frac{w^2}{r} = -\frac{1}{\rho}\frac{\partial p}{\partial r}+\nu\left(\frac{\partial^2 v}{\partial r^2}+\frac{1}{r}\frac{\partial v}{\partial r}-\frac{v}{r^2}+\frac{\partial^2 v}{\partial x^2}\right)$$

$$u\frac{\partial w}{\partial x}+v\frac{\partial w}{\partial r}+\frac{vw}{r} = \nu\left(\frac{\partial^2 w}{\partial r^2}+\frac{1}{r}\frac{\partial w}{\partial r}-\frac{w}{r^2}+\frac{\partial^2 w}{\partial x^2}\right)$$

$$\frac{\partial(ru)}{\partial x}+\frac{\partial(rv)}{\partial r} = 0$$

Treating the jet region as a boundary layer, the transverse dimension of which is small at large Reynolds numbers, we shall assume the radial velocity v is small as compared to the longitudinal u and transverse w. We shall also discard $\partial^2 u/\partial x^2$ and $\partial^2 w/\partial x^2$ in the brackets on the right as compared with radial derivatives. Then we obtain the following equations of axisymmetric jet propagation which are general for the cases of nonswirled and swirled jets

$$u\frac{\partial u}{\partial x}+v\frac{\partial u}{\partial r} = -\frac{1}{\rho}\frac{\partial p}{\partial x}+\nu\left(\frac{\partial^2 u}{\partial r^2}+\frac{1}{r}\frac{\partial u}{\partial r}\right) \qquad \frac{\partial p}{\partial r} = \frac{\rho w^2}{r}$$

$$u\frac{\partial w}{\partial x}+v\frac{\partial w}{\partial r}+\frac{vw}{r} = \nu\left(\frac{\partial^2 w}{\partial r^2}+\frac{1}{r}\frac{\partial w}{\partial r}-\frac{w}{r^2}\right)$$

$$\frac{\partial(ru)}{\partial x}+\frac{\partial(rv)}{\partial r} = 0 \qquad (170)$$

In the case considered the flow will occur in a meridian plane, so that

$$w = 0 \qquad \frac{\partial p}{\partial r} = 0 \qquad p = p(x) = \text{const} \qquad (171)$$

the latter is the consequence of the equality of pressure in an infinite space surrounding a jet. When equalities (171) are satisfied Equations 170 are simplified and reduced to the following

$$u\frac{\partial u}{\partial x}+v\frac{\partial u}{\partial r} = \nu\left(\frac{\partial^2 u}{\partial r^2}+\frac{1}{r}\frac{\partial u}{\partial r}\right) \qquad \frac{\partial(ru)}{\partial x}+\frac{\partial(rv)}{\partial r} = 0 \qquad (172)$$

As it directly follows from the considerations of dimensionality, the solution of Equations 172 for the case of a nonswirled jet issuing from an infinitely thin hole with zero flow rate and finite momentum will be self-similar. In fact, in the case of extremely large Reynolds numbers the momentum flux which is equal for all cross-sections is determined as

$$J_0 = \rho\int_0^\infty 2\pi r u^2 dr = \text{const} \qquad (173)$$

and will have the order ($\mathbf{Re} = UL/\nu$)

$$\rho U^2 \cdot L^2/\mathbf{Re} = \mu UL$$

and the order of the stream function $\psi = \int_0^r rudr$ will be

$$UL^2/\mathbf{Re} = \nu L$$

Consequently, the solution of the problem should have the form

$$\psi = \nu Lf\left(\frac{x}{L}, \frac{r}{L}\sqrt{\frac{UL}{\nu}}\right) = \nu Lf\left(\frac{x}{L}, \frac{r}{L}\sqrt{\frac{J_0}{\rho\nu^2}}\right)$$

and by the condition of lack of dependence on L

$$\psi = \nu L \cdot \frac{x}{L}\Theta\left(\frac{r/L\sqrt{J_0/\rho\nu^2}}{x/L}\right) = \nu x\Theta\left(\sqrt{\frac{J_0}{\rho\nu^2}}\frac{r}{x}\right) \qquad (174)$$

thus proving self-similarity of the problem.

To simplify calculations we seek the stream function in the form

$$\psi = \nu x a(\eta) \qquad \eta = \frac{r}{x\sqrt{\nu}} \qquad (175)$$

We find (the prime denotes the derivative with respect to η)

$$u = \frac{1}{x}\frac{a'}{\eta} \qquad v = \frac{\sqrt{\nu}}{x}\left(a' - \frac{a}{\eta}\right)$$

$$\frac{\partial u}{\partial r} = \frac{1}{x^2\sqrt{\nu}}\left(\frac{a'}{\eta}\right)' \qquad \frac{\partial^2 u}{\partial r^2} = \frac{1}{\nu x^3}\left(\frac{a'}{\eta}\right)'' \qquad \frac{\partial u}{\partial x} = -\frac{1}{x^2}\left[\eta\left(\frac{a'}{\eta}\right)' + \frac{a'}{\eta}\right]$$

and substituting (172) into the first equation of the system, we obtain the unsolved ordinary differential equation of the third order to determine $a(\eta)$

$$\left(\frac{a'}{\eta}\right)'' + \frac{1+a}{\eta}\left(\frac{a'}{\eta}\right)' + \left(\frac{a'}{\eta}\right)^2 = 0 \qquad (176)$$

The boundary conditions will be

$$a = 0, \quad a' = 0 \;\; \text{at} \;\; \eta = 0$$

with the second condition following from the condition of the boundedness of longitudinal velocity $u = a'/(x\eta)$ on the axis. Moreover, we have the condition of the boundedness of a when $\eta \to \infty$, since according to Equation 175 a is proportional to fluid volumetric flow rate through a given jet cross-section.

Equation 176 is once integrated directly and gives

$$\eta a'' - a' + aa' = 0$$

Introducing a new variable $\zeta = \ln \eta$ instead of η, we obtain the equation

$$\frac{d^2 a}{d\zeta^2} - 2\frac{da}{d\zeta} + a\frac{da}{d\zeta} = 0$$

which being once integrated under the adopted boundary conditions is reduced to the following

$$\frac{da}{d\zeta} = \frac{4a - a^2}{2}$$

Integrating once again, we obtain

$$a(\eta) = \frac{a^2\eta^2}{1 + \frac{1}{4}a^2\eta^2} \tag{177}$$

To determine the constant of integration α it remains only to use the integral condition (173), which with the aid of Equation 175, is rewritten in the form

$$\int_0^{\infty} [a'(\eta)]^2 \frac{d\eta}{\eta} = \frac{J_0}{2\pi\mu}$$

Having calculated by Equation 177

$$a' = \frac{2\alpha^2\eta}{\left(1 + \frac{1}{4}\alpha^2\eta^2\right)^2}$$

and substituted it into the previous relation, we obtain

$$\alpha = \sqrt{\frac{3}{16}\frac{J_0}{\pi\mu}}$$

Reverting to Equation 175 and to subsequent formulas for u and v, we obtain the unknown solution of the problem

$$\psi = \nu x \frac{\alpha^2\eta^2}{1 + \frac{1}{4}\alpha^2\eta^2} \qquad u = \frac{2\alpha^2}{x}\frac{1}{\left(1 + \frac{1}{4}\alpha^2\eta^2\right)^2}$$

$$v = \frac{\alpha\sqrt{\nu}}{x}\frac{\alpha\eta\left(1 - \frac{1}{4}\alpha^2\eta^2\right)}{\left(1 + \frac{1}{4}\alpha^2\eta^2\right)^2} \tag{178}$$

The shape of stream lines and the profiles of longitudinal velocity in the considered case of an axisymmetric flow is generally similar to that in the plane case.

According to Equation 178 the fluid mass flow rate through the given jet cross-section is equal to

$$M = 2\pi\rho\int_0^\infty urdr = 2\pi\rho\psi(\infty) = 8\pi\rho\nu x = 8\pi\mu x \tag{179}$$

It is interesting to note that this flow rate does not depends on the momentum flux that characterizes the given jet, but rather on the fluid viscosity. The momentum grows proportionally to the distance from the jet source.

Distribution of maximum velocity on the axis ($\eta = 0$) by Equation 178 is presented by the expression

$$u_{\max} = \frac{2\alpha^2}{x} = \frac{3}{8}\frac{J_0}{\pi\mu x} \tag{180}$$

Determining the jet boundary as a geometrical position of the points where the ratio $u/u_{\max}$ retains some small but constant value, we shall satisfy ourselves that a right circular cone with the semivertex angle proportional to $\sqrt{\nu^2\rho/J_0}$ will serve as such a conditional boundary of an axisymmetric jet. Noting that the dimensionless quantity $J_0/(\rho\nu^2)$ plays the role of the Reynolds number $\mathbf{Re} = UL/\nu$, we verify that the conditional width of the jet diminishes with the growth of the Reynolds number by the law $1/\sqrt{\mathbf{Re}}$, thus confirming the possibility of applying the boundary layer equations in this case.

As an example of more complex nonself-similar flow we consider the problem of the propagation of a laminar swirled axisymmetric jet in the space submerged into the same but quiescent fluid.[56] In this case we manage to obtain the solution in the form of an asymptotic series constructed by the inverse powers of the jet cross-section distance from the jet source.

The equations of the viscous fluid motion in a boundary layer – a swirled jet – are presented by the system (170). They involve, along with the longitudinal u and the lateral v components, the transversal velocity component w that characterizes jet twisting. Though in the outer flow, according to the problem conditions, pressure is equal everywhere, there is still radial pressure drop in the jet itself that counterbalances centrifugal forces caused by jet twisting. This drop is related to the transversal velocity by the second equality of the system (170). Its presence also causes pressure variability along the axis, thus making it impossible to neglect the term $\partial p/\partial x$ in the first equation of this system.

Two main conservation laws may be derived from the system of equations (170): momentum and moment of momentum conservation, from which two integral conditions of nontriviality of solutions will follow.

Making use of the last equation of the system (170) (the continuity equation), after simple transformations we rewrite the first equation of this syntem in the form

$$\frac{\partial}{\partial x}[r(p+\rho u^2)] + \frac{\partial}{\partial r}\left[r\left(\rho uv - \mu\frac{\partial u}{\partial r}\right)\right] = 0$$

and, integrating across a jet, we obtain

$$\frac{d}{dx}\int_0^\infty r(p+\rho u^2)\,dr + \left[r\left(\rho uv - \mu\frac{\partial u}{\partial r}\right)\right]_{r=0}^{r=\infty} = 0$$

---

[56] Loitsyanskiy, L. G., Propagation of a swirled jet in an infinite space submerged into the same fluid, *Prikl. Mat. Mech.*, 17(1), 1953.

Here p means the pressure at the jet point relative to the pressure outside the jet. In this case the existence of the integral on the left side is assumed. If we consider (this may be verified by the solution obtained later) that with the growth of r, u and $\partial u/\partial r$ decrease rather quickly, then the substitution vanishes and the previous equality leads to the first integral condition of the problem

$$2\pi\int_0^\infty r(p+\rho u^2)\,dr = \text{const} = J_0 \tag{181}$$

which generalizes the condition (173) to the case of a swirled jet. The left side of this equality will be called the flow of total momentum through the jet cross-ection. When $x \to \infty$, $p \to 0$ and the constant in Equation 181 coincides with the constant of a nonswirled jet.

Note that the third equation of the system (170), with both sides being multiplied into r, and the continuity equation used, may be rewritten in the form

$$\frac{\partial(ruw)}{\partial x} + \frac{\partial(rvw)}{\partial r} + vw = \nu r\frac{\partial}{\partial r}\left[\frac{1}{r}\frac{\partial}{\partial r}(rw)\right]$$

or being remultiplied into r it is

$$\frac{\partial}{\partial x}(r^2uw) + \frac{\partial}{\partial r}(r^2vw) = \nu\left\{\frac{\partial}{\partial r}\left[r\frac{\partial}{\partial r}(rw)\right] - 2\frac{\partial}{\partial r}(rw)\right\}$$

We shall integrate both sides of the last equality across the jet. Then, with the asumptions on the rate of the decrease of u and w with the growth of r, that are similar to the previous ones, we obtain

$$\frac{d}{dx}\int_0^\infty r^2uw\,dr = 0$$

or multiplying into the constant $2\pi\rho$ and integrating

$$2\pi\rho\int_0^\infty r^2uw\,dr = \text{const} = L_0 \tag{182}$$

The left side is the transfer of the main moment of momentum through the jet cross-section. This moment L_0 is the same along the entire jet. Consequently, it may serve as a measure of jet twisting.

The presence of the two constant quantities J_0 and L_0 along a swirled jet, that ambiguously determine characteristic length and velocity, serves as a barrier to the possibility of reducing Equations 170 to one ordinary equation, making the problem nonself-similar.

In this case we introduce the stream function ψ of the meridian flow and will seek it in the form of the expansion

$$\psi = \nu\left(ax + a_0 + \frac{a_1}{x} + \ldots\right) \tag{183}$$

where a, a_0, a_1 ... are the unknown functions of the same variable η as that used in the case of a nonswirled jet. Then, the velocity components in meridional cross-sections are (the prime again stands for the derivative with respect to η)

$$u = \frac{1}{r}\frac{\partial\psi}{\partial r} = \frac{a'}{\eta}\frac{1}{x} + \frac{a_0'}{\eta}\frac{1}{x^2} + \frac{a_1'}{\eta}\frac{1}{x^3} + \dots$$

$$v = -\frac{1}{r}\frac{\partial\psi}{\partial x} = \frac{\sqrt{\nu}}{x}\left[a' - \frac{a}{\eta} + a_0'\frac{1}{x} + \left(a_1' + \frac{a_1}{\eta}\right)\frac{1}{x^2} + \dots\right] \tag{184}$$

We shall specify the peripheral speed w by the series

$$w = \frac{b_1}{x} + \frac{b_2}{x^2} + \dots \tag{185}$$

and, finally, introduce also the expaneion for the pressure p relative to the pressure of the quiescent fluid surrounding the jet

$$\frac{p}{\rho} = \frac{c_1}{x} + \frac{c_2}{x^2} + \dots \tag{186}$$

Here, b_1, b_2, ..., c_1, c_2, ... are the unknown functions of the variable η.

Substituting the expansions mentioned into the system of equations (170) and equating the coefficients at equal powers of x, we obtain the system of ordinary differential equations for determining the unknown coefficients.

We obtain from the first equation of the system (170)

$$c_1 + \eta c_1' = 0$$

$$\left(\frac{a'}{\eta}\right)'' + \frac{1+a}{\eta}\left(\frac{a'}{\eta}\right)' + \left(\frac{a'}{\eta}\right)^2 + 2c_2 + \eta c_2' = 0$$

$$\left(\frac{a_0'}{\eta}\right)'' + \frac{1+a}{\eta}\left(\frac{a_0'}{\eta}\right)' + \frac{3a'a_0'}{\eta\,\eta} + 3c_3 + \eta c_3' = 0$$

$$\dots\dots\dots\dots\dots\dots\dots\dots \tag{187}$$

From the second equation of this system we find

$$\eta c_1' = 0 \qquad \eta c_2' = b_1^2 \qquad \eta c_3' = 2b_1 b_2 \qquad \eta c_4' = b_2^2 + 2b_1 b_3, \quad \dots \tag{188}$$

and from the third

$$b_1'' + \frac{1+a}{\eta}b_1' - \frac{1-a}{\eta^2}b_1 = 0$$

$$b_2'' + \frac{1+a}{\eta}b_2' - \frac{1-a-\eta a'}{\eta^2}b_2 = 0$$

$$\dots\dots\dots\dots\dots\dots\dots\dots \tag{189}$$

Moreover, similar operation with the integral conditions (181) and (182) will lead to the following system of integral conditions

$$\int_0^\infty \eta c_1 d\eta = 0 \qquad \int_0^\infty \left(\eta c_2 + \frac{a'^2}{\eta}\right) d\eta = \frac{J_0}{2\pi\mu}$$

$$\int_0^\infty \left(\eta c_3 + \frac{2a'a_0'}{\eta}\right) d\eta = 0 \qquad \int_0^\infty \left(\eta c_4 + \frac{a_0'^2 + 2aa_1'}{\eta}\right) d\eta = 0, \quad \dots$$

$$\int_0^\infty \eta a' b_1 d\eta = 0 \qquad \int_0^\infty \eta\,(a'b_2 + a_0'b_1)\,d\eta = \frac{L_0}{2\pi\mu\sqrt{\nu}}, \quad \dots \tag{190}$$

Taking the jet axis $\eta = 0$ as the zero streamline and using the condition of the velocity on tne axis, we shall find the boundary conditions

$$a = a_0 = a_1 = \dots = 0 \qquad a' = a_0' = a_1' = \dots = 0 \quad \text{at} \quad \eta = 0 \tag{191}$$

and from the condition of finiteness of the volumetric flow rate through the cross-section or the values of ψ at $\eta = \infty$, we obtain the condition of the boundedness of the quanties

$$a(\infty), \quad a_0(\infty), \quad \dots \tag{192}$$

Further, by the definition of the twisting velocity we have

$$b_1 = b_2 = \dots = 0 \quad \text{at} \quad \eta = 0$$

$$b_1 = b_2 = \dots = 0 \quad \text{at} \quad \eta = \infty \tag{193}$$

and, finally, by the definition of pressure

$$c_1 = c_2 = \dots = 0 \quad \text{at} \quad \eta = \infty \tag{194}$$

Note first of all that from the first equation of the system (189) there directly follows the existence of the solution $b_1 \equiv 0$ satisfying the boundary conditions (193) and the integral conditions (190). Then, it follows from the first three equations of the system (188) and from the boundary conditions (194) that

$$c_1 = c_2 = c_3 = 0$$

In this case it is easy to see that the function $a(\eta)$ satisfies the same equation and the same boundary conditions and integral conditions as that in the case of a nonswirled jet. Accurate up to small values of the order of $1/x^2$ the difference is reduced to the terms involving the functions $a_0(\eta)$ and $b_2(\eta)$ which can be easily found.

Assuming in the third equation of the system (187)

$$\frac{a_0'}{\eta} = A \tag{195}$$

and passing over from the argument η to the new argument $\xi = a/4$ related to that previously determined by

$$\xi = \frac{\frac{1}{4}\alpha^2\eta^2}{1 + \frac{1}{4}\alpha^2\eta^2}$$

we come to the hypergeometric equation

$$\xi(\xi - 1)\frac{d^2A}{d\xi^2} - \frac{dA}{d\xi} - 6A = 0$$

the regular soluticn of which under the evident boundary conditions

$$\xi = 0 \quad \text{at} \quad \eta = 0 \qquad \xi = 1 \quad \text{at} \quad \eta = \infty$$

will be

$$A(\xi) = C(1 - \xi)^2(1 - 4\xi)$$

Referring then to Equation 195 and repeating integration, we obtain

$$a_0(\eta) = \beta\xi(2\xi - 1) = -\beta\frac{\frac{1}{4}\alpha^2\eta^2\left(1 - \frac{1}{4}\alpha^2\eta^2\right)}{\left(1 + \frac{1}{4}\alpha^2\eta^2\right)^2}$$

where β is the integration constant.

The function $a_0(\eta)$ satisfies the conditions (191) and (192).

To determine $b_2(\eta)$ we integrate the second equation of the system (189) that is presented in the form

$$\eta^2 b_2'' + \eta b_2' - b_2 + \eta a b_2' + \eta a' b_2 + a b_2 = (\eta^2 b_2' - \eta b_2 + \eta a b_2)' = 0$$

We obtain

$$\eta b_2' = (1 - a)b_2$$

where it follows

$$b_2(\eta) = \gamma\frac{\alpha\eta}{\left(1 + \frac{1}{4}\alpha^2\eta^2\right)^2} \tag{196}$$

with the integration constant γ being determined from the last integration of system (190), which at $b_1 = 0$ is rewritten in the form

$$\int_0^\infty \eta a' b_2 d\eta = \frac{L_0}{2\pi\mu\sqrt{\nu}}$$

Having integrated, we find

$$\gamma = \frac{3\sqrt{3}}{64\pi\sqrt{\pi}} \frac{L_0\sqrt{\rho J_0}}{\mu^2} \tag{197}$$

The pressure is characterized by the function $c_4(\eta)$, which satisfies the last equation of system (188) at $b_1 = 0$ and is equal to

$$c_4 = -\frac{2}{3}\gamma^2 \frac{1}{\left(1+\frac{1}{4}\alpha^2\eta^2\right)^3}$$

Making use of the obtained expressions of the unknown functions, we arrive at the following approximate formulas of the theory of a swirled jet (u_{max} stands for the velocity on the jet axis)

$$\psi = \nu\left[\frac{\alpha^2\eta^2}{1+\frac{1}{4}\alpha^2\eta^2}x - \beta\frac{\frac{1}{4}\alpha^2\eta^2\left(1-\frac{1}{4}\alpha^2\eta^2\right)}{\left(1+\frac{1}{4}\alpha^2\eta^2\right)^2}\right]$$

$$u = \frac{2\alpha^2}{\left(1+\frac{1}{4}\alpha^2\eta^2\right)^2}\frac{1}{x} - \frac{1}{2}\beta\alpha^2\frac{1-\frac{3}{4}\alpha^2\eta^2}{\left(1+\frac{1}{4}\alpha^2\eta^2\right)^3}\frac{1}{x^2}$$

$$v = \sqrt{\nu}\left[\frac{\alpha^2\eta\left(1-\frac{1}{4}\alpha^2\eta^2\right)}{\left(1+\frac{1}{4}\alpha^2\eta^2\right)^2}\frac{1}{x} - \frac{1}{2}\beta\alpha^2\frac{\eta\left(1-\frac{3}{4}\alpha^2\eta^2\right)}{\left(1+\frac{1}{4}\alpha^2\eta^2\right)^3}\frac{1}{x^2}\right]$$

$$w = \gamma\frac{\alpha\eta}{\left(1+\frac{1}{4}\alpha^2\eta^2\right)^2}\frac{1}{x^2}$$

$$p = -\frac{2}{3}\gamma^2\frac{1}{\left(1+\frac{1}{4}\alpha^2\eta^2\right)^3}\frac{1}{x^4} \qquad u_{max} = \frac{2\alpha^2}{x} - \frac{1}{2}\beta\alpha^2\frac{1}{x^2}$$

$$Q = 2\pi\mu(4x+\beta) \tag{198}$$

The constants α and γ are expressed in terms of the quantities characteristic of the given swirled jet: momentum J_0, moment L_0, and the physical constants ρ, μ. As for the constant β, its appearance is attributed not to the jet twisting but to the verification of the solution for a nonswirled jet due to the terms of the order of $1/x^2$ in the expressions for the projections of velocity u and v and for the free term in the stream function. The previous solution of the problem for a nonswirled jet was exact for the source with infinitely small diameter of the outlet section. As can be easily concluded by the last formula of the system (198) the terms involving β give the correction to the finiteness of the jet initial flow rate.[57] Making use of this formula and assuming $x = 0$, we obtain in the approximation adopted (Q_0 is the flow rate in the initial section)

$$\beta = \frac{Q_0}{2\pi\mu} = \frac{\pi d^2}{4}\frac{\rho U_0}{2\pi\mu} = \frac{U_0 d^2}{8\nu} \tag{199}$$

Note some qualitatively important conclusions. While longitudinal and lateral velocities decrease in inverse proportion to the first degree of the distance from the jet source, the preripheral velocity (twisting) decreases in inverse proportion to the square of the same distance. This explains the fact that at issuing from the hole, initial twisting does not remain noticeable at a distance from the jet source.

The maximum value of the peripheral velocity

$$w_{max} = \frac{3\sqrt{3}}{8}\frac{\gamma}{x^2}$$

is attained on a cone[58]

$$\eta = \frac{2\sqrt{3}}{3\alpha}$$

The problem of laminar jets was systematically analyzed by Goldshtik.[59]

12. Plane Nonstationary Boundary Layer

The cause of comparative complexity of the solution of the nonstationary boundary layer problems is, first, in the presence in its equations, along with the terms expressing convective acceleration, of the additional term local acceleration, and, second, in the appearance of initial conditions along with boundary ones.

It is known that the relation between the values of local and convective accelerations is characterized by the order of magnitude of the Strouhal number, which is equal to the quotient of the length characteristic for the given flow and the product of characteristic velocity and time. Esential specific features of nonstationary flows manifest themselves rather strongly at comparatively large values of the Strouhal number. At small values of this parameter it is enough to make use of quasi-stationary methods, i.e., to consider nonstationary phenomena at each moment as if they are stationary, but have their instantaneous values as determining parameters.

In this volume it is impossible to go into detail about this complex section of the boundary layer theory. The discussion is restricted to consideration of one simple problem which is interesting from the viewpoint of the comprehension of the mechanism of vorticity diffusion from the place of origin on the streamlined body surface. This is the problem of instantaneous (pulse) setting of the body, submerged into quiescent infinite viscous incompressible fluid, into translational uniform rectilinear motion.

If by the problem condition the body acquires steady-state motion instantly, this cannot be said of the fluid surrounding it. Of natural interest are the processes of fluid flow evolving with time: origination and development of a boundary layer on the body surface, appearance

---

[57] The solution corresponding to the finite initial diameter of nonswirled jet issuing was first obtained by Rumer in the paper The problem on a submerged jet, *Prikl. Mat. Mekh.*, 16(2) 1952.

[58] For the theoretical study of more complex laminar jets see Korobko, V. I., *Teoriya neavtomodelnykh struy vyazkoy zhidkosti (Theory of Non self similar Jets of Viscous Fluid)*, Saratov University Press, 1977, pt. 1, pp. 3–138.

[59] Goldshtik, M. A., *Vikhrevye potoki (Vortical Flows)*, Nauka Press, Novosibirsk, 1981, 235 259.

of separation and its shift upstream, and boundary layer transition to its steady-state form that corresponds to a stationary flow past a body.

The possibility of a comparatively simple solution of this problem in explained by the fact that the outer free irrotational flow is stationary in the case of translational rectilinear and uniform motion of the body and the velocity on the body surface is determined by the function of one variable x only. Such a simple solution takes place before the onset of separation origination and until the separation acquires its full development, i.e., at the beginning of the acceleration section.

An inverse effect of the boundary layer on the outer flow at the stage of developing, and moving separation will become noticeable and lead to the appearance of time in the number of velocity and outer flow arguments. Using comparative smallness of the acceleration duration and introducing time into the definition of the boundary layer thickness, we can seek the problem solution in the form of the series in time powers whose convergence at rather small t is ensured. This circumstance also facilitates solution.

The nonstationary boundary layer equations will differ by the presence of local acceleration $\partial u/\partial t$ on the left side of Equation 11, by the boundary and initial conditions. In the case under consideration these equations will have the form

$$\frac{\partial u}{\partial t}+u\frac{\partial u}{\partial x}+v\frac{\partial u}{\partial y}=U\frac{dU}{dx}+\nu\frac{\partial^2 u}{\partial y^2} \qquad \frac{\partial u}{\partial x}+\frac{\partial v}{\partial y}=0$$

$$u=U(x) \qquad v=0 \quad \text{at} \quad y=0, \quad t=0,$$

$$u=0 \qquad v=0 \quad \text{when} \quad y=0, \quad t>0,$$

$$u\to U(x) \quad \text{for} \quad y\to\infty,$$

$$u=U(0) \quad \text{when} \quad x=0, \quad y\geq 0, \quad t\geq 0. \tag{200}$$

The first line of the conditions given expresses the fact that there is no boundary layer at the initial instant $t = 0$ and fluid slides over the contour of a cylindrical body. At small values of time t the boundary layer is very thin, velocities u are close to their outer value $U(x)$ and v slightly differs from zero. Then, the first of the differential equations of the system (200) may be reduced to a linear form coinciding in type with the known equation of the theory of heat conduction in a solid body (subscripts at u, v, f, and F indicate their belonging to the corresponding approximation)

$$\nu\frac{\partial^2 u_1}{\partial y^2}-\frac{\partial u_1}{\partial t}=0 \tag{201}$$

Assuming[60]

$$u_1=U(x)f_1(\eta) \qquad \eta=y/(2\sqrt{\nu t}) \tag{202}$$

we obtain for the determination of $f_1(\eta)$ an ordinary differential equation of the second kind (prime is the derivative with respect to η)

$$f_1''+2\eta f_1'=0$$

[60] Blasius, H., Grenzschichten in Flüssigkeiten mit kleiner Reibung, Z. Math. Phys., 56, 1–37, 1908.

which under the adopted conditions (200) has the solution in the form of the known Gaussian error function

$$f_1 = \mathrm{Erf}\,\eta = \frac{2}{\sqrt{\pi}}\int_0^{\eta} e^{-\alpha^2}d\alpha$$

thus giving the first approximation for longitudinal velocity

$$u_1 = U(x)\,\mathrm{Erf}\,\eta \tag{203}$$

Making use of the continuity equation, we introduce the value of the lateral velocity

$$v_1 = -2\sqrt{\nu t}\,\frac{dU}{dx}\left[\eta\mathrm{Erf}\,\eta - \frac{1}{\sqrt{\pi}}(1 - e^{-\eta^2})\right] \tag{204}$$

corresponding to this first approximation.

To find the second approximation we take in the first equation of system (200) $u = u_1 + u_2$, $v = v_1 + v_2$.

Then, the correction u_2 will be determined from the nonuniform "thermal"-type equation

$$\nu\frac{\partial^2 u_2}{\partial y^2} - \frac{\partial u_2}{\partial t} = -U\frac{dU}{dx} + u_1\frac{\partial u_1}{\partial x} + v_1\frac{\partial u_1}{\partial y} =$$

$$= U\frac{dU}{dx}\left[\mathrm{Erf}^2\eta - \frac{2}{\sqrt{\pi}}\eta e^{-\eta^2}\mathrm{Erf}\,\eta - 1 + \frac{2}{\pi}(e^{-\eta^2} - e^{-2\eta^2})\right] \tag{205}$$

seeking u_2 in the form

$$u_2 = tU\frac{dU}{dx}f_2(\eta) \tag{206}$$

to determine the function $f_2(\eta)$ we obtain a nonuniform equation

$$f_2'' + 2\eta f_2' - 4f_2 = 4\left[\mathrm{Erf}^2\eta - \frac{2}{\sqrt{\pi}}\eta e^{-\eta^2}\mathrm{Erf}\,\eta - 1 + \frac{2}{\pi}(e^{-\eta^2} - e^{-2\eta^2})\right] \tag{207}$$

the solution of which is

$$f_2(\eta) = \frac{1}{2}(2\eta^2 - 1)\,\mathrm{Erf}^2\eta + \frac{3}{\sqrt{\pi}}\eta e^{-\eta^2}\mathrm{Erf}\,\eta + 1 - \frac{4}{3\pi}e^{-\eta^2} + \frac{2}{\pi}e^{-2\eta^2} +$$

$$+ \alpha(2\eta^2 + 1) + \beta\left[\frac{\sqrt{\pi}}{2}(2\eta^2 + 1)\,\mathrm{Erf}\,\eta + \eta e^{-\eta^2}\right] \tag{208}$$

The constants α and β entering into this equation are determined from the adopted boundary conditions of $f_2(\eta)$ vanishing at $\eta = 0$ and $\eta = \infty$; we find

$$\alpha = -\left(1 + \frac{2}{3\pi}\right) = -1.212 \qquad \beta = \frac{1}{\sqrt{\pi}}\left(1 + \frac{4}{3\pi}\right) = 0.804 \tag{209}$$

Making again use of the continuity equation, we determine the lateral velocity v in the second approximation

$$v_2 = 2\sqrt{\nu t^3}\left[U\frac{d^2U}{dx^2} + \left(\frac{dU}{dx}\right)^2\right]\left\{\frac{1}{3\sqrt{\pi}}\left(1+\frac{4}{3\pi}\right) - \right.$$

$$-\frac{1}{6}\,(2\eta^3 - 3\eta)\,\mathrm{Erf}^2\eta + \frac{1}{6\sqrt{\pi}}\,[4 - (4\eta^2 - 1)\,e^{-\eta^2}]\,\mathrm{Erf}\,\eta - \eta -$$

$$-\frac{1}{3\pi}\eta e^{-\eta^2} - \frac{4\sqrt{2}}{3\sqrt{\pi}}\,\mathrm{Erf}\,(\eta\sqrt{2}) + \left(1+\frac{2}{3\pi}\right)\left(\frac{2}{3}\eta^3 + \eta\right) -$$

$$\left. -\frac{1}{\sqrt{\pi}}\left(1+\frac{4}{3\pi}\right)\left[\frac{\sqrt{\pi}}{6}\,(2\eta^3 + 3\eta)\,\mathrm{Erf}\,\eta + \frac{1}{3}\,(\eta^2+1)\,e^{-\eta^2}\right]\right\} \tag{210}$$

Proceeding similarly, we can then find the third approximation.[61] In this case we have to solve a nonuniform equation

$$\nu\frac{\partial^2 u_3}{\partial y^2} - \frac{\partial u_3}{\partial t} = u_1\frac{\partial u_2}{\partial x} + u_2\frac{\partial u_1}{\partial x} + v_1\frac{\partial u_2}{\partial y} + v_2\frac{\partial u_1}{\partial y} \tag{211}$$

The function u_3 should involve t^2 as a multiplier and have the form

$$u_3 = t^2\left[U^2\frac{d^2U}{dx^2}f_3(\eta) + U\left(\frac{dU}{dx}\right)^2 F_3(\eta)\right] \tag{212}$$

where the functions f_3 and F_3 satisfy the following nonuniform equations

$$f_3'' + 2\eta f_3' - 8f_3 = 4\theta(\eta) \qquad F_3'' + 2\eta F_3' - 8F_3 = 4\theta(\eta) + 4\Phi(\eta)$$

We shall not write the right sides in the developed form, referring the reader to the paper by Goldstein and Rosenhead.[61] In this paper the authors spared no effort to calculate the value of the correction u3 in a closed form.[62] Unfortunately, the problem of the convergence range of the thus obtained series of functions

$$u = u_1 + u_2 + u_3 + \ldots \qquad v = v_1 + v_2 + v_3 + \ldots$$

remains unsolved.

To find the relation between the abscissa of the separation point xs and the time t_s corresponding to it, we determine the root of the equation $(\partial u/\partial y)_{y=0} = 0$, or $(\partial u/\partial \eta)_{\eta=0} = 0$. Then, we obtain in the second approximation

$$1 + \left(1+\frac{4}{3\pi}\right)\left(\frac{dU}{dx}\right)_s t_s = 0 \tag{213}$$

[61] Goldstein, S. and Rosenhead, L., Boundary-layer growth, *Proc. Cambr. Phil. Soc.*, 32, 392–401, 1936.

[62] In this calculation the authors made mistakes which were then indicated in Wundt, H., Wachstum der laminaren Grenzschichtan schräg angeströmten Zylinder bei Anfahrt aus der Ruhe, *Ing. Arch.*, 23(3), 218, 1955, and which we took into account.

Separation may occur only in the range of a negative value of the longitudinal derivative of the velocity on the outer edge of the layer ($dU/dx < 0$) and will begin at the point with maximum in magnitude value of this derivative. The time interval t_s from the onset of motion until the moment of separation origination is determined by the formula

$$t_s = \frac{1}{\left(1 + \frac{4}{3\pi}\right)\left(-\frac{dU}{dx}\right)_{max}} = \frac{0.702}{\left(-\frac{dU}{dx}\right)_{max}} \tag{214}$$

It is remarkable that the moment until the onset of separation origination on the body surface is independent of the physical properties of fluid (density and. viscosity) but it depends only on the maximum inclination of the curve of outer velocity distribution.

In the case of a round cylinder with the radius a it is (x is the arc of a circle reckoned from the front critical point)

$$U(x) = 2U_\infty \sin\left(\frac{x}{a}\right) \qquad \frac{dU}{dx} = 2\frac{U_\infty}{a}\cos\left(\frac{x}{a}\right) \tag{215}$$

Maximum in magnitude value of this derivative is attained at $x = \pi a$ at the rear critical point.

The time after the lapse of which from the moment of the motion onset originates boundary layer separation at a rear critical point, according to Equation 214 will be equal to

$$t_s = \frac{a}{2\left(1 + \frac{4}{3\pi}\right)U_\infty} = 0.351\frac{a}{U_\infty} \tag{216}$$

Let σ, which is the distance traversed by the cylinder moving with the velocity U_∞ from the onset of the motion until the moment of separation origination in the rear edge, be determined by the equality

$$\sigma = U_\infty t_s = 0.351a \tag{217}$$

and be equal to about one third of the cylinder radius. The third approximation gives somewhat smaller value

$$\sigma = 0.32a \tag{218}$$

In Figure 190 a dashed line indicates the dependence of the angular location θ_s^0 of the separation point (in degrees), reckoned from the front critical point of a round cylinder, on the dimensionless time $\tau = tU_\infty/d$ (d is the cylinder diameter). This curve is calculated by the stated method of successive approximations. Within the time interval from $\tau = 0$ to $\tau_s = t_sU_\infty/d = 0.16$ the angle θ_s^0 does not change and is equal to $180°$. Then, starting from this moment, the angle θ_s^0 decreases, thus corresponding to the shift of the separation point upstream around the cylinder contour. For the Sake of comparison a solid line presents the same dependence determined by the approximate method similar to that of Pohlhausen.[63] In contrast to the previous one this solid line asymptotically ($\tau \to \infty$) tends to a stationary value (overestimated, as explained earlier) of the separation point angle ($107°5'$).

Neither method takes into account the inverse effect of the developing separation on the outer theoretical irrotational flow, but moreover in the Blasius power-law method the rate of

[63] Schuh, H., Calculation of unsteady boundary layer in two-dimensional laminar flow, *Z. Flugwiss.*, 1, 122–131, 1953.

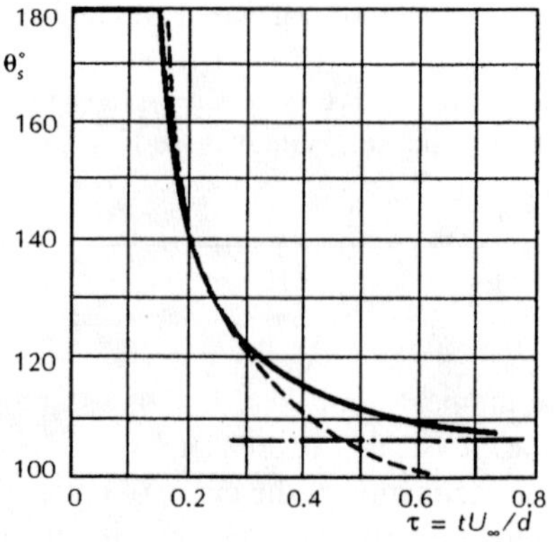

Fig. 190

convergence of the series with time τ is lost, thus explaining the absence of asymptotic tendency to a stationary value of the separation point angular coordinate. Due to the reason mentioned, the power-law solution (in the third approximation) is probably suitable only when $\tau \le 0.3$.

At present not only are cases of impulsive setting of the body into uniform translational flow but also equally accelerated cases with power-law and exponential grows of velocity are studied in detail. Solution of these and many other problems of an unsteady boundary layer may be found in the earlier-mentioned monographs on the boundary layer theory and original papers cited in them.

The problem on the conditions of an unsteady laminar boundary layer on a moving surface is considered by Williams.[64]

In view of the considerable grows of the number of similarity parameters the application of the method of generalized similarity in the theory of an unsteady boundary layer becomes complex and with up-to-date possibilities of electronic computers, feasible only in local approximations[65] or under additional simplifying assumption.[66]

The solution presented vividly shows the origin of the process of viscous effect distribution on a flow past a body set into motion. Experiments confirm a theoretical description of the phenomenon. It is enough to thoroughly examine the known photographs by Titiens[67] which describe the onset of round cylinder motion in a water chute, to convince oneself in the validity of this statement. It is vividly seen in these photographs that at first the absent bound-

---

[64] *Vikhrevye dvizheniya zhidkosti: Ustoychivost i otryv pogranichnogo sloya, svobodnye i kvantovye vikhri, Mekhanika. Novoye v zarubezhnoy nauke* (*Vertical Notions of Fluid: Stability and Separation of a Boundary Layer, Mechanics. New in Foreign Science*), Mir Press, Moscow, 1979, 58–100.

[65] Bushmarin, O. N. and Sarayev, Yu. V., Parametric method in the theory of an unsteady boundary layer, *J. Eng. Phys*, 27(1), 1974; Bushmarin, O. N. Parametric method for calculating an unsteady laminar boundary layer in incompressible fluid with suction or injection, *J. Eng. Phys*, 31(4), 1976.

[66] Djuric, M., On the universal form of unsteady incompressible boundary layer equation and its solving. Publ. de l'Instit. Math., Belgrad, 9, 1969; Ascovic, R., Quelques contributions a l'etude de la couche limite laminaire tridimensionelle en regime non stationaire, Rap. A-16 du Labor, d'Aerod. de l'Universite Laval, Québec, Canada, Juin, 1970; Saljnikov, V., Sur une forme possible d'equations universelles de la couche limite laminaire instationaire, *C. R. Acad. Sci.*, Paris, 272, 830–833, 1971; Saljnikov, V. and Djukic, D., L'universalisation des equations de la couche limite laminaire instationaire, *Recentes Recherches sur les Couches Limites Instationaires, Jutam. Symposium* 1971, Vol 1, Eichelbrenner, E. A., Ed., Québec, Canada, 1972, 168–205.

[67] Schlichting, G., *Teoriya pogranichnogo sloya* (*Boundary Layer Theory*): (translated from German), Nauka Press, Moscow, 1974, 42; Goldstein, S., *Sovremennoye sostoyaniye gidroaerodinamiki vyazkoy zhidkosti* (*Up-to-date Hydrodynamics of Viscous Fluid*), Vol. 1 (translated from English), Inostrannaya Literature Press, Moscow, 1948, photo 7–8.

ary layer becomes gradually thicker until at some maximum thickness in the vicinity of a rear critical point separation of the layer takes place. This separation develops and propagates, tending to occupy its limiting position that corresponds to a steady-state flow past a cylinder.

The results of the experiments on measuring pressure distributions over a surface of a round cylinder at different stages of its motion from rest performed by Schwabe[68] confirm that at the beginning of motion pressure distribution is close to the theoretical one corresponding to an irrotational flow of ideal fluid past a cylinder. This also speaks of the fact that at the beginning of motion a boundary layer even on such a blunt body, as a round cylinder, is very thin, it completely envelopes body surface and, therefore, does not exert a noticeable inverse effect on the outer flow. Only after separation origination and its shift upstream from the rear edge of the cylinder there appears sharp change of pressure distribution curve that ends by transition to that ordinary distribution which is observed in a real steady-state flow past a cylinder.

Similarly, if a body undergoes a steady-state motion and at some instant this motion is disturbed, for example, a wing angle of attack suddenly changes, then transition to a new steady-state flow corresponding to a new position of a wing in a flow does not take place so rapidly as the angle of attack changes, but it is delayed. Reconstruction of streamlining requires some finite time due to the effect of viscous forces in a boundary layer. By means of extending a smooth flow past a wing onto postcritical angles of attack one can obtain, for a short period of time a noticeable increase in the coefficient of maximum wing lift force (a dynamic coefficient of lift force).

Mathematical difficulties arising when considering the problems of the theory of an unsteady laminar boundary layer are displayed in the review paper by Stuart.[69] An example of a periodic layer[70] is considered elsewhere in the volume.

13. Temperature and Concentration Boundary Layers in Incompressible Fluid

This section considers the simplest case of incompressible viscous fluid with constant physical characteristics (density, coefficients of viscosity, thermal conductivity, diffusion) that is quite admissible if velocities of motion are much smaller than the sound velocity and differences of temperature and concentrations of admixtures are small. Moreover, we neglect dissipation of mechanical energy and inner sources of heat and substance origination. In Chapter 15 the concern is with dynamics and thermodynamics of gas at large velocities, and the restrictions on the generality of the formulation of problems on heat and mass transfer will be lifted.

As was noted at the beginning of the chapter, at large values of the flow Reynolds number $\mathbf{Re}$ and the number $\mathbf{Pe}$ and $\mathbf{Pe}_d$, there will form, along with velocity boundary layers, temperature and concentration boundary layers. In these thin layers, as compared with the linear dimension, which is characteristic of flows, the rate of temperature or concentration variation in a direction lateral to the flow will considerably exceed variations in longitudinal direction.

This important fact makes it possible to present the equations of heat (174) and substance

[68] Schwabe, M. S., Über Druckermittlung in der nichtstationären ebenen, *Stromung. Ing.-Arch.*, 6, 1935.

[69] Stuart, J. T., Unsteady boundary layers (earlier cited Proceedings of the Canadian symposium Vol. 1; *ibid*, results are presented of different studies into unsteady laminar and turbulent boundary layers.

[70] Lin, C. C., Motion in the boundary layer with a rapidly oscillating external flow, Proc. 9th Int. *Congr. Appl. Mech.*, Vol. 4, Brussels, 1957, 155–167.

(182) distribution (Chapter 10) at large values of thermal and diffusional Peclet numbers in the form (we restrict ourselves to a plane stationary flow)

$$u\frac{\partial T}{\partial x}+v\frac{\partial T}{\partial y} = a\frac{\partial^2 T}{\partial y^2} \qquad u\frac{\partial c}{\partial x}+v\frac{\partial c}{\partial y} = D\frac{\partial^2 c}{\partial y^2} \tag{219}$$

We recall that here a and D are the coefficients of thermal conductivity and diffusion. These equations of thermal and diffusional boundary layer should be solved together with the equations of a velocity boundary layer

$$u\frac{\partial u}{\partial x}+v\frac{\partial u}{\partial y} = -\frac{1}{\rho}\frac{dp}{dx}+\nu\frac{\partial^2 u}{\partial y^2} \qquad \frac{\partial u}{\partial x}+\frac{\partial v}{\partial y} = 0 \tag{220}$$

which in the formulation adopted are noninteracting, independent of the solution of Equations 219. Conversely, to be solved Equations 219 require the knowledge of the velocity field, i.e., the preliminary solution of the dynamic problem (220).

We shall take

$$T = T_w \qquad c = c_w \quad \text{at} \quad y=0$$

$$T \to T_\infty \qquad c \to c_\infty \quad \text{when} \quad y \to \infty$$

$$T = T_0(y) \qquad c = c_0(y) \quad \text{when} \quad x = x_0 \tag{221}$$

as the boundary conditions for Equations 219. In this case we assume that the temperature of a streamlined surface T_w and the temperature T_∞ of a free stream are independent of x, i.e., they are constant along a temperature boundary layer. Similarly, we take concentrations of substance admixture on the surface, c_w, and at a distance from it, c_∞, to be constant along the flow.

Under the conditions of the existence of a self-similar solution for a velocity boundary layer (Section 7) Equations 219 will also have self-similar solutions. It is enough to show this for the first of equations (219).

Let us introduce the dimensionless temperature θ, relating the temperature difference $T_w - T$ to the scale characteristic of it, $T_w - T_\infty$. We have

$$\theta(\xi) = \frac{T_w - T}{T_w - T_\infty} \qquad \Phi(\xi) = \frac{\psi}{U\delta} \qquad \xi = \frac{y}{\delta}$$

$$T = T_w - (T_w - T_\infty)\,\theta(\xi) \tag{222}$$

Recalling expressions (86) for the derivatives $\partial\psi/\partial y = u$, $\partial\psi/\partial x = -v$ and having directly calculated the derivatives (point is the sign of differentiation with respect to ξ, prime – with respect to x)

$$\frac{\partial T}{\partial x} = (T_\infty - T_w)\,\dot{\theta}\left(-\xi\frac{\delta'}{\delta}\right) = -\frac{1}{2}\frac{z'}{z}(T_\infty - T_w)\,\xi\dot{\theta}$$

$$\frac{\partial T}{\partial y} = (T_\infty - T_w)\,\dot{\theta}\frac{1}{\delta} \qquad \frac{\partial^2 T}{\partial y^2} = (T_\infty - T_w)\,\ddot{\theta}\frac{1}{\delta^2}$$

we substitute these expressions into the first of Equations 219. Then, after obvious reduction of terms involving the product $\Phi\dot{\theta}$, we obtain the equation to be solved in the similarity variables

$$\ddot{\theta} + \sigma\left(\beta + \frac{1}{2}\bar{\beta}\right)\Phi\dot{\theta} = 0 \tag{223}$$

where for brevity the Prandtl number **Pr** denoted by σ is equal to the ratio ν/a of the kinematic coefficient of viscosity ν to the thermal diffusivity a.

Equation 223 is integrated and gives the solution in the form of a quadrature. Under the adopted boundary conditions (221) this solution $[\bar{\beta} = 2(1-\beta)]$ has the form

$$\theta(\xi;\beta,\sigma) = \frac{\int_0^{\xi} \exp\left(-\sigma\int_0^{\xi}\Phi d\xi\right) d\xi}{\int_0^{\infty} \exp\left(-\sigma\int_0^{\xi}\Phi d\xi\right) d\xi} \tag{224}$$

In this case the values of the function $\Phi(\xi;\beta)$ that enter into this solution may be obtained by integration with respect to ξ of the values of $\dot{\Phi}(\xi;\beta)$ given in Table 14.

As can be easily concluded by Equations 219 and 221, the same quadrature (224) will also determine self-similar concentration distribution

$$\theta_d = \frac{c_w - c}{c_w - c_\infty} = \theta(\xi;\ \beta, \sigma_d) \tag{225}$$

where the subscript d refers to the distribution of concentrations and to the Schmidt number corresponding to it (**Sc** = $\mathbf{Pr}_d = \sigma_d$).

Having no possibility to present the tables and graphs, which occupy a large volume of the distributions $\theta(\xi;\beta,\sigma)$, of the distributions $\theta(\xi;\beta,\sigma_d)$, which are equivalent in form, and even of the distributions $\dot{\theta}(0;\beta,\sigma)$, $\dot{\theta}(0;\beta,\sigma_d)$ less cumbersome and most important for calculating temperature and concentration layers. We shall refer the reader to the monograph of H. L. Evans (see above) where these distributions are presented and analyzed in detail.

Recalling the considerations of the comparative thickness of velocity, temperature, and concentration boundary layers depending on the values of the Prandtl and Schmidt numbers, we complement them with similar information related to the same value of these numbers, which is equal to unity, but to different values of β.

If $\beta > 0$, the fact corresponds to a converging section of a boundary layer with the accelerated fluid flow on its outer edge, then the temperature (as well as the concentration) boundary layer is thicker than the velocity one. If $\beta < 0$ (a diverging section with the decelerated fluid flow on the outer edge of the boundary layer), then the temperature (concentration) layer is thinner than the velocity layer. This regularity, strictly applicable in self-similar boundary layers in incompressible fluid, retains its validity in gas flows with large velocities and pressure drops.

We examine two cases of heat transfer corresponding to the values $\beta = 0$ and $\beta = 1$.

In the first case ($\beta = 0$) heat transfer takes place from the surface of the plate in a longitudinal flow.[71] The function Φ is recognized as its normalized value Φ_0^* prescribed by a standard form of the Blasius equation (34). Then, noting that according to this equation it is

$$\Phi_0^* = -\frac{\dddot{\Phi}_0^*}{\ddot{\Phi}_0^*} \qquad \int_0^\xi \Phi_0^* d\xi = -\int_0^\xi \frac{\dddot{\Phi}_0^*}{\ddot{\Phi}_0^*} d\xi = -\ln\frac{\ddot{\Phi}_0^*(\xi)}{\ddot{\Phi}_0^*(0)}$$

$$\exp\left(-\sigma\int_0^\xi \Phi_0^* d\xi\right) = \left[\frac{\ddot{\Phi}_0^*(\xi)}{\ddot{\Phi}_0^*(0)}\right]^\sigma$$

we shall find the unknown solution (224) in the following simple form

$$\theta(\xi) = \frac{\int_0^\xi [\ddot{\Phi}_0^*(\xi)]^\sigma d\xi}{\int_0^\infty [\ddot{\Phi}_0^*(\xi)]^\sigma d\xi} \tag{226}$$

At the Prandtl number $\sigma = 1$ it follows from this equation

$$\theta(\xi) = \frac{T_w - T}{T_w - T_\infty} = \dot{\Phi}_0^*(\xi) = \frac{u}{U_\infty} \tag{227}$$

thus expressing at $\sigma = 1$ the condition of similarity of the distributions of: (1) the difference between the wall temperature and that at the given point of the boundary layer and (2) flow velocity in the same section.

In a general case ($\sigma \neq 1$) the distribution $\theta(y)$ in a temperature boundary layer is not similar to the distribution $u(y)$ in a hydrodynamic layer. For most gases σ differs comparatively slightly from unity and the difference between the curves $\theta(\xi)$ and $\dot{\Phi}_0^*(\xi)$ is small. For liquids, σ changes within wide limits ($\sigma \gg 1$ for viscous oils, glycerine; $\sigma \ll 1$ for liquid metals) and the difference becomes very noticeable. In Figure 191 the curves of the distribution $\theta(\xi)$ are presented for several values of σ that confirm the above; the curve for $\sigma = 1$ presents simultaneously the distribution of the dimensionless velocity u/U_∞.

In Figure 191, the regularity found at the beginning of this chapter is vividly seen; i.e., in the fact that the thickness of the temperature boundary layer grows with the decrease of the Prandtl number σ coinciding with the thickness of the velocity boundary layer only at $\sigma = 1$. When $\sigma > 1$ the temperature layer is thinner then the velocity layer and when $\sigma < 1$, conversely, the velocity boundary layer is thinner then the temperature layer.

We shall determine the local coefficient of heat transfer $\alpha(x)$, which is equal to

$$\alpha(x) = \frac{q_w}{|T_w - T_\infty|} = -\lambda\left(\frac{\partial T}{\partial y}\right)_w \frac{1}{|T_w - T_\infty|} =$$

$$= \frac{1}{2}\lambda\left(\frac{d\theta}{d\xi}\right)_{\xi=0}\sqrt{\frac{U_\infty}{\nu x}} = \frac{1}{2x}\lambda f(\sigma)\sqrt{\mathbf{Re}_x} \tag{228}$$

---

[71] Pohlhausen, E., Wärmeaustausch zwischen festen Körpern und Flüssigkeiten mit kleiner Reibung und kleiner Warmeleitung, Z. *Angew. Math. Mech.*, 1, 115, 1921.

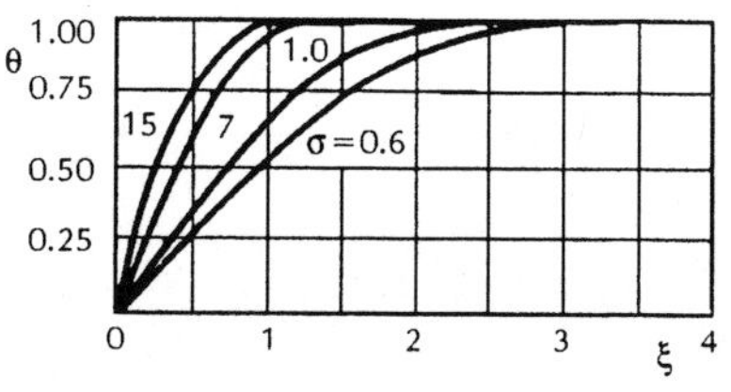

Fig. 191

where q_w is the projection of the heat flux vector $\boldsymbol{q}^*$ (Equation 173, Chapter 10) onto the normal to the surface, i.e., the quantity of heat per second transferred through the unit of this surface area, and

$$f(\sigma) = \left(\frac{d\theta}{d\xi}\right)_{\xi=0} = [\ddot{\Phi}_0^*(0)]^\sigma \left\{\int_0^\infty [\ddot{\Phi}_0^*(\xi)]^\sigma d\xi\right\}^{-1} \qquad \mathbf{Re}_x = \frac{U_\infty x}{\nu} \tag{229}$$

We find the local Nusselt number

$$\mathbf{Nu}_x = \frac{\alpha x}{\lambda} = \frac{1}{2} f(\sigma)\sqrt{\mathbf{Re}_x} \tag{230}$$

that characterizes heat transfer at the given point of the plate surface. The function $f(\sigma)$ calculated by Pohlhausen within the range $0.6 \le \sigma \le 15$ is[72]

$$f(\sigma) \approx 0.664\sqrt[3]{\sigma} \tag{231}$$

so that within this range of Prandtl numbers the local equation of heat transfer

$$\mathbf{Nu}_x \approx 0.332\sqrt[3]{\sigma}\sqrt{\mathbf{Re}_x} \tag{232}$$

may be used.

Performing integration over the entire length of the plate L, we obtain the expression for a total Nusselt number **Nu** of the plate

$$\mathbf{Nu} = \frac{Q}{\lambda|T_w - T_\infty|} = \frac{\int_0^L q_w dx}{\lambda|T_w - T_\infty|} = \frac{1}{2} f(\sigma)\sqrt{\frac{U_\infty}{\nu}}\int_0^L \frac{dx}{\sqrt{x}} = f(\sigma)\sqrt{\frac{U_\infty L}{\nu}} \tag{233}$$

or in an approximate form for the above-mentioned range of the Prandtl number

$$\mathbf{Nu} \approx 0.664\sqrt[3]{\sigma}\sqrt{\mathbf{Re}_\infty} \qquad \mathbf{Re}_\infty = \frac{U_\infty L}{\nu} \tag{234}$$

Similar formulas also will be obtained for the coefficient of local mass transfer (q_d is the quantity of substance transferred through the unit of surface area per second)

---

[72] The experiments of F. Elias (*Z. Angew. Math. Mech.*, 9, 434–453, 1929. and in the same publication, 10, 1–14, 1930) confirmed the theoretical results of Pohlhausen.

$$\alpha_d = \frac{q_d}{|c_w - c_\infty|} = \frac{D\left|\frac{\partial c}{\partial y}\right|_{y=0}}{|c_w - c_\infty|}$$

and, respectively, the local diffusion Nusselt number

$$\mathbf{Nu}_{xd} = \frac{\alpha_d x}{D} = \frac{1}{2} f(\sigma_d)\sqrt{\mathbf{Re}_x} \tag{235}$$

and the total diffusion Nusselt number for the entire plate surface

$$\mathbf{Nu}_d = \frac{Q_d}{D|c_w - c_\infty|} = \frac{\int_0^L D\left|\frac{\partial c}{\partial y}\right|_{y=0} dx}{D|c_w - c_\infty|} = f(\sigma_d)\sqrt{\mathbf{Re}_\infty} \tag{236}$$

In the case of an arbitrary β the local Nusselt number is determined by the equation

$$\mathbf{Nu}_x = K(\sigma;\beta)\,\mathbf{Re}_x^{1/2} \tag{237}$$

where it is assumed that

$$K(\sigma;\beta) = (2-\beta)^{-1/2}\left\{\int_0^\infty \exp\left[-\sigma\int_0^\xi \Phi^*(\xi)\,d\xi\right]d\xi\right\}^{-1} \tag{238}$$

To the temperature boundary layer near the front critical point on a cylindrical body corresponds $U = cx$, $\beta = 1$. In this case the local Nusselt number will be equal to

$$\mathbf{Nu}_x = \frac{\alpha x}{\lambda} = K(\sigma;1)\,\mathbf{Re}_x^{1/2} = K(\sigma;1)\sqrt{\frac{c}{\nu}}\,x \tag{239}$$

where the coefficient $K(\sigma;1)$ is found from the following table:

σ	0.6	0.7	0.8	0.9	1.0	1.1	7.0	10	15
$K(\sigma;1)$	0.466	0.495	0.521	0.546	0.570	0.592	1.18	1.34	1.54

For a round cylinder with the diameter d near the front critical point

$$U = 2U_\infty \sin\frac{2x}{d} \approx 4U_\infty x/d = cx$$

so that $c = 4U_\infty/d$ and Equation 239 will be reduced to the form

$$\mathbf{Nu}_x = 2K(\sigma;1)\left(\frac{U_\infty d}{\nu}\right)^{1/2}\frac{x}{d} \tag{240}$$

The theory of heat transform from the surface of a rotating disk in infinite and finite regions of viscous fluid flow under laminar conditions presented in the monograph by Dorfman (cited earlier). Heat transfer in non-Newtonian fluids is discussed in collected papers compendium published in 1968.[73]

To calculate thermal and concentration boundary layers with arbitrary velocity distribution on the outer edge of the velocity boundary layer the method of generalized similarity may be used. The noninteracting hydrodynamic equation (220) is by familiar means reduced to the universal form. Introducing additional succession of thermal parameters we can also reduce Equations 219 to the universal form.[74]

The application of the method of generalized similarity will be shown in Chapter 15 for a more general case of a gas flow of large velocities and great temperature drops.

The problem on free laminar convection[75] of incompressible fluid near a surface of an infinite plate but with bounded lower edge can serve as an example of the problem for which the equation of a velocity boundary layer will not be noninteracting, but will be connected with the equation of a temperature boundary layer. The plate temperature T_w is kept constant; the surrounding temperature at a distance from the plate is equal to T_∞.

In this case the motion in a boundary layer is caused by the presence of buoyancy force, the specific value of which (related to the mass unit) may be presented in the form

$$g\frac{T-T_\infty}{T_\infty} = g\frac{T-T_\infty}{T_w-T_\infty}\,\frac{T_w-T_\infty}{T_\infty} = g\Theta\frac{T_w-T_\infty}{T_\infty} \tag{241}$$

The equations of motion and heat propagation in velocity and temperature boundary layers will be

$$u\frac{\partial u}{\partial x}+v\frac{\partial u}{\partial y} = \nu\frac{\partial^2 u}{\partial y^2}+g\frac{T_w-T_\infty}{T_\infty}\Theta$$

$$\frac{\partial u}{\partial x}+\frac{\partial v}{\partial y} = 0 \qquad u\frac{\partial \Theta}{\partial x}+v\frac{\partial \Theta}{\partial y} = a\frac{\partial^2 \Theta}{\partial y^2}$$

$$u = v = 0 \qquad \Theta = 1 \quad \text{at} \quad y = 0$$

$$u = 0 \qquad \Theta = 0 \quad \text{when} \quad y = \infty . \tag{242}$$

Having introduced new variables (ψ is the stream function)

$$\xi = \left[\frac{g\,(T_w-T_\infty)}{4\nu^2 T_\infty}\right]^{1/4}\frac{y}{x^{1/4}} = C\frac{y}{x^{1/4}}$$

$$\psi\,(x, y) = 4\nu C x^{3/4}\Phi\,(\xi) \qquad \Theta\,(x, y) = \vartheta\,(\xi)$$

---

[73] *Teplomassoobmen v neniyutonovskh zhidkostyakh* (*Heat Transfer in Non-Newtonian Fluids*), collected papers edited by A. V. Luikov and B. M. Smolskiy, Energiya Press, Moscow, 1968.

[74] The paper by Saljnikov, V. and Djordjevic, V., Universalisferung der Gleichung vom Temperaturengrenzschichtproblem, *Z. Angew. Math. Mech.*, 48(8), 237–241,1968, deals with this problem.

[75] The concern of the monograph by Gershuni, G. Z. and Zhukhovitskiy, E. M., *Konvektivnaya ustoychivost neszhimayemoy zhidkosti* (*Convective Stability of Incompressible Fluid*), Nauka Press, Moscow, 1972, is a vast range of problems of laminar free convection, The authors of the monograph are the leaders of the well-known school at the Perm University and Pedagogical Institute. Various mathematical methods for solving the problems of free convection near a vertical surface are presented in the recently published monograph: Sokovishin, Yu. A. and Martynenko O. G., *Vvedeniye v teoriyu svobodno-konvektivnogo teploobmena* (*Introduction into the Theory of Free-Convective Heat Transfer*), Leningrad State University Press, Leningrad, 1982.

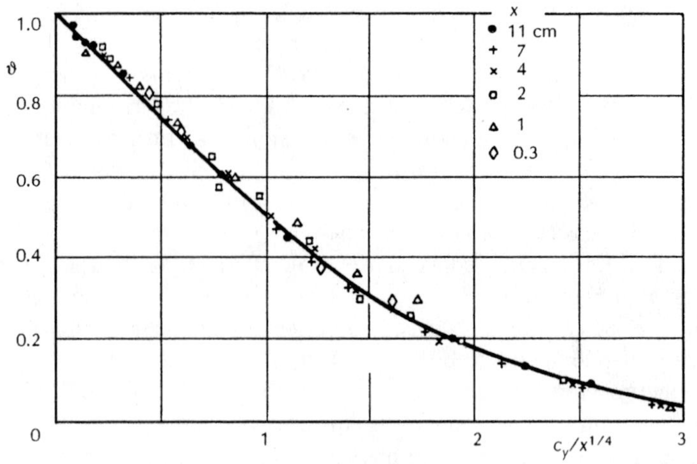

Fig. 192

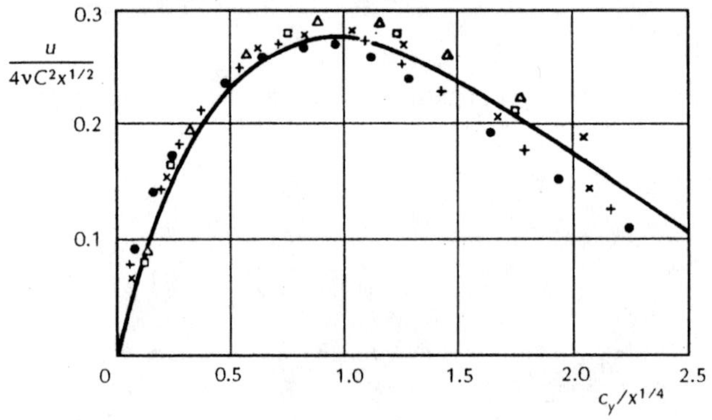

Fig. 193

$$C = \left[\frac{g\,(T_w - T_\infty)}{4\nu^2 T_\infty}\right]^{1/4} \tag{243}$$

the choice of which may be made based on the consideration of dimensions or from the condition of problem self-similarity, we come to the following system of ordinary differential equation

$$\dddot{\Phi} + 3\Phi\ddot{\Phi} - 2\dot{\Phi}^2 + \vartheta = 0 \qquad \ddot{\vartheta} + 3\sigma\Phi\dot{\vartheta} = 0$$

$$\Phi(0) = \dot{\Phi}(0) = 0 \qquad \vartheta(0) = 1 \qquad \dot{\Phi}(\infty) = \vartheta(\infty) = 0 \tag{244}$$

Placing the coordinate origin ($x = 0$) on the lower edge of the plate, we also have the conditions $u = 0$, $\vartheta = 0$ at $x = 0$.

Equations 244 were integrated by Pohlhausen[76] using expansions into series for air (σ = 0.733).

Figures 192 and 193 show the dimensionless profiles of temperatures and velocities in the layer cross-sections that are calculated theoretically; in the same figures experimental points[77] are presented that are in quite satisfactory agreement with the theory. A theoretical value of the local Nusselt number at the point lying on the height h from the lower edge of the plate is equal to

$$\mathbf{Nu}_h = \frac{\alpha h}{\lambda} = 0.508\left[\frac{gh^3(T_w - T_\infty)}{4\nu^2 T_\infty}\right]^{1/4} = 0.359\,(\mathbf{Gr})^{1/4} \tag{245}$$

where the Grashof number equal to

$$\mathbf{Gr} = \frac{gh^3(T_w - T_\infty)}{\nu^2 T_\infty}$$

is used (see Chapter 10).

A nonself-similar case of free convection on a surface of a vertical plate with arbitrarily prescribed temperature or heat flux on its surface was studied by Zolotov,[78] a postgraduate student of the Department of Hydroaerodynamics of the M. I. Kalinin Leningrad Polytechnical Institute, who used the method of generalized similarity in his study.

---

[76] Pohlhausen, E. Wärmemaustausch zwischen festen Körpern und Flussigkeiten mit kleiner Wärmeleiturg, *Z. Angew. Math. Mech.*, 1, 115, 1921.

[77] The Elias experiments (*Z. Angew. Math. Mech.*, 9, 434–453, 1929).

[78] Zolotov, N. L., The method of generalized similarity in the problems of free convection with arbitrary distribution of temperature or heat flux on a vertical wall, *Mekh, Zhidk. Gaza*, No.3, 1980; Zolotov, N. L., Two-term solution of the problem on free-convection near a vertical non-isothermal plate, *Mekh. Zhidk. Gaza*, No.5, 490–492, 1980.

13

Turbulent Flows of Incompressible Viscous Fluid

1. Instability of Laminar Flows and Occurrence of Turbulence

Viscous fluid flows, considered in the two previous chapters; related to a number of laminar flows. Trajectories of particles, streamlines, fields of velocities, and pressures in these flows were of quite definite, "regular" character. The fact that a general picture of laminar flow observed in reality and many of their details were rather well described by the solutions of the Navier-Stokes equations under the corresponding, also "regular", initial and boundary conditions, serves as the expression of this regularity of a laminar flow. One can recall, for example, a Poiseuille flow of a viscous fluid in a round tube; the agreement between theoretically calculated characteristics of this flow (parabola of velocities, formulas of flow rate, and resistance) and the test data was successfully confirmed long ago. The same relates to numerous examples of viscous fluid laminar flows: to the motion of lubricant in narrow clearances between a shaft and a bearing journal that is quite satisfactorily described by the hydrodynamic theory of bearing lubrication, to the motions in laminar boundary layers that are rather accurately calculated by the theory stated in the previous chapter, etc.

The noted coincidence between the results of calculations of laminar flows with experiments serves as the basis making for conclusion on the validity of the Navier-Stokes equations and their applicability for theoretical description of viscous fluid flows. One should not, however, think that the absence of the possibility in a number of cases to make such a conclusion may serve as a foundation for an argument on the discrepancy between the theory and reality.

The presence of various, very often small in magnitude, random deviations or "disturbances" may either very slightly change the flow considered (this will indicate flow stability relative to small disturbances), or distort it completely (this takes place in the case of flow instability). Thus, in reality only those solutions of the Navier-Stokes equations are observed that are stable with respect to possible disturbances.

In stable flows, small disturbances that originated randomly or are introduced into a flow by a researcher do not develop with time but, conversely, they attenuate without noticeably affecting the processes occurring in a liquid flow. In contrast to this, in unstable flow the disturbances, which at first are small, grow substantially alternating the character of the initial flow and facilitating its transition either to a new stable flow, if this exists among the possible solution of the Navier-Stokes equations, or to some chaotic flow formed by liquid masses irregularly moving and interacting with each other. The processes of origination and development of such flows as well of their destruction are of random character and require statistic approaches to be studied.

This form of viscous fluid flow widely, found in nature and engineering devices, has the appellation "turbulent flow". Air flows in atmosphere, water flows in seas, oceans, rivers, and channels, in water-conducting pipes, in gas pipelines, turbines, pumps and compressors, in nozzles of rocket and jet engines are all examples of turbulent flow.

Specific features of turbulent motion can be easily observed when, for example, looking from a bridge onto a water surface in a channel that is covered with small particles of sweepings or a petroleum film. One can see that separate bodies participating in a medium of water flow in a channel perform complicated transverse motions, while near the sides they take part in retrograde flows. Similar motions may be observed from about ship, especially near the stern portion.

Historically, the first scientific observations of turbulent motions were the experiments that were performed by the English physicist Reynolds in 1883 in which he studied water flows in a round cylindrical tube. As the velocity of a laminar liquid flow increases waves begin to superimpose upon a uniformly colored and well-defined jet, and propagation of these waves along the jet indicates origination of disturbances in a previously quiet rectilinear motion. Gradually, as water velocity grows, the quantity of these waves and their amplitude increase until, finally, the jet is divided into irregular small jets which mix with each other and whose chaotic character allows one to judge the occurrence of laminar-to-turbulent flow transition. The described picture of transition fully corresponds to the earlier-mentioned cause of this transition. As velocity grows, laminar flow loses its stability; in this case random perturbations, which previously evoked only oscillations of jets about their rectilinear position, quickly develop and lead to a new form of fluid flow – turbulent flow.

In these experiments Reynolds revealed that laminar-to-turbulent flow transition is characterized by the attainment of a critical value of some dimensionless number or criterion which then was named after him. By the experiments of Reynolds himself the critical value was equal to $\mathbf{Re}_{cr} = (u_{mean}d/\nu)_{cr} = 1.3 \cdot 10^4$; here, u_{mean} is the velocity mean over the flow rate, d is the tube diameter, and ν is the coefficient of kinematic viscosity. Subsequently he discovered the presence of such a low critical value of $\mathbf{Re}_{cr}$, which is approximately equal to 2300, that for $\mathbf{Re} < \mathbf{Re}_{cr}$ a flow in a tube remained laminar, irrespective of the disturbances introduced into the flow.[1] Moreover, it was noted that by removing disturbances at the tube inlet or by reducing their initial intensity one can artificially maintain a laminar flow at considerably larger values of the Reynolds number. In particular, the value $1.3 \cdot 10^4$ obtained by Reynolds was explained by the presence of a smooth tube inlet in his experiments. However, he did not manage to obtain a definite value for the upper boundary of the critical number; this boundary was extended many times by more thorough experiments[2] and was increased to about $5 \cdot 10^4$. Of course, this laminar flow does not permit the occurrence even of very small perturbations without directly passing over to a turbulent one.

---

[1] Reynolds, O., On the dynamical theory of incompressible viscous fluids and the determination of the criterion, *Phil. Trans. R. Soc.*, 1895.

[2] See Chapter 5 of the monograph by Shiller, L., *Dvizheniye zhidkosti v trubakh (Fluid Flow in Tubes)*, ONTI Press, Moscow, 1936.

Leaving aside the problem on test values of the critical Reynolds number for cylindrical tubes with different shapes of their cross-section,[3] we note that the critical Reynolds number $\mathbf{Re}_{cr}$ is strongly affected by the deviation from the tube cylindricality, i.e., the tube diverging or converging character. Thus, in converging tubes $\mathbf{Re}_{cr}$ exceeds the corresponding number for a cylindrical tube, and, conversely, in diverging channels $\mathbf{Re}_{cr}$ is comparatively small.

Note and confirm in what follows from test results that wall roughness does not affect the value of the critical Reynolds number. This is natural because the low Reynolds number Requirement is attributed to the flow stability rather than to the presence or absence of disturbances in it.

The mathematical theory of stability of laminar flows is well developed. Betchov and Criminale in their monograph, "The Problems of Hydrodynamical Stability" ("Voprosy gidrodinamicheskoy ustoychivosti", Mir Press, Moscow, 1971), simply and clearly present the state of this problem giving a great number of different applications including magneto-hydrodynamic and visco-elastic flows and also flows of gas with solid particles suspended in it. This monograph to a great extent may fill in the deficiency of this course.

The theory of laminar flow stability relative to small disturbances is the simplest section of the general theory of stability of laminar flows. This linear theory acquired the greatest development and is stated in a great number of special courses and monographs.[4]

The nonlinear theory of hydrodynamic stability was recently developed. Its foundations are formulated at the end of the review composed by Stuart and published in the manual edited Rosenhead (pp. 562 to 578).[4] This branch of the stability theory also used the methods of the oscillation theory but studies the development of disturbances of finite amplitude.[5] The problem of the stability of a laminar flow refers only to the initial stage of the phenomenon of laminar-to-turbulent flow transition. Of major importance are the transition phenomena proper that present the development of those random disturbances which existed in the flow until it lost its stability or, at the moment of stability loss noticeably increased and led finally to the developed turbulent flow.

The studies in this direction that have appeared in the last few decades show that nonlinear effects in viscous flows are rather peculiar. Extremely specific in this sense are the phenomena originating in a round tube on increase of the Reynolds number beyond the critical value. These phenomena are typical also of other cases of viscous fluid motion, in particular of a Couette flow between moving parallel planes, between the surfaces of rotating coaxial cylinders and in boundary layers.

At the inlet to the tube initial section, the flow carries disturbances of varying nature. These may be either disturbances from the outside, for example, from the enclosure in which

---

[3] Shiller, L., Chapter 4.

[4] Kochin, N. E., Kibel, I. A., and Roze, N. V., *Teoreticheskaya gidromekhanika* (*Theoretical Gydromechanics*), Part II, 4th ed., Fizmatgiz Press, Moscow, 1963, 658–686; Slyozkin, N. A., *Dinamika vyaskoy neszhimayemoy zhidkosti* (*Dynamics of Viscous Incompressible Fluid*), Gostekhizdat Press, Moscow, 1955, 385–432; Schlichting, G., *Teoriya pogranichnogo sloya* (*Boundary Layer Theory*), Nauka Press, Moscow, 1974, chap.16 and 17 (in Russian); Lin, C. C., *Teoriya gidrodinamicheskoy ustoychivosti* (*Theory of Hydrodynamical Stability*), Inostrannaya Literature Press, Moscow, 1958 (in Russian); *Laminar Boundary Layers*, Rosenhead, S., Ed., Clarendon Press, Oxford, 1963, 492–578; Monin, A. S and Yaglom, A. M., *Statisticheskaya gidromekhanika* (*Statistical Hydromechanics*), Part 1, Section "Gidrodinamicheskaya neustoychivost i vozniknoveniye turbulentnosti" ("Hydrodynamic Instability and Origination of Turbulence"). Nauka Press, Moscow, 1965, 77–161. This list is complemented by the Betchov, R. and Criminale, V. monograph. See also Reshotko, E., *Ustoychivost laminarnogo pogranichnogo sloya i ego perekhod v turbulentnyy* (*Stability of a Laminar Boundary Layer and Its Transition to a Turbulent One*), Vol. 21, Mir Press, Moscow, 1979, 11–57.

[5] A short review of this trend with detailed bibliography can be found in the report by H. Görtler and W. Welte to the Symposium on Turbulence, Kyoto, Japan, 19–24 September 1966 (Görtler, H. and Welte, W., Recent mathematical treatment of laminar flow transition problems, in *Physics of Fluids*, Vol. 10, Part 2 1966); Stuart, J. T., Nonlinear stability theory, *Ann. Rev. Fluid Mech.*, 3, 347–370, 1971.

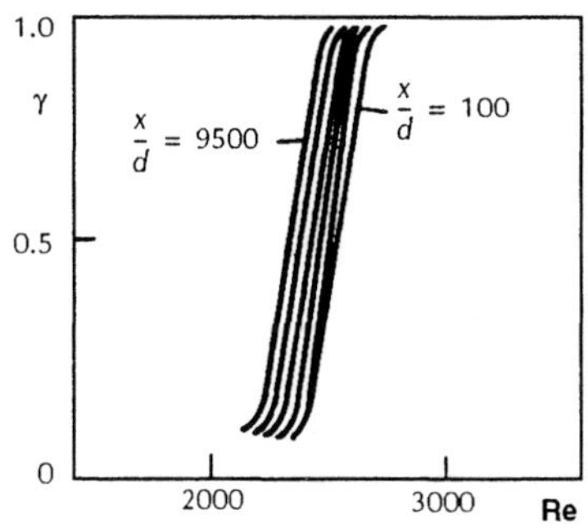

Fig. 194

the air intake tube is positioned, or from the tank with water escaping through a tube, or disturbances formed due to nonsmoothness of the tube inlet. The last reason is usually dominant. As was mentioned above, it was Reynolds who noted in his first experiments that at the values of **Re**, which are far from critical, discrete waves or groups of waves, decaying downstream, pass over rectilinear jets of dye at the tube initial section. Approaching its critical value, these disturbances, superposable on laminar flow, become even more intense until they fill in the entire flow region and the flow becomes fully turbulent.

It was noted by other researchers (Schiller et al.) that primary originations of these disturbances that are comparatively rare by the frequency of occurrence do not affect either velocity profiles in tube cross-sections or general tube resistance. Only in direct vicinity to the crisis does the influence of these waves becomes noticeable: velocity profiles distort, the resistance law changes.

A thorough investigation of the flow in a tube at Reynolds numbers close to critical showed that in one and the same fixed tube cross-section and at the same value of the Reynolds number **Re** $= u_{mean} d/\nu$ alternation of laminar and turbulent regimes may take place. This phenomenon is called intermittency. The cause of intermittency of flow regimes lie in the fact that turbulence, as shown by detailed experiments, originates first in discrete flow regions in the form of "clouds" or "spots" which, in the case of a tube, fill in the tubs cross-section with "plugs" that may have a length along the tube of the order of several tens of tube diameters, this length is dependent on the flow Reynolds number.

The fraction of time of turbulent regime existence in the given tube section serves as the main quantitative characteristic of the phenomenon. This dimensionless quantity, which is equal to zero when the flow is always laminar, and to unity when the flow retains a turbulent form, is called "the coefficient of intermittency" and has the notation γ. This quantity depends on both the flow Reynolds number and the distance x from the tube inlet. In Figures 194 and 195 (experimental points are omitted) the change of the intermittency coefficient is shown in a round tube depending on the Reynolds number **Re** and the relative distance from the tube inlet x/d by the Rotta[6] experiments in air and water and the Coles[7] experiments in water. The intermittence coefficient sharply increases in the zone of the critical Reynolds number, noting that in the sections close to the inlet it increases later than in the far ones. In Figure 196 in the circles not joined by a curve show the profile of velocities at **Re** = 2550 and $\gamma = 0.7$. In the same figure the velocity profiles relating to the same Reynolds number are presented also under laminar conditions ($\gamma = 0$, a solid line is the Poiseuille parabola, light circles) and under

---

[6] Rotta, J. C., Experimenteller Beitrag zur Entstehung turbulenter Strömung in Rohr, *Ing. Arch.*, No. 24, 258–281, 1956.

[7] Coles, D., Interfaces and Intermittency in turbulent Shear Flow, Mecanique de la Turbulence, Colloque Internationale du Centre Nationale de la Recherche Scientifique, Marseilles, France, 229–251.

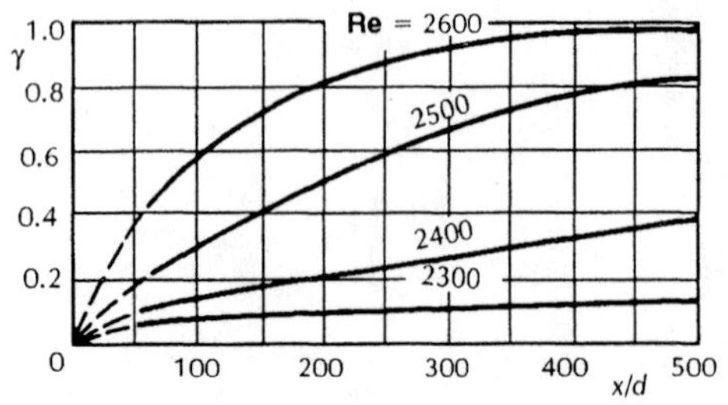

Fig. 195

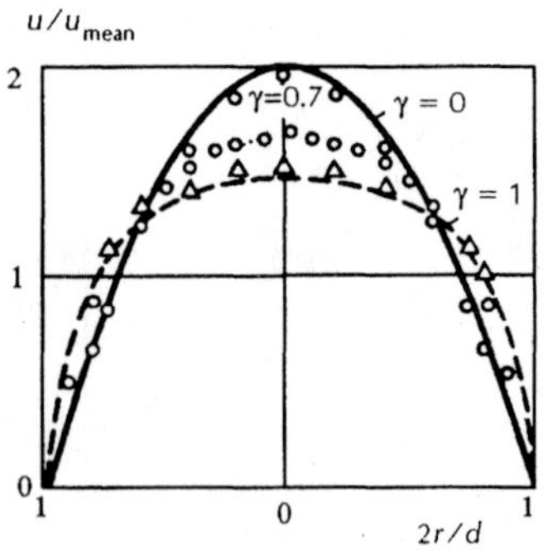

Fig. 196

turbulent conditions ($\gamma = 1$, a dashed line, triangles).

We note the fact of flattering of the velocity profile of laminar-to-turbulent flow transition. In this case, velocity increases on the tube axis and decreases at some fixed distance from the tube wall. Placing the velocimeter at a certain small distance from the wall we can judge by the growth of the velocity head on the laminar-to-turbulent flow transition. This technique was successfully applied in experimental studies of transition in a boundary layer. More accurate investigation of intermittency is attributed to the study into origination and development of velocity fluctuations in a flow by oscillographic recording of hotwire anemometer readings.

In Figure 197 the data on velocities of turbulent "plugs" motion in a tube are presented. Curves I and II represent the velocity of the front and rear walls of the "plug", respectively, expressed in terms of the portions of the flow mean velocity. As can be easily concluded from these two curves, the front edges of plugs in postcritical regimes move more quickly than the rear ones. By virtue of this the "plugs" stretch occupying (when moving) greater tube volumes. Moreover, the leading edge of one "plug" overtakes the rear edge of the neighboring "plug". All this leads to the fact that at postcritical values of **Re** at a distance from the tube inlet, a turbulent flow is established. In the same figure, for the sake of comparison, curves III and IV of the velocities of turbulent and laminar flows on the tube axis are given, respectively, that are also related to the flow mean velocity.

The velocity of the front wall of a "plug" is at first smaller, but as the Reynolds number grows it becomes greater than the velocity of a turbulent flow on the tube axis, and the rear wall velocity is much smaller than this velocity. Also note that below the critical Reynolds number the velocity of the "plug" front edge is, on the contrary, smaller than the velocity of the rear edge; this leads to the reduction of lengths of forming "plugs" and to their disappear-

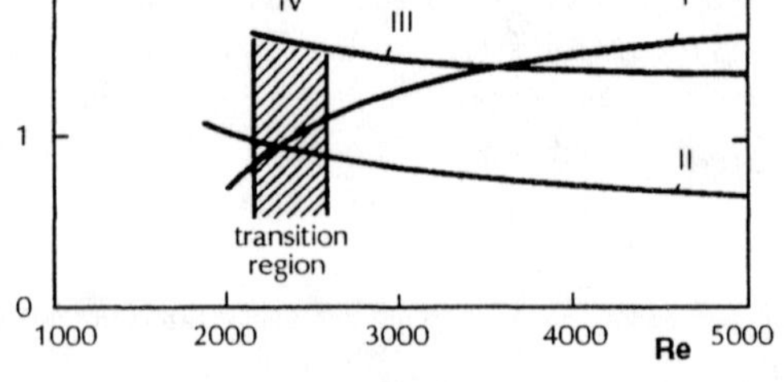

Fig. 197

ance in a laminar flow.

Along with viscous fluid flow in circular cylindrical tubes, Coles[7] also studied transition flows in a space between coaxial rotating cylinders. When passing over some Reynolds number, a first stable circular flow of fluid particles in planes normal to the axis of rotation is replaced by a flow with a cellular structure of closed secondary flows that is periodical in the direction parallel to the axis of rotation. This flow, usually called a Taylor flow, is formed in the case of a dominant flow of the inner cylinder. In a case of a dominant value of the outer cylinder rotation a stable rotational flow of particles changes over to a spiral mixed laminar-turbulent flow. These periodically spaced spirals, retaining their forms and interposition, rotate as the integral about a common axis of cylinders with angular velocity close to the arithmetic mean of angular velocity of cylinders.

Distribution of turbulent "plugs" along spirals obeys qualitatively the same laws as in the case of a mixed laminar-turbulent motion in a cylindrical tube. The intermittency coefficient γ on the central radius takes on values ranging from 0.3 to 0.7.

Changing the angular velocities of rotation of inner and outer cylinders, one can vividly observe the processes of origination and destruction of different regimes of viscous fluid flows between rotating cylinders, covering the range from the periodical Taylor flows to doubly periodical spiral structures. Of great interest is the relation between the turbulence characteristics in "plugs" and the Taylor secondary structures that thus serve as finite disturbances facilitating the laminar-to-turbulent flow transition.[8]

At present transient phenomena in laminar-turbulent flows attract the attention of theorists and experimenters. In addition to those already noted, we should mention successes achieved in the comprehension of the processes of laminar-to-turbulent flow transition which are studied from the viewpoint of nonlinear mechanics. The wave approach to the study into disturbances developing in a laminar flow that lost its stability was widely applied. Theoretical studies refer mainly to a classical so-called circular Couette between two coaxial cylinders (predominantly with motionless outer and rotating inner ones). The succession of conversions of flow from a purely azimuthal to a transverse wavy, vortical flow, its modulations and, finally, transition to a nonperiodical chaotic one, i.e., to a properly turbulent flow have been studied. Also analyzed are free convective flows (Rayleigh-Benard vortices), jet flows, etc. The most characteristic of the modern studies into the phenomena in a transient region and in a subsequent developed turbulent motion are the application of the notions related to discrete formation of vortices of a complex structure, whose origination and interaction are studied both experimentally and by numerical methods.

The notions of nonlinear oscillations form the basis of a general theory of transient phenomena, with the method of characteristics, which is in comparison of an oscillatory regime with the motion of a "representing point" and which substantially facilitates comprehension of the process, being successfully used for their study. The change in spectra of the frequencies of fluctuations that accompany the transition phenomena and is characteristic of them is studied theoretically and experimentally. A growth of the frequency of oscillations from the very rare discrete at the initial state of the transition to the continuous spectrum at the stationary developed process of turbulent flow is observed in the course of transition from an ordered laminar flow to a chaotic turbulent one. At present this phenomenon is subjected to a thorough analysis[9] and is confirmed by numerous experiments.[10] In spite of these successes, the problem of the transition of laminar flow conditions to turbulent ones is far from being

---

[8] Coles, D., Transition in circular Couette flow, *J. Fluid Mech.*, 21, 385–482, 1965. In this paper the problem of nonstationary phenomena in circular Couette flow is studied in detail and numerous photographs of visualized flows are supplied.

[9] *Nonlinear Dynamics*, ed. by R. Helleman, *Annals of the New York Academy of Science*, 357, 1980.

[10] IUTAM Symposium on Turbulence and Chaotic Phenomena in Fluids, Kyoto, September 5 to 10, 1983.

solved. Betchov, the well-known specialist in the field, notes that in contrast to fully laminar and turbulent flows which allow the familiar "first approximations", nobody has yet suggested a fruitful "first approximation" for the transient regime. In view of this, experimental results are very important (see below).[11]

2. Transient Phenomena in a Boundary Layer. Resistance Crisis of Blunt Bodies

A similarity of the phenomena of laminar-to-turbulent flow transition in a cylindrical tube with those in a circular Couette flow carries over to viscous fluid flow in boundary layers on surfaces of solid bodies, in jets, and wakes behind bodies. If we agree, using a comparatively rough approach, to quantitatively compare the velocity on the outer edge of a boundary layer with the velocity on the tube axis and the boundary layer thickness with the tube radius, we should introduce into consideration the Reynolds number for a turbulent layer

$$\mathbf{Re}_\delta = \frac{U\delta}{\nu}$$

that characterize a flow in the given layer section.

Numerous experiments on the determination of a critical number, $\mathbf{Re}_{\delta\,cr}$, for a boundary layer on a plate led to the values close to the tube critical number. The same order of $\mathbf{Re}_{\delta\,cr}$ was found in the case of a flow past a round cylinder, sphere, and wing profiles. Here it was revealed that relative position of a boundary layer critical section in which a laminar boundary layer passes over to a turbulent one is substantially dependent on the degree of disturbance of an outer flow running against the body. The critical Reynolds number of a boundary layer changes with this factor.

In contrast to the transient phenomena considered in the previous section, the presence of any flow mode in a boundary layer is stipulated by flow development along a boundary layer. Therefore, the initial portion of a layer is usually laminar, a transient region is positioned behind it where turbulent and laminar zones of a flow exist simultaneously and, finally, the zone of a developed turbulent flow that consists of a turbulent core and a thin viscous sublayer adjacent to a solid wall.

At small intensity of disturbances in an outer flow in the experiments with both plates and wings it was possible to transition to values $\mathbf{Re}_{\delta\,cr}$ much higher than those in strongly disturbed flows. Thus, for example, in a boundary layer on a plate, placed in a slightly turbulent wind tube a laminar flow was observed up to a critical boundary layer section where $\mathbf{Re}_{\delta\,cr} = 6290$, and on polished metallic airplane wings in a flight $\mathbf{Re}_{\delta\,cr}$ was brought up to 9300. This indicates that a relative size of a boundary layer laminar portion on a wing, especially in a quiet free stream, is dependent on wing surface roughness in the vicinity of its leading edge or on the presence of production shortcomings of surface treatment in this wing region. This difference of fluid flow in a boundary layer from the tube flow may be explained by the fact that near the wing tip the boundary layer is very thin, roughness asperities pene-

[11] The mechanism of laminar-to-turbulent boundary layer transition has been insufficiently studied. Its connection with the presence of local nonstationary separations in a viscous sublayer that are the consequence of pressure fluctuations formed the basis for a semiempirical method for determining the transition point that was suggested by A. A. Dorodnitsyn and L. G. Loitsyanskiy in the paper "On the theory of laminar-to-turbulent layer transition", *Prikl. Mat. Mekh.*, 9(4), 1945. A modern presentation of the problem on transition of laminar boundary layer to a turbulent one is found in the review paper by S. A. Gaponov and V. Ya. Levchenko, Modern problems of boundary layer transition, in *Uspekhi Mekhaniki*, No.4, VINITI Pess, Moscow, 1981.

trate through a boundary layer, and become the sources of disturbances in an outer flow which will pass inside a boundary layer through its outer edge.

Instead of $\mathbf{Re}_\delta$, which contains the inaccurate quantity δ, the numbers

$$\mathbf{Re}^* = \frac{U\delta^*}{\nu} \qquad \mathbf{Re}^{**} = \frac{U\delta^{**}}{\nu}$$

may be considered that are composed by more accurately determined quantities: by the displacement thickness and momentum loss thickness. Their corresponding values may be found directly from the measurements of velocities in layer sections and from recalculation. At present wide use is made of $\mathbf{Re}^{**}$. The value of $\mathbf{Re}^{**}_{cr}$ by the experiments on different wings in different wind tubes varies from 600 in strongly turbulent tubes to 1300 in slightly turbulent ones (by some data relating to the tubes with very small turbulence, $\mathbf{Re}^{**}_{cr}$ attained the value of 2300).

The observed difference in values of $\mathbf{Re}^{**}$ for different wings has one more cause. Similar to that in the case of a tube with alternating cross-section, a critical value $\mathbf{Re}^{**}_{cr}$ in a boundary layer depends also on the fact whether a critical section is in converging or diverging portions of a boundary layer. In the region of an accelerated flow (a converging portion of a layer) $\mathbf{Re}^{**}_{cr}$ has higher values than those in the region of a delayed flow (a diverging portion of a layer). In the case of a free boundary layer as, for example, in a jet or in a wake at a distance from the body, critical values of the Reynolds number are very small and practically always one has to deal with turbulent jets and wakes behind a body.

Theoretical determination of the lower critical Reynolds number of a boundary layer $\mathbf{Re}^{**}_{cr}$, by which we mean the value $\mathbf{Re}^{**}$ in that boundary layer section, where the flow stability is lost, may be fulfilled with satisfactory accuracy by one-term approximation. Thus, designating $\mathbf{Re}^{**}$ and the form parameter f in the layer section, where a laminar flow loses its stability, in terms of

$$\mathbf{Re}^{**}_{cr} = \left(\frac{U\delta^{**}}{\nu}\right)_{cr} \qquad f_{cr} = \left(\frac{U'\delta^{**2}}{\nu}\right)_{cr}$$

one may, using the stability theory, obtain an approximate relation between $\mathbf{Re}^{**}_{cr}$ and f_{cr}, which is presented by a solid line in Figure 198. Within the framework of the stability theory this plot is common for all plane boundary layers, i.e., it is independent of the velocity distribution $U(x)$ on the outer edge of a boundary layer.

It follows directly from the graph that critical Reynolds numbers $\mathbf{Re}^{**}_{cr}$ corresponding to positive value of f_{cr}, i.e., to the converging portion of a boundary layer, considerable exceed critical numbers in the zone of a retarded flow in a diverging region. This fact is conventionally expressed, stating that a laminar flow in a converging portion of a boundary layer is more stable then in a diverging portion. In this case the value of the critical Reynolds number $\mathbf{Re}^{**}_{cr}$ is taken as the quantitative measure of stability.

Having calculated, by the one-term theory, the quantities $f = f(\bar{x})$ and $\mathbf{Re}^{**} = \mathbf{Re}^{**}(\bar{x})$ for a flow with a prescribed distribution of velocities $U(\bar{x})$, we shall eliminate the dimensionless abscissa $\bar{x}$ and construct the curve of relation between current values of f and $\mathbf{Re}^{**}$ in the same coordinates as in Figure 198. This will lead to the curve (a dashed line in Figure 198) whose inclination has a positive sign and that intersects the basic curve. In fact, while the critical Reynolds number $\mathbf{Re}^{**}_{cr}$ in a diverging portion of a boundary layer ($f < 0$) is smaller than in a converging one ($f > 0$), the current Reynolds number $\mathbf{Re}^{**}$, on the contrary, grows downstream along the body surface, i.e., on transition from positive to negative f. The intersection point of solid and dashed lines will determine the unknown f_{cr} for the given specific case of prescribing the outer velocity $U(\bar{x})$ and thus the abscissa $\bar{x}_{cr}$ of

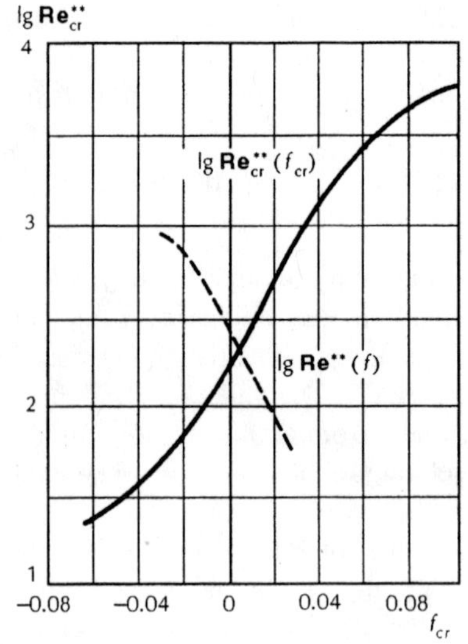

Fig. 198

the point of stability loss.

This point of laminar boundary layer stability loss should not be confused either with the beginning of the transition region or with that point of laminar-to-turbulent flow transition that is of interest for practice. By the beginning of the transition region we usually mean the point (the boundary layer section) at which the developing disturbances grow so noticeably that they begin to change a laminar character of flow in a boundary layer, and by the transition point we mean such an intermediate point of the transition region where a turbulent character of flow has substantially manifested itself, for example, in distortion of a velocity profile in boundary layer cross-sections. In those cases when the length of a transition region is small compared with the body size or high accuracy in determining the transition position is not required, the use of the "transition point" concept is quite acceptable.

Experimental determination of the transition point in a boundary layer is usually carried out in the following way. A pilot tube whose opening is directed toward the flow is made to move along boundary layer, maintaining the tube tip D (a dynamic opening) at the same small distance h (Figure 199) from the wing surface. Substracting from the total head, registered by the tube opening D, the pressure in the corresponding section of a boundary layer that is measured by means of the opening on the wing surface, positioned directly below the microtube tip, we may determine velocity at a chosen fixed distance from the surface in different sections of a boundary layer.

In view of the boundary layer thickening downstream, the dimensionless velocity u/U, which is measured at the same distance from the wing surface inside a layer, should decrease. Actually, in this case the relative coordinate h/δ of the measuring point decreases and the point itself, as it goes deeper into the boundary layer, passes over to relatively smaller velocities.

Figure 199 shows the velocity profiles in subsequent cross-sections 1 to 6 of laminar (solid lines), transient, and turbulent (dashed lines) boundary layers on a wing at one and the same value of **Re** of a free stream. As shown in Figure 196 these profiles greatly differ from each other. A vertical line corresponds to the chosen distance $y = h$ of the microtube tip from the wing surface. The points A_1, A_2, A_3 give the values of u/U registered by a microtube in a laminar portion of a boundary layer. When the tube tip gets into the region of transient or turbulent modes, the value of u/U sharply jumps from relatively small value (the point A_3) to

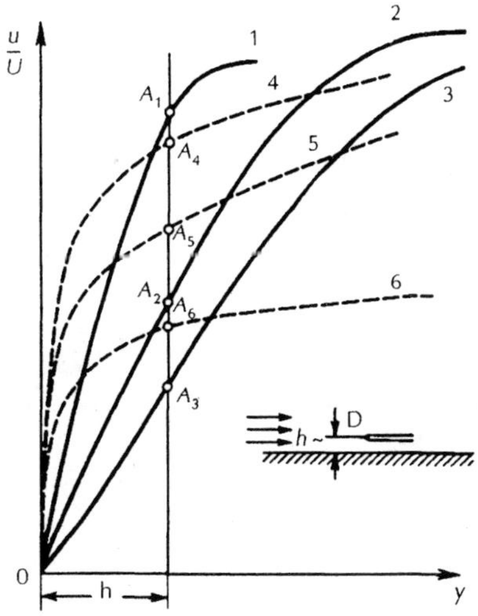

Fig. 199

the value A_4 and then it decreases, passing the values A_5 and A_6. If we lay off u/U on the axis of ordinates (Figure 200) and relative distances along the wing surface on the abscissa, the result of such a construction is that the curves may be obtained which are similar to those given in Figure 200. The region to the left from the vertical dashed line corresponds to a laminar boundary layer, transient region lies between dashed line and vertical lines, and, finally, to the right from vertical lines a turbulent flow takes place. In Figure 200 several curves are given that refer to different Reynolds numbers $\mathbf{Re}_\infty = U_\infty b/\nu$ within the range from $1.7 \cdot 10^6$ to $5.1 \cdot 10^6$. It is seen from the consideration of these curves that the length of the transient region decreases with the increase in the Reynolds number of a free stream, but it still has the values quite comparable with the wing chord. Experimental determination of the transition point exhibits some arbitrariness; some authors determine the transition point as a center of the transition region, others as the minimum point on the curve u/U, and get others as the maximum point.

Numerous experimental studies, both laboratory and natural (an airplane, ship body), showed that the position of the transition point is greatly dependent on many flow parameters. These are, first of all, the Reynolds number **Re** and quantitative characteristics of a free stream turbulent structure. Among these characteristics the following three are of basic importance: (1) the degree, or intensity of turbulence ε determined by the ratio of the time-averaged amplitude of velocity fluctuations in a free stream to its mean velocity; (2) the turbulence scale L characterizing spatial extension of fluid volumes in an outer flow, which are subjected to disturbances statistically interrelated in the sense that the correlation coefficient of velocity fluctuations at any two points of these volumes is other than zero, and (3) frequency of fluctuations in an outer flow n taken by the maximum value of the distribution function of kinetic energy of fluctuations by frequencies.[12]

The transition phenomenon depends on pressure distribution in the outer flow: in a converging portion of a boundary layer, where the outer flow accelerates transition is delayed shifting downstream, while in a diverging portion with a delaying flow it conversely is hastened, passing over to the upstream portions. Finally, of great importance is the state of the

[12] We shall restrict ourselves to these qualitative estimates of the degree, scale, and turbulence frequency of a free stream; their quantitative determination is given later.

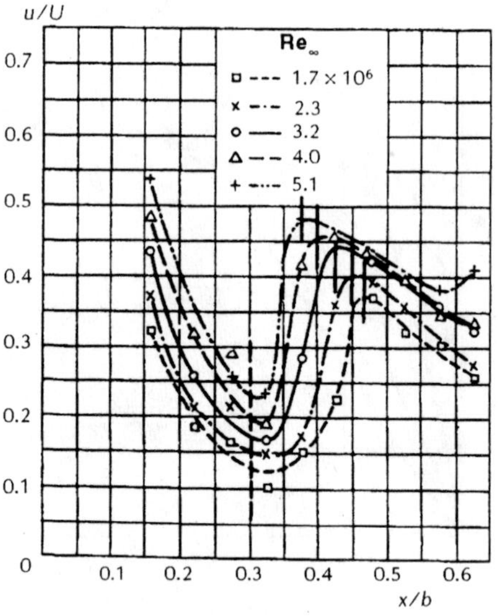

Fig. 200

surface of a streamlined body: the degree of its roughness, waviness, heating, etc.

Recently, experiments have been conducted to reveal the effect of admixtures introduced in small concentrations into fluid in a boundary layer on delaying the extension of a laminar portion or on reducing the rate of the already-developed turbulence. Especially useful in this case are high-polymer admixtures,[13] the introduction of which even in small portions influences flow mode (the Thoms effect).

We shall pause on illustrating the influence of some just-mentioned factors on the position of the point of laminar-to-turbulent flow transition in a boundary layer.

The effect of the Reynolds number on the position of the transition point on the surface of a smooth wing is expressed in a shift of this point, with the growth of the Reynolds number toward the leading edge. For different wing profiles this shift occurs in different ways, also being dependent on test conditions, i.e., on free stream turbulence, etc. Some general notes may be taken of this fact.

If, on the wing surface behind the point of minimum pressure, there exist the separation point of a laminar layer. Then this points farthest (downstream) possible transition point, because a stalling layer nearly instantly changes over to a turbulent state. As the Reynolds number grows the transition point shifts upstream and turn upstream of the separation point. In this case laminar separation ceases and is replaced by a turbulent one which is either firmed, but considerably further downstream, or is completely. The separation point shifts toward the point of pressure minimum and then goes over to the layer converging portion. This is schematically shown in Figure 201 for the upper surface of a wing profile with the delayed converging portion (the point of minimum pressure occurs at approximately 45% of the chord); in the same figure, for comparison, the curve of stability loss point shifting is also given.

As is seen from the graph, a laminar portion of the boundary layer on this profile spreads

---

[13] Ioselevich, V. A. And Pilipenko, V. N., Turbulent boundary layer of fluid with polymer admixtures, in *Metody raschyota turbulentnogo pogranichnogo sloya. Ser. Itogi nauki i tekhniki* (*Methods for Calculation of a Turbulent Boundary Layer. Series: Results of Science and Technology*), VINITI Press, Moscow, 1977, chap.5; Lumley, J. L., Drag reduction by additives, Annu. Rev. Fluid Mech., 1, 367–384, 1969.

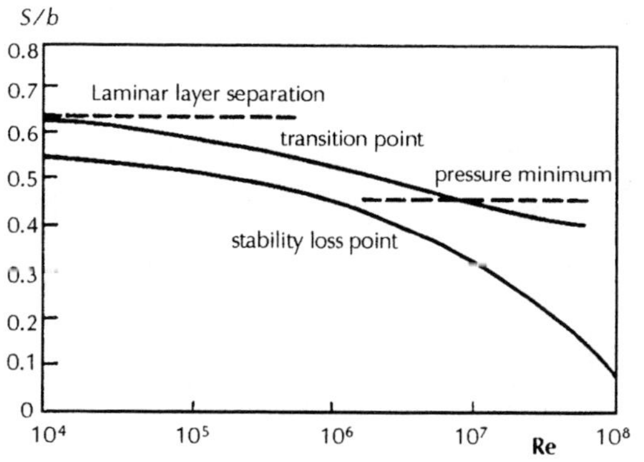

Fig. 201

over about the entire front region of the wing profile even at large values of the Reynolds number. Such wing profiles are called laminarized. On ordinary wing profiles the point of minimum pressure lies considerably closer to the profile nose, thus reducing the laminar layer portion.

To roughly estimate the position of the transition point on a wing profile with a smooth surface in the virtually most interesting range of the Reynolds numbers 10^6 to 10^7, it may be recommended to take the point of pressure minimum as the position of the transition point.

The experiments by Schubauer and Skramstad[14] conducted in a wind tube with the intensity of flow turbulence in a working portion varying within a wide range, starting from very small values on the order of 0.02%, showed the existence of two main reasons for the origination of laminar-to-turbulent layer transition, *viz.*, the loss of laminar layer stability and the presence of intensive disturbances in the outer flow.

Figure 202 presents the curves of the effect of the intensity of the outer flow turbulence ε on the local Reynolds number $\mathbf{Re}_x = U_\infty x/\nu$ composed for the abscissas of the points separating a laminar portion of a boundary layer on a plate in a longitudinal flow from the transient zone and the zone of a developed turbulent flow in a boundary layer. As can be judged by this curve, at the intensity of outer flow turbulence not exceeding 0.1%, the boundaries of laminar and turbulent portion of a boundary layer are independent of it. In this case the transition phenomenon is explained by the first of the reasons mentioned – the loss of laminar layer stability. To the beginning of the transition zone corresponds the value of $\mathbf{Re}_x$ somewhat smaller than $3 \cdot 10^6$; to the end of the transition zone, i.e., to the beginning of the zone of a developed turbulent flow corresponds the value of $\mathbf{Re}_x$ close to $4 \cdot 10^6$.

As the intensity of the outer flow turbulence grows, the second reason, the effect of outer disturbances becomes dominant. It is vividly seen from. Figure 202 that with the increase in the intensity of the outer flow turbulence the values of the Reynolds numbers on the transient zone boundaries begin to sharply decrease, the size of a laminar portion reduces nearly two-fold with the intensity being conserved within a comparatively narrow range (up to 0.36%). In ordinary wind tubes the turbulence intensity may attain 1% and in other cases, for example, in a circulating portion of a turbine or a compressor, it may reach considerably greater values. Here, laminar portions on the surface of streamlined bodies (wing profiles, impeller vanes) are insufficient. Conversely, in the case of body motion through a quiescent fluid (an airplane in a quiet atmosphere, etc.) intensity of free stream turbulence may be very small and the length of a laminar portion is large. The presence of the noticeable zone of transition, which

[14] Schubauer, G. B. and Skramstad, H. K., Laminar Boundary-Layer Oscillations and Transition on a Flat Plate, NACA Rep., 909, 1948.

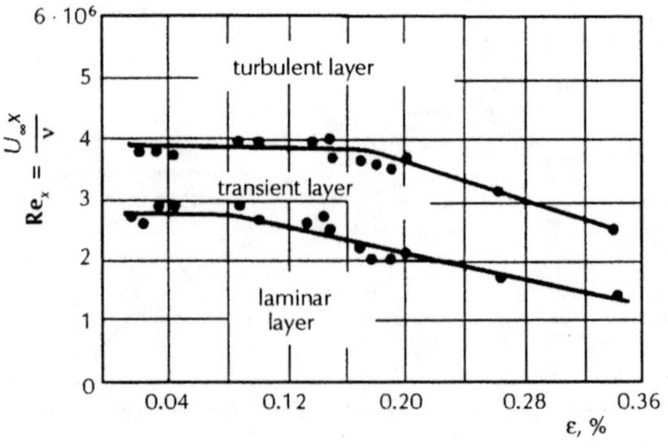

Fig. 202

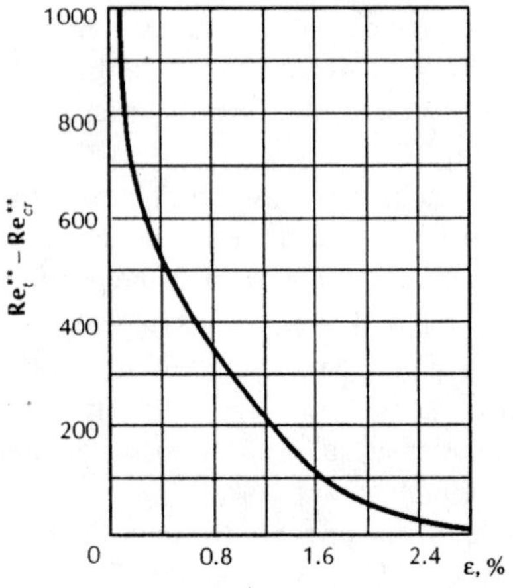

Fig. 203

should not be neglected in detailed calculation of a boundary layer, also attracts attention.

The influence of the intensity of a free stream turbulence on the length of a transition portion may be judged by the Greenwill empirical curve (Figure 203)[15] composed from the data of the experiments of Schubauer-Skramstad, Hall-Hislope and Driden (test points are omitted) on a plate in a longitudinal flow. In contrast to Figure 203, the difference, $\mathbf{Re}_t^{**} - \mathbf{Re}_{cr}^{**}$, of the Reynolds numbers involving the thickness of momentum loss at the points of transition and stability loss as the length is plotted on the ordinate axis. Judging by this curve the length of the transition zone rapidly decreases with the growth of turbulence intensity ε, attaining small values at ε on the order of 2.5 to 2.8%.

To the less studied factors one should refer the effect of the scale of free stream turbulence (see Section 4) on the position of the transition point. The results of the experiments[16] with a boundary layer on an elliptical cylinder positioned at the zero angle in an air flow, which is made turbulent by grids placed in front of the cylinder at some distance from it (the dimensions of grid cells are given in the figure), may serve as an example of this effect (Figure 204).

---

[15] See the frequently cited monograph by Schlichting, p. 445, Figure 16.21.

[16] Dryden, H. L., Modern development of boundary layer mechanics, in *Problemy mekhaniki* (*Problems in Mechanics*), Mises, R. and Karman, T., Eds., IL Press, Moscow, 1955.

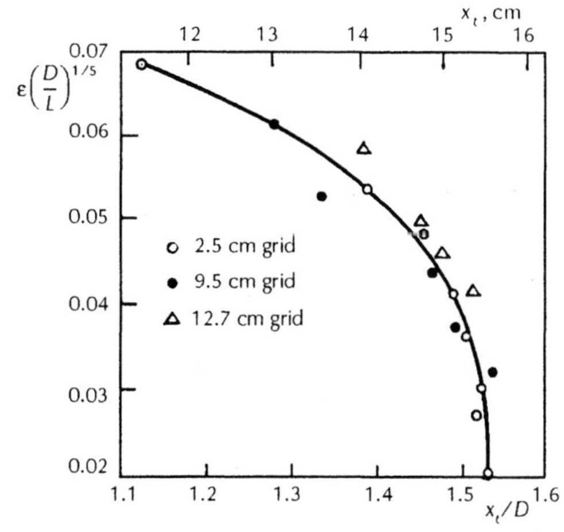

Fig. 204

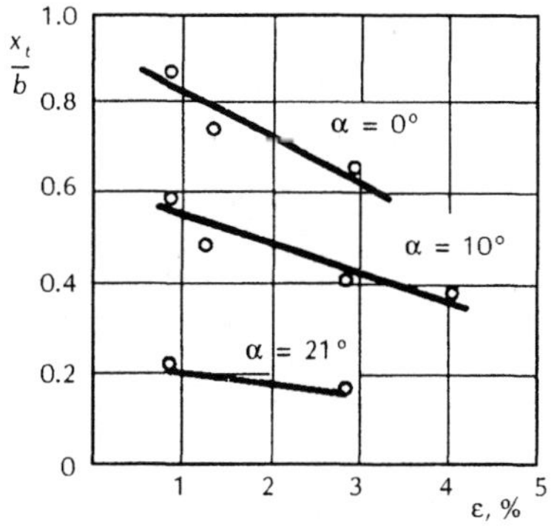

Fig. 205

While moving downstream the vortices, created by grid rods, are destroyed thus forming smeared regions of a disturbed flow dimensions represent the turbulence scale. The turbulence scale L may be measured and its ratio to the linear size of a streamlined body (in our case, to the smaller diameter of the ellipse D) along with turbulence intensity ε serves as the characteristic of free stream turbulence.

The graph in Figure 204 expresses the relation between a dimensionless abscissa of the point of laminar-to-turbulent layer transition on the surface of an elliptical cylinder and the Taylor parameters,[17] which represents the product of turbulent intensity by the fifth root of the ratio of the body characteristic size D to the scale of turbulence L. It is seen from the graph that at small values of the Taylor parameter outer disturbances exert a weak effect on the size of the layer laminar portion. Here, everything is determined by the inner stability of motion in a layer. At comparatively large values of the parameter this influence sharply increases and the length of the laminar portion rapidly reduces.

To qualitatively illustrate the effect of the streamwise pressure drop on the position of the transition point in a boundary layer at different intensities of free stream turbulence we may use the results of the experiments conducted by Minskiy[18] (Figure 205). The relative position of the transition point on the upper surface of the 14%-thick wing profile is plotted on the axis of ordinate, and the degree of turbulence ε is laid off on the axis of abscissas. As is shown by the graph, with the grows of the free stream turbulence intensity, a significant shift of the transition to the wing tip is observed.

The length of a laminar portion in a wing frontal section also sharply decreases with increase in the angle of attack (the curves of Figure 205 refer to different values of the angle of attack α marked on them). This is natural because with the increase in the angle of attack follows an intensification of pressure recovery, thus leading to the increase in boundary layer diffusivity, which in turn causes weakening of stability of a boundary layer laminar portion.

As mentioned in the previous chapter, the curvature of a surface at small ratios of the laminar boundary layer thickness to the curvature radius appears only in the outer flow velocity. It may be noted that on concave surfaces, even comparatively small curvature exerts a considerable effect on transition. On convex surfaces this effect is unnoticeable; the transition

[17] Theoretical substantiation of the choice of this parameter may be found in Taylor, G. I., Statistical theory of turbulence, *Proc. R. Soc. London*, A 151, 421, 1935.

[18] Minskiy, E. M., The effect of free stream turbulence on transition, *Tr. TsAGI*, 4–15, 1939.

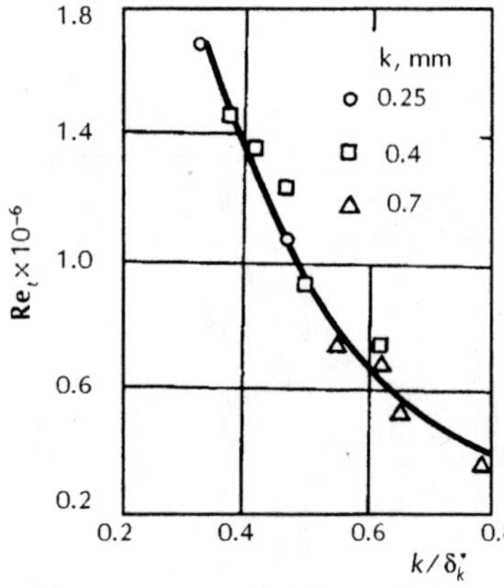

Fig. 206

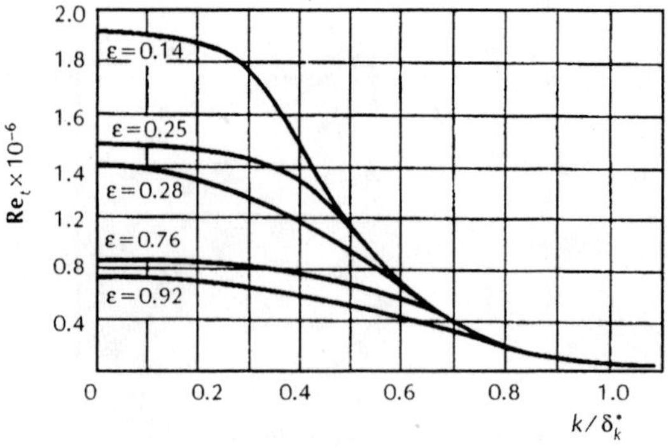

Fig. 207

phenomenon occurs in the same manner as on the plate with the corresponding distribution of pressures.

Of great practical interest is the problem on the effect of the condition (treatment) of the surface on the position of the point of laminar-to-turbulent layer transition.

Roughness asperities of the surface play the role of sources for origination of disturbances in a boundary layer. These disturbances add to those introduced into a boundary layer from the outer flow.

The ratio of k/δ_k^* of the height k of the roughness asperity to the displacement thickness δ_k^* at the point of this asperity dislocation on the body surface is taken as the quantitative estimation of the effect of the single element (asperity) on the origination of transition in a wake behind it or on the position of the transition point. The Dryden curve (Figure 206), which expresses the dependence of the Reynolds number at the transition point $\mathbf{Re}_t = x_t U_\infty/\nu$ on the relative roughness k/δ_k^*, may be taken as the justification of the choice of this characteristic. Application of this characteristic allows one to combine on one curve the results of the experiments relating to different absolute roughnesses k. The generality of the mechanism of the effect on the position of the transition point of disturbances, originating due to the presence of separate roughness asperities and then developing in the wake region behind the asperity and existing in the free stream itself, may be judged by the plot presented in Figure 207.[19] It follows from the curves of this figure that at small relative roughness k/δ_k^* the position of the transition point is determined by the intensity of free stream turbulence ε and at comparatively large relative roughness is independent of this intensity.

In Dryden's paper one may find different experimental data on the problem of the surface roughness effect on laminar-to-turbulent boundary layer transition.

Experimental studies of the boundary layer structure conducted on the basis of modern methods of measurements showed great complexity of the mechanism of laminar-to-turbulent flow mode transition. Referring those interested to the detailed report of Tani,[20] we note only some (the most peculiar) features of the phenomenon considered.

First, of interest is the instability of the form of disturbances. If by means of a special device, for example, of an oscillating filament stretched along the plate span, disturbances are

[19] Dryden, H. L., Laminar-to-turbulent flow transition (a review paper), in *Turbulentnye techeniya i teplopere-dacha* (*Turbulent Flows and Heat Transfer*), Inostrannaya Literature Press, Moscow, 1963.

[20] Tani, I., Review of some experimental results on boundary-layer transition, *Phys. Fluids*, 10(9), 11–16, 1967.

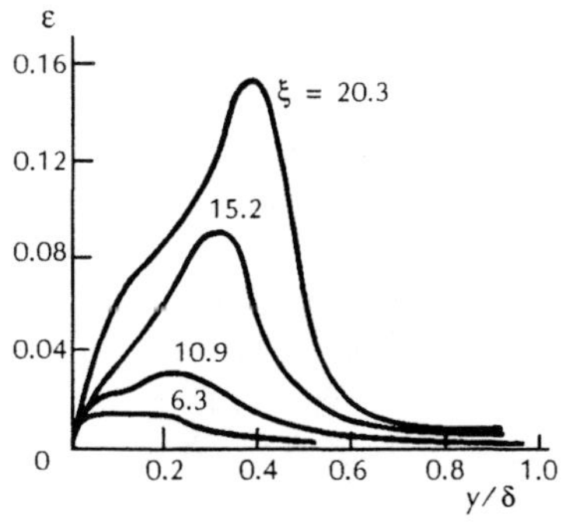

Fig. 208

generated equal in amplitude. They very quickly lose their plane character and acquire a wavy structure with peaks and dips in the distributions of longitudinal (along the flow) and transverse (along the span) averaged velocities, and with mean intensity of longitudinal velocity fluctuations, periodically alternating along the span.

Concentration of flow vorticity corresponds to the peaks in these distributions. High-speed filming vividly shows origination of vortex loops of the type Π-shaped vortices with vortical "ropes" distributing along the flow (Section 10, Chapter 9). These "free" vortices, coinciding in direction with streamlines, induce velocities in planes normal to the flow direction, thus facilitating the interlacement of these vortical ropes and even greater complication of the flow structure.

General flow instability in the transition region leads to the increase in the amplitude of flow longitudinal fluctuations and thus to the growth of the turbulence local intensity ε. In this case, as may be judged by the test curves of Klebanoff, Tidstream, and Sargent[20] given in Figure 208, the intensity ε grows sharply at a distance from the source of disturbances (in the figure the values of the distance ξ from the source of disturbances given in centimeters are presented by the markers on the curves).

As the distance from the source of disturbances grows the maximum of ε shifts from the wall attaining $\xi = 20.3$ cm, the position corresponding to two fifths of the boundary layer thickness. These curves refer to the crests of the wavy distribution of disturbances transverse to the flow. Disturbances of small intensity, whose frequencies lead one to assume that laminar modes persist in these regions for a long periods of time, are observed in the valleys of this distribution.

Thus, in a boundary layer, as well as in tube flows, turbulence originates in limited regions coexisting with the laminar flow regions. These turbulent "clouds" or "spots" similar to turbulent plugs in tube flows propagate downstream in a boundary layer and form the phenomenon of intermittency of laminar and turbulent flow modes in a transition region.

Along with this intermittency, which is typical of the transition region (i.e., the consequence of formation of closed turbulent zones, "clouds" and "spots"), the intermittency of another type is observed near the outer edge of a boundary layer that is characterized by mutual penetration of fluid volumes through the boundary layer edge from a comparatively weakly disturbed outer flow into the boundary layer region filled with disturbances of a greater amplitude.

It should be noted that in general one can speak of the outer edge of a turbulent layer, as well as of the boundary between a viscous sublayer and a turbulent core only as some thin transverse regions filled either with laminar or turbulent inner structure, with "protuberances" penetrating the boundary layer from the side of the outer flow and viscous sublayer, and imparting an intermittent character to the entire flow. In the outer zone, the so-called "sub-layer", sharp changes take place in the degree of flow turbulence from the substantial value, as compared with the outer flow, to a small degree of turbulence in the inner flow.

In spite of such complex inner motions, the boundary layer region is rather simple if considered in a comparatively rough-averaged form corresponding to the indications of such commonly used measuring devices as Pitot tubes connected with standard micromanometers.

Thus, making use of experimental profiles of velocities in the cross-sections of the boundary layer transient region, Persh[21] calculated conditional thickness δ^* and δ^{**} as well as their ratio H. Persh could show that independent of the longitudinal pressure drop in a boundary layer (and even at variable Mach numbers) the value of H constantly falls from its laminar value (about 2.6) to a turbulent one (1.3 to 1.4). On the basis of processing of a great amount of experimental data, he formulated the law of dimensionless velocity variation in conformable points of transition layer cross-sections depending on the fall of the parameter H and showed that the velocity profiles in the boundary layer sections of the transition zone form a one-term family with the parameter H.

The phenomena in the transition zone of a boundary layer on a plate in a longitudinal flow were considered by Dhawan and Narasimha[22] in a quantitative formulation from the viewpoint of the scheme of "intermittency" of the origination of turbulent "spots " in a boundary layer. The coefficient of intermittency γ, whose determination was presented in the previous section, was found experimentally by processing; oscillograms of velocity fluctuations measured by a low-inertia hotwire anemometer. The following exponential function

$$\gamma = 1 - \exp(-A\xi^2) \tag{1}$$

may serve as an analytical expression of the variation of γ in the function of the longitudinal coordinate x, where A = const = 0.412, and $\xi = (x - x_t)/\lambda$ is the dimensionless difference between the longitudinal coordinates: x, inside the transition zone; x_t, origin of the transition zone; and λ, the conditional length of the transition zone determined by the curve $\gamma(x)$ in the form

$$\lambda = x_{\gamma = 0.75} - x_{\gamma = 0.25}$$

This length entering into the main relation (Equation 1) is the characteristic linear scale of the transition zone.

Comparison of the profiles of averaged velocities in the transition zone u_P that are determined experimentally by the Pitot tube, with the theoretical laminar u_L and turbulent u_T profiles, shows that a simple relationship exists:

$$u_P^2 = (1-\gamma)\,u_L^2 + \gamma u_T^2 \tag{2}$$

For a time-mean velocity $\bar{u}$ obtained as a result of processing of oscillograms this relationship is much simpler and is reduced to the linear

$$\bar{u} = (1-\gamma)\,u_L + \gamma u_T \tag{3}$$

Equation 2 may serve for approximate determination of the intermittency coefficient γ by the readings of the Pitot tube.

---

[21] Persh, J., The behaviour of the boundary layer in the region of transition from laminar to turbulent flow, *J. Aeron. Sci.*, 22(6), 443, 444, 1955.

[22] Dhawan, S. and Narasimha, E., Some properties of boundary layer flow during the transition from laminar to turbulent motion, *J. Fluid Mech.*, 3(4), 418–436, 1958.

The relationship

$$H = H_T + (H_L - H_T)\, e^{-A\xi^2} \tag{4}$$

was determined for the ratio of boundary layer conditional thickness $H = \delta^*/\delta^{**}$.

The empirical theory of fluid motion in a transition zone of a boundary layer makes it possible to accurately quantitatively describe not only kinematic (velocity profiles, ratio of layer thicknesses), but also dynamic (local and overall coefficient of friction) aspects of the phenomenon. However, in this case the basic quantity – the abscissa x_t of the origin of the transition zone entering into the determination of the variable ξ – remains unknown.

It may be considered at a very small intensity of the turbulence ε of the outer flow that the origin of the transition zone coincides with the point of the loss of laminar flow stability in a boundary layer, the calculation of whose position by the graph $\mathbf{Re}^{**}_{cr}(f_{cr})$ was explained earlier.

The phenomenon of laminar-to-turbulent flow transition is allied to the reverse of a turbulent flow to a laminar one that is observed in converging portions ($dp/dx < 0$) of boundary layers. Such a "relaminarization" of a boundary layer may be determined based on the existing model schemes of a turbulent motion at comparatively small values of the Reynolds number. The found local criteria of relaminarization that are expressed by "critical" values of the parameters determining motion[23] are

$$K_{cr} = \frac{\nu}{U^2}\frac{dU}{dx} = 3.7 \cdot 10^{-6}, ^{24}$$

$$\Delta_{p.\,cr} = -\frac{\nu}{\rho v_*^3}\frac{dp}{dx} = 0.025, ^{25}$$

where $U(x)$ and $p(x)$ are the distributions of velocity along a boundary layer on its outer edge and of pressure in boundary layer cross-sections; $v_* = \sqrt{\tau_w/\rho}$ is the "dynamic" velocity which is considered later.

We shall pause on the consideration of some effects related to the transition phenomenon. If we analyze the curves (Figure 209) of the dependence of the coefficient of resistance c_x of a sphere on the Reynolds number, then we may note a sharp decrease of c_x in the range of **Re** of the order $(1 \text{ to } 3)\cdot 10^5$. This phenomenon is called the drag crisis. The critical Reynolds number $\mathbf{Re}_{cr}$ corresponding to it, defined as the value of **Re** at $c_x = 0.3$, is strongly dependent on those parameters that exert an effect on the laminar-to-turbulent boundary layer transition, in particular, on turbulence intensity.

The less turbulent the flow, the higher the value of $\mathbf{Re}_{cr}$ attained at measurements of sphere resistance in this tube. Thus, the curve V ($\mathbf{Re}_{cr} \approx 270{,}000$) corresponds to the experiments in a wind tunnel with the turbulence intensity 0.5%, the curve I ($\mathbf{Re}_{cr} \approx 125{,}000$) corresponds to the flow with the turbulence intensity of 2.5%; the other curves correspond to the tunnels with intermediate values of turbulence degree.

To understand the cause of the noted phenomenon of sharp decrease in sphere resistance we shall resort to the consideration of curves of pressure distribution over its surface (Figure

[23] See Chapter 1, Section 1 (item 4), composed by A. N. Sekundov, in the review Ginevskiy, A. S., Ioselevich, V. A., Kolesnikov, A. V., Lapin, Yu. V., Pilipenko, V. N., and Sekundov, A. N., Methods for calculation of a turbulent boundary layer, *Results Sci. Tecnol. Fluid Mech.*, 11, 166–170, 1978.

[24] Kline, S. J., Reynolds, W. C., Schraub, F. A., and Runstadler, P. W., The structure of turbulent boundary layers, *J. Fluid Mech.*, 3P, 741–773, 1967.

[25] Patel, V. C. and Head, M. R., Reversion of turbulent to laminar flow, *J. Fluid Mech.*, 34, 371, 1968.

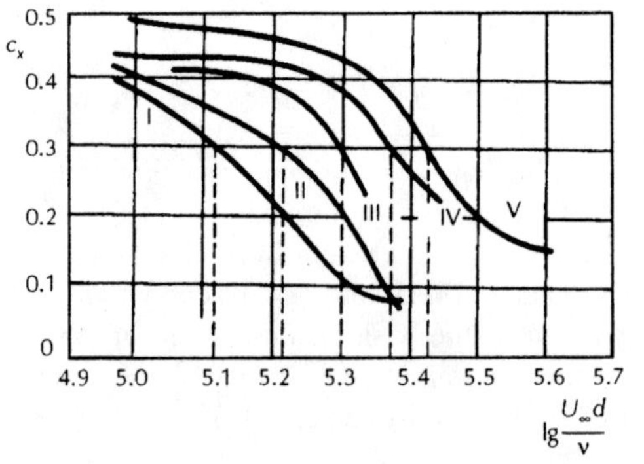

Fig. 209

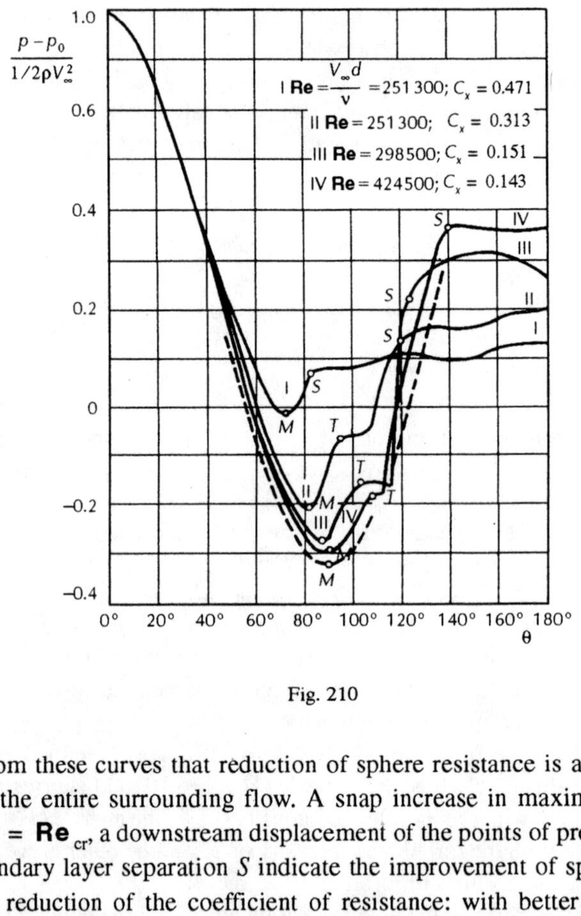

Fig. 210

210) It follows from these curves that reduction of sphere resistance is attributed to radical reconstruction of the entire surrounding flow. A snap increase in maximum pressure drop observed near **Re** = **Re** $_{cr}$, a downstream displacement of the points of pressure minimum *M* and points of boundary layer separation *S* indicate the improvement of sphere streamlining. This explains the reduction of the coefficient of resistance: with better enveloping of the sphere surface by a flow, the pressure distribution approaches the ideal (in Figure 210 shown by dashes) at which, according to the d'Alembert paradox, resistance should be equal to zero.

It should be noted that visual observations (Figure 211) confirm the described picture of the improvement of sphere streamlining in the range of Reynolds number mentioned. This phenomenon, called the flow crisis, is explained by the change in the position of the laminar-to-turbulent boundary layer transition on a sphere. At **Re** smaller than $1.5 \cdot 10^5$ separation of a laminar boundary layer on a sphere surface takes place that changes over to a tur-

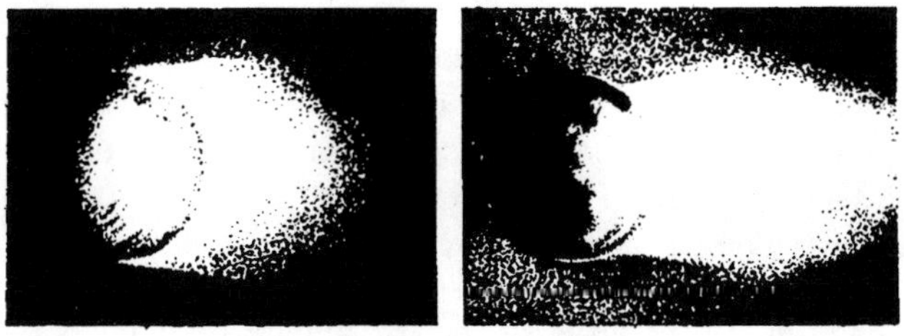

Fig. 211

bulent one outside the sphere in a separated layer.

As the Reynolds number grows the transition point T located in the wake behind a sphere moves toward the flow, first to the sphere surface. As the point T reaches the point of laminar layer separation S, the flow in a separated layer near the separation point becomes turbulent, agitation of fluid flowing from the separation zone along the sphere surface becomes stronger, the boundary layer entrains the fluid, the flow is improved and the separation point (Figure 210) begins to shift downstream. The separation point S corresponds to turbulent boundary layer separation since the transition point precedes the point of turbulent boundary layer separation.

It will be proved that the separation point of a turbulent boundary layer, with the same velocity distribution in the outer flow, is usually located downstream of the point of laminar boundary layer separation. Taking place at the transition point T is local separation of a laminar layer that is not then developed and that is accompanied by reverse sticking of the already-turbulent boundary layer to the sphere surface. Such a turbulent bubble of separation in its developed form was observed on front portions of wing profiles. Its appearance and disappearance led to "enigmatic" variations of lift force and resistance of wings at large attack angles, and to the "hysteresis" of the lift force coefficient at initial increase and subsequent decrease of the angle of attack, among others. One of the first descriptions of this phenomenon may be found in the collected papers edited by Goldstein.[26]

The essence of the phenomenon of bubble origination is in the fact that at comparatively large values of the Reynolds number the separated laminar layer is extremely unstable and immediately changes to a turbulent state. In this case the near-wall edge of the layer, separated from the body surface, thanks to the origination of the intensive exchange between fluid masses of this boundary and those of the separation zone where fluid flow has a retrograde character, is smeared, sucked in, and sticks to the body surface, thus forming a closed separation zone which is the separation bubble itself. This bubble, similar to the developed separation, distorts the outer flow, but much weaker than the latter, leads to the so-called strong interaction between a boundary layer and the outer irrotational flow.

Analogous closed separation zones were observed in the vicinity of a leading edge of wing profiles at comparatively large angles of attack[27] and on the surface of an elliptic cylinder.[28] It should be noted that bubble formation was observed only in wind tunnels with low turbulence levels. Increase in the free stream turbulence leads to the disappearance of the separation bubble. It may be supposed that the phenomenon described usually coincides with the drag crisis if the effect of outer disturbances is small.

---

[26] *Sovremennoe sostoyaniye gidrodinamiki vyazkoy zhidkosti (Modern State of Viscous Fluid Hydrodynamics)*, Vol.2, Goldstein, S., Ed., Inostrannaya Literature Press, Moscow, 1948.

[27] Gault, D. E., Boundary-Layer and Stalling Characteristics of the NACA 63–009 Airfoil Section, NACA Tech. Note, No.1894, 1949.

[28] Schubauer, G. B., Air Flow in Boundary Layer of an Elliptic Cylinder, NACA Rep. 652, 1939.

The drag crisis may occur at the Reynolds numbers considerable smaller than the critical if the boundary layer is artificially made turbulent by introducing special separate roughnesses-turbulizers into it. Thus, for example, in his classical experiments Prandtl obtained improvement of the flow past a sphere and reduction of its resistance when a thin wire ring was placed onto the sphere surface (Figure 211).

In ship model towing basins turbulizers are used such that their effect is to substitute for the high Reynolds numbers otherwise inaccessible in the basins and thus to converge laboratory and natural conditions. Not always, of course, the increase in the degree of flow turbulence leads to the same change of drag or lift force as would accompany the growth of the Reynolds number.[29] This especially refers to wing profiles near the front point of which crisis phenomena, similar to those occurring on the surface of a round cylinder, develop.

In particular, the flow crisis phenomenon explains the fact of sharp difference between maximum values $c_{y\,\max}$ of the coefficient of wing lifting force obtained in laboratory studies in wind tunnels (comparatively small Reynolds numbers) and on an airplane (large Reynolds numbers). It is known that the coefficient of lift force c_y grows with the angle of attack α to some critical value α_{cr} at which it attains its maximum. Deviation of c_y from the linear dependence on α is explained by boundary layer thickening in a rear (diverging) portion of the layer and, thus, by the increase of the boundary layer inverse effect on the outer irrotational flow. This influence results in considerable distortion of the outer flow and, as a consequence, to the violation of the linear dependence $c_y(\alpha)$ theoretically predicted in substantially wider range of attack angles.

Laminar boundary layer thickening on the face of a wing profile leads to early separation in the front edge region where the layer is laminar and easily separates under the effect of the inverse pressure drop. In this case, if observation is performed in low-turbulent tunnels or under natural conditions of flow in a low-turbulent atmosphere, formation of the separation bubble, i.e., of the closed separation zone, is probable, with the growth of the attack angle, which will lead to developed open-flow stalling from the wing surface and a sharp disruption of circulation and reduction of the coefficient of lifting force.

As the Reynolds number **Re** grows at the fixed angle of attack, in full conformity with the above-described crisis phenomenon, wing streamlining is improved and the possibility arises to pass over to large angles of attack and to obtain higher values of c_y and, consequently, of $c_{y\,\max}$. In this case both cymax and the critical angle α_{cr} increases. Continuing to increase the Reynolds number, one can obtain high values of $c_{y\,\max}$.

All the above refers, of course, to such wing profiles on the frontal portion of which at large angles of attack are created the conditions for flow crisis origination (i.e., to the profiles; the shape of the edges provides the presence of a laminar layer on the profile upper surface and boundary layer separation at laminar mode of flow in it. Example of such are symmetric and slightly bent profiles with a comparatively substantial distance from the leading edge to maximum thickness ("laminarized" profiles).

The available ("supporting" profiles, which usually have considerable curvature, do not possess this property. From the surface of these wing profiles at large angles of attack a turbulent layer stalls. On these profiles the growth of the Reynolds number does not lead to the increase in the critical angle of attack α_{cr}; conversely, it may even lead to their decrease. This is explained by the decrease of a laminar portion on the wing upper surface due to the displacement of the transition point upstream and, consequently, to turbulent layer thickening, thus resulting in the shift of the point of turbulent layer separation toward the wing tip, i.e., to flow deterioration.[30]

---

[29] Loitsyanskiy, L. G., *Aerodinamika pogranichnogo sloya* (*Aerodynamics of a Boundary Layer*), Gostekhizdat Press, Moscow, 1941, 257–260.

The position of the transition point exerts a great effect not only on the lifting force, but also on the wing resistance. As will be revealed later, skin friction drag is much smaller with the laminar boundary layer than with the turbulent one. In view of this, it is natural to delay the position of the transition point downstream as much as possible and to increase a relative length of the boundary layer laminar portion. This is achieved first of all by the improvement of laminar flow stability in a boundary layer by choosing a shape of the wing profile such that the maximum profile thickness was positioned at 40 to 60% of the chord from the wing leading edge.

Due to the delay of the maximum thickness the length of the boundary layer converging portion is enlarged in which laminar flow retains its stability at considerably greater local Reynolds numbers $\mathbf{Re}^{**}$ than in a diverging one. Moreover, the frontal portion of the wing surface is thoroughly polished in order to reduce disturbances to minimum that are caused by wing roughness or by separate asperities on its surface in the wing frontal portion where the boundary layer is still relatively thin.

The experiment shows that these wings with the delayed boundary layer really possess rather small resistances, but they easily lose their advantage with the least amount of dust, rain droplets, or insects on the wing surface.

At present many other methods of drag reduction are being developed. For example, disturbances accumulated in air when passing through the frontal portion of a boundary layer are suction off from the wing surface through slots or a porous surface. This free of external disturbances which arrive from outside and retain a laminar flow mode in a boundary layer. Economy of such "control" of a boundary layer is achieved thanks to the fact that only small amounts of air need be suction off at small rates through boundary layer cross-sections.

3. The Reynolds Equations of the Averaged Turbulent Motion

To describe turbulent motion Reynolds[31] suggested the following technique which has since been widely used. Registering in time velocities and pressures at a fixed point it may be assumed that

$$u = \bar{u} + u' \qquad v = \bar{v} + v' \qquad w = \bar{w} + w' \qquad p = \bar{p} + p' \tag{5}$$

where u, v, w, and p are the real instantaneous – we shall call them actual – velocity projections and pressure, $\bar{u}$, $\bar{v}$, $\bar{w}$, and $\bar{p}$ are the mean (with respect to time) velocity and pressure components, and u', v', w', and p' are the fluctuating values of velocity and pressure, or briefly, the fluctuations. By a mean value of actual quantity φ, denoted as $\bar{\varphi}$ or $\langle\varphi\rangle$, we mean the usual average over the time t, taken over the time interval T, called the period of averaging:

$$\bar{\varphi}(x, y, z; t) = \frac{1}{T}\int_{t-T/2}^{t+T/2} \varphi(x, y, z; \tau)\, d\tau \tag{6}$$

We assume that in every turbulent flow exists a constant period of averaging T, which is

---

[30] Krasilshchikov, P. P., The influence of the Reynolds number and flow turbulence on the wing lifting force, *Tr. TsAGI*, 268, 1936; Loitsyanskiy, L. G., *Aerodinamika pogranichnogo sloya* (*Aerodynamics of a Boundary Layer*), Gostekhizdat, Leningrad-Moscow, 1941, 256–262.

[31] Reynolds, O., On the dynamic theory of incompressible viscous fluids and the determination of the criterion, *Phil. Trans. R. Soc.*, 1895.

sufficiently large as compared with the period of turbulent fluctuations, but small in comparison with a characteristic time of the mean turbulent motion (the periodic time of an oscillatory flow, the time in which a body moves through a distance equal to its length, etc.). Thus, a further averaging of a quantity already averaged according to Equation 6 leads to the same result.

This means that

$$\bar{\bar{\varphi}} = \bar{\varphi} \tag{7}$$

If the averaging process (Equation 6) is carried out at a fixed point at different times t and produces the same value of $\bar{\varphi}$, then such a mean flow is known as steady and the turbulent flow itself is quasi-steady.

The assumption (7) is equivalent to claiming that the mean of a fluctuation $\varphi' = \varphi - \bar{\varphi}$ of the quantity φ is zero. Indeed, by virtue of the linearity of (6) and (7) we have

$$\overline{\varphi'} = \bar{\varphi} - \bar{\bar{\varphi}} = 0 \tag{8}$$

In what follows we deal exclusively with quasi-steady turbulent flows. In such flows a mean value $\bar{\varphi}$ will be a function of position only. For, if ψ denotes another fluctuating function of time and position then it follows from (6) that

$$\overline{\bar{\varphi}\psi} = \bar{\varphi}\bar{\psi} \tag{9}$$

(the bar placed over a group of quantities signifies that the operation (6) is to be applied to the whole group).

If a turbulent flow is not quasi-steady, Equation 9 has to be regarded as a further property of the time average (6).

From the definition of the process of averaging (6) it immediately follows that the mean of a space derivative of a function is equal to the derivative of the mean

$$\overline{\frac{\partial \varphi}{\partial x}} = \frac{\partial \bar{\varphi}}{\partial x}, \ldots, \tag{10}$$

since differentiation with respect to a space coordinate and integration with respect to time are independent processes. This is also true of the time derivative. Indeed, by the well-known rule for the differentiation of an integral with variable limits

$$\frac{\partial \bar{\varphi}}{\partial x} = \frac{\partial}{\partial t}\frac{1}{T}\int_{t-T/2}^{t+T/2} \varphi(x, y, z; \tau)\, d\tau =$$

$$= \frac{1}{T}\left[\varphi\left(x, y, z; t + \frac{T}{2}\right) - \varphi\left(x, y, z; t - \frac{T}{2}\right)\right] = \frac{1}{T}\int_{t-T/2}^{t+T/2} \frac{\partial \varphi}{\partial \tau} d\tau$$

and therefore

$$\frac{\partial \bar{\varphi}}{\partial t} = \overline{\frac{\partial \varphi}{\partial t}} \tag{11}$$

The adopted law of averaging (6) is not the only possible one. Not departing from non-linearity of time-averaging (6), we can introduce some weight function under the integral sign whose distribution will correspond to a predominant role of one time instant relative to others in averaging with respect to the period T (see the end of Chapter 15).

Besides time averaging we can also carry out averaging over the results of multiply repeated experiments – ensemble averaging. In what follows the simplest time averaging (6) and validity of rules (Equations 7 to 11) corresponding to it are utilized.

The quantity $\overline{\varphi\psi}$ obtained as a result of averaging the product of two fluctuating functions φ and ψ has the name of a single-point (by virtue of the fact that the values of φ and ψ in integration are taken at the same space-time point) double correlation and the relationship

$$R = \frac{\overline{\varphi\psi}}{\sqrt{\overline{\varphi^2}}\sqrt{\overline{\psi^2}}}$$

is called the coefficient of correlation between two statistically related quantities. The equality of the coefficient of correlation either to plus or minus unity indicates a full deterministic relation between the phenomena described by the functions of coordinates and time φ and ψ. The "minus" sign, the negative value of the correlation coefficient, means opposite phases of oscillations, and the equality of the coefficient of correlation to zero indicates statistical independence of the phenomena. To intermediate degrees of statistical correlation of fluctuating quantities correspond absolute values of the coefficient of correlation ranging from zero to unity. The coefficient of correlation between the fluctuations occurring at two different points of space and, generally speaking, at different time instants, will be called the coefficient of two-point space-time correlation and, depending on the number of correlated fluctuating functions, double, triple, etc., correlation.

The concept of correlation (reflection of statistical correlation of phenomena) and of the coefficient of correlation (the measure, normalized by the range (–1, 1) of this correlation) forms the basis of statistical theories of turbulence. The correlations $\overline{\varphi'\psi'}$, $\overline{\varphi'\psi'\chi'}$, etc. also have the names "correlation moments",[32] or simply, the "moments" of the second, third, etc. orders, adding the epithets "one-point", "two-point" etc.

The notion of "the correlation moment" of the second and higher orders presupposes the presence of at least two correlated quantities. For generality we introduce the term "the first order moments", meaning the averaged quantities $\bar{u}$, $\bar{v}$, $\bar{w}$, and $\bar{p}$.

Making use partly of postulated properties and partly of those derived from the law of averaging (6), we may obtain differential equations of the averaged motion of incompressible fluid. It should only be assumed, as was done by Reynolds, that actual flow, in spite of being irregular and being affected by random circumstances attributed to the previous history of the flow, is still strictly described by the Navier-Stokes equations. The basic idea of the general approach to the description of turbulent flows proposed by Reynolds consists in this simple, but far from obvious, assumption. It should be noted that attempts thus far develop to a purely statistical theory of turbulent flows, that does not rely on the Navier-Stokes equations, have not led to any substantial results.

Using this principal thesis as a basis, we refer to the formulation of equations for velocities and pressures averaged by the previous technique. For this, the Navier-Stokes equations in actual velocities and pressures (volumetric forces are omitted and summing up is meant by

[32] The term "correlation moment" was first introduced by A. A. Fridman and H. B. Keller in their basic paper "Differential Equations of Incompressible Fluid Turbulent Flow" presented at the First International Congress for Applied Mechanics in Delft in 1924 (see *Proc. 1st Int. Cong. Appl. Mech.*, Delft, 1924, 395–405.

the subscript j doubly repeated in monomials)

$$\rho \frac{\partial v_i}{\partial t} + \rho v_j \frac{\partial v_i}{\partial x_j} = -\frac{\partial p}{\partial x_i} + \mu \frac{\partial^2 v_i}{\partial x_j \partial x_j}$$

$$\frac{\partial v_j}{\partial x_j} = 0 \quad (i = 1, 2, 3)$$

Making use of the above equation, rewrite in the form

$$\rho \frac{\partial v_i}{\partial t} + \rho \frac{\partial (v_i v_j)}{\partial x_j} = -\frac{\partial p}{\partial x_i} + \mu \frac{\partial^2 v_i}{\partial x_j \partial x_j}$$

$$\frac{\partial v_j}{\partial x_j} = 0 \quad (i = 1, 2, 3) \tag{12}$$

Substitute actual velocities v_i and pressures for the averaged $\bar{v}_i$, $\bar{p}$ and fluctuational v_i', p' by (5)

$$v_i = \bar{v}_i + v_i' \quad (i = 1, 2, 3) \qquad p = \bar{p} + p'$$

Then, noting that

$$v_i v_j = (\bar{v}_i + v_i')(\bar{v}_j + v_j') = \bar{v}_i \bar{v}_j + \bar{v}_i v_j' + \bar{v}_j v_i' + v_i' v_j'$$

and averaging both parts of the equation over time according to (6), we have in conformity with the adopted rules (Equations 7 to 11)

$$\overline{v_i v_j} = \bar{v}_i \bar{v}_j + \overline{v_i' v_j'} \tag{13}$$

averaging both sides of the Navier-Stokes equations by (6) and using (13), we obtain

$$\rho \frac{\partial \bar{v}_i}{\partial t} + \rho \frac{\partial (\bar{v}_i \bar{v}_j)}{\partial x_j} = -\frac{\partial \bar{p}}{\partial x_i} + \mu \frac{\partial^2 \bar{v}_i}{\partial x_j \partial x_j} + \frac{\partial}{\partial x_j}\left(-\rho \overline{v_i' v_j'}\right)$$

$$\frac{\partial \bar{v}_j}{\partial x_j} = 0 \tag{14}$$

or, applying the last equation of system (14)

$$\rho \frac{\partial \bar{v}_i}{\partial t} + \rho \bar{v}_j \frac{\partial \bar{v}_i}{\partial x_j} = -\frac{\partial \bar{p}}{\partial x_i} + \mu \frac{\partial^2 \bar{v}_i}{\partial x_j \partial x_j} + \frac{\partial}{\partial x_j}\left(-\rho \overline{v_i' v_j'}\right)$$

$$\frac{\partial \bar{v}_j}{\partial x_j} = 0 \tag{15}$$

Note that from the last equations of systems (12) and (15) there directly follows the equality

$$\frac{\partial v_j'}{\partial x_j} = 0 \qquad (16)$$

which expresses the condition of fluid incompressibility in a fluctuating flow.

The Reynolds equations (15) may be treated as the first in the general system of equations of the transfer of flow turbulent characteristics, *viz.*, as the equations of momentum transfer $\rho \bar{v}_i$, and accordingly, Equations 15 involving $\bar{v}_i$ $(i = 1, 2, 3)$ and $\bar{p}$ as the unknowns may be called the "first-order momental equations".

Note the thesis which will be confirmed in a more general case Equations 15 serving for $\bar{v}_i$ determination contain the "second-order moments" $\overline{v_i'v_j'}$ which make this system indeterminate. The equations, which should determine the "second-order moments" entering into Equations 15, will turn to involve the "third-order moments", etc.

So, from the very beginning of the formulation of the equations of the transfer of flow turbulent characteristics the indeterminancy of the system of transfer equations is revealed. The problem of closing these equations has not acquired a rational solution as yet and it requires, in every separate case, the introduction of additional assumptions which have an approximate character only.

Recalling general equations of fluid dynamics "in stresses"

$$\rho \frac{\partial v_i}{\partial t} + \rho v_i \frac{\partial v_i}{\partial x_j} = \frac{\partial p_{ji}}{\partial x_j} = \mathrm{Div}_i P \quad (i = 1, 2, 3)$$

we may say that the averaged turbulent flow equations (15) differ from the Navier-Stokes equations by the presence of the additional tensor of turbulent stress

$$\Pi = \begin{pmatrix} -\rho\overline{v_1'^2} & -\rho\overline{v_1'v_2'} & -\rho\overline{v_1'v_3'} \\ -\rho\overline{v_1'v_2'} & -\rho\overline{v_2'^2} & -\rho\overline{v_2'v_3'} \\ -\rho\overline{v_1'v_3'} & -\rho\overline{v_2'v_3'} & -\rho\overline{v_3'^2} \end{pmatrix} \qquad (17)$$

with the components

$$\pi_{ij} = -\rho\,\overline{v_i'v_j'} \qquad (18)$$

The tensor 17 is symmetric; its components (18) are called the Reynolds stresses. The Reynolds stresses have a physical sense of the components of the averaged momentum transfer by fluctuating velocities. That is why the first equation in (15), involving in its right side the derivatives of the Reynolds stresses, also was called the momentum equation of velocity fluctuations. In what follows we shall briefly call the first equation of (15) the "momentum equation".

We shall call the tensor *P*, which is equal to

$$P = -\bar{p}E + 2\mu\bar{\dot{S}} + \Pi \qquad (19)$$

and which has the components

$$p_{ij} = -\bar{p}\delta_{ij} + \mu\left(\frac{\partial \bar{v}_i}{\partial x_j} + \frac{\partial \bar{v}_j}{\partial x_i}\right) + \left(-\rho\,\overline{v_i' v_j'}\right) \tag{20}$$

the tensor of total stress.

Omitting the derivation analogous to the previous, we write the equations of momentum transfer in cylindrical coordinates

$$\rho\left(\frac{\partial \bar{v}_r}{\partial t} + \bar{v}_r\frac{\partial \bar{v}_r}{\partial r} + \frac{\bar{v}_\varepsilon}{r}\frac{\partial \bar{v}_r}{\partial \varepsilon} + \bar{v}_z\frac{\partial \bar{v}_r}{\partial z} - \frac{\bar{v}_\varepsilon^2}{r}\right) =$$

$$= -\frac{\partial \bar{p}}{\partial r} + \mu\left(\frac{\partial^2 \bar{v}_r}{\partial r^2} + \frac{1}{r}\frac{\partial \bar{v}_r}{\partial r} + \frac{1}{r^2}\frac{\partial^2 \bar{v}_r}{\partial \varepsilon^2} + \frac{\partial^2 \bar{v}_r}{\partial z^2} - \frac{\bar{v}_r}{r^2} - \frac{2}{r^2}\frac{\partial \bar{v}_\varepsilon}{\partial \varepsilon}\right) +$$

$$+\frac{1}{r}\frac{\partial}{\partial r}\left(-r\rho\,\overline{v_r'^2}\right) + \frac{1}{r}\frac{\partial}{\partial \varepsilon}\left(-\rho\,\overline{v_r' v_\varepsilon'}\right) + \frac{\partial}{\partial z}\left(-\rho\,\overline{v_r' v_z'}\right) - \frac{1}{r}\left(-\rho\,\overline{v_\varepsilon'^2}\right)$$

$$\rho\left(\frac{\partial \bar{v}_\varepsilon}{\partial t} + \bar{v}_r\frac{\partial \bar{v}_\varepsilon}{\partial r} + \frac{\bar{v}_\varepsilon}{r}\frac{\partial \bar{v}_\varepsilon}{\partial \varepsilon} + \bar{v}_z\frac{\partial \bar{v}_\varepsilon}{\partial z} + \frac{\bar{v}_r \bar{v}_\varepsilon}{r}\right) =$$

$$= -\frac{1}{r}\frac{\partial \bar{p}}{\partial \varepsilon} + \mu\left(\frac{\partial^2 \bar{v}_\varepsilon}{\partial r^2} + \frac{1}{r}\frac{\partial \bar{v}_\varepsilon}{\partial r} + \frac{1}{r^2}\frac{\partial^2 \bar{v}_\varepsilon}{\partial \varepsilon^2} + \frac{\partial^2 \bar{v}_\varepsilon}{\partial z^2} - \frac{\bar{v}_\varepsilon}{r^2} + \frac{2}{r^2}\frac{\partial \bar{v}_r}{\partial \varepsilon}\right) +$$

$$+\frac{\partial}{\partial r}\left(-\rho\,\overline{v_r' v_\varepsilon'}\right) + \frac{1}{r}\frac{\partial}{\partial \varepsilon}\left(-\rho\,\overline{v_\varepsilon'^2}\right) + \frac{\partial}{\partial z}\left(-\rho\,\overline{v_\varepsilon' v_z'}\right) + \frac{2}{r}\left(-\rho\,\overline{v_r' v_\varepsilon'}\right)$$

$$\rho\left(\frac{\partial \bar{v}_z}{\partial t} + \bar{v}_r\frac{\partial \bar{v}_z}{\partial r} + \frac{\bar{v}_\varepsilon}{r}\frac{\partial \bar{v}_z}{\partial \varepsilon} + \bar{v}_z\frac{\partial \bar{v}_z}{\partial z}\right) =$$

$$= -\frac{\partial \bar{p}}{\partial z} + \mu\left(\frac{\partial^2 \bar{v}_z}{\partial r^2} + \frac{1}{r}\frac{\partial \bar{v}_z}{\partial r} + \frac{1}{r^2}\frac{\partial^2 \bar{v}_z}{\partial \varepsilon^2} + \frac{\partial^2 \bar{v}_z}{\partial z^2}\right) +$$

$$+\frac{1}{r}\frac{\partial}{\partial r}\left(-\rho\,\overline{v_r' v_z'}\right) + \frac{1}{r}\frac{\partial}{\partial \varepsilon}\left(-\rho\,\overline{v_\varepsilon' v_z'}\right) + \frac{\partial}{\partial z}\left(-\rho\,\overline{v_z'^2}\right)$$

$$\frac{\partial}{\partial r}(r\bar{v}_r) + \frac{1}{r}\frac{\partial}{\partial \varepsilon}(r\bar{v}_\varepsilon) + \frac{\partial}{\partial z}(r\bar{v}_z) = 0 \tag{21}$$

The components of total stresses in cylindrical coordinates are

$$p_{rr} = -\bar{p} + 2\mu\frac{\partial \bar{v}_r}{\partial r} - \rho\overline{v_r'^2} \qquad p_{\varepsilon\varepsilon} = -\bar{p} + 2\mu\left(\frac{1}{r}\frac{\partial \bar{v}_\varepsilon}{\partial \varepsilon} + \frac{\bar{v}_r}{r}\right) - \rho\overline{v_\varepsilon'^2}$$

$$p_{zz} = -\bar{p} + 2\mu\frac{\partial \bar{v}_z}{\partial z} - \rho\overline{v_z'^2}$$

$$p_{r\varepsilon} = \mu\left(\frac{\partial \bar{v}_\varepsilon}{\partial r} + \frac{1}{r}\frac{\partial \bar{v}_r}{\partial \varepsilon} - \frac{\bar{v}_\varepsilon}{r}\right) - \rho\overline{v_r' v_\varepsilon'}$$

$$p_{rz} = \mu\left(\frac{\partial \bar{v}_r}{\partial z} + \frac{\partial \bar{v}_z}{\partial r}\right) - \rho\overline{v_r' v_z'} \qquad p_{\varepsilon z} = \mu\left(\frac{1}{r}\frac{\partial \bar{v}_z}{\partial \varepsilon} + \frac{\partial \bar{v}_\varepsilon}{\partial z}\right) - \rho\overline{v_\varepsilon' v_z'} \tag{22}$$

The same technique of averaging, that was carried out when deriving Equation 16, but applied to Equation 174 of Chapter 10, makes it possible to obtain the following equation of heat distribution in a turbulent flow

$$\rho c_p\left(\frac{\partial \bar{T}}{\partial t} + \bar{u}\frac{\partial \bar{T}}{\partial x} + \bar{v}\frac{\partial \bar{T}}{\partial y} + \bar{w}\frac{\partial \bar{T}}{\partial z}\right) = \frac{\partial}{\partial x}\left(\lambda\frac{\partial \bar{T}}{\partial x} - \rho c_p\overline{u'T'}\right) +$$

$$+\frac{\partial}{\partial y}\left(\lambda\frac{\partial \bar{T}}{\partial y} - \rho c_p\overline{v'T'}\right) + \frac{\partial}{\partial z}\left(\lambda\frac{\partial \bar{T}}{\partial z} - \rho c_p\overline{w'T'}\right) \tag{23}$$

Here, the actual temperature T is presented by the sum of the averaged, $\bar{T}$, and fluctuating, T', temperatures. The physical constants density ρ, heat capacity c_p, and thermal conductivity λ are assumed to be constant.

The vector $\boldsymbol{q}$ with the components

$$q_x = -\rho c_p\overline{u'T'} \qquad q_y = -\rho c_p\overline{v'T'} \qquad q_z = -\rho c_p\overline{w'T'} \tag{24}$$

expresses the turbulent heat flux.

In cylindrical coordinates Equation 13 has the form

$$\rho c_p\left(\frac{\partial \bar{T}}{\partial t} + \bar{v}_r\frac{\partial \bar{T}}{\partial r} + \frac{\bar{v}_\varepsilon}{r}\frac{\partial \bar{T}}{\partial \varepsilon} + \bar{v}_z\frac{\partial \bar{T}}{\partial z}\right) = \frac{1}{r}\frac{\partial}{\partial r}\left(\lambda r\frac{\partial \bar{T}}{\partial r} - \rho c_p r\overline{v_r'T'}\right) +$$

$$+\frac{1}{r}\frac{\partial}{\partial \varepsilon}\left(\frac{\lambda}{r}\frac{\partial \bar{T}}{\partial \varepsilon} - \rho c_p\overline{v_\varepsilon'T'}\right) + \frac{\partial}{\partial z}\left(\lambda\frac{\partial \bar{T}}{\partial z} - \rho c_p\overline{v_z'T'}\right) \tag{25}$$

Substituting in Equation 23 for the specific heat $c_p T$ the concentration of the admixture of some substance c, we obtain the equation of turbulent diffusion of this substance

$$\rho\left(\frac{\partial \bar{c}}{\partial t} + \bar{u}\frac{\partial \bar{c}}{\partial x} + \bar{v}\frac{\partial \bar{c}}{\partial y} + \bar{w}\frac{\partial \bar{c}}{\partial z}\right) = \frac{\partial}{\partial x}\left(\rho D\frac{\partial \bar{c}}{\partial x} - \rho\overline{u'c'}\right) +$$

$$+\frac{\partial}{\partial y}\left(\rho D\frac{\partial \bar{c}}{\partial y} - \rho\overline{v'c'}\right) + \frac{\partial}{\partial z}\left(\rho D\frac{\partial \bar{c}}{\partial z} - \rho\overline{w'c'}\right) \tag{26}$$

where D is the coefficient of ordinary molecular diffusion of substance in the mixture, and the vector $\boldsymbol{m} = -\rho\boldsymbol{V}'c'$ with the components

$$m_x = -\rho\overline{u'c'} \qquad m_y = -\rho\overline{v'c'} \qquad m_z = -\rho\overline{w'c'} \tag{27}$$

is the vector of a turbulent flux of substance.

It may be noted that Equation 26 differs from (23) by the substitution of $c_p\bar{T}$ for $\bar{c}$ and λ for ρD; by the same rule we obtain from Equation 25 the equation of turbulent diffusion in cylindrical coordinates.

Returning to the system of continuity and momentum equations (15), we note that there are ten unknowns in the combination of these four equations ($\bar{v}_i$, $\bar{p}$, π_{ij}; $i, j = 1, 2, 3$). Thus, Equations 15 do not present a closed (determinant) system of equations, and since the excess of unknowns refers to the Reynolds stresses $\pi_{ij} = -\rho\overline{v_i'v_j'}$ it is quite natural to make an attempt to derive the equations of the transport of Reynolds stresses from the Navier-Stokes equations and thus close the system of equation (15). Formulation of the transport equations π_{ij} is not difficult. We shall refer to the first equation of system (12) and express each actual quantity in it as a sum of the averaged and fluctuating quantities. The result of this substitution may be presented in the form:

$$\rho\frac{\partial\bar{v}_i}{\partial t} + \rho\frac{\partial}{\partial x_j}(\bar{v}_i\bar{v}_j) + \frac{\partial\bar{p}}{\partial x_i} - \mu\frac{\partial^2\bar{v}_i}{\partial x_j\partial x_j} + \frac{\partial}{\partial x_j}\left(\rho\overline{v_i'v_j'}\right) +$$

$$+\rho\frac{\partial v_i'}{\partial t} + \rho\frac{\partial}{\partial x_j}(\bar{v}_iv_j' + \bar{v}_jv_i') + \frac{\partial p'}{\partial x_i} - \mu\frac{\partial^2 v_i'}{\partial x_j\partial x_j} -$$

$$-\frac{\partial}{\partial x_j}(-\rho v_i'v_j') - \frac{\partial}{\partial x_j}\left(\rho\overline{v_i'v_j'}\right) = 0$$

Noting that according to the first equation of (14), the first line is equal to zero, we come to the following equation of velocity fluctuations

$$\rho\frac{\partial v_i'}{\partial t} + \rho\frac{\partial}{\partial x_j}(\bar{v}_iv_j' + \bar{v}_jv_i') + \frac{\partial p'}{\partial x_i} =$$

$$= \frac{\partial}{\partial x_j}\left(\mu\frac{\partial v_i'}{\partial x_j} - \rho v_i'v_j' + \rho\overline{v_i'v_j'}\right) \quad (i = 1, 2, 3) \qquad (28)$$

Replacing in this system of equations the "free" subscript i by n, we obtain the equation, similar to Equation 28, for v_n'

$$\rho\frac{\partial v_n'}{\partial t} + \rho\frac{\partial}{\partial x_j}(\bar{v}_nv_j' + \bar{v}_jv_n') + \frac{\partial p'}{\partial x_n} =$$

$$= \frac{\partial}{\partial x_j}\left(\mu\frac{\partial v_n'}{\partial x_j} - \rho v_n'v_j' + \rho\overline{v_n'v_j'}\right) \quad (n = 1, 2, 3) \qquad (29)$$

We multiply both sides of Equation 28 into v_n' and both sides of Equation 29 into v_i' and sum the results. In this case we take into account the equality

$$\frac{\partial}{\partial x_j}(v_i'v_n'\bar{v}_j) = \bar{v}_j\frac{\partial}{\partial x_j}(v_i'v_n')$$

which follows directly from the last equation of system (15), and also note that

$$v_i' \frac{\partial}{\partial x_j}(v_n' v_j') + v_n' \frac{\partial}{\partial x_j}(v_i' v_j') = \frac{\partial}{\partial x_j}(v_i' v_n' v_j')$$

Combining the terms in the equations obtained and dividing both sides by ρ yields

$$\frac{\partial}{\partial t}(v_i' v_n') + \bar{v}_j \frac{\partial v_i' v_n'}{\partial x_j} = -\left(v_i' v_j' \frac{\partial \bar{v}_n}{\partial x_j} + v_n' v_j' \frac{\partial \bar{v}_i}{\partial x_j} \right) -$$

$$-2\nu \frac{\partial v_i'}{\partial x_j} \frac{\partial v_n'}{\partial x_j} + \frac{p'}{\rho}\left(\frac{\partial v_n'}{\partial x_i} + \frac{\partial v_i'}{\partial x_n} \right) - \frac{\partial}{\partial x_j}(v_i' v_n' v_j') - \frac{1}{\rho}\frac{\partial}{\partial x_i}(v_n' p') -$$

$$-\frac{1}{\rho}\frac{\partial}{\partial x_n}(v_i' p') + \nu \frac{\partial^2 v_i' v_n'}{\partial x_j \partial x_j} + v_n' \frac{\partial}{\partial x_j}\left(\overline{v_i' v_j'} \right) + v_i' \frac{\partial}{\partial x_j}\left(\overline{v_n' v_j'} \right) \tag{30}$$

where j is the "dummy" subscript, $i, n = 1, 2, 3$.

Let us introduce the notation for the double single-point correlation or the second-order moment

$$\overline{v_i' v_n'} = \tau_{in} = -\frac{1}{\rho}\pi_{in} \tag{31}$$

We shall carry out averaging (6) over both sides of the equations of system (30), making use of the rules of averaging (Equations 7 to 11); we then obtain the unknown system of equations

$$\frac{\partial \tau_{in}}{\partial t} + \bar{v}_j \frac{\partial \tau_{in}}{\partial x_j} = -\left(\tau_{ij} \frac{\partial \bar{v}_n}{\partial x_j} + \tau_{nj} \frac{\partial \bar{v}_i}{\partial x_j} \right) - 2\nu \overline{\frac{\partial v_i'}{\partial x_j} \frac{\partial v_n'}{\partial x_j}} +$$

$$+ \overline{\frac{p'}{\rho}\left(\frac{\partial v_n'}{\partial x_i} + \frac{\partial v_i'}{\partial x_n} \right)} - \frac{\partial}{\partial x_j}\overline{(v_i' v_n' v_j')} - \frac{1}{\rho}\frac{\partial}{\partial x_i}\overline{(v_n' p')} -$$

$$-\frac{1}{\rho}\frac{\partial}{\partial x_n}\overline{(v_i' p')} + \nu \frac{\partial^2 \tau_{in}}{\partial x_j \partial x_j} \tag{32}$$

or, combining the last four terms on the right side and applying the evident equalities (δ_{ij} are the Kronecker symbols)

$$\frac{\partial}{\partial x_i}\overline{(v_n' p')} = \frac{\partial}{\partial x_j}\left(\delta_{ij} \overline{v_n' p'} \right)$$

$$\frac{\partial}{\partial x_n}\overline{(v_i' p')} = \frac{\partial}{\partial x_j}\left(\delta_{nj} \overline{v_i' p'} \right)$$

we shall reduce Equations 32 to the form

$$\frac{\partial \tau_{in}}{\partial t} + \bar{v}_j \frac{\partial \tau_{in}}{\partial x_j} = -\left(\tau_{ij} \frac{\partial \bar{v}_n}{\partial x_j} + \tau_{nj} \frac{\partial \bar{v}_i}{\partial x_j} \right) - 2\nu \overline{\frac{\partial v_i'}{\partial x_j} \frac{\partial v_n'}{\partial x_j}} +$$

$$+\overline{\frac{p'}{\rho}\left(\frac{\partial v_n'}{\partial x_i}+\frac{\partial v_i'}{\partial x_n}\right)}-\frac{\partial}{\partial x_j}\left(\overline{v_i'v_n'v_j'}+\frac{\delta_{ij}}{\rho}\overline{v_n'p'}+\frac{\delta_{nj}}{\rho}\overline{v_i'p'}-\nu\frac{\partial\tau_{in}}{\partial x_j}\right) \tag{33}$$

Pausing on this form of the equation of the transfer of correlation of velocity fluctuations $\tau_{in} = \overline{v_i'v_n'}$, we explain a physical meaning of the terms entering into Equation 33. On the left side there is an individual derivative with respect to time τ_{in}, which is composed of the local and convective derivatives with the latter corresponding to the transfer of τ_{in} by the averaged flow. An ordinary phenomenon expressed by this side of Equation 33 is called advection. Equation 33 is the balance between advection and the sum of terms on the right side. The first of them involves the components, related to the mass unit, of tensor products of the tensor of turbulent stresses Π by the differential tensor of the field of averaged velocities. It is natural to treat the combination of these products as the power spent by the field of averaged velocities to form the Reynolds stresses. In conformity with this, the first term on the right side is called turbulence generation.

These two terms of Equation 33 have an evident value and can be measured. The second term to the right, involving the kinematic coefficient of molecular viscosity and the averaged sum of coupled products of the components of the differential tensor of the field of fluctuating velocities, presents the "dissipation rate". Less obvious and almost unmeasurable is the term

$$\overline{\frac{p'}{\rho}\left(\frac{\partial v_n'}{\partial x_i}+\frac{\partial v_i'}{\partial x_n}\right)}$$

which has a sense of the correlation between pressure fluctuations and the tensor of the deformation rate of the field of fluctuating velocities. Finally, the last group of terms, under the common differentiation sign $\partial/\partial x_j$, is the "diverging" portion, meaning diffusion both due to velocity and pressure fluctuations and to viscosity.

The equation of the transport of the averaged kinetic energy of a fluctuating flow per unit mass

$$k = \frac{1}{2}\overline{v_i'v_i'} = \frac{1}{2}\tau_{ii} \tag{34}$$

is easily derived from Equation 33. To obtain this equation we assume in (33) $n = i$ and preserve the rule of summing by the subscript doubly repeating in monomials. We directly obtain

$$\frac{\partial\tau_{ii}}{\partial t}+\bar{v}_j\frac{\partial\tau_{ii}}{\partial x_j} = -2\tau_{ij}\frac{\partial\bar{v}_i}{\partial x_j}-2\nu\overline{\frac{\partial v_i'}{\partial x_j}\frac{\partial v_i'}{\partial x_j}}+$$

$$+2\overline{\frac{p'}{\rho}\frac{\partial v_i'}{\partial x_i}}-\frac{\partial}{\partial x_j}\left(\overline{v_i'v_i'v_j'}+\frac{2}{\rho}\overline{v_j'p'}-\nu\frac{\partial\tau_{ii}}{\partial x_j}\right)$$

Taking into account the notation of (34) and noting that, according to Equation 16 the third term to the right is equal to zero, we come (reducing by the multiplier 2) to the unknown equation of the transport of turbulence kinetic energy

$$\frac{\partial k}{\partial t}+\bar{v}_j\frac{\partial k}{\partial x_j} = -\tau_{ij}\frac{\partial \bar{v}_i}{\partial x_j}-\nu\,\overline{\frac{\partial v_i'}{\partial x_j}\frac{\partial v_i'}{\partial x_j}}-\frac{\partial}{\partial x_j}\left[\overline{v_j'\left(\frac{v_i'v_i'}{2}+\frac{p'}{\rho}\right)}-\nu\frac{\partial k}{\partial x_j}\right] \tag{35}$$

Equations 33 and 35 will be used in the Chapter 14, and together with the equation of transport dissipation rate, the derivation of which will be omitted, they will serve as the basis for modern methods of calculating a turbulent boundary layer.

All of the above mentioned transfer equations as well as the others (the equations of vortex fluctuations transfer, of the coefficient of turbulent viscosity, etc.) possess the basic drawback: besides the unknowns they also involve the additional characteristics of turbulence (triple correlations of velocity fluctuations, correlations of these fluctuations with the fluctuations of pressure, and others). In this case, as new equations are added to the system of transfer equations, there arise new, unknown correlation terms. Thus, when formulating the equations of transfer of triple correlations of velocity fluctuations by the previous method, there appear quadrupole correlations and so on. It is proved experimentally that double, triple, etc. in order correlations of velocity fluctuations grow in magnitude, thus preventing convergence of techniques.

Thus, there appeared the basic problem in the turbulence theory of closing the system of turbulent flow equations that for a long time has attracted the attention of researchers, but is still far from being solved. Numerous approximate methods of the statistical approach to the solution of the problems of turbulent flow theory and, in particular, rather reliable methods for calculating turbulent boundary layers are developed to solve this fundamental problem. These pursuits are reduced to the expression of complex correlations entering into the transfer equations in terms of more simple ones, relative to which it is easier to make additional considerations closing the problem formulation. Great successes are achieved here that are supported by wide possibilities of the methods of numerical calculation on computers.

This book is not aimed at the presentation of these extremely various methods, whose enumeration would occupy much space. We refer the reader to some monographs and reviews of the problem.[33]

4. Some Data on the Inner Structure of Turbulent Flows

The equations of transfer presented in the previous section are of a general statistical character, but to some extent they reflect the physical processes occurring in reality. Comprehension of these processes requires penetration to the inner structure of turbulent motions which will be now briefly considered.

As has been already mentioned in Sections 2 and 3, a turbulent flow possesses a complex vortical structure involving various vortical formations which differ both in forms and in sizes. Speaking of the dimensions of vortices one implies their mean sizes and distinguishes small- and large-scale vortical structures.

[33] The theory of turbulent flows was thoroughly formulated in a fundamental monograph Monin, A. S. and Yaglom, A. M., *Statisticheskaya gidromekhanika. Mekhanika turbulentnosti* (*Statistical Hydromechanics. Mechanics of Turbulence*), Part 1, Nauka Press, Moscow, 1965; Part 2, Nauka Press, Moscow, 1967. Still of great interest is the review by O. M. Phillips, Shear flow turbulence, California, *Annu. Rev. Fluid Mec.*, 1, 245–264, 1969. The most easily understood is the monograph Bradshaw, P., Cebeci, T., and Whitelaw, J., *Engineering Calculation Methods for Turbulent Flow*, Academic Press, London, 1981. Among the most detailed reviews are *Turbulentnost, printsipy i primeneniya* (*Turbulence, Principles and Application*), W. Frost and T. Moulden, Eds., Mir Press, Moscow, 1980; *Turbulentnost* (*Turbulence*), P. Bradshaw, Ed., Mashinostroeniye Press, Moscow, 1980. Note also a brief but thorough review by I. A. Belov, *Modeli turbulentnosti* (*Turbulence Models*), Leningrad Mechanical Institute Press, Leningrad, 1986.

In the transport processes of momentum and kinetic energy of fluctuations, and also of Reynolds stresses and for other turbulent characteristics vortices of different scales take place. Of special importance are large-scale vortices. A comparatively new branch of the turbulence theory is concerned with the study of their interaction. This new branch deals with the so-called coherent turbulence[34] that is formed in turbulent jets, wakes behind streamlined bodies, in preseparation regions of "near-wall" turbulent boundary layers and also under natural conditions (tornados, cyclones).[35]

To the main processes evoking the formation, development, and disappearance of vortices belong convection (advection), generation, diffusion, and, at the last stage of degeneration, dissipation (transition to heat) that were mentioned earlier.

Decomposition of large vortices to smaller ones, in which inertia phenomena are still predominant over viscous, takes place in the process of turbulent diffusion. These vortices, which are said to be within an "inertia range of scales", take part in convection and turbulent diffusion, but to a negligibly small extent are subjected to the effect of viscosity. A general process of further degradation of vortices leads them finally to small vortices with a small scale which are affected by viscous diffusion and subsequent viscous dissipation of kinetic energy to heat. This "cascade" scheme[36] rather roughly, of course, reproduces real processes occurring in turbulent flows, but correctly describes general tendencies. The small-scale vortices, rapidly decaying, are notable to long maintain and transport downstream the information on disturbances obtained by them upstream. This role is played by the vortices of large and partially inertial scales thanks to whose presence the appearance of the "memory effect" is possible in turbulent flows.

Kinetic energy of large vortices has the order of specific energy of a local averaged flow, but reduces with the scale, and in small vortices it becomes negligible.

Quantitative determination of the "turbulence scale" is closely connected with the statistical correlation of velocity fluctuations in the studied region of the disturbed flow. The coefficient of correlation between the velocity fluctuations at the points of a fluid volume carrying the traces of that initial vortical disturbance, which is gradually transformed from the volumes of one scale to other smaller scales, serves as the measure of this correlation. Having determined spatial determination of the coefficient of correlation, we thus can evaluate the spatial structure of turbulent disturbances and find the scale of the vortex at each stage of its destruction.

Modern measuring devices show different ways for studying the inner structure of turbulent flows. Among these is direct filming of water flows which are made visible with the aid of solid admixtures or gas bubbles[37] that allows one to obtain the picture of streamlines, and

---

[34] Ginevskiy, A. S. and Vlasov E. V., Coherent structures in turbulent jet flows, in *Modeli mekhaniki sploshnoy sredy* (*Models of Continuum Mechanics*), Siberian Branch of the USSR Academy of Sciences, Institute of Theoretical and Applied Mechanics Press, Novosibirsk, 1983, 91–117.

[35] Specific features of vortical structures of turbulent flows are presented in the monograph Albring, W., *Elementarvorgänge fluider Wirbelbewegungen*, Academie Verlag, Berlin, 1981. Sect. 305.

[36] The origin of this scheme goes back to L. Richardson (*Weather Prediction of Numerical Process*, Cambridge University Press, London, 1922) and served as a basis for the known familiar works by A. N. Kolmogorov in which this general scheme was applied to the determination of specific quantitative regularities in the flows with large Reynolds numbers (Richardson himself did not make any quantitative conclusions from the scheme suggested, though in a subsequent (1926) work he intuitively used it; see the historical essay in the introduction to the first part of the monograph Monin, A. S. and Yaglom, A. M. *Statisticheskaya gidromekhanika* (*Statistical Hydromechanics*), Nauka Press, Minsk, 1965.

[37] Minskiy, E. M., On velocity fluctuations in a free stream, *Tekhn. Zam. TsAGI*, No. 105, 1936; Minskiy E. M. and Fridman, B. A., On experimental determination of some statistical characteristics of turbulent flows, *Izv. Energetich. Inst. G. M. Krzhizhanovskogo*, 9; Fridman, B. A. Application of high-speed filming to the study of the velocity field of a turbulent flow, *Izv. AN SSSR*, Ser. Geograf. Geofiz., 12(2), 1946, and other works of the same author (*Izv. AN SSSR, OTN*, No. 11, 1953, etc.).

to measure mean intensity of velocity fluctuations and other statistical characteristics such as the coefficient of correlation and scale.

In air flows the method of thermal anemometry is successfully used which is based on the effect of cooling an electrically heated, thin, short platinum wire. In stable equilibrium the flow averaged velocity may be judged by the filament electric resistance. The intensity of velocity fluctuations in a flow may be determined by the deviations from the equilibrium in the compensation scheme (fluctuations of an oscillograph loop placed in the zero branch of the bridge). These fluctuations may be related to scale.[38]

Making use of various optical, electric, thermal, electrodynamic, and laser[39] devices, we can directly record mean quadratic values of velocity fluctuations, i.e. mean values of their products at the same or two different points of the flow. This gives the value of the turbulence intensity, turbulent skin friction, and the coefficient of correlation between velocity fluctuations at two points of the flow. As shown later, it is not difficult to find a mean scale of turbulence as well. For the details of the experiment we refer the reader to special manuals.[40]

Then, we denote the projections of the averaged velocity by the capital letters U, V, W and the projections of the fluctuating velocity component by the corresponding lower case letters u, v, and w. Retaining the adopted sign of time averaging in the form of the overbar, we compose the three basic averaged characteristics of fluctuating velocities – the square roots of the mean quadratic fluctuations of velocity

$$\sqrt{\overline{u^2}} \qquad \sqrt{\overline{v^2}} \qquad \sqrt{\overline{w^2}}$$

These quantities, related to the local or total averaged velocity, determine the first statistical characteristics of a turbulent flow – turbulence intensities in different directions.

At the given time instant t we take two points in space $M(x, y, z)$ and $M'(x', y', z')$ and designate the projections of fluctuating velocities of fluid at these two points of space by u, v, w and u', v', w', correspondingly. In a general case the asymmetric tensor of the second kind with the table

$$\begin{pmatrix} \overline{uu'} & \overline{uv'} & \overline{uw'} \\ \overline{vu'} & \overline{vv'} & \overline{vw'} \\ \overline{wu'} & \overline{wv'} & \overline{ww'} \end{pmatrix} \tag{36}$$

will present the tensor of moments of a two-point correlation between fluctuating velocities at the points M and M'.

Tending the point M' toward the point M, we obtain in the limit the symmetric tensor of the moments of single-point correlation

---

[38] The first experimental studies in this direction were conducted by the Dutch physicist J. Bürgers and van der Hegge Zyinen in 1924 in Delft (see Bürgers, J. M., the motion of fluid in the boundary layer along smooth surface, *Proc. 1st Int. Cong. Appl. Mech.*, Delft, 1924, 116–120 and a number of papers mentioned by the author in the *Transactions of the Dutch Academy of Sciences*).

[39] Laser methods for measuring local velocities are stated in the reviews Rinkevichyus, B. S., Doppler method for measuring local velocities by lasers. Usp. Fiz. nauk, 11(2), 305–330, 1973. Experimental technique of laser measurements in jets with solid admixtures was developed and introduced into engineering practice by a group of Tallinn scientists: M. K. Laats, F. A. Frishman, A. Z. Rosenshtein, in *M. K. Laats, Edited Collected Papers of the Proceedings of the 3rd All-Union Conference on Theoretical and Applied Aspects of Turbulent Flows "Turbulentnye dvukhfaznye techeniye" (Turbulent Two-Phase Flows)*, Tallinn, 1979. See also other collected papers of the Institute of Thermal and Electrical Physics of the Academy of Sciences of Estonian SSR that are published under the same title, and the works by M. K. Laats from the years 1979 to 1982.

$$\begin{pmatrix} \overline{u^2} & \overline{uv} & \overline{uw} \\ \overline{vu} & \overline{v^2} & \overline{vw} \\ \overline{wu} & \overline{wv} & \overline{w^2} \end{pmatrix} \tag{37}$$

whose components differ from the components of the tensor Π of turbulent stresses (17) only by constant multipliers ($-\rho$). Moving the point M' off the point M, we obtain the moments of two-point correlation between fluctuating velocities of the flow points that are less statistically correlated. The components of tensor (36) will tend in this case to zero.

Let us introduce the coefficients of spatial correlation between velocity fluctuations at two points: $u_i(x_1, x_2, x_3)$ and $u_j(x_1', x_2', x_3')$, where $x_1' = x_1 + \delta_{1m} r$, $x_2' = x_2 + \delta_{2m} r$, $x_3' = x_3 + \delta_{3m} r$ (δ_{km} is the Kronecker symbol, $m = 1, 2, 3$), having assumed

$$R_{ij}^{(1)}(r, 0, 0) = \frac{\overline{u_i(x_1, x_2, x_3)\, u_j(x_1 + r, x_2, x_3)}}{\sqrt{\overline{u_i^2}(x_1, x_2, x_3)}\,\sqrt{\overline{u_j^2}(x_1 + r, x_2, x_3)}}$$

$$R_{ij}^{(2)}(r, 0, 0) = \frac{\overline{u_i(x_1, x_2, x_3)\, u_j(x_1, x_2 + r, x_3)}}{\sqrt{\overline{u_i^2}(x_1, x_2, x_3)}\,\sqrt{\overline{u_j^2}(x_1, x_2 + r, x_3)}}$$

$$R_{ij}^{(3)}(r, 0, 0) = \frac{\overline{u_i(x_1, x_2, x_3)\, u_j(x_1, x_2, x_3 + r)}}{\sqrt{\overline{u_i^2}(x_1, x_2, x_3)}\,\sqrt{\overline{u_j^2}(x_1, x_2, x_3 + r)}} \tag{38}$$

Here, the subscripts at R on the left side refer to the numbers of the projections of the correlated fluctuating velocities, the superscript – to the number of the coordinate axis in whose direction the points M and M' are positioned.

In a general case of a turbulence spatial structure, we have to deal with a great number of the coefficients of two-point correlation.

---

[40] Povkh, I. L., *Aerodinamicheskiy eksperiment v mashinostroyenii* (*Aerodynamic Experiments in Machine Building*), Mashinostroyeniye Press, Leningrad, 1974, Chapters X to XII; Loitsyanskiy, L. G., *Aerodinamika pogranichnogo sloya* (*Aerodynamics of a Boundary Layer*), Gostekhizdat Press, Leningrad–Moscow, 1941, 382–402. Among the works on thermoanemometry the recently published monograph is distinguished: Yarin, L. P., Genkin, A. L., and Kukes, V. I., *Termoanemometriya gazovykh potokov* (*Thermoanemometry of Gas Flows*), Mashinostroyeniye Press, Leningrad, 1983. See also Zakharov, Yu. G. and Minskiy, E. M., Study of turbulence by a hot wire anemometer, *Tekh. Zam. TsAGI*, No. 172, 1978; Zakharov, Yu. G., Minskiy, E. M., and Filippov, M. S., On the technique for turbulence measurements by a hotwire anemometer, *Tr. TsAGI*, No. 402, 1939. The analysis of errors of measurements with the aid of a hotwire anemometer can be found in the paper: Zakharov, Yu. G., Measuring mean and fluctuating velocities of an air flow by a hotwire anemometer, *Tr. TsAGI*, No. 599, 1946. Some useful schemes are described in: Smirnov, G. V., A hot wire anemometer with filament heating by alternate current, Tr. LPI, No. 217, Leningrad, 1961, 168–175; Some electrical-radio-engineering methods for measuring aerodynamic quantities. *Tr. LPI*, No. 280, Leningrad, 1967, 107–116. Foreign literature on thermoanemometry is also voluminous. We note only some: Dryden, H., Turbulence investigations at the National Bureau of Standards. *Proc. 5th Int. Congr. Appl. Mech.*, 1938; Dryden, H., Schubauer, G., Mock, W., and Skramstad, H., Measurements of Intensity and Scale of Wind Tunnel Turbulence and Their Relation to the Critical Reynolds Number of Spheres, Tech. Rev. NACA, 581, 1937; Hall, A., Measurements of the intensity and scale of turbulence, ARC R & M, No. 1842, 1938; Corrsin, S., Extended applications of the hot wire anemometer, *Rev. Sci. Instr.*, 18, 469, 1947; Townsend, A., The measurement of double and triple correlation derivatives in isotropic turbulence, *Proc. Cambr. Phil. Soc.*, 43, 560, 1947. Conte-Bellot, J., *Turbulentnoe techeniye v kanale s parallelnymi stenkami* (*Turbulent Flow in a Channel with Parallel Walls*), (translated from French), Mir Press, Moscow, 1968; collected papers: *Beiträge zur Turbulenzforschung und Messtechnik/ Herausgegeben von Hoffmeister, M. und Förste, J.*, Academie Verlag, Berlin, 1973.

The integrals

$$L_{ij}^{(k)} = \int_0^\infty R_{ij}^{(k)}\, dr \tag{39}$$

where $R_{ij}^{(k)}$ is one of the correlation coefficients introduced by the system of equations (38), give the system of "turbulence scales" $L_{ij}^{(k)}$, which in a general case is very complicated.

Usually we restrict ourselves to the consideration of a "longitudinal" and two "lateral", relative to the direction of flow, scales, prescribing them by the integrals

$$L_x = \int_0^\infty R_{11}^{(1)}\, dr \qquad L_y = \int_0^\infty R_{11}^{(2)}\, dr \qquad L_z = \int_0^\infty R_{11}^{(3)}\, dr \tag{40}$$

which are among the most simple to determine experimentally. The turbulence scale is the second statistical characteristic of turbulence. Depending on the value of the ratio of turbulence scales to the characteristic dimensions of the flow (tube radius, "nominal" thickness of a boundary layer, etc.), "small-scale" and "large-scale" turbulences are identified.

Finally, we introduce the third characteristic of turbulence – turbulence spectrum – the function $F(k)$ of the distribution of kinetic energy of fluctuations by frequencies k of these fluctuations in time. An infinitely small quantity $F(k)dk$ determines the portion, relative to unit, of the energy of fluctuations with frequencies lying within the range $(k, k + dk)$ of the total spectrum of the averaged energy of a fluctuation flow. Omitting the numerical factor 1/2, we determine this mean by frequencies energy by the expression (u is the fluctuating velocity)

$$\overline{u^2} = \frac{\int_0^\infty \overline{u^2}(k)\, F(k)\, dk}{\int_0^\infty F(k)\, dk} = \int_0^\infty \overline{u^2}(k)\, F(k)\, dk \tag{41}$$

since from the function determination itself it follows that

$$\int_0^\infty F(k)\, dk = 1 \tag{42}$$

We present some quantitative results referring to the characteristics of a turbulent flow (see above). We start with consideration of free turbulence. The picture of turbulence intensity reduction is observed in a jet surrounded by a cocurrent flow (Figure 212). The curves[41] are presented of the distribution $\sqrt{\overline{u^2}}$ in the cross-sections normal to the flow and located at different relative distances x/d from the nozzle exit section. The velocity at the nozzle exit is 10.2 m/s, the cocurrent flow velocity is 8.5 m/s. It follows from the curves that near the

[41] Kobashi, Y., Experimental studies on compound jets, in *Proc. 2nd Jpn. Natl. Congr. Appl. Mech.*, 1952, 223–226.

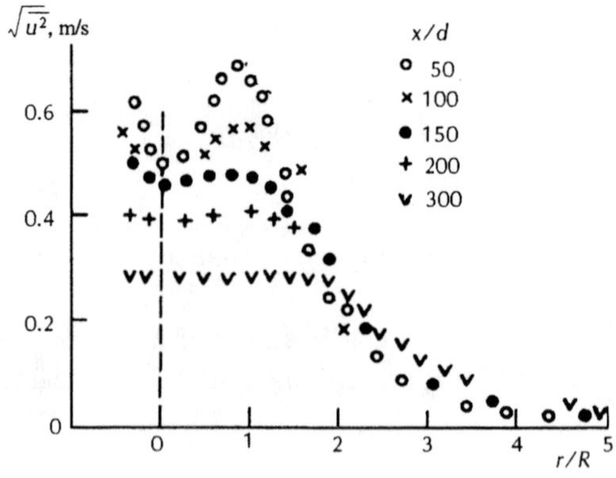

Fig. 212

boundary between a jet and the cocurrent flow at the nozzle exit ($r/R = 1$) there originate disturbances of high intensity that decrease downstream. The intensity of a turbulent flow escaping from the nozzle also decreases in this case.

A similar, but more detailed study was conducted by Ginevskiy et al.[42] Making use of the method of thermoanemometry the authors measured a number of turbulent characteristics of a jet in a cocurrent flow: distribution of intensities in a longitudinal $\varepsilon_u = \sqrt{\overline{u^2}}/U$ and lateral $\varepsilon_v = \sqrt{\overline{v^2}}/U$, $\varepsilon_w = \sqrt{\overline{w^2}}/U$ fluctuations of velocity, the two-point coefficient of correlation $R_{11}^{(2)}(0, y) = R_{uu}(y)$ and the turbulence scale $L_{11}^{(2)} = L(y)$ corresponding to it, and also the one-point coefficient of correlation r between the longitudinal, u and lateral, v, fluctuations.[43] The measurements were made at different ratios $m = U_\infty/U_0$ of the cocurrent flow velocity, U_∞, to the velocity at the exit section of the nozzle, U_0, from which the jet issued.

Figure 213a presents a summary graph which should be considered. In this figure, referring to the jet cross-section $x/d = 20$ (x is the distance between the cross-section and the nozzle exit section, d is the nozzle exit section diameter), all of the single-point characteristics: ε_u, ε_v, ε_w; r and a separately presented graph of the quantity $\Delta U = (U - U_\infty)/(U_m - U_\infty)$ of the excess averaged velocity $U - U_\infty$ related to the difference of velocity on the axis U_m and in the cocurrent flow U_∞ are given. Plotted on the abscissa axis is the dimensionless distance $\eta_{1/2} = y/\delta_{1/2}$ from the jet axis related to a value of $y = \delta_{1/2}$ which corresponds to the point where $\Delta U = 1/2$. In addition to Figure 212 where the decrease in the intensity of the longitudinal fluctuation with the distance from the nozzle exit section is shown, in Figure 213 data are presented on the various fluctuation quantities as they depend on the jet velocity at the nozzle exit section relative to the cocurrent flow velocity.

The attention is first attracted to the presence of maxima of fluctuation intensities and then of rapid lessening of intensities when approaching the jet "boundary". Second, it is attracted to the presence of the concept on the "jet boundary" as the point of the given section at which the excess velocity is equal to zero ($\eta_{1/2} = 2.2$) and an imaginary point at which

[42] Ginevskiy, A. S., Ilizarova, L. I., and Shubin, Yu. M., Study of turbulent jet microstructure in a cocurrent flow, *Mekh. Zhidk. Gaza*, 81–88, 1966.

[43] The usual notation ε_u, ε_v, ε_w; r of the turbulence intensity and of the coefficient of correlation should not be confused with the notation ε for the dissipation rate and r for the radial coordinate used later.

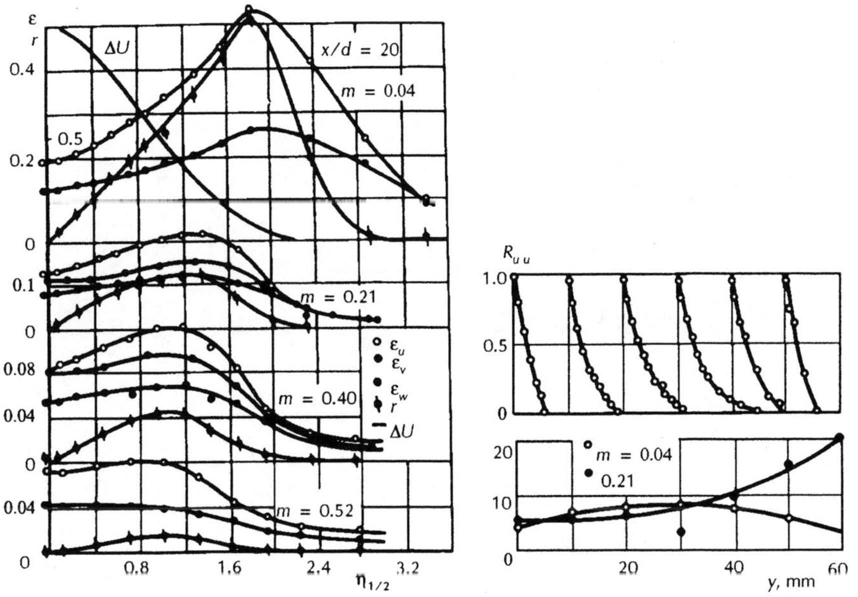

(a) (b)

Fig. 213

all of the disturbances produced by a jet in a cocurrent flow were equal to zero.

It may be noted that the concept on this second physically imagined boundary would be difficult to determine quantitatively because there is a boundary layer between the jet and a cocurrent flow where a smooth transition from the jet to a cocurrent flow takes place. At small values of the parameter, the second boundary turned out to be noticeably different for various turbulent characteristics. The graph of the jet-section-distribution of the coefficient of a two-point correlation between longitudinal fluctuation velocities $R_{uu}(y)$ (upper curves in Figure 213) and the lateral scale $L(y)$ determined by the change in the correlation coefficient according to the first integral (40) is borrowed from the just cited paper. We note the scale constancy in the jet central region and its change in the different ways on the jet boundary at different values of the parameter m.

Numerous experimental data on the flow turbulent structure in a plane tube may also be found in the book by Conte-Bellot.[44] We shall consider some of the data. In Figure 214a, b distributions of the intensity of longitudinal fluctuations, relative to the dynamic velocity $v_* = \sqrt{\tau_w/\rho}$ are presented as functions of the dimensionless distance from the wall differently composed for a near-wall and central portions of the flow. These distribution refer to different cross-sections at different relative distances x/D from the tube inlet (D is the semidistance between the walls of a plane tube; experimental dots are omitted).

Figure 214a describes a "near-wall" region. On the axis of abscissa laid off is the so-called "universal" coordinate (the dimensionless distance from the wall) $\eta = yv_*/\nu$. In a viscous sublayer ($\eta < 11.5$) the intensity of longitudinal fluctuations grows sharply with the distance from the wall attaining maximum nearly in the center of the region of transition from a viscous sublayer to a logarithmic zone. Then, intensity smoothly decreases up to its value on the tube axis, as is shown in Figure 214b, differing from Figure 214a by the abscissa scale.

[44] Conte-Bellot, J., *Turbulentnoye techeniye v kanale s parallelnymi stenkami* (*Turbulent Flow in a Channel with Parallel Walls*), Mir Press, Moscow, 1968.

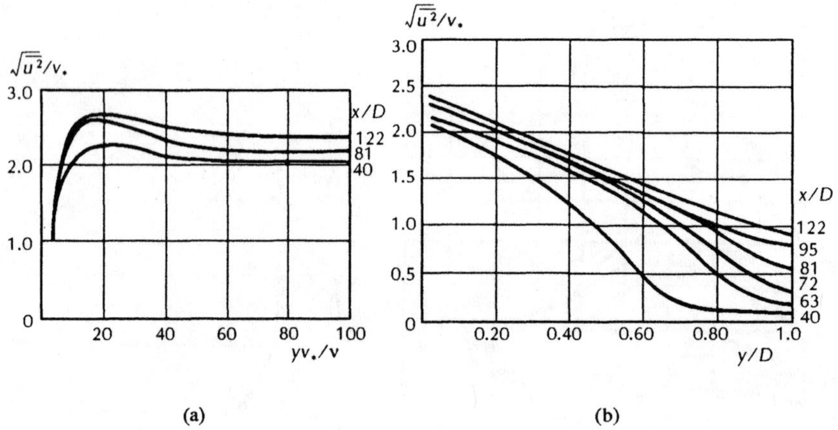

Fig. 214

The graphs presented refer to the flow with the Reynolds number $\mathbf{Re} = U_{ave}D/\nu$ = 12,000. As is shown by graphs given in the cited book, intensities of longitudinal and lateral fluctuations related to the dynamic velocity are weakly dependent on the Reynolds number of the flow.

Omitting the graphs of spectra of frequencies and of the correlation coefficients (available in the book by Conte-Bellot), we give the only common graph (Figure 215) (experimental points are omitted; their scatter, as compared with other graphs, is substantial) of the distributions of the longitudinal L_x and two lateral L_y and L_z turbulence scales over the tube section $x/D = 118$ at the Reynolds number $\mathbf{Re} = 120{,}000$. The longitudinal scale L_x considerably exceeds both lateral ones in magnitude.

For the scales L_x and L_y the presence of maxima is characteristic at about 30% of the semidistance D between the tube walls, measured from the tube wall.

Studies of turbulence in the boundary layer region are also numerous. Both the distributions of intensities and correlation between scales and frequency spectra have been investigated.

The reader is referred to the review by Dryden,[45] which is available in Russian, where

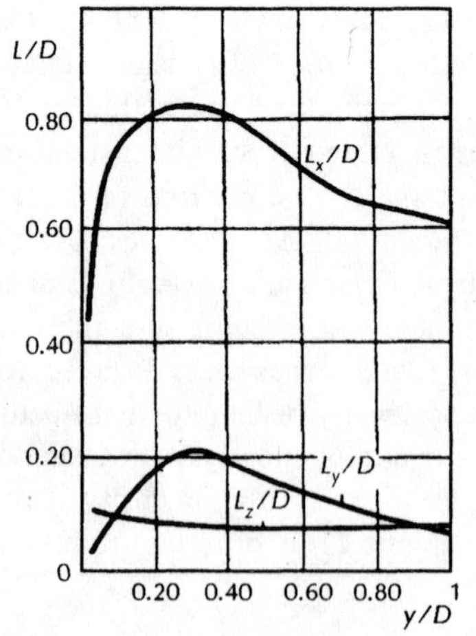

Fig. 215

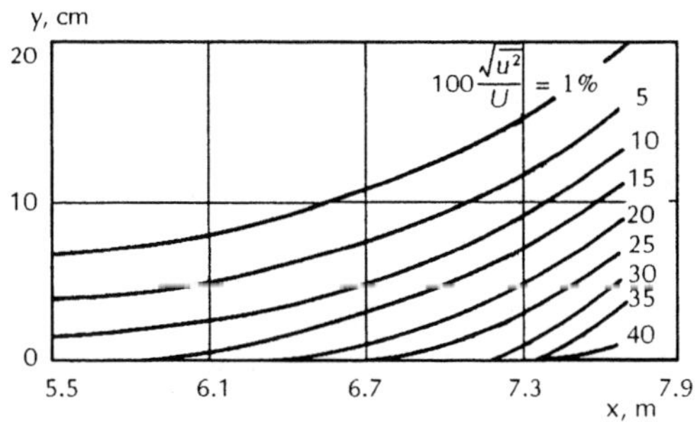

Fig. 216

systematic information on the structure of a turbulent boundary layer in the presence of the longitudinal negative and positive pressure drops is collected. As an example, we present the diagram of the distribution of the local intensity of turbulence $\left(\sqrt{\overline{u}^2}\right)/U$ in a diverging region of a boundary layer (Figure 216). The curves are loci of points of equal turbulence intensity shown in percent on the right.

The growth of turbulence intensity in the given lateral cross-section of a boundary layer when approaching the wall is explained by the influence of two factors: the increase in the absolute quadratic mean fluctuation of the velocity $\overline{u}^2$ similar to that shown in Figure 214a, and the simultaneous decrease in the velocity of the averaged flow U. The distribution of the local turbulence intensity along a boundary layer at the given value of the ordinate (y = const) shows a sharp growth of the intensity in a diverging portion of a layer ($x > 5.5$ m). Extremely high intensity (40%) takes place near the point of boundary layer separation ($x_s = 7.84$ m).

Frequency (spectral) characteristics of a turbulent boundary layer show that inside a layer the oscillations of low frequency, as compared with the frequencies of oscillation outside the boundary layer, are predominant. Thus, in a boundary layer on a plate in the section corresponding to the Reynolds number $\mathbf{Re}_x = 650{,}000$ prevail the frequencies up to 40 to 50 Hz; at $\mathbf{Re}_x = 1.6$ million, up to 20 Hz; and outside a boundary layer, of about 100 Hz. The portion of high frequencies (about 100 Hz) is small. Reduction of frequencies on transition to greater Reynolds numbers $\mathbf{Re}_x$ indicate the growth of the turbulence scale with the boundary layer thickness.

A simple correlation between the frequency of oscillations and turbulence scale may be observed. As a large-scale vortex passes by a motionless pick-up of a frequency meter the greater period of time will be registered as compared with passage of a small-scale vortex. Hence, there follows a general principle: to the vortices greater in scale corresponds smaller frequency, and vice versa. Using the wavelength instead of the frequency, we shall convince ourselves in full conformity of the previously noted difference in the process of vortex scattering to the familiar general law on a more rapid scattering of short waves as compared with long ones.[46]

Greatly developed is the theory of "coherent" turbulence that explains turbulent interaction on interfaces formed, for example, in free jets or jets in cocurrent flows and wakes behind streamlined bodies by the presence of discrete vortical structures (annular, toroidal, Π-shaped, etc.). For a great number of applied problems and especially the theory of a turbu-

[45] Dryden, H., Modern development of boundary layer mechanics, in *Problemy mekhaniki* (*Problems of Mechanics*), Inostrannaya Literatura Press, Moscow, 1955, 277–290.

[46] To get acquainted with the general theory of turbulence we suggest the most accessible monograph: Hinze, J. O., *Turbulentnost* (*Turbulence*), Fizmatgiz Press, Moscow, 1963.

lent boundary layer, the existence of "coherent" structures serves as an explanation of such a specific phenomenon as separation of a turbulent boundary layer from a smooth surface. A similar problem was considered in Chapter 12 for the case of a laminar boundary layer in which the separation phenomenon was subjected to simple mathematical modeling. Such affairs are different with a turbulent boundary layer. According to modern concepts, large-scale turbulence develops in the outer portion of a boundary layer. The structure of this turbulence presents a dynamic equilibrium process of the interaction of large vortices, which have the order of the size of a boundary layer thickness. On the one hand, they disintegrate to smaller vortices and, on the other hand, they coagulate to larger ones, thus forming the so-called "coherent" structure. Being entrained by a separated boundary layer, these vortices fill in the "wake" behind a body that also has a coherent structure. We emphasize that in these regions the theory of turbulent transport loses its validity and is replaced by the theory of interaction of vortical systems that has not been fully developed.

5. Transfer of Momentum, Heat, and Admixtures in Turbulent Flows. Hypotheses of Boussinesq, Fourier, and Fick

The separation of some simpler averaged flow from the general turbulent motion, suggested by Reynolds, does not eliminate the necessity of taking into account those physical processes which in reality take place in turbulent flows. Streamlines of the averaged flow that are impermeable for this conditionally introduced motion, are permeable for a fluctuating flow that transports momentum, heat, matter, and other types of physical substances from layer to layer across the streamlines of an averaged flow. This transport is similar to that which takes place in the case of molecular transfer in laminar flows, and determines turbulent friction between the layers in an averaged flow, heat and mass transfer between them, and other various transport phenomena.

Here, the difference from laminar (molecular) transfer is in the fact that substance in turbulent transfer is carried not by separate molecules that are comparatively negligible in mass but by finite volumes of fluid, as they are sometimes called, "moles". In accordance with this the processes of turbulent transfer themselves are called "molar" in contrast to "molecular" processes of transfer in laminar flows.

From this point of view the technique of averaged flow separation may be presented as follows. An actual turbulent flow with serpentine chaotically interwoven streamlines and trajectories that are characteristic of it, is substituted by some ordered laminated (but not laminar) flow. This substitution may be expressed by the term "stratification" which is adopted in meteorological applications of the turbulence theory (from the Latin "stratus", meaning layer). Stratification may occur in terms of different flow characteristics: velocity, density, temperature, etc.

Considering, for example, a steady-state averaged turbulent flow in a plane tube (Figure 217), one presents the streamlines of an averaged motion in the form of straight lines parallel to the tube axis. This is velocity stratification. In the case of a steady-state flow an identical profile of averaged velocities $\bar{u}(y)$ takes place in all cross-sections of the tube. The profile shape depends on the properties of a turbulent flow and will be determined later. Streamlines of a fluctuating flow intersect streamlines of an averaged motion, penetrate from one its layer to another, and create mixing of fluid through small areas located along streamlines of an averaged flow.

Such mixing, called turbulent or molar mixing, is accompanied by the transporting through the boundaries between layers, of momentum, energy, and other mechanical or thermodynamical parameters of an averaged motion of fluid, heat, and of admixtures involved in

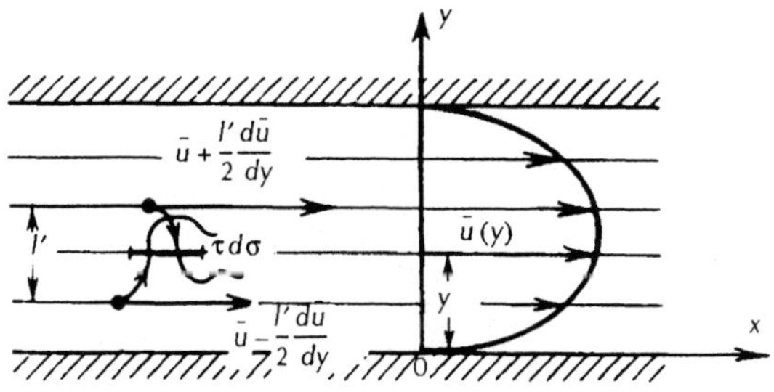

Fig. 217

fluid (e.g., smoke or dust in air, silt or sand in water, etc.).

Momentum transfer produces turbulent friction between the layers, heat transfer stimulates turbulent heat conduction, transport of admixtures – turbulent diffusion of these admixtures. Mechanism of turbulent mixing is similar both for friction and for heat conduction or diffusion; the difference is only in specific features of substance – momentum, heat, or admixture – transported by a fluctuating flow.

In 1877, Boussinesq,[47] using the analogy between a laminar flow of a viscous fluid and its turbulent averaged motion as a basis, suggested the equation for tangent stress τ_t between the layers of an averaged flow, that became widely popular:

$$\tau_t = A\frac{d\bar{u}}{dy} \tag{43}$$

This equality resembles the Newton equation (Equation 1, Chapter 10). The difference is only in the fact that in Equation 43 the dynamic coefficient of ordinary (molecular) viscosity μ is substituted by a substantially greater dynamic coefficient of turbulent (molar) viscosity A which was first called the coefficient of turbulent mixing ("Austauschgrösse" according to Prandtl)and by $\bar{u}$ we mean the averaged velocity of a turbulent fluid flow.

Just as the dynamic coefficient of viscosity presents the physical constant of liquid or gas at their given thermodynamic states, the coefficient of proportionality A was also first treated as a constant coefficient coinciding in dimensional representation with μ. The calculations of turbulent fluid flow, performed under this simplest assumption on the constancy of A, showed that in order for the results of these calculations to agree with experiment the value of A should by many orders exceed μ. It may be noted as an historical curiosity that Richardson, a well-known specialist in the dynamics of atmospheric air motions, compared the dynamic coefficient of air turbulent viscosity A with analogous coefficient of syrup viscosity μ, and the kinematic coefficient A/ρ with the kinematic coefficient of shoe polish ν. It was Lorentz[48] who first explained this abnormal fact. He showed that in turbulent flows friction, turbulent viscosity, in contrast to ordinary viscosity, is stipulated by transport through the layers of moving fluid of momenta, and not of microscopic particles, but of mixing finite volumes of media.

This idea of Lorentz was then developed in the works of Taylor, Richardson, and Schmidt,[49] who created the model of turbulent mixing and indicated the role of the "mixing

---

[47] Boussinesq, J., Esai sur la théorie des eaux courantes, *Mémoires Préssentées par Diverses Savants á l'Academe du Science*, Paris, 23, 1877.

[48] Lorentz, H. A., Vorschlagen der Akad. von Wetenschapen, Amsterdam, 6, 1897.

[49] Keller, L. V., The theory of convection in turbulent atmosphere according to Taylor, Richardson, and Schmidt, *Izv. Gl. Geofiz. Observ.*, No. 4, 3, 1931; *ibid*, voluminous bibliography.

length" concept.

We derive a general formula of turbulent friction in the simplest case of a developed plane rectilinear averaged motion presented in Figure 217. Consider an elementary small area $d\sigma = dx \cdot 1$, which is parallel to the streamline of an averaged flow and is positioned at a distance y from a solid wall. The streamlines of a fluctuating flow pass through this area and transport momenta of adjacent layers located both above and behind the area at some distance $l'/2$ with the fluctuating velocity v' which is transverse to the averaged flow serving as the transport velocity. The tangent stress of turbulent friction $\pi_{xy} = \pi_{yx}$ will be denoted for brevity in terms of τ_t and be determined as the time-mean projection onto the axis $0x$ of the quantity of averaged turbulent flow through the area located along the streamline per area unit. Meaning by $\tau_t d\sigma$ the friction force applied from the upper layer, ware averaged velocities are higher to the lower one, we consider momentum transferred from the upper layer to the lower one as acquired, i.e., positive, and momentum transported from the lower layer to the upper one as lost, i.e., negative. Then, designating the time-averaged quantity by the overbar, we find

$$\tau_t d\sigma = \overline{\rho v'\left[\bar{u}\left(y+\frac{l'}{2}\right)-\bar{u}\left(y-\frac{l'}{2}\right)\right]}d\sigma$$

or substituting approximately

$$\bar{u}\left(y \pm \frac{l'}{2}\right) = \bar{u}(y) \pm \frac{l'}{2}\frac{d\bar{u}}{dy}$$

and performing evident reduction, we shall obtain

$$\tau_t = -\rho\overline{u'v'} = \rho\overline{v'l'}\frac{d\bar{u}}{dy} \qquad (44)$$

Comparing equalities (44) and (43) yields the expression for the coefficient of turbulent mixing A (ρ is the constant)

$$A = \rho\overline{v'l'} \qquad (45)$$

which, thanks to the presence of velocity v' and l', imparts to this empirical quantity the inner sense, relating it to the mechanism of turbulent mixing. The Boussinesq equation (43), specified by expression (45), which is derived for the simplest rectilinear shear flow, in full conformity with the consideration stated at the beginning of Chapter 10, will be generalized to the case of any spatial motion.

We introduce an assumption on a local isotropy of the mechanism of turbulent mixing. This means that the coefficient A is a scalar quantity. An assumption on anisotropy of this mechanism would lead to a sharp complication of the mathematical aspect of the problem caused by the fact that the quantity A would become in this case the fourth-order tensor.

We have, assuming the correlation between the Reynolds stress tensor Π (see Equations 17 and 18) and the tensor of velocities of deformations in an averaged flow $\bar{\dot{S}}$ to be linear,

$$\Pi = a\bar{\dot{S}} + bE \qquad (46)$$

where a and b are the scalars dependent on the turbulent flow structure, E is the tensor unity, whose component is equal to the Kronecker symbols δ_{ij}. A comparison between (46) and (43) for the simplest form of an averaged flow gives

$$a = 2A \tag{47}$$

and equating linear invariants on the left and right sides of Equation 46 makes it possible to conclude that

$$b = -\frac{1}{3}\rho\left(\overline{v_1'^2} + \overline{v_2'^2} + \overline{v_3'^2}\right) = -\frac{1}{3}\rho\,\overline{(v_1'^2 + v_2'^2 + v_3'^2)} = -\frac{2}{3}\rho k \tag{48}$$

where, by the previous k is the kinetic energy of a fluctuating flow referred to the mass unit

$$k = \frac{1}{2}\overline{v_i'v_i'} \tag{49}$$

Further, we use the commonly adopted designations

$$A = \mu_t \qquad \frac{A}{\rho} = \nu_t \tag{50}$$

where μ_t and ν_t are the dynamic and kinematic coefficients of "turbulent viscosity", respectively. Then, we come to the following form of the generalized Boussinesq hypothesis

$$\Pi = 2\mu_t\bar{\bar{S}} - \frac{2}{3}\rho k E \tag{51}$$

and in the rectangular Carterian coordinates

$$\pi_{ij} = -\rho\overline{v_i'v_j'} = 2\mu_t\bar{S}_{ij} - \frac{2}{3}\rho k\delta_{ij} = \mu_t\left(\frac{\partial\bar{v}_i}{\partial x_j} + \frac{\partial\bar{v}_j}{\partial x_i}\right) - \frac{1}{3}\rho\,\overline{(v_m'v_m')}\,\delta_{ij} \tag{52}$$

The quantity

$$\frac{\rho}{3}\overline{v_m'v_m'} = p_t \tag{53}$$

may be called "turbulent pressure", and equality (52) changed to the form

$$\pi_{ij} = -\rho\overline{v_i'v_j'} = \mu_t\left(\frac{\partial\bar{v}_i}{\partial x_j} + \frac{\partial\bar{v}_j}{\partial x_i}\right) - p_t\delta_{ij} \tag{54}$$

whose right side is similar to the generalized Newton law for viscous fluid (Equation 10, Chapter 10).

The Boussinesq equation in its initial form (43), and then in a generalized form (51 or 52), acquired wide application in hydraulics, meteorology, oceanography and other branches of technology and natural science. In this case different assumptions were made on changing the character of the quantities μ_t and ν_t in a flow, and their constancy was shortly renounced.[50] As Boussinesq himself mentioned, and later confirmed experimentally, μ_t and ν_t change over the section normal to the flow in a tube or in a boundary layer, they increase from zero on a streamlined solid surface up to some maximum and then decrease to the minimum value on the tube axis or on the outer edge of a turbulent boundary layer. On a sharp growth of $\mu_t = A$ at a distance frm a solid surface one may judge by Figure 218 which pre-

[50] See Forchheimer, F., *Gidravlika (Hydraulics)*, ONTI Press, Moscow-Leningrad, 1935 (Russian translation).

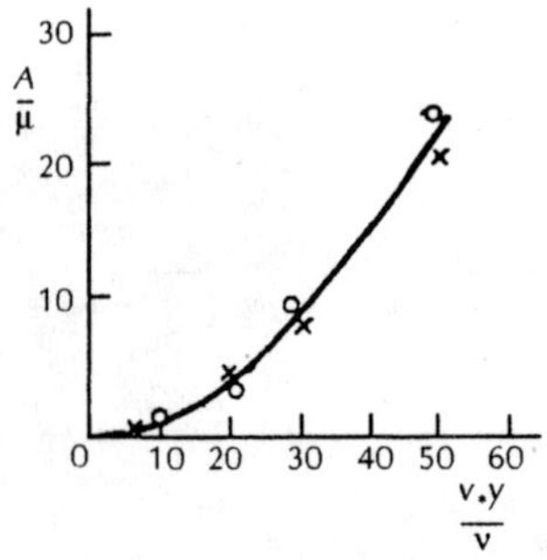

Fig. 218

sents the results of the experiments by Laufer in tubes (circles) and Schubauer in boundary layers (crosses).[51] The distance from the solid wall related to the quantity ν/v_* having the dimensions of length, where $v_* = \sqrt{\tau_w/\rho}$, and τ_w is the skin friction on a wall, is plotted on the abscissa axis. On the ordinate axis, as compared to ordinary molecular viscosity, the ratio $\mu_t/\mu = A/\mu$ is plotted. It may be noted that on approaching the solid wall the coefficient of turbulent viscosity rapidly decreases. In meteorology and oceanography the quantity $A = \mu_t$ is usually treated as an increasing function of the distance between the measurement point and the Earth's surface or ocean bottom.

Historically the Boussinesq equation was a large step on the way of closing the momentum equation (16), having made it possible to express six unknown Reynolds stresses in terms of one unknown μ_t, but it did not give complete solution of the closure problem. The Boussinesq equation, composed analogously to the generalized determining Newton relation for a laminar viscous fluid flow, forms the basis of the overwhelming majority of the methods for calculating turbulent fluid flows in hydraulics, hydrodynamic, and their technological applications. At the same time this equation deserves some critical notes. Similarly to the Newton equation, the Boussinesq equation refers to the number of the so-called local equations in the sense that it determines the values of turbulent (Reynolds) stresses at the given point of fluid flow depending on the averaged rates of deformations in the vicinity of the same point, i.e., it does not involve the effect on the determining relation of the flow regions at a distance from this point. This limitedness of the Boussinesq equation deprives it of a number of important applications, for instance, in asymmetric flows in channels with different roughness of walls. In such channels, in full nonconformity with the Boussinesq equations, the points of zero turbulent friction do not coincide with the points of maximum velocity ($\partial\bar{u}/\partial y = 0$). In general, it may be noted that the Boussinesq equation (43) does not reflect a vortical structure of a flow and the dimensions of vortical masses exchanging momentum, though allowance for these factors may be very crucial in describing some phenomena specific to turbulence. As will be shown later attenuation of turbulent friction stresses in the wake behind the body disturbing a flow, calculated based on the local Boussinesq equation, reflects only a general tendency of this phenomenon but it cannot give an accurate quantitative description of this "hereditary" process. Other examples of the inadequacy of the Boussinesq equation may be found in the survey part of the thesis by the Dutch scientist Builtjes.[52]

When considering turbulent flows in real fluids, along with momentum transfer, one often has to deal with simultaneous transport of heat and substance (mass). Practically important problems of heat and mass transfer in turbulent flows usually allow simple temperature

---

[51] Cited in: Hinze, I. O., *Turbulentnost (Turbulence)*, Fizmatgiz Press, Moscow, 1963 (Russian translation).

[52] Builtjes, P. J. H., Memory Effects in Turbulent Flows Department of Mechanical Engineering, Delft, The Netherlands, 1977, 5–8.

and concentration stratification which coincides with velocity stratification. Making use of the Boussinesq idea on imparting the form of the Newton "laminar" law to the equation of turbulent friction, the form formally generalizing the Fourier and Fick laws, which we know from previous chapters, may be given to turbulent fluxes of heat and substance (mass).

We designate the coefficients of turbulent transfer of momentum, heat and admixture concentration in terms of $A_\tau = \mu_t$, A_q and A_m, respectively. Recalling the expressions for heat and substance flows (24) and (27) and restricting ourselves to the simplest case of a steady-state shear flow near a wall where the averaged velocities $\bar{u}$, temperature $\bar{T}$, and concentration of the admixture in fluid $\bar{c}$ will be the functions of the distance from the wall y only, we find the following expressions for turbulent flows of momentum, heat and substance

$$\tau_t = -\rho\overline{u'v'} = A_\tau \frac{d\bar{u}}{dy}$$

$$q_t = -\rho c_p \overline{v'T'} = -c_p A_q \frac{d\bar{T}}{dy}$$

$$m_t = -\rho\overline{v'c'} = -A_m \frac{d\bar{c}}{dy} \tag{55}$$

where the primes distinguish the fluctuations of relevant quantities from their averaged values denoted by an overbar.[53]

To emphasize the role of the transverse fluctuating velocity v' as similar for all of the considered processes of "substance" (momentum, heat, matter) transfer from a layer to a layer of moving fluid, we express the coefficients of transfer A_τ, A_q, and A_m in terms of v' and some "transport lengths" l'_τ, l'_q and l'_m typical of each separate substance, having assumed analogously to (45) that

$$A_\tau = \rho\overline{v'l'_\tau} \qquad A_q = \rho\overline{v'l'_q} \qquad A_m = \rho\overline{v'l'_m} \tag{56}$$

The sense of the definitions of these "transport lengths", given by Prandtl, is explained in the next section. Sometimes, instead of the coefficients of transfer A_τ, A_q and A_m, the ratios of them to the medium density are introduced

$$\varepsilon_\tau = \frac{A_\tau}{\rho} = \frac{\mu_\tau}{\rho} = \nu_\tau \qquad \varepsilon_q = \frac{A_q}{\rho} \qquad \varepsilon_m = \frac{A_m}{\rho} \tag{57}$$

Along with the expressions of turbulent (molar) transfer of the substances mentioned we compose the expressions of complete transfers combining laminar (molecular) and turbulent (molar) transfers of momentum, heat, and substance

$$\tau = \mu\frac{d\bar{u}}{dy} + \tau_t = (\mu + A_\tau)\frac{d\bar{u}}{dy} = (\mu + \rho\varepsilon_\tau)\frac{d\bar{u}}{dy}$$

$$q = -\lambda\frac{d\bar{T}}{dy} + q_t = -(\lambda + c_p A_q)\frac{d\bar{T}}{dy} = -(\lambda + \rho c_p \varepsilon_q)\frac{d\bar{T}}{dy}$$

[53] Considering it impossible to abandon commonly adopted notation c for concentration and c_p for heat capacity at constant pressure, we warn against confusing these quantities similar in designations.

$$m = -\rho D\frac{d\bar{c}}{dy} + m_t = -(\rho D + A_m)\frac{d\bar{c}}{dy} = -(\rho D + \rho\varepsilon_m)\frac{d\bar{c}}{dy} \quad (58)$$

To the laminar (molecular) coefficients of transfer-coefficients of viscosity μ, thermal conductivity λ, and diffusion ρD (introduced in Section 3, Chapter 12), correspond the coefficients of turbulent viscosity A_τ, thermal conductivity A_q, and diffusion A_m.

Equations 56 give the dependence of the quantities A_τ, A_q and A_m on v', l'_τ, l'_q, and l'_m which remain unknown. Credit for the reduction in the number of these unknowns and thus of the approach to the solution of the problem of closing the equations of momentum, heat, and substance transfer belongs to Prandtl, who, having retained the role of the transverse fluctuating velocity in the mechanism of transfer, expressed this velocity in terms of the analogues of the quantities l'_τ, l'_q, and l'_m and the transverse derivative of the averaged velocity $d\bar{u}/dy$.

6. The Prandtl "Mixing Length" Theory

As mentioned above, the development of the problem of closing the equations of momentum, heat, and matter transfer did not reduced to the determination of Equations 46 to 48 only, which are far from being the reflection of the real mechanism of transfer of these substances. In the first 25 years of the 20th century there appeared the transfer theory, which considerably deepened the concepts of this field and virtually closed the equations of momentum, heat, and matter transfer.

This theory was developed by Prandtl.[54] Proceeding from the analogy between the transfer of finite in volume vortical masses in a turbulent flow and transport of molecules in a laminar flow, Prandtl suggested a simple and obvious model of substance transfer between the layers of an averaged turbulent flow. According to his idea a vortical mass emanated in the given "initial" layer as some individual volume, moves during the transfer process normally to the averaged flow and then moves in this direction, having lost its individual character – the difference in a longitudinal averaged velocity – or at the constant density, its momentum, within some adjacent layer. To this point there were no differences between the picture of transfer described above and that stated in the previous section. As earlier, use is made of the concept on a "vortical volume", which moves from one layer to another, passing some distance between the layers that determines the typical of this volume "mixing length". The quantitative determination of the mixing length was unambiguous until the appearance of the Prandtl theory.

New to the Prandtl model is the assumption of the fact that a vortical mass retains its difference in momentum during the whole period of moving from the initial layer to the final one, and only at the moment of mixing does it immediately lose its individual character, having thus evoked perturbation of the averaged velocity in the final layer. This perturbation is taken proportional to the distance l' between the initial and final layers and to the value of difference between the averaged velocities in these layers $\Delta\bar{u} \approx l' d\bar{u}/dy$. By the Prandtl idea this perturbation in the layer averaged velocity is the cause of origination of a fluctuating velocity with the projections u' and v' in it, so it is natural to assume that ($\backsim$ is the sign of proportionality)

[54] Prandtl, L., Untersuchungen zur ausgebildete Turbulenz, Z. *Angew. Math Mech.*, 5, 1925. See also Results of recent works on turbulence investigation, in *Problemy turbulentnosti* (*Problems of Turbulence*), ONTI Press, Moscow, 1936, 14–16 (Russian translation).

$$u' \backsim l'\frac{d\bar{u}}{dy} \qquad v' \backsim l'\frac{d\bar{u}}{dy} \tag{59}$$

This equation, which is central to the Prandtl theory, leads to the expression of the tangent component of the Reynolds stress in the case of the simplest rectilinear shear flow under consideration

$$\tau = -\rho\overline{u'v'} = -\rho\overline{l'^2}\left(\frac{d\bar{u}}{dy}\right)^2$$

If we agree to understand by τ the skin friction applied to the upper edge of the layer from the side of the region of higher velocities ($\tau > 0$) and, moreover, to introduce the averaged length $l^2 = \overline{l'^2}$, then the previous expression for τ will take on the form

$$\tau = \rho l^2\left(\frac{d\bar{u}}{dy}\right)^2 \tag{60}$$

Prandtl called the quantity l, which remained the only undetermined quantity here, the mixing length (Mischungsweg). This is why the entire theory acquired the name of the mixing length theory.

In order to avoid mentioning the sign at τ, Prandtl suggested that Equation 60 should have the form

$$\tau = \rho l^2\left|\frac{d\bar{u}}{dy}\right|\frac{d\bar{u}}{dy} \tag{61}$$

For the dynamic coefficient of turbulent friction $A_\tau = \mu_t$ and its kinematic value $A_\tau/\rho = \mu_t/\rho = \nu_t$ the equations

$$A_\tau = \mu_t = \rho l^2\left|\frac{d\bar{u}}{dy}\right| \qquad \nu_t = l^2\left|\frac{d\bar{u}}{dy}\right| \tag{62}$$

are valid. Taking into account generality of the mechanisms of turbulent transfer of momentum, heat, and matter (mass), we compose the expressions for the coefficients of mixing (56) in the form

$$A_\tau = \mu_t = \rho l_\tau^2\left|\frac{d\bar{u}}{dy}\right| \qquad A_q = \rho l_q^2\left|\frac{d\bar{u}}{dy}\right| \qquad A_m = \rho l_m^2\left|\frac{d\bar{u}}{dy}\right| \tag{63}$$

For the kinematic coefficients the equalities

$$\frac{A_\tau}{\rho} = \nu_t = \varepsilon_\tau = l_\tau^2\left|\frac{d\bar{u}}{dy}\right| \qquad \frac{A_q}{\rho} = \varepsilon_q = l_q^2\left|\frac{d\bar{u}}{dy}\right| \qquad \frac{A_m}{\rho} = \varepsilon_m = l_m^2\left|\frac{d\bar{u}}{dy}\right| \tag{64}$$

are valid, where the difference in the expressions of mixing lengths for momentum, heat, and matter, determined by the equations

$$l_\tau^2 = \overline{l_\tau'^2} \qquad l_q^2 = \overline{l_\tau' l_q'} \qquad l_m^2 = \overline{l_\tau' l_m'} \tag{65}$$

is admitted.

Recalling the Prandtl and Schmidt numbers (the Prandtl diffusion number) introduced in the theory of molecular (laminar) heat and mass transfer

$$\mathbf{Pr} = \frac{\mu c_p}{\lambda} \qquad \mathbf{Sc} = \mathbf{Pr}_d = \frac{\mu}{\rho D} \tag{66}$$

we shall generalize these definitions to the case of turbulent heat and mass transfer, having assumed

$$\mathbf{Pr}_t = \frac{c_p A_\tau}{c_p A_q} = \frac{A_\tau}{A_q} = \frac{\varepsilon_\tau}{\varepsilon_q} \qquad \mathbf{Sc}_t = \frac{A_\tau}{A_m} = \frac{\varepsilon_\tau}{\varepsilon_m} \tag{67}$$

Naturally, these turbulent criteria differ principally from their laminar prototypes first of all in the fact that they are dependent on the form of motion and not just on the physical constant of the medium. Moreover, characterizing the relations between molar transfers, which by many orders (by final values) exceed molecular transfers, the turbulent numbers of Prandtl and Schmidt are only slightly dependent on their molecular values. If, for example, the Prandtl numbers (molecular) for viscous oils and liquid metals differ by 1 million times, then turbulent Prandtl numbers, in similar flows of so physically differing media, are close to each other in value.

"The carriers" of transport of various substances are the same "vortical" masses, but the processes of their mixing with a surrounding medium may differ for different substances. Depressing the activity of this process on the way of the "carrier" motion from the initial to final layer, i.e., considering the substance passive, we may assume that the coefficients of transfer should be equal to each other and, consequently, according to Equation 67, turbulent Prandtl and Schmidt numbers are equal to unity. In fact, the transferred substance is not completely passive. Some difference in the activity of heat and mass transfer takes place and is compared with momentum transfer (friction). This results in the difference (not large) of the turbulent Prandtl and Schmidt number from unity to this or that side.

There is voluminous literature on the experimental determination of the turbulent Prandtl number, especially for such a "limiting" case, as turbulent flows of liquid metals in which molecular thermal conductivity is very large as compared with viscosity (the Prandtl numbers in a laminar flow have the order of 10^{-3} to 10^{-4}). However, in these cases the turbulent Prandtl number is also close to unity, remaining within the limits $0.5 < \mathbf{Pr}_t < 2$.[55] In the studies of turbulent heat transfer in gases it is more often assumed that $\mathbf{Pr}_t = 1$ or 0.7.

Unfortunately, the results of experiments on the determination of the mean values of the turbulent Prandtl number are slightly accurate and sometimes even contradictory due to a great number of complicating circumstances (determination of derivatives of velocities and temperatures prescribed by inaccurate experimental curves, contamination of surfaces, etc.). Thus, for example, via the Ludwieg experiment (1956),[56] $\mathbf{Pr}_t$ grows from the wall to the tube axis, while via the Sleicher tests (1957)[57] it decreases. A theoretical aspect of this problem is difficult even in a semiempirical formulation. Nevertheless some steps in this direction

[55] The survey of the studies in this trend may be found in the monograph by Kutateladze, S. S., Borishanskiy, V. M., Novikov, I. I., and Fedynskiy, O. S., *Zhidkometallicheskiye teplonositeli (Liquid-Metal Coolants)*, Atomizdat Press, Moscow, 1967.

[56] Ludwieg, H., Bestimmung des Verhaltnisses der Austauschkoeffizien ten für Wärme und Impuls bei turbulenten Grenzschichten, *Z. Flugwiss.*, 4, (1/2), 1956.

[57] Sleicher, C., Experimental Velocity and Temperature Profiles for Air in Turbulent Pipe Flow, Paper N 57–HT–9, American Society of Mechanical Engineers, 1957.

have been made both in Russia[58] and abroad.[59]

Simplest, and practically sufficient, is the assumption of the passivity of the substance transferred and, consequently, of the equality of the numbers $\mathbf{Pr}_t$ and $\mathbf{Sc}_t$ to unity. This assumption was successfully used in the studies of heat transfer from the walls of tubes and in other cases of near-wall turbulence, but became ineffective for the flows under the conditions of free turbulence (jets, wakes behind a body).

Coming back to the Prandtl theory, one can note the similarity between the concept of the "mixing length" introduced into it and the free path length of a molecule of the kinetic theory of gases.

Doing justice to the "mixing length" theory, one cannot but note its limitation to applicability only to flow regions where $d\bar{u}/dy \neq 0$. When approaching the tube axis or the boundary layer outer edge, where $d\bar{u}/dy = 0$, the role of the first derivative of the averaged velocity disappears. Its equality to zero, according to the Prandtl equations (59), would mean the equality of the fluctuating velocities u' and v' to zero, thus not corresponding to reality. The necessity for substituting the first derivative entering the Prandtl equation by the higher-order derivatives was shown, in particular, by Prandtl himself, but it was not developed from the practical viewpoint.[60] This leads us to restrict ourselves in the application of the mixing length theory only to near-wall regions of turbulent fluid flows or, in any case, to the regions at a distance from the flow points where $d\bar{u}/dy = 0$ (axis of a jet or wake).

The Prandtl equations (60) and (61), published in 1926, have stood the test of time and until now have been successfully applied in calculations of various turbulent flows. Under conditions of extremely complicated chaotic flows and interactions of separate "vortical" volumes the Prandtl equations could distinguish the basic determining factors of the turbulent transfer phenomenon. These extremely simple and clearly formulated equations are still fundamental to the theory of turbulent flows.

As to the only uncertain quantity, the mixing length, in the Prandtl formulas, it is subjected, as was mentioned by Prandtl himself, to persuasive intuitive determination, at least in a number of the simplest cases (discussed later). This makes it possible to recognize the Prandtl "mixing length" theory as that closing the equations of momentum, heat, and matter (mass) transfer.

The Prandtl theory, similar to the Boussinesq theory that grounds it, belongs to the class of local theories of turbulent motions based on the assumption that turbulent mixing at the given flow point is completely determined by physical constants ρ, μ, ν, and by the averaged flow in direct vicinity to this point. Analytically this is expressed in the dependence of the transport coefficients on the combination of the derivatives $d\bar{u}/dy$, $d^2\bar{u}/dy^2$, etc. The velocity $\bar{u}(y)$ itself should not enter into the number of these arguments since, combining a fluid particle moving at the constant velocity along the given layer $\bar{u}(y)$ with the translationally moving system of coordinates, we should observe in this inertial Galileoan system of reckoning the same dynamic phenomena as in a motionless system irrespective of the translational flow velocity $\bar{u}$.

The derivation of the Prandtl equation (60), and of other formulas related to it, conserved the style of presentation of the mixing length theory close to the style of the original Prandtl paper cited above. If the local character of the mixing length theory forms the basis of derivation, then, based on the dimensional method, one can come to the same Prandtl equation (60). By the evidence of persons collaborating with Prandtl, he himself first arrived at his formula

---

58 Voskresenskiy, K. D. and Turilina, Ye. S., in *Teploperedacha i teplovoye modelirovaniye* (*Heat Transfer and Thermal Modeling*), Izd. AN SSSR Press, Moscow, 1958, 87.

59 Dwyer, Ye. 0. Eddy transport in liquid-metal heat transfer, *AIChE J.*, 9(2), 261–268, 1963.

60 Loitsyanskiy, L. G., On some applications of the similarity method in the turbulence theory, *Prikl. Mat. Mekh.*, 2(2), 180–206, 1935.

from the considerations of dimensions and only then took that physical image to it.

We shall assume that even at a small distance from the wall the coefficient of turbulent mixing A so greatly exceeds μ that the effect of ordinary molecular viscosity on turbulent viscosity may be neglected. In other words, we consider the process of turbulent mixing disregarding the effect of molecular viscosity on it, although in fact this effect takes place; this issue is discussed later. Then, this coefficient of purely turbulent mixing A may be presented by a one-term power-law relation ($\backsim$ is the sign of proportionality; the coefficient of proportionality is dimensionless)

$$A \backsim \rho^a \left(\frac{d\bar{u}}{dy}\right)^b \left(\frac{d^2\bar{u}}{dy^2}\right)^c \ldots \tag{68}$$

We shall try to restrict ourselves first to the dependence on the density ρ and to the first derivative of the velocity $d\bar{u}/dy$, having assumed that

$$A \backsim \rho^a \left(\frac{d\bar{u}}{dy}\right)^b \tag{69}$$

Noting that in the physical system of dimensions mass–length–time (M, L, T)

$$[A] = \left[\frac{\tau}{d\bar{u}/dy}\right] = \left[\frac{M}{T^2 L} T\right] = [ML^{-1}T^{-1}]$$

$$[\rho] = [ML^{-3}] \qquad \left[\frac{d\bar{u}}{dy}\right] = [T^{-1}]$$

we shall convince ourselves that Equation 69, when comparing the indices at M, L, and T, yields the unfulfilled system of equations

$$a = 1 \qquad 3a = 1 \qquad b = 1;$$

i.e., the formula of the (69) type is impossible. Complementing it with the multiplier containing some degree of length, i.e., assuming

$$A \backsim \rho^a l^m \left(\frac{d\bar{u}}{dy}\right)^b \qquad [l] = [L] \tag{70}$$

we shall obtain, equating the indices at M, L, and T, the system of equations

$$a = 1 \qquad -1 = 3a + m \qquad b = 1$$

which has the solution

$$a = 1 \qquad m = 2 \qquad b = 1$$

that, according to Equation 70, leads to the formulas (the coefficient of proportionality is included in the determination of l)

$$A = \rho l^2 \frac{d\bar{u}}{dy} \qquad \tau = A\frac{d\bar{u}}{dy} = \rho l^2 \left(\frac{d\bar{u}}{dy}\right)^2 = \rho l^2 \left|\frac{d\bar{u}}{dy}\right| \frac{d\bar{u}}{dy}$$

which coincide with the Prandtl equations (60) and (61).

Now we shall take into account the role of the second derivative $d^2\bar{u}/dy$; for this sake it will be assumed

$$A \backsim \rho^a \left(\frac{d\bar{u}}{dy}\right)^b \left(\frac{d^2\bar{u}}{dy^2}\right)^c \tag{71}$$

In this case there is no necessity to additionally apply the length, because the relation $(d\bar{u}/dy) / (d^2\bar{u}/dy^2)$ has the dimensions of length and, consequently, it should be

$$l = -\frac{æ\,(d\bar{u}/dy)}{d^2\bar{u}/dy^2} \tag{72}$$

here, æ is some dimensionless constant, and the sign on the right side is selected so that near a solid wall, where $d\bar{u}/dy > 0$, $d^2\bar{u}/dy^2 < 0$, the value of 1 was positive. Substituting the value l from Equation 72 into (62) and (61) yields

$$A = æ^2\rho \frac{(d\bar{u}/dy)^3}{(d^2\bar{u}/dy^2)^2} \qquad \tau = æ^2\rho \frac{(d\bar{u}/dy)^4}{(d^2\bar{u}/dy^2)^2} \tag{73}$$

Equations 73 were first derived by von Kàrmàn[61] from considerably more complex consideration. von Kàrmàn proceeded from the assumption of the similarity of fields of fluctuating velocity in different layers of a turbulent shear flow.

The von Kàrmàn equations (72) and (73) are typical of the local approach to the study of turbulent motions. The Prandtl equation (60) is in this sense less typical because the quantity of mixing length 1 which remains unknown, presents the possibility of its use for determining other approaches on which one can judge by comparison with applications based on the Prandtl and von Kàrmàn equations.

We shall pause on two examples containing direct use of the Prandtl theory with some additions of intuitive character relative to the law of the mixing length variation. Reference to these intuitive considerations presupposes the introduction of some constants, the values of which are determined by generalization of experimental data. This deprives the Prandtl theory of strict inner closure and causes one to refer this theory to the class of the so-called semiempirical theories. It should be emphasized that the quantity of such theories, not only in fluid mechanics but in other branches of science, is much larger than is often realized. In the present and following chapters we shall widely use the term a "semiempirical theory".

7. "Free" Turbulence. Submerged Jets. Far Wake

Passing over to applications of semiempirical methods of calculating turbulent flows, we shall distinguish two classes of motions: (1) free flows occurring at a distance from solid surfaces and obeying the laws governing the so called free turbulence, and (2) near-wall flows which in contrast to previous develop near solid surfaces and are described by the laws governing near-wall turbulence.

Various cases of propagation of turbulent jets in quiescent fluid and in cocurrent flows, formation of a wake behind a body, and others refer to the first class.

The absence of interactions between molecular and molar processes of transfer in the

[61] von Kàrmàn, Th., Mechanische Ähnlichkeit und Turbulenz, *Nachr Ges. Wiss. Göttingen, Math. Phys. Kl.*, 1930.

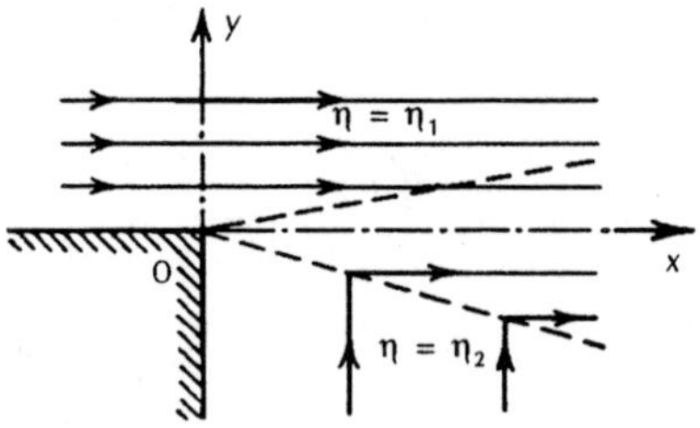

Fig. 219

motions of this class is a specific feature of free turbulence. In such cases we have to deal with purely turbulent flows and with molar transfer processes only, thus greatly simplifying calculation. The simplest in this case are the techniques of assigning the coefficients of transfer and mixing length.

In modern technology a practical value of the theory of turbulent jets (jet and rocket engines, combustion chambers of steam generators, heating and ventilating systems, apparatuses of chemical industry, etc.) is so large that this theory became an independent branch of applied hydrodynamics many years ago and it is the concern of voluminous monographs.[62] Therefore, we shall restrict ourselves to the presentation of some of the most interesting results, from the theoretical viewpoint, in order to illustrate the application of semiempirical methods to this branch of general hydrodynamics.

We shall start from the direct application of the Prandtl equation (60). Consider an averaged turbulent flow in a boundary layer formed in the region of mixing of a jet of a very large diameter with surrounding fluid of the same density.

The choice of the coordinate axes is shown in Figure 219. The jet diameter is taken infinitely large so that the flow in the boundary region may be considered as plane.

The equations of an averaged turbulent flow in a jet boundary layer may be derived from the Reynolds equations (15) by a procedure similar to that of deriving the equation of laminar boundary layer from the Navier-Stokes equations. We shall obtain, discarding the dashes as averaging designations,

$$u\frac{\partial u}{\partial x} + v\frac{\partial u}{\partial y} = \frac{1}{\rho}\frac{\partial \tau}{\partial y} \qquad \frac{\partial u}{\partial x} + \frac{\partial v}{\partial y} = 0 \tag{74}$$

In these equations, in view of the absence of solid surfaces bounding the flow (the tube wall effect downstream behind the point 0 is not taken into account), viscous terms are omitted; moreover, we neglect the derivative $\rho\partial \overline{u'^2}/\partial x$ as compared with $\partial\tau/\partial y$ and pressure variation is neglected through the entire region. Following Prandtl, we shall assume the quantity l constant over the cross-section of the region of jet mixing with the surrounding fluid and varying from section to section. Equation 74 will take on the form

$$u\frac{\partial u}{\partial x} + v\frac{\partial u}{\partial y} = l^2(x)\frac{\partial u}{\partial y}\frac{\partial^2 u}{\partial y^2} \qquad \frac{\partial u}{\partial x} + \frac{\partial v}{\partial y} = 0$$

or, if use is made of the stream function $\psi(x, y)$ of an averaged motion,

[62] Abramovich, G. N., *Teoriya turbulentnykh struy* (*Theory of Turbulent Jets*), Fizmatgiz Press, Moscow, 1960 and a new edition of this book published in 1984; Ginevskiy, A. S., *Teoriya turbulentnykh struy i sledov* (*Theory of Turbulent Jets and Wakes*), Mashinostroeniye Press, Moscow, 1969; Korobko, V. N., *Teoriya neavtomodelnykh struy vyazkoy zhidkosti* (*Theory of Non-self-similar Jets of Viscous Fluid*), Part II, Saratov University Press, 1977.

$$\frac{\partial\psi}{\partial y}\frac{\partial^2\psi}{\partial x\,\partial y}-\frac{\partial\psi}{\partial x}\frac{\partial^2\psi}{\partial y^2} = l^2(x)\frac{\partial^2\psi}{\partial y^2}\frac{\partial^3\psi}{\partial y^3} \tag{75}$$

We shall formulate the boundary conditions. Assuming the velocity in a jet undisturbed portion equal to U_0, we have[63]

$$\psi \to U_0 y, \quad \frac{\partial\psi}{\partial y} \to U_0 \quad \text{when} \quad y \to \infty, \quad x > 0,$$

$$\frac{\partial\psi}{\partial y} = U_0 \quad \text{when} \quad x = 0, \quad y > 0. \tag{76}$$

The boundary shape remains unknown. The missing third boundary condition is formulated on the lower edge ($y \to -\infty$) and it will probably have the form

$$\frac{\partial\psi}{\partial y} \to 0 \quad \text{when} \quad y \to -\infty, \quad x > 0,$$

$$\frac{\partial\psi}{\partial y} = 0 \quad \text{when} \quad x = 0, \quad y < 0, \tag{77}$$

since the mixing region borders from below with fluid which has no longitudinal velocity ($u = 0$). Note that involvement of the surrounding fluid into a jet (the so-called entrainment) occurs and is implemented thanks to the presence of the transverse velocity v on the lower edge (Figure 219).

To determine $l(x)$ and to reveal whether the formulated problem is self-similar, i.e., whether in the case considered the partial differential equation (75) with the boundary conditions (76) and (77) is reduced to the ordinary differential equation, we shall apply the consideration which is similar to that often used in the laminar boundary layer theory.

We shall designate a completely arbitrary scale of lengths in terms of L, not determined by the problem conditions. Contrary to a laminar layer where the scale of transverse lengths was obtained from the scale of longitudinal lengths by dividing the latter into $\sqrt{\mathbf{Re}}$ in the turbulent layer theory, on neglecting the viscous terms and consequently the Reynolds number, the longitudinal and transverse lengths have equal scales.

Moving on to dimensionless quantities, we shall convince ourselves of the fact that Equation 75 does not give any relation between the scales of the stream function Ψ and the lengths L since it is uniform relative to ψ. From the first boundary condition (76) we obtain the relation between the scales of the stream function, velocities, and lengths

$$\Psi = U_0 L$$

Consequently, the solution of Equation 75 should have general form (the primes are temporary taken to denote the dimensionless coordinates and the stream function)

$$\psi' = \psi'(x', y')$$

or

$$\psi = U_0 L \psi'\left(\frac{x}{L}, \frac{y}{L}\right)$$

[63] Tollmien, W., Berechung turbulenter Ausbreitungsvorgänge. *Z. Angew. Math. Mech.*, 6, 468–478, 1926.

The scale L is absent in the problem formulation, however, is also absent in the problem solution. Hence, we conclude that the function ψ' should have the form

$$\psi' = x'\varphi\left(\frac{y'}{x'}\right)$$

In fact, in this case the scale L is omitted and the dimensional function will have the form

$$\psi = U_0 x\varphi\left(\frac{y}{x}\right) = U_0 x\,\varphi(\eta) \qquad \eta = \frac{y}{x} \tag{78}$$

Now we shall determine the form of the function $l(x)$ at which this solution of Equation 75 is possible. Calculating the derivatives (the prime is the symbol of the derivative with respect to η)

$$\frac{\partial\psi}{\partial y} = U_0\varphi'(\eta) \qquad \frac{\partial^2\psi}{\partial y^2} = U_0\frac{1}{x}\varphi''(\eta) \qquad \frac{\partial^3\psi}{\partial y^3} = U_0\frac{1}{x^2}\varphi'''(\eta)$$

$$\frac{\partial\psi}{\partial x} = U_0[\varphi(\eta) - \eta\varphi'(\eta)] \qquad \frac{\partial^2\psi}{\partial x\partial y} = -\frac{U_0}{x}\eta\,\varphi''(\eta)$$

and substituting them into Equation 75, we obtain after simple transformations

$$-\varphi\varphi'' = \frac{l^2}{x^2}\varphi''\varphi''' \tag{79}$$

According to Equation 78 the quantity φ is the stream function of η only; consequently, in the previous equation the variable x should be absent. This yields

$$l = cx \tag{80}$$

where c is the empirical constant dependent, as is shown by the experiments, on the turbulent structure of a boundary layer, i.e., on the previous history of the flow.

Equation 79 is then reduced to the ordinary differential equation

$$\varphi''(c^2\varphi''' + \varphi) = 0$$

which breaks down into

$$c^2\varphi''' + \varphi = 0 \tag{81}$$

$$\varphi'' = 0 \tag{82}$$

Equation 82 corresponds to the motion in an undisturbed jet core, i.e., to the first two equalities (76), which, according to Equation 78, now may be rewritten in the form

$$\varphi(\eta) = \eta \qquad \varphi'(\eta) = 1 \tag{83}$$

We shall designate the smallest positive root of the equation $\varphi'(\eta) = 1$ in terms of η_1 and the smallest in magnitude negative root of the equation $\varphi'(\eta) = 0$ in terms of η_2. Then, according to Equations 76, 77, and 78, the equalities

$$\eta = \frac{y}{x} = \eta_1 \quad \text{and} \quad \eta = \frac{y}{x} = \eta_2 \tag{84}$$

will represent the equations of rectilinear edges of the boundary layer of jet mixing with the surrounding fluid, and the jet width $b(x)$ will be determined by the difference

$$b(x) = (\eta_1 - \eta_2)x \tag{85}$$

Comparing Equation 80 and 85, we convince ourselves that the earlier adopted assumption on the constancy of the mixing length over the jet cross-section equivalent to the assumption on the proportionality of the mixing length to the width of a jet in its given cross-section. This assumption along with that of the dependence of the introduced constant c on the flow previous history (i.e., on the motion prior to the jet escape into the infinite space flooded with the same fluid), is deprived of local character as is the Prandtl formula (60) itself.

Integrating the ordinary (the problem is self-similar!) linear differential equation of the third order (81), we obtain its general integral in the form

$$\varphi(\eta) = C_1 e^{-\alpha\eta} + e^{\frac{1}{2}\alpha\eta}\left(C_2 \cos\frac{\sqrt{3}}{2}\alpha\eta + C_3 \sin\frac{\sqrt{3}}{2}\alpha\eta\right)$$

where $\alpha = c^{-2/3}$ is the only empirical constant of the problem. To determine the integration constants C_1, C_2, and C_3 and the bounding values of η_1 and η_2 we have four equations

$$\varphi''(\eta_1) = 0 \qquad \varphi'(\eta_1) = 1 \qquad \varphi(\eta_1) = \eta_1 \qquad \varphi'(\eta_2) = 0$$

and one additional

$$\varphi''(\eta_2) = 0$$

which express the condition of smoothness of transition of velocities to the prescribed values on the mixing region boundaries.

Omitting simple computations related to the calculation of these constants, we present the results

$$C_1 = -0.0062 \qquad C_2 = 0.987 \qquad C_3 = 0.577$$

$$\alpha\eta_1 = 0.981 \qquad \alpha\eta_2 = -2.04$$

Figure 220 shows the curves of the distribution of longitudinal and transverse velocities in the mixing region. The experiments conducted in Prandtl's laboratory confirmed the theory and provided the opportunity to determine the values of the constant c or α; they were equal: $c = 0.0246$, $\alpha = 11.8$. According to Equation 85 the width b of the mixing region in this case is equal to

$$b = \frac{3.021}{11.8}x = 0.255x$$

The calculation stated may be used in the design of a wind tunnel with the opened operation section on the boundary of which the above-described mixing region is formed.[64]

Consider the submerged axisymmetric straight jet.[65] Take the axis $0x$ as the jet axis. Referring to the Reynolds equations in the cylindrical coordinates (21), we make simplifica-

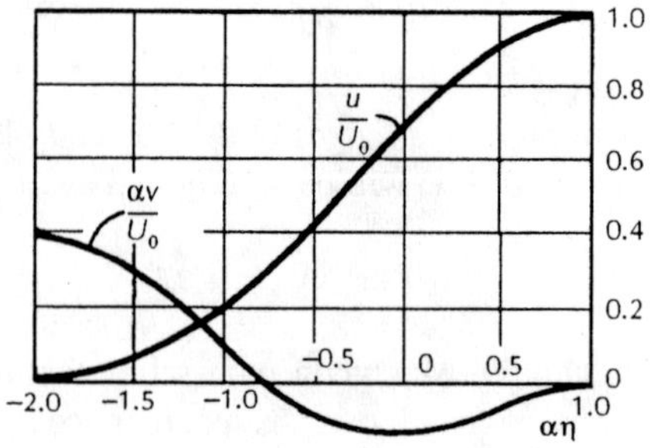

Fig. 220

tions corresponding to general concepts of a boundary layer. Note that in this case $\bar{v}_\varepsilon = 0$ and the derivatives with respect to ε are also equal to zero. Moreover, from the considerations of symmetry follows the equality to zero of the tangent turbulent stresses $\pi_{\varepsilon r} = -\rho \overline{v'_\varepsilon v'_r}$ and $\pi_{\varepsilon x} = -\rho \overline{v'_\varepsilon v'_x}$. According to considerations similar to those of the laminar jet theory (the submerged space infinity), we shall discard the term involving pressure; we shall also neglect variation of the quantity $\overline{v'^2_x}$ along the axis.

We denote for brevity

$$\pi_{rx} = \tau = -\rho \overline{v'_r v'_x} \qquad \bar{v}_x = u \qquad \bar{v}_r = v$$

Then the equations of turbulent jet propagation, as they follow from the last two lines of system (21), will be

$$u\frac{\partial u}{\partial x} + v\frac{\partial u}{\partial r} = \frac{1}{\rho}\left(\frac{\partial \tau}{\partial r} + \frac{\tau}{r}\right) \equiv \frac{1}{\rho}\frac{1}{r}\frac{\partial}{\partial r}(r\tau) \qquad \frac{\partial}{\partial x}(ru) + \frac{\partial}{\partial r}(rv) = 0$$

Using the Prandtl equation for turbulent friction that, as can be easily comprehended, in the present case ($\partial u/\partial r < 0$ at any r) will be written in the form

$$\tau = \rho l^2 \left|\frac{\partial u}{\partial r}\right| \frac{\partial u}{\partial r} = -\rho l^2 \left(\frac{\partial u}{\partial r}\right)^2$$

we shall obtain the equation of jet propagation in the form

$$u\frac{\partial u}{\partial x} + v\frac{\partial u}{\partial r} = -\frac{1}{r}\frac{\partial}{\partial r}\left[rl^2\left(\frac{\partial u}{\partial r}\right)^2\right] \qquad \frac{\partial}{\partial x}(ru) + \frac{\partial}{\partial r}(rv) = 0 \tag{86}$$

Introducing the stream function ψ related to the velocities u, v by the formulas

$$u = \frac{1}{r}\frac{\partial \psi}{\partial r} \qquad v = -\frac{1}{r}\frac{\partial \psi}{\partial x} \tag{87}$$

[64] Abramovich, G. N., Aerodynamics of flow in the opened operation section of a wind tunnel, *Tr. TsAGI*, 223, 236, 1935.

[65] See the paper by W. Tollmien.

we reduce the problem to the solution of a third-order equation

$$\frac{1}{r}\frac{\partial\psi}{\partial r}\frac{\partial^2\psi}{\partial x\partial r} - \frac{\partial\psi}{\partial x}\frac{\partial}{\partial r}\left(\frac{1}{r}\frac{\partial\psi}{\partial r}\right) = -\frac{\partial}{\partial r}\left\{rl^2\left[\frac{\partial}{\partial r}\left(\frac{1}{r}\frac{\partial\psi}{\partial r}\right)\right]^2\right\} \tag{88}$$

To this equation the condition of momentum conservation is added

$$2\pi\rho\int_0^\infty ru^2dr = 2\pi\rho\int_0^\infty \frac{1}{r}\left(\frac{\partial\psi}{\partial r}\right)^2 dr = J_0 \tag{89}$$

and the boundary conditions

$$\psi = 0 \qquad \frac{\partial}{\partial r}\left(\frac{1}{r}\frac{\partial\psi}{\partial r}\right) = 0 \quad \text{at} \quad r = 0$$

$$\frac{1}{r}\frac{\partial\psi}{\partial r} \to 0 \quad \text{when} \quad r \to \infty \tag{90}$$

We shall introduce the scale of the stream function Ψ and the scale of lengths L. Going over to the dimensionless quantities we convince ourselves that Equation 88 and the boundary conditions (90), similar to the first problem of this section, do not impose any restrictions to Ψ and L. From condition (89) we obtain one correlation between the scales Ψ and L:

$$\Psi = \sqrt{\frac{J_0}{2\pi\rho}}L$$

Hence, it follows that the solution for ψ should have a general form (the prime is the symbol of the dimensionless coordinates and the stream function)

$$\psi = \Psi\psi'(x', r') = \sqrt{\frac{J_0}{2\pi\rho}}L\psi'\left(\frac{x}{L}, \frac{r}{L}\right)$$

However, since the scale L not involved in the problem conditions it should not be present in the solution either. To fulfill this condition we demand that the function $\psi'(x', r')$ should have the form

$$\psi' = x'\varphi\left(\frac{r'}{x'}\right)$$

Then the scale L falls out and the dimensional stream function ψ will be determined by the expression

$$\psi = \sqrt{\frac{J_0}{2\pi\rho}}x\varphi(\eta) \qquad \eta = \frac{r}{x} \tag{91}$$

Assuming, as before, the mixing length l to be a function x only and substituting expression (91) into Equation 88, we shall convince ourselves that in the presence of the stream functions of the type of (91) the equality $l = cx$ should be valid.

Introducing a new variable η into Equation 88, according to (91), and taking the formula for l into account, we shall find that the problem is self-similar; we have an ordinary differential equation (the prime is the derivative with respect to η)

$$\frac{\partial}{\partial\eta}\left(\frac{\varphi\varphi'}{\eta}\right) = \frac{\partial}{\partial\eta}\left[\frac{c^2}{\eta}\left(\varphi'' - \frac{\varphi'}{\eta}\right)^2\right]$$

After termwise integration of both sides of this equation we come to the second-order equation

$$\varphi\varphi' = c^2\left(\varphi'' - \frac{\varphi'}{\eta}\right)^2 + \text{const}\cdot\eta \tag{92}$$

Here, the integration constant is equal to zero. In fact, when $\eta \to 0$ $(r \to 0)$, i.e., on the jet axis, the velocity u remains finite, and v tends to zero. Calculating these velocities, we shall obtain ($\backsim$ is the proportionality sign)

$$u \backsim \frac{1}{x}\frac{\varphi'}{\eta} \qquad v \backsim \frac{1}{x}\left(\varphi' - \frac{\varphi}{\eta}\right)$$

consequently, φ'/η tends to the finite value, and $\varphi \to 0$ when $\eta \to 0$. Constructing then the derivative $\partial u/\partial r$ and demanding that it should vanish when $\eta \to 0$, we obtain

$$\frac{\partial u}{\partial r} \backsim \frac{1}{\eta x^2}\left(\varphi'' - \frac{\varphi'}{\eta}\right) \to 0 \quad \text{for} \quad \eta \to 0$$

Consequently,

$$\frac{1}{\eta}\left(\varphi'' - \frac{\varphi'}{\eta}\right) \to 0 \quad \text{for} \quad \eta \to 0$$

where there follows the equality of the constant to zero in Equation 92.

Selecting in Equation 92 the variable $\eta/\sqrt[3]{c^2}$ instead of η, we obtain the equation and the boundary conditions independent of partial values of the constants. This equation is integrated numerically.

The above formulated problem of round jet propagation may be solved considerably simpler if instead of the Prandtl theory stated in the previous section, use is made of another, also semiempirical Prandtl theory from 1942.[66] When solving the problems of free turbulence in which the profiles of longitudinal averaged velocities usually have an inflection (Figure 220), we replace expression (62) for the coefficient of turbulent friction A by some simplified one based on the following considerations. We shall assume that in the cross-section M_1M_2 (Figure 221) the velocity continuously changes over from some value $u = u_1$ to the value $u = u_2$. Thus, in a jet, propagating through the flooded infinite space, the velocity u_1 on the jet outer edge is equal to zero, and u_2 is the maximum velocity u_{max} on the axis. In the case of an aerodynamic wake the velocity u_1 at a distance behind the body corresponds to the minimum velocity on the wake axis, and $u_2 = U_\infty$ to the velocity of an undisturbed outer flow falling onto the body.

The derivative $\partial u/\partial y$ on the edges of the range M_1M_2 vanishes both on transition to uniform flows and in the cases when the points M_1 and M_2 correspond to the maximum and minimum of the velocity. Here, the graph of velocities has the inflection point in the range

---

[66] Prandtl, L., Bemerkungen zur Theorie des freien Turbulenz, Z. *Angew. Math. Mech.*, 22(5), 241, 1942; Görtler, H., Berechung von Aufgaben der freien Turblenz auf Grund eines neues Näherusansatzes, *ihid.*, 244.

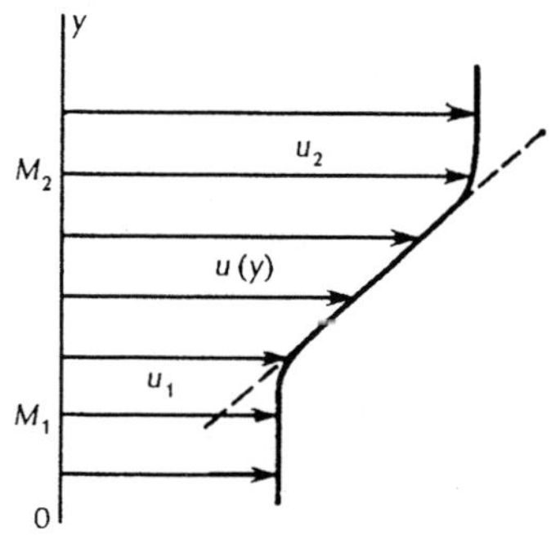

Fig. 221

considered where $\partial^2 u/\partial y^2 = 0$ and approaches a straight line everywhere excluding the regions adjacent to the range ends. Making use of this observation, we can make an approximate substitution in expression (62) of the coefficient of turbulent transfer A

$$\left|\frac{\partial u}{\partial y}\right| \approx \frac{|u_2 - u_1|}{b}$$

and assume

$$A \approx \rho l^2 \frac{|u_2 - u_1|}{b}$$

where $b = M_1 M_2$ is the width of the turbulent mixing region. The error, appearing in this case on the region boundaries, is not substantial because in the expression of turbulent friction (62) the quantity A is multiplied by the derivative $\partial u/\partial y$ which vanishes at the region boundaries. Thus, the coefficient of turbulent transfer in the problems of free turbulence may be taken constant over the cross-section, i.e., it is independent of y (but, generally speaking, dependent on x, i.e., variable along the flow). Assuming the mixing length which is constant over the layer section to be proportional to the width of the transfer region b, we obtain the following, general for the majority of problems in the turbulence theory, formula for the coefficient of turbulent friction

$$A = k\rho b|u_2 - u_1| \tag{93}$$

where k is some constant coefficient of proportionality; the quantities b and $|u_2 - u_1|$ change from section to section and represent the unknown functions of the coordinate reckoned along the flow.

We shall prove that in the case of an axisymmetric propagation of a turbulent straight jet the coefficient of turbulent friction A is constant over the entire flow region, i.e., it is not dependent either on x, or on r. We shall recall that the mixing length l is proportional to the jet width b and, as has been already proved, $b \backsim x$.

On the other hand, in the case given

$$u_1 = 0 \qquad u_2 = u_{max}(x) = u(x, 0)$$

and it follows from Equations 87 and 91 that

$$u_{max}(x) = u|_{r=0} = \frac{1}{r}\frac{\partial\psi}{\partial r}\bigg|_{r=0} = \sqrt{\frac{J_0}{2\pi\rho}}\frac{1}{x}\left[\frac{\varphi'(\eta)}{\eta}\right]_{\eta=0}$$

i.e.,

$$u_{max} \backsim \sqrt{\frac{J_0}{\rho}}\frac{1}{x}$$

Consequently, we have according to Equation 93 (σ is a new empirical constant)

$$A = \text{const} = \sigma\sqrt{\rho J_0} \qquad \varepsilon_\tau = \frac{A}{\rho} = \text{const} = \sigma\sqrt{\frac{J_0}{\rho}} \tag{94}$$

The hypothesis on the constancy of the coefficient of turbulent mixing has been applied in the problems of turbulent motion in free atmosphere, oceans, and rivers in aerodynamic and thermal wakes. In 1938 Trubchikov[67] formulated the same hypothesis for the case of a turbulent fluid flow in aerodynamic and thermal wakes.

Now, Equations 86 may be rewritten in the form

$$u\frac{\partial u}{\partial x} + v\frac{\partial u}{\partial r} = \sigma\sqrt{\frac{J_0}{\rho}}\frac{1}{r}\frac{\partial}{\partial r}\left(r\frac{\partial u}{\partial r}\right) \qquad \frac{\partial}{\partial x}(ru) + \frac{\partial}{\partial r}(rv) = 0$$

which differs from the equations of laminar jet propagation only by the fact that the constant kinematic coefficient of molecular viscosity ν is substituted by the kinematic coefficient of turbulent viscosity $\varepsilon_\tau = \sigma\sqrt{J_0/\rho}$, where σ is some empirical constant dependent on the flow turbulent structure.

Thus, there is no necessity to solve the problem anew because from the mathematical viewpoint a similar problem was solved in an earlier section. To compare the solutions we note that in turbulent jet calculation one should make substitution in Equation 178 of Chapter 12

$$\alpha = \sqrt{\frac{3}{16\pi\sigma}\sqrt{\frac{J_0}{\rho}}} = \sqrt{\frac{3}{16\pi\sigma}}\sqrt[4]{\frac{J_0}{\rho}}$$

$$\alpha\eta = \frac{1}{\sigma}\sqrt{\frac{3}{16\pi}}\frac{r}{x} \qquad \alpha\sqrt{\nu} = \sqrt{\frac{3}{16\pi}}\sqrt{\frac{J_0}{\rho}} \tag{95}$$

We shall have

$$\psi = \frac{3}{16\pi\sigma}\sqrt{\frac{J_0}{\rho}}\,x\,\frac{\left(\frac{r}{x}\right)^2}{1 + \frac{3}{64\pi\sigma^2}\left(\frac{r}{x}\right)^2}$$

---

[67] Trubchikov, B. Ya., A thermal method for measuring turbulence in wind tunnels, *Tr. TsAGI*, 372, 16, 1938. See also the previously cited papers by Prandtl and Görtler.

$$u = \frac{3}{8\pi\sigma}\sqrt{\frac{J_0}{\rho}}\frac{1}{x}\frac{1}{\left[1+\frac{3}{64\pi\sigma^2}\left(\frac{r}{x}\right)^2\right]^2}$$

$$v = \frac{3}{16\pi\sigma}\sqrt{\frac{J_0}{\rho}}\frac{r}{x^2}\frac{\left[1-\frac{3}{64\pi\sigma^2}\left(\frac{r}{x}\right)^2\right]}{\left[1+\frac{3}{64\pi\sigma^2}\left(\frac{r}{x}\right)^2\right]^2} \tag{96}$$

By equality (179) of Chapter 12 we find the per second mass flow rate through the jet sections

$$M = 8\pi\sigma\rho\sqrt{\frac{J_0}{\rho}}\,x \tag{97}$$

and by Equation 180 of Chapter 12 the formula of maximum velocity on the jet axis

$$u_{\max} = \frac{3}{8\pi\sigma}\sqrt{\frac{J_0}{\rho}}\frac{1}{x} \tag{98}$$

To promote experimental verification of the validity of the formulas obtained, we shall transform the formula of the longitudinal velocity (the second equality of Equation 96) to the form

$$\frac{u}{u_{\max}} = F\left(\frac{r}{R}\right)$$

where R means the value of r at which $u = \frac{1}{2}u_{\max}$. From the equation

$$\frac{1}{2} = \frac{1}{\left(1+\frac{3}{64\pi\sigma^2}\frac{R^2}{x^2}\right)^2}$$

we shall obtain

$$\frac{3}{64\pi\sigma^2}\frac{1}{x^2} = \frac{\sqrt{2}-1}{R^2} = \frac{0.414}{R^2}$$

and, consequently

$$\frac{u}{u_{\max}} = \frac{1}{\left(1+0.414\frac{r^2}{R^2}\right)^2}$$

In Figure 222 the graph of this function is given. In the same figure test points are also plotted.[68] The agreement may be considered rather satisfactory.

---

[68] Kuethe, A., Investigations of the turbulent mixing regions, formed by jets, *J. Appl. Mec.*, 11 (3), 1935.

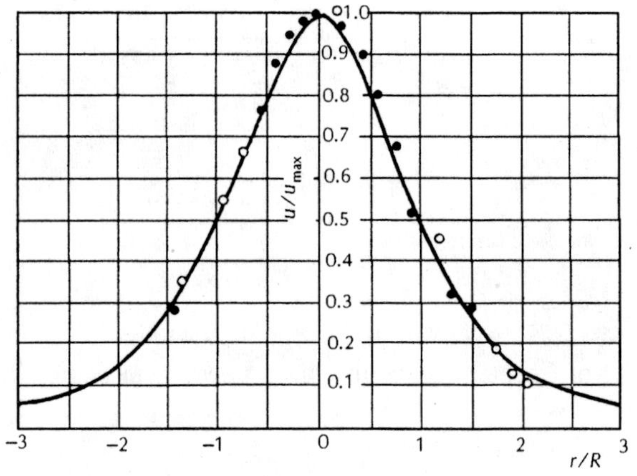

Fig. 222

In the case of a straight turbulent jet issuing from the hole of a finite diameter with finite initial flow rate and also of a turbulent swirled jet, we shall assume that the formulas of dynamic and kinematic coefficients of turbulent friction (94) remain valid.[69]

Generalizing the earlier-derived formulas for a laminar jet to the case of a turbulent jet and substituting ν into $\sigma\sqrt{J_0/\rho}$, we have instead of equality (199) of Chapter 12

$$\beta = \frac{U_0 d^2}{8\sigma\sqrt{\dfrac{J_0}{\rho}}}$$

Or, assuming the momentum in the jet initial cross-section to be equal to

$$J_0 = \rho\frac{\pi d^2}{4}U_0^2$$

we have the following expression for the constant β:

$$\beta = \frac{d}{4\sqrt{\pi}\,\sigma}$$

The formula of the maximum velocity on the jet axis, according to the last but one relation of system (198) of the previous chapter and the adopted expressions (95) for α and αη, is

$$u_{max} = \frac{3}{8\pi\sigma}\sqrt{\frac{J_0}{\rho}}\frac{1}{x}\left(1 - \frac{1}{16\sqrt{\pi}\,\sigma}\frac{d}{x}\right)$$

Noting that the coefficient in front of the brackets is the maximum velocity $u_{max}^{(1)}$, which in earlier approximation ($d = 0$) is determined by Equation 98, we shall rewrite the previous equality in the form

[69] Loitsyanskiy, L. G., Swirled jet propagation in an infinite space flooded with the same fluid, *Prikl. Mat. Mekh.*, 17 (1), 14, 1953.

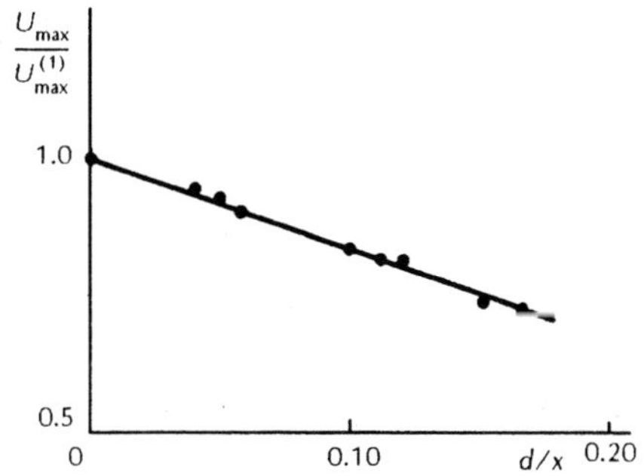

Fig. 223

$$\frac{u_{max}}{u^{(1)}_{max}} = 1 - \frac{1}{16\sqrt{\pi}\sigma}\frac{d}{x} \tag{99}$$

This formula may be treated as the expression of the correction for the finiteness of the outlet section diameter in the distribution of maximum velocity on the jet axis. Having used the test materials of Dubov,[70] we plotted them (Figure 223) on a straight line (99). The agreement is rather satisfactory if it is assumed that $\sigma = 0.21$.[71] The essence of the correlation is evident: at $d/x = 0.18$ it attains 30%.

Substituting in formulas (198) of Chapter 12 α, $\alpha\eta$, μ, and ν into their turbulent analogues, we obtain the solution of the problem on a swirled turbulent jet. Not writing the transformed formulas we present the results of the comparison between the data of theoretical calculation and experiment given in the work of Dubov. The experiments confirmed the correctness of the assumption of the constancy of the coefficient of turbulent transfer in a straight jet with the finite outlet diameter as well as in a swirled jet.

To give an insight into this fact, Figure 224 a, b present the distributions of longitudinal u and also of circumferential w speeds at different distances x from the jet outlet. The intensity of swirling was characterized by the dimensionless parameter

$$æ = \frac{4L_0}{J_0 d}$$

and in the cases given in Figure 224 it corresponded to the value $æ = 0.52$. The absence of depressions in the graphs of the longitudinal velocity near the axis at small distances from the outlet shows that to this value of the parameter corresponds a comparatively weak jet twisting. Along with the experimental points are theoretical curves calculated by Dubov. As it follows from the graphs, as a distance from the outlet grows the agreement between the theory and the experiment improves.

To show the effect of twisting increase, Figure 225 gives graphs of the longitudinal velocity with comparatively strong twisting ($æ - 1.25$). At small values of x the depressions, that decrease at a distance from the jet outlet, are vividly seen in the graphs of the longitudinal velocity.

In direct vicinity to the nozzle from which the jet issues we may observe even reverse flows. The lower curve confirms the fact that at a sufficient distance from the nozzle a swirled jet behaves as a straight one.

[70] Dubov, V. S., Propagation of a free swirled jet in a flooded space, *Tr. LPI*, No. 176, 137–145, 1955.
[71] The line and data shown in Figure 223 suggest the value of $\sigma = 0.021$ for use in Equation 99.

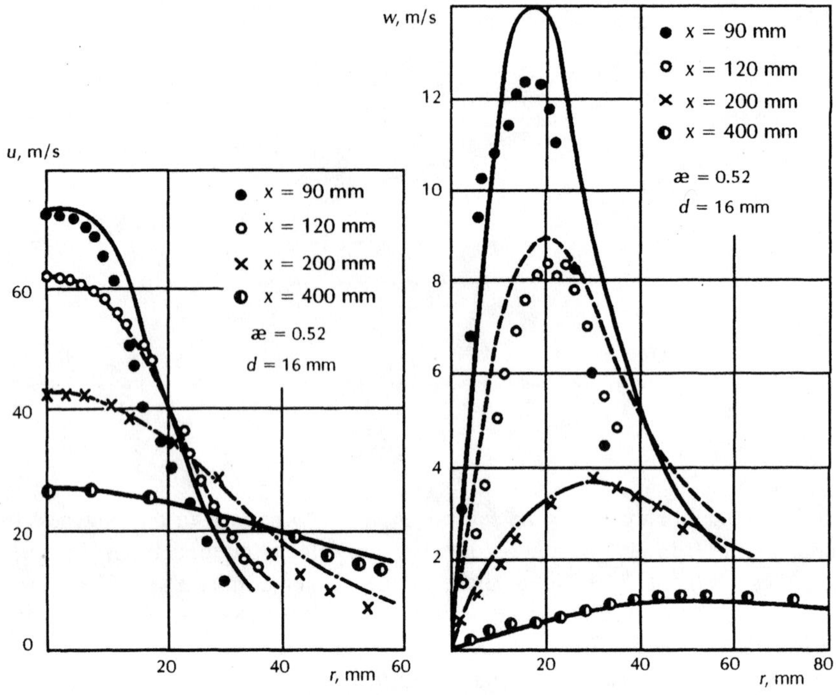

(a) (b)

Fig. 224

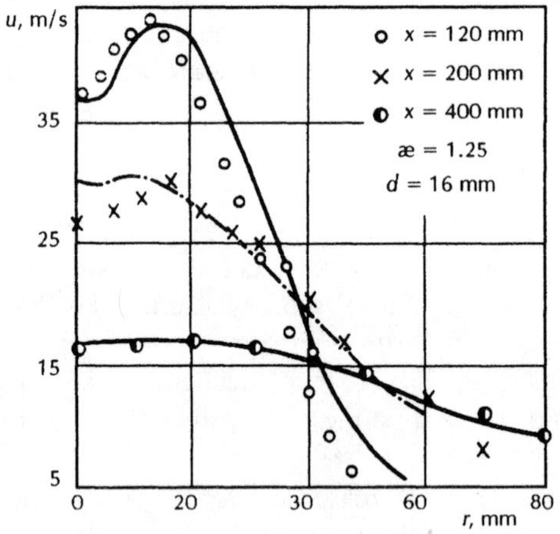

Fig. 225

The problem of velocity distribution in a turbulent wake far behind the body in a case of a plane flow is fully similar to the problem on a laminar wake solved in an earlier section. One can easily convince oneself if again a small difference of velocities is introduced outside and inside a wake $u_1 = U_\infty - u$ the mixing length l is assumed to be proportional to the wake width $2b$, and the modulus of the derivative $|du/dy|$ in the expression of the kinematic coefficient of turbulent viscosity ε_τ is approximately substituted by the ratio $u_{1\,max}/b$, where

$u_{1\,max}$ is the maximum value of u_1 in the wake section that takes place on its axis. In this case we obtain

$$\varepsilon_\tau = l^2 \left|\frac{du}{dy}\right| \backsim b^2 \frac{u_{1\,max}}{b} = b u_{1\,max}$$

On the other hand, having assumed that $u_1 = u_{1\,max}\varphi(y/b)$, similar to the laminar wake, we have

$$\rho U_\infty \int_{-b}^{b} u_1 dy = \rho U_\infty b u_{1\,max} \int_{-1}^{1} \varphi\left(\frac{y}{b}\right) d\left(\frac{y}{b}\right) = W$$

Consequently,

$$b u_{1\,max} = \text{const} \cdot \frac{W}{\rho U_\infty} \tag{100}$$

Hence, we conclude that (k is the empirical constant dependent on a wake turbulent structure)

$$\varepsilon_\tau = \frac{kW}{\rho U_\infty} = \text{const} \tag{101}$$

The Reynolds equation, if the squares and the higher powers of small quantities are discarded, will be reduced to the equation

$$U_\infty \frac{\partial u_1}{\partial x} = \varepsilon_\tau \frac{\partial^2 u_1}{\partial y^2} \tag{102}$$

which is quite similar to Equation 43 of Chapter 12 for a laminar wake. Solving this equation under the same boundary conditions, we obtain

$$u_1 = U_\infty - u = \frac{1}{2\sqrt{\pi}} \sqrt{\frac{W}{k\rho x}}\, e^{-\frac{\rho U_\infty^2 y^2}{4kWx}}$$

Introducing the quantity Y as the ordinate y at which $u_1 = \frac{1}{2} u_{1\,max}$, we find

$$\frac{u_1}{u_{1\,max}} = e^{-0.7(y/Y)^2} \tag{103}$$

In Figure 226 this equality is presented in the form of a curve in the coordinates ($u_1/u_{1\,max}$, y/Y); in the same figure are the results of the experimental examples at a considerable distance from the body. As is seen the coincidence is quite satisfactory everywhere with the exception of direct vicinity to the wake edge where the experimental points lie somewhat below; this fact is probably explained by the "intermittency" phenomena on the wake edge.

Jets in cocurrent flows and spatial wakes behind the body and a system of bodies are calculated by analogous methods, Those problems are stated in special literature.[72]

The problem of the distribution of temperature and admixture concentration in the flows

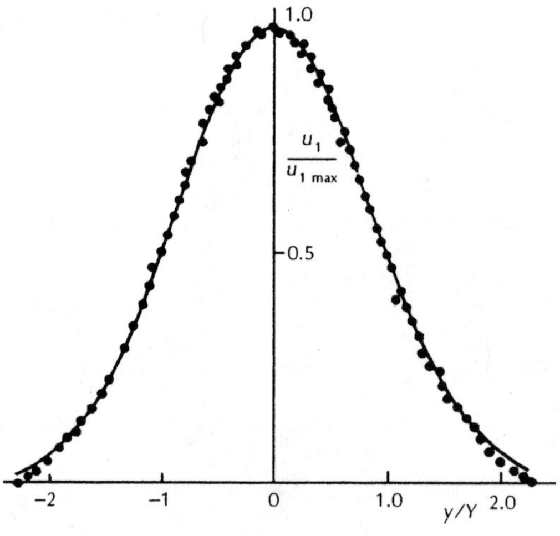

Fig. 226

under the conditions of free turbulence leads us to the necessity of the correct choice of the corresponding coefficients of heat transfer ε_q and admixture concentration ε_m. At present we have to restrict ourselves to the consideration of experimental materials which cannot yet be called exhaustive.

If the turbulent Prandtl and Schmidt numbers (67) were equal to unity, i.e., $\varepsilon_\tau = \varepsilon_q = \varepsilon_m$, then the profiles of velocities, relative excessive temperatures, and concentrations in turbulent jets and wakes would be similar to each other. Experiments confirm the presence of the similarity of the profiles of excessive temperatures and concentrations, but vividly indicate the absence of similarity between the profiles of velocities and excessive temperatures, and consequently, concentrations.

We shall present, as an example, the graph (Figure 227), borrowed from the oft-cited monograph by Abramovich, of the results of the experiments performed by the author of the monograph and by Borodachyov on a plane heated flooded air jet which contains the CO_2 admixture. The curves of the velocity (a dashed line) and of an excessive temperature (a dashed and dotted line) are given without experimental points; the distribution of the dimensionless concentration is presented by the experimental points corresponding to different dis-

---

[72] We refer the reader to the monographs by G. N. Abramovich cited earlier, as well as to: *Sovremennoe sostoyaniye gidroaerodinamiki vyazkoy zhidkosti* (*Up-to-date State of Hydroaerodynamics of Viscous Fluid*), Vol. 2, Goldstein, S., Ed., Inostrannaya Literatura Press, Moscow, 1948 (Russian translation); Loitsyanskiy, L. G., *Aerodinamika pogranichnogo sloya* (*Aerodynamics of a Boundary Layer*), Gostekhizdat Press, Leningrad–Moscow, 1941; Schlichting, G., *Teoriya pogranichnogo sloya* (*Boundary Layer Theory*), Nauka Press, Moscow, 1974 (Russian translation); Vulis, L. A. and Kashkarov, V. P., *Teoriya struy vyazkoy zhidkosti* (*The Theory of Viscous Fluid Jets*), Nauka Press, Moscow, 1965; Ginevskiy, A. S., *Teoriya turbulentnykh struy i sledov* (*The Theory of Turbulent Jets and Wakes*), Mashinostroeniye Press, Moscow, 1969; Korobko, V. N, *Teoriya neavtomodelnykh struy vyazkoy zhidkosti* (*The Theory of Non-self-similar Jets of Viscous Fluid*), Part II, Saratov University Press, 1977. Grodzovskiy, G. L., Solution of axisymmetric problems of free turbulence by the theory of turbulent diffusion, *Prikl. Mat. Mekh.*, 14 (4), 1950; Bushmarin, O. N., A turbulent axisymmetric jet of incompressible fluid issuing into a cocurrent uniform flow of the same fluid, *Tr. LPI*, Energomashinostroeniye, Tekh. Gidromekh., No. 5, 15–23, 1953 and A swirled jet in a cocurrent flow of the fluid of the same density, *Tr. LPI*, No. 176, 1955; Loitsyanskiy, L. G., On the theory of plane laminar and turbulent jets, *Tr. LPI*, No. 176, 1955; Ginevskiy, A. S., Turbulent wake and a jet in a cocurrent flow in the presence of the longitudinal pressure gradient, *Izv. AN SSSR, Mekh. Mashinostr.* No. 2, 1959, and Approximate equations of motion in the problems of the theory of turbulent jets, *ibid*, No. 5, 1963, and a great number of works by L. A. Vulis and colleagues both in the just cited monograph and in the collected papers: *Issledovaniye fizicheskikh osnov rabochego protsessa topok i pechey* (*Study of Physical Fundamentals of the Operation of Furnaces and Fire Chambers*), Alma-Ata, 1956.

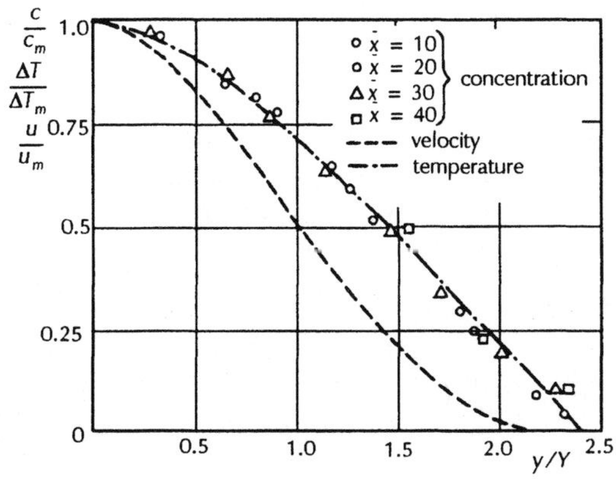

Fig. 227

tances from the jet outlet made nondimensional by the jet mouth width. The subscript m denotes the values of the quantities on the axis ΔT is the temperature difference at the given point of a jet and in the outer space.

The graph vividly shows that the profiles of the excessive temperature and concentrations in a plane jet are alike (the dimensionless profiles coincide), and the velocity profile is not similar to them. In the case considered the numbers $\mathbf{Pr}_t$ and $\mathbf{Sc}_t$ are close to 0.5.

Not pausing on details we note that Taylor[73] suggested another semiempirical theory of turbulent flow, called "the theory of vorticity transport". According to this theory, in the case of rectilinear averaged flow with the velocity distribution $u = u(y)$ there will be ($\tau = -\rho u'v'$ is the purely turbulent skin friction that is independent of viscosity)

$$\frac{\partial \tau}{\partial y} = A_\omega \frac{\partial \omega}{\partial y} = A_\omega \frac{\partial^2 u}{\partial y^2} = \rho l_{\mathrm{Ta}}^2 \left|\frac{\partial u}{\partial y}\right| \frac{\partial^2 u}{\partial y^2} \qquad A_\omega = \rho l_{\mathrm{Ta}}^2 \left|\frac{\partial u}{\partial y}\right| \tag{104}$$

where $\omega = \partial u/\partial y$, and l_{Ta} is the Taylor mixing length which differs from the Prandtl. As Taylor noted, and this is also confirmed by experiment, the coefficient of turbulent transfer of averaged vorticity A_ω under the conditions of free turbulence coincides in magnitude with the coefficients of heat A_q and substance A_m transfer, but it is not equal to the coefficient of momentum transfer A_τ.

The details of the problem of heat and mass transfer in flows with free turbulence may be found elsewhere.

8. Two-Layer Scheme of "Near-Wall" Turbulence. Logarithmic Velocity Profile

The laws governing mixing length and the coefficient of turbulent transfer in fluid motion at a solid wall differ principally from the laws governing free turbulent flows (see above) at a distance from solid surfaces. The presence of a considerable effect of the molecular viscosity on the processes of turbulent transfer greatly complicates the study of near-wall turbulence.

To emphasize the main specific feature of a turbulent motion near a solid wall, we consider the following idealized case[74] which simply and vividly explains the matter. Assume

[73] Taylor, G. I., The transport of vorticity and heat through fluids in turbulent motion, *Proc. R. Soc. Ser. A*, 5, 1932.

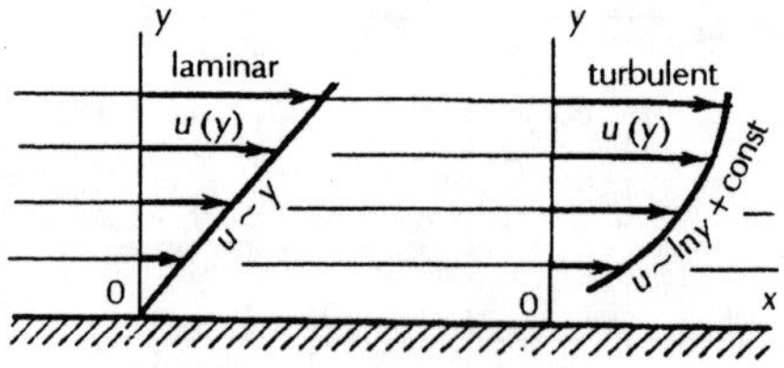

Fig. 228

that fluid occupying the upper semiplane executes a plane stationary averaged motion (Figure 228) which is parallel to an infinite solid wall coinciding with the axis $0x$ with the volumetric forces being absent. In the case of this stratification by averaged velocities any two cross-sections which are transverse to streamlines are similar kinetically and dynamically, i.e., all the derivatives with respect to x are equal to zero and the elements of motion may be dependent only on the ordinate y. This motion is called "developed".

We compare laminar and averaged turbulent motions of the kind. Noting that $u = u(y)$, $v = 0$, $p = p(y)$, we reduce the equations of a laminar flow to the form

$$\mu \frac{d^2u}{dy^2} = 0 \qquad \frac{dp}{dy} = 0$$

where it follows that

$$u = C_1 y + C_2 \qquad p = \text{const}$$

In the motion considered the only boundary condition is

$$u = 0 \quad \text{at} \quad y = 0$$

thus giving $C_2 = 0$. To determine C_1 we take the skin friction on a wall as the adopted constant

$$\tau_w = \mu \left(\frac{du}{dy} \right)_{y=0}$$

Then, the velocity distribution $u(y)$ in a laminar flow is

$$u = \frac{\tau_w}{\mu} y \tag{105}$$

and the velocity profile will become linear (Figure 228, on the left), and the skin friction between any two layers in an averaged motion will be constant and equal to the skin friction in a wall

$$\mu \frac{du}{dy} = \text{const} = \tau_w$$

Now we go to a turbulent flow described in the given case by the Reynolds equation (τ is a purely turbulent friction)

---

[74] Prandtl, L., Neuere Ergebnisse der Turbulenzforschung, VDI, No. 5, 1933.

$$\mu \frac{d^2u}{dy^2} + \frac{d\tau}{dy} = 0$$

Its integration yields

$$\mu \frac{du}{dy} + \tau = C_3$$

or noting that the turbulent skin friction $\tau = -\rho \overline{u'v'}$ on a wall is equal to zero, since fluctuation v' normal to it cannot exist on the wall, we obtain $C_3 = \tau_w$ and, consequently,

$$\mu \frac{du}{dy} + \tau = \tau_w \tag{106}$$

In direct vicinity to the wall turbulent friction τ is much smaller than the summand $\mu du/dy$ corresponding to molecular friction; in this region Equation 116 coincides with the equation of a laminar flow and its integration will lead to the linear profile of velocities (105).

Conversely, in the region at a distance from the wall the term $\mu du/dy$ is small compared with the turbulent friction τ and may be omitted. If the turbulent skin friction is prescribed by the Prandtl formula, Equation 106 is

$$\tau = \rho l^2 \left(\frac{du}{dy}\right)^2 = \tau_w \tag{107}$$

Noting that the distance y of the given point from a solid wall is the only length typical of this point in an infinite flow, Prandtl assumed that in this simplest case the mixing length l is proportional to y

$$l = æy \tag{108}$$

where the coefficient of proportionality æ is some numerical constant determined from the experiment.[75]

Equation 107 in the presence of relation (108) may be integrated rather simply and its integral expressed in a logarithmic form

$$u = \frac{1}{æ}\sqrt{\frac{\tau_w}{\rho}}\ \ln y + C \tag{109}$$

Comparison of this velocity distribution with the earlier obtained "laminar" distribution (105) shows a deep difference between them, from the mathematical viewpoint this difference is expressed in the fact that the linear velocity profile in a laminar flow becomes logarithmic in a case of a turbulent flow. It is essential that this specific feature of a turbulent flow survives near the wall in the cases of more complex flows than the simplified scheme considered above.

One cannot use the boundary condition on the wall for the determination of the integration constant C, since in a near-wall region Equation 107 is invalid. A thin "viscous sublayer" with the linear velocity profile should be distinguished near a solid boundary and then the

[75] In the literature, especially on engineering applications of the turbulence theory, this simplest formula (108) which was introduced by Prandtl is taken as a general law valid for all of the turbulent boundary layers and is even named "the Prandtl law". Moreover, this formula takes place only in a near-wall region. The importance of this formula as well as of the logarithmic velocity law directly derived from it, expressing a specific feature of a turbulent flow near a wall, is irrefutable and should be emphasized.

logarithmic solution should be matched with a linear one (105).

We take the following simplified scheme. Let the flow be divided into two regions which drastically differ in structure: a thin near-wall region of a purely viscous flow [a viscous (laminar) sublayer] and the region of a fully developed turbulent flow independent of viscosity (a flow turbulent core). The division adopted is schematic. In fact, at a distance from the wall the viscosity effect decreases constantly and is not concentrated in some sharply bounded region. To specify this scheme one can also introduce an intermediate region in which along with turbulent friction there is molecular friction. The introduction of such a transition region turns out to be especially useful when studying heat and mass transfer and will be formulated as will the theory of continuous decrease of the viscosity effect. We shall restrict ourselves to the scheme of two regions: of a viscous sublayer and a turbulent core.

Write δ_v for the viscous sublayer thickness and u_v for the velocity on the boundary between a viscous sublayer and a flow turbulent core that is common for both regions. The flow in a viscous sublayer is characterized by the value of the skin friction on the wall τ_w and by physical constants of fluid μ and ρ. Considering these quantities from the viewpoint of dimensional theory, we shall compose two possible combinations

$$v_* = \sqrt{\frac{\tau_w}{\rho}} \qquad l_* = \frac{\nu}{\sqrt{\tau_w/\rho}} = \frac{\nu}{v_*} \tag{110}$$

The first of these quantities v_* has the dimensions of velocity, though by its nature it consists of dynamic quantities: stress and density; therefore, we call it dynamic velocity. The second has the units of length, and for the same reason it may be called dynamic length.[76] To facilitate memorizing these expressions, which are important for further presentation, one may note that if the dynamic length and dynamic velocity are recognized as the scales of lengths and velocities, then the Reynolds number $v_* l_*/\nu$, composed with their aid, will be equal to unity.

One can easily convince oneself that the expressions of dynamic velocity and length determined by equalities (110) may differ only by dimensionless factors from any other one-term combinations of the quantities τ, ρ, μ, which have the unity of velocity or length. Actually, we shall assume that the length δ may be presented in tile form of a power-law monomial dependent on physical constants μ, p, and skin friction on a wall τ_w

$$\delta = \alpha \mu^a \rho^b \tau_w^c$$

where α is some dimensionless constant.

Formulating the equation for correlating dimensions

$$[L] = \left[\frac{M}{TL}\right]^a \left[\frac{M}{L^3}\right]^b \left[\frac{M}{LT^2}\right]^c$$

and comparing the power indices at M, L, and T on the left and on the right, we shall obtain the system of equations

$$a + b + c = 0 \qquad -a - 3b - c = 1 \qquad -a - 2c = 0$$

[76] In non-Russian literature, for these quantities use is made of the terms *Schubspannunggeschwindigkeit*, *friction-velocity*, and *friction-length*, which are difficult to translate into Russian. We have avoided this difficulty by introducing the epithet "dynamic" for "reduced" velocity and length because they are expressed in terms of the dynamic quantities τ_w and ν.

which has the only solution

$$a = 1 \qquad b = -\frac{1}{2} \qquad c = -\frac{1}{2}$$

Hence, it follows

$$\delta = \alpha\mu\rho^{-1/2}\tau_w^{-1/2} = \alpha\frac{\nu}{\sqrt{\tau_w/\rho}} = \alpha l_*$$

Similarly, we shall convince ourselves that any one-term combination of ρ, μ, and τ_w which has the dimensions of velocity may differ from v_* only by the dimensionless factor.

It follows that the viscous sublayer thickness δ_v and the velocity on the sublayer boundary uv should be proportional to the dynamic length l_* and dynamic velocity v_*, correspondingly.

We shall prove that they have equal coefficients of proportionality. We shall assume

$$\delta_v = \alpha l_* = \alpha\frac{\nu}{v_*} = \alpha\frac{\nu}{\sqrt{\tau_w/\rho}} \tag{111}$$

where α is the dimensionless constant. Then, according to Equation 105 we shall have

$$u_v = \frac{\tau_w}{\rho}\delta_v = \alpha\frac{\tau_w}{\mu}\frac{\nu}{\sqrt{\tau_w/\rho}}$$

In fact,

$$u_v = \alpha\sqrt{\frac{\tau_w}{\rho}} = \alpha v_* \tag{112}$$

Making use of the formulas obtained, we shall determine the constant C in equality (109), which now may be rewritten in the form

$$u = \frac{v_*}{æ}\ln y + C \tag{113}$$

We have on the viscous sublayer boundary

$$u_v = \frac{1}{æ}v_*\ln\delta_v + C$$

so that Equations 111 and 112 yield

$$C = \alpha v_* - \frac{1}{æ}v_*\ln\left(\alpha\frac{\nu}{v_*}\right) = v_*\left(\alpha - \frac{1}{æ}\ln\alpha\right) - \frac{1}{æ}v_*\ln\frac{\nu}{v_*}$$

Substituting this value of the constant into Equation (113), we find

$$\varphi = \frac{1}{æ}\ln\eta + \alpha - \frac{1}{æ}\ln\alpha \tag{114}$$

where

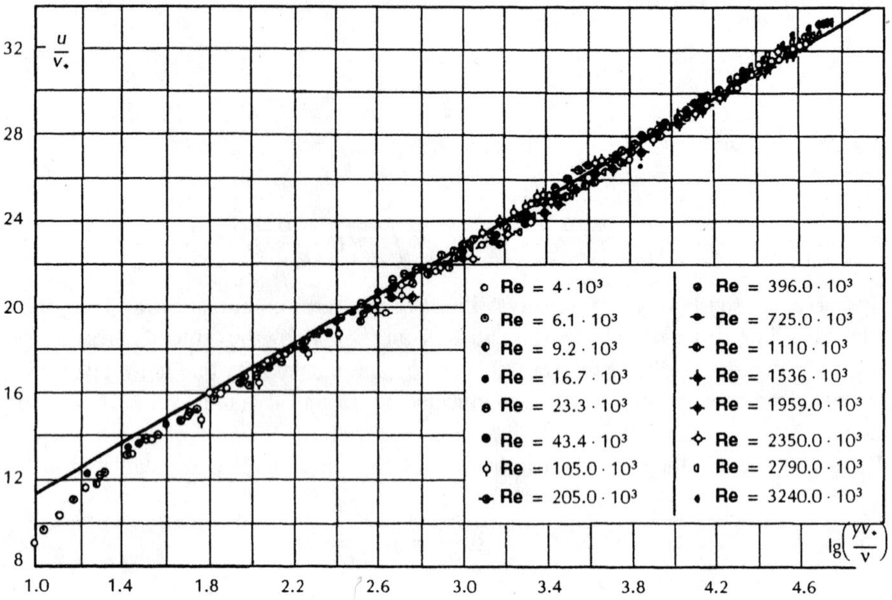

Fig. 229

$$\varphi = \frac{u}{v_*} \qquad \eta = \frac{yv_*}{\nu} = \frac{y}{l_*}$$

or, moving from natural logarithms to common ones (lg)

$$\varphi = \frac{2.303}{æ}\lg\eta + \alpha - \frac{2.303}{æ}\lg\alpha \tag{115}$$

Nikuradze[77] conducted the experiments with turbulent water flow in long cylindrical round tubes with smooth walls within a wide range of Reynolds numbers **Re** $= u_{\text{mean}}d/\nu$ (u_{mean} is the velocity mean over the flow rate, d is the tube diameter, and ν is the kinematic coefficient of viscosity) from their critical value to **Re** $\approx 3.24 \cdot 10^6$.

In Figure 229 presents the results of Nikuradze's accurate systematic experiments on measuring velocities in tube cross-section. As follows from the graph, experimental points agree rather satisfactory with the curve corresponding to the logarithmic velocity profile

$$\frac{u}{v_*} = \varphi = A\lg\eta + B = 5.75\lg\eta + 5.5 = 5.75\lg\frac{yv_*}{\nu} + 5.5 \tag{116}$$

Of principal importance is the fact that the logarithmic formula (116) retains its form for all the Reynolds numbers of flow, or, as it is common to say, it is universal. Further, power-law equations for the velocity will be introduced that do not possess a universal property. A structure of logarithmic formulas is so that the effect of the Reynolds number, i.e., viscosity is

---

[77] Nikuradze, I., Gesetzmässigkeiten der turbulenten Strömung in glatten Rohren. VDI, Forschungsheit, 1932, Bd. 356.

completely included into the scales of lengths l_* and velocities v_*; this makes formula (116) universal. From the physical viewpoint the mentioned property of logarithmic equations is explained by the presence of a viscous sublayer in which the total viscosity effect is concentrated and which is distinguished by proportionality of the scales l_* and v_* to the sublayer thickness δ_v and velocity on its outer boundary u_v. The considerations presented may serve as justification for naming the scales $l_* = \nu / v_*$ and $v_* = \sqrt{\tau_w / \rho}$ universal and the quantities $\varphi = u/v_*$ and $\eta = y/l_* = yv_*/\nu$ the universal nondimensional velocity and coordinate, respectively.

According to Equation 105 in a laminar sublayer in universal variables there will exist the equality

$$\varphi = \eta \tag{117}$$

and on the sublayer boundary, by Equation 111,

$$\eta_v = \frac{\delta_v}{l_*} = \frac{\delta_v v_*}{\nu} = \alpha$$

Comparing Equation 116 with the logarithmic distribution of velocities (115), we convince ourselves of the fact that it is necessary to assume for quantitative coincidence that

$$æ \approx 0.4 \qquad \alpha \approx 11.5$$

Experimental data deviate from the curve (116) only in the region of comparatively small values of $\eta = yv_*/\nu$ corresponding to the points close to the tube wall where the viscosity effect becomes noticeable (the transition region); to the extreme left point of a horizontal scale $\lg \eta \approx 1$ there approximately corresponds the viscous sublayer boundary. Making use of the coordinates φ, η we rewrite the linear profile of velocities (117) in a viscous sublayer in the form

$$\varphi = \eta = 10^{\lg \eta}$$

With a logarithmic scale of abscissas adopted in Figure 229 to this profile will correspond a connecting line which is not shown in the figure, nor is the transition to this line from the straight line (116) is shown at small values of $\lg \eta$. The position of this connecting line will be shown further on. We also note the shown deviation of the experimental points downward from the straight line (116) in its right side that is dependent on the effect of the difference of the flow in a round tube from the considered simplified case of a plane flow near an infinite plane.

A problem may arise of the legitimacy of the comparison of theoretical results related to the idealized flow in an infinite region (Figure 228) with the results of experiments in a round tube conducted by Nikuradze. The consideration that the plane flow studied may be presented as a limiting case of a tube flow, if at a fixed distance y of the tube flow point from the wall we let the distance between the planes in a plane tube or a radius in a round cylindrical tube approach infinity, serves as an answer. The correctness of such a treatment is proven below.

Implying some difference between a theoretical logarithmic profile of velocities (114) and the Nikuradze curve (Figure 229) near the tube axis, we shall apply equality (114) to the tube axis, where $y = a$, $u = u_{max}$, and from the result obtained we subtract (114) term by term. Then, we come to the following formula of the "velocity defect" as the difference $u_{max} - u$ is called

$$\frac{u_{max} - u}{v_*} = \frac{1}{æ} \ln \frac{a}{y} \tag{118}$$

which contains the only empirical constant æ; substituting æ = 0.4, we obtain

$$\frac{u_{max} - u}{v_*} = 5.75 \ln \frac{a}{y} \tag{119}$$

As shown in Figure 229, this simple equation gives good agreement with the experiment.

The constants $æ \approx 0.4$ and $\alpha \approx 11.5$ represent the two basic empirical constants characterizing a turbulent flow. Sometimes it is convenient to introduce the third constant f, which is expressed in terms of the first two and which determines the slope of the logarithmic curve of velocities on the boundary of a viscous sublayer but from the side of a turbulent core.

Having taken the derivative of both sides of Equation 114, we obtain

$$f = \left(\frac{d\varphi}{d\eta}\right)_{\eta = \eta_\nu} = \frac{1}{æ\alpha} = 0.218 \tag{120}$$

Thus, this simplified theoretical analysis makes it possible to trace the basic laws governing the distribution of velocities during a turbulent flow in a round tube. The necessity of using the two empirical constants is retained in further presentation, though the nature of the constants themselves changes.

The noted correlation between the flows about an infinite plane and near the round tube wall that leads to the coincidence of simple theoretical formulas of velocity (115) and "the velocity defect" (118) with the test data, indicates the identity of local properties as turbulent flow in these regions. It may be noted that the "velocity defect" calculated for a plane tube with a finite distance $2h$ between the planes gives worse agreement with the experiment. We shall use the actual distributions of full skin friction τ over the cross-section of a plane tube of the height $2h$

$$\tau = \tau_w\left(1 - \frac{y}{h}\right) = \tau_w \frac{z}{h} \tag{121}$$

where y is the distance of the point to the wall, and z is the distance of the point from the axis. These are easily derived in the case of a steady flow from the condition of equilibrium of the elementary fluid volume between two tube cross-sections and the von Kármán formula for turbulent skin friction (73). This formula valid outside a viscous sublayer ($y \gg \delta_\nu$) (the prime denotes the derivative with respect to y)

$$\rho æ^2 \frac{u'^4}{u''^2} = \tau_w\left(1 - \frac{y}{h}\right)$$

This equation may be rewritten in the form (the minus sign on the right side was chosen in view of the fact that $u'' < 0$)

$$\frac{u''}{u'^2} = -\frac{æ}{v_*}\frac{1}{\sqrt{1 - \frac{y}{h}}}$$

it is easily integrated and gives the first integral

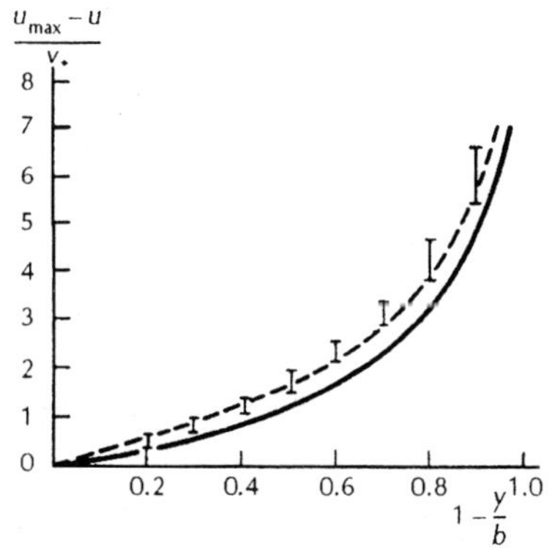

Fig. 230

$$-\frac{1}{u'} = 2\frac{æ}{v_*}h\sqrt{1-\frac{y}{h}} + C$$

Following von Kármán,[78] we adopt

$$u' = \infty \quad \text{at} \quad y = 0$$

thus, approximately corresponding to a very high value of the velocity curve inclination near a wall. Hence, it follows

$$C = -\frac{2æh}{v_*} \qquad u'(y) = \frac{v_*}{2æh}\frac{1}{1-\sqrt{1-y/h}}$$

Integration of the last equation leads to distribution of the velocity "defect" that contains the only constant æ

$$\frac{u_{max} - u}{v_*} = -\frac{1}{æ}\left[\ln\left(1-\sqrt{1-\frac{y}{h}}\right)+\sqrt{1-\frac{y}{h}}\right] \qquad (122)$$

Figure 230 presents, for the sake of comparison, the theoretical curves (122) (a solid line) and (118) (a dashed line) at æ = 0.4; in the same figure, in the form of vertical sections, the boundaries of the Nikuradze experimental data are drawn within the range of **Re** from $4 \cdot 10^4$ to $3.24 \cdot 10^6$. As is seen from the graph, a simple equation (118) at æ = 0.4 satisfactorily represents an actual distribution of velocities. At the same æ Equation 122 gives somewhat underestimated values of the "velocity defect".

Making use of relation (122) one can interpret Equation 118 in the following way. We fix the value of the ordinate y in a plane or round tube and let the width h or the diameter d approach the infinity. At small y/h we obtain in this case instead of (122)

$$\frac{u_{max} - u}{v_*} \approx \frac{1}{æ}\ln\frac{h}{y}$$

i.e., at $h = a$ (a is the tube radius) – Equation 118.

[78] von Kármán, Th., Mechanische Ähnlichkeit und Turbulenz, Nachrichten der Geselschaft der Wissenschlaften zu Göttingen, *Math. Phys.*, Kl, 1930.

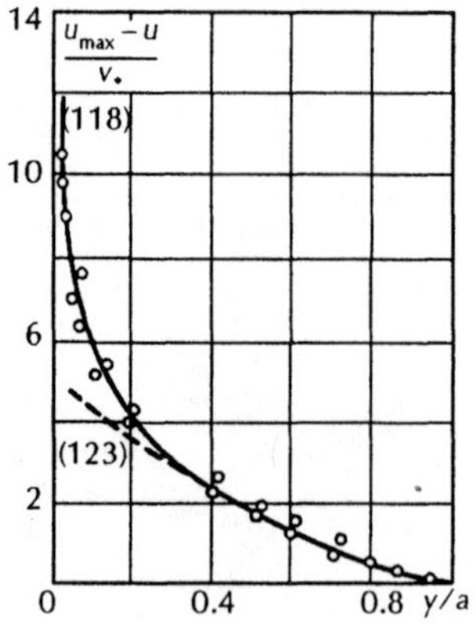

Fig. 231

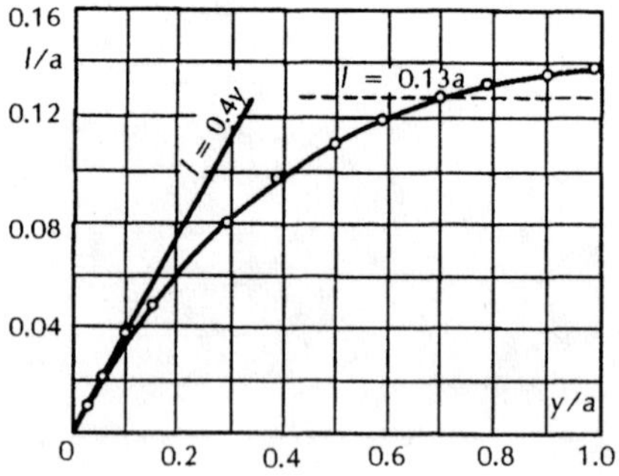

Fig. 232

Thus, the flow considered by Prandtl is the limiting case of plane or round tubes if we let the half-width h or the tube radius a approach infinity, and the distance of the point from the tube wall is fixed. The profile of velocity defect (118) in the tube central portion is close to that experimentally determined by Darcy[79] and shown in Figure 231 by dashes,

$$\frac{u_{\max} - u}{v_*} = 5.08\left(1 - \frac{y}{a}\right)^{3/2} = 5.08\left(\frac{r}{a}\right)^{3/2} \tag{123}$$

and adopted in hydraulic courses. We shall show that this profile may be obtained on the basis of the Prandtl semiempirical theory.[80] For this purpose we note that the mixing length determined by the Nikuradze experimental data from the Prandtl equation

$$l = \frac{\sqrt{\frac{\tau}{\rho}}}{du/dy} = \frac{\sqrt{\frac{\tau_w}{\rho}\frac{r}{a}}}{du/dy} = \frac{v_*\sqrt{1 - \frac{y}{a}}}{du/dy} \tag{124}$$

may be presented by the graph (Figure 232) from which it directly follows that near the axis the ratio l/a is almost constant. Assuming $l = \beta a$ and τ, according to Equation 121, equal to $\tau = (\tau_w r)/a$, we obtain from the Prandtl equation (60)

$$\tau = \rho l^2\left(\frac{du}{dr}\right)^2 = \beta^2\rho a^2\left(\frac{du}{dr}\right)^2 = \tau_w\frac{r}{a}$$

where it follows (the minus sign from the condition $\frac{du}{dr} < 0$)

$$-\frac{d(u/v_*)}{d(r/a)} = \frac{1}{\beta}\sqrt{\frac{r}{a}}$$

After integration and with the boundary condition

[79] Darcy, H., Recherches expérimentales relatives au mouvement de l'eaux; dans les tuyaux, Mem. Diversed Davsnted Entrangers, 15, 1858.

[80] Loitsyanskiy, L. G., *Aerodynamika pogranichnogo sloya (Aerodynamics of a Boundary Layer)*, Gostekhizdat Press, Moscow-Leningrad, 1941, 296, 297.

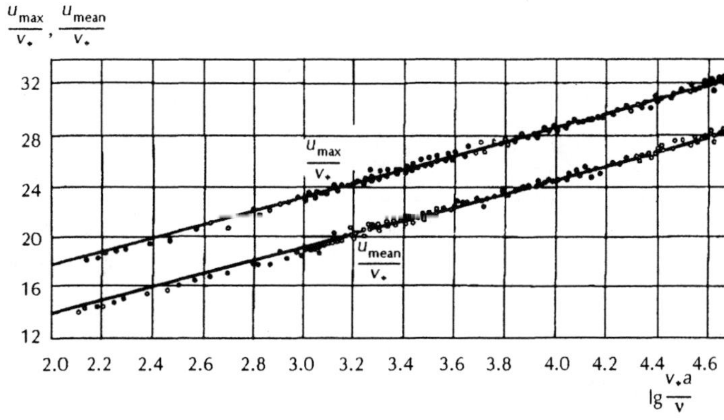

Fig. 233

$$u = u_{max} \quad \text{at} \quad r = 0$$

taken into account, we obtain

$$\frac{u_{max} - u}{v_*} = \frac{2}{3\beta}\left(\frac{r}{a}\right) = \frac{2}{3\beta}\left(1 - \frac{y}{a}\right)$$

that at $\beta = 0.13$ is given by the Darcy law (123). As is seen from Figure 232, the given value of β slightly differs from the experimental one $\beta = 0.14$ at the point $\tau = 0$ where the determination of l by Equation 124 in which at $r = 0$ both the numerator and denominator vanish, is inaccurate.

We shall determine the mean velocity in a tube u_{mean} as

$$u_{mean} = \frac{1}{\pi a^2}\int_0^a u \cdot 2\pi(a - y)\,dy$$

Performing the mentioned averaging over both sides of formula (118), we obtain at $æ = 0.4$

$$\frac{u_{max} - u_{mean}}{v_*} = -\frac{1}{æ\pi a^2}\int_0^a \ln\left(\frac{y}{a}\right) \cdot 2\pi(a - y)\,dy =$$

$$= -\frac{2}{æ}\int_0^1 \ln\left(\frac{y}{a}\right)\left(1 - \frac{y}{a}\right)d\left(\frac{y}{a}\right) \approx 4.08 \qquad (125)$$

The numerical value of this relation, 4.08, taken on the right side somewhat exceeds the value 3.75 determined by the integral, but it is closer to the experiment.

This formula of correlation between the maximum (on the tube axis) and the velocity mean over the tube cross-section is well confirmed experimentally as is seen from Figure 233. In contrast to the laminar flow in a round tube when $u_{max}/u_{mean} = 2$, in a turbulent

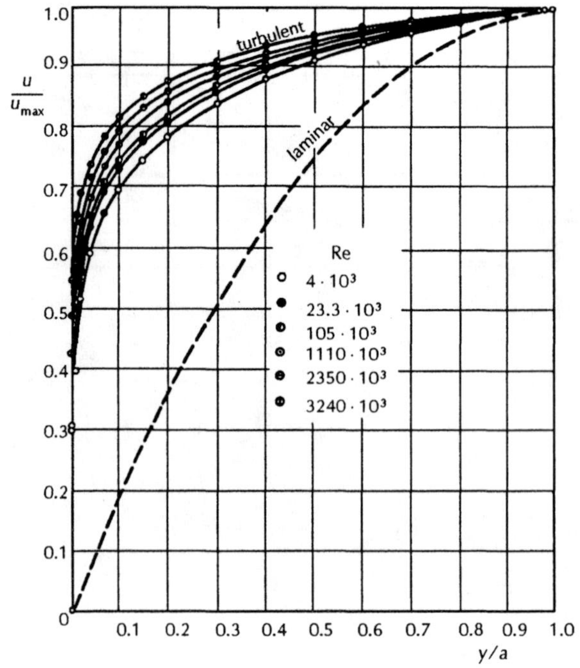

Fig. 234

flow this ratio decreases with the growth of the Reynolds number from 1.3 at its small values ($\mathbf{Re} \approx 5000$) to 1.5 at comparatively large ones ($\mathbf{Re} \approx 3 \cdot 10^6$). This indicates a sharp difference between the form of the velocity profile in a turbulent flow and the parabola of velocities in a laminar flow.

In Figure 234 the velocity profiles are drawn in the coordinates u/u_{max}, y/a and this difference is vividly seen. It may be noticed that turbulent profiles lie considerably higher, or are "fuller" than the laminar ones, the degree of their fullness increasing with the Reynolds number. Experimental points correspond to the Nikuradze tests cited above.

9. Logarithmic and Power-Law Formulas for the Velocity Profiles and the Resistance of Smooth Tubes

The velocity distribution laws derived above contain the unknown quantity v_*, related to the skin friction stress of the pipe. To complete the calculation we need an additional relation between v_* and u_{max} or u_{mean}. Such a relation constitutes the resistance law[81] of the turbulent tube flow.

The required resistance laws can be easily derived using the velocity distributions and the expressions for the thickness of the laminar sublayer and the velocity at its outer edge. The problem reduces to the determination of the relation between the resistance coefficient λ, defined by the formula known from Chapter 10

$$\Delta p = \lambda \frac{L}{d} \frac{\rho u_{mean}^2}{2} \tag{126}$$

[81] Also often called a "constitutive" law.

(where Δp is the pressure drop in a length of tube, L, d is the tube diameter, and umean is the mean velocity), and the Reynolds number $\mathbf{Re} = u_{mean}d/\nu$.

In uniform steady developed tube flow the pressure force $\Delta p \cdot \pi d^2/4$ is balanced by the friction resistance $\tau_w \pi d \cdot L$, so that from the equality $\Delta p \cdot \pi d^2/4 = \tau_w \pi d \cdot L$ it follows that

$$\Delta p = \frac{4L}{d}\tau_w$$

Substituting the obtained expression for Δp into Equation 126

$$\tau_w = \lambda \frac{\rho u_{mean}^2}{8}$$

or, recalling the definition of the dynamic velocity $v_* = \sqrt{\tau_w/\rho}$,

$$v_*^2 = \frac{\lambda}{8} u_{mean}^2$$

where

$$\frac{u_{mean}}{v_*} = \frac{2\sqrt{2}}{\sqrt{\lambda}} \tag{127}$$

To derive the resistance law, i.e., a relation between λ and $\mathbf{Re} = u_{mean}d/\nu$, we either calculate the velocity at the axis ($u = u_{max}$ at $y = a$) from Equation 116, which was obtained by matching the velocity distributions in the turbulent core and in the laminar sublayer, or we evaluate Equation 119, which satisfies the boundary condition at the edge of the laminar sublayer. In either case, we obtain the equation

$$\frac{u_{max}}{v_*} = 5.75 \lg \frac{a v_*}{\nu} + 5.5$$

which may be written in the equivalent form (with $d = 2a$)

$$\frac{u_{max} - u_{mean}}{v_*} + \frac{u_{mean}}{v_*} = 5.75 \lg\left(\frac{u_{mean}d}{\nu}\frac{v_*}{2u_{mean}}\right) + 5.5$$

or, making use of Equations 125 and 127

$$\frac{1}{\sqrt{\lambda}} = C \lg(\mathbf{Re}\sqrt{\lambda}) + D$$

Results of numerous experiments (Stanton and Pannel, Ombeck, Nusselt, Jacob and Erk, Shiller and Hermann, Nikuradze) are in good agreement with the formula containing the rounded-off coefficients[77]

$$\frac{1}{\sqrt{\lambda}} = 2 \lg(\mathbf{Re}\sqrt{\lambda}) - 0.8 \tag{128}$$

Equation 128 represents the desired relation $\lambda(\mathbf{Re})$ in an implicit form. Nikuradze suggested the explicit relation

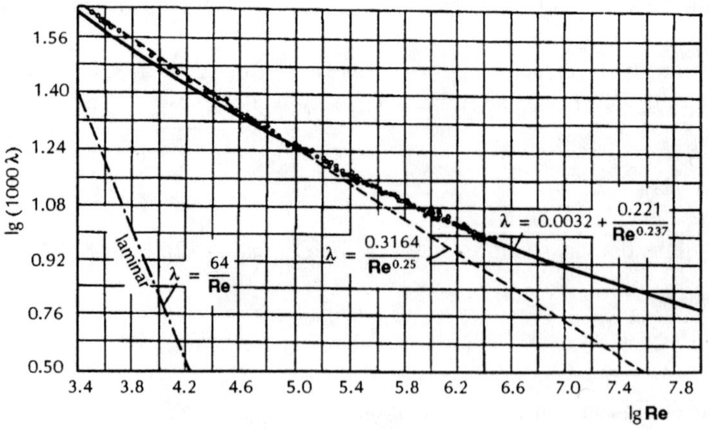

Fig. 235

$$\lambda = 0.0032 + \frac{0.221}{\mathbf{Re}^{0.237}} \tag{129}$$

compared with experimental results in Figure 235. Also plotted in the figure is the Blasius formula

$$\lambda = \frac{0.3164}{\mathbf{Re}^{0.25}} \tag{130}$$

(often used in hydraulics) whose validity is limited to $\mathbf{Re} < 10^5$. The dashed-dotted straight line appearing in the lower left corner of the figure corresponds to the laminar resistance law $\lambda = 64/\mathbf{Re}$.

The procedure for calculating the fully developed turbulent flow in a round tube is as follows. Given the tube diameter d, the kinematic viscosity of the fluid ν, and the required volume flow, we find the mean velocity u_{mean} and the Reynolds number $\mathbf{Re} = u_{\text{mean}}d/\nu$. From Equation 129 we calculate the resistance coefficient λ and then the pressure drop Δp in a given length of tube L

$$\Delta p = \lambda \frac{L}{d} \frac{\rho u_{\text{mean}}^2}{2}$$

We determine by Δp

$$\tau_w = \frac{\lambda}{8} \rho u_{\text{mean}}^2$$

and

$$v_* = \sqrt{\frac{\tau_w}{\rho}} = \frac{\sqrt{\lambda}}{2\sqrt{2}} u_{\text{mean}}$$

Finally, Equation 116 for the velocity distribution is used to complete the solution of the problem.

Besides the semiempirical logarithmic formulas for the velocity distribution and the resistance, an important role is still played by purely empirical power-law expressions. To this class belongs the Blasius formula (130), which represents the particular case of the general power-law resistance formula

$$\lambda = c/\mathbf{Re}^m \tag{131}$$

As shown by experiments c and m depend on the Reynolds number; the magnitude of m decreases with increasing **Re**. Using the Blasius formula, von Kármán showed[82] by dimensional arguments that to the resistance law (130) corresponds the velocity profile

$$u/u_{\text{max}} = (y/a)^{1/7}$$

which later became known as the one-seventh power law.

Similarly, the general expression (131) corresponds to the velocity profile

$$u/u_{\text{max}} = (y/a)^n \tag{132}$$

To find the relation between the exponents m and n we follow the same method as was used to derive the logarithmic resistance law from the logarithmic velocity profile. Using (132), we find

$$\frac{u_{\text{mean}}}{u_{\text{max}}} = \frac{1}{\pi a^2}\int_0^a 2\pi(a-y)\left(\frac{y}{a}\right)^n dy = 2\int_0^1\left(1-\frac{y}{a}\right)\left(\frac{y}{a}\right)^n d\left(\frac{y}{a}\right) = \frac{2}{(n+1)\,(n+2)}$$

Next, we evaluate (132) at the edge of the laminar sublayer, putting

$$y = \delta_\nu = \alpha\frac{\nu}{v_*} \qquad u = u_\nu = \alpha v_*$$

We have then

$$\frac{\alpha v_*}{u_{\text{max}}} = \left(\frac{\alpha\nu}{a v_*}\right)^n \qquad \frac{v_*}{u_{\text{mean}}}\cdot\frac{u_{\text{mean}}}{u_{\text{max}}} = 2^n\alpha^{n-1}\left(\frac{\nu}{u_{\text{mean}}d}\right)^n\left(\frac{u_{\text{mean}}}{v_*}\right)^n$$

where it follows, using Equation 127, that

$$\lambda = \frac{2^{\frac{5n+1}{n+1}}\alpha^{\frac{2(n-1)}{n+1}}\left[(n+1)\,(n+2)\right]^{\frac{2}{n+1}}}{\mathbf{Re}^{\frac{2n}{n+1}}} \tag{133}$$

Comparing this expression with Equation 131, we find

$$m = \frac{2n}{n+1} \tag{134}$$

$$c = 2^{\frac{5n+1}{n+1}}\alpha^{\frac{2(n-1)}{n+1}}\left[(n+1)\,(n+2)\right]^{\frac{2}{n+1}} \tag{135}$$

---

[82] von Kármán, Th., Über laminare und turbulente Reibung, *Z. Angew. Math. Mech.*, 1, 1921.

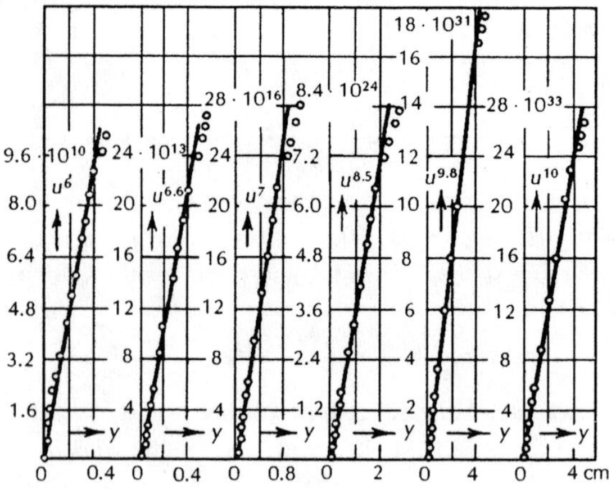

Fig. 236

It is immediately apparent from these equations that the Blasius resistance law (130), where $m = 1/4$, corresponds to the one seventh power law for the velocity profile.

In Figure 236 reproduces some further results of Nikuradze's experiments which illustrating the reduction of m with increasing Reynolds number.

It is interesting to express the power-law velocity profile in the coordinates u/v_*, $y/l_* = yv_*/\nu$, i.e.,

$$\frac{u}{v_*} = A\left(\frac{yv_*}{\nu}\right)^n \tag{136}$$

The expressions

$$\frac{u}{v_*} = 8.74\left(\frac{yv_*}{\nu}\right)^{1/7} \quad \text{for} \quad 40 < \frac{yv_*}{\nu} < 700$$

$$\frac{u}{v_*} = 9.60\left(\frac{yv_*}{\nu}\right)^{1/8} \quad \text{for} \quad 70 < \frac{yv_*}{\nu} < 1100$$

are in good agreement with experiment, as are other similar expressions with a decreasing exponent and an increasing factor of proportionality.

The main reason for the superiority of the logarithmic velocity profile is the fact that Equation 116 holds over in an extraordinary wide range of yv_*/ν, from about 40 up to a value which according to the available experimental data, is certainly not less than 100,000, but is probably considerably higher. This means that the logarithmic velocity profile is universal.

For future use, we need the power-law resistance formula in a form easily deduced from Equation 136

$$\tau_w = \zeta \mathbf{Re}_{max}^{-\frac{2n}{n+1}} \cdot \frac{1}{2}\rho u_{max}^2 \qquad \mathbf{Re}_{max} = \frac{u_{max}\alpha}{\nu} \tag{137}$$

(where ζ is the resistance coefficient dependent on n). In particular, the one seventh power

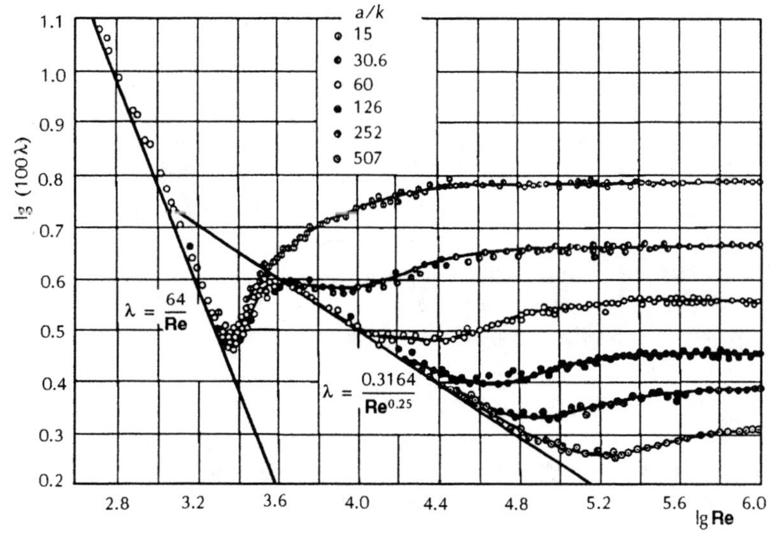

Fig. 237

law ($n = 1/7, A = 8.7$) gives

$$\frac{\tau_w}{\rho u_{max}^2} = 0.0225\mathbf{Re}_{max}^{-1/4} \tag{138}$$

The above considerations relate only to the flow in smooth tubes with a strictly cylindrical surface. In practice, we have to deal with pipes whose inner surfaces have some degree of roughness and waviness.

The study of the effects of various kinds of roughness on the resistance of tubes comes within the province of hydraulics, which has at its disposal a great many engineering formulas for estimating the resistance of tubes used in technological processes.

To idealize and, at the same time, to generalize the concept of roughness we suppose that the inner surface of the tube is covered with roughness elements in the form of particles of approximately uniform size. We shall denote by k the height of the particles (in practice this is taken to be the average height) and shall refer to the quantity k expressed in millimeters as the absolute roughness and to the ratio k/a (where a is the tube radius) as the relative roughness. We shall assume that the relative roughness is small (between 0.2 and 7%).

Examination of typical experimental data on the resistance of tubes with such grain or sand roughness[83] (Figure 237) leads to the following conclusions (on the curves, the quantity a/k inverse to the relative roughness is taken as the parameter):

1. The magnitude of the relative roughness has no effect on the critical Reynolds number, $\mathbf{Re}_{cr}$, at which transition from laminar to turbulent flow begins; in Figure 237 all the curves with different values of a/k run into the straight line $\lambda = 64/\mathbf{Re}$, corresponding to laminar flow, at the same value of $\mathbf{Re}_{cr}$, approximately equal to $2 \cdot 10^3$ ($\lg\mathbf{Re}_{cr} \approx 3.3$).
2. The transition regime is almost unaffected by the relative roughness.
3. The lower the magnitude of the relative roughness, the wider the range of Reynolds num-

[83] Nikuradze, J. *Strömungsgesetze in rauhen Rohren*, VDI Forschungshert, No. 361, 1933.

bers in which ordinary, smooth pipe types of turbulent flow are observed; thus, with relative roughness on the order of 0.2%, the experimental resistance curve agrees with the Blasius formula $\lambda = 0.3164/\mathbf{Re}^{0.25}$ up to $\mathbf{Re} = 5 \cdot 10^4$; by contrast, the results for k/a between 3 to 7% are markedly different from the smooth tube data.

4. At sufficiently high Reynolds numbers (which increase with decreasing relative roughness), the coefficient of resistance ceases to depend on **Re** and is determined solely by the degree of roughness, its value increasing with the relative roughness.

A satisfactory explanation of these fundamental results can be given if the magnitudes of the roughness height k and the laminar sublayer thickness δ_v are compared.

The following three regimes of flow can then be distinguished:

1. The first limiting case: roughness elements are submerged in the viscous sublayer ($k \ll \delta_v$). The presence of the roughness does not disturb the sublayer laminar character and the flow is free from separation and vortices. There being no differences between this case and the flow in smooth tubes, such rough tube can be regarded as hydrodynamically smooth.
2. The second limiting case: roughness grains protrude above the viscous sublayer ($k \gg \delta_v$). The flow past the grains separate, and the resistance of the tube is made up of the drag of the grains, behaving like bluff bodies, which is independent of the Reynolds number and proportional to the velocity head of the flow. This case may be called the regime of developed roughness.
3. The intermediate case: k and δ_v are of the same order; (1) and (2) are limiting cases of this more general regime.

We shall now describe the semiempirical basis of the theory of turbulent flow in rough tubes. In view of the fact that, in general, the resistance τ_w represents the sum of the drags of roughness elements distributed over a unit area of the tube surface, we assume that the resistance law is of the form

$$\frac{\tau_w}{\rho u_k^2} = f\left(\frac{ku_k}{\nu}\right)$$

where u_k is the velocity at the roughness height, k is some mean roughness height, and the quantity ku_k/ν plays the part of the Reynolds number of the flow past the roughness elements. Therefore,

$$\frac{v_*^2}{u_k^2} = f\left(\frac{u_k}{v_*} \cdot \frac{kv_*}{\nu}\right)$$

or, solving for u_k/v_*,

$$\frac{u_k}{v_*} = \Phi\left(\frac{kv_*}{\nu}\right)$$

Φ is an unknown function whose form is to be determined by experiment. It is important to note that the above arguments imply that this function depends neither on the Reynolds number nor on the degree of roughness.

We use the condition $u = u_k$ at the roughness height $y = k$ to determine the constant of integration C in the expression (113) for the logarithmic velocity profile. The resulting equation for the velocity distribution in a rough tube is

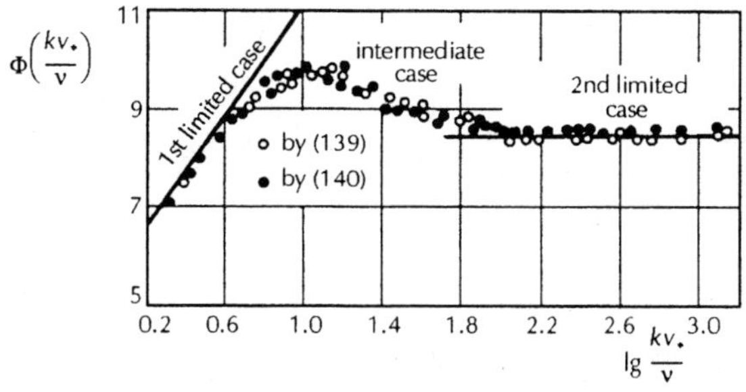

Fig. 238

$$\frac{u}{v_*} = 5.75\lg\frac{y}{k} + \frac{u_k}{v_*} = 5.75\lg\frac{y}{k} + \Phi\left(\frac{kv_*}{\nu}\right) \tag{139}$$

With the aid of this equation we derive the resistance law in the usual way. We evaluate (139) at the tube axis and proceed as in the smooth tube case:

$$\frac{u_{\max}}{v_*} = 5.75\lg\frac{a}{k} + \Phi\left(\frac{kv_*}{\nu}\right) \qquad \frac{u_{\max} - u_{\text{mean}}}{v_*} + \frac{u_{\text{mean}}}{v_*} - 5.75\lg\frac{a}{k} = \Phi\left(\frac{kv_*}{\nu}\right)$$

Substituting, as before

$$\frac{u_{\max} - u_{\text{mean}}}{v_*} = 3.75^{84} \qquad \frac{u_{\text{mean}}}{v_*} = \frac{2\sqrt{2}}{\sqrt{\lambda}}$$

we obtain the resistance law

$$3.75 + \frac{2\sqrt{2}}{\sqrt{\lambda}} - 5.75\lg\frac{a}{k} = \Phi\left(\frac{kv_*}{\nu}\right) \tag{140}$$

Comparing Equations 139 and 140, it can be seen that there are two independent ways of determining the unknown function $\Phi(kv_*/\nu)$, first, from the experimental velocity profiles and second, from measurements of tube resistance:

$$\Phi\left(\frac{kv_*}{\nu}\right) = \frac{u}{v_*} - 5.75\lg\frac{y}{k} = 3.75 + \frac{2\sqrt{2}}{\sqrt{\lambda}} - 5.75\lg\frac{a}{k}$$

Two sets of values of Φ obtained from Nikuradze's experimental results are compared in Figure 238. The possibility of such a comparison makes the determination of the function $\Phi(kv_*/\nu)$ more reliable.

In the first limiting case, when the velocity profile should have the ordinary logarithmic form of the smooth tube profile, we have

$$\Phi\left(\frac{kv_*}{\nu}\right) = 5.75\lg\frac{kv_*}{\nu} + 5.5$$

and in the second limiting case

[84] Previously, instead of this semiempirical quantity 4.08 was adopted (see Equation 125).

$$\Phi\left(\frac{kv_*}{\nu}\right) = \text{const} = 8.48$$

As seen from Figure 238, the first regime extends up to

$$\lg\frac{kv_*}{\nu} < 0.47 \qquad \frac{kv_*}{\nu} < 3$$

so that the limit of applicability of the smooth tube law is

$$\frac{k}{\delta_\nu} < 0.25$$

Using the definition of the thickness of the laminar sublayer we obtain from Equations 111, 127, and 129

$$\frac{\delta_\nu}{a} = \alpha\frac{\nu}{av_*} = \alpha\frac{\nu}{2a \cdot u_{\text{mean}}}\,\frac{2u_{\text{mean}}}{v_*} = \frac{\alpha}{\mathbf{Re}}\,\frac{4\sqrt{2}}{\sqrt{\lambda}} = \frac{65}{\mathbf{Re}\sqrt{0.0032 + 0.221\,\mathbf{Re}^{0.237}}}$$

which leads to the following estimate of the limit of applicability of the smooth tube formula:

$$\frac{k}{a} < \frac{16.3}{\mathbf{Re}\sqrt{0.0032 + 0.221\,\mathbf{Re}^{-0.237}}}$$

According to Figure 238 the second limiting regime extends from

$$\lg\frac{kv_*}{\nu} > 1.8 \qquad \frac{kv_*}{\nu} > 60$$

or

$$\frac{k}{\delta_\nu} > 6$$

so that the tube surface is hydraulically rough if

$$\frac{k}{a} > \frac{390}{\mathbf{Re}\sqrt{0.0032 + 0.221\,\mathbf{Re}^{-0.237}}}$$

Thus, to each value of the tube Reynolds number **Re** correspond definite ranges of the relative roughness *k/a* in which one or the other set of fomulas may be used.

Also of importance is the second limiting case when

$\Phi\,(kv_*/\nu)$ is constant and equal to 8.48. By Equation 139, the corresponding velocity profile is

$$\frac{u}{v_*} = 5.75\,\lg\frac{y}{k} + 8.48 \tag{141}$$

and the resistance law (140) reduces to

$$\lambda = \frac{1}{\left(2\lg\frac{a}{k} + 1.74\right)^2} \tag{142}$$

It should be noted that the formulas derived from the theory of the "idealized roughness" can be applied to practical cases if one determines experimentally the equivalent sand roughness k_e/a of the actual surface.[85]

The semiempirical theory of steady-state turbulent flow in plane and cylindrical tubes with smooth and rough surfaces, formulated in this and previous sections, is about a half century old and has become commonly accepted. One cannot but indicate some of its disadvantages: the impossibility of applying the logarithmic velocity profile near the tube edge, some required "play" of constants in transition from the logarithmic velocity profile to the logarithmic resistance law, etc.

10. Empirical Method for Calculating a Turbulent Boundary Layer on a Plate with Smooth and Rough Surfaces in a Longitudinal Flow

Application of general equations of transfer (Section 3) to the region of a turbulent boundary layer is still of a normal character. Deriving the equations of a turbulent boundary layer by the technique similar to the procedure given in Chapter 12 for the laminar boundary layer equations we neglect longitudinal derivatives of the turbulence characteristics as compared with transverse ones.

Restricting ourselves to a plane stationary flow, we present the equation of momentum transfer in the turbulent boundary region in the following simplified form

$$u\frac{\partial u}{\partial x} + v\frac{\partial u}{\partial y} = U\frac{dU}{dx} + \frac{1}{\rho}\frac{\partial \tau}{\partial y} \qquad \frac{\partial u}{\partial x} + \frac{\partial v}{\partial y} = 0 \tag{143}$$

where τ denotes the total tangent skin friction between the layers of an averaged flow and includes both turbulent and ordinary viscous friction

$$\tau = \mu\frac{\partial u}{\partial y} - \rho\overline{u'v'} = \mu\frac{\partial u}{\partial y} + \tau_t$$

In formulation of the first equation in (143) the terms $\frac{\partial}{\partial x}(-\rho\overline{u'^2})$ and $\mu(\partial^2 u/\partial x^2)$ are discarded as small compared with the main term $\partial\tau/\partial y$ characterizing the friction effect. This discarding is commonly adopted though there is no certainty that it equally holds over all the regions of a boundary layer, in particular, near separation.

Turbulent boundary layer equations (143) represent an indeterminate system of equations, since in contrast to the case of a laminar layer, τ involves the unknown summand τ_t. We shall dwells on the simplest techniques of turbulent boundary layer calculation that are

[85] A more detailed treatment of the theory of turbulent flow past rough walls can be found in Schlichting, H., *Pogranichnyy sloy* (*Boundary Layer*), Izd. Nauka, Moscow, 1974 (in Russian), in some papers written by L. G. Loitsyanskiy, On Universal Formulae in the Theory of Resistance of Rough Pipes, *TsAGI Rep.* No. 250, 1936 and K. K. Fedyayevskiy, Approximate Calculation of Friction Intensity and "Admissible" Heights of Roughness for an Airfoil, *idem.* and also Schlichting, H., Experimentelle Untersuchungen zum Rauhigkeitsproblem. *Ing.-Arch.*, 7, 1936. (The concept of the equivalent sand roughness was probably first introduced in this work.)

widely used in practice and suit it to some extent. These techniques are based on the application of the integral relation of moments

$$\frac{d\delta^{**}}{dx} + \frac{U'\delta^{**}}{U}(2+H) = \frac{\tau_w}{\rho U^2} \qquad H = \frac{\delta^*}{\delta^{**}} \tag{144}$$

which coincides in form with the similar relation (107) from Chapter 12 for a laminar boundary layer and which is derived from the system of equations (143).

Three quantities δ^{**}, H, and τ_w are unknown in Equation 144. To determine them we should add two more equations relating these quantities and allow some simplifications which as a rule are of an intuitive character.

These empirical methods being the oldest among the existing ones (they are dated as early as the 1940s to 1950s) are based on the momentum relation (144) and the following empirical laws found as a result of processing a great number of experiments:

1. Falkner's equation[86]

$$c_{f_0} = \frac{2\tau_w}{\rho U_\infty^2} = 0.0131\,\mathbf{Re}^{**-1/6} \qquad \mathbf{Re}^{**} = \frac{U_\infty \delta^{**}}{\nu} \tag{145}$$

which is, generally speaking, suitable for calculating a turbulent layer on a plate in a longitudinal flow (the subscript zero on c_f corresponds to this; U_∞ is the free stream velocity).

2. Lüdwieg-Tillman's equation[87]

$$c_f = 0.246\,\mathbf{Re}^{**-0.268} \cdot 10^{-0.678H} \qquad \mathbf{Re}^{**} = \frac{U\delta^{**}}{\nu} \qquad H = \frac{\delta^*}{\delta^{**}} \tag{146}$$

suggested for turbulent boundary layers with the arbitrary distribution of velocity $U(x)$ at the outer edge of a boundary layer.

Comparatively simple is the solution of a turbulent boundary layer problem in a case of smooth and rough plates in a longitudinal flow, when $U = \text{const} = U_\infty$, $U' = 0$. In this case the integral relation of momentum (144) is reduced to the simpler

$$\frac{d\delta^{**}}{dx} = \frac{\tau_w}{\rho U_\infty^2} \tag{147}$$

acquiring a definite form if we apply the Falkner equation (145) to the right side and on the left side we pass over from δ^{**} and x to the quantities

$$\mathbf{Re}^{**} = \frac{U_\infty \delta^{**}}{\nu} \qquad \mathbf{Re}_x = \frac{U_\infty x}{\nu}$$

We have

$$\frac{d\mathbf{Re}^{**}}{d\mathbf{Re}_x} = 0.00655\,\mathbf{Re}^{**-1/6}$$

[86] Falkner, V. M., The resistance of a smooth flat plate with turbulent boundary layer, *Aircr. Eng.*, 15, 65, 1943.

[87] Lüdwieg, H. and Tillman, W., Untersuchungen über die Wandschubspannung in turbulenten Reibungsschichten, *Ing.-Arc.*, 17, 288, 1949.

and after direct integration

$$\mathbf{Re}^{**\,7/6} = \frac{7}{6}\cdot 0.00655\,\mathbf{Re}_x + C \tag{148}$$

We first assume that a laminar portion of a boundary layer is negligibly small and a turbulent layer is formed directly on the plate leading edge.

Then, $\delta^{**} = 0$ at $x = 0$ or, the equivalent, $\mathbf{Re}^{**} = 0$ at $\mathbf{Re}_x = 0$, thus meaning $C = 0$. With such an assumption we have, according to Equation 148

$$\mathbf{Re}^{**} = 0.0153\,\mathbf{Re}_x^{6/7} \tag{149}$$

or, proceeding from the Reynolds numbers $\mathbf{Re}^{**}$ and $\mathbf{Re}_x$ to the momentum loss thickness δ^{**} and the abscissa x,

$$\delta^{**} = 0.0153\left(\frac{\nu}{U_\infty}\right)^{1/7} x^{6/7} \tag{150}$$

Thus, the momentum loss thickness in a turbulent boundary layer on a plate grows proportionally to the abscissa in the six sevenths power; this law slightly differs from the linear one. Let us recall that in the case of a laminar layer on a plate the momentum loss thickness increased in proportion to the square root of the abscissa, i.e., much slower than in a turbulent layer.

The ratio of the momentum loss thickness to the abscissa is a weak function of the Reynolds number

$$\delta^{**}/x = 0.0153\,\mathbf{Re}_x^{-1/7}$$

All these relations are in good agreement with test data obtained in wind tunnels and towing basins.

By the Falkner formula and Equation 149 we find (the subscript 0 at the coefficient of resistance means that a longitudinal flow past a plate is considered when dp/dx and dU/dx are equal to zero)

$$c_{f0} = \frac{\tau_w}{\frac{1}{2}\rho U_\infty^2} = 0.0131\,\mathbf{Re}^{**-1/6} = 0.0131\cdot 0.0153^{-1/6}\mathbf{Re}_x^{-1/7} \tag{151}$$

The final form of the formula for a local coefficient of friction is

$$c_{f0} = 0.0263\,\mathbf{Re}_x^{-1/7} \tag{152}$$

Hence, the expression of the coefficient of the total resistance of a plate with the length L can be easily determined

$$C_{f0} = \frac{W}{\frac{1}{2}\rho U_\infty^2 L}$$

We have

$$C_{f0} = \frac{\int_0^L \tau_w dx}{\frac{1}{2}\rho U_\infty^2 L} = \int_0^1 c_{f0}\, d\left(\frac{x}{L}\right) = \frac{1}{\mathbf{Re}} \int_0^{\mathbf{Re}} c_{f0}\, d\mathbf{Re}_x$$

and by virtue of Equation 152

$$C_{f0} = 0.0307\,\mathbf{Re}^{-1/7} \tag{153}$$

where **Re** is the Reynolds number of the flow past the plate

$$\mathbf{Re} = \frac{U_\infty L}{\nu}$$

The empirical equations (152) and (153) are in good agreement with experiments conducted at large Reynolds numbers and may be successfully used for calculating the resistance of smooth plates under the same conditions of streamlining when a laminar portion is small enough.

In Figure 239 a master graph of the values of C_f is presented (the subscript 0 is omitted) on which the experimental points are plotted related to different test conditions in air and water on the plates, both completely smooth and with roughnesses near the leading edge, that serve for early formation of a turbulent boundary layer; the tests were conducted within a wide range of the Reynolds number.[88]

Within a wide range of the Reynolds number values the suggested power-law formula (153) virtually does not differ from the logarithmic Prandtl-Schlichting formula[89] (a solid line in Figure 239)

$$C_{f0} = 0.455\,(\lg \mathbf{Re})^{-2.58} \tag{154}$$

and it is in good agreement with the test points of purely turbulent flow past a plate without a laminar segment in the nose portion. The power-law relation shown by a dashed-dotted line

$$C_{f0} = 0.074\,\mathbf{Re}^{-1/6} \tag{155}$$

is suitable only for comparatively small **Re**, approximately up to $\mathbf{Re} = 5 \cdot 10^6$. At large **Re** this straight line deviates from test points as is seen in the lower portion of Figure 239. Empirical substantiation of formula (155) is attributed to the application of the one seventh law to the velocity profile.

It follows from the upper graph of Figure 239 that the resistance coefficient of the plate with a completely laminar layer is much smaller than that of the plate with a completely turbulent r layer. Thus, for example, if by any means we could obtain a flow past a plate with a fully laminar layer at $\mathbf{Re} = 500{,}000$, then its coefficient of resistance would be $C_{f\,\mathrm{lam}} = 0.0018$; with a fully turbulent layer and at the same **Re** we have $C_{f\,\mathrm{turb}} = 0.005$, i.e., it is three times as large for a laminar layer. At high Reynolds numbers this difference becomes more pronounced. Hence, the conclusion can be regarding the profit of laminar layer preser-

---

[88] *Sovremennoye sostoyaniye gidroaerodinamiki vyazkoy zhidkosti* (*The Modern State of Hydroaerodynamics of Viscous Fluid*), Vol. 2, Inostrannaya Literatura Press, Moscow, 1958, 40–42.

[89] See, for instance, our monograph: *Aerodinamika pogranichnogo sloya* (*Boundary Layer Aerodynamics*), Gostekhizdat Press, Moscow, 1941, 313.

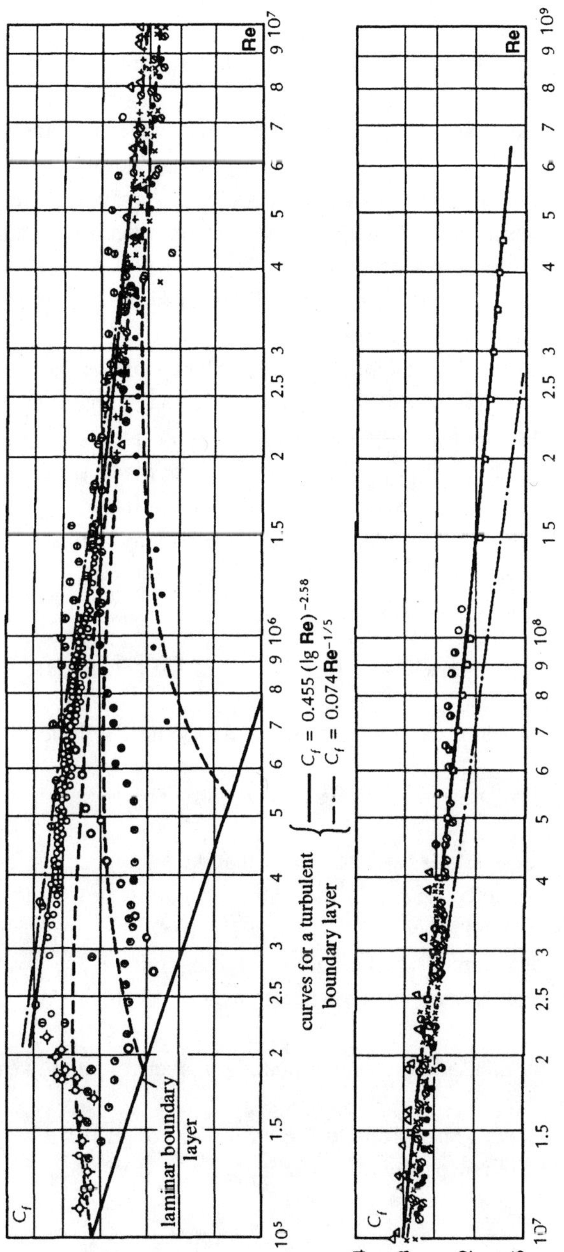

C_f
Re
laminar boundary layer
curves for a turbulent boundary layer
$C_f = 0.455\,(\lg \mathbf{Re})^{-2.58}$
$C_f = 0.074\,\mathbf{Re}^{-1/5}$
0.006
0.005
0.004
0.003
0.0025
0.002
0.0015
0.004
0.003
0.002
0.0015
10^5
10^6
10^7
10^8
10^9

Fig. 239

vation on a streamlined body by one of the means of displacement of the transition point that were mentioned at the beginning of the present chapter.

To calculate the resistance of a plate with a segment of a laminar boundary layer in a nose portion, one should find the constant C, entering into equality (148), that in this case differs from zero.

Substituting the transition region by one point we should agree on the means of joining the solutions on the boundary between the regions of laminar and turbulent flows. The most natural from the viewpoint of the techniques adopted in the previous and the present chapters is the use of the assumption on the identity of the momentum thickness loss in the section in which laminar and turbulent portions join; in this case δ^{**} or $\mathbf{Re}^{**}$ at the initial point of a turbulent boundary layer are equated to their values at the end of a laminar portion which are calculated by the laminar boundary layer theory.

Integrating both sides of momentum equations with respect to x from the plate leading edge to the transition point, we shall conclude that the adopted matching condition is a physically natural requirement stemming from continuity in the growth of total resistance $W_x = \rho U_\infty^2 \delta^{**}(x)$ of the plate portion from $x = 0$ to the given x on the transition of the abscissa of the portion end over the abscissa of the transition point.

Designating the values of the corresponding Reynolds numbers, which are common for both portions of a boundary layer, in terms of $\mathbf{Re}_{xt}$ and $\mathbf{Re}_t^{**}$ we obtain according to Equation 138

$$\mathbf{Re}^{**\,7/6} - \mathbf{Re}_t^{**\,7/6} = 0.00765\,(\mathbf{Re}_x - \mathbf{Re}_{xt}) \tag{156}$$

and according to the laminar boundary layer theory there will be

$$\mathbf{Re}_t^{**} = 0.664\sqrt{\mathbf{Re}_{xt}}$$

Not dwelling on details, we shall show that the allowance for the effect of $\mathbf{Re}_{xt}$ value on the plate total resistance results in the transition curves presented in Figure 239 by dashed lines. For flows in wind tunnels or other artificial flows the position and shape of these transition curves are determined by the value of the parameter $\mathbf{Re}_{xt}$ which is dependent on the incident flow turbulence, on the surface roughness near the leading edge, and on other conditions.

The empirical method of turbulent boundary layer calculation (presented above) refers to a smooth plate only.

The method was generalized to the case of a rough plate by Droblenkov,[90] who suggested the use of the power law of resistance

$$\frac{\tau_w}{\rho U_\infty^2} = 0.0031\left(\frac{\delta^{**}}{k}\right)^{-1/6} \tag{157}$$

for the case of developed roughness when $\tau_w/\rho U_\infty^2$ is independent of the Reynolds number but is dependent on the local relative roughness δ^{**}/k only.

Reapplying the momentum equation, we obtain the equation

[90] Droblenkov, V. F., Turbulent boundary layer on a rough curvilinear surface, *Izv. AN SSSR, OTN*, 8, 17–21, 1955.

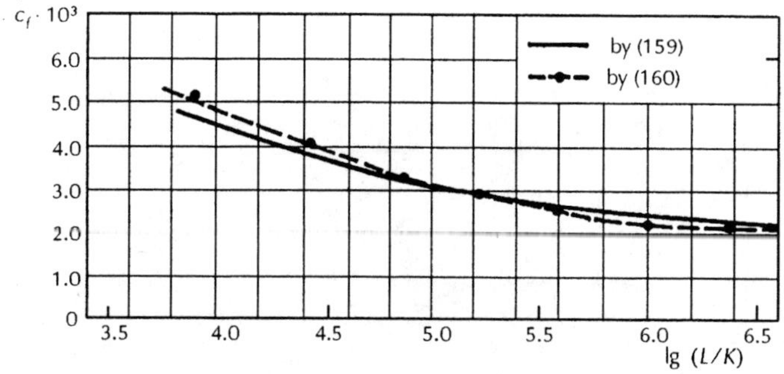

Fig. 240

$$\frac{d\delta^{**}}{dx} = 0.0031\left(\frac{\delta^{**}}{k}\right)^{-1/6}$$

which after integration at the boundary condition $\delta^{**} = 0$ at $x = 0$ will give the formula of the change in the thickness δ^{**} along the plate

$$\delta^{**} = 0.008\,x\left(\frac{x}{k}\right)^{-1/7}$$

Now, using Equation 157, we can easily find the distribution of the local coefficient of friction

$$c_{f0} = \frac{\tau_w}{\frac{1}{2}\rho U_\infty^2} = 0.0139\left(\frac{x}{k}\right)^{-1/7} \tag{158}$$

and, after the integration over the plate length L, also the coefficient of the plate total resistance

$$C_{f0} = \frac{\int_0^L \tau_w dx}{\frac{1}{2}\rho U_\infty^2 L} = 0.0162\left(\frac{L}{k}\right)^{-1/7} \tag{159}$$

Figure 240 gives, for the sake of comparison, the results of the calculation by Equation 159 and by the Schlichting formula[91]

$$C_{f0} = \left(1.89 + 1.62\,\mathrm{lg}\frac{L}{k}\right)^{-2.5} \tag{160}$$

derived from more complex theoretical considerations.

We shall not pause on the problem of the calculation of plate resistance in an intermediate flow mode in a boundary layer corresponding to the roughness range $1/4 < k/\delta_v < 6$. We indicate only that Prandtl and Schlichting thoroughly studied this problem and developed

[91] Schlichting, H., Experimentelle Untersuchungen zum Rauhigkeits-problem. *Ing.-Arch.*, 7, 1936.

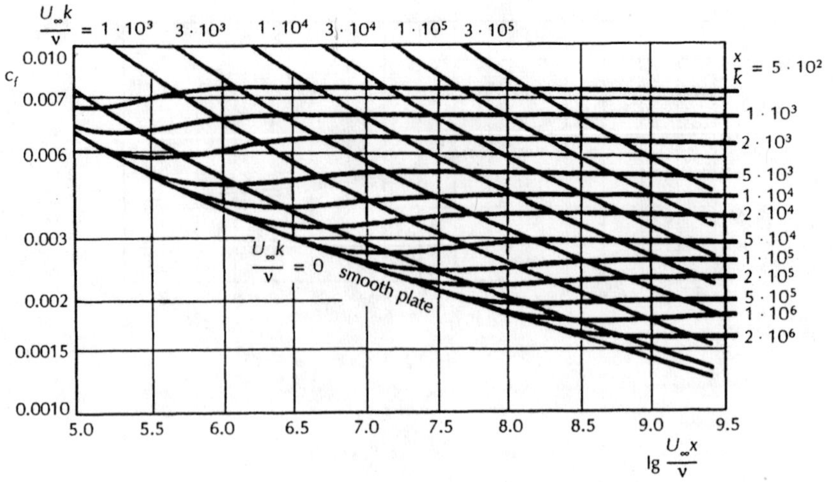

Fig. 241

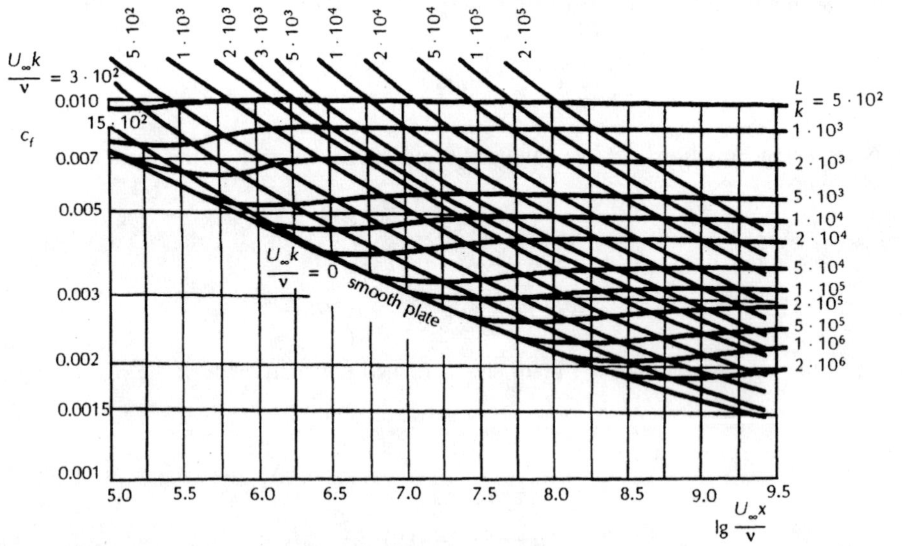

Fig. 242

common nomograms for calculating c_{f0} and C_{f0} under any roughness conditions.[92] They are presented in Figures 241 and 242. Making use of these nomograms one can perform different calculations to evaluate various effects. Thus, for example, if we assign an absolute size of a roughness element k, then the curves of the family $x/k = \text{const}$ in Figure 241 or $L/k = \text{const}$ in Figure 242 will represent the dependence of the local coefficient of resistance on the local or general Reynolds numbers at fixed x or on the coefficient of the total resistance at fixed plate length L. The curves of the family $U_\infty k/\nu = \text{const}$ at the given absolute roughness k made it possible to determine the change of the same coefficients with variable x or L. We can behave conversely, assigning x or L and varying k.

All the above refer, of course, to the case of the so-called general roughness, which is

[92] Prandtl, L. and Schlichting, H., Der Winderstandgesetz rauher Platten. Werft., Reederei, Hafen, 1934, Bd. 14.

equal and uniformly distributed over the entire plate surface. The problems become complicated if roughness has the character of isolated elements, even though they are regularly encountered on the surfaces (welded seams, rivets, etc.). Studying the effect of such local asperities is essential for aviation, shipbuilding, and other branches of technology.[93] To the number of practically important there belongs the problem of a turbulent boundary layer on a disk or a round cylinder rotating in infinite fluid or that bounded by a motionless shell. These problems are thoroughly considered in the monograph by Dorfman.[94]

11. Empirical Method for Calculating Turbulent Boundary Layer with Arbitrary Velocity Distribution on the Outer Edge

An empirical method, which is similar in simplicity to that presented in the previous section, also may be constructed for the calculation of a turbulent boundary layer with arbitrary distribution of flow velocity on its outer edge.

We shall consider one of these methods analogous to the one-term method of laminar boundary layer calculation.[95] Recall that for one-term calculation of a laminar boundary layer the following parameters were introduced

$$f = \frac{U'\delta^{**2}}{\nu} \qquad \zeta = \frac{\tau_w \delta^{**}}{\mu U} \qquad H = \frac{\delta^*}{\delta^{**}}$$

the first two of which may be rewritten in the form

$$f = \frac{U'\delta^{**}}{U}\mathbf{Re}^{**} \qquad \zeta = \frac{\tau_w}{\rho U^2}\mathbf{Re}^{**} \qquad \mathbf{Re}^{**} = \frac{U\delta^{**}}{\nu}$$

The presence of $\mathbf{Re}^{**}$ in the expressions for f and ζ with the dimensionless quantities $U'\delta^{**}/U$ and $\tau_w/\rho U^2$ characterizing the longitudinal pressure on the conventional boundary layer thickness δ^{**}

$$-\delta^{**}\frac{dp/dx}{\rho U^2} = -\delta^{**}\frac{-\rho U U'}{\rho U^2} = \frac{U'\delta^{**}}{U}$$

and the skin friction on a solid wall τ_w, related to the velocity head, leads to the fact that in a laminar boundary layer the functional relationships $\zeta(f)$ and $H(f)$ retain their forms at any Reynolds numbers of a boundary layer. Strictly speaking, in the case of a turbulent boundary layer it is impossible to find such a common multiplier $G(\mathbf{Re}^{**})$ for the quantities $U'\delta^{**}/U$ and $\tau_w/\rho U^2$, so that, similar to the laminar boundary layer, the quantities

$$f = \frac{U'\delta^{**}}{U}G(\mathbf{Rc}^{**}) \qquad \zeta = \frac{\tau_w}{\rho U^2}G(\mathbf{Re}^{**}) \tag{161}$$

[93] Fedyayevskiy, K. K., Approximate calculation of friction intensity and "admissible" heights of roughness for an airfoil, *Tr. TsAGI*, No. 250, 1936, Fedyayevskiy, K. K. and Fomina, N. N., Study of roughness effect on drag and state of a boundary layer, *Tr. TsAGI*, No. 441, 1939.

[94] Dorfman, L. A., *Gidrodinamicheskoye soprotivleniye i teplootdacha vrashchayushchikhsya tel (Hydrodynamics and Heat Transfer of Bodies of Revolution)*. Fizmatgiz Press, Moscow, 1960, 68–92, 135–160, 193–206.

[95] Loitsyanskiy, L. G., An approximate method for calculation of a turbulent boundary layer on the airfoil profile, *Prikl. Mat. Mekh.*, 9, 433–448, 1945.

which are a natural generalization of the Buri parameters,[96] are connected by the relation

$$\zeta = \zeta(f) \tag{162}$$

not involving the Reynolds number. However, judging by the available test data, the quantity $G(\mathbf{Re}^{**})$ may be chosen in such a way that the effect of the Reynolds on relationship (162) is comparatively small. A weak function of the Reynolds number is also presented by the relation $\delta^*/\delta^{**} = H$. Thus, by the Hama experiments[97] for smooth and rough plates the general relationship

$$H_0 = \left(1 - 6.8\sqrt{\frac{c_f}{2}}\right)^{-1} \tag{163}$$

is valid that in the case of a smooth plate and, with resistance Equation 152 used, yields

$$H_0 = (1 - 0.78\,\mathbf{Re}_x^{-1/14})^{-1}$$

The power index of $\mathbf{Re}_x$ indicates that H_0 is a weak function of $\mathbf{Re}_x$. Actually, the quantity H_0 on transition from $\mathbf{Re}_x = 10^6$ to $\mathbf{Re}_x = 10^7$ reduces from 1.41 to 1.33, i.e., by approximately 6%. On a rough plate the role of the Reynolds number is even smaller and in the case of developed roughness it is simply absent. Taking account of this weak variability of H_0 with the growth of the Reynolds number on the plate ($dp/dx = 0$), we shall either consider this quantity independent of $\mathbf{Re}_x$ and assume

$$H = H(f)$$

or allow for the effect of the Reynolds number from the equations for a plate.

Not restricting ourselves to the form of the function $G(\mathbf{Re}^{**})$ and having taken the quantity f, determined by the first equation of (161), as a form parameter, we derive from the integral relation (144) the equation for determining the form parameter f. For this purpose we multiply both sides of (144) into $G(\mathbf{Re}^{**})$. Then we obtain, according to Equation 161

$$G(\mathbf{Re}^{**})\frac{d\delta^{**}}{dx} + (2 + H)f = \zeta \tag{164}$$

The first term on the left side may be transformed so (the prime denotes the derivative with respect to x)

$$G(\mathbf{Re}^{**})\frac{d\delta^{**}}{dx} = \frac{d}{dx}\left[G(\mathbf{Re}^{**})\frac{U'\delta^{**}}{U}\frac{U}{U'}\right] - \delta^{**}\frac{dG}{d\mathbf{Re}^{**}}\frac{d\mathbf{Re}^{**}}{dx} =$$

$$= \frac{d}{dx}\left(f\frac{U}{U'}\right) - \delta^{**}\frac{dG}{d\mathbf{Re}^{**}}\left(\frac{U}{\nu}\frac{d\delta^{**}}{dx} + \frac{U'\delta^{**}}{\nu}\right) =$$

$$= \frac{d}{dx}\left(f\frac{U}{U'}\right) - \mathbf{Re}^{**}\frac{dG}{d\mathbf{Re}^{**}}\frac{d\delta^{**}}{dx} - \mathbf{Re}^{**}\frac{dG/d\mathbf{Re}^{**}}{G(\mathbf{Re}^{**})}f =$$

[96] Buri, A., Eine Berechnungsgrundlage für die Turbulente Grenzschicht bei Beschleunigter und Verzögerter Grundströmung, Dissertation, Zürich, 1931; see also Nikuradze, J., *Untersuchungen über die Strömungen des Wassers in Konvergenten und divergenten Kanälen*, VDI Forschungsheft, 1929, No. 289.

[97] Hama, F. R., Boundary layer characteristics for smooth and rough surfaces. *Trans. Soc. Nav. Architects Mar. Eng.*, 62, 333–358, 1954.

$$= \frac{d}{dx}\left(f\frac{U}{U'}\right) - \frac{\mathbf{Re}^{**}dG/d\mathbf{Re}^{**}}{G(\mathbf{Re}^{**})} G(\mathbf{Re}^{**})\frac{d\delta^{**}}{dx} - \frac{\mathbf{Re}^{**}dG/d\mathbf{Re}^{**}}{G(\mathbf{Re}^{**})} f$$

Introducing the quantity

$$m = \frac{\mathbf{Re}^{**}dG/d\mathbf{Re}^{**}}{G(\mathbf{Re}^{**})} = \frac{d\lg G(\mathbf{Re}^{**})}{d\lg\mathbf{Re}^{**}} \tag{165}$$

which in its structure is weakly dependent on $\mathbf{Re}^{**}$, we find from the previous equation

$$(1+m)G(\mathbf{Re}^{**})\frac{d\delta^{**}}{dx} = \frac{d}{dx}\left(\frac{U}{U'}\right) - mf$$

Making use of equality (164), we eliminate the quantity $G(\mathbf{Re}^{**})\frac{d\delta^{**}}{dx}$; this yields

$$\frac{d}{dx}\left(\frac{U}{U'}\right) = (1+m)\zeta - [2+m+(1+m)H]f$$

or, having expanded the derivative on the left side,

$$\frac{df}{dx} = \frac{U'}{U}F(f) + \frac{U''}{U'}f \tag{166}$$

where for brevity the designation

$$F(f) = (1+m)\zeta - [3+m+(1+m)H]f \tag{167}$$

is adopted.

Equation 166 does not differ in form from its laminar analogue (Equation 111, Chapter 12) that was basic to laminar boundary layer calculation. The difference is only in the form of the functional relation $F(f)$ which is presented by equality (113) (Chapter 12) for a laminar layer and by equality (167) of this chapter for a turbulent layer. If it is assumed in Equation 167 that $m = 1$ (the fact corresponds to a laminar layer), then the right sides of Equation 167 and Equation 113 of Chapter 12 will coincide.

The function $G(\mathbf{Re}^{**})$ for the region at a distance from the separation point will be determined, taking into account the fact that the multiplier $\mathbf{Re}^{**}$, which ensures the ζ independence of the Reynolds number in the case of a laminar layer, remained the same in the presence and absence of the longitudinal pressure variation in a layer. Let us assume that in the region under consideration this property is valid also in the case of a turbulent layer. We take the value of the universalizing factor $G(\mathbf{Re}^{**})$ inversely proportional to the local coefficient of the plate skin friction ($U' = 0$), having assumed that ($\backsim$ is the proportionality sign)

$$G(\mathbf{Re}^{**}) \backsim \frac{2}{c_{f_0}} \tag{168}$$

It is easily seen that the numerical value of the coefficient of proportionality is of no importance because the change of this coefficient will cause proportional variation of ζ and f, but it will not affect m and, consequently, result in the same variation of $F(f)$. In this case the form of Equation 166 will not be disturbed.

We assume, following Equations 168 and 145,

$$G(\mathbf{Re}^{**}) = 153.2\,\mathbf{Re}^{**\,1/6} \tag{169}$$

then, according to Equation 165, we find $m = 1/6$ and the function $F(f)$ will be equal to

$$F(f) = \frac{7}{6}\zeta - \left(\frac{19}{6} + \frac{7}{6}H\right)f \tag{170}$$

We linearize the function $F(f)$, having assumed similar to the plate

$$\zeta = (\zeta)_{f=0} = 1 \qquad H = H_0 = 1.3 - 1.4$$

Taking the values of ζ and H, we find the linear presentation for $F(f)$

$$F(f) = a - bf \tag{171}$$

that is analogous to Equation 115 of Chapter 12, but in the case of turbulent layer with the constants a and b these values are equal to

$$a = 1.17 \qquad b = 4.7 - 4.8 \tag{172}$$

Equation 166 admits here the solution in the form of a simple quadrature

$$f(x) = \frac{U'(x)}{U^b(x)}\left[a\int_0^x U^{b-1}(\xi)\,d\xi + C\right]$$

If the laminar layer on the surface of an airfoil profile is assumed to be absent, then we have

$$f(x) = \frac{aU'(x)}{U^b(x)}\int_0^x U^{b-1}(\xi)\,d\xi \tag{173}$$

with (similar to a laminar layer) the point of flow bifurcation ($x = 0$) being singular. Evaluating the limit, we find

$$f(0) = aU'(0)\frac{U^{b-1}(0)}{bU^{b-1}(0)\,U'(0)} = \frac{a}{b} = 0.24$$

If the presence of a laminar layer within the range of abscissas ($0 < x < x_t$) is taken into account, then the expression for f will become somewhat complex and will take on the form

$$f(x) = \frac{U'(x)}{U^b(x)}\left[a\int_{x_t}^x U^{b-1}(\xi)\,d\xi + \frac{U_t^b}{U_t'}f_t\right]$$

Here, U_t, U_t' and f_t represent the values of U, U', and f at the transition point $x = x_t$, with f_t being calculated by the first formula of (161)

$$f_t = \frac{U_t'\delta_t^{**}}{U_t}G(\mathbf{Re}_t^{**}) = \frac{\nu U_t'}{U_t^2}\mathbf{Re}_t^{**}\,G(\mathbf{Re}_t^{**})$$

Finally, we have

$$f(x) = \frac{U'(x)}{U^b(x)}\left[a\int_{x_t}^{x} U^{b-1}(\xi)\,d\xi + \nu U_t^{b-2}\mathbf{Re}_t^{**}\,G(\mathbf{Re}_t^{**})\right] \tag{174}$$

According to the earlier adopted condition of laminar and turbulent boundary layers the quantity $\mathbf{Re}_t^{**}$ should be calculated by the theory of a laminar boundary layer.

Making use of formulas (173) and (174), we find $f(x)$. Then, according to the first equality of system (161), we obtain the equation for $\mathbf{Re}^{**}(x)$

$$\mathbf{Re}^{**}G(\mathbf{Re}^{**}) = \frac{U^2}{\nu U'}f(x)$$

Thus, for example, for a fully turbulent layer over the entire profile surface we have, according to Equation 173

$$\mathbf{Re}^{**}G(\mathbf{Re}^{**}) = 153.2\,\mathbf{Re}^{**7/6} = \frac{a}{\nu U^{b-2}(x)}\int_0^x U^{b-1}(\xi)\,d\xi \tag{175}$$

Having performed quadrature and determined $\mathbf{Re}^{**}(x)$ and $\delta^{**}(x)$, in the approximation adopted ($\zeta = 1$), following the second equality of system (161), we find the formula for the local coefficient of friction

$$c_f = \frac{\tau_w}{\frac{1}{2}\rho U^2} = \frac{2}{G(\mathbf{Re}^{**})} = 0.0131\,\mathbf{Re}^{**-1/6} \tag{176}$$

which is similar to the equation for a plate but at $\mathbf{Re}^{**}$ calculated for the given distribution of the outer flow velocity $U(x)$.

Having thus determined τ_w or c_f as a function of x and summed up, over an airfoil surface, the projections of friction elementary forces $\tau_w dx$ onto the direction of the free stream, we find the total skin friction of the airfoil.

The determination of the displacement thickness $\delta^*(x)$ follows. In the approximation adopted this quantity is equal to

$$\delta^*(x) = H_0\delta^{**}(x) = (1.3 - 1.4)\,\delta^{**}(x) \tag{177}$$

The bracketed numerals indicate the limits of the values of H_0 at different values of the flow Reynolds number: the first corresponds to higher values, the second to lower ones. However approximate the above empirical approach (resting to a large extent on the analogy with the problem of a turbulent boundary layer on a plate, i.e., on the case of the velocity constancy on the boundary layer outer edge), results of calculation of the quantity $\delta^{**}(x)$ and consequently of $\mathbf{Re}^{**}(x)$ by this method led to other, more perfect methods[98] with improved accuracy. As to the value of c_f, within the framework of the above-described method, it may

---

[98] See p. 19 of the monograph by K. K. Fedyayevskiy, A. S. Ginevskiy, and A. V. Kolesnikov, *Raschyot turbulentnogo pogranichnogo sloya neszhimayemoy zhidkosti* (*Calculation of a Turbulent Boundary Layer of an Incompressible Fluid*), Sudostroeniye Press, Leningrad, 1973.

be determined by the Lüdwieg-Tillmann equation (146), specifying the values of $H = H_0$ from the previous section or using more complex empirical methods (for example, the method of Truckenbrodt presented in the monograph by G. Schlichting[99]) as well as more modern techniques that are discussed in the following chapter.

Empirical methods are far from being exhausted and they may be still useful, especially in view of the considerable development of experimental methods and accumulation of new test data on one hand and, on the other, in view of exploration of the unlimited possibilities of computer calculation. It also should be taken into account that more perfect methods may be of limited value to a practicing engineer due to their complexity and law visuality.

Statistical processing of accumulated test data on flows in turbulent boundary layers might play a significant role in the improvement of empirical methods. Such an investigation was conducted at the end of the 1960s by Kamenetskiy,[100] who suggested the following empirical formulas for the dependence of the basic characteristics of a turbulent boundary layer on the parameter f:

$$\zeta = \frac{c_f}{c_{f_0}} = 1 + 0.1367f + 0.015f^2 + 0.00333f^3$$

$$H = H_0 - 0.0701f + 0.02913f^2 + 0.01083f^3 + 0.001606f^4$$

$$F = F_0 - 4.694f + 0.0238f^2 - 0.0246f^3 \tag{178}$$

Further development of the mentioned approach on the basis of the use of new experimental materials[101] seems very promising.

The existing empirical methods present especially noticeable errors in a diverging region of a boundary layer ($dp/dx > 0$, $dU/dx < 0$) when approaching the separation point. Here, the effect is especially strong of adopted assumption that turbulent layer characteristics with arbitrary pressure distribution are similar to their values in a boundary layer on a plate in a longitudinal flow ($dp/dx = 0$, $dU/dx = 0$). Note that the calculation of a turbulent boundary layer near the separation point is still a weak spot in all, even modern, theories.r

The separation region is characterized by a developed outer portion, the velocity profile of which may be likened to the profile in an aerodynamic wake behind the body or on a jet edge. Near a separation point and behind it in a stalled jet the effect of the Reynolds number becomes negligible. This admits a simple solution which serves for the determination of δ^{**}, the momentum thickness loss. According to the condition of the independence of the Reynolds number over the entire preseparation region, we shall take $m = 0$; consequently,

$$G(\mathbf{Re}^{**}) = \text{const} = G_s$$

and moreover,

$$\zeta = 0 \qquad H = H_s$$

Then, according to Equation 166 we come to the equation

---

[99] Schlichting, G., *Teoriya pogranichnogo sloya* (*Boundary Layer Theory*), Nauka Press, Moscow, 1969, 624–638 (Russian translation).

[100] Kamenetskiy, A. I., A new empirical method for calculating a turbulent boundary layer in an incompressible fluid, *Tr. LPI*, No. 313, 62–67, 1970.

[101] Computation of turbulent boundary layer, 1968, *Proc. AFOSR-JFR-Stanford Conference*, Vol. 2, Coles, D. E. and Hirst, E. A., Eds. 1969.

$$\frac{df}{dx} = \left[\frac{U''}{U'} - (3 + H_s)\frac{U'}{U}\right]f \tag{179}$$

the integral of which is

$$f = C\frac{U'}{U^{3 + H_s}}$$

If it is assumed that in the previous region whose boundary with the second preseparation region is characterized by the quantity f_1, calculation was made by the above technique. Then, matching solutions at $f = f_1$, we shall determine the integration constant C as

$$C = \frac{U_1^{3 + H_s} f_1}{U_1'}$$

and obtain the following equation for calculation of f or $\bar{f} = f/f_s$ near separation:

$$\frac{f}{f_1} = \frac{\bar{f}}{\bar{f}_1} = \frac{U'}{U_1'}\left(\frac{U_1}{U}\right)^{3 + H_s} \tag{180}$$

Noting that near separation the relation

$$f = \frac{U'\delta^{**}}{U}G_s$$

will be valid, we shall find the expression, corresponding to Equation 180, for the ratio of the thicknesses of the momentum loss ln the separation region

$$\frac{\delta^{**}}{\delta_1^{**}} = \left(\frac{U_1}{U}\right)^{2 + H_s} \tag{181}$$

Proceeding from other theoretical considerations and making use of the empirical data Ross and Robertson[102] suggested the equation

$$\frac{\delta^{**}}{\delta_1^{**}} = \left(\frac{U_1}{U}\right)^{4.8} \tag{182}$$

which is similar to Equation 181 at $H_s = 2.8$. It should be noted that the choice of the value $H_s = 2.8$ is not contradictory to the experimental data for the value of H_s at the separation point. Proceeding from the considerations of dimensionality and some assumptions, which are general for the boundary layer theory, Bam-Zelikovich arrived at the same result.[103]

The above empirical technique was generalized to the case of a rough surface by V. F. Droblenkov.[104]

---

[102] Ross, D. and Robertson, I., An empirical method for calculation of the growth of a turbulent boundary layer, *J. Aeron. Sci.*, 21 (5), 355–358, 1954.

[103] Bam-Zelikovich, G. M., Calculation of boundary layer separation, *Izv. AN SSSR, OTN*, No.12, 1954.

[104] Droblenkov, V. F., Turbulent boundary layer on a rough curvilinear surface, *Izv. AN SSSR, OTN*, No. 8, 1955.

12. Semiempirical Method for Calculating Turbulent Boundary Layer on a Plate in a Longitudinal Flow

Moving from the empirical methods to semiempirical ones which are based on the use of the semiempirical theories of Prandtl and von Kármán, we emphasize that the creators of these theories applied them directly to the calculation of a turbulent boundary layer on a plate in a longitudinal flow.[105] We present the von Kármán theory, along with the idea of derivation being completely preserved and the mathematical aspect being somewhat changed.[106] Having taken the integral relation of momentum (147) as a basis and expressing the thickness of momentum loss δ^{**} in universal variables ($\varphi = u/v_*$, $\eta = yv_*/\nu$, $h = U_\infty/v_*$):

$$\delta^{**} = \int_0^\infty \frac{u}{U_\infty}\left(1 - \frac{u}{U_\infty}\right)dy = \frac{\nu h}{U_\infty}\int_0^\infty \frac{\varphi}{h}\left(1 - \frac{\varphi}{h}\right)d\eta \tag{183}$$

we move from the independent variable η to the independent variable φ/h, having obtained as a result the following relationship for δ^{**} (the dot above the letter denotes the derivative with respect to φ):

$$\delta^{**} = \frac{\nu h^2}{U_\infty}\int_0^1 \frac{\varphi}{h}\left(1 - \frac{\varphi}{h}\right)\dot{\eta}\, d\left(\frac{\varphi}{h}\right) \tag{184}$$

The value of $\dot{\eta} = d\eta/d\varphi$ in a viscous sublayer is equal to unity. In a turbulent core, to determine $\dot{\eta}$ use is made of the von Kármán relation (the second formula in Equation 73) (the prime denotes the derivative with respect to y)

$$\tau = \frac{\rho æ^2 u'^4}{u''^2}$$

$æ$ is the turbulence constant equal to 0.4.

Adopting the assumption of the constancy of the skin friction across a layer $\tau = \tau_w$ and moving on to universal coordinates, we obtain (now the prime denotes the derivative with respect to η)

$$\frac{\varphi''}{\varphi'^2} = -æ \tag{185}$$

The minus sign appears when the root is taken in view of the fact that $\varphi'' < 0$. In equality (185) we interchange the argument and the function; then, we obtain the equation $\ddot{\eta}/\dot{\eta} = æ$, from which it follows

$$\dot{\eta} = Ce^{æ\varphi} \tag{186}$$

[105] A free presentation of these computational methods may be found in the monograph Loitsyanskiy, L. G., *Aerodinamika pogranichnogo sloya* (*Aerodynamics of a Boundary Layer*). Leningrad-Moscow, 1941, 310–313 and 335–338.

[106] Lapin, Yu. V. and Loitsyanskiy, L. G., Application of the von Kármán method to the calculation of a turbulent boundary layer on a plate in a gas flow. *Tr. LPI*, No. 217, 1961.

If we denote the value of the derivative $d\varphi/d\eta$ on the viscous sublayer edge $\eta = \eta_v = \alpha$ (120) in terms of f, then equality (186) takes on the form

$$\dot{\eta} = \frac{e^{-æ\alpha}}{f} e^{æ\varphi} \tag{187}$$

where is α the turbulence constant equal to 11.5.

Reverting to equality (184), we note that strictly speaking, the integral on the right side of this equality should be divided into two portions: $0 \le \varphi \le \alpha$ and $\alpha \le \varphi \le h$ and in each of them its own value of η should be substituted. However, in view of the relative thickness of a viscous sublayer, which slightly affects the value of the integral (184), the first portion may be omitted, having extended the second to the wall. As is shown by calculations, at large h there is no great difference in the values of the integral.

Thus, eliminating $\dot{\eta}$ from equalities (184) and (187), we obtain the following expression for the loss of momentum thickness

$$\delta^{**} = \frac{\nu h^2 e^{-æ\alpha}}{f U_\infty} \int_0^1 \bar{u}(1-\bar{u})\, e^{æh\bar{u}} d\bar{u} \tag{188}$$

where $\bar{u} = \varphi/h = u/U_\infty$.

The value of $æh$, as seen from its determination, noticeably exceeds unity. In these cases it is convenient to make use of the general presentation of the integral, involving the exponential function,

$$\int e^{æh\psi} f(\psi)\, d\psi = \frac{e^{æh\psi}}{æh}\left[f(\psi) - \frac{f'(\psi)}{æh} + \frac{f''(\psi)}{æ^2h^2} - \ldots\right] \tag{189}$$

that is easily derived by integration in parts. In cases in which the series in square brackets is not truncated, this form may be treated as the series expressing the value of the integral at large $æh$.

Performing in (188) integration with the aid of the series (189), we obtain the expression for the thickness of momentum loss

$$\delta^{**} \approx \frac{\nu e^{-æ\alpha}}{f æ^2 U_\infty} e^{æh}\left(1 - \frac{2}{æh}\right) \tag{190}$$

We also find the expression for the form parameter $H = \delta^*/\delta^{**}$. By the technique analogous to that shown above one can find the following expression for the displacement thickness:

$$\delta^* \approx \frac{\nu e^{-æ\alpha}}{f æ^2 U_\infty} e^{æh}$$

Composing the ratio δ^*/δ^{**}, we obtain the expression for H

$$H \approx \frac{1}{1 - 2/(æh)} = \left(1 - 5\sqrt{\frac{c_f}{2}}\right)^{-1}$$

which at the Reynolds numbers 10^6 to 10^7 gives $H = 1.3$, which is in good agreement with the experiment.

Reverting to Equation 190, we note that in the calculation of the thickness of momentum loss the second term in the brackets may be neglected as compared with the first. As is shown by calculations, at large $æh$ the error obtained in computation of the friction coefficient with such an approximation is insignificant.

Having taken such approximation for the thickness of momentum loss we substitute it into the integral relation of momentum. After simple transformations we obtain the following equation

$$\frac{e^{-æ\alpha}}{fæ^2} h^2 d\left(e^{æh}\right) = d\mathbf{Re}_x \qquad \mathbf{Re}_x = \frac{U_\infty x}{\nu}$$

Taking the integral of both sides of this equation and using the boundary condition $\mathbf{Re} = 0$ at $v_* = \infty$, i.e., $h = 0$, we have in the same approximation

$$\frac{e^{-æ\alpha}}{fæ^2} h^2 e^{æh} = \mathbf{Re}_x$$

Taking logs of both sides of this equality and reverting from the variable h to c_f by the equality $h = \sqrt{2/c_f}$, we obtain the von Kármán formula

$$\frac{1}{\sqrt{c_f}} = A + B\lg(\mathbf{Re}_x c_f) \tag{191}$$

where A and B are expressed in terms of the constants $æ$ and α. With the test data corresponds the values $A = 1.7$, $B = 4.15$, which are close to them.

The semiempirical von Kármán equation represents the implicit dependence between the local coefficient of resistance c_f and the Reynolds number $\mathbf{Re}_x$. For the coefficient of the total friction resistance of the plate with the length L the logarithmic Prandtl-Schlichting formula (154) was mentioned earlier.

13. Integral Methods for Calculation of Turbulent Boundary Layers

Among the semiempirical methods of turbulent boundary layer calculation that are discussed later, an important role is played by integral semiempirical methods based on the application of the Prandtl mixing length theory.

We shall pause on the basic techniques of integral methods, referring the reader for the details to a thoroughly composed monograph,[107] the first part of which is concerned with the detailed presentation of integral methods.

In its initial form the idea of the integral method was in the determination of velocity profiles in the sections of a turbulent boundary layer by assigning polynomial approximations of turbulent skin friction τ and the mixing length l in the formula

$$U - u = \int_y^\delta \sqrt{\frac{\tau}{\rho}} \frac{dy}{l} \tag{192}$$

[107] Fedyayevskiy, K. K., Ginevskiy, A. S., and Kolesnikov, A. V., *Raschyot turbulentnogo pogranichnogo sloya neszhimayemoy zhidkosti* (*Calculation of Turbulent Boundary Layer of Incompressible Fluid*), Sudostroeniye Press, Leningrad, 1973, Chap. 1 and 2.

which can be easily derived from the Prandtl law.

Proceeding in a way similar to the method of Pohlhausen, who assigned velocity profiles in the laminar boundary layer cross-sections by the fourth power polynomial (Equation 117 in Chapter 12), which satisfied the boundary conditions (Equation 118 of Chapter 12), use is made in the integral method of turbulent boundary layer calculation of the analogous polynomial approximations for the skin friction

$$\frac{\tau}{\tau_w} = \sum_{i=0}^{n} \alpha_i \eta^i \qquad \eta = \frac{y}{\delta} \tag{193}$$

where δ is the finite boundary layer thickness, or of the polynomial representation which is more convenient for calculations

$$\sqrt{\frac{\tau}{\tau_w}} = \sum_{i=0}^{n} b_i \eta^i \tag{194}$$

The coefficients a_i and b_i in these formulas are selected from the corresponding boundary conditions, the examples of which are

$$\frac{\tau}{\tau_w} = 1 \qquad \frac{d}{d\eta}\left(\frac{\tau}{\tau_w}\right) = \beta \qquad \frac{d^2}{d\eta^2}\left(\frac{\tau}{\tau_w}\right) = 0 \quad \text{at} \quad \eta = 0$$

$$\frac{\tau}{\tau_w} = 0 \qquad \frac{d}{d\eta}\left(\frac{\tau}{\tau_w}\right) = 0 \qquad \text{at} \quad \eta = 1 \tag{195}$$

where

$$\beta = \frac{\delta}{\tau_w}\frac{dp}{dx} \tag{196}$$

The first conditions in both lines are evident, the second in the first line follows from the boundary layer equations, the third one in the first line is derived by differentiation of the both sides of the boundary layer equation with respect to η (taking account of the continuity equation), and the last condition in the second line is complementary, requiring the smoothness of the transition of τ/τ_w to the zero value on the boundary layer outer edge.

Using the five boundary conditions (195), we obtain the profile of turbulent skin friction in the form of the fourthpower polynomial, correspondingly,

$$\frac{\tau}{\tau_w} = 1 - 4\eta^3 + 3\eta^4 + \beta\eta\,(1 - 3\eta^2 + 2\eta^3) \tag{197}$$

Taking two conditions at $\eta = 0$ and one at $\eta = 1$,

$$\frac{\tau}{\tau_w} = 1 - \eta^2 + \beta\eta\,(1 - \eta) \qquad (-2 < \beta \leq \infty) \tag{198}$$

Many other profiles of τ, including two- and three-term ones, are presented in the cited monograph by Fedyaevskiy, Ginevskiy, and Kolesnikov.

The mixing length l is also presented by different approximate expressions from which we note the following:

1. According to Prandtl

$$l = æy \quad \text{and} \quad \frac{l}{\delta} = 0.14 - 0.08\left(1 - \frac{y}{\delta}\right)^2 - 0.06\left(1 - \frac{y}{\delta}\right)^4$$

2. According to Kutateladze and Leontiyev

$$\frac{l}{\delta} = \frac{æy}{\delta} - (2æ - 3\gamma)\left(\frac{y}{\delta}\right)^2 + (æ - 2\gamma)\left(\frac{y}{\delta}\right)^3 \qquad æ = 0.39 \qquad \gamma = 0.14$$

3. According to the paper of the Stanford Conference (see the earlier reference)

$$\frac{l}{\delta} = \left(\frac{l_\delta}{l}\right) \operatorname{th}\left[\frac{æ\,(y/\delta)}{l_\delta/l}\right] \qquad æ = 0.41 \qquad \frac{l_\delta}{l} = 0.085$$

4. According to Chi and Chang

$$\frac{l}{\delta} = 0.4\frac{y}{\delta} - 0.5\left(\frac{y}{\delta}\right)^2 + 0.2\left(\frac{y}{\delta}\right)^3$$

5. According to Escudier and Spalding:

$$\frac{l}{\delta} = \frac{æy}{\delta} \quad \left[0 \le \frac{y}{\delta} \le \left(\frac{1}{æ}\right)\frac{l_\delta}{\delta}\right] \qquad \frac{l}{\delta} = \frac{l_\delta}{\delta} \quad \left[\left(\frac{1}{æ}\right)\left(\frac{l_\delta}{\delta}\right) \le 1\right]$$

There is some arbitrariness in the selection of the profiles of τ and l for the boundary layer cross-sections, which is sometimes moderated by additional intuitive considerations.

With the distributions τ/τ_w and l/δ being assigned, they are substituted into the main Prandtl equation (192). After integration the distribution of velocities u/U in the function of y/δ and of the parameter β are obtained. Then δ^*, δ^{**}, and H are found as the functions of β. Searching for $\beta(x)$ is the last stage in which the integral von Kármán relation (144) is used. In the case in which for the ratio τ/τ_w multiterm representations are taken, one has to use, along with Equation 144, the integral relations (of energy and others) following it. That is why the integral method (see above) is called the "method of integral relations".

We shall restrict ourselves to a short review, referring the reader for details and examples to the cited monograph in which much weight is given to the integral method.

14. Drag of a Body Moving in Fluid. Profile Drag, Friction Drag, Pressure Drag

Calculation of a turbulent boundary layer forms the basis for the determination of body drag in its motion in a viscous fluid. Examine the following terms. Total (frontal) drag, composed of the profile drag, which in turn consists of friction drag and pressure drag, and of the induced drag. The last type of drag is associated with finite character of a body span (a non-plane character of a flow past a body), which results in a local lifting force that gives a non-zero projection onto the direction of a total free stream. The nature of induced drag is different from the first two components and that is why induced drag is not considered here.

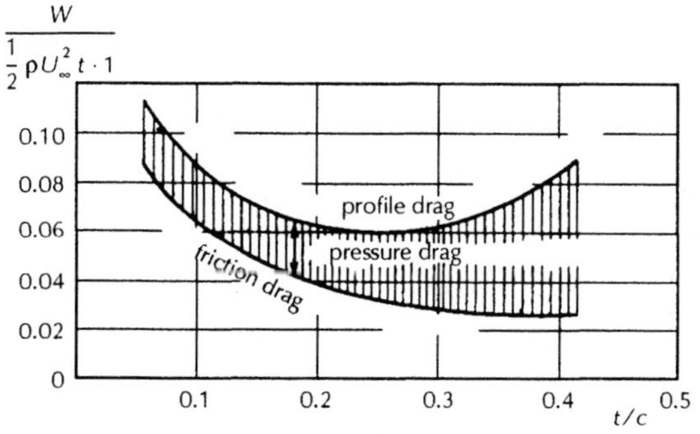

Fig. 243

Friction drag is determined as the projection onto the direction of motion of the resultant of tangential forces applied from the liquid side to the body surface. Pressure drag, correspondingly, is determined by an analogous projection of the result of normal forces.

We further discuss this second component of the profile drag. According to the concept, generally for both laminar and turbulent boundary layers, a flow outside the boundary layer region may be considered an irrotationally moving ideal (i.e., free of viscous friction) liquid. With a thin enough boundary layer and its familiar property to transmit, by the layer sections, the pressure of the outer flow with respect to the boundary layer, to the body surface, the result of normal forces is essentially equal to zero (following the D'Alembert paradox (Section 73)), and consequently pressure drag should also differ only slightly from zero. This would be close to reality if the boundary layer did not disturb the outer irrotational flow. In fact, stream lines, due to the wall retarding effect, are displayed away from the wing surface. Such a distortion of the flow pattern leads to the violation of the ideal pressure distribution over the wing profile.

The boundary layer is not only controlled by an outer flow but also exerts an inverse effect on it. Strictly speaking, it is impossible to *a priori* prescribe the distribution of pressures or velocities in the outer flow because this distribution is, in turn, dependent on the boundary layer development. Consequently, it is the function of the Reynolds number, the surface roughness, and other factors. However, if the body is streamlined without stallings and Reynolds numbers are rather large, then neglect of the boundary layer inverse effect on the distribution of pressures and velocities in the outer flow is permitted. The boundary layer inverse effect on the outer flow is especially strong in those portions of the boundary layer where it is the thickest, as, for example, near the separation point.

Figure 243, for comparison, shows the curves of the dependence of profile drag and friction drag coefficients of the Zhukovskiy symmetric profiles on their relative thickness. On the diagram the resistance force is related to the wing middle section rather than to the area in the plane. This explains why the coefficients of profile drag and friction drag grow sharply with the reduction of the relative thickness. The difference between the coefficients of profile drag and friction drag, shown by vertical dashed lines, determines the coefficient of pressure drag. Studying the diagram composed at the fixed Reynolds number ($U_\infty c/\nu = 4 \cdot 10^5$) leads to the obvious conclusion concerning the growth of the role of the pressure drag with the relative thickness of the profile and, conversely, on the growth of the friction drag importance wing pass over to thin profiles.[108]

[108]This problem is considered more thoroughly in the book *Sovremennoe sostoyaniye gidroaerodinamiki vyazkoy zhidkosti* (*Modern State of Viscous Fluid Hydroaerodynamics*), Vol. 2, Goldstein, S., Ed., Inostrannaya Literatura Press, Moscow, 1948, 78–85 (in Russian).

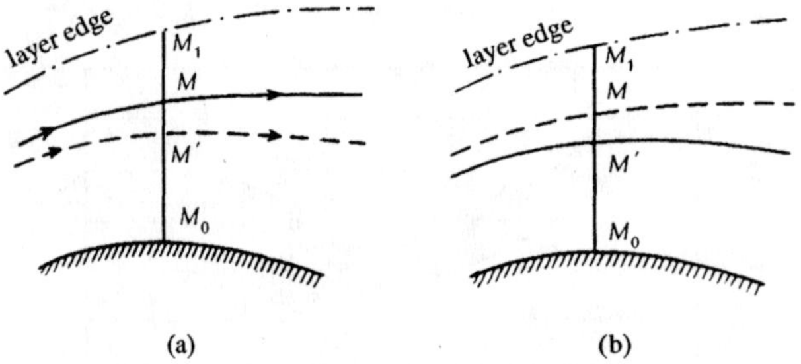

Fig. 244

As is shown by the experiments, pressure drag of a perfectly streamlined wing profile in the presence of a fully laminar and fully turbulent boundary layer on the wing surface decreases with the growth of the Reynolds number. This is natural because as the Reynolds number grows the boundary layer thickness decreases and the outer flow approaches the irrotational flow of an ideal fluid past the profile.

To elucidate the causes of the distortion of a theoretical pressure distribution in the irrotational flow of an ideal fluid past the wing profile, compare any real streamline (Figure 244 a, a solid line) which comes to the point M of the given cross-section M_0M_1 of the boundary layer and a streamline of the irrotational flow of an ideal fluid (shown in Figure 244 a by a dashed line) that coincides with the previous one at a distance in front of a body. The Section MM' represents the displacement of a real streamline relative to an ideal one. From the condition of the equality of the fluid volumetric flow rate in the compared motions through the cross-sections $M_0M = y$ and $M_0M' = y - MM'$, i.e., the consequence of the coincidence of both streamlines at a distance upstream of the body, we conclude that (U stands for the longitudinal velocity on the layer outer edge)

$$\int_0^y u\,dy = U(y - MM')$$

When formulating the right side of this equality, we take into account the fact that through a small thickness of the layer the velocity in an irrotational flow of an ideal fluid may be taken to be constant and equal to U. According to the last equality the mentioned displacement of a streamline at the point M with the coordinate y is

$$MM' = \int_0^y \left(1 - \frac{u}{U}\right) dy$$

On the surface of a streamlined body ($y = 0$) the streamline displacement disappears; the zero streamline, which is common for both flows, real and ideal, coincides with the wing surface. At a distance from the wing surface displacements of actual streamlines increase with respect to ideal ones. Finally, at a boundary layer edge ($y = \delta$) the value of displacement attains its maximum[109]

[109] It may be assumed that $\delta = \infty$.

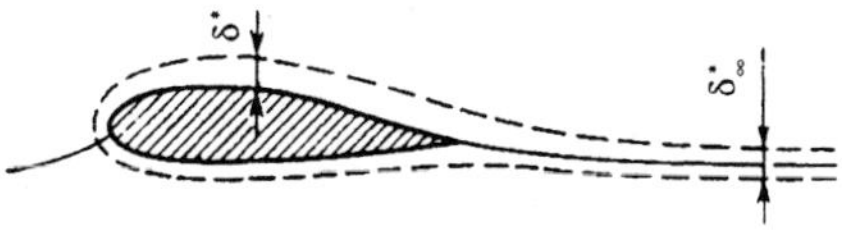

Fig. 245

$$(MM')_{y=\delta} = \int_0^\delta \left(1 - \frac{u}{U}\right) dy = \delta^*$$

Thus, the displacement of an actual streamline relative to that of an irrotational flow of an ideal fluid past a body on the boundary layer outer edge is equal to the displacement thickness δ^*.

We shall prove that the real distribution of pressure over the wing profile surface with a plane viscous fluid flow past this profile coincides with the pressure distribution in the case of an irrotational flow of an ideal fluid past a semi-body (Figure 245) formed by building up the displacement thickness onto the wing profile and on both sides from the zero streamline in its wake that was calculated by the actual pressure distribution; the contour of this semibody will be called "effective".

Assume that there is prescribed a plane flow of a real (viscous) fluid past a wing profile that is accompanied by the formation of a boundary layer on a body and of an aerodynamic wake behind the body. Along with this real flow in a boundary layer we consider a fictitious potential flow which would be the continuous extension of a real outer potential flow onto the region occupied by a boundary layer.

According to the known property of a boundary layer, pressures and longitudinal velocities in the potential flow (above) should coincide with pressures and velocities in the flow on the outer edge of the boundary layer. Instead of the decrease in the velocity from some value on the layer outer edge to the zero value on the wing surface that is typical of the flow in a boundary layer, in the potential flow (equivalent by pressures) the velocity will be identical everywhere on the given normal that is equal to the velocity on the layer outer edge. Hence, the considered potential flow possessing a volumetric flow rate through the cross-section of the considered jet similar to that of a real flow in a boundary layer cannot fill in the entire boundary layer region (with an aerodynamic wake included in the concept of a boundary layer).

To determine a new flow region we consider (Figure 244b) some point M of the boundary layer cross-section. We shall designate the real streamline passing through the point M by a solid line, and the streamline of the potential flow coinciding with the real one (see above) at a distance upstream of the body by a dashed line coming into some point M' on the same normal.

We emphasize the difference between the fictitious potential flow coinciding with the real one everywhere outside the boundary layer, and the earlier considered potential flow possessing a common zero streamline with the real one.

As seen from Figure 244, real streamlines lie in one case above the ideal ones, and in the other behind them.

Formulating the condition of the equality of the flow rate in real and fictitious potential flows through the cross-sections M_1M and M_1M' reckoned from the outer edge of a boundary layer,

$$\int_y^\delta u\,dy = U(\delta - y - MM')$$

so that the distance between the compared streamlines in real and fictitious flows will be equal to

$$MM' = \int_y^\delta \left(1 - \frac{u}{U}\right)dy$$

At the edge of a boundary layer ($y = \delta$) $MM' = 0$ the streamlines will coincide. Going deeper into the boundary layer the value of MM' will increase and the fictitious streamlines will be displaced from the wing surface. When the real streamline finally coincides with the surface of a wing profile, the streamlines of the fictitious irrotational flow will be displaced from the surface at a distance

$$(MM')_{y=0} = \int_0^\delta \left(1 - \frac{u}{U}\right)dy = \delta^*$$

which is equal to the displacement thickness.[110] Thus, the main zero streamline of a real flow (Figure 245), bifurcating at a leading stagnation point of the body contour, coinciding with the body contour, and passing through an aerodynamic wake into the infinity behind the body, in a fictitious irrotational flow should be replaced by some infinite semibody formed by building up the displacement thickness which is calculated by the actual pressure distribution along the normal to the zero streamline.

In Figure 245 the main profile and the zero streamline behind it are shown by a straight line and the "effective" contour whose streamlining by a potential flow is equivalent in pressure distribution to that by a real fluid. A fictitious irrotational flow entering the boundary layer through its outer edge (not shown in the figure) with the velocities equal to those of the real flow (this fictitious flow is then not affected by the frictional drag) has inside the boundary layer the velocities higher than in a real one. In this case the fictitious flow cannot occupy the entire boundary layer region. A portion of the plane between the zero stream line in the real flow and the semibody boundary in the fictitious flow remains free of fluid, and the line $y = \delta^*$ is the boundary streamline.

The given concepts on the simplest technique of allowing for the inverse effect of the boundary layer on the outer irrotational flow were presented assuming fluid incompressibility. We can easily convince ourselves that a similar technique also may be applied in the case of the boundary layer in a compressible fluid, in a gas flow with sub- and supersonic velocities. It is enough to repeat all these considerations, substituting only volumetric flow rates $\int_0^y u\,dy$ or $\int_0^\delta u\,dy$ by the corresponding mass flow rates $\int_0^y \rho u\,dy$ or $\int_0^\delta \rho u\,dy$ and having taken the quantity

[110] This may be treated as the origin of the term "the displacement thickness".

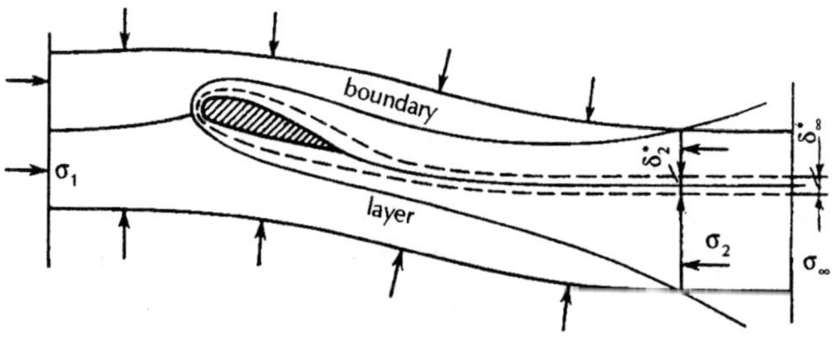

Fig. 246

$$\delta^* = \int_0^\delta \left(1 - \frac{\rho u}{\rho_\delta U}\right) dy$$

where ρ_δ is the density at the layer outer edge, as the displacement thickness. We also emphasize that all the formulations in the present section refer both to turbulent and laminar boundary layers.

Practical determination of the semibody shape and subsequent calculation of the pressure drag as the result of pressure forces of the irrotational flow of an ideal fluid on the semibody surface onto the free stream direction is very difficult. Therefore, it is preferable to determine pressure drag as the result of the subtraction of frictional drag from the profile drag. A simple approximate way of profile drag calculation will be given now. As to the frictional drag, it is found by the summation of friction elementary forces over the surface.

15. Approximate Formulae for the Profile Drag[111]

Consider an aerofoil profile (Figure 246) in an unbounded stream of a fluid of density ρ whose velocity at infinity is U_∞. Again compare two equivalent (with regard to the pressure distribution) flows: (1) the actual flow, including the boundary layer and the wake, and (2) the fictitious irrotational flow of an ideal fluid past the equivalent half-body, shown by a dotted line in Figure 246; the two flows are identical outside the boundary layer.

Take any section σ_2 spanning the wake and perpendicular to the free stream direction and draw the external flow streamlines passing through the points where the wake boundaries intersect σ_2; these streamlines form a streamtube enclosing the body.

Let σ_1 be a cross-section of this tube far ahead of the body and parallel to σ_2. Then, applying the momentum theorem (Chapter 4) in the Euler form to the segment of the streamtube between σ_1 and σ_2 in actual and fictitious flows, we have

1. For the actual flow:

$$\rho \int_{\sigma_1} u^2 dy - \rho \int_{\sigma_2} u^2 dy - R_x + X_p = 0$$

[111] It is assumed that the flow past an aerofoil profile occurs without separation of a boundary layer.

2. For the fictitious flow:

$$\rho \int_{\sigma_2} u^2 dy - (\sigma_2 - \delta_2^*)\, \rho u_2^2 - R_{ix} + X_p = 0$$

In these equations R_x denotes the drag of the aerofoil profile in the actual flow; i.e., the required profile drag, R_{ix} the pressure drag of the half-body up to the σ_2 section, X_p the x-component of the result of pressure forces acting on the surface of the stream tube (indicated by arrows in Figure 246), u_2 the velocity in the x-direction in the potential flow across the section σ_2, and δ_2^* the displacement thickness of the wake at σ_2. Subtracting termwise the left sides of the two equations, we have

$$(\sigma_2 - \delta_2^*)\, \rho u_2^2 - \rho \int_{\sigma_2} u^2 dy + R_{ix} - R_x = 0$$

Noting that from the definition of the displacement thickness δ^*

$$\sigma_2 - \delta_2^* = \sigma_2 - \int_{\sigma_2} \left(1 - \frac{u}{u_2}\right) dy = \int_{\sigma_2} \frac{u}{u_2}\, dy$$

we find from the preceding equation that

$$R_x = \rho \int_{\sigma_2} u\,(u_2 - u)\, dy + R_{ix}$$

Suppose now that the section σ_2 is infinitely far downstream of the body. Then, as was shown in Chapter 9, the pressure drag of the half-body shown dotted in Figure 246 is zero, because its thickness at infinity tends to the finite value of δ_∞^*. Taking the limit of the last equation we find that

$$R_x = \lim_{\sigma_2 \to \sigma_\infty} \rho \int_{\sigma_2} u\,(u_2 - u)\, dy + \lim_{\sigma_2 \to \sigma_\infty} R_{ix} = \rho \int_{-\infty}^{\infty} u\,(U_\infty - u)\, dy = \rho U_\infty^2 \delta_\infty^{**}$$

where δ_∞^{**} is the momentum thickness of the wake infinitely far downstream of the body. Therefore, the profile drag coefficient C_{xp} in terms of the momentum thickness at infinity δ_∞^{**} is

$$C_{xp} = \frac{R_x}{\frac{1}{2}\rho U_\infty^2 c} = 2\frac{\delta_\infty^{**}}{c} \qquad (199)$$

where c is the wing chord.

It is not possible to directly determine the value of δ_∞^{**} and we shall, therefore, develop an approximate relation between δ_∞^{**} and the momentum thickness at the trailing edge of the aerofoil[112] which can easily be determined theoretically or experimentally. For this purpose we recall the derivation of the von Kármán integral relation (107) given in Chapter 12. Similar derivation applied to equations of a turbulent boundary layer (143) led to the same integral relation (144). Employing this relation to the case of a wake, we shall integrate within the limits $(-\infty, \infty)$. Then, the term τ_w does not appear on the right side of the equation, and the von Kármán integral relation will take on the form

$$\frac{d\delta^{**}}{dx} + \frac{2}{U}\frac{dU}{dx}\delta^{**} + \frac{1}{U}\frac{dU}{dx}\delta^{*} = 0 \tag{200}$$

Dividing both sides of this equation by δ^{**} and integrating with respect to x from the trailing edge (subscript k) to infinity (subscript ∞) we get

$$\ln\frac{\delta_\infty^{**}}{\delta_k^{**}} = \ln\frac{U_k^2}{U_\infty^2} - \int_{x_k}^{\infty}\frac{\delta^*}{\delta^{**}}\frac{U'}{U}dx$$

Restricting ourselves to low values of the pressure gradient outside the boundary layer (i.e., assuming the aerofoil to be thin, of small camber, and at low incidence), we replace the ratio $\delta^*/\delta^{**} = H$ under the integral sign by its mean value

$$H_{\text{mean}} = \frac{1}{2}(H_k + H_\infty)$$

Completing the integration we have

$$\ln\frac{\delta_\infty^{**}}{\delta_k^{**}} = \ln\frac{U_k^2}{U_\infty^2} - \frac{1}{2}(H_k + H_\infty)\ln\frac{U_\infty}{U_k}$$

or,

$$\frac{\delta_\infty^{**}}{\delta_k^{**}} = \left(\frac{U_k}{U_\infty}\right)^{2+\frac{1}{2}(H_k+H_\infty)} \tag{201}$$

Substituting this value of δ_∞^{**} into Equation 199 we obtain the following approximate formula for the profile drag coefficient

$$C_{xp} = 2\left(\frac{U_k}{U_\infty}\right)^{2+\frac{1}{2}(H_k+H_\infty)}\frac{\delta_k^{**}}{c} \tag{202}$$

Measurements of pressure distributions on wing profiles show that the ratio U_k/U_∞ is little different from unity if the profile is relatively thin, of low camber, and at a small angle of incidence (so that the lift coefficient c_y is small). The last condition is satisfied during flight at speeds near maximum. Thus, for example, on a 14% thick profile (NACA 2414) at $c_y = 0.18$, U_k/U_∞ is equal to 0.9, i.e., by 10% is smaller than unity; even on a 25% thick profile at $c_y = 0.25$ this ratio departs from unity by not more than 25%.

The value of H_∞ can be shown to be equal to unity by substituting $u = U - u' = U_\infty - u'$ into the expressions for δ^* and δ^{**} and neglecting the second and higher powers of the velocity perturbation u'. As has already been stated, the value of H near the point of

---

[112] Squire, H., and Young, A., Calculation of the aerofoil profile drag, in *K voprosu o maksimalnoy skorosti samolyota* (*On Maximum Velocity of a Plane*), Oborongiz, Moscow, 1941.

minimum pressure is between 1.3 and 1.4; in the absence of separation this increases to $H =$ 1.8 to 2.3 at the trailing edge.

At first sight it would appear that one is faced with the difficult task of determining very accurately the value of U_k/U_∞, since this ratio is raised to a fairly high power (between 3.2 and 3.4) in Equation 202. However, it can be easily shown from Equation 175 that δ_k^{**}/c is inversely proportional to U_k/U_∞ raised to a power which is little different from the power of U_k/U_∞ in Equation 202. Thus, C_{xp} is proportional to a low power of U_k/U_∞. This explains why Equation 202, which for $H_\infty = 1$ and $H_k = 1.4$ is identical to the well-known Squire and Young formula,

$$C_{xp} = 2\left(\frac{U_k}{U_\infty}\right)^{3.2}\frac{\delta_k^{**}}{c} \tag{203}$$

is in good agreement with the experiment.

The two empirical constants appearing in Equation 174 are found to be $a = 1.17$ and $b =$ 4.7 to 4.8. We shall retain this value of a and put $b \approx 5$ which considerably simplifies the calculation. Then, taking $G(\mathbf{Re}^{**}) = 153.2\mathbf{Re}^{**1/6}$ we get

$$\frac{\delta^{**}}{c} = \frac{0.0153}{\mathbf{Re}^{1/7}}\frac{1}{(U/U_\infty)^{25/7}}\left[\int_{x_t/c}^{x/c}\left(\frac{U}{U_\infty}\right)^4 d\left(\frac{x}{c}\right) + 131\mathbf{Re}^{1/6}\left(\frac{U_t}{U_\infty}\right)^{25/6}\left(\frac{\delta_t^{**}}{c}\right)^{7/6}\right]^{6/7}$$

where

$$\mathbf{Re} = \frac{U_\infty c}{\nu}$$

and c is the wing chord.

To find δ_k^{**}/c we apply this formula, with $x/c = x_k/c$, to the top and bottom surfaces of the aerofoil separately and add the results. Substituting the value of $(\delta^{**}/c)_{x = x_k}$ so obtained into the preceding formula, we can cancel the factor $(U_k/U_\infty)^{25/7} \approx (U_k/U_\infty)^{3.6}$ in the denominator with the factor $(U_k/U_\infty)^{2 + (1/2)(H_k + H_\infty)}$ in the numerator, since $2 + \frac{1}{2}(H_k + H_\infty) \approx 3.2$ to 3.4. This leads to the following formula for the profile drag coefficient

$$C_{xp} = \frac{0.0307}{\mathbf{Re}^{1/7}}\left[\int_{x_t/c}^{1}\left(\frac{U}{U_\infty}\right)^4 d\left(\frac{x}{c}\right) + 131\mathbf{Re}^{1/6}\left(\frac{U_t}{U_\infty}\right)^{25/6}\left(\frac{\delta_t^{**}}{c}\right)^{7/6}\right]^{6/7}$$

It should be remembered that this must be calculated separately for the two surfaces and the results added. It will be seen that the velocity distribution $U(x)$ occurs in this expression only under the integral sign. At high Reynolds numbers, the value of U at the transition point (U_t) differs little from the value of U at the point of maximum velocity (U_m). The fact that U_k and H_k, the values of U and H at the trailing edge, are not required, makes this formula very convenient to use. We note that in the case of a flat plate parallel to the stream ($U = \text{const} = U_\infty$) with a fully turbulent boundary layer, the above equation agrees with formula (153),

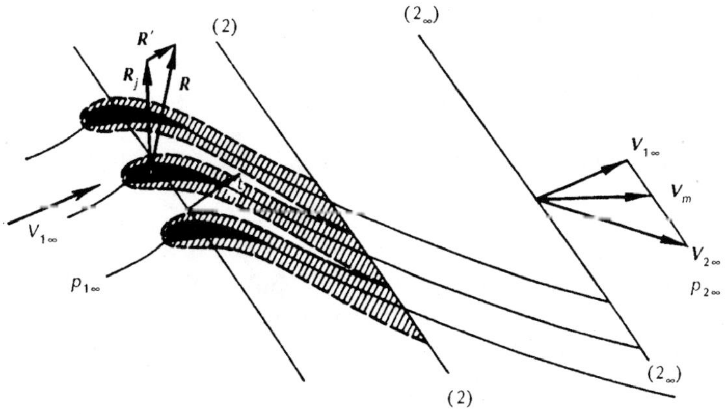

Fig. 247

provided one takes into account the difference between the definitions C_f and C_{xp}; the former is the average skin friction coefficient, or the friction drag coefficient of one side of the plate, while the latter is the profile drag coefficient of both sides of the plate.

The formula just derived is valid at high Reynolds numbers. At lower Reynolds numbers ($\mathbf{Re} < 5 \cdot 10^6$ approximately), when the skin friction law for smooth flat plates is given by Equation 155, the formula

$$C_{xp} = \frac{0.074}{\mathbf{Re}^{1/5}}\left[\int_{xt/c}^{1}\left(\frac{U}{U_\infty}\right)^{7/2} d\left(\frac{x}{c}\right) + 62.5\mathbf{Re}^{1/4}\left(\frac{U_t}{U_\infty}\right)^{15/4}\left(\frac{\delta_t^{**}}{c}\right)^{5/4}\right]^{4/5}$$

derived by Speidel,[113] should be used; as before the contributions of the top and bottom surfaces should be summed. Charts are available from which the profile drag coefficient can be determined, given the geometric parameters of the aerofoil profile, the position of the transition point and the Reynolds number of the flow.[114]

The method described above can be extended to the case of aerofoils in cascades.[115] We shall consider the plane flow past a cascade of aerofoils (Figure 247), with the pressure $p_{1\infty}$ and the velocity $V_{1\infty}$ at infinity upstream and the corresponding values of $p_{2\infty}$ and $V_{2\infty}$ at infinity downstream. Let ρ denote the density of the fluid and t the gap vector. Applying the momentum theorem, we find that the equation for the resultant force acting on an aerofoil in the cascade in viscous flow has the same form as the ideal flow case (Chapter 7)

$$\boldsymbol{R} = (p_{1\infty} - P_{2\infty})\boldsymbol{t} + \rho(\boldsymbol{t} \cdot \boldsymbol{V}_{1\infty})(\boldsymbol{V}_{1\infty} - \boldsymbol{V}_{2\infty})$$

However, because of the presence of the viscous dissipation the total pressure downstream of the case is no longer equal to the total pressure in the free stream; the loss of the total pressure across the cascade is equal to

$$p' = \left(p_{1\infty} + \frac{1}{2}\rho V_{1\infty}^2\right) - \left(p_{2\infty} + \frac{1}{2}\rho V_{2\infty}^2\right)$$

---

[113] See *H. Schlichting's Boundary Layer Theory*, Pergamon Press, London, 1955, 514.

[114] See Squire and Young, *loc. cit.*

[115] Lointsyanskiy, L. G., The drag of a cascade in a viscous incompressible flow, *Prikl. Mat. Mekh.*, 9(4), 1947.

The expression for the resultant force vector $\boldsymbol{R}$ may be written as

$$\boldsymbol{R} = \frac{1}{2}\rho\,(V_{2\infty}^2 - V_{1\infty}^2)\,\boldsymbol{t} + \rho\,(\boldsymbol{t}\cdot\boldsymbol{V}_{1\infty})\,(\boldsymbol{V}_{1\infty} - \boldsymbol{V}_{2\infty}) + p'\boldsymbol{t}$$

or, using the notation of Chapter 7

$$\boldsymbol{R} = \rho\,(\boldsymbol{V}_m\cdot\boldsymbol{V}_d)\,\boldsymbol{t} - \rho\,(\boldsymbol{t}\cdot\boldsymbol{V}_m)\,\boldsymbol{V}_d + p'\boldsymbol{t} = \rho\boldsymbol{V}_m\times(\boldsymbol{t}\times\boldsymbol{V}_d) + p'\boldsymbol{t}$$

The first term on the right side represents the "Zhukovskiy force" and may be denoted by $\boldsymbol{R}_j$, the second term is the profile drag which will be denoted by $\boldsymbol{R}'$. Thus,

$$\boldsymbol{R} = \boldsymbol{R}_j + \boldsymbol{R}'$$

where the drag $\boldsymbol{R}'$ is simply related to the total pressure loss p'

$$\boldsymbol{R}' = p'\boldsymbol{t}$$

The determination of the cascade profile, as compared to that of an isolated aerofoil, is complicated by the fact that at some distance downstream of the cascade the individual wakes join together (see Figure 247) to form a region of flow which cannot be described by the equations of a boundary layer type. We shall denote the section in which this occurs by the single subscript 2 (without ∞) and assume that the velocity perturbations at that section are small. It can be easily shown[115] that the total pressure loss is then given by

$$p' = \rho V_{2\infty}^2\,\frac{\delta_2^{**}}{t\cos\beta_{2\infty}}$$

where δ_2^{**} is the momentum thickness at the section 2 and $\beta_{2\infty}$ is the angle between the directions of $\boldsymbol{V}_{2\infty}$ and $\boldsymbol{t}$. Relating the conditions at the section 2 to those at the trailing edges ($\delta^{**} = \delta_k^{**}$, $U = U_k$) by the same method as was used in the case of an isolated aerofoil, we obtained the following formulas for the total pressure loss p' and the drag force $\boldsymbol{R}'$

$$p' = \rho V_{2\infty}^2\left(\frac{U_k}{V_{2\infty}}\right)^{3.2}\frac{\delta_k^{**}}{t\cos\beta_{2\infty}} \qquad R' = \rho V_{2\infty}^2\left(\frac{U_k}{V_{2\infty}}\right)^{3.2}\frac{\delta_k^{**}}{\cos\beta_{2\infty}}$$

These expressions contain $\boldsymbol{V}_{2\infty}$, the velocity at infinity downstream, rather than $\boldsymbol{V}_m$, the mean velocity vector commonly used in the theory of cascades. Noting that $V_{2\infty}\cos\beta_{2\infty} = V_m\cos\beta_m$, where β_m is the angle between the directions of $\boldsymbol{V}_m$ and $\boldsymbol{t}$, we have

$$p' = \rho V_m^2\left(\frac{U_k}{V_m}\right)^{3.2}\left(\frac{V_m}{V_{2\infty}}\right)^{0.2}\frac{\delta_k^{**}}{t\cos\beta_m} \tag{204}$$

and a corresponding equation for the drag.

The mean velocity $\boldsymbol{V}_m$ may be regarded as an effective free stream velocity so that the drag force is then the component $\boldsymbol{D}$ of $\boldsymbol{R}'$ in the direction of $\boldsymbol{V}_m$

$$D = R'\cos\beta_m = \rho V_m^2\left(\frac{U_k}{V_m}\right)^{3.2}\left(\frac{V_m}{V_{2\infty}}\right)^{0.2}\delta_k^{**} \tag{205}$$

and the corresponding drag coefficient becomes

$$C_D = \frac{D}{\frac{1}{2}\rho V_m^2 c \cdot 1} = 2\left(\frac{U_k}{V_m}\right)^{3.2}\left(\frac{V_m}{V_{2\infty}}\right)^{0.2}\frac{\delta_k^{**}}{c} \tag{206}$$

Equation 206 differs from the corresponding expression (203) for an isolated aerofoil only by the factor $(V_m/V_{2\infty})^{0.2}$, whose value in practice is close to unity.

A comparative simplicity of the method just presented for calculation of profile drag is attributed to the integral character of the latter.[116] Separate components of profile drag, the skin friction drag and pressure drag, cannot be calculated so easily.

As follows from the above, the determination of profile drag by approximate formulas requires only the knowledge of momentum loss thickness δ_k^{**} at the aerofoil trailing edge. This makes it possible to restrict ourselves to approximate empirical and semiempirical methods for calculating a turbulent boundary layer presented in this section, which provide an integral value of the momentum loss thickness δ^{**} with accuracy sufficient for practice. Note that these approximate computational methods are insufficient for the determination of the local coefficient of friction on the surface of a streamlined aerofoil profile and consequently, the total frictional resistance.

One should bear in mind that in the formulation of empirical methods for calculating a turbulent boundary layer with arbitrary distribution of pressure use was made of the quantities applied for calculations in the case of constant pressure. This drawback is especially critical on strongly diverging ones ($dp/dx \gg 0$) and, in particular, on preseparation zones of a boundary layer.

This fact is confirmed by experiments on the determination of losses in turbine and compressor cascades.[117] The calculations of drag of turbine cascades, in which a role played by diverging segment is small made according to Equation 206, agree rather satisfactorily with test data. Difficulties arose when attempts were made to calculate drags of compressor cascades by the same approximate formulas in which great longitudinal pressure gradients are observed. In these cases we required the more accurate methods for calculation of turbulent boundary layers that are considered in the following chapter.

Hydrodynamic and hydraulic calculations of some turbulent flows that are encountered in engineering practice (curvilinear channels, cascades, divergent ducts, vortex chambers, etc.) can be found in special monographs.[118]

---

[116] Another approach to the calculation of flow parameters downstream the cascade is presented in the book by G. Yu. Stepanov, *Gidrodinamika reshyotok turbomashin* (*Hydrodynamics of Cascades in Turbines*), Fizmatgiz, Moscow, 1962, Sections 52 and 54.

[117] Zysina-Molozhen, L. M., Calculations of losses in turbine cascades, in *Aerogidrodinamika* (*Aerohydrodynamics*), TsKTI Book 27, Mashgiz, Moscow, 1954. The results of investigations of losses in turbine rotors and guide vanes, including some results obtained by the present author, are collected in I. I. Kirillov's *Gas Turbines and Gas Turbine Plants*, Mashgiz, Moscow, 1956, 134 166.

[118] Idelchik, I. E., *Aerodinamika tekhnologicheskikh apparatov* (*Aerohydrodynamics of Engineering Apparatuses*), Mashinostroenie, Moscow, 1983; Goldshtik, M. A., *Vikhrevye potoki* (*Vortex Flows*), Nauka, Novosibirsk, 1981, chap. 4; Bradshaw, P., Cebeci, T., and Whitelaw, J., *Engineering Calculation Methods for Turbulent Flow*, Academic Press, London, 1981, 210.

14

Some Modern Methods for Calculation of a Turbulent Boundary Layer in an Incompressible Fluid

1. "Moment" Methods of the Turbulent Boundary Layer Theory

The methods of turbulent boundary layer calculation presented in the previous chapter cannot satisfy modern practice in full measure because they make it possible to calculate mainly the boundary layer integral quantities such as the displacement and momentum thickness. The most important boundary layer characteristics – primarily skin friction distribution on a solid boundary – are determined by these methods with insufficient authenticity.

The methods which have appeared recently are, by their potential, substantially superior to those stated in Chapter 13, thus making it possible to calculate not only surface friction, but also characteristics of turbulent boundary layers. It is quite natural that the possibilities of these methods can be limited by certain conditions. Many of the methods were developed as applied to specific classes of problems so that they are not universal.

At present the number of publications concerned with the methods of turbulent boundary layer calculation is so great that we have no possibility to consider them thoroughly within this book. In this section common properties of a wide class of the methods based on "moment" equations of the two first orders are briefly discussed, while specific applications of some of them are reviewed.

The momental equations of this or that order were introduced in Chapter 13 as the momentum equations of averaged motion of Reynolds stresses and of the kinetic energy of velocity fluctuations, but they also admit different content corresponding to other characteristics of turbulent motions (dissipation velocity, coefficient of turbulent friction, etc.). All of them are based on the averaging of the Navier-Stokes equations of incompressible viscous fluid flow, reduced to different forms and in this conditional sense (because the techniques of averaging themselves are conditional) they can be considered rigorous consequences of the

Navier-Stokes equations. Thus, it is no wonder that momental equations of different orders are considered a reliable basis for modern methods of calculation of turbulent boundary layers.

The structure of momental equations is complex. Equations, used for determining the first order moments, involve unknown moments of the second order and equations for the second order moments the moments of the third order, and so on. This makes the systems of momental equations unclosed; thus, they require additional assumptions – "connections". Besides, the momental equations include different correlations of fluctuating velocities and pressures which require additional empirical and sometimes simply intuitive "model presentations".

Comparatively simple are the momental equations of the first order. These are the equations of momentum transfer (Equation 16, Chapter 13). The main unknown quantities in them are averaged velocities and pressures, but they also include the second order moments, i.e., Reynolds stresses. To close the system of the first order momental equations one has to resort, within the entire boundary layer region, to the Boussinesq hypothesis of "turbulent viscosity" (Equation 51, Chapter 13). In the vicinity of the solid surface the Prandtl "mixing length" theory is used with the correction for a damping factor, while at a distance from the wall use is made of the Clauser hypothesis (see below). More complex is the matter with the equations for the second order moments involving the third order moments, thus making one use for their determination the so-called "model relations", which take the initial equations into the sphere of intuitive conjectures and assumptions based on the considerations of the dimensional theory (Chapter 10). Such are the equations of the transfer of Reynolds stresses (32) or (33) and also of the transfer of kinetic energy of fluctuations (Equation 35, Chapter 13).

The justification of the possibility of using these and other, like methods based on the equations of moments of different orders is the confirmation of their applicability for solving the more complex problems.

Variety and complexity of the modern methods for calculating turbulent boundary layers make us restrict ourselves to a short, nearly review presentation only of some of them and to recommend the following collected papers and monographs[1] for expanding the knowledge of these and other methods.

2. Method of the Moments of the First Order. Two-Layer Procedure

By the method of the first order moments for calculation of a turbulent boundary layer in a plane stationary flow we understand the problem of solving the system of equations

$$u\frac{\partial u}{\partial x} + v\frac{\partial u}{\partial y} = -\frac{1}{\rho}\frac{dp}{dx} + \frac{\partial}{\partial y}\left(\nu\frac{\partial u}{\partial y} - \overline{u'v'}\right)$$

$$\frac{\partial u}{\partial x} + \frac{\partial v}{\partial y} = 0 \tag{1}$$

[1] *The proceedings of Soviet workshops* in: *Turbulentnye techeniya (Turbulent Flows)*, V. V. Struminskiy, Ed., Izd. Nauka, Moscow,1974 and 1977; *Mekhanika turbulentnykh potokov (Mechanics of Turbulent Flows)*, Izd. Nauka, Moscow, 1980; *Metody raschyota turbulentnykh techeniy (Methods for Calculation of Turbulent Flows)*, Kollman, Ed., Mir Publishers, Moscow, 1984 (in Russian); *Turbulentnye sdvigovye techeniya (Turbulent Shear Flows)*, Mashinostroeniye, Moscow (Collected Papers No.1), 1982; (Collected Papers No.2), 1983; *Turbulentnost, printsipy i primeneniya (Turbulence, Principles and Applications)*, W. Frost and T. Moulden, Eds., Mir Publishers, Moscow, 1980 (in Russian); *Turbulentnost (Turbulence)*, P. Bradshaw, Ed., Mashinostroeniye, Moscow, 1982 (in Russian); Bradshaw P., Cebeci T., and Whitelaw J., *Engineering Calculation Methods for Turbulent Flow*, Academic Press, New York, 1981.

which is the system of Equations 15 of Chapter 13 simplified by ordinary assumptions of the boundary layer theory with the boundary condition) typical of the boundary layer where u, υ, p are the averaged values of the projections of actual velocity and pressure, u' and v' are the fluctuational velocities and the only Reynolds stress $(-\rho\overline{u'v'})$ is supposed to be expressed in terms of the main unknowns.

In the recent literature the method of the first order is often considered comparatively rough and does not allow one to determine internal characteristics of a turbulent boundary layer, but it approximately gives the profiles of averaged velocities, skin friction on a solid boundary, and conditional layer thickness. However, this method has been widely applied for several decades and, probably, being used skilfully, it can give satisfactory results. The method is most suitable for those flows which are determined by one scale of lengths and by one scale of velocities. "Two layer" calculation procedures adopted in the method of moments of the first order comply with these requirements. According to this procedure the entire boundary layer region in the direction perpendicular to the flow is divided into two subregions: (1) a "near-wall" one, bounding with the streamlined body surface, and (2) an "outer" one, occupying the remainder of the layer. The "near-wall" subregion is thinner than the outer one; thus, in the case of a nongradient outer inviscid flow (longitudinal flow past a plate) each of them accounts for about 20 and 80% of a nominal boundary layer thickness.

The "near-wall" subregion consists of a viscous sublayer, transitional section, and a logarithmic portion (Chapter 13) which obey the "wall law"; the "outer" subregion is controlled by the so-called "wake law".

For each of these subregions separately their own scales of lengths and velocities are typical. In a nongradient boundary layer these are (see Chapter 13) "universal" scales of lengths $l_* = \nu/v_*$ and velocities $v_* = \sqrt{\tau_w/\rho}$ for the "near-wall" subregion and δ^* or δ^{**} and U (the velocity on the outer edge of the boundary layer) for the "outer" subregion. The choice of scales in a general case of the boundary layer with converging $(dp/dx < 0)$ and diverging $(dp/dx > 0)$ portions requires special consideration (see Section 6).

3. "Near-Wall" Subregion of a Turbulent Boundary Layer. Interaction between Molecular and Turbulent Viscosities

The main problem in the theory of a "near-wall" subregion is the establishment of a continuous smooth profile of the averaged velocity in the boundary layer cross-sections instead of the division of a "near-wall" subregion into a viscous sublayer and a turbulent flow core. This continuous distribution of velocities should be valid through the entire "near-wall" subregion of a solid wall up to the lower limit of the "outer" subregion of the boundary layer.

One of possible constructions of such a velocity profile is the following.[2] Remaining within the framework of the problem of a plane shear flow (Chapter 13) with the typical constancy of a skin friction τ over the entire cross-section, i.e., equal to its value on the wall $(\tau = \tau_w)$, from the main equality

$$\tau = \mu \frac{du}{dy} - \rho\overline{u'v'} \tag{2}$$

[2]Loitsyanskiy, L. G., Turbulent fluid flow and the internal problem, *Izv. NIIG*, 9, 1933, and Semi-empirical theories of the interaction between the processes of molecular and molar exchanges in a turbulent fluid flow, *Tr. Vsesoyuznogo Teor. Prikl. Mekh.*, AN SSSR, Moscow-Leningrad, 1962, 145–166. Modern presentation of this problem is given in Damping factor to the Prandtl equation for a transition portion of a turbulent boundary layer, *Inzh.-Fiz. Zh.*, 45(6), 924–932, 1983.

we obtain according to the Boussinesq hypothesis (Equation 43, Chapter 13)

$$\tau = \mu\frac{du}{dy} + \mu_t\frac{du}{dy} = \mu\left(\frac{du}{dy}\right)\left(1 + \frac{\nu_t}{\nu}\right) \tag{3}$$

Here, by μ_t and $\nu_t = \mu_t/\rho$ we respectively understand dynamic and kinematic coefficients of "turbulent viscosity", which, unlike the Prandtl μ_{tP} and ν_{tP}, allow for mutual influence of the processes of molecular and molar (turbulent) momentum transfer. We shall retain for τ_t and μ_t the Prandtl form of the law of turbulent friction with the mixing length l being indeterminate for the present

$$\tau_t = \mu_t\frac{du}{dy} = \rho l^2\left(\frac{du}{dy}\right)^2 \qquad \mu_t = \rho l^2\frac{du}{dy} \tag{4}$$

However, we distinguish them from the Prandtl one, which do not depend on molecular viscosity and which assume a linear law for l_P:

$$\tau_{tP} = \rho\, l_P^2\left(\frac{du}{dy}\right)^2 \qquad \mu_{tP} = \rho\, l_P^2\frac{du}{dy} \qquad l_P = æy \tag{5}$$

It directly follows from Equations 4 and 5 that

$$\tau_t = \rho\left(\frac{l}{l_P}\right)^2 l_P^2\left(\frac{du}{dy}\right)^2 = \left(\frac{l}{l_P}\right)^2\tau_{tP} \tag{6}$$

The first cofactor on the right side of the last equality

$$D = \left(\frac{l}{l_P}\right)^2 \tag{7}$$

expresses the some correction, which should be adopted in order to pass over from the classical Prandtl equation (5), disregarding the influence of molecular viscosity on turbulent exchange, to Equation 4. This correction (7), which represents the damping influence of viscosity on turbulent exchange (friction) in the form of a factor equal to the square of the ratio of the new mixing length l, involving the viscosity effect, to the old Prandtl l_P neglecting this effect, we shall take to be a "damping factor".

To find the dependence of the "damping factor" on the distance to a solid wall we shall resort to the theory of interaction between molecular and turbulent exchanges stated in the just-cited works. In these works the relation between a "transition" function

$$f(\mathbf{R}) = 1 + \frac{\nu_t}{\nu} \tag{8}$$

monotonically changing, from the value $f = 1$ directly on a solid wall to the value in a turbulent flow core

$$f = 1 + \frac{\nu_{tP}}{\nu} = 1 + l_P^2\frac{du/dy}{\nu} \tag{9}$$

with the "local Reynolds number"[3]

$$\mathbf{R} = l_P^2\frac{du/dy}{\nu} = æ^2y^2\frac{du/dy}{\nu} \tag{10}$$

is found.

In Equation 9 the fact is expressed that on the outer edge of the "transition portion" and outside it the kinematic coefficient of turbulent friction ν_t ceases to depend on viscosity and becomes equal to ν_{tP}. It follows from Equations 9 and 10 that the equality

$$f_\infty = 1 + \mathbf{R} \tag{11}$$

will be valid in this subregion. Equations 4, 7, and 8 yield

$$f - 1 = \frac{\nu_t}{\nu} = l^2 \frac{du/dy}{\nu} = \left(\frac{l}{l_P}\right)^2 l_P^2 \frac{du/dy}{\nu} = D\mathbf{R} \tag{12}$$

Introducing the universal variables known from the previous chapter

$$\frac{yv_*}{\nu} = \eta \qquad \frac{u}{v_*} = \varphi \qquad v_* = \sqrt{\frac{\tau_w}{\rho}} \tag{13}$$

we transform Equation 10 to the form

$$\mathbf{R} = æ^2\eta^2 \frac{d\varphi}{d\eta} \tag{14}$$

and according to the definition of the "transition" function f (Equations 3 and 8) we shall find in terms of these universal variables

$$\frac{\tau}{\tau_w} = f\frac{d\varphi}{d\eta} \tag{15}$$

Dividing on the basis of terms both sides of Equation 14 by the corresponding parts of Equation 15, we obtain after simple transformations

$$\mathbf{R}f(\mathbf{R}) = æ^2\eta^2\tau/\tau_w \tag{16}$$

We shall restrict ourselves to a particular case of purely shear quasi-rectilinear averaged motion in which $\tau = \text{const} = \tau_w$.[4] Then, returning to Equations 12 and 16 at $\tau = \tau_w$, we shall have the following parametric (in terms of the parameter **R**) determination of the "damping factor" $D(\eta)$ in the function of the universal coordinate η

$$D(\mathbf{R}) = \frac{f(\mathbf{R}) - 1}{\mathbf{R}} \qquad \mathbf{R}f(\mathbf{R}) = æ^2\eta^2 \tag{17}$$

Thus, the matter is reduced to the assignment of the form of the "transition" function f(**R**). As yet there is no simple rationally substantiated model of the phenomenon of transition

---

[3] The quantities f, **R**, and the concepts of "transition function" and "local Reynolds number" related to them were later used in the paper by G. L. Mellor, The effects of pressure gradients on turbulent flow near a smooth wall, *J. Fluid Mech.*, 24 (2), 255–274, 1966 without references to the report of the present author made at the 10th International Congress of Applied Mechanics in Stresa (Italy) in 1960 and which was published in the proceedings of the congress (Loitsyanskiy, L., Sur l'action reciproque de la transmission moleculaire et molaire dans l'ecoulement turbulent, in Proc. *10th Int. Congr. of Appl. Mech. Italy*, Stresa, 1960, Elsevier, Amsterdam, 1962, 202–204) 6 years before Mellor's work was published.

[4] To this case there will also correspond the flow in a transition portion in a nongradient ($dp/dx = 0$) turbulent boundary layer.

from a laminar motion in a "viscous sublayer" to a purely turbulent one in the flow "core". Remaining within a framework of a semiempirical approach to the description of this process, we determine it by two quantities: the position of the conditional "transition point" $\mathbf{R} = \mathbf{R}_{cr}$ and the measure of the conditional from the quantitative viewpoint concept on the extent of the "transition region" in the given cross-section of a turbulent boundary layer.[5]

We shall denote the value of the "transition" function in the transition region ($\mathbf{R} < \mathbf{R}_{cr}$) in terms of $f_0(\mathbf{R})$ and its value in the flow "core", i.e., outside the transition region ($\mathbf{R} > \mathbf{R}_{cr}$), in terms of $f_\infty(\mathbf{R})$.

$f(\mathbf{R})$ may be determined as a rough approximation based only on the assignment of the "transition point" position $\mathbf{R} = \mathbf{R}_{cr}$ with the aid of a discontinuity function

$$\operatorname{sgn} x = \begin{cases} 1 \text{ when } x > 0 \\ 0 \text{ when } x = 0 \\ -1 \text{ when } x < 0 \end{cases}$$

assuming

$$f(\mathbf{R}) = \frac{1}{2}(f_0 + f_\infty) + \frac{1}{2}(f_\infty - f_0)\operatorname{sgn}(\mathbf{R} - \mathbf{R}_{cr}) \tag{18}$$

In fact, in this case

$$f(\mathbf{R}) = f_0(\mathbf{R}) \quad \text{when} \quad \mathbf{R} < \mathbf{R}_{cr} \quad \text{and} \quad \operatorname{sgn}(\mathbf{R} - \mathbf{R}_{cr}) = -1$$

$$f(\mathbf{R}) = f_\infty(\mathbf{R}) \quad \text{when} \quad \mathbf{R} > \mathbf{R}_{cr} \quad \text{and} \quad \operatorname{sgn}(\mathbf{R} - \mathbf{R}_{cr}) = 1 \tag{19}$$

From the discontinuous determination of the "transition" function $f(\mathbf{R})$ by Equation 18 we can pass over to a continuous one recalling that by the definition of the function $\operatorname{sgn} x$, the equations will exist[6]

$$\operatorname{sgn} x = \lim_{\alpha \to \infty}\left(\frac{2}{\pi}\operatorname{arctg}\alpha x\right)$$

$$\operatorname{sgn} x = \lim_{\alpha \to \infty}\operatorname{erf}(\alpha x) \qquad \operatorname{erf} t = \frac{2}{\sqrt{\pi}}\int_0^t e^{-\xi^2}d\xi$$

$$\operatorname{sgn} x = \lim_{\alpha \to \infty}\operatorname{th}(\alpha x), \text{ etc.} \tag{20}$$

Having omitted the limit sign, we shall obtain approximate asymptotic expressions of the function $\operatorname{sgn} x$ at large α in the form of continuous functions. Thus, having retained the previous designations for the functions $f_0(\mathbf{R})$ and $f_\infty(\mathbf{R})$ and having used, for example, the last equation in the system (20), we have the following continuous expression for the "transition"

[5]Loitsyanskiy, L. G., Damping factor in the Prandtl equation for the transition portion of a turbulent boundary layer, *Inzh. Fiz. Zh.*, 45 (6), 924–932, 1983.

[6]Korn, G. and Korn, T., *Spravochnik po matematike dlya nauchnykh rabotnikov i inzhenerov* (*Reference Book on Mathematics for Scientific Workers and Engineers*), Nauka, Moscow, 1977 (in Russian).

function:[7]

$$f(\mathbf{R}) = \frac{1}{2}[f_0(\mathbf{R}) + f_\infty(\mathbf{R})] + \frac{1}{2}[f_\infty(\mathbf{R}) - f_0(\mathbf{R})]\,\mathrm{th}\,[\alpha(\mathbf{R} - \mathbf{R}_{cr})] \qquad (21)$$

This expression contains two free parameters $\mathbf{R}_{cr}$ and α. Dealing with them we can obtain different approximate expressions for a "transition" function. For $\mathbf{R}_{cr}$ we shall retain the magnitude $\mathbf{R}_{cr} = 11$ stated by Mellor. The increase of α will lead to narrowing of the "transition" portion. We shall take $\alpha = 100$ and retain for large $\mathbf{R}$ the previously mentioned asymptotics $f_\infty(\mathbf{R})$ (Equation 11). As to the asymptotics $f_0(\mathbf{R})$, for small $\mathbf{R}$, determine it by the expression

$$f_0(\mathbf{R}) = 1 + \gamma\mathbf{R}^2 \qquad (22)$$

justified experimentally by the fourth power law of decrease in the kinematic coefficient of turbulent viscosity ν_t as approaching the solid wall that will be prescribed by the equation

$$\frac{\nu_t}{\nu} = \gamma æ^4\eta^4 \qquad (23)$$

where $\gamma = 0.0092$ from Deissler's[8] experiments and $\gamma = 0.0125$ from Hanratty's[9] experiments, $æ = 0.4$ and η the universal ordinate equal to yv_*/ν.

We shall take for γ the rounded off value of $\gamma = 0.01$. According to Equation 21 the "transition" function $f(\mathbf{R})$ is determined by the interpolation expression

$$f(\mathbf{R}) = \frac{1}{2}(2 + \mathbf{R} + 0.01\mathbf{R}^2) + \frac{1}{2}(\mathbf{R} - 0.01\mathbf{R}^2)\,\mathrm{th}\,\alpha(\mathbf{R} - 11) \qquad (24)$$

and the damping factor $D(\mathbf{R})$ by the first equality of system (17) is equal to

$$D(\mathbf{R}) = \frac{1}{2}(1 + 0.01\mathbf{R}) + \frac{1}{2}(1 - 0.01\mathbf{R})\,\mathrm{th}\alpha(\mathbf{R} - 11) \qquad (25)$$

The set of the last two equations in combination with the second equality of system (17) is the parametric determination of the transition function f and of the damping factor D as the functions of η. Figure 248 present, the graph $D(\eta)$ in the form of the curve (first from the left). This curve, judging by the closeness of the "outer boundary" of the transition portion to the experimentally justified magnitude $\eta_1 \approx 35$, can be taken as the damping factor which is in satisfactory agreement with the test data. This is confirmed by the calculation of the velocity profile, the experimental shape of which is well known (see Figure 249).

Some difficulty in the parametric assignment of the transition function and of the damping factor is not usually an obstacle for their numerical determination. Nevertheless, an explicit form of these dependencies, which simplified practical calculations, is worth mentioning.

The greatest ground has been gained with regard to the van Driest formula

---

[7] In the paper by L. G. Loitsyanskiy, (see footnote 5) the first function $(2/\pi)$arctg of the system (20) was used instead of th, which did not noticeably affect the calculations.

[8] Deissler, R., Analysis of Turbulent Heat Transfer, Mass Transfer and Friction in Smooth Tubes at High Prandtl and Schmidt Numbers, NACA Rep. 1210, 1959.

[9] Hanratty, Th., Study of turbulence close to a solid wall, *Phys. Fluids, Suppl.*, 126–133, 1967.

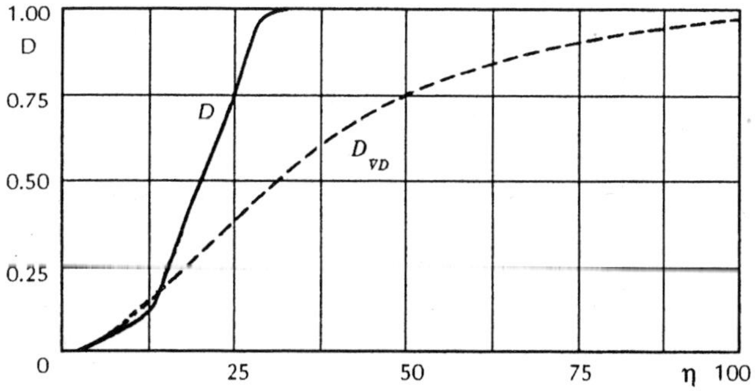

Fig. 248

$$D = [1 - \exp(-\eta/A_*)]^2 \qquad A_* = 26 \tag{26}$$

which is the exponential expression of the damping factor; here, A_* is the empirical coefficient and $\eta = yv_*/\nu$, $v_* = \sqrt{\tau_w/\rho}$.

The fact that this equation is derived on the basis of the considerations which have little in common with semiempirical theories of turbulence deserves our attention. The model adopted by van Driest rests on the analogy between two phenomena absolutely different in nature: damping of a turbulent oscillatory flow by viscosity and a laminar nonstationary viscous fluid flow near an infinite plate oscillating, harmonically. In this case, based on simple considerations of dimensionality, van Driest compared the "dynamic velocity" v_*, typical of a turbulent flow, with the square root of the product of the kinematic coefficient of viscosity of the angular frequency of plate oscillations. We suggest a new derivation of the van Driest equation (26) based on the above notion of the viscosity influence on the turbulent friction. We emphasize that in this derivation the necessity to introduce a new empirical constant A_* is removed. It is expressed in terms of the agreed-upon constances of turbulence æ and γ.

We shall take the transition function $f(\mathbf{R})$ and the damping factor $D(\mathbf{R})$, expressed in

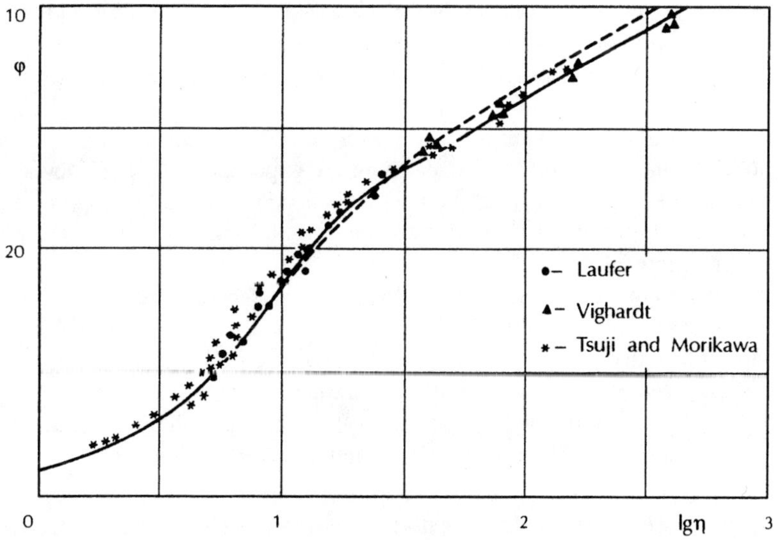

Fig. 249

terms of it, in the form

$$f(\mathbf{R}) = 1+\mathbf{R}\{1-\exp[-(\gamma\mathbf{R})^{1/n}]\}^n$$
$$D(\mathbf{R}) = \{1-\exp[-(\gamma\mathbf{R})^{1/n}]\}^n \qquad (27)$$

which at any n satisfies asymptotic equations (11) and (22).

If the second line of Equation 17 is substituted by the relation $\mathbf{R} = æ^2\eta^2$, which follows from it at $f(\mathbf{R}) = 1$, then, according to Equation 27, at $n = 2$ we obtain the expression of the damping factor

$$D(\eta) = [1-\exp(-æ\sqrt{\gamma}\eta)]^2 \qquad (28)$$

coinciding with the van Driest equation (26), if it is assumed in it that

$$A_* = 1/(æ\sqrt{\gamma}) \qquad (29)$$

At $æ = 0.4$ and $\gamma = 0.0092$ the value of the constant $A_* = 26$ adopted by van Driest is confirmed. This consideration throws new light on the van Driest equation and makes it possible to express the empirical constant A_* in terms of the main constants of turbulence æ and γ. Note that at $n = 1$ the transition function has the form

$$f(\mathbf{R}) = 1+\mathbf{R}[1-\exp(-\gamma\mathbf{R})] \qquad (30)$$

which leads with $\mathbf{R}$ to the expression of the damping factor

$$D_0(\eta) = 1-\exp(-\eta^2/A_*^2) \qquad (31)$$

satisfactorily describing velocity distribution in the inner portion of the transition region, but greatly distorting the transition to the logarithmic straight line in the region of the turbulent core.

Figure 248 shows the comparison graphs of the functions $D(\eta)$ and $D_{\text{V.D.}}(\eta)$. The difference between them has a slight affect on the velocity profile in the transition region thanks to integral determination of this velocity (Equation 34).

In spite of the physical weakness of the van Driest model and a certain arbitrariness in the choice of the form of the transition function (27), and especially of the relation $\mathbf{R} = æ^2\eta^2$ which is associated with the assumption of $f(\mathbf{R}) = 1$, the calculations by the van Driest equation are in satisfactory agreement with the experiment. A small deviation of the curve (26) may be noted from the experimental points inside the transition region and a somewhat delayed reaching of the logarithmic curve situated a bit higher the experimental curves (dashed curve in Figure 249).

The van Driest equation is used in the calculation of both nongradient and gradient boundary layers, substituting for the skin friction on the wall τ_w, and entering the definition of v_*, its current quantity in the boundary layer cross-section $\tau(y)$. This generalization of the van Driest equation or previously formulated parametric equations follows directly from equality (16), in which the term on the right side $\tau(y)$ can be replaced by the available approximate relations of friction distribution over the boundary layer cross-section.[10]

[10] Coles, D., The law of the wall in turbulent shear flow, 50 Jahre Grenzschichtforschung, Braunschweig, 1955, S. 154–163; Rotta, I. K., *Turbulentnyy pogranichnyy sloy v neszhimaemoy zhidkosti* (*Turbulent Boundary Layer in Incompressible Fluid*), Mir Publisher, Moscow, 1967, 75–78; Loitsyanskiy, L. G., On one common formula of the theory of heat and mass transfer in a wall region of a turbulent boundary layer, in *Collected Papers Gidroaerodinamika* (*Hydroaerodynamics*), Leningrad Politekhnika Institute, Leningrad, 1983 3–10.

Making use of the introduced damping factor, instead of considering the velocity distributions separately in the regions of the viscous sublayer and the flow turbulent core, we can obtain a continuous distribution of velocity in the cross-sections of the entire near-wall subregion from the solid wall to the outer boundary of this subregion. For this purpose, recalling the expression for complete skin friction τ, we take account of Equations 3 and 4 and restrict ourselves in the near-wall subregion to the approximate equality $\tau = \tau_w$. The τ is presented in the form of the sum

$$\tau = \tau_w = \mu \frac{du}{dy} + \rho l^2 \left(\frac{du}{dy}\right)^2 \tag{32}$$

or, passing over from l to $l_p = æy$ and using the damping factor (7)

$$\tau_w = \mu \frac{du}{dy} + \rho æ^2 y^2 D(y) \left(\frac{du}{dy}\right)^2 \tag{33}$$

The introduction of universal quantities (Equation 13) into this equation transforms it to

$$1 = \frac{d\varphi}{d\eta} + æ^2 \eta^2 D(\eta) \left(\frac{d\varphi}{d\eta}\right)^2$$

Solving this equation, which is quadratic in $d\varphi/d\eta$, we find

$$\frac{d\varphi}{d\eta} = \frac{2}{1 + \sqrt{1 + 4æ^2\eta^2 D(\eta)}}$$

$$\varphi = \frac{u}{v_*} = \int_0^\eta \frac{2d\eta}{1 + \sqrt{1 + 4æ^2\eta^2 D(\eta)}} \tag{34}$$

Designating the universal ordinate of the boundary layer section point that corresponds to the end of the transition portion (in terms of η_1) we present the second equation of system (34) in the form

$$\varphi(\eta) = \int_0^{\eta_1} \frac{2d\eta}{1 + \sqrt{1 + 4æ^2\eta^2 D(\eta)}} + \int_{\eta_1}^{\eta} \frac{d\eta}{æ\eta} =$$

$$= \int_0^{\eta_1} \frac{2d\eta}{1 + \sqrt{1 + 4æ^2\eta^2 D(\eta)}} - \frac{1}{æ}\ln\eta_1 + \frac{1}{æ}\ln\eta \tag{35}$$

Comparing the last equation with the logarithmic profile of velocities in a turbulent flow core ($\eta \geq \eta_1$) represented by Equation 116 of Chapter 13

$$\varphi = \frac{u}{v_*} = A\lg\eta + B$$

we determine the constants A and B

$$A = \frac{2.303}{æ} \qquad B = \int_0^{\eta_1} \frac{2\,d\eta}{1+\sqrt{1+4æ^2\eta^2 D(\eta)}} - \frac{2.303}{æ}\lg \eta_1$$

where the numerical factor 2.303 is the modulus of decimal logarithms.

The calculations performed using the values of the constants $æ = 0.4$, $\gamma = 0.0092$, $\eta_1 = 30$, and Equations 35 made it possible to construct a graph of velocity distribution in a "near-wall" subregion (Figure 249) and to determine constants A and B. In the same figure the graph of velocity distribution, based on the van Driest damping factor, is shown by a dashed line.

The value of the coefficient B is equal to 4.7 which is close to the experimental value 4.8. According to van Driest $B_{\text{v.D.}} = 5.4$.

The test data correspond to the most reliable experiments of Laufer,[11] Vighardt,[12] and Tsuji and Morikawa.[13] The recent and most accurate calculations[14] led to the values $\mathbf{R}_{cr} = 13$, $B = 5.1$. The general case[14] of a "gradient" boundary layer is considered and parabolic approximations of the damping factor which compare favorably with the van Driest formula (26) in simplicity and accuracy are suggested.

4. Heat and Mass Transfer in a "Near-Wall" Subregion

Presentation of the existing method of heat and mass transfer calculations is not the aim of this monograph, but it has been the concern of numerous publications.[15]

In the present section we shall dwell only on some principal problems closely connected with turbulent flow and turbulent heat transfer accompanying it. As to the turbulent transfer of substance, a semiempirical theory of these processes coincides with a similar theory of the processes of heat propagation, so everything given in the present section relates to both processes equally.

The solution of the general problem of transfer in turbulent flows hinges on (see Chapter 13) a sufficient knowledge of the coefficients of transfer ε_τ, ε_q, ε_m. If for the first of these coefficients we can construct a rather satisfactory semiempirical expression involving the mixing length concept, then for the rest we have to use the assumption of a passive character of the transferred substance, of the equality of Prandtl and Schmidt numbers to unity, specification of some empirical mean values of these numbers, or accounting for their empirical distributions over the flow.

In this section we shall go into the details of the heat and mass transfer phenomena under the conditions of "near-wall" turbulence exclusively.

---

[11] Laufer, J., The Structure of Turbulence in Fully Developed Flow, NACA Rep. 1174, 1954.

[12] Computation of turbulent boundary layer, in *Proc. AFOSR-IFR-Stanford Conference*, 1968, Vol. 2, University of Stanford, Stanford, CA, 1969, 98–123, (Experiment 1400).

[13] Tsuji, J. and Morikawa, G., Turbulent boundary layer with pressure gradient alternating in sign, *Aeron. Q.*, 27 (1), 15–28, 1976.

[14] Loitsyanskiy, L. G. and Zyabrikov, V. V., Damping factor in the theory of Prandtl mixing length, Rep. VI All-Union Meet. Theoretical and Applied Mechanics, Tashkent, September 24 to 30, 1986.

[15] Zhukauskas, A. A., *Konvektivnyy perenos v teploobmennikakh* (*Convective Transfer in Heat Exchangers*), Nauka, Moscow, 1982; Groeber, H., Erk, S., and Grigull, U., *Osnovy ucheniya o teploobmene* (*Principles of Heat Transfer*), Inostr. Literatura, Moscow, 1958 (in Russian); Patankar, S. V. and Spalding, D. B., *Heat and Mass Transfer in Boundary Layers*, Intertext Books, London, 1970; Luikov, A. V. and Mikhailov, Yu. A., *Teoriya teplo- i massoperenosa* (*Theory of Heat and Mass Transfer*), Gosenergoizdat, Moscow-Leningrad, 1963; Kutateladze, S. S. and Leontiyev, A. I., *Teploobmen i treniye v turbulentnom pogranichnom sloe* (*Heat Transfer and Friction in a Turbulent Boundary Layer*), Energia, Moscow, 1972.

Of principal importance is the problem of whether in the heat transfer phenomena there is retained the division of flow into a viscous sublayer with molecular nature of transfer (temperature sublayer) and a turbulent core where the processes of transfer are purely molar, independent of liquid molecular structure, and what should be the relation between the thickness of a viscous and temperature sublayers.

Similar to the fact mentioned in the theory of a laminar boundary layer (Chapter 13), thickness of viscous and temperature sublayers can coincide only in the case of the equality of the molecular Prandtl number to unity (**Pr** = 1), because only in this case are the profiles of velocity and temperature distribution in a sublayer similar.

If the molecular Prandtl number is less than unity (**Pr** < 1), thus indicating the increased role of liquid thermal conductivity as compared with viscosity ($\lambda > \mu c_p$), molecular processes of heat conduction retain their importance in the turbulent core region, where molecular viscosity can be neglected. Hence, it follows that when **Pr** < 1 the thickness of a temperature sublayer will exceed that of a viscous sublayer. So, for example, in liquid metals (mercurium, metal melts), for which **Pr** ≪ 1, the processes of molecular heat conduction will prevail in the most part of a turbulent core.

Conversely, at the molecular Prandtl number greater than unity (**Pr** > 1), a turbulent (molar) character of heat transfer predominates over the molecular one, i.e., over ordinary heat conduction. This leads to the situation when turbulent heat transfer develops in some outer portion of a viscous layer and, consequently, a temperature sublayer becomes thinner than a viscous one. Such a relation between the thickness of viscous and turbulent sublayers is especially sharply exhibited in the flows of extremely viscous liquids (lubricating oils, glycerine, etc.) with **Pr** ≫ 1.

Among the classical results in this region we should mention the Reynolds analogy,[16] which establishes a simple relation between friction and heat transfer in a turbulent flow with laminar and turbulent Prandtl numbers being equal to unity and in the absence of a longitudinal pressure drop in the flow.

Under these conditions ($\mathbf{Pr} = \mu c_p/\lambda = 1$, $\mathbf{Pr}_t = \varepsilon_\tau/\varepsilon_q = 1$) the equations of plane velocity and temperature layers, according to Equations 1, 23, 55, and 57 of Chapter 13, will have the form

$$u\frac{\partial u}{\partial x} + v\frac{\partial u}{\partial y} = \frac{\partial}{\partial y}\left[(\nu + \varepsilon_\tau)\frac{\partial u}{\partial y}\right]$$

$$u\frac{\partial T}{\partial x} + v\frac{\partial T}{\partial y} = \frac{\partial}{\partial y}\left[(\nu + \varepsilon_\tau)\frac{\partial T}{\partial y}\right] \qquad (36)$$

and the boundary conditions, as applied to the plate situated along the axis $0x$, are

$$u = v = 0 \qquad T = T_w \quad \text{at} \quad y = 0$$

$$u = U_\infty \qquad T = T_\infty \quad \text{at} \quad y = \infty \qquad (37)$$

where T_w is the surface temperature, T_∞ is the temperature at a distance from the surface.

Considering the system of equations (36) and the boundary conditions (37), we shall make sure that the second equation may be satisfied assuming that

---

[16] Reynolds, O., On the extent and action of the heating surface for steam boilers, *Proc. Manchester Lit. Phil. Soc.*, 14, 1874.

$$\frac{T-T_w}{T_\infty - T_w} = \frac{u}{U_\infty} \tag{38}$$

This expresses the fact that distributions of velocities and temperature drops in the flow sections normal to a solid wall are similar.

Differentiating both sides of Equation 38 with respect to y and then assuming that $y = 0$, we have

$$\frac{1}{U_\infty}\left(\frac{\partial u}{\partial y}\right)_{y=0} = \frac{1}{T_\infty - T_w}\left(\frac{\partial T}{\partial y}\right)_{y=0}$$

Introducing the skin friction stress $\tau_w = \mu(\partial u/\partial y)_{y=0}$ and the heat flux $q_w = -\lambda(\partial T/\partial y)_{y=0}$ on the body surface into this equation, we obtain the relation

$$\frac{\tau_w}{\rho U_\infty^2} = \frac{q_w}{\rho c_p U_\infty (T_w - T_\infty)} \tag{39}$$

expressing the Reynolds analogy (mentioned above) and relating friction to heat transfer on a solid impermeable wall in a viscous fluid flow.

The left side of the last equation is equal to $c_f/2$, while the right side is the Stanton number **St**. In case when the quantities U_∞ and T_∞ corresponding to the free stream are taken as the scales of velocity and temperature, the Stanton number is designated as $\mathbf{St}_\infty$, assuming that

$$\frac{q_w}{\rho c_p U_\infty (T_w - T_\infty)} = \mathbf{St}_\infty \tag{40}$$

(The expression for the Stanton number constructed by other quantities is introduced later.)

The Reynolds analogy (39) for the case of the flow along impermeable wall and of the absence of a longitudinal pressure gradient will be written in the short form

$$\frac{c_f}{2} = \mathbf{St}_\infty \tag{41}$$

In the developed turbulent flow in a plane or round cylindrical tube, as well as in boundary layers, the scales U_∞ and T_∞ are replaced by the liquid velocity and temperature on the tube axis or on the outer edge of the boundary layer. In the case of a tube it is preferable to deal with section-mean values of the velocity u_{mean} and temperature T_{mean}, and the coefficient of tube resistance λ, entering into the equation

$$\Delta p = \lambda \frac{L}{d}\frac{\rho u_{\text{mean}}^2}{2}$$

is introduced instead of c_f. Noting that in the developed (same in laminar or turbulent) flow the relation

$$\tau_w = \Delta p \quad \text{if} \quad L = \frac{d}{4}$$

is valid, we find

$$\frac{\tau_w}{\rho u^2_{\text{mean}}} = \frac{\lambda}{8} \tag{42}$$

Determining the section-mean values by Equation 38, we obtain

$$\frac{u_{\text{mean}}}{U_\infty} = \frac{T_w - T_{\text{mean}}}{T_w - T_\infty} \tag{43}$$

Then, having rewritten Equation 39 in the form

$$\frac{\tau_w}{\rho u^2_{\text{mean}}}\left(\frac{u_{\text{mean}}}{U_\infty}\right)^2 = \frac{q_w}{\rho c_p u_{\text{mean}}(T_w - T_{\text{mean}})} \frac{T_w - T_{\text{mean}}}{T_w - T_\infty} \frac{u_{\text{mean}}}{U_\infty}$$

and using Equation 43, we come to the Reynolds analogy for a tube in the form

$$\frac{\lambda}{8} = \frac{q_w}{\rho c_p u_{\text{mean}}(T_w - T_{\text{mean}})} = \mathbf{St} \tag{44}$$

Here, **St** denotes the Stanton number constructed by the mean values of u_{mean} and T_{mean}.

Extension of the Reynolds analogy (Equation 39 and 41) to the case of molecular Prandtl numbers, which are not equal but are close to unity, was suggested by Taylor[17] and Prandtl.[18] Assuming that in this case a small difference between the thickness of viscous and temperature sublayers may be neglected and the velocity and temperature distributions inside these thin layers coinciding in thickness may be taken to be linear, we have ($\tau = \text{const} = \tau_w$, $q = \text{const} = q_w$, T_{out} is the temperature of the viscous sublayer outer edge)

$$\tau = \tau_w = \mu\left(\frac{\partial u}{\partial y}\right)_{y=0} = \mu\frac{u_{\text{out}}}{\delta_{\text{out}}}$$

$$q = q_w = -\lambda\left(\frac{\partial T}{\partial y}\right)_{y=0} = \lambda\frac{T_w - T_{\text{out}}}{\delta_{\text{out}}} \tag{45}$$

According to the adopted condition that the turbulent Prandtl number is equal to unity, in a turbulent flow core independent of the condition $\mathbf{Pr} \neq 1$, exists the similarity between the distributions of averaged velocities and temperatures, so that

$$\frac{T - T_{\text{out}}}{T_\infty - T_{\text{out}}} = \frac{u - u_{\text{out}}}{U_\infty - u_{\text{out}}} \tag{46}$$

($U_\infty > u_{\text{out}}$, $T_\infty < T_{\text{out}} < T_w$ is the velocity and temperature at a distance from the solid wall at the outer edge of the turbulent boundary layer or on the tube axis) and, consequently, outside the viscous (or temperature that for $\mathbf{Pr} \approx 1$ is the same) sublayer the relation

[17] Taylor, G. J., Conditions at the Surface of a Hot Body Exposed to the Wind, Tech. Rep. Adv. Com. Aeronautics, 2(272), 1916.

[18] Prandtl, L., Bemerkung über den Wärmeübergang im Rohr, *Phys. Z.*, 29, 1928.

$$\frac{1}{T_\infty - T_{\text{out}}}\frac{\partial T}{\partial y} = \frac{1}{U_\infty - u_{\text{out}}}\frac{\partial u}{\partial y}$$

takes place which can be identically rewritten in the form ($A_q = A_\tau$, $\varepsilon_q = \varepsilon_\tau$)

$$\frac{c_p A_q (\partial T/\partial y)}{\rho c_p (T_\infty - T_{\text{out}})(U_\infty - u_{\text{out}})} = \frac{A_\tau (\partial u/\partial y)}{\rho (U_\infty - u_{\text{out}})^2}$$

or [$c_p A_q (\partial T/\partial y) = q = \text{const} = q_w$; $A_\tau (\partial u/\partial y) = \tau = \text{const} = \tau_w$],

$$\frac{q_w}{\rho c_p (U_\infty - u_{\text{out}})(T_{\text{out}} - T_\infty)} = \frac{\tau_w}{\rho (U_\infty - u_{\text{out}})^2} \tag{47}$$

that is the generalization of Equation 39.

Eliminating the two quantities δ_{out} and T_{out} from the system of Equations 45 and 47, we have

$$\delta_{\text{out}} = \frac{\mu u_{\text{out}}}{\tau_w} \qquad c_p T_{\text{out}} = c_p T_w - \mathbf{Pr}\, u_{\text{out}} \frac{q_w}{\tau_w}$$

and Equation 48 changes over to

$$\frac{q_w}{\rho c_p U_\infty (T_w - T_\infty)} = \frac{\tau_w}{\rho U_\infty^2 [1 + (\mathbf{Pr} - 1) u_{\text{out}}/U_\infty]} \tag{48}$$

or, according to the previously adopted notation,

$$\mathbf{St}_\infty = \frac{c_f/2}{1 + (\mathbf{Pr} - 1) u_{\text{out}}/U_\infty} \tag{49}$$

Passing over in the case of a plane or round cylindrical tube from U_∞ and T_∞ to mean flow rate quantities u_{mean} and T_{mean} (when calculating these quantities we can, at **Pr** close to unity, neglect the small dissimilarity in the distributions of velocities and temperatures in a viscous sublayer which constituted the negligible portion of the integration interval) we shall have instead of Equation 49,

$$\mathbf{St} = \frac{\lambda/8}{1 + (\mathbf{Pr} - 1) u_{\text{out}}/u_{\text{mean}}} \tag{50}$$

The obtained relations Equation 49 and 50 give the generalizations of the Reynolds analogies (Equations 41 and 44) to the case of small difference of **Pr** from unity. Rewriting them in the form ($\alpha \approx 11.5$)

$$\frac{1}{\mathbf{St}_\infty} = \frac{2}{c_f} + (\mathbf{Pr} - 1)\frac{u_{\text{out}}}{U_\infty}\frac{2}{c_f} = \frac{2}{c_f} + (\mathbf{Pr} - 1)\frac{u_{\text{out}}}{v_*}\frac{v_*}{U_\infty}\frac{2}{c_f} =$$

$$= \frac{2}{c_f} + (\mathbf{Pr} - 1)\alpha\sqrt{\frac{2}{c_f}} = \frac{2}{c_f} + \sqrt{\frac{2}{c_f}}\, g(\mathbf{Pr})$$

$$g(\mathbf{Pr}) = \alpha(\mathbf{Pr} - 1) \tag{51}$$

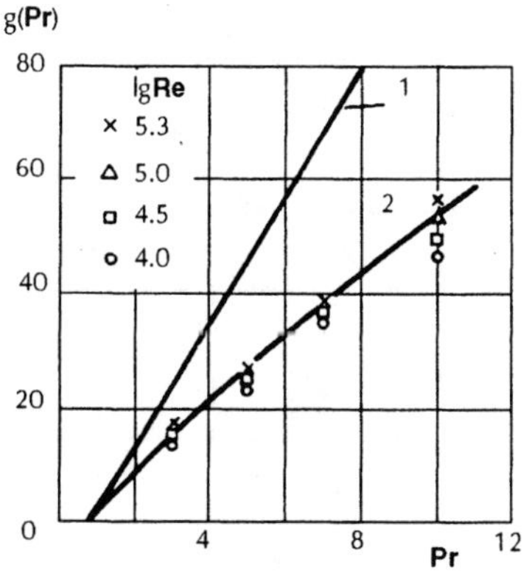

Fig. 250

or

$$\frac{1}{\mathbf{St}} = \frac{8}{\lambda} + (\mathbf{Pr} - 1)\frac{u_{\text{out}}}{u_{\text{mean}}}\frac{8}{\lambda} = \frac{8}{\lambda} + (\mathbf{Pr} - 1)\frac{u_{\text{out}}}{v_*}\frac{v_*}{u_{\text{mean}}}\frac{8}{\lambda} = \frac{8}{\lambda} + g(\mathbf{Pr})\sqrt{\frac{8}{\lambda}} \qquad (52)$$

we find that in the considered case of small difference of **Pr** from unity the matter is reduced to the correction term in the classical formulation of the Reynolds analogy (41) and to its transformed expression (44). A new approach, as compared with (41) and (44) is the introduction of the intermediate velocity on the outer edge of a viscous sublayer u_{out} and the relation from Chapter 13 $u_{\text{out}}/v_* = \alpha \approx 11.5$. In this case appears the von Kármán function g(**Pr**) (Equation 51), which retains its value (see below) and the form of which, as will be now shown, changes with the extension of the range of allowable values of the Prandtl number **Pr**.

Figure 250 presents the graph (straight line 1) of the function g(**Pr**) entering into Equation 51. For comparison this figure contains the experimental points of Eagle and Fergusson.[19] The coincidence of the Prandtl equation (51) with the experiment can be taken satisfactory only at **Pr** close to unity.

von Kármán[20] abandoned the "two-layer" scheme and placed between a viscous sublayer and a turbulent core, in which

$$\varphi = \eta \quad \text{when} \quad 0 \le \eta \le 5 \qquad \varphi = 5.75\,\lg\eta + 5.5 \quad \text{when} \quad \eta \ge 30$$

Correspondingly, a new "transition" ("buffer") layer with the distribution of velocities (Figure 251)

$$\varphi = 11.5\,\lg\frac{\eta}{5} + 5 \quad \text{for} \quad 5 \le \eta \le 30$$

Introduction of the region of transition from a viscous sublayer to a turbulent core extended the applicability range of the generalized Reynolds analogy (52). Curve 3 corresponding to the equation of this "three-layer" scheme of von Kármán (Figure 251)

[19] Eagle, A. and Fergusson, R., On the coefficient of heat transfer from the internal surface of tube walls, *Proc. R. London Soc. Ser. A*, 127, 1930.

[20] von Kármán, Th. The analogy between fluid friction and heat transfer, *Trans. Am. Soc. Mech. Eng.*, 61, 1939.

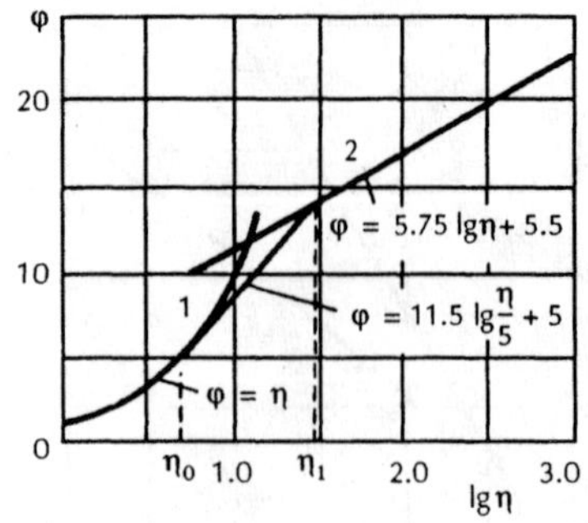

Fig. 251

$$g(\mathbf{Pr}) = 5\left\{(\mathbf{Pr}-1) + 2.3\lg\left[1 + \frac{5}{6}(\mathbf{Pr}-1)\right]\right\} \tag{53}$$

was in good agreement with the data of Eagle and Fergusson out to the values of the **Pr** number of about 10.

To move further into the range of large Prandtl numbers "multilayer" schemes were tested, thus leading to complex computational techniques. The theory of continuously distributed (over the entire region) interaction between molecular and molar processes of transfer, representing the extension of the theory formulated in the previous section to the case of heat and mass transfer, is deprived of this drawback.[21]

Following the notation of Equations 8 and 10 adopted in Section 3, we shall rewrite the system of equations (45) in the following form, having added the expression for the mass flow density m:

$$\tau = \mu\frac{du}{dy}\left(1 + \frac{\varepsilon_\tau}{\nu}\right) = \mu\frac{du}{dy}f(\mathbf{R})$$

$$q = \lambda\frac{dT}{dy}\left(1 + \mathbf{Pr}\frac{\varepsilon_\tau}{\nu}\right) = \lambda\frac{dT}{dy}\,[1 - \mathbf{Pr} + \mathbf{Pr}f(\mathbf{R})]$$

$$m = \rho D\frac{dc}{dy}\left(1 + \mathbf{Sc}\frac{\varepsilon_\tau}{\nu}\right) = \rho D\frac{dc}{dy}\,[1 - \mathbf{Sc} + \mathbf{Sc}f(\mathbf{R})] \tag{54}$$

Retaining the approximate condition $\tau = \text{const} = \tau_w$ which is taken to be a principle of the theory of interaction between molecular and molar exchanges in the region of transition from a viscous sublayer to a turbulent core and adding the analogous assumptions $q = \text{const} = q_w$ and $m = \text{const} = m_w$ to it, we complement the previously introduced universal scales of velocities $v_* \approx \sqrt{\tau_w/\rho}$ and lengths $l_* = \nu/v_*$ with analogous scales of temperatures and concentrations of substance, having adopted

$$T_* = \frac{q_w}{\rho c_p v_*} \qquad c_* = \frac{m_w}{\rho v_*} \tag{55}$$

[21] Loitsyanskiy, L. G., Semi-empirical theories of interaction between the processes of molecular and molare exchange in a turbulent liquid flow, in *Proc. All-Union Meet.* Theoretical and Applied Mechanics, AN SSSR, Moscow-Leningrad, 1962, 145–166.

The ordinates y, velocities u, temperatures T, and concentrations of substance c expressed in terms of fractions of these scales are further denoted as

$$\frac{y}{l_*} = \frac{yv_*}{\nu} = \eta \qquad \frac{u}{v_*} = \varphi \qquad \frac{T - T_w}{T_*} = \psi \qquad \frac{c - c_w}{c_*} = \chi \tag{56}$$

Dividing both sides of Equations 54 by τ_w, q_w, and m_w, respectively, and passing over to universal quantities (56), to determine $\varphi(\eta)$, $\psi(\eta)$, and $\chi(\eta)$ we obtain the following system of equations

$$1 = f(\mathbf{R})\frac{d\varphi}{d\eta} \qquad 1 = \frac{1 - \mathbf{Pr} + \mathbf{Pr}f(\mathbf{R})}{\mathbf{Pr}}\frac{d\psi}{d\eta} \qquad 1 = \frac{1 - \mathbf{Sc} + \mathbf{Sc}f(\mathbf{R})}{\mathbf{Sc}}\frac{d\chi}{d\eta} \tag{57}$$

In this system of equations four unknown quantities are included – φ, ψ, χ and $\mathbf{R}$ – and the transition function $f(\mathbf{R})$ is determined by Equation 24. Recalling also equality (17), according to which

$$\eta = \frac{1}{æ}\sqrt{\mathbf{R}f(\mathbf{R})} \qquad d\eta = \frac{1}{2æ}\frac{\mathbf{R}f'(\mathbf{R}) + f(\mathbf{R})}{\sqrt{\mathbf{R}f(\mathbf{R})}}d\mathbf{R}$$

we obtain the expressions of the functions $\varphi(\eta)$, $\psi(\eta)$, and $\chi(\eta)$ in the quadratures in the parametric dependence on $\mathbf{R}$

$$\varphi = \frac{1}{2æ}\int_0^{\mathbf{R}}\frac{\mathbf{R}f'(\mathbf{R}) + f(\mathbf{R})}{\sqrt{\mathbf{R}f^3(\mathbf{R})}}d\mathbf{R} \qquad \psi = \frac{1}{2æ}\int_0^{\mathbf{R}}\frac{\mathbf{R}f'(\mathbf{R}) + f(\mathbf{R})}{\left[\frac{1}{\mathbf{Pr}} + f(\mathbf{R}) - 1\right]\sqrt{\mathbf{R}f(\mathbf{R})}}d\mathbf{R}$$

$$\chi = \frac{1}{2æ}\int_0^{\mathbf{R}}\frac{\mathbf{R}f'(\mathbf{R}) + f(\mathbf{R})}{\left[\frac{1}{\mathbf{Sc}} + f(\mathbf{R}) - 1\right]\sqrt{\mathbf{R}f(\mathbf{R})}}d\mathbf{R} \qquad \eta = \frac{1}{æ}\sqrt{\mathbf{R}f(\mathbf{R})} \tag{58}$$

Direct consideration of system (58) reveals that the second and third equations of this system at $\mathbf{Pr} = \mathbf{Sc}$ become identical. This corresponds to the above note the quantitative equivalence of heat and mass processes, Moreover, in the same direct manner we can make a conclusion on the similarity of distributions of velocities and excess temperatures and, consequently, excess concentrations at $\mathbf{Pr} = \mathbf{Sc} = 1$.

On assignment of $f(\mathbf{R})$ in the form of Equation 24 the quadratures involved in system (58) can be determined by a numerical method. We leave aside possible simplifications by transition to approximate formulas for a damping factor.

We now resort to the problem of seeking after the approximate explicit analytical expression of the function $g(\mathbf{Pr})$ and Stanton and Schmidt numbers at arbitrary large values of the Prandtl number $\mathbf{Pr}$ that corresponds to the location of temperature and concentration boundary layers deep inside the velocity boundary layer.

Making use of relation (22) to express $f(\mathbf{R})$, i.e., the more valid the greater the Prandtl number, we substitute the variables in system (58)

$$\gamma\mathbf{R}^4 = t^4 \tag{59}$$

Then, the system of equations (58) will change over to

$$\varphi = \frac{1}{æ \sqrt[4]{\gamma}} \int_0^t \frac{1 + 3t^4}{(1 + t^4)^{3/2}} dt$$

$$\psi = \frac{1}{æ \sqrt[4]{\gamma}} \int_0^t \frac{1 + 3t^4}{(1/\mathbf{Pr} + t^4)(1 + t^4)^{1/2}} dt$$

$$\eta = \frac{1}{æ \sqrt[4]{\gamma}} t (1 + t^4)^{1/2} \tag{60}$$

The equation for χ omitted because it differs by nothing from the equation for ψ in case the Schmidt number **Sc** is substituted by the Prandtl number **Pr**.

The integrals entering the right sides of system (60) are calculated most easily by expanding the integrands in powers of t with the extraction of the transcendental part

$$B(x) = \int_0^x \frac{dx}{1 + x^4} = \frac{1}{4\sqrt{2}} \ln \frac{1 + x\sqrt{2} + x^2}{1 - x\sqrt{2} + x^2} + \frac{1}{2\sqrt{2}} \operatorname{arctg} \frac{x\sqrt{2}}{1 - x^2} \tag{61}$$

tabulated by Bakhmetiyev and widely applied in hydraulics.[22] We obtain

$$\varphi = 8.06 (t + 0.3t^5 - 0.291t^9 + 0.264t^{13} - \ldots)$$

$$\psi = 8.06 \Big[(2.5t - 0.225t^5 + 0.09t^9 - \ldots) +$$

$$+ \frac{1}{\mathbf{Pr}} (1.125t - 0.1625t^5) + \frac{0.08125\, t}{\mathbf{Pr}^2} + \left(1 - \frac{2.5}{\mathbf{Pr}} - \frac{1.125}{\mathbf{Pr}^2}\right) \mathbf{Pr}^{3/4} B(\mathbf{Pr}^{1/4} t) \Big]$$

$$\eta = 8.06 (t + 0.5t^5 - 0.125t^9 + \ldots) \tag{62}$$

We shall use this solution within the range $0 \le \eta \le \eta_0$, $0 \le t \le t_0$ corresponding to a viscous sublayer, the upper limit of which η_0 will be found from the condition that the tangent drawn at this point in the direction to the velocity profile φ [the first equation of system (60)] could pass through the point $\eta_1 = 30$, $\varphi_1 = 13.63$ of the origin of the logarithmic velocity profile taken by the verified Nikuradze equation (Chapter 13) in the form

$$\varphi = 5.75 \lg \eta + 5.24 \tag{63}$$

For this purpose we should assume $\eta_0 = 6.07$, $t_0 = 0.68$. Then the equation of the tangent will take on the form

$$\varphi = 11.3 \lg \eta - 3.09 \qquad (6.07 \le \eta \le 30) \tag{64}$$

When $\eta \ge 30$ profile (63) will serve as the velocity profile. To the selected universal velocity profile corresponds simple analytical expressions for the universal temperature profile

[22] Esman, I. E., *Gidravlika (Hydraulics)*, Gonti NKTR SSSR Press, Moscow 1938, 365 (Table 58).

$$\psi = 11.3\ \lg\left(\eta + 4.98 - \frac{4.98}{\mathbf{Pr}}\right) - 11.3\ \lg\left(11.05 - \frac{4.98}{\mathbf{Pr}}\right) + \psi_0$$

$$\text{for} \quad 6.07 \le \eta \le 30\,,$$

$$\psi = 5.75\ \lg\eta - 8.49 + \psi_1 \quad \text{for} \quad 30 \le \eta\,, \tag{65}$$

when ψ_0 means the value of ψ calculated by the second equation of system (62) at $t = t_0 = 0.68$, and ψ_1 means ψ calculated by the first equation of system (65) at $\eta = \eta_1 = 30$. We shall give the values of these quantities

$$\psi_0 = 13.46 + \frac{5.976}{\mathbf{Pr}} + \frac{4.453}{\mathbf{Pr}^2} +$$

$$+\left(8.06 - \frac{20.15}{\mathbf{Pr}} - \frac{9.068}{\mathbf{Pr}^2} - \frac{6.549}{\mathbf{Pr}^3}\right)\mathbf{Pr}^{3/4}B\,(0.68\mathbf{Pr}^{1/4})$$

$$\psi_1 = 11.3\lg\left(25.09 + \frac{4.91}{\mathbf{Pr}}\right) - 11.3\lg\left(1.16 + \frac{4.91}{\mathbf{Pr}}\right) + \psi_0 \tag{66}$$

We introduce into consideration the Stanton number $\mathbf{St}_m$ constructed by the values of the velocity and excess temperature U_m and $\theta_m = T_w - T_m$ on the tube axis. Retaining the determination of the Stanton number in the previously found form, we shall have

$$\mathbf{St}_m = \frac{q_w}{\rho c_p U_m (T_w - T_m)} \tag{67}$$

Recalling the notation of Equation 56, we obtain

$$\frac{1}{\mathbf{St}_m} = \frac{U_m}{v_*}\frac{\rho c_p v_*}{q_w}(T_w - T_m) = \varphi_m \psi_m \tag{68}$$

In this case according to (63) and the last equation of system (65)

$$\psi_m = 5.75\ \lg\eta_m - 8.49 + \psi_1 = \varphi_m - 13.73 + \psi_1$$

moreover,

$$\varphi_m = \left(\frac{2}{c_f}\right)^{1/2} \qquad c_f = \frac{2\tau_w}{\rho U_m^2}$$

Hence, it follows

$$\frac{1}{\mathbf{St}_m} = \frac{2}{c_f} + \left(\frac{2}{c_f}\right)^{1/2}(\psi_1 - 13.73) \tag{69}$$

Introducing the von Kármán function $g(\mathbf{Pr})$ by the equality which generalizes the right side of the first equation of (51) and comparing Equations 51 and 69, we obtain the expression of function $g(\mathbf{Pr})$

$$g(\mathbf{Pr}) = \psi_1 - 13.73$$

and, using the system of equations (66), we have its final expression

$$g(\mathbf{Pr}) = 11.31\mathrm{g}\left(25.09 + \frac{4.91}{\mathbf{Pr}}\right) - 11.31\mathrm{g}\left(1.16 + \frac{4.91}{\mathbf{Pr}}\right) - 0.27 + \frac{5.976}{\mathbf{Pr}} +$$

$$+ \frac{4.453}{\mathbf{Pr}^2} + \left(8.06 - \frac{20.15}{\mathbf{Pr}} - \frac{9.068}{\mathbf{Pr}^2} - \frac{6.549}{\mathbf{Pr}^3}\right)\mathbf{Pr}^{3/4} B\,(0.68\mathbf{Pr}^{1/4}) \tag{70}$$

which is valid for any $\mathbf{Pr} \geq 1$.

The last equation includes a more simple asymptotic equation for $g(\mathbf{Pr})$ at high Prandtl number **Pr**

$$g(\mathbf{Pr}) \sim 8.97\,(\mathbf{Pr})^{3/4} + 6.26 - \frac{22.4}{\mathbf{Pr}^{1/4}} + O\left(\frac{1}{\mathbf{Pr}}\right) \tag{71}$$

We note in conclusion that Stanton number **St** constructed by mean values of velocity and excess temperature u_{mean} and $T_w - T_{\text{mean}}$ will be, similar to Equation 68, determined by the equation

$$\mathbf{St} = \frac{1}{\varphi_{\text{mean}}\psi_{\text{mean}}} = \frac{1}{\varphi_m\psi_m}\,\frac{\varphi_m}{\varphi_{\text{mean}}}\,\frac{\psi_m}{\psi_{\text{mean}}} = \mathbf{St}_m\,\frac{\varphi_m}{\varphi_{\text{mean}}}\,\frac{\psi_m}{\psi_{\text{mean}}} \tag{72}$$

with

$$c_f = \frac{2\tau_w}{\rho U_m^2} = 2\frac{1}{\varphi_{\text{mean}}^2}\left(\frac{\varphi_{\text{mean}}}{\varphi_m}\right)^2 \qquad \varphi_{\text{mean}} = \frac{2\sqrt{2}}{\sqrt{\lambda}}$$

so that, passing in (72) to mean characteristics, we shall have

$$\mathbf{St} = \frac{\lambda\psi_m/\psi_{\text{mean}}}{8\varphi_m/\varphi_{\text{mean}} + 2\sqrt{2}\sqrt{\lambda}\,g(\mathbf{Pr})} \tag{73}$$

As can be judged by the curves of the dependence of the dimensionless temperature $\psi/\psi_m = (T - T_w)/(T_m - T_w)$ on the dimensionless distance from the tube wall y/a (Figure 252), with the growth of **Pr** the temperature profiles become "fuller" so that the ratio

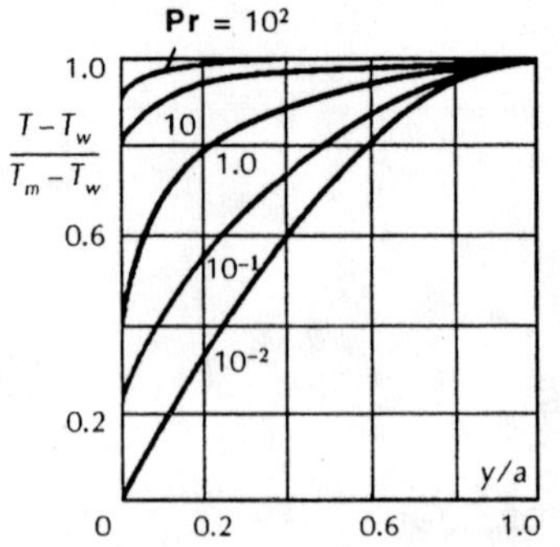

Fig. 252

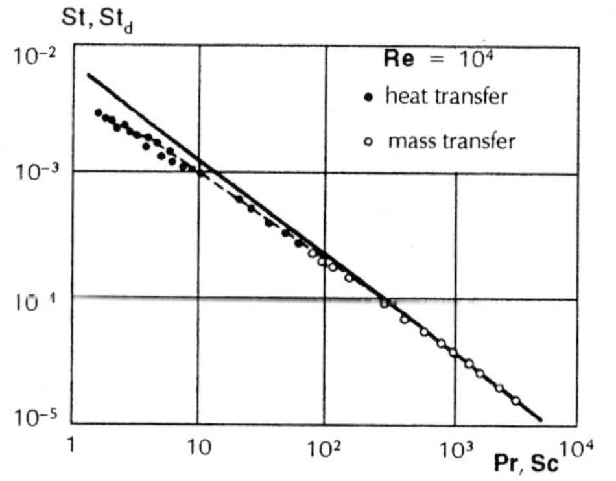

Fig. 253

$\psi_m/\psi_{\text{mean}}$ approaches unity. Assuming approximately $\psi_m/\psi_{\text{mean}} \approx 1$, we shall use Equation 125 of Chapter 13, rewritten as

$$\frac{\varphi_m}{\varphi_{\text{mean}}} = 1 + \frac{4.08}{\varphi_{\text{mean}}} \approx 1 + 1.44\sqrt{\lambda}\ . \tag{74}$$

Based on these approximate relations and the asymptotic representation (71) of the function $g(\mathbf{Pr})$, we obtain from Equation 73 the following asymptotic expression (at high **Pr**) for the Stanton number

$$\mathbf{St} \sim 0.04\lambda^{1/2}\mathbf{Pr}^{-3/4} - (0.0124 + 0.0175\lambda^{1/2})\,\mathbf{Pr}^{-5/4} + $$
$$+ 0.0985\lambda^{1/2}\mathbf{Pr}^{-7/4} + O(\mathbf{Pr}^{-9/4}) \tag{75}$$

Figure 253 presents the graph of the dependence of the numbers **St** and $\mathbf{St}_d$ (which is determined in terms of concentration) on the Prandtl number **Pr** of the Schmidt number **Sc** which is equivalent by the value. The test points are borrowed from the work by Deissler.[23] A dashed curve is constructed by Equation 75 and checked by numerical integration of the function $\psi(\eta)$ by the second and third formulas of system (60). A solid line corresponds to the limiting Deissler equation

$$\mathbf{St} = 0.04\lambda^{1/2}\mathbf{Pr}^{-3/4} \tag{76}$$

which is valid for very high Prandtl and Schmidt numbers. The graph in the figure corresponds to the value of the Reynolds number $\mathbf{Re} = 10^4$ at which λ was determined by the Blasius equation (Equation130 of Chapter 13).

The above consideration related only to heat and mass transfer at high **Pr** and at $\mathbf{Pr}_t = 1$. The case of small Prandtl numbers **Pr** corresponding to metal melts and to strongly ionized gases by its mechanism drastically differs from the considered case of high **Pr** values.

At small **Pr** the temperature boundary layer is thicker than the velocity one, the temperature is "smeared" over a wide (as compared with the velocity boundary layer) region. As seen from Figure 252 the temperature profiles at small **Pr** become similar to the "laminar" analogue of velocities (see Figure 177 in Chapter 12) The literature is voluminous in this field of

---

[23] Deissler, R., Analysis of Turbulent Heat Transfer, Mass Transfer and Friction in Smooth Tubes at High Prandtl and Schmidt Numbers, NACA Rep. 1210, 1959.

heat and mass transfer.[24]

The material discussed in this section elucidated the main concepts of the modern study of heat and mass transfer in turbulent flows briefly and to a small extent (see the earlier mentioned monographs in this field). To enrich the section we shall give·some additional information on temperature distribution over the tube of boundary layer sections, in this case we shall drop the abovemade assumption that $\mathbf{Pr}_t = 1$.

From the second equation of system (58) (Chapter 13), rewritten in the form (the sign is included in the definition of the derivative dT/dy)

$$q = (\lambda + \rho c_p \varepsilon_q)\frac{dT}{dy} = \left(\lambda + \frac{\rho c_p \varepsilon_\tau}{\mathbf{Pr}_t}\right)\frac{dT}{dy}$$

where ε_q is replaced by $\varepsilon_\tau = \nu_t$, according to Equation 67, Chapter 13

$$\varepsilon_q = \frac{\varepsilon_\tau}{\mathbf{Pr}_t} = \frac{\nu_t}{\mathbf{Pr}_t} \tag{77}$$

we find, making use of the simplification ($q = \text{const} = q_w$), that

$$q_w = \left[\lambda + \frac{\rho c_p æ^2}{\mathbf{Pr}_t} y^2 D(y) \frac{du}{dy}\right]\frac{dT}{dy}$$

Passing over to universal variable (56) having divided both sides of the last equation by q_w, we obtain

$$1 = \left[\frac{1}{\mathbf{Pr}} + \frac{1}{\mathbf{Pr}_t} æ^2 \eta^2 D(\eta) \frac{d\varphi}{d\eta}\right]\frac{d\psi}{d\eta}$$

where

$$\frac{d\psi}{d\eta} = \left[\frac{1}{\mathbf{Pr}} + \frac{1}{\mathbf{Pr}_t} æ^2 \eta^2 D(\eta) \frac{d\varphi}{d\eta}\right]^{-1}$$

$$\psi = \int_0^\eta \frac{d\eta}{\dfrac{1}{\mathbf{Pr}} + \dfrac{1}{\mathbf{Pr}_t} æ^2 \eta^2 D(\eta) \dfrac{d\varphi}{d\eta}} \tag{78}$$

These equalities together with Equations 34 solve the formulated problem of temperature distribution over the tube and boundary layer sections in the integral form based on the adoption of the previously mentioned approximate expression of the damping factor (Equations 30 and 32).

The equations of the form (78) usually present in the monographs[25] are distinguished, like Equations 34, by the fact that they involve the quantities æ and $D(\eta)$ typical of the Prandtl theory and its modification given at the beginning of this chapter. If, for example, velocity distribution is given by the test curve of it is known from the preliminary calculation

---

[24] Borishanskiy, V. M., Kutateladze, S. S., Novikov, I. I., and Fedynskiy, O. S., *Zhidkometallicheskie teplonositely (Liquid Metal Heat Carriers)*, Atomizdar, Moscow, 1967.

[25] See the monograph by A. A. Zhukauskas, p.106, equation (7.39).

according to some theory, to determine temperature we can use the integral equation which does not include æ and $D(\eta)$. In other words, the integral equation is not related to the mixing length theory. This equation can easily be obtained,[26] having eliminated from the system of equations

$$\left(1+\frac{\varepsilon_t}{\nu}\right)\frac{d\varphi}{d\eta} = 1$$

$$\frac{1}{\mathbf{Pr}}\left(1+\frac{\mathbf{Pr}}{\mathbf{Pr}_t}\frac{\varepsilon_t}{\nu}\right)\frac{d\psi}{d\eta} = 1 \tag{79}$$

which was used in the derivation of Equations 78, the quantity $\varepsilon_\tau/\nu = \nu_t/\nu$.

If, for brevity we introduce the notations

$$\frac{d\varphi}{d\eta} = \varphi', \qquad \frac{d\psi}{d\eta} = \psi' \tag{80}$$

the following simple relation will be obtained as a result of this elimination

$$\frac{1}{\psi'} - \frac{1}{\mathbf{Pr}_t}\frac{1}{\varphi'} = \frac{1}{\mathbf{Pr}} - \frac{1}{\mathbf{Pr}_t} \tag{81}$$

As is easily seen the known condition of similarity between velocity and excess temperature profiles

$$\psi' = \varphi' \qquad \psi = \varphi$$

that takes placc at $\mathbf{Pr}$ and $\mathbf{Pr}_t$ equal to unity, is contained in this equality.

In a general case ($\mathbf{Pr} \neq 1$, $\mathbf{Pr}_t \neq 1$) Equation 81 establishes a linear relation between the inverse values of the derivatives φ' and ψ' which is a generalization of the Reynolds analogy to this general case. From Equation 81 directly follows the integral equation for determination of temperature by the prescribed velocity profiles and by the Prandtl numbers $\mathbf{Pr}$ and $\mathbf{Pr}_t$:

$$\psi = \int_0^\eta \frac{d\eta}{\frac{1}{\mathbf{Pr}_t}\frac{1}{\varphi'} + \frac{1}{\mathbf{Pr}} - \frac{1}{\mathbf{Pr}_t}} \tag{82}$$

We shall not discuss the general consequences of from the two last equations, such as their possible generalization to the case of the presence of a longitudinal pressure gradient which is not so obvious as expressions (81) and (82).[26]

The considerations made are illustrated by numerical calculation of the relative excess temperature distribution by Equation 82.

To find the velocity distribution $\varphi(\eta)$ we shall make use of the second equation of (34) in which the damping factor is determined by the set of Equation 25 and the second relation of system (17). To this velocity distribution corresponds a solid line in Figure 249 which is in good agreement with experimental points.

---

[26] Loitsyanskiy, L. G., On one general formula in the theory of heat and mass transfer in a near-wall region of a turbulent boundary layer, in *Gidroaerodinamika* (*Hydroaerodynamics*), LPI, Leningrad, 1983, 3–10.

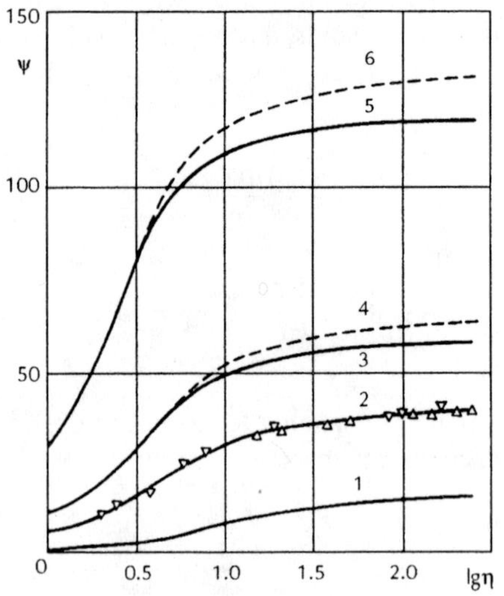

Fig. 254

The Prandtl number $\mathbf{Pr}_t$ was taken as 0.8. The numbers $\mathbf{Pr}$ were prescribed as 5.4, 10, and 30. To these values of $\mathbf{Pr}$ in Figure 254 correspond curves 2, 3, and 5. Curve 1 conforms to $\mathbf{Pr}_t = 1$, $\mathbf{Pr} = 1$, thus reflecting the adopted velocity distribution $\varphi(\eta) = \psi(\eta)$. It is admissible to use the value $\mathbf{Pr}_t = 1$ in calculations instead of the real one, which is shown by dashed lines 4 and 6. According to these curves the difference in temperature distributions at $\mathbf{Pr}_t = 0.8$ and $\mathbf{Pr}_t = 1$ for $\mathbf{Pr} = 10$ and $\mathbf{Pr} = 30$ is quite significant, but it decreases as $\mathbf{Pr}$ approaches unity.

Curve 2 ($\mathbf{Pr}_t = 0.8$, $\mathbf{Pr} = 5.4$) is similar to the upper curve in Fig. 7.10 given in the Zhukauskas monograph. Solid line corresponds to our computational values, triangles, to the experimental data of Zhukauskas and Shanchyauskas.[27]

5. "Outer" Subregion of a Turbulent Boundary Layer. Clauser Hypothesis

The method of the first order moment is trustworthy because it is accompanied by the scheme of two subregions in a turbulent boundary layer: the "near-wall" one governed by the "wall law", implying application of the Prandtl mixing length theory with correction for the damping factor effect and the outer one described by the "wake law". The near wall subregion was rather thoroughly discussed in two previous sections.

We shall now consider the outer subregion, which occupies a greater portion of the boundary layer, up to 80% of its thickness and, in the near-separation zone, virtually the entire boundary layer.

It is useful to look for the origin of a "two-layer" model of a turbulent boundary layer in the works by Clauser,[28] though the idea of a two-layer model, as applied to a laminar boundary layer with the problem of "joining" the internal and external solutions typical of this

[27] Zhukauskas, A. A. and Shlanchyaushas, A. A., *Teplootdacha v turbulentnom potoke zhidkosti* (*Heat Transfer in a Turbulent Fluid Flow*) Mintis, Vilnius, 1973, 107–111 (cited by the previously mentioned monograph of Zhukauskas).

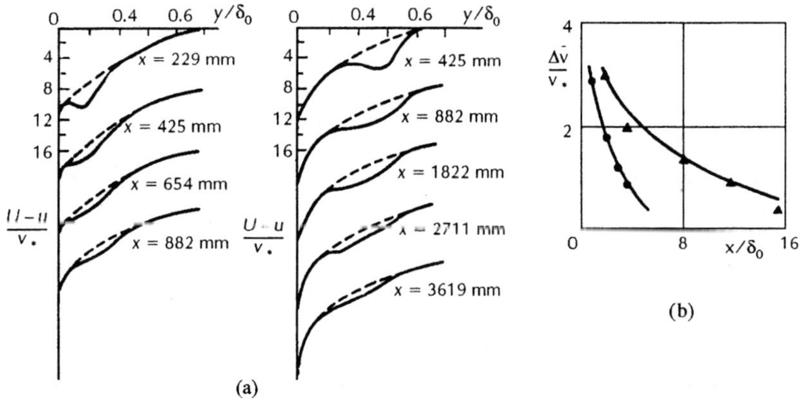

Fig. 255

model, was formulated much earlier by von Kármán.[29]

The outer subregion of a turbulent boundary layer is located between the outer edge of a near-wall subregion and the outer edge (nominally meant as a finite quantity) of a boundary layer. In this region a fluid flow is determined by the "law of the wake" whose main specific feature, in accordance with the Clauser hypothesis, is the constancy (independence of y) of the coefficient of turbulent viscosity ν_t across a boundary layer. A real character of ν_t variation along this region is determined by the "modified Clauser hypothesis", suggested by Lapin and Strelets[30] and stated below. As a turbulent boundary layer develops and the flow passes over to a diverging portion ($dp/dx > 0$), the entire boundary layer and its outer subregion become thicker and, approaching the separation point, the outer subregion tends to fully occupy the entire boundary layer section.

The difference in structures of turbulence of these two subregions was evidently shown by the following experiment, conducted by Clauser, which played a very important role and made it possible not only to better understand the vortex structure of these subregions, but also to call into question the applicability of the Boussinesq hypothesis (Equation 43) of Chapter 13 to the outer subregion of the boundary layer.

A rod with the diameter $d = 12.7$ mm was placed near a solid wall normally to the flow in the section of a boundary layer with the thickness $\delta_0 = 236$ mm at two distances y from the wall, which are defined by the quantities $y/\delta_0 = 16\%$ and $y/\delta_0 = 59\%$. The velocity profiles of this disturbed flow are shown by solid lines in the both parts of Figure 255 a on the left for $y/\delta_0 = 16\%$ (i.e., the rod is placed closer to the wall); on the right for $y/\delta_0 = 59\%$ (i.e., the rod is more remote). The velocity distributions undisturbed by the rod are shown by dashed lines. It can be concluded that the inner portion of a boundary layer

[28] Clauser, F. H., Turbulent boundary layers in adverse pressure gradients, *J. Aeron. Sci.*, 21, 91–108, 1954, and the survey by the same author: Turbulent boundary layer, in *Advances in Applied Mechanics*, Vol.4, Academic Press, New York, 1956, which was translated into Russian and published as *Problemy mekhaniki* (*Problems of Mechanics*), vol.2, Inostrannaya Literatura, Moscow, 1959. The terms the "wall law" and the "wake law" were introduced by D.Coles in The law of the wake in the turbulent boundary layer, *J. Fluid Mech.*, 1(2), 1956. He also constructed a one-term form of velocity profiles in boundary layer sections in the "wake law" region.

[29] von Kármán, Th. and Millikan, C. B., On the Theory of Laminar Boundary Layers Involving Separation, NACA Rep. 504, 1934, and L. G. Loitsyanskiy, *Aerodinamika pogranichnogo sloya* (*Aerodynamics of a Boundary Layer*), Gostekhizdat, Moscow-Leningrad, 1941, 191–204.

[30] Lapin, Yu. K. and Strelets, M. Kh., Modification of the Clauser hypothesis for equilibrium and non-equilibrium turbulent boundary layers. *Teplofiz. Vysok. Temp.*, 23(3), 522–529, 1985.

returns to the undisturbed state quicker than the outer one. The relaxation lengths of disturbances can be estimated as 483 mm at $y/\delta_0 = 16\%$ and 1930 mm at $y/\delta_0 = 59\%$. Some stages of this process are shown in Figure 255a in the form of the dependencies of velocity defects related to dynamic velocity on the distances x between the given boundary layer section and the section with the rod. In Figure 255b the same is shown in the form of a graph of the maximum disturbance $\Delta\bar{v}$ vs. x/δ_0 for the two mentioned rod locations. The cause of this difference is in the rate of disturbance in attenuation Clauser attributed to the difference in sizes of those "vortex masses" which, moving along with the flow and taking part in the motion lateral to the flow, form a turbulent flow structure in the boundary layer sections. Near a solid surface, the vortices formed on the outer edge of the viscous sublayer (of, more precisely, due to the loss of the laminar flow stability on its boundary) have the dimensions of the order of the viscous sublayer thickness (of about 1 to 2% of this thickness), i.e. they are very small compared with the total boundary layer thickness. Turbulence originating in this zone can be determined as "small scale". Conversely, on reaching the outer edge of the boundary layer the vortices attain comparatively large dimensions of the order of the boundary layer thickness. In this case in the outer subregion appears "large-scale" turbulence.

As was noted at the end of Section 1 of Chapter 10 the "lifetime" of large vortices and, consequently, of disturbances transferred by them considerably exceeds the "lifetime" of small vortices, thus explaining the difference in the rate of attenuation of disturbances noted by Clauser (Figure 255). The influence of this peculiarity of the turbulence vortex structure is in no way taken into account by the Boussinesq hypothesis in which turbulence is determined by local isotropy of some imaginary "turbulent fluid" with the law, which is usual for viscous fluid, of the stress tensor dependence on the tensor of deformation velocities in an averaged flow and only with strongly increased value of the variable coefficient of turbulent viscosity. The Clauser experiment offers an example of a turbulent flow with a sharply pronounced phenomenon of the role of prehistory, including this effect and the influence of inhomogeneity of flow vortex structure. Other examples are given at the end of this chapter.

Clauser emphasized this specific effect of flow vortex structure on the relaxation of disturbances introducing, as the parameter allowing for the external aspect of the phenomenon described, the quantity

$$\beta = \frac{\delta^* dp/dx}{\tau_w} \tag{83}$$

which has the sense of the ratio of pressure gradient over the length δ^* characteristic of the boundary layer to the stress of local friction τ_w on a solid edge of the near-wall boundary layer.

Further, it will be shown that it is not sufficient to have one parameter for the description of the whole variety of turbulent flows in boundary layers; the Clauser parameter β is only the first among the parameters, but it is of principal importance in the phenomenon description.

Boundary layers in which all sections the parameter β has the same value are called equilibrium. In these boundary layers velocity profiles in the sections are similar. In this sense "equilibrium" turbulent boundary layers present self-similar solutions for the equations of a turbulent boundary layer. This analogy with a laminar boundary layer occurs strictly in the presence of power-law velocity distribution on the outer edge of the boundary layer[31] with the exponent in this distribution being considered as the parameter related to it unambigously;[32] thus, at $\beta = 0$, 0.9, and 5.4 this parameter is 0, −0.15, and −0.255, respectively.

---

[31] Alber, J. E., Similar Solutions for a Family of Separated Turbulent Boundary Layers, *AIAA Paper*, 1971, 71–203.

[32] Bradshaw, P., The turbulence structure of equilibrium boundary layers, *J. Fluid Mech.*, 29, 625–645, 1967.

By a universal profile of averaged velocities in a turbulent boundary layer we agree to mean the profile prescribed in such universal variables that the shape of this profile is independent of the local coefficient of friction c_f and, consequently, on local Reynolds number. In a near-wall region (i.e., in a viscous sublayer, transitional region, and the logarithmic law region) the profile of the "universal" variables $\varphi = u/v_*$ and $\eta = yv_*/\nu$ will serve as this profile. As g was noted by Ludweig and Tillmann,[33] at not very great values of dp/dx in the near-wall region this profile possesses, besides universality by c_f (the local Reynolds number), also universality by the longitudinal change in pressure (i.e., is independent of β). As is shown by experiments the "universality" by the local Reynolds number is retained in the profiles of velocity defect $(U-u)/v_*$ expressed as a function of y/δ. This universality extends to the entire region of the boundary layer thus making it possible to separate the effect of the longitudinal pressure gradient proper on the distribution of velocities over the entire boundary layer section. With this purpose in mind, as was suggested by Clauser, we shall take instead of δ the quantity

$$\Delta = \int_0^\infty \frac{U-u}{v_*} dy \tag{84}$$

as the conditional boundary layer thickness and introduce into consideration the ratio of two integrals

$$G = \frac{\int_0^\infty \left(\frac{U-u}{v_*}\right)^2 dy}{\int_0^\infty \frac{U-u}{v_*} dy} = \int_0^\infty \left(\frac{U-u}{v_*}\right)^2 d\left(\frac{y}{\Delta}\right) \tag{85}$$

which is independent of the Reynolds number. The thicknesses δ^* and δ^{**} are related to Δ by

$$\frac{\delta^*}{\Delta} = \frac{v_*}{U} = \sqrt{\frac{c_f}{2}} \qquad \frac{\delta^{**}}{\Delta} = \left(1 - G\sqrt{\frac{c_f}{2}}\right)\sqrt{\frac{c_f}{2}} \tag{86}$$

and the parameter H is determined by the equation

$$H = \frac{\delta^*}{\delta^{**}} = \left(1 - G\sqrt{\frac{c_f}{2}}\right)$$

The parameter G is directly dependent on β; at $\beta = 0$ it is equal to 6.8. The graph of the function $G(\beta)$ is given at the end of this chapter due to its use as the index of nonequilibrium of the boundary layer considered there. The role of G as the parameter of boundary layer nonequilibrium is illustrated by the graphs of the velocity defect $(U-u)/v_*$ as the function of y/δ (Figure 256), where δ is the so-called nominal boundary layer thickness, denoted earlier (Equation 40, Chapter 12) by δ_{99}, and the parameter H in the function of c_f (Figure 257). On both graphs the quantity G is adopted as the parameter. The experimental points in Figure

---

[33] Ludweig, H. and Tillmann, W., Investigation of the Wall Shearing Stress in Turbulent Boundary Layers, NACA Tech. Memo., No.1285, 1950.

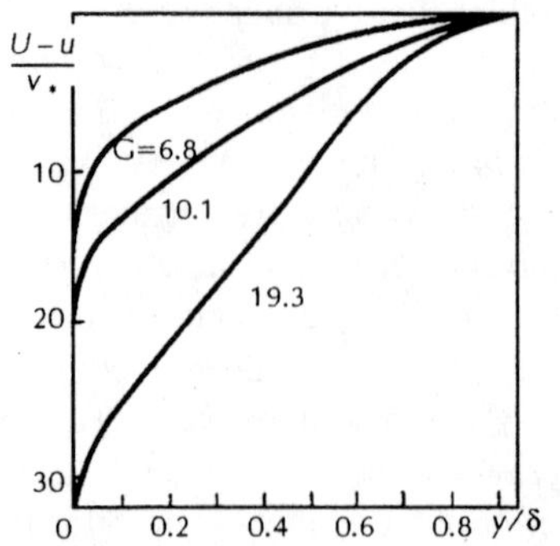

Fig. 256

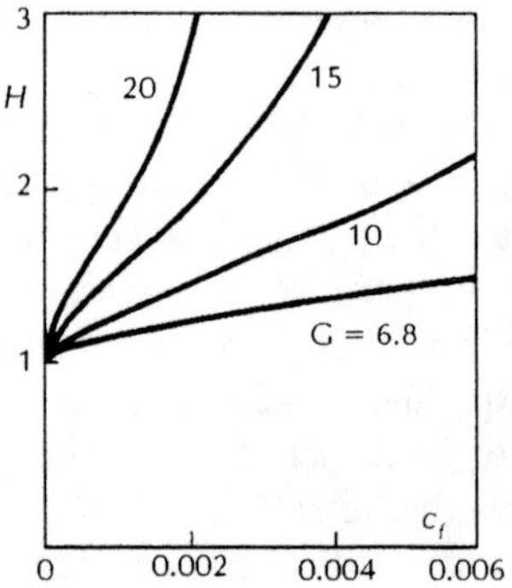

Fig. 257

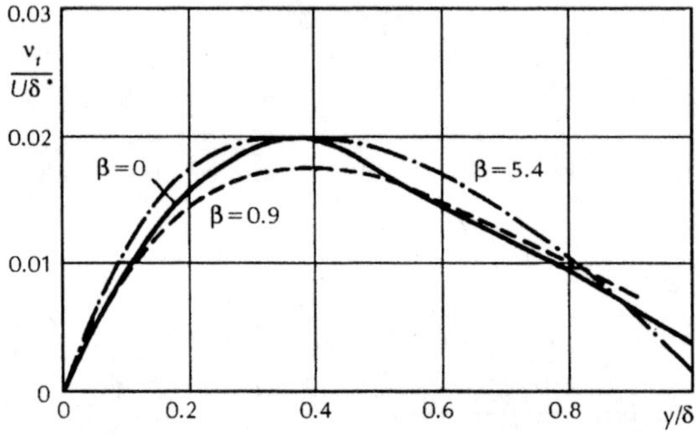

Fig. 258

256 are omitted; they lie closely grouped on the curves given.

The model of the "wake law" also belongs to Clauser. The assumption of the constancy of the kinematic coefficient of turbulent viscosity ν_t in the outer portion of boundary layer sections with its possible change from section to section forms the basis of this model. The experiments do not justify this assumption. As is indicated by modern experimental data,[34] the relationship $\nu_t/(U\delta^*)$ in the function of the dimensionless coordinate y/δ (Figure 258) constructed for three values of $\beta = 0$, 0.9, and 5.4 (the last value of β corresponds to considerable divergence), decreases at all values of β upon approaching the outer edge of the boundary layer. However, as Bradshaw notes "tolerable approaching to the profile of averaged velocity is obtained on the assumption on the constancy of this quantity" (ν_t is meant). This Clauser assumption with these or other corrections is widely used in modern methods of calculation of turbulent boundary layers.[35]

As to the change of ν_t along the outer region of the boundary layer, in this case Clauser made great simplifications. Having noted that the value of ν_t is determined by the product of velocity by the length, Clauser took the dynamic velocity $v_* = \sqrt{\tau_w/\rho}$, which is connected with friction and which has a determining role within the entire region of the boundary layer, as the characteristic velocity, and the size of large vortices occupying the outer region of the boundary layer that is proportional to the boundary layer thickness, as the

[34] Cited by the Report of Bradshaw: The turbulence structure of equilibrium boundary layers. NPL, Teddington; see also the previously cited paper of the same title in *J. Fluid Mech.*

[35] Cebeci, T. and Smith, A.M.O., Analysis of turbulent boundary layer, *Appl. Math. Mech.*, 15, 1974.

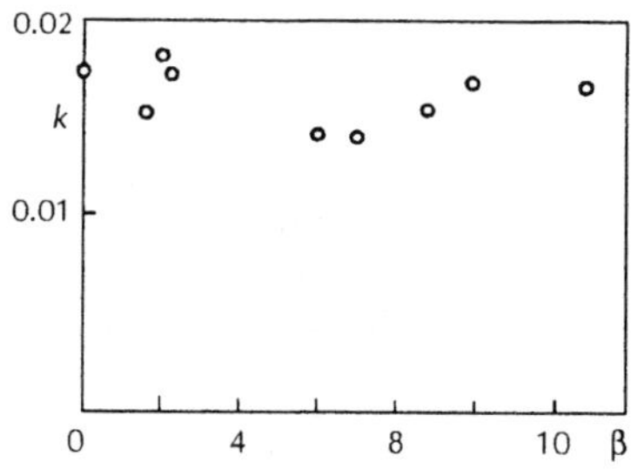

Fig. 259

length introduced by Δ (Equation 85). This led Clauser to the equation for the kinematic coefficient of viscosity ν_t

$$\nu_t = k v_* \Delta$$

or to the following expression of the Clauser hypothesis

$$\nu_t = kU\delta^* = k\int_0^\infty (U-u)\,dy \tag{87}$$

where the coefficient of proportionality k can be, generally speaking, the function of the parameter β (Equation 83) that is equivalent to it by the first equality of Equation 85. The processing of the experiments available at that time gave scattering of points shown in Figure 259, which express the dependence $k(\beta)$; this scattering attains the value of about 25%. Clauser attributed this scattering to inaccuracies in the experiments and suggested that k be considered as the constant value equal to approximately 0.018 over the entire length of the boundary layer. Modern authors, using the assumption of the constancy of k, take $k = 0.0168$.

Further experiments showed that k is a greatly varying function of the parameter β, especially in a nonequilibrium region, so that the diagram in Figure 259 should be extended to both the greater β and smaller k.

Experimental data of Simpson, Chew, and Shivaprasad[36] can serve as an evident confirmation of this fact. The main characteristics of the turbulent boundary layer studied by these researchers are given in Figure 260 Judging by the velocity distributions $U(x)$ on the outer edge on the boundary layer, of pressure gradient $(1/\rho) \cdot (dp/dx)$, and particularly, by the rate of β variation along the boundary layer that is represented by the dimensionless quantity $(d\beta/dx)\,\delta^*$, a conclusion can be made on the considerable nonequilibrium of the boundary layer studied by the authors.

Experimental data of these authors were processed by Lapin and Strelets.[30] The most interesting of their results is the change in the Clauser coefficient k along the boundary layer shown in Figure 261 by a dashed line. This sharp decrease of k in a preseparation nonequilibrium region ($dp/dx > 0$) makes it possible to assume that k and, consequently, ν_t tend to zero. Thickening of the boundary layer in a preseparation region indicates the growth of vortex masses occupying the boundary layer and attaining by their mean dimensions the thickness of this layer. Interaction of these "large vortices" is described by the laws of "coherent turbu-

---

[36] Simpson, R. L., Chew, Y. T., and Shivaprasad, B. G., Structure of separating turbulent boundary layer. I. Mean flow and Reynolds stresses, *J. Fluid Mech.*, 113, 23–51, 1981.

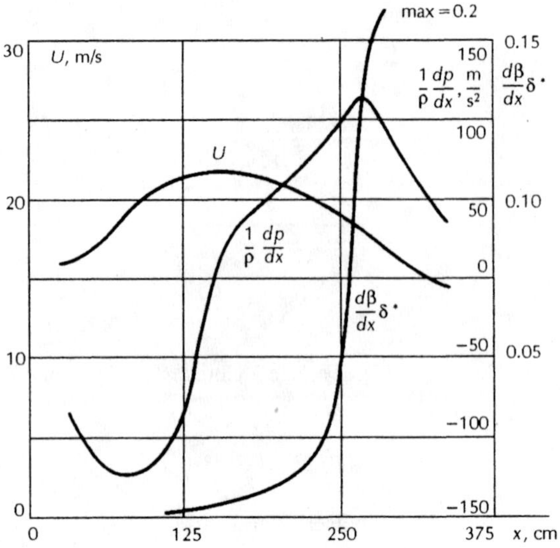

Fig. 260

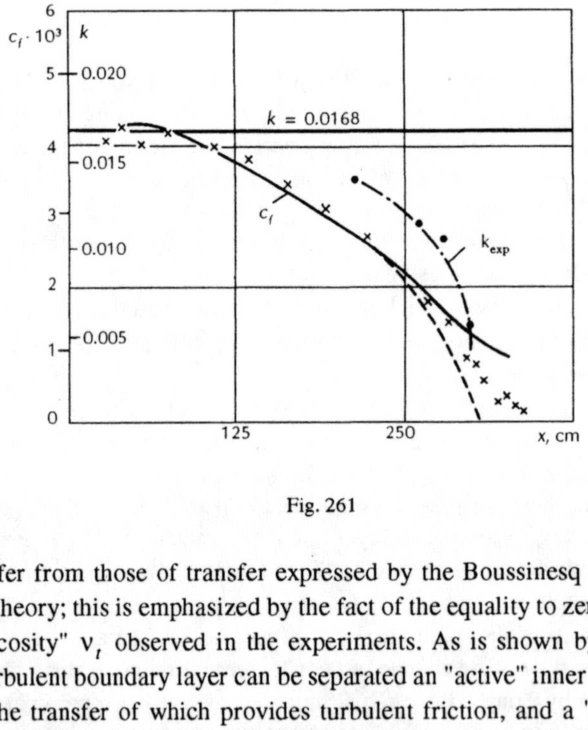

Fig. 261

lence" which differ from those of transfer expressed by the Boussinesq hypothesis and the "mixing length" theory; this is emphasized by the fact of the equality to zero of the coefficient of "turbulent viscosity" ν_t observed in the experiments. As is shown by Bradshaw,[32] in a fluid flow in a turbulent boundary layer can be separated an "active" inner portion which generates vortices, the transfer of which provides turbulent friction, and a "passive" outer one containing developed "large vortices" which are subjected to pressure fluctuations, but which do not take part in friction formation. The mechanism of interaction of these "large" vortices has not been defined as yet, but in some cases it allows study by numerical methods. These problems are the concern of the modern theory of "coherent turbulence".

Bradshaw emphasizes the importance of skin friction τ_w on a solid boundary (i.e., on the lower boundary of the near-wall subregion) for the character of flow over the entire boundary layer, including both its inner and outer subregions. This also determines a great role for the dynamic velocity $v_* = \sqrt{\tau_w/\rho}$ as the main scale of velocities in boundary layer sections, a fact frequently mentioned in this and previous chapters.

The Clauser hypothesis on the constancy of k, as shown by the comparison of calculations with test material, retains its value in converging ($dp/dx < 0$) and nongradient ($dp/dx = 0$) portions of a boundary layer and, with certain approximation, in a diverging portion at small deviations from an equilibrium pressure distribution. In preseparation regions of nonequilibrium boundary layers, Lapin and Strelets successfully used a semiempirical formula which they suggested for the Clauser coefficient k

$$k = 0.0168\exp\left(-A\beta + B\frac{d\beta}{dx}\delta^* - \dots\right) \tag{88}$$

where $\beta = (dp/dx)(\delta^*/\tau_w)$.

This formula gives good results in the case of equilibrium layers ($d\beta/dx = 0$) and may be successfully applied to nonequilibrium layers in its general form (88). Moreover, Equation 88 is deprived of the drawback of the drawback of the initial Clauser equation (87) which does not satisfy the conditions near a separation point

$$k \to 0 \quad \text{with} \quad \tau_w \to 0 \qquad \beta \to \infty \tag{89}$$

To determine constants A and B the authors of relationship (88) used the series of experiments (1100, 1200, 1300, 2100, 2200, 2300, 3300) presented in the previously cited *Proceedings of the Stanford Conference*; in this case at each step of the numerical calculation there were chosen such values of A and B leading to the agreement with the experiment that then were averaged by the method of least squares. Following are the recommended values of the constants

$$A = 0.177 \qquad B = 7.0 \tag{90}$$

6. Results of Numerical Calculations of Turbulent Boundary Layers by the Method of First Order Moments

By its complexity the problem of calculating turbulent boundary layers cannot be compared with an analogous problem for laminar flows. The matter here is in principal difficulties of the formulation of the problems on turbulent flows rather than in the volume of calculations; with the modern level of development of computers this would not be difficult. The absence of full clarity in the concepts of the mechanism of turbulent flows manifest itself, first of all, by the nonclosed character of the system of equations of momentum transfer, Reynolds stresses, and other turbulent characteristics and also by insufficiently rational justification of additional relations which are called "connectives" and which are meant for approximate closure of transport equations. For the method of first order moments, stated in this section, such "connectives" are the Prandtl theory of mixing length with the correction for a damping factor in the inner near-wall subregion of a turbulent boundary layer and the Clauser hypothesis (for the outer subregion) with the modification made by Lapin and Strelets mentioned at the end of the previous section.

In spite of the existence of a great number of methods for turbulent boundary layer calculation (the method stated is nearly the simplest among them) the problem of selecting a practically acceptable method has always been related to the degree of agreement between calculation results and test data. Even when calculating involves the experimental distribution of the outer velocity, thus removing the problem on a separate account of the inverse effect of the boundary layer on the external irrotational flow, there remains doubt about the correctness of the computational determination of the separation point. The fact remains that, similar to the case of a laminar boundary layer (recall the end of Section 2, Chapter 12), the turbulent boundary layer equations (1), which are the result of rejection in Equations 15 of the Chapter 13 of the terms $\partial^2 u/\partial x^2$, $\partial\tau/\partial x$, and $\partial p/\partial y$, in accordance with the "boundary layer approximation", cannot be used in the separation region. In the case of a turbulent boundary layer the problem is aggravated by the fact that the flow structure in the separation region precludes the application of the Boussinesq model and of Equation 43 of Chapter 13 because here the mechanism of turbulent transfer typical of the near-wall flows is replaced by the interaction of "large vortices" which is described by the model of "coherent" turbulence (see the end of Section 1). The usual technique of determining the position of the separation point becomes even more "conventional" as compared with the case of a laminar boundary layer. A general analysis of the phenomenon of turbulent separation can be found in the monograph by Gogish and Stepanov.[37] The works by Stratford[38] deal with the experimental study of turbulent boundary layer separation. The available literature on moment methods of turbulent boundary layer calculation is so voluminous that the bibliography on the topic would take too much place. We shall restrict ourselves to some recommendations given in monographs and reviews.[39]

The method of moments of the first order presented in this section is reduced to numerical integration of the system of equations (1) in which it is assumed

$$-\overline{u'v'} = \nu_t \frac{\partial u}{\partial y} \tag{91}$$

$$\nu_t = \begin{cases} æ^2 y^2 D(y) \dfrac{\partial u}{\partial y} & \text{for} \quad 0 \le y \le y_m \\ kU\delta^* \gamma & \text{for} \quad y_m < y < \infty \end{cases} \tag{92}$$

Here, besides the notation adopted in Equation 1, the following is available: ν_t is the kinematic coefficient of "eddy viscosity", $D(y)$ is the damping coefficient determined by the van Driest formula (26) or by Equation 28, æ is the von Kármán constant, k is the multiplier in the Clauser modified equation (88), U is the velocity on the outer edge of the boundary layer, δ^* is the displacement thickness; $y_m = y_m(x)$ is the interface between outer and inner subregions of a boundary layer; and γ is the coefficient of intermittency of laminar and turbulent boundary layers adopted to be equal to ($b \approx 6$ is the empirical coefficient)[40]

---

[37] Gogish, L. U. and Stepanov, G. Yu., *Turbulentnye otryvnye techeniya (Turbulent Separation Flows)*, Nauka, Moscow, 1979.

[38] Stratford, B. S., The prediction of separation of the turbulent boundary layer, *J. Fluid Mech.*, 5(1), 1, 1959. An experimental flow with zero skin friction throughout its region of pressure rise, *J. Fluid Mech.*, 17–35, 1959.

[39] Ginevskiy, A. S., Ioselevich, V. A., Kolesnikov, A. V., Lapin, Yu. V., Pilipenko, V. N., and Sekundov, A. N. Itogi nauki i tekhniki. Mekhanika zhidkosti i gasa, (Methods for calculation of a turbulent boundary layer in *Review of Science and Technology Fluid Mechanics*, Vol.11, VINITI Press, Moscow, 1978, 155–304. This review includes about 700 references to Soviet and foreign papers. Of great interest are the monographs Fedyayevskiy, K. K., Ginevskiy, A. S., and Kolesnikov, A. V., *Calculation of a Turbulent Boundary Layer of Incompressible Fluid*, Sudostroyeniye Press, Leningrad, 1973; *Turbulence, Principles, Application*, W. Frost and T. Moulden, Eds., Mir Press, Moscow, chap. 7 to 10; Belov I. A., *Modeli turbulentnosti (Turbulence Models)*, Leningradskiy Mekhanicheskiy Institut Press, Leningrad, 1986; Bradshaw, P., Cebeci, T., and Whitelaw, J., *Engineering Calculation Methods for Turbulent Flow*, Academic Press, London, 1981, chap. 1 to 7.

[40] Klebanoff, P. S., Characteristics of Turbulence in a Boundary Layer With Zero Pressure Gradient, NACA Rep. No.1247, 1955.

$$\gamma = \left[1 + 5.5\left(\frac{y}{\delta}\right)^b\right]^{-1}$$

Cebeci and Smith,[39] having replaced in assumption (29) the dynamic velocity v_* by the velocity $v_v = 11.8v_*$, which is equal to the velocity at the outer edge of a viscous sublayer, substituted $A_* = 26$ in the van Driest equation by a new constant $A = 26(1 + 11.8p_*)^{-1/2}$, where $p_* = \dfrac{dp}{dx}\dfrac{\nu}{\rho v_*^3}$, and introduced into the Clauser equation on the right side of the last equation of system (92) the additional multiplier, which takes into account the Reynolds number effect.

Neglecting these corrections but accounting for the effect of the coefficient of intermittency, γ, Lapin and Strelets, using Equations 88 and 90, calculated a boundary layer that was experimentally studied by Simpson, Chew, and Shivaprasad (see the earlier reference). The most important boundary layer characteristic was calculated, i.e., the coefficient of local friction on a solid surface c_f. Having a desire to find a comparative value of the second term in the exponent index of Equation 88, Lapin and Strelets made two separate calculations: the first by the complete Equation 88 (shown by a solid curve in Figure 261), and the second by the same equation, but without the second term in the exponent (shown by a dashed curve in Figure 261). These curves bounded from two sides the test points which are shown by crossess; this indicates a relative closeness of the calculation results by the complete equation to the experiment, at least within the region $0 < x < 300$ cm. Extrapolating the curves to the zero value of c_f we obtain some scattering in the position of the separation point. It hardly can be attributed to the imperfection of formula (88), which for further improvement requires resorting either to the next derivatives of β and U (an operation hardly accurately realizable), or to the general failure of the Boussinesq hypothesis-based calculation in this transient (toward the "coherent turbulence") region of the boundary layer which far overshoots the limits of equilibrium.

The results of calculations of two characteristics, c_f and $H = \delta^*/\delta^{**}$, performed by Lapin and Strelets (Figure 262) can serve as the confirmation of the credibility of the coefficient k expressed by formulas (88) and (90) in the case of a boundary layer slightly differing from the equilibrium value (experiment "1100" by the classification used in the Proceedings of the Stanford Conference). The solid lines in Figure 262 refer to calculations with the aid of the formula modified by these authors for k (Equations 88 and 90) and dashed lines refer to calculations by the classical Clauser formula with the constant $k = 0.0168$. Experimental values are shown by crosses for c_f and by dots for H. The curves in Figure 262 confirm the assumptions made earlier of the satisfactoriness of the modified Clauser formula application to about the entire boundary layer, provided that its nonequilibrium is small. The deviation of the results of calculations by the method based on the initial Clauser hypothesis about the coefficient k constancy are noticeable even in this case of boundary layer small nonequilibrium.

A priori judgments on the fitness or unfitness of the method of first-order moments for separate problems with arbitrary velocity distribution on the outer edge of the boundary layer can hardly be made. The necessity to validate the calculation results based on different experimental data still remains.

Of some difficulty for numerical calculation by a two-layer scheme is the necessity of matching "inner" and "outer" solutions. This problem is thoroughly considered by Lapin.[41] In

---

[41] Lapin, Yu. V., *Turbulentnyi pogranichyi sloy v sverkhzvukovykh potokakh gaza* (*Turbulent Boundary Layer in Supersonic Gas Flows*), 2nd ed., Nauka Press, Moscow, 1982.

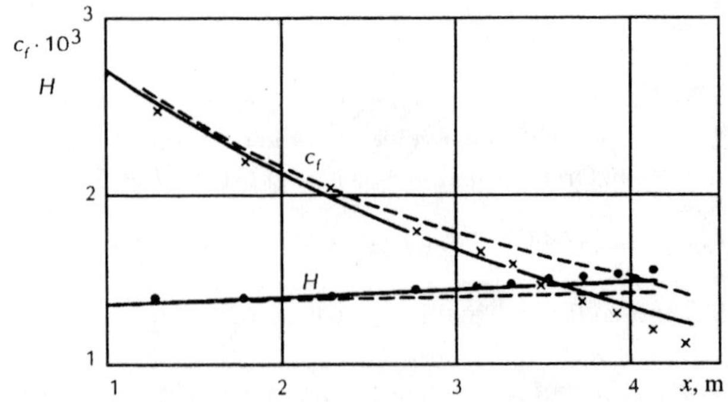

Fig. 262

these calculations matching was made by the coefficient of eddy viscosity ν_t and, because the method is based on the Boussinesq hypothesis (43) of Chapter 13, also by the derivative $\partial u/\partial y$. Note that in the case when the point of matching of subregions in the cross-sections of boundary layers that approach the separation cross-section gets first into the transition portion of a near-wall subregion and then into a viscous sublayer, the determination of the ordinate y_m of the point of matching is reduced to the solution of a complex transcendental equation; it is solved in the course of numerical calculation and confirms the y_m tendency to zero, thus corresponding to the following-in of the entire separation cross-section by the outer subregion subjected to the wake law.

Estimating the validity of method of first order moments we show its advantages compared with empirical and semiempirical methods stated in Chapter 13. The calculations performed demonstrated its higher accuracy in comparison, even with the best of the "old" methods – the method of "integral" relations.

To the shortcomings of the method of first order moments can be attributed its one-parameter nature, which takes place in the initial formulation of the method, based on the Clauser hypothesis on the k constancy, and also the absence of rational justification for the choice of the quantity β as the only parameter – a feature that is unconvincing for the majority of known nonequilibrium boundary layers. In this sense, of special value is the modification of the Clauser formula (88), suggested by Lapin and Strelets, by introducing the multiplier involving two parameters β and $(d\beta/dx)\,\delta^*$. This modification alleviates the mentioned shortcoming of the method of first order moments.

Among the publications dealing with detailed numerical calculations of turbulent boundary layers by the above method is the paper by Yemeliyanova[42] who used the implicit differential scheme and the factorization method (discussed later). The Clauser formula was applied in its initial nonmodified form, thus making it impossible for the author to approach a preseparation region and to determine the separation point. The results are compared with the experimental data given in Proceedings of the Stanford Conference.

The method of generalized similarity in application layers rests on the same two semiempirical bases as the formulated method of the first order. These bases are the "mixing length" theory, refined by the introduction of the damping factor (for the inner subregion of the boundary layer), and the Clauser hypothesis in its initial or modified form (for the "outer subregion"). The method of generalized similarity includes nothing new from the viewpoint of the verification of the turbulence mechanism. However, the role of the method of generalized

---

[42] Yemeliyanova, G. N., Numerical method for solving equations of a plane turbulent boundary layer of an incompressible fluid with longitudinal pressure gradient, *Tr. TsAGI*, No. 1543, 1973.

similarity in rationalizing the techniques for the determination of the basic parameters together with the known calculational advantages allows one to consider it promising.

The first, which were not general searches for the application of the method of generalized similarity in the theory of a turbulent boundary layer introduced little into a wide formulation of the problem.[43]

In contrast to a laminar boundary layer for which the method of generalized similarity was formulated in Sections 12 and 13 of Chapter 12, in the case of a turbulent boundary there is much freedom in the choice of main scales of the basic quantities used in the method, with these scales being different for near-wall and outer subregions. In other respects the algorithm remains as before, and, as in the case of a laminar boundary layer, leads to the transformation of the boundary layer differential equation to "the variables of generalized similarity", i.e., to the "universal" form for all boundary layer problems which require only single numerical integration and composition of tables of used solutions when solving separate concrete problems.

For the inner (near-wall) subregion of a turbulent boundary layer the following universal scales (Section 10 of Chapter 13) (we shall call them "inner") are typical: (1) of lengths $l_* = \nu/v_*$, (2) of velocities $v_* = \sqrt{\tau_w/\rho}$ (the dynamic velocity), (3) of skin friction $\tau_w = \mu(\partial u/\partial y)_{y=0}$, and (4) of stream function $v_* l_* = \nu$.

The outer scales for the outer subregion are δ^*, δ^{**} or any other conventional boundary layer thickness, the length scales; the velocity on the outer edge of the boundary layer $U(x)$, the velocity scale; the fitting by the dimensions quantity ρU^2, the skin friction scale; the products $U\delta^*$, $U\delta^{**}$ or similar, the stream function scale.

Along with these, directly relating to the corresponding subregions of the boundary layer, other "mixed" scales are also possible. Thus, for instance, in the outer region, as the velocity scale one can use the same velocity scale v_*, that in the inner region, instead of $U(x)$ and the skin friction scale τ_w or any other value of skin friction in the boundary layer sections can be taken instead of ρU^2.

If we obey natural restrictions in the choice of scales, having required that they are not equal to zero or infinity within the limits of their application, the set of "inner" scales should be divided into two portions. To the first zone of the boundary layer that involves the point of minimum pressure ($dp/dx = 0$), but that is a rather large distance from the separation point ($\tau_w = 0$) the system of "inner" scales (mentioned above) may be related. The second subregion of the boundary layer, following the first, does not suit the previous group of scales because the used quantity of skin friction on a solid wall τ_w and, consequently, $v_* = \sqrt{\tau_w/\rho}$ vanish at the separation point while l_* becomes infinity.

In this second zone of the "inner" subregion, the value of the longitudinal pressure gradient dp/dx, which does not vanish throughout the zone may be taken as the basis for the determination of scales, having assigned

$$l_S = \frac{\nu}{v_S} \qquad v_S = \sqrt{\left(\frac{\nu}{\rho}\right)\left(\frac{dp}{dx}\right)} \qquad \tau_S = \rho v_S^2 = \rho\sqrt{\left[\left(\frac{\nu}{\rho}\right)\left(\frac{dp}{dx}\right)\right]^2}$$

and the stream function scale $v_S l_S = \nu$ similar to the first zone.[44]

The following is worth mentioning. If in the first zone outside a viscous sublayer and the

---

[43] Loitsyanskiy, L. G., Similarity parameters in the boundary layer theory, in *Problemy gidrodinamiki i mekhaniki sploshoy sredy* (*Problems of Hydrodynamics and Mechanics of Continuum* (*Collected Papers Commemorating the 60th Anniversary of Academician L.I.Sedov*) Nauka Press, Moscow, 1969, 301–310, sect. 4; Mekhanika i Energomashinostroyeniye, Leningrad, Trudy LPI, 1–76, No. 352, 4–9.

[44] Stratford, B. S., The prediction of separation of the turbulent boundary layer, *J. Fluid Mech.*, 1959, v. 5, part 1, p. 5.

transition zone there occurs a logarithmic velocity profile, then in the second zone it is replaced by a power-law velocity profile. In fact, in the familar expression for the distribution of friction τ near a solid wall, i.e., at small y,

$$\tau = \tau_w + \left(\frac{dp}{dx}\right)y$$

when approaching the separation point the first term is small and may be omitted as compared with the second one. Retaining the Prandtl equation and the linear law of mixing length with the same von Kármán constant æ in this second zone outside the viscous sublayer and the transition zone, we shall have the equation

$$\rho æ^2 y^2 \left(\frac{du}{dy}\right)^2 = \frac{dp}{dx} y$$

whose integral, in contrast to the logarithmic law in the first zone of the boundary layer acquires the form of the power law of the "root square"

$$u = \frac{2}{æ}\sqrt{\frac{1}{\rho}\frac{dp}{dx}}\sqrt{y} + \text{const}$$

which is confirmed by the experiments.[45]

A general problem of a rational choice of the scales of length of velocities and other quantities for different zones along and across a boundary layer acquired an interesting treatment in a number of papers by Kader and Yaglom[46] that are based on the dimensional theory.

Choice of this or that set of scales determines the form of a "universal" equation and of its solution. At present in the Department of Hydroaerodynamics of the Leningrad Polytechnic Institute studies on possible applications of the method of generalized similarity to the calculation of a turbulent boundary layer are conducted only within the framework of the first order moment equation.

Complete calculation of velocity profiles in the inner subregion of the first stage has been made by the method of generalized similarity.[47] The corresponding numerical calculation for the outer region is impeded by the effect of a great number of outer parameters, c_f, $H = \delta^*/\delta^{**}$, $\mathbf{Re}^{**}$, and these are presently programmed by Strelets.

7. Review of the Application of Second Order Models. Models "$k - \varepsilon$", "$\overline{u'v'} - k - \varepsilon$"

Use of even more powerful computers in computational practice attracts the researchers' attention to the utilization of the equations of transfer of turbulence characteristics (Section 6 of Chapter 13) involving second order moments as the unknown quantities but having also third order moments (both velocity and pressure fluctuations). To reduce these equations

[45] This "half-power" law was first mentioned by E. E. Solodkinn and I. I. Mezhirov in 1950 (see the review edited by A.S.Ginevskiy, cited earlier, p.183). It was experimentally confirmed in the paper Stratford, B. S. An experimental flow with zero skin friction throughout its region of pressure rise, *J. Fluid Mech.*, 5(1), 17–35, 1959.

[46] Kader, B. A. and Yaglom, A. M., Application of similarity considerations to the calculation of retarding turbulent boundary layers, *Dokl. AN SSSR*, 233(1), 52–55, 1977. By the same authors: Similarity laws for near-wall turbulent flows, Itogi nauki i tekhniki, *VINITI, Mekh. Zhidk. Gaza*, 15, 81–155, 1980.

[47] Loitsyanskiy, L. G., Method of generalized similarity in the calculations of the inner subregion of a turbulent boundary layer, *Mekh. Zhidk. Gaza*, No. 5, 1990.

(second order models) to closed form the necessity arises to apply "determining relations" to separate terms of these equations that simplify the formulation of the problem. As is known by recent investigations[48] the first order models are less accurate than the second order, though the latter require much more computer time. We briefly elucidate these models.

At present the number of methods that are based on the use of the second-order models and which differ from each other in the number of involved transfer equations and unknown quantities expressing the turbulent flow characteristics is large. Chapter 8 of the collected papers edited by Frost and Moulden gives the classification of the existing methods and their short description.[49]

The most favorable referee comments in the literature refer to the methods containing the equations of kinetic energy transfer. Among them the method based on combined solution of equations of transfer of momentum, kinetic energy, and dissipation energy, the so-called $k-\varepsilon$ method, is distinguished. In this method the set of equations is subjected to the combined solution of

$$\frac{\partial u}{\partial x}+\frac{\partial v}{\partial y} = 0$$

$$u\frac{\partial u}{\partial x}+v\frac{\partial u}{\partial y} = -\frac{1}{\rho}\frac{dp}{dx}+\frac{\partial}{\partial y}\left(\nu\frac{\partial u}{\partial y}-\overline{u'v'}\right)$$

$$u\frac{\partial k}{\partial x}+v\frac{\partial k}{\partial y} = \frac{\partial}{\partial y}\left(\frac{\nu_t}{\sigma_k}\frac{\partial k}{\partial y}\right)+\nu_t\left(\frac{\partial u}{\partial y}\right)^2-\varepsilon$$

$$u\frac{\partial \varepsilon}{\partial x}+v\frac{\partial \varepsilon}{\partial y} = \frac{\partial}{\partial y}\left(\frac{\nu_t}{\sigma_\varepsilon}\frac{\partial \varepsilon}{\partial y}\right)+C_1\frac{\varepsilon}{k}\nu_t\left(\frac{\partial u}{\partial y}\right)^2-C_2\frac{\varepsilon^2}{k} \tag{93}$$

where it is assumed that

$$k = \frac{1}{2}\overline{(u'^2+v'^2)} \qquad \varepsilon = \nu\overline{\left(\frac{\partial u'}{\partial x}\right)^2+\left(\frac{\partial v'}{\partial y}\right)^2}$$

$$-\overline{u'v'} = \nu_t\frac{\partial u}{\partial y} \tag{94}$$

To close the system of equations use is made of the Prandtl-Kolmogorov "jointing"

$$\nu_t = C_\nu\frac{k^2}{\varepsilon} \tag{95}$$

based on local analogy with the theory of homogeneous and isotropic turbulence.

On the right sides of the last two equations of system (93) the coefficients ν_k for kinetic energy k and ν_ε for dissipation velocity ε are substituted by the Prandtl numbers $\sigma_k = \nu_t/\nu_k$ and $\sigma_\varepsilon = \nu_t/\nu_\varepsilon$.

---

[48] Chang, K. C. and Cebeci, T., Numerical aspects of turbulence models, in *Int. Symp. Comput. Fluid. Dynamics, Tokyo*, 1985, preprint Vol. 2, Tokyo, 1985, 538–549.

[49] See the earlier cited monograph by P. Bradshaw, pp.45 to 49, in which some of these methods are presented in the form easily understood by engineers; critical treatment by A. N. Sekundov can be found in Section 1, Chapter 1 of the review written by A. S. Ginevskiy et al mentioned earlier.

The system of equations of the $k-\varepsilon$ method contains a number of empirical constants: $C_v = 0.09$, $C_1 = 1.44$, $C_2 = 1.92$ $\sigma_k = 1.0$, and σ_ε determined from

$$\sigma_\varepsilon = \frac{æ^2}{\sqrt{C_v}\,(C_2 - C_1)} \qquad æ = 0.4 \tag{96}$$

A detailed analysis of the method illustrated by a great number of boundary layer calculations, experimental data for which are given in Proceedings of the Stanford Conference, can be found in the paper by Singhal and Spalding.[50] Note that judging by the second line of Equations 94, the $k-\varepsilon$ method is based on the Boussinesq hypothesis which, as is shown in the next section, does notsatisfactorily reflect the properties of turbulent flow in those regions of a boundary layer where the flow structure is determined by large-scale turbulence. These regions refer to that located near the outer edge of the boundary layer and nearly the entire boundary layer in the preseparation portion. This shortcoming is eliminated in the $\overline{u'v'} - k - \varepsilon$ method in which the Boussinesq hypothesis is replaced by the equation of Reynolds stress transfer that involves turbulent skin friction as the unknown dependent variable. We shall be content with the presentation of the system of equations of the $\overline{u'v'} - k - \varepsilon$ method:

$$u\frac{\partial \overline{u'v'}}{\partial x} + v\frac{\partial \overline{u'v'}}{\partial y} = C_s \frac{\partial}{\partial y}\left(\frac{k^2}{\varepsilon}\frac{\partial \overline{u'v'}}{\partial y}\right) - C_{\varphi 1}\left(\frac{\overline{u'v'}\varepsilon}{k} + C_\mu k \frac{\partial u}{\partial y}\right)$$

$$u\frac{\partial k}{\partial x} + v\frac{\partial k}{\partial y} = 0.9\,C_s \frac{\partial}{\partial y}\left(\frac{k^2}{\varepsilon}\frac{\partial k}{\partial y}\right) - \overline{u'v'}\frac{\partial u}{\partial y} - \varepsilon$$

$$u\frac{\partial \varepsilon}{\partial x} + v\frac{\partial \varepsilon}{\partial y} = C_\varepsilon \frac{\partial}{\partial y}\left(\frac{k^2}{\varepsilon}\frac{\partial \varepsilon}{\partial y}\right) - C_{\varepsilon 1}\frac{\overline{u'v'}\,\varepsilon}{k}\frac{\partial u}{\partial y} - C_{\varepsilon 2}\frac{\varepsilon^2}{k} \tag{97}$$

In contrast to the system of equations (93) of the $k-\varepsilon$ method the Prandtl numbers for $\overline{u'v'}$, k, and ε are supposed to be equal to unity. The constants in Equations 97 are taken to be[51]

$$C_S = 0.1 \quad C_{\varphi 1} = 2.8 \quad C_\mu = 0.09 \quad C_\varepsilon = 0.09 \quad C_{\varepsilon 1} = 1.40 \quad C_{\varepsilon 2} = 1.95.$$

Frost and Moulden[39] state different moment methods and the results of computations by these methods are compared.

A conclusion can be reached that the problem of turbulent boundary layer calculation, including the calculation of the separation point position, is far from being completed.

The complexity of the existing turbulent boundary layer models is attributed to the operation of averaging the Navier-Stokes equations and to the difficulties arising due to it in closing the system of the obtained equations with the aid of different "jointings" and determining relationship, that lead the equation to a numerically integrated form. A recently developed method (Polezhayev et al.) which is based on direct (without preliminary averaging) integration of the Navier-Stokes equations is deprived of all these shortcomings.

In view of the development of effective numerical methods for solving the Navier-Stokes

---

[50] Singhal, A. K. and Spalding, D. B., Prediction of two-dimensional boundary layers with the aid of the "$k-\varepsilon$" model of turbulence. in *Computer Methods in Applied Mechanics and Engineering*, Vol. 25, North-Holland, Amsterdam, 1981, 305–383.

[51] Frost, W. and Moulden, T., Eds., *Turbulence, Principles and Application*, Mir Press, Moscow, 1980.

equations of viscous incompressible fluid flow and of the appearance of powerful computers there arose the possibility to conduct "numerical experiments" which directly model turbulent flows. We mention the fundamental study of Rozhdestvenskiy and Simakin,[52] who studied two- and three-dimensional nonstationary flows of incompressible viscous fluid in an infinite plane channel with parallel walls. The authors note that the results of their calculations perfectly reflect characteristic features of turbulent flows, with the calculations of three-dimensional flows being in the best agreement with the experiment. Characteristics of turbulent flows such as the coefficient of channel resistance, mean velocity profile, and distribution of square-mean velocity fluctuations are calculated with high accuracy.

8. Relaxation Phenomena in Turbulent Boundary Layers. Hinze Relaxation Equation

Just as the specific features of the molecular structure of a viscoelastic fluid (Section 2, Chapter 10) stipulated a number of its special properties (finiteness of the velocity of disturbance propagation, influence of flow prehistory on subsequent motion) which justified its name as a "hereditary" fluid, so the specific features of the "vortical" structure of turbulent flows provide the latter with the properties similar to those mentioned above of viscoelastic fluids. This remarkable fact is warranted by numerous experiments, among which the Clauser experiment (Section 5) on the slowed down relaxation of disturbances in a large-scale turbulence flow was historically the first. There also exist theoretical proofs of the hereditary nature of small-scale turbulence flows,[53] where heredity shows itself to be comparatively weak.

Note that the approach to a turbulent viscous fluid, flow for a laminar flow of some imaginary "turbulent" fluid, which possesses specific features taken by Boussinesq (see Section 5, Chapter 13), cannot be fully justified because it by no means transmits the entire variety of processes occurring in turbulent flows. This, of course, also refers to the above-mentioned possibility of considering the turbulent flow as a motion of some hypothetical viscoelastic fluid. However, if we restrict ourselves only to a partial analogy between separate specified properties, then this approach is quite acceptable.[54] We shall make use of such an analogy, confronting it with a more strict, theoretically warranted approach.

We shall first address the interesting and widely developed experiments of Japanese scientists Tsuji and Morikawa.[55] In contrast to "smooth" velocity distributions on the outer edge of the boundary layer $U(x)$ with one maximum portion and one converging and diverging portion that are usually in the experiments on boundary layers, these authors considered the case of a more complex peculiar distribution $U(x)$ (Figure 263), with alternating sign of dp/dx in the presence of two diverging ducts.

In the inlet portion I, which is not shown in the figure, $U = \text{const}$, $U' = 0$. Then, as can be seen from Figure 263, diverging portion II ($U' < 0$), converging portion III ($U' > 0$), diverging portion IV ($U' < 0$) follow and converging portion V ($U' > 0$) begins. In the converging

---

[52] Rozhdestvenskiy, B. L. and Simakin, I. N., Simulation of turbulent flows in a plane channel, *Zh. Vych. Mat. Mat. Fiz.*, 25(1), 86–121, 1985 (contains a voluminous bibliography). See also Shumann, V. V., Gretzbach, G., and Kleiser, L., Direct methods of numerical simulation of turbulent flows, in *Methods for Calculation of Turbulent Flows*, Kollman, W., Ed., Mir press, Moscow, 1984; Orzag, S., Numerical simulation of turbulent flows, in *Turbulence, Principles and Applications*, (Frost, W. and Moulden, T., Eds., Mir Press, Moscow, 1980.

[53] Crow, S. C., Viscoelastic properties of finegrained incompressible turbulence, *J. Fluid Mech.*, 33(1), 1–20, 1968.

[54] Rivlin, R. S., *Q. Appl. Math.*, 15(212), 1957.

[55] Tsuji, I. and Morikawa, G., Turbulent boundary layer with pressure gradient alternating its sign, *Aeron. Q.*, 27(1), 15–28, 1976.

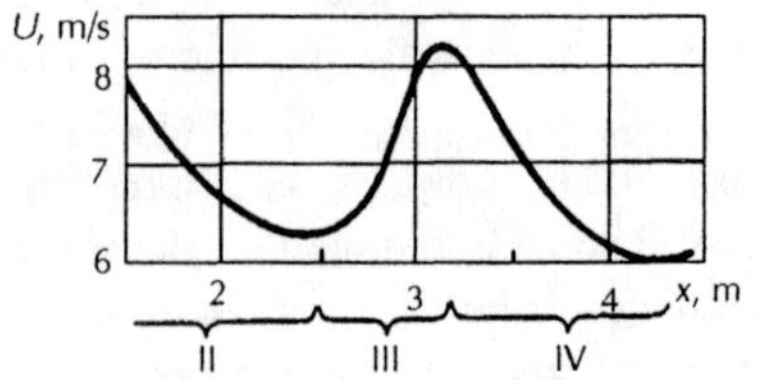

Fig. 263

portions the condition of relaminarization was provided, while in the diverging ones the condition of boundary layer separation was provided.

The study includes the Clauser diagram (Section 5) by which a conclusion can be made (Figure 264) on nonequilibrium of the boundary layer considered with the exception of the initial portion only. The solid line on the diagram corresponds to $G(\beta)$ for equilibrium boundary layers with different but constant along the flow values of β. The dashed spiral curve expresses the deviation from the equilibrium state on subsequent portions of a boundary layer.

Using the thermal anemometry method, the authors accurately measured the quantities $\overline{u'v'}$ over the boundary layer cross-sections, which correspond to different values of the abscissa x (Figure 265).

The straight lines on the left indicate the slopes of tangent lines to the curves of full skin friction $\tau/\rho = \nu du/dy - \overline{u'v'}$ referred to as density. At some distance from the wall this slope can be referred to the distribution of turbulent friction only.

As follows from the equations of first order moments (Equation 1)

$$\left[\frac{\partial(\tau/\rho)}{\partial y}\right]_{y=0} = \frac{dp}{dx} \tag{98}$$

or, neglecting usual viscosity,

$$\left[\frac{\partial(-\overline{u'v'})}{\partial y}\right]_{y=0} = \frac{dp}{dx} \tag{99}$$

In conformity with this the straight lines characterizing the slopes of tangent lines to the curves of skin friction distribution should represent positive slopes in diverging portions $(dp/dx > 0)$ and negative slopes in converging portions of a boundary layer correspondingly (see Figure 265). As follows from the graphs of Figure 265 this condition is observed only on

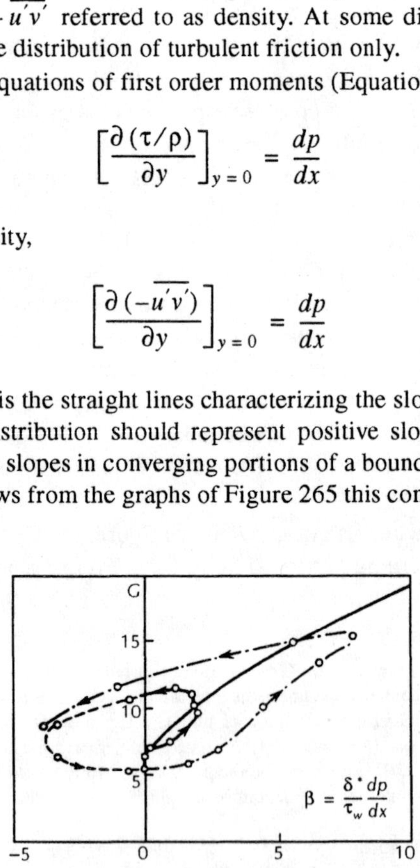

Fig. 264

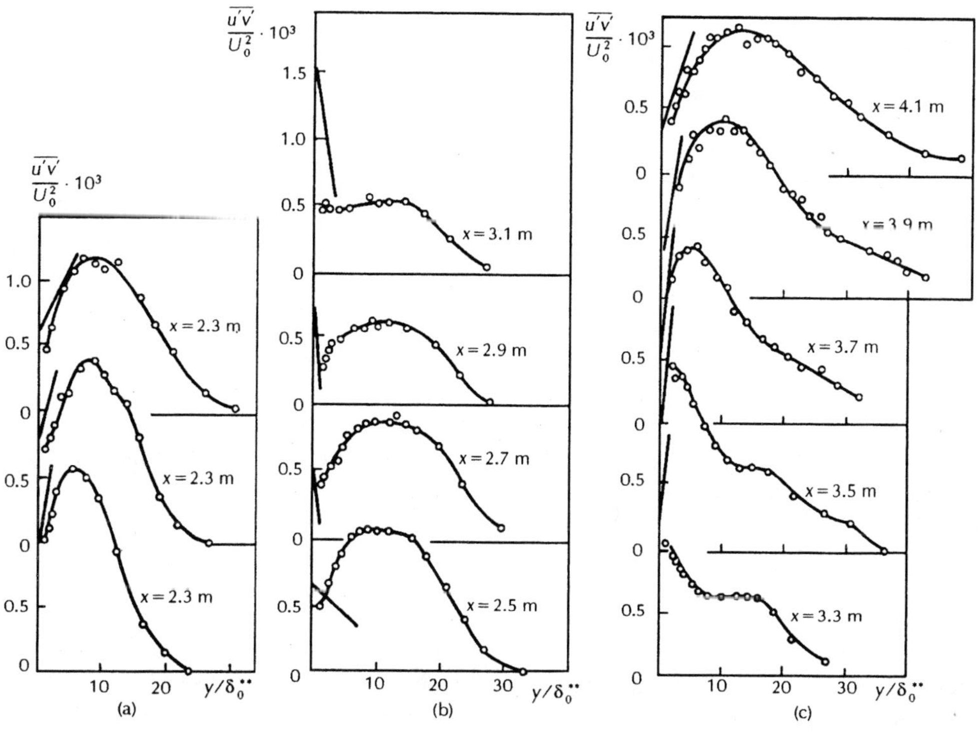

Fig. 265

portion II (Figure 265a). On converging portion III ($2.5 \leq x \leq 3.1$) (Figure 265b) the nearest to wall experimental points are situated on the curves which evidently have a diverging type of turbulent skin friction distribution. Finally, in the last diverging portion IV (Figure 265c) the condition of a slope's positive nature is fully recovered. Only in the transient sections $x = 3.1$, $x = 3.3$ is some uncertainty in the fulfillment of near-wall conditions observed. Considering Figure 265 as a whole we can conclude that in spite of the presence of both diverging and converging portions in the flow (Figure 263), skin friction distribution preserves the diverging character obtained in portion II.

Not dwelling on other interesting results stated in the paper cited we emphasize that Figure 265 gives an extremely obvious example of the presence of hereditary phenomena in non-equilibrium turbulent boundary layers, with the main manifestation of this effect being referred to the flow points at a distance from the wall where large scale turbulence is observed. As is shown by the near-wall velocity profiles expressed in "universal" scales (i.e., referred to the dynamic velocity v) the influence of heredity in this region of small-scale turbulence is slightly appreciable. To conclude the description of this experiment we give the graph of the local coefficient on the wall $c_f(x)$, having compared it with that calculated by the Ludwieg-Tillman equation (146) known from Section 10 of Chapter 13 (Figure 266). As is seen from the figure the experimental points (clear dots), starting from the left boundary of converging portion III noticeably deviate from the Ludwieg-Tillman distribution c_f shown by a solid line.

Two main conclusions can be made from the above: (1) "hereditary" phenomena in turbulent flows are determined by a "vortical" flow structure and (2) large vortices (large-scale turbulence) provide intensive effects of "heredity", whereas small vortices (small-scale turbulence) weakly reflect the effects of heredity.

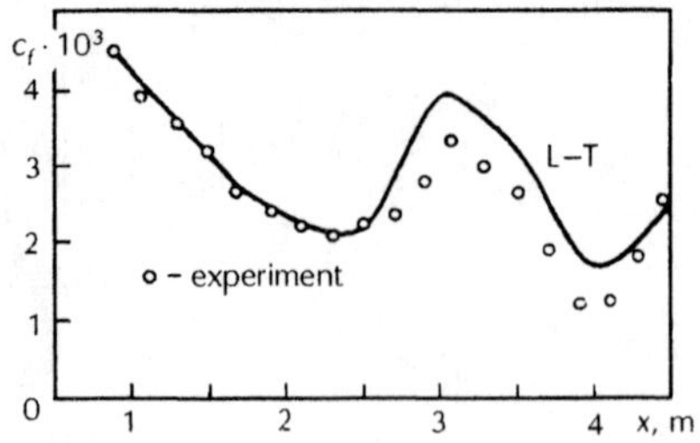

Fig. 266

The absence of similar vortical structures in laminar boundary layers deprive the latter of the "heredity" property. Of course, a conventional molecular diffusion of perturbations (taking place, for instance, in the problem of the initial tube portion) is reserved for the layers.

The methods of boundary layer calculation that are based on the Boussinesq hypothesis are not suitable to account for heredity because in essence it expresses the model of a laminar flow, with the coefficient of viscosity that is more variable in space. This hypothesis can be used only in the regions of small-scale turbulence where the effects of heredity do not show up essentially.

To calculate heredity in turbulent boundary layers the Boussinesq equation should be replaced by a new equation of turbulent skin friction that would take account of relaxation processes. The relaxation equation derived by Hinze[56] and its integral, which should replace the Boussinesq equation (Equation 43, Chapter 13), fall into the class of the first though rather imperfect attempts in this direction.

Among several different ways of deriving relaxation relationships preference should be given to the most natural, based on the use of the equation of transfer of tangential component of turbulent stress tensor.

Hinze based this conclusion on the equation

$$u\frac{\partial \overline{u'v'}}{\partial x} + v\frac{\partial \overline{u'v'}}{\partial y} = -\overline{v'^2}\frac{\partial u}{\partial y} - 2\nu\left(\overline{\frac{\partial u'}{\partial x}\frac{\partial v'}{\partial x}} + \overline{\frac{\partial u'}{\partial y}\frac{\partial v'}{\partial y}}\right) +$$

$$+\overline{\frac{p'}{\rho}\left(\frac{\partial u'}{\partial y} + \frac{\partial v'}{\partial x}\right)} - \frac{\partial \overline{u'v'^2}}{\partial y} - \frac{1}{\rho}\frac{\partial \overline{u'p'}}{\partial y} + \nu\frac{\partial^2 \overline{u'v'}}{\partial y^2} \tag{100}$$

which, in the case of the stationary plane turbulent boundary layer is the equation of the Reynolds stress transfer (Equations32 or 33, Chapter 13). This equation was simplified repeatedly. First, the terms containing ordinary viscosity were omitted. Then the derivatives of the quantities $\overline{u'v'^2}$ and $\overline{u'p'}$ with respect to y were discarded. The remaining third term on the right side was expanded into two summands, the first of which was connected with the averaged deformation motion and taken to be equal to

$$c_1 k^2 \frac{\partial u}{\partial y}$$

and the second was determined by the presence of turbulent fluctuations

[56] Hinze, J. O., Gedächniseffekte in der Turbulenz, *Z. Angew. Math. Mech.*, 56, 403–415, 1976.

$$-c_2\varepsilon\,(k^2)^{-1}\,\overline{u'v'}$$

where k is the double-averaged kinetic energy of velocity fluctuations, ε is the dissipation rate, and c_1 and c_2 are the constant coefficients.

Having adopted these assumptions we can obtain the equation

$$u\frac{\partial \overline{u'v'}}{\partial x} + v\frac{\partial \overline{u'v'}}{\partial y} = -\left(\overline{v'^2} - c_1 k^2\right)\frac{\partial u}{\partial y} - c_2\frac{\varepsilon}{k^2}\overline{u'v'} \tag{101}$$

Further, it was assumed from the considerations of dimensional representation that

$$\frac{c_2\varepsilon}{k^2} = \frac{1}{\lambda_t} \qquad \left(\overline{v'^2} - c_1 k^2\right)\lambda_t = \nu_t \tag{102}$$

where λ_t, which has the dimension of time, was taken to be equal to the characteristic time of turbulent relaxation. Having made in Equation 101 the substitution multiplying both sides of the thus-obtained equation by λ_t and introducing the notation

$$\lambda_t u - L_x \qquad \lambda_t v = L_y \tag{103}$$

we have the Hinze equation

$$L_x\frac{\partial \overline{u'v'}}{\partial x} + L_y\frac{\partial \overline{u'v'}}{\partial y} + \overline{u'v'} = -\nu_t\frac{\partial u}{\partial y} \tag{104}$$

which replaces the Boussinesq equation but does not contain it as a particular case at $L_x = L_y = 0$ because in this case it would be $\lambda_t = 0$ and $\nu_t = 0$.

The quantities L_x and L_y, which have the dimension of length and are proportional to the relaxation time, are considered by Hinze to be longitudinal and transverse relaxation lengths, respectively.

Analyzing the deviation of the Hinze equation from the general equation of transfer of Reynolds stresses, attention is directed to the physical meaning of the summands in the transfer equations discarded when derived (Section 3, Chapter 13). These are the terms involving the coefficient of ordinary (molecular) viscosity. Further, and very important, the author neglects the terms, as small terms, expressing turbulent diffusion. Hence, it can be concluded that the Hinze equation allows the application only in those regions of a turbulent boundary layer where large-scale turbulence is dominant, i.e., in the outer subregion of a boundary layer.

This conclusion confirms the earlier remark concerning the difference in the applicability ranges of the Boussinesq and Hinze equations: the first in the inner near-wall subregion, and the second in the outer subregion of a turbulent boundary layer. This statement is further confirmed by an example of calculation and by a comparison with the experiments.

Using the considerations made as a background and paying attention to the fact that the basic property of large-scale turbulence is to "remember" the previous history of the flow, we shall present a simple derivation of the Hinze equation (104) that is based on the analogy between a large-scale turbulent flow and the flow of Maxwell viscoelastic fluid.[57] We reem-

---

[57] Loitsyanskiy, L. G., Hereditary phenomena in turbulent boundary layers, in *Symp. Numerical Methods in Hydraulics*, Telavi, April, 1980. Vodnye Resursy, No.3, 1981 and by the same author Hereditary phenomena in turbulent motions, in Report to the 5th All-Union Conf. on Theoretical and Applied Mechanics (Alma-Ata, 1981), *Mekh. Zhidk. Gaza*, No.9, 5–19, 1982.

phasize that identity of these two absolutely different in nature flows is not stated here but their common property – heredity – is used.

Similar to the Boussinesq equation (43 of Chapter 13), treated as the generalization of the Newton equation for the simplest plane shear laminar flow to the case of the Boussinesq turbulent fluid which does not possess hereditary properties, the Maxwell viscoelastic fluid with its pronounced hereditary properties can form the basis of the image of the Hinze "turbulent fluid". This requires only the substitution made in the determining relation (Equation 21, Chapter 10): λ into λ_t, μ into μ_t, ν into ν_t, τ into $\tau_t = -\rho \overline{u'v'}$, u into the averaged velocity $\bar{u}$ (the bar over u is later omitted); the derivative with respect to time $\dot{\tau}$ on the left side is replaced by the individual derivative, which in the given plane stationary flow is equal to

$$\dot{\tau} = u\frac{\partial \tau}{\partial x} + v\frac{\partial \tau}{\partial y}$$

With this substitution Equation 21 of Chapter 10 is rewritten as

$$\lambda_t\left(u\frac{\partial \tau_t}{\partial x} + v\frac{\partial \tau_t}{\partial y}\right) + \tau_t = \mu_t\frac{\partial u}{\partial y} \tag{105}$$

and after substitution of $\tau_t = -\rho \overline{u'v'}$ and introduction of notation (103) for relaxation "lengths" coincide with Equation 104.

In the outer subregion of a turbulent boundary layer longitudinal relaxation dominates transverse relaxation, thus making it possible to neglect the relaxation length L_y as compared with L_x. In this case Equation 104 ($L_x = L$) changes to

$$L\frac{\partial \overline{u'v'}}{\partial x} + \overline{u'v'} = -\nu_t\frac{\partial u}{\partial y} \tag{106}$$

Having assumed that L is the function of x only we can obtain the solution of the last equation in the quadratic form

$$\overline{u'v'}(x) = \overline{u'v'}(x_0)\exp\left(-\int_{x_0}^{x}\frac{d\xi}{L(\xi)}\right) - \int_{x_0}^{x}\frac{(\nu_t\partial u/\partial y)_\xi}{L(\xi)}\exp\left(-\int_{\xi}^{x}\frac{d\zeta}{L(\zeta)}\right)d\xi \tag{107}$$

In the second integral the expression within brackets $(\nu_t\partial u/\partial y)_\xi$ indicates that when integrating with respect to x (or ξ), the coordinate y takes a fixed value (i.e., is the parameter).

Calculations of some examples[58] and comparison of the results of these calculations with experiment showed that the linear dependence $L = ax$, where a is the constant, which in different examples ranges from 0.3 to 0.4, can be taken for the function $L(x)$. In this case the previous solution changes to

$$\overline{u'v'} = (\overline{u'v'})_{x=x_0} - \frac{x^{-1/a}}{a}\int_{x_0}^{x}\left(\nu_t\frac{\partial u}{\partial y}\right)_\xi \xi^{1/a-1}d\xi \tag{108}$$

Builtjes[58] cites the examples of the Hinze theory and relaxation equation (104) application. These examples (a flow past a semispherical obstacle in a boundary layer on a plate, a plane

[58] Builtjes, P. J., Memory Effects in Turbulent Flows, Ph. D. thesis, Delft, Universitat Teknologie, 1977.

aerodynamic wake behind a transverse flow past a round cylinder, a flow behind a cascade) were aimed at illustrating the possibility of using the Hinze equation rather than the Boussinesq equation, which in all of these cases does not work. The thesis also presents the examples of the unsuitability of the Boussinesq equation, in particular in the problems of plane tubes with walls of different roughness, on near-wall jets, and in other asymmetric flows. The thesis also contains a survey of theories that differ from the Hinze theory.

In conclusion we shall note that Zyabrikov[59] studied the problem of the applicability of the computation techniques based on the Boussinesq hypothesis and on the Hinze relaxation equation to different ranges of a boundary layer.

Recall that in a near-wall subregion a small-scale flow structure corresponds to very weak properties of heredity, while in the outer subregion portion close to the outer edge of the boundary layer, conversely, a large-scale character of the turbulent flow structure shows itself in strong hereditary effects. It is natural to assume the applicability of the Boussinesq theory in the first of these boundary layer subregions, whereas the Hinze relaxation equation should be used in the second subregion.

Zyabrikov confirmed the validity of these statements, having made comparative calculations of the nonequilibrium boundary layer experimentally studied by Tsuji and Morikawa.[55] The author used two computational techniques: the first mentioned at the end of Section 6, based on the generalized similarity considerations and the Boussinesq hypothesis, and the second using the Hinze relaxation theory. Figures 267 and 268 depict the results of turbulent skin friction calculation by these methods: dashed lines 1 refer to the calculation performed with the aid of the first method, solid lines 2 refer to the second method; clear dots correspond to the experiment. Figure 268 shows the distributions of the dimensionless friction $\tau/(\rho U^2)$ along the boundary layer at a small distance from the wall $y = 0.1\delta$ (δ is the "nominal" boundary layer thickness). Along this straight line at the end of region I (recall Figure 263), over region III entirely, and at the beginning of region IV the calculation made by the Boussinesq hypothesis is closer to the experiment than the calculation made by the Hinze relaxation equation. Figure 268 presents friction distributions that correspond to the layer at a distance from the wall $y = 0.7\delta$. On this straight line the distribution of skin friction over the entire boundary layer length closely corresponds to the relaxation theory. Though the results presented are preliminary, thus explaining some deviations of experimental points from the corresponding curves, the general tendency is expressed rather definitely.

The theory of heredity in turbulent boundary layers is at the initial stage of its development. Along with the stated Hinze theory are the attempts made at 2 simple empirical account of the influence of heredity. We shall comment on this trend,[60] which uses the standard relaxation equation

$$\nu_t = \nu_0 + (\nu_p - \nu_0)\left[1 - \exp\left(-\frac{\Delta x}{l}\right)\right]$$

where ν_0 is the turbulent viscosity in the cross-section of the relaxation beginning, ν_p is the equilibrium value of the coefficient ν_t in the given cross-section, Δx is the distance from the initial section, l is the relaxation length expressed by the parts of layer thicknesses d in the initial cross-section. As shown by the experiments l is on the order of about 5 to 10 δ.

The scientists from Kaunas[61] systematically studied relaxation phenomena in a turbulent boundary layer on a plate heated from below in plane velocity and temperature boundary lay-

---

[59] Zyabrikov, V. V., Relaxation of the Reynolds skin friction in a non-equilibrium turbulent boundary layer, *Inzh.-Fiz. Zh.*, 44(4), 1983.

[60] Hortsmann, C. C., Turbulence model for non-equilibrium adverse pressure gradient flows, *AIAA J.*, 15(2), 131, 132, 1977.

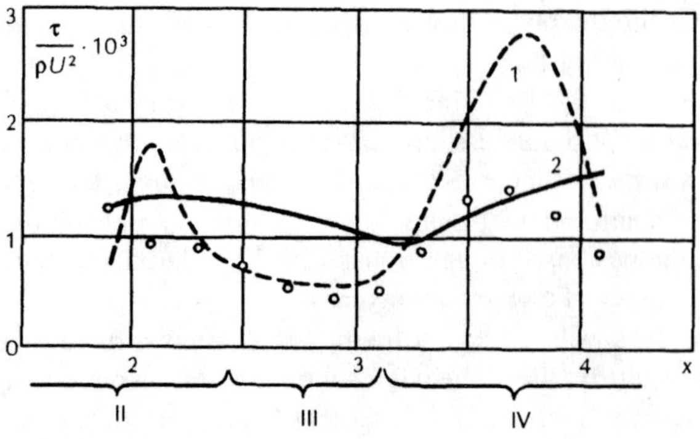

Fig. 267

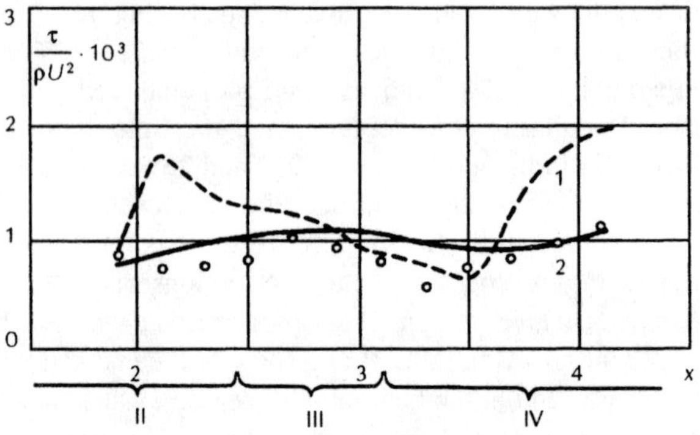

Fig. 268

ers with disturbances in the form of obstacles as rectangular parallelepiped. The authors made a detailed analysis of flow regions behind the point of sticking of boundary layers, stalled from the obstacles, to the plate surface. Similar to previous studies the growth in the intensity relaxation of Reynolds stresses and, correspondingly, of turbulent heat transfer were found at a distance from the plate surface. Complex interaction of large-scale with small-scale vortices, which are generated near the surface, was noticed.

---

[61] Zhlanchyauskas, A. A., Pyadishyus, A. A., and Zigmantas, G. P., Heat transfer in a turbulent boundary layer in the presence of disturbances and the relaxation, in *Teplomassoobmen-IV* (*Heat and Mass Transfer IV*), Vol.1, Part 2, Minsk, 1980 185–196.

15

Dynamics of a Viscous Gas

1. Basic Equations of Viscous Gas Flows

The following general assumptions will be taken as the basis for the study of viscous gas flows:

1. Gas is perfect, i.e., pressure p, density ρ and absolute temperature T satisfy the Clapeyron equation of state

$$\frac{p}{\rho} = RT \tag{1}$$

which on particular assumption about the constancy of specific heat under the constant pressure c_p may also be written in the form

$$\frac{p}{\rho} = \frac{R}{c_p}(c_p T) = \frac{R}{c_p}h \tag{2}$$

where h is the enthalpy (thermal function), which in a general case of the temperature-dependent cp is expressed by the integral

$$h = \int_0^T c_p(T)\, dT \tag{3}$$

2. Gas is a "Newtonian" fluid which follows the generalized Newton's law on the linear relation between strain tensor and rate of strain tensor (see Chapter 10, Section 1). In contrast

to incompressible fluid, in the case of gas, which similar to Section 2 of Chapter 10, will be taken an isotropic medium, the scalar coefficient b entering the basic linear law

$$P = a\dot{S} + bE \tag{4}$$

depends not only on the linear invariant of the strain tensor $p_{11} + p_{22} + p_{33}$, but also on the linear invariant of the rate of strain tensor $\partial V_1/\partial x_1 + \partial V_2/\partial x_2 + \partial V_3/\partial x_3 = \text{div}\, \boldsymbol{V}$ which in the case of high speed gas flow will differ from zero.

Similar to Section 1, Chapter 10 we assume that $a = 2\mu$ and, to find the coefficient b, equate linear invariants of both sides of Equation 4. This yields

$$p_{11} + p_{22} + p_{33} = 2\mu\, \text{div}\boldsymbol{V} + 3b \tag{5}$$

Generalizing the definition of pressure for incompressible fluid employed in Section 1 of Chapter 10 we assume

$$\frac{1}{3}(p_{11} + p_{22} + p_{33}) = -p + \mu'\text{div}\boldsymbol{V} \tag{6}$$

where the coefficient μ' will be termed the coefficient of volumetric viscosity or the second coefficient of viscosity.

Substitution of expression (6) into Equation 5 gives

$$b = -p - \left(\frac{2}{3}\mu - \mu'\right)\text{div}\boldsymbol{V} \tag{7}$$

and the formula of the generalized Newton's law for gas (4) takes on the form

$$P = 2\mu\dot{S} + \left[-p - \left(\frac{2}{3}\mu - \mu'\right)\text{div}\boldsymbol{V}\right]E \tag{8}$$

The possibility of assumption (6) was mentioned by Stokes and then by Kirchhoff in his lectures on the theory of heat. The second viscosity considerably affects high speed processes in gas, such as detonation, gas passage through a shock wave, etc.

Further, we shall restrict ourselves only to allowance for the effect of the conventional first viscosity and neglect the second viscosity assuming

$$\mu' = 0 \tag{9}$$

In this case

$$P = 2\mu\dot{S} + \left(p + \frac{2}{3}\mu\,\text{div}\boldsymbol{V}\right)E \tag{10}$$

or in the matrix form

$$\begin{pmatrix} p_{xx} & p_{xy} & p_{xz} \\ p_{yx} & p_{yy} & p_{yz} \\ p_{zx} & p_{zy} & p_{zz} \end{pmatrix} = 2\mu \begin{pmatrix} \dfrac{\partial u}{\partial x} & \dfrac{1}{2}\left(\dfrac{\partial u}{\partial y} + \dfrac{\partial v}{\partial x}\right) & \dfrac{1}{2}\left(\dfrac{\partial u}{\partial z} + \dfrac{\partial w}{\partial x}\right) \\ \dfrac{1}{2}\left(\dfrac{\partial v}{\partial x} + \dfrac{\partial u}{\partial y}\right) & \dfrac{\partial v}{\partial y} & \dfrac{1}{2}\left(\dfrac{\partial v}{\partial z} + \dfrac{\partial w}{\partial y}\right) \\ \dfrac{1}{2}\left(\dfrac{\partial w}{\partial x} + \dfrac{\partial u}{\partial z}\right) & \dfrac{1}{2}\left(\dfrac{\partial w}{\partial y} + \dfrac{\partial v}{\partial z}\right) & \dfrac{\partial w}{\partial z} \end{pmatrix} -$$

$$-\left(p+\frac{2}{3}\mu\operatorname{div}\boldsymbol{V}\right)\begin{pmatrix}1&0&0\\0&1&0\\0&0&1\end{pmatrix} \tag{11}$$

3. Coefficient of dynamic viscosity μ is the function of absolute temperature T only. In what follows different laws of this dependence are used. First, we shall give the Sutherland formula

$$\frac{\mu}{\mu_0} = \left(\frac{T}{T_0}\right)^{3/2}\frac{T_0+T_S}{T+T_S} \tag{12}$$

where T_S is the Sutherland constant, having for air the value close to 122 K and T_0 and μ_0 are the absolute temperature and the coefficient of viscosity corresponding to some initial state of gas. The power law equation

$$\mu/\mu_0 = (T/T_0)^n \tag{13}$$

is also widely used, where for comparatively high temperatures $n = 1/2$ and for lower ones $n = 1$. According to von Kármán $n = 0.76$ on the average. Sometimes it is assumed that

$$n = 8/9 \quad \text{when} \quad 90 < T < 300 \text{ K}$$

$$n = 3/4 \quad \text{when} \quad 250 < T < 600 \text{ K} \tag{14}$$

Chapman and Rubesin[1] suggested that in the boundary layer theory a simple linear law of relation between μ and T

$$\frac{\mu}{\mu_\infty} = C\frac{T}{T_\infty} = C\frac{h}{h_\infty} \tag{15}$$

should be used with the constant C which is taken from the condition of the possible approximation to the Sutherland formula (12) and is equal to (T_w is the wall temperature, T_∞ is the temperature on the boundary layer outer edge)

$$C = \left(\frac{T_w}{T_\infty}\right)^{1/2}\frac{T_\infty+T_S}{T_w+T_S} \tag{16}$$

4. The specific heats c_p and c_v and consequently their ratio k are independent of the gas absolute temperature and are physical constants of the gas.
5. The thermal conductivity of gas λ is proportional to the coefficient of dynamic viscosity μ, so that the Prandtl number **Pr** $= \mu c_p/\lambda$, which is for brevity denoted σ, is considered as the gas physical constant

$$\mu c_p/\lambda = \sigma = \text{const} \tag{17}$$

To derive equations for the dynamics of a viscous gas we shall use the momentum equation (Equation 6, Chapter 4)

---

[1] Chapman, D. R. and Rubesin, M. W., Temperature and velocity profiles in the compressible laminar boundary layer with arbitrary distribution of surface temperature, *J. Aeron. Sci.*, 16, 547, 1949.

$$\rho\frac{dV}{dt} = \rho F + \mathrm{Div} P \tag{18}$$

and substituting P from Equation 10 into it obtain the following basic Navier Stokes equation for the dynamics of a viscous gas

$$\rho\frac{dV}{dt} = \rho F - \mathrm{grad}\left(p + \frac{2}{3}\mu\,\mathrm{div} V\right) + 2\mathrm{Div}\,(\mu\dot{S}) \tag{19}$$

Resolved into components, Equation 19 has the form

$$\begin{aligned}
\rho\frac{du}{dt} &= \rho F_x - \frac{\partial p}{\partial x} + 2\frac{\partial}{\partial x}\left(\mu\frac{\partial u}{\partial x}\right) + \frac{\partial}{\partial y}\left[\mu\left(\frac{\partial u}{\partial y} + \frac{\partial v}{\partial x}\right)\right] + \\
&\quad + \frac{\partial}{\partial z}\left[\mu\left(\frac{\partial u}{\partial z} + \frac{\partial w}{\partial x}\right)\right] - \frac{2}{3}\frac{\partial}{\partial x}(\mu\,\mathrm{div} V) \\
\rho\frac{dv}{dt} &= \rho F_y - \frac{\partial p}{\partial y} + \frac{\partial}{\partial x}\left[\mu\left(\frac{\partial u}{\partial y} + \frac{\partial v}{\partial x}\right)\right] + 2\frac{\partial}{\partial y}\left(\mu\frac{\partial v}{\partial y}\right) + \\
&\quad + \frac{\partial}{\partial z}\left[\mu\left(\frac{\partial v}{\partial z} + \frac{\partial w}{\partial y}\right)\right] - \frac{2}{3}\frac{\partial}{\partial y}(\mu\,\mathrm{div} V) \\
\rho\frac{dw}{dt} &= \rho F_z - \frac{\partial p}{\partial z} + \frac{\partial}{\partial x}\left[\mu\left(\frac{\partial u}{\partial z} + \frac{\partial w}{\partial x}\right)\right] + \frac{\partial}{\partial y}\left[\mu\left(\frac{\partial v}{\partial z} + \frac{\partial w}{\partial y}\right)\right] + \\
&\quad + 2\frac{\partial}{\partial z}\left(\mu\frac{\partial w}{\partial z}\right) - \frac{2}{3}\frac{\partial}{\partial z}(\mu\,\mathrm{div} V)
\end{aligned} \tag{20}$$

In addition, we have the continuity equations

$$\frac{\partial \rho}{\partial t} + \mathrm{div}\,(\rho V) = 0 \tag{21}$$

or

$$\frac{\partial \rho}{\partial t} + \frac{\partial(\rho u)}{\partial x} + \frac{\partial(\rho v)}{\partial y} + \frac{\partial(\rho w)}{\partial z} = 0 \tag{22}$$

The presence of the variable quantity T makes the obtained system of equations incomplete; the number of unknowns (u, v, w, p, ρ, μ, T) exceeds by one the number of equation. To complete the system of equations we add the equation of energy balance in the moving gas. For this purpose we use the differential form of the energy equation (Equation 21, Chapter 4)

$$\rho\frac{d}{dt}\left(c_v T + \frac{V^2}{2}\right) = \rho F \cdot V + \mathrm{div}\,(PV) + \rho q \tag{23}$$

and, as was done in Chapter 5, we pass over from the internal energy $c_v T$ to the enthalpy h, related by the formula

$$h = c_p T = c_v T + \frac{p}{\rho}$$

Moreover, we shall restrict ourselves to consideration of heat transfer by conduction only, i.e., we assume

$$\rho q = \operatorname{div}(\lambda \operatorname{grad} T)$$

where, according to Equation 17, the coefficient of thermal conductivity λ will be regarded as the function of the temperature T and proportional to μ.

Substituting in Equation 23 the stress tensor P by its expression (10), we obtain

$$\rho \frac{d}{dt}\left(h + \frac{V^2}{2}\right) = \rho \boldsymbol{F} \cdot \boldsymbol{V} + \rho \frac{d}{dt}\left(\frac{p}{\rho}\right) + 2\operatorname{div}(\mu \boldsymbol{V}\dot{S}) -$$

$$- \operatorname{div}\left[\left(p + \frac{2}{3}\mu \operatorname{div}\boldsymbol{V}\right)\boldsymbol{V}\right] + \operatorname{div}\left(\frac{\lambda}{c_p}\operatorname{grad} h\right)$$

or making use of obvious transformation

$$\rho \frac{d}{dt}\left(\frac{p}{\rho}\right) = \frac{dp}{dt} - \frac{p}{\rho}\frac{d\rho}{dt} = \frac{\partial p}{\partial t} + \boldsymbol{V} \cdot \operatorname{grad} p + p \operatorname{div}\boldsymbol{V} = \frac{\partial p}{\partial t} + \operatorname{div}(p\boldsymbol{V})$$

and of Equation 17 we obtain the following form of the energy equation

$$\rho \frac{d}{dt}\left(h + \frac{V^2}{2}\right) = \rho \boldsymbol{F} \cdot \boldsymbol{V} + \frac{\partial p}{\partial t} + \operatorname{div}\left(2\mu \boldsymbol{V}\dot{S} - \frac{2}{3}\mu \boldsymbol{V} \operatorname{div}\boldsymbol{V} + \frac{\mu}{\sigma}\operatorname{grad} h\right) \tag{24}$$

The product of the velocity vector $\boldsymbol{V}$ and the tensor $\dot{S}$, of which the divergence is taken on the right side of Equation 24 can be expressed via differential vector operations on the vector $\boldsymbol{V}$ in the form

$$\boldsymbol{V}\dot{S} = \operatorname{grad}\left(\frac{V^2}{2}\right) - \frac{1}{2}\boldsymbol{V} \times \operatorname{rot}\boldsymbol{V} \tag{25}$$

For future reference, we note the particular case of steady flow without body forces. Making use of Equation 21 we obtain

$$\rho \frac{d}{dt}\left(h + \frac{V^2}{2}\right) = \rho \frac{\partial}{\partial t}\left(h + \frac{V^2}{2}\right) + \rho \boldsymbol{V} \cdot \operatorname{grad}\left(h + \frac{V^2}{2}\right) = \operatorname{div}\left[\rho \boldsymbol{V}\left(h + \frac{V^2}{2}\right)\right]$$

and Equation 24 is transformed to the following, the so called "divergent" form

$$\operatorname{div}\left[\rho \boldsymbol{V}\left(h + \frac{V^2}{2}\right) - 2\mu \boldsymbol{V}\dot{S} + \frac{2}{3}\mu \boldsymbol{V} \operatorname{div}\boldsymbol{V} - \frac{\mu}{\sigma}\operatorname{grad} h\right] = 0 \tag{26}$$

Using expression (25), Equation 26 can be put into the form

$$\operatorname{div}\left[\rho \boldsymbol{V}\left(h + \frac{V^2}{2}\right) - \mu \operatorname{grad}\left(\frac{h}{\sigma} + V^2\right) - \mu \operatorname{rot}\boldsymbol{V} \times \boldsymbol{V} + \frac{2}{3}\mu \boldsymbol{V} \operatorname{div}\boldsymbol{V}\right] = 0 \tag{27}$$

It is possible to derive other useful forms of the energy equation. We recall the equation for change in the kinetic energy (16) of Chapter 4

$$\rho \frac{d}{dt}\left(\frac{V^2}{2}\right) = \rho \boldsymbol{F} \cdot \boldsymbol{V} + \operatorname{div}(P\boldsymbol{V}) + N_{\text{in}} \tag{28}$$

Substituting into expression (17) of Chapter 4 the rate of work done by internal forces the tensor P from Equation 10 we obtain the following relation

$$N_{\text{in}} = -2\mu\dot{S}^2 + p\,\text{div}\,\boldsymbol{V} + \frac{2}{3}\mu\,(\text{div}\,\boldsymbol{V})^2 \tag{29}$$

Comparing this with the corresponding expression for N_{in} in an incompressible fluid, we satisfy ourselves that in the presence of compressibility two new terms appear in the expression for N_{in}

$$p\,\text{div}\,\boldsymbol{V} + \frac{2}{3}\mu\,(\text{div}\,\boldsymbol{V})^2$$

The first of them stands for work done by pressure and the second for work of internal friction during the compression of the gas. The rate of dissipation, i.e., irreversible part of the rate of work done by internal forces with its sign reversed, is equal to

$$N_{\text{dis}} = 2\mu\dot{S}^2 - \frac{2}{3}\mu\,(\text{div}\,\boldsymbol{V})^2 \tag{30}$$

The expression for the rate of dissipation (30) in the form of the difference of two quadratic terms does not allow one to directly judge the sign of this quantity and the conditions that make it vanish. We shall show that the rate of dissipation can be presented as a sum of quadratic terms. From Equation 30 we have

$$N_{\text{dis}} = 2\mu\left(\dot{S}_{11}^2 + \dot{S}_{22}^2 + \dot{S}_{33}^2 + 2\dot{S}_{12}^2 + 2\dot{S}_{23}^2 + 2\dot{S}_{31}^2\right) -$$

$$-\frac{2}{3}\mu\,(\dot{S}_{11}^2 + \dot{S}_{22}^2 + \dot{S}_{33}^2 + 2\dot{S}_{11}\dot{S}_{22} + 2\dot{S}_{22}\dot{S}_{33} + 2\dot{S}_{33}\dot{S}_{11}) =$$

$$= 4\mu\,(\dot{S}_{12}^2 + \dot{S}_{23}^2 + \dot{S}_{31}^2) + \frac{2}{3}\mu\,(2\dot{S}_{11}^2 + 2\dot{S}_{22}^2 + 2\dot{S}_{33}^2 - 2\dot{S}_{11}\dot{S}_{22} - 2\dot{S}_{22}\dot{S}_{33} - 2\dot{S}_{33}\dot{S}_{11})$$

or combining the terms within the second bracket

$$N_{\text{dis}} = 4\mu\,(\dot{S}_{12}^2 + \dot{S}_{23}^2 + \dot{S}_{31}^2) + \frac{2}{3}\mu\,[(\dot{S}_{11} - \dot{S}_{22})^2 + (\dot{S}_{22} - \dot{S}_{33})^2 + (\dot{S}_{33} - \dot{S}_{11})^2] \tag{31}$$

From this equation it follows directly that in addition to the trivial case of quasi solid motion of a gas already considered in Section 10 of Chapter 10 dealing with the flow of a viscous incompressible fluid, the mechanical energy of a viscous gas is not dissipated into heat in an isotropic radial expansion or compression of the gas when shear velocities are equal to zero and the linear rates of strain are the same in all directions: $\dot{S}_{11} = \dot{S}_{22} = \dot{S}_{33}$. It should be noted that if the second coefficient of viscosity is taken into account, the dissipation of molecular energy into heat is no longer zero in a radial expansion or compression.

Substracting both sides of Equation 28 from Equation 3 and passing over from the internal energy to enthalpy, we get, after some simple transformations, the following form of the energy equation

$$\rho\frac{dh}{dt} = \frac{dp}{dt} + 2\mu\dot{S}^2 - \frac{2}{3}\mu\,(\text{div}\,\boldsymbol{V})^2 + \text{div}\left(\mu\,\text{grad}\,\frac{h}{\sigma}\right) \tag{32}$$

Comparing this equation with Equation 43 of Chapter 5 and noting the arguments presented there, we conclude that in the presence of viscosity the flow cannot be barotropic. If it is adiabatic it cannot be isentropic.

Equation 24 or 32 completes the system of equations for the dynamics of a viscous gas, at least within the formulation given at the beginning of this chapter. Wider considerations allowing heat transfer by radiation, the phenomena of dissociation, ionization, etc., that are essential in supersonic motion, require special study.

As in the case of an incompressible viscous fluid the condition of zero velocity on a solid motionless boundary and the condition of the coincidence between the velocity of the fluid particles adjacent to the surface of the body moving in the fluid with the corresponding velocities of the points on the body surface (the "no slip" condition) are among the boundary conditions.

The no slip condition does not hold in rarefied gases; at a solid boundary there is then a slip velocity which may be taken to be proportional to the gradient, in the direction normal to the surface of the velocity component parallel to the surface. The no slip condition completely loses its significance in highly rarefied gases when the molecular mean free path is comparable to the linear dimensions of the body. In this case gas cannot be considered a continuous medium. These flows fall outside the scope of mechanics in the narrow sense of the word. Rather, they constitute the subject of the kinetic theory of gases. Note that recently the problems of rarefied gas flows past bodies acquired a practical significance in connection with the flights of bodies to high altitudes.

In the case of unbounded external flows past bodies, one of the boundary conditions is the assignment of the free stream velocity or the velocity of the body relative to undesturbed gas at a distance from the body; in channel flows the value of the mass flow per second is usually given, in the study of the spreading of jets – the momentum flux, etc.

Various boundary conditions can arise in the case of temperature. The most frequent is the assignment of the temperature distribution along the body surface or channel walls as well as the free stream temperature. In other cases the distribution of the heat transfer, i.e., the quantity of heat transferred per second through a surface area unit, "heat flux vector", is specified.

According to the Fourier law this is equivalent to prescribing the temperature gradient normal to the body surface or the channel flow. Boundary conditions of this kind are based on the assumption that there is no temperature jump on the channel wall. This is well confirmed by experiments in nonrarefied gases (or, more precisely, when the mean free path of the gas is small compared with the body or channel dimensions). In the case of rarefied and extremely rarefied gases the boundary conditions mentioned lose their meaning. In rarefied gas flows, the slip velocity is associated with a temperature jump which can be taken to be proportional to the normal gradient of the gas temperature near the wall. In highly rarefied gases, the whole concept of temperature (as well as velocity) requires elucidation which is given in the kinetic theory of gases.

Another boundary condition is provided by the knowledge of the pressure at a point, usually far from the body, at channel entry, etc.

The initial conditions specify, as in an incompressible fluid the velocity and the temperature field at the initial instant of time and the corresponding value of pressure at some point of the flow.

The problem of the condition for the existence and uniqueness of solutions of Navier-Stokes equations for the dynamics of a viscous gas rests on a variety of problem formulations and has not been completely solved.

The simplest example of an exact solution of Navier-Stokes equations for the dynamics of a viscous gas is given by the problem of a one dimensional stationary gas flow in an unbounded plane that is accompanied by transition from a super to subsonic flow when passing through a shock wave (see Section 110 of the 5th edition). However, this solution has no physical sense because the shock wave density is of the order of the molecular free path and, consequently, in the shock wave region the model of a continuous media is incorrect, thus making the use of Navier-Stokes equations impossible. As a result of this formal solution we obtain an underestimated value of the wave thickness as compared with the experimental one. Note that in spite of the facts mentioned the theory of a shock wave in a continuous gaseous medium was the object of many studies.[2]

Among the many searches conducted for solutions of this problem, which get further into the phenomenon but nevertheless are based on a model of a continuous medium and on the Navier-Stokes equations, one should mention the works accounting for the effect of a thermodynamic nonequilibrium state of a gas flow in a shock wave region.[3] The time of gas residence within narrow limits of the shock is so small that the increment of the kinetic energy of gas molecules (gas heating) cannot be uniformly distributed over all degrees of freedom of molecules of two or polyatomic gases.

The relaxation time of translational and reciprocal portions of molecule energy is comparable to the time of gas passage through a shock wave while the oscillatory portion has a greater relaxation time. This has an effect on the values of gas physical constant and considerably changes the process of flow affecting both the shock wave thickness and distribution of velocities and temperature in it. The theory based on kinetic considerations was developed by Lipman. The shock wave thickness calculated by this theory is in agreement with the experimental results.[4]

There was a small number of exact solutions of the problems of gas dynamics. Worth noting is an exact solution of the problem on a viscous gas plane flow around a rotating cylinder and in a gap between rotating cylinders derived by Stepanyants,[5] without limiting assumptions on flow isothermicity and without reduction of inertial terms, and a numerical study of gas-free convection in a plane rectangular cavity published by Polezhaev.[6]

The development and use of computers in scientific practice has resulted in the appearance of a great number of numerical solutions of dynamic and thermal problems on a gas flow at relatively small Reynolds numbers. Systematically studied were the problems of a one-dimensional, nonstationary flow of a real gas, which is assumed to be perfect, possessing viscosity, thermal conductivity, diffusion, and experiencing the effect of chemical reactions. Accurate numerical solutions of the problems of one-dimensional nonstationary flows of this gas can be found in a special monograph by Demiyanov et al.[7]

Of great technical interest is the dynamics of a viscous gas, the theory of "gas lubrication", which is a generalization of the hydrodynamic theory of lubricating with an incompressible viscous fluid (see Sections 4 and 5 of Chapter 11) to the case of gas "lubrication". In

---

[2]The first solutions of this problem ($\sigma = 0, 3/4, \infty$; independence of m and l of temperature) were long ago given by Rankine, Rayleigh, Prandtl, Taylor, and Hamel. A more accurate solution belongs to Becker (Stosswelle und Detonation, *Z. Phys.*, 8, 321–322, 1921–1922). Becker's solution was verified ($\sigma = 3/4$, $\mu \sim \sqrt{T}$) by L. G. Thomas (Note on Becker's theory of shock front, *J. Chem. Phys.*, 12, 449–453, 1944) and by a number of other scientists. See Mordukhov, M. and Libi, P., On complete solution of equations on a one-dimensional flow of a viscous heat conducting compressible gas, in *Mekhanika*, Vol. 1, Inostrannaya Literatura, Moscow, 1950; Liber, A., Romano, F., and Lev, G., Approximate solutions for a shock wave in a steady state one dimensional flow of a viscous compressible gas, in *Mekhanika*, Vol. 1, Inostrannaya Literatura, Moscow, 1952.

[3]Lunkin, Yu. P., On the structure of shock waves, *Zh. Tekhn, Fiz.*, 27(6), 1957.

[4]Schmidt, B., Electron beam density measurements in shock waves in argon, *Int. Congr. Appl. Mech.*, Abstr. 103, 1968.

[5]Stepanyants, L. G., Some problems of compressible gas flow, Tr. LPI, *Energomashinostroeniye, Tekhnicheskaya Gidromekhanika*, No. 5, 111–128, 1953.

[6]Polezhaev, V. I., Numerical solution of the system of two-dimensional nonstationary Navier-Stokes equations for a compressible gas in an enclosure, *Mekh. Zhidk. Gasa*, No. 2, 109–11, 1967.

[7]Demiyanov, Yu. A., Sekrieru, G. V., Igoshin, L. I., Kireyev, V. I., and Pinskiy, V. L., *Odnomernye nestatsionarnye techeniya realnogo gaza* (*One-Dimensional Nonstationary Flows of a Real Gas*), Shtiintsa, Kishinev, 1980 (has extensive bibliography).

this theory one deals with the nonlinear Reynolds equation whose solution is determined mainly by approximate numerical methods. This is currently a large, independent branch of technical gas dynamics. The basic mathematical difficulty is in a nonlinearity of Reynolds equations generalized to the case of "lubricating" gas compressibility. Even under the assumption of working medium isothermicity ($\rho \sim p$) the generalized Reynolds equation includes the pressure p along with its square, thus making the equation nonlinear. The boundary conditions, which take into account various shapes of bushings and seats of bearings, are also complex. Those interested in details are referred to the primary sources.[8]

Among the possible approaches to the numerical solutions of the problems of viscous gas dynamics, the method of finite differences, which is thoroughly considered in Chapter 16, is practiced on a large scale.

An essential property of the problems on the dynamics of a viscous gas is the fact that pressure is related to density and temperature (internal energy) by the equation of state. Consequently, the pressure gradient can be expressed in terms of these quantities, with the conventional elimination of pressure by the vortex operation and the formulation of the problem on vortex transfer being unnecessary.

2. Conditions for the Similarity of Viscous Gas Flows

Proceeding with the question of dynamical similarity in incompressible viscous fluid flows, we shall set up a system of nondimensional equations of motion of a viscous gas. We shall restrict ourselves to the case of a stationary body immersed in an unbounded stream which is uniform at infinity and has there a velocity V_∞, density ρ_∞, pressure p_∞, temperature T_∞, enthalpy h_∞, coefficient of viscosity μ_∞; the quantities c_p, c_v, and their ratio $c_p/c_v = k$ will be regarded as constant throughout the flow. We shall denote the length scale as L and take the values of other quantities at infinity as their scales. In the case of steady flow without body forces, Equations 1, 13, 20, 22, and 27 will take the form

$$\frac{\rho_\infty V_\infty^2}{L}\left(\rho' u'\frac{\partial u'}{\partial x'} + \ldots\right) = -\frac{\rho_\infty}{L}\frac{\partial p'}{\partial x'} + \frac{\rho_\infty V_\infty}{L^2}\left[2\frac{\partial}{\partial x'}\left(\mu'\frac{\partial u'}{\partial x'}\right) + \ldots\right]$$

$$\frac{\rho_\infty V_\infty}{L}\left[\frac{\partial(\rho' u')}{\partial x'} + \frac{\partial(\rho' v')}{\partial y'} + \frac{\partial(\rho' w')}{\partial z'}\right] = 0$$

[8]Sheinberg, S.A., Gas lubrication of sliding bearings, in *Treniye i iznos v mashinakh* (*Friction and Wear in Machines*), Collected Papers VII, Institute of Science of Machines of the USSR Academy of Sciences, Moscow, 1953; Stepanyants, L. G., Some methods of gas dynamic methods of lubrication, *Tr. LPI*, No. 280, 1967; Zablotskiy, N. D. and Sipenkov, I. E., One method for formulation of the problem on forced gas lubrication of sliding bearings, *Tr. LPI*, No. 265, 1966; Kotlyar, Ya. M., On the theory of spheres suspended by air, *AN SSSR, OTN*, 6, 1959; Stepanyants, L. G., Zablotskiy, N. D., and Sipenkov, I. E., Method of theoretical investigation of externally pressurized gas-lubricated bearings, *Trans. ASME*, No. 68, LUBS–44, 1969; Sheinberg, S. A., Zhed, V. P., and Shisheyev, M. D., *Opory skolzheniya s gasovoy smazkoy* (*Sliding Base with Gas Lubrication*), Mashinostroyeniye, Moscow, 1968; Konstantinesku, V. N., *Gasovaya smaska* (*Gas Lubrication*), Mashinostroyeniye, Moscow, 1968; Pinkus, O. and Steenlicht, B., *Theory of Hydrodynamic Lubrication*, McGraw Hill, New-York, 1961; Saibel, E. A. and Macken, N. A., The fluid mechanics of lubrication, *Annu. Rev. Fluid Mech.*, 5, 185–212, 1973; Pinegin, S. V., Tabachnikov, Yu. B., and Sipenkov, I. E., *Statisticheskiye i dinamicheskiye kharakteristiki gazostaticheskikh opor* (*Statictical and Dynamic Charaoteristics of Gas Static Seats*), Nauka, Moscow, 1982. The problems of gas lubrication are the object of attention of the scientific workers of the Department of Hydroaerodynamics of the Leningrad Polytechnical Institute, guided by N. D. Zablotskiy, and the scientific workers of the Institute of Science of Machines of the USSR Academy of Sciences, led by S. V. Pinegin.

$$\frac{1}{L}\operatorname{div}'\left[\rho_\infty V_\infty \rho' V'\left(h_\infty h' + \frac{1}{2}V_\infty^2 V'^2\right) - \frac{\mu_\infty}{L}\mu' \operatorname{grad}'\left(\frac{h_\infty}{\sigma}h' + V_\infty^2 V'^2\right) - \right.$$

$$\left. - \frac{\mu_\infty V_\infty^2}{L}\mu'\left(\operatorname{rot}'\boldsymbol{V}' \times \boldsymbol{V}' - \frac{2}{3}\boldsymbol{V}'\operatorname{div}'\boldsymbol{V}'\right)\right] = 0$$

$$\frac{p_\infty p'}{\rho_\infty \rho'} = \frac{k-1}{k}h_\infty h' \qquad \mu' = f(h')$$

Let us divide both sides of the first three projections of the equation of motion by $\rho_\infty V_\infty^2/L$ by the coefficient of the nondimensional convective acceleration. In the fourth equation the scale factor is cancelled. Divide both sides of the fifth equation by $\rho_\infty V_\infty h_\infty/L$. We have, then, in terms of the nondimensional quantities

$$\rho' u' \frac{\partial u'}{\partial x'} + \ldots = -\frac{p_\infty}{\rho_\infty V_\infty^2}\frac{\partial p'}{\partial x'} + \frac{\mu_\infty}{\rho_\infty V_\infty L}\left[2\frac{\partial}{\partial x'}\left(\mu' \frac{\partial u'}{\partial x'}\right) + \ldots\right]$$

$$\frac{\partial(\rho' u')}{\partial x'} + \frac{\partial(\rho' v')}{\partial y'} + \frac{\partial(\rho' w')}{\partial z'} = 0$$

$$\operatorname{div}'\left[\rho' \boldsymbol{V}'\left(h' + \frac{1}{2}\frac{V_\infty^2}{h_\infty}V'^2\right) - \frac{\mu_\infty}{\rho_\infty V_\infty L}\mu' \operatorname{grad}'\left(\frac{h'}{\sigma} + \frac{V_\infty^2}{h_\infty}V'^2\right) - \right.$$

$$\left. - \frac{\mu_\infty}{\rho_\infty V_\infty L}\frac{V_\infty^2}{h_\infty}\mu'\left(\operatorname{rot}'\boldsymbol{V}' \times \boldsymbol{V}' - \frac{2}{3}\boldsymbol{V}'\operatorname{div}'\boldsymbol{V}'\right)\right] = 0$$

$$p' = \frac{k-1}{k}\frac{\rho_\infty h_\infty}{p_\infty}h'\rho' \qquad \mu' = f(h')$$

The primed differential operators indicate that the operations are carried out in nondimensional coordinates.

Infinitely far from the body the velocity field is uniform, the rates of strain are absent, and the flow of a viscous gas is the same as that of the ideal gas. Therefore, at infinity we can use the ideal gas relations. In particular, we have

$$\frac{p_\infty}{\rho_\infty V_\infty^2} = \frac{1}{k}\frac{a_\infty^2}{V_\infty^2} = \frac{1}{k\mathbf{M}_\infty^2} \qquad \frac{V_\infty^2}{h_\infty} = (k-1)\frac{V_\infty^2}{a_\infty^2} = (k-1)\,\mathbf{M}_\infty^2$$

$$\frac{p_\infty}{\rho_\infty h_\infty} = \frac{p_\infty}{\rho_\infty V_\infty^2}\frac{V_\infty^2}{h_\infty} = \frac{1}{k\mathbf{M}_\infty^2}(k-1)\,\mathbf{M}_\infty^2 = \frac{k-1}{k}$$

In addition, we shall introduce the notation of the flow Reynolds number

$$\mathbf{Re}_\infty = \frac{\rho_\infty V_\infty L}{\mu_\infty}$$

We thus obtain the following set of the nondimensional equations of steady flow of a viscous gas

$$\rho'\left(u'\frac{\partial u'}{\partial x'}+v'\frac{\partial u'}{\partial y'}+w'\frac{\partial u'}{\partial z'}\right) = -\frac{1}{k\mathbf{M}_\infty^2}\frac{\partial p'}{\partial x'}+\frac{1}{\mathbf{Re}_\infty}\left\{2\frac{\partial}{\partial x'}\left(\mu'\frac{\partial u'}{\partial x'}\right)+\right.$$

$$\left.+\frac{\partial}{\partial y'}\left[\mu'\left(\frac{\partial u'}{\partial y'}+\frac{\partial v'}{\partial x'}\right)\right]+\frac{\partial}{\partial z'}\left[\mu'\left(\frac{\partial u'}{\partial z'}+\frac{\partial w'}{\partial x'}\right)\right]-\frac{2}{3}\frac{\partial}{\partial x'}(\mu'\operatorname{div}'\boldsymbol{V}')\right\}$$

$$\rho'\left(u'\frac{\partial v'}{\partial x'}+v'\frac{\partial v'}{\partial y'}+w'\frac{\partial v'}{\partial z'}\right) = -\frac{1}{k\mathbf{M}_\infty^2}\frac{\partial p'}{\partial y'}+\frac{1}{\mathbf{Re}_\infty}\left\{\frac{\partial}{\partial x'}\left[\mu'\left(\frac{\partial u'}{\partial y'}+\frac{\partial v'}{\partial x'}\right)\right]+\right.$$

$$\left.+2\frac{\partial}{\partial y'}\left(\mu'\frac{\partial v'}{\partial y'}\right)+\frac{\partial}{\partial z'}\left[\mu'\left(\frac{\partial v'}{\partial z'}+\frac{\partial w'}{\partial y'}\right)\right]-\frac{2}{3}\frac{\partial}{\partial y'}(\mu'\operatorname{div}'\boldsymbol{V}')\right\}$$

$$\rho'\left(u'\frac{\partial w'}{\partial x'}+v'\frac{\partial w'}{\partial y'}+w'\frac{\partial w'}{\partial z'}\right) = -\frac{1}{k\mathbf{M}_\infty^2}\frac{\partial p'}{\partial z'}+\frac{1}{\mathbf{Re}_\infty}\left\{\frac{\partial}{\partial x'}\left[\mu'\left(\frac{\partial u'}{\partial z'}+\frac{\partial w'}{\partial x'}\right)\right]+\right.$$

$$\left.+\frac{\partial}{\partial y'}\left[\mu'\left(\frac{\partial v'}{\partial z'}+\frac{\partial w'}{\partial y'}\right)\right]+2\frac{\partial}{\partial z'}\left(\mu'\frac{\partial w'}{\partial z'}\right)-\frac{2}{3}\frac{\partial}{\partial z'}(\mu'\operatorname{div}'\boldsymbol{V}')\right\}$$

$$\frac{\partial(\rho' u')}{\partial x'}+\frac{\partial(\rho' v')}{\partial y'}+\frac{\partial(\rho' w')}{\partial z'} = 0$$

$$\operatorname{div}'\left\{\rho'\boldsymbol{V}'\left(h'+\frac{k-1}{2}\mathbf{M}_\infty^2 V'^2\right)-\frac{1}{\mathbf{Re}_\infty}\mu'\operatorname{grad}'\left[\frac{h'}{\sigma}+(k-1)\,\mathbf{M}_\infty^2 V'^2\right]-\right.$$

$$\left.-\frac{k-1}{\mathbf{Re}_\infty}\mathbf{M}_\infty^2\mu'\left(\operatorname{rot}'\boldsymbol{V}'\times\boldsymbol{V}'-\frac{2}{3}\boldsymbol{V}'\operatorname{div}'\boldsymbol{V}'\right)\right\} = 0$$

$$p' = h'\rho' \qquad \mu' = f(h') \tag{33}$$

To this system of equations are added nondimensional boundary conditions which have already been discussed in general terms. In the case under consideration, these boundary conditions amount to specifying, in a nondimensional form, the equations of the body surface (the velocity at the surface is equal to zero) prescribing the distribution of the nondimensional temperature of its normal derivative and putting the nondimensional values of velocity, pressure, and temperature equal to unity at infinity.

The nondimensional system of equations and boundary conditions describing the flow of a viscous gas is of interest in itself, for it allows the study of not only a single flow but also the whole class of motions which differ from the given one in their scales of body dimensions and in the scales of velocity, temperature, etc. At the same time, the nondimensional system of equations makes it possible to deduce the conditions necessary for the similarity of two gas flows. Suppose, for example, that we are studying two geometrically, kinematically, and dynamically similar steady flows of viscous gases past a body, or a system of bodies, with negligible body forces. The boundaries of the streamlined bodies in the two flows must be geometrically similar and similarly disposed with respect to free streams; this condition is involved in the definition of the geometrical similarity which is a part of the general similarity.

In the presence of geometrical similarity, the non dimensional coordinates (i.e., expressed in terms of the respective length scales) of the corresponding points will be identical pure numbers. The nondimensional boundary conditions will also be identical and so will the nondimensional values of velocity, pressure, and temperature at the corresponding flow points. These values are the solutions of nondimensional system (33). Consequently, the equations themselves should also be identical. As is seen from the structure of system (33), this requires that two similar systems must have identical values of $\mathbf{Re}_\infty$, $\mathbf{M}_\infty$, k, and σ; if the body temperature is specified, then from the temperature nondimensional boundary conditions it follows that the ratio of dimensional temperatures at any reference points on the wall to the temperature at infinity also should be identical. This ratio, T_w/T_∞, of the temperature on the body wall T_w to the free stream temperature T_∞ is known as the temperature factor.

From this follows the direct theorem of similarity: if two steady flows of a viscous gas without body forces and radiation are similar, then the corresponding numbers $\mathbf{Re}_\infty$, $\mathbf{M}_\infty$, k, σ, and T_w/T_∞ are identical for the two flows. The question arises of the determination of the sufficient conditions, i.e., the conditions for the similarity of two hydroaerodynamic phenomena. However, the solution of this problem requires a rigorous proof of the theorem on the existence and uniqueness of solutions of the equations.

In a more general case that includes the existence of body forces, for example, of weight forces, one should introduce the Froude number and in the case of an unsteady flow the Strouhal number becomes important.

In the problems of the dynamics of a viscous gas, along with the similarity theory, the theory of dimensional analysis is applied in all the cases discussed in Chapters 10 and 11. Here, one should of course take into account the growing number of parameters which determine the medium flow.

3. The Laminar Boundary Layer in High-Speed Gas Flows

To derive the basic equations of the laminar flow of a viscous gas we shall use a procedure which does not differ in essence from that previously employed in the case of an incompressible fluid.

We shall consider only the case of the plane steady flow without body forces. Momentum equations (20) and energy equations (27) in this case are reduced to the form

$$\rho u\frac{\partial u}{\partial x}+\rho v\frac{\partial u}{\partial y} = -\frac{\partial p}{\partial x}+\frac{4}{3}\frac{\partial}{\partial x}\left(\mu\frac{\partial u}{\partial x}\right)-\frac{2}{3}\frac{\partial}{\partial x}\left(\mu\frac{\partial v}{\partial y}\right)+\frac{\partial}{\partial y}\left(\mu\frac{\partial u}{\partial y}\right)+\frac{\partial}{\partial y}\left(\mu\frac{\partial v}{\partial x}\right)$$

$$\rho u\frac{\partial v}{\partial x}+\rho v\frac{\partial v}{\partial y} = -\frac{\partial p}{\partial y}+\frac{\partial}{\partial x}\left(\mu\frac{\partial u}{\partial y}\right)+\frac{\partial}{\partial x}\left(\mu\frac{\partial v}{\partial x}\right)+\frac{4}{3}\frac{\partial}{\partial y}\left(\mu\frac{\partial v}{\partial y}\right)-\frac{2}{3}\frac{\partial}{\partial y}\left(\mu\frac{\partial u}{\partial x}\right)$$

$$\frac{\partial(\rho u)}{\partial x}+\frac{\partial(\rho v)}{\partial y} = 0$$

$$\frac{\partial}{\partial x}\left[\rho u\left(h+\frac{u^2}{2}+\frac{v^2}{2}\right)-\mu\frac{\partial}{\partial x}\left(\frac{h}{\sigma}+\frac{4}{3}\frac{u^2}{2}+\frac{v^2}{2}\right)-\mu v\frac{\partial u}{\partial y}+\frac{2}{3}\mu u\frac{\partial v}{\partial y}\right]+$$

$$+\frac{\partial}{\partial y}\left[\rho v\left(h+\frac{u^2}{2}+\frac{v^2}{2}\right)-\mu\frac{\partial}{\partial y}\left(\frac{h}{\sigma}+\frac{u^2}{2}+\frac{4}{3}\frac{v^2}{2}\right)-\mu u\frac{\partial v}{\partial x}+\frac{2}{3}\mu v\frac{\partial u}{\partial x}\right] = 0$$

$$\frac{p}{\rho} = \frac{k-1}{k}h \qquad \frac{\mu}{\mu_1} = f\left(\frac{h}{h_1}\right) \tag{34}$$

We shall go over to the nondimensional form of these equations, expressing all the quantities in terms of characteristic scales, but unlike the previously cases, we shall take into consideration the difference between the transverse and longitudinal scales of coordinates and velocities. Therefore, we shall retain for x and u scales: L (some dimension characteristic for the streamlined body, e.g., the wing chord) and U_∞ (the free stream velocity), while y and v will have their own as yet unspecified scales Y and V. With the remaining notation (Section 2) and the scales removed, we shall obtain using primes to denote the nondimensional quantities

$$\begin{aligned}
&\frac{\rho_\infty U_\infty^2}{L}\rho' u'\frac{\partial u'}{\partial x'} + \frac{\rho_\infty U_\infty V}{Y}\rho' v'\frac{\partial u'}{\partial y'} = -\frac{p_\infty}{L}\frac{\partial p'}{\partial x'} + \frac{4}{3}\frac{\mu_\infty U_\infty}{L^2}\frac{\partial}{\partial x'}\left(\mu'\frac{\partial u'}{\partial x'}\right) - \\
&\quad - \frac{2}{3}\frac{\mu_\infty V}{LY}\frac{\partial}{\partial x'}\left(\mu'\frac{\partial v'}{\partial y'}\right) + \frac{\mu_\infty U_\infty}{Y^2}\frac{\partial}{\partial y'}\left(\mu'\frac{\partial u'}{\partial y'}\right) + \frac{\mu_\infty V}{LY}\frac{\partial}{\partial y'}\left(\mu'\frac{\partial v'}{\partial x'}\right)
\end{aligned}$$

$$\begin{aligned}
&\frac{\rho_\infty U_\infty V}{L}\rho' u'\frac{\partial v'}{\partial x'} + \frac{\rho_\infty V^2}{Y}\rho' v'\frac{\partial v'}{\partial y'} = -\frac{p_\infty}{Y}\frac{\partial p'}{\partial y'} + \frac{\mu_\infty U_\infty}{LY}\frac{\partial}{\partial x'}\left(\mu'\frac{\partial u'}{\partial y'}\right) + \\
&\quad + \frac{\mu_\infty V}{L^2}\frac{\partial}{\partial x'}\left(\mu'\frac{\partial v'}{\partial x'}\right) + \frac{4}{3}\frac{\mu_\infty V}{Y^2}\frac{\partial}{\partial y'}\left(\mu'\frac{\partial v'}{\partial y'}\right) - \frac{2}{3}\frac{\mu_\infty U_\infty}{LY}\frac{\partial}{\partial y'}\left(\mu'\frac{\partial u'}{\partial x'}\right)
\end{aligned}$$

$$\frac{\rho_\infty U_\infty}{L}\frac{\partial(\rho' u')}{\partial x'} + \frac{\rho_\infty V}{Y}\frac{\partial(\rho' v')}{\partial y'} = 0$$

$$\begin{aligned}
&\frac{1}{L}\frac{\partial}{\partial x'}\Bigg[\rho_\infty U_\infty \rho' u'\left(h_\infty h' + U_\infty^2\frac{u'^2}{2} + V^2\frac{v'^2}{2}\right) - \\
&\quad - \frac{\mu_\infty}{L}\mu'\frac{\partial}{\partial x'}\left(\frac{h_\infty}{\sigma}h' + \frac{4}{3}U_\infty^2\frac{u'^2}{2} + V^2\frac{v'^2}{2}\right) - \frac{\mu_\infty U_\infty V}{Y}\mu' v'\frac{\partial u'}{\partial y'} + \\
&\quad + \frac{2}{3}\frac{\mu_\infty U_\infty V}{Y}\mu' u'\frac{\partial v'}{\partial y'}\Bigg] + \frac{1}{Y}\frac{\partial}{\partial y'}\Bigg[\rho_\infty V \rho' v'\left(h_\infty h' + U_\infty^2\frac{u'^2}{2} + V^2\frac{v'^2}{2}\right) - \\
&\quad - \frac{\mu_\infty}{Y}\mu'\frac{\partial}{\partial y'}\left(\frac{h_\infty}{\sigma}h' + U_\infty^2\frac{u'^2}{2} + \frac{4}{3}V^2\frac{v'^2}{2}\right) - \frac{\mu_\infty U_\infty V}{L}\mu' u'\frac{\partial v'}{\partial x'} + \frac{2}{3}\frac{\mu_\infty U_\infty V}{L}\mu' v'\frac{\partial u'}{\partial x'}\Bigg] = 0
\end{aligned}$$

$$p' = \rho' h' \qquad \mu' = f(h') \tag{35}$$

We divide both sides of the first three equations by the constant group of scales, which appears at the first terms on the left side, and divide the fourth equation by $\rho_\infty U_\infty h_\infty / L$. Then, taking into account the definitions of the nondimensional complexes of scales in terms of $\mathbf{M}_\infty$, $\mathbf{Re}_\infty$, and k already introduced in Section 2, we obtain the following system of nondimensional equations

$$\rho' u'\frac{\partial u'}{\partial x'} + \frac{LV}{YU_\infty}\rho' v'\frac{\partial u'}{\partial y'} = -\frac{1}{k\mathbf{M}_\infty^2}\frac{\partial p'}{\partial x'} + \frac{4}{3}\frac{1}{\mathbf{Re}_\infty}\frac{\partial}{\partial x'}\left(\mu'\frac{\partial u'}{\partial x'}\right) -$$

$$-\frac{2}{3}\frac{LV}{YU_\infty}\frac{1}{\mathbf{Re}_\infty}\frac{\partial}{\partial x'}\left(\mu'\frac{\partial v'}{\partial y'}\right)+\frac{1}{\mathbf{Re}_\infty}\left(\frac{L}{Y}\right)^2\frac{\partial}{\partial y'}\left(\mu'\frac{\partial u'}{\partial y'}\right)+\frac{LV}{YU_\infty}\frac{1}{\mathbf{Re}_\infty}\frac{\partial}{\partial y'}\left(\mu'\frac{\partial v'}{\partial x'}\right)$$

$$\rho'u'\frac{\partial v'}{\partial x'}+\frac{LV}{YU_\infty}\rho'v'\frac{\partial v'}{\partial y'} = -\frac{1}{k\mathbf{M}_\infty^2}\frac{LU_\infty}{YV}\frac{\partial p'}{\partial y'}+\frac{1}{\mathbf{Re}_\infty}\frac{LU_\infty}{YV}\frac{\partial}{\partial x'}\left(\mu'\frac{\partial u'}{\partial y'}\right)+$$

$$+\frac{1}{\mathbf{Re}_\infty}\frac{\partial}{\partial x'}\left(\mu'\frac{\partial v'}{\partial x'}\right)+\frac{4}{3}\frac{1}{\mathbf{Re}_\infty}\left(\frac{L}{Y}\right)^2\frac{\partial}{\partial y'}\left(\mu'\frac{\partial v'}{\partial y'}\right)-\frac{2}{3}\frac{1}{\mathbf{Re}_\infty}\frac{LU_\infty}{YV}\frac{\partial}{\partial y'}\left(\mu'\frac{\partial u'}{\partial x'}\right)$$

$$\frac{\partial(\rho'u')}{\partial x'}+\frac{LV}{YU_\infty}\frac{\partial(\rho'v')}{\partial y'} = 0$$

$$\frac{\partial}{\partial x'}\left\{\rho'u'\left[h'+(k-1)\,\mathbf{M}_\infty^2\left(\frac{u'^2}{2}+\frac{V^2}{U_\infty^2}\frac{v'^2}{2}\right)\right]-\right.$$

$$-\frac{1}{\mathbf{Re}_\infty}\mu'\frac{\partial}{\partial x'}\left[\frac{h'}{\sigma}+(k-1)\,\mathbf{M}_\infty^2\left(\frac{4}{3}\frac{u'^2}{2}+\frac{V^2}{U_\infty^2}\frac{v'^2}{2}\right)\right]-$$

$$\left.-\frac{1}{\mathbf{Re}_\infty}\frac{LV}{YU_\infty}(k-1)\,\mathbf{M}_\infty^2\left(\mu'v'\frac{\partial u'}{\partial y'}-\frac{2}{3}\mu'u'\frac{\partial v'}{\partial y'}\right)\right\}+$$

$$+\frac{L}{Y}\frac{\partial}{\partial y'}\left\{\frac{V}{U_\infty}\rho'v'\left[h'+(k-1)\,\mathbf{M}_\infty^2\left(\frac{u'^2}{2}+\frac{V^2}{U_\infty^2}\frac{v'^2}{2}\right)\right]-\right.$$

$$-\frac{1}{\mathbf{Re}_\infty}\frac{L}{Y}\mu'\frac{\partial}{\partial y'}\left[\frac{h'}{\sigma}+(k-1)\,\mathbf{M}_\infty^2\left(\frac{u'^2}{2}+\frac{4}{3}\frac{V^2}{U_\infty^2}\frac{v'^2}{2}\right)\right]-$$

$$\left.-\frac{1}{\mathbf{Re}_\infty}(k-1)\,\mathbf{M}_\infty^2\frac{V}{U_\infty}\left(\mu'u'\frac{\partial v'}{\partial x'}-\frac{2}{3}\mu'v'\frac{\partial u'}{\partial x'}\right)\right\} = 0$$

$$p' = \rho'h' \qquad \mu' = f(h') \tag{36}$$

We now choose the scales Y and V so that the transverse and longitudinal nondimensional velocities and distances, as well as the velocity derivatives, are all of the zeroth order in the Reynolds number, i.e., they remain finite as $\mathbf{Re}_\infty \to \infty$. This leads to two conditions which, if we include in Y and V the factors independent of $\mathbf{Re}_\infty$, may be written in the form

$$\frac{LV}{YU_\infty} = 1 \qquad \frac{1}{\mathbf{Re}_\infty}\left(\frac{L}{Y}\right)^2 = 1 \quad \text{when} \quad Y = \frac{L}{\sqrt{\mathbf{Re}_\infty}} \qquad V = \frac{U_\infty}{\sqrt{\mathbf{Re}_\infty}} \tag{37}$$

Making use of these expressions for Y and V we shall reduce the previous system of equations to the form

$$\rho'u'\frac{\partial u'}{\partial x'}+\rho'v'\frac{\partial u'}{\partial y'} = -\frac{1}{k\mathbf{M}_\infty^2}\frac{\partial p'}{\partial x'}+\frac{4}{3}\frac{1}{\mathbf{Re}_\infty}\frac{\partial}{\partial x'}\left(\mu'\frac{\partial u'}{\partial x'}\right)-$$

$$-\frac{2}{3}\frac{1}{\mathbf{Re}_\infty}\frac{\partial}{\partial x'}\left(\mu'\frac{\partial v'}{\partial y'}\right)+\frac{\partial}{\partial y'}\left(\mu'\frac{\partial u'}{\partial y'}\right)+\frac{1}{\mathbf{Re}_\infty}\frac{\partial}{\partial y'}\left(\mu'\frac{\partial v'}{\partial x'}\right)$$

$$\rho' u'\frac{\partial v'}{\partial x'}+\rho' v'\frac{\partial v'}{\partial y'} = -\frac{1}{k\mathbf{M}_\infty^2}\mathbf{Re}_\infty\frac{\partial p'}{\partial y'}+\frac{\partial}{\partial x'}\left(\mu'\frac{\partial u'}{\partial y'}\right)+$$

$$+\frac{1}{\mathbf{Re}_\infty}\frac{\partial}{\partial x'}\left(\mu'\frac{\partial v'}{\partial x'}\right)+\frac{4}{3}\frac{\partial}{\partial y'}\left(\mu'\frac{\partial v'}{\partial y'}\right)-\frac{2}{3}\frac{\partial}{\partial y'}\left(\mu'\frac{\partial u'}{\partial x'}\right)$$

$$\frac{\partial(\rho' u')}{\partial x'}+\frac{\partial(\rho' v')}{\partial y'} = 0$$

$$\frac{\partial}{\partial x'}\left\{\rho' u'\left[h'+(k-1)\mathbf{M}_\infty^2\left(\frac{u'^2}{2}+\frac{1}{\mathbf{Re}_\infty}\frac{v'^2}{2}\right)\right]-\right.$$

$$-\frac{1}{\mathbf{Re}_\infty}\mu'\frac{\partial}{\partial x'}\left[\frac{h'}{\sigma}+(k-1)\mathbf{M}_\infty^2\left(\frac{4}{3}\frac{u'^2}{2}+\frac{1}{\mathbf{Re}_\infty}\frac{v'^2}{2}\right)\right]-$$

$$\left.-\frac{1}{\mathbf{Re}_\infty}(k-1)\mathbf{M}_\infty^2\left(\mu' v'\frac{\partial u'}{\partial y'}-\frac{2}{3}\mu' u'\frac{\partial v'}{\partial y'}\right)\right\}+$$

$$+\frac{\partial}{\partial y'}\left\{\rho' v'\left[h'+(k-1)\mathbf{M}_\infty^2\left(\frac{u'^2}{2}+\frac{1}{\mathbf{Re}_\infty}\frac{v'^2}{2}\right)\right]-\right.$$

$$-\mu'\frac{\partial}{\partial y'}\left[\frac{h'}{\sigma}+(k-1)\mathbf{M}_\infty^2\left(\frac{u'^2}{2}+\frac{4}{3}\frac{1}{\mathbf{Re}_\infty}\frac{v'^2}{2}\right)\right]-$$

$$\left.-\frac{1}{\mathbf{Re}_\infty}(k-1)\mathbf{M}_\infty^2\left(\mu' u'\frac{\partial v'}{\partial x'}-\frac{2}{3}\mu' v'\frac{\partial u'}{\partial x'}\right)\right\} = 0$$

$$p' = \rho' h' \qquad \mu' = f(h') \tag{38}$$

We shall now go over to the limit of $\mathbf{Re}_\infty \to \infty$ or take the equivalent step of discarding in this set of equations all the terms of the first and higher orders in the small quantity $1/\mathbf{Re}_\infty$. This gives

$$\rho' u'\frac{\partial u'}{\partial x'}+\rho' v'\frac{\partial u'}{\partial y'} = -\frac{1}{k\mathbf{M}_\infty^2}\frac{\partial p'}{\partial x'}+\frac{\partial}{\partial y'}\left(\mu'\frac{\partial u'}{\partial y'}\right)$$

$$\frac{\partial p'}{\partial y'} = 0 \qquad \frac{\partial(\rho' u')}{\partial x'}+\frac{\partial(\rho' v')}{\partial y'} = 0$$

$$\frac{\partial}{\partial x'}\left[\rho' u'\left(h'+\frac{k-1}{2}\mathbf{M}_\infty^2 u'^2\right)\right]+\frac{\partial}{\partial y'}\left[\rho' v'\left(h'+\frac{k-1}{2}\mathbf{M}_\infty^2 u'^2\right)\right] =$$

$$= \frac{\partial}{\partial y'}\left[\mu'\frac{\partial}{\partial y'}\left(\frac{h'}{\sigma}+\frac{k-1}{2}\mathbf{M}_\infty^2 u'^2\right)\right]$$

$$p' = \rho' h' \qquad \mu' = f(h') \tag{39}$$

Note that by virtue of the second equation we may replace $\partial p'/\partial x'$ by dp'/dx'; moreover, using the third equation (the continuity equation), the left side of the fourth equation (the energy equation) may be rewritten in the form

$$\rho' u' \frac{\partial}{\partial x'}\left(h' + \frac{k-1}{2}\mathbf{M}_\infty^2 u'^2\right) + \rho' v' \frac{\partial}{\partial y'}\left(h' + \frac{k-1}{2}\mathbf{M}_\infty^2 u'^2\right) \tag{40}$$

Thus, the following first basic form of the nondimensional equations of a plane, steady, high-speed, laminar boundary layer in a viscous gas is obtained:

$$\rho' u' \frac{\partial u'}{\partial x'} + \rho' v' \frac{\partial u'}{\partial y'} = -\frac{1}{k\mathbf{M}_\infty^2}\frac{dp'}{dx'} + \frac{\partial}{\partial y'}\left(\mu' \frac{\partial u'}{\partial y'}\right)$$

$$\frac{\partial(\rho' u')}{\partial x'} + \frac{\partial(\rho' v')}{\partial y'} = 0$$

$$\rho' u' \frac{\partial}{\partial x'}\left(h' + \frac{k-1}{2}\mathbf{M}_\infty^2 u'^2\right) + \rho' v' \frac{\partial}{\partial y'}\left(h' + \frac{k-1}{2}\mathbf{M}_\infty^2 u'^2\right) =$$

$$= \frac{\partial}{\partial y'}\left[\mu' \frac{\partial}{\partial y'}\left(h' + \frac{k-1}{2}\mathbf{M}_\infty^2 u'^2\right)\right] + \left(\frac{1}{\sigma} - 1\right)\frac{\partial}{\partial y'}\left(\mu' \frac{\partial h'}{\partial y'}\right)$$

$$p' = \rho' h' \qquad \mu' = f(h') \tag{41}$$

Introduce the enthalpy of adiabatically and isentropically slowed gas, or the stagnation enthalpy

$$h_0 = h + \frac{u^2}{2} \tag{42}$$

In a viscous gas h_0 is no longer a constant, as it is for an ideal gas, but represents an unknown function of coordinates. Referring h_0 to the free stream enthalpy h_∞, we obtain

$$h_0' = \frac{h_0}{h_\infty} = \frac{h}{h_\infty} + \frac{u^2}{2h_\infty} = \frac{h}{h_\infty} + \frac{U_\infty^2}{2h_\infty}\frac{u^2}{U_\infty^2}$$

Allowing for $h_\infty = a_\infty^2/(k-1)$, we find

$$h_0' = h' + \frac{k-1}{2}\mathbf{M}_\infty^2 u'^2 \tag{43}$$

Therefore, the first basic form of the system of nondimensional equations of the boundary layer (41) becomes

$$\rho' u' \frac{\partial u'}{\partial x'} + \rho' v' \frac{\partial u'}{\partial y'} = -\frac{1}{k\mathbf{M}_\infty^2}\frac{dp'}{dx'} + \frac{\partial}{\partial y'}\left(\mu' \frac{\partial u'}{\partial y'}\right)$$

$$\frac{\partial(\rho' u')}{\partial x'} + \frac{\partial(\rho' v')}{\partial y'} = 0$$

$$\rho' u' \frac{\partial h_0'}{\partial x'} + \rho' v' \frac{\partial h_0'}{\partial y'} = \frac{\partial}{\partial y'}\left(\mu' \frac{\partial h_0'}{\partial y'}\right) + \left(\frac{1}{\sigma} - 1\right)\frac{\partial}{\partial y'}\left(\mu' \frac{\partial h'}{\partial y'}\right)$$

$$p' = \rho' h' \qquad \mu' = f(h') \tag{44}$$

The second basic form of these equations is obtained from (41) by simple transformations of the third equation of this system. We have

$$\rho' u' \frac{\partial h'}{\partial x'} + \rho' v' \frac{\partial h'}{\partial y'} + (k-1)\,\mathbf{M}_\infty^2 u' \left(\rho' u' \frac{\partial u'}{\partial x'} + \rho' v' \frac{\partial u'}{\partial y'}\right) =$$

$$= \frac{1}{\sigma}\frac{\partial}{\partial y'}\left(\mu' \frac{\partial h'}{\partial y'}\right) + (k-1)\,\mathbf{M}_\infty^2 \frac{\partial}{\partial y'}\left(\mu' u' \frac{\partial u'}{\partial y'}\right) =$$

$$= \frac{1}{\sigma}\frac{\partial}{\partial y'}\left(\mu' \frac{\partial h'}{\partial y'}\right) + (k-1)\,\mathbf{M}_\infty^2 u' \frac{\partial}{\partial y'}\left(\mu' \frac{\partial u'}{\partial y'}\right) + (k-1)\,\mathbf{M}_\infty^2 \mu' \left(\frac{\partial u'}{\partial y'}\right)^2 \tag{45}$$

which, together with the first equation of system (41), leads to

$$u'\left(\rho' u' \frac{\partial u'}{\partial x'} + \rho' v' \frac{\partial u'}{\partial y'}\right) = -\frac{1}{k\mathbf{M}_\infty^2} u' \frac{dp'}{dx'} + u' \frac{\partial}{\partial y'}\left(\mu' \frac{\partial u'}{\partial y'}\right)$$

In this case the preceding equation becomes

$$\rho' u' \frac{\partial h'}{\partial x'} + \rho' v' \frac{\partial h'}{\partial y'} = \frac{k-1}{k} u' \frac{dp'}{dx'} + (k-1)\,\mathbf{M}_\infty^2 \mu' \left(\frac{\partial u'}{\partial y'}\right)^2 + \frac{1}{\sigma}\frac{\partial}{\partial y'}\left(\mu' \frac{\partial h'}{\partial y'}\right) \tag{46}$$

Thus, we obtain the second basic form of the system of the nondimensional equations of a plane steady, high speed, laminar boundary layer in a viscous gas

$$\rho' u' \frac{\partial u'}{\partial x'} + \rho' v' \frac{\partial u'}{\partial y'} = -\frac{1}{k\mathbf{M}_\infty^2} \frac{dp'}{dx'} + \frac{\partial}{\partial y'}\left(\mu' \frac{\partial u'}{\partial y'}\right)$$

$$\frac{\partial(\rho' u')}{\partial x'} + \frac{\partial(\rho' v')}{\partial y'} = 0$$

$$\rho' u' \frac{\partial h'}{\partial x'} + \rho' v' \frac{\partial h'}{\partial y'} = \frac{k-1}{k} u' \frac{dp'}{dx'} + (k-1)\,\mathbf{M}_\infty^2 \mu' \left(\frac{\partial u'}{\partial y'}\right)^2 + \frac{1}{\sigma}\frac{\partial}{\partial y'}\left(\mu' \frac{\partial h'}{\partial y'}\right)$$

$$p' = \rho' h' \qquad \mu' = f(h') \tag{47}$$

The third equation of this system is an explicit statement of the balance between the nondimensional convective rate of enthalpy change (the left side of the equation), the rate of work done by the pressure forces (the first term on the right), the heat produced by dissipation of the mechanical energy (the second term on the right), and the heat transferred by conduction (the third term on the right).

It is not difficult to return to the dimensional form of these equations. Recalling the scales adopted at the beginning of this section, we obtain the systems of equations in terms of dimensional quantities. The first basic form of the equations is

$$\rho u \frac{\partial u}{\partial x} + \rho v \frac{\partial u}{\partial y} = -\frac{dp}{dx} + \frac{\partial}{\partial y}\left(\mu \frac{\partial u}{\partial y}\right) \qquad \frac{\partial(\rho u)}{\partial x} + \frac{\partial(\rho v)}{\partial y} = 0$$

$$\rho u \frac{\partial h_0}{\partial x} + \rho v \frac{\partial h_0}{\partial y} = \frac{\partial}{\partial y}\left(\mu \frac{\partial h_0}{\partial y}\right) + \left(\frac{1}{\sigma} - 1\right)\frac{\partial}{\partial y}\left(\mu \frac{\partial h}{\partial y}\right)$$

$$p = \frac{k-1}{k}\rho h \qquad \frac{\mu}{\mu_\infty} = f\left(\frac{h}{h_\infty}\right) \tag{48}$$

The second basic form is

$$\rho u \frac{\partial u}{\partial x} + \rho v \frac{\partial u}{\partial y} = -\frac{dp}{dx} + \frac{\partial}{\partial y}\left(\mu \frac{\partial u}{\partial y}\right) \qquad \frac{\partial(\rho u)}{\partial x} + \frac{\partial(\rho v)}{\partial y} = 0$$

$$\rho u \frac{\partial h}{\partial x} + \rho v \frac{\partial h}{\partial y} = u\frac{dp}{dx} + \mu\left(\frac{\partial u}{\partial y}\right)^2 + \frac{1}{\sigma}\frac{\partial}{\partial y}\left(\mu \frac{\partial h}{\partial y}\right)$$

$$p = \frac{k-1}{k}\rho h \qquad \frac{\mu}{\mu_\infty} = f\left(\frac{h}{h_\infty}\right) \tag{49}$$

If desired, then we may go over from the enthalpy h directly to the absolute temperature T. Then, the third equation of system 49, the energy equation, will take the form

$$\rho c_p\left(u\frac{\partial T}{\partial x} + v\frac{\partial T}{\partial y}\right) = u\frac{dp}{dx} + \mu\left(\frac{\partial u}{\partial y}\right)^2 + \frac{c_p}{\sigma}\frac{\partial}{\partial y}\left(\mu \frac{\partial T}{\partial y}\right) \tag{50}$$

Carrying out in the system of equations (38) the limiting transition $\mathbf{Re}_\infty \to \infty$, we neglected the fact that the energy equation involves the terms of the order of $\mathbf{M}_\infty^2/\mathbf{Re}_\infty$ which at relatively small $\mathbf{Re}_\infty$ for which, of course, the boundary layer is still valid ($\mathbf{Re}_\infty > 100$), and great $\mathbf{M}_\infty$ (for example $\mathbf{M}_\infty > 10$) are no longer small and discarding these terms is unjustified.

In this connection we shall estimate the order of magnitude of $\mathbf{M}_\infty^2/\mathbf{Re}_\infty$ from the physical viewpoint. For this purpose, along with the characteristic dimensions of the body L and the boundary layer δ, which were adopted earlier, we shall consider the main molecular dimension – the mean free path of gas molecules l between two subsequent collisions. By the known formula for the dynamic coefficient of viscosity

$$\mu = 0.499\rho\bar{v}l$$

where $\bar{v}$ is the mean velocity of free path of molecules which is expressed in terms of pressure and density or the speed of sound in the form ($k = c_p/c_v$)

$$\bar{v} = \sqrt{\frac{8}{\pi}}\sqrt{\frac{p}{\rho}} = \sqrt{\frac{8}{\pi k}}\,a$$

we shall obtain

$$l = 1.255\sqrt{k}\,\frac{\nu}{a}$$

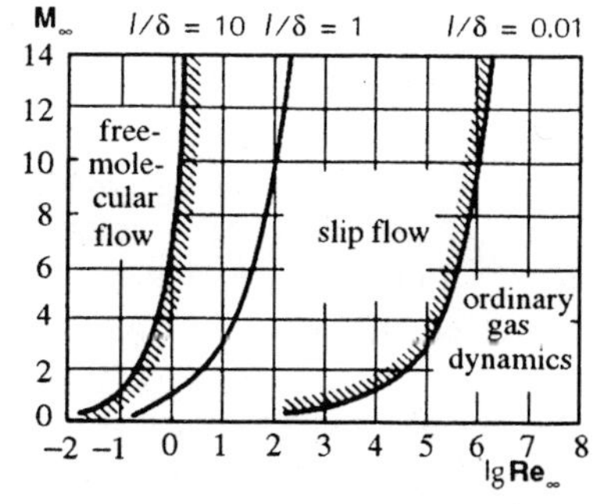

Fig. 269

Let us compose the relation[9]

$$\frac{l}{\delta} = \frac{l}{L}\frac{L}{\delta} = 1.255\sqrt{k}\,\frac{v_\infty}{a_\infty L}\,\frac{L}{\delta} = 1.255\sqrt{k}\,\frac{v_\infty}{U_\infty L}\,\frac{U_\infty}{a_\infty}\,\frac{L}{\delta}$$

which is typical of the phenomenon considered. At great Reynolds numbers $\mathbf{Re}_\infty = U_\infty L/\nu$ the relation $\delta/L = 1/\sqrt{\mathbf{Re}_\infty}$ is warranted, so that

$$\frac{l}{\delta} = 1.255\sqrt{k}\,\frac{\mathbf{M}_\infty}{\sqrt{\mathbf{Re}_\infty}}$$

Hence, it follows that the quantity $\mathbf{M}_\infty/\sqrt{\mathbf{Re}_\infty}$ has the order of the free path of gas molecules to the boundary layer thickness ratio.

Thus, we can conclude that the boundary layer equations derived in the previous section are justifiable (1) at sufficiently great Reynolds numbers and (2) under the condition of the comparatively small free path of molecules as compared with the boundary layer thickness. The latter means that the gas should not be rarefied.

If gas moves at small Reynolds numbers[10], then the idea of the boundary layer is lost. It can be emphasized that if the boundary layer thickness becomes commensurate with the flow as a whole ($\delta \approx L$). Then, in contrast to the above equation for l/δ at great Reynolds numbers, at small $\mathbf{Re}_\infty$ it will be

$$\frac{l}{\delta} \approx \frac{l}{L} = 1.255\sqrt{k}\,\frac{\mathbf{M}_\infty}{\mathbf{Re}_\infty}$$

With the ratio l/δ being used as a basic criterion of boundary layer equation validity, the ranges of the Reynolds to Mach numbers ratio can be approximately found for which at the given k being slightly different for various gases, one should use a method for viscous gas flow calculation. Figure 269 in a semilogarithmic scale, plots the lines of the relation between

[9]Tsien, H. S., Aerodynamics of rarefied gases, in *Gas Dynamics*, Inostrannaya Literatura, Moscow, 1950, 310–357.

[10]Th. von Kármán (*Abh. Aerodyn. Aachen*, No.4, 1925) noted that the Reynolds number, as it directly follows from kinetic considerations used in deriving previous formulas can be presented in the form $\mathbf{Re}_\infty = (U_\infty/\bar{v})(L/l)$. The smallness of this quantity can take place in the following cases:(1) $L \approx l$, $U_\infty \ll \bar{v}$ (the Brownian motion of small particles suspended in a liquid or a gas); (2) $L \gg l$, $U_\infty \ll \bar{v}$ (conventional conditions of a "slow" motion of a body in a viscous liquid); (3) $L \ll l$, $U_\infty \approx \bar{v}$ (motion of bodies in highly rarefied gases).

$\mathbf{M}_\infty$ and $\mathbf{Re}_\infty$ under the three prescribed values of the parameter l/δ: 0.01, 1, 10.[9] These lines, rather conventionally divide the validity ranges of different methods for studying gas flows.

The right side limiting range characterizes the set of $\mathbf{Re}_\infty$ and $\mathbf{M}_\infty$ values, for which the Navier-Stokes equations are justifiable. At great Reynolds numbers in this region use can be made of the boundary layer equations in a high speed gas, if the Mach numbers $\mathbf{M}_\infty$ greatly differ from zero, and of the boundary layer equations in an incompressible fluid, if the Mach numbers $\mathbf{M}_\infty$ slightly differ from zero. At very small Reynolds numbers, the asymptotic behavior of the curve bounding the region considered shows that under these conditions only at negligible values of $\mathbf{M}_\infty$ (i.e., at very small absolute flow velocities) is the use of hydrodynamic equations admissible. This corresponds to the classic range of "slow flows", the Stokes problem on a sphere, etc.

The left side limiting range of $\mathbf{M}_\infty$ and $\mathbf{Re}_\infty$ values holds for strongly rarefied gases when one cannot already speak of gas as of a continuous medium. This is a range of a free molecular gas flow which is described by statistical methods of the kinetic theory of gases. At present these methods are used to calculate power and heat effects of rarefied atmosphere on a body flying at very great heights (superaerodynamics).[11]

An intermediate region is a matter of special difficulty. Until recently one could speak of the established methods for calculating boundary layer flows within this range of $\mathbf{Re}_\infty$ and $\mathbf{M}_\infty$ values, though such problems for general motions of a viscous gas have been the concern of investigations conducted by Maxwell in the second half of the 19th century and by Knudsen, Millikan, and others at the beginning of the 20th century. If we consider that part of the intermediate region which borders on the right side limiting range of Navier-Stokes equations applicability, then we can restrict ourselves to introducing some corrections into conventional methods of fluid mechanics. These corrections have two leads. First, it becomes essential to introduce additional terms into the Navier-Stokes equations that indicate the necessity to use in these cases certain nonlinear laws which substitute the Newton, Fourier, and Fick linear laws.

It follows from kinetic considerations that in the studied portion of the transitional region corresponding rarefied gases, the components of viscous stresses, tensor, vectors of heat, and substance flows should involve, along with ordinary linear terms, nonlinear combinations of velocity derivatives with respect to coordinates.[12] The ratio of these additional terms to the main terms, which correspond to the linear laws, has the order of magnitude $\mathbf{M}_\infty^2/\mathbf{Re}_\infty$ or of the square of l/δ, the free path to the boundary layer thickness ratio.

Moreover, and this is probably of greatest importance, both the velocity and temperature boundary conditions on a solid body surface are subject to radical changes. As early as 1875, Kundt and Warburg (conducting experiments with a disk oscillating in a rarefied gas) noted that attenuation amplitudes reduce with pressure in a surrounding medium. This fact, which falls away from the laws of dynamics of Newtonian viscous fluid, can be explained only by giving up the main property of viscous gases ("sticking" of gas particles to a solid wall). It was assumed that rarefied gas "slips" over the disk surface, with the slip velocity u_w for an isothermal gas flow being equal to

[11] An idea of the problems and methods of this branch of aerodynamics is given in Chapter 10 of Hayes, W. D. and Probstein, R. F., *Teoriya giperzvukovykh techeniy* (*The Theory of Hypersonic Flows*), Inostrannaya Literatura, Moscow, 1962.

[12] Barnett, D., The distribution of molecular velocities and the mean motion in a nonuniform gas, *Proc. London Math. Soc.*, 40, 382–435, 1935; see also Chapman, S. and Cowling, T., *Mathematicheskaya teoriya neodnorodnykh gazov* (*Mathematical Theory of Non Uniform Gases*) (translated from English), Inostrannaya literatura, Moscow, 1960.

$$u_w = \gamma \frac{\partial u}{\partial y}$$

The coefficient γ, which enters into this formula, is called the slip coefficient of a gas. Based on their experiments, Kundt and Warburg showed that the slip coefficient, having the dimension of length, is proportional to the free path of molecules and is even close to it numerically; this coefficient is in inverse proportion to the absolute pressure in a gas. The theory of the "slip coefficient" was given by Maxwell in 1879,[13] who suggested the formula

$$\gamma = 0.998 \frac{2-f}{f} l$$

where f expressed the portion of tangential momentum of molecules lost during the impact of molecules against the body surface. This portion is close to unity.

Gas molecules, in running into a solid body, get into a molecularly complex "rough" body surface. In this case only a small portion of molecules is directly reflected from the body surface, and the overwhelming number of molecules that "stick" are absorbed by the surface and only then diffusely emitted (i.e., independently of the direction of their falling onto the surface). As shown by the experiments, in an air flow above the solid metallic surface or mercury the values of the coefficient f are close to unity and somewhat differ from unity in the case of flowing above glass or oil.

In nonisothermal flow of a rarefied gas the velocity boundary conditions become more complex. The necessity arises to change the boundary condition for wall temperature. Similar to velocity, the temperature near the body surface also has a gradient proportional to the free path of a molecule l, *viz.*,

$$T - T_p = 1.996 \frac{2-\alpha}{\alpha} \frac{k}{k+1} \frac{\lambda}{\mu c_p} l \frac{\partial t}{\partial y}$$

Here, α is the "accommodation" coefficient introduced by Knudsen and indicates the portion of those molecules that after the contact with the surface and being re-emitted acquire a mean energy accommodated to the energy of molecules with the surface temperature T_p; $k = c_p/c_v$, λ is the coefficient of gas thermal conductivity. The accommodation coefficients α and the slip coefficients γ are tabulated by Tsien[9] and in special guides of the kinetic theory of gases.

Application of the equations of motion of rarefied gases (the Barnett equations) to calculation of specific flows, in particular, to a boundary layer is for the present a matter of great difficulty. In the works of this new approach of gas physical mechanics use is still made of the Navier-Stokes equations, although slip and accommodation conditions are taken as boundary conditions. At present there exist special guides on the dynamics of a rarefied gas.[14]

4. The Laminar Boundary Layer on a Plate in a Parallel High Speed Gas Flow

To study the laminar boundary layer formed on a plate parallel to a high speed flow of a viscous gas we use the second basic form (47) of the nondimensional equations. Taking into

[13] Maxwell, C., *Phil., Trans. R Soc.* London, 170, 231–256, 1879.

[14] Kogan, M. N., *Dinamika razrezhennogo gaza (Dynamics of a Rarefied Gas)*, Nauka, Moscow, 1967; Shidlovskiy, V. P., *Vvedeniye v dinamiku razrezhennykh gazov (Introduction to the Dynamics of Rarefied Gases)*, Nauka, Moscow, 1965.

consideration that in this case $dp'/dx' = 0$, $p' = \text{const} = 1$, we shall use the power law of viscosity. The problem is reduced to integration of the system of equations

$$\rho' u' \frac{\partial u'}{\partial x'} + \rho' v' \frac{\partial u'}{\partial y'} = \frac{\partial}{\partial y'}\left(\mu' \frac{\partial u'}{\partial y'}\right) \qquad \frac{\partial(\rho' u')}{\partial x'} + \frac{\partial(\rho' v')}{\partial y'} = 0$$

$$\rho' u' \frac{\partial h'}{\partial x'} + \rho' v' \frac{\partial h'}{\partial y'} = (k-1)\mathbf{M}_\infty^2 \mu' \left(\frac{\partial u'}{\partial y'}\right)^2 + \frac{1}{\sigma}\frac{\partial}{\partial y'}\left(\mu' \frac{\partial h'}{\partial y'}\right)$$

$$\rho' = \frac{1}{h'} \qquad \mu' = h'^n \tag{51}$$

Dorodnitsyn[15] discovered a general coordinate transformation which reduces the equations for boundary layer in a compressible gas to a form very close to that for the boundary layer in an incompressible fluid. This transformation is defined as

$$\xi = \int_0^x \frac{p}{p_0} dx \qquad \eta = \int_0^y \frac{\rho}{\rho_0} dy \tag{52}$$

where p_0 and ρ_0 are the main stream stagnation values of pressure and density.

Using for the plate ($p = p_\infty$) p_∞ and ρ_∞ instead of p_0 and ρ_0 we have (prime stands for a nondimensional quantity)

$$\xi' = x' \qquad \eta' = \int_0^{y'} \rho' dy' \tag{53}$$

To simplify the form of subsequent equations, for the time being we shall discard the primes denoting nondimensional quantities; the reintroduction of the dimensional quantities will be emphasized wherever necessary.

The equations of going over from differentiation with respect to x, y to differentiation with respect to ξ, η will become

$$\frac{\partial}{\partial x} = \frac{\partial}{\partial \xi} + \frac{\partial \eta}{\partial x}\frac{\partial}{\partial \eta} \qquad \frac{\partial}{\partial y} = \rho \frac{\partial}{\partial \eta}$$

as ρ and η are the functions not only of y but also of x.

The first equation of system (51) is transformed to

$$\rho u \frac{\partial u}{\partial \xi} + \rho u \frac{\partial \eta}{\partial x}\frac{\partial u}{\partial \eta} + \rho v \cdot \rho \frac{\partial u}{\partial \eta} = \rho \frac{\partial}{\partial \eta}\left(\mu \rho \frac{\partial u}{\partial \eta}\right) \tag{54}$$

which, dividing by ρ and using the last two equations of system (51), becomes

$$u \frac{\partial u}{\partial \xi} + \left(u \frac{\partial \eta}{\partial x} + \rho v\right)\frac{\partial u}{\partial \eta} = \frac{\partial}{\partial \eta}\left(h^{n-1}\frac{\partial u}{\partial \eta}\right) \tag{55}$$

The second equation of this system (the continuity equation) implies the existence of a

---

[15] Dorodnitsyn, A. A., Boundary layer in a compressible gas, *Prikl. Math. Mekh.*, 6(6), 1942.

stream function ψ, such that

$$\rho u = \frac{\partial \psi}{\partial y} = \rho \frac{\partial \psi}{\partial \eta} \qquad \rho v = -\frac{\partial \psi}{\partial x} = -\frac{\partial \psi}{\partial \xi} - \frac{\partial \eta}{\partial x}\frac{\partial \psi}{\partial \eta} = -\frac{\partial \psi}{\partial \xi} - \frac{\partial \eta}{\partial x} u$$

Hence, the conclusion can be made of the validity of the relations

$$u = \frac{\partial \psi}{\partial \eta} \qquad u\frac{\partial \eta}{\partial x} + \rho v = -\frac{\partial \psi}{\partial \xi} \tag{56}$$

Comparing this with Equation 55, it is seen that if we introduce

$$u\frac{\partial \eta}{\partial x} + \rho v = \tilde{v} \tag{57}$$

Equations 55 and 56 will take the form

$$u\frac{\partial u}{\partial \xi} + \tilde{v}\frac{\partial u}{\partial \eta} = \frac{\partial}{\partial \eta}\left(h^{n-1}\frac{\partial u}{\partial \eta}\right) \qquad u = \frac{\partial \psi}{\partial \eta} \qquad \tilde{v} = -\frac{\partial \psi}{\partial \xi} \qquad \frac{\partial u}{\partial \xi} + \frac{\partial \tilde{v}}{\partial \eta} = 0 \tag{58}$$

Transforming similarly the third equation of system (51) (the energy equation) we have

$$\rho u\frac{\partial h}{\partial \xi} + \rho u\frac{\partial \eta}{\partial x}\frac{\partial h}{\partial \eta} + \rho v \rho\frac{\partial h}{\partial \eta} = \frac{1}{\sigma}\rho\frac{\partial}{\partial \eta}\left(\mu\rho\frac{\partial h}{\partial \eta}\right) + (k-1)\,\mathbf{M}_{\infty}^2\mu\rho^2\left(\frac{\partial u}{\partial \eta}\right)^2$$

or, dividing both sides by ρ, using definition (57) and the last two relations in (51)

$$u\frac{\partial h}{\partial \xi} + \tilde{v}\frac{\partial h}{\partial \eta} = \frac{1}{\sigma}\frac{\partial}{\partial \eta}\left(h^{n-1}\frac{\partial h}{\partial \eta}\right) + (k-1)\,\mathbf{M}_{\infty}^2 h^{n-1}\left(\frac{\partial u}{\partial \eta}\right)^2 \tag{59}$$

Similar to the problem of the boundary layer on a plate in an incompressible fluid flow, we seek the expression for the longitudinal velocity $u(\xi, \eta)$ and the enthalpy $h(\xi, \eta)$ as functions only of the single variable ζ equal to

$$\zeta = \frac{\eta}{2\sqrt{\xi}} \tag{60}$$

Then, according to the second equation of system (58), we shall have

$$\psi = \int_0^{\eta} u\left(\frac{\eta}{2\sqrt{\xi}}\right)d\eta = 2\sqrt{\xi}\int_0^{\eta/(2\sqrt{\xi})} u\left(\frac{\eta}{2\sqrt{\xi}}\right)d\left(\frac{\eta}{2\sqrt{\xi}}\right) = 2\sqrt{\xi}\int_0^{\zeta} u(\zeta)\,d\zeta$$

For brevity, introduce the notation

$$2\int_0^{\zeta} u(\zeta)\,d\zeta = \varphi(\zeta)$$

Then we shall have the following expressions for the stream function ψ, velocities u and $\tilde{v}$, the velocity u, and enthalpy, h, derivatives with respect to ζ. (Following they will be primed)

$$\psi = \sqrt{\xi}\,\varphi(\zeta) \qquad u = \frac{1}{2}\varphi'(\zeta) \qquad \tilde{v} = \frac{1}{2\sqrt{\xi}}(\zeta\varphi' - \varphi)$$

$$\frac{\partial u}{\partial \xi} = -\frac{1}{4\xi}\zeta\varphi''(\zeta) \qquad \frac{\partial u}{\partial \eta} = \frac{1}{4\sqrt{\xi}}\varphi''(\zeta) \qquad \frac{\partial^2 u}{\partial \eta^2} = \frac{1}{8\xi}\varphi'''(\zeta)$$

$$\frac{\partial h}{\partial \xi} = -\frac{1}{2\xi}\zeta h' \qquad \frac{\partial h}{\partial \eta} = \frac{1}{2\sqrt{\xi}}h'$$

Substituting these expressions into the first equation of system (58) and into Equation 59, we obtain the following two ordinary differential equations for determination of the unknown functions φ and h

$$(h^{n-1}\varphi'')' + \varphi\varphi'' = 0$$

$$(h^{n-1}h')' + \frac{\sigma}{4}(k-1)\,\mathbf{M}_\infty^2\, h^{n-1}\varphi''^2 + \sigma\varphi h' = 0 \tag{61}$$

Note that the problem of the laminar boundary layer on a plate in a parallel flow of a uniform gas is, analogous to the case of an incompressible fluid, self-similar, thus resulting in the use of a single independent variable over ζ according to (60).

The boundary conditions for φ will be the same as in the case of an incompressible fluid

$$\varphi = 0 \quad \varphi' = 0 \ \text{at} \ \zeta = 0$$

$$\varphi' = 2 \ \text{at} \ \zeta = \infty \tag{62}$$

Various sets of boundary conditions may apply to the enthalpy h. If the constant nondimensional wall temperature T_w (referred to T_∞) is assigned, then the nondimensional boundary conditions are

$$h = h_w \ \text{at} \ \zeta = 0$$

$$h = 1 \ \text{at} \ \zeta = \infty \tag{63}$$

For a heat-insulated plate the enthalpy boundary conditions will be reduced to

$$\frac{dh}{d\zeta} = 0 \ \text{at} \ \zeta = 0$$

$$h = 1 \ \text{at} \ \zeta = \infty \tag{64}$$

In a general case, Equations 61 can only be integrated by numerical methods.

We shall consider the simplest case when the viscosity-temperature relation is linear ($n = 1$). In this case, instead of (61), we shall obtain the following system of equations

$$\varphi''' + \varphi\varphi'' = 0$$

$$h'' + \sigma\varphi h' + \frac{\sigma}{4}(k-1)\,\mathbf{M}_\infty^2\,\varphi''^2 = 0 \tag{65}$$

The first of these equations, solved with the boundary conditions (62), does not differ formally from the corresponding equation of the problem on the boundary layer on a plate in an incompressible fluid. Integrating the second equation of system (65), we find an expression for $h(\zeta)$ in the form

$$h(\zeta) = \frac{1}{8}(k-1)\mathbf{M}_\infty^2 \vartheta(\zeta) - \frac{C}{8}\int_\zeta^\infty [\varphi''(\zeta)]^\sigma d\zeta + C_1 \tag{66}$$

where the notation has been introduced

$$\vartheta(\zeta) = 2\sigma\int_\zeta^\infty [\varphi''(\zeta)]^\sigma d\zeta \int_0^\zeta [\varphi''(\zeta^*)]^{2-\sigma} d\zeta^* \tag{67}$$

and the constants of integration C and C_1 should be determined from the boundary conditions (63) or (64). Assuming $\zeta = \infty$, we shall find the value of the constant $C_1 = 1$; and assuming $\zeta = 0$ we obtain, according to Equation 63,

$$C = \frac{1 - h_w + \frac{1}{8}(k-1)\mathbf{M}_\infty^2 \vartheta(0)}{\frac{1}{8}\int_0^\infty [\varphi''(\zeta)]^\sigma d\zeta} \tag{68}$$

Denote by h_t and T_t the values of the enthalpy and temperature of the plate under the conditions of the absence of heat transfer (64). In this case the plate can play a role of a temperature measuring device, i.e., a plate thermometer.

Differentiating (66) and noting that from Equation 67 $\vartheta'(0) = 0$, we find that $C = 0$, i.e., when $h_w = h_t$ we shall obtain from Equation 68

$$h_t = 1 + \frac{1}{8}(k-1)\mathbf{M}_\infty^2 \vartheta(0) \tag{69}$$

or passing over to absolute temperatures

$$T_t = T_\infty\left[1 + \frac{1}{8}\vartheta(0)(k-1)\mathbf{M}_\infty^2\right] \tag{70}$$

or, according to (67)

$$\vartheta(0) = 2\sigma\int_0^\infty [\varphi''(\zeta)]^\sigma d\zeta \int_0^\zeta [\varphi''(\zeta^*)]^{2-\sigma} d\zeta^* \tag{71}$$

The constant C, in the general case of the presence of heat transfer from the plate surface, can be presented according to (68) and (69) as

$$C = \frac{h_t - h_w}{\frac{1}{8}\int\limits_0^\infty [\varphi''(\zeta)]^\sigma d\zeta} \tag{72}$$

Let us now analyze the obtained results. It is easy to see that as $\mathbf{M}_\infty \to 0$ the relations (66) and (72) agree with formula (226) derived previously in terms of the variable ζ of Chapter 12 for the incompressible case; the equation thus obtained

$$\frac{T_w - T}{T_w - T_\infty} = \frac{\int\limits_0^\zeta [\varphi''(\zeta)]^\sigma d\zeta}{\int\limits_0^\infty [\varphi''(\zeta)]^\sigma d\zeta} \tag{73}$$

gives the temperature distribution in the boundary layer on a plate in an incompressible fluid with account for the linear law of relation between the coefficient of viscosity and temperature.

Returning to the case of the high speed flow of a gas where the effects of compressibility (the number $\mathbf{M}_\infty$) cannot be neglected, we shall suppose that the function $\vartheta(\zeta)$ has been tabulated for various σ. It is especially important to know the values of $\vartheta(0)$, some of which we give below illustrating their variation with σ(for gases σ does not exceed unity):

$$\sigma = 0.6,\ 0.8,\ 1.0$$

$$\vartheta(0) = 3.08,\ 3.58,\ 4.00$$

Turning to Equation 70, we see that when $n = 1$ it represents the solution of the problem of measuring the temperature of a gas flow $T = T_\infty$ by means of a direct measurement of the surface temperature $T = T_w = T_t$ of the plate in a parallel gas stream under the condition that heat is not transferred from the plate (there is no heat transfer through the holder and wires of a thermocouple). As can be seen from Equation 70, such a plate thermometer will not indicate the temperature T_∞, but the higher temperature T_t, which increases with the Mach number $\mathbf{M}_\infty$. This is natural because the plate decelerates the flow, and as a result of the dissipation of its mechanical energy into heat the plate should be heated. This process of dissipation is related to the increase of entropy. However, as it directly follows from Equation 70, when $\sigma = 1$ and $\vartheta(0) = 4$ the thermometer will indicate the temperature which is equal to the main stream stagnation temperature:

$$(T_t)_{\sigma=1} = T_\infty\left(1 + \frac{k-1}{2}\mathbf{M}_\infty^2\right) = T_0$$

This is no longer so for $\sigma < 1$, when $T_t < T_0$.

Equation 70 may be used for calculation of corrections to the readings of the plate thermometer in a gas with a known value of σ, different from unity. The quantity $(1/4)\vartheta(0)$ is called the recovery factor, within a wide range of $(k-1)\mathbf{M}_\infty^2$ the recovery factor is a function of σ very little different from $\sqrt{\sigma}$, as can be easily verified directly from the table of $\vartheta(0)$ seen above.

The surface of a projectile traveling at supersonic speeds will have approximately the same temperature (70). At high $\mathbf{M}_\infty$ this temperature may exceed the maximum temperature permissible from the viewpoint of the structural strength. At high temperatures radiation becomes significant and helps to cool the surface.[16]

In order to determine the coefficient of heat transfer from plate whose surface temperature is $T = T_w$ we shall calculate the absolute value of the normal temperature gradient, $\partial T / \partial y$, at the plate surface. Going over to dimensional quantities on the right side, we have

$$\frac{dh}{d\zeta} = \frac{1}{T_\infty} 2\sqrt{\xi}\,\frac{\partial T}{\partial \eta} = \frac{2}{T_\infty}\sqrt{\frac{\nu_\infty x}{U_\infty}}\,\frac{\rho_\infty}{\rho}\,\frac{\partial T}{\partial y}$$

where

$$\left(\frac{\partial T}{\partial y}\right)_{y=0} = \frac{T_\infty}{2}\sqrt{\frac{U_\infty}{\nu_\infty x}}\,\frac{\rho_w}{\rho_\infty}\left(\frac{dh}{d\zeta}\right)_{\zeta=0}$$

However, from Equation 66:

$$\left(\frac{dh}{d\zeta}\right)_{\zeta=0} = \frac{C}{8}\,[\varphi''(0)]^\sigma$$

Moreover, in the case of the plate ($p = p_\infty$) this yields by the equation of state

$$\frac{\rho_w}{\rho_\infty} = \frac{T_\infty}{T_w}$$

Using the expression for C (Equation 72), which in terms of dimensional quantities takes on the form

$$C = \frac{1}{T_\infty}\,\frac{T_t - T_w}{\frac{1}{8}\int_0^\infty [\varphi''(\zeta)]^\sigma d\zeta}$$

we shall have

$$\left(\frac{\partial T}{\partial y}\right)_{y=0} = \frac{1}{2}(T_t - T_w)\cdot\frac{T_\infty}{T_w} f(\sigma)\sqrt{\frac{U_\infty}{\nu_\infty x}} \tag{74}$$

whereas in Equation 231 of Chapter 12, it is assumed that

$$f(\sigma) = \frac{1}{\int_0^\infty \left[\frac{\varphi''(\zeta)}{\varphi''(0)}\right]^\sigma d\zeta} \approx 0.664\sqrt[3]{\sigma} \tag{75}$$

Introducing the Nusselt number **Nu**, we obtain, similar to Equation 233 of Chapter 12

[16] Kibel, I. A., The boundary layer in a compressible fluid with account of radiation, *Dokl. AN SSSR*, 25(4), 1939.

$$\mathbf{Nu} = -\frac{Q}{\lambda_\infty (T_w - T_\infty)} = -\frac{1}{\lambda_\infty (T_w - T_\infty)} \int_0^L \lambda_w \left(\frac{\partial T}{\partial y}\right)_{y=0} dx =$$

$$= \frac{T_w - T_t}{T_w - T_\infty} \frac{\lambda_w}{\lambda_\infty} \frac{T_\infty}{T_w} f(\sigma) \sqrt{\frac{U_\infty}{\nu_\infty}} \int_0^L \frac{dx}{2\sqrt{x}}$$

Taking into account that the coefficient of heat conduction is proportional to the coefficient of dynamic viscosity we have for $n = 1$,

$$\frac{\lambda_w}{\lambda_\infty} = \frac{T_w}{T_\infty}$$

so that carrying out the integration we get

$$\mathbf{Nu} = \frac{T_w - T_t}{T_w - T_\infty} f(\sigma) \sqrt{\mathbf{Re}_\infty} \tag{76}$$

Here, the quantity T_t is determined from Equation 70.

In the absence of heat transfer $T_w = T_t$ and $\mathbf{Nu} = 0$. With heat transfer at small $\mathbf{M}_\infty$, i.e., at low speeds, when the effects of compressibility are insignificant, $T_t = T_\infty$, and consequently,

$$\mathbf{Nu} = f(\sigma) \sqrt{\mathbf{Re}_\infty} = 0.664 \sqrt[3]{\sigma} \sqrt{\mathbf{Re}_\infty} \tag{77}$$

This coincides with Equation 234 considered in Chapter 12.

Finally, when $\sigma = 1$, we have

$$\mathbf{Nu} = 0.664 \frac{T_w - T_0}{T_w - T_\infty} \sqrt{\mathbf{Re}_\infty} \tag{78}$$

Returning to the question of the drag of the plate, we shall first find the local skin friction τ_w. In terms of dimensional variables we have

$$\varphi''(\zeta) = 4 \sqrt{\frac{\nu_\infty x}{U_\infty}} \frac{1}{U_\infty} \frac{\rho_\infty}{\rho} \frac{\partial u}{\partial y}$$

where

$$\tau_w = \mu_w \left(\frac{\partial u}{\partial y}\right)_{y=0} = \frac{1}{4} \sqrt{\frac{U_\infty}{\nu_\infty x}} U_\infty \frac{T_\infty}{T_w} \mu_\infty \frac{T_w}{T_\infty} \varphi''(0)$$

$$c_f = \frac{\tau_w}{\frac{1}{2}\rho_\infty U_\infty^2} = \frac{1}{2} \sqrt{\frac{\mu_\infty}{\rho_\infty U_\infty x}} \varphi''(0)$$

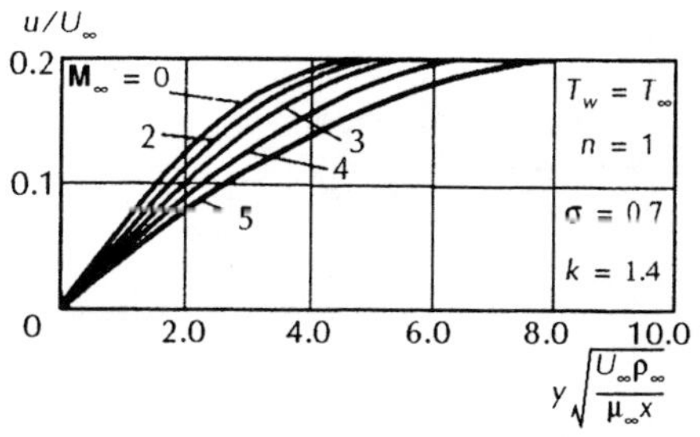

Fig. 270

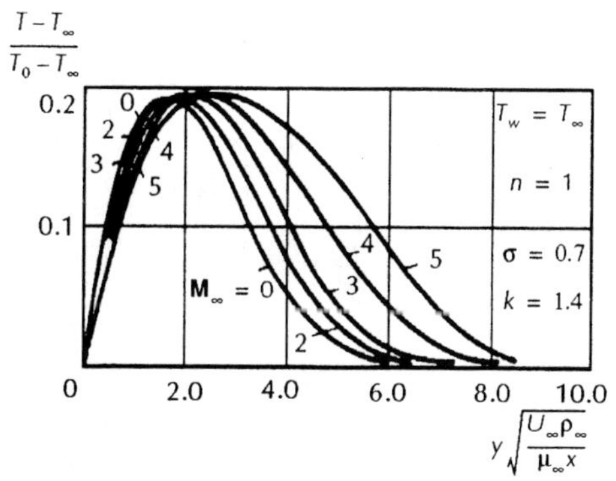

Fig. 271

where $\varphi''(0)$ has the value 1.328, the same value as for an incompressible fluid. Therefore, if for $n = 1$ the values of μ and ρ in the formula for the local skin friction coefficient are calculated at the main stream temperature, local skin friction coefficient remains the same as in incompressible flow:

$$\tau_w = 0.332\sqrt{\frac{\mu_\infty \rho_\infty U_\infty^3}{x}} \qquad c_f = \frac{0.664}{\sqrt{\mathbf{Re}_x}} \qquad C_f = \frac{1.328}{\sqrt{\mathbf{Re}_\infty}} \tag{79}$$

As can be concluded from the analysis, to be able to calculate the coefficients of friction and heat transfer it is not necessary to know the explicit relation between the new variables ζ and $y/\sqrt{x}$, since the final expressions involve only conditions at $y = 0$ or at $y = \infty$. The question of velocity and temperature distribution across the boundary layer is more involved because the distributions of velocity $u/U_\infty = (1/2)\varphi'(\zeta)$ and temperature (Equation 92) are expressed in terms of the independent variable ζ which is related to the physical coordinates by the formula

$$\zeta = \frac{1}{2}\sqrt{\frac{U_\infty}{\nu_\infty x}} \int_0^y \frac{T_\infty}{T} dy$$

and, in turn, depends on the temperature distribution. Differentiating with respect to y, we shall get

$$\frac{\partial \zeta}{\partial y} = \frac{1}{2}\sqrt{\frac{U_\infty}{\nu_\infty x}} \frac{h_\infty}{h(\zeta)}$$

The relation between ζ and $y/\sqrt{x}$ is determined by the integral expression

$$\frac{1}{h_\infty}\int_0^\zeta h(\zeta)\, d\zeta = \frac{1}{2}\sqrt{\frac{U_\infty}{\nu_\infty}} \frac{y}{\sqrt{x}}$$

Figures 270 and 271[17] illustrate the effect of $\mathbf{M}_\infty$ on the velocity and temperature pro-

[17] Hantzsche, W. and Wendt, H., Die laminare Grenzschicht der ebener Platte mit und ohne Warmeubertragung, Jahrbuch 1942 der Deutschen Luftfahrtforschung, 40–51. Figures 270 and 271and a number of others are reproduced from this paper, although the method of analysis described above differs from that of Hantzsche and Wendt.

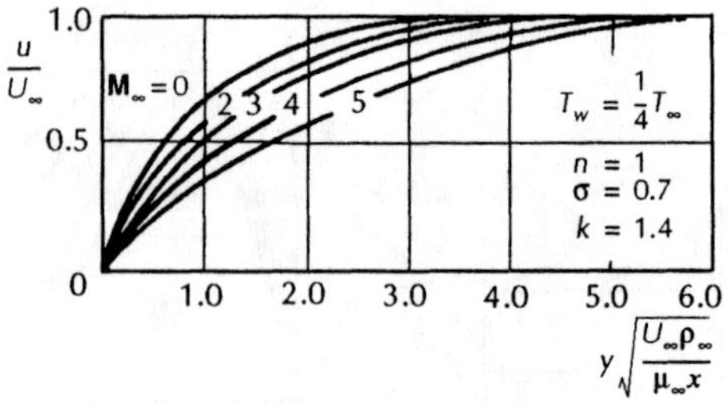

Fig. 272

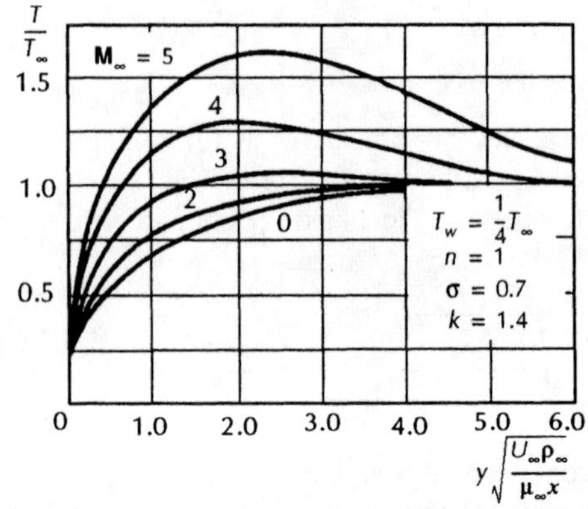

Fig. 273

files on a plate cooled so that its temperature remains equal to the free stream temperature, for the case $n = 1$, $\sigma = 0.7$ and $k = 1.4$. The growth of the thickness of the velocity and temperature boundary layers with $\mathbf{M}_\infty$ attracts the reader's attention in both figures.

The velocity profile becomes flatter at high Mach numbers. The temperature at first increases at a distance from a wall and then returns to the initial value, with the maximum of the ratio $(T - T_\infty)/(T_0 - T_\infty)$ (T_0 is the free stream stagnation temperature) moving farther away from the wall, as the boundary layer becomes thicker.

Comparison of these curves with those shown in Figures 272 and 273 and corresponding to the case of a cooled plate $[T_w = (1/4)T_\infty]$ shows that cooling slightly thickness boundary layers.

Figure 274 demonstrates, in the absence of heat transfer from the plate, the tendency of the velocity profiles in the coordinates u/U_∞, $y\sqrt{U_\infty \rho_w/(\mu_w x)}$ to become nearly linear as $\mathbf{M}_\infty$ increases.

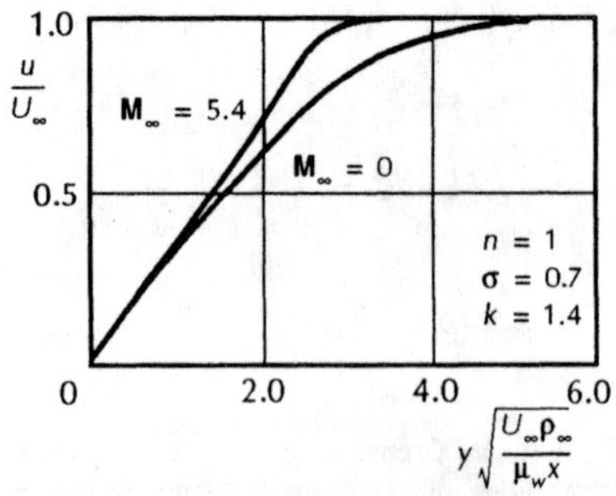

Fig. 274

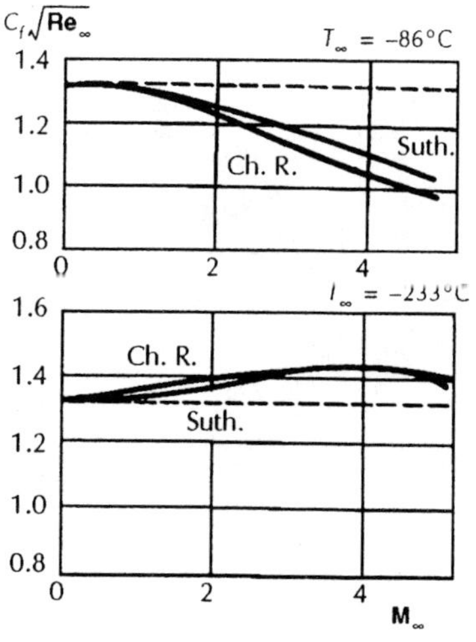

Fig. 275

The solution above, which is valid at $n = 1$, is suitable only at comparatively small temperatures of the plate surface. It is not difficult in this case to improve the accuracy of calculations if a more real linear Chapman-Rubesin formula (15), which approximates Sutherland's law (12), is taken instead of the simplest nondimensional law $\mu' = h'$. They differ only in the right sides of Equation 54, where the nondimensional product $\mu\rho$ is not replaced by unity, but rather by the Chapman Rubesin nondimensional constant C

$$C = \frac{1 + h_S}{h_w + h_S} h_w^{1/2}$$

where $h_S = T_S/T_\infty$, and $T_S = 122$ K. Everything remains the same if the constant multiplier $\sqrt{C}$ is introduced into the scales of the transverse coordinate and the stream function.

Then, for the drag coefficients, instead of (79) the formulas

$$c_f = \frac{0.664}{\sqrt{\mathbf{Re}_x}}\sqrt{C} \qquad C_f = \frac{1.328}{\sqrt{\mathbf{Re}_\infty}}\sqrt{C} \tag{80}$$

will be valid, which at high Mach numbers approach the solutions based on the Sutherland formula. This is confirmed by the graphs in Figure 275. On the two graphs applied to the low temperatures of a flow $T_\infty = -86$°C (for flight of 50 km) and $T_\infty = -233$°C (an experiment in a wind tunnel at supersonic speeds) but to high wall temperatures which increase with the Mach number, the dashed line indicates the results of calculation by the simplest linear formula for viscosity (79), while the solid curves with the notations "Ch. R." and "Suth." correspond to the results from formulas (15) (Chapman Rubesin) and (12) (Sutherland). The use of the Chapman Rubesin formula is warranted.

Simple empirical formulas for calculation of the skin friction drag of the plate are suggested by Young.[18] These expressions

$$c_f\sqrt{\mathbf{Re}_x} = 0.664\,[0.45 + 0.55\,(T_w/T_\infty) + 0.09\,(k-1)\,\mathbf{M}_\infty^2\sqrt{\sigma}\,]^{-(1-n)/2} \tag{81}$$

---

[18] Young, A. D., Boundary layers, in *Sovremennoe sostoyaniye aerodinamiki bolshikh skorostey (State of the Art of Hight-Speed Aerodynamics)*, Vol.1, Howarth, L., Ed., Inostrannaya Literatura, Moscow, 1955, 433

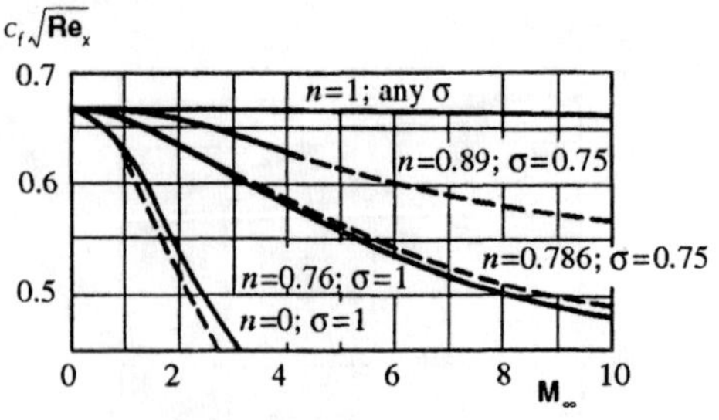

Fig. 276

for a general case of prescribing the temperatures of a plate and a gas free stream

$$c_f\sqrt{\mathbf{Re}_x} = 0.664\,[1 + 0.365\,(k-1)\,\mathbf{M}_\infty^2\sqrt{\sigma}\,]^{-\frac{1-n}{2}} \tag{82}$$

and in the absence of heat transfer (a plate thermometer) slightly differ (within 1%) from theoretical solutions for the real values of σ and n which are of practical significance, for $\mathbf{M}_\infty < 10$.

Figure 276 compares some theoretical calculation[19] of drag in the absence of heat transfer with the empirical formula (82). Solid lines correspond to theoretical calculations; dashed lines correspond to Equation 82. The upper straight line refers to the case $n = 1$ when the product $c_f\sqrt{\mathbf{Re}_x}$ is independent either of the Mach number or of σ.

To conclude we shall give one more graph of temperature distribution in the layer sections with $\mathbf{M}_\infty \le 10$ for the case of the absence of heat transfer at $\sigma = 1$, $n = 0.76$ (Figure 277). A sharp increase of the wall temperature and the temperature layer thickness with $\mathbf{M}_\infty$ that is natural in the case of the absence of heat transfer (cooling) from the plate surface is vividly seen in the figure. Comparing this graph with the distribution of velocities given in the same figure we shall note that in the case of σ =1 the thickness of velocity and thermal layers are equal. The velocity profiles have a bend which is especially noticeable at high $\mathbf{M}_\infty$.[20]

We restricted ourselves only to considering one of the simplest integration of the set of equations (61), *viz.*, the case of $n = 1$, when the first equation of this set becomes independent of and is integrated separately from the second. Of interest is another simple case – when the Prandtl number σ is taken to be unity (for air σ = 0.72). We have to return to this case in consideration of the more complex problem on a laminar boundary layer in the presence of a longitudinal pressure gradient. We shall restrict ourselves only to the following general analysis of the case in considering the first form of the boundary layer equations, which is presented by the system (48). Assuming in the third equation σ =1, we shall get a linear, nondimensional equation with respect to h_0

[19] Numerical methods for solving the boundary layer equations for the plate were used by H. W. Emmons and J. G. Brainerd (see *J. Appl. Mech.*, 8, A–105, 1941 and 9, A–1, 1942). With the present high development of computers, such problems do not offer difficulties.

[20] The details of the solutions of the problems for a laminar boundary layer in a gas can be found in the extended review Kuerti, G., Laminar boundary layer in a compressible fluid, in *Problemy mekhaniki* (*Problems of Mechanics*), Mises, R. and von Kármán, Th., Eds., Inostrannaya Literatura, Moscow, 1955. Unfortunately, @ papers predominantly published outside Russia up to 1948 are included in this review. The papers of Soviet authors published approximately within the same period are surveyed in Loitsyanskiy, L. G., Boundary layer, in *Mekhanika v SSSR za XXX Let* (*Mechanics in the USSR During 30 Years*), Gostekhizdat, Moscow, 1950, 300–320; and *Laminarnyi pogranichnyi sloy* (*Laminar Boundary Layer*), Fizmatgiz, Moscow, 1962, 319–352.

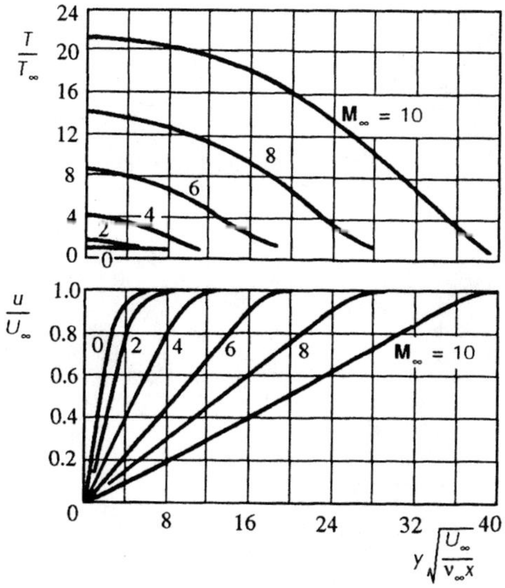

Fig. 277

$$\rho u \frac{\partial h_0}{\partial x} + \rho v \frac{\partial h_0}{\partial y} = \frac{\partial}{\partial y}\left(\mu \frac{\partial h_0}{\partial y}\right)$$

which for $dp/dx = 0$, by virtue of the first equation of the system (48), has an obvious particular integral (a and b are the constants)

$$h_0 = au + b$$

Going over to dimensional quantities this integral, known as the Crocco integral, may be rewritten in the form

$$\frac{T}{T_\infty} = -\frac{k-1}{2}\mathbf{M}_\infty^2\left(\frac{u}{U_\infty}\right)^2 + a\frac{u}{U_\infty} + b \tag{83}$$

Using the obvious conditions

$$T = T_w \text{ at } u = 0 \qquad T = T_\infty \text{ at } u = U_\infty \tag{84}$$

we shall determine a and b; finally, we find the following relation between the temperature and velocity distributions (the Crocco integral)

$$\frac{T}{T_\infty} = -\frac{k-1}{2}\mathbf{M}_\infty^2\left(\frac{u}{U_\infty}\right)^2 + \left(1 - \frac{T_w}{T_\infty} + \frac{k-1}{2}\mathbf{M}_\infty^2\right)\frac{u}{U_\infty} + \frac{T_w}{T_\infty} \tag{85}$$

Using as earlier, the subscript zero at the stagnation temperature, we shall have for any point of the boundary layer

$$T_0 = T\left(1 + \frac{k-1}{2}\mathbf{M}^2\right) = T\left(1 + \frac{k-1}{2}\frac{u^2}{kRT}\right) =$$

$$= T\left(1 + \frac{k-1}{2}\frac{U_\infty^2}{kRT_\infty}\frac{T_\infty}{T}\frac{u^2}{U_\infty^2}\right) = T\left(1 + \frac{k-1}{2}\mathbf{M}_\infty^2\frac{T_\infty}{T}\frac{u^2}{U_\infty^2}\right)$$

and consequently, its values at the outer edge of the boundary layer ($u = U_\infty$) on the plate surface ($u = 0$) are

$$T_{\infty 0} = T_\infty\left(1 + \frac{k-1}{2}\mathbf{M}_\infty^2\right) \qquad T_{w0} = T_w$$

Rewriting Equation 85 in the form

$$\frac{T + \frac{k-1}{2}\mathbf{M}_\infty^2 T_\infty\left(\frac{u}{U_\infty}\right)^2 - T_w}{T_\infty\left(1 + \frac{k-1}{2}\mathbf{M}_\infty^2\right) - T_w} = \frac{u}{U_\infty}$$

we get

$$\frac{T_0 - T_{w0}}{T_{\infty 0} - T_{w0}} = \frac{u}{U_\infty} \tag{86}$$

This extends the relation (227) obtained in Chapter 12 to the case of high speed flow of an ideal gas. Equation 86 shows that for $\sigma = 1$ and arbitrary value of the exponent n in the viscosity-temperature power law the distribution of the difference between the local stagnation (or total) temperature and the wall temperature is similar to the velocity distribution across the boundary layer.

The calculation of the velocity profiles, to which the temperature profiles are related by (85), presents some difficulties. This difficulty should be eliminated before numerical integration of the ordinary nonlinear differential equation to which the problem is reduced.

To construct the above-mentioned equation we shall eliminate φ from the first equation of system (61) and from the result of this very equation differentiation with respect to ζ

$$(h^{n-1}\varphi'')' + \varphi\varphi'' = 0 \qquad (h^{n-1}\varphi'')'' + \varphi\varphi''' + \varphi'\varphi'' = 0$$

For this purpose we shall multiply the first of these equations by φ''', the second equation by φ'' and subtract one from the other. We have

$$(h^{n-1}\varphi'')''\varphi'' + \varphi'\varphi''^2 - (h^{n-1}\varphi'')'\varphi''' = 0$$

We shall take into account that the nondimensional enthalpy h, according to the Crocco integral (85), is the well known function of the nondimensional velocity u; in addition we shall recall that the function $\varphi(\zeta)$ was taken from the condition $u = \frac{1}{2}\varphi'(\zeta)$. Assuming $h^{n-1}u' = s$, we rewrite the previous equation in the form

$$s\frac{d^2 s}{du^2} = -2uh^{n-1}(u) \tag{87}$$

It is assumed in this case that $h(u)$ is replaced by its expression according to (85). The study of the case $\sigma = 1$ is thus connected with the integration of Equation 87. Here, the Dorodnitsyn transform is not used because the coordinate y does not appear in the arguments.

From the first equation of system (61) rewritten in the transformed form

$$\left(\frac{ds}{du}+\varphi\right)\varphi'' = 0$$

and obvious relations $\varphi'' \neq 0$, $\varphi = 0$ at $u = 0$ there follows the boundary condition

$$\frac{ds}{du} = 0 \quad \text{at} \quad u = 0 \tag{88}$$

Moreover, it follows from the definition of s that

$$s = 0 \quad \text{at} \quad u = 1 \tag{89}$$

Consequently, $s = 0$, $u = 1$ is the singular point for the second order nonlinear equation (87). Assuming on the right side of the equation $u = 1$ and emphasizing that by Equation 85 $h = 1$, we shall construct the approximate equation

$$s\frac{d^2s}{du^2} = -2$$

the integral curves of which coincide with the integral curves of the exact equation 87 in the vicinity of the singular point $u = 1$, $s = 0$. Determination of the integral curves of the approximate equation offers no difficulty. We have (A is the integration constant)

$$\frac{ds}{du}\frac{d^2s}{du^2} = -\frac{2}{s}\frac{ds}{du} \qquad \left(\frac{ds}{du}\right)^2 = -4\ln(As)$$

where from the substitution $-\ln(As) = z^2$ and with the use of the boundary condition (89) we can obtain (erf is the notation for the error integral known from the previous sections)

$$u = 1 - \frac{\sqrt{\pi}}{2A}\text{erf}\left[\sqrt{1-\ln(As)}\right] + \frac{\sqrt{\pi}}{2A}$$

Assigning different values for the arbitrary constant of integration A, we shall choose an integral curve in such a way that passing from the singular point along this curve and then extending it by numerical integration of the exact equation (87), we can come, according to the boundary condition (88), to the point where $u = 0$, $ds/du = 0$. This method is frequently applied to solutions of the boundary value problems of a boundary layer. The results of the above computation in the absence of heat transfer and with the exponent $n = 0.76$ in the law of the viscosity coefficient dependence on the wall temperature have been already given and discussed in Figure 277. The high wall temperatures (at $\mathbf{M}_\infty = 10$ they exceed the absolute temperature of the external flow by 20 fold) were attributed to the absence of plate surface cooling. If, for example, it is demanded that at the expense of strong cooling $T_w = \frac{1}{4}T_\infty$, then computation will lead to the curves depicted in Figure 278. It can be seen that in this case the temperature maxima will shift from the plate surface to inside the boundary layer and reduce greater as compared with the case of a thermally insulated plate. The thicknesses of velocity and temperature boundary layers will be equal again because $\sigma = 1$.

When solving the problems of a laminar boundary layer in a high speed gas flow, in par-

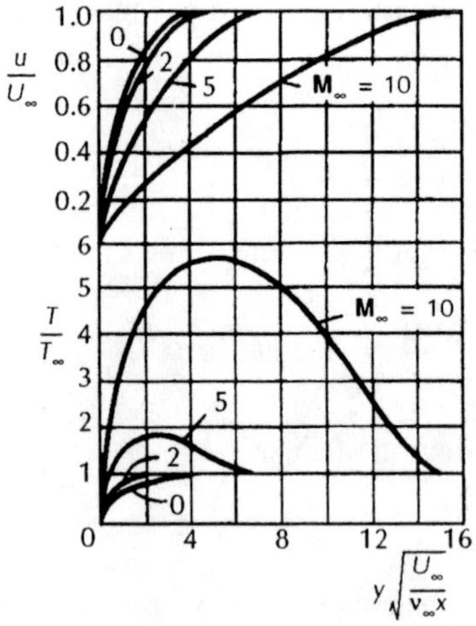

Fig. 278

ticular, for the plate ($dp/dx = 0$), use can be successfully made of the Mises and the Crocco variables described in Chapter 12 (Section 6). Here, no possibility exists to discuss this problem and refer the reader to the previously quoted special reviews on the theory of the laminar boundary layer in high speed gas flow or to our monograph.[21] Note that the stated investigation of the behavior of integral curves of Equation 87 in the vicinity of a singular point was conducted by the method of transition to the velocity as an independent variable. This technique is close to the use of the Crocco variables.

As shown in footnote 21 use of the Crocco variables makes it possible to reduce the order of the set of equations (61), in particular, in the case of the coefficient of viscosity being linearly dependent on temperature. Instead of the first equation of system (65) that is the third order equation, we can obtain the second order equation

$$KK'' + 2\eta = 0$$

where $K = 2\tau\sqrt{\mathbf{Re}_\infty \xi}/(\rho_\infty U_\infty^2)$, and the primes denote differentiation with respect to the variable $\eta = u/U_\infty$. In spite of the apparent simplicity this equation cannot be solved by quadratures and requires the use of integration methods. The first case of applying the Mises variables to the problem of a laminar boundary layer on a plate parallel to a gas flow was published by von Kármán and Tsien.[22]

5. Laminar Boundary Layer on a Cone Parallel to a Supersonic Flow

A certain analogy can be seen between the boundary layer on a plate and that on a cone in a parallel flow.[23] Assume that the semivertex angle θ_0 corresponds, with the Mach number $\mathbf{M}_\infty$ being set in a free stream, to the case of a conical shock wave attached to the cone apex.

---

[21] Loitsyanskiy, L. G., *Laminarnyi pogranichnyi sloy (Laminar Boundary Layer)*, Fizmatgiz, Moscow, 1962, 334–346.

[22] von Kármán, Th. and Tsien, H. S., Boundary layer in compressible fluid, *J. Aeron. Sci.*, 5, 1938.

[23] Hantzsche, W. and Wendt, H., Die laminare Grenzschicht an einem mit Überschallgeschwindigkeit angeströmten nicht angestellten Kegel, Jahrbuch 1941 der Deutschen Luftfartforschung, S. 1–74.

Upstream of the wave the flow of an ideal (nonviscous) gas will be potential and conical (i.e., all the gas parameters should remain constant along any conical surface which is coaxial with a streamlined cone) has a common apex with it, and lies between the cone and a shock wave. In particular, in the motion of an ideal gas the pressure should be constant on the cone surface and consequently, by the known boundary layer property, the pressure will be constant through the entire boundary layer in a viscous gas. This fact brings together the cases of the flow in a boundary layer on a cone and a parallel flow past a plate. A simple correspondence can be shown between the two flows.

For this purpose we shall compare the boundary layer equations on a cone in spherical coordinates

$$\rho\left(u\frac{\partial u}{\partial r}+\frac{v}{r}\frac{\partial u}{\partial \theta}\right)=\frac{1}{r^2}\frac{\partial}{\partial \theta}\left(\mu\frac{\partial u}{\partial \theta}\right)$$

$$\frac{\partial(\rho u)}{\partial r}+\frac{1}{r}\frac{\partial(\rho v)}{\partial \theta}+\frac{2\rho u}{r}=0$$

$$\rho\left(u\frac{\partial h}{\partial r}+\frac{v}{r}\frac{\partial h}{\partial \theta}\right)=\frac{1}{\sigma}\frac{1}{r^2}\frac{\partial}{\partial \theta}\left(\mu\frac{\partial h}{\partial \theta}\right)+\frac{\mu}{r^2}\left(\frac{\partial u}{\partial \theta}\right)^2 \tag{90}$$

and the corresponding boundary conditions ($\theta_\infty > \theta_0$ is the conventional angle representing the layer edge)

$$u=0 \qquad v=0 \qquad h=h_w \quad \left(\text{or } \frac{\partial h}{\partial \theta}=0\right) \quad \text{at } \theta=\theta_0$$

$$u \to u_e \qquad h \to h_e \quad \text{for } \theta \to \theta_\infty \tag{91}$$

with the boundary layer equations and the boundary conditions on a plate in the same gas

$$\rho\left(\bar{u}\frac{\partial \bar{u}}{\partial x}+\bar{v}\frac{\partial \bar{u}}{\partial y}\right)=\frac{\partial}{\partial y}\left(\mu\frac{\partial \bar{u}}{\partial y}\right) \qquad \frac{\partial(\rho\bar{u})}{\partial x}+\frac{\partial(\rho\bar{v})}{\partial y}=0$$

$$\rho\left(\bar{u}\frac{\partial \bar{h}}{\partial x}+\bar{v}\frac{\partial \bar{h}}{\partial y}\right)=\frac{1}{\sigma}\frac{\partial}{\partial y}\left(\mu\frac{\partial \bar{h}}{\partial y}\right)+\mu\left(\frac{\partial \bar{u}}{\partial y}\right)^2$$

$$\bar{u}=0 \quad \bar{v}=0 \quad \bar{h}=\bar{h}_w \quad \left(\text{or } \frac{\partial \bar{h}}{\partial y}=0\right) \quad \text{at } y=0$$

$$\bar{u} \to \bar{u}_e \quad \bar{h} \to \bar{h}_e \quad \text{for } y \to \infty \tag{92}$$

Noting that both problems are self similar (there is no characteristic length scale in the formulations of the problems), we shall seek the solution of the problem of the longitudinal flow past a plate in an ordinary form

$$\bar{u}=\bar{u}(\zeta) \qquad \bar{v}=\frac{\tilde{V}(\zeta)}{\sqrt{x}} \qquad \bar{h}=\bar{h}(\zeta) \qquad \zeta=\frac{y}{\sqrt{x}} \tag{93}$$

and for the boundary layer on a cone, in an analogous form,

$$u = u(\zeta) \qquad v = \frac{v(\zeta)}{\sqrt{r}} \qquad h = h(\zeta)$$

$$\zeta = \frac{r(\theta - \theta_0)}{\sqrt{r}} = (\theta - \theta_0)\sqrt{r} \tag{94}$$

Then the set of equations (90) passes over to the system of ordinary differential equations

$$\left(\frac{\rho u \zeta}{2} + \rho \tilde{v}\right)\frac{du}{d\zeta} = \frac{d}{d\zeta}\left(\mu \frac{du}{d\zeta}\right)$$

$$\frac{\zeta}{2}\frac{d(\rho u)}{d\zeta} + \frac{d(\rho \tilde{v})}{d\zeta} + 2\rho u = 0$$

$$\left(\frac{\rho u \zeta}{2} + \rho \tilde{v}\right)\frac{dh}{d\zeta} = \frac{1}{\sigma}\frac{d}{d\zeta}\left(\mu \frac{dh}{d\zeta}\right) + \mu\left(\frac{du}{d\zeta}\right)^2 \tag{95}$$

and the system (95), to the following:

$$\left(-\frac{1}{2}\rho \bar{u}\zeta + \rho \tilde{V}\right)\frac{d\bar{u}}{d\zeta} = \frac{d}{d\zeta}\left(\mu \frac{d\bar{u}}{d\zeta}\right)$$

$$-\frac{1}{2}\zeta \frac{d(\rho \bar{u})}{d\zeta} + \frac{d}{d\zeta}(\rho \tilde{V}) = 0$$

$$\left(-\frac{1}{2}\rho \bar{u}\zeta + \rho \tilde{V}\right)\frac{d\bar{h}}{d\zeta} = \frac{1}{\sigma}\frac{d}{d\zeta}\left(\mu \frac{d\bar{h}}{d\zeta}\right) + \mu\left(\frac{d\bar{u}}{d\zeta}\right)^2$$

$$\bar{u} = 0 \quad \tilde{V} = 0 \quad \bar{h} = \bar{h}_w \quad \left(\text{or } \frac{d\bar{h}}{d\zeta} = 0\right) \text{ at } \zeta = 0$$

$$\bar{u} \to \bar{u}_e \quad \bar{h} \to \bar{h}_e \text{ for } \zeta \to \infty \tag{96}$$

Additional transformation of the system (95)

$$\zeta_1 = \sqrt{3\zeta} \qquad \tilde{v}(\zeta) = \sqrt{3}\left[W(\zeta_1) - \frac{2}{3}\zeta_1 u\right] \tag{97}$$

gives (the conditions in the last lines of Equations 96 and 98 are as usual asymptotic)

$$\left(-\frac{\rho u \zeta_1}{2} + \rho W\right)\frac{du}{d\zeta_1} = \frac{d}{d\zeta_1}\left(\mu \frac{du}{d\zeta_1}\right)$$

$$-\frac{\zeta_1}{2}\frac{d(\rho u)}{d\zeta_1} + \frac{d(\rho W)}{d\zeta_1} = 0$$

$$\left(-\frac{\rho u \zeta_1}{2} + \rho W\right)\frac{dh}{d\zeta_1} = \frac{1}{\sigma}\frac{d}{d\zeta_1}\left(\mu \frac{dh}{d\zeta_1}\right) + \mu\left(\frac{du}{d\zeta_1}\right)^2$$

$$u = 0 \quad W = 0 \quad h = h_w \quad \left(\text{or} \quad \frac{dh}{d\zeta_1} = 0\right) \text{ at } \zeta_1 = 0$$

$$u \to u_e \quad h \to h_e \text{ for } \zeta_1 \to \infty \tag{98}$$

which is absolutely similar to the equations and the boundary conditions (96) for a plate, so that $\tilde{u} = u$, $\tilde{V} = W$.

Hence, it can be concluded that if the functions of the argument ζ

$$\bar{u} = u^*(\zeta) \qquad \bar{v} = \frac{\tilde{v}^*(\zeta)}{\sqrt{x}} \qquad \bar{h} = h^*(\zeta) \tag{99}$$

are the solution for a plate, then according to Equation 97 the functions of the argument ζ_1

$$u = u^*(\zeta_1) \qquad v = \frac{\sqrt{3}}{\sqrt{r}}\left[W(\zeta_1) - \frac{2}{3}\zeta_1 u^*(\zeta_1)\right] \qquad h = h^*(\zeta_1) \tag{100}$$

will give the solution for a cone.

Comparing these solutions, we can draw the following conclusions. At the same values of the abscissa $r = x$, conventional thicknesses of the layer δ^*, δ^{**} on a cone are by a factor of $\sqrt{3}$ smaller than on the plate ($\bar{\delta}^*$, $\bar{\delta}^{**}$). Actually we have, for example,

$$\overline{\delta^*} = \int_0^\infty \left[1 - \frac{u^*(\zeta)}{u_e^*}\right] dy = \sqrt{x}\int_0^\infty \left[1 - \frac{u^*(\zeta)}{u_e^*}\right] d\zeta$$

$$\delta^* = \int_{\theta_0}^{\theta_\infty} \left[1 - \frac{u^*(\zeta_1)}{u_e^*}\right] r\,d\theta = \frac{\sqrt{r}}{\sqrt{3}}\int_0^\infty \left[1 - \frac{u^*(\zeta_1)}{u_e^*}\right] d\zeta_1$$

so that

$$\delta^* = \frac{\overline{\delta^*}}{\sqrt{3}}$$

and analogous relationships for the momentum thickness of a boundary layer.

Also compare the skin friction τ_w on the cone and $\overline{\tau_w}$ on the plate. We have

$$\tau_w = \left(\frac{\mu}{r}\frac{\partial u}{\partial \theta}\right)_{\theta = \theta_0} = \left(\frac{\mu}{r}\frac{du^*}{d\zeta_1}\frac{\partial \zeta_1}{\partial \theta}\right)_{\substack{\zeta_1 = 0 \\ \theta = \theta_0}} = \left(\frac{\mu}{r}\frac{du^*}{d\zeta_1}\sqrt{3}\sqrt{r}\right)_{\zeta_1 = 0} =$$

$$= \sqrt{3}\left(\mu\frac{\partial \bar{u}}{\partial y}\right)_{y = 0} = \sqrt{3}\,\bar{\tau}_w$$

so that at the same value of the abscissa $r = x$ the skin friction on the cone is by a factor of $\sqrt{3}$ greater than in the plate. Similarly, we shall convince ourselves that the coefficient of total friction resistance on the cone, under other conditions being equal, is by a factor of

$\frac{2}{3}\sqrt{3}$ greater than the corresponding coefficient on the plate. In fact, the first of these coefficients is

$$\left(\frac{1}{2}\rho_e u_e^2 \pi L^2\right)^{-1}\int_0^L 2\pi r \tau_w dr = \left(\frac{1}{2}\rho_e u_e^2 \pi L^2\right)^{-1}\int_0^L 2\pi r \sqrt{3}\,\bar{\tau}_w dr =$$

$$= \frac{(4/\sqrt{3})\,\bar{\tau}_w(L)}{\frac{1}{2}\rho_e u_e^2}$$

and the second

$$\left(\frac{1}{2}\rho_e \bar{u}_e^2 L\right)^{-1}\int_0^L \bar{\tau}_w dx = \frac{2\bar{\tau}_w(L)}{\frac{1}{2}\rho_e \bar{u}_e^2}$$

which, with $u_e = \bar{u}_e$, leads to the above conclusion. In the same relations $\sqrt{3}$ and $\frac{2}{3}\sqrt{3}$ are the local and mean coefficients of heat transfer.

These results can be also obtained from a more general view-point, applying the Mangler transform[24] to the equations of a boundary layer on a cone parallel to the flow that stems from transition from the coordinates x, y, reckoned along the meridional section of the surface of the body of revolution and normal to it, to the coordinates $\bar{x}$, $\bar{y}$ in a corresponding plane boundary layer by the formulas ($r_0(x)$ is the radius of the transverse curvature of the body of revolution)

$$\bar{x} = \int_0^x r_0^2(\xi)\,d\xi \qquad \bar{y} = r_0(x)\,y \tag{101}$$

All the other parameters of the flow are preserved

$$\bar{u} = u \quad \bar{u}_e = u_e \quad \bar{\psi} = \psi$$

$$\bar{p} = p \quad \bar{\rho} = \rho \quad \bar{h} = h \quad \bar{\mu} = \mu \tag{102}$$

excluding the transverse velocity v, which is transformed by the equation

$$\bar{v} = \frac{v}{r_0} + \frac{r_0'(x)}{r_0^2}yu \tag{103}$$

Let us compose the equations of the boundary layer on the body of revolution in the case of a parallel gas flow in the coordinates x and y, using the transform mentioned in Section 11 of Chapter 12 in derivation of the set of equations (165) of Chapter 12. We shall obtain (for

---

[24] Mangler, W., Zusammenhang zwischen ebenen und rotationssymmetrischen Grenzschichten in kompressiblen Flüssigkeiten, *Z. Angew. Math. Mech.*, 28, 97, 1948. Transforms (101) and (103) coincide with those by E. I. Stepanov (Section 1, Chapter 13).

the generality, in the first equation the term defined by the pressure gradient is preserved)

$$\rho\left(u\frac{\partial u}{\partial x}+v\frac{\partial u}{\partial y}\right) = \rho u_e\frac{du_e}{dx}+\frac{\partial}{\partial y}\left(\mu\frac{\partial u}{\partial y}\right)$$

$$\frac{\partial(r_0\rho u)}{\partial x}+\frac{\partial(r_0\rho v)}{\partial y} = 0$$

$$\rho\left(u\frac{\partial h}{\partial x}+v\frac{\partial h}{\partial y}\right) = \frac{1}{\sigma}\frac{\partial}{\partial y}\left(\mu\frac{\partial h}{\partial y}\right)+\mu\left(\frac{\partial u}{\partial y}\right)^2 \tag{104}$$

We introduce the stream function $\psi(x, y)$ by the equations

$$r_0 u = \frac{1}{\rho}\frac{\partial\psi}{\partial y} \qquad r_0 v = -\frac{1}{\rho}\frac{\partial\psi}{\partial x} \tag{105}$$

which satisfy the second equation of the system (104). Applying the transform (101), according to which

$$\frac{\partial}{\partial x} = r_0^2\frac{\partial}{\partial\bar{x}}+r_0'(x)\,y\frac{\partial}{\partial\bar{y}} \qquad \frac{\partial}{\partial y} = r_0\frac{\partial}{\partial\bar{y}} \tag{106}$$

to the system of equations (105), we have

$$r_0 u = \frac{1}{\rho}r_0\frac{\partial\psi}{\partial\bar{y}} \qquad r_0 v = -\frac{1}{\rho}r_0^2\frac{\partial\psi}{\partial\bar{x}}-r_0'(x)\,y\frac{1}{\rho}\frac{\partial\psi}{\partial\bar{y}} \tag{107}$$

where

$$u = \bar{u} = \frac{1}{\rho}\frac{\partial\psi}{\partial\bar{y}} \qquad \bar{v} = \frac{v}{r_0}+\frac{r_0'(x)}{r_0^2}yu = -\frac{1}{\rho}\frac{\partial\psi}{\partial\bar{x}} \tag{108}$$

Applying the transforms (106) to the first and third equations of the system (104) and using Equations 108 for the second equation of this system, we obtain from (102) the following system of equations in new variables:

$$\bar{\rho}\left(\bar{u}\frac{\partial\bar{u}}{\partial\bar{x}}+\bar{v}\frac{\partial\bar{u}}{\partial\bar{y}}\right) = \bar{\rho}\bar{u}_e\frac{d\bar{u}_e}{d\bar{x}}+\frac{\partial}{\partial\bar{y}}\left(\bar{\mu}\frac{\partial\bar{u}}{\partial\bar{y}}\right)$$

$$\frac{\partial(\bar{\rho}\bar{u})}{\partial\bar{x}}+\frac{\partial(\bar{\rho}\bar{v})}{\partial\bar{y}} = 0$$

$$\bar{\rho}\left(\bar{u}\frac{\partial\bar{h}}{\partial\bar{x}}+\bar{v}\frac{\partial\bar{h}}{\partial\bar{y}}\right) = \frac{1}{\sigma}\frac{\partial}{\partial\bar{y}}\left(\bar{\mu}\frac{\partial\bar{h}}{\partial\bar{y}}\right)+\bar{\mu}\left(\frac{\partial\bar{u}}{\partial\bar{y}}\right)^2 \tag{109}$$

Coinciding with the boundary layer equations in a plane gas flow; the transformed boundary conditions also will be in agreement.

Making use of this general transformation in the particular case of the cone $[r_0(x) = ax]$ considered at the beginning of this section, we shall have from the first equation of the system (101)

$$\bar{x} = \frac{1}{3}a^2x^3 \qquad \bar{y} = axy \tag{110}$$

where $a = \sin\theta_0$ (θ_0 is the half of the cone apex). We shall compare, for example, skin friction for the cone τ_w and for the plate $\bar{\tau}_w$, According to the second equation of the system (101) we obtain at the corresponding points (x, y) and $(\bar{x}, \bar{y})$ the general relation

$$\tau_w = \mu_w\left(\frac{\partial u}{\partial y}\right)_{y=0} = \bar{\mu}_w\left(\frac{\partial \bar{u}}{\partial \bar{y}}\right)_{\bar{y}=0}\left(\frac{\partial \bar{y}}{\partial y}\right)_{y=0} = \bar{\tau}_w r_0 = \bar{\tau}_w ax \tag{111}$$

We shall recall that Hantzsche and Wendt compared τ_w and $\bar{\tau}_w$ at the same values of abscissa on the cone generatrix and along the plate; this means that according to Equation 110 it should be assumed

$$x = \frac{1}{3}a^2x^3 \quad \text{or} \quad ax = \sqrt{3}$$

Returning then to Equation 111 we have

$$\tau_w = \sqrt{3}\,\bar{\tau}_w$$

in full conformity with the previous result. The remaining relations can be obtained similarly.

In the cited monograph,[25] on the theory of a laminar boundary layer, many examples are given on the calculation of a boundary layer in a gas in the absence of a longitudinal pressure gradient.

6. The Laminar Boundary Layer in High-Speed Flow with External Pressure Gradients

First, we examine a simple approximate method of calculating the development of the laminar boundary layer in a compressible gas with given external pressure gradients.[25] which is valid at moderate Mach numbers, for example, $\mathbf{M}_\infty < 2$.

We shall consider the case of a heat insulated surface, i.e., we shall take the heat transfer through the surface to be zero so that

$$\frac{\partial T}{\partial y} = 0 \quad \text{at} \quad y = 0 \tag{112}$$

We shall also assume that $\sigma = 1$. Then it can be easily verified that a particular integral of the energy equation (the third formula in Equation 48), satisfying the above boundary condition, is

$$h_0 = \text{const}$$

or, remembering (42),

[25] Dorodnitsyn, A. A. and Loitsyanskiy L. G., On the theory of transition from laminar to turbulent flow in the boundary layer, *Prikl. Mat. Mekh.*, 9(4), 1945.

$$h + \frac{u^2}{2} = h_0 \tag{113}$$

where h_0 is the constant for all the boundary layer points. The last equation is equivalent to (T_0, similar to h_0, is constant throughout the layer):

$$T = T_0\left(1 - \frac{u^2}{2h_0}\right) \tag{114}$$

Noting that outside the boundary layer there is potential flow, and that the pressure is constant across the layer, so that we obtain, for all points of the layer, the well known isentropic relation (Chapter 5)

$$p = p_0\left(1 - \frac{u_e^2}{2h_0}\right)^{\frac{k}{k-1}} \tag{115}$$

where u_e is the velocity at the outer edge of the boundary layer and p_0 is the mainstream stagnation pressure constant throughout the outer edge of the boundary layer. From the equation of state the corresponding density relation is (with ρ_0 denoting the main stream stagnation density)

$$\rho = \frac{\rho_0\left(1 - \frac{u_e^2}{2h_0}\right)^{\frac{k}{k-1}}}{1 - \frac{u^2}{2h_0}} \tag{116}$$

Making use of the powerlaw relation between the coefficient of viscosity and the absolute temperature, we obtain

$$\mu = \mu_0\left(\frac{T}{T_0}\right)^n = \mu_0\left(1 - \frac{u^2}{2h_0}\right)^n \tag{117}$$

Because a solution of the energy equation already has been found (Equation 113), we are left with the first two boundary layer equations (in a dimensional form)

$$\rho\left(u\frac{\partial u}{\partial x} + v\frac{\partial u}{\partial y}\right) = -\frac{dp}{dx} + \frac{\partial}{\partial y}\left(\mu\frac{\partial u}{\partial y}\right) \qquad \frac{\partial(\rho u)}{\partial x} + \frac{\partial(\rho v)}{\partial y} = 0 \tag{118}$$

and, introducing for brevity the definitions

$$\alpha = \frac{u}{\sqrt{2h_0}} \qquad \bar{\alpha} = \frac{u_e}{\sqrt{2h_0}} \tag{119}$$

we apply to Equations 118 the Dorodnitsyn transformation,[26] which in this case takes the form

$$\xi = \int_0^x \frac{p}{p_0}dx = \int_0^x (1-\bar{\alpha}^2)^{\frac{k}{k-1}}dx \qquad \eta = \int_0^y \frac{\rho}{\rho_0}dy = \int_0^y \frac{(1-\bar{\alpha}^2)^{\frac{k}{k-1}}}{(1-\alpha^2)}dy \tag{120}$$

[26] Dorodnitsyn, A. A., The boundary layer in a compressible gas, *Prikl. Mat. Mekh.*, 6(6), 1942.

We note that

$$\frac{\partial}{\partial x} = \frac{p}{p_0}\frac{\partial}{\partial \xi} + \frac{\partial \eta}{\partial x}\frac{\partial}{\partial \eta} = (1-\bar{\alpha}^2)^{\frac{k}{k-1}}\frac{\partial}{\partial \xi} + \frac{\partial \eta}{\partial x}\frac{\partial}{\partial \eta}$$

$$\frac{\partial}{\partial y} = \frac{\rho}{\rho_0}\frac{\partial}{\partial \eta} = \frac{(1-\bar{\alpha}^2)^{\frac{k}{k-1}}}{(1-\alpha^2)}\frac{\partial}{\partial \eta} \tag{121}$$

and that from the first of these equations and the relations (115)

$$\frac{dp}{dx} = (1-\bar{\alpha}^2)^{\frac{k}{k-1}}\frac{dp}{d\xi} = -p_0\frac{k}{k-1}(1-\bar{\alpha}^2)^{\frac{k+1}{k-1}}\cdot 2\bar{\alpha}\bar{\alpha}' =$$

$$= -\frac{p_0}{h_0}\frac{k}{k-1}(1-\bar{\alpha}^2)^{\frac{k+1}{k-1}}u_e u_e' = -\rho_0(1-\bar{\alpha}^2)^{\frac{k+1}{k-1}}u_e u_e' \tag{122}$$

where primes denote differentiation with respect to ξ.

Substituting these expressions into the first of Equations 118 we shall reduce it to the form

$$u\frac{\partial u}{\partial \xi} + \tilde{v}\frac{\partial u}{\partial \eta} = \frac{1-\alpha^2}{1-\bar{\alpha}^2}u_e u_e' + \nu_0\frac{\partial}{\partial \eta}\left[(1-\alpha^2)^{n-1}\frac{\partial u}{\partial \eta}\right] \tag{123}$$

where it is assumed for brevity

$$\tilde{v} = \frac{v}{1-\alpha^2} + \frac{u}{(1-\bar{\alpha}^2)^{\frac{k}{k-1}}}\frac{\partial \eta}{\partial x} \tag{124}$$

follows from the continuity equation that

$$\rho u = \frac{\partial \psi}{\partial y} = \frac{\rho}{\rho_0}\frac{\partial \psi}{\partial \eta} \qquad \rho v = -\frac{\partial \psi}{\partial x} = -(1-\bar{\alpha}^2)^{\frac{k}{k-1}}\frac{\partial \psi}{\partial \xi} - \frac{\partial \eta}{\partial x}\frac{\partial \psi}{\partial \eta}$$

or

$$u = \frac{1}{\rho_0}\frac{\partial \psi}{\partial \eta} \qquad \frac{v}{1-\alpha^2} = -\frac{1}{\rho_0}\frac{\partial \psi}{\partial \xi} - \frac{u}{(1-\bar{\alpha}^2)^{\frac{k}{k-1}}}\frac{\partial \eta}{\partial x}$$

It terms of the velocity $\tilde{v}$ defined in (124) we rewrite the last two equations in the form

$$u = \frac{1}{\rho_0}\frac{\partial \psi}{\partial \eta} \qquad \tilde{v} = -\frac{1}{\rho_0}\frac{\partial \psi}{\partial \xi}$$

where the continuity equation in the Dorodnitsyn coordinates follows directly

$$\frac{\partial u}{\partial \xi} + \frac{\partial \tilde{v}}{\partial \eta} = 0 \tag{125}$$

Making use of Equations 123 and 125 in a manner analogous to that already described in

Chapter 12, we derive the momentum integral equation in terms of the Dorodnitsyn variables

$$\frac{d}{d\xi}\int_0^{\infty,\,\delta_\eta} u\,(u_e - u)\,d\eta + u_e'\int_0^{\infty,\,\delta_\eta}\left(\frac{1-\alpha^2}{1-\bar{\alpha}^2}\,u_e - u\right)d\eta = v_0\left(\frac{\partial u}{\partial \eta}\right)_{\eta = 0} \tag{126}$$

where δ_η is the finite thickness of the boundary layer defined as the value of η for which $u \approx u_e$. Introducing the analogous conventional boundary layer thicknesses

$$\delta_\eta^* = \int_0^{\infty,\,\delta_\eta}\left(1 - \frac{u}{u_e}\right)d\eta \qquad \delta_\eta^{**} = \int_0^{\infty,\,\delta_\eta}\frac{u}{u_e}\left(1 - \frac{u}{u_e}\right)d\eta \tag{127}$$

Then, using the identity

$$\frac{1-\alpha^2}{1-\bar{\alpha}^2}\,u_e - u = \frac{1}{1-\bar{\alpha}^2}\,[u_e - u + \alpha\bar{\alpha}\,(u_e - u)]$$

we transform Equation 126 into the form

$$\frac{d\delta_\eta^{**}}{d\xi} + \left(2\frac{u_e'}{u_e} + \frac{\bar{\alpha}\bar{\alpha}'}{1-\bar{\alpha}^2}\right)\delta_\eta^{**} + \frac{u_e'}{u_e(1-\bar{\alpha}^2)}\delta_\eta^* = \frac{v_0}{u_e^2}\left(\frac{\partial u}{\partial \eta}\right)_{\eta = 0} \tag{128}$$

This is the momentum equation. When $\bar{\alpha} \to 0$, $\xi \to x$, $\eta \to y$, $v_0 \to v$ (i.e., going over to low speeds), we obtain the momentum equation for an incompressible fluid derived in Chapter 12.

We now assume that the velocity profiles across the boundary layer are of the one parameter family

$$\frac{u}{u_e} = \varphi\left(\frac{\eta}{\delta_\eta^{**}}; f\right) \tag{129}$$

which does not depend explicitly on $\mathbf{M}_\infty$, i.e., it is the same as in the incompressible case ($\mathbf{M}_\infty = 0$). Here, the implicit effect of the Mach number $\mathbf{M}_\infty$ in terms of the quantities u_e, δ_η^{**}, η, and f is, of course, preserved.

This assumption is valid at moderate $\mathbf{M}_\infty$. Thus, for example, for a plate ($f = 0$) the quantity

$$\zeta = \left[\frac{\partial(u/u_e)}{\partial(y/\delta^{**})}\right]_{y=0} = \left[\frac{\partial(u/u_e)}{\partial(\eta/\delta_\eta^{**})}\right]_{\eta=0} = \varphi'(0;0)$$

should be independent of $\mathbf{M}_\infty$ and equal to 0.221. Exact calculations show that in fact this quantity grows with the Mach number $\mathbf{M}_\infty$ and takes the values of 0.222, 0.227, and 0.234 at $\mathbf{M}_\infty$ values of 0.65, 1.59, and 3.05, respectively. Therefore, the present solution, whose merit is simplicity, can be used only at relatively small Mach numbers $\mathbf{M}_\infty$. Making use of Equation 129, we have

$$\frac{\delta_\eta^*}{\delta_\eta^{**}} = \int_0^{\infty}\left[1 - \varphi\left(\frac{\eta}{\delta_\eta^{**}}; f\right)\right]d\left(\frac{\eta}{\delta_\eta^{**}}\right) = H(f)$$

$$\left(\frac{\partial u}{\partial \eta}\right)_{\eta=0} = \frac{u_e}{\delta_\eta^{**}}\varphi'(0;f) = \frac{u_e}{\delta_\eta^{**}}\zeta(f) \tag{130}$$

After substitution of these expressions into Equation 128 we get

$$\frac{1}{2}\frac{d}{d\xi}\left(\frac{\delta_\eta^{**2}}{\nu_0}\right) + \left(2\frac{u_e'}{u_e} + \frac{\bar{\alpha}\bar{\alpha}'}{1-\bar{\alpha}^2}\right)\frac{\delta_\eta^{**2}}{\nu_0} + \frac{u_e'}{u_e(1-\bar{\alpha}^2)}\frac{\delta_\eta^{**2}}{\nu_0}H(f) = \frac{\zeta(f)}{u_e}$$

Multiplying both sides of this equation by $1-\bar{\alpha}^2$, we shall reduce it to the form

$$\frac{1}{2}\frac{d}{d\xi}\left[\frac{(1-\bar{\alpha}^2)\,\delta_\eta^{**2}}{\nu_0}\right] = \frac{1-\bar{\alpha}^2}{u_e}\left\{\zeta(f) - \frac{u_e'\delta_\eta^{**2}}{\nu_0(1-\bar{\alpha}^2)}[2+H(f)]\right\}$$

It can be seen that the quantity

$$f = \frac{u_e'\delta_\eta^{**2}}{\nu_0(1-\bar{\alpha}^2)} \tag{131}$$

plays the role of a form parameter because the expression within the curly brackets turns out to be a function of a form parameter only. The preceding equation may be written in the form

$$\frac{d}{d\xi}\left[\frac{(1-\bar{\alpha}^2)^2}{u_e'}f\right] = \frac{1-\bar{\alpha}^2}{u_e}F(f)$$

or, carrying out differentiation on the left side

$$\frac{df}{d\xi} = F(f)\frac{d}{d\xi}\ln\frac{u_e}{\sqrt{1-\bar{\alpha}^2}} + f\frac{d}{d\xi}\ln\frac{u_e'}{(1-\bar{\alpha}^2)^2} \tag{132}$$

This equation is a generalization of the theory of Chapter 12 to the considered case of a high speed gas flow. The function entering into Equation 132

$$F(f) = 2\{\zeta(f) - [2+H(f)]f\} \tag{133}$$

does not differ from the corresponding function $\bar{f}$ for an incompressible fluid.

Retransforming to the basic variable x, we have (with dashes denoting differentiation with respect to x)

$$\frac{df}{dx} = F(f)\frac{d}{dx}\ln\frac{u_e}{\sqrt{1-\bar{\alpha}^2}} + f\frac{d}{dx}\ln\frac{u_e'}{(1-\bar{\alpha}^2)^{2+\frac{k}{k-1}}} \tag{134}$$

From the previous assumption that the velocity profile (129) is independent of $\mathbf{M}_\infty$ it follows that the form of the function $F(f)$ should be the same as in incompressible flow ($\mathbf{M}_\infty = 0$). This makes it possible, as before, to replace the function $F(f)$ with its linear approximation

$$F(f) = a - bf$$

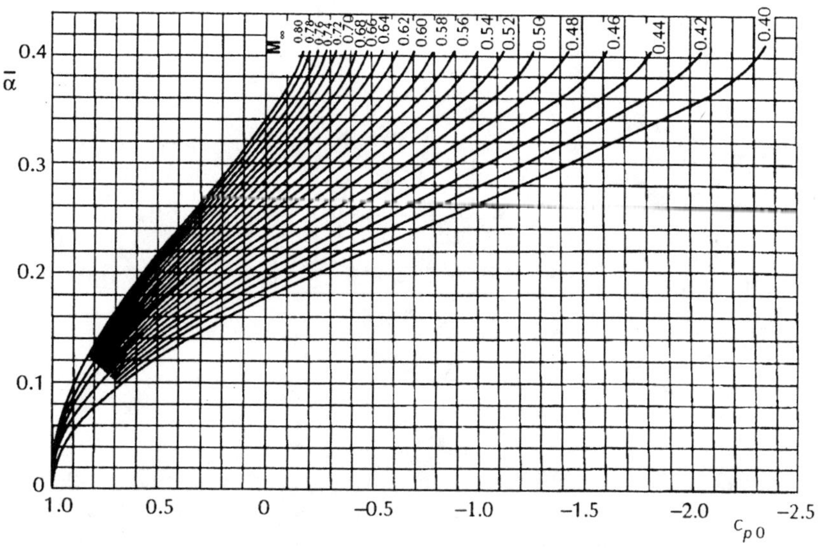

Fig. 279

and after simple integration of Equation 134 to obtain

$$f = \frac{au_e'}{u_e^b(1-\bar{\alpha}^2)^m} \int_0^x u_e^{b-1}(1-\bar{\alpha}^2)^{m-1}dx \tag{135}$$

where

$$m = 2 + \frac{k}{k-1} - \frac{b}{2} \tag{136}$$

For air ($k = 1.4$), with $b = 5.75$, m is approximately 2.63. Equation 135 can be rewritten in the form

$$f = a\frac{\bar{\alpha}'}{\bar{\alpha}^b(1-\bar{\alpha}^2)^m} \int_0^x \bar{\alpha}^{b-1}(1-\bar{\alpha}^2)^{m-1}dx \tag{137}$$

which is more convenient for numerical calculations. To convert the given distribution of the pressure coefficient $c_{p0}(x)$ along the profile surface into the distribution $\bar{\alpha}(x)$ use may be made of the set of curves reproduced in Figure 279, for $\mathbf{M}_\infty < \mathbf{M}_{cr}$.

The assumption adopted earlier leads us to conclude that at large subcritical velocities fs, the value of f at separation, is independent of $\mathbf{M}_\infty$. Remembering that in the subcritical region an increase of $\mathbf{M}_\infty$ implies increased expansion of the flow, and, consequently, larger $u_e'(x)$ beyond the point of minimum pressure, we conclude from Equation 135 that the value of xs decreases, i.e., the "point" of separation moves upstream. Hence, it follows that at subcritical speeds compressibility of the fluid accelerates the onset of boundary layer separation and so worsens the flow past the body. Calculations confirm this argument. Thus, for example, the laminar separation point on the upper surface of NACA 4412 aerofoil, which is situated approximately 11% of the chord downstream of the leading edge at $\mathbf{M}_\infty = 0$ and $c_y = 0.146$,

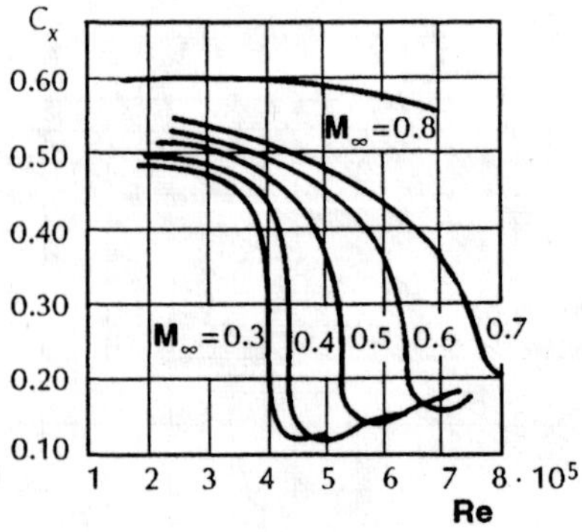

Fig. 280

moves forward to the 5% chord position at $\mathbf{M}_\infty = 0.4$.

An evident confirmation of the separation point upstream displacement with the increase of $\mathbf{M}_\infty$ is provided by the results of Ferri's experimental investigation of the sphere drag crisis. Because of the deterioration of the flow past the sphere with increasing $\mathbf{M}_\infty$, one would expect that to improve this flow associated with the drag crisis the critical Reynolds number should increase with increasing $\mathbf{M}_\infty$. The results of Ferri's studies into the flow past the sphere at different $\mathbf{M}_\infty$, depicted in Figure 280 provide good confirmation of this assumption. As $\mathbf{M}_\infty$ increases from 0.3 to 0,7, the conventional Reynolds number $\mathbf{Re}_{cr}$, corresponding to $c_x = 0.3$, grows from 400,000 to 740,000.

The phenomenon of earlier separation due to gas compressibility is also responsible for the observed reduction of $c_{y\,\max}$ with increasing $\mathbf{M}_\infty$ under subcritical conditions.

Now we consider a general case of arbitrary Mach, $\mathbf{M}_\infty$, and Prandtl numbers, which differ from unity,[27] but in fact are not so great that the phenomena of gas dissociation and ionization, typical of its high temperatures, should be taken into account.

In this case there is no longer integral (113) of the motion and energy equations and it is necessary to solve the complete system of equations (48), which after simple transformations of the third equation of the system, can be rewritten as

$$\rho u \frac{\partial u}{\partial x} + \rho v \frac{\partial u}{\partial y} = -\frac{dp_e}{dx} + \frac{\partial}{\partial y}\left(\mu \frac{\partial u}{\partial y}\right) \qquad \frac{\partial(\rho u)}{\partial x} + \frac{\partial(\rho v)}{\partial y} = 0$$

$$\rho u \frac{\partial h_0}{\partial x} + \rho v \frac{\partial h_0}{\partial y} = \frac{1}{\sigma}\frac{\partial}{\partial y}\left(\mu \frac{\partial h_0}{\partial y}\right) - \left(\frac{1}{\sigma} - 1\right)\frac{\partial}{\partial y}\left(\mu u \frac{\partial u}{\partial y}\right)$$

$$\frac{k-1}{k}\rho h = p = p_e \qquad \mu = \mu_{e0} a(\chi) \tag{138}$$

where it is assumed that

$$\chi = \frac{T}{T_{e0}} = \frac{h}{h_{e0}} \tag{139}$$

and h_0 is the variable total enthalpy (42); in the case under consideration $h_0 \neq h_{e0}$. The func-

[27] Dorodnitsyn, A. A., Laminar boundary layer in a compressible gas, in *Teoreticheskiye raboty po aerodinamike* (*Theory of Aerodynamics*), TsAGI Oborongiz, Moscow, 1957, 140–173.

tion $a(\chi)$ which enters into the last equation of the system (138) expresses some general dependence of the viscosity coefficient on temperature (enthalpy).

From the definition of the stagnation parameters of gas, we get

$$p = p_e = p_{e0}(1-\alpha_e^2)^{\frac{k}{k-1}} \qquad \rho = \rho_{e0}\frac{(1-\alpha_e^2)^{k/(k-1)}}{\chi} \qquad \alpha_e = u_e/\sqrt{2h_{e0}} \tag{140}$$

Applying the Dorodnitsyn transform to the first three equations of the system (138)

$$\xi = \int_0^x \frac{p_e}{p_{e0}}dx \qquad \eta = \int_0^y \frac{\rho}{\rho_{e0}}dy \tag{141}$$

we find

$$\frac{\partial}{\partial x} = \frac{p_e}{p_{e0}}\frac{\partial}{\partial \xi} + \frac{\partial \eta}{\partial x}\frac{\partial}{\partial \eta} \qquad \frac{\partial}{\partial y} = \frac{\rho}{\rho_{e0}}\frac{\partial}{\partial \eta} \tag{142}$$

In addition, from Equation 140 (prime denotes the derivative with respect to ξ)

$$\frac{dp_e}{dx} = \frac{p_e}{p_{e0}}\frac{dp_e}{d\xi} = -\frac{k}{k-1}\frac{p_e}{h_{e0}}u_e u_e'(1-\alpha_e^2)^{\frac{1}{k-1}}$$

and, consequently,

$$\frac{dp_e}{dx} = -\rho_{e0}(1-\alpha_e^2)^{\frac{k+1}{k-1}}u_e u_e' \tag{143}$$

Applying the transformation (142) to the first equation of the system (138) and making use of (143), we obtain

$$\rho\frac{p_e}{p_{e0}}u\frac{\partial u}{\partial \xi} + \rho\left(\frac{\partial \eta}{\partial x}u + \frac{\rho}{\rho_{e0}}v\right)\frac{\partial u}{\partial \eta} =$$

$$= \rho_{e0}(1-\alpha_e^2)^{\frac{k+1}{k-1}}u_e u_e' + \frac{\rho}{\rho_{e0}}\mu_{e0}\frac{\partial}{\partial \eta}\left[a(\chi)\frac{\rho}{\rho_{e0}}\frac{\partial u}{\partial \eta}\right]$$

where, dividing both sides of the equation by the quantity

$$\rho\frac{p_e}{p_{e0}} = \rho_{e0}\frac{(1-\alpha_e^2)^{\frac{2k}{k-1}}}{\chi}$$

we get the following expression for the first (dynamic) equation

$$u\frac{\partial u}{\partial \xi} + \tilde{v}\frac{\partial u}{\partial \eta} = \frac{\chi}{\chi_e}u_e u_e' + \nu_{e0}\frac{\partial}{\partial \eta}\left[b(\chi)\frac{\partial u}{\partial \eta}\right] \tag{144}$$

where it is assumed that

$$\tilde{v} = \frac{p_{e0}}{p_e}\frac{\partial \eta}{\partial x}u + \frac{\rho}{\rho_{e0}}\frac{p_{e0}}{p_e}v = (1-\alpha_e^2)^{-\frac{k}{k-1}}u\frac{\partial \eta}{\partial x} + \frac{v}{\chi}$$

$$b(\chi) = \frac{a(\chi)}{\chi} \qquad \nu_{e0} = \frac{\mu_{e0}}{\rho_{e0}} \qquad \chi_e = \frac{T_e}{T_{e0}} = 1-\alpha_e^2 \tag{145}$$

Introducing the stream function ψ by the equations

$$\rho u = \frac{\partial \psi}{\partial y} = \frac{\rho}{\rho_{e0}}\frac{\partial \psi}{\partial \eta}$$

$$\rho v = -\frac{\partial \psi}{\partial x} = -\frac{p_e}{p_{e0}}\frac{\partial \psi}{\partial \xi} - \frac{\partial \eta}{\partial x}\frac{\partial \psi}{\partial \eta} = -\frac{p_e}{p_{e0}}\frac{\partial \psi}{\partial \xi} - \rho_{e0}u\frac{\partial \eta}{\partial x}$$

we have

$$u = \frac{1}{\rho_{e0}}\frac{\partial \psi}{\partial \eta} \qquad \tilde{v} = -\frac{1}{\rho_{e0}}\frac{\partial \psi}{\partial \xi} \tag{146}$$

so that

$$\frac{\partial u}{\partial \xi} + \frac{\partial \tilde{v}}{\partial \eta} = 0\,. \tag{147}$$

Finally, we shall analogously transform the third equation of the system (138) (the energy equation). We shall introduce the nondimensional enthalpy instead of its full value h_0

$$\vartheta = \frac{h_0}{h_{e0}} = \frac{h}{h_{e0}} + \frac{u^2}{2h_{e0}} = \chi + \alpha^2 \tag{148}$$

Applying the transformation (142) to the third equation of the system (138), we obtain

$$\rho u\frac{p_e}{p_{e0}}\frac{\partial \vartheta}{\partial \xi} + \rho\left(u\frac{\partial \eta}{\partial x} + \frac{\rho v}{\rho_{e0}}\right)\frac{\partial \vartheta}{\partial \eta} =$$

$$= \frac{\mu_{e0}}{\sigma}\frac{\rho}{\rho_{e0}}\frac{\partial}{\partial \eta}\left[a(\chi)\frac{\rho}{\rho_{e0}}\frac{\partial \vartheta}{\partial \eta}\right] - \left(\frac{1}{\sigma}-1\right)\frac{\rho\mu_{e0}}{\rho_{e0}h_{e0}}\frac{\partial}{\partial \eta}\left[a(\chi)\frac{\rho}{\rho_{e0}}u\frac{\partial u}{\partial \eta}\right]$$

Then, having divided both sides by $\rho p_e/p_{e0}$, we determine finally the following form of the energy equations in terms of the Dorodnitsyn variables

$$u\frac{\partial \vartheta}{\partial \xi} + \tilde{v}\frac{\partial \vartheta}{\partial \eta} = \frac{\nu_{e0}}{\sigma}\frac{\partial}{\partial \eta}\left[b(\chi)\frac{\partial \vartheta}{\partial \eta}\right] - \left(\frac{1}{\sigma}-1\right)\nu_{e0}\frac{\partial}{\partial \eta}\left[b(\chi)\frac{\partial(\alpha^2)}{\partial \eta}\right] \tag{149}$$

The set of Equations 144, 147, and 149, with the boundary conditions

$$u = \tilde{v} = 0 \quad \text{at} \quad \eta = 0$$

$$u = u_e(\xi) \quad \text{at} \quad \eta = \infty$$

$$\frac{\partial \vartheta}{\partial \eta} = 0 \quad \text{at} \quad \eta = 0 \quad \text{(if the body surface is thermally insulated)}$$

$$\vartheta = \vartheta_w(\xi) \quad \text{at} \quad \eta = 0 \quad \text{(if the surface temperature is fixed)}$$

$$\vartheta = 1 \quad \text{at} \quad \eta = \infty$$

represents the formulation of the problem of a plane laminar boundary layer on an aerofoil in a high speed gas flow under the given pressure distribution on the outer edge of the boundary layer.

Analyzing the equations obtained we can see that the left sides of the first three equations of the problem, which are written in terms of the Dorodnitsyn variables, coincide with the corresponding equations for the plane laminar boundary layer in an incompressible fluid flow. However, the right sides of these equations involve the explicit influence of compressibility in terms of the quantities χ/χ_e and $b(\chi)$.

Using the dependence of the viscosity coefficient on temperature (enthalpy) in the Chapman Rubezin form t (15), we shall have $b(\chi) = 1$. In fact, according to (15)

$$\frac{\mu}{\mu_e} = \frac{Ch}{h_e} \qquad \frac{\mu_{e0}}{\mu_e} = \frac{Ch_{e0}}{h_e}$$

so that

$$\frac{\mu}{\mu_{e0}} = a(\chi) = \chi b(\chi) = \frac{h}{h_{e0}} = \chi$$

In the paper by Dorodnitsyn,[27] along with the approximate formula (15), another assumption is made which is also close to the Sutherland equation, but is nonlinear and corresponds to the value of the function $b(\chi)$

$$b(\chi) = 1 + 0.3(1 - \chi)$$

The dashed line in Figure 281 corresponding to this law, lies between the Sutherland curves for $T_e = 330$ K and $T_e = 660$ K, which are presented by solid lines.

In the presence of such a relation between μ and T, for $\sigma = 14/19$ (valid for two atomic gases), and on the assumption about the absence of heat transfer, Dorodnitsyn numerically

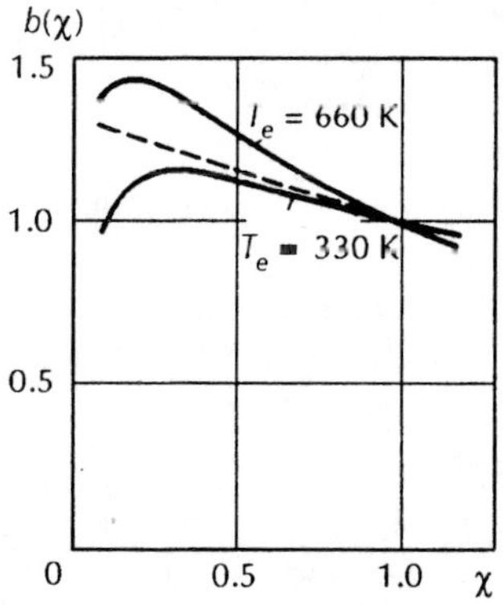

Fig. 281

determined the unknown functions u and ϑ for three cases of prescribing $u_e(\xi)$:

1. For a parabolic distribution of $u_e(\xi) = c_1\xi + c_3\xi^3$
2. For a longitudinal flow past the plate $u_e = \text{const}$
3. For a one slope profile $u_e(\xi) = b_0 - b_1\xi$

Applying these classes of solutions sequentially to a converging portion of the boundary layer, the region of pressure minimum and to a diverging portion of the boundary layer, Dorodnitsyn constructed an approximate one term solution of the problem considered which is a generalization of the solution (discussed at the beginning of this section) to the case of limited values of the Mach number for a free stream.

We shall form two basic integral relations (prime denotes the derivative with respect to ξ)

$$\delta^{**\prime} + \frac{u_e'}{u_e(1-\alpha_e^2)}(2-\alpha_e^2 + H - H_\vartheta)\,\delta^{**} = \frac{\nu_{e0}}{u_e^2}\, b(\chi_w)\left(\frac{\partial u}{\partial \eta}\right)_{\eta=0}$$

$$(u_e\delta_\vartheta^{**})' = \frac{\nu_{e0}}{\sigma}\, b(\chi_w)\left(\frac{\partial \vartheta}{\partial \eta}\right)_{\eta=0} \tag{150}$$

where the following notations are taken

$$\delta^* = \int_0^\infty\left(1-\frac{u}{u_e}\right)d\eta \qquad \delta^{**} = \int_0^\infty\frac{u}{u_e}\left(1-\frac{u}{u_e}\right)d\eta$$

$$\delta_\vartheta^* = \int_0^\infty(1-\vartheta)\,d\eta \qquad \delta_\vartheta^{**} = \int_0^\infty\frac{u}{u_e}(1-\vartheta)\,d\eta$$

$$H = \frac{\delta^*}{\delta^{**}} \qquad H_\vartheta = \frac{\delta_\vartheta^*}{\delta_\vartheta^{**}} \tag{151}$$

Under the condition of a thermally insulated surface of a body ($\partial\vartheta/\partial\eta = 0$ at $\eta = 0$) the second integral relation is reduced to the equation

$$(u_e\delta_\vartheta^{**})' = 0$$

Integrating and assuming the condition at the front stagnation point $u_e = 0$, we conclude that in general, $\delta_\vartheta^{**} = 0$.

Now in contrast to the previously adopted assumption which restricts the argument to small values of $\mathbf{M}_\infty$ and $\mathbf{M}_e$, the velocity and temperature profiles in the boundary layer cross sections may be presented by the expressions

$$\frac{u}{u_e} = \varphi_1\left(\frac{\eta}{\delta^{**}}, f, \mathbf{M}_e, \sigma\right) \qquad \vartheta = \varphi_2\left(\frac{\eta}{\delta^{**}}, f, \mathbf{M}_e, \sigma\right) \tag{152}$$

where the role of parameters is played by the following quantities:

1. The form parameter

$$f = \frac{u_e' \delta^{**2}}{\nu_{e0} \chi_e} \tag{153}$$

analogous to that considered at the beginning of this section;
2. The local Mach number $\mathbf{M}_e$, related to α_e as

$$\alpha_e^2 = \frac{\frac{k-1}{2}\mathbf{M}_e^2}{1 + \frac{k-1}{2}\mathbf{M}_e^2} \tag{154}$$

3. The Prandtl number σ.

Then, substituting expressions (152) into integral Equations 151 and the first of the integral relations (150), we shall come to the same equations (134) and (135) for determining the form parameter f as was done in the preceding simplest case. The difference will be only in the function F which will be dependent on the parameters α_e and σ and will take the form

$$F(f; \alpha_e, \sigma) = 2\zeta - 2f(2 + H - H_\vartheta) \tag{155}$$

where the reduced drag coefficient ζ will also be different:

$$\zeta = \frac{\delta^{**}}{u_e} b(\chi_w) \left(\frac{\partial u}{\partial \eta} \right)_{\eta = 0} \tag{156}$$

As calculations show, within the range of $\mathbf{M}_e$ from zero to 2.378 and for $\sigma = 14/19 \approx 0.74$, which is close to the Prandtl number for air, the form of the function $F(f; \alpha_e, \sigma)$ is weakly dependent on α_e. This confirms the simplyfying assumption, made at the beginning of this section, on the F independence of $\mathbf{M}_e$, at moderate Mach numbers, and makes it possible to again use the linear presentation of the function $F(f)$ and to obtain the value of $f(x)$ in the form of (137) at the same constants a and b.

Similarly, $\zeta(f; \alpha_e, \sigma)$ at $\sigma = 14/19$ weakly depends on α_e and can be determined according to the conventional graph or the table as for an incompressible fluid. From determination of ζ in Equation 156, the skin friction τ_w is expressed as

$$\tau_w = \mu_w \left(\frac{\partial u}{\partial y} \right)_{y = 0} = \frac{\mu_{e0} u_e}{\delta^{**}} \zeta(f) \, (1 - \alpha_e^2)^{\frac{k}{k-1}} \tag{157}$$

with δ^{**} being calculated *a priori* by the already known $f(x)$ and Equation 153, which is rewritten for the variable x in the form

$$f = \frac{\delta^{**2} \frac{du_e}{dx}}{\nu_{e0} (1 - \alpha_e^2)^{\frac{2k-1}{k-1}}} \tag{158}$$

Thus for Mach numbers $\mathbf{M}_\infty$ up to approximately 2.4, the dynamic quantities are easily calculated. Figures 282 and 283 demonstrate the dependences of $\chi_w = T_w/T_{e0}$ and H_ϑ on f

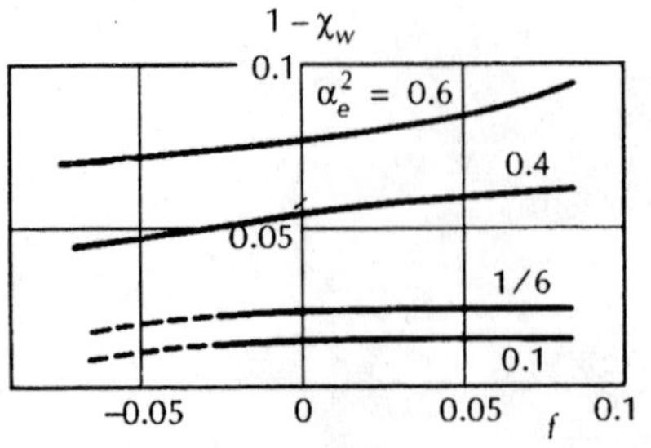

Fig. 282

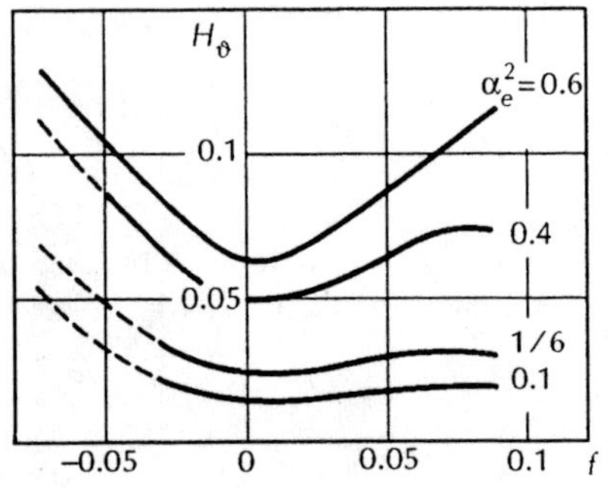

Fig. 283

at different α_e^2, ranging from 0.1 to 0.6. As seen from the graphs presented, the influence of the parameter $\mathbf{M}_e$ on these quantities is substantial.

7. Modification of the Laminar Boundary Layer Equations to Equations for an Incompressible Fluid

Some modifications of the Dorodnitsyn transform can bring, under certain limiting assumptions, an exact coincidence between the first (dynamic) equation of the system (138) and the corresponding equation for an incompressible fluid.[28]

First, we shall consider the case of $\sigma = 1$, $n = 1$ and the absence of heat transfer (thermally insulated surface).

With a denoting the speed of sound we replace the previous index $e0$, which corresponds to the stagnation value at the outer edge of the boundary layer, with 1. Then, from the formulas given in Chapter 5, we have

$$\chi_e = \frac{T_e}{T_1} = \left(\frac{a_e}{a_1}\right)^2 = \left(1 + \frac{k-1}{2}\mathbf{M}_e^2\right)^{-1} = \left(1 + \frac{k-1}{2}\frac{U_e^2}{a_1^2}\right)^{-1} \tag{159}$$

where U_e is

$$U_e = \frac{a_1}{a_e} u_e \tag{160}$$

Moreover, the same formulas will yield

$$\frac{p_e}{p_1} = \left(\frac{T_e}{T_1}\right)^{\frac{k}{k-1}} = \chi_e^{\frac{k}{k-1}} \qquad \frac{\rho_e}{\rho_1} = \chi_e^{\frac{1}{k-1}} \tag{161}$$

Transformation of the coordinates and velocities in the first two equations of the system (138) suggested by Stewartson[28] gives

[28] Stewartson, K., Correlated incompressible and compressible boundary layers, *Proc. R..Soc. London*, A200, 84–100, 1949.

$$X = \int_0^x \frac{p_e}{p_1}\frac{a_e}{a_1}dx \qquad Y = \frac{a_e}{a_1}\int_0^y \frac{\rho}{\rho_1}dy \qquad U = \frac{a_1}{a_e}u \qquad V = \frac{a_1}{a_e}\tilde{v} \tag{162}$$

($\tilde{v}$ is presented in Equation 167). Using Equations 159 and 161 we can rewrite Equation 162 as

$$X = \int_0^x \chi_e^{\frac{3k-1}{2(k-1)}}dx \qquad Y = \chi_e^{\frac{k+1}{2(k-1)}}\int_0^y \frac{\rho}{\rho_e}dy \qquad U = \chi_e^{-1/2}u \qquad V = \chi_e^{-1/2}\tilde{v} \tag{163}$$

Under the adopted limiting conditions the third equation of system (138) (the energy equation) can be replaced by its obvious integral

$$\frac{T}{T_e} = 1 + \frac{k-1}{2}\mathbf{M}_e^2 - \frac{k-1}{2}\mathbf{M}_e^2\left(\frac{u}{u_e}\right)^2$$

which in the notation given will be written in the form

$$\frac{T}{T_e} = 1 + \frac{k-1}{2a_1^2}\frac{u_e^2}{\chi_e} - \frac{k-1}{2a_1^2}\frac{u^2}{\chi_e} = 1 + \frac{k-1}{2}\frac{U_e^2}{a_1^2} - \frac{k-1}{2}\frac{U^2}{a_1^2} = \frac{1}{\chi_e} - \frac{k-1}{2}\frac{U^2}{a_1^2} \tag{164}$$

The formulas of transformation to new variables will be

$$\frac{\partial}{\partial x} = \chi_e^{\frac{3k-1}{2(k-1)}}\frac{\partial}{\partial X} + \frac{\partial Y}{\partial x}\frac{\partial}{\partial Y} \qquad \frac{\partial}{\partial y} = \chi_e^{\frac{k+1}{2(k-1)}}\frac{\rho}{\rho_e}\frac{\partial}{\partial Y} \tag{165}$$

Note that equations

$$\frac{\mu}{\mu_1} = \frac{T}{T_1} \qquad \frac{p}{p_1} = \frac{\rho T}{\rho_1 T_1} = \frac{p_e}{p_1} = \chi_e^{\frac{k}{k-1}}$$

give

$$\frac{\mu\rho}{\mu_1\rho_1} = \chi_e^{\frac{k}{k-1}}$$

By virtue of Equations 165 and the above relation, we shall rewrite the first equation of system (138) in the form

$$\rho\chi_e^{\frac{3k-1}{2(k-1)}}u\frac{\partial u}{\partial X} + \left[\rho u\frac{\partial Y}{\partial x} + \rho v\chi_e^{\frac{k+1}{2(k-1)}}\frac{\rho}{\rho_e}\right]\frac{\partial u}{\partial Y} =$$

$$= \rho_e\chi_e^{\frac{3k-1}{2(k-1)}}u_e\frac{du_e}{dX} + \frac{\rho}{\rho_e}\chi_e^{\frac{2k+1}{k-1}}\frac{\mu_1\rho_1}{\rho_e}\frac{\partial^2 u}{\partial Y^2}$$

and then, dividing both sides by $\chi_e^{\frac{3k-1}{2(k-1)}}\rho$

$$u\frac{\partial u}{\partial X} + \tilde{v}\frac{\partial u}{\partial Y} = \frac{\rho_e}{\rho}u_e\frac{du_e}{dX} + \chi_e^{1/2}\nu_1\frac{\partial^2 u}{\partial Y^2} \tag{166}$$

where the previously derived quantity $\tilde{v}$ is determined by the expression

$$\tilde{v} = \chi_e^{-\frac{3k-1}{2(k-1)}}\left[u\frac{\partial Y}{\partial x} + \frac{\rho v}{\rho_e}\chi_e^{\frac{k+1}{2(k-1)}}\right] \tag{167}$$

We shall transform the velocity components. According to the second line in Equation 163, we shall have instead of Equation 166

$$\chi_e^{1/2}U\frac{\partial}{\partial X}(\chi_e^{1/2}U) + \chi_e V\frac{\partial U}{\partial Y} = \frac{T}{T_e}\chi_e^{1/2}U_e\frac{d}{dX}(\chi_e^{1/2}U_e) + \chi_e\nu_1\frac{\partial^2 U}{\partial Y^2}$$

Making differentiation and using Equation 164, we have (the prime stands for derivatives with respect to X)

$$\chi_e\left(U\frac{\partial U}{\partial X} + V\frac{\partial U}{\partial Y}\right) + \frac{1}{2}\chi_e' U^2 =$$

$$= \left(\frac{1}{\chi_e} - \frac{k-1}{2}\frac{U^2}{a_1^2}\right)\left(\chi_e U_e U_e' + \frac{1}{2}\chi_e' U_e^2\right) + \chi_e\nu_1\frac{\partial^2 U}{\partial Y^2} \tag{168}$$

With the use of the relation between χ_e' and U_e' that can be easily obtained from Equation 159

$$\chi_e' = -\frac{k-1}{a_1^2}\chi_e^2 U_e U_e' \tag{169}$$

we shall find the following two forms for the expression between the second round brackets in the first term of the right side of Equation 168:

1. $\chi_e U_e U_e' + \frac{1}{2}\chi_e' U_e^2 = \chi_e U_e U_e' - \frac{k-1}{2a_1^2}\chi_e^2 U_e^3 U_e' = \chi_e^2 U_e U_e'\left(\frac{1}{\chi_e} - \frac{k-1}{2}\frac{U_e^2}{a_1^2}\right) = \chi_e^2 U_e U_e'$

2. $\chi_e U_e U_e' + \frac{1}{2}\chi_e' U_e^2 = -\frac{a_1^2}{k-1}\frac{\chi_e'}{\chi_e} + \frac{1}{2}U_e^2\chi_e' = -\frac{a_1^2}{k-1}\left(\frac{1}{\chi_e} - \frac{k-1}{2}\frac{U_e^2}{a_1^2}\right)\chi_e' = -\frac{a_1^2}{k-1}\chi_e'$

Multiplying the first form by the first summand between the first brackets of the right side of Equation 168 and the second one by the second summand between the same brackets, we shall obtain instead of Equation 168

$$\chi_e\left(U\frac{\partial U}{\partial X} + V\frac{\partial U}{\partial Y}\right) + \frac{1}{2}\chi_e' U^2 = \chi_e U_e U_e' + \frac{1}{2}\chi_e' U^2 + \chi_e\nu_1\frac{\partial^2 U}{\partial Y^2}$$

and after obvious reductions we shall come to the equation

$$U\frac{\partial U}{\partial X} + V\frac{\partial U}{\partial Y} = U_e\frac{dU_e}{dX} + \nu_1\frac{\partial^2 U}{\partial Y^2} \tag{170}$$

which completely corresponds to the boundary layer equation for an incompressible fluid.

To show that the transformed components U and V satisfy the condition of fluid incompressibility we shall introduce the stream function ψ, having assumed that

$$\rho u = \rho_1 \frac{\partial \psi}{\partial y} \qquad \rho v = -\rho_1 \frac{\partial \psi}{\partial x} \tag{171}$$

Making use of the transformation formulas (162) and differential relations (165) we shall have

$$U = \chi_e^{-1/2} u = \chi_e^{-1/2} \frac{\rho_1}{\rho} \chi_e^{1/2} \frac{\rho}{\rho_1} \frac{\partial \psi}{\partial Y} = \frac{\partial \psi}{\partial Y}$$

and, similarly, by Equation 167 for the reduced velocity $\tilde{v}$

$$V = \chi_e^{-1/2} \tilde{v} = \chi_e^{-1/2} \chi_e^{-\frac{3k-1}{2(k-1)}} \left(u \frac{\partial Y}{\partial x} + \chi_e^{1/2} \frac{\rho v}{\rho_1} \right) =$$

$$= \chi_e^{-1/2} \chi_e^{-\frac{3k-1}{2(k-1)}} \left\{ \chi_e^{1/2} \frac{\partial \psi}{\partial Y} \frac{\partial Y}{\partial x} - \chi_e^{1/2} \left[\chi_e^{\frac{3k-1}{2(k-1)}} \frac{\partial \psi}{\partial X} + \frac{\partial Y}{\partial x} \frac{\partial \psi}{\partial Y} \right] \right\} = -\frac{\partial \psi}{\partial X}$$

Thus, we have

$$U = \frac{\partial \psi}{\partial Y} \qquad V = -\frac{\partial \psi}{\partial X} \qquad \frac{\partial U}{\partial X} + \frac{\partial V}{\partial Y} = 0 \tag{172}$$

i.e., the reduced velocity components U and V actually correspond to some "fictitious" flow of an incompressible fluid and the stream function for this flow in the plane (X, Y) is simultaneously the stream function for the considered gas flow in the plane (x, y).

The boundary layer equations in the plane (X, Y) can be written in the form analogous to Equation 15 of Chapter 12:

$$\frac{\partial \psi}{\partial Y} \frac{\partial^2 \psi}{\partial X \partial Y} - \frac{\partial \psi}{\partial X} \frac{\partial^2 \psi}{\partial Y^2} = U_e \frac{dU_e}{dX} + \nu_1 \frac{\partial^3 \psi}{\partial Y^3} \tag{173}$$

The boundary conditions for the set of Equations 170 and 172 or for Equation 173 only also coincide with the ordinary conditions for an incompressible fluid

$$U = V = 0 \text{ at } Y = 0 \qquad U \to U_e(X) \quad \text{for } Y \to \infty$$

or

$$\psi = \frac{\partial \psi}{\partial Y} = 0 \quad \text{at} \quad Y = 0 \qquad \frac{\partial \psi}{\partial Y} \to U_e(X) \quad \text{for} \quad Y \to \infty$$

Skin friction on the body surface is determined by the formula

$$\tau_w = \mu_w \left(\frac{\partial u}{\partial y} \right)_{y=0} = \mu_1 \chi_e^{\frac{2k-1}{k-1}} \left(\frac{\partial U}{\partial Y} \right)_{Y=0} = \mu_1 \chi_e^{\frac{2k-1}{k-1}} \left(\frac{\partial^2 \psi}{\partial Y^2} \right)_{Y=0} \tag{174}$$

The main difficulty, which arises in calculations by the above method, is that in the presence of a simple relationship between the actual velocity of the outer flow u_e and the longitudinal coordinate x the relation $U_e(X)$, which corresponds to this relationship, in a "fictitious" flow of an incompressible fluid can be rather complex and not appropriate to the known class of exact solutions. In other cases, on the contrary, the simple relation $U_e(X)$ will be bound up with the complex distribution $u_e(X)$.

We shall dwell, finally, on a more general case when the Prandtl number σ differs from unity and the body surface transfers heat, but restrict ourselves to the assumption of the linearity of the relation between gas viscosity, and its temperature and enthalpy. We shall express this relation quantitatively using the aforementioned Chapman-Rubezin formula with the constant C, which enters into it, being introduced into the first of transforms (163) as a multiplier.

Introduction of this constant will not upset the above transformations so that they remain applicable. A principal difference is only in the fact that integral (164) in the case of $\sigma \neq 1$ and in the presence of surface transferring heat will not take place and in the general system of equations one had to consider also a differential energy equation.

We shall consider a thermal function S, having assumed

$$S = \frac{h_0}{h_1} - 1 \qquad S \to 0 \quad \text{when} \quad Y \to \infty \tag{175}$$

Making in the system of equations (138) the transformation mentioned and passing over to the reduced velocities U and V, we shall come to the following general system of laminar boundary layer equations

$$U\frac{\partial U}{\partial X} + V\frac{\partial U}{\partial Y} = U_e\frac{dU_e}{dX}(1+S) + \nu_1\frac{\partial^2 U}{\partial Y^2} \qquad \frac{\partial U}{\partial X} + \frac{\partial V}{\partial Y} = 0$$

$$U\frac{\partial S}{\partial X} + V\frac{\partial S}{\partial Y} = \nu_1\left\{\frac{1}{\sigma}\frac{\partial^2 S}{\partial Y^2} - \frac{1-\sigma}{\sigma}\frac{\frac{k-1}{2}\mathbf{M}_e^2}{1+\frac{k-1}{2}\mathbf{M}_e^2}\frac{\partial^2}{\partial Y^2}\left[\left(\frac{U}{U_e}\right)^2\right]\right\} \tag{176}$$

with the boundary conditions

$$U = V = 0 \qquad S = S_w \quad \text{for} \quad Y = 0$$

$$U \to U_e(X) \qquad S \to 0 \quad \text{for} \quad Y \to \infty$$

$$U = U_0(Y) \qquad S = S_0(Y) \quad \text{for} \quad X = X_0 \tag{177}$$

The quantity

$$S_w = \frac{h_w}{h_1} - 1 = \frac{T_w}{T_1} - 1 = \frac{T_w}{T_\infty}\left(1 + \frac{k-1}{2}\mathbf{M}_\infty^2\right)^{-1} - 1$$

which enters into boundary conditions (177) at the a priori given Mach number for a free stream $\mathbf{M}_\infty$ and the adiabatic constant k plays, in the case of a high speed gas flow, the role of a "temperature factor" considered in Chapter 12.

It can be noted that in the previously studied specific case, $\sigma = 1$, and in the absence of heat transfer from the body surface the integral $h_0 = h_1$ takes place or, according to Equation 175, $S = 0$. Then the system of Equations 176 will be simplified and reduced to the system (170) and (172).

Cohen and Reshotko[29] used the system of Equations 176 for seeking a self similar solution corresponding to a power-law assignment of $U_e = cX^m$ and representing the generalization of the Falkner Skan Hartree solution, known from Chapter 12, to the case of a gas flow. Based on the solution obtained, Cohen and Reshotko developed an approximate one term method for calculation of a laminar boundary layer in a gas at arbitrary distribution of external velocity;[30] this method is the analogue of the Kochin-Loitsyanskiy method (Chapter 12).

As stated at the end of Chapter 12, the latter is a local one term method which gives underestimated drag and too early separation; this very drawback is also characteristic of the Cohen Reshotko method. Therefore, we content ourselves with the given short remarks concerning this method and pass over to presentation of a more exact one.

8. The Method of Generalized Similarity in the Theory of a Laminar Boundary Layer in a High Speed Gas Flow

Let us introduce, according to Equation 172, the stream function $\psi(X, Y)$ into Equations 176 and rewrite them in the form

$$\frac{\partial \psi}{\partial Y}\frac{\partial^2 \psi}{\partial X \partial Y} - \frac{\partial \psi}{\partial X}\frac{\partial^2 \psi}{\partial Y^2} = U_e \frac{dU_e}{dX}(1+S) + \nu_1 \frac{\partial^3 \psi}{\partial Y^3}$$

$$\frac{\partial \psi}{\partial Y}\frac{\partial S}{\partial X} - \frac{\partial \psi}{\partial X}\frac{\partial S}{\partial Y} = \frac{\nu_1}{\sigma}\left\{\frac{\partial^2 S}{\partial Y^2} - (1-\sigma)\,\text{æ}\,\frac{\partial^2}{\partial Y^2}\left[\left(\frac{1}{U_e}\frac{\partial \psi}{\partial Y}\right)^2\right]\right\}$$

$$\psi = \frac{\partial \psi}{\partial Y} = 0 \qquad S = S_w \quad \text{at} \quad Y = 0$$

$$\frac{\partial \psi}{\partial Y} \to U_e \qquad S \to 0 \quad \text{for} \quad Y \to \infty \tag{178}$$

where $S_w = \text{const}$ and an additional definition is used for the "compressibility parameter"

$$\text{æ} = 1 - \chi_e = \frac{\frac{k-1}{2}\mathbf{M}_e^2}{1 + \frac{k-1}{2}\mathbf{M}_e^2} \tag{179}$$

As to the boundary conditions in the initial cross section $X = X_0$, which are placed on the last line of system (177), they will be now substituted in full conformity with the technique considered in Chapter 12 for the case of an incompressible fluid, by some integral conditions.

Making use of motion and energy equations in the form of Equation 176 and the ordinary technique described in Chapter 12, we shall derive the following integral relations

[29] Cohen, C. B. and Reshotko, E., Similar Solution for the Compressible Laminar Boundary Layer with Heat Transfer and Pressure Gradient, NACA Rep. 1293, 1956.

[30] Cohen, C. B. and Reshotko, E., The Compressible Laminar Boundary Layer with heat Transfer and Arbitrary Pressure Gradient, NACA Rep. 1294, 1956.

$$\frac{dZ^{**}}{dX} = \frac{F}{U_e} \qquad \frac{dZ_s}{dX} = \frac{F_s}{U_e} \tag{180}$$

with the notation

$$Z^{**} = \frac{\Delta^{**2}}{\nu_1} \qquad Z_s = \frac{\Delta_s^2}{\nu_1}$$

$$F = 2[\zeta - (2+H)f] \qquad F_s = -2(\zeta_s/\sigma + f_s)$$

$$\zeta = \left[\frac{\partial(U/U_e)}{\partial(Y/\Delta^{**})}\right]_{Y=0} \qquad \zeta_s = \left[\frac{\partial S}{\partial(Y/\Delta_s)}\right]_{Y=0}$$

$$H = \frac{\Delta^*}{\Delta^{**}} \qquad f = U_e' Z^{**} \qquad f_s = U_e' Z_s$$

$$\Delta^* = \int_0^\infty \left(1 + S - \frac{U}{U_e}\right) dY \qquad \Delta^{**} = \int_0^\infty \frac{U}{U_e}\left(1 - \frac{U}{U_e}\right) dY \qquad \Delta_s = \int_0^\infty \frac{U}{U_e} S \, dY \tag{181}$$

The relationship

$$\zeta^* = \left[\frac{\partial S}{\partial(Y/\Delta^{**})}\right]_{Y=0} = \zeta_s \cdot \frac{\Delta^{**}}{\Delta_s} \qquad f = f_s \cdot \left(\frac{\Delta^{**}}{\Delta_s}\right)^2 \tag{182}$$

take place.

For the conditions in the initial cross-section ($X = X_0$) we shall take the following integral conditions

$$Z^{**} = Z_0^{**} \qquad Z_s = Z_{s0} \quad \text{at} \quad X = X_0 \tag{183}$$

These conditions should be prescribed a priori for the solutions of the system of two ordinary differential equations of the first order (180) with the unknown quantities Z^{**} and Z_s are determined.

By analogy with the method stated in Chapter 12, Section 9, let us introduce a system of parameters

$$f_n = U_e^{n-1} \frac{d^n U_e}{dX^n} Z^{**n} \quad (n = 1, 2, ..., \infty) \tag{184}$$

which in the case of arbitrary assignment of the function $U_e(X)$ in the class of analytical functions may be treated as independent variables. As regards the compressibility parameter, æ, which is found from equality (179), it is assumed that this parameter exerts a weak quantitative effect on motion and heat transfer and therefore is not included in the basic parameters. This is justified by the assumption that the main portion of the compressibility effect has been taken into account in the Dorodnitsyn-Stewartson transform. With the terminology of Section 9 of Chapter 12 used, we may say that the influence of this additional parameter of compressibility æ is allowed for locally (i.e., the parameter æ itself is retained in Equations 178 and in subsequent "universal" equations), but the derivatives taken with respect to it are omitted. A complete account for the parameter influence somewhat complicates the form of the universal equations and their integration, but it also can be made.[31]

The process of reducing Equations 178 to a "universal" form containing the function $U_e(X)$ does not differ from that made previously (Section 9, Chapter 12) for the case of an incompressible fluid.

Change over in Equations 178 from the independent variables X, Y and the stream function ψ to the new variables

$$\xi = B\frac{Y}{\Delta^{**}} \qquad f_n \ (n = 1, 2, \ldots) \quad \text{and} \quad \Phi = B\frac{\psi}{U_e\Delta^{**}} \tag{185}$$

Then, following exactly the transformations made in Chapter 12, we obtain the following system of "universal" equations:

$$\frac{\partial^3\Phi}{\partial\xi^3} + \frac{F + 2f_1}{2B^2}\Phi\frac{\partial^2\Phi}{\partial\xi^2} + \frac{f_1}{B^2}\left[1 + S - \left(\frac{\partial\Phi}{\partial\xi}\right)^2\right] =$$

$$= \frac{1}{B^2}\sum_{n=1}^{\infty}\{[(n-1)f_1 + nF]f_n + f_{n+1}\}\left(\frac{\partial\Phi}{\partial\xi}\frac{\partial^2\Phi}{\partial\xi\partial f_n} - \frac{\partial\Phi}{\partial f_n}\frac{\partial^2\Phi}{\partial\xi^2}\right)$$

$$\frac{\partial^2 S}{\partial\xi^2} + \sigma\frac{F + 2f_1}{2B^2}\Phi\frac{\partial S}{\partial\xi} + 2(\sigma - 1)\,æ\frac{\partial}{\partial\xi}\left(\frac{\partial\Phi}{\partial\xi}\frac{\partial^2\Phi}{\partial\xi^2}\right) =$$

$$= \frac{\sigma}{B^2}\sum_{n=1}^{\infty}\{[(n-1)f_1 + nF]f_n + f_{n+1}\}\left(\frac{\partial\Phi}{\partial\xi}\frac{\partial S}{\partial f_n} - \frac{\partial\Phi}{\partial f_n}\frac{\partial S}{\partial\xi}\right)$$

$$\Phi = \frac{\partial\Phi}{\partial\xi} = 0 \qquad S = S_w \quad \text{at} \quad \xi = 0$$

$$\frac{\partial\Phi}{\partial\xi} \to 1 \qquad S \to 0 \quad \text{for} \quad \xi \to \infty$$

$$\Phi = \Phi_0(\xi) \qquad S = S_0(\xi) \quad \text{at} \quad f_1 = 0,\ f_2 = 0,\ \ldots \tag{186}$$

Here $\Phi_0(\xi)$ and $S_0(\xi)$ are the solution of the problem of a longitudinal flow past a plate (see Section 4, Chapter 12), and B is a normalizing multiplier chosen from the condition that the first of the equations for the problem should have an ordinary "Blasius" form (a dot denotes the derivative with respect to ξ)

$$\dddot{\Phi}_0 + \Phi_0\ddot{\Phi}_0 = 0$$

$$\ddot{S}_0 + \sigma\Phi_0\dot{S}_0 + 2(\sigma - 1)\,æ_0\left(\ddot{\Phi}_0^2 + \dot{\Phi}_0\dddot{\Phi}_0\right) = 0$$

---

[31] This is fulfilled by Lyubenov, a postgraduate student in the Department of Hydroaerodynamics of the Leningrad Polytechnical Institute. As shown by his studies the local approach is rather accurate (Lyubenov, V., Two-parametric solution of boundary layer equations for gas, *Tr. LPI*, No. 3, 28 35, 1970).

$$\Phi_0 = \dot{\Phi}_0 = 0 \qquad S_0 = S_w \quad \text{at} \quad \xi = 0$$

$$\dot{\Phi}_0 \to 1 \qquad S_0 \to 0 \quad \text{for} \quad \xi \to \infty \tag{187}$$

Leaving behind the problem of solving the system (186) in a form of power-law series,[32] we shall consider the results of the numerical solution of the system (186) in a uniparametric approximation obtained by Kapustyanskiy[33] following the program worked out by Simuni and Terentiyev.[34]

Especially simple is the solution of the system considered in the case of one-term approximation at the Prandtl number equal to unity ($\sigma = 1$). In this case, system (186) is reduced to the form

$$\frac{\partial^3\Phi}{\partial\xi^3} + \frac{F+2f_1}{2B^2}\Phi\frac{\partial^2\Phi}{\partial\xi^2} + \frac{f_1}{B^2}\left[1 + S - \left(\frac{\partial\Phi}{\partial\xi}\right)^2\right] =$$

$$= \frac{1}{B^2}Ff_1\left(\frac{\partial\Phi}{\partial\xi}\frac{\partial^2\Phi}{\partial\xi\partial f_1} - \frac{\partial\Phi}{\partial f_1}\frac{\partial^2\Phi}{\partial\xi^2}\right)$$

$$\frac{\partial^2 S}{\partial\xi^2} + \frac{F+2f_1}{2B^2}\Phi\frac{\partial S}{\partial\xi} = \frac{1}{B^2}Ff_1\left(\frac{\partial\Phi}{\partial\xi}\frac{\partial S}{\partial f_1} - \frac{\partial\Phi}{\partial f_1}\frac{\partial S}{\partial\xi}\right)$$

$$\Phi = \frac{\partial\Phi}{\partial\xi} = 0 \qquad S = S_w \quad \text{at} \quad \xi = 0$$

$$\frac{\partial\Phi}{\partial\xi} \to 1 \qquad S \to 0 \quad \text{for} \quad \xi \to \infty$$

$$\Phi = \Phi_0(\xi) \qquad S = S_0(\xi) \quad \text{at} \quad f_1 = 0 \tag{188}$$

where f_1 is the analogue of the form parameter in uniparametric methods

$$f_1 = \frac{U_e'\Delta^{**2}}{\nu_1}$$

We shall go into a detailed analysis of the results of numerical calculation of Equations 188.

Figure 284 gives the curves of the dimensionless velocity $U/U_e = \partial\Phi/\partial\xi$ (a) and of the thermal function S (b) at three values of the parameter f_1: $f_1 = 0$, $f_1 = -0.05$, $f_1 = -0.0646$ and at the temperature factor $S_w = 0.4$ (the crossed arrows show the scales by which reckoning should be made).

Figure 285 shows graphs of the main computational quantities ζ and H as functions of f_1. Numbers 1 and 2 indicate the curves $\zeta(f_1)$ at $S_w = -0.4$ by the presented method and by the Cohen-Reshotko method, correspondingly. Attention is concentrated on underesti-

---

[32] Kapustyanskiy, S. M., Laminar boundary layer in a high-speed gas flow, *Tr. LPI*, No. 248, 59–64, 1965.

[33] Kapustyanskiy, S. M., One-parametric solution of a laminar boundary layer in a gas flow with arbitrary external velocity and temperature gradient, *Inzh.-Fiz. Zh.*, 9(6), 768–774, 1965.

[34] Simuni, L. M. and Terentiyev, N. M., Numerical solution of a one-parametric equation of the theory on a laminar boundary layer, *Tr. LPI*, No. 248, 56–58, 1965.

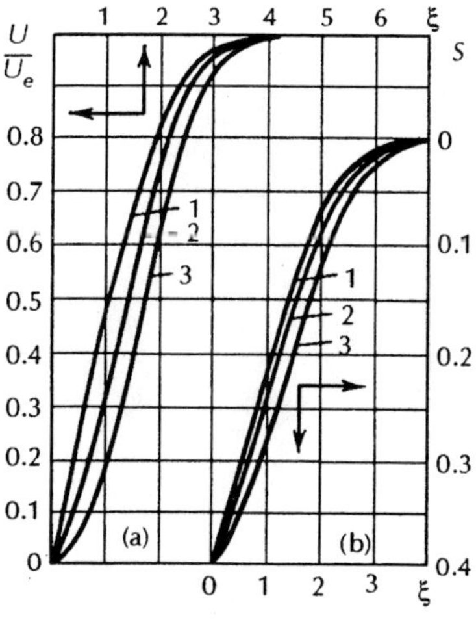

Fig. 284

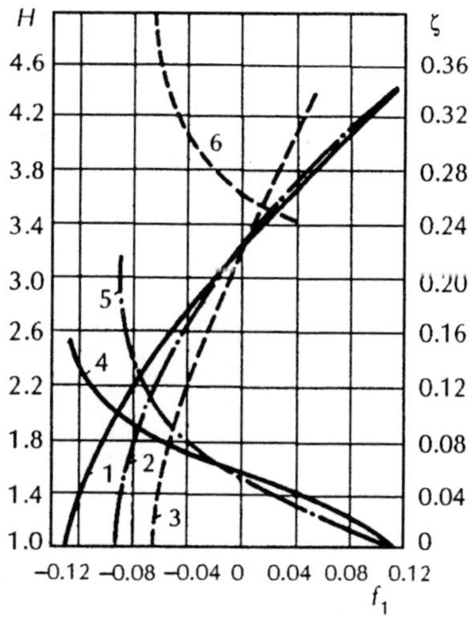

Fig. 285

mated values of ζ and the absolute separation value of the parameter f_{1s}, calculated by the Cohen-Reshotko method, in comparison with the Kapustyanskiy uniparameter solution, which is the analogue of the more exact solution of Howarth in an incompressible fluid. Number 3 denotes the curve $\zeta(f_1)$ for $S_w = 0.4$. It can be seen that the parameter S_w exerts a strong effect on the separation value of the parameter $f_1 = f_{1s}$ as is vividly shown in Figure 286.

Figure 285 makes a comparison of the Kapustyanskiy and Cohen-Reshotko curves (numbers 4 and 5, respectively) for $H(f_1)$ at $S_w = -0.4$; number 6 stands for the same value according to Kapustyanskiy at $S_w = 0.4$.

In Figure 287 the graphs of the function $F(f_1)$ and the values of this curve deviations from the straight line $\varepsilon(f_1)$ are numbered analogously to Figure 285: 4 – at $S_w = -0.4$, 5 – at $S_w = 0.4$.

The graphs of the reduced heat transfer coefficient $\zeta^*(f_1)$ at $S_w = -0.4$ are shown in Figure 288: 1 and 2, according to Kapustyanskiy and Cohen-Reshotko, respectively; 3,

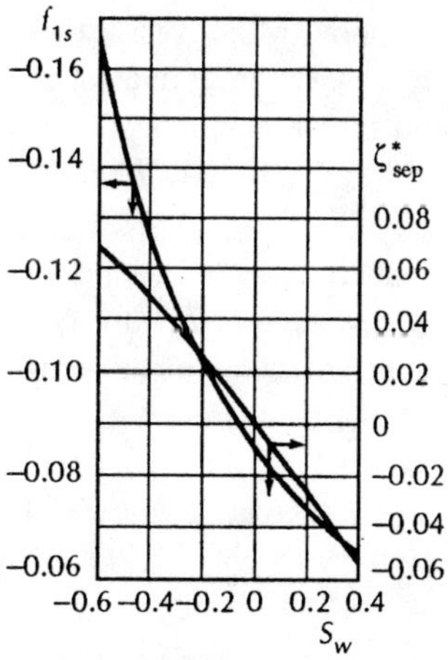

Fig. 286

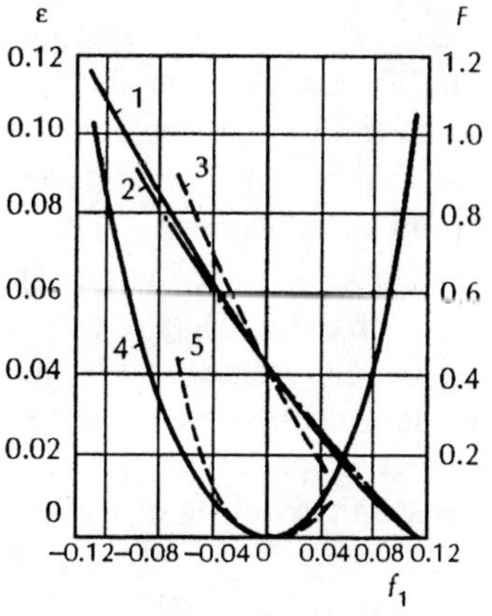

Fig. 287

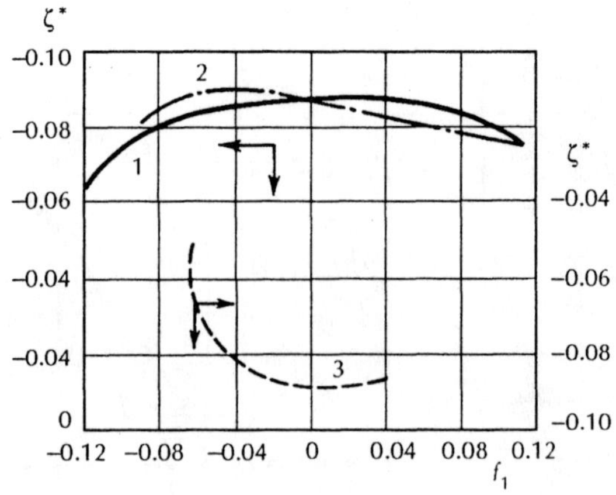

Fig. 288

according to Kapustyanskiy for $S_w = 0.4$. The values of ζ^* at the separation point depending on the temperature factor S_w are given in Figure 286.

The previous results related to the simplest case of the Prandtl number equal to unity ($\sigma = 1$). Subsequently, Kapustyanskiy Numerically integrated the system of Equations 186 for the case $\sigma \neq 1$.[35] In the paper cited the author studies the influence of the change in æ and σ on the relations $\zeta^*(f_1)$ and $\zeta_s(f_1)$ and also on the reduced coefficient of friction ζ. The fact is confirmed that disregard of the derivatives with respect to æ, i.e., use of æ as a local parameter, exerts a slight effect on friction, but a somewhat stronger one on heat transfer. The influence of æ and σ on ζ in a converging region of a boundary layer is negligible, while in a diverging one it is more pronounced; the influence of the same parameters on F is, conversely, noticeable in the converging region and negligible in the diverging one; the behavior of the quantity H is analogous.

An appreciable quantitative effect is exerted by the parameters æ and σ on the reduced coefficients of heat transfer ζ^* and ζ_s. We content ourselves with the presentation of the graphs $F_s(f_{1s})$ (solid lines) and $\zeta_s(f_{1s})$, (dashed lines) (Figure 289). These quantities were determined by Equations 181 and 182. The graphs correspond to the temperature factor $S_w = -0.4$.

The heat transfer coefficient ζ_s diminishes in magnitude at a distance from the front stagnation point and when approaching the separation point. Moreover, at the given value of f_s the magnitude of ζ_s diminishes sharply as æ increases. Of great interest is the study of a specific flow at different Mach number $\mathbf{M}_\infty$ and wall temperatures T_w. Table 20 gives the results of the calculation of the separation point abscissa x_{sep} in the Cohen-Reshotko example (I by Kapustyanskiy, II by Cohen-Reshotko).

It is seen from the table presented that in the considered case of a supersonic flow at the constant Mach number $\mathbf{M}_\infty$ the increase in the body surface temperature facilitates separation; conversely, at fixed temperature the growth of the Mach number leads to separation delay. The table also confirms the earlier-mentioned fact of the underestimation of the separation point abscissas calculated by the Cohen-Reshotko method; the difference between these values and more exact ones obtained by Kapustyanskiy increases with the Mach number. Readers interested in the details of method application, the tables of "universal" functions,

[35] Kapustyanskiy, S. M., One-parametric solution of the equations of the boundary layer in a high-speed gas flow at the Prandtl number different from unity, *Tr. LPI*, No. 265, 24–34, 1966.

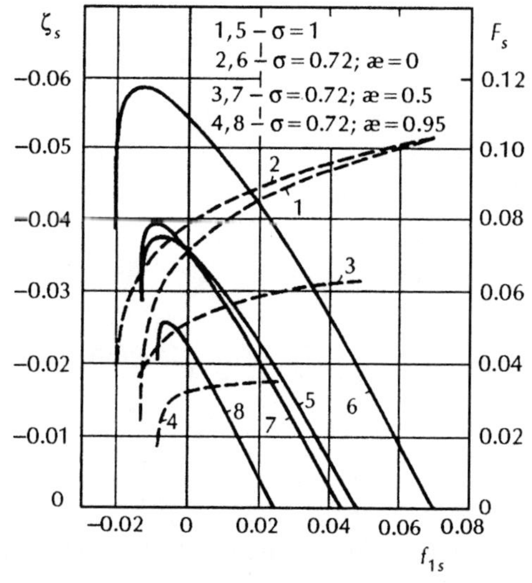

Fig. 289

Table 20

T_w		1045 K		1465 K		
M_∞		3.606	5	3.606	4.5	6.08
x_s	I	0.404	0.800	0.300	0.407	0.867
	II	0.345	0.540	0.260	0.335	0.560

and corresponding graphs are referred to the thesis by Kapustyanskiy.[36]

At velocities of gas motion comparable or slightly exceeding the value of the velocity of the distribution of small perturbations in it (sound velocity), there originate phenomena, specific for these motions. The theoretical analysis of these phenomena is a matter of computational rather then fundamental difficulties. The methods for integration of boundary layer equations and the programs of their numerical integration on computers have been developed for these cases. More serious difficulties arise when considering gas flows in boundary layers at very high supersonic, or as it is sometimes called, hypersonic velocities. Physicochemical phenomena, whirl accompany such flows, are very complex and many of them have been insufficiently studied as yet. Of basic importance are the phenomena accompanied by the mechanical-to-thermal energy transition. This is heating of gas during its pass through shock waves and especially through a powerful "bow wave" formed on blunt bodies. Of great importance is also dissipation of mechanical energy to heat that takes place in boundary layers.

The temperatures in these cases reach great values. Thus, for example, on return of a spaceship into dense atmospheric layers of Earth with the second cosmic velocity (of about

[36] Kapustyanskiy, S. M., Parametric Method for Solving Laminar Boundary Layer Equations in a High-Speed Gas Flow, tesis, LPI, 1965; Saljnikov, V. and Boričič, Z., Die universelle Grenzschichtleichungen für den Fall der kompressiblenen laminaren Strömung, *Z. Math. Mech. Sonderheft-T,* 74, 146–148, 1974.

Table 21 c_p (cal/mol · K) **as a Function of Absolute Temperature (*T*)**

Gas	Temperature (*T*)					
	1000	3000	5000	7000	9000	11,000
C	4.970	5.170	5.470	5.588	5.810	7.247
CO	7.930	8.895	9.100	9.362	10.110	12.160
CO_2	12.960	14.880	15.320	14.840	14.870	14.880
O	4.997	5.002	5.208	5.410	5.513	5.569
O_2	8.333	9.552	10.200	10.340	9.882	9.098

11 km/s) the temperature at the surface may attain 12,000 to 13,000° (on return with the first cosmic velocity this temperature has the order of 8000°). Even greater velocities and consequently temperatures may be reached in the case of meteorite entering Earth's atmosphere.

Under such heating conditions air cannot be considered a homogeneous gas. At the temperatures of about 1000 K, molecular nitrogen N_2 is the main gas in the air content rather than molecular oxygen O_2 (the content of other components is small). This makes it possible to consider air, with a good approximation, to be a uniform medium. As temperature increases the collision energy of molecules becomes so great that at first a phenomenon of dissociation develops (for oxygen, the reaction $O_2 = O + O$ noticeable at a temperature of about 3000 K; for nitrogen, $N_2 = N + N$ at temperatures of about 6000 K), and then ionization (the onset of formation of nitrogen positive ions N^+ and of liberation of electrons e^- at 1000 K and O^+ at 8500 K). Such chemical reactions lead to the fact that at high temperatures air changes to a multicomponent mixture of gases (N_2, N, O, O_2, N^+, O^+, e^-, etc.), dynamics and thermodynamics of which require a more complex analysis as compared with the case of a homogeneous gas.

This is complicated by the fact that heating a gas takes place in such a thin region (the shock wave thickness has the order of a molecule free path) so that within this small length the kinetic energy imparted to molecules during heating has no time to distribute over all internal degrees of freedom of molecules and gas is not in a complete thermodynamic state of equilibrium. In these cases gas is said to relax and the time required to reach an equilibrium state as well as the length equivalent to this time are called the relaxation time and the relaxation length, respectively.

Relaxation is determined by the quantity of collisions required for the acquisition of equilibrium energy in motions of a molecule with certain degrees of freedom. Thus, for example, it is known that to establish equilibrium motion with translational degrees of freedom only few collisions of molecules are enough, while for rotational degrees these amount to tens of collisions and for oscillatory many thousands. Complete equilization of the energy of molecules by all the degrees of freedom requires tens of thousands of collisions.

Nonequilibrium of the phenomena of gas dissociation and ionization makes the computation of these flows much more complicated.

The presence of internal degrees of freedom in multiatomic molecules greatly affects the coefficients of heat capacity, which at rather high temperatures cannot be considered physical constants of a gas. Table 21[37] shows the importance of the effect of temperature on heat

[37] Alekseyev, B. V., *Pogranichnyi sloy s khimicheskimi reaktsiyami* (*Boundary Layer with Chemical Reactions*), Vychislitelnyi Tsentr AN SSSR, Moscow, 1967, 19.

capacity c_p for different gases. With this dependence taken into account gas enthalpy is determined by integral (3).

9. Laminar Boundary Layer in a Supersonic Flow of Reacting Gases

In a number of fields of modern technology (e.g., aviation, space, power, laser, chemical, etc.) nonuniform, multicomponent flows of mixtures of chemically reacting gases occur and find application. Nonuniform mixtures, all components of which belong to the same liquid or gaseous state, are called homogeneous or single phase. Depending on the number of components entering these media we distinguish two-component, three-component, etc., media. Nonuniform media, including substances in different states of aggregation (phases), are called heterogeneous or according to the quantity of phases entering these media, two-phase, three-phase, etc.

Mechanics and thermodynamics of heterogeneous media are based on the assumption of the possibility to pass over from the systems of discrete particles to continuous media. This is achieved by particle ensemble averaging. The difference from the kinetic theory of gases lies in that these particles greatly exceed molecules in size. Such a statistical transition naturally requires the presence of a rather great number of particles in an elementary volume of averaging. The principal difficulties arising in the study of dynamics and thermodynamics of medium are attributed to complexity of the mechanisms of intercomponent and especially interphase interactions which accompany the motion of these media.

In this section attention is paid only to multicomponent mixtures of gases.[38]

Using a general technique of statistical mechanics as a basis we postulate the existence of the so-called distribution functions of particles, forming a system via their location in space and velocities. Characterizing the position of separate particles by their vector-radii $\boldsymbol{r}$ at the given time instant t end the velocity vectors $\boldsymbol{V}$, we shall introduce the distribution function $f(\boldsymbol{r}, \boldsymbol{V}, t)$ as the coefficient of proportionality in the expression of the probable number of particles δN which are situated in the vicinity $(\boldsymbol{r}, \boldsymbol{r} + \delta\boldsymbol{r})$ of the given point of space and moves at the velocities within range $(\boldsymbol{V}, \boldsymbol{V}+\delta\boldsymbol{V})$

$$\delta N = f(\boldsymbol{r}, \boldsymbol{V}, t)\ (\delta\boldsymbol{r})\ (\delta\boldsymbol{V}) \tag{189}$$

where the symbols $(\delta\boldsymbol{r})$ and $(\delta\boldsymbol{V})$ stand for elementary volumes in a geometrical space of the coordinates x, y, z and the space of velocities u, v, w which will be taken equal to $(\delta\boldsymbol{r}) = \delta x\delta y\delta z$ and $(\delta\boldsymbol{V}) = \delta u\delta v\delta w$. These volumes will be further denoted as $(\delta\boldsymbol{r})$ and $(\delta\boldsymbol{V})$ as it has been adopted in the kinetic theory of gases.

From the determination of the distribution function (189) follows the expression for the total number of particles in the system

$$N = \int f(\boldsymbol{r}, \boldsymbol{V}, t)\ (\delta\boldsymbol{r})\ (\delta\boldsymbol{V}) \tag{190}$$

---

[38] A general problem of the flow of heterogeneous media is the concern of the following monographs: Nigmatulin, R. I., *Osnovy mekhaniki geterogennykh sred* (*The Principles of Mechanics of Heterogeneous Media*), Nauka, Moscow, 1978 ; Sou, S., *Gidrodinamika mnogofaznykh sistem* (*Hydrodynamics of Multiphase Systems*) (translated from English), Mir Publishers, Moscow; Nigmatulin, R. I., Kraiko, A. N., Starkov, V. K., and Sternin, L. E., Mekhanika mnogofaznykh sred (Mechanics of Multiphase Media), Itogi nauki i tekhniki, Gidromekhanika (Results of Science and Technology, Hiydromechanics), VINITI, 1972, v. 6. Some ideas on the difficulties arising in the study of dynamics and thermodynamics of heterogeneous media can be obtained when studying Section 13 of the 5th edition of this book.

as the volumetric integral over the regions of change in positions and velocities of system particles. Equality (190) denotes the adopted normalizing of the distribution function. Relating the probable number of particles (189) in the volume $(\delta \boldsymbol{r})$ $(\delta \boldsymbol{V})$ to the total number of particles (190), the sense of the probability of particles being in the above-mentioned elementary volume can be imparted to the distribution function. Thus, the normalization condition (190) will be substituted by the ordinary condition of the equality to unity of the sum of probabilities of all possible positions and velocities of the system particles.

Relating the probable number (189) to the volume $(\delta \boldsymbol{r})$, we shall obtain the probable density of the velocity distribution of particles within the range $(\boldsymbol{V}, \boldsymbol{V} + \delta \boldsymbol{V})$

$$\delta n = \frac{\delta N}{(\delta \boldsymbol{r})} = f(\boldsymbol{r}, \boldsymbol{V}, t)\,(\delta \boldsymbol{V}) \tag{191}$$

and then the probable density of the distribution of particles in space at the given moment

$$n = \int f(\boldsymbol{r}, \boldsymbol{V}, t)\,(\delta \boldsymbol{V}) \tag{192}$$

Multiplying this density by the mass m of the particle situated at the given point of space yields mass density or, briefly, density ρ of a continuous medium at the given point and given time instant

$$\rho = nm \tag{193}$$

Considering a continuous heterogeneous medium as a system of particles composed of a certain number of subsystems (components of the ith kind) with the particles of equal masses we shall take for each of them their own distribution function $f^{(i)}(\boldsymbol{r}^{(i)}, \boldsymbol{V}^{(i)}, t)$. Substituting velocities of separate particles of the ith kind $\boldsymbol{V}^{(i)}$ by means of velocities $\bar{\boldsymbol{V}}^{(i)}$ at the given point of space occupied by the particles of the ith kind and taking the distribution function $f^{(i)}$ as the "weight" of averaging, we find

$$\bar{\boldsymbol{V}}^{(i)} = \frac{1}{n^{(i)}} \int f^{(i)} \boldsymbol{V}^{(i)} (\delta \boldsymbol{V}^{(i)}) = \frac{\int f^{(i)} \boldsymbol{V}^{(i)} (\delta \boldsymbol{V}^{(i)})}{\int f^{(i)} (\delta \boldsymbol{V}^{(i)})} \tag{194}$$

To determine mean values at the points of a continuous medium as a whole (the mixture of components) we shall assume general "formulas of mixing" by mass for density

$$\rho = \sum_{(i)} n^{(i)} m^{(i)} = \sum_{(i)} \rho^{(i)} \tag{195}$$

and by momentum for velocity

$$\rho \bar{\boldsymbol{V}} = \sum_{(i)} \rho^{(i)} \bar{\boldsymbol{V}}^{(i)} \qquad \bar{\boldsymbol{V}} = \frac{1}{\rho} \sum_{(i)} \rho^{(i)} \bar{\boldsymbol{V}}^{(i)} \tag{196}$$

The ratio of densities at the given point

$$\frac{\rho^{(i)}}{\rho} = \frac{n^{(i)}m^{(i)}}{\sum_{(i)} n^{(i)}m^{(i)}} = c^{(i)} \tag{197}$$

determines the "concentration" $c^{(i)}$ of the particles of the ith kind at the given point of the continuous heterogeneous medium. According to (197) we have

$$\sum_{(i)} c^{(i)} = \frac{1}{\rho}\sum_{(i)} \rho^{(i)} = 1$$

It remains only to deal with mean velocities $\bar{V}^{(i)}$ and $\bar{V}$ in this connection we shall omit the bar over V. The difference of velocities

$$V^{*(i)} = V^{(i)} - V \tag{198}$$

can be treated as a mean velocity of ith component propagation through a continuous heterogeneous medium at the given point and can be called the velocity of diffusion of the ith component.

The continuity equation of the ith component will have the same form (Equations 12 or 13 of Chapter 3) as in the case of uniform medium, the only difference being that $\rho^{(i)}$ and $V^{(i)}$ will replace ρ and V, and by $J^{(ji)}$ we mean the rate of increment, related to the unit volume, of the ith component mass due to the reactions of transition from the jth component to the ith one. It is evident that

$$J^{(i)} = \sum_{(j)} J^{(ji)} \qquad \sum_{(i)} J^{(i)} = \sum_{(i,j)} J^{(ji)} = 0 \tag{199}$$

Equation 13 of Chapter 3 in this case passes over to the continuity equation of the ith component:

$$\frac{\partial \rho^{(i)}}{\partial t} + \operatorname{div}(\rho^{(i)} V^{(i)}) = J^{(i)} = \sum_{(j)} J^{(ji)} \tag{200}$$

Summing up both sides of this equation by all the components we obtain, according to Equations 195 and 196, the continuity equation for a mixture as a whole

$$\frac{\partial \rho}{\partial t} + \operatorname{div}(\rho V) = 0 \tag{201}$$

In Equation 200 we can eliminate the mean velocity $V^{(i)}$ of separate component motion, having expressed it (according to Equation 198), in terms of the mean velocity of medium as a whole V and the velocity of component diffusion $V^{*(i)}$; this yields

$$\frac{\partial \rho^{(i)}}{\partial t} + \operatorname{div}[\rho^{(i)}(V^{*(i)} + V)] = J^{(i)} \tag{202}$$

or, introducing by (197) the concentration $c^{(i)} = \rho^{(i)}/\rho$,

$$\frac{\partial (\rho c^{(i)})}{\partial t} + \operatorname{div}[\rho c^{(i)}(V^{*(i)} + V)] = J^{(i)} \tag{203}$$

Combining the terms

$$c^{(i)}\left[\frac{\partial \rho}{\partial t}+\operatorname{div}(\rho \boldsymbol{V})\right]+\rho\frac{\partial c^{(i)}}{\partial t}+\rho \boldsymbol{V}\cdot \operatorname{grad} c^{(i)} = J^{(i)}-\operatorname{div}(\rho c^{(i)}\boldsymbol{V}^{*(i)})$$

and making use of the continuity equation for mixture (201), we finally have

$$\rho\frac{\partial c^{(i)}}{\partial t}+\rho \boldsymbol{V}\cdot \operatorname{grad} c^{(i)} = J^{(i)}-\operatorname{div}(\rho c^{(i)}\boldsymbol{V}^{*(i)}) \tag{204}$$

This equation is called the equation of component concentration. The left side of this equation is an aggregate of local and convective change in the concentration of the ith component in the mixture flow. The right side characterizes the origin of the *i*th component due to physicochemical transitions of *j*th components to *i*th ones and diffusion (propagation) of the *i*th component in a mixture. In this equation the mean velocity of mixture $\boldsymbol{V}$, the quantities $J^{(ji)}$ and their sum $J^{(i)} = \sum_{(j)} J^{(ji)}$, and velocity of diffusion $\boldsymbol{V}^{*(i)}$ are unknown as well as concentration $c^{(i)}$. In the so called diffusion approximation the mixture velocity in Equation 204 is determined from the equation of homogeneous medium dynamics with the values of physical constants corresponding to the mixture. As shown later, this is equivalent to elimination in the exact equation of mixture dynamics of a nonlinear term characterizing the total effect of diffusion motions of components. This elimination is admissible when diffusion velocities are comparatively small. The diffusion velocity $\boldsymbol{V}^{*(i)}$ is related to the gradient, $\operatorname{grad} c^{(i)}$, of the *i*th component concentration by Fick's law and the velocity of this component formation $J^{(i)} = \sum_{(j)} J^{(ji)}$ is expressed in terms of physicochemical parameters entering into the kinetic equations of the process of component formation. This approximate "diffusion" approach which linearizes Equation 204 with respect to the concentration $c^{(i)}$ turns out in many cases, to be sufficient for the study of homogeneous gas media.

In a more strict approach (this refers in the first place to heterogenous media) which is required in practice, one must consider the dynamic stress equations and the equations of energy balance for both separate components and their mixture as a whole.

Derivation of these equations as applied to separate components of mixture is based on the same principles as in the case of a homogeneous continuous medium, being different only in the interaction of components with each other and in the possible presence of physicochemical transitions of one mixture of components to others. The equations of dynamics and balance of energy of the mixture are derived from the equations for mixture components by numerical term by term summation of these equations.

In this case the problem arises of expressing the means over the entire mixture of the mechanical and thermodynamical characteristics of the total flow and also of the mean values of physical constants in terms of the given assembly of their values for separate components. Sometimes, for example, for densities (195) and velocities (196), this relation is very simple; in other cases this refers, for example, to the determination of the mean special analysis which exceeds the bounds of purely physicochemical consideration of the problem and causes one to take account of medium motion.

Let us form the equations of a plane stationary boundary layer ln a flow of a mixture of reacting gases, assuming that all the processes have thermodynamic equilibrium. We shall preserve the notation ρ for density, p for pressure, $V(u, v)$ for velocity, h for enthalpy, and T for absolute temperature of gas mixture, and agree to denote the corresponding values of these quantities for separate components entering into this mixture by the index i. $J^{(i)}$ stands for mass velocity of the ith component formation at the given point of the flow per unit volume.

Restricting ourselves to the stationary case, we shall have, according to Equation 200, the continuity equation for the ith component of the mixture

$$\frac{\partial}{\partial x}(\rho^{(i)}u^{(i)}) + \frac{\partial}{\partial y}(\rho^{(i)}v^{(i)}) = J^{(i)} \tag{205}$$

and according to Equation 201 for the mixture as a whole

$$\frac{\partial}{\partial x}(\rho u) + \frac{\partial}{\partial y}(\rho v) = 0 \tag{206}$$

The dynamic equation of the boundary layer for a mixture of gases will be taken as similar to that for a homogeneous gas, with the density and the coefficient of viscosity corresponding to the mixture of gases, *viz.*,

$$\rho u\frac{\partial u}{\partial x} + \rho v\frac{\partial u}{\partial y} = -\frac{dp}{dx} + \frac{\partial}{\partial y}\left(\mu\frac{\partial u}{\partial y}\right) \tag{207}$$

One can roughly estimate the coefficient of mixture viscosity by the equation

$$\frac{1}{\mu} = \sum_{(i)}\frac{c^{(i)}}{\mu^{(i)}}\sqrt{\frac{m}{m^{(i)}}} \qquad \frac{1}{m} = \sum_{(i)}\frac{c^{(i)}}{m^{(i)}}$$

where ρ, μ, and m are the density, the dynamic coefficient of viscosity, and the molar mass at the given point of mixture, respectively, and $\rho^{(i)}$, $\mu^{(i)}$, and $m^{(i)}$ the same for the ith component ($\rho^{(i)}$ is the density at the partial pressure $p^{(i)}$ of the component); $c^{(i)} = \rho^{(i)}/\rho$ is the mass, or weight, concentration of the ith component at the given point.

We shall take the equation of state of the mixture ith component in the form

$$p^{(i)} = R^{(i)}\rho^{(i)}T^{(i)} \tag{208}$$

The temperatures $T^{(i)}$ of separate mixture components are taken to be alike and equal to the total equilibrium temperature T of the mixture at the given point. Then, determining the mixture pressure p at the given point as a sum of partial pressures $p^{(i)}$, according to the Dalton law we have

$$p = \sum_{(i)}p^{(i)} = \left(\sum_{(i)}R^{(i)}\rho^{(i)}\right)T = R\rho T \tag{209}$$

where the gas constant R of the mixture is found from the equalities

$$R = \frac{1}{\rho}\sum_{(i)}R^{(i)}\rho^{(i)} = \sum_{(i)}\frac{\rho^{(i)}}{\rho}R^{(i)} = \sum_{(i)}c^{(i)}R^{(i)} \tag{210}$$

In such a choice of the mixture gas constant the equation of its state preserves the same form as the equation of state of a homogeneous gas.

The main specific features of the processes of motion and heat and mass transfer in mixtures of reacting gases lie in a thermodynamic, or more precisely, in a thermochemical region. Consider the derivation of the equation of energy balance of a moving mixture of gases and the diffusion equations of its separate components.

Note that convective heat supply in a mixture differs in no way from that in a homogeneous gas; when forming the left side of the energy balance equation one should only take into account that the mixture enthalpy h is determined as a sum of the products of enthalpies $h^{(i)}$ of separate components into the concentrations $c^{(i)}$ of these components

$$h = \sum_{(i)} c^{(i)} h^{(i)} \tag{211}$$

The right side of the energy balance equation as well as Equation 32 for a homogeneous gas, will include the terms standing for the work of pressure forces, the quantity of mechanical energy dissipating to heat per second, and heat transferred by heat conduction. Except for these heat sources, which are common for a mixture and a homogeneous gas, there are two sources of heat generation that are specific for a moving gas mixture.

First, in formation of the ith component of mixture with the mass velocity $J^{(i)}$ the amount of heat $J^{(i)} h^{*(i)}$ is liberated per second where $h^{*(i)}$ means some enthalpy, characteristic of each component, that is "latent" heat of the ith component formation. This leads to the necessity of adding to the earlier mentioned terms of the right side of the balance equation one more term:

$$\sum_{(i)} J^{(i)} h^{*(i)}$$

which is equal, according to Equation 205 and to the relation $\rho^{(i)} = \rho c^{(i)}$, to

$$\sum_{(i)} h^{*(i)} \left[\frac{\partial}{\partial x} (\rho c^{(i)} u^{(i)}) + \frac{\partial}{\partial y} (\rho c^{(i)} v^{(i)}) \right] \tag{212}$$

Table 22 gives the "enthalpies of formation" of some most frequently encountered gas components.[39]

Secondly, the diffusion of components in a mixture serves as an additional cause of heat transfer in a mixture. Recalling hat the diffusion velocity of the particles of the *i*th component in the mixture is determined by the difference of the velocities of the *i*th component and of the mixture itself at the given point, we shall obtain the expression for the last additional term in the form of a sum of divergences of the vectors of heat fluxes, taken with the sign "–", of separate components $\rho^{(i)} (V^{(i)} - V) h^{(i)} = \rho c^{(i)} (V^{(i)} - V) h^{(i)}$, which is equal to

$$-\frac{\partial}{\partial x}\left[\rho \sum_{(i)} c^{(i)} (u^{(i)} - u) h^{(i)}\right] - \frac{\partial}{\partial y}\left[\rho \sum_{(i)} c^{(i)} (v^{(i)} - v) h^{(i)}\right] \tag{213}$$

Thus, the equation of the energy balance for a moving mixture of gases in a boundary layer will be reduced to the form ($h^{*(i)} = \text{const}$)

---

[39] Alekseyev, B. V., *Pogranichnyi sloy s khimicheskimi reaktsiyami (Boundary Layer with Chemical Reactions)*, Vychislitelnyi Tsentr AN SSSR, Moscow, 1967, 21.

Table 22 Enthalpies of Formation of Frequently Encountered Gas Components

Component	C	CO	CO_2	O	O_2	C^+	CO^+	O^+	O_2^+
Enthalpy $h^{*(i)}$ (kcal/mol)	264	66.77	0	58.97	0	523.8	390.4	372.9	282.4

$$\rho u \frac{\partial}{\partial x}\left(\sum_{(i)} c^{(i)} h^{(i)}\right) + \rho v \frac{\partial}{\partial y}\left(\sum_{(i)} c^{(i)} h^{(i)}\right) =$$

$$= u\frac{dp}{dx} + \mu\left(\frac{\partial u}{\partial y}\right)^2 + \frac{\partial}{\partial y}\left(\lambda\frac{\partial T}{\partial y}\right) + \frac{\partial}{\partial x}\left(\sum_{(i)} \rho u^{(i)} c^{(i)} h^{*(i)}\right) + \frac{\partial}{\partial y}\left(\sum_{(i)} \rho v^{(i)} c^{(i)} h^{*(i)}\right) -$$

$$- \frac{\partial}{\partial x}\left[\sum_{(i)} \rho (u^{(i)} - u) c^{(i)} h^{(i)}\right] - \frac{\partial}{\partial y}\left[\sum_{(i)} \rho (v^{(i)} - v) c^{(i)} h^{(i)}\right] \tag{214}$$

The basic properties of the boundary layer, which make it possible to reduce the right side of the energy balance equation relative to the additional terms, have not been taken into account as yet, before these simplifications, we shall make an identity transformation in the second line

$$\frac{\partial}{\partial x}\left(\sum_{(i)} \rho u^{(i)} c^{(i)} h^{*(i)}\right) + \frac{\partial}{\partial y}\left(\sum_{(i)} \rho v^{(i)} c^{(i)} h^{*(i)}\right) \equiv \frac{\partial}{\partial x}\left[\sum_{(i)} \rho (u^{(i)} - u) c^{(i)} h^{*(i)}\right] +$$

$$+ \frac{\partial}{\partial y}\left[\sum_{(i)} \rho (v^{(i)} - v) c^{(i)} h^{*(i)}\right] + \frac{\partial}{\partial x}\left(\sum_{(i)} \rho u c^{(i)} h^{*(i)}\right) + \frac{\partial}{\partial y}\left(\sum_{(i)} \rho v c^{(i)} h^{*(i)}\right)$$

Transposing the last two terms into the left side of the equality and taking into account the continuity equation for the mixture of gases (201), we shall rewrite the balance equation (214) in the form containing the difference of enthalpies $h^{(i)} - h^{*(i)}$ as a main unknown quantity

$$\rho u \frac{\partial}{\partial x}\left[\sum_{(i)} c^{(i)} (h^{(i)} - h^{*(i)})\right] + \rho v \frac{\partial}{\partial y}\left[\sum_{(i)} c^{(i)} (h^{(i)} - h^{*(i)})\right] =$$

$$= u\frac{dp}{dx} + \mu\left(\frac{\partial u}{\partial y}\right)^2 + \frac{\partial}{\partial y}\left(\lambda\frac{\partial T}{\partial y}\right) - \frac{\partial}{\partial x}\left[\sum_{(i)} \rho (u^{(i)} - u) c^{(i)} (h^{(i)} - h^{*(i)})\right] -$$

$$- \frac{\partial}{\partial y}\left[\sum_{(i)} \rho (v^{(i)} - v) c^{(i)} (h^{(i)} - h^{*(i)})\right]$$

The next to last term from the right may be (as usual) in a boundary layer, neglected as compared to the other ones, and thus the balance equation may be rewritten in the form

$$\rho u \frac{\partial \bar{h}}{\partial x} + \rho v \frac{\partial \bar{h}}{\partial y} = u \frac{dp}{dx} + \mu \left(\frac{\partial u}{\partial y} \right)^2 +$$

$$+ \frac{\partial}{\partial y} \left[\lambda \frac{\partial T}{\partial y} - \sum_{(i)} \rho \left(v^{(i)} - v \right) c^{(i)} \left(h^{(i)} - h^{*(i)} \right) \right] \tag{215}$$

where, for brevity, the notation

$$\bar{h} = \sum_{(i)} c^{(i)} \left(h^{(i)} - h^{*(i)} \right) \tag{216}$$

is introduced.

Restricting ourselves to the scheme of a binary mixture ($i = 1, 2$; $D^{(1)} = D^{(2)} = D$), we shall take the generalized law of diffusion according to which the diffusion velocity of the ith component is determined in terms of the gradients of concentration and temperature by the equation

$$\mathbf{V}^{(i)} - \mathbf{V} = -\frac{D}{c^{(i)}} \operatorname{grad} c^{(i)} - \frac{D^{T(i)}}{T c^{(i)}} \operatorname{grad} T \tag{217}$$

The terms on the right side express the velocities of mass diffusions and thermodiffusions, respectively; i.e., those velocities of the ith component propagation relative to the mixture that arise due to nonuniformity of the fields of the concentration of this component and also of the temperature common for all the components of the mixture.

The quantities D and D^T are the coefficients of mass diffusion (they are usually called the coefficients of diffusion) and thermodiffusion, respectively.

Noting that the difference of velocities $v^{(i)} - v$, which enters into the last term of the right side of Equation 215, according to (217) can be rewritten in the form

$$v^{(i)} - v = -\frac{D}{c^{(i)}} \frac{\partial c^{(i)}}{\partial y} - \frac{D^{T(i)}}{T c^{(i)}} \frac{\partial T}{\partial y} \tag{218}$$

we shall finally obtain the next form of the heat balance equation in a boundary layer formed in a flow of a mixture of reacting gases

$$\rho u \frac{\partial \bar{h}}{\partial x} + \rho v \frac{\partial \bar{h}}{\partial y} = u \frac{dp}{dx} + \mu \left(\frac{\partial u}{\partial y} \right)^2 + \frac{\partial}{\partial y} \left(\lambda \frac{\partial T}{\partial y} \right) +$$

$$+ \frac{\partial}{\partial y} \left[\rho \sum_{(i)} \left(h^{(i)} - h^{*(i)} \right) \left(D \frac{\partial c^{(i)}}{\partial y} + \frac{D^{T(i)}}{T} \frac{\partial T}{\partial y} \right) \right] \tag{219}$$

Let us transform the continuity equation of the *i*th component (Equation 200). Separating the diffusion velocities from the velocities u_i and v_i and expressing them by Equation 217 in the boundary layer approximation, we obtain the equation

$$\rho u \frac{\partial c^{(i)}}{\partial x} + \rho v \frac{\partial c^{(i)}}{\partial y} = \frac{\partial}{\partial y}\left[\rho\left(D\frac{\partial c^{(i)}}{\partial y} + \frac{D^{T(i)}}{T}\frac{\partial T}{\partial y}\right)\right] + J^{(i)} \tag{220}$$

which is usually called the equation of diffusion (sometimes called concentration) of the ith component.

Not dwelling on further transformations and possible simplifications[40] of the obtained system of equations, which does not meet the problems presented in this volume, we note some difficulties connected with the determination of boundary conditions for concentrations. The conditions, which the concentrations of mixture components on a solid surface of the body streamlined by a gas mixture are subject to, depend on catalytic properties of this surface. If the question concerns the reaction of dissociation, then the body surface may to this or that extent facilitate the recombination of atoms to molecules. On an absolutely catalytic surface concentrations of atoms are usually $c_A = 0$; on an absolutely noncatalytic wall $\partial c_A/\partial y = 0$. Usually one has to deal with an intermediate case and to introduce the condition of the balance of flow of atoms onto a wall and their absorption on the wall with the rate proportional to some degree of concentration of atoms:

$$\rho_w D_w\left(\frac{\partial c_A}{\partial y}\right)_w = æ_w (c_{Aw}\rho_w)^n$$

where the exponent n determines the order of the reaction of absorption of atoms and the coefficient $æ_w$ determines the catalytic ability of the body surface.

With modern possibilities of numerical calculation on computers the problems of a laminar boundary layer in a dissociating gas are analyzable. Especially simple are the solutions of two limiting problems: (1) the case of the "frozen" flow ($J^{(i)} = 0$), in which all the processes in a dissociated gas are determined only by convection and diffusion of separate components (the velocity of reactions is equal to zero) and (2) the case of "equilibrium" dissociation, in which concentration is the prescribed function of temperature and pressure and the diffusion equation is absent. In the latter case the solution can be obtained in a "universal" form by the method of generalized similarity (Section 8).[41]

Having restricted ourselves to these general remarks, readers are referred to another publication by the author,[42] in which integration of the equations of a laminar boundary layer in a dissociating and recombinating gas near the front surface of the body of revolution[43] is studied in detail, and to the available review papers in this field.[44]

Of special complexity are the problems of a boundary layer in a hypersonic flow in the presence of body surface destruction due to the high temperatures of a gas flow past the surface. The processes of melting or direct evaporation (sublimation) of the body surface, which

---

[40] We shall refer to the monograph Hayes, W. D. and Probstein, R. F., *Teoriya giperzvukovykh techeniy* (*The Theory of Hypersonic Flows*), Inostrannaya Literatura, Moscow, 1962, in which self similar solutions of these equations are given in detail (pp. 376 to 400). Equation 219 above is transformed to the equation of relatively complete enthalpy $h_0 = \sum_{(i)} c^{(i)} (h^{(i)} - h^{*(i)}) + \frac{1}{2}u^2$ and the term accounting for thermodiffusion is omitted.

[41] Krivtsova, N. V., Laminar boundary layer in a gas dissociated in an equilibrium state with arbitrary distribution of external velocity, *Mekh. Zhidk, Gaza*, No.6, 1966.

[42] Loitsyanskiy, L. G., *Laminare Grenzschichten*, Academie Verlag, Berlin, 1967.

[43] For more detail see Fay, J. and Ridell, F., Theory of stagnation point heat transfer in dissociated air, *J. Aeron. Sci.*, 25(2), 73–85, 1958; in Russian see: *Problemy dvizheniya golovnoy chasti raket dalnego deystviya* (*Problems of Motion of a Nose Tip of Missiles*), Inostrannaya Literatura, Moscow, 1959, 217–256.

[44] Ludweig, H. and Heil, M., *Teoriya pogranichnogo sloya s dissotsiatsiyey i ionizatsiyey* (*The Theory of a Boundary Layer with Dissociation and Ionization*), Vol. 4, Inostrannaya Literatura, Moscow, 1963, 37–99.

originate under these conditions, generate multicomponent mixtures in a boundary layer that contains along with dissociated and ionized particles, complex molecules of the destructing surface material. This leads to the necessity of establishing boundary conditions on a deforming surface, of accounting for heat phenomena in a solid body bounded by the destructing surface, and of considering the conditions of its mechanical destruction (formation of cracks and their "weathering").

Despite the extreme complexity of these phenomena, which require data from different fields of mechanics for study, computational methods[45] in thermodynamics and thermal chemistry are available that are quite satisfactory for practice.

In motion of bodies with high velocities close to the second cosmic velocity, the influence of gas radiation on aerothermodynamic parameters of the boundary layer becomes essential. This radiation flow affects the heat flux to the surface of the body near its nose tip and is comparable to the heat flux originating due to the mechanical energy dissipation in a boundary layer. To become acquainted with the existing studies in this field see the book by Pai.[46]

There is great interest in the problems of the flow of ionized gases, the so called cold plasma (temperature to 15,000 K), which possess electric conductivity and increased thermal conductivity (the flow at small Prandtl numbers). Of special engineering importance are the problems of plasma motion in a magnetic field.

10. Interaction between a Laminar Boundary Layer and an External Nonviscous Supersonic Flow

The methods for calculation of a laminar boundary layer in a gas flow of high sub and supersonic velocities stated in previous sections of this chapter followed the style of the classical boundary layer theory: the pressure distribution in an outer irrotational nonviscous flow was assumed to be set a priory, and an inverse effect of the boundary layer on the external flow was not taken into account, even in the cases of such evident violations of the premises in the Prandtl theory that took place in a preseparation region where transverse dimensions and velocities in a boundary layer lose their comparative smallness.

Recently, a new and important for practice branch of the boundary layer theory has been developed: the theory of interaction between a boundary layer and an external nonviscous flow. This theory extends the limits of the classical theory to the case of viscous medium (incompressible and compressible) motion in the regions that border some specific points of flows, e.g., the point of layer separation from a solid surface and of its subsequent "adhesion" to the surface, the point of destruction of the body "smoothness", of motions in the "bottom" region behind the shell edge, in the "near" wake behind the body, etc.

These problems acquire special importance under the conditions of supersonic and even

---

[45] Tirskiy, G. A., Blunt body sublimation in the vicinity of the stagnation point in the plane and axisymmetric flow of a gas mixture, *Zh. Vych. Mat. Mech. Fiz.*, 1(5), 884–902, 1961; Shchennikov, V. V., Calculation of a laminar boundary layer near a sublimating surface, *Zh. Vyc. Mat. Mech. Fiz.*, 1(5), 1961; Anfimov, N. A., Laminar boundary layer on a chemically active surface, *Izv. AN SSSR, Mekh. Mashin.*, No. 3, 3–12, 1962 and Burning of graphite in an air flow at high temperatures, *Izv. AN SSSR, Mekh. Mashin.*, No. 5,1964; Ovsyannikov, V. M., Destruction of an axisymmetric body of revolution made of a chemically complex material in a high enthalpy air flow, *Mekh. Zhidk. Gasa*, No. 5, 1965; the review paper in *Mekhanika v SSSR za 50 let* (*Mechanics in the USSR Over 50 Years*); *Dinamika vyazkikh zhidkostey i gazov, teoriya laminarnykh i pogranichnykh sloyov* (*Dynamics of Viscous Liquids and Gases, the Theory of Laminar and Turbulent Boundary Lyers*), Loitsyanskiy, L. G., Ed.; *The Problem of Heat Protection of Body Surface in a Hypersonic Flow* (compiled by G. A. Tirskiy), Nauka, Moscow, 1970, 552–559.

[46] Pai, S. I., *Dinamika izluchlayushchego gaza* (*Dynamics of Radiatlng Gas*), Mir Publishers, Moscow, 1968 (Russian translation).

more of hypersonic flows, where the growth of the role of the boundary layer inverse effect on the external nonviscous flow and, consequently, the increase of interaction between them are stipulated by a comparatively great thickness of the hypersonic boundary layer region. The cause of this specific feature of the hypersonic boundary layer is gas expansion at high temperatures (characteristic of flows with large Mach numbers), with accompanying decrease of gas density, and the decrease in the Reynolds number, thus leading to the growth of the role of viscous friction on the body surface.

As noted, direct integration of the Navier-Stokes equations at the Reynolds numbers characteristic of the boundary layer theory of the first approximation (the Prandtl equation), in the cases considered is impossible not only for analytical but also numerical solutions on a computer. The asymptotic methods (the methods of small perturbations) come to their aid. We are already familiar with a specific case of application of these methods in considering the standard boundary layer theory procedure of "joining" the solution of the Prandtl equations with an external nonviscous flow (Section 2, Chapter 12).

The essence of asymptotic methods is in the following. The entire flow region under consideration is divided to some subregions with inner scales characteristic of each of them: linear longitudinal and transverse dimensions, velocities, pressure, and other mechanical and thermodynamical quantities. Together with these inner scales are considered external scales common for all regions: the size of the flow region as a whole, velocity U_∞, and others. By these external scales the Reynolds number **Re** $= U_\infty L/\nu$ is constructed and assumed to be large and the reciprocal 1/**Re** small. In each subregion changes are made from the dimensional quantities to dimensionless ones referred to the scales characteristic of the subregion; the main assumption is made that at the limiting transition **Re** $\to \infty$ all the thus-normalized quantities in the subregions tend to finite values, or in other words, they have the order of unity.

As a result of this limiting transition the Navier Stokes equations formulated for all subregions are simplified, taking on different forms depending on specific features of flow in this subregion (the Euler and Prandtl equations, equations of a "slow" viscous flow). The solutions of these simplified equations found for every two conjugate subregions are "joined" with each other. The Prandtl classical theory of a boundary layer serves as an obvious example. The limiting transition **Re** $\to \infty$, corresponding to the "disappearance" of viscosity ($\nu \to 0$), converts the Navier-Stokes equations to the Euler. However, the Euler equations do not admit integration under the boundary conditions corresponding to medium adhesion to the solid body surface (zero relative velocity on the solid surface). Naturally, the necessity arises to divide the entire flow region to two subregions: to the outer one described by the Euler equations with the boundary condition of surface "impermeability" only, i.e., the normal component of the relative velocity on the surface is equal to zero, and to the inner thin near wall region (boundary layer) in which the "adhesion" condition is valid. Due to a small thickness of this region the Navier Stokes equations are simplified and pass over to the Prandtl equations. We shall recall that the Prandtl equations are derived from the Navier Stokes equations by the limiting transition **Re** $\to \infty$ only when all of the quantities in a boundary layer are referred to their characteristic scales – longitudinal ones which have the order of unity and transverse ones with the order $1/(\sqrt{\mathbf{Re}})$.

In the asymptotic theories in which we are interested, along with the subregions of the classical boundary layer type other subregions arise whose orders in longitudinal and transverse dimensions, velocities, pressure drops, etc. differ from $1/(\sqrt{\mathbf{Re}})$. Estimation of the orders of these subregions by the Reynolds number of extent scales, and of mechanical and thermodynamic characteristics of medium motions in them, is the main stage of the construction of asymptotic solutions. The second stage is the formation of a series by the parameters, the smallness of which is provided by the tendency of the external Reynolds number to infinity, and determination of the coefficients of these series in this or another simplest approximation. In this case asymptotic solutions in conjugate subregions are "joined".

Having no possibility to go into the details of the use of asymptotic methods, we shall only point out that the problem of the determination of the orders of characteristic scales of flow subregions requires, in each separate case, the application of, in many respects intuitive, physical considerations that particularly refer to complex super and hypersonic problems.

The importance of asymptotic methods is not limited to the account of the boundary layer inverse effect on the external nonviscous flow that is expressed in the distortion of the external flow due to displacement of streamlines from a solid surface caused by the retarding effect of a solid wall (Section 3, Chapter 13). Of significance is the fact that these methods reveal the nature of other very important physical phenomena in a supersonic boundary layer, one of the most essential of which is possibility for perturbations to propagate upstream. this is initially inconsistent with the hyper- and parabolic characters of the equations of motion in the outer and inner regions of the boundary layer. The mechanism of this propagation later becomes clear and quantitatively determined thanks to the consideration of the sublayer of small velocities, which is located directly on the solid surface and which transmits the perturbation waves upstream. This effect is termed "free interaction" and the boundary layer region where it takes place is called the "region of free interaction".

Pausing to discuss the simple problem of the interaction between a laminar boundary layer and the external hypersonic flow on a longitudinally streamlined plate, we shall restrict ourselves only to accounting for the effect of pushing aside the streamlines and we digress from the above-mentioned effect of free interaction.[47] This problem was considered in a more general form. We suggest review of the problem of laminar boundary, layer interaction with a supersonic flow by V. Ya. Neiland.[48]

As a basis of the quantitative analysis of the phenomenon we shall assume the known relation between pressure in a hypersonic flow and the angle of deviation of streamlines from the direction of the incident unperturbed flow.

We agree to distinguish between the "local angle of attack" θ_w formed by the tangent to the contour of the streamlined body at a given point with the direction of an unperturbed flow and the effective local angle of attack θ, constructed in the same way, for the effective contour, i.e., that obtained by building up the displacement thickness $\delta^*(x)$, found from (221) along the normal to the streamlined surface. The angles θ and θ_w compared with each other for the points belonging to the same section of the boundary layer. The difference between these angles $\theta^* = \theta - \theta_w$ in the case of their smallness can be approximately equated to the value of the derivative $d\delta^*/dx$ at the point of the same section of the boundary layer. In the case of a compressible medium by the "displacement thickness" we mean the quantity

$$\delta^* = \int_0^\delta \left(1 - \frac{\rho u}{\rho_\delta u_\delta}\right) dy \tag{221}$$

where ρ_δ and u_δ denote the values of the density and velocity on the outer edge of the boundary layer calculated by the pressure distribution with a correction which has been already made for pushing aside the streamlines due to the boundary layer effect.

---

[47] Hayes, W. D. and Probstein, R. F., *Teoriya giperzvukovykh techeniy* (*The Theory of Hypersonic Flows*), Inostrannaya Literatura, Moscow, 1962, 430 (Russian translation).

[48] Neiland, V. Ya., Assymptotic theory of separation and interaction between a boundary layer with supersonic flow of gas, *Adv. Mech.*, 4(2), 1981; This review possesses numerous references. Moreover, see the paper by Neiland in *Mekh. Zhidk. Gaza*, No.4, 1971, and others).

Note that in a hypersonic flow ($\mathbf{M}_\infty \gg 1$) the density ρ is small and it can be assumed that $\delta^* \approx \delta$. Then we shall have the equalities

$$\theta = \theta_w + \theta^* \approx \theta_w + \frac{d\delta^*}{dx} \approx \theta_w + \frac{d\delta}{dx}$$

$$K = \mathbf{M}_\infty \theta = \mathbf{M}_\infty (\theta_w + \theta^*) \approx \mathbf{M}_\infty \left(\theta_w + \frac{d\delta^*}{dx} \right) \approx \mathbf{M}_\infty \left(0_w + \frac{d\delta}{dx} \right) \tag{222}$$

In the case of a longitudinal flow past a flat plate ($\theta_w = 0$) when the dependence of layer thickness δ^*, δ on the longitudinal coordinate is a one term power law one, the previous equalities will change over to

$$\theta = \frac{d\delta^*}{dx} \approx \frac{\delta^*}{x} \approx \frac{\delta}{x} \qquad K = \mathbf{M}_\infty \frac{d\delta^*}{dx} \approx \mathbf{M}_\infty \frac{\delta^*}{x} \approx \mathbf{M}_\infty \frac{\delta}{x} \tag{223}$$

Assuming that the bow shock wave is formed in the external nonviscous flow with respect to the boundary layer, we shall take the following model of a hypersonic viscous gas flow past a thin aerofoil profile with a sharp leading edge.

The external homogeneous flow with the velocity U_∞, density ρ_∞, pressure p_∞, and temperature T_∞ passes through the bow shock wave on the leading sharp edge of the profile and only then does it meet the outer boundary layer edge $y = \delta(x)$ where the values of the velocity density, pressure, and temperature are equal to u_δ, ρ_δ, p_δ, and T_δ. The flow region between the front shock wave and the outer boundary layer edge is usually called a shock layer. To estimate the orders of the boundary layer thickness δ and ratio of pressures $p_\delta/p_\infty = p/p_\infty$ we shall leave aside the flow vorticity in a shock layer and, according to the conventional Prandtl scheme, assume (the bar over the symbol denotes mean over the cross sectional value of the quantity)

$$\delta^2 \approx \delta^{*2} \backsim \frac{\bar{\mu} x}{\bar{\rho} u_\delta} = \frac{\bar{\mu}}{\mu_\infty} \frac{\rho_\infty}{\bar{\rho}} \frac{\mu_\infty x}{\rho_\infty u_\delta} \tag{224}$$

We shall take the linear law (15) for the dependence of the viscosity coefficients on temperature and note that in a hypersonic flow ($\mathbf{M}_\infty \gg 1$) an oblique shock makes such a small angle with the incident flow that u_δ can be easily replaced by U_∞. Besides taking the body surface to be thermally insulated we assume $\bar{T}/T_\infty \approx ((k-1)/2)\mathbf{M}_\infty^2$. Then, instead of (224) we obtain $[(\rho_\infty/\bar{\rho}) = (p_\infty/p)(\bar{T}/T_\infty)]$

$$\delta^2 \approx \delta^{*2} \backsim \left(\frac{k-1}{2} \mathbf{M}_\infty^2 \right)^2 C \frac{p_\infty}{p} \left(\frac{\mu_\infty x}{\rho_\infty U_\infty} \right)$$

or, recalling (223)

$$\theta \approx \frac{\delta^2}{x} \approx \frac{\delta^{*2}}{x} \backsim \frac{k-1}{2} \mathbf{M}_\infty^2 \sqrt{C} \frac{\sqrt{p_\infty/p}}{\sqrt{\mathbf{Re}_{x\infty}}} \qquad \mathbf{Re}_{x\infty} = \frac{\rho_\infty U_\infty x}{\mu_\infty} \tag{225}$$

According to Equation 106 of Chapter 8, we find the following expression for the pressure coefficient c_p:

$$c_p = \frac{p - p_\infty}{\frac{1}{2}\rho_\infty U_\infty^2} = \frac{2}{k\mathbf{M}_\infty^2}\left(\frac{p}{p_\infty} - 1\right) = \left[\frac{k+1}{2} + \sqrt{\left(\frac{k+1}{2}\right)^2 + \frac{4}{K^2}}\right]\theta^2$$

where

$$\frac{p}{p_\infty} = 1 + k\left[\frac{k+1}{4} + \sqrt{\left(\frac{k+1}{4}\right)^2 + \frac{1}{K^2}}\right]K^2 \tag{226}$$

In the considered case of a hypersonic flow past the "effective" profile the similarity criterion $K = \mathbf{M}_\infty\theta$ known from Chapter 8, satisfies, as follows from Equation 225, the proportionality condition

$$K \backsim \chi\sqrt{\frac{p_\infty}{p}} \qquad \chi = \mathbf{M}_\infty^3\frac{\sqrt{C}}{\sqrt{\mathbf{Re}_{x\infty}}} \tag{227}$$

Comparing Equations 226 and 227, we conclude that the real value of the ratio of pressures p/p_∞ is determined as the function of the parameter χ, which thus becomes a generalized similarity criterion of hypersonic flows past bodies which accounts for the portion of the above-mentioned interaction between a boundary layer with the external nonviscous flow.

The considerations presented refer to the case of a longitudinal flow past a thin plate but allow the generalization to the case of a thin wedge in the case of an axial flow and in the presence of a small angle of attack. Here, one should also take account, along with the parameter χ, of the effect of the parameter $K_w = \mathbf{M}_\infty\theta_w$ because in this case it is not justifiable to reject θ_w as compared with $\theta^* = d\delta^*/dx$.

The two limiting cases of interaction are

1. Weak interaction ($\chi < 1$ or of the order of unity): This limiting case means comparatively small hypersonic velocities ($K < 1$) and great Reynolds numbers ($\mathbf{Re}_\infty \gg 1$). The pressure distribution in this case scarcely differs from that unperturbed by a boundary layer. The study of weak interaction with the aid of asymptotic expansions by the powers of the wall parameter χ is not very difficult and uses the flow without interaction as an initial approximation.

A detailed analysis of the effect of weak interaction on the pressure distribution leads[47] to the equation (for air $k = 1.4$)

$$\frac{p}{p_\infty} = 1 + 0.31\chi + 0.05\chi^2 \qquad (0 \le \chi < 4) \tag{228}$$

which is in a good agreement with the experimental results in this range of χ values.

2. Strong interaction ($\chi \gg 1$): In this case corresponding to very high Mach numbers ($\mathbf{M}_\infty \gg 1$) and finite values of $\mathbf{Re}_\infty$, the effects of interaction are very substantial. When $K \gg 1$, the right side of Equation 226 may be replaced by an asymptotic (at great K) series

$$\frac{p}{p_\infty} = \frac{k(k+1)}{2}K^2 + \frac{3k+1}{k+1} - \frac{8k}{(k+1)^3}\frac{1}{K^2} + \ldots \tag{229}$$

where the estimation of the quantity p_∞/p by K when $K \gg 1$ follows:

$$\frac{p_\infty}{p} \approx \frac{2}{k(k+1)}\frac{1}{K^2} = \frac{2}{k(k+1)}\frac{1}{\mathbf{M}_\infty^2\theta^2} \approx \frac{2}{k(k+1)\mathbf{M}_\infty^2}\left(\frac{x}{\delta^*}\right)^2 \tag{230}$$

According to this estimation we have from (225)

$$\frac{\delta^*}{x} \approx \frac{k-1}{2} \mathbf{M}_\infty^2 \sqrt{C} \sqrt{\frac{2}{k(k+1)\mathbf{M}_\infty^2}} \, \frac{x}{\delta^*} \, \frac{1}{\sqrt{\mathbf{Re}_{x\infty}}} \tag{231}$$

so that

$$\theta \approx \frac{\delta^*}{x} \approx \frac{\delta}{x} \approx \left[\frac{k-1}{k(k+1)} \frac{\mathbf{M}_\infty \sqrt{C}}{\sqrt{\mathbf{Re}_{x\infty}}} \right]^{1/2} \tag{232}$$

Return to Equation 231 we obtain the estimation for the ratio of pressures

$$\frac{p}{p_\infty} \approx \frac{k-1}{2} \frac{\mathbf{M}_\infty^3 \sqrt{C}}{\sqrt{\mathbf{Re}_{x\infty}}} = \frac{k-1}{2} \chi \tag{233}$$

Without derivation,[47] the formula

$$\frac{p}{p_\infty} = 0.514\chi + 0.759 + O\left(\frac{1}{\chi}\right) \qquad (4 \leq \chi \leq 7.5)$$

is in good agreement with the experimental results for air ($k = 1.4$).

Unlike weak interaction, when a relative thickness of the boundary layer has the order of $\delta/L \backsim \mathbf{Re}_L^{-1/2}$ (L is the plate length, $\mathbf{Re}_L = U_\infty L/\nu_\infty$), in the case of strong interaction $\delta/L \backsim \mathbf{Re}_L^{-1/4}$. In the case of weak interaction development of the boundary layer thickness along the plate is usually determined by a conventional regularity $\delta \backsim x^{1/2}$ while in the opposite case a more rapid development of $\delta \backsim x^{3/4}$ takes place; the ratio of pressures p/p_∞ in both cases is presented by the functions of the same parameter $\chi = \mathbf{M}_\infty^3 \sqrt{C}/\sqrt{\mathbf{Re}_{x\infty}}$; in the case of strong interaction the relation $p/p_\infty \backsim x^{-1/2}$ will be valid.

The mechanism of boundary layer interaction with the external nonviscous flow is greatly complicated in the case of thin bodies with a blunted leading edge. In such cases bow shock waves of a complex curvilinear shape are formed at very high Mach numbers. When passing through this wave an incident uniform isentropic flow becomes vortical and nonisentropic. In this case, under the conditions corresponding to the notion of strong interaction, the vorticity and gradient of entropy in the region between the front wave and the outer edge of the boundary layer may be quite intense.

The classical theory of a laminar boundary layer does not take into account the external flow vorticity but allows only for the velocity at the outer edge of the boundary layer. Attempts to extend the Prandtl theory to this case were not sufficiently developed. The considered effect of pushing aside the streamlines in the presence of "vortical interaction" may be substantially altered, especially near the front blunted edge of the body. We shall also note that in a hypersonic viscous gas flow past slender bodies of revolution, with the exception of the above-mentioned effects, the effect of the body transverse curvature is also important in that in the case of the flows of small velocities in manifests itself only on very slender bodies.

The study of separation phenomena concerned with the existence of shock waves in the external flow which interact with the boundary layer comprises a large portion of the theory of boundary layer interaction with the external nonviscous flow.

Having no possibility to go into a detailed analysis of special problems, we note some general facts concerned with the study of local phenomena of interaction between a boundary layer and shock waves falling onto it and the resulting local separations.

According to classical notions of the theory of a boundary layer in a supersonic flow the propagation of perturbations upstream is impossible because this propagation would be inconsistent with a parabolic character of the equations of motions in the boundary layer region and to a hyperbolic character of the equation of motion in the outer region. Nevertheless, "anticipation" of the external flow influence on the flow in a boundary layer is observed.

In the vicinity of the separation point, as well as near any other point of sharp, longitudinal change in the boundary layer parameters, the main assumption is violated. This assumption was used in the derivation of the boundary layer equations: the slow character of the change in the quantities along the flow as compared with their sharp variation across the flow. Recovery of the role of longitudinal derivatives results in the Navier Stokes equations, which in the case of stationary motions have an elliptical character. Besides assigning the boundary condition, normal for stationary parabolic equations of the boundary layer in the initial sections, on the wall, and on the outer boundary layer edge, it is necessary to prescribe the boundary condition somewhere downstream, as otherwise the elliptical equations fail to give a definite solution.

The methods based on matching asymptotic solutions (when $\mathbf{Re} \to \infty$) of the equations of gas flow in regions that differ in the character of motion, are practiced on a large scale. It was found when developing these methods that in contrast to the classical boundary layer theory with two typical regions (a boundary layer and an external nonviscous flow), in the asymptotic theory, conforming to the considered problem of a gas flow near a specific point with a sharp longitudinal change in the laminar boundary layer external characteristics, one has to deal with the problem of matching the solutions in the three zones situated near the specific point of the boundary layer. According to Neiland,[48] the longitudinal dimension, common to all these zones, is the order of $\mathbf{Re}^{-3/8}$.

The external zone has the same cross sectional dimension and the flow in it ln the first approximation can be described by the linear theory of supersonic flows. In the second "intermediate" region, which is directly adjacent to the first and has the cross sectional dimension on the order of $\mathbf{Re}^{-1/2}$, the flow is retained so that the first approximation coincides with one observed in an undisturbed boundary layer at a distance from the considered specific point. Disturbances in the intermediate zone are small and in the first approximation do not affect pressure distribution.

Finally, the third "near wall" region, which in asymptotic theories plays an especially important role because the change in the wall layer thickness is the very reason for the origin of the longitudinal pressure gradient in the external flow; it also stipulates flow stalling from the body surface. The thickness of this region is comparatively small: $\mathbf{Re}^{-5/8}$. Though the flow here is described by the equations of the boundary layer in an incompressible fluid that are ordinary in the form, these equations have a specific fundamental feature – the term on the right side dp/dx is no longer assigned a priori and it should be determined in the course of solution from the condition of matching the flow in a near wall region with an outer supersonic flow. This condition retains an elliptical character of the equation of motion in the near wall region because together with the Ackeret equation for the pressure coefficient (given in Chapter 8) calculated by the linear theory of supersonic flows, it makes it possible to express the quantity dp/dx in terms of the second derivative of the displacement thickness in a viscous portion of the flow region by the formula

$$\frac{dp}{dx} = k\mathbf{M}_\infty p_\infty \frac{d\theta}{dx} \approx \frac{\rho_\infty U_\infty^2}{\mathbf{M}_\infty} \frac{d^2\delta^*}{dx^2} \tag{234}$$

which follows directly from the previous content of Chapter 8. In the equations of the asymptotic theory the term appears involving the "senior" second derivative with respect to x of the unknown function $\delta^*(x)$, this fact requires assignment of the second boundary condition by the variable x at the point downstream of the specific point of the boundary layer.

This leads to the fact that disturbances arising behind the separation point exert an effect on the boundary layer regions situated upstream. This is where the elliptical character of the initial Navier Stokes equations is felt.

The calculations of pressure distributions over the body surface in a stalling zone and in its vicinity lead, according to the asymptotic theories, to good agreement with the available experimental data.

In this book we have no possibility to go into a detailed analysis of different methods of asymptotic presentations of the solutions of the Navier Stokes equations, which is why we refer readers to the corresponding review papers.[49]

The formulated considerations explaining the mechanism of transfer of abrupt changes in the external flow shock wave (falling onto a boundary layer, a boundary layer separation, or a deflection point on the body surface) upstream along the laminar boundary layer describe a wide class of phenomena both in the superhypersonic and in the subsonic gas flows. Similar in form processes have (see Chapter 12) a place in small velocity flows. With the desire to emphasize a specific feature of the entire group of these phenomena, which are capable of transferring disturbances in the boundary layer both upstream and downstream and in their reverse effect on the external flow, for these kind of flows we use the above-mentioned "flow with free interaction".[50]

The contribution of Russian scientists to this new, quickly developing area of general boundary layer theory is indisputably great. The names of the researchers who made much progress and also references to the papers published by them can be found in footnote 49.

11. Turbulent Boundary Layer on a Plate in a Gas Flow Parallel to It

The problem of the effect of gas compressibility on the origin of turbulence as well as on the mechanism of steadystate turbulence has been little studied. Theoretical works on the stability of a laminar boundary layer at high velocities show that, with other conditions being equal, laminar layer stability becomes weaker with the increase in the number $\mathbf{M}_\infty$; this should be understood in the sense that as $\mathbf{M}_\infty$ grows the lower critical Reynolds number Recr, above which disturbance in the layer ceases to decay, should decrease.

However, the experiments[51] confirming the correctness of this thesis approximately up to $\mathbf{M}_\infty = 3.5$ show that with further growth of $\mathbf{M}_\infty$ the phenomenon develops in an opposite direction.

Figure 290 presents experimental data on the change, with the growth of $\mathbf{M}_\infty$, of the critical Reynolds number corresponding to the point of the loss of laminar layer stability, i.e., to the beginning of the transition region, and of the Reynolds number at the point behind which the developed turbulent flow (the end of the transition region) takes place.

As follows from the graphs presented at the $\mathbf{M}_\infty$ value of about 3.5, maximum upstream

[49] Brown, S. N. and Stewartson, K., Laminar separation, Annu. Rev. Fluid Mech., 1, 45–72, 1969; *Annu. Rev. Ind.*, Mikhailov, V. V., Neiland, V. Ya., and Sychev, V. V., The theory of viscous hypersonic flow, *Annu. Rev. Ind.*, 371–396, 1971; Chang, P. K., *Otryvnye techeniya (Separation Flows)*, Vol. 1 to 3, 3rd ed., Mir Publishers, Moscow, 1972–1973 with the review paper by Golubinskiy, A. I., Maykapara, G. I., and Neiland, V. Ya., published in the Volume 3.

[50] The theory of flows with free interaction in the case of a turbulent boundary layer was less developed. The state of the problem is presented in the book Gogish, L. V. and Stepanov, G. Yu., *Turbulentnye otryvnye techeniya (Turbulent Separation Flows)*, Nauka, Moscow, 1979.

[51] Korkegi, R., Transition studies and skin friction measurements on an insulated flat plate at a Mach number 5.8, *J. Aeron. Sci.*, 23(2), 97–107, 1956.

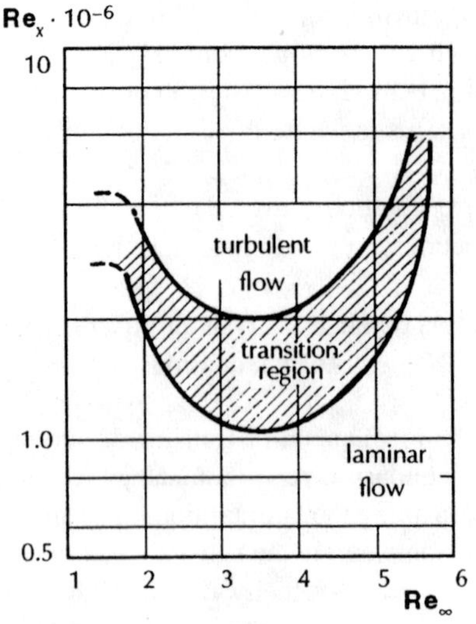

Fig. 290

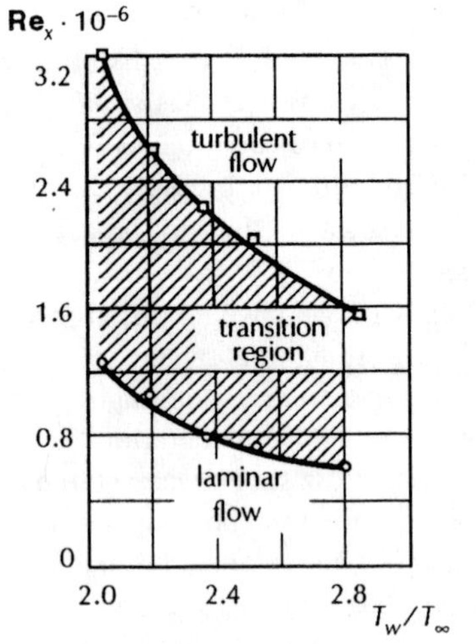

Fig. 291

displacement of the transition region boundaries and the maximum width of the region itself are observed. With further growth of $\mathbf{M}_\infty$ (up to $\mathbf{M}_\infty = 5.8$) the boundaries shift downstream and probably have the tendency to continue displacement in the same direction at greater $\mathbf{M}_\infty$. This fact is very important from the viewpoint of solving the problem of the boundary layer structure on the surface of bodies moving with great $\mathbf{M}_\infty$ numbers; otherwise, any accurate calculation of the resistance and heat transfer of bodies is impossible.

The reason for the divergence between theoretical studies and the experiment is in the incompleteness of the theory of the boundary layer stability in compressible fluid at high velocities. Doubt has been expressed in the literature[51] concerning the exact nature of this theory, in particular due to the absence of accounting for the effect of $\mathbf{M}_\infty$ number on the wavelengths of propagation of perturbations and on the growth of their amplitude at high supersonic velocities. As is shown by experiments, perturbations produced in the flow by turbulizing cascades at high values of $\mathbf{M}_\infty$ attenuate almost instantly, thus indicating the increased stability of laminar flows at these $\mathbf{M}_\infty$. One may think that under such conditions the above considerations of the gradual nature of the phenomenon of laminar to turbulent layer transition becomes invalid. Probably, in this case only the loss of flow stability in a laminar layer, which at high $\mathbf{M}_\infty$ begins in the region of great Reynolds numbers, serves as the cause of transition. The considerations given require methodical experimental testing.

Heat supply through a streamlined surface destabilizes a laminar boundary layer and, on the contrary, heat transfer from the body surface exerts a stabilizing effect. This indisputable fact is proved both theoretically and experimentally.[52] In Figure 291 the graph is given which illustrates this postulate with the help of the experiments of Higgins and Pappas[53] with the plate at $\mathbf{M}_\infty = 2.40$.

At present the theory of gas flow stability in a laminar boundary layer at high sub and supersonic velocities is still under development. Having no possibility to expound the results of this special area of study and give voluminous literature in the field, we refer the reader to

---

52 Van Driest, E., Calculation of the stability of the laminar boundary layer in a compressible fluid on a flat plate with heat transfer, *J. Aeron. Sci.*, 19(2), 801, 1952; Shapiro, A., Effects of pressure gradient and heat transfer on a stability of the compressible laminar boundary layer, *J. Aeron. Sci.*, 23(1), 81–83, 1956.

53 Shapiro, A., The dinamics and thermodynamics of compressible fluid flow, Vol. 2, New York, 1953, 1077.

the review paper published by S. A. Gaponov and V. Ya. Levchenko in Advances in Mechanics (see Section 1, Chapter 13). The results of extensive theoretical and experimental studies of the scientists at the Siberian Branch of the USSR Academy of Sciences performed in this field were generalized in the doctoral thesis of S. A. Gaponov, "Development of Disturbances in a Supersonic Boundary Layer".

Gas compressibility for $\mathbf{M}_\infty \ll 1$ slightly affects the mechanism of turbulent exchange. The 1933 to 1938 experiments of Russian scientists Varshavskiy, Gukhman, Ilyukhin, Lelchuk, and Tarasova, as well as later experiments performed by foreign scientists Frössel, Young, Keenan, and Neumann, showed that at subsonic velocities one can successfully use the same power law and logarithmic equations of resistance as was done for an incompressible fluid, if for the velocity and density their values averaged in a certain manner over the tube section are meant.[54] Theoretically, in a shockless flow this possibility is retained with sufficient approximation for not very high supersonic velocities ($\mathbf{M} \leq 1.7$);[55] however, supersonic motions in tubes are accompanied by the formation of a complex system of shock waves which do not allow one to consider flow as one-dimensional or to use the concept of a steady state turbulent flow.

Gas flow in a turbulent boundary layer with high sub and supersonic velocities is a crucial problem of modern aerodynamics. The means of its practical solution lie in the generalization of empirical and semiempirical theories of turbulence in an incompressible fluid to the case of high-speed gas flow. It should be noted that the experimental test of the feasibility to apply such theories to high speed gas flows became quite possible due to extensive development of experimental techniques in supersonic wind tunnels and during flight. The comparison of theoretical and experimental results undertaken, have shown that making use of empirical and semiempirical methods one can find formulas which answer the requirements of practical application.

We shall go into details of the study of a turbulent boundary layer on a smooth plate parallel to a gas flow. Restricting ourselves to the case of a thermally insulated plate and leaving aside the problem on the form of velocity and temperature profiles in the layer sections, we compose the empirical formula of the dependence of the coefficient of local resistance c_f on the local Reynolds number $\mathbf{Re}_x$. For this purpose we shall use the familiar empirical relations between cf and $\mathbf{Re}_x$ in an isothermal flow of an incompressible fluid. Unlike a flow where the constants μ and ρ are the same through the entire flow, in this case the quantities μ and ρ change with temperature over the layer cross section. There are no grounds to take μ and ρ corresponding to the free stream temperature T_∞ because near the plate surface the gas temperature is probably T_w, which at high $\mathbf{M}_\infty$ considerably exceeds T_w. Relating μ and ρ to the surface temperature T_w seems to be more substantiated,[56] although it is clear that in this case the effect of temperature on gas viscosity and density will be overestimated. So, it is logical to relate the gas physical constants to some mean temperature T_m which is greater than T_∞, but smaller than T_w. Tucker[57] made the simplest assumption, having taken T_m equal to the arithmetic mean value of T_∞ and T_w

---

[54] Gukhman, A. A. and Ilyukhin, N. V., *Osnovy ucheniya o teploobmene pri techenii gaza s bolshoy skorostiyu* (*Foundations of the Teaching on Heat Transfer in High Speed Gas Flow*), Mashgis, Moscow, 1951. One can find a critical survey of the majority of Russian and foreign studies on this problem and detailed references. Also presented are power-law formulas of resistance and heat transfer of tubes convenient for practical calculations.

[55] Deissler, R., Analytical and Experimental Investigation of Adiabatic Turbulent Flow in Smooth Tubes, NACA Tech. Note, No.2138, 1950.

[56] This assumption was first made by Th. von Kármán in the report to the Volt Congress in Rome in 1935: Th. von Kármán, The problem of resistance in a compressible fluid, in *Gazovaya dinamika* (*Gas Dynamics*), GONTI, Moscow, 1939, 58.

[57] Tucker, M., Approximate Calculation of Turbulent Boundary Layer Development in Compressible Flow, NACA Tech. Note, No.2337, 1951.

$$T_m = \frac{1}{2}(T_\infty + T_w)$$

Making use of the resistance equation (152 of Chapter 13), we shall have, putting the subscript m at the gas constants for $T_\infty = T_m$

$$\frac{\tau_w}{\frac{1}{2}\rho_m U_\infty^2} = 0.0263\left(\frac{\rho_m U_\infty x}{\mu_m}\right)^{-1/7} \tag{235}$$

with (p = const)

$$\frac{\rho_m}{\rho_\infty} = \frac{T_\infty}{T_m} \qquad \frac{\mu_m}{\mu_\infty} = \left(\frac{T_m}{T_\infty}\right)^{n} \tag{236}$$

Emphasizing that we are searching for the relation between c_f and $\mathbf{Re}_x$, which is expressed in terms of the a priori given characteristics of the free flow falling onto the plate U_∞, ρ_∞, μ_∞, i.e.,

$$c_f = \frac{\tau_w}{\frac{1}{2}\rho_\infty U_\infty^2} \qquad \mathbf{Re}_x = \frac{\rho_\infty U_\infty x}{\mu_\infty} \tag{237}$$

we shall eliminate ρ_m and μ_m from Equation 235 based on Equation 236. First, we obtain

$$\frac{\tau_w}{\frac{1}{2}\rho_\infty U_\infty^2}\frac{\rho_\infty}{\rho_m} = 0.0263\left(\frac{\rho_\infty U_\infty x}{\mu_\infty}\right)^{-1/7}\left(\frac{\rho_\infty}{\rho_m}\right)^{1/7}\left(\frac{\mu_m}{\mu_\infty}\right)^{1/7} \tag{238}$$

and then

$$c_f \mathbf{Re}_x^{1/7} = 0.0263\left(\frac{T_\infty}{T_m}\right)^{6/7}\left(\frac{T_m}{T_\infty}\right)^{n/7} = 0.0263\left(\frac{T_\infty}{T_m}\right)^{\frac{6-n}{7}} \tag{239}$$

The exponent n in the law of the dependence of the coefficient of viscosity on temperature within a practical range of temperatures from 1 to 1/2 is close to 3/4; it will not be a great error if, for simplicity, we assume $n = 1$. Further, we shall take the coefficient of temperature recovery on the plate surface to be equal to unity, i.e., under the condition of the absence of heat transfer from the plate surface we shall consider the plate temperature T_w equal to the temperature of adiabatically and isentropically stagnated gas flowing onto the plate with the velocity U_∞. Then, we have

$$\frac{T_\infty}{T_m} = \frac{2T_\infty}{T_\infty + T_w} = \frac{2T_\infty}{T_\infty + T_\infty\left(1 + \frac{k-1}{2}\mathbf{M}_\infty^2\right)} = \frac{2}{2 + \frac{k-1}{2}\mathbf{M}_\infty^2}$$

Substituting this expression into expression (239), we find

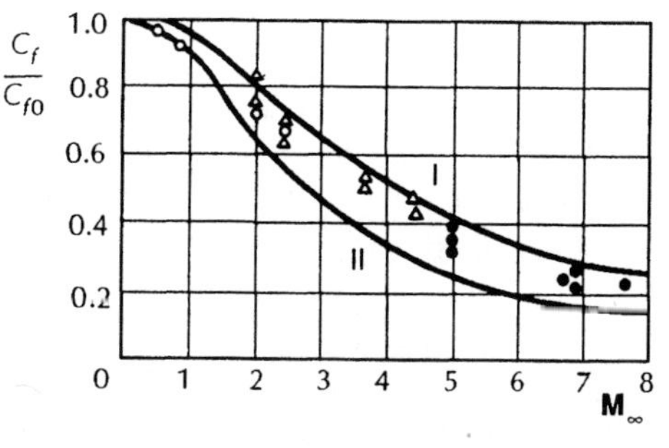

Fig. 292

$$c_f \mathbf{Re}_x^{1/7} = 0.0263 \left(\frac{2}{2 + \frac{k-1}{2} \mathbf{M}_\infty^2} \right)^{5/7} \tag{240}$$

Integrating over the entire length of the plate, we obtain the formula for the coefficient of total resistance

$$C_f \mathbf{Re}_x^{1/7} = 0.0307 \left(\frac{2}{2 + \frac{k-1}{2} \mathbf{M}_\infty^2} \right)^{5/7} \tag{241}$$

which is the generalization of Equation 153, Chapter 13 to the case of gas.

As follows from these empirical formulas, the influence of the number $\mathbf{M}_\infty$ on the coefficients of local and total resistance is independent of the effect of the Reynolds number $\mathbf{Re}_x$ or $\mathbf{Re}$, so that taking the conventional notation by the subscript 0 at $\mathbf{M}_\infty = 0$, we have

$$\frac{C_f}{C_{f0}} = \frac{c_f}{c_{f0}} = \left(\frac{2}{2 + \frac{k-1}{2} \mathbf{M}_\infty^2} \right)^{5/7} \tag{242}$$

In Figure 292 formula (242), curve I, is compared with the data of different authors.[58] The upper limit of the numbers $\mathbf{M}_\infty$ extends to about 8. The coincidence of the Tucker equation with the experimental data, as it follows from Figure 292 is quite satisfactory at these comparatively high values of $\mathbf{M}_\infty$. The same figure also gives curve II, which corresponds to the use of the plate surface temperature T_∞, as the characteristic mean temperature T_w, i.e., the curve

$$\frac{C_f}{C_{f0}} = \frac{c_f}{c_{f0}} = \frac{1}{\left(1 + \frac{k-1}{2} \mathbf{M}_\infty^2 \right)^{5/7}}$$

[58] In Figure 292 clear dots stand for the experimental results of Chapman, D. R. and Kester, P. H., *J. Aeron. Sci.*, 20(7), 441–449, 1953; triangles indicate the data of Coles, D., *J. Aeron. Sci.*, 21(7), 443–449, 1954; black dots represent the experiments of Lobb, R. K., Winkler, E. M., and Persh, J., *J. Aeron. Sci.*, 22(1), 1–10, 1955.

This curve confirms the earlier assumption that use of the plate surface temperature as the mean temperature in a boundary layer should lead to overestimation of the influence of air compressibility (the number $\mathbf{M}_\infty$) on the coefficient of plate resistance.

Another, also empirical method of calculation of a turbulent boundary layer in a high-speed gas flow parallel to a plate was published by Spalding and Chi.[59] This method, unlike previous methods, can be applied also in the presence of heat transfer from the plate surface. In the method use is made of tabulated empirical functions, thus simplifying the application of the method. The presence of the two cited well known references allows avoidance of the details of the method.

Passing over to semiempirical methods, we note that they are based on the generalization of the methods, suggested by the founders of semiempirical theories, Prandtl and von Kármán, for an incompressible fluid, to the case of high speed gas flows (Section 1, Chapter 13).

Historically, a semiempirical theory of a turbulent boundary layer in a compressible gas with high velocities has two trends. The first, opened by the work of Frankl and Voishel,[60] was based on the direct transfer of the equations of von Kármán semiempirical theory to gas dynamics. Further, van Driest[61] followed the same method.

In the works of the second trend the Dorodnitsyn variables were used and it was assumed that the formulas of semiempirical theories of turbulence remain valid in these variables. The answer to the question as to which trend gives results closer to reality can be given only by experiments conducted at high Mach number. At the time when both trends appeared, there were no such experimental data. At Mach numbers slightly exceeding unity, both trends gave results that virtually coincided with each other. Only when the data on the friction resistance of plane surfaces at high Mach number (approximately reaching 8) were obtained, it became clear that only the first trend gives correct results.

In their work Frankl and Voishel generalized the von Kármán method directly to the case of a gas flow, having simplified it only by the assumption of the constancy of skin friction across a boundary layer. Following this course they first found a shape of velocity profiles in the layer sections and then by usual means they obtained the so called law of resistance, i. e., the relation between the local coefficient of friction and the Reynolds number for a boundary layer, excluding use of the Reynolds number constructed by the free stream velocity and abscissa of the given point on the plate.

The degree of approximation, assumed by Flankl and Voishel, allowed them to make calculations only up to Mach numbers slightly exceeding unity. Transition to higher Mach numbers would require either greater complication of the already complex computational technique or direct application of numerical methods.

Using the von Kármán equation and assuming constancy of the skin friction across a boundary layer a simpler way exists of solution construction that does not require the notions of the boundary layer Reynolds number and of the law of resistance to be introduced *a priori*.[62] This greatly simplifies the study of the problems on a turbulent boundary layer in a gas flow. The use of a simple asymptotic expansion which has been already applied in Chap-

---

[59] Spalding, D. B. and Chi, S. W., Resistance in a compressible boundary layer on a smooth flat plate with and without heat transfer, *Mekhanika* (*Mechanics*), Coll. Pap. No.6, 1964; See also the monograph by Lapin, Yu. V., (pp. 161–170) cited earlier.

[60] Frankl, F. I. and Voishel, V. V., Friction in a turbulent boundary layer near a plate in a plane parallel flow of a compressible gas at high velocities, *Tr. TsAGI*, No.321 1937.

[61] Van Driest, E. R., Turbulent boundary layer in compressible fluids, *J. Aeron. Sci.*, 18(3), 1951, Turbulent boundary layer at different Prandtl numbers, in *Problema pogranichnogo sloya i voprosy teploperedachi* (*The Problem of a Boundary Layer and Heat Transfer*), Energoizdat, Moscow, 1960, 216–229. In the latter paper an approximate account is taken of the effect of the recovery coefficient.

[62] Lapin, Yu. V. and Loitsyanskiy, L.G., Use of the von Kármán method for calculation of the turbulent boundary layer on a plate in a gas flow, *Tr. LPI*, No.217, 1961.

ter 13 for an incompressible fluid makes it possible to generalize the von Kármán theory of the plate resistance in an incompressible fluid to the case of a gas flow with high Mach numbers.

We shall restrict ourselves to the case when the Prandtl number is equal to unity ($\sigma = 1$) and consider a flow past a plate both as thermally insulated ("plate thermometer") and with heat transfer.

Let the technique, which was used in Chapter 13 for the case of an incompressible fluid, be applied to the problem of a turbulent boundary layer on a plate in a gas flow. In the notation of this section

$$\varphi = \frac{u}{v_*} \qquad h = \frac{U_\infty}{v_*} \qquad \frac{u}{U_\infty} = \frac{\varphi}{h} \qquad \eta = \frac{v_* y}{\nu_w}$$

The Crocco equation (85) can be easily transformed to

$$\frac{T}{T_w} = 1 - \omega\left(\frac{\varphi}{h}\right) - \gamma\left(\frac{\varphi}{h}\right)^2 \tag{243}$$

where

$$\omega = 1 - \frac{T_t}{T_w} \quad (\omega > 0 \text{ for surface heating and } \omega < 0 \text{ for cooling})$$

$$\gamma = \frac{k-1}{2}\mathbf{M}_\infty^2 \frac{T_\infty}{T_w}$$

$T_t = T_\infty\left(1 + \frac{k-1}{2}\mathbf{M}_\infty^2\right)$ is the thermometer temperature which is equal to that of plate surface T_w in the absence of heat transfer ($\omega = 0$).

Noting that with constant pressure in the layer section the equality

$$\frac{\rho}{\rho_w} = \frac{T_w}{T} = \frac{1}{1 - \omega\left(\frac{\varphi}{h}\right) - \gamma\left(\frac{\varphi}{h}\right)^2} \tag{244}$$

takes place, we shall compose the main von Kármán relation (Equation 185, Chapter 13), which in the presence of medium compressibility will have the form

$$\frac{\varphi''}{\varphi'^2} = \frac{-æ}{\sqrt{1 - \omega\left(\frac{\varphi}{h}\right) - \gamma\left(\frac{\varphi}{h}\right)^2}} \tag{245}$$

Replacing, as in Chapter 13, the argument and the function, we obtain instead of Equation 245

$$\frac{\ddot{\eta}}{\dot{\eta}} = \frac{æ}{\sqrt{1 - \omega\left(\frac{\varphi}{h}\right) - \gamma\left(\frac{\varphi}{h}\right)^2}} \tag{246}$$

the first integral of which is the equality

$$\dot{\eta} = C\exp\left(\frac{æh}{\sqrt{\gamma}}\arcsin\frac{\sqrt{\gamma}\dfrac{\varphi}{h}+\dfrac{\omega}{2\sqrt{\gamma}}}{\sqrt{1+\dfrac{\omega^2}{4\gamma}}}\right) \tag{247}$$

To determine the constant C we have to prescribe the value of the derivative $d\varphi/d\eta$ at the boundary of a laminar sublayer from the side of a turbulent core. The simplest assumption is the requirement for this quantity to have the same value $f = 1/(æ\alpha)$ as in the case of small velocities, i.e., for the derivative $\varphi'(\alpha+0)$ to be independent of the coefficients γ and ω. This assumption is equivalent to the disregard of the change in the density along the laminar sublayer; it leads to the following result ($\bar{u} = \varphi/h$):

$$\dot{\eta} = \frac{1}{f}\exp\left[\frac{æh}{\sqrt{\gamma}}\left(\arcsin\frac{\sqrt{\gamma}\,\bar{u}+\dfrac{\omega}{2\sqrt{\gamma}}}{\sqrt{1+\dfrac{\omega^2}{4\gamma}}}-\arcsin\frac{\sqrt{\gamma}\dfrac{\alpha}{h}+\dfrac{\omega}{2\sqrt{\gamma}}}{\sqrt{1+\dfrac{\omega^2}{4\gamma}}}\right)\right] \tag{248}$$

It can be easily seen that at $\omega = \gamma = 0$ the right side of this equation coincides with the corresponding expression (187) for an incompressible fluid (Chapter 13).

Express the displacement thicknesses and momentum losses of a boundary layer in the case of gas by the equations

$$\delta^* = \int_0^\infty \frac{\rho}{\rho_\infty}\left(1-\frac{u}{U_\infty}\right)dy \qquad \delta^{**} = \int_0^\infty \frac{\rho u}{\rho_\infty U_\infty}\left(1-\frac{u}{U_\infty}\right)dy \tag{249}$$

Using Equations 244 and 248 and making transformations similar to those in Chapter 13, we obtain the following expression for the momentum loss thickness

$$\delta^{**} = \frac{(1-\omega-\gamma)\,h^2\nu_w}{fU_\infty}\int_0^1 \frac{\bar{u}(1-\bar{u})}{1-\omega\bar{u}-\gamma\bar{u}^2}\times$$

$$\times\exp\left[\frac{æh}{\sqrt{\gamma}}\left(\arcsin\frac{\sqrt{\gamma}\,\bar{u}+\dfrac{\omega}{2\sqrt{\gamma}}}{\sqrt{1+\dfrac{\omega^2}{4\gamma}}}-\arcsin\frac{\sqrt{\gamma}\dfrac{\alpha}{h}+\dfrac{\omega}{2\sqrt{\gamma}}}{\sqrt{1+\dfrac{\omega^2}{4\gamma}}}\right)\right]d\bar{u} \tag{250}$$

Introducing a new variable

$$\psi = \frac{1}{\sqrt{\gamma}}\arcsin\frac{\sqrt{\gamma}\,\bar{u}+\dfrac{\omega}{2\sqrt{\gamma}}}{\sqrt{1+\dfrac{\omega^2}{4\gamma}}} \tag{251}$$

change (250) over to the form

$$\delta^{**} = \frac{(1-\omega-\gamma)\,\nu_w h^2 \sqrt{1+\dfrac{\omega^2}{4\gamma}}}{fU_\infty \gamma} \int\limits_{\psi_w}^{\psi_\infty} \frac{\exp\,[\text{æ}h\,(\psi-\psi_B)\,]}{\cos\sqrt{\gamma}\,\psi} \times$$

$$\times\,(\sin\sqrt{\gamma}\,\psi - \sin\sqrt{\gamma}\,\psi_w)\,(\sin\sqrt{\gamma}\,\psi_\infty - \sin\sqrt{\gamma}\,\psi)\,d\psi \qquad (252)$$

where it is assumed that

$$\psi_\infty = \frac{1}{\sqrt{\gamma}}\arcsin\frac{\sqrt{\gamma}+\dfrac{\omega}{2\sqrt{\gamma}}}{\sqrt{1+\dfrac{\omega^2}{4\gamma}}} \qquad \psi_w = \frac{1}{\sqrt{\gamma}}\arcsin\frac{\dfrac{\omega}{2\sqrt{\gamma}}}{\sqrt{1+\dfrac{\omega^2}{4\gamma}}}$$

$$\psi_B = \frac{1}{\sqrt{\gamma}}\arcsin\frac{\sqrt{\gamma}\,\dfrac{\alpha}{h}+\dfrac{\omega}{2\sqrt{\gamma}}}{\sqrt{1+\dfrac{\omega^2}{4\gamma}}} \qquad (253)$$

To calculate a certain integral on the right side of Equation 252, we again apply the asymptotic expansion (Equation 189 of Chapter 13). Making integration with the accuracy up to the term involving æ^3h^3 in the denominator and presenting ψ_B approximately in the form

$$\psi_B = \psi_w + \frac{\alpha}{h} \qquad (254)$$

we obtain the following expression for the momentum loss thickness

$$\delta^{**} = \frac{e^{-\text{æ}\alpha}\,(1-\omega-\gamma)\,\nu_w}{f\text{æ}^2U_\infty}\,e^{\text{æ}h\,(\psi_\infty-\psi_w)}\left(1-\frac{2-1.5\omega-\gamma}{\text{æ}h\sqrt{1-\omega-\gamma}}\right) \qquad (255)$$

By a similar technique the expression can be obtained for the displacement thickness (249)

$$\delta^{*} = \frac{e^{-\text{æ}\alpha}\,(1+\gamma)\,\nu_w}{f\text{æ}^2U_\infty}\,e^{\text{æ}h\,(\psi_\infty-\psi_w)}\left[1-\frac{\gamma\left(3-1.5\omega-\gamma+\dfrac{\omega}{2\gamma}\right)}{\text{æ}h\,(1+\gamma)\sqrt{1-\omega-\gamma}}\right] \qquad (256)$$

Constructing, according to the last two equations, the relation $\delta^*/\delta^{**} = H$, we find

$$H = \frac{(1+\gamma)\left[1-\dfrac{\gamma\left(3-1.5\omega-\gamma-\dfrac{\omega}{2\gamma}\right)}{\text{æ}h\,(1+\gamma)\sqrt{1-\omega-\gamma}}\right]}{(1-\omega-\gamma)\left[1-\dfrac{2-1.5\omega-\gamma}{\text{æ}h\sqrt{1-\omega-\gamma}}\right]} \qquad (257)$$

In the limiting case of the flow of an incompressible gas ($\gamma = 0$) and of the absence of heat transfer ($\omega = 0$) we obtain the relation for H which completely coincides with the corresponding equality in Chapter 13.

Having taken a rougher approximation for the momentum loss thickness (neglecting the second term within the last round bracket of Equation 255), we substitute its expression

$$\delta^{**} = \frac{e^{-æ\alpha}\nu_w(1-\omega-\gamma)}{fæ^2 U_\infty} e^{æh(\psi_\infty - \psi_w)} \tag{258}$$

into the integral equation of momentum. After simple transformations we have

$$\frac{e^{-æ\alpha}}{fæ^2} h^2 d[e^{æh(\psi_\infty - \psi_w)}] = d\mathbf{Re}_{xw} \qquad \mathbf{Re}_{xw} = \frac{U_\infty x}{\nu_w} \tag{259}$$

Taking the integral of both sides of this equation and using the boundary condition $\mathbf{Re}_{xw} = 0$ at $h = 0$, we obtain in the same approximation

$$\frac{e^{-æ\alpha}}{fæ^2} h^2 e^{æh(\psi_\infty - \psi_w)} = \mathbf{Re}_{xw} \tag{260}$$

Within the limit Equation 260 changes over to the corresponding equation for an incompressible fluid (subscript 0)

$$\frac{e^{-æ\alpha}}{fæ^2} h_0^2 e^{æh_0} = \mathbf{Re}_{x0} \tag{261}$$

Further, divide both sides of Equation 260 by both sides of Equation 261, correspondingly; then we have

$$\left(\frac{h}{h_0}\right)^2 e^{æh(\psi_\infty - \psi_w) - æh_0} = \frac{\rho_w \mu_\infty}{\rho_\infty \mu_w} \tag{262}$$

Passing over in the last equality from h to c_f and taking logs, we obtain

$$-\lg\frac{c_f}{c_{f0}} + \frac{æ\sqrt{2}\lg e}{\sqrt{c_{f0}}}\left[\frac{(\psi_\infty - \psi_w)\sqrt{1-\omega-\gamma}}{\sqrt{\frac{c_f}{c_{f0}}}} - 1\right] = \lg\frac{\mu_\infty}{\mu_w} \tag{263}$$

Taking for c_{f0} the power-law dependence on $\mathbf{Re}_{x0}$

$$c_{f0} = 0.0263\mathbf{Re}_{x0}^{-1/7} \tag{264}$$

and introducing the notations

$$F = \frac{0.25}{\sqrt{c_{f0}}} = 1.55\mathbf{Re}_{x0}^{1/14} \qquad G = \lg\frac{\mu_\infty}{\mu_w}$$

$$K = \sqrt{\frac{1-\omega-\gamma}{\gamma}}\left(\arcsin\frac{\sqrt{\gamma} + \frac{\omega}{2\sqrt{\gamma}}}{\sqrt{1+\frac{\omega^2}{4\gamma}}} - \arcsin\frac{\frac{\omega}{2\sqrt{\gamma}}}{\sqrt{1+\frac{\omega^2}{4\gamma}}}\right) \tag{265}$$

yields the dependence of the friction coefficient on the free stream parameters and conditions on the wall in the following implicit form

$$-\lg\frac{c_f}{c_{f0}} + FK\sqrt{\frac{c_{f0}}{c_f}} = F + G \tag{266}$$

This equation can be reduced to the transcendental equation with one parameter

$$\lg N + N = L \tag{267}$$

if it is assumed that

$$L = \lg\frac{FK}{2} + \frac{F+G}{2} \qquad N = \frac{FK}{2\sqrt{c_f/c_{f0}}} \tag{268}$$

Thus, the expression for the friction coefficient will be written in the form

$$\frac{c_f}{c_{f0}} = \frac{F^2K^2}{4N^2} \tag{269}$$

where F and K are the known functions and N is found from the solution of Equation 267.

As is shown by calculations, Equation 267 can be substituted by an approximate simpler equality

$$N = 0.123 + 0.820L \tag{270}$$

which is rather accurate within a wide range of variation of all parameters. In this case Equation 269 takes on the form

$$\frac{c_f}{c_{f0}} = \frac{F^2K^2}{4\left[0.123 + 0.820\left(\lg\frac{FK}{2} + \frac{F+G}{2}\right)\right]^2} \tag{271}$$

To find G one should specify the law of viscosity change with temperature. Either the Sutherland equation or power-law dependence of viscosity on temperature can be taken as this law. In the latter case the expression for G will be

$$G = n\lg\frac{T_\infty}{T_w}$$

where n is the exponent in the law $\mu \sim T^n$. It is interesting to compare the technique presented with the van Driest method based on the transition of the Prandtl equation to gas dynamics. After the reduction of the van Driest equation to the form of Equation 266 it becomes clear that the difference is in the quantity G, which, with the power-law change of viscosity being adopted, is expressed by the equality

$$G = \left(n + \frac{1}{2}\right)\lg\frac{T_\infty}{T_w}$$

When considering Equation 271, we find that at high Reynolds numbers the magnitude of the relation c_f/c_{f0} weakly depends on $\mathbf{Re}_{x0}$. Actually, tending the Reynolds number to

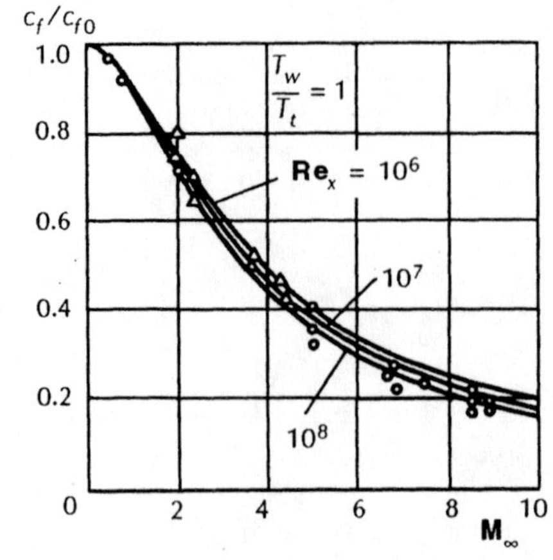

Fig. 293

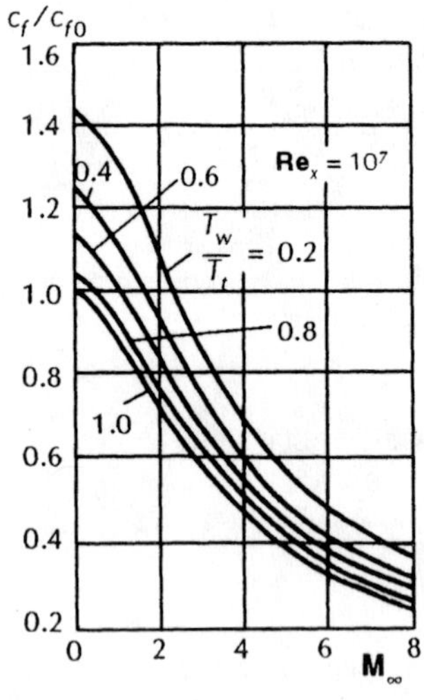

Fig. 294

infinity ($\mathbf{Re}_{x0} \to \infty$) or F to infinity, we obtain the following limit equation for the relation c_f/c_{f0}

$$\frac{c_f}{c_{f0}} = K^2$$

This formula may be useful for approximate calculations of friction resistance at high Reynolds numbers.

The local coefficient of friction was calculated by the method stated. The results of calculation for the case of the absence of heat transfer ($T_w/T_t = 1$) and comparison with experimental data of different authors are shown in Figure 293. Three curves in this figure correspond to three different values of the Reynolds number, 10^6, 10^7, and 10^8.

As follows from Figure 294, the calculation and experimental data are in good agreement up to high $\mathbf{M}_\infty$ numbers. The influence of the temperature factor on the coefficient of friction are shown in Figure 294.

Let us consider two specific cases which are of practical importance.

1. Thermally insulated plate. In this case the wall temperature T_w is equal to the stagnation temperature

$$T_w = T_\infty\left(1 + \frac{k-1}{2}\mathbf{M}_\infty^2\right) \tag{272}$$

and the quantities ω and γ are, correspondingly,

$$\omega = 0 \qquad \gamma = \frac{\frac{k-1}{2}\mathbf{M}_\infty^2}{1 + \frac{k-1}{2}\mathbf{M}_\infty^2} \tag{273}$$

The function K, which determines the process will have the form

$$K = \sqrt{\frac{1-\gamma}{\gamma}} \arcsin\sqrt{\gamma} \tag{274}$$

2. Plate heated or cooled at small velocities. In this case, passing over to the limit at $\gamma = 0$, we obtain

$$K = \frac{2(\sqrt{1-\omega} + \omega - 1)}{\omega} \tag{275}$$

In the limiting case of an incompressible fluid flow in the absence of heat transfer ($\omega = \gamma = 0$) $K = 1$.

The solution given is among semiempirical ones. Recall that the von Kármán equation used as a basis for the derivation of Equation 245 becomes incorrect near the outer edge of the boundary layer, where all consecutive derivatives of the averaged velocity along the coordinate normal to the plate surface tend to zero. An assumption on the constancy of the shear stress over the entire boundary layer region is incorrect. As it follows directly from the Reynolds equations for the turbulent boundary layer

$$\rho u \frac{\partial u}{\partial x} + \rho v \frac{\partial u}{\partial y} = -\frac{dp}{dx} + \frac{\partial \tau}{\partial y} \qquad \frac{\partial(\rho u)}{\partial x} + \frac{\partial(\rho v)}{\partial y} = 0 \tag{276}$$

under the condition $dp/dx = 0$ corresponding to the considered case of a flow parallel to the plate, it will be

$$\left(\frac{\partial \tau}{\partial y}\right)_{y=0} = 0 \tag{277}$$

Differentiating both sides of the first equation in system (276) with respect to y and expanding the derivatives in the second equation we shall satisfy ourselves that at $y = 0$ the system of equations

$$\left[\rho\left(\frac{\partial u}{\partial x}\frac{\partial u}{\partial y} + \frac{\partial v}{\partial y}\frac{\partial u}{\partial y}\right)\right]_{y=0} = \left(\frac{\partial^2 \tau}{\partial y^2}\right)_{y=0}$$

$$\left(\rho\frac{\partial u}{\partial x} + \rho\frac{\partial v}{\partial y}\right)_{y=0} = 0$$

will take place, from which it follows directly that

$$\left(\frac{\partial^2 \tau}{\partial y^2}\right)_{y=0} = 0 \tag{278}$$

The presence of Equations 277 and 278 allows one to consider that at small distances from the plate surface it can be approximately assumed that $\tau = \text{const} = \tau_w$. This to a certain extent justifies the above assumption but, of course, limits its validity by a "wall region". The substantiation of the stated semiempirical method and its discussion can be found in more detail in Section 14 (pp. 135 to 161) of the monograph by Lapin.[63]

As follows from the comparison of the computational results with the experimental data

[63] Lapin, Yu. V., *Turbulentnyi pogranichnyi sloy v sverkhzvukovykh potokakh gaza* (*Turbulent Boundary Layer in Supersonic Gas Flows*), Nauka, Moscow, 1982.

the approximation taken describes rather well the influence of free stream Mach number on a characteristic of the boundary layer such as the ratio of the coefficient of skin friction in a high-speed gas flow to the corresponding value of the same coefficient in the absence of the compressibility effect ($\mathbf{M}_\infty = 0$). We shall emphasize that the errors, which appear when numerator and denominator are determined separately, may be to a certain extent concealed; this fact, apparently explains the good agreement between computational results and experimental data shown in Figure 293.

The problem of thermal characteristics of a boundary layer was left aside, though it can be easily studied thanks to Equation 243, whose validity is intuitively justified by the assumption of the equality of the Prandtl number to unity. The absence of this assumption substantially complicates the problem due to the neccesity of introducing the coefficient of recovery. Recall (Section 7) that in the case of a laminar boundary layer on a plate in a parallel gas flow with high supersonic velocities the coefficient of recovery $\vartheta_0(0)/4$ was determined by the analytical expression (71). There is no simple solution to this problem in the case of a turbulent boundary layer, and one is usually restricted to experimental data, according to which the coefficient of recovery lies within the range from 0.88 to 0.90 and is independent of the Reynolds and Mach numbers.

12. Turbulent Boundary Layer in a Gas Flow in the Presence of the Longitudinal Pressure Gradient

The problem of calculation of a turbulent boundary layer in a gas flow in the general case with longitudinal pressure gradient is very complicated. As compared to the problem for an incompressible fluid, the fact of density variability and origin of the new unknown temperature dependence affecting the physical "constants" – coefficients of viscosity and thermal conductivity – points to additional difficulties in the formulation and solution of this problem. If one could have doubts as to the completeness of the problem of calculation of a turbulent boundary layer in an incompressible fluid (Sections 6 to 8 of Chapter 14), then it becomes clear in the considered case of a high-speed gas flow that a similar problem is far from being solved. That is why we must restrict ourselves here to the indication of those ways by which the solution of this urgent problem is sought . With the practical aspect of possible engineering applications in mind, we shall dwell only on the methods based on the "two-layer scheme" (known from Sections 3 and 5 of Chapter 14) with flow division into the inner region subjected to the "wall law" and the outer one subjected to the "wake law". Further, we restrict ourselves to the case of a plane stationary near-wall turbulent boundary layer in a uniform gas flow in the absence of volumetric forces.

Let us recall that the derivation of general equations describing a turbulent flow started from the determination of the laws of velocity averaging and of other characteristics of the real flow with subsequent application of these laws to averaging the Navier-Stokes equations and with simplification of the obtained equations conforming to the problem of a turbulent boundary layer.

The equations of viscous gas dynamics were derived in Section 1. They were averaged by the ordinary technique of preliminary expansion of equalities in the flow to averaged and fluctuating terms with subsequent averaging of the presented equations of gas dynamics. We shall omit this derivation as well as further simplifications of equations by the method of a boundary layer.[63]

Note only that to simplify a final form of equations, instead of the direct averaging of real quantities φ, fluctuating in time, by the law (see Section 3 of Chapter 13)

$$\bar{\varphi} = \frac{1}{T}\int_{t-T/2}^{t+T/2} \varphi dt \qquad (279)$$

we shall make use of the more complicated "weighted in density" averaging (which appeared to simplify equations)

$$\tilde{\varphi} = \frac{\int_{t-T/2}^{t+T/2} \rho\varphi dt}{\int_{t-T/2}^{t+T/2} \rho dt} \qquad (280)$$

which according to Equation 279 also can be defined as the relation

$$\tilde{\varphi} = \frac{\overline{\rho\varphi}}{\bar{\rho}} \qquad (281)$$

Representing the real (oscillating) quantity φ as a sum of $\tilde{\varphi}$, averaged by Equations 280 and 281, and of the fluctuation that in contrast to the sum corresponding to averaging by (279) will be denoted by two primes φ'', we obtain

$$\varphi = \tilde{\varphi} + \varphi'' \qquad (282)$$

Note that averaging by (279) yields

$$\overline{\rho\varphi} = \overline{(\bar{\rho}+\rho')(\tilde{\varphi}+\varphi'')} = \overline{\bar{\rho}\tilde{\varphi}+\rho'\tilde{\varphi}+\rho\varphi''} = \bar{\rho}\tilde{\varphi}+\overline{\rho'\tilde{\varphi}}+\overline{\rho\varphi''}$$

with the second term on the right being $\overline{\rho'\tilde{\varphi}} = \overline{\rho'}\tilde{\varphi} = 0$, and by the definition of (281) we find that

$$\overline{\rho\varphi} = \bar{\rho}\tilde{\varphi}$$

Then the previous equality will pass over to

$$\overline{\rho\varphi''} = 0 \qquad (283)$$

which substitutes the equality $\overline{\varphi'} = 0$ inherent in averaging by (279), and is the most important for simplification of transfer equations. Note that for the oscillations φ'' the condition

$$\overline{\varphi''} \neq 0$$

takes place.

In the above notation, omitting the sign of averaging everywhere except for the terms with products of oscillations we shall have the following system of equations for a plane stationary turbulent boundary layer in a gas flow

$$\frac{\partial(\rho u)}{\partial x} + \frac{\partial(\rho v)}{\partial y} = 0$$

$$\rho u \frac{\partial u}{\partial x} + \rho v \frac{\partial u}{\partial y} = -\frac{dp}{dx} + \frac{\partial \tau}{\partial y}$$

$$\rho u \frac{\partial h_0}{\partial x} + \rho v \frac{\partial h_0}{\partial y} = \frac{\partial}{\partial y}(q + u\tau) \qquad (284)$$

where $h_0 = h + u^2/2$ is the total enthalpy, τ and q mean total skin friction and heat flux involving both laminar (molecular) and turbulent (molar) components and accounting for the interaction of molecular and molar exchanges.

The last equation in system (284) can be replaced by the equation relative to the enthalpy h

$$\rho u \frac{\partial h}{\partial x} + \rho v \frac{\partial h}{\partial y} = u\frac{dp}{dx} + \frac{\partial q}{\partial y} + \tau \frac{\partial u}{\partial y} \qquad (285)$$

Assuming that the external flow (subscript e) is isentropic and on the outer edge of the boundary layer the Bernoulli equation is valid (U is the velocity on the outer edge of the boundary layer, the subscript e is unnecessary)

$$\frac{dp}{dx} = -\rho_e U \frac{dU}{dx} \qquad (286)$$

we shall write the boundary conditions in an ordinary form

$$u = v = 0 \qquad h_0 = h_w \quad \text{at} \quad y = 0$$

$$u \to U \qquad h_0 \to h_e + \frac{1}{2}U^2 \quad \text{at} \quad y \to \infty \qquad (287)$$

To close the system of Equations 284 we use the following formulas of turbulent components of skin friction and of heat flux

$$\tau_t = -\rho\overline{u''v''} = \mu_t \frac{\partial u}{\partial y} \qquad (288)$$

$$q_t = -\rho\overline{v''h''} = \lambda_t \frac{\partial T}{\partial y} \qquad (289)$$

The first of these is the Boussinesq equation, the second is the Fourier equation.

We shall consider the inner and outer subregions of a turbulent boundary layer separately.

As it follows from the foregoing, semiempirical hypotheses of turbulence retain their applicability at Mach numbers slightly exceeding unity. We also assume that these hypotheses are independent of the choice of the technique of averaging, in other words, making use of the Prandtl theory of a "mixing length" both for turbulent skin friction τ_t and for turbulent heat flux q_t, we can assume

$$\mu_t = \rho l^2 \frac{\partial u}{\partial y} = \rho æ^2 y^2 [1 - e^{-yv_*/(vA)}]^2 \frac{\partial u}{\partial y}$$

$$\lambda_t = \rho c_p l_q^2 \frac{\partial u}{\partial y} = \rho c_p æ_q^2 y^2 \left[1 - e^{-yv_*/(\nu A_q)}\right]^2 \frac{\partial u}{\partial y} \tag{290}$$

In the last equality $æ_q$ and A_q are the constants characterizing a turbulent heat flux q_t differing from æ = 0.4 and A = 26. According to Meier and Rotta[64] the constants $æ_q$ and A_q are deter mined by the equalities

$$\left(\frac{æ}{æ_q}\right)^2 = 0.86 \qquad \frac{A_q}{A} = 1.25 \tag{291}$$

The quantity A for a gas flow is suggested to be found from the equations.[65]

$$A = 26\left(\frac{\rho}{\rho_w}\right)^{1/2}\left[1 - 11.8\left(\frac{\mu_w}{\mu_e}\right)\left(\frac{\rho_e}{\rho_w}\right)^2 \frac{\mu_e U}{\rho_e v_*^3}\frac{dU}{dx}\right]^{-1/2}$$

The problem of numerical calculation of the inner region of a turbulent boundary layer, based on the considerations given, requires verification by the comparison with test data which are scanty.

More complicated is the problem of outer region calculation where the use of the Clauser equation (even with modification of Lapin and Strelets) remains disputable and is awaiting new studies. The method of matching inner and outer regions is also vague.

The two-layer scheme was used in the fundamental work of Reews,[66] who abandoned an application of the Clausius hypothesis in the outer region, having replaced it by three integral relations: momentum, the first moment of velocity, and energy. The results of calculation by the Reews method showed good agreement with the experimental velocity profiles in the boundary layer cross-sections and with the distribution of local friction over the body surface. A thorough analysis of the state of the theory of a turbulent boundary layer in gas flows can be found in Section 1 of the review by A. S. Ginevskiy et al. (cited earlier). The section written by A. N. Sekundov presents the characteristic of the existing computational techniques which gives a rather obvious picture of the state of the art of this problem. We also suggest Section 5, dealing with the effect of gas compressibility on the processes of turbulent boundary layer development. The problem of the calculation of a turbulent boundary layer in a gas flow is far from being completed, thus, making it impossible to give a detailed account here.

---

[64] Meier, H. V. and Rotta, J. C., Experimental and Theoretical Investigations of Temperature Distribution in Supersonic Boundary Layer, AIAA Paper No.744, 1970.

[65] Cebeci, T. and Smith, A. M. 0., *Analysis of Turbulent Boundary Layers*, Academic Press, New York, 1974.

[66] Reews, B. L., Two-layer model of turbulent boundary layer, *AIAA J.*, 12(7), 932–939, 1971.

16

Some Numerical Methods for Solving Equations of Hydrogas Dynamics[1]

1. Introduction: Basic Concepts of Theory of Difference Schemes

The methods of approximate numerical solution of the equations of hydrogas dynamics have recently undergone intensive development. It is these methods that present now, along with a physical experiment, the main instruments for studying problems of fluid mechanics. To understand the reasons of swift expansion of numerical methods in the considered branch of mechanics, it is enough to pay attention to familiar specific features of the basic equations of motion of solid flowing media. The majority of hydroaerodynamics problems, interesting from the practical viewpoint, are described by the systems of nonlinear partial equations. One often has to set the boundary-value conditions for these equations on the *a priori* unknown limits. In the solution there may occur singularities and singular surfaces, discontinuities of gas dynamic functions and the equations may change their type in the calculational region (for example, when $\mathbf{M}$ becomes equal to unity). The circumstances mentioned nearly eliminate the possibility of analytical solutions. Therefore, it is quite natural that general ideas, relating to the search for approximate solutions of equations, have immediately found a rich soil in the problems of hydrogas dynamics. A wide distribution of numerical methods became possible due to the progress in the field of numerical technique and to the development of numerical methods themselves with the interrelation between these trends becoming more noticeable during recent years.

Numerical methods are widely applied to solve ordinary differential equations as well as one-dimensional integral and integro-differential ones to which separate problems in fluid mechanics are reduced. However, the most substantial contribution to hydrogas dynamics is connected with the application of numerical methods to direct integration of partial equations

[1] By request of the author, this chapter has been written by Dr. Yu P. Golovachyov and S.B. Koleshko of the Leningrad M.I. Kalinin Polytechnical Institute (Chair of Hydroaerodynamics).

describing motion, heat/mass transfer, and more complex physical phenomena in fluids. In a number of cases, due to the difficulties of representing actual conditions in physical experiments, numerical modeling is the only source of information on the flow parameters (motion of bodies with cosmic velocities, flows of aggressive media, etc.).

It should be emphasized that the wide application of numerical modeling did not deprecate traditional analytical approaches but somewhat changed their role. Thus, the asymptotic methods, being the means for studying limiting flow modes, give the information on the orders of magnitudes of unknown functions, scales of their variation in these or those portions of calculational regions. This information is useful for formulating problems of numerical modeling and realizing numerical methods. Analytical solutions which usually refer to simplified particular cases are very valuable as the "standards" for estimating the properties of calculational schemes and accuracy of numerical solutions.

Development of numerical methods aimed at solving the problems of fluid mechanics led to the appearance of a new scientific discipline – calculational hydrogas dynamics. Many of its propositions have been already clarified themselves and have become classical. Nevertheless, there are a number of unresolved problems in this field. Among the central ones, there is the problem of the adequacy of numerical results to the solution of the initial problem that is especial acute in those cases when in the equation small parameters are present at higher derivatives (e.g., 1/**Re**). Nonlinear analysis of stability of numerical schemes is another problem. Great attention has been paid to the means of processing and visual presentation of information obtained from the numerical solution of multidimensional problems.

In view of the incompleteness of the modern theory of numerical methods and limitations of computer resources, the development of numerical hydrogas dynamics is directly based on a numerical experiment, in the course of which elements of a numerical method, algorithms, and computer programs may be corrected. As a result of calculations there sometimes arises the necessity to change the form of a physicomathematical model of the phenomenon studied.

To solve the problems of fluid mechanics use is made of a number of various numerical methods. Among them are the methods of integral relations, finite differences, finite elements, spectral methods, methods of particles, and others. The methods of finite differences are the most wide spread; they differ according to a universal character: simplicity of realization and the possibility to obtain highly accurate results. In view of these reasons we shall restrict ourselves to the consideration of these methods. Of course, within the framework of this rather small chapter we can only give the main concepts of the methods and their applications, having referred the reader to special manuals and original papers in order to study these problems in detail.[2]

The present chapter, besides being a brief presentation of the concepts of the theory of difference schemes, also gives a description of some widespread methods of numerical inte-

---

[2]The literature on the method of finite differences, in general and as applied to the problems of fluid mechanics, is voluminous. We shall note only some monographs and teaching guides which possess numerous references to other sources: Samarskiy, A. A., *Teoriya raznostnykh skhem* (*The Theory of Difference Schemes*), Nauka, Moscow, 1983; Samarskiy, A. A. and Nikolayev, Ye. S., *Metody resheniya setochnykh uravneniy* (*Methods for Solving Grid Equations*), Nauka, Moscow, 1980; Marchuk, G. I., *Metody vychislitelnoy matematiki* (*Methods of Computational Mathematics*), Nauka, Moscow, 1980; Godunov, S. K. and Ryabenkiy, V. S., *Raznostnye skhemy: Vvedeniye v teoriyu* (*Difference Schemes: Introduction into the Theory*), Nauka, Moscow, 1973; Richtmeier, R. and Morton, K., *Raznostnye metody resheniya kraevykh zadach* (*Difference Techniques for Solving Boundary-Value Problems*), Mir Press, Moscow, 1972 (in Russian); Belotserkovskiy, O. M., *Chislennoye modelirovaniye v mekhanike sploshnykh sred* (*Numerical Modelling in Continuum Mechanics*), Nauka, Moscow, 1984; Kovenya, V. M. and Yanenko, N. N., *Metod rasschepleniya v zadachakh gasovoy dinamiki* (*Method of Splitting in Problems of Gas Dynamics*), Nauka, Novosibirsk, 1981; Godunov, S. K. et al., *Chislennoye resheniye mnogomernykh zadach gasovoy dinamiki* (*Numerical Solution of Multi-Dimensional Problems of Gas Dynamics*), Nauka, Moscow, 1976.

gration of the equations of nonviscous gas motions of boundary layers, and of the Navier-Stokes equations for incomnpressible fluids. Examples of numerical solution of the corresponding problems are presented.

We shall refer to the main concepts of the theory of difference schemes. The essence of finite-difference methods is as follows.

The region of continuous variation of the argument is substituted by a discrete one, i.e., by that consisting of separate points (nodes). The distances between the nodes are called mesh widths. A combination of nodes composes a grid which may be uniform (with mesh width constant in all directions) or nonuniform.

An approximate substitution (approximation) of differential equations, of boundary, initial, and other conditions is performed by their difference analogues. A combination of nodes, to which the quantities entering into separate approximating relations belong, form a difference model. The collection of thus-obtained difference relations gives a difference scheme. The content of the first two stages composes the process of digitization.

The solution of the corresponding difference problem – the grid function in the form of a uni- and multi-dimensional table – is taken as an approximate solution of the difference problem.

The difference scheme and its solution depend on the grid pitches as well as on the parameters.

Each of the above-mentioned stages may not be realized uniquely. The method of realization determines of the possibility itself of obtaining an approximate solution, its accuracy, and the efficiency of the corresponding algorithm.

We shall pause on the basic concepts and definitions of the theory of difference schemes, illustrating in a number of cases our presentation with simple examples.

1.1. Concerning the Construction of Difference Grids

When solving the problems of fluid flows in regions of a complex shape the coordinates that reflect the initial region in a physical space are often transformed onto the computational region of a standard shape (depending on the problem dimensions – a single section, rectangle, parallelepiped). A uniform grid is usually introduced in the computational region. As a result, construction and realization of the finite-difference scheme as well as the realization of boundary conditions are simplified because the region boundaries may coincide with coordinate lines or surfaces.

Transformation of the given physical region to a certain standard form may be carried out with the aid of different shapes, and consequently may differently correlate the points of initial and computational regions. Determining the coordinates of the physical region points which correspond to the nodes of a uniform grid in a space of new (transformed) variables is the problem of grid plotting.

It is clear that the transformation mentioned should, in any case, provide a mutual, unambiguous correlation between the coordinates and the nodes of grids in physical and computational regions. Additional demands to the grids obtained in a physical region as a result of transformation may be dictated by the limitations to the shape of meshes, nonuniformity of fields of gas dynamic functions, the form of boundary conditions, etc. We shall emphasize that application of an optimal difference grid, the nodes of which are located in a physical plane in conformity with specific features of the flow (for example, the nodes crowd together in the region of high gradients of gas dynamic quantities), substantially reduces consumption of electronic computer resources and sometimes determines the feasibility of numerical modeling.

A description of modern approaches to the construction of grids may be found in monographs[3] and reviews.[4] Not pausing any more on this problem we shall consider approximation of differential equations on uniform difference grids. It is assumed that as needed, one may perform the corresponding transformation of coordinates.

1.2. Difference Approximation

Several means of obtaining algebraic relations by which the initial differential problem is approximated. The most simple and illustrative approach lies in the fact that to each derivative there corresponds an analogue. At this stage use is often made of different polynomial approximations of functions and of other, more formal methods. Note, moreover, that rather than differential equations, integral forms of laws of conservation of physical quantities (Sections 1 to 4 of Chapter 4) written for the meshes of a difference grid may serve as initial objects of approximation.

First we shall consider the simplest problem of approximation of the derivative du/dx at some node $x_m = m\Delta x$ of a uniform difference grid with the mesh width Δx. Thus, the expressions

$$\frac{u_{m+1}-u_m}{\Delta x} \qquad \frac{u_m-u_{m-1}}{\Delta x} \qquad \frac{u_{m+1}-u_{m-1}}{2\Delta x}$$

(u_m is the value of the grid function at the node x_m) are clear since their relation with the definition of the derivative concept itself is obvious.

By the theorem on the average, these algebraic expressions give an accurate value of the continuous function derivative $u(x)$ at some inner point of the interval $[x_m, x_m + \Delta x]$, $[x_m - \Delta x, x_m]$, and $[x_m - \Delta x, x_m + \Delta x]$. It is natural to expect that they are a good approximation for $(du/dx)_{x=x_m}$ at finite but rather small values of steps Δx.

The above expressions differ both in adopted standards and values of error.

To determine the approximation error one uses the expansion of the function into a Taylor series. Taking into account that

$$u_{m\pm1} = u_m \pm \left(\frac{du}{dx}\right)_{x_m}\Delta x + \frac{1}{2}\left(\frac{d^2u}{dx^2}\right)_{x_m}\Delta x^2 \pm \dots \tag{1}$$

we obtain

$$\frac{u_{m+1}-u_m}{\Delta x} = \left(\frac{du}{dx}\right)_{x_m} + \frac{\Delta x}{2}\left(\frac{d^2u}{dx^2}\right)_{x_m} + O(\Delta x^2)$$

$$\frac{u_m-u_{m-1}}{\Delta x} = \left(\frac{du}{dx}\right)_{x_m} - \frac{\Delta x}{2}\left(\frac{d^2u}{dx^2}\right)_{x_m} + O(\Delta x^2)$$

$$\frac{u_{m+1}-u_{m-1}}{2\Delta x} = \left(\frac{du}{dx}\right)_{x_m} + \frac{\Delta x^2}{6}\left(\frac{d^3u}{dx^3}\right)_{x_m} + O(\Delta x^4) \tag{2}$$

---

[3]Godunov, S. K. et al., *Chislennoye resheniye mnogomernykh zadach gasovoy dinamiki* (*Numerical Solution of Multi-Dimensional Problems of Gas Dynamics*), Nauka, Moscow, 1976; Kovenya, V. M. and Yanenko, N. N., *Metod rasshchepleniya v zadachakh gazovoy dinamiki* (*Method of Splitting in Gas Dynamics Problems*), Nauka, Novosibirsk, 1981; Anderson, D. A., Tannehill, J. C., and Pletcher, R. H., *Computational Fluid Mechanics and Heat Transfer*, Hemisphere Washington, 1984.

[4]Tompson, J. E., A survey of dynamically-adaptive grids in the numerical solution of partial differential equations, *Appl. Numer. Math.*, 1(1), 3–27, 1985.

i.e., the first two approximating expressions have an error of the order of Δx and the third of the order Δx^2.

When constructing not so obvious approximating expressions use can be made of the method of uncertain coefficients. We approximate, for example, du/dx with the error of the order of Δx^2 in the node m on a three-point standard containing the nodes m, $m+1$, $m+2$. Having written the expansion into a Taylor series for u_{m+2}

$$u_{m+2} = u_m + \left(\frac{du}{dx}\right)_{x_m} \cdot 2\Delta x + \frac{1}{2}\left(\frac{d^2u}{dx^2}\right)_{x_m}(2\Delta x)^2 + \ldots \tag{3}$$

we shall compose a linear combination of equality (1) for $m+1$ and expression (3) multiplied by some coefficient c:

$$u_{m+1} + cu_{m+2} - (1+c)\,u_m = (1+2c)\left(\frac{du}{dx}\right)_{x_m}\Delta x + \frac{1}{2}(1+4c)\left(\frac{d^2u}{dx^2}\right)_{x_m}\Delta x^2 + O(\Delta x^3)$$

Assuming $c = -1/4$, thus providing vanishing of the second term on the right side, yields the equality

$$\left(\frac{du}{dx}\right)_{x_m} = \frac{-3u_m + 4u_{m+1} - u_{m+2}}{2\Delta x} + \frac{\Delta x^2}{3}\left(\frac{d^3u}{dx^3}\right)_{x_m} + O(\Delta x^3)$$

The second derivative in the node x_m is usually presented by the expression

$$\left(\frac{d^2u}{dx^2}\right)_{x_m} \sim \frac{1}{\Delta x}\left(\frac{u_{m+1} - u_m}{\Delta x} - \frac{u_m - u_{m-1}}{\Delta x}\right) = \frac{u_{m-1} - 2u_m + u_{m+1}}{\Delta x^2}$$

which has the second order of accuracy.

The above difference approximations of the first and second derivatives written on two- and three-point standards, have the accuracy of $O(\Delta x)$ or $O(\Delta x^2)$. An order of accuracy may be increased by extending a standard. Another way is in the use of the so-called compact approximations.

As an example we shall consider derivation of a symmetric approximating equation for du/dx on a three-point standard. For this we shall refer to the last of relations (2). Designating $P = du/dx$, we shall present the main term of the error in the form

$$\frac{\Delta x^2}{6}\left(\frac{d^3u}{dx^3}\right)_{x_m} = \frac{\Delta x^2}{6}\left(\frac{d^2P}{dx^2}\right)_{x_m} = \frac{\Delta x^2}{6}\left(\frac{P_{m-1} - 2P_m + P_{m+1}}{\Delta x^2} + O(\Delta x^2)\right)$$

As a result the relation mentioned will be reduced to the compact approximating formula accurate to $O(\Delta x^4)$:

$$\frac{P_{m-1} + 4P_m + P_{m+1}}{6} = \frac{u_{m+1} - u_{m-1}}{2\Delta x} \tag{4}$$

Similarly, a compact approximating equation may be derived for $Q = d^2u/dx^2$ that also has the accuracy $O(\Delta x^4)$:

$$\frac{Q_{m-1} + 10Q_m + Q_{m+1}}{12} = \frac{u_{m-1} - 2u_m + u_{m+1}}{\Delta x^2} \tag{5}$$

A characteristic feature of Equations 4 and 5 is in the fact that each of them contains in the standard nodes not one, but several values of unknown approximating quantities, i.e., the equation data are implicit. In view of this the transition to compact approximations noticeably complicates the structure of algebraic relations to which the initial problem will be reduced as a result of approximation. Equations 4 or 5, written for the inner nodes of the grid, form the system relative to P or Q that is solved together with finite-difference approximations of initial equations.

Note that equations of compact numerical differentiation may be obtained by other means,[5] for example, with the aid of splines, interpolation Hermite polynomials, etc.

The considered means for approximating the derivatives of the first and second orders make it possible to construct difference schemes for hydrogas dynamic equations by approximating separate terms of these equations. Another approach to the construction of difference schemes is in approximation of the entire equation, e. g., a widespread integro-interpolation method. We shall demonstrate the essence of this method in the example of the simple one-dimensional equation of the first order

$$\frac{du}{dx} = f(x) \tag{6}$$

where $f(x)$ is the known function. We shall integrate Equation 6 within the interval $x_{m-1} \leq x \leq x_{m+1}$

$$u_{m+1} - u_{m-1} = \int_{x_{m-1}}^{x_{m+1}} f(x)\, dx \tag{7}$$

and then approximately express the integral in terms of f_{m-1}, f_m, and f_{m+1} on the basis of the quadratic interpolation of the function $f(x)$

$$f(x) \sim f_{m-1} + \frac{-3f_{m+1} + 4f_m - f_{m-1}}{2}\xi + \frac{f_{m-1} - 2f_m + f_{m+1}}{2}\xi^2$$

where $\xi = (x - x_{m-1})/\Delta x$, which has the error

$$(-\xi^3 + 3\xi^2 - 2\xi)\left(\frac{d^3 f}{dx^3}\right)_{x_{m-1}} \cdot \frac{\Delta x^3}{6} + O(\Delta x^4)$$

Then

$$\int_{x_{m-1}}^{x_{m+1}} f(x)\, dx = \Delta x \int_0^2 f d\xi = \frac{\Delta x}{3}(f_{m-1} + 4f_m + f_{m+1}) + O(\Delta x^5)$$

Thus, after substitution into (7) and division by $2\Delta x$, leading to the compact equation

[5] Tolstykh, A. I., *Raznostnye metody povyshennoy tochnosti dlya zadach aerogidrodinamiki (Difference Methods of Improved Accuracy For the Problems of Aerohydrodynamics)*, MFTI Press, Moscow, 1985.

$$\frac{u_{m+1} - u_{m-1}}{2\Delta x} = \frac{f_{m-1} + 4f_m + f_{m+1}}{6}$$

which coincides, with allowance for the designation $P = du/dx = f$, with Equation 4.

If the initial differentiation problem involves not only the first but also the second derivatives, then the integro-interpolation method gives compact schemes with much more complicated structures of relations between unknown and known quantities. We shall refer to the model equation

$$\frac{d^2u}{dx^2} + a(x)\frac{du}{dx} = f(x) \tag{8}$$

Designating

$$\frac{du}{dx} = P \tag{9}$$

and integrating Equations 8 and 9 within the interval $x_{m-1} \le x \le x_{m+1}$, we shall come to the relations

$$P_{m+1} - P_{m-1} + \frac{\Delta x}{3}[(aP)_{m-1} + 4(aP)_m + (aP)_{m+1}] =$$

$$= \frac{\Delta x}{3}(f_{m-1} + 4f_m + f_{m+1}) + O(\Delta x^5) \tag{10}$$

$$u_{m+1} - u_{m-1} = \frac{\Delta x}{3}(P_{m-1} + 4P_m + P_{m+1}) + O(\Delta x^5) \tag{11}$$

It is evident that to derive a finite approximating equation which relates only the values of u and f in the grid nodes it is necessary to enlist additional relations, allowing one to eliminate P_{m-1}, P_m, and P_{m+1}. These relations may be obtained having integrated Equations 8 and 9 from x_{m-1} to x_m with the use of the same interpolations

$$P_m - P_{m-1} + \frac{\Delta x}{12}[5(aP)_{m-1} + 8(aP)_m - (aP)_{m+1}] =$$

$$= \frac{\Delta x}{12}(5f_{m-1} + 8f_m - f_{m+1}) + O(\Delta x^4) \tag{12}$$

$$u_m - u_{m-1} = \frac{\Delta x}{12}(5P_{m-1} + 8P_m - P_{m+1}) + O(\Delta x^4) \tag{13}$$

If the system (11) to (13) is now solved relative to P_{m-1}, P_m, and P_{m+1} and the relations obtained are substituted into the left side of (10), then we shall have as a result the scheme of the third order of accuracy on a three-point pattern (with allowance for dividing by Δx).

A more detailed presentation of the method as applied to the boundary layer equations is given in Section 3.

The presented examples, which refer to separate derivatives or ordinary differential equations, demonstrate a typical specific feature of the method of finite differences on the stage of approximation – a great number of possible types of approximating equations. When approximating partial differential equations a number of possible versions multiply. In what follows it is convenient to explain some specific features of difference schemes by means of the equations

$$\frac{\partial u}{\partial t} + a\frac{\partial u}{\partial x} = f(x, t) \tag{14}$$

and

$$\frac{\partial u}{\partial t} + a\frac{\partial u}{\partial x} - \nu\frac{\partial^2 u}{\partial x^2} = f(x, t) \tag{15}$$

where $a(x, t)$ and $f(x, t)$ are the known functions of the spatial coordinate x and time t, ν is the positive coefficient. The left side of Equation 14, as follows from Section 6 of Chapter 2, is the individual (Lagrangian) derivative of u on a particle motion with the velocity along the axis $0x$. The function $f(x, t)$ has a sense of the prescribed source of u variation. Equation 15 supplemented with the term $\nu\partial^2 u/\partial x^2$, which represents diffusion phenomena (diffusion proper, heat conduction, viscosity depending on the physical content of the function u) in a generalized form. The first equation type is hyperbolic and the second is parabolic.

The presented model equations are similar in type to the equations of unsteady unidimensional flow of ideal (Equation 14) and viscous (Equation 15) fluids, with the considerable simplification that the multiplier a in a convective term is adopted to be known *a priori* (sometimes even constant). The simplification mentioned, removing principal difficulties related to the nonlinearity of hydrodynamics equations, allows the utility of applying difference schemes to model equations from the viewpoint of using them to describe various properties adherent to convection and diffusion phenomena.

It is more worthy that model equations have known accurate analytical solutions which turn out to be very useful when studying approximate numerical solutions.

Let the Cauchy problem with the initial condition $u(x, 0) = u^{(0)}(x)$ be formulated for Equations 14 and 15 within the region $-\infty < x < \infty$, $0 < t \le T$. We shall introduce in this region a rectangular uniform grid with the mesh widths Δx and Δt and designate the values of the grid functions in the grid node $x_m = m\Delta x$, $t_n = n\Delta t$ as u_m^n, a_m^n, f_m^n. We shall assume that the derivative $\partial u/\partial t$ in the node x_m, t_n is approximated by the expression $(u_m^{n+1} - u_m^n)/\Delta t$. Then, to the differential equation (14) there may correspond the following algebraic relations

$$\frac{u_m^{n+1} - u_m^n}{\Delta t} + a_m^n \frac{u_m^n - u_{m-1}^n}{\Delta x} = f_m^n$$

$$\frac{u_m^{n+1} - u_m^n}{\Delta t} + a_m^n \frac{u_{m+1}^n - u_m^n}{\Delta x} = f_m^n$$

$$\frac{u_m^{n+1} - u_m^n}{\Delta t} + a_m^n \frac{u_{m+1}^n - u_{m-1}^n}{2\Delta x} = f_m^n \tag{16}$$

($m = 0, \pm 1, \pm 2, \ldots$; $n = 0, 1, \ldots, M-1$, where M is the smallest portion of the relation $T/\Delta t$), complemented by the initial condition $u_m^0 = u^{(0)}(x_m)$. The patterns corresponding to these schemes are shown in Figure 295.

For Equation 15, the term corresponding to the second derivative $\partial^2 u/\partial x^2$ will also enter into the left sides of approximating expressions. As a result the difference equations

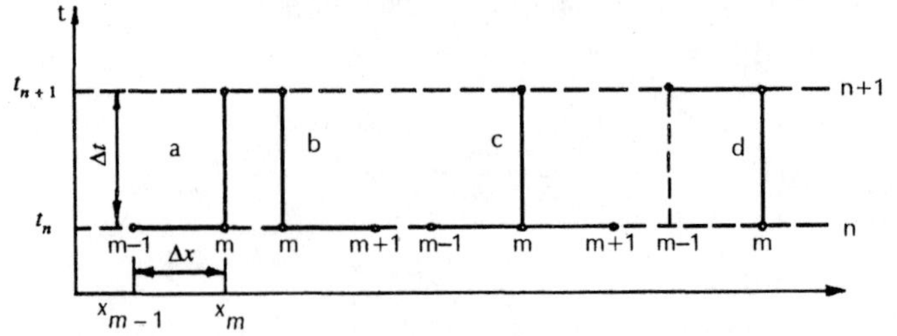

Fig. 295

$$\frac{u_m^{n+1} - u_m^n}{\Delta t} + a_m^n \frac{u_m^n - u_{m-1}^n}{\Delta x} - \nu \frac{u_{m-1}^n - 2u_m^n + u_{m+1}^n}{\Delta x^2} = f_m^n \tag{17}$$

are obtained.

It is obvious that a number of the versions of difference schemes can be easily increased using at least other variants of the approximation of the derivative $\partial u/\partial x$. Moreover, spatial derivatives may be referred not to the "lower" time layer t_n, but to the "upper" one t_{n+1} or to some intermediate by introducing the weighting coefficients α ($0 \le \alpha \le 1$): for example,

$$\frac{\partial u}{\partial x} \sim \alpha \frac{(u_m^n - u_{m-1}^n)}{\Delta x} + (1 - \alpha) \frac{(u_m^{n+1} - u_{m-1}^{n+1})}{\Delta x}$$

Such a variety of types of approximating relations makes a preliminary evaluation of difference schemes obligatory. With this purpose in mind we shall consider the approximation feature from more general viewpoints and also determine two other fundamental concepts: convergence and stability.

1.3. Convergence, Approximation Stability

It is adopted in the theory of difference schemes to briefly and generally write a mathematical formulation of a differential problem, with initial and boundary conditions included, in the form

$$Lu = f \tag{18}$$

where L is the differentiation operator, u and f are the unknown function of the continuous argument and the prescribed right side, correspondingly. The notation L represents those operations on the function u which are assigned by differential equations and additional conditions of the problem. In that part, which refers to the equations, the operator L for Equation 14 has the form $\partial/\partial t + a\,\partial/\partial x$, and for Equation 15 $\partial/\partial t + a\partial/\partial x - \nu\partial^2/\partial x^2$.

The corresponding difference problem is written in the form

$$L_h u^{(h)} = f^{(h)} \tag{19}$$

where L_h is the difference operator, $u^{(h)}$ is the grid function (the solution of Equation 19), $f^{(h)}$ is the prescribed right side, h is the mesh width of the grid (a combination of spatial-temporal steps). The concept of a difference operator unites all algebraic operations (which refer also to initial and boundary conditions), the aid of which the grid function $u^{(h)}$ is transformed to $f^{(h)}$.

The most natural demand on the approximate solution of $u^{(h)}$ is its convergence with u

as the mesh width h becomes smaller. However, a direct comparison between $u^{(h)}$ and u is prevented by the difference in their natures: one of the functions is determined at a discrete set of points, the other is continuous. In view of this, one has to introduce one more function $[u]_h$, a grid analogue of u. The rules, according to which $[u]_h$ is introduced, are set depending on the specific features of the function u. Thus, for the continuous u it is natural to determine $[u]_h$ as a table compiled from the values of u in the grid nodes ("projection" of u onto the grid), and for the discontinuous one – as a table involving the integral-mean values of u in the vicinity of each node. It is $[u]_h$ that is then considered as an accurate solution ("pattern table") with which the solution of Equation 19 compared.

One can speak of the convergence of the solution $u^{(h)}$ to the accurate solution $[u]_h$ with the kth order of accuracy with respect to the step h, if the norm[6] of the error $[u]_h - u^{(h)}$ of an approximate solution tends to zero for $h \to 0$, as the value of the order h^k

$$\left\| [u]_h - u^{(h)} \right\|_{h \to 0} \leq ch^k$$

As is seen the determination of convergence has an asymptotic character, whereas real calculations are conducted with small but finite steps. Because the value of the constant c depends on the type of the scheme, in some cases transition to the scheme of formally higher order may not give a desired effect at finite h.

As a rule, direct verification of convergence is impossible, especially the function u is unknown. Therefore, convergence is determined by two stages: verification of approximation and analysis of scheme stability.

Approximation is investigated in the following way. We shall assume that an accurate solution, i.e., the grid function $[u]_h$, is substituted into the difference scheme (as it will become clear later, to evaluate the approximation one need not possess this function). As a result of this substitution the discrepancy (error) of approximation is obtained

$$\delta f^{(h)} = L_h [u]_h - f^{(h)}$$

If with $h \to 0$ the discrepancy norm tends to zero, being in this case the value of the order of h^{k_1} then the scheme is called approximating the initial problem (18) on the solution u, with the order k_1 relative to the step h.

To practically investigate approximation it is enough to present the solution of the initial problem in the form of the expansion into the Taylor series in terms of the powers h near the corresponding grid nodes.

As an example we shall elucidate the character of the approximation of the differential equation (17) in the vicinity of the node x_m, t_n. Using the expansions

$$u(x_m, t_{n+1}) = u(x_m, t_n) + \Delta t\left(\frac{\partial u}{\partial t}\right)_{x_m, t_n} + \frac{\Delta t^2}{2}\left(\frac{\partial^2 u}{\partial t^2}\right)_{x_m, t_n} + \frac{\Delta t^3}{6}\left(\frac{\partial^3 u}{\partial t^3}\right)_{x_m, t_n} + \ldots$$

[6]By the norm $\|z^{(h)}\|$ of the grid function $z^{(h)}$ we shall mean a non-negative number characterizing the deviation of this function from zero at all of the points. Depending on the type of the problem use is made of various definitions of norms. Thus, the greatest by the modulus component of $z^{(h)}$ or the quadratic mean value of $z^{(h)}$ over all the grid nodes may be taken as the norm.

$$u(x_{m\pm1}, t_n) = u(x_m, t_n) \pm \Delta x\left(\frac{\partial u}{\partial x}\right)_{x_m, t_n} + \frac{\Delta x^2}{2}\left(\frac{\partial^2 u}{\partial x^2}\right)_{x_m, t_n} \pm \frac{\Delta x^3}{6}\left(\frac{\partial^3 u}{\partial x^3}\right)_{x_m, t_n} + \ldots \quad (20)$$

determining $[u]_h$ by the "design" technique and taking into account that $u(x,t)$ satisfies Equation 15, we shall obtain after substitution of (20) into scheme (17)

$$(\delta f^{(h)})_{x_m, t_n} = \frac{\Delta t}{2}\left(\frac{\partial^2 u}{\partial t^2}\right)_{x_m, t_n} - a_m^n \frac{\Delta x}{2}\left(\frac{\partial^2 u}{\partial x^2}\right)_{x_m, t_n} + O(\Delta t^2, \Delta x^2)$$

Hence, it is seen that local (in the node vicinity) approximation has the first order in both variables. It is also easy to find that the scheme with symmetric approximation of the derivative $\partial u/\partial x$ by the expression $(u_{m+1}^n - u_{m-1}^n)/(2\Delta x)$ gives the discrepancy of the order $O(\Delta t, \Delta x^2)$.

Verification of approximation is not usually difficult. In the case of uniform grids it is enough to convince oneself in the approximation at any arbitrary point.

It may seem that if the approximation error tends to zero with $h \to 0$, then it ensures the convergence of the solution $u^{(h)}$ of the corresponding difference problem to the accurate solution $[u]_h$. However, in a general case this statement is incorrect.

The difficulty is that the basic data of the problem (right sides, coefficients, and initial and boundary conditions) are prescribed with some error. For the same initial problem, depending on the structure of a difference scheme, initial errors may behave differently. In stable schemes the errors mentioned do not grow. Conversely, that are schemes unstable by the initial data increase random errors and in view of the avalanche-type character of this phenomenon they are not suitable for operation.

A stable scheme should possess the properties of unambiguous solvability and continuous dependence on the initial data. The second demand means that small variations in the problem conditions should result in small changes of the solution.

It may be assumed that for schemes written in the form $L_h u^{(h)} = f^{(h)}$, which involves, initial and boundary conditions of the problem, the values of initial data are determined by the right side $f^{(h)}$. The condition of continuous dependence of the solution $u^{(h)}$ on the initial data $u^{(h)}$ for linear difference problems means that $u^{(h)}$ should be on the same order as $f^{(h)}$, i.e.,

$$\|u^{(h)}\| \le c\|f^{(h)}\|$$

This inequality should be valid for any right side f with c being independent of h.

A fundamental statement that approximation and stability result in convergence, with the order of the convergence accuracy k being equal to the order of approximation k_1, is valid.

We shall emphasize that if approximation establishes the agreement between the difference scheme and initial differential equations, then stability is a purely inner property of a difference scheme not connected with its origin.

The methods developed up to the present, and strictly substantiated for investigating stability, belong to linear problems. As to nonlinear problems, for them *a priori* estimation of stability substituted by the corresponding one for linear analogues of these problems is of course of limited value. Authentic determination of stability boundaries for real nonlinear problems is possible only with the help of numerical experiments.

Of the difference schemes for solving a vast class of the so-called evolution problems described by hyperbolic and parabolic equations, the most widely used is the spectral method of stability investigation. The procedure of solving the corresponding difference problems may be considered a successive application of the operator of transition on a new layer to the solution on each previous layer. In the simplest case this succession of actions may be presented by:

$$u^{n+1} = R_h u^n \qquad u^{n+2} = R_h u^{n+1} = R_h\left(R_h u^n\right) = R_h^2 u^n, \quad \ldots$$

$$u^{n+N} \equiv R_h^N u^n$$

For stability of difference schemes it is necessary that the norms of the powers of the transition operator R_h be limited. We shall take the spectral radius of transition operator, i.e., its greatest-in-magnitude eigenvalue as the norm of the transition operator. Then, the necessary condition of stability is formulated as: the spectral radius $\rho(R_h)$ of the transition operator should not exceed unity. (A detailed analysis leads to a weaker condition $\rho \le 1 + O(\Delta t)$, where Δt is the mesh width of a grid along the evolution coordinate: however, in practice, one usually restricts oneself to the condition $\rho \le 1$.)

For linear difference operators with constant coefficients the harmonic functions (harmonics) serve as eigenfunctions. In the case of two continuous arguments x, t (t is the evolution coordinate) the harmonic has the form

$$u = e^{\omega t + ikx}$$

where ω is the frequency, k is the wave number, and $i^2 = -1$. For the functions of a discrete argument that are determined on the grid $x_m = m\Delta x$, $t_n = n\Delta t$, the corresponding harmonic is

$$u_m^n = e^{\omega n\Delta t + ikm\Delta x} = \lambda^n e^{i\alpha m} \tag{21}$$

where $\lambda = e^{\omega\Delta t}$ is the transition multiplier and $\alpha = k\Delta x$ is the phase angle. The stability condition takes on the form $|\lambda| \le 1$.

It should be valid for all harmonics with phase angles $0 \le \alpha \le 2\pi$ (the value $\alpha = 2\pi$ corresponds to the most short-wave harmonic resolved on the grid with the mesh width Δx).

The stability condition is verified by the analysis of the characteristic equation which results from substitution of harmonic (21) into the difference scheme.

Thus, for the scheme with the "back" difference over x

$$\frac{u_m^{n+1} - u_m^n}{\Delta t} + a\frac{u_m^n - u_{m-1}^n}{\Delta x} = 0 \qquad a = \text{const} > 0 \tag{22}$$

(the first of schemes (16) with the right side f omitted) the characteristic equation has the form

$$\lambda = 1 - æ + æ e^{-i\alpha} \qquad æ = a\Delta t/\Delta x > 0$$

The scheme is stable when $æ \le 1$. Such a stability is called conditional because it takes place when a certain condition of the inequality type is fulfilled.

For an analogous scheme with the "forward" difference over

$$\frac{u_m^{n+1} - u_m^n}{\Delta t} + a\frac{u_{m+1}^n - u_m^n}{\Delta x} = 0 \qquad a = \text{const} > 0 \tag{23}$$

$\lambda = 1 + æ - æ e^{i\alpha}$ and the scheme is unconditionally (i.e., at any ratios between Δt and Δx)

unstable.

The presented examples show that changes in the approximation type may result in the scheme instability and, consequently, in its inefficiency.

The above spectral method for studying stability is applicable to homogeneous difference equations with constant coefficients. On its application to the estimation of stability of difference schemes for solving gas dynamics problems, preliminary linearization of equations and "freezing" of the coefficients of these equations at the values equal to their local values at the point under consideration.

1.4. Amplitude and Phase Errors

The analysis given above provides a guide to the amplitude and phase errors of an approximate solution, i.e., to the effects of scheme dissipation and dispersion. We shall consider this problem of the example of scheme (22) for Equation 14 with $f = 0$.

We shall compare the changes experienced per time step Δt by the amplitude and phase of a harmonic with the wavenumber k in an accurate solution and in the difference scheme mentioned. We shall first consider an accurate solution. Substituting $u = e^{\omega t + ikx}$ into differential equation (14) yields the dispersion relation

$$\omega = -iak$$

where

$$u = e^{ik(x-at)}$$

The multiplier of the transition from the layer n to the layer $n + 1$ has the form

$$\lambda_e = \frac{u(x, t+\Delta t)}{u(x, t)} = e^{-ika\Delta t} = e^{i\Delta\varphi_e} \tag{24}$$

where $\Delta\varphi_e = -ka\Delta t = -æ\alpha$, $æ = a\Delta t/\Delta x$, $\alpha = k\Delta x$.

It is seen from expression (24) that in the accurate solution the harmonic amplitude does not change ($|\lambda_e| = 1$) and the phase changes by the value $\Delta\varphi_e = -æ\alpha$ during the time Δt.

Now we shall refer to the difference scheme (22). For this scheme, the transition multiplier is determined by the relation

$$\lambda(\alpha) = 1 - æ + æe^{-i\alpha} = 1 - æ + æ\cos\alpha - iæ\sin\alpha$$

so that

$$|\lambda|^2 = 1 - 2æ(1-æ)(1-\cos\alpha) \tag{25}$$

It is seen from (25) that in the difference scheme the amplitudes of all harmonics (with those corresponding to $\alpha = 0$ and 2π excluded) decrease when, $æ < 1$ since $|\lambda| < 1$. The dependence of the amplitude error on the phase angle α is shown in Figure 296.

Now we shall evaluate the phase error of the solution in the difference scheme considered. We shall present the transition multiplier in the form $\lambda(\alpha) = |\lambda|e^{i\Delta\varphi}$, where $\Delta\varphi$ is the shift in the phase solution by the step Δt. We shall determine this value approximately, assuming Δt, and consequently $\Delta\varphi$ small:

$$\Delta\varphi \approx \sin\Delta\varphi = \frac{\operatorname{Im}\lambda}{|\lambda|} = \frac{-æ\sin\alpha}{\sqrt{1-2æ(1-æ)(1-\cos\alpha)}}$$

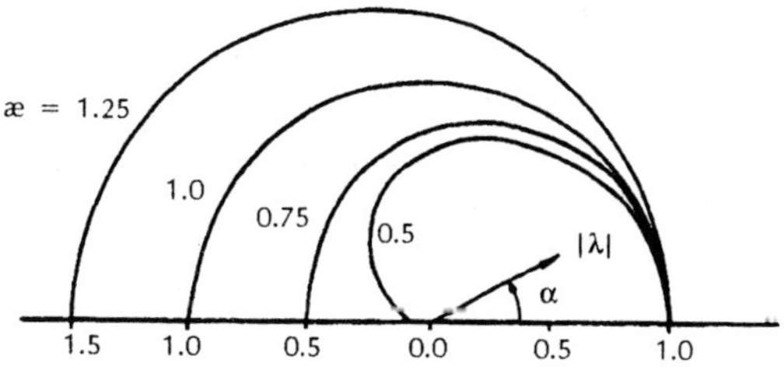

Fig. 296

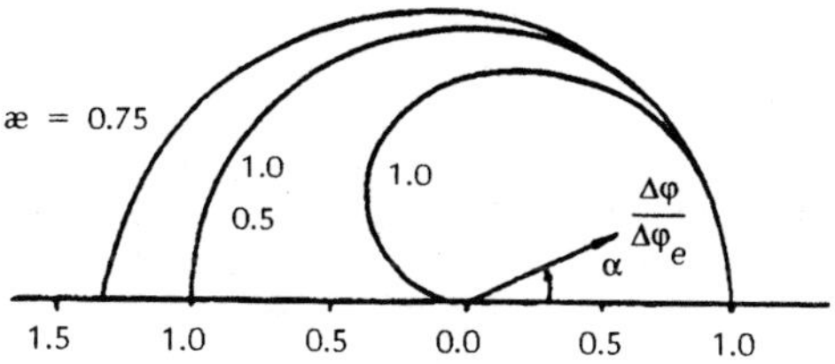

Fig. 297

Comparing the result obtained with the phase variation in an accurate solution $\Delta\varphi_e = -æ\alpha$, the expression of the phase error $\Delta\varphi/\Delta\varphi_e$ of an approximate solution for the difference scheme (22) may be obtained (Figure 297).

The existence of phase errors may, in the case of multiple transition to subsequent time layers, noticeably deform spatial distribution of an approximate solution because its different Fourier components propagate over the grid with different velocities. However, this effect is of no importance if it is necessary to find a stationary (developed) solution of the problem.

1.5. Differential (Approximation of the Difference Scheme) Scheme Viscosity

To study the properties of difference scheme use is widely made, besides spectral analysis, of the method of differential approximation. If the expansion of the continuous argument into the Taylor series is substituted into the difference scheme (assuming that this function coincides with the solution $u^{(h)}$ in the grid nodes), and the grid step is allowed to tend to zero, then for the approximate scheme all the terms containing the grid step will vanish in the limit, and the remaining form is the initial differential equation. If after the substitution mentioned we reduce the main terms of approximation error (i.e., the terms with minimum powers of grid meshes), then we obtain the equation which is called the first differential approximation of the difference scheme. On retaining the terms with subsequent powers of grid, the combination of all of the retained terms gives the second differential approximation, etc.

Thus, differential approximations of difference schemes are the differential equations which involve both the terms of the initial equations and additional terms in the form of the products of derivatives by definite powers of grid meshes. By the properties of the solutions of these differential equations one can judge the features of the difference scheme. Though the last argument is proved only for some classes of problems,[7] the thus-obtained informa-

---

[7] Shokin, Yu. N., *Method differentialnogo priblizheniya* (*Method of Differential Approximation*), Nauka, Novosibirsk, 1981.

tion on the properties of difference schemes is usually in conformity with the results of numerical experiments.

As an example we shall consider the first differential approximation of scheme (22). Having carried out the above procedure, we shall obtain

$$\frac{\partial u}{\partial t} + a\frac{\partial u}{\partial x} + \frac{\Delta t}{2}\frac{\partial^2 u}{\partial t^2} - \frac{a\Delta x}{2}\frac{\partial^2 u}{\partial x^2} = 0 \qquad (26)$$

The hyperbolic equation (26) is called the first differential approximation in the hyperbolic form for the considered difference scheme. More convenient for analysis is the first differential approximation in a parabolic form obtained as a result of the elimination of the term with the second derivative with respect to time from Equation 26. Having differentiated (26) over t and assuming that Δt is proportional to Δx, we obtain with the accuracy of the quantities of the order of Δx: $\partial^2 u/\partial t^2 = a^2 \partial^2 u/\partial x^2$.

Then Equation 26 takes on the form:

$$\frac{\partial u}{\partial t} + a\frac{\partial u}{\partial x} = \nu_e \frac{\partial^2 u}{\partial x^2} \qquad (27)$$

where $\nu_e = a\Delta x(1 - æ)/2$, $æ = a\Delta t/\Delta x > 0$.

It is seen that the first differential approximation differs from the initial equation by the term $\nu_e \partial^2 u/\partial x^2$ on the right side. It is natural to give the name of the coefficient of schematic (grid) viscosity to the multiplier ν_e. On fulfilling the condition of stability ($æ < 1$, $\nu_e > 0$) the additional term of Equation 27 exerts a smoothing effect on the problem solution. This circumstance is of substantial importance. We shall emphasize that though the difference equation (22) was the result of approximation of the equation of inviscid flow $\partial u/\partial t +$ $+ a\partial u/\partial x = 0$, the behavior of the difference solution, as it follows directly from the differential approximation derived above, reflects the properties of some imaginary viscous flow with the coefficient of viscosity ν_e. This circumstance explains a great number of specific features of a difference solution which do not take place in a flow of a physically inviscid medium and which sometimes lead to incorrect treatment of computational results. Besides the mentioned smoothing effect which is the most important in practice, we shall also mention the increase of entropy and other more complex manifestations of schematic dissipation effects.

Note that in case the difference equation accounted for the physical viscosity of medium ν (as, for example, Equation 17 approximating the differential equation (15) of a viscous flow), the first differential approximation at $f = 0$ would have the form

$$\frac{\partial u}{\partial t} + a\frac{\partial u}{\partial x} = (\nu + \nu_e)\frac{\partial^2 u}{\partial x^2}$$

It is obvious that if $\nu_e \approx \nu$ the use of such a scheme is equivalent to the calculation of a flow of considerably more viscous fluid. The account for the effect of schematic viscosity becomes predominant at large Reynolds numbers. It also should be born in mind that when solving the equations with non linear convective terms the unknown velocity u rather than the prescribed a will enter into the expression for ν_e, and in this connection the quantitative estimation of the role of schematic viscosity becomes possible only after the solution of the difference problem.

1.6. Conservativeness

Differential equations of hydrogas dynamics may be written both in ordinary and in divergent forms. In the latter case all of the spatial derivatives enter into these equations in the form of the divergence of the corresponding vectors and tensors. The application of such a form of differential equations, as was observed in the case of the integral form of conservation the laws, provides the fulfillment of the following property: after the summation of the equations written for several adjoining volumes there remain only the flows through the outer surface of a total volume, and all the rest of the flows through the contacting surfaces of adjoining volumes mutually cancel out.

An analogous property also may be imparted to an approximating system obtained on the basis of the above-mentioned forms of initial equations presentation.

If after the summation of difference equations over all the grid meshes the expressions are obtained which involve only grid quantities on the region boundaries, then this scheme is called conservative. The value of this property is in the fact that, in the conservative scheme, difference analogues of the conservation laws are valid accurate to the errors of rounding-off, independent of grid meshes. In nonconservative schemes these laws are fulfilled only accurate to approximation errors.

The importance of the conservativeness property is especially substantial in the case of unobstructed computation of a discontinuous flow.

1.7. Monotonicity

Difference schemes which, when passing over to a new time layer, preserve a monotonic character of solutions are called monotonous. Such schemes play an important role on obtaining nonsmooth solutions and calculating the flows with large gradients of functions.

For linear schemes which may be reduced to the form

$$u_m^{n+1} = \sum_{(k)} \alpha_k u_{m+k}^n \tag{28}$$

the conditions of monotonicity are prescribed by the Godunov theorem: in order for the difference scheme (28) to transform all the monotonous functions into monotonous functions with the same direction of growth, it is necessary and sufficient that all the coefficients α_k were non-negative. Note that the conditions of this theorem coincide with necessary and sufficient conditions of stability.

It is proved that in the class of linear homogeneous difference schemes for hyperbolic equations monotonous schemes with the approximation order higher than the first do not exist. An increase of approximation order in these schemes leads to the occurrence of oscillations of gas dynamic functions in the regions of high gradients. To construct monotonous schemes of an elevated order of approximation use is made of nonlinear and inhomogeneous algorithms in which different smoothing operators are applied that depend on local properties of the solution. Some approaches to the construction of such schemes are considered in the papers by Harten, Yee, and others.[8]

---

[8] Harten, A., High resolution schemes for hyperbolic conservation laws. *J. Comput. Phys.*, 49, 357–393, 1983; Yee, H. C., Construction of explicit and implicit symmetric TVD schemes and their applications, *J. Comput. Phys.*, 68(1), 151–179, 1987; Kopchenov, V. I. and Kraiko, A. N., A monotonous difference scheme of the second order for hyperbolic systems with two independent variables, *Zh.Vych. Mat. Mat. Fis.*, 23(4), 848–859, 1983.

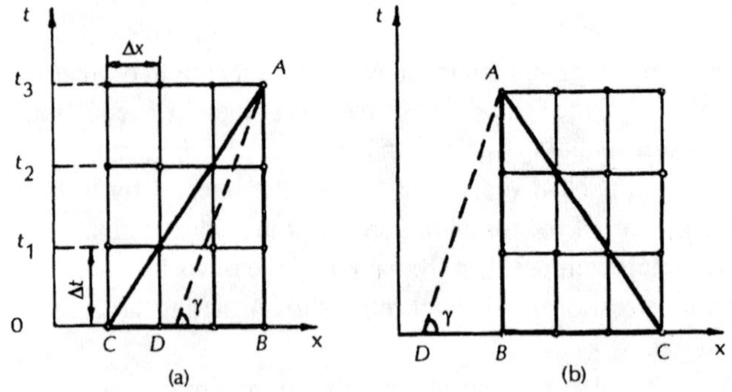

Fig. 298

1.8. Explicit and Implicit Schemes

The above-considered schemes (16) and (17) contained only one point on the upper ($t = t_{n+1}$) time layer, which is why each unknown value u^{n+1} was directly calculated by the known values of u^n on the lower ($t = t_n$) layer. These schemes are called explicit.

The explicit schemes for hyperbolic equations may possess only conditional stability. The necessary condition of the Courant-Friedrichs-Levi (CFL) stability requires that the region of the dependence of the solution to the difference problem (i.e., the combination of nodes containing the quantities determining the solution at this point) include the region of the dependence of the differential problem solution.

We shall consider the explicit scheme that is prescribed by the first relation (16) and that approximates the differential problem

$$\frac{\partial u}{\partial t} + a\frac{\partial u}{\partial x} = f(x, t) \qquad u(x, 0) = u^{(0)}(x)$$

For the case of $a = \text{const}$ we shall elucidate what is the region of the dependence of the solution of the difference problem in the node $A(x_m, t_{n+1})$ of the plane x, t. In accordance with a pattern shown in Figure 295a, the value of u_m^{n+1} at the point A is determined only by the quantities u_m^n, u_{m-1}^n, and f_m^n on a layer t_n. In turn, u_m^n is determined by u_m^{n-1}, u_{m-1}^{n-1}, f_m^{n-1}, and u_{m-1}^n by u_{m-1}^{n-1}, u_{m-2}^{n-1}, f_{m-1}^{n-1} on a layer t_{n-1}, etc. Following this consideration we shall come to the conclusion that the region of the solution dependence at the point A is composed of the nodes which lie in the region of a rectangular triangle ABC with the hypotenuse AC, passing by the diagonals of the cells, with the vertical cathetus AB ($x = x_m$) and the horizontal cathetus BC – the section of the axis $t = 0$ of the length $(n+1)\Delta t$ – on which the initial condition is formulated. Such a region for $n = 2$ is shown in Figure 298a.

The initial differential equation pertains to a hyperbolic type and its characteristics at $a = \text{const}$ are straight lines with the angular coefficients $dt/dx = a^{-1}$. The solution to the differential problem is written in the form

$$u(x, t) = u^{(0)}(x_0) + \int_0^t f(x_0 + at', t')\,dt' \qquad x = x_0 + at$$

Hence, the solution at any point of a semiplane $-\infty < x < \infty$, $t > 0$ is unambiguously determined by values of f on the characteristic $x - at = \text{const}$ passing through this point and by the initial value of the function at the point of the characteristic intersection with the line of initial data $t = 0$. Thus, the region of the dependence of the differential problem solution at the point A is the segment of the corresponding characteristic between the axis $t = 0$ and this point. In Figure 298a the characteristic AD is shown by the dashed line. It is directly seen that the CFL condition (AD lies within the limits of the ABC triangle) is valid for the considered scheme when $\text{tg}\gamma \geq \Delta t/\Delta x$ (conventional stability), which is in full conformity with the condition $æ = a\Delta t/\Delta x \leq 1$. The ABC triangle of the dependence of the difference problem solution for an analogous scheme with the difference "forward" over x (the second of relations (16) with the pattern shown in Figure 295b is given in Figure 298b. As is seen, for $a > 0$ the CFL condition is not fulfilled at any relations between Δt and Δx (unconditional instability). If $a < 0$, then, conversely, the previous scheme will become unconditionally stable.

We shall take for simplicity $f \equiv 0$. Then at $a = \text{const}$ the differential equation (14) describes the propagation of such a wave when the fixed form of the initial disturbance $u^{(0)}(x)$ is displaced as a whole with the velocity a along the axis $0x$, while the values of the function $u(x,t)$ are constant along the characteristics. It is interesting to elucidate the extent to which the conventionally stable scheme (22) reflects this property of the differential problem solution.

If we assume in the scheme $æ = 1$, then the diagonal of the grid mesh AC coincides with the characteristic AD (Figure 298a), and it will follow from relation (22) that $u_m^{n+1} = u_{m-1}^n = \ldots = u_{m-(n+1)}^0$ i.e., the difference solution exactly reflects the wave propagation. However, it is difficult to perform calculations on the upper limit of stability. Thus, if the coefficient a and $æ = a\Delta t/\Delta x$ are variable, then nodes where $æ < æ_{max} = 1$ inevitably exist. Then, as it is clear from the above, the region of the dependence of the difference problem solution in the given node includes more than one point at $t = 0$ that does not correspond to the properties of the differential problem.

The condition of the stability of explicit schemes ($\Delta t \leq \Delta x/a$) often imposes more rigid limitations to the value of the time step than those that follow from the considerations of the required accuracy of the numerical solution. As a result this may unjustifiably increase the time of calculation on a computer. On the one hand, the stability condition involves the digitization parameters Δt, Δx. Hence, it follows that the limitation mentioned is stipulated exclusively by the structure of a difference scheme rather than the essence of the initial problem. In view of these considerations schemes were developed for which the stability conditions to a lesser extent limit the value of the step Δt or are even absent (in the latter case the schemes are called unconditionally stable). They belong to the class of implicit schemes.

To illustrate this we again refer the reader to the equation $\partial u/\partial t + a\partial u/\partial x = 0$, $a = \text{const}$. We approximate it by the difference equation

$$\frac{u_m^{n+1} - u_m^n}{\Delta t} + a\frac{u_m^{n+1} - u_{m-1}^{n+1}}{\Delta x} = 0$$

which differs from Equation 22 only by the fact that in the second term the differences are taken on the upper layer t_{n+1} (the pattern is shown in Figure 295b. The spectral method gives $|\lambda|^{-2} = 1 + 2æ\sin^2(\alpha/2) + æ^2\sin^2\alpha$. Since $|\lambda| \leq 1$ at any $æ > 0$, then the scheme for $a > 0$ is unconditionally stable.

In implicit schemes each difference equation contains several unknowns which are algebraically interrelated. High stability of such schemes is obtained by complicating the algebraic aspect of the problem. This complication usually has two aspects: nonlinearity and high dimensionality of the systems of difference equations. As for the first aspect, to solve nonlinear systems use is, as a rule, made of the iteration method. However, the solution of nonlinear (or of those which became so as a result of this iteration) systems of difference equation of high dimensionality may be a complex problem which to be solved requires a great number of arithmetic operations. The solution of this problem is promoted by the fact that in many cases the matrix of the coefficients of the system of differential equations contains a great number of zero moments.

First, we consider one of the most widely used direct methods for solving a boundary-value problem for a unidimensional system of differential equations with a tridiagonal matrix of the coefficients – the factorization method.

Let an unknown grid function u_m be determined in the nodes x_m ($m = 0, 1, ..., M$), which in the inner nodes satisfies the approximating three-point relations

$$a_m u_{m-1} + b_m u_m + c_m u_{m+1} = f_m$$

($m = 1, 2, ..., M-1$; a_m, b_m, c_m are the known coefficients, fm are the known right sides of equations) and the boundary conditions $u_0 = U_0$, $u_M = U_M$. The solution of this problem as a general system at large M is rather labor consuming. With the form of the system being taken into account we shall present its solution in the form of two-point relations

$$u_m = A_m u_{m+1} + B_m \qquad m = 0, 1, \ldots, M-1$$

thus, leading after the substitution into initial equations, to recurrent relations between factorization coefficients A_m and B_m

$$A_m = -\frac{c_m}{a_m A_{m-1} + b_m} \qquad B_m = \frac{f_m - a_m B_{m-1}}{a_m A_{m-1} + b_m}$$

Since it follows from the condition $u_0 = U_0$ that $A_0 = 0$, $B_0 = U_0$, then, based on these formulas, we can consecutively find A_1, B_1, ..., A_{m-1}, B_{m-1} (the stage of "direct" factorization). Then, knowing the factorization coefficients and using the second boundary condition $u_M = U_M$, we shall calculate (in decreasing order of the subscripts) u_{M-1}, u_{M-2}, ..., u_1 (the stage of "reverse" factorization).

A sufficient condition of nongrowing errors in the course of factorization has the form $|b_m| \geq |a_m| + |c_m|$, with at least at one of the m values of this inequality being strict.

If, as it was assumed, in each node x_m one number u_m is determined, then factorization is called scalar. In case the quantities u_m, a_m, b_m, c_m, f_m denote the matrices of a dimensionality, then the analogous in sense operations are called "matrix factorization".

Computational problems arising as a result of the application of implicit techniques to multidimensional problems will be explained in the example of the equation

$$\frac{\partial u}{\partial t} = \frac{\partial^2 u}{\partial x^2} + \frac{\partial^2 u}{\partial y^2} \tag{29}$$

We shall assume that the simplest-in-forum implicit scheme is written for this equation where all the elements of the spatial operator are referred to the "upper" time layer $n + 1$:

$$\frac{u^{n+1} - u^n}{\Delta t} = \frac{\delta^2 u^{n+1}}{\delta x^2} + \frac{\delta^2 u^{n+1}}{\delta y^2}$$

where $\delta^2 u/\delta x^2$, $\delta^2 u/\delta y^2$ are the difference analogues of the corresponding derivatives (spatial indices are omitted). In this case, to determine an unknown grid function u^{n+1} we obtain the system of algebraic equations with a pentadiagonal matrix. The solution of such a system for a great number of grid nodes is very labour-consuming. Therefore, in similar situations one prefers to resort to such schemes which, on the one hand, preserve rather high stability, and, on the other hand, are similar to implicit schemes for spatially one-dimensional problems.

One of these is the scheme of alternate directions or of longitudinal-lateral factorizations (LLF). In this scheme u^{n+1} is determined by u^n as a result of two "semi integer" or "fractional" time steps.

On the first fractional step (from t_n to $t_{n+1/2} = t_n + \Delta t/2$) the derivative with respect to x in Equation 29 is approximated implicitly while the derivative with respect to y explicitly. On the second fractional step (from $t_{n+1/2}$ to t_{n+1}), conversely, the derivative with respect to x is expressed explicitly and the derivative with respect to y implicitly. Thus, the scheme has the form:

$$\frac{u^{n+1/2} - u^n}{\tau} = \frac{\delta^2 u^{n+1/2}}{\delta x^2} + \frac{\delta^2 u^n}{\delta y^2}$$

$$\frac{u^{n+1} - u^{n+1/2}}{\tau} = \frac{\delta^2 u^{n+1/2}}{\delta x^2} + \frac{\delta^2 u^{n+1}}{\delta y^2}$$

or

$$(I - \tau\Lambda_1)\, u^{n+1/2} = (I + \tau\Lambda_2)\, u^n \tag{30}$$

$$(I - \tau\Lambda_2)\, u^{n+1} = (I + \tau\Lambda_1)\, u^{n+1/2} \tag{31}$$

where I is the unit operator, Λ_1, Λ_2 are the difference analogues of the operators of the second derivatives with respect to x and y, respectively, $\tau = \Delta t/2$; spatial indices are omitted. On each of the fractional steps Equation 29 is approximated over time with the order Δt and calculations are realized by factorizations along the corresponding coordinate lines. Moreover, as a result of these steps the second order of approximation over time is attained. To confirm this fact we shall write the scheme in the integer steps having eliminated an intermediate grid function $u^{n+1/2}$ from Equations 30 and 31

$$\frac{u^{n+1} - u^n}{\Delta t} = (\Lambda_1 + \Lambda_2)\frac{u^n + u^{n+1}}{2} - \frac{\Delta t^2}{4}\Lambda_1\Lambda_2\left(\frac{u^{n+1} - u^n}{\Delta t}\right)$$

Expanding the function u into the Taylor series shows that the error of approximation is of the second order with respect to Δt.

Study of the stability of the LLF scheme by spectral methods shows that in spite of conventional stability of each of the fractional steps (due to the absence of explicit approximations) the scheme as a whole turns to be unconditionally stable.

One more widespread method of approximate numerical solution of spatially multidimensional problems is the method of splitting. The essence of this method will be explained in the example of the same two-dimensional equation (29).

Similar to the LLF scheme, the grid function u^{n+1} is determined by the known u^n in two steps. On the first step the auxiliary grid function $\tilde{u}$ is found that obeys the equation

$$\frac{\tilde{u} - u^n}{\Delta t} = \frac{\delta^2 \tilde{u}}{\delta x^2}$$

On the second step the function u^{n+1} is determined from the equation

$$\frac{u^{n+1} - \tilde{u}}{\Delta t} = \frac{\delta^2 u^{n+1}}{\delta y^2}$$

Calculations are performed by ordinary factorization along the coordinate lines $y = \text{const}$ and $x = \text{const}$, correspondingly.

In the given method the spatial operator of the initial problem is split into one-dimensional operators and on each step the corresponding equation, "truncated" with respect to spatial variables, is approximated by the implicit scheme. Equation 29 is approximated only by combination of two steps. Actually, writing down the scheme on each step with the help of the notation Λ_1 and Λ_2

$$(I - \Delta t \Lambda_1)\, \tilde{u} = u^n$$

$$(I - \Delta t \Lambda_2)\, u^{n+1} = \tilde{u}$$

and eliminating the intermediate quantity $\tilde{u}$, we shall obtain the scheme in integer steps

$$(I - \Delta t \Lambda_1)\,(I - \Delta t \Lambda_2)\, u^{n+1} = u^n$$

or

$$\frac{u^{n+1} - u^n}{\Delta t} = (\Lambda_1 + \Lambda_2)\, u^{n+1} - \Delta t \Lambda_1 \Lambda_2 u^{n+1}$$

The presented scheme of splitting has the first order of accuracy over time.

Studying by spectral methods indicates unconditional stability of separate steps and, consequently, of the scheme as a whole.

1.9. Schemes Based on the Time-Dependent Technique

The previous schemes refer to the equations describing unsteady processes. However, to solve the problems in stationary formulations $Lu = f$ (the operator L contains only the derivatives with respect to spatial variables) the use of the technique developed for nonstationary problems is preferred. For this purpose partial derivatives with respect to time of unknown quantities are added to the equations passing the problem over to the class of hyperbolic or parabolic ones. A stationary field is obtained as a limit to which the solution of a nonstationary problem tends at large t under stationary boundary conditions. Such an approach, which is iterative in essence, is called the time-dependent technique. Examples of application of this method are given later.

Construction of time-dependent techniques presupposes a greater freedom in the treatment of derivatives with respect to time, as compared to the solution of nonstationary problems proper, since in the time-dependent technique these terms will vanish. The value of the step Δt is determined by the rate of convergence to the stationary solution rather than by the accuracy of the determination of time-dependent relations. In view of this the application of implicit schemes becomes more justifiable.

2. Methods for Solving the Euler Equations

A system of Euler's equations describes the flows of inviscid nonheat-conducting liquid or gas. In the majority of cases the inviscid fluid flow can be considered irrotational, thus its calculation is reduced to the solution of the Laplace equation for the velocity potential. The methods to solve such problems were discussed in Chapters 7 and 9. In this section we shall consider some methods for solving the system of Euler's equations for gas.

2.1. Some Specific Features of Numerical Simulation of Inviscid Gas flows

Nonstationary Euler's equations form a system of hyperbolic equations. Stationary equations retain hyperbolic properties in the case of supersonic flows and become elliptical at subsonic velocities. We shall pause now on the formulation of initial and boundary conditions of the corresponding problems.

The values of gas dynamic functions on the lines or surfaces that are not characteristic may be prescribed as initial conditions. When solving nonstationary equations the values of the functions over the entire computational region at the initial time instant are usually specified.

The formulation of boundary conditions depends on the type of the computational region limits. The limit may lie in the region with the prescribed values of gas dynamic functions or coincide with the surface on both sides of which gas parameters are related by the familiar functional dependence, for example, with the surface of a shock wave or contact discontinuity. On solid surfaces the condition of impermeability is usually prescribed. When studying the flows possessing the properties of symmetry or periodicity in space use is made of the boundary conditions which express the mentioned properties in a mathematical form.

The boundary conditions on opened permeable surfaces, for example, at the inlet and outlet of the computational region are to a great extent dependent on the character of the flow in the vicinity of these boundaries. On the determination of the quantity and type of conditions, which should be prescribed on the surface σ limiting the considered flow region Ω, we can use the following approach.[9]

In the vicinity of the surface element σ with the outer normal n to Ω a one-dimensional nonstationary flow is considered, the parameters of which depend only on time and distance from the surface. A weakly disturbed flow of such a type is a combination of three plane waves propagating with the velocities q_n, $q_n \pm c$, where q_n is the projection of the gas velocity vector onto the normal n, c is sound velocity. The first wave is completely determined by the three quantities, for example, by entropy and the components of gas velocity that are tangential to the surface σ. Each of the two other waves is characterized by one quantity, for example, by pressure or by the corresponding Riemann invariant. The number of conditions, set on the surface element σ, should be taken equal to the number of the parameters determining those waves which propagate from the considered portion of the boundary into the region Ω. In this case, among the prescribed quantities there cannot be the parameters which determine the waves not coming into the computational region. In accordance with this rule, at supersonic velocities all the boundary conditions are set at the computational region inlet. At subsonic velocities the $N + 1$ conditions should be assigned at the inlet and one more at the outlet, here N is the number of spatial coordinates. At the outlet boundary pressure is

[9] Godunov, S. K., Zabrodin, A. V., Ivanov, M. Ya., Kraiko, A. N., and Prokopov, G. P., *Chislennye resheniya mnogomernykh zadach gasovoy dinamiki* (*Numerical Solution of Multidimensional Problems of Gas Dynamics*), Nauka, Moscow, 1976.

usually prescribed.

Application of numerical methods to the integration of equations leads to some specific features in the formulation of boundary conditions. To them belong the possibility to construct computational boundary conditions and difference approximations for calculating the functions of those boundaries of the computational region where they are not determined by physical boundary conditions. The necessity in additional boundary conditions arises when use is made of difference equations of higher order. Different ways for determining gas dynamic functions on the outer edges of supersonic flows serve as examples for application of boundary approximations. Here, the difference equations with one-side approximation of derivatives are used, or extrapolation inside the computational region is applied that may be carried out with account for characteristic properties of equations.

The possibility for the occurrence of shock waves is an important feature of gas flows. Two approaches are used to calculate flows with shock waves and other disrupters. The first is based on the discrimination of specific features. Here, differential equations are used only in the regions of continuous variation of the functions and on the distinguished discontinuity surfaces the conditions of compatibility are used. The approach mentioned may provide high accuracy, though its realization becomes difficult in the presence of complex systems of gas dynamic discontinuities.

Another approach is based on the application of homogeneous finite-difference schemes of unobstructed computation. Use of such schemes is the most expedient in the cases when the studied flow pattern is not known *a priori* or it qualitatively changes with time. In the methods of unobstructed computation the Euler equations are written in the conservative form which allows the existence of discontinuity solutions. In the case of unobstructed computation the discontinuity surfaces manifest themselves as the regions of large gradients of gas dynamic functions which, depending on the properties of the difference scheme, are "smeared" to some number of cells of the difference grid. The homogeneity of the algorithm and simplicity of its realization is the advantage of the approach considered, while the decrease of accuracy in the vicinity of discontinuities is its disadvantage.

In many cases it is expedient to combine the above two approaches to the calculation of gas flows. Thus, for example, in the problems of supersonic flows past bodies it is reasonable to distinguish the leading shock wave, and to perform unobstructed computation of subsequent shock waves, if they exist at all.

In calculations of inviscid gas flows, explicit methods have acquired a relatively greater use. Their application is especially expedient in the study of flows described by hyperbolic equations because in this case the value of the step of integration over the hyperbolic coordinate is first limited by the requirements of accuracy.

2.2. The Godunov Method

We shall consider the basic ideas of this method on the example of the integration of acoustic equations. The system of differential equations that describes the propagation of plane sound waves has the form

$$\frac{\partial u}{\partial t} + \frac{1}{\rho_0}\frac{\partial p}{\partial x} = 0 \qquad \frac{\partial p}{\partial t} + \rho_0 c_0^2 \frac{\partial u}{\partial x} = 0 \tag{32}$$

Here, u, p are small deviations of velocity and pressure from their values in an undisturbed medium that are caused by the propagation of sound waves; the subscript 0 denotes the constant parameters of undisturbed state of gas.

Equations 32 are the system of hyperbolic equations. Integrating them over an arbitrary region with the boundary σ in the plane (x, t) and passing over to contour integrals, we shall

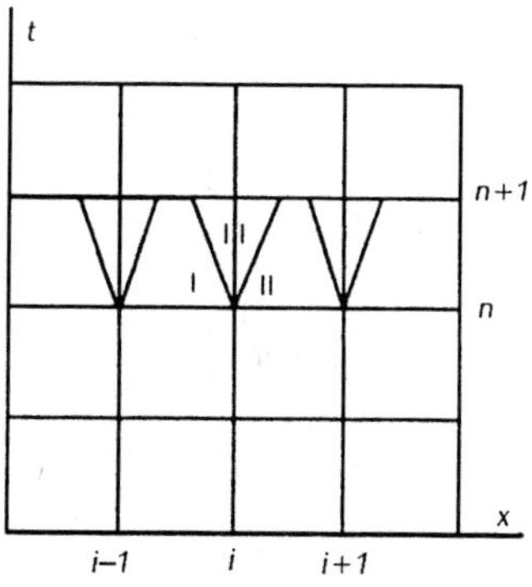

Fig. 299

obtain the integral equalities

$$\oint_{\sigma}(\rho_0 u dx - p dt) = 0 \qquad \oint_{\sigma}\left(\frac{p}{c_0^2}dx - \rho_0 u dt\right) = 0 \tag{33}$$

We shall introduce the difference scheme in the plane (x,t) (see Figure 299) and assume that the gas parameters on the nth time layer may be presented by step functions with discontinuities on the boundaries of cells. We shall take in Equations 33 the boundary of the cell as the contour σ and designate

$$\frac{1}{\Delta x}\int_{x_{i-1}}^{x_i} u(x,t)\,dx = u_{i-1/2} \qquad \frac{1}{\Delta x}\int_{x_{i-1}}^{x_i} u(x,t+\Delta t)\,dx = u^{i-1/2}$$

$$\frac{1}{\Delta t}\int_{t_n}^{t_{n+1}} u(x_{i-1},t)\,dt = U_{i-1} \qquad \frac{1}{\Delta t}\int_{t_n}^{t_{n+1}} u(x_i,t)\,dt = U_i$$

Similar designations also will be used for the integrals for pressure. Then, based on the integral equalities (33) the following difference scheme is obtained

$$u^{i-1/2} = u_{i-1/2} - \frac{\Delta t}{\Delta x\,\rho_0}(P_i - P_{i-1})$$

$$p^{i-1/2} = p_{i-1/2} - \frac{\Delta t}{\Delta x}\rho_0 c_0^2 (U_i - U_{i-1}) \tag{34}$$

The "principle" values P_i and U_i, which are the mean values of the functions on the side edges of the cell, are obtained from the solution of the self-similar problem of the decomposition of gas dynamic discontinuity which occurs at the point $x = x_i$ that separates two different states of gas in neighboring cells on the layer n. As a result of such a decomposition sound waves propagating to the left and right originate with the sound velocity c_0. We shall consider this problem.

By simple transformations the system of Equations 32 may be written in the invariants

$$\frac{\partial}{\partial t}\left(u+\frac{p}{\rho_0 c_0}\right)+c_0\frac{\partial}{\partial x}\left(u+\frac{p}{\rho_0 c_0}\right) = 0$$

$$\frac{\partial}{\partial t}\left(u-\frac{p}{\rho_0 c_0}\right)-c_0\frac{\partial}{\partial x}\left(u-\frac{p}{\rho_0 c_0}\right) = 0 \qquad (35)$$

It is seen from Equations 35 that the values of Riemann's invariants $Y = u+p/(\rho_0 c_0)$ and $Z = u-p/(\rho_0 c_0)$ remain unchangeable along straight lines with the inclination angle $dx/dt = \pm c_0$ called characteristics. With the Riemann invariants known we can easily find the quantities we are interested in

$$u = \frac{1}{2}(Y+Z) \qquad p = \frac{1}{2}\rho_0 c_0 (Y-Z) \qquad (36)$$

Now, we shall consider the case when the initial conditions of the problem at $t = 0$ have the form:

$$u_0(x) = u_{\mathrm{I}} \qquad p_0(x) = p_{\mathrm{I}} \qquad \text{for} \quad x<x^*$$

$$u_0(x) = u_{\mathrm{II}} \qquad p_0(x) = p_{\mathrm{II}} \qquad \text{for} \quad x>x^*$$

where $U_{\mathrm{I}}, p_{\mathrm{I}}, U_{\mathrm{II}}, p_{\mathrm{II}}$ are some constant values. Allowing for the fact that the values of the invariants Y, Z remain unchangeable on the characteristics $x \mp c_0 t = \text{const}$, we come to the structure presented in Figure 300. In region I ($x<x^*-c_0 t$):

$$Y = Y_{\mathrm{I}} \quad Z = Z_{\mathrm{I}}$$

where

$$u = u_{\mathrm{I}} \quad p = p_{\mathrm{I}}$$

In region II ($x>x^*+c_0 t$):

$$Y = Y_{\mathrm{II}} \quad Z = Z_{\mathrm{II}}$$

where

$$u = u_{\mathrm{II}} \quad p = p_{\mathrm{II}}$$

In region III ($x^*-c_0 t<x<x^*+c_0 t$):

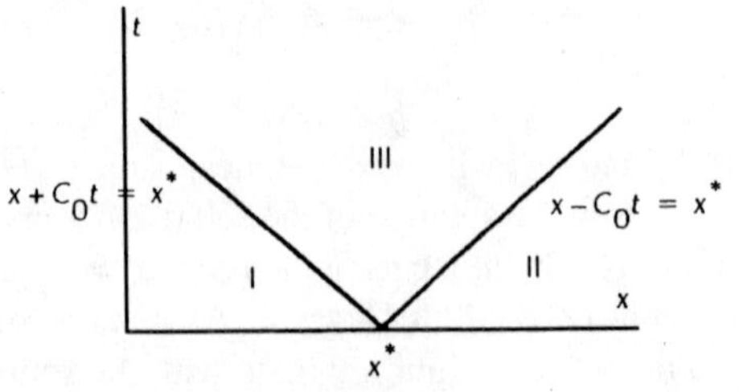

Fig. 300

$$Y = Y_{\mathrm{I}} \quad Z = Z_{\mathrm{II}}$$

where in accordance with the general solution (36)

$$u = \frac{u_{\mathrm{I}} + u_{\mathrm{II}}}{2} - \frac{p_{\mathrm{II}} - p_{\mathrm{I}}}{2\rho_0 c_0} \qquad p = \frac{p_{\mathrm{I}} + p_{\mathrm{II}}}{2} - \rho_0 c_0 \frac{u_{\mathrm{II}} - u_{\mathrm{I}}}{2}$$

Thus, in regions I and II gas parameters retains the undisturbed values of, and along the lines separating regions I, II, and III the constructed functions $u(x,t)$, $p(x,t)$ suffer discontinuity. These two lines are formed from the initial discontinuity at the point $x = x^*$.

The solution of the problem considered determines the technique for calculating "high" values in the difference scheme (34):

$$U_i = \frac{u_{i-1/2} + u_{i+1/2}}{2} - \frac{p_{i+1/2} - p_{i-1/2}}{2\rho_0 c_0}$$

$$P_i = \frac{p_{i-1/2} + p_{i+1/2}}{2} - \rho_0 c_0 \frac{u_{i+1/2} - u_{i-1/2}}{2} \tag{37}$$

If we consider that the decomposition of the discontinuity occurs on the boundaries of all the cells, then during some period of time the flow structure will have the form schematically shown in Figure 299. This structure remains until sound waves from the neighboring nodes reach the considered node of the difference grid. Hence, the limitation to the step of integration follows with respect to time: $\Delta t \le \Delta x / c_0$, which coincides with the CFL condition.

When solving nonlinear problems of gas dynamics the initial equations are also integrated over the volume of an elementary cell. Finite-difference approximation is similar to that considered above for a linear problem. The difference is only in the determination of "high" values. With the use of a coarse difference grid great differences in the values of gas dynamic functions in neighboring cells are possible, especially in the vicinity of strong discontinuities. In this case the linear theory and Equations 37 following from it become nonapplicable. In these cases "high" values are calculated using the solutions of a nonlinear problem of the decomposition of an arbitrary discontinuity. The problem also has a self-similar solution. The decomposition of an arbitrary discontinuity is accompanied by the origination of shock waves, centralized waves of rarefaction, and contact discontinuities. Here, one configuration is realized (Figure 301). "High" values on the cell boundary are calculated by different formulas depending on which of the five possible situations is realized on the decomposition of discontinuity. These formulas are presented in books by Rozhdestvenskiy and Yanenko[10] and Godunov.[11]

The above-described version of the Godunov method has the first order of approximation over the spatial coordinate and time. Generalization of the method to multidimensional and stationary supersonic flows and also numerous examples of the solution of difference problems of continuum mechanics are given in Godunov.[11] Further, Godunov's method was improved to obtain the second order of accuracy over spatial coordinates[12] and time.[13]

---

[10] Rozhdestvenskiy, B. L. and Yanenko, N. N., *Sistemy kvazilineynykh uravneniy (Systems of Quasilinear Equations)*, Nauka, Moscow, 1978.

[11] Godunov, S. K. et al., *Chislennoye reshenie mnogomernykh zadach gazovoy dinamiki (Numerical Solution of Multi-Dimensional Problems of Gas Dynamics)*, Nauka, Moscow, 1976.

[12] Kolgan, V. P., Application of the principle of minimum values of the derivative to the construction of finite-difference schemes to calculate discontinuity solutions of gas dynamics, *Uch. Zap. TsAGI*, 3(6), 68–77, 1972.

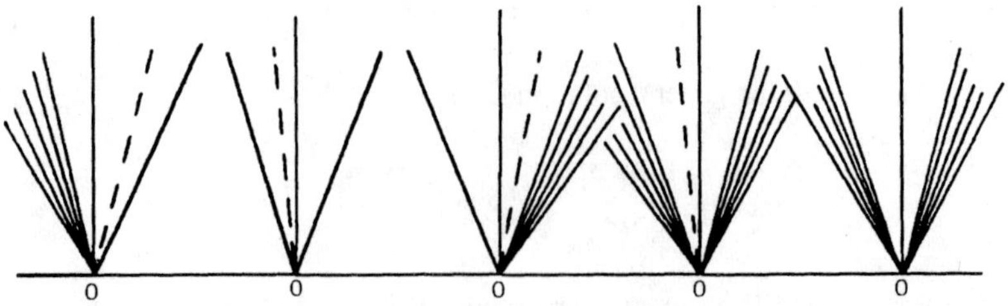

Fig. 301

2.3. Method of Large Particles

In this method developed by Belotserkovskiy and Davydov[14] use is made of the basic elements of the method of particles in cells PIC previously suggested by Harlow. In the region considered an immovable Euler difference scheme fixed in space is introduced. Continuum is treated as a combination of particles of a fixed mass (the Lagrange grid of particles) which move through the Euler grid of cells. Particles serve to determine the parameters of the fluid itself (mass, momentum, energy), and the Euler grid is used to calculate field characteristics (pressure, density, temperature).

Along with great possibilities for studying complex hydrodynamic phenomena the PIC method has a number of drawbacks. The main one lies in the fact that due to the representation of continuum by a great number of particles, the flow parameters in each cell change step-wise. As soon as a particle crosses the boundary of the Euler cell, its mass, momentum, and energy are subtracted from the corresponding values of this cell and are added to the neighboring cell. This leads to nonphysical fluctuations of the calculated quantities. Some difficulties also arise in calculation of strongly rarefied regions from which practically all particles go away. At the same time a substantial increase in the number of particles is impossible due to the limitation of computer power.

In view of the above-mentioned difficulties it is expedient to discard the discrete model of the medium and to proceed from the concept of continuity considering a mass flow through the boundaries of Euler cells instead of particles. In this case gas density is determined by the mass conservation law for the given Euler cell rather than by dividing the total mass of all the particles in a cell into its volume. Thus, instead of the combination of particles in cells the mass of the entire cell is considered as a whole (a large particle). Flows of these large particles through the Euler grid are described on the basis of finite-difference concepts of conservation laws.

The main idea of the method of large particles is in splitting the initial system of equations by physical processes. The modeled medium is substituted by the system of large particles occupying at the given time instant the volumes which coincide with the cells of the Euler grid. Calculation of each time layer is divided into three stages:

1. Intermediate values of the functions, on the assumption of the absence of mass flow through the boundaries of cells, are determined on a fixed Euler grid
2. The density of mass flow through the boundaries of cells is calculated
3. Finite values of functions in each cell of the Euler grid are found on the basis of the con-

[13] Rodionov, A. V., Improvement of the order of the approximation of Godunov's scheme, *Zh. Vych. Mat. Mat. Fiz.*, 27(12), 1978.

[14] Belotserkovskiy, O. M. and Davydov, Yu. M., *Metod krupnykh chastits v gazovoy dinamike* (*Method of Large Particles in Gas Dynamics*), Nauka, Moscow, 1982.

servation laws

Essentially, in the first stage the change in momentum and energy of the Lagrange elementary volume of gas, enclosed inside the Euler cell, during the time Δt is considered. In the second stage the motion of particles through the boundaries of the Euler cells is modeled, i.e., the change of the position of the volume mentioned is taken into account. On the third stage, with the use of the results obtained, the variation of gas parameters in the initial Euler stage during the time Δt is determined. In fact, the calculation is made in the local Lagrange coordinates with recalculation on each time step in terms of the Euler grid.

Thus, the evolution of the entire system during the time Δt is described by the following: at first, the change of the inner state of subsystems which occupy the cells in the assumption of their stationarity (the Euler stage); then, displacement of all large particles in proportion with their velocity and time Δt without inner state variation (the Lagrange state), and, finally recalculation of gas parameters in terms of the initial Euler grid (the final stage).

We shall consider the realization of this procedure in the case of two spatial coordinates. The Euler equations in the divergent form in the Cartesian coordinates are written as

$$\frac{\partial \rho}{\partial t} + \mathrm{div}(\rho \mathbf{V}) = 0$$

$$\frac{\partial \rho u}{\partial t} + \mathrm{div}(\rho u \mathbf{V}) + \frac{\partial p}{\partial x} = 0$$

$$\frac{\partial \rho v}{\partial t} + \mathrm{div}(\rho v \mathbf{V}) + \frac{\partial p}{\partial y} = 0$$

$$\frac{\partial \rho E}{\partial t} + \mathrm{div}(\rho E \mathbf{V}) + \mathrm{div}(p \mathbf{V}) = 0 \tag{38}$$

Here after E is the specific total gas energy.

We shall introduce the stationary Euler grid which consists of rectangular cells with the sides Δx, Δy into the integration region (Figure 302).

In the first stage, in accordance with the assumption on the absence of transfer through the edges of cells, equations are solved which do not contain the terms of the type $\mathrm{div}(\rho\varphi\mathbf{V})$, where $\varphi = (1, u, v, E)$. It follows from the continuity equation that in the first stage the gas density does not vary. The rest of the equations take on the form

$$\rho\frac{\partial u}{\partial t} + \frac{\partial p}{\partial x} = 0$$

$$\rho\frac{\partial v}{\partial t} + \frac{\partial p}{\partial y} = 0$$

$$\rho\frac{\partial E}{\partial t} + \mathrm{div}(p\mathbf{V}) = 0 \tag{39}$$

Intermediate values of the velocity components and specific energy E may be found by the simplest finite-difference approximations

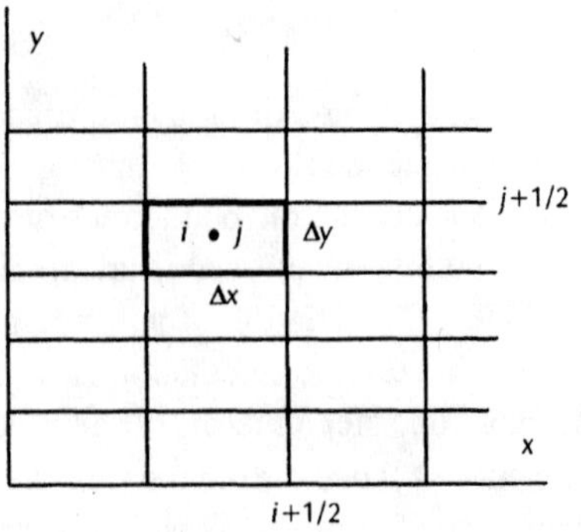

Fig. 302

$$\tilde{u}_{ij} = u^n_{ij} - \frac{p^n_{i+1/2,j} - p^n_{i-1/2,j}}{\Delta x} \cdot \frac{\Delta t}{\rho^n_{ij}} \qquad \tilde{v}_{ij} = v^n_{ij} - \frac{p^n_{i,j+1/2} - p^n_{i,j-1/2}}{\Delta y} \cdot \frac{\Delta t}{\rho^n_{ij}}$$

$$\tilde{E}_{ij} = E^n_{ij} - \left[\frac{(pu)^n_{i+1/2,j} - (pu)^n_{i-1/2,j}}{\Delta x} + \frac{(pv)^n_{i,j+1/2} - (pv)^n_{i,j-1/2}}{\Delta y}\right] \cdot \frac{\Delta t}{\rho^n_{ij}} \tag{40}$$

Here, the quantities with integral indices refer to the centers of cells and the quantities with half-integral indices to their edges (see Figure 302). The quantities with half-integral indices are calculated as half-sums of the values of the corresponding parameters in the neighboring cells, for example, $p^n_{i+1/2,j} = (p^n_{ij} + p^n_{i+1,j})/2$. The difference scheme (40) is known to be insufficiently stable. Its use is possible thanks to the corresponding compensation on subsequent stages that provides stability of the total time step.

In the second stage, mass flows of particles through the boundaries of Euler cells are modeled. Thus, for instance

$$\Delta M_{i+1/2,j} = \langle\rho\rangle_{i+1/2,j}\langle u\rangle_{i+1/2,j}\Delta y\Delta t \tag{41}$$

Within the angular brackets written are the values of gas density and velocity on the cell boundary. They are calculated with account for the velocity vector direction by the formulas of the first or second orders of accuracy. The formulae of the first order of accuracy have the form

$$\Delta M_{i+1/2,j} = \begin{cases} \rho^n_{ij}\dfrac{\tilde{u}_{ij} + \tilde{u}_{i+1,j}}{2}\Delta y\Delta t, & \text{when } \tilde{u}_{ij} + \tilde{u}_{i+1,j} > 0 \\ \rho^n_{i+1,j}\dfrac{\tilde{u}_{ij} + \tilde{u}_{i+1,j}}{2}\Delta y\Delta t, & \text{when } \tilde{u}_{ij} + \tilde{u}_{i+1,j} < 0 \end{cases} \tag{42}$$

The equations of the third stage represent the laws of mass, M, momentum, p and total specific energy, E, conservation written for the considered cell in the form

$$F^{n+1} = F^n + \sum \Delta F_k \quad \text{where} \quad F = (M, p, E) \tag{43}$$

summation is carried out over all cell boundaries. It is assumed upon application of formulas

(43) that mass flows ΔM through the boundaries of cells, determined on the second stage, have intermediate values of velocity and specific energy calculated on the first stage. Proceeding from the above, final values of the unknown functions ρ, $X = (u, v, E)$ on a new time layer are found by the following formulas

$$\rho_{ij}^{n+1} = \rho_{ij}^{n} + \frac{\sum \Delta M_k}{\Delta x \Delta y} \qquad X_{ij}^{n+1} = \frac{\rho_{ij}^{n}}{\rho_{ij}^{n+1}} \tilde{X}_{ij} + \frac{\sum \tilde{X}_k \Delta M_k}{\rho_{ij}^{n+1} \Delta x \Delta y} \tag{44}$$

To solve the equations of the first stage and calculate the mass flows in the second stage, use can be made of other techniques.[15] In the cells adjacent to the boundaries of the computational region, on all three stages, calculations are performed, accounting for boundary conditions. When studying flow in the regions of an arbitrary shape the contour of the streamlined surface may intersect the rectangular Euler grid in different ways. In such cases fractional cells are introduced.

The properties of the method of large particles were studied by differential approximations. It was shown that over the entire flow region the method has the first order of approximation over time or the second order of approximation over spatial coordinates depending on the formulas for ΔM.

In a one-dimensional case, on using the formulas of the first order of accuracy (42) the first differential approximation of the method of large particles has the form

$$\frac{\partial \rho}{\partial t} + \frac{\partial \rho u}{\partial x} = \frac{\partial}{\partial x}\left(\varepsilon \frac{\partial \rho}{\partial x}\right) + O(\Delta t, \Delta x^2)$$

$$\frac{\partial \rho u}{\partial t} + \frac{\partial (\rho u^2 + p)}{\partial x} = \frac{\partial}{\partial x}\left(\varepsilon \frac{\partial \rho u}{\partial x}\right) + O(\Delta t, \Delta x^2)$$

$$\frac{\partial \rho E}{\partial t} + \frac{\partial}{\partial x}[u(\rho E + p)] = \frac{\partial}{\partial x}\left(\varepsilon \frac{\partial \rho E}{\partial x}\right) + O(\Delta t, \Delta x^2) \tag{45}$$

where $\varepsilon = |u|\Delta x/2$ is the coefficient of approximation viscosity. Written the left sides of (45) are the terms of initial differential equations, and on the right sides, the terms arising due to problem digitization.

It is seen that the necessary condition of the considered scheme stability, that is, in the positiveness of the coefficient of approximation viscosity, is fulfilled independently of the relation between the mesh width for both time and spatial coordinates. Similar analysis shows that calculation of ΔM by the equations of the second order of accuracy take place only when artificial dissipation is introduced on the first stage. Numerical investigations, conducted for a number of one- and two-dimensional problems, showed that use of the second order formulas do not give considerable increase in the accuracy of solutions. Therefore, the authors suggest the use of the first order scheme which is simpler in practice.

2.4. The Monotonized McCormack Scheme

The method of Voynovich et al.[16] is an explicit homogeneous scheme of unobstructed calculation intended for studying multidimensional nonstationary discontinuity flows. The method

[15] Belotserkovskiy, O. M. and Davydov, Yu. M., *Metod krupnykh chastits v gazovoy dinamike (The Method of Large Particles in Gas Dynamics)*, Nauka, Moscow, 1982.

is constructed on the basis of the laws of mass, momentum, and energy conservation written in the integral form that in the Euler coordinates are

The discontinuity equation:

$$\int\limits_V \rho dV \Big|_{t_0}^{t} + \int\limits_{t_0}^{t}\int\limits_{\sigma} \rho u_n ds dt = 0 \tag{46}$$

The equation of momentum balance:

$$\int\limits_V \rho \mathbf{u} dV \Big|_{t_0}^{t} + \int\limits_{t_0}^{t}\int\limits_{\sigma} (\rho u_n \mathbf{u} + p\mathbf{n})\, ds dt = 0 \tag{47}$$

or in the projections on the axis of the Cartesian rectangular system of coordinates

$$\int\limits_V \rho u_x dV \Big|_{t_0}^{t} + \int\limits_{t_0}^{t}\int\limits_{\sigma} (\rho u_n u_x + p n_x)\, ds dt = 0 \tag{47a}$$

$$\int\limits_V \rho u_y dV \Big|_{t_0}^{t} + \int\limits_{t_0}^{t}\int\limits_{\sigma} (\rho u_n u_y + p n_y)\, ds dt = 0 \tag{47b}$$

$$\int\limits_V \rho u_z dV \Big|_{t_0}^{t} + \int\limits_{t_0}^{t}\int\limits_{\sigma} (\rho u_n u_z + p n_z)\, ds dt = 0 \tag{47c}$$

the equation of total energy balance

$$\int\limits_V \rho E dV \Big|_{t_0}^{t} + \int\limits_{t_0}^{t}\int\limits_{\sigma} (\rho E + p)\, u_n ds dt = 0 \tag{48}$$

Equations 46 to 48 are written for the gas volume V bounded by a closed surface σ with the outer normal n. In this case Equations 47a, b, and c are written relative to the Cartesian components of momentum, thus making it possible to construct conservative schemes of arbitrary difference schemes.

The studied region is divided into a great number of "elementary" volumes (spatial cells) inside which gas dynamic parameters for each time instant t are considered to be equal over the entire cell volume. The position of a specific cell on a grid is characterized by specification of the indices j_1, j_2, j_3 ($j_1 = 1, ..., N_1, j_2 = 1,..., N_2, j_3 = 1, ..., N_3$). A combination of the cells in a computational region, which differ in the value of one of the indices, is a grid

[16] Voynovich, P. A., Zhmakin, A. I., and Fursenko, A. A., The Program of Calculating Multi-Dimensional Non-Stationary Discontinuity Flows of an Ideal Gas, Preprint No. 1268, Leningrad Institute of Physical Engineering, USSR Academy of Sciences, Leningrad, 1988.

radial line. Then, along with the directions prescribed by the directions of the axes of the Cartesian system of coordinates, use will be made of the concept of "grid" directions 1, 2, or 3 which correspond to the direction of the outer normal to the surface of the given cell edge (j_1, j_2, j_3) from its side to a neighboring cell with a greater value of the corresponding index.

We shall consider the procedure of determining gas dynamic functions on the layer $t = t^{n+1}$ by the known values on the layer $t = t^n$, based on the example of one cell with the indices (j_1, j_2, j_3). The integration surface σ in Equations 46 to 48 in this case includes the edges that separate the cells in the grid direction 1, $S_{j_1-1/2}$ and $S_{j_1+1/2}$; in the grid direction 2, $S_{j_2-1/2}$ and $S_{j_2+1/2}$; and in the grid direction 3, $S_{j_3-1/2}$ and $S_{j_3+1/2}$. Thus, the integrals over the surface consist of six summands according to the number of cell edges. Each of these summands with its sign will be called the flow of the corresponding quantity through the given edge of the cell. The algebraic sums of mass, momentum, and energy flows over all six edges of the cell determine the rates of the change in gas dynamic parameters inside a cell and allows one to determine, by their known values at the time instant t^n, the values at the moment $t^{n+1} = t^n + \Delta t$ of the gas dynamic parameters for calculating flows through the cell edges are known.

The described version of the method is based on the decomposition of the time step operator $L(\Delta t)$ along spatial direction into the three operators $L_1(\Delta t)$, $L_2(\Delta t)$, and $L_3(\Delta t)$, each of which is realized by a two-step explicit McCormack scheme in the concept of finite volumes in combination with matched conservative smoothing.[17] The McCormack scheme has the second order of approximation over spatial coordinates and time. Decomposition of the operator $L(\Delta t)$ makes it possible to reduce the solution of a three-dimensional problem to the sequence of one-dimensional problems.

We shall consider realization of the one-dimensional operator $L_1(\Delta t)$ (the operators $L_2(\Delta t)$ and $L_3(\Delta t)$ are realized similarly).

We introduce the following notation

$$U = \begin{bmatrix} \rho \\ \rho u_x \\ \rho u_y \\ \rho u_z \\ \rho E \end{bmatrix} \qquad F_\alpha = \begin{bmatrix} \rho u_\alpha \\ \rho u_\alpha u_x + \delta_{\alpha x} p \\ \rho u_\alpha u_y + \delta_{\alpha y} p \\ \rho u_\alpha u_z + \delta_{\alpha z} p \\ (\rho E + p) u_\alpha \end{bmatrix} \qquad H = \begin{bmatrix} 0 \\ \delta_{\alpha x} p \\ \delta_{\alpha y} p \\ \delta_{\alpha z} p \\ 0 \end{bmatrix}$$

$$\alpha = x, y, z \qquad \delta_{\alpha\beta} = \begin{cases} 1, & \text{if } \alpha = \beta \\ 0, & \text{if } \alpha \neq \beta \end{cases}$$

To shorten the notation we shall designate the variable index of the grid radial line $j_1 = i$, and the indices j_2 and j_2, which are constant for all of the cells of the given radial line, will be omitted.

The operator $L_1(\Delta t)$ is realized in three stages:

1. The predictor (the first step of the McCormack scheme)

---

[17] Zhmakin, A. I. and Fursenko, A. A., On one monotonous difference scheme of unobstrated calculation, *Zh. Vych. Mat. Mat Fiz.*, 20(1), 1980.

$$\tilde{U}_i = U_i^n - \frac{\Delta t}{V_i} \sum_{\alpha = x, y, z} [((F_\alpha)_{i+1}^n (S_\alpha)_{i+1/2} - (F_\alpha)_i^n (S_\alpha)_{i-1/2}) - -H_i^n ((S_\alpha)_{i+1/2} - (S_\alpha)_{i-1/2})] \quad (49)$$

Here, U_i^n is the vector of unknowns on the nth layer over the time ($t = t_n$) in a cell $j_1 = i$, j_2, j_3; V_i the volume of the calculated cell; S_α is the area of projection of the corresponding cell edge onto the plane normal to the direction $\alpha(x, y, \text{or } z)$.

2. The corrector (the second step of the McCormack scheme):

$$\tilde{\tilde{U}}_i = \frac{1}{2}\left\{U_i^n + \tilde{U}_i - \frac{\Delta t}{V_i} \sum_{\alpha = x, y, z} \left[\left((\tilde{F}_\alpha)_i (S_\alpha)_{i+1/2} - (\tilde{F}_\alpha)_{i-1} (S_\alpha)_{i-1/2}\right) - -\tilde{H}_i ((S_\alpha)_{i+1/2} - (S_\alpha)_{i-1/2})\right]\right. \quad (50)$$

Here, the quantities $\tilde{F}_\alpha$, $\tilde{H}$ are calculated using the values of gas dynamic parameters obtained on the predictor step.

3. Matched conservative smoothing:

$$U_i^{n+1} = \tilde{\tilde{U}}_i + \frac{1}{V_i}(\varphi_{i+1/2}^* - \varphi_{i-1/2}^*)$$

$$\varphi_{i+1/2}^* = \begin{cases} \varphi_{i+1/2}, & \text{if } (\delta\tilde{\tilde{\rho}}_{i+1/2} \cdot \delta\tilde{\tilde{\rho}}_{i+3/2} < 0) \vee (\delta\tilde{\tilde{\rho}}_{i+1/2} \cdot \delta\tilde{\tilde{\rho}}_{i-1/2} < 0) \\ 0 & \text{in an opposite case} \end{cases}$$

$$\varphi_{i+1/2} = Q\delta U_{i+1/2} / (1/V_i + 1/V_{i+1})$$

$$\delta f_{i+1/2} = f_{i+1} - f_i \quad (51)$$

Here, Q is the constant coefficient. It is usually assumed in calculations that $Q \approx 0.1$. The mentioned algorithm of smoothing makes it possible to effectively suppress nonphysical oscillations originating in the vicinity of discontinuities without reducing the order of scheme approximation on smooth solutions. The source term H, not distorting the approximation of the initial system of equations, provides an accurate description of stationary gas flows at constant pressure of arbitrary grids.

In view of nonlinearity, one-dimensional operators $L_1(\Delta t)$, $L_2(\Delta t)$, and $L_3(\Delta t)$ do not commute, thus leading to the reduction of the order of approximation of $L(\Delta t) = L_1(\Delta t)\, L_2(\Delta t)\, L_3(\Delta t)$ as compared with the order of approximation of one-dimensional operators. A symmetric succession of one-dimensional operators

$$L(2\Delta t) = L_1(\Delta t) L_2(\Delta t) L_3(\Delta t) L_3(\Delta t) L_2(\Delta t) L_1(\Delta t)$$

allows the provision of the second orders of approximation.

The scheme given is conventionally stable. The maximum possible value of Δt is determined by the CFL condition of the stability of one-dimensional operators $L_1(\Delta t)$, $L_2(\Delta t)$, $L_3(\Delta t)$:

$$\Delta t \leq \min \{\Delta t_1, \Delta t_2, \Delta t_3\}$$

$$\Delta t_1 = \min_{j_1, j_2, j_3} \left\{ \frac{V_{j_1 j_2 j_3}}{\sum\limits_{\alpha = x, y, z} (u_\alpha S_\alpha)_{j_1 j_2 j_3} + cS_{j_1 j_2 j_3}} \right\}$$

where c is the sound velocity,

$$(S_\alpha)_{j_1 j_2 j_3} = \frac{1}{2}\left[(S_\alpha)_{j_1 + 1/2, j_2, j_3} + (S_\alpha)_{j_1 - 1/2, j_2, j_3}\right]$$

$$S_{j_1 j_2 j_3} = \sqrt{\sum_{\alpha = x, y, z} (S_\alpha)^2_{j_1 j_2 j_3}}$$

Δt_2 and Δt_3 are determined similarly. It is assumed in calculations that

$$\Delta t = C \cdot \min \{\Delta t_1, \Delta t_2, \Delta t_3\}$$

where $C = 0.8 \ldots 0.9$.

2.5. Computation Example

To illustrate the possibilities of numerical methods we shall consider the problem on the motion of a shock wave and of a cocurrent flow behind it in the vicinity of conjugation of two rectangular channels.[18] Figure 303 depicts the flow patterns by an interferogram obtained in a shock tube at the Mach number **M** = 1.7 for the time instant when a shock wave has already passed the region of channels conjugation. Interference bands are lines of equal density of gases. They correspond to the step $\Delta\rho/\rho_0 = 0.083$. On the interferogram one can see the position of shock waves, the presence of a vortex formed as a result of cocurrent flow passing an angle, and specific features of the interaction between a shock wave reflected from the wall and this vortex.

The calculation of the considered flow imposes rather strict demands on the numerical method. The results of calculations performed using the above-considered numerical techniques are given in Figure 304 to 306, where, with the step $\Delta\rho/\rho_0 = 0.166$, the lines of constant densities are constructed for the time instant corresponding to Figure 303. All the calculations are made on a uniform grid with the number of nodes 103×54. Figure 304 presents the results obtained by the method of large particles, Figure 305 presents results obtained by the Godunov method, and by the monotonized McCormack method in Figure 306. Calculations via the method of large particles were conducted in two versions: without artificial viscosity (Figure 304a) and with artifical "viscous" pressure of the type $q = -\eta\, \partial u/\partial x$ being introduced on the Euler stage (Figure 304b). In the case of calculation without artificial viscosity the solution in a considerable portion of the computational region is of an oscillating character, with the oscillations in the cross-hatched portion of the figure

[18] Zhmakin, A. I., Komissaruk, V. A., Mende, N. P., Fursenko, A. A., and Tsimbalov, V. V., On the Propagation of Shock Waves in the Vicinity of Conjugation of Plane Channels, Preprint No. 950, Leningrad Institute of Physical Engineering, USSR Academy of Sciences, Leningrad, 1985.

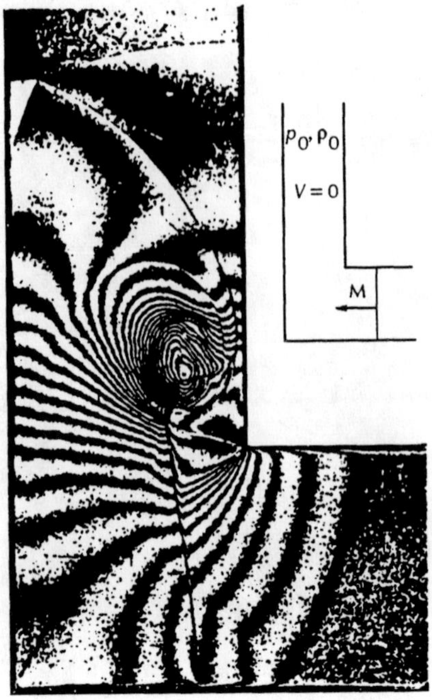

Fig. 303

being so great that construction of isolines here was not expedient. The introduction of artificial viscosity with $\eta = 0.5$ nearly completely avoids oscillations, but leads to large "smearing" of gas dynamic discontinuities. It is seen from the comparison of the results that the structure of the considered flow is most accurately described by the monotonized McCormack scheme which has a higher (second) order of approximation.

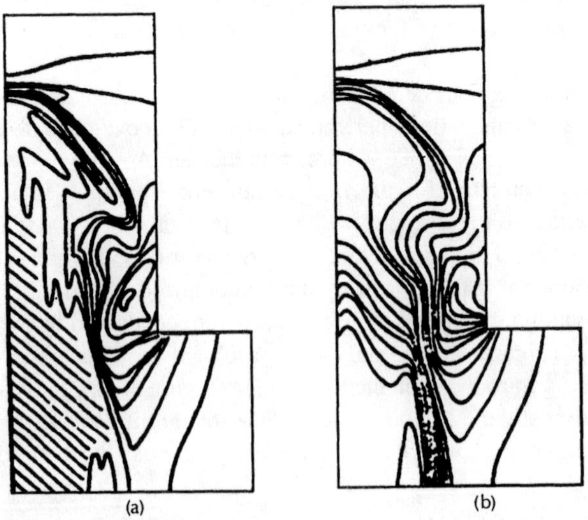

Fig. 304

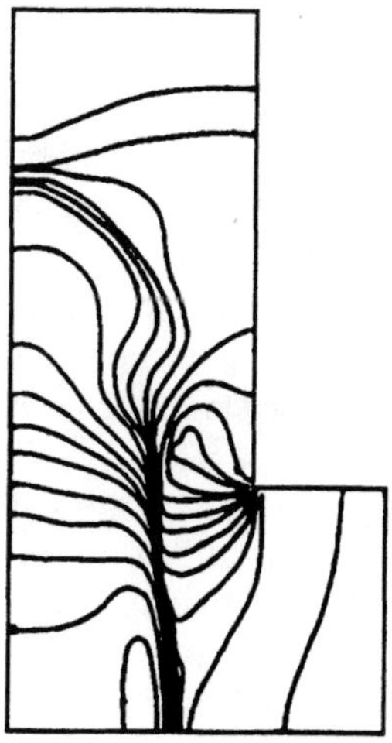

Fig. 305

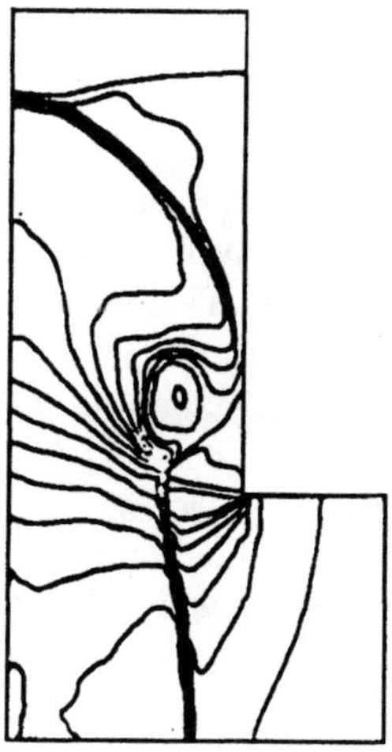

Fig. 306

3. Methods for Solving the Boundary Layer Equations

Chapters 12 to 15 present the formulations of various problems on near-wall and "free" boundary layers. These formulations in some cases admit self-similar solutions to which correspond ordinary differential equations. However, in real situations as a rule one has to deal with nonself-similar flows.

Numerical methods of solving such problems of a boundary layer (at least for the traditional problems of this class) form a more complete picture as compared with the numerical methods for the equations of an inviscid medium and complete Navier-Stokes equations. The explanation of this is in the fact that the initial formulations of boundary layer problems do not involve many of the prerequisites that offer great difficulties when solving the Euler and Navier-Stokes equations. Thus, the problem of accounting for compressibility and, in particular, for the transition through sonic lines and surfaces at $\mathbf{M} = 1$ is not insurmountable boundary layer problems. Pressure distribution in a boundary layer is usually assigned *a priori*, and the problem of the presence of the small parameter $1/\mathbf{Re}$ at senior derivatives, that is acute for viscous fluid dynamics, is in fact solved by the familiar transformation of the coordinate transverse to flow (Chapter 12).

We shall restrict ourselves to the consideration of the system describing a laminar stationary plane ($s = 0$) or axisymmetric ($s = 1$) flow of a homogeneous gas in a boundary layer and involving the continuity, momentum, and energy equations

$$\frac{\partial}{\partial x}(\rho u r^s) + \frac{\partial}{\partial y}(\rho v r^s) = 0 \tag{52}$$

$$\rho u \frac{\partial u}{\partial x} + \rho v \frac{\partial u}{\partial y} = -\frac{dp}{dx} + \frac{\partial}{\partial y}\left(\mu \frac{\partial u}{\partial y}\right) \tag{53}$$

$$\rho u \frac{\partial h}{\partial x} + \rho v \frac{\partial h}{\partial y} = u\frac{dp}{dx} + \frac{\partial}{\partial y}\left(\frac{\mu}{\mathbf{Pr}} \frac{\partial h}{\partial y}\right) + \mu\left(\frac{\partial u}{\partial y}\right)^2 \tag{54}$$

The notation corresponds to that adopted in Chapter 12. Gas is taken to be ideal with the state equation $p/\rho = RT/m$.

The system (52) to (54) includes the first order equation (52) and two quasi-linear second order equations (53) and (54). The equations are related to each other only by the coefficients dependent on the sought functions and also, thanks to the presence of the term $\mu\,(\partial u/\partial y)^2$ in the energy equation, by the first derivative of u with respect to y.

In accordance with the general parabolic type of the system, the boundary conditions at $y = 0$, on a wall, and at $y = \delta(x)$ (or for $y \to \infty$ in asymptotic approach), on the outer edge of the boundary layer, as well as the initial conditions (at the origin of the coordinate x) should be formulated for it.

The boundary layer equations allow the "layered" calculation along the coordinate x, which in this sense is similar to the variable t in the model equation (29). However, in Equations 53 and 54 the longitudinal velocity u, which vanishes at $y = 0$, is one of the factors with a derivative with respect to x. In view of this the use of ordinary explicit schemes becomes difficult because the stability condition of these schemes would require a practically impossible division of Δx (Section 1). One explicit scheme, the three-layer Dufort-Frankel[19] scheme has value for the boundary layer problems. As a rule, calculations of flows in boundary layers are based on implicit schemes. In what follows we shall consider three rather typical difference schemes.

3.1. Brailovskaya and Chudov Scheme

We shall give this scheme[20] for a plane ($s = 0$) stationary flow described by Equations 52 to 54.

Each of the quasi-linear second-order equations may be presented in the following form

$$a\frac{\partial\varphi}{\partial x} + b\frac{\partial\varphi}{\partial y} - \frac{\partial}{\partial y}\left(c\frac{\partial\varphi}{\partial y}\right) = d \tag{55}$$

For the motion equation $\varphi = u$, $c = \mu$, $d = -dp/dx$; for the energy equation $\varphi = h$, $c = \mu/\mathbf{Pr}$, $d = u\dfrac{dp}{dx} + \mu\left(\dfrac{\partial u}{\partial y}\right)^2$; for both equations $a = \rho u$, $b = \rho v$.

The boundary conditions in a general form are

$$y = 0:\quad \alpha\frac{\partial\varphi}{\partial y} + \beta\varphi = \gamma \quad v = 0 \tag{56}$$

$$y = \delta(x):\quad \varphi = \Phi(x) \tag{57}$$

Equation 56 makes it possible to express the condition both on a thermally insulated surface and on the surface of the prescribed temperature.

In the plane (x,y) we shall introduce a uniform difference grid with the meshes Δx, Δy. The values of the sought functions $\varphi_m^n = \varphi(x_n, y_m)$ will be determined in the nodes of the main grid $x_n = x_0 + n\Delta x$, $y_m = m\Delta y$ ($m, n = 0, 1, \ldots$) with integer indices. Moreover, we shall introduce an auxiliary grid with the nodes $\{x_n, y_{m+1/2} = (m+1/2)\Delta y\}$ and $\{x_{n+1/2} = x_0 + (n+1/2)\Delta x, y_m\}$ which have half-integer indices. We shall find the coef-

---

[19] Roache, P., *Vychislitelnaya gidrodinamika* (*Computational Hydrodynamics*), Mir Publishers, Moscow, 1980, 95–101 (Russian translation).

[20] Brailovskaya, I. Yu. and Chudov, L. A., Solution of boundary layer equations by the difference technique, in *Vychislitelnye metody i programmirovaniye*, MGU Press, Moscow, 1962, 167–182.

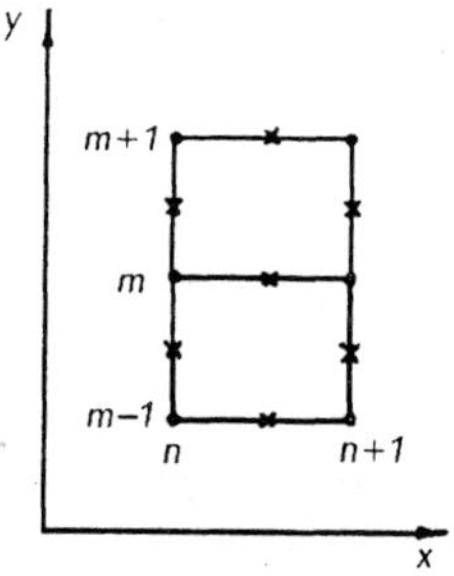

Fig. 307

ficients of equations in these nodes. Because these coefficients are dependent on the sought functions, we must use iterations.

We approximate Equation 55 by an implicit two-layer scheme on a six-point pattern (Figure 307):

$$a_m^{n+1/2}\frac{\varphi_m^{n+1}-\varphi_m^n}{\Delta x}+b_m^{n+1/2}\left(sK_m^{n+1}+(1-s)K_m^n\right)-$$

$$-sL_m^{n+1}-(1-s)L_m^n=d_m^{n+1/2} \qquad (58)$$

where

$$K_m^n=\frac{\varphi_{m+1}^n-\varphi_{m-1}^n}{2\Delta y} \qquad L_m^n=\frac{c_{m+1/2}^n(\varphi_{m+1}^n-\varphi_m^n)-c_{m-1/2}^n(\varphi_m^n-\varphi_{m-1}^n)}{\Delta y^2}$$

$1/2\le s\le 1$ is the weight coefficient. The approximation error is the value of the order $(s-1/2)\,\Delta x+\Delta x^2,\ \Delta y^2$ (at $s = 1/2$ the scheme has the second order of accuracy along the coordinate x).

The boundary conditions for φ^{n+1} in the difference form will be written as

$$\alpha^{n+1}\frac{4\varphi_1^{n+1}-3\varphi_0^{n+1}-\varphi_2^{n+1}}{2\Delta y}+\beta^{n+1}\varphi_0^{n+1}=\gamma^{n+1} \qquad (59)$$

(the error $O(\Delta y^2)$).

$$\varphi_{M(n+1)}^{n+1}=\Phi^{n+1} \qquad (60)$$

Here, $M(n + 1)$ is the number of the last (upper) node for the layer $n + 1$.

Calculations are performed in the following order. Let a, b, c, d be known in some approximation. Then the values of φ_m^{n+1} ($m = 0, 1, ..., M(n + 1)$) will be found from the system, which is composed of $M(n + 1) - 1$ three-point equations for φ_m^{n+1} (55) and the boundary conditions (59) and (60), by simple scalar factorization. The number of nodes on the layer $n + 1$ will be determined from the condition of smooth conjugation $|\varphi_{M(n+1)}^{n+1}-\varphi_{M(n+1)-1}^{n+1}|\le\varepsilon$ ($\varepsilon > 0$ is a small number), reflecting an asymptotic character of the transition of the sought functions to their values on the outer edge of a boundary layer.

After the determination of all the functions φ^{n+1} the transverse velocity $v^{n+1/2}$ is calculated, on an intermediate layer $x_{n+1/2}$ from the two-point difference continuity equation for $v^{n+1/2}$ approximated with the second order of accuracy over x and y:

$$\frac{1}{2\Delta x}[(\rho u)_m^{n+1} - (\rho u)_m^n] + \frac{1}{2\Delta x}[(\rho u)_{m+1}^{n+1} - (\rho u)_{m+1}^n] +$$

$$+\frac{1}{\Delta y}[(\rho v)_{m+1}^{n+1/2} - (\rho v)_m^{n+1/2}] = 0$$

The density ρ depends on p and T; therefore, in the given iteration it is the known function. Proceeding from the boundary condition $v_0^{n+1/2} = 0$, the rest of the values of $v_m^{n+1/2}$ are found.

The coefficients a, b, c depend on φ only, and on the right side in the energy equation d depends on $\partial u/\partial y$ also. The values of φ in the nodes of the auxiliary grid that are necessary for the calculation of the coefficients, are determined with the help of linear interpolation

$$\varphi_m^{n+1/2} = \frac{1}{2}(\varphi_m^n + \varphi_m^{n+1}) \qquad \varphi_{m+1/2}^n = \frac{1}{2}(\varphi_m^n + \varphi_{m+1}^n)$$

and the derivative $(\partial\varphi/\partial y)_m^{n+1/2}$ is expressed by the formula of second order of accuracy

$$\left(\frac{\partial\varphi}{\partial y}\right)_m^{n+1/2} = \frac{1}{2}\frac{\varphi_{m+1}^n - \varphi_{m-1}^n}{2\Delta y} + \frac{1}{2}\frac{\varphi_{m+1}^{n+1} - \varphi_{m-1}^{n+1}}{2\Delta y}$$

Because the initial problem is nonlinear (the coefficients a, b, c and the right side d in the energy equation depend on the unknown φ^{n+1}), iterations are used. The iteration process is finished as the condition $\|(\varphi^{n+1})^{l+1} - (\varphi^{n+1})^l\| \le \varepsilon_1$ is attained, where the notations $l + 1$ and l refer to two subsequent iterations.

The stability of this difference scheme was studied by a spectral technique. Stability takes place with the arbitrary path of Δx, Δy tending to zero. Numerous computations showed that the difference between the real problem and the model one (the boundary conditions, non-linearity, variable step over y) does not invalidate the scheme stability.

3.2. Compact Integro-Interpolation Scheme

This scheme is written, similar to that in Section 3.1, on a six-point pattern, however, the former has greater accuracy. First, in system (52) to (54) we shall pass over from the variables x, y to the Dorodnitsyn-Liez variables

$$\xi(x) = \int_0^x \rho_w \mu_w u_e r^{2s}\,dx \qquad \eta(x, y) = \frac{u_e r}{\sqrt{2\xi}}\int_0^y \rho\,dy \tag{61}$$

where the indices w and e refer to the body surface and outer edge of the boundary layer, respectively. The use of the variables ξ, η brings the equations closer to the form characteristic of an incompressible fluid, and also thanks to a self-similar construction of the coordinate η (of the type y/d), makes it possible to have a unique computational value of η_e over all of the boundary layer sections.

In the variables ξ, η the motion and energy equations take on the form

$$\frac{\partial}{\partial\eta}\left(l\frac{\partial\bar{u}}{\partial\eta}\right) + f\frac{\partial\bar{u}}{\partial\eta} + 2\xi\left(\frac{\partial f}{\partial\xi}\frac{\partial\bar{u}}{\partial\eta} - \bar{u}\frac{\partial\bar{u}}{\partial\xi}\right) = \beta\left(\bar{u}^2 - \frac{\rho_e}{\rho}\right) \tag{62}$$

$$\frac{\partial}{\partial\eta}\left(\frac{l}{\mathbf{Pr}}\frac{\partial\bar{h}}{\partial\eta}\right)+f\frac{\partial\bar{h}}{\partial\eta}+2\xi\left(\frac{\partial f}{\partial\xi}\frac{\partial\bar{h}}{\partial\eta}-\bar{u}\frac{\partial\bar{h}}{\partial\xi}\right) =$$

$$= \beta\frac{u_e^2}{h_0}\frac{\rho_e}{\rho}\bar{u}-l\frac{u_e^2}{h_0}\left(\frac{\partial\bar{u}}{\partial\eta}\right)^2 \qquad (63)$$

Here, $\bar{u} = u/u_e$, $\bar{h} = h/h_0$ (h_0 is the stagnation enthalpy in the outer flow), $l = \rho\mu/(\rho_w\mu_w)$, $f = \int_0^\eta \bar{u}d\eta$ is the reduced stream function $\beta = 2d\ln u_e/d\ln\xi$.

As a result of the introduction of the stream function the continuity equation is satisfied automatically.

Taking into account similarity of the left sides of Equations 62 and 63, we shall present each of them in the form

$$\frac{\partial}{\partial\eta}\left(a\frac{\partial\varphi}{\partial\eta}\right)+f\frac{\partial\varphi}{\partial\eta}+2\xi\left(\frac{\partial f}{\partial\xi}\frac{\partial\varphi}{\partial\eta}-\bar{u}\frac{\partial\varphi}{\partial\xi}\right) = R$$

or

$$\frac{\partial}{\partial\eta}\left(a\frac{\partial\varphi}{\partial\eta}\right)+f\frac{\partial\varphi}{\partial\eta}+2\xi\left[\frac{\partial}{\partial\eta}\left(\frac{\partial f}{\partial\xi}\varphi\right)-\frac{\partial}{\partial\xi}(\bar{u}\varphi)\right] = R \qquad (64)$$

where $\varphi = \bar{u}$, $a = l$ for the motion equation, $\varphi = \bar{h}$, $a = l/\mathbf{Pr}$ for the energy equation; R, the right sides of the corresponding equations.

The boundary-value conditions have the form: at $\eta = 0$: $\bar{u} = 0, f = 0$ (or $f = f_w(\xi)$ with injection of gas through the surface), $\bar{h} = \bar{h}_w(\xi)$ (the surface with prescribed temperature) or $\partial\bar{h}/\partial\eta = 0$ (thermally insulated surface); at $\eta = \eta_e$: $\bar{u} = 1$, $\bar{h}_e = \bar{h}_e(\xi)$; at $\xi = \xi_0$: $\bar{u} = \bar{u}^{(0)}(\eta)$, $\bar{h} = \bar{h}^{(0)}(\eta)$

The value of η_e is chosen rather great to manifest an asymptotic character of the behavior of the sought functions near the outer edge of the boundary layer. The "initial" profiles $\bar{u}^{(0)}(\eta)$ and $\bar{h}^{(0)}(\eta)$ are usually determined from the solution of the corresponding self-similar problem.

In the computational region we shall introduce the difference grid $\xi_n = \xi_0 + n\Delta\xi$, $\eta_m = m\Delta\eta$ ($n = 0, 1, ...; m = 0, 1, ..., M$) with the meshes $\Delta\xi$, $\Delta\eta$. Following the work by Shchennikov,[21] we shall integrate both sides of Equation 64 with respect to area S of the rectangular cell of division

$$\iint_S \frac{\partial}{\partial\eta}\left(a\frac{\partial\varphi}{\partial\eta}\right)d\xi d\eta + \iint_S f\frac{\partial\varphi}{\partial\eta}d\xi d\eta + 2\iint_S \xi\frac{\partial}{\partial\eta}\left(\frac{\partial f}{\partial\xi}\varphi\right)d\xi d\eta -$$

$$-2\iint_S \xi\frac{\partial}{\partial\xi}(\bar{u}\varphi)\,d\xi d\eta = \iint_S R d\xi d\eta \qquad (65)$$

[21] Shchennikov, V. V., Calculation of a laminar boundary layer along the side of the subliming body of revolution, *Zh. Vych. Mat. Mat. Fiz.*, 5(5), 142–144, 1965.

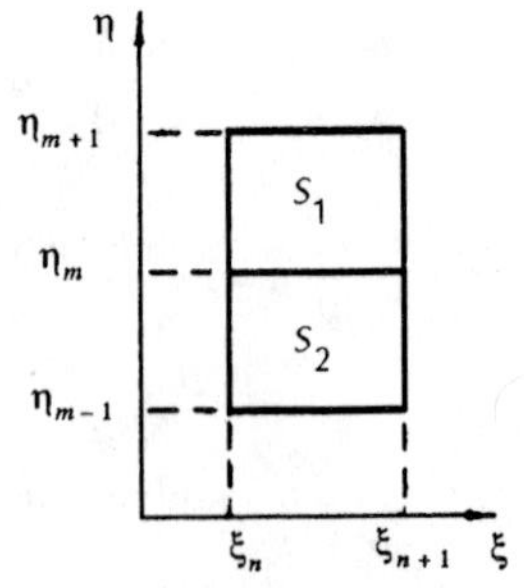

Fig. 308

We approximate the functions under the integrals by the quadratic relation over η and the linear relation over ξ and calculate the integrals in (65) for $S = S_1 + S_2$ and $S = S_1$ (Figure 308).

Then, with allowance for the corresponding results and estimates presented in Section 1, we shall have after the division into $\Delta\xi\Delta\eta$:

for $S = S_1 + S_2$:

$$\frac{3}{\Delta\eta}(P_{m+1} - P_{m-1})^{n+1/2} + \left[\left(f\frac{P}{a}\right)_{m-1} + 4\left(f\frac{P}{a}\right)_m + \left(f\frac{P}{a}\right)_{m+1}\right]^{n+1/2} +$$

$$+ 6\frac{\xi^{n+1/2}}{\Delta\eta}[(f_\xi\varphi)_{m+1} - (f_\xi\varphi)_{m-1}]^{n+1/2} -$$

$$- 2\frac{\xi^{n+1/2}}{\Delta\xi}\{[(\bar{u}\varphi)_{m-1} + 4(\bar{u}\varphi)_m + (\bar{u}\varphi)_{m+1}]^{n+1} -$$

$$-[(\bar{u}\varphi)_{m-1} + 4(\bar{u}\varphi)_m + (\bar{u}\varphi)_{m+1}]^n\} = (R_{m-1} + 4R_m + R_{m+1})^{n+1/2}$$

$$(\text{accuracy} \quad O(\Delta\xi^2 + \Delta\eta^4)) \tag{66}$$

for $S = S_2$:

$$\frac{12}{\Delta\eta}(P_m - P_{m-1})^{n+1/2} + \left[5\left(f\frac{P}{a}\right)_{m-1} + 8\left(f\frac{P}{a}\right)_m - \left(f\frac{P}{a}\right)_{m+1}\right]^{n+1/2} +$$

$$+ 24\frac{\xi^{n+1/2}}{\Delta\eta}[(f_\xi\varphi)_m - (f_\xi\varphi)_{m-1}]^{n+1/2} -$$

$$- 2\frac{\xi^{n+1/2}}{\Delta\xi}\{[5(\bar{u}\varphi)_{m-1} + 8(\bar{u}\varphi)_m - (\bar{u}\varphi)_{m+1}]^{n+1} -$$

$$-[5(\bar{u}\varphi)_{m-1} + 8(\bar{u}\varphi)_m - (\bar{u}\varphi)_{m+1}]^n\} = (5R_{m-1} + 8R_m - R_{m+1})^{n+1/2}$$

$$(\text{accuracy} \quad O(\Delta\xi^2 + \Delta\eta^3)) \tag{67}$$

Here, $P = a\partial\varphi/\partial\eta$, $f_\xi = \partial f/\partial\xi$; the quantities with the index n + 1/2 are determined as half sums on the layers n and $n + 1$.

Similarly, integrating the equalities $\partial\varphi/\partial\eta = P/a$ on the portions $\eta_{m-1} \le \eta \le \eta_{m+1}$ and $\eta_{m-1} \le \eta \le \eta_m$, we obtain the relations valid at any ξ:

$$\frac{3}{\Delta\eta}(\varphi_{m+1} - \varphi_{m-1}) = \left(\frac{P}{a}\right)_{m-1} + 4\left(\frac{P}{a}\right)_m + \left(\frac{P}{a}\right)_{m+1} \quad \text{(accuracy } O(\Delta\eta^4)\text{)} \tag{68}$$

$$\frac{12}{\Delta\eta}(\varphi_m - \varphi_{m-1}) = 5\left(\frac{P}{a}\right)_{m-1} + 8\left(\frac{P}{a}\right)_m - \left(\frac{P}{a}\right)_{m+1} \quad \text{(accuracy } O(\Delta\eta^3)\text{)} \tag{69}$$

Having written Equations 68 and 69 for the layer $n + 1/2$, we express P_{m-1}, P_m, P_{m+1} from systems (67) to (69). This system is linear and inhomogeneous, with respect to the sought quantities

$$\sum_{j=1}^{3} c_{ij} P^{n+1/2}_{m-2+j} = d_i \quad (i = 1, 2, 3) \tag{70}$$

with the coefficients

$$c_{11} = 5\left(\frac{f}{a}\right)^{n+1/2}_{m-1} - \frac{12}{\Delta\eta} \qquad c_{12} = 8\left(\frac{f}{a}\right)^{n+1/2}_{m} + \frac{12}{\Delta\eta} \qquad c_{13} = -\left(\frac{f}{a}\right)^{n+1/2}_{m+1}$$

$$c_{21} = \left(\frac{1}{a}\right)^{n+1/2}_{m-1} \qquad c_{22} = \left(\frac{4}{a}\right)^{n+1/2}_{m} \qquad c_{23} = \left(\frac{1}{a}\right)^{n+1/2}_{m+1}$$

$$c_{31} = \left(\frac{5}{a}\right)^{n+1/2}_{m-1} \qquad c_{32} = \left(\frac{8}{a}\right)^{n+1/2}_{m} \qquad c_{33} = -\left(\frac{1}{a}\right)^{n+1/2}_{m+1}$$

$$d_1 = -24\frac{\xi^{n+1/2}}{\Delta\xi\Delta\eta}\left[((f^{n+1} - f^n)\varphi^{n+1/2})_m - ((f^{n+1} - f^n)\varphi^{n+1/2})_{m-1}\right] +$$

$$+ 2\frac{\xi^{n+1/2}}{\Delta\xi}\left[(5(\bar{u}\varphi)_{m-1} + 8(\bar{u}\varphi)_m - (\bar{u}\varphi)_{m+1})^{n+1} -\right.$$

$$\left. - (5(\bar{u}\varphi)_{m-1} + 8(\bar{u}\varphi)_m - (\bar{u}\varphi)_{m+1})^n\right] + (5R_{m-1} + 8R_m - R_{m+1})^{n+1/2}$$

$$d_2 = \frac{3}{\Delta\eta}(\varphi_{m+1} - \varphi_{m-1})^{n+1/2} \qquad d_3 = \frac{12}{\Delta\eta}(\varphi_m - \varphi_{m-1})^{n+1/2}$$

Here, with the accuracy of $O(\Delta\xi^2)$ $f^{n+1/2}_{\xi}$ is replaced by the expression $(f^{n+1} - f^n)/\Delta\xi$. The solution of system (70) may be presented in the form

$$P^{n+1/2}_{m-1} = \frac{\Delta_1}{\Delta} = d_1 X_{m-1} + d_2 Y_{m-1} + d_3 Z_{m-1}$$

$$P^{n+1/2}_{m} = \frac{\Delta_2}{\Delta} = d_1 X_m + d_2 Y_m + d_3 Z_m$$

$$P^{n+1/2}_{m+1} = \frac{\Delta_3}{\Delta} = d_1 X_{m+1} + d_2 Y_{m+1} + d_3 Z_{m+1} \tag{71}$$

where $\Delta_1, \Delta_2, \Delta_3$ are the determinants obtained by the substitution of the first, second, and third columns, respectively, of the determinant Δ by the column of the right sides d; X, Y, Z are the coefficients at d_i $(i = 1, 2, 3)$ in the decomposition of the corresponding determinants by the elements of the columns of d. It is evident that X, Y, Z depend on f and a only.

Substituting (71) into (66), we come to the following expression relating the values of φ in the nodes of a six-point pattern and approximating the equations of motion and energy on this pattern with accuracy $O(\Delta\xi^2, \Delta\eta^3)$:

$$(\gamma_1)_m \varphi_{m-1}^{n+1} + (\gamma_2)_m \varphi_m^{n+1} + (\gamma_3)_m \varphi_{m+1}^{n+1} = \delta_m \quad (m = 1, \ldots, M-1) \tag{72}$$

Here,

$$(\gamma_1)_m = -\frac{3}{2\Delta\eta}B_Y - \frac{6}{\Delta\eta}B_Z + \frac{\xi^{n+1/2}}{\Delta\xi}\left\{B_x\left[10\bar{u}_{m-1}^{n+1} + \frac{12}{\Delta\eta}(f_{m-1}^{n+1} - f_{m-1}^{n})\right] - \right.$$
$$\left. -2\bar{u}_{m-1}^{n+1} - \frac{3}{\Delta\eta}(f_{m-1}^{n+1} - f_{m-1}^{n})\right\}$$

$$(\gamma_2)_m = \frac{6}{\Delta\eta}B_Z + \frac{\xi^{n+1/2}}{\Delta\xi}\left\{B_x\left[16\bar{u}_m^{n+1} - \frac{12}{\Delta\eta}(f_m^{n+1} - f_m^{n})\right] - 8\bar{u}_m^{n+1}\right\}$$

$$(\gamma_3)_m = \frac{3}{2\Delta\eta}B_Y + \frac{\xi^{n+1/2}}{\Delta\xi}\left[-2(1 + B_x)\bar{u}_{m+1}^{n+1} + \frac{3}{\Delta\eta}(f_{m+1}^{n+1} - f_{m+1}^{n})\right]$$

$$\delta_m = (\delta_1)_m \varphi_{m-1}^{n} + (\delta_2)_m \varphi_m^{n} + (\delta_3)_m \varphi_{m+1}^{n} + (R_{m-1} + 4R_m + R_{m+1})^{n+1/2} -$$
$$-(5R_{m-1} + 8R_m - R_{m+1})^{n+1/2}$$

$$(\delta_1)_m = \frac{3}{2\Delta\eta}B_Y + \frac{6}{\Delta\eta}B_Z + \frac{\xi^{n+1/2}}{\Delta\xi}\left\{B_x\left[10\bar{u}_{m-1}^{n} - \frac{12}{\Delta\eta}(f_{m-1}^{n+1} - f_{m-1}^{n})\right] - \right.$$
$$\left. -2\bar{u}_{m-1}^{n} + \frac{3}{\Delta\eta}(f_{m-1}^{n+1} - f_{m-1}^{n})\right\}$$

$$(\delta_2)_m = -\frac{6}{\Delta\eta}B_Z + \frac{\xi^{n+1/2}}{\Delta\xi}\left\{B_x\left[16\bar{u}_m^{n} - \frac{12}{\Delta\eta}(f_m^{n+1} - f_m^{n})\right] - 8\bar{u}_m^{n}\right\}$$

$$(\delta_3)_m = -\frac{3}{2\Delta\eta}B_Y + \frac{\xi^{n+1/2}}{\Delta\xi}\left[-2(1 + B_x)\bar{u}_{m+1}^{n} + \frac{3}{\Delta\eta}(f_{m+1}^{n+1} - f_{m+1}^{n})\right]$$

$$B_X = \frac{3}{\Delta\eta}(X_{m+1} - X_{m-1}) + \left(\frac{f}{a}\right)_{m-1}^{n+1/2} X_{m-1} + 4\left(\frac{f}{a}\right)_m^{n+1/2} X_m + \left(\frac{f}{a}\right)_{m+1}^{n+1/2} X_{m+1}$$

$$B_Y = \frac{3}{\Delta\eta}(Y_{m+1} - Y_{m-1}) + \left(\frac{f}{a}\right)_{m-1}^{n+1/2} Y_{m-1} + 4\left(\frac{f}{a}\right)_m^{n+1/2} Y_m + \left(\frac{f}{a}\right)_{m+1}^{n+1/2} Y_{m+1}$$

$$B_Z = \frac{3}{\Delta\eta}(Z_{m+1} - Z_{m-1}) + \left(\frac{f}{a}\right)_{m-1}^{n+1/2} Z_{m-1} + 4\left(\frac{f}{a}\right)_m^{n+1/2} Z_m + \left(\frac{f}{a}\right)_{m+1}^{n+1/2} Z_{m+1}$$

Equations 72 for $\varphi^{n+1} = \bar{u}^{n+1}$ and $\varphi^{n+1} = \bar{h}^{n+1}$ along with the corresponding boundary conditions are solved by scalar factorization and, in view of the nonlinear character of the equations, their solution requires iterations (the coefficients $\gamma_1, \gamma_2, \gamma_3, \delta_1, \delta_2, \delta_3$ depend on $\bar{u}^{n+1}$ and $\bar{h}^{n+1}$, and the right sides also depend on the derivatives $\partial\bar{u}/\partial\eta$).

Calculations are performed as follows:

1. The coefficients $(\gamma_{1,2,3})_m$, $(\delta_{1,2,3})_m$ and the values of the right sides R are calculated by the values $(\bar{u}^{n+1}, \bar{h}^{n+1})^l$ and $(\partial\bar{u}^{n+1}/\partial\eta)^l$ on the $(n + 1)$th layer in the lth iteration. The velocity derivatives are calculated by Equations 71.
2. Based on factorizations $(\bar{u}^{n+1}, \bar{h}^{n+1})^{l+1}$ in the $(l + 1)$th iteration are found. The presented stream function $f = \int_0^\eta \bar{u}d\eta$ is determined by some quadrature formula (for instance, by the Simpson formula).
3. In the absence of the coincidence with the required accuracy of functions in the two last subsequent iterations the iteration process should be continued.

To accelerate the convergence of iterations damping may be applied, i.e., calculation of the function $\left(\tilde{\varphi}^{n+1}\right)^{l+1}$, which replaces $(\varphi^{n+1})^{l+1}$ at the beginning of the next iteration cycle:

$$\left(\tilde{\varphi}^{n+1}\right)^{l+1} = (\varphi^{n+1})^l + \alpha[(\varphi^{n+1})^{l+1} - (\varphi^{n+1})^l]$$

(α is the damping coefficient).

3.3. Difference Schemes with Exponential Operator

One scheme, suggested by El-Mistikawy and Werle,[22] is especially useful for calculation of flows containing the regions of sharp variation of functions with a priori unknown locations. Under these conditions the difference schemes with the approximation of convective terms by central differences that are typical of the boundary layer may give nonmonotonous and even oscillating velocity profiles.

As an example illustrating the main idea of the scheme we shall consider a self-similar solution of the problem on a stationary flow of an incompressible fluid in a boundary with injection through a surface. The corresponding equation has the form (all notations are from the previous section)

$$\frac{d^2\bar{u}}{d\eta^2} + f\frac{d\bar{u}}{d\eta} + \beta(1 - \bar{u}^2) = 0 \tag{73}$$

where $f = \int_0^\eta \bar{u}d\eta + f_w$, f_w is the reduced stream function on a wall that characterizes injection. In accordance with this the boundary conditions have the form: $\bar{u} = 0, f = f_w$ at $\eta = 0$;

---

[22] El-Mistikawy, T. M. and Werle, M. J., Numerical method for boundary layers with blowing – the exponential box scheme, *AIAA J.*, 16(7), 749–750, 1978.

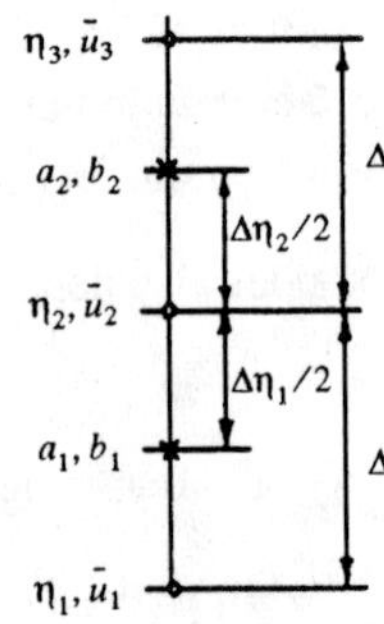

Fig. 309

$\bar{u} = 1$ at $\eta = \eta_e$.

Taking into consideration the iteration solution of the problem, we shall write expression (73) in the form

$$\frac{d^2\bar{u}}{d\eta^2} - (a+b)\frac{d\bar{u}}{d\eta} + ab\bar{u} = -\beta \tag{74}$$

where the coefficients $a = -f/2 + \sqrt{(f/2)^2 + \beta\bar{u}}$, $b = -f/2 - \sqrt{(f/2)^2 + \beta\bar{u}}$ are assumed to be known from the previous iteration.

Note that the term $\beta\bar{u}^2$, which in the previous scheme was referred to the right side of the equation, i.e., it was taken from the previous iteration, is now presented in the form $ab\bar{u}$ (a, b are known from the previous iteration) and added to other parts of the left (iterated) side.

We shall introduce the difference grid, which for the sake of generality, is nonuniform, with the nodes η_m. Consider two neighboring intervals between the nodes: $\eta_{m-1} \le \eta \le \eta_m$ and $\eta_m \le \eta \le \eta_{m+1}$; for brevity we shall designate η_{m-1}, η_m, η_{m1} by η_1, η_2, η_3 correspondingly, and the values of $\bar{u}$ in these nodes are $\bar{u}_1$, $\bar{u}_2$, $\bar{u}_3$; $\eta_2 - \eta_1 = \Delta\eta_1$, $\eta_3 - \eta_2 = \Delta\eta_2$ (Figure 309). Coefficients a and b are taken to be step functions equal to their values in the midpoints of the intervals. Then, the general solutions of Equation 74 may be written in the form

$$\bar{u}(\eta) = A_1 e^{a_1\eta} + B_1 e^{b_1\eta} + g_1 \quad \text{for} \quad \eta_1 \le \eta \le \eta_2 \tag{75}$$

$$\bar{u}(\eta) = A_2 e^{a_2\eta} + B_2 e^{b_2\eta} + g_2 \quad \text{for} \quad \eta_2 \le \eta \le \eta_3 \tag{76}$$

where $g_1 = -\beta/(a_1 b_1)$, $g_2 = -\beta/(a_2 b_2)$, and the parameters A, B are the integration constants. The parameters A_1, B_1 may be expressed from relation (75) in terms of the values of $\bar{u}_1$ and $\bar{u}_2$

$$\bar{u}_1 = A_1 e^{a_1(\eta_2 - \Delta\eta_1)} + B_1 e^{b_1(\eta_2 - \Delta\eta_1)} + g_1$$

$$\bar{u}_2 = A_1 e^{a_1\eta_2} + B_1 e^{b_1\eta_2} + g_1$$

where

$$A_1 = \frac{e^{b_1\eta_2}}{\Delta_1}\begin{vmatrix} \bar{u}_1 - g_1 & e^{-b_1\Delta\eta_1} \\ \bar{u}_2 - g_1 & 1 \end{vmatrix} \qquad B_1 = \frac{e^{a_1\eta_2}}{\Delta_1}\begin{vmatrix} e^{-a_1\Delta\eta_1} & \bar{u}_1 - g_1 \\ 1 & \bar{u}_2 - g_1 \end{vmatrix}$$

$$\Delta_1 = e^{(a_1+b_1)\eta_2}(e^{-a_1\Delta\eta_1} - e^{-b_1\Delta\eta_1}) \tag{77}$$

Passing over to solution on the outer range (76), we demand the coincidence between the values of the functions $\bar{u}$ and its derivative $\partial\bar{u}/\partial\eta$ on the lower limit $\eta = \eta_2$ of this range and the corresponding values on the upper limit of the previous range, i.e., the fulfillment of the equalities

$$\bar{u}_2 = A_2 e^{a_2\eta_2} + B_2 e^{b_2\eta_2} + g_2$$

$$A_1 a_1 e^{a_1\eta_2} + B_1 b_1 e^{b_1\eta_2} = A_2 a_2 e^{a_2\eta_2} + B_2 b_2 e^{b_2\eta_2}$$

Finding A_2 and B_2 from these equalities, we obtain

$$A_2 = \frac{e^{b_2\eta_2}}{\Delta_2}\begin{vmatrix} \bar{u}_2 - g_2 & 1 \\ A_1 a_1 e^{a_1\eta_2} + B_1 b_1 e^{b_1\eta_2} & b_2 \end{vmatrix}$$

$$B_2 = \frac{e^{a_2\eta_2}}{\Delta_2}\begin{vmatrix} 1 & \bar{u}_2 - g_2 \\ a_2 & A_1 a_1 e^{a_1\eta_2} + B_1 b_1 e^{b_1\eta_2} \end{vmatrix} \qquad \Delta_2 = e^{(a_2+b_2)\eta_2}(b_2 - a_2) \tag{78}$$

Substituting now into Equation 75 at $\eta = \eta_3$

$$\bar{u}_3 = A_2 e^{a_2(\eta_2+\Delta\eta_2)} + B_2 e^{b_2(\eta_2+\Delta\eta_2)}$$

relations (77) for A_2, B_2 and expressing A_1, B_1, entering into them, in terms of $\bar{u}_1$, $\bar{u}_2$ (according to formulas Equation 76 we shall have after reductions

$$r_1\bar{u}_1 - (s_1 + s_2)\bar{u}_2 + r_2\bar{u}_3 = (r_1 - s_1)g_1 + (r_2 - s_2)g_2 \tag{79}$$

where

$$r_1 = (a_1 - b_1)e^{b_1\Delta\eta_1}/t_1 \qquad r_2 = (a_2 - b_2)e^{-a_2\Delta\eta_2}/t_2$$

$$s_1 - [a_1 - (1 - t_1)b_1]/t_1 \qquad s_2 = [a_2(1 - t_2) - b_2]/t_2$$

$$t_1 = 1 - e^{-(a_1 - b_1)\Delta\eta_1} \qquad t_2 = 1 - e^{-(a_2 - b_2)\Delta\eta_2}$$

Similar three-point relations also may be derived for other ranges. As a result we obtain the system of difference equations with a tridiagonal matrix that is solved on each iteration by ordinary factorization.

The condition of diagonal predominance of the matrix $|s_1 + s_2| \geq |r_1| + |r_2|$ is fulfilled at

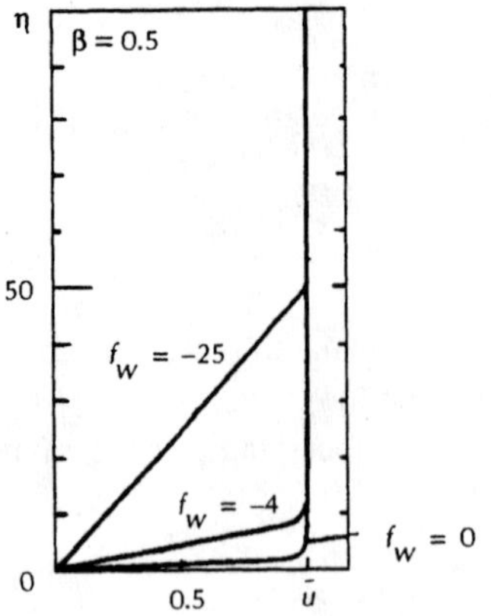

Fig. 310

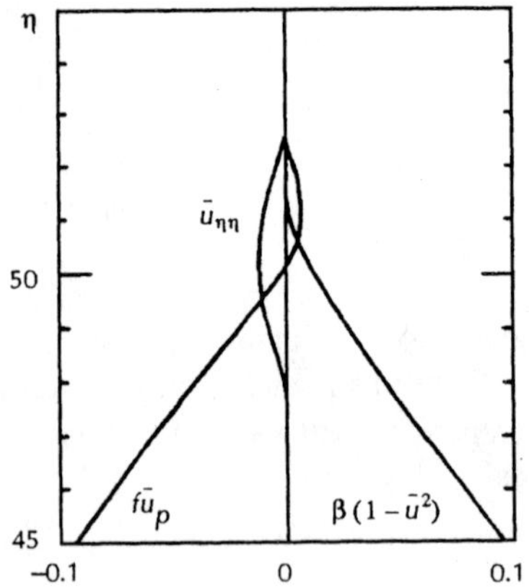

Fig. 311

$ab = -\beta\bar{u} < 0$. The values of $f = \int_0^{\eta} \bar{u}d\eta + f_w$ are calculated on each iteration after the determination of grid values of $\bar{u}$ by analytical integration of relations of the type of (74) and (75) within the corresponding ranges.

We shall take profiles of the longitudinal velocity $\bar{u}$ that correspond to a self-similar solution for the vicinity of the critical point on an axisymmetric body ($\beta = 0.5$) at $f_w = 0, -4$, and -25 (Figure 310), as an example demonstrating specific features of a flow in the presence of intensive injection. Figure 311 gives an idea of the relative contribution of separate terms of the equation of momentum balance for the layer portion $45 \le \eta \le 55$ ($f_w = -25$). It is seen that for $\eta < 48$ the convective terms and the pressure gradient counterbalance each other and the effect of viscous forces becomes commensurate with them only above this portion.

The considered difference scheme automatically changes over smoothly from the boundary layer-type scheme ($\eta > 48$) to that with the approximation of convective terms by the differences opposite to the flow (virtually, inviscid region, $\eta < 48$). In view of this no difficulties or instabilities arise in calculations, even on uniform grids. In both subregions of the flow this scheme provides the second order of accuracy.

4. Methods for Solving the Navier-Stokes Equations of Incompressible Fluid

The Navier-Stokes equation for an isothermal flow of an incompressible fluid has the form (Equation 29, Chapter 10):

$$\frac{\partial \boldsymbol{V}}{\partial t} + (\boldsymbol{V} \cdot \nabla)\boldsymbol{V} = -\frac{1}{\rho}\operatorname{grad} p + \nu\nabla^2\boldsymbol{V} \tag{80}$$

(the term corresponding to volumetric forces is omitted). The system consisting of the equation of momentum balance (80) and the continuity (incompressibility) equation

$$\operatorname{div} \boldsymbol{V} = 0 \tag{81}$$

is usually called the system of Navier-Stokes equations. After passing over to the dimensionless variables (the lengths and velocities are related to the scales L and U, correspondingly, the time to L/U, the pressure to ρU^2) Equation 80 is reduced to the form

$$\frac{\partial V}{\partial t} + (V \cdot \nabla) V = -\operatorname{grad} p + \frac{1}{\mathbf{Re}} \nabla^2 V \tag{82}$$

where the previous notation is left for the dimensionless quantities and spatial operators.

Use is often made of the conservative form of the motion equation

$$\frac{\partial V}{\partial t} + \operatorname{Div}(VV) = -\operatorname{grad} p + \frac{1}{\mathbf{Re}} \nabla^2 V \tag{83}$$

which differs from the initial one only by the presentation of the convective term.

We emphasize an essential specific feature of the system of Equations 81, 80 or 82, 80: Pressure p, entering into the motion equation under the gradient sign, is free of any other mathematical relations. This is one of the differences of the Navier-Stokes equations for an incompressible fluid from the corresponding equations for gas, where pressure is related to density and temperature in terms of the state equation $p/\rho = RT/m$. Hence, it follows that in the system of Equations 81, 80 or 82, 80 the quantity p plays the only role of the potential of the vector of nonviscous surface forces per unit volume.

The distribution of this potential p at each time instant should be such that the constant solenoidality of the velocity field V ($\operatorname{div} V = 0$, the sources or sinks are absent) is provided. In this sense the problem of the field $p(r, t)$ determination is quasistationary, whereas to determine a nonstationary velocity field $V(r, t)$ from Equations 82 or 83, including $\partial V/\partial t$, one should solve the problem with the initial conditions.

In a stationary case the system of the Navier-Stokes equations for incompressible fluid possesses elliptical properties and in a nonstationary case, parabolic-elliptical properties (the latter is due to the presence of the pressure p with a quasi-stationary character of determination in the momentum equation).

The formulation of initial and boundary equations for the sought functions V and p is determined by the mentioned mathematical type of equations.

The initial (at $t = t_0$) velocity distribution $V_0(r)$ should be prescribed for the problem of a nonstationary flow. (The function $V_0(r)$ should naturally satisfy the condition $\operatorname{div} V_0 = 0$.) The conditions for the components of velocity vector are formulated on the region boundary. These conditions depend both on the problem character and on the type of the corresponding boundary portion and were discussed in Chapter 10. Therefore, we shall discuss here only those conditions whose formulations in the numerical solution of the Navier-Stokes equations have various specific features.

For the problems of the flow past bodies, which is homogeneous at infinity from them, the condition $V|_{r\to\infty} \to V_\infty$ takes place. In numerical calculations a similar condition is usually prescribed at large but finite distance $r = R \gg L$ (with the case of a transformation of coordinates such that physical infinity corresponds to a finite value of the new coordinates excluded). In view of this, to reduce the error conditions are used that follow from asymptotic estimates of the solution behavior at large r. Such estimates, for instance, for a longitudinal flow past a cylinder may be found in the paper by Babenko et al.[23]

For inner problems (for example, flows in tubes, channels) the conditions for velocity in inlet and outlet section should be formulated.

The inlet conditions ($x = x_1$) are determined by fluid supply in a channel and are sometimes reduced to the formulation of all of the velocity components: $V|_{x=x_1} = V_1(y, z, t)$. In

[23] Babenko, K. I., Vedenskaya, N. D., and Orlova, M. G., Calculation of stationary viscous fluid flow past a round cylinder, *Zh. Vych. Mat. Mat. Fiz.*, 15(1), 183–196, 1975.

other cases only the velocity component normal to the initial section is fixed, while for other components the "mild" conditions of the type $\partial^k V_i/\partial x^k = 0$ $(k = 1, 2)$ are written.

A more complicated case is the formulation of the conditions at the channel outlet ($x = x_2$). Two approaches have been developed here. In the first of them the solutions corresponding to the steady-state flow are used as the boundary conditions. For example, for a rather long cylindrical tube this is the Poiseuille velocity profile (Chapter 10). It is possible to assign velocity profiles from the parameter family of solutions of the boundary layer equations, etc.

In the other approach "mild" conditions are formulated. We note that in a real flow at $x = x_2$ the conditions of the type $\partial^k V_i/\partial x^k = 0$ may not be fulfilled. However, as is shown by calculations, in many cases the effect of such conditions extends upstream along a relatively small (as compared to the channel length $x_2 - x_1$) portion, thus making their application nonexpedient in the absence of a priori information on the flow parameters at the outlet. To quantitatively estimate a "deforming" influence of these conditions on the calculation results it is preferable to vary the coordinate of the section x_2 where these conditions are prescribed.

In principle, pressure does not require any conditions with the exception of the assignment of $p(t)$ at some point of the region to determine the level of its reckoning.

Returning to the Navier-Stokes equations, we shall note that determination of pressure p entering into the initial forms of these equations requires special techniques in the process of numerical integration. The potential vector grad p can be easily eliminated from Equations 82 and 83 operating on all of their terms by rot. As a result we come to the Helmholtz equation of vortex transfer Ω in the viscous fluid flow (Equation 161, Chapter 10)

$$\frac{\partial \Omega}{\partial t} + (V \cdot \nabla)\,\Omega = (\Omega \cdot \nabla)\,V + \nu\nabla^2\Omega \tag{84}$$

or in the conservative form

$$\frac{\partial \Omega}{\partial t} + \mathrm{Div}\,(V\Omega - \Omega V) = \nu\nabla^2\Omega \tag{85}$$

The use of the system consisting of the vortex transfer equation (84) or (85) and the continuity equation (81) is the most effective for two-dimensional flows (for example, plane or axisymmetric), when two of the three components of the vector Ω are equal to zero and the stream function may be introduced. For simplicity we shall consider a plane flow. Then, only the component $\Omega = \Omega_z = \partial v/\partial x - \partial u/\partial y$ may differ from zero, so that the scalar equation of the zth vortex component transfer

$$\frac{\partial \Omega}{\partial t} + (V \cdot \nabla)\,\Omega = \nu\nabla^2\Omega \tag{86}$$

in combination with the continuity equation (81) remains to be solved. Introducing the stream function ψ, we shall obtain, according to the determination of Ω,

$$\nabla^2\psi = -\Omega \tag{87}$$

Thus, the system includes the parabolic equation of vortex transfer (86) and the elliptic Poisson equation for the stream function (87).

Now we shall consider the problem of initial and boundary conditions for the Navier-Stokes equations written in the variables vortex-stream function. It should be emphasized that relative simplification of the type of the system of Equations 86 and 87, as compared to the initial system of Equations 80 and 81 is accompanied by the appearance of the problem of formulation and numerical expression of the boundary conditions for a vortex on a solid surface.

The solution of the parabolic equation (86) requires the initial (at $t = t_0$) vortex distribution $\Omega_0(\boldsymbol{r})$ to be prescribed. Then, the stream function ψ_0 for $t = t_0$ will be determined as a result of the integration of the Poisson equation (87) $\nabla^2\psi_0 = -\Omega_0$. Here, we pay attention to the following. When solving Equation 87 with the boundary conditions for ψ, which correspond to a viscous flow (for example, in an enclosure with impermeable walls $\psi_w = 0$, $(\partial\psi/\partial n)_w = 0$), formally there arises the redefinition or overdetermination of the boundary conditions. Thus, having integrated the impermeability condition $\psi_w = 0$, the Poisson equation with the arbitrary right side $\Omega_0(\boldsymbol{r})$ we obtain some function $\psi_0(\boldsymbol{r})$, which does not generally satisfy the adhesion condition $(\partial\psi_0/\partial n)_w = 0$. It ia clear that the fulfillment of the adhesion condition at $t = t_0$ (similar to each subsequent time layer) may be provided only at the expense of the certain agreement between this condition and the vortex distribution Ω on the right side of Equation 87. This problem will be considered on the derivation of the difference analogue of the boundary condition for the vortex on a solid surface.

In the physical formulation the initial velocity distribution $V_0(\boldsymbol{r})$ is usually known. From it $\psi_0(\boldsymbol{r})$ and $\Omega_0(\boldsymbol{r})$ may be calculated, and in this case Equation 87 will be automatically fulfilled.

Besides the initial condition the boundary conditions for the functions Ω and ψ should also be assigned. The formulations of the majority of hydrodynamic problems are such that on the boundaries either the velocities themselves or their derivatives over spatial variables are prescribed. The passage to the conditions for the stream function is rather obvious. As to the boundary conditions for a vortex, the greatest difficulties arise due to the derivation of the conditions for a vortex on solid surfaces that are absent in the physical formulation. These conditions are discussed separately; however, first we shall discuss other conditions for a vortex:

1. At an infinite distance from a streamlined body in a homogeneous flow the condition $\Omega|_{r\to\infty} \to 0$ takes place. If the condition of transition to the homogeneous flow is formulated at finite distance $R \gg L$, then use is made of the corresponding asymptotic expressions.[23]
2. The condition $\Omega = 0$ on the symmetry axis $y = \text{const}$ follows from the equalities $v = 0$, $\partial u/\partial y = 0$ at $y = \text{const}$.
3. For the problems of internal flows the forms of initial (for the velocity components) conditions at the inlet and outlet are rather various. On this basis, the conditions for a vortex should be obtained. This problem is most easily solved if the profile of a longitudinal velocity $u(y)$ and a "mild" condition for a transverse velocity $\partial v/\partial x = 0$ are prescribed, thus directly giving $\Omega = -\partial u/\partial y$. The analysis of other possible cases may be found in the book by Roache.

To numerically integrate the two-dimensional equation of vortex transfer (86) and the Poisson equation for stream function (87) a great number of different schemes is suggested. We shall consider the methods of finite differences and various types of the conditions for a vortex on a solid surface that are characteristic for these problems.

4.1. An Implicit Scheme of Alternate Directions

We shall write Equations 86 and 87, having reduced them to a dimensionless form, in a rectangular Cartesian coordinate system

$$\frac{\partial\Omega}{\partial t}+\frac{\partial\psi}{\partial y}\frac{\partial\Omega}{\partial x}-\frac{\partial\psi}{\partial x}\frac{\partial\Omega}{\partial y}=\frac{1}{\mathbf{Re}}\left(\frac{\partial^2\Omega}{\partial x^2}+\frac{\partial^2\Omega}{\partial y^2}\right) \tag{88}$$

$$\frac{\partial^2\psi}{\partial x^2}+\frac{\partial^2\psi}{\partial y^2}=-\Omega \tag{89}$$

Let us introduce a uniform difference grid with the meshes Δx and Δy. Let the values of Ω^n, ψ^n in the grid nodes on a time layer tn be known. To determine the grid function Ω^{n+1} on the layer $t_{n+1}=t_n+\Delta t$ we shall write the following scheme of alternate directions:[24]

$$\frac{\Omega^{n+1/2}-\Omega^n}{\Delta t/2}=\frac{1}{\mathbf{Re}}\left(\frac{\partial^2\Omega^{n+1/2}}{\partial x^2}+\frac{\partial^2\Omega^n}{\partial y^2}\right)-\frac{\partial\psi^n}{\partial y}\frac{\partial\Omega^{n+1/2}}{\partial x}+\frac{\partial\psi^n}{\partial x}\frac{\partial\Omega^n}{\partial y}$$

$$\frac{\Omega^{n+1}-\Omega^{n+1/2}}{\Delta t/2}=\frac{1}{\mathbf{Re}}\left(\frac{\partial^2\Omega^{n+1/2}}{\partial x^2}+\frac{\partial^2\Omega^{n+1}}{\partial y^2}\right)-\frac{\partial\psi^n}{\partial y}\frac{\partial\Omega^{n+1/2}}{\partial x}+\frac{\partial\psi^n}{\partial x}\frac{\partial\Omega^{n+1}}{\partial y} \tag{90}$$

The quantities $\partial\Omega/\partial x$, $\partial^2\Omega/\partial x^2$, etc. in Equation 90 denote the difference analogues of the corresponding derivatives; spatial indices are omitted.

The scheme approximates Equation 88 with the order $O(\Delta t)$ because the derivatives of ψ in the convective terms are taken on the layer t_n. $\Omega^{n+1/2}$ is calculated by factorizations along the portions y = const, and Ω^{n+1} along the portions x = const.

The determination of the stream function ψ on each time layer requires solution of the Poisson equation (89), with the values of Ω of the right side corresponding to this layer. For the difference equation approximating (89) this solution may be obtained, for example, by the time-dependent technique also following the scheme of alternate directions:

$$\frac{\psi^{(l+1/2)}-\psi^{(l)}}{\tau}=\frac{\partial^2\psi^{(l+1/2)}}{\partial x^2}+\frac{\partial^2\psi^{(l)}}{\partial y^2}+\Omega$$

$$\frac{\psi^{(l+1)}-\psi^{(l+1/2)}}{\tau}=\frac{\partial^2\psi^{(l+1/2)}}{\partial x^2}+\frac{\partial^2\psi^{(l+1)}}{\partial y^2}+\Omega \tag{91}$$

where τ is the iteration parameter and l and $l+1$ are the iteration indices. The solution obtained by the time-dependent technique will correspond to the same time layer (not assigned in Equations 91) to which corresponds the function Ω on the right side.

The study of stability will be conducted for the linear equations

$$\frac{\partial\Omega}{\partial t}=\frac{1}{\mathbf{Re}}\nabla^2\Omega+A\frac{\partial\Omega}{\partial x}+B\frac{\partial\Omega}{\partial y} \tag{92}$$

$$\nabla^2\psi=-\Omega \tag{93}$$

where A, B are the constants, with $|A|, |B| \le 1$. In the dimensionless equation (92) the latter condition are provided due to the choice of the velocity scale. We shall assume that all the derivatives with respect to x, y will be approximated by symmetric differences, for instance, $(\partial\Omega/\partial x)_{m,s}\sim(\Omega_{m+1,s}-\Omega_{m-1,s})/(2\Delta x)$.

[24] Kuskova, T. V., Numerical investigations of two-dimensional viscous incompressible fluid flows, in *Nekotorye primeneniya setok v gazovoy dinamike* (*Some Applications of Grids in Gas Dynamics*), Vol. 3, MGU Press, Moscow, 1971, 7–136.

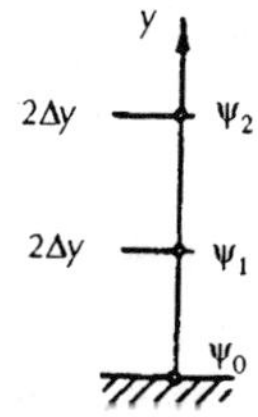

Fig. 312

Then the scheme will have the error over the spatial variables $O(\Delta x^2, \Delta y^2)$. In accordance with the spectral method we shall present the solution in the form

$$\Omega^n_{ms} = \lambda^n e^{i(\alpha m + \beta s)} \tag{94}$$

Substituting (94) into the scheme of alternate directions written for Equation 92, we obtain

$$\lambda = \frac{1 - 2æ_1 \sin^2\frac{\alpha}{2} + ic_1 \sin\alpha}{1 + 2æ_1 \sin^2\frac{\alpha}{2} - ic_1 \sin\alpha} \cdot \frac{1 - 2æ_2 \sin^2\frac{\beta}{2} + ic_2 \sin\beta}{1 + 2æ_2 \sin^2\frac{\beta}{2} - ic_2 \sin\beta}$$

where

$$æ_1 = \frac{1}{\mathbf{Re}}\frac{\Delta t}{\Delta x^2} \qquad æ_2 = \frac{1}{\mathbf{Re}}\frac{\Delta t}{\Delta y^2} \qquad c_1 = A\frac{\Delta t}{\Delta x} \qquad c_2 = B\frac{\Delta t}{\Delta y}$$

It is evident that $|\lambda| \le 1$ at any ratios between Δt, Δx, and Δy, i.e., the scheme in linear approximation is unconditionally stable.

Now we shall consider the derivation of approximate conditions for a vortex on a solid surface providing, with certain accuracy, fulfillment of the sticking conditions. The expansion of the stream function to the Taylor series in the wall vicinity ($y = 0$, Figure 312) forms the basis of this derivation

$$\psi_1 = \psi_0 + \left(\frac{\partial\psi}{\partial y}\right)_0 \Delta y + \frac{1}{2}\left(\frac{\partial^2\psi}{\partial y^2}\right)_0 \Delta y^2 + \frac{1}{6}\left(\frac{\partial^3\psi}{\partial y^3}\right)_0 \Delta y^3 + O(\Delta y^4) \tag{95}$$

Note that $(\partial\psi/\partial y)_0 = u_0 = 0$, $(\partial^2\psi/\partial y^2)_0 = (\partial u/\partial y)_0 = -\Omega_0$, $(\partial^3\psi/\partial y^3)_0 = -(\partial\Omega/\partial y)_0$. (In the latter two equalities it is assumed that $v_0(x) \equiv 0$, i.e., the wall is impermeable.) If we retain in the expansion (95) the terms of the order Δy^2 then we shall come to the Thome condition with the error $O(\Delta y^2)$:

$$\Omega_0 = \frac{\psi_0 - \psi_1}{\Delta y^2/2} \tag{96}$$

and if the term on the order of Δy^3 is retained and $(\partial\Omega/\partial y)_0 = (\Omega_1 - \Omega_0)/\Delta y + O(\Delta y)$, then we shall obtain the Woods condition with the error $O(\Delta y^2)$

$$2\Omega_0 + \Omega_1 = \frac{\psi_0 - \psi_1}{\Delta y^2/6} \tag{97}$$

The second order condition may be obtained in the form, which does not contain Ω_1, if the Taylor series is written for ψ_2

$$\psi_2 = \psi_0 + \left(\frac{\partial\psi}{\partial y}\right)_0 2\Delta y + \frac{1}{2}\left(\frac{\partial^2\psi}{\partial y^2}\right)_0 (2\Delta y)^2 + \frac{1}{6}\left(\frac{\partial^3\psi}{\partial y^3}\right)_0 (2\Delta y)^3 + O(\Delta y^4) \tag{98}$$

and equalities (95) and (98) are then combined so that the terms with the third derivative may be eliminated. As a result the condition

$$\Omega_0 = \frac{7\psi_0 - 8\psi_1 + \psi_2}{2\Delta y^2} \tag{99}$$

is obtained. The given conditions can be easily generalized to the cases of moving and permeable walls. Other variants of approximating expressions, including those of the increased order of accuracy,[25] are possible.

The difference conditions for a vortex relate one or two values of Ω (on a wall and in a node nearest to the vortex) with some near-wall values of ψ. Therefore, satisfaction of such a boundary condition within a framework of an implicit scheme requires an additional iteration cycle. In fact, in view of their form the boundary conditions for Ω may be directly realized only when the necessary near-wall values of ψ on the same time layer are known. However, to determine ψ in terms of the Poisson equation solution it is necessary to know the values of Ω in the inner nodes.

Application of various versions of boundary conditions for a vortex is dictated both by accuracy and the convenience of the combination with the difference scheme used. As a test for the analysis of the influence of conditions (96), (97), and (99) on the stability of calculation and the speed of convergence the Poiseuille flow in a plane channel with the symmetry axis ($y = 0$) and in a solid wall ($y = 1/2$) was calculated.[24] When the section-mean velocity is taken as a corresponding scale the exact solution has the form

$$u = \frac{3}{2} - 6y^2 \qquad \psi = \frac{3}{2}y - 2y^3 \qquad \Omega = 12y$$

In the formulation of the accurate condition $\Omega = 6$ on the wall convergence at all the values of $\Delta t, \tau$ was observed within the range of 0.005 to 2.5. If any of the mentioned approximate conditions for Ω was taken, then stability sharply decreases and one could obtain convergence only for $\Delta t, \tau \le 0.005$. The use of the formulas of the increased order of accuracy may lead to instability of calculation with growing **Re** (see, for example, the book by Roache).

4.2. Explicit Scheme of Second Order of Accuracy in Terms of Time

To numerically solve the Navier-Stokes equations in the variables Ω, ψ both implicit and explicit schemes are used. As an example we present an explicit three-layered scheme for whose construction use is made of the expansion of the function Ω into the Taylor series in terms of time:

$$\Omega^{n+1} = \Omega^n + \left(\frac{\partial\Omega}{\partial t}\right)^n \Delta t + \frac{1}{2}\left(\frac{\partial^2\Omega}{\partial t^2}\right)^n \Delta t^2 + O(\Delta t^3) \tag{100}$$

---

[25] Tarunin, E. L., On the choice of the approximating formula for a vortex on a solid boundary when solving the problems of viscous fluid dynamics, in *Chislennye metody mekhaniki sploshnoy sredy* (*Numerical Methods of Continuum Mechanics*), Vol. 9, Novosibirsk, 1979.

Approximating $(\partial^2\Omega/\partial t^2)^n$ in the form

$$\left(\frac{\partial^2\Omega}{\partial t^2}\right)^n = \frac{1}{\Delta t}\left[\left(\frac{\partial\Omega}{\partial t}\right)^n - \left(\frac{\partial\Omega}{\partial t}\right)^{n-1}\right] + O(\Delta t)$$

and substituting this expression into (100), we obtain, after division by Δt, the relation

$$\frac{\Omega^{n+1} - \Omega^n}{\Delta t} = \frac{3}{2}\left(\frac{\partial\Omega}{\partial t}\right)^n - \frac{1}{2}\left(\frac{\partial\Omega}{\partial t}\right)^{n-1} + O(\Delta t^2)$$

Paying attention to the equation of vortex transfer

$$\frac{\partial\Omega}{\partial t} + \Lambda\Omega = 0$$

(Λ, the spatial operator), we come to the three-layered Adams-Bashfort[26] scheme of the second order of accuracy over time

$$\frac{\Omega^{n+1} - \Omega^n}{\Delta t} = -\frac{3}{2}(\Lambda\Omega)^n + \frac{1}{2}(\Lambda\Omega)^{n-1} \tag{101}$$

Having chosen difference approximations for the elements of the spatial operator Λ, we can directly calculate Ω^{n+1} by the values of the functions known from the two previous time layers. The "starting" (on the first time step) calculation should be made by any two-layered scheme.

The techniques for approximating the convective terms in the Navier-Stokes equations are very diverse. In those cases when special attention is paid to the conservativeness, there may be used approximations suggested by Arakawa.[27] The convective terms are in this case approximated in the central node m, s of the nine-point pattern by the formula

$$\begin{aligned}
&\frac{1}{3}\left[\frac{\Psi_{m,s+1} - \Psi_{m,s-1}}{2\Delta y}\cdot\frac{\Omega_{m+1,s} - \Omega_{m-1,s}}{2\Delta x} - \frac{\Psi_{m+1,s} - \Psi_{m-1,s}}{2\Delta x}\cdot\frac{\Omega_{m,s+1} - \Omega_{m,s-1}}{2\Delta y} - \right.\\
&-\frac{1}{2\Delta x}\left(\Psi_{m+1,s}\frac{\Omega_{m+1,s+1} - \Omega_{m+1,s-1}}{2\Delta y} - \Psi_{m-1,s}\frac{\Omega_{m-1,s+1} - \Omega_{m-1,s-1}}{2\Delta y}\right) +\\
&+\frac{1}{2\Delta y}\left(\Psi_{m,s+1}\frac{\Omega_{m+1,s+1} - \Omega_{m-1,s+1}}{2\Delta x} - \Psi_{m,s-1}\frac{\Omega_{m+1,s-1} - \Omega_{m-1,s-1}}{2\Delta x}\right) -\\
&-\frac{1}{2\Delta y}\left(\Omega_{m,s+1}\frac{\Psi_{m+1,s+1} - \Psi_{m-1,s+1}}{2\Delta x} - \Omega_{m,s-1}\frac{\Psi_{m+1,s-1} - \Psi_{m-1,s-1}}{2\Delta x}\right) +\\
&\left.+\frac{1}{2\Delta x}\left(\Omega_{m+1,s}\frac{\Psi_{m+1,s+1} - \Psi_{m+1,s-1}}{2\Delta y} - \Omega_{m-1,s}\frac{\Psi_{m-1,s+1} - \Psi_{m-1,s-1}}{2\Delta y}\right)\right]
\end{aligned} \tag{102}$$

We can easily convince ourselves that the terms in each line of square brackets are approximated with the second order of accuracy over x, y, respectively

[26] See, for instance, the cited book by P. Roache.

[27] Arakawa, A., Computational design of long-term numerical integration of the equations of fluid motion. I. Two-dimensional incompressible flow, *J. Comput. Phys.*, 1(2), 230–236, 1966.

$$\frac{\partial\psi}{\partial y}\frac{\partial\Omega}{\partial x}-\frac{\partial\psi}{\partial x}\frac{\partial\Omega}{\partial y} \qquad -\frac{\partial}{\partial x}\left(\psi\frac{\partial\Omega}{\partial y}\right) \qquad \frac{\partial}{\partial y}\left(\psi\frac{\partial\Omega}{\partial x}\right)$$

$$-\frac{\partial}{\partial y}\left(\Omega\frac{\partial\psi}{\partial x}\right) \qquad \frac{\partial}{\partial x}\left(\Omega\frac{\partial\psi}{\partial y}\right)$$

thus, indicating approximation of convective vortex transfer by the entire difference equation (102).

Approximation (102) possesses the property of conservativeness with respect not only to the vortex Ω, but also to the quantities Ω^2 and the kinetic energy $[(\partial\psi/\partial y)^2 + (\partial\psi/\partial x)^2]/2$. The viscous terms are introduced into the scheme by adding the ordinary difference expressions (on a five-point pattern) into $(\Lambda\Omega)^{n-1}$ and $(\Lambda\Omega)^n$.

Having determined the vortex Ω^{n+1} and the stream function ψ^{n+1} by either technique, pressure p may be calculated on a new time layer. Expressing by numerical differentiation velocities and their derivatives in terms of ψ, pressure may be found directly from the motion equation

$$\text{grad}\, p = \nu\nabla^2 \boldsymbol{V} - (\boldsymbol{V}\cdot\nabla)\boldsymbol{V} - \frac{\partial \boldsymbol{V}}{\partial t}$$

Having prescribed pressure at some point M_0 we present pressure at any point M in the form

$$p_M = p_{M_0} + \int_{M_0}^{M}\left[\nu\nabla^2 \boldsymbol{V} - (\boldsymbol{V}\cdot\nabla)\boldsymbol{V} - \frac{\partial \boldsymbol{V}}{\partial t}\right] d\boldsymbol{r}$$

However, due to computational errors the vector $\nu\nabla^2 \boldsymbol{V} - (\boldsymbol{V}\cdot\nabla)\boldsymbol{V} - \frac{\partial \boldsymbol{V}}{\partial t}$ is not exactly potential and, therefore, pressure may depend on the choice of the integration technique. To eliminate this it is preferable to find the field of p from the Poisson equation obtained by applying the operation div to Equations 82 or 83. Because as a result of differentiation the order of the pressure equation increases, to close the problem we should formulate boundary conditions for p. This problem is discussed below.

Refer to another group of methods for numerical integration of the Navier-Stokes equations of incompressible fluid. These methods are based on the use of the primitive forms of (82) or (83) of the motion equations, i.e., the variables "velocity-pressure". With such an approach the basic problem is in the development of the technique for pressure field determination.

Now we shall formulate the equation which should subordinate pressure p at each time instant so that a solenoidal character of the velocity profile is provided.

Assume that the initial velocity distribution $\boldsymbol{V}_0(\boldsymbol{r})$ (at $t = t_0$ with the property $\text{div}\, \boldsymbol{V}_0 = 0$ is known. It is obvious that in case we have for one and the same time instant, in addition to $\boldsymbol{V}_0(\boldsymbol{r})$, some distribution $p(\boldsymbol{r})$, then the field $\partial \boldsymbol{V}/\partial t$ (at $t = t_0$) is calculated directly from the equation of motion. The latter means that one may determine the velocity field for the next (infinitely close) time instant and evaluate whether it is solenoidal. To preserve a solenoidal character it is necessary and sufficient that

$$\frac{\partial}{\partial t}\text{div}\, \boldsymbol{V} = \text{div}\,\frac{\partial \boldsymbol{V}}{\partial t} = 0 \tag{103}$$

Substituting $\partial V/\partial t$ from Equation 83 into 103, we obtain (substituting here and below for consistency $1/\mathbf{Re} = \nu$):

$$\nabla^2 p = \mathrm{div}\ [\nu\nabla^2 \boldsymbol{V} - \mathrm{Div}\ (\boldsymbol{VV})] = \nu\nabla^2 (\mathrm{div}\ \boldsymbol{V}) - \mathrm{div}\ \mathrm{Div}\ (\boldsymbol{VV})$$

Noting that the first term on the right is zero, we come to the Poisson equation

$$\nabla^2 p = -\mathrm{div}\ \mathrm{Div}\ (\boldsymbol{VV}) \tag{104}$$

for which the time t is a parameter.[28]

On the formulation of the corresponding difference problem the boundary relations for p, which close the difference analogue of the Poisson equation, may differ depending on the type of the difference grid and on the approximation used on it.

4.3. Problems of the Determination of the Pressure Field on the Example of the Explicit Difference Scheme for Equation 82

To comprehend a general idea we shall write this scheme for the inner nodes of the grid not specifying the grid construction and the type of spatial operators related to it:

$$\frac{\boldsymbol{V}^{n+1} - \boldsymbol{V}^{n}}{\Delta t} + \mathrm{Div}\,(\boldsymbol{V}^n\boldsymbol{V}^n) = -\mathrm{grad}\ p + \nu\nabla^2 \boldsymbol{V} \tag{105}$$

$$\mathrm{div}\ \boldsymbol{V}^{n+1} = 0 \tag{106}$$

The scheme has the first order of accuracy in time; the notation of the difference operators are taken to be the same as the differential ones. Transposing all the known (relating to the nth layer) quantities to the right side of Equation 105 obtain

$$\boldsymbol{V}^{n+1} + \Delta t\ \mathrm{grad}\, p = \boldsymbol{f}^n \tag{107}$$

where

$$\boldsymbol{f}^n = \boldsymbol{V}^n + \Delta t \nu\nabla^2 \boldsymbol{V}^n - \Delta t \mathrm{Div}\,(\boldsymbol{V}^n\boldsymbol{V}^n) \tag{108}$$

It can be easily seen that $\boldsymbol{f}^n$ may be interpreted as the velocity for the moment $t_{n+1} = t_n + \Delta t$ in the case of a homogeneous pressure field and that $\boldsymbol{f}^n$ accurately describes the vortical properties of the unknown grid function $\boldsymbol{V}^{n+1}$.

The linear algebraic system of Equations 106 and 107 in combination with the boundary conditions for $\boldsymbol{V}^{n+1}$ should determine $\boldsymbol{V}^{n+1}$ and p (accurate to the additive constant) in the grid nodes. Usually the equations entering into this system are combined so that the group of relations involving only unknown p is distinguished. Substituting (107) into (106), we have

$$\nabla^2 p = \frac{1}{\Delta t}\mathrm{div}\,\boldsymbol{f}^n \tag{109}$$

the difference Poisson equation for pressure.[29]

Note that it is assumed in expression $\mathrm{div}\,\boldsymbol{f}^n$ that $\mathrm{div}\,\boldsymbol{V}^n = 0$, then

[28] Here, as in the previous expression, the pressure term on the left should be divided by the constant density p. Alternately, consider unit density.

[29] Again, Equations 107 and 109 are only valid if we read p/ρ for p.

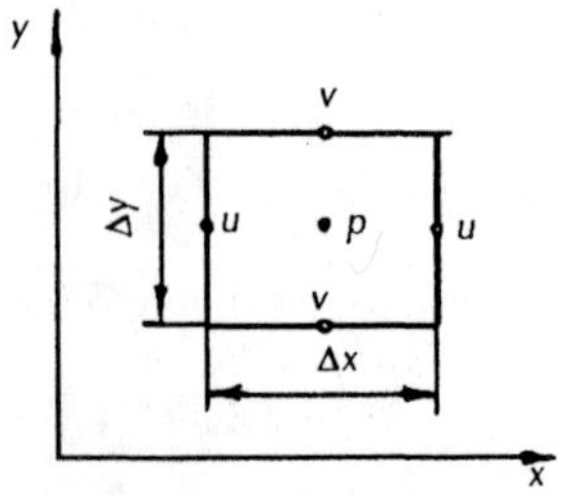

Fig. 313

$$\operatorname{div} f^n = \operatorname{div} V^n + \Delta t \nu \nabla^2 \operatorname{div} V^n - \Delta t \operatorname{div} \operatorname{Div}(V^n V^n) = -\Delta t \operatorname{div} \operatorname{Div}(V^n V^n)$$

However, in calculations, due to different errors, the condition $\operatorname{div} V^n = 0$, is not satisfied accurately. Therefore, application of the "truncated" expression $\operatorname{div} f^n = -\Delta t \operatorname{div} \operatorname{Div}(V^n V^n)$ (disregarding the real value of the velocity divergence on the nth layer) may become the cause of calculation instability.

Formulation of additional (closing) conditions for p to the difference Poisson equation (109), as was mentioned previously, is directly connected with the grid used. On utilization of the variables V, p, especially convenient are the "hybrid" difference grids named also "MAC grids", in which separate velocity components and pressure are determined in the nodes shifted relative to each other. The difference grid in the form of a rectangular parallelepiped (in a plane case, rectangular; Figure 313) is the basis of this grid. The velocity components normal to the edges are determined in the middle of them, and pressure is found in the center of the cell. The continuity equation is naturally written for each cell in terms of fluid flows through the edges. The protection of the motion equation on the axis of the Cartesian system of coordinates are approximated in the nodes by the determination of the corresponding velocity components.

Now we shall consider the construction of the difference scheme for a rectangular two-dimensional region with the boundary Γ that is divided into $N_1 \times N_2$ equal cells of the MAC grid (Figure 314). For simplicity we shall consider the velocities on Γ as prescribed. To determine at the nodes the unknowns $(N_1 - 1) \times N_2$ values of u, $N_1 \times (N_2 - 1)$ values of v, and $N_1 \times N_2$ values of p we shall write $N_1 \times N_2$ relations approximating the continuity equation (for each cell, or in the nodes of p), $(N_1 - 1) \times N_2$ relations approximating the equation of motion (107) in the projection on the axis x (in the inner nodes of u), and $N_1 \times (N_2 - 1)$ analogous relations for the projections on the axis y (in the inner nodes of v), Thus, the total

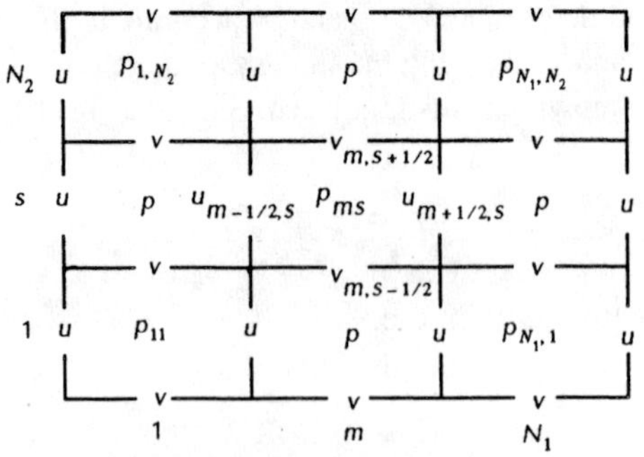

Fig. 314

number of unknowns is equal to the number of approximating relations. It should be noted that on summation of the difference continuity equations over all of the cells only the velocity components which refer to the boundary Γ and are normal to it will remain. This means that one of the $N_1 \times N_2$ difference continuity equations is the consequence of the rest equations and the boundary conditions. This requires the arbitrary setting of one of the values of p, i.e., to select the additive constant for pressure.

The listed approximating relations compose a closed algebraic system not resorting to any additional considerations with respect to pressure on the boundary Γ. These relations are absent in the physical formulation of the problem. This is the manifestation of one of the advantages of hybrid grids that is caused by the location of the node of pressure determination in the cell center. However, the structure of hybrid grids possesses a shortcoming: the absence of the nodes with the tangential velocity components on the boundaries of cells and thus on the boundary of the entire region (the grid is assumed to be in agreement with the region boundary, i.e., the sides of cells serve as the elements of Γ) Because of this we have to use special techniques to approximate separate terms of the motion equations in near-boundary nodes and to satisfy the sticking conditions.

To approximate spatial operators entering into Equations 106 to 108, we shall take the following expressions with the error $O(\Delta x^2, \Delta y^2)$:

$$(\operatorname{div} V)_{m,s} \sim D_{ms}(V) = \frac{u_{m+1/2,s} - u_{m-1/2,s}}{\Delta x} + \frac{v_{m,s+1/2} - v_{m,s-1/2}}{\Delta y}$$

$$(1 \le m \le N_1 \quad 1 \le s \le N_2) \tag{110}$$

$$u^{n+1}_{m+1/2,s} + \frac{\Delta t}{\Delta x}(p_{m+1,s} - p_{m,s}) = (f_x)^n_{m+1/2,s}$$

$$(1 \le m \le N_1 - 1 \quad 1 \le s \le N_2) \tag{111}$$

$$v^{n+1}_{m,s+1/2} + \frac{\Delta t}{\Delta y}(p_{m,s+1} - p_{m,s}) = (f_y)^n_{m,s+1/2}$$

$$(1 \le m \le N_1 \quad 1 \le s \le N_2 - 1) \tag{112}$$

where

$$(f_x)_{m+1/2,s} = u_{m+1/2,s} + \frac{\Delta t \nu}{\Delta x^2}(u_{m-1/2} - 2u_{m+1/2} + u_{m+3/2})_s +$$

$$+ \frac{\Delta t \nu}{\Delta y^2}(u_{s-1} - 2u_s + u_{s+1})_{m+1/2} - \frac{\Delta t}{\Delta x}(u^2_{m+1} - u^2_m)_s -$$

$$- \frac{\Delta t}{\Delta y}[(uv)_{s+1/2} - (uv)_{s-(1/2)}]_{m+1/2} \tag{113}$$

$$(f_y)_{m,s+1/2} = v_{m,s+1/2} + \frac{\Delta t \nu}{\Delta x^2}(v_{m-1} - 2v_m + v_{m+1})_{s+1/2} +$$

$$+ \frac{\Delta t \nu}{\Delta y^2}(v_{s-1/2} - 2v_{s+1/2} + v_{s+3/2})_m -$$

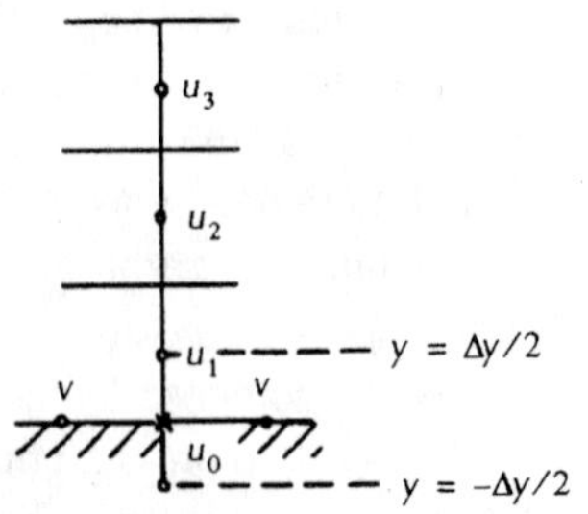

Fig. 315

$$-\frac{\Delta t}{\Delta x}[(uv)_{m+1/2}-(uv)_{m-1/2}]_{s+1/2}-\frac{\Delta t}{\Delta y}(v_{s+1}^2-v_s^2)_m \tag{114}$$

The quantities u, v with two integer or two fractional indices, entering into the convective terms, are determined as half sums of the nearest values corresponding to u, v in the nodes of the grid. Thus,

$$u_{m+1,s} = \frac{1}{2}(u_{m+3/2,s}+u_{m+1/2,s})$$

$$u_{m+1/2,s+1/2} = \frac{1}{2}(u_{m+1/2,s+1}+u_{m+1/2,s}) \quad \ldots$$

In view of the computation of $(f_x)^n$ and $(f_y)^n$ in near-boundary (i.e., at a half-mesh distance from the wall) nodes this should be especially noted. The terms entering into f_x and approximating $\partial(uv)/\partial y$ and $\partial^2 u/\partial y^2$ and those entering into f_y and approximating $\partial(uv)/\partial x$ and $\partial^2 v/\partial x^2$, in view of the mentioned properties of a hybrid grid cannot be directly calculated by Equations 113 and 114. For near-boundary nodes (for example, for the node of u1 determination in Figure 315, where the second index is omitted due to its insignificance), the approximation of the derivatives of the tangential velocity components over the normal requires special consideration. If, in particular, the boundary is a solid surface, then it is simultaneously connected with the realization of the adhesion conditions in the node $y = 0$ not entering into the pattern of the difference grid.

The adhesion condition may be exactly satisfied if additional nodes for the tangential velocity component on solid surfaces are introduced (the node $y = 0$ with $u_{1/2} = 0$ marked by a cross in Figure 315). In this case we have to use nonstandard approximations in the near-wall nodes, for instance,

$$\left(\frac{\partial^2 u}{\partial y^2}\right)_1 = \frac{-25u_1+10u_2-u_3}{5\Delta y^2}+O(\Delta y^2)$$

It would be simpler to extend the pattern half-mesh downward from the line $y = 0$, i.e., to introduce the node with u_0 at $y = -\Delta y/2$. Then the condition of adhesion may be fulfilled with the second order of accuracy, having assumed $u_0 + u_1 = 0$, and $(\partial^2 u/\partial y^2)_1$ approximated as usual in terms of u_0, u_1, u_2 at equally spaced points. However, the approximate satisfaction of the adhesion condition may cause a noticeable error in the character of the balance of forces on a solid surface.

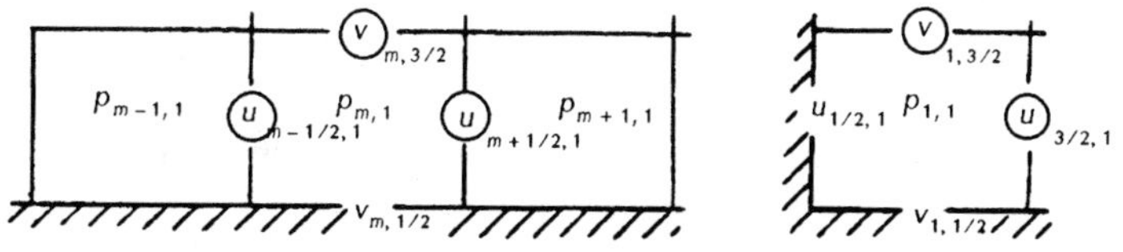

a b

Fig. 316

Below we shall consider one more method to achieve adhesion condition realization with controlled error.

Formulate, proceeding from relations (110) to (114), the difference analogue of the Poisson equation for pressure and additional conditions to it.

Having substituted the expressions u^{n+1}, v^{n+1} from (111) and (112) into the continuity equation (110) written for the inner cells $(1 < m < N_1,\ 1 < s < n_2)$, we obtain $(N_1 - 2) \times (N_2 - 2)$ difference approximations of the Poisson equation on a five-point patterns

$$\frac{p_{m-1,s} - 2p_{m,s} + p_{m+1,s}}{\Delta x^2} + \frac{p_{m,s-1} - 2p_{m,s} + p_{m,s+1}}{\Delta y^2} = \frac{1}{\Delta t} D_{ms}(f^n) \tag{115}$$

where

$$D_{ms}(f) = \frac{(f_x)_{m+1/2,s} - (f_x)_{m-1/2,s}}{\Delta x} + \frac{(f_y)_{m,s+1/2} - (f_y)_{m,s-1/2}}{\Delta y}$$

With the components V^{n+1} being eliminated from the equations $D_{ms}(V^{n+1}) = 0$ for the cells adjacent to the region boundary, we should note that the motion equation is not approximated at the points of the boundary itself, though the boundary conditions exist. Thus, for a cell $m, 1$, which is not angular but adjacent to a solid wall (Figure 316a), we shall have, having substituted the values of u^{n+1}, v^{n+1} (marked by circles) from (111) and (112) and the value $v_{m,1/2} = 0$ (the boundary condition) into (110), the four-point relation

$$\frac{p_{m-1,1} - 2p_{m,1} + p_{m+1,1}}{\Delta x^2} + \frac{p_{m,2} - p_{m,1}}{\Delta y^2} = \frac{1}{\Delta t}\tilde{D}_{m,1}(f^n) \tag{116}$$

where $\tilde{D}_{m,1}(f^n)$ is obtained from a general expression of $D_{m,s}(f^n)$ by replacing $(f_y)_{m,1/2} = = v_{m,1/2} = 0$.

Even more "truncated" is an analogous relation for an angular cell 1,1 (Figure 316b) involving pressure at three points:

$$\frac{p_{2,1} - p_{1,1}}{\Delta x^2} + \frac{p_{1,2} - p_{1,1}}{\Delta y^2} = \frac{1}{\Delta t}\tilde{D}_{1,1}(f^n) \tag{117}$$

where $\tilde{D}_{1,1}(f)$ is obtained by substitution in a general expression of $D_{m,s}(f)$: $(f_x)_{1/2,1} = = u_{1/2,1} = 0$, $(f_y)_{1,1/2} = v_{1,1/2} = 0$.

Having solved Equations (115) with the closing relations of the type of (116) and (117), we calculate u^{n+1}, v^{n+1} from Equations (111) and (112).

Another approach exists to the approximation of the motion equations in the near-wall nodes, with the adhesion conditions been taken into account.[30] Instead of using in these nodes expressions (111) and (112), which are common in form, with partially modified right sides $(f_x)^n$, $(f_y)^n$, we expand the tangential velocity components near the wall. For example, we have for the point $y = \Delta y/2$ (Figure 315):

$$u_{1/2}^{n+1} = 0 = u_1^{n+1} - \frac{\Delta y}{2}\left(\frac{\partial u}{\partial y}\right)_1^{n+1} + \frac{\Delta y^2}{8}\left(\frac{\partial^2 u}{\partial y^2}\right)_1^{n+1} + O(\Delta y^3) \tag{118}$$

Further transformations of (117) refer to the derivatives $(\partial u/\partial y)_1^{n+1}$ and $(\partial^2 u/\partial y^2)_1^{n+1}$. The derivative $(\partial u/\partial y)_1^{n+1}$ is replaced by $(\partial u/\partial y)_1^n$ with the error $O(\Delta t)$, the latter derivative is approximated over the values $u_{1/2}^n = 0$, u_1, u_2 with the second order of accuracy

$$\left(\frac{\partial u}{\partial y}\right)_1^{n+1} = \left(\frac{\partial u}{\partial y}\right)_1^n + O(\Delta t) = \frac{3u_1 + u_2}{3\Delta y} + O(\Delta t, \Delta y^2) \tag{119}$$

Then, the projection of the motion equation onto the axis x at the same point for the expression $\partial^2 u/\partial y^2$ in terms of $\partial p/\partial x$. In view of $u_1^{n+1} = O(\Delta y)$ we have

$$\left(\frac{\partial^2 u}{\partial y^2}\right)_1^{n+1} = \frac{1}{\nu}\left(\frac{\partial p}{\partial x}\right)_1^{n+1} + O\left(\frac{\Delta y}{\nu}\right) \tag{120}$$

If we relate expressions (118) to (120) to the value $x_{m+1/2}$ and substitute

$$\left(\frac{\partial p}{\partial x}\right)_{m+1/2,1} = \frac{p_{m+1,1} - p_{m,1}}{\Delta x} + O(\Delta x^2) \tag{121}$$

then (118) with allowance for (119) to (121) is reduced to the approximate equality with the common error $O\left(\Delta y^3, \frac{\Delta y^3}{\nu}, \Delta y \cdot \Delta t, \Delta x^2\right)$:

$$u_{m+1/2,1}^{n+1} + \frac{\Delta\tilde{t}}{\Delta x}(p_{m+1,1} - p_{m,1}) = \left(\tilde{f}_u\right)_{m+1/2,1} \tag{122}$$

where

$$\Delta\tilde{t} = \Delta y^2/(8\nu) \qquad \left(\tilde{f}_u\right)_{m+1/2,1} = (3u_1^n + u_2^n)/6 \tag{123}$$

and the index $n + 1$ for pressure is omitted (pressure providing the solenoidal character of the velocity field in the transition from t_n to t_{n+1}. Relations of the type (Equations 122 and 123) and analogous ones for the transverse velocity replace expressions (111) and (112) with the modified right sides in near-wall nodes.

The stability of the presented difference scheme was estimated us applied to the explicit stage of (108). The function f^n may be treated as the velocity for the moment t_{n+1} calculated by the explicit scheme for the fictitious isobaric flow.

---

[30] Belotserkovskiy, O.M., Gushchin, V.A. and Shchennikov, V. V., The release method as applied to the solution of the problems of viscous incompressible fluid dynamics, *Zh. Vych. Mat. Mat. Fiz.*, 15(1), 197–207, 1975; Gushchin, V. A. and Konshin, V. N., *Chislennoye modelirovaniye volnovykh dvizheniy zhidkosti* (*Numerical Simulation of Wave Fluid Flows*), AN SSSR Press, Moscow, 1985.

Estimation of stability is based on the requirement of the positiveness of the coefficients at second derivatives in a parabolic form of the first differential approximation and it has the form:

$$\nu > \frac{\Delta t}{2}\max\,(u^2, v^2)$$

$$\nu > \max\left(\frac{\Delta t}{2}u^2 + \frac{\Delta x^2}{4}\frac{\partial u}{\partial x}, \frac{\Delta t}{2}v^2 + \frac{\Delta y^2}{4}\frac{\partial v}{\partial y}\right) \tag{124}$$

4.4. Implicit Time-Dependent Scheme

This scheme calculates stationary flows in the variables V, p that uses the "relaxation" form of the continuity equation or the model of "artifical compressibility":[31]

$$A\frac{\partial p}{\partial t} + \operatorname{div}V = 0 \tag{125}$$

($A > 0$ is the iteration parameter). Equation 125 involves the direct mechanism correcting the p field in order to suppress the sources or sinks (in the case of a source pressure decreases whereas in the case of a sink it increases).

A completely nonstationary form of the system including Equation 125 and the equation of motion (83) allows the application of the schemes of splitting over the spatial variables.

Write Equations 83 and 125 in the matrix form

$$\frac{\partial U}{\partial t} + \Lambda U = 0 \tag{126}$$

We have for a plane flow in a rectilinear Cartesian system of coordinates

$$U = \begin{bmatrix} u \\ v \\ p \end{bmatrix} \qquad \Lambda = \begin{bmatrix} \frac{\partial u\cdot}{\partial x} + \frac{\partial v\cdot}{\partial y} - \nu\left(\frac{\partial^2}{\partial x^2} + \frac{\partial^2}{\partial y^2}\right) & 0 & \frac{\partial}{\partial x} \\ 0 & \frac{\partial u\cdot}{\partial x} + \frac{\partial v\cdot}{\partial y} - \nu\left(\frac{\partial^2}{\partial x^2} + \frac{\partial^2}{\partial y^2}\right) & \frac{\partial}{\partial y} \\ \frac{1}{A}\frac{\partial}{\partial x} & \frac{1}{A}\frac{\partial}{\partial y} & 0 \end{bmatrix}$$

$\nu = 1/\mathbf{Re}$; the dot in the expressions of the type $(\partial u\cdot)/\partial x$ means multiplication under the sign of the derivatives by the corresponding velocity component when multiplying the matrix Λ by the column of the unknowns U.

The implicit time-dependent schema in the so-called delta form[32]

---

[31] Vladimirova, N. N., Kuznetsov, B. G., and Yanenko, N. N., Numerical calculations of the symmetric plane flow of a viscous incompressible fluid past a plane, in *Nekotorye voprosy vychislitelnoy i prikladnoy matematiki* (*Some Problems of Computational and Applied Mathematics*), Nauka, Novosibirsk, 1966, 186–192; Chorin, A. J., A numerical method for solving incompressible viscous flow problems, *J. Comput. Phys.*, 2(1), 12–26, 1967.

[32] Yegorov, Yu. E. and Koleshko, S. B., Application of the method of fractional steps to numerical solution of the equations of incompressible viscous fluid in natural variables, in *Dinamika neodnorodnykh i szhimaemykh sred* (*Dynamics of Nonhomogeneous and Comressible Media*), Vol. 8, LGU Press, Leningrad, 1984, 80–92.

$$\frac{\Delta U^{n+1}}{\Delta t} + (\Lambda U)^{n+1} = 0 \tag{127}$$

where $\Delta U^{n+1} = U^{n+1} - U^n$ forms the basis of the stationary problem $\Lambda U = 0$ with the prescribed conditions for velocity on the boundary of the region Γ. For simplicity we preserve the notation of the difference operators that coincides with the notation of differential operators. Extract from the spatial part $(\Lambda U)^{n+1}$ the unknown time (iteration) increment ΔU^{n+1}, having presented

$$(\Lambda U)^{n+1} = (\Lambda U)^n + L^n (\Delta U)^{n+1} + O(\Delta t^2)$$

$$L^n = \left(\frac{\partial \Lambda}{\partial U}\right)^n = \begin{bmatrix} 2u^n \frac{\partial}{\partial x} + v^n \frac{\partial}{\partial y} - \nu\left(\frac{\partial^2}{\partial x^2} + \frac{\partial^2}{\partial y^2}\right) & u^n \frac{\partial}{\partial y} & \frac{\partial}{\partial x} \\ v^n \frac{\partial}{\partial x} & u^n \frac{\partial}{\partial x} + 2v^n \frac{\partial}{\partial y} - \nu\left(\frac{\partial^2}{\partial x^2} + \frac{\partial^2}{\partial y^2}\right) & \frac{\partial}{\partial y} \\ \frac{1}{A}\frac{\partial}{\partial x} & \frac{1}{A}\frac{\partial}{\partial y} & 0 \end{bmatrix}$$

As a result we come to the system of linear equations relative to ΔU^{n+1}

$$(I + \Delta t L^n)\, \Delta U^{n+1} = \Delta f^n \tag{128}$$

where I is the unit operator, $\Delta f^n = -\Delta t (\Lambda U)^n$ is the residual of the stationary problem solution on the time layer t^n accurate to the factor $(-\Delta t)$.

To approximately solve the two-dimensional problem (128) over the spatial variables, we shall use the splitting method (see Section 1). Having represented the operator $I + \Delta t L^n$, with the error $O(\Delta t^2)$ by the factorized operator $(I + \Delta t L_1^n) \cdot (I + \Delta t L_2^n)$, we obtain instead of (128)

$$(I + \Delta t L_1^n)\,(I + \Delta t L_2^n)\, \Delta U^{n+1} = \Delta f^n \tag{129}$$

where the one-dimensional operators L_1^n and L_2^n have the form

$$L_1^n = \begin{bmatrix} 2u^n \frac{\partial}{\partial x} - \nu \frac{\partial^2}{\partial x^2} & 0 & \frac{\partial}{\partial x} \\ v^n \frac{\partial}{\partial x} & u^n \frac{\partial}{\partial x} - \nu \frac{\partial^2}{\partial x^2} & 0 \\ \frac{1}{A}\frac{\partial}{\partial x} & 0 & 0 \end{bmatrix}$$

$$L_2^n = \begin{bmatrix} v^n \frac{\partial}{\partial y} - \nu \frac{\partial^2}{\partial y^2} & u^n \frac{\partial}{\partial y} & 0 \\ 0 & 2v^n \frac{\partial}{\partial y} - \nu \frac{\partial^2}{\partial y^2} & \frac{\partial}{\partial y} \\ 0 & \frac{1}{A}\frac{\partial}{\partial y} & 0 \end{bmatrix}$$

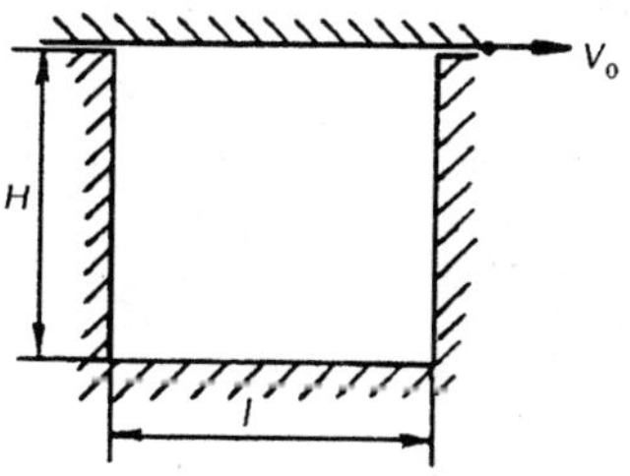

Fig. 317

The difference scheme (129) is easily divided into two subsystems. Introducing the intermediate function $\Delta\tilde{U} = \begin{bmatrix} \Delta\tilde{u} \\ \Delta\tilde{v} \\ \Delta\tilde{p} \end{bmatrix}$ (the time index is omitted), we obtain

$$(I + \Delta t L_1^n)\,\Delta\tilde{U} = \Delta f^n \tag{130}$$

$$(I + \Delta t L_2^n)\,\Delta U^{n+1} = \Delta\tilde{U} \tag{131}$$

Equations 130, supplemented with the boundary conditions for $\Delta\tilde{U}$, are easily solved by longitudinal factorizations (along the portions y = const) and Equations 131 with their boundary conditions, along the portions x = const. The structure of the matrices L_1^n, L_2^n does not require vector factorizations because it allows the elimination of $\Delta\tilde{p}$, Δp^{n+1} from the motion equation with the aid of the split continuity equation.

Application of the spectral method of the stability analysis to the linear analogue of scheme (129) has shown that for $A > 0$ the scheme is unconditionally stable.

4.5. Calculation of Viscous Incompressible Fluid Flow in a Rectangular Cavity with the Upper Wall Moving in Its Plane with the Velocity V_0 (Figure 317)

The fluid, filling the entire cavity, is entrained in motion by viscous forces. This problem illustrates characteristic features of many flows described by the Navier-Stokes equations: difference relations between inertia and viscous forces, simultaneous existence of regions with large and small gradients, and formation of zones with a recirculation flow.

Let the length l and the velocity V_0 be taken as the scales entering into the problem formulation. Then lV_0 and V_0/l are the scales of the stream function and inverse time. The Reynolds number **Re** $= lV_0/\nu$ is the only similarity criterion and the dimensionless depth of the cavity $\bar{H} = H/l$ is the problem geometrical parameter.

In Figure 318a to c, taken from the work of Belov and Isayev[33] (variables Ω,ψ, using Equation 101 with the approximation of convective terms (102), stream lines for a square

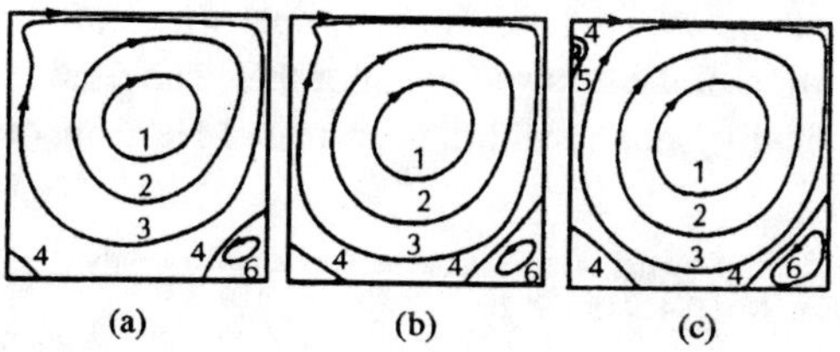

Fig. 318

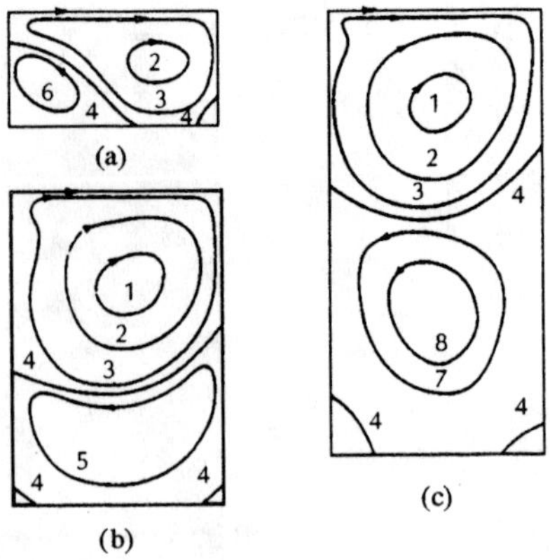

Fig. 319

cavity are shown for Reynolds numbers 400, 1000, and 2500, respectively. The following values of the dimensionless stream function ψ correspond to the figures: 1 (–0.10), 2 (–0.06), 3 (–0.01), 4 (0), 5 (0.0001), and 6 (0.001). A vast central region of a circulation flow (a "central vortex") and secondary "angular vortices" in the lower portion of the cavity are distinguished. (Here, the concept "vortex" is used to denote the bounded region of a circulating flow.) At first, the dimensions of angular vortices grow with **Re**, but at **Re** > 1000 their values virtually stabilize. Starting from **Re** ≈ 1500 one more secondary vortex appears near a moving wall.

Figure 319a to c, also borrowed from Belov and Isayev, show the flow evolution at **Re** = = 1000 as a result of $\bar{H}$ variation which is correspondingly equal to 0.5, 1.4, and 2.0. Additional numerical notations have the following sense: 7 – $\psi = 0.001$, 8 – $\psi = 0.02$. Two large regions of a circulation flow are observed at small depths (Figure 319a): of the basic one and one of less intensity on the left at the bottom. There is also one more secondary vortex near a rear wall. With the growth of $\bar{H}$ the dimensions and intensity of the central vortex increase, the size of the vortex at a front wall decreases, and that at a rear wall grows. With further ($\bar{H} > 1$) increase of the cavity height the angular vortices converge, thus forming a single secondary vortex over the entire width of the cavity that rotates oppositely to the main one and is less intense.

For the cavities with $\bar{H} = 1.4$ and 2.0 new vortices are formed at the lower corners that grow with the depth. This corresponds to the cellular structures in the form of the system of vertical vortices, that are generated in deep cavities, of the intensity decreasing in the bottom direction.

In Figure 320 taken from the cited work by Yegorov and Koleshko the isobars for the flow in a quadrangular cavity at **Re** = 400 are shown. The numbers denote the values of the dimensional pressure $(p - p_1)/(\rho V_0^2) \cdot 100$, were p_1 is the pressure at the lower right angle of the cavity. The pressure distribution has a rather complex character. The regions of the noticeable pressure growth in the vicinity of the upper right angle, where the fluid is entrained by a moving boundary, is carried to a vertical wall, and the region of relative rarefaction in the upper portion of the left wall is noticeable. The flow near the bottom of a quadrangular cavity is practically isobaric within the range of **Re** from 400 to 2500.

[33] Belov, I. A. and Isayev, S. A., Circulation fluid flow in a rectangular cavity at mean and large Reynolds numbers, *Zh. Prikl. Mekh. Tekh. Fiz.*, No. 1, 41–45, 1982.

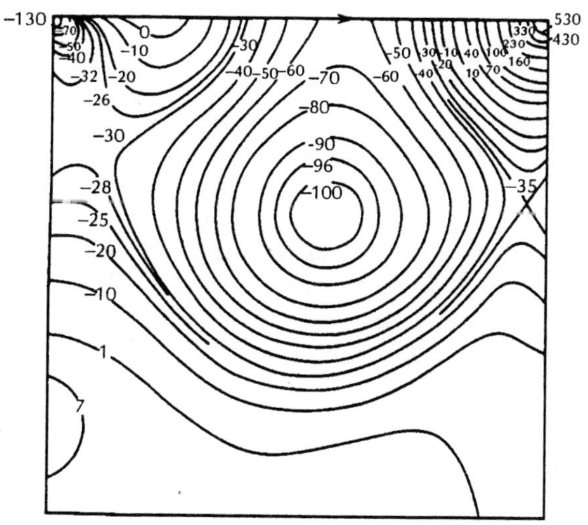

Fig. 320

5. Methods for Solving the Navier-Stokes Equations for a Gas

In this section the methods of the numerical integration of the balance equations of mass, momentum, and energy of a viscous heat conducting gas are considered. The corresponding system of equations, which is called the system of the Navier-Stokes equations for a gas, is usually applied to the calculation of laminar flows. The methods given below are also applied to the solution of the Reynolds equations describing variation in the averaging of the characteristics of turbulent flows.

First, we consider some specific features of problem formulation and numerical solution of the system of equations of viscous gas dynamics.

The nonstationary Navier-Stokes equations possess hyper- and parabolic properties. For stationary flows the system of the Navier-Stokes equations is elliptical. In accordance with this, either initial boundary value or boundary value problems are formulated for it.

In the use of nonstationary equations the values of the functions at the reference time instant are prescribed at the initial conditions. The formulation of the boundary value substantially differs from that mentioned in Section 2 for the inviscid gas model.

On solid surfaces at high Reynolds numbers use is made of the conditions of impermeability, adhesion, and temperature. In the presence of injection through a streamlined surface the parameters of the injected gas are specified. At small Reynolds numbers the Navier-Stokes equations are integrated with the boundary conditions of sliding and of the temperature jump.[34]

The problem of the correct formulation of boundary conditions for the Navier-Stokes equations at the inlet to and outlet from the computational region in a general case remains unsolved. The greatest difficulty is the formulation of conditions of these boundaries at sub-

[34] Kogan, M. N., *Dinamika razrezhennogo gaza (Dynamics of Rarefied Gas)*, Nauka, Moscow, 1967.

sonic velocities, especially in calculation of inner flows. In these cases the essential role is played by the effects caused by the reflection of acoustic waves from the boundaries. To eliminate these undesirable effects the boundary conditions are used that provide free emergence of acoustic perturbations.[35] The correctness of the formulation of boundary conditions and selection of boundary approximations when calculating viscous gas flows is usually verified by numerical experiments.

As the Reynolds number grows a small parameter at higher derivatives appears in the Navier-Stokes equations. In this case regions with large gradients of functions are formed in the flow field, the description of which requires application of difference grids with small steps over the spatial coordinates. Under these conditions, due to strict limitation of the value of the time step, the use of the explicit schemes results in unjustified increase in the time of computation, especially in stationary problems. Therefore, at present, to solve the Navier-Stokes equations use is mainly made of the implicit techniques possessing greater stability.

5.1. Implicit Factorized Scheme

This numerical method is presented based on the work of Fedorchenko.[36] For a two-dimensional flow the Navier-Stokes equations in the Cartesian coordinates are written in the following conservative form:

$$\frac{\partial U}{\partial t} + \frac{\partial F(U)}{\partial x} + \frac{\partial G(U)}{\partial y} = \frac{\partial V_1(U, U_x)}{\partial x} + \frac{\partial V_2(U, U_y)}{\partial x} + \frac{\partial W_1(U, U_x)}{\partial y} + \frac{\partial W_2(U, U_y)}{\partial y} \tag{132}$$

Here, U is the vector of the conservative variables; F, G, V, W are the vectors of the flows:

$$U = \begin{bmatrix} \rho \\ \rho u \\ \rho v \\ \rho E \end{bmatrix} \qquad F = \begin{bmatrix} \rho u \\ \rho u^2 + p \\ \rho u v \\ (\rho E + p)\, u \end{bmatrix} \qquad G = \begin{bmatrix} \rho v \\ \rho u v \\ \rho v^2 + p \\ (\rho E + p)\, v \end{bmatrix}$$

$$V = V_1 + V_2 = \begin{bmatrix} 0 \\ \frac{4}{3}\mu u_x - \frac{2}{3}\mu v_y \\ \mu(v_x + u_y) \\ \mu v(u_y + v_x) - \frac{2}{3}\mu u(u_x + v_y) + 2\mu u u_x + \lambda T_x \end{bmatrix}$$

---

[35] Fedorchenko, A. T., On the problems of numerical simulation of nonstationary spatial flows of inviscid gas in nozzles, *Zh.Vych. Mat. Mat. Fiz.*, 22(1), 178–196, 1982.

[36] Beam, R. M. and Warming, R. F., An implicit factored scheme for the compressible Navier-Stokes equations, *AIAA J.*, 16(4), 393–402, 1978.

$$W = W_1 + W_2 = \begin{bmatrix} 0 \\ \mu\,(u_y + v_x) \\ \frac{4}{3}\mu v_y - \frac{2}{3}\mu u_x \\ \mu u\,(v_x + u_y) - \frac{2}{3}\mu v\,(v_y + u_x) + 2\mu v v_y + \lambda T_y \end{bmatrix}$$

Consider the following two-parametric family of difference schemes:

$$(1+\xi)\frac{U^{n+1} - U^n}{\Delta t} - \xi\frac{U^n - U^{n-1}}{\Delta t} + \theta\Lambda^{n+1}U^{n+1} + (1-\theta)\,\Lambda^n U^n = 0 \tag{133}$$

where Λ is the matrix difference operator approximating the derivatives over the spatial coordinates; ξ, θ are the numerical parameters. At different values of these parameters ($\xi = -1/2$, 0, 1/2; $\theta = 0$, 1) Equation 133 represents the family of the familiar explicit and implicit difference schemes which have the second order of approximation over time at $\theta = 1/2 + \xi$ and the first order of approximation in other cases.

Write this scheme for the increment of the unknown functions. The use of this formulation provides the splitting schemes with the independence of the stationary solution on the value of the time step. Denote

$$\Delta U^n = U^{n+1} - U^n \qquad \Delta F^n = F^{n+1} - F^n = F(U^{n+1}) - F(U^n) \quad \ldots$$

Passing over to the increments, we obtain

$$(1+\xi)\frac{\Delta U^n}{\Delta t} - \xi\frac{\Delta U^{n-1}}{\Delta t} + \theta\left(\frac{\partial F}{\partial x} + \frac{\partial G}{\partial y} - \frac{\partial V}{\partial x} - \frac{\partial W}{\partial y}\right)^{n+1} +$$

$$+ (1-\theta)\left(\frac{\partial F}{\partial x} + \frac{\partial G}{\partial y} - \frac{\partial V}{\partial x} - \frac{\partial W}{\partial y}\right)^{n} + O\left[\left(\theta - \frac{1}{2} - \xi\right)\Delta t + \Delta t^2\right] = 0$$

or

$$(1+\xi)\frac{\Delta U^n}{\Delta t} - \xi\frac{\Delta U^{n-1}}{\Delta t} + \theta\left[\frac{\partial}{\partial x}(\Delta F^n - \Delta V^n) + \frac{\partial}{\partial y}(\Delta G^n - \Delta W^n)\right] +$$

$$+ \frac{\partial}{\partial x}(F^n - V^n) + \frac{\partial}{\partial y}(G^n - W^n) + O\left[\left(\theta - \frac{1}{2} - \xi\right)\Delta t + \Delta t^2\right] = 0 \tag{134}$$

Solving Equation 134 relative to the vector ΔU^n we obtain

$$\Delta U^n = \frac{\theta\Delta t}{1+\xi}\left[\frac{\partial}{\partial x}(-\Delta F^n + \Delta V^n) + \frac{\partial}{\partial y}(-\Delta G^n + \Delta W^n)\right] + \frac{\Delta t}{1+\xi}\left[\frac{\partial}{\partial x}(-F + V)^n +\right.$$

$$\left. + \frac{\partial}{\partial y}(-G + W)^n\right] + \frac{\xi}{1+\xi}\Delta U^{n-1} + O\left[\left(\theta - \frac{1}{2} - \xi\right)\Delta t^2 + \Delta t^3\right] \tag{135}$$

Having replaced in (135) the derivatives with respect to x, y by the finite-difference analogues we can obtain the system of algebraic difference equations relative to the vector ΔU^n.

The increments of the vectors of flows ΔF^n, ΔG^n, ΔV^n, ΔW^n are the nonlinear functions of the vector of the conservative variables U and the vectors of the derivatives U_x, U_y. To obtain the linear system the expansion of the vectors of flows into the Taylor series

$$F^{n+1} = F^n + \left(\frac{\partial F}{\partial U}\right)^n (U^{n+1} - U^n) + O(\Delta t^2) = F^n + \left(\frac{\partial F}{\partial U}\right)^n \Delta U^n + O(\Delta t^2)$$

$$G^{n+1} = G^n + \left(\frac{\partial G}{\partial U}\right)^n \Delta U^n + O(\Delta t^2)$$

$$V_1^{n+1} = V_1^n + \left(\frac{\partial V_1}{\partial U}\right)^n \Delta U^n + \left(\frac{\partial V_1}{\partial U_x}\right)^n \Delta U_x^n + O(\Delta t^2)$$

$$W_2^{n+1} = W_2^n + \left(\frac{\partial W_2}{\partial U}\right)^n \Delta U^n + \left(\frac{\partial W_2}{\partial U_y}\right)^n \Delta U_y^n + O(\Delta t^2) \tag{136}$$

As a result, accurate to the terms of the order $O(\Delta t^2)$, the increments of the vectors of flows are written in the form

$$\Delta F^n = A^n \Delta U^n \qquad \Delta G^n = B^n \Delta U^n$$

$$\Delta V_1 = P^n \Delta U^n + R^n \Delta U_x^n = (P - R_x)^n \Delta U^n + \frac{\partial}{\partial x}(R\Delta U)^n$$

$$\Delta W_2^n = (Q - S_y)^n \Delta U^n + \frac{\partial}{\partial y}(S\Delta U)^n \tag{137}$$

where A, B, P, R, Q, S are the Jacobi matrices:

$$A = \frac{\partial F}{\partial U} \quad B = \frac{\partial G}{\partial U} \quad P = \frac{\partial V_1}{\partial U_x} \quad R = \frac{\partial V_1}{\partial U_x} \quad Q = \frac{\partial W_2}{\partial U} \quad S = \frac{\partial W_2}{\partial U_y}$$

The use of the expansions of the type of (136) or the terms of the equations with the mixed derivatives V_2, W_1 complicates the organization over spatial coordinates in the event of further splitting. Therefore, instead of the mentioned expansions the explicit approximation is used by the following formulas:

$$\Delta V_2^n = \Delta V_2^{n-1} + O(\Delta t^2) \qquad \Delta W_1^n = \Delta W_1^{n-1} + O(\Delta t^2) \tag{138}$$

After the substitution of (137) and (138) into equation (135), this equation takes on the form

$$\left[I + \frac{\theta \Delta t}{1+\xi}(L_x + L_y)\right]\Delta U^n = \frac{\Delta t}{1+\xi}\left[\frac{\partial}{\partial x}(-F+V)^n + \frac{\partial}{\partial y}(-G+W)^n\right] +$$

$$+ \frac{\theta \Delta t}{1+\xi}\left[\frac{\partial}{\partial x}(\Delta V_2)^{n-1} + \frac{\partial}{\partial y}(\Delta W_1)^{n-1}\right] +$$

$$+ \frac{\xi}{1+\xi}\Delta U^{n-1} + O\left[\left(\theta - \frac{1}{2} - \xi\right)\Delta t^2 + \Delta t^3\right] \tag{139}$$

where

$$L_x = \frac{\partial}{\partial x}(A - P + R_x)^n - \frac{\partial^2}{\partial x^2}R^n$$

$$L_y = \frac{\partial}{\partial y}(B - Q + S_y)^n - \frac{\partial^2}{\partial y^2}S^n$$

The operator on the left side of (139) is replaced by the factorized operator

$$I + \frac{\theta\Delta t}{1+\xi}(L_x + L_y) = \left(I + \frac{\theta\Delta t}{1+\xi}L_x\right)\left(I + \frac{\theta\Delta t}{1+\xi}L_y\right) + O(\Delta t^2)$$

Then, the difference scheme is written in the form

$$\left(I + \frac{\theta\Delta t}{1+\xi}\Lambda_x^n\right)\left(I + \frac{\theta\Delta t}{1+\xi}\Lambda_y^n\right)\Delta U^n =$$

$$= \frac{\Delta t}{1+\xi}\left[\frac{\partial}{\partial x}(-F+V)^n + \frac{\partial}{\partial y}(-G+W)^n\right] +$$

$$+ \frac{\theta\Delta t}{1+\xi}\left[\frac{\partial}{\partial x}(\Delta V_2)^{n-1} + \frac{\partial}{\partial y}(\Delta W_1)^{n-1}\right] + \frac{\xi}{1+\xi}\Delta U^{n-1} \quad (140)$$

where Λ_x, Λ_y are the difference operators approximating L_x, L_y.

The difference scheme (139) is realized by two fractional steps

$$\left(I + \frac{\theta\Delta t}{1+\xi}\Lambda_x^n\right)\Delta U^* = \text{r.h. side}$$

$$\left(I + \frac{\theta\Delta t}{1+\xi}\Lambda_y^n\right)\Delta U^n = \Delta U^* \quad (141)$$

where the notation "r.h. side" is used for the right side of Equation 140 calculated by the values of the functions on the previous time layers. The values of the functions on a new time layer are found by the formula $U^{n+1} = U^n + \Delta U^n$.

The derivatives with respect to spatial coordinates in the scheme under consideration are approximated by central difference relations of the second order of accuracy

$$\left(\frac{\partial f}{\partial x}\right)_{i,j} \sim \frac{f_{i+1,j} - f_{i-1,j}}{2\Delta x} \qquad \left(\frac{\partial f}{\partial y}\right)_{i,j} \sim \frac{f_{i,j+1} - f_{i,j-1}}{2\Delta y} \quad (142)$$

$$\left(\frac{\partial^2 f}{\partial x^2}\right)_{i,j} \sim \frac{f_{i+1,j} - 2f_{i,j} + f_{i-1,j}}{\Delta x^2} \qquad \left(\frac{\partial^2 f}{\partial y^2}\right)_{i,j} \sim \frac{f_{i,j+1} - 2f_{i,j} + f_{i,j-1}}{\Delta y^2} \quad (143)$$

where f is any of the unknown functions. The systems of difference equations obtained on each fractional step have block-tridiagonal matrices of the coefficients and are solved by vector factorization.

The authors studied stability of the considered method on the model equations with the constant coefficients

$$\frac{\partial u}{\partial t} + C_1\frac{\partial u}{\partial x} + C_2\frac{\partial u}{\partial y} = 0 \tag{144}$$

$$\frac{\partial u}{\partial t} = a\frac{\partial^2 u}{\partial x^2} + b\frac{\partial^2 u}{\partial x \partial y} + c\frac{\partial^2 u}{\partial y^2} \quad (a, c > 0,\ b^2 < 4ac) \tag{145}$$

The first of them is the hyperbolic equation, the second is parabolic. It was shown by the spectral method that the factorized scheme of the second order of accuracy over time ($\theta = 1/2 + \xi$) written for Equation 144 in the form

$$\left(I + \frac{\theta\Delta t}{1+\xi}C_1\frac{\partial}{\partial x}\right)\left(I + \frac{\theta\Delta t}{1+\xi}C_2\frac{\partial}{\partial y}\right)\Delta u^n = -\frac{\Delta t}{1+\xi}\left(C_1\frac{\partial u}{\partial x} + C_2\frac{\partial u}{\partial y}\right)^n + \frac{\xi}{1+\xi}\Delta u^{n-1}$$

is unconditionally stable when $\xi > 0$. This Scheme for Equation 145 written in the form

$$\left(I - \frac{\theta\Delta t}{1+\xi}a\frac{\partial^2}{\partial x^2}\right)\left(I - \frac{\theta\Delta t}{1+\xi}c\frac{\partial^2}{\partial y^2}\right)\Delta u^n = \frac{\Delta t}{1+\xi}\left(a\frac{\partial^2 u}{\partial x^2} + b\frac{\partial^2 u}{\partial x \partial y} + c\frac{\partial^2 u}{\partial y^2}\right)^n +$$

$$+ \frac{\theta\Delta t}{1+\xi}b\left(\frac{\partial^2 \Delta u}{\partial x \partial y}\right)^{n-1} + \frac{\xi}{1+\xi}\Delta u^{n-1}$$

is unconditionally stable with more strict limitation $\xi \geq 0.385$.

The considered numerical method possesses the properties of conservativeness, complete approximation, and unconditional stability in linear approximation. The absence of iterations on transition to a new time layer is an essential advantage of the method that improves its effectiveness. The following remarks should be made relative to the practical realization of the method:

1. Approximation of convective transfer by central difference relations of the type of (142) leads to the origin of oscillations in a numerical solution for whose elimination one requires artificial dissipation.
2. When calculating the elements of the Jacobi matrices P and Q it appears possible in many cases to disregard the dependence of the transport coefficients on the components of the vector of state U. This greatly simplifies the expressions for the elements of the matrices mentioned. If we assume the coefficient of transfer by local constants, then $-P + R_x = 0$, $-Q + S_y = 0$ and in the difference scheme only the matrices A, B, R, S remain.The simplifications mentioned are quite justified in the calculations of the time-dependent technique of stationary flows because the stationary solution (if it exits) is independent of simplifications adopted in the calculation of the Jacobi matrices.
3. The considered difference scheme in a general case is three-layered, i.e., it contains the values of the functions on three time layers: $n - 1$, n, $n + 1$. However, calculations may be organized so that for each node of the difference grid only two data files, U and ΔU, will be stored in the computer memory.
4. Because the scheme in three-layered, to begin calculation we should have the initial data on the two time layers. The values of the vector of state in all the nodes of one layer are known from the initial condition. The data of the second layer may be obtained by the two-layered scheme which is obtained from the general equation (133) at $\xi = 0$.

As an example of method application we shall consider the problem on the interation between a shock wave and a laminar boundary layer on a plate in a supersonic flow. The flow

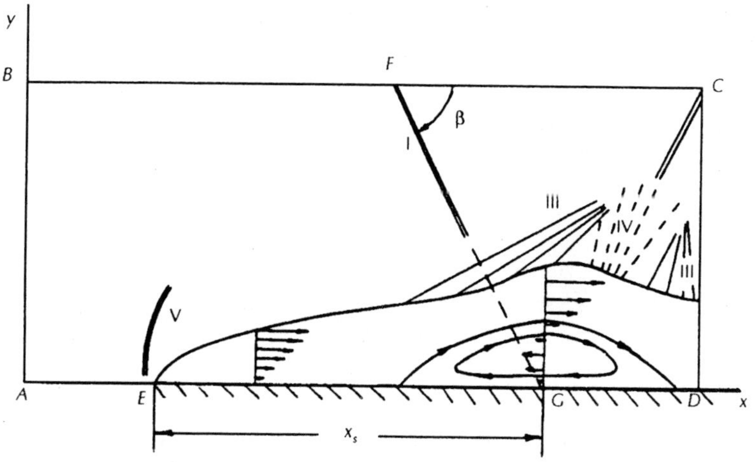

Fig. 321

pattern is schematically presented in Figure 321. Because of the rather large intensity of the impinging shock I the increase in pressure behind it results in boundary layer separation. The pressure disturbance induced by the impinding shock propagates upstream in the sublayer portion of a boundary layer. Therefore, separation occurs to the left of the point of impingement. Due to the separation and subsequent flow attachment near the plate surface regions with concave streamlines originate where compression waves III are generated that form the reflected wave II. A zone of rarefaction IV lies between the regions of compression. Also shown in the figure is shock wave V originating near the plate tip.

The problem is formulated in the following way. The computational region is the rectangular *ABCD* in the plane of the Cartesian coordinates (x, y). A supersonic flow with the prescribed values of gas dynamic parameters flows in through the boundary *AB*. The condition of flow symmetry is used on the portion of the lower boundary *AE* in front of the plate. The conditions of adhesion, impermeability, and the condition for temperature are specified on the plate surface *ED*. The condition on the upper boundary *BC* corresponds to the presence of the oblique shock *FG* at the point *F*. In this case in the portion *BF* the parameters of an unperturbed free stream are prescribed and the gas parameters in the portion *FC* are determined by the formulas of transition in terms of the oblique shock with the assigned inclination angle β. On the outlet boundary *CD* use is made of "mild" conditions of the form $\partial^2 f/\partial x^2 = 0$.

Calculations were performed for the following conditions: $\mathbf{M}_\infty = 2$, $\mathbf{Re}_\infty = (\rho_\infty V_\infty x_s)/\mu_\infty = 2.96 \cdot 10^5$, $\beta = 32.6°$, $x_s = 0.047$ m. The difference scheme contained 32x45 nodes positioned with a uniform step over x and narrowing toward the plate surface. In Figure 322, the computed change in pressure and the coefficient of friction over

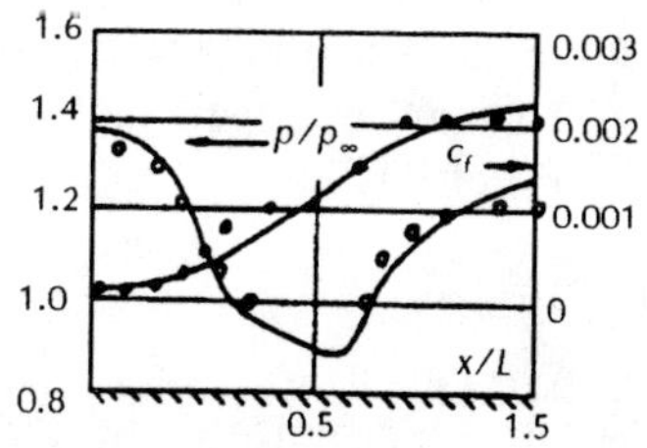

Fig. 322

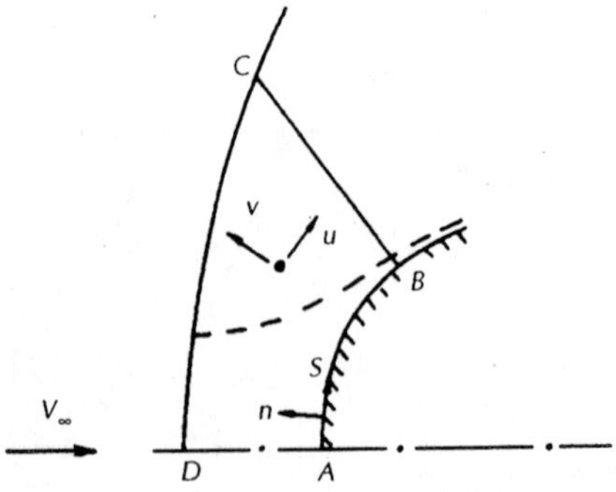

Fig. 323

the plate surface are compared with the experimental data. The results of calculations are plotted by solid lines; the results of measurements are marked by circles.

5.2. Scheme of Constant Direction

In implicit schemes of constant direction the functions on a new time layer, similar to the splitting schemes, are calculated on several stages. On each, one-dimensional systems of equations are solved by the factorization method. The difference from the schemes of splitting and of alternate directions lies in the fact that the direction of factorizations remains unchangeable. In some cases the transition to a new time layer in the schemes of constant direction may be fulfilled without iterations. In such cases the stationary problems, when solving, may be treated as explicit-implicit methods of relaxation in lines. These methods are conditionally stable. However, the limitation on the value of the time layer that follows from the stability condition may be quite acceptable in the calculation of flows with a rather smooth variation of the functions in the direction corresponding to explicit approximation. It should be noted that the methods of constant direction are very convenient for relaxation on computers with parallel processors because in their application there does not arise the necessity to change the structure of data during calculation. With the change in the direction of factorizations the data should be redistributed among the processors, thus considerably reducing the effectiveness of the parallel algorithm.

Consider application of an implicit scheme of constant direction to the example of the problem of supersonic axisymmetric flow past a blunt body.[37] Based on this example we shall illustrate the technique of shock wave separation and use of movable grids.

At rather large Reynolds numbers a supersonic flow past a blunt body is accompanied by the formation of a separated shock wave in front of the body. The flow pattern in one of meridian sections is shown in Figure 323, where the contours of the body and of the separated shock wave are presented by the solid lines *AB* and *CD*. The flow between a frontal shock wave and the body surface is of a mixed character. In the vicinity of the stagnation line *AD* it is subsonic. Further downstream the gas, with the exception of a near-wall region, is accelerated up to a supersonic velocity. The boundary separating sub- and supersonic regions is represented by a dashed line. The form of this line corresponds to large regions of supersonic flow velocities.

[37] Golovachyov, Yu. P. and Popov, F. D., Calculation of a supersonic flow of a viscous gas past blunt bodies at large Reynolds numbers, *Zh. Vych. Mat. Mat. Fiz.*, 12(5), 1292–1303, 1972; Golovachyov, Yu. P., Kuzmin, A. M., and Popov, F. D., On the calculation of a flow past blunt bodies by full and simplified Navier-Stokes equations, *Zh. Vych. Mat. Mat. Fiz.*, 13(4), 1973.

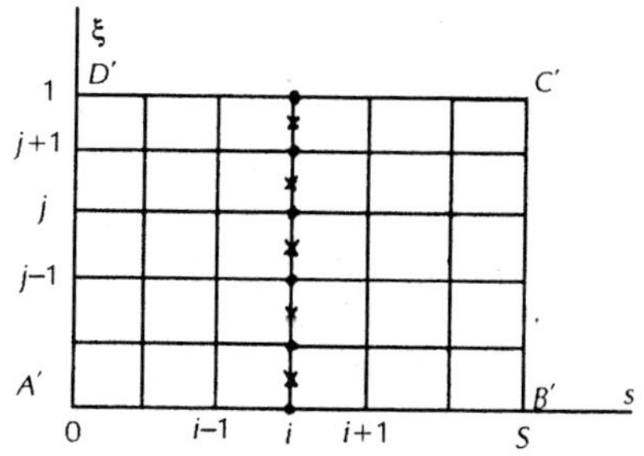

Fig. 324

The flow is considered in the region $ABCD$ which is bounded by a frontal shock wave, the body surface, the symmetry axis, and by some radial line BC. The boundary BC lies at a large enough distance from the symmetry axis so that gas velocity component normal to it is supersonic everywhere with the near-wall region excluded. To describe the flow use is made of the system of Navier-Stokes equations written in curvilinear orthogonal coordinates (s, n) that are related to the body surface (see Figure 323). The initial conditions of the problem are described by the assignment of the field of gas dynamic functions at the time instant $t = 0$. The boundary value conditions are formulated as follows. On the frontal shock wave use is made of the Rankine-Hugoniot relations, on the body surface the conditions $u = v = 0$, $T = T_w$ are valid, on the flow axis the symmetry conditions

$$u = 0 \qquad \partial v/\partial s = 0 \qquad \partial \rho/\partial s = 0 \qquad \partial T/\partial s = 0$$

are applied, and on the outlet boundary BC the "mild" conditions of the type $\partial^2 f/\partial s^2 = 0$ are used where f is any of the unknown functions.

The coordinates of the separated shock wave are not known a priori. They are determined simultaneously with the solution of equations. To close the problem we shall assume that in the momentum equation in the projection on n the terms with the second derivatives with respect to n do not exceed, in accordance with the concepts of the boundary layer theory, $\delta = \mathbf{Re}^{-1/2}$ by the order of magnitude. Accounting for this fact, whose satisfaction in the difference scheme is considered below, it is quite enough to have three conditions on the body surface and four Rankine-Hugoniot relations for integration across the shock layer and determination of the frontal shock wave position.

To numerically solve the problem, the distance from the body surface n is normalized by the value of the frontal shock wave separation ε, i.e., a new independent variable $\xi = n/\varepsilon(s, t)$ is introduced instead of n. As a result, the considered flow region is transformed into the rectangular $A'B'C'D'$ (see Figure 324). In the region $0 \le s \le S$, $0 \le \xi \le 1$, $t \ge 0$ the difference grid with the following coordinates of the nodes

$$s_i = i\Delta s \quad \Delta s = S/N \quad i = 0, 1, \ldots, N$$

$$\xi_j = j\Delta\xi \quad \Delta\xi = 1/M \quad j = 0, 1, \ldots, M$$

$$t^n = n\Delta t \quad n = 0, 1, \ldots \tag{146}$$

is introduced. Due to the fact that the value of the frontal shock wave separation from the body surface changes with time, the movable grid whose nodes move along the radial lines s = const corresponds to the fixed grid (146) in the computational region.

In new variables the Navier-Stokes equations are written in the matrix form as

$$E\frac{\partial X}{\partial t}+\frac{1}{\rho}\frac{\partial}{\partial \xi}\left(A\frac{\partial X}{\partial \xi}\right)+B\frac{\partial X}{\partial \xi}+D=0 \tag{147a}$$

$$e\frac{\partial X}{\partial t}+b\frac{\partial X}{\partial \xi}+d=0 \tag{147b}$$

Here, $X=\{u\ T\ v\ p\}^{\mathrm{T}}$ is the vector column of the unknown functions. The first of the above equations expresses the momentum balance in the projection on s and the energy balance, the second contains the equation of momentum balance in the projection on n and the continuity equation. The elements of the vectors D, d, besides the unknown functions themselves, also involve the terms with the derivatives.

For the difference approximation of Equations 147 use is made of the implicit two-layered scheme with the weight α. On each radial line $s=s_i$ Equation 147a is approximated in the grid nodes with the integral values of the index j, and Equation 147b in the nodes with half-integral values of the same index (see Figure 324).

We denote $X^n_{ij}=X(s_i,\xi_j,t^n)$, $X^\alpha_{ij}=\alpha X^n_{ij}+(1-\alpha)X^{n-1}_{ij}$. The difference relations used in the inner nodes of the grid for Equation 147a have the form

$$\frac{\partial X}{\partial t}\sim\frac{1}{\Delta t}(X^n-X^{n-1})_{ij}$$

$$\frac{\partial}{\partial \xi}\left(A\frac{\partial X}{\partial \xi}\right)\sim\frac{1}{\Delta\xi^2}\left[A_{j+1/2}(X_{j+1}-X_j)-A_{j-1/2}(X_j-X_{j-1})\right]^\alpha_i$$

$$\frac{\partial X}{\partial \xi}\sim\frac{1}{2\Delta\xi}(X_{j+1}-X_{j-1})^\alpha_i \qquad \frac{\partial X}{\partial s}\sim\frac{1}{2\Delta s}(X_{i+1}-X_{i-1})^\alpha_j$$

$$\frac{\partial}{\partial s}\left(\gamma\frac{\partial X}{\partial \xi}\right)\sim\frac{1}{4\Delta\xi\Delta s}\left[\gamma_{i+1,j}(X_{j+1}-X_{j-1})_{i+1}-\gamma_{i-1,j}(X_{j+1}-X_{j-1})_{i-1}\right]^\alpha$$

$$\frac{\partial}{\partial s}\left(\gamma\frac{\partial X}{\partial s}\right)\sim\frac{1}{\Delta s^2}\left[\gamma_{i+1/2}(X_{i+1}-X_i)-\gamma_{i-1/2}(X_i-X_{i-1})\right]^\alpha_j$$

For Equation 147b in the inner grid nodes use is made of the difference relations

$$\frac{\partial X}{\partial t}\sim\frac{1}{2\Delta t}\left[(X_j+X_{j-1})^n-(X_j+X_{j-1})^{n-1}\right]_i$$

$$\frac{\partial X}{\partial \xi}\sim\frac{1}{\Delta\xi}(X_j-X_{j-1})^\alpha_i \qquad \frac{\partial X}{\partial s}\sim\frac{1}{4\Delta s}\left[(X_j+X_{j-1})_{i+1}-(X_j+X_{j-1})_{i-1}\right]^\alpha$$

$$\frac{\partial}{\partial s}\left(\gamma\frac{\partial X}{\partial \xi}\right)\sim\frac{1}{2\Delta\xi\Delta s}\left[\gamma_{i+1,j-1/2}(X_j-X_{j-1})_{i+1}-\gamma_{i-1,j-1/2}(X_j-X_{j-1})_{i-1}\right]^\alpha$$

$$\frac{\partial}{\partial \xi}\left(\gamma\frac{\partial X}{\partial \xi}\right)\sim\frac{1}{2\Delta\xi^2}\left[\gamma_j(X_{j+1}-X_{j-1})-\gamma_{j-1}(X_j-X_{j-2})\right]^\alpha_i$$

$$\frac{\partial}{\partial s}\left(\gamma\frac{\partial X}{\partial s}\right)\sim\frac{1}{2\Delta s^2}\left[\gamma_{i+1/2,j-1/2}(X_{i+1,j}+X_{i+1,j-1}-X_{ij}-X_{i,j-1})\ -\right.$$

$$-\gamma_{i-1/2,j-1/2}(X_{ij}+X_{i,j-1}-X_{i-1,j}-X_{i-1,j-1})\,]^{\alpha}$$

On the boundaries and in near-boundary nodes asymmetric difference approximations of the second order of accuracy are applied.

The coefficients of the difference equations are calculated by the values of their arguments X_{ij}^{α} for (147a) and $X_{i,j-1/2}^{\alpha} = (X_j + X_{j-1})_i^{\alpha}/2$ for Equation 147b.

The values of the functions on the layer n are determined from the known values on the layer $n-1$ by the method of subsequent approximations. We denote the number of iteration by q, the vector of the unknown functions on the layer n in this iteration by $X^{n(q)}$

$$X^{n(0)} = X^{n-1} \qquad X^{\alpha(q)} = \alpha X^{n(q)} + (1-\alpha)X^{n-1}$$

All the coefficients of the difference equations are calculated by the values of $X^{\alpha(q)}$, where q is the number of the previous iteration. In this case the system of the difference equations is decomposed into independent linear subsystems on the radial lines $s = s_i$. Each of these subsystems has the form

$$P_j^{\alpha(q)} X_{j+1}^{n(q+1)} + R_j^{\alpha(q)} X_j^{n(q+1)} + Q_j^{\alpha(q)} X_{j-1}^{n(q+1)} + \Phi_j = 0$$

$$(j = 1, 2, ..., M-1) \qquad (148a)$$

$$r_{j-1/2}^{\alpha(q)} X_j^{n(q+1)} + q_{j-1/2}^{\alpha(q)} X_{j-1}^{n(q+1)} + \varphi_{j-1/2} = 0$$

$$(j = 1, 2, ..., M) \qquad (148b)$$

The coefficients P, R, Q, Φ, r, q, φ are expressed in terms of the matrices of the coefficients of the initial system of Equations 147.

On each radial line the solution of the difference equations (148) supplemented with the boundary conditions on the frontal shock wave and the body surface is found by vector factorization. Omitting the indices $n(q+1)$ and $\alpha(q)$, we describe the method for solving the difference equations on the radial line. The boundary conditions on the body surface for the tangential velocity components and temperature in combination with Equations 148b at $j = 1$ are written in the form

$$\begin{bmatrix} 0\,0\,0\,0 \\ 0\,0\,0\,0 \\ r_{1/2} \end{bmatrix} X_1 + \begin{bmatrix} 1\,0\,0\,0 \\ 0\,1\,0\,0 \\ q_{1/2} \end{bmatrix} X_0 + \begin{bmatrix} 0 \\ -T_w \\ \varphi_{1/2} \end{bmatrix} = 0$$

Hence, it follows

$$X_0 + \Omega_1 X_1 + \Pi_1 = 0 \qquad (149)$$

where

$$\Omega_1 = \begin{bmatrix} 1\,0\,0\,0 \\ 0\,1\,0\,0 \\ q_{1/2} \end{bmatrix}^{-1} \begin{bmatrix} 0\,0\,0\,0 \\ 0\,0\,0\,0 \\ r_{1/2} \end{bmatrix} \qquad \Pi_1 = \begin{bmatrix} 1\,0\,0\,0 \\ 0\,1\,0\,0 \\ q_{1/2} \end{bmatrix}^{-1} \begin{bmatrix} 0 \\ -T_w \\ \varphi_{1/2} \end{bmatrix}$$

The relations of the type of (149) are easily obtained for all $j = 1, 2, ..., M$. In fact, Equations 148 may be written as

$$\begin{bmatrix} P_j \\ r_{j+1/2} \end{bmatrix} X_{j+1} + \begin{bmatrix} R_j \\ q_{j+1/2} \end{bmatrix} X_j + \begin{bmatrix} Q_j \\ 0\,0\,0\,0 \\ 0\,0\,0\,0 \end{bmatrix} X_{j-1} + \begin{bmatrix} \Phi_j \\ \varphi_{j+1/2} \end{bmatrix} = 0$$

Eliminating X_{j-1} with the aid of the relation

$$X_{j-1} + \Omega_j X_j + \Pi_j = 0 \tag{150}$$

we obtain

$$\begin{bmatrix} P_j \\ r_{j+1/2} \end{bmatrix} X_{j+1} + \begin{bmatrix} R_j - Q_j\Omega_j \\ q_{j+1/2} \end{bmatrix} X_j + \begin{bmatrix} \Phi_j - Q_j\Pi_j \\ \varphi_{j+1/2} \end{bmatrix} = 0$$

or

$$X_j + \Omega_{j+1} X_{j+1} + \Pi_{j+1} = 0 \tag{151}$$

where

$$\Omega_{j+1} = \begin{bmatrix} R_j - Q_j\Omega_j \\ q_{j+1/2} \end{bmatrix}^{-1} \begin{bmatrix} P_j \\ r_{j+1/2} \end{bmatrix} \qquad \Pi_{j+1} = \begin{bmatrix} R_j - Q_j\Omega_j \\ q_{j+1/2} \end{bmatrix}^{-1} \begin{bmatrix} \Phi_j - Q_j\Pi_j \\ \varphi_{j+1/2} \end{bmatrix}$$

The above recurrent formulas determine the coefficients of factorization.

Now, consider the procedure of determining the velocity of shock wave motion and of calculating the values of gas dynamic functions behind its front. The condition of impermeability of a streamlined body surface that has not been used as yet, is written in the form

$$\omega_0 X_0 + \pi_0 = 0 \quad \text{where} \quad \omega_0 = [0\,0\,1\,0] \quad \pi_0 = 0$$

Using (149), such relations may be obtained for all $j = 0, 1, ..., M$:

$$\omega_j X_j + \pi_j = 0$$

with

$$\omega_{j+1} = \omega_j \Omega_{j+1} \qquad \pi_{j+1} = \omega_j \Pi_{j+1} - \pi_j$$

On fulfilling direct factorization the coefficients Ω_j, Π_j are calculated simultaneously with the coefficients ω_j, π_j. As a result, the relation

$$\omega_M X_M + \pi_M = 0$$

will be obtained on a shock wave. Having added it to the four Rankine-Hugoniot relations, we shall have the closed system of equations from which all of the vector components X_m and the velocity of the shock wave front motion ε_t' are found.

The values of the functions in all the inner nodes of the grid and on the body surface on the considered radial line $s = s_i$ are determined as the result of inverse factorization carried out according to Equation 151. A new value of shock wave separation from the body surface is calculated by the equation

$$\varepsilon^{n(q+1)} = \varepsilon^{n-1} + \Delta t\left[\alpha(\varepsilon_t')^{n(q+1)} + (1-\alpha)(\varepsilon_t')^{n-1}\right]$$

The above computations are carried out sequentially on all the radial lines $s = s_i$ ($i = 0, 1, \ldots, N$). If the resulting values of the unknown functions coincide with the prescribed accuracy with the corresponding values of the previous iteration, then they are taken as the solution on the time layer n. If not, a new iteration is performed.

It is not difficult to show that this difference scheme generally has the first order of approximation over time and the second order of approximation over the spatial coordinates. At $\alpha = 1/2$ the approximation order over time becomes second. The study of scheme stability will be conducted on the model of homogeneous differential equations with the constant coefficients

$$\frac{\partial u}{\partial t} + B\frac{\partial u}{\partial x} - \frac{\partial}{\partial x}\left(A\frac{\partial u}{\partial x}\right) = 0 \quad A > 0 \tag{152}$$

$$\frac{\partial u}{\partial t} + b\frac{\partial u}{\partial x} = 0 \tag{153}$$

Making use of the standard procedure, we shall write a parabolic form of the first differential approximation of the difference scheme for these equations. In the case of $\alpha = 1$ we have

$$\frac{\partial u}{\partial t} + B\frac{\partial u}{\partial x} - \left(A + B^2\frac{\Delta t}{2}\right)\frac{\partial^2 u}{\partial x^2} = 0 \tag{154}$$

$$\frac{\partial u}{\partial t} + b\frac{\partial u}{\partial x} - b^2\frac{\Delta t}{2}\cdot\frac{\partial^2 u}{\partial x^2} = 0 \tag{155}$$

A positive nature of the coefficients at the second derivative with respect of x indicates unconditional stability of the scheme. In the analysis presented the problem of the convergence of iterations used for the transition to a new time layer, requires special consideration, is not solved. Calculations show that the admitted value of the time layer depends on the number of iterations.

The problem on a monotonous character of the scheme under consideration is studied based on the example of the model stationary equation with constant coefficients

$$\alpha\frac{d^2u}{dx^2} + \beta\frac{du}{dx} + \gamma u = 0 \qquad \alpha > 0 \qquad \gamma \le 0 \tag{156}$$

The finite-difference equation corresponding to the considered scheme has the form

$$\alpha\frac{u_{j-1} - 2u_j + u_{j+1}}{\Delta x^2} + \beta\frac{u_{j+1} - u_{j-1}}{2\Delta x} + \gamma u_j = 0 \tag{157}$$

Compare the behavior of the solutions of (156) and (157).

A general solution of (156) is

$$u = c_1 e^{\mu_1 x} + c_2 e^{\mu_2 x} \tag{158}$$

where c_1, c_2 are the constants determined from the boundary conditions; μ_1, μ_2 are the roots of the characteristic equations

$$\alpha\mu^2 + \beta\mu + \gamma = 0$$

In the nodes of the difference schemes the solution of (158) is written as follows

$$u(x_j) = c_1\Lambda_1^j + c_2\Lambda_2^j$$

where

$$\Lambda_{1,2} = \exp(\mu_{1,2}\Delta x) = \exp\left[(-\beta \pm \sqrt{\beta^2 - 4\alpha\gamma})\,\Delta x/2\alpha\right]$$

It is easy to see that for $\alpha > 0$, $\gamma \le 0$ we have

$$0 < \Lambda_1 \le 1 \qquad 1 \le \Lambda_2 < \infty$$

The general solution of the difference equation (157) has the form

$$u_j = c_1\lambda_1^j + c_2\lambda_2^j \tag{159}$$

where $\lambda_{1,2}$ are the roots of the characteristic equations

$$\psi(\lambda) = \left(\alpha + \beta\frac{\Delta x}{2}\right)\lambda^2 + (\gamma\Delta x^2 - 2\alpha)\,\lambda + \left(\alpha - \beta\frac{\Delta x}{2}\right) = 0$$

We can easily convince ourselves that the roots $\lambda_{1,2}$ are real. To determine their position on the actual axis we shall write the values of the function $\psi(\lambda)$ at some values of the argument found in Table 1. Comparison of the signs at $\psi(\lambda)$ at the chosen points yields

$$0 < \lambda_1 < 1 \qquad 1 \le \lambda_2 < \infty \quad \text{when} \quad \alpha \pm \beta\frac{\Delta x}{2} > 0$$

$$-1 < \lambda_1 < 0 \qquad 1 < \lambda_2 < \infty \quad \text{when} \quad \alpha - \beta\frac{\Delta x}{2} < 0$$

$$0 < \lambda_1 \le 1 \qquad -\infty < \lambda_2 < -1 \quad \text{when} \quad \alpha + \beta\frac{\Delta x}{2} < 0$$

Thus, the positions of roots of the characteristic equations of the differential and difference problems coincide only when $\alpha \pm \beta\Delta x/2 > 0$, or when $\Delta x < 2\alpha/|\beta|$. If this condition is violated, then one of the roots of the characteristic equation in the difference problem becomes negative. In accordance with (159), to it corresponds a notched solution which is not even in qualitative agreement with the solution of the differential problem. The obtained limitation to the value of the difference grid step becomes substantial at small α, i.e., in the regions of large gradients. Allowing for the order of magnitudes of the coefficients of model equation (156), we can easily see the limitation to the grid Reynolds number is $\mathbf{Re}_h = \rho V\Delta x/\mu < 2$.

As an example of the application of this numerical method we shall consider the problem of a supersonic flow past a spherical blunt body in a wake of another body.[38] A supersonic far wake behind an axisymmetric body streamlined at the zero contact angle is a parallel axisym-

[38] Golovachyov, Yu. P. and Leontiyeva, N. N., A supersonic flow past a sphere in a wake of a blunt cylinder, *J. Eng. Phys.*, 51(6), 955–959, 1986.

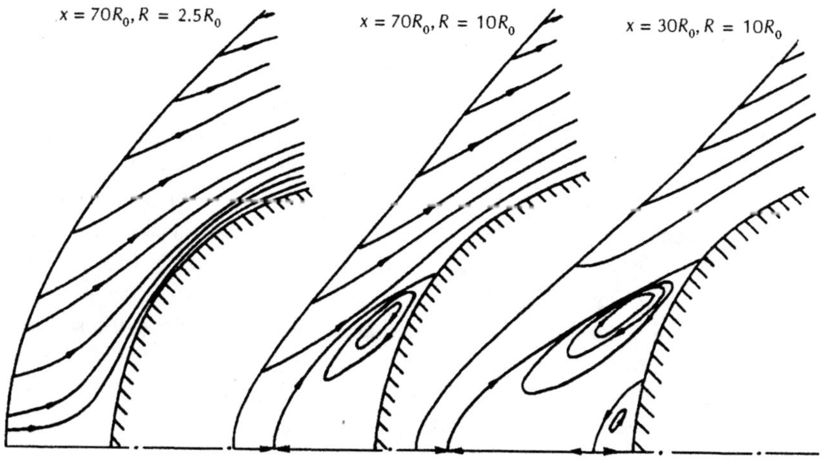

Fig. 325

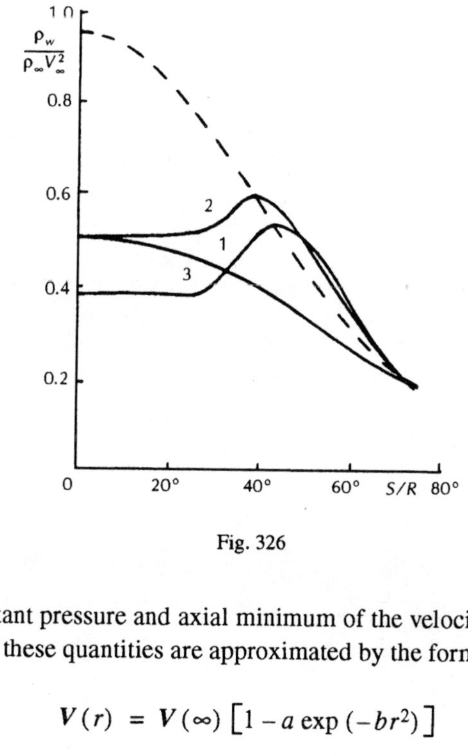

Fig. 326

metric flow with constant pressure and axial minimum of the velocity and temperature maximum. Distributions of these quantities are approximated by the formulas

$$V(r) = V(\infty)\left[1 - a\exp(-br^2)\right]$$

$$T(r) = T(0)\left[1 + c\left(1 - \frac{V^2(r)}{V^2(0)}\right)\right] \tag{160}$$

where r is the distance from the wake axis related to the radius of blunting; V is the modulus of the velocity vector; a, b, c are parameters whose values were taken from comparison with the results of calculation of the flow in a wake.

A stationary flow past a sphere whose center lies on the wake axis is considered. The ratio of gas-specific heat capacities is $\gamma = 1.4$, the dependence of the viscosity coefficient on temperature is approximated by the function $\mu \sim \sqrt{T}$. The flow in a shock wake was calculated on a uniform difference grid with the steps $\Delta\theta = 2.5$, $\Delta\xi = 0.02$.

In Figure 31 and 32 the results of the calculation of a flow past a thermally insulated sphere in a wake behind a spherical blunt cylinder at $\mathbf{M}_\infty = 3$, $\mathbf{Re}_\infty = 200$, $k = 0.257$, $L = 3.6R_0$ are presented. Here, $\mathbf{M}_\infty$, $\mathbf{Re}_\infty$ are the Mach and Reynolds numbers of a uniform flow impinging on a cylinder; L and R_0 are the length and radius of the blunting cylinder; k is the ratio of cylinder surface temperature to that of the free stream stagnation. Figure 325 shows the position of the frontal shock wave and stream lines in a shock layer at different values of the distance between the bodies x and the sphere radius R. The manner of flow structure variation depending on the geometric parameters of the considered pair of bodies is seen from this figure. Figure 326 presents the change of pressure along the sphere surface. Curves 1 to 3 correspond to: (1) $x = 70R_0$; $R = 2.5R_0$; (2) $x = 70R_0$, $R = 10R_0$; (3) $x = 30R_0$, $R = 10R_0$. A dashed line shows, for comparison, pressure distribution in the case of a uniform flow with $\mathbf{M}_\infty = 3$, $\mathbf{Re}_\infty = 10^3$ past a sphere.

Author's Index

Index